# Fault-related Rocks

## A Photographic Atlas

# Fault-related ROCKS

## A Photographic Atlas

*Edited by*

Arthur W. Snoke, Jan Tullis, and

Victoria R. Todd

PRINCETON UNIVERSITY PRESS

PRINCETON, NEW JERSEY

Published by Princeton University Press, 41 William Street,
Princeton, New Jersey 08540
In the United Kingdom: Princeton University Press, Chichester, West Sussex



*Library of Congress Cataloging-in-Publication Data*

Fault-related rocks : a photographic atlas / edited by Arthur W. Snoke,
Jan Tullis, and Victoria R. Todd.
p. cm.
Includes bibliographical references and index.
ISBN 0-691-01220-2 (alk. paper)
1. Mylonite—Atlases. 2. Petrofabric analysis—Atlases.
I. Snoke, Arthur W. II. Tullis, Jan. III. Todd, Victoria R. (Victoria Roy)
QE475 M95P75 1998
552'.4—DC21 97-19775 CIP Rev.

This book has been composed in Times Roman and Helvetica

Princeton University Press books are printed on acid-free paper and meet the guidelines for permanence and durability of the Committee on Production Guidelines for Book Longevity of the Council on Library Reserces

http://pup.princeton.edu

Printed in the United States of America

10 9 8 7 6 5 4 3 2 1

# ERRATA

**Page 4:** In Figure 0.2, the scattered dot pattern meant to depict crush breccias was printed out as solid gray. Below is the corrected figure.

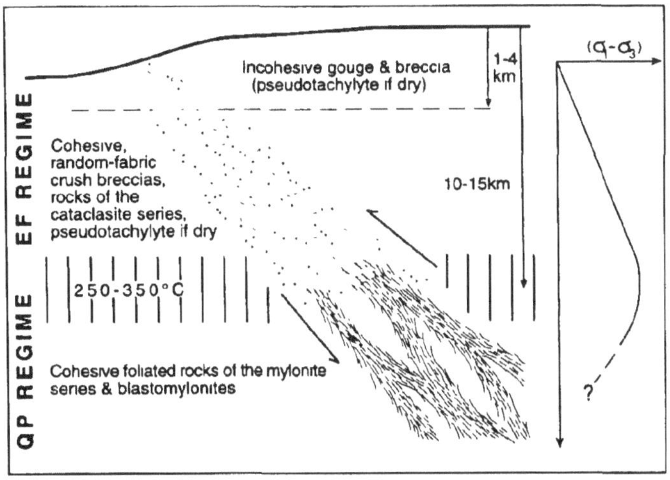


**Page 116:** In Figure 32.2, the scale bar should read 0.2 mm.

**Page 117:** In Plates 32H and 32I, the scale is 0.2 mm to the inch.

**Page 393:** In Plate 119A, the line to the right of the one labeled $S_m$ should be labeled $S_1$.

**Page 395:** In Plate 119D, the line at the lower left should be labeled $S_m$.

**Page 590:** In Figure 176.1, the portions depicting crust were printed very faintly. Below is a better reproduction.

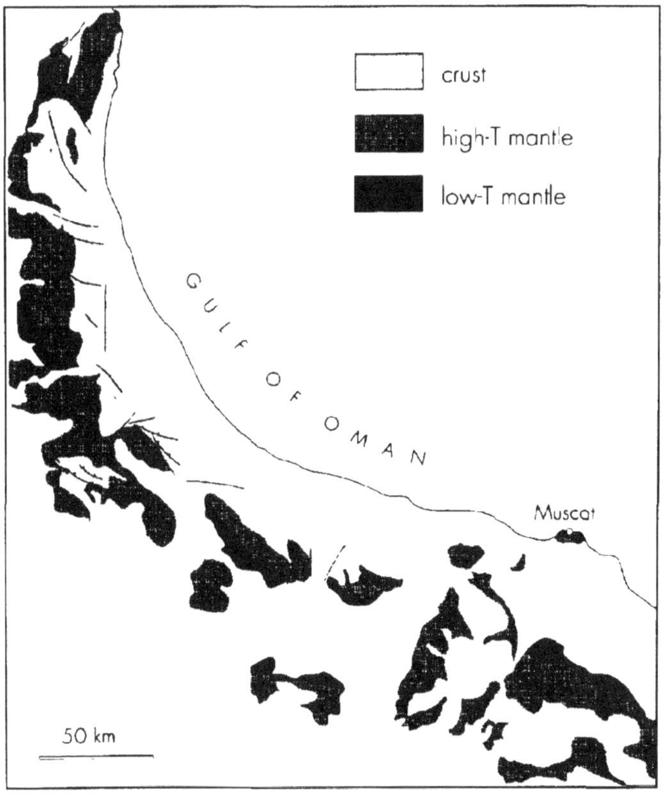

# Preface

The initial concept of a photographic atlas on fault rocks was developed during informal discussions at the Geological Society of America Penrose Conference "Significance and Petrogenesis of Mylonitic Rocks" held at San Diego, California, in spring 1981, and coconvened by the editors of this book. Many participants felt that a new consensus about mylonites and ductile shear zones had been achieved at the conference, based largely upon the contributions of geoscientists who brought perspectives from the laboratory as well as from the field. In part motivated by the publication (1982) of the *Atlas of Deformational and Metamorphic Rock Fabrics* (and its predecessor, the *Atlas of Rock Cleavage*) both edited by G. J. Borradaile, M. B. Bayly, and C. McA. Powell, we felt that a collection of photographs of features characteristic of faults and shear zones, ranging in scale from mesoscopic to microscopic to even submicroscopic, together with carefully documented locations and regional (or experimental) information, would provide a valuable reference as well as a teaching device.

A call for contributions was issued in fall 1981 by the editors for the proposed volume. All participants of the Penrose Conference were solicited as well as many other individuals who the editors judged would be interested in the project. Contributions were submitted to the editors during 1982 and 1983. However, a variety of professional and personal circumstances intervened toward the end of that period, and the project was left uncompleted. Over ten years passed, until January 1994, when the editors made a commitment to revitalize and complete the project. We felt that the original goal of a photographic atlas of features found in fault- and shear-zone rocks was still important and had not been met by any other publication. In addition, it seemed to us that a number of major advances in our understanding of faults and shear zones had been made in the intervening years. In February 1994, a letter was sent out to all former contributors advising them of our plan to complete the atlas and asking them if they would like to continue to participate in the project. Also, we requested that they modify their earlier submitted contributions, if necessary. In the intervening years, the personal computer had evolved into a standard research tool and all previously submitted contributions were scanned and formatted by Snoke to a common standard. These manuscripts were sent to the original contributors for their additions and modifications. Nearly all of the former contributors continued on with the project and virtually all made modifications to their ten-year-old-plus manuscripts. At the same time, the editors actively sought new contributions from many other scientists who had become active in fault-rocks research during the past ten years. During 1994 and 1995 we had an excellent response, and many new contributions were submitted for consideration for the atlas. The editors evaluated all contributions, striving for quality of photographs and for a degree of uniformity of style and terminology. By mid-July 1995, the manuscript of the book was completed for formal review and submitted to Princeton University Press for consideration for publication. The outside reviewers selected by the Press were John M. Christie and Robert J. Twiss, who performed the Herculean task of critiquing our gigantic manuscript; we deeply appreciate their editorial comments and useful suggestions for improving the atlas. Between September 1995 and June 1996, editing of the atlas continued until the final submittal of the manuscript to Princeton University Press in late June. During this period, Snoke and Tullis reworked their introductory overview on fault rocks many times and benefited from constructive review comments by Barbara John, Bruno Lafrance, Win Means, Ernie Rutter, Rick Sibson, Carol Simpson, John Spray, and Vicki Todd. We also owe special thanks to Rick Law who provided us with a detailed and exceptionally helpful critique of our overview essay.

Several individuals helped us during the preparation of this book. During the early stages of the project, M. B. Bayly provided us with much useful advice about the production of such a book. Gloria Correra helped assemble the massive "References Cited" section. Judy Snoke helped with many details of the preparation of both the review and final versions of the manuscript. Sara Van Rheenen was our initial editor at Princeton University Press, and Jack Repcheck was our final editor. We thank all these individuals for their collective help in bringing this long-lived project to completion.

*June 1996* ARTHUR W. SNOKE, JAN TULLIS, AND VICTORIA R. TODD

# Contributors' addresses

Susan M. Agar
Department of Geological Sciences
Northwestern University
Evanston, IL 60208
U.S.A.

I. Allison
Department of Geology and Applied Geology
University of Glasgow
Glasgow, Scotland G12 8QQ
United Kingdom

Håkon Austrheim
Mineralogisk-Geologisk Museum
Sarsgate 1
N-0562, Oslo
Norway

Hassan A. Babaie
Department of Geology
Georgia State University
Atlanta, GA 30303
U.S.A.

R. L. Barnett
RR#2
Lambeth, ON N6A 4B6
Canada

Rodey Batiza
Department of Geology and Geophysics
University of Hawaii
Honolulu, HI 96822
U.S.A.

J. M. Bedford
Department of Geosciences
The University of Arizona
Tucson, AZ 85721
U.S.A.

Nicholas M. Beeler
U.S. Geological Survey
345 Middlefield Road
Mail Stop 977
Menlo Park, CA 94025
U.S.A.

T. H. Bell
Department of Geology
James Cook University of North Queensland
Townsville, Queensland 4811
Australia

Michael L. Blanpied
U.S. Geological Survey
345 Middlefield Road
Mail Stop 977
Menlo Park, CA 94025
U.S.A.

T. G. Blenkinsop
Department of Geology
University of Zimbabwe
P.O. Box MP 167
Mount Pleasant, Harare
Zimbabwe

Andy R. Bobyarchick
Department of Geography and Earth Sciences
The University of North Carolina at Charlotte
Charlotte, NC 28223
U.S.A.

Paul D. Bons
Victorian Institute of Earth and Planetary Sciences
Department of Earth Sciences
Monash University
Clayton, Victoria 3168
Australia

Jean Luc Bouchez
URA CNRS n° 67
Laboratoire de Pétrophysique et Tectonique
Université Paul Sabatier
38 rue des Trente-Six-Ponts
31400, Toulouse
France

Françoise Boudier
Laboratoire de Tectonophysique
ISTEEM
Université Montpellier II
34095 Montpellier cedex 5
France

Anne-Marie Boullier
LGIT-IRIGM
BP 53
38041 Grenoble cedex 9
FRANCE

Theresa M. Boundy
Department of Geology
Ball State University
Muncie, IN 47306
U.S.A.

K. H. Brodie
Department of Geology
University of Manchester
Manchester M13 9PL
United Kingdom

William L. Brown
C.N.R.S.-C.R.P.G.
BP 20
54501 Vandoeuvre-lès-Nancy cedex
France

L. Burlini
Consiglio Nazionale delle Ricerche
Dipartimento di Scienze della Terra
Università di Milano
I-20133 Milano
Italy

Jay P. Busch
Department of Geological Sciences
University of Michigan
Ann Arbor, MI 48109
U.S.A.

David L. Carmichael
Department of Geological Sciences
San Diego State University
San Diego, CA 92182
U.S.A.

J. Carreras
Departament de Geologia
Universitat Autònoma de Barcelona
08193 Bellaterra (Cerdanyola)
Spain

J. M. Casas
Departament de Geologia, Dinamica, Geofisica i Paleontologia
Universitat de Barcelona
Zona Universitaria de Pebrables
08071 Barcelona
Spain

Martin Casey
Department of Earth Sciences
The University
Leeds LS2 9JT
United Kingdom

Georges Ceuleneer
UPR CNRS n° 234
Dynamique Terrestre et Planétaire
14 avenue Edouard Belin
31400, Toulouse
France

Frederick M. Chester
Center for Tectonophysics
Texas A&M University
College Station, TX 77843-3113
U.S.A.

Clint H. Christensen
Department of Geoscience
University of Nevada
Las Vegas, NV 89154
U.S.A.

K. N. Constenius
Department of Geosciences
The University of Arizona
Tucson, AZ 85721
U.S.A.

Angela J. Coulton
Department of Geological Sciences, ES 351
State University of New York at Albany
Albany, NY 12222
U.S.A.

Darrel S. Cowan
Department of Geological Sciences, AJ-20
University of Washington
Seattle, WA 98195
U.S.A.

L. J. Cox
U.S. Geological Survey
Tucson, AZ 85721
U.S.A.

S. F. Cox
Department of Geology
The University of Newcastle
University Drive
Callaghan, New South Wales 2308
Australia

George H. Davis
Department of Geosciences
The University of Arizona
Tucson, AZ 85721
U.S.A.

Lisa N. Dell'Angelo
Geologisches Institut
ETH-Zentrum
CH-8032 Zürich
Switzerland

H. G. Dong
Department of Geological Sciences
State University of New York at Albany
1400 Washington Avenue
Albany, NY 12222
U.S.A.

M. R. Drury
Institute of Earth Sciences
P.O. Box 90.021
3508 TA Utrecht
The Netherlands

Ernest M. Duebendorfer
Department of Geology
Box 4099
Northern Arizona University
Flagstaff, AZ 86011
U.S.A.

W. James Dunlap
Research School of Earth Sciences
The Australian National University
Canberra, A.C.T. 0200
Australia

Phillip D. Ellzey
Department of Geology
Arizona State University
Tempe, AZ 85287
U.S.A.

Bradley G. Erskine
Department of Geology and Geophysics
University of California
Berkeley, CA 94720
U.S.A.

M. A. Etheridge
Etheridge & Henley Geoscience Consultants
P.O. Box 3778
Manuka, A.C.T. 2603
Australia

James P. Evans
Department of Geology
Utah State University
Logan, UT 84322-4505
U.S.A.

Charles J. Ferranti
Department of Geoscience
New Mexico Institute of Mining and Technology
Socorro, NM 87801
U.S.A.

David M. Fountain
Department of Geology and Geophysics
University of Wyoming
Laramie, WY 82071-3006
U.S.A.

M. Friedman (deceased)
Center for Tectonophysics
Texas A&M University
College Station, TX 77843-3113
U.S.A.

Paul D. Fullagar
Department of Geology, CB# 3315
The University of North Carolina at Chapel Hill
Chapel Hill, NC 27599-3315
U.S.A.

D. Gapais
Géosciences Rennes
(CNRS, UPR 4661)
Campus de Beaulieu
Université de Rennes
35042 Rennes Cedex
France

Jane A. Gilotti
Geological Survey
New York State Museum
Albany, NY 12230
U.S.A.

Gary H. Girty
Department of Geological Sciences
San Diego State University
San Diego, CA 92182
U.S.A.

Gayle Gleason
Institute of Geophysics
University of California, Riverside
Riverside, CA 92521
U.S.A.

Laurel B. Goodwin
Department of Geoscience
New Mexico Institute of Mining and Technology
Socorro, NM 87801
U.S.A.

Liang-Yan Gong
Department of Earth Sciences
The University
Leeds LS2 9JT
United Kingdom

Robert T. Gregory
Department of Geological Sciences
Southern Methodist University
Dallas, TX 75275
U.S.A.

D. Gritti
Dipartimento di Scienze della Terra
Università di Milano
I-20133 Milano
Italy

Yves Guéguen
Géosciences Rennes
Campus de Beaulieu
Université de Rennes
35042 Rennes Cedex
France

Bradley R. Hacker
Department of Geological Sciences
University of California at Santa Barbara
Santa Barbara, CA 93106
U.S.A.

Mark R. Handy
Institut für Geowissenschaften
Justus-Liebig-Universität Giessen
Senckenbergstrasse 3
D-35390 Giessen
Germany

Simon Hanmer
Geological Survey of Canada
601 Booth Street
Ottawa, ON K1A 0E8
Canada

Robert D. Hatcher, Jr.
Department of Geological Sciences
University of Tennessee
Knoxville, TN 37996-1410
U.S.A.

J. F. Hippertt
Departamento de Geologia
Morrodo Cruzeiro
Universidade Federal de Ouro Preto
CEP 35.400-000, Ouro Preto, MG
Brazil

Greg Hirth
Woods Hole Oceanographic Institution
Woods Hole, MA 02543
U.S.A.

E. H. Hoogerduijn Strating
Shell International Petroleum Company
P.O. Box 162
2501 AN The Hague
The Netherlands

Robert J. Hooper
Conoco, Inc.
P.O. Box 2197
Houston, TX 77252-2197
U.S.A.

Benoît Ildefonse
Laboratoire de Tectonophysique
ISTEEM
Université Montpellier II
34095 Montpellier cedex 5
France

M. P. A. Jackson
Bureau of Economic Geology
University of Texas at Austin
University Station, Box X
Austin, TX 78713
U.S.A.

P. Jégouzo
Géosciences Rennes
(CNRS, UPR 4661)
Campus de Beaulieu
Université de Rennes
35042 Rennes Cedex
France

Mark W. Jessell
Victorian Institute of Earth and Planetary Sciences
Department of Earth Sciences
Monash University
Clayton, Victoria 3168
Australia

Shaocheng Ji
Département de Géologie
Université de Montréal
C.P. 6128, succursale Centre-ville
Montréal, QC H3C 3J7
Canada

Dazhi Jiang
Department of Geology
The University of New Brunswick
Fredericton, NB E3B 5A3
Canada

J. A. Johnson
Department of Geosciences
The University of Arizona
Tucson, AZ 85721
U.S.A.

Ramon Julià
Institut de Ciències de la Terra "Jaume Almera" (C.S.I.C.)
Martí Franqués s/n,
08028 Barcelona
Spain

Thomas J. Kalakay
Department of Geology and Geophysics
University of Wyoming
Laramie, WY 82071-3006
U.S.A.

J. A. Karson
Department of Geology
Box 90277
Duke University
Durham, NC 27708-0277
U.S.A.

Patricia B. Kaygi
Harding Lawson Assoc.
707 17th Street, Suite 2400
Denver, CO 80202-3424
U.S.A.

R. W. Kerrich
Department of Geological Sciences
University of Saskatchewan
Saskatoon, SK S7N 0W0
Canada

David L. Kirschner
Department of Earth and Atmospheric Sciences
3507 Laclede Avenue
Saint Louis University
St. Louis, MO 63103
U.S.A.

Robert J. Knipe
Department of Earth Sciences
The University
Leeds LS2 9JT
United Kingdom

F. Kong
Department of Geosciences
The Unversity of Arizona
Tucson, AZ 85721
U.S.A.

Hemin Koyi
Hans Ramberg Tectonic Laboratory
Institute of Earth Sciences
Uppsala University
S-752 36 Uppsala
Sweden

Andreas Kronenberg
Center for Tectonophysics and Department of Geology and Geophysics
Texas A&M University
College Station, TX 77843
U.S.A.

Bruno Lafrance
Saskatchewen Energy and Mines
P.O. Box 5000
LaRonge, SK S0J 1L0
Canada

Larry S. Lane
Geological Survey of Canada
3303 33rd St., NW
Calgary, AB T2L 2A7
Canada

Timothy E. La Tour
Department of Geology
Georgia State University
Atlanta, GA 30303
U.S.A.

R. D. Law
Department of Geological Sciences
Virginia Polytechnic Institute and State University
Blacksburg, VA 24061
U.S.A.

J. Lee
Department of Geology
Central Washington University
Ellensburg, WA 98926
U.S.A.

Aiming Lin
Department of Earth and Planetary Sciences
Faculty of Science
Kobe University
Kobe 657
Japan

Gordon S. Lister
Victorian Institute of Earth and Planetary Sciences
Department of Earth Sciences
Monash University
Clayton, Victoria 3168
Australia

G. E. Lloyd
Department of Earth Sciences
The University
Leeds LS2 9JT
United Kingdom

J. M. Logan
P.O. Box 1776
Bandon, OR 97411
U.S.A.

M. Losantos
Servei Geològic de Catalunya
Barcelona
Spain

Jean Macaudière
E.N.S.G.-C.N.R.S.
BP 452
54001 Nancy cedex
France

Robert H. Maddock
30 Kingsway
Woking GU21 1NT
United Kingdom

Jerry F. Magloughlin
Department of Earth Resources
Colorado State University
Fort Collins, CO 80523
U.S.A.

C. K. Mawer
RIFTEC Consultant Geoscientists
P.O. Box 1113
Belconnen, A.C.T. 2617
Australia

Andrew McCaig
Department of Earth Sciences
The University
Leeds LS2 9JT
United Kingdom

Allen J. McGrew
Department of Geology
University of Dayton
Dayton, OH 45469-2364
U.S.A.

W. D. Means
Department of Geological Sciences
State University of New York at Albany
1400 Washington Avenue
Albany, NY 12222
U.S.A.

Martin G. Miller
Department of Geology
University of Wisconsin, Eau Claire
Eau Claire, WI 54702
U.S.A.

Robert B. Miller
Department of Geology
San Jose State University
San Jose, CA 95192-0102
U.S.A.

Gautam Mitra
Department of Geological Sciences
University of Rochester
Rochester, NY 14627
U.S.A.

Sharon Mosher
Department of Geological Sciences
University of Texas at Austin
Austin, TX 78712
U.S.A.

Mary Karen Nelis
Department of Geological Sciences
University of Texas at Austin
Austin, TX 78712
U.S.A.

Steven L. Netto
Department of Geological Sciences
San Diego State University
San Diego, CA 92182
U.S.A.

Julie Newman
Institute of Earth Sciences
P.O. Box 90.021
3508 TA Utrecht
The Netherlands

Adolphe Nicolas
Laboratoire de Tectonophysique
ISTEEM
Université Montpellier II
34095 Montpellier cedex 5
France

Niels Ø. Olesen
Department of Earth Sciences
Aarhus University
C. F. Møllers Allé
DK-8000 Århus C
Denmark

David L. Olgaard
Exxon Production Research Company
Room N-343
P.O. Box 2189
Houston, TX 77252
U.S.A.

R. Ornelas
Department of Geosciences
The University of Arizona
Tucson, AZ 85721
U.S.A.

Cees W. Passchier
Johannes Gutenberg-Universität
Institut für Geowissenschaften
55099 Mainz
Germany

Scott R. Paterson
Department of Earth Sciences
University of Southern California
Los Angeles, CA 90089-0740
U.S.A.

M. Pecha
Department of Geosciences
The University of Arizona
Tucson, AZ 85721
U.S.A.

Keith S. Pollman
Department of Geological Sciences
University of Texas at Austin
Austin, TX 78712
U.S.A.

Patrice Rey
Department of Earth Sciences
Monash University
Clayton, Victoria 3168
Australia

Stephen J. Reynolds
Department of Geology
Arizona State University
Tempe, AZ 85287
U.S.A.

E. P. Rodriguez
Department of Geosciences
The University of Arizona
Tucson, AZ 85721
U.S.A.

David W. Rowe
P.O. Box 2708
Crested Butte, CO 81224
U.S.A.

E. H. Rutter
Rock Deformation Laboratory
University of Manchester
Manchester M13 9PL
United Kingdom

Pere Santanach
Departament de Geologia, Dinàmica, Geofísica i Paleontologia
Universitat de Barcelona
Zona Universitària de Pedralbes
08071 Barcelona
Spain

Paul Segall
Department of Geophysics
Stanford University
Stanford, CA 94305
U.S.A.

Zachary D. Sharp
Université de Lausanne
Institut de Minéralogie et Pétrographie BFSH-2
CH-1015 Lausanne
Switzerland

T. Shimamoto
Earthquake Research Institute
University of Tokyo
No. 1-1, Yayoi 1-chome
Bunkyo-ku
Tokyo 113
Japan

Carol Simpson
Department of Earth Sciences
675 Commonwealth Avenue
Boston University
Boston, MA 02215
U.S.A.

John P. Sims
School of Earth Sciences
University of Melbourne
Parkville, Victoria 3052
Australia

Arthur W. Snoke
Department of Geology and Geophysics
University of Wyoming
Laramie, WY 82071-3006
U.S.A.

John G. Spray
Department of Geology
The University of New Brunswick
Fredericton, NB E3B 5A3
Canada

Mel R. Stauffer
Department of Geological Sciences
University of Saskatchewan
Saskatoon, SK S7N 0W0
Canada

Jürgen E. Streit
Department of Geology
The University of Newcastle
University Drive
Callaghan, New South Wales 2308
Australia

Holger Stünitz
Geologisches Institut
Universität Basel
Bernoullistrasse 32
CH-4056 Basel
Switzerland

Mark T. Swanson
Department of Geosciences
7 College Avenue
University of Southern Maine
Gorham, ME 04038
U.S.A.

Hideo Takagi
Institute of Earth Science
School of Education
Waseda University
1-6-1 Nishiwaseda, Shinjuku-ku
Tokyo 169-50
Japan

Christopher J. Talbot
Hans Ramberg Tectonic Laboratory
Institute of Earth Sciences
Uppsala University
S-752 36 Uppsala
Sweden

Kirsten S. Techmer
Institut für Geologie und Dynamik der Lithosphäre
Universität Göttingen
Goldschmidtstrasse 3
37077 Göttingen
Germany

Coen E. ten Brink
Institute of Earth Sciences
Utrecht University P.O. Box 80.021
3508 TA Utrecht
The Netherlands

Christian Teyssier
Department of Geology and Geophysics
University of Minnesota
Minneapolis, MN 55455
U.S.A.

Celeste N. Thomson
Department of Geological Sciences
San Diego State University
San Diego, CA 92182
U.S.A.

Victoria R. Todd
Department of Earth Sciences
University of California, Riverside
Riverside, CA 92521
U.S.A.

Tsuyoshi Toyoshima
Department of Geology
Faculty of Science
Niigata University
8050 Ikarashi-2-nocho
Niigata 950-21
Japan

Jan Tullis
Department of Geological Sciences
Brown University
Providence, RI 02912
U.S.A.

Ben A. van der Pluijm
Department of Geological Sciences
University of Michigan
Ann Arbor, MI 48109
U.S.A.

Dirk van der Wal
Application Laboratory
Philips Electron Optics BV, Building AAE
P.O. Box 218
NL-5600 MD Eindhoven
The Netherlands

David A. Vanko
Department of Geology
Georgia State University
Atlanta, GA 30303
U.S.A.

R. H. Vernon
School of Earth Sciences
Macquarie University
Sydney, New South Wales 2109
Australia

R. L. M. Vissers
Department of Geology
Utrecht University Institute of Earth Sciences
P.O. Box 80021
3508 TA Utrecht
The Netherlands

Rene Wagemakers
Department of Geological Sciences
San Diego State University
San Diego, CA 92182
U.S.A.

H.-R. Wenk
Department of Geology and Geophysics
University of California
Berkeley, CA 94720
U.S.A.

Joseph Clancy White
Department of Geology
The University of New Brunswick
Fredericton, NB E3B 5A3
Canada

S. H. White
Department of Geology
Institute of Earth Sciences
Utrecht University
P.O. Box 80.021
3508 TA Utrecht
The Netherlands

Chris J. L. Wilson
School of Earth Sciences
University of Melbourne
Parkville, Victoria 3052
Australia

Robert P. Wintsch
Department of Geological Sciences
Indiana University
Bloomington, IN 47405
U.S.A.

W. Adolph Yonkee
Department of Geology
Weber State University
Ogden, UT 84408-2507
U.S.A.

J. Douglas Yule
Division of Geological and Planetary Science
California Institute of Technology, 170-25
Pasadena, CA 91125
U.S.A.

Richard A. Yund
Department of Geological Sciences
Brown University
Providence, RI 02912
U.S.A.

# Fault-related Rocks

A Photographic Atlas

# AN OVERVIEW OF FAULT ROCKS

Arthur W. Snoke and Jan Tullis



The characteristics of faults and shear zones are strikingly variable. One goal of this *Atlas* is to provide a visual overview of the characteristics of the rocks associated with faults and shear zones. Such rocks are broadly called *fault rocks* (Sibson, 1977) and range from brittle, incohesive breccia or gouge at the highest structural levels in the Earth's crust to high-temperature, crystal-plastic mylonites in the deep crust or upper mantle.

In this overview, we include (1) a brief discussion of fault-rock nomenclature and classification, (2) a summary of the important grain-scale deformation mechanisms operative within fault and shear zones, (3) comments on crustal-scale fault- and shear-zone models, and (4) some important questions regarding fault rocks and fault and shear zones. Although this essay thus touches upon some of the major topics that have dominated research on fault rocks and fault and shear zones during the past twenty or more years, our treatment is not meant to be exhaustive.

At the outset of this overview, it is important to discuss the commonly used terms *brittle* and *ductile*. Some authors have used these terms genetically, but as Rutter (1986) points out, they are best used to describe the macroscopic nature of deformation. If the deformation involves a through-going discontinuity (a fault), it is brittle; if the deformation is continuous and homogeneously distributed, it is ductile. Brittle deformation involves cracking and frictional sliding, whereas ductile deformation can be accomplished by a variety of deformation mechanisms ranging from cataclastic flow to crystal-plastic processes (see the later section on deformation mechanisms). Thus, the use of the terms brittle and ductile depends on the scale that one is referring to: for example, cataclastic flow involves cracking on the grain scale, but because the cracks remain distributed the deformation is macroscopically ductile. However, cataclastic flow may occur within the gouge or cataclasite of a (brittle) fault zone.

We recognize the problems in terminology highlighted by Rutter (1986), the importance of specific deformation mechanisms as argued by Schmid and Handy (1991), and the evidence of both brittle and crystal-plastic deformation in some low-grade, polyphase mylonitic rocks. However, in an attempt to organize a diverse group of fault rocks, we have chosen a classification based upon macroscopic features rather than genetic factors. Thus, in the organization of this *Atlas*, we have grouped the plates into three categories: brittle behavior, semi-brittle behavior, and ductile behavior.

Major fault zones are commonly rooted in the middle to lower crust (e.g., see Ramsay, 1980a; Coward, 1984; Figure O.1) and evolve over a considerable period of time. Consequently, a fault zone with a significant dip-slip component may consist of a wide variety of fault rocks representing a range in P, T, and fluid conditions (e.g., Sibson, 1977; Figure O.2). Fault rocks from different depths in the crust will record different deformation mechanisms. Fault zones commonly exhibit a complex overprinting of fault-rock microstructures, recording changes in deformation mechanisms during progressive deformation or deformation during exhumation. It is common to observe crystal-plastic microstructures overprinted by brittle fracturing. The reverse of this situation, in which brittlely deformed rocks were overprinted by high-temperature metamorphic processes (e.g., recrystallization and grain growth), may also be common, but the early brittle history is difficult, if not impossible, to recognize after the superposition of a higher-temperature event.

In the case of large-displacement, normal-sense fault zones, crystal-plastic deformation microstructures in the footwall block are juxtaposed against brittle deformation microstructures of the hanging-wall block (e.g., Cordilleran-type metamorphic core complexes; also see Passchier, 1984). For large-displacement, reverse-sense fault zones, crystal-plastic deformation in the hanging-wall block (e.g., mylonitic rocks) is juxtaposed above brittle deformation of the footwall block. Schmid and Handy (1991) noted the common juxtaposition of fault rocks of different character across major dip-slip fault zones and concluded that such a distribution indicates that the rate of displacement and shearing far exceeded the rate of thermal equilibration across the fault zone. In other cases, fault rocks have developed over a protracted period of time, as when an old shear zone was reactivated by a significantly younger tectonic event. Many authors have inferred that long-established ductile shear zones influence the location of younger, brittle fault systems. Older brittle faults can also influence the location of younger brittle faults, e.g., in the Sevier fold-and-thrust belt of the U.S. Cordillera, late listric normal faults are developed near ramps of older thrust faults (Royse and others, 1975; Royse, 1993).

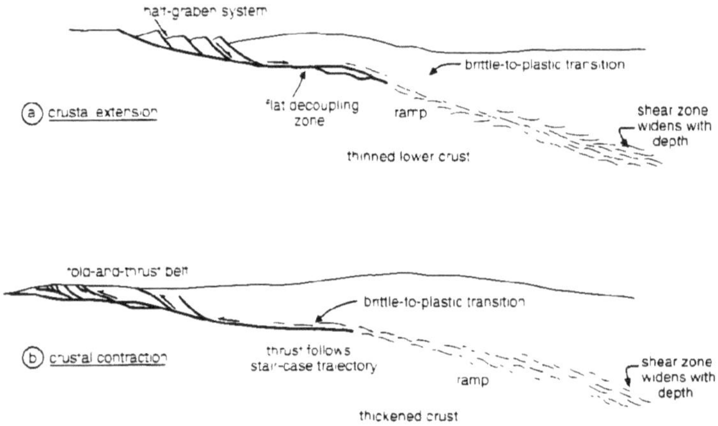


**Figure 0.1.** Brittle and ductile shear zones in (a) crustal extension and (b) crustal contraction (modified from Coward (1984)). The inferred relative position of the brittle-to-plastic transition is indicated for both cases.

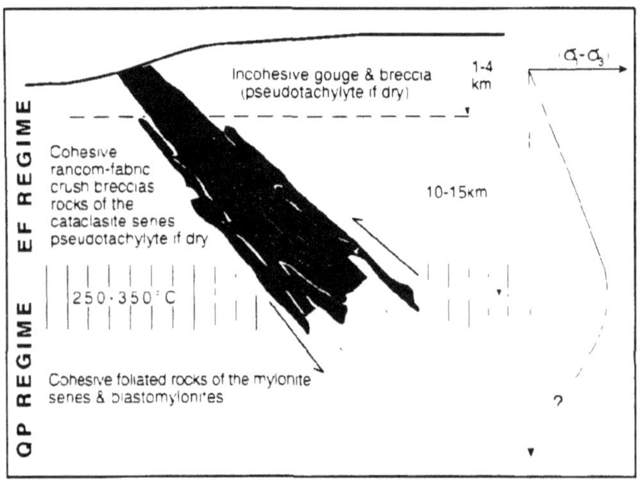


**Figure 0.2.** Sibson's conceptional model of a major fault zone (1977; reproduced with permission, the Geological Society of London. Copyright © 1977). EF is elastico-frictional, QP is quasi-plastic; inset on right shows trend of strength as a function of depth. See text for further discussion.

## A brief discussion of nomenclature and classification

Nomenclature associated with fault rocks remains controversial (Tullis and others, 1982; Schmid and Handy, 1991) despite many attempts at systematic classifications (e.g., Spry, 1969; Higgins, 1971; Zeck, 1974; Sibson, 1977; White, 1982; Wise and others, 1984; Heitzmann, 1985). A problem that has plagued virtually all classification schemes is the use of genetic terms in describing or naming fault rocks when the deformation mechanisms that produced the texture and fabric of the rocks are poorly known or incorrectly inferred. On the other hand, devising a completely nongenetic classification has proven difficult (Schmid and Handy, 1991). We will return to this issue of fault-rock nomenclature in a later part of this overview, but first we present the historic usage of several important terms commonly associated with fault rocks. It should be noted that most of the terms and concepts were based on observations of quartzo-feldspathic rocks.

The term *mylonite* was introduced by Lapworth (1885) to describe rocks found within the Moine thrust zone (specifically at Arnaboll Hill, Eriboll) in northwest Scotland. Lapworth described the field setting and overall characteristics of mylonite as follows (1885, p. 559):

> The most intense mechanical metamorphism occurs along the grand dislocation (thrust) planes, where the gneisses and pegmatites resting on those planes are crushed, dragged, and ground out into a finely-laminated schist (*Mylonite*, Gr. *mylon*, a mill) composed of shattered fragments of the original crystals of the rock set in a cement of secondary quartz, the lamination being defined by minute inosculating lines (fluxion lines) of kaolin or chloritic material and secondary crystals of mica. Whatever rock rests immediately upon the thrust-plane, whether Archæan, igneous, or Palæozoic, &c., is similarly treated, the resulting mylonite varying in colour and composition according to the material from which it is formed. The variegated schists which form the transitional

zones between the Arnaboll gneiss and Sutherland mica-schists are all essentially mylonites in origin and structure, and appear to have been formed along many dislocation planes, some of which still show between them patches of recognisable Archæan and Palæozoic rocks. These variegated schists (Phyllites or Mylonites) differ locally in composition according to the material from which they have been derived, and in petrological character according to the special physical accidents to which they have been subjected since their date of origin—forming frilled schists, veined schists, glazed schists, &c., &c. The more highly crystalline flaggy mica-schists, &c., which lie generally to the east of the zones of the variegated schists, appear to have been made out of similar materials to those of the variegated schists, but to have been formed under somewhat different conditions. They show the fluxion-structure of the mylonites; but the differential motion of the component particles seems to have been less, while the chemical change was much greater. In some of these crystalline schists (the augen-schists) the larger crystals of the original rock from which the schist was formed, are still individually recognisable, while the new matrix containing them is a secondary crystalline matrix of quartz and mica arranged in the fluxion-planes. While the *mylonites* may be described as microscopic pressure-breccias with fluxion structure, in which the interstitial dusty, siliceous, and kaolinitic paste has only crystallised in part; the *augen-schists* are pressure-breccias, with fluxion-structure, in which the whole of the interstitial paste has crystallised out. The *mylonites* were formed along the thrust-planes, where the two superposed rock-systems moved over each other as solid masses; the *augen-schists* were probably formed in the more central parts of the moving system, where the all-surrounding weight and pressure forced the rock to yield somewhat like a plastic body. Between these augen-schists there appears to be every gradation, on the one hand to the mylonites, and on the other to the typical mica-schists composed of quartz and mica. Like the mylonites, the crystalline augenites and micalites present us with local differences in chemical composition (calcareous, hornblendic, quartzose, &c.), suggestive of Archæan, igneous, or Palæozoic origin. They also show similar structural varieties due to secondary physical changes (frilled, veined, glazed, &c.), as well as others due to the presence of special minerals (garnet, actinolite, &c., &c.).

Although Lapworth did not include illustrations with his initial description of mylonite, Teall provided an informative set of photomicrographs of mylonitic rocks from the Moine thrust zone, including an example of Lapworth's type mylonite (Teall's plate 1, figure 1). Teall wrote the following in regard to this sample (1918, p. 2):

> During our first day's excursion in the Eribol district he [Lapworth] led us to Arnabol Hill, showed us Archæan gneiss resting almost horizontally on Cambrian quartzite, explained that this was due to an overthrust fault, and, pointing to the lower portion of the gneiss, used words to this effect: If you want to study the microscopic structure of rolled out gneiss take a specimen from there. That was my first introduction to a special type of dynamic metamorphism, and here is the specimen I collected on that occasion.
>
> It is a compact, cherty-looking rock, showing a streaky aspect on a joint surface. It possesses no marked cleavage, but has a tendency to break along certain planes which have a faint silky appearance suggestive of slickensides. Looked at simply as a hand specimen, and in the light of the knowledge previously available, it would have been difficult if not impossible to determine its mode of origin. But the field evidence settled this beyond a doubt, for it could be traced, in a few feet, into rocks which could be identified as portions of the Lewisian gneiss. This is the kind of rock for which Prof. Lapworth subsequently proposed the term *mylonite*. Under the microscope . . . the streaky structure becomes still more marked. Colourless lenticles are seen to be separated by sinuous lines of darker material. Small grains of felspar may sometimes be recognised in these lenticles, which consist for the most part of irresolvable crypto-crystalline material. The dark streaks are also irresolvable.

At the type locality of mylonite, nonfoliated structureless fault rocks are scarce (White, 1982), and were not recognized by Lapworth (1885) as part of the fault-rock suite at Arnabol Hill, Eriboll. Subsequently, Grubenmann and Niggli (1924) introduced the term *cataclasite* which they described as an "aphanitic, structureless rock" (White, 1982). Although mylonite and cataclasite were originally coined as distinct microstructural terms based on field and thin section examination, their distinctiveness became blurred when the term *cataclastic rocks* was introduced by Waters and Campbell (1935) as a collective name for all rocks of the gouge–breccia–cataclasite–mylonite kindred. This collective association led to the misleading implication that such rocks have developed chiefly by frictional processes or cataclasis. The pervasiveness of this misconception among geologists is perhaps best demonstrated in Higgins' (1971) extensive review paper entitled "Cataclastic Rocks." For example, Higgins defines mylonite as follows (1971, p. 7):

> A coherent microscopic pressure-breccia with fluxion structure that may be megascopic or visible only in thin section and with porphyroclasts generally larger than 0.2 mm. These porphyroclasts make up from about 10 to about 50 percent of the rock. Mylonites generally show recrystallization and even new mineral formation (neomineralization) to a limited degree, but the dominant texture is cataclastic.

Although Lapworth's (1885) original definition of mylonite may seem to focus on brittle deformational aspects because he used terms such as "crushed," "ground out," and "shattered," a complete reading of Lapworth's article as well as Teall's (1918) description, as pointed out by White (1982), indicates that crystal-plastic processes were also recognized as essential in mylonite formation. Teall clearly recognized the difference in behavior of quartz and feldspar (1918, p. 3): plagioclase "illustrate[s] the cataclastic effects in a very beautiful manner," whereas quartz "yields more readily to the deforming stresses." Christie (1960, p. 86; 1963) recognized the importance of neomineralization in the development of mylonites in association with the Moine thrust zone in the Assynt region. He referred to such mylonitic rocks as *blastomylonites* and suggested that they graded into the regionally metamorphosed Moine schists. He referred to these mylonitic rocks as "primary mylonitic rocks" and argued that they were formed during the progressive metamorphism of the Moine schists in a discrete "movement-zone" associated with the Moine thrust. Christie (1960, 1963) also recognized another group of fault rocks, which he referred to as "secondary mylonitic rocks"; these fault rocks were clearly cataclastic and he reasoned that they were related to a separate, later phase of brittle deformation associated with the thrust zone.

In spite of the recognition of crystal-plastic deformation processes and neomineralization in mylonites of the Moine thrust zone, for about fifty years after Teall (1918), the brittle (cataclastic) aspects of the mylonite-generating process were generally emphasized (e.g., Hsu, 1955; Spry, 1969, table 8; Higgins, 1971, table 1). This was due in part to the lack of success in producing crystal-plastic deformation of silicates in the laboratory. However, Carter and others (1964) included a few pictures (e.g., their plate 9) of quartzites experimentally deformed at high pressures and temperatures and compared the textures in these samples to naturally deformed mylonitic quartzites from the Moine thrust zone. The authors concluded that analogous textures in both the experimentally and naturally deformed quartzites originated by recrystallization rather than by "crushing" or "granulation." Somewhat later, Tullis and others (1973) experimentally deformed quartzites over a wide range of conditions within the crystal-plastic field; their photomicrographs show flattened original grains progressively replaced with increasing strain and temperature by small, new syntectonically recrystallized grains. In that same year, Bell and Etheridge (1973) and White (1973) recognized that syntectonic recrystallization during relatively high-temperature ductile deformation was responsible for the grain-size reduction in many natural mylonites.

Although this revised view of mylonite genesis was not immediately accepted by the geologic community at large, an indication of change in the prevailing view on mylonite petrogenesis among structural geologists was reflected in the widely used textbook by Hobbs, Means, and Williams (1976, p. 303):

> Blastomylonites have not necessarily suffered any of the brittle fragmentation or "milling" envisioned for ordinary microbreccias. Their fine-grain size and distinctive microstructure may be due entirely to ductile deformation accompanied by recrystallization [Bell and Etheridge (1973)].

Significant progress in promoting a better understanding of mylonite petrogenesis among geologists was achieved in 1981, when a Geological Society of America Penrose Conference entitled "Significance and Petrogenesis of Mylonitic Rocks" was convened by Arthur W. Snoke, Jan Tullis, and Victoria R. Todd (see Tullis and others, 1982). This conference brought together structural geologists, petrologists, field geologists, experimentalists, and theoreticians from Australia and Europe as well as North America. The meeting generated significant interest in mylonitic rocks as well as in faults and shear zones in general, and stimulated fruitful collaborations among scientists of diverse backgrounds. One of the goals of this international conference was to clarify the terminology of mylonites. Although no specific definition of mylonite emerged from the meeting, most conference participants agreed that three features were usually implied: (1) grain-size reduction; (2) occurrence in a tabular zone, although the zone could be as wide as tens of kilometers; and (3) enhanced foliation (fluxion structure) and/or lineation.

After the publication of the report that summarized the highlights of the Penrose Conference on mylonitic rocks (Tullis and others, 1982), Wise and others (1984) proposed a broad framework for the terminology of fault-related rocks. This framework in part focused on a rate-of-strain versus rate-of-recovery diagram (Figure O.3). These authors argued that (p. 391):

> Competition between rate of strain and rate of recovery/ recrystallization is a major determinant of the texture of fault-related rocks, the rates being functions of composition, grain size, temperature, fluids, and the stress field.

The basis of Wise and others' (1984) framework for fault-rock terminology requires further explanation. Recovery is the term for the processes that reduce the stored strain energy associated with dislocations in material that has deformed by crystal plasticity; it involves the development of subgrains and/or recrystallized grains. This type of recovery does not occur at all in a rock undergoing microcracking, frictional sliding, or pressure solution (although frictional melting and precipitation from fluids could be considered processes of "recovery" of intact rock properties for materials undergoing brittle deformation). Recovery is a necessary phenomenon in minerals undergoing dislocation creep, and its rate is chiefly temperature dependent but is also influenced by grain size and presence of fluids. Strain, on the other hand, can occur by any of the previously mentioned mechanisms, as well as by dislocation glide. Thus, the classification of Wise and others (1984) compares the rate of strain by any deformation mechanism to the rate of recovery during dislocation creep.

Wise and others also proposed a formal definition of mylonite that states (1984, p. 393):

> *mylonite* is a general term for *coherent rocks with at least microscopic foliation, with or without porphyroclasts, characterized by intense syntectonic crystal-plastic grain-size reduction of the country rock to an average diameter less than 50 microns (0.5 mm* [this should have read 0.05 mm]) *and invariably showing at least minor syntectonic recovery/recrystallization.*

An important aspect of this revised definition of mylonite is that all grain-size reduction is attributed to crystal-plastic deformation (i.e., dynamic recrystallization). However, some recent studies of mylonites, particularly those employing transmission electron microscopy (TEM), have demonstrated that grain-size reduction in some polymineralic mylonitic rocks involves cataclasis of strong minerals (e.g., amphibole or feldspar at greenschist- to lower amphibolite-facies conditions)

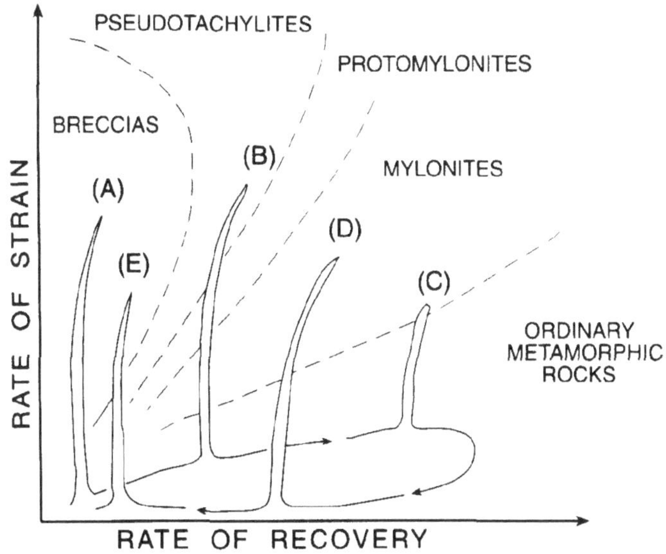


**Figure 0.3.** Wise and others' rate-of-recovery versus rate-of-strain diagram summarizing the terminology of fault-related rocks (1984, reproduced with permission of the publisher, the Geological Society of America, Boulder, Colorado U S A Copyright © 1984) A-B-C-D-E represent possible high-strain-rate events on a general P-T-t path See text for discussion

and synchronous, crystal-plastic deformation and resultant dynamic recrystallization of weaker minerals (e.g., quartz) (Simpson, 1985; Nyman and others, 1992). Thus, although the Wise and others' (1984) definition reflects a recognition of the importance of crystal-plastic deformation processes in the textural evolution of mylonites, it neglects the role of cataclasis in the grain-size reduction of some minerals during mylonitization. Furthermore, various chemical reactions involving intergranular fluids also play a key role in the textural evolution of many mylonites (e.g., Knipe and McCaig, 1994), and these reactions are commonly essential for the strain softening and textural/mineralogical evolution of polymineralic mylonites.

Subsequently, another attempt to define mylonite was initiated at a 1984 meeting of the Canadian Tectonics Group. Based upon replies to a questionnaire that asked the question: "What is a mylonite?," Mawer devised the following definition (1986):

> Mylonites occur in shear zones that is, zones across which one block of rock is displaced with respect to another, and which show no evidence of major loss of material continuity. They develop primarily by ductile deformation processes. They are well-foliated and commonly well-lineated, and show an overall reduction in grain size when compared to their hosts. Mylonites commonly contain a suite of diagnostic asymmetric structures, such as rotated porphyroclasts/blasts and S-C textures. Mylonite zones can be developed at any scale, in rocks of any original composition and fabric.

The definition of mylonite offered by Mawer (1986) reflects the relatively recent recognition that composite foliations as well as various other kinematic indicators are common features of mylonitic rocks. The widely cited paper by Ramsay and Graham (1970) is often considered the beginning of modern analysis of the kinematic evolution of shear zones. In the same year, a paper by Eisbacher (1970) presented examples of microstructural features (especially "mica fish") that could be used to determine the sense-of-shear in mylonitic rocks. This paper was apparently before its "time," because it was generally overlooked for over a decade until it was rediscovered in the early 1980s. Between 1979 and 1986, a series of papers by various authors laid the groundwork for kinematic analysis of ductile shear zones based on mesoscopic and microstructural observations (e.g., Berthé and others, 1979a,b; Platt and Vissers, 1980; Bouchez and others, 1983; Simpson and Schmid, 1983; Lister and Snoke, 1984; Passchier and Simpson, 1986). Excellent summaries and critiques of shear-sense indicators have been compiled by Hanmer and Passchier (1991) and Passchier and Trouw (1996).

The preceding comments on the usage of "mylonite" in the geological literature indicate that grain-size reduction commonly has been recognized as a fundamental aspect of the mylonitization process. However, deformation at high homologous temperatures can promote grain growth via grain-boundary migration recrystallization when grain boundaries are not pinned by other phases. Thus, "mylonites" in zones of concentrated shear strain in some monomineralic materials such as glacial ice (Hudleston, 1980) or rock salt (Talbot, 1981; Urai and others, 1986) may actually increase in grain size during shear strain (see also Schmid and Handy, 1991). The products of recrystallization and grain growth that accompany mylonitization in high-temperature, polymineralic mylonitic rocks are sometimes called *mylonitic gneiss* or *blastomylonite*. It is important to note that high-temperature mylonites may exhibit little evidence of crystal-plastic strain at the grain scale, e.g., the grains are polygonal with no undulatory extinction, although they may show abundant mesoscopic evidence of high strain, e.g., strongly developed foliation and lineation, kinematic indicators, and sheath folds. It can be difficult to determine whether true static annealing has followed a high-temperature deformation.

The nonfoliated character of cataclasites has served as a major distinguishing characteristic in virtually all modern classification schemes for brittle fault rocks (e.g., Spry, 1969; Higgins, 1971; Sibson, 1977; Wise and others, 1984; Heitzmann, 1985). However, Chester and others (1985) have demonstrated that cataclasis associated with brittle, upper crustal faulting can produce well-foliated fault gouge. According to Chester and others (1985, p. 143), microstructures within foliated gouges associated with the Punchbowl fault, Los Angeles County, California "show that cataclastic mechanisms of fracture, rigid-rotation of grain fragments, and frictional sliding along grain boundaries were dominant" in the textural development of these fault rocks. Textural and compositional layering has also been produced in experimentally deformed polymineralic, simulated fault gouge (examples cited in Chester and others, 1985) as well as in feldspar-rich rocks (Tullis and Yund, 1987). Foliated cataclasites, there-

**Table O.1.** Textural classifications of fault rocks

<table>
<tr><td colspan="3"></td><td>Random fabric</td><td colspan="2"></td><td>Foliated</td><td></td><td></td></tr>
<tr><td rowspan="2">Incohesive</td><td colspan="3">Fault breccia<br>(visible fragments > 30% of rock mass)</td><td colspan="5">?</td></tr>
<tr><td colspan="3">Fault gouge<br>(visible fragments < 30% of rock mass)</td><td colspan="5">Foliated gouge</td></tr>
<tr><td rowspan="6">Cohesive</td><td rowspan="6">Nature of matrix</td><td>Glass–devitrified glass</td><td colspan="2">Pseudotachylyte</td><td colspan="3">?</td><td></td></tr>
<tr><td rowspan="4">Tectonic reduction in grain size dominates grain growth by recrystallization and neomineralization</td><td colspan="4">Crush breccia (fragments > 0.5 cm)<br>Fine crush breccia (0.1 < fragments < 0.5 cm)<br>Crush microbreccia (fragments < 0.1 cm)</td><td>0-10</td><td rowspan="4">Percent of matrix</td></tr>
<tr><td rowspan="3">Cataclasite series</td><td>Protocataclasite</td><td rowspan="3">Mylonite series</td><td colspan="2">Protomylonite</td><td>10-50</td></tr>
<tr><td>Cataclasite</td><td>Mylonite</td><td rowspan="2">Phyllonite varieties</td><td>50-90</td></tr>
<tr><td>Ultracataclasite</td><td>Ultramylonite</td><td>90-100</td></tr>
<tr><td>Grain growth pronounced</td><td colspan="2">?</td><td colspan="4">Blastomylonite</td></tr>
</table>

*Note* Sibson's (1977) textural classification of fault rocks as modified in Scholz (1990, table 3.1), reproduced with permission of the publisher, Cambridge University Press. Copyright © 1990. The only significant departure from the original Sibson (1977) classification is the addition of *foliated gouge*, which has been produced in laboratory experiments and recognized in natural fault zones (Chester and others, 1985). This important addition implies that both random- and foliated-fabric fault rocks can occur in the upper crustal, brittle frictional regime, whereas Sibson (1977) used fabric to distinguish fault rocks of the frictional regime (random fabric) from those of the quasi-plastic regime (foliated fabric).

fore, are a rock type that must be included in any fault-rocks classification.

When Sibson (1977) proposed his widely cited classification (see Table O.1 for a modified version after Scholz, 1990, table 3.1), the presence or absence of foliation was considered a useful criterion to distinguish mylonitic rocks from cataclastic rocks. However, with the demonstration by Chester and others (1985) that foliation can develop during cataclasis at shallow crustal depths, it is the mechanism of foliation development that now distinguishes mylonitic and cataclastic rocks rather than its presence or absence (Schmid and Handy, 1991).

In a significant departure from traditional attempts at establishing a systematic, "nongenetic" classification of fault rocks, Schmid and Handy (1991) proposed a genetic classification of fault rocks into cataclastic and mylonitic rocks (their table 16.2). These authors argue that all previous classifications are *not* truly nongenetic, because they incorporate obvious or hidden genetic connotations. The transition from frictional to viscous flow is the fundamental criterion upon which Schmid and Handy (1991) based their subdivision of fault rocks. Such a classification requires identification of the mineral(s) and deformation mechanisms that accommodated most of the strain in the rock and thus controlled the rheology of the rock. This classification would be difficult to apply at the hand sample level during geologic mapping or other field-based studies, although it may be practical and useful at the microscopic level (see table 16.3 in Schmid and Handy, 1991, for petrographic criteria).

In summary, earth scientists undoubtedly will continue to study fault rocks using diverse approaches (mesoscopic to submicroscopic scale) and with a great variety of research objectives. Any systematic classification of fault rocks is destined to be inadequate in some respect because of variations in the scale of specific studies. At one extreme is the field geologist, who is attempting to develop useful mesoscopic descriptions of fault rocks in his/her study area. On the other hand, an experimentalist must consider the role of various deformation mechanisms at a submicroscopic level, commonly with the aid of transmission electron microscopy. Such contrasting levels of examination of fault rocks are not amenable to the development of a single, all-purpose classification. In view of this reality, we suggest that a modified version of Sibson's (1977) classification, as employed by Scholz (1990, table 3.1; Table O.1), is probably the best compromise at present and is particularly useful for the field geologist. In regard to this classification, the random-fabric cataclasite series (protocataclasite-cataclasite-ultracataclasite) and the foliated mylonite series (protomylonite-mylonite-ultramylonite) can be viewed as end-member series, and gradational transitions occur between them (R. H. Sibson, personal comm., 1996). If, during a research project, microstructural analysis is undertaken, then the classification scheme of Schmid and Handy (1991) might be more useful.

We conclude our comments on fault-rock nomenclature and classification with a brief discussion of *pseudotachylyte*—a rock with a controversial past and the subject of much recent interest (e.g., Spray, 1995). In this *Atlas*, we have employed the spelling as originally used by Shand (1916). He used "pseudotachylyte" for branching, glassy, dike-like rocks occurring within granitic rocks of the Vredefort structure as exposed in the Parijs region of South Africa and favored a melt origin for these rocks. However, there has been controversy about the role of frictional melting (e.g., Sibson, 1975) versus cataclasis (e.g., Wenk, 1978) in the origin of pseudotachylyte. Clearly, part of this controversy has resulted from the broad and improper use of "pseudotachylyte" for rocks that superficially resemble it or are intimately associated with it (e.g., ultramylonite or ultracataclasite), but whose origin involved solid-state deformation mechanisms. However, perhaps even more significant in regard to this discussion are the results of recent high-speed slip experiments that indicate that comminution is an essential precursor to melting by friction (Spray, 1995). In this light, the fracture or fusion controversy is thus a nonproblem, because as stated by Spray (1995, p. 1122): "pseudotachylyte is the result of both fracture *and* fusion."

Magloughlin and Spray summarized the characteristics of pseudotachylyte as (1992, p. 198):

> a particular assemblage of mesoscale and microscale characteristics, including typically dark, aphanitic veins showing intrusive relations, sharp boundaries, included clasts and crystals of the host rock and, critically, an association with fault or shear zones. The veins may contain glass, microlites, spherulites, vesicles, amygdules and embayed lithic fragments and show chilled margins.

An important aspect of pseudotachylyte is that it is widely acknowledged as providing evidence of seismicity along ancient, now exhumed, faults. Pseudotachylyte is thus one of the few criteria for the seismic nature of ancient faults, although it also has been found in some landslides (Masch and others, 1985) and impact structures (e.g., Wilshire, 1971; Thompson and Spray, 1994). It should be noted that the pseudotachylyte at Shand's (1916) type locality is now equated with the 2.0 Ga Vredefort impact event (Martini, 1992; Spray and others, 1995). Some workers attribute impact-related pseudotachylyte formation to shock melting (e.g., Reimold and Colliston, 1994), whereas others believe it is primarily the product of frictional melting incurred during gravitational collapse of the impact-generated transient cavity, and is thus formed by the same mechanism as endogenic examples (Spray and Thompson, 1995). Regardless of origin, all pseudotachylytes are high-strain-rate features. Most pseudotachylyte developed in the upper crust in massive rocks with low porosity such as granitoids, gabbroic rocks, or high-grade gneisses. An interesting occurrence of eclogite-facies pseudotachylytes, apparently developed at very deep crustal levels, has been described recently from the Bergen Arcs of western Norway (Austrheim and Boundy, 1994). This locality provides important geologic evidence of brittle faulting, frictional melting, and presumed seismic activity at a depth of 60 km or more in an ancient collision zone.

Some pseudotachylytes are associated with fault rocks that have formed dominantly via crystal-plastic deformation processes, such as ultramylonite. Passchier (1982a,b) described variable pseudotachylyte-ultramylonite relationships in the Saint-Barthélemy massif, French Pyrenees, where ultramylonite is localized in and overprints pseudotachylyte layers, whereas younger pseudotachylyte veins at the same locality indicate recurrent brittle faulting and frictional melting. Such cyclic pseudotachylyte-ultramylonite formation suggests periodic, seismogenic brittle faulting during a continuum of steady-state flow (e.g., Sibson, 1980; Hobbs and others, 1986; John and Howard, 1995). Pseudotachylyte thus is an excellent example of how the careful interpretation of fault rocks can provide unique information about rheology, strain rate, fluid activity, and paleoseismicity in exhumed, ancient fault and shear zones.

### Deformation mechanisms

As implied in the previous section, a key problem with classification schemes for fault rocks is inferring the processes responsible for the textural characteristics visible in a hand sample or a thin section. Historically, this involved a lack of understanding of the specific processes by which ductile deformation was achieved. More recently, our knowledge of possible deformation mechanisms has improved greatly, in part by analogy with metals, in part by development of apparatus and techniques capable of inducing ductile deformation in refractory silicate minerals within relatively short laboratory times, and in part by the use of TEM to identify the types of submicroscopic defects involved in the deformation.

There are relatively few grain-scale deformation mechanisms that combine to allow the large-scale deformation of heterogeneous rock masses (see reviews by Schmid, 1983; Knipe, 1989; Williams and others, 1994). These deformation mechanisms include cracking and frictional sliding, pressure solution, dislocation creep, (dry) diffusion creep, and grain-boundary sliding. Each of these mechanisms is characterized by a distinct form of flow law, a characteristic set of microstructures, and a specific range of pressure–temperature–fluid conditions over which it is dominant.

Deformation may involve steady-state flow (e.g., the flow stress remains constant with increasing strain), or it may involve strain hardening or strain weakening (e.g., the stress increases or decreases, respectively, with increasing strain). The flow law for a given deformation mechanism relates the strain rate to steady-state flow stress, temperature, and other material parameters. For some deformation mechanisms, the form of the flow law can be derived theoretically (e.g., diffusion creep), whereas for others it has been determined empirically (e.g., dislocation creep). For different minerals deforming by the same mechanism, the material constants will differ and thus the flow strengths will differ for a given temperature and strain rate. Natural deformation commonly involves polymineralic aggregates, several different deformation mechanisms, and non-steady-state flow. However, experimental studies of steady-state flow by single deformation mechanisms in monomineralic aggregates have provided information that is useful for interpreting the microstructures of naturally deformed rocks. In addition, at present only steady-state flow laws can be extrapolated from experimental to natural conditions; such flow laws at least provide constraints on the stress levels in natural deformations.

For a given mineral, the fields of dominance for different deformation mechanisms over a range of crustal conditions are well illustrated on a *deformation-mechanism map* (Figure O.4) having axes of temperature (commonly normalized by the melting temperature) and flow stress (commonly normalized by the shear modulus). Boundaries separating fields of dominance for the different deformation mechanisms are obtained by finding the temperature-stress conditions where the flow laws for two mechanisms yield equal strain rates. Away from these boundaries, a single mechanism contributes dominantly to the aggregate strain rate. The location of the mechanism field boundaries will obviously vary for different minerals; for example, feldspar requires a significantly higher temperature than quartz to activate dislocation creep. In addition, because the flow laws for some deformation mechanisms depend on grain size and fluids, the location of the mechanism field boundaries may change with changing grain size or fluid activity. Deformation-mechanism maps can be contoured for strain rate; this is helpful for showing that certain mechanisms (e.g., dislocation creep) can be activated both at fast experimental strain rates as well as at slower geologic strain rates (Figure 0.4).

Laboratory experiments have provided information about the microstructures produced by the operation of most of these deformation mechanisms and how they evolve with progressive strain for different minerals over a range of conditions (e.g., Tullis, 1990). Experiments must be done at faster strain rates and higher temperatures than natural deformation, but similarities in experimental and natural microstructures can be used to verify that the processes actually are the same. Once this verification has been made, the microstructures of naturally deformed rocks can be used to provide useful information about the conditions and rheology of the deformation.

One difficulty with interpreting the microstructures in naturally deformed rocks is that the external conditions of deformation (e.g., temperature, strain rate, and fluid availability)

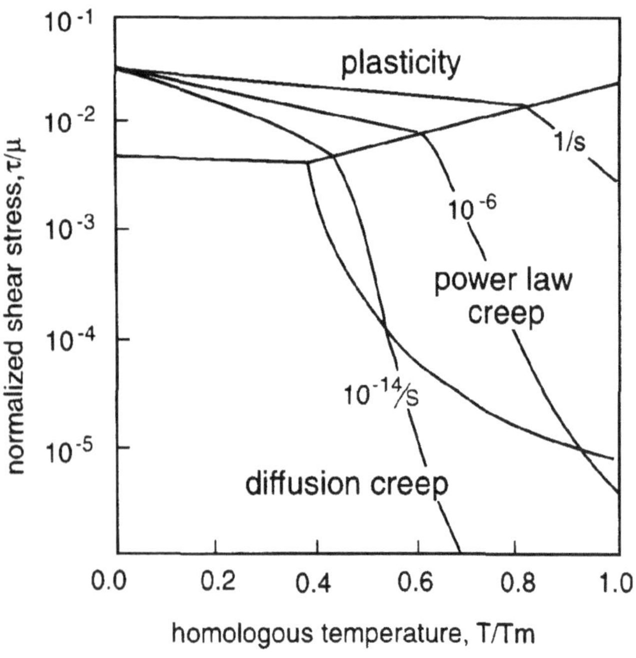


**Figure 0.4.** Schematic deformation-mechanism map, with axes of shear stress normalized by shear modulus and temperature normalized by melting temperature, adapted from maps for olivine by Frost and Ashby (1982) for grain size of ~50 mm Fields of dominance for plasticity, power-law creep, and diffusion creep are shown, together with an experimental strain-rate contour ($10^{-6}s^{-1}$) and a natural strain-rate contour ($10^{-14}s^{-1}$) From Tullis (1990, reproduced with permission of the publisher, Unwin Hyman Copyright © 1990)

almost certainly changed over time. Similarly, some of the internal material parameters (e.g., grain size, mineralogy, and arrangement of the various phases) may have changed with progressive strain. Thus, the deformation microstructures that are preserved in a rock largely reflect the latest increment, and it may be difficult to say what portion of the total rock strain they represent. Experimental studies are needed to determine the kinetics of response of deformation microstructures to changing conditions. However, useful information about the deformation history of a polymineralic rock can be obtained from observations of coexisting deformation microstructures in different phases, as well as metamorphic reactions, fluid inclusions, and overprinting relationships.

Whether or not deformation microstructures give an accurate idea of bulk rock (or mineral) strain depends on the rock (mineral) composition and the dominant deformation mechanism. In general, for rocks deformed at lower temperatures, the microstructures are more indicative of the rock strain. However, in rocks deformed at higher temperatures, the rate of recovery keeps pace with the rate of deformation, so that the rock may appear statically "annealed" or unstrained. This was one of the early difficulties with the terminology for mylonites. The original mylonite was a moderate-grade quartzo-feldspathic rock in which the feldspars were semi-brittle and the quartz deformed by climb-accommodated dislocation creep. A pure quartzite deformed at the same, or at higher, temperature might not appear to be as highly strained, at least on the scale of a thin section, because it would be completely recrystallized to an aggregate of polygonal grains. However, one indication of high strain would be provided by a strong crystallographic preferred orientation (CPO).

By definition, rocks deformed in fault and shear zones have undergone greater strain than the surrounding rocks, so an important question concerns the process that caused the localization of strain. It is important to assess the deformation mechanisms that operated in the rocks of fault and shear zones in order to determine the reasons for strain localization as well as the strength of the zone once it has formed. Was strain localized along some original inhomogeneity in rock composition, or in a zone characterized by higher fluid content? Or was the zone initiated due to an instability in the bulk rock deformation? Because the rocks in ductile shear zones are characterized by finer grain size than that in the host rocks, it is commonly assumed that grain-size reduction has induced strain weakening. In the following sections we briefly review the important grain-scale deformation mechanisms that may contribute to fault- and shear-zone rocks, with a particular emphasis on the microstructures they produce and on whether they are capable of producing strain weakening.

### *Cracking and frictional sliding*

Faults characterized by brittle or semi-brittle deformation form in the upper 10–15 km of the crust where (1) pressures are not high enough to inhibit crack opening, (2) temperatures are not high enough for thermally activated processes to be competitive, and (3) most rocks already have an abundance of microcracks, voids, and other flaws. Many faults tend to follow a preexisting weakness (although some important exceptions have been noted, e.g., Wernicke and others, 1985). Rather than forming by the propagation of a single sharp shear fracture, faults form by the propagation of a process zone of high crack density behind which there is coalescence into a through-going fault (e.g., Hallbauer and others, 1973; Blenkinsop and Rutter, 1986; Lloyd and Knipe, 1992; Moore and Lockner, 1995).

Once formed, faults undergo frictional sliding. This involves adhesive and abrasive wear of the wall rocks, which eventually produces a zone of comminuted material. At shallow depths, this material consists of incohesive fine-grained fault gouge, whereas at somewhat greater depths, cohesive cataclasite is produced. Both gouge and cataclasite are characterized by angular rock fragments in a wide range of sizes; some amorphous material is produced even after small offsets on experimental faults (Yund and others, 1990). Within experimental fault gouge, localized slip surfaces ("Riedel shears") commonly develop both parallel and oblique to the fault plane (e.g., Mandl and others, 1977; Logan and others, 1992), and some evidence of these features has been found in natural fault zones (Chester and others, 1993). Such slip surfaces can be used as shear-sense indicators (e.g., Petit, 1987).

Experimental studies have shown that frictional sliding is velocity dependent (e.g., Dieterich, 1979). If the resistance to sliding decreases with increasing velocity, then the behavior is termed velocity weakening, and it is likely to produce earthquake instabilities. If the resistance to sliding increases with increasing velocity, the behavior is termed velocity strengthening, and sliding is stable (aseismic). The processes responsible for this switch in frictional behavior remain poorly understood, but it appears that increasing temperature promotes a transition from velocity weakening to velocity strengthening (e.g., Shimamoto, 1989). Thus the seismic-aseismic transition on faults such as the San Andreas may correlate with a change in the frictional properties (e.g., Tse and Rice, 1986), rather than indicating a transition to crystal plasticity.

The dilatancy associated with brittle deformation allows fluid influx, and fluids are of crucial importance to fault behavior. Fluids may weaken a fault zone in at least three ways: mechanically, by reducing the effective stress (e.g., Rice, 1992), or chemically, by causing stress corrosion of crack tips (e.g., Atkinson, 1984; Kerrich, 1986) and/or allowing reaction of the finely comminuted material with water to form weaker hydrous phases (e.g., Mitra, 1978). Fluids may indirectly strengthen a fault zone by precipitating quartz or calcite (e.g., Fredrich and Evans, 1992). Fluids are directly related to the earthquake cycle on faults (e.g., Sibson, 1990); during interseismic periods precipitation causes decreasing permeabilty and increasing fluid pressure, leading eventually to rupture, which increases the permeability and decreases the fluid pressure. Evidence for these cycles is preserved in vein systems and fluid inclusions (e.g., Cox, 1995; Bouillier and Robert, 1992).

Some segments of active faults tend to undergo creep, producing only small earthquakes, whereas other portions become locked and eventually produce large earthquakes. At present there are no good microstructural criteria for distinguishing between cataclasites that have resulted from "slow" or "fast" fault slip, and pseudotachylytes remain the only reliable criterion for seismicity on ancient faults (e.g., Sibson, 1975; Spray, 1995). The coefficient of friction for most rocks is remarkably constant (Byerlee, 1978); however, a few rocks, especially those that contain substantial amounts of montmorillonite and chrysotile serpentine, have a much lower coefficient, and these rock types may contribute to the existence of stably creeping regions (e.g., Reinen and others, 1991). Recently, it was recognized that some active major faults, such as the San Andreas fault, are anomalously weak; that is, slip occurs almost normal to the direction of maximum compression (e.g., Zoback and others, 1987). Most hypotheses to explain this involve high fluid pressure within the fault zone (e.g., Rice, 1992; Sleep and Blanpied, 1992). It remains to be determined whether there are microstructures that correlate with this weakness that, if found in ancient faults, would allow inference of similar behavior (e.g., Evans and Chester, 1995).

### *Pressure solution*

Pressure solution is an important deformation mechanism at shallow to intermediate depths in the crust. It involves (1) the dissolution of material at relatively high-stress sites in a deforming aggregate, (2) diffusive transfer of the solutes through an interconnected grain-boundary fluid down a chemical potential gradient, and (3) precipitation of these materials at lower-stress sites (e.g., Robin, 1978; Rutter, 1983). Diffusive transfer may occur on the grain scale or over considerable distances; in the latter case, pressure solution may accomplish significant volume loss (e.g., Wright and Platt, 1982). The simplest case involves dissolution and reprecipitation of a single phase, such as quartz, but it is possible for one phase to dissolve and its components to precipitate at different sites, singly or in combination with other components, as new minerals ("incongruent pressure solution"; Beach, 1979). Thus, pressure solution is intimately related to metamorphism.

There is abundant microstructural evidence for the operation of pressure solution in rocks that were naturally deformed at moderately shallow levels of the crust, e.g., ~200–300°C. These include dissolution features (dissolution seams and stylolites, truncated grains and fossils, and pitted pebbles) as well as features indicative of precipitation (fibers precipitated in pressure shadows and in extensional veins). Because the precipitated material is added progressively during deformation, it records information about the strain path, indicating whether the deformation accumulated coaxially or noncoaxially (e.g., Durney and Ramsay, 1973; Beutner and Diegel, 1985). It is likely that pressure solution may also contribute substantially to deformation at somewhat higher temperatures, but that the microstructural evidence for it is destroyed by the concurrent operation of other processes such as dislocation creep.

The efficiency of pressure solution depends on the availability of relatively soluble minerals, such as calcite and quartz. It also depends on grain size: smaller grains require shorter diffusion distances. Furthermore, the existence of interconnected fluid channels within the aggregate to provide fast diffusivity pathways is critical for the mechanism's efficiency. There is abundant evidence that interphase boundaries are effectively wider and/or "wetter" than grain boundaries between like phases. In naturally deformed rocks, pressure solution is more extensive in micaceous quartz aggregates than in pure quartz aggregates (e.g., Houseknecht, 1988; Hippertt, 1994d). Recent experiments have shown that the addition of clay minerals along grain boundaries produces higher rates of compaction creep (e.g., Dewers and Ortoleva, 1991; Hickman and Evans, 1995). Differential stress is another factor which may increase the interconnectedness of grain-boundary fluids, either melts (e.g., Bulau and others, 1979; Jin and others, 1994) or aqueous fluids (e.g., Tullis and others, 1996), and thus significantly increase the rate of pressure solution.

Pressure solution has not generally been considered as a mechanism that may contribute to strain localization in the formation of faults or ductile shear zones. However, pressure solution is probably enhanced within the deeper parts of fault zones, where cataclasis decreases the grain size and porosity allows fluid influx. The operation of pressure solution will further modify the properties of faults and ductile shear zones: material dissolved at greater depth may travel upward and be precipitated at higher levels within fault zones, helping to heal and strengthen them (Fredrich and Evans, 1992). Within ductile shear zones, the dissolution of more soluble minerals may change the composition of the shear-zone material over time (e.g., Power and Tullis, 1989), causing either weakening or strengthening (e.g., O'Hara, 1988).

### *Mechanical twinning*

Mechanical twinning is a common form of low-temperature crystal plasticity that occurs in minerals such as calcite, plagioclase, and pyroxene. It involves a specific crystallographic transformation accomplished under resolved shear stress, and it accommodates only a limited amount of strain which is determined by the particular crystal structure (e.g., for calcite a maximum shear strain of 0.35). In calcite aggregates, mechanical twins can be used to infer the stress orientation (e.g., Turner, 1962) as well as the strain magnitude in relatively low-strain rocks (e.g., Groshong and others, 1984). Some authors suggested that mechanical twins might provide a useful indication of stress magnitude (e.g., Tullis, 1980); however, recently the resistance to twinning has been shown to be strongly dependent on grain size (e.g., Rowe and Rutter, 1990). Mechanical twins provide internal boundaries that may be important in hindering or enhancing deformation by other processes (e.g., Rutter, 1995), and the twinning process can provide an extra degree of freedom that facilitates homogeneous rock deformation. In calcite, mechanical twinning produces a strong CPO at relatively low strains (e.g., Burkhard, 1993).

### *Dislocation creep*

Perhaps the dominant deformation mechanism in the mid- to lower crust is dislocation creep (crystal plasticity). Steady-state deformation (implied by the term "creep") involves both a strain-producing process and a recovery process. Strain is produced by the glide of line defects known as dislocations through a grain, which accomplishes a unit cell shear offset. Dislocations gliding on intersecting planes tend to become tangled or pinned at obstacles, and this produces strain hardening (e.g., Nicolas and Poirier, 1976; Twiss and Moores, 1992). Recovery processes, which include dynamic recrystallization and dislocation climb, act to offset strain hardening by allowing the material to continuously "recover" a state of lower internal energy. (It is important to note that dynamic recrystallization by itself is not a deformation mechanism.) Both dislocation glide and recovery are thermally activated, and they depend little on pressure.

Semi-brittle flow involves a combination of dislocation glide and distributed microcracking (e.g., Hirth and Tullis, 1994). This mode of deformation occurs at temperatures somewhat lower than those that allow steady-state dislocation creep, where dislocations are generated and undergo limited glide, but recovery is not rapid enough to offset strain hardening. Locally the internal stresses rise high enough to produce microcracks.

The dislocation creep field can be divided into a lower-temperature regime within which the recovery process is recrystallization by grain-boundary migration, and two higher-temperature regimes within which it is dislocation climb. Dynamic recrystallization accompanies the deformation in all three regimes, although the processes and microstructures are distinctively different (e.g., Hirth and Tullis, 1992). In addition, there are important differences in the mechanical behavior.

In the lower-temperature regime, termed *recrystallization-accommodated dislocation creep*, dislocations can glide but dislocation climb is limited. High densities of tangled dislocations are developed, and local contrasts in dislocation density provide the driving force for small-scale grain-boundary migration. This process produces very small strain-free recrystallized grains. These strain-free recrystallized grains are able to undergo an increment of "easy" dislocation glide, although the dislocation density rapidly builds up to the same level as that in the host grain, initiating a new episode of grain-boundary migration recrystallization. In this low-temperature dislocation creep regime, original grains are little changed in shape, but they are progressively replaced by very fine (1–5 μm) recrystallized grains which form along original grain boundaries and along internal deformation bands and twin boundaries (e.g., Tullis and Yund, 1985). Progressive strain produces augen separated by fine-grained recrystallized tails; this texture is commonly seen in the feldspars of low- to medium-grade mylonites (e.g., Simpson, 1986; Stauffer and Lewry, 1993).

One of the important aspects of recrystallization-accommodated dislocation creep is that it produces significant strain weakening of initially coarse-grained aggregates, and tends to partition strain into finer-grained zones. In a coarse-grained aggregate, an imposed constant strain rate will result in an early stress peak followed by progressive weakening, whose onset corresponds to the development of a layer of "soft" recrystallized grains along original grain boundaries. Strain weakening continues until the aggregate is completely recrystallized, because mechanical steady state requires microstructural steady state. In an aggregate of varying grain size, this process of recrystallization-accommodated dislocation creep will cause strain to be partitioned into finer-grained zones, because they will more rapidly become completely recrystallized (e.g., Tullis and others, 1990). For example, if fine-grained cataclastic rocks are brought to higher temperatures, they will be much weaker than coarser-grained wall rocks, and strain will be highly partitioned into the preexisting fine-grained zone. Zones of concentrated strain involving low-temperature grain-boundary migration recrystallization are observed on all scales, from macroscopic (e.g., Simpson, 1985) down through finer-scale S and C structures (e.g., Berthé and others, 1979a; Gapais, 1989). These zones can be distinguished from fine-grained fault gouge by their uniform grain size and absence of porosity. They can be distinguished from the products of grain-boundary sliding and/or diffusion creep by the presence of a high dislocation density in some grains and a strong CPO.

At temperatures high enough to allow climb of dislocations, there is a transition to *climb-accommodated dislocation creep* (Hirth and Tullis, 1992). Climb allows the free dislocation density to be lowered by annihilation of dislocations of opposite sign and arrangement of many of the remaining dislocations into low-energy subgrain boundaries (i.e., low-angle internal boundaries with misorientations <10°; see Poirier, 1985). Climb increases the mobility of dislocations, and allows them to continue gliding without strain hardening, thus producing steady-state creep.

At lower temperatures within the climb-accommodated dislocation creep regime, the characteristic texture is very different from that in the lower-temperature recrystallization-accommodated dislocation creep regime. Original grains tend to become homogeneously flattened and to develop a core and mantle structure (White, 1976). The central "core" regions of grains have larger subgrains, defining a gentle sweeping undulatory extinction. The margins of the grains, where the local stresses and strains are higher, have smaller subgrains as well as recrystallized grains that have developed by progressive subgrain misorientation (e.g., Guillope and Poirier, 1979). The size of the recrystallized grains is only slightly larger than that of the subgrains, due to the combination of low driving force and somewhat limited grain-boundary mobility. The recrystallized grains have a similar free dislocation density, and thus a similar strength, as the deformed original grains from which they originated. Thus recrystallization in this regime produces only minor strain weakening, unless a change to a grain-size-sensitive deformation mechanism is precipitated.

At somewhat higher temperatures within the climb-accommodated dislocation creep regime, the characteristic texture is distinctly different. Climb is very easy, and thus the average free dislocation density is very low. Grain-boundary mobility is high, even though the driving force (dislocation density contrast) is low. Thus, although subgrains are common, they tend not to accumulate enough misorientation (>~10°) to become recrystallized grains. Instead, recrystallization results from the migration of segments of original grain boundaries; even after low strain the original grains are completely replaced by recrystallized grains which tend to be much larger than coexisting subgrains. In this high-temperature regime, the grain-shape fabric appears "unstrained" throughout the deformation, although a strong CPO develops. Because recovery (climb plus recrystallization) keeps pace with deformation, the microstructure may appear to be "annealed" even though the recrystallization is dynamic rather than static. There is no easy way to tell whether high-temperature mylonites have been statically annealed other than by the presence of random-fabric metamorphic minerals, or perhaps from the distribution of subgrain sizes about the mean (e.g., Jin and others, 1989).

Within the dislocation creep regime, several microstructures may be useful as "paleopiezometers," e.g., measures of the flow stress. Experiments have shown that the steady-state flow stress is directly related to the dislocation density and inversely related to the subgrain size and recrystallized grain size (e.g., Poirier, 1985). The extent of recrystallization in a rock may reflect the total strain and/or the temperature, but the *size* of the recrystallized grains reflects the flow stress. Similarly, the dislocation density within grains does not reflect the strain (after the first few percent), but rather the flow stress. The usefulness of these microstructures as "paleopiezometers," in allowing estimates of the magnitudes of the ductile flow stresses that have operated in nature, depends on accurate laboratory calibration. Experimental calibrations of stress-microstructure relations have been made for some microstructures in some minerals (e.g., olivine—van der Wal and others, 1993; calcite—Schmid and others, 1980; de Bresseur, 1996), but they have not yet been determined for quartz and feldspar. For the latter minerals, the theoretical relation of Twiss (1977) for recrystallized grain size has commonly been used. However, recently it has been realized that different mechanisms of dynamic recrystallization may have different relations with the flow stress (e.g., Derby, 1991; Rutter, 1995).

There are a number of caveats in applying laboratory-derived piezometers to naturally deformed rocks (e.g., White, 1979). First, any deformed rock has passed through a stress peak during exhumation, and thus higher-temperature (and lower-stress) dislocation creep microstructures may have been overprinted by lower-temperature (and higher-stress) microstructures. A second problem results from the fact that piezometers are calibrated for steady-state flow; this may be rare in nature, and there are few data on the rates of microstructural response to changing conditions. Third, the recrystallized grain size for one phase might be pinned at less than its equilibrium value by other phases, or it might increase during a post-deformation anneal. However, even if one can never derive accurate measures of the flow stress from these dislocation creep microstructures, in a polyphase rock they can provide very useful qualitative information about the deformation history. For a given phase, dislocation density and subgrain size have different recovery kinetics; thus, the former is more likely to reflect a later, lower-temperature, higher-stress part of the deformation. In addition, different phases undergo deformation mechanism transitions at different absolute temperatures. Thus the dislocation density in quartz may equilibrate with the flow stress down to a lower temperature than that recorded by feldspar, and quartz would show recrystallized grains formed by subgrain misorientation at the same conditions where feldspar would show recrystallized grains formed by low-temperature grain-boundary migration (e.g., Dell'Angelo and Tullis, 1996). Qualitative estimates of stress variations within a deforming polyphase aggregate were made by Lister and Price (1978); they documented strong variations in recrystallized grain size of quartz around more rigid feldspar augen, with smaller grains on the high-stress sides.

It has long been known that trace amounts of water can profoundly affect the strength of silicates experimentally deformed by dislocation creep (e.g., "hydrolytic weakening"; Griggs, 1967; Kronenberg, 1994). Recently, it has been shown that for quartz this reflects the dependence of the dislocation creep strength on water fugacity (Post and others, 1996), which must be incorporated in the flow law as a water fugacity term with an exponent of about 1 (Gleason and Tullis, 1995; Kohlstedt and others, 1995). If rocks are naturally deformed in the presence of a free-water phase, then they cannot be further weakened. However, some rocks might be dried out and thus strengthened by high-grade metamorphism, and then hydrolytically weakened by a later influx of fluids; this might contribute to the formation of a ductile shear zone (e.g., Kronenberg and others, 1990).

One of the hallmarks of deformation by dislocation creep that can be ascertained optically is the production of a CPO (see Wenk, ed., 1985; and the review by Law, 1990). The pattern of preferred orientation may provide useful information concerning the geometry of the strain, because noncoaxial strain produces asymmetric CPOs; it may also provide useful information on the relative temperature of the deformation, because higher temperatures activate different slip systems. Many authors have assumed that the fine grain size characterizing most mylonites and ductile shear zones implies that the rocks deformed by some combination of grain-boundary sliding and diffusion creep. However, most fine-grained shear-zone rocks have a strong CPO (e.g., Law and others, 1986; Rutter and others, 1994), implying that dislocation creep contributed a significant component of the strain. There has been some debate as to whether the progressive development of a CPO might produce geometric hardening or weakening (e.g., Lister and Williams, 1983; Schmid and others, 1987; Takeshita and Wenk, 1988), and, therefore, whether it would tend to promote or prevent further strain localization in a mylonite zone. Glide alone will continually rotate crystal axes with respect to the stress, but dynamic recrystallization may continually "reset" the grain orientations and thus allow a steady-state CPO as well as steady-state flow (e.g., Law, 1986; Herwegh and Handy, 1996). For the case of simple shear, it is commonly thought that the easy slip plane will align with the shear plane, producing an initial strain weakening and subsequently allowing steady-state flow (e.g., Schmid and Casey, 1986; Schmid, 1994), although there has been some debate on this point (e.g., Wenk and Christie, 1991).

### *Diffusion creep and grain-boundary sliding*

Diffusion creep and grain-boundary sliding commonly have been invoked for the strain softening believed to be involved in the development of ductile shear zones. These processes are treated together because neither can operate alone; each requires an accommodation process (e.g., Poirier, 1985).

Diffusion creep involves the atom-by-atom transfer of material down a chemical potential gradient from high- to low-stress sites; it may occur by volume diffusion (Nabarro-Herring creep) or by dry grain-boundary diffusion (Coble creep). As grains change shape by this process, they must undergo some motion relative to their neighbors; thus a component of nonfrictional grain-boundary sliding is required to avoid gaps and overlaps between grains. The rates of volume and grain-boundary diffusion in most minerals are very slow; thus this process is only competitive with dislocation creep for very fine grain sizes (e.g., short diffusion paths) or very high temperatures. However, because high temperature promotes rapid grain growth, this mechanism may only be important over a restricted intermediate depth range (e.g., Karato and Wu, 1993).

Under certain high-temperature conditions, aggregates may undergo dominantly grain-boundary sliding (nonfrictional, isovolumetric, i.e., not pressure dependent). In this case, original grains remain equant, and the aggregate changes shape mostly by grain neighbor switching (e.g., Ashby and Verall, 1973). However, there must be an accommodation mechanism to prevent gaps and overlaps at grain boundaries; this may involve diffusion or local dislocation motion.

Both diffusion creep and grain-boundary sliding are processes whose flow laws have a very low stress exponent (i.e., 1 to 2, compared to 3 to 8 for dislocation creep). In addition, the strain rate has a strong dependence on grain size, because smaller grains involve shorter diffusion paths and a greater area of grain boundaries is more efficient in allowing strain of an aggregate. Another important aspect of both these deformation mechanisms is that they do not produce a CPO provided that they are not accompanied by any intracrystalline plasticity. However, after large strains even a minor component of crystal plasticity can result in a measurable CPO (e.g., Rutter and others, 1994).

Probably the best-known example of a natural mylonite whose deformation appears to have involved a significant component of grain-boundary sliding and/or diffusion creep is the Lochseiten calc-mylonite, a ~1 m thick layer of extremely fine-grained limestone underlying the Glarus thrust (Schmid, 1975). This layer has undergone high internal strain but maintained equant grains and a very weak CPO, consistent with grain-boundary sliding and diffusion creep, although M. Burkhard (personal comm., 1995) has observed microstructural evidence for other mechanisms in some places within the mylonite. The likely importance of grain-boundary sliding and diffusion creep in naturally deformed limestones has been confirmed by the results of experimental deformation (Schmid and others, 1977) and diffusion (Farver and Yund, 1996) studies of calcite aggregates. Other criteria that have been used to infer grain-boundary sliding in macroscopically "superplastic" mylonites include evidence of phase mixing on a fine scale (e.g., Bouillier and Guéguen, 1975).

There are several processes that might cause an initial grain-size reduction, allowing a change to diffusion creep and/or grain-boundary sliding; these include cataclasis, dynamic recrystallization, and neomineralization. In a polyphase aggregate, progressive recrystallization of several of the phases may cause pinning of grain boundaries at a grain size smaller than the equilibrium size for dislocation creep, thus allowing a switch to dominantly grain-boundary sliding and/or diffusion creep. This has been elegantly demonstrated by Wenk and Pannetier (1990), who showed that quartz in the Santa Rosa granitic mylonites has a strong CPO, whereas quartz in ultramylonite bands (which appear to represent more extreme strain) has an almost random CPO. An additional factor here may be that boundaries between unlike phases, especially those involving micas, may have faster diffusion rates and greater ease of sliding than boundaries between the same phase. There are some indications of this effect both in experiments (Hickman and Evans, 1995) and in naturally deformed rocks (e.g., Kronenberg, 1981).

## Comments on fault- and shear-zone models

A conceptual model is a useful way to graphically summarize the textural variations, P-T conditions, deformation processes, and rheology spanning the brittle-to-plastic transition along a crustal fault and shear zone. An especially influential model was developed by Sibson (1977, his figure 8); this diagrammatic cross section for a fault and shear zone in quartzo-feldspathic crust is reproduced as Figure O.2. Sibson's (1977) model delineated two major deformational regimes in the Earth's crust: his EF (elastico-frictional) regime and QP (quasi-plastic) regime. Subsequently, Sibson (1982) referred to his EF regime as the "frictional regime." The frictional regime is separated from the underlying quasi-plastic regime by a transition zone whose position in the crust depends on T, P, fluid content and pressure, rock type, and strain rate. The frictional regime is brittle and seismogenic; deformation involves cataclasis and frictional sliding on discrete fault surfaces, and produces incohesive breccia or gouge at shallower levels and cohesive cataclasite at greater depths. The quasi-plastic regime is ductile and aseismic; deformation takes place largely by crystal-plastic flow, and produces mylonites and blastomylonites. This model is somewhat oversimplified, because frictional sliding can be either stable (aseismic) or unstable (seismic) and because other deformation processes such as diffusion creep are important in the quasi-plastic regime of fault and shear zones.

In his conceptual model of a fault and shear zone, Sibson (1977) included an indication of how rock strength would be expected to vary with depth (inset to Figure O.2). Within the brittle regime, the strength increases linearly with depth, due to the dependence of frictional sliding on normal stress. Within the quasi-plastic regime, strength decreases with depth, due to the dependence of crystal plasticity on temperature. Microearthquake studies in seismically active areas with moderate heat flow (e.g., San Andreas fault; see figure 5.7 in Hill and others, 1990) support this model and show that the base of the seismogenic zone has an apparent "roughness" along the strike of the fault zone (Sibson, 1984). This roughness was inferred to indicate variations in the depth of the transition from the seismic frictional regime to the aseismic quasi-plastic regime, and thus may reflect variations in rock composition and fluid pressure, or variations in other parameters such as geometry and mode of faulting, strain rate, and geothermal gradient.

Sibson (1984) augmented his original fault-zone model by using laboratory data on the rheological properties of quartz-dominated crust in order to construct profiles showing the limits to the shear resistance as a function of depth for different modes of faulting (Figure O.5). These profiles incorporated a number of simplifying assumptions, including Byerlee's (1978) simple friction law ($\mu \sim 0.75$), hydrostatic fluid pressure, a fixed geothermal gradient ($dT/Dz = 25°C\ km^{-1}$), a constant strain rate, and the laboratory-determined flow law for quartz available at that time. However, Sibson (1984) indicated the general effects of increasing fluid pressure on the strength in the frictional regime, and the effects of strain rate, temperature, water content, and bulk composition in the quasi-plastic regime. Furthermore, on his profile for strike-slip faults, he indicated qualitatively the expected gradual transition in strength in the complex brittle-plastic transition region.

Scholz (1988) somewhat modified Sibson's models, postulating that for quartzo-feldspathic crust there is a broad transition zone of semi-brittle behavior between the onset of quartz crystal plasticity at about 300°C and feldspar crystal plasticity at about 450°C (Figure O.6). He also incorporated regimes of unstable and stable frictional sliding (A–B values on

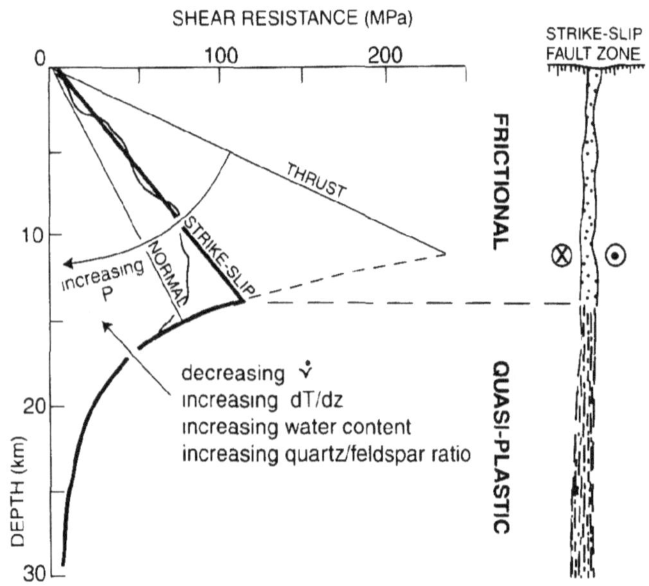


**Figure O.5.** Sibson's synoptic diagram illustrating shear resistance profiles for different types of faults, including the effects of increased fluid pressure (P) (1986, *reproduced with permission, from the Annual Review of Earth and Planetary Sciences*, volume 14, © 1986, by Annual Reviews, Inc.). For the strike-slip profile, Sibson schematically shows the expected subsidiary "roughness" in the frictional regime and the reduction of the peak shear resistance. For the quasi-plastic regime, the arrow shows that strength is decreased by decreasing shear strain rate, and increasing geothermal gradient, fluid pressure, and ratio of quartz to feldspar.

**Figure O.6.** Scholz's synoptic shear zone model (1988, reproduced with permission from the publisher, Springer-Verlag. Copyright © 1988). In this model, Scholz defined a broad transition zone between brittle, frictional behavior and crystal-plastic deformation. For quartzo-feldspathic crust, this transition zone was defined by the onset of quartz crystal-plasticity at about 300°C and feldspar crystal-plasticity at about 450°C. However, note that the strength profile for the model (right-hand frame) continues to follow the friction law into the brittle-to-plastic transition zone (i.e., there is no strength discontinuity at $T_1$). The depth scale on the left is dependent on the geothermal gradient; in this model, the gradient is based on a heat flow model for the region of the San Andreas fault (Lachenbruch and Sass, 1980, model B)

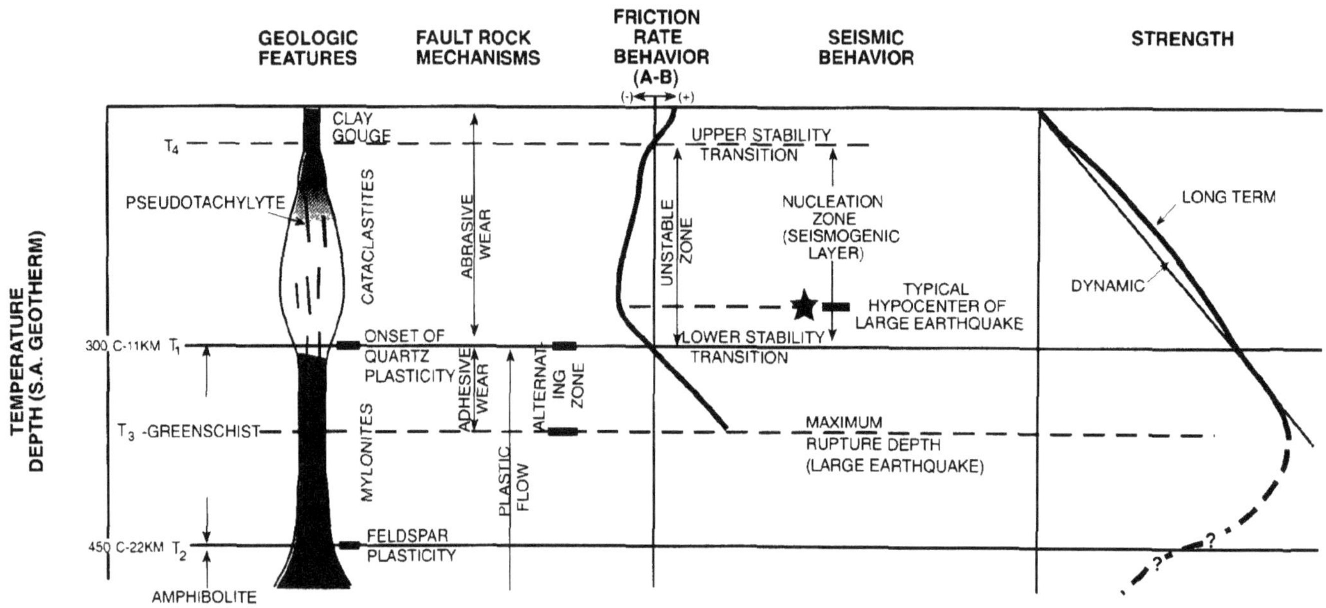


Figure O.6 negative and positive, respectively), and indicated schematically the depth interval that would undergo alternating brittle and ductile deformation. Scholz (1990) adopted the terminology suggested by Rutter (1986) and cast his model in terms of the "brittle-to-plastic transition." In contrast, Schmid and Handy (1991) advocated the use of "frictional-viscous transition," because they viewed the term plastic as virtually synonymous with crystal plasticity and thereby excluding the role of such important deformation mechanisms as diffusional mass transfer and grain-boundary sliding. Schmid and Handy (1991) suggested that the transition from frictional to viscous flow may occur at shallower crustal levels than predicted by Scholz (1988) for his brittle-to-plastic transition, because viscous flow processes depend on grain size and on local and regional fluid flux as well as on geothermal gradient and strain rate. Schmid and Handy (1991) also noted that the depth interval for the frictional-viscous transition will narrow with strain, because grain-size reduction due to cataclasis in the frictional regime and dynamic recrystallization in the viscous regime will induce a change to grain-size-sensitive deformation mechanisms.

Whether the "best" conceptual model involves a brittle-plastic or a frictional-viscous transition, compositional heterogeneity within the lithosphere will produce rheological stratifications, and hence a number of such transitions (e.g., Ranalli and Murphy, 1987; Carter and Tsenn, 1987). The absolute temperature required for crystal-plastic behavior in intermediate-mafic and ultramafic rocks is significantly higher than that for quartzo-feldspathic rocks, and in areas of low heat flow (e.g., old cratons), seismic rupture is observed in the lower crust or even the upper mantle (e.g., Chen and Molnar, 1983). Early rheological profiles of continental lithosphere (reviewed by Kirby, 1983) assumed a uniform crust characterized by a quartz rheology underlain by an upper mantle characterized by an olivine rheology (e.g., Goetze and Evans, 1979). More recent rheological models commonly employ multiple layers of different composition for the continental crust, such as a quartz/granitic upper crust and an intermediate/basic lower crust (e.g., Smith and Bruhn, 1984; Rutter and Brodie, 1991). Even the more homogeneous oceanic lithosphere is probably not rheologically uniform (Kirby and Kronenberg, 1987; Molnar, 1988). Laboratory experiments as well as observations of naturally deformed rocks provide qualitative information on the relative strengths of quartz, feldspar, pyroxene, and olivine, and thus the relative strengths of granitic and gabbroic rocks. However, it should be noted that, at present, quartz and olivine are the only silicates for which there are flow law data that can be accurately extrapolated to construct stress versus depth profiles (Kohlstedt and others, 1995).

Although rheological stratification of the lithosphere is a useful concept for better understanding the mechanical behavior and depth distribution of seismicity, there are some significant limitations and complexities that must be considered when rheological aspects are incorporated into a general fault- and shear-zone model. First, stress versus depth profiles based on the brittle-to-plastic transition (i.e., Figure O.5) provide only an upper bound for the stress at a given depth. As Rutter and Brodie (1988) pointed out, such profiles apply to situations in which deformation occurs at constant stress, strain rate, temperature, fluid pressure, and microstructure. However, fault and shear zones develop over time, and the deformation mechanisms and consequently the stress within these zones will change progressively as the fault and shear zone matures. For example, grain-size reduction in both the frictional and quasi-plastic regimes can promote a change to pressure solution, diffusion creep, and grain-boundary sliding, which in turn would cause a decrease in strength. In addition, increasing pore fluid pressure will decrease the strength within the frictional regime (e.g., Hubbert and Rubey, 1959), and it can allow transient brittle behavior even within the high-temperature plastic regime by effectively reducing normal stresses (e.g., Secor, 1965; Wilshire and Kirby, 1989). Partial melt may play a similar role (i.e., melt-enhanced embrittlement, see Davidson and others, 1994; Keleman and Dick, 1995). Furthermore, both field and experimental studies (Hollister and Crawford, 1986; Dell'Angelo and Tullis, 1988) have recognized the potential overall "weakening effect" that melt can have on rocks in deep crustal shear zones. As a final example, pore fluids in fault and shear zones may drive hydrothermal mineral reactions which can lead to a decrease in strength, if the reaction products are weaker than the reactants (e.g., White and Knipe, 1978; Evans, 1990; Wintsch and others, 1995).

Recently there has been an increased focus on periodic variations in fluid pressure, fault strength, and permeability in fault zones and the relationship to the earthquake cycle (e.g., Sibson, 1990; Hickman and others, 1995). Analyses of vein arrays and fluid inclusions in ancient fault zones indicate a complex, time-varying behavior (e.g., Cox, 1995). During the interseismic period of a fault zone, static fluids facilitate crack healing and pore-space sealing, and this decreases the permeability, increases fluid pressure, and decreases fault strength. Eventually the fluid pressure reaches lithostatic and causes rupture, which in turn increases permeability, decreases fluid pressure, and increases fault strength. Sibson (1990) argued that this *seismic pumping* and *fault-valve behavior* is especially likely to occur near the base of the seismogenic zone on high-angle reverse faults, because the low permeability cap produced at the brittle-to-plastic transition (Etheridge and others, 1983) will allow metamorphic fluids to develop a high fluid pressure. The fact that faults commonly act as fluid conduits, with alternating rupture and sealing, has important economic consequences; for example, many gold deposits are located in such settings (e.g., Sibson and others, 1988).

Rocks within the frictional-plastic transition zone of faults will perhaps be subject to the most complex, time-varying deformation. During interseismic periods these rocks may deform in a fully visco-plastic fashion. However, earthquake ruptures will propagate brittle deformation, and perhaps fluids, downward. Thus there is a depth interval within which deformation will be alternately brittle and ductile. Considering all of the possible complexities in deformation likely to occur in fault and shear zones, it may be difficult to know what aspect of a rock's deformational history is represented by the preserved microstructures.

## Concluding remarks

Enormous advances have been made in the understanding of fault rocks during the past twenty-five years. These advances have covered all aspects of fault rocks, from field analysis using kinematic indicators, to experimental studies involving polyphase aggregates and simple shear, to new and improved techniques for imaging the submicroscopic structures of minerals. Nevertheless, as is characteristic of science in general, advances invariably lead to new research problems and questions. Furthermore, despite an improved understanding of the petrogenesis of fault rocks, earth scientists still struggle with problems of nomenclature and classification. In this light, we conclude our review of fault rocks by listing ten important questions that require further research. Unquestionably, other researchers would prepare a somewhat different list; such lists, although useful as an overview of important problems, clearly reflect the research prejudices of the authors.

1) What are the relationships among pseudotachylyte, cataclasite, and ultramylonite in places where these fault rocks occur together, and what are their implications for the earthquake cycle

and seismic-aseismic transition? Can one distinguish cataclasites resulting from fast fault slip from those resulting from slow fault creep? Besides pseudotachylyte, are there other microstructural indicators of seismicity along ancient faults? Is the scarcity of pseudotachylyte real or apparent?

2) What processes are responsible for the extremely low resolved shear stresses on certain large-displacement faults such as the San Andreas fault? Is a thick, clay-rich gouge zone required? Is a significant fluid-pressure difference between the fault zone and host rock required? Are there distinctive mesoscopic features (e.g., vein systems diagnostic of fluid overpressure) or microstructural signatures of this weakness that could be recognized in ancient exhumed faults?

3) What processes of brecciation operate within and around fault zones, and how are they affected by fluids? Are vein systems diagnostic of fluid overpressure?

4) What are the volumes, fluxes, and compositions of fluids that move through fault and shear zones? What is the source of these fluids? What techniques (e.g., fluid inclusion analysis, stable isotopic analysis) can be used to obtain these data?

5) What microstructural criteria can be used to recognize the operation of grain-size-sensitive deformation mechanisms (grain-boundary sliding and/or diffusional creep) in mylonites, and how common are these processes in crustal and upper mantle shear zones? Are they common in polymineralic rocks?

6) How do the mechanical characteristics of a fault or shear zone change with progressive strain, and what are the resulting modifications of the microstructures (and, possibly, mineral assemblage)? What conditions promote strain weakening, steady-state flow, and strain hardening?

7) What geobarometric techniques can be used to accurately estimate pressure conditions during the formation of various fault rocks?

8) What role does melt play in the evolution of high-grade shear zones? What role does pressure solution play?

9) What mesoscopic and microscopic aspects can be used to assess the strain history of high-grade shear zones in which recovery and recrystallization have kept pace with deformation? Are there diagnostic microstructures that will allow us to identify shear zones that have been statically annealed following deformation?

10) What radiometric dating techniques can be employed on the rocks of fault and shear zones to yield meaningful geologic ages for the deformation?

# BRITTLE BEHAVIOR

## Geometrical characteristics and microstructures



Common abbreviations used in regard to the plates in this section include PPL = plane polarized light, NX = crossed nicols, SEM = scanning electron microscope, and TEM = transmission electron microscope.

This set of plates illustrates some mesoscopic to microscopic characteristics of brittle fault zones in the upper crust. Many of the examples are in granitic rocks, although characteristics of brittle deformation in other rock types are also illustrated in several of the plates, including mafic rocks of the oceanic crust. Fractures at all scales from the single mineral grain to mesoscopic fracture arrays are the focus of many of the plates. Although mesoscopic and microscopic observations are emphasized in many of the plates, other more sophisticated techniques, e.g., the SEM (scanning electron microscope) or TEM (transmission electron microscope), are exceptionally useful in identifying brittle microstructures in fault rocks. Kinematic indicators in brittle fault zones are discussed in several of the plates, and additional examples are illustrated in the next section, "Cataclasis and gouge development." As an additional reference, a useful paper by Petit (1987) summarizes the characteristics of various small-scale structures that commonly occur in fault zones and whose recognition and proper interpretation are essential to a complete understanding of the kinematic history.

# 1. FAULT ZONE IN GRANODIORITE

PAUL SEGALL

Exposure of a strike-slip fault located south of Seven Gables Lakes, Mount Abbot quadrangle, Sierra Nevada, California, U.S.A. The fault left-laterally offsets a series of mafic dikes as well as pluton contacts by approximately 100 m (Lockwood and Lydon, 1975). The host rock is Late Cretaceous granodiorite of Lake Edison.

**1A.** View to the east-northeast along the northern margin of the fault zone (FZ). The fault is exposed as the prominent notch in the skyline ridge and in the right foreground. Numerous fractures related to the faulting are exposed in the wall rock adjacent to the fault zone (left foreground). Fractures of the primary set (P) parallel the fault zone and may exhibit left-lateral displacements (see Figure 1.1). Fractures of the second set (S) strike obliquely to the fault zone and cross-cut the primary fractures. This demonstrates that the oblique fractures postdated formation of the primary fractures.

**1B.** Close-up of the northern boundary of the fault zone. The transition between nearly unfractured wall rock (WR) and highly fractured fault-zone rocks (FZ) is extremely sharp. Granodiorite within the fault zone is broken into fist-sized and smaller fragments, the surfaces of which are coated with chlorite and are commonly slickensided. Rock hammer (arrow) for scale.

**1C.** Concentrated deformation along the northern boundary of fault zone (between arrows). Unfractured wall rock (WR) is visible on left side of photograph and highly fractured granodiorite within the fault zone (FZ) is visible toward the upper right. Smaller fault zones exposed in the area, of the order of several meters in width, display similar intense internal fracturing. In some cases, it has been documented that most of the net slip through the fault zone is localized at the boundaries of the zones (Segall and Pollard, 1983).

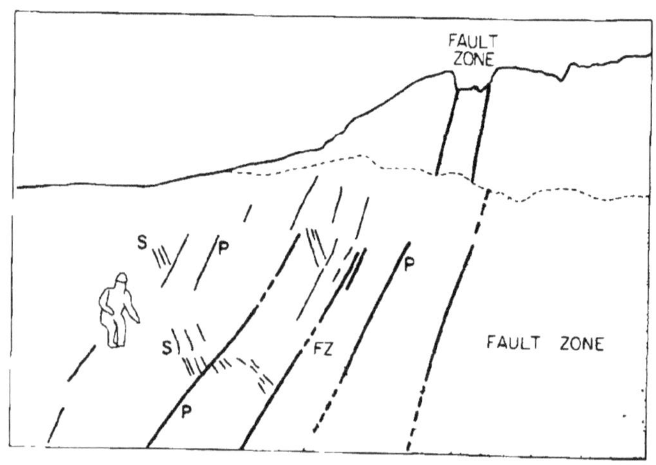


**Figure 1.1.** Sketch of critical field relationships described in Plate 1A. Note geologist for scale.

1 A

1 B

1 C

# 2. BRITTLE DEFORMATION OF A GRANITE

MARY KAREN NELIS AND SHARON MOSHER

The Marble Falls fault is a northeast-striking, Late Pennsylvanian normal fault with an estimated throw of 3000 feet (915 m) (Warner, 1961) located in Marble Falls, Texas, U.S.A. (Figure 2.1). The fault puts homogeneous Late Proterozoic Town Mountain Granite in contact with Carboniferous Marble Falls Limestone. The amount of deformation within the granite decreases away from the fault trace; however, the amount of observed displacement between individual fragments of granite of all sizes is minor throughout the fault zone. Two features characterize the deformed granite throughout the fault zone: (1) the deformation is entirely brittle, and (2) the texture at all scales is characterized by fracture-bounded fragments surrounded by a matrix of smaller, angular fragments.

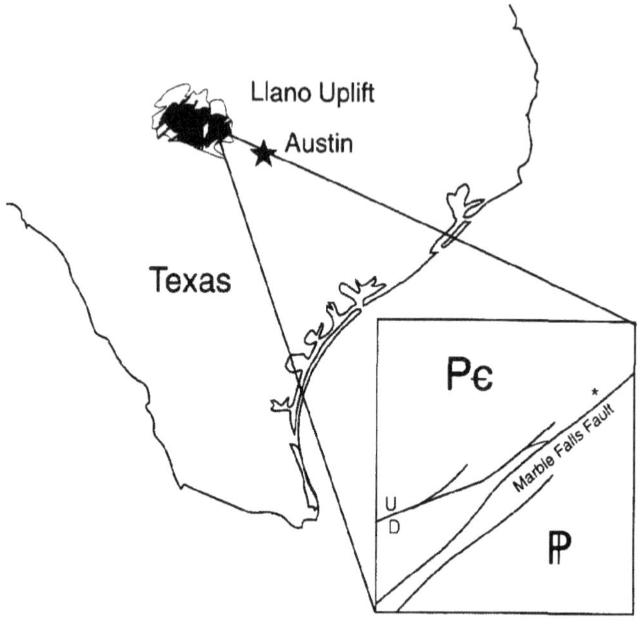


**Figure 2.1.** Index map showing location of the Marble Falls fault, Llano uplift, central Texas, U.S.A.

### Geologic map

**2A.** Detailed map of the fracture pattern in the granite along a 2 m wide traverse across the fault zone. (Width of fracture indicated by line thickness.) Close to the fault trace, linear to curvilinear fractures mostly parallel the trace and form a closely spaced network of interconnecting fractures filled with angular granite fragments and rock flour. In the region 1 to 9 m from the fault trace only major fractures are shown, because the granite between these fractures is cut by numerous fine fractures which commonly branch into very fine splay fractures at their tips. Further away from the fault trace, there is a transition to fractures that are both parallel and perpendicular to the trace, although the longest fractures parallel the fault trace. Many fractures are filled with angular granite fragments and rock flour, and they bound blocks of undeformed, unfractured granite (shown as wide black lines). At distances greater than 30 m from the fault, there are fractures parallel and perpendicular to the fault trace, but these are more widely spaced and are not filled with granite fragments.

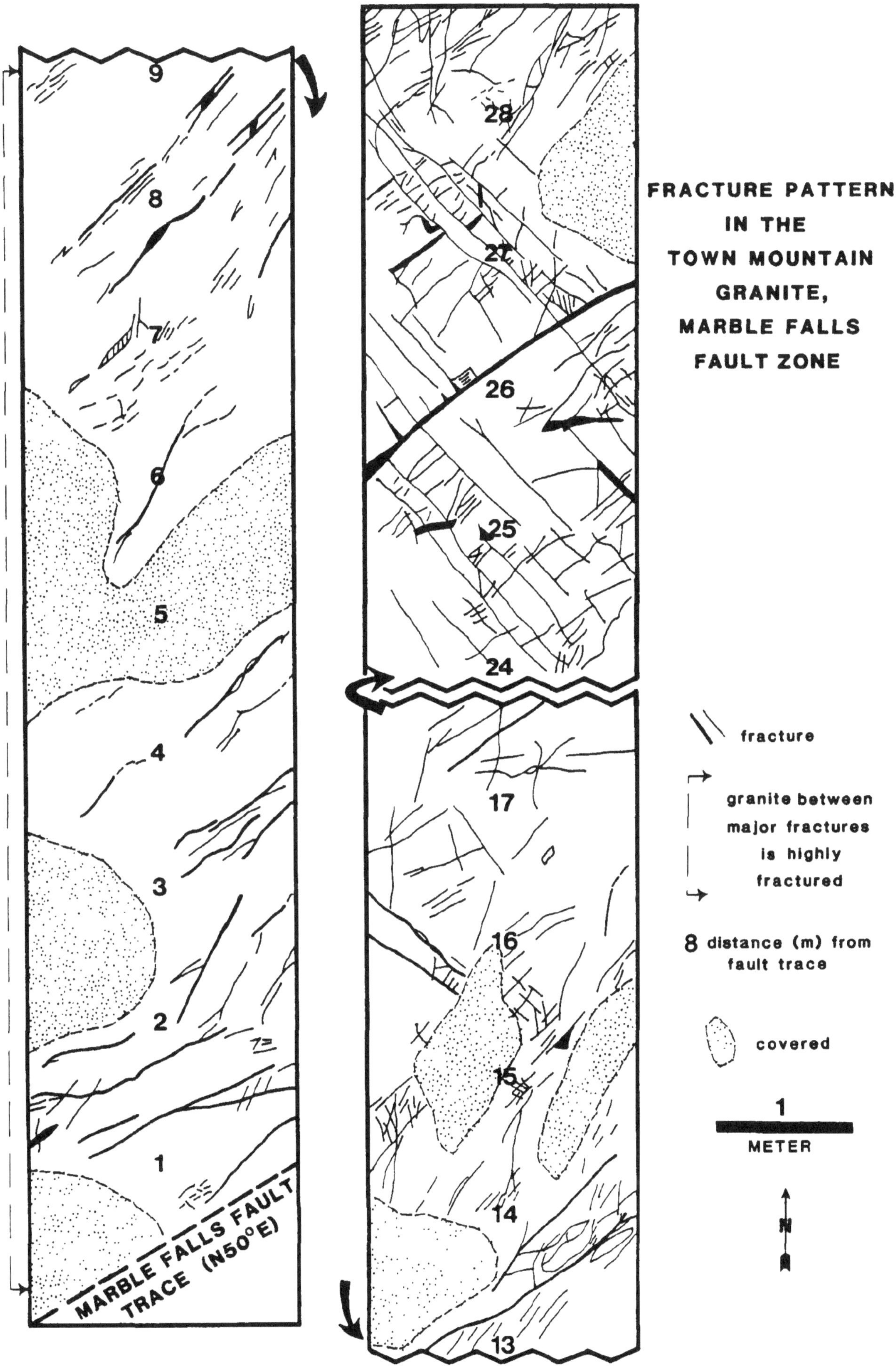
FRACTURE PATTERN
IN THE
TOWN MOUNTAIN
GRANITE,
MARBLE FALLS
FAULT ZONE
MARBLE FALLS FAULT
TRACE (N50°E)
fracture
granite between
major fractures
is highly
fractured
8 distance (m) from
fault trace
covered
1
METER
N

2 A

### Outcrop scale

**2B.** Angular to subrounded blocks of undeformed granite bounded by fractures filled with brecciated rock and grain fragments ranging in size from several centimeters to submicroscopic. The general lack of rounding of fragments, as well as minimal displacements (less than 3 cm offset) of mafic xenoliths elsewhere within the granite, indicate that offset along the fractures was minimal. Photograph taken 25 m from fault trace.

**2C.** Detail from 2B showing the range in breccia fragment sizes and shapes. Note the sharp boundaries of the fractures and breccia fragments filling the fractures. The visible granite fragments are suspended in a matrix of smaller granite fragments. Compare the texture at this scale to the identical texture seen in the photomicrographs of 2D–2G.

2 B

2 C

### Microscopic scale

**2D.** Photomicrograph of a small fracture in the granite 2 m from the fault trace. Note the sharp boundaries with undeformed wall rock and angular fragments of granite and individual crystals floating in a matrix of submicroscopic particles. NX

**2E.** Photomicrograph at the edge of a small fracture showing the sharp boundary between deformed and undeformed granite. Fractured fragments of single crystals are surrounded by a submicroscopic matrix which consists of extremely fine rock flour stained red by iron oxides. NX

**2F.** Photomicrograph showing brittle deformation of a single quartz grain. The three segments of the grain show little displacement or rotation. Note the sharp boundaries of the fractured fragments and the very fine breccia filling the fractures as well as surrounding the grain. Sample taken 20 m from fault trace. NX

**2G.** Photomicrograph showing brittle deformation of a single feldspar grain. Some fractures parallel cleavage planes; note, however, a large fracture filled with fine breccia cutting across cleavage planes. The grain is surrounded by a fine breccia. Sample taken 20 m from fault trace. NX

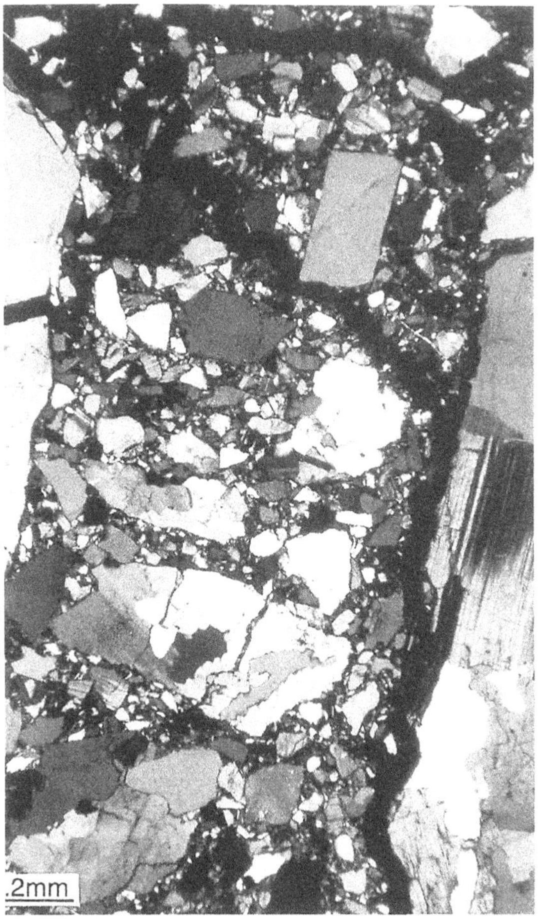


2 D

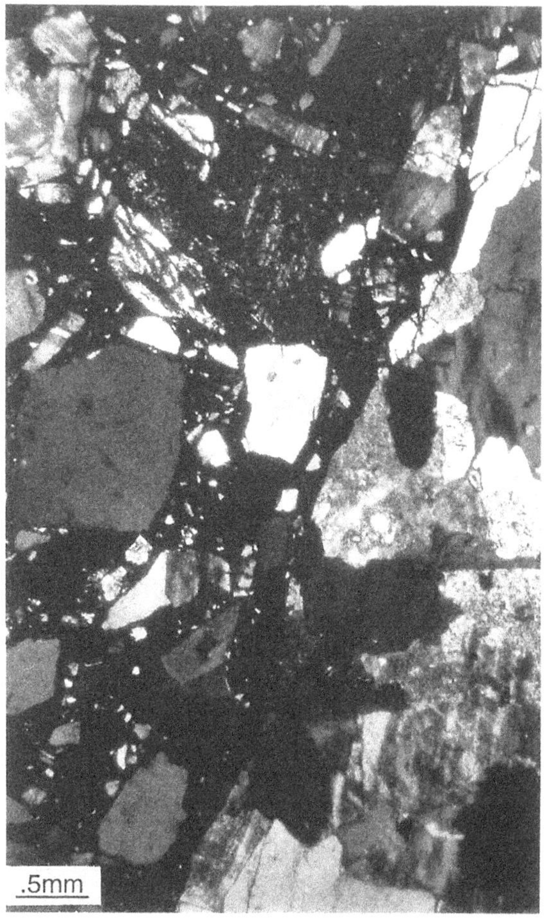


2 E

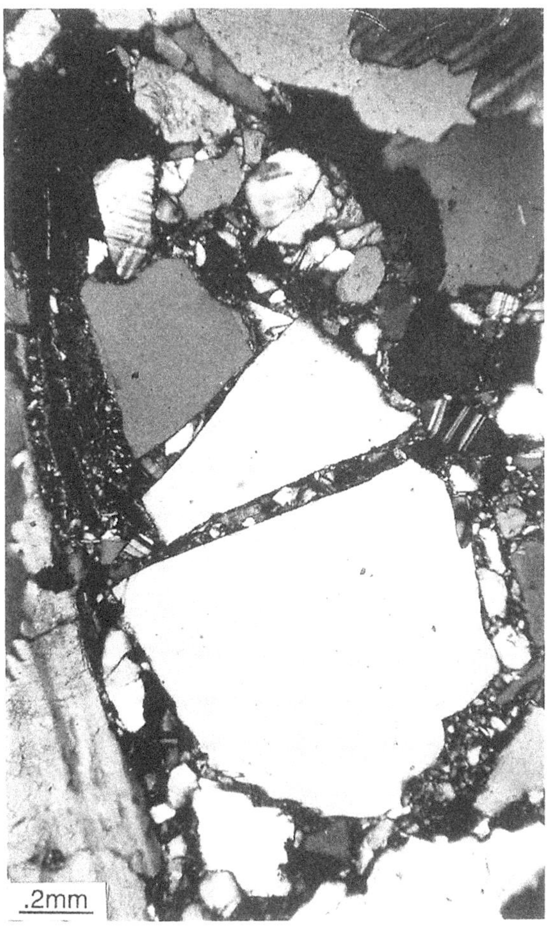


2 F

2 G

# 3. BRITTLE STRUCTURES IN A NORMAL-SENSE FAULT ZONE

LARRY S. LANE

The Columbia River fault zone (CRF, Figure 3.1) is a shallowly eastward-dipping zone of middle Eocene brittle fracturing and normal-sense displacement (Lane, 1984) which is superimposed on and disrupts the Jurassic(?) to early Eocene ductile Monashee décollement (MD, Figure 3.1) (Brown and others, 1986; Lane and others, 1989). The fault zone was temporarily well exposed in a damsite excavation near Revelstoke (Figure 3.1). Brittle faulting and associated clay alteration were confined to a 100 m thick zone within metasedimentary rocks of the Monashee complex (Lane, 1984; Lane and Brown, 1987).

**3A.** Structures in clay gouge, Revelstoke damsite. This view of the principal displacement zone of CRF (typically 0.5–2 m thick) shows the upper two-thirds of the gouge-breccia zone, here 5 m thick (Lane, 1984, figure 8). The accompanying sketch (Figure 3.2) labels gouge-zone features. Rock hammer gives scale. Local thickening and internal deformation are presumed to relate to slip over a large-scale irregularity on the fault. The finest gouge is at the upper gouge-rock interface and is correlated with the zone of highest strain (Logan and others, 1979). Also visible are folded color bands and Riedel shears (Tchalenko, 1970), which are locally rotated and branched. These structures, combined with prominent down-dip striations and grooves on the upper wall-rock surface, indicate normal dip-slip displacement (Lane, 1984).

**3B.** Internal structure of subsidiary shears in CRF, Revelstoke damsite. Steep reverse faults (R) are conjugate structures, active simultaneously with shallow-dipping layer-parallel shears (N), as shown by mutually offsetting relationships in the early stage of brittle fault-zone evolution (Lane, 1984, figure 9). Reverse motion is demonstrated by 1.5 m offset of a distinctive mylonitic carbonate marker bed (M). Note rock hammer at left center of the photograph for scale.

**3C.** Inset view of reverse fault (R) shown in 3B. Internal structure shows elongate breccia fragments derived from layered wall rock along the fault zone. Rotation of fragments is consistent with the known reverse slip. Intensity of brecciation and rotation increase toward margins of the fault zone. Lens cap is 6 cm in diameter.

(GSC contribution no. 36594)

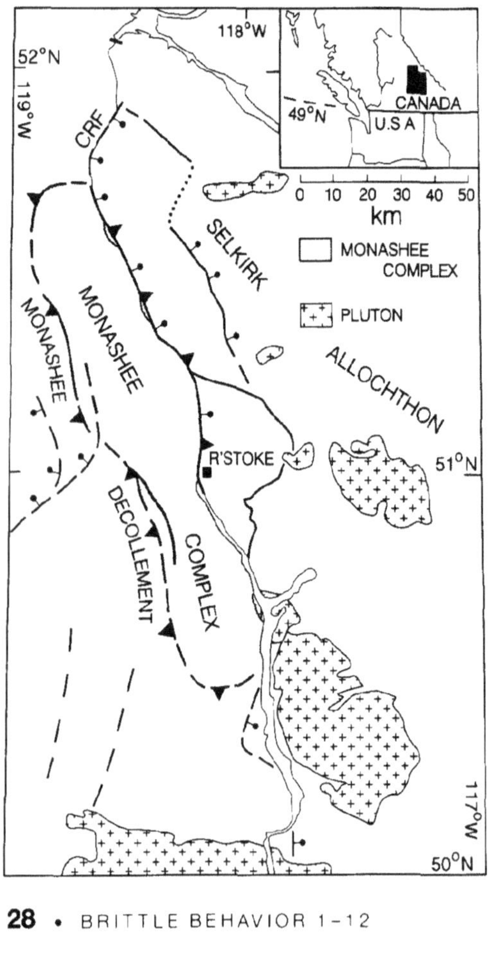


**Figure 3.1** *(left)*. Geologic index map showing major tectonic elements near Revelstoke, British Columbia, Canada. The Monashee complex forms the footwall of the Monashee décollement (MD). CRF = Columbia River fault.

**Figure 3.2** *(right)*. Sketch of Plate 3A. Locally folded color bands are offset by Riedel shears (heavy lines).

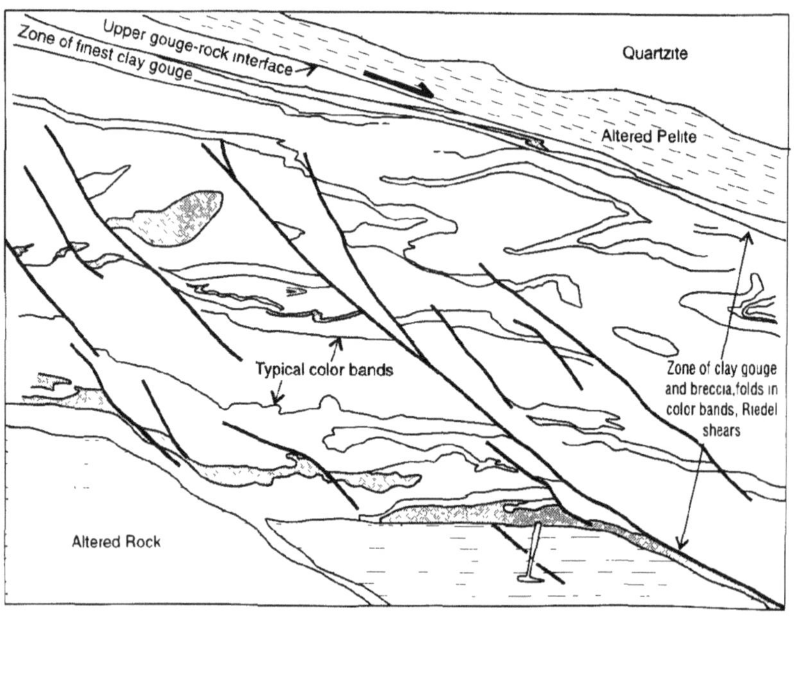

3 A

3 B

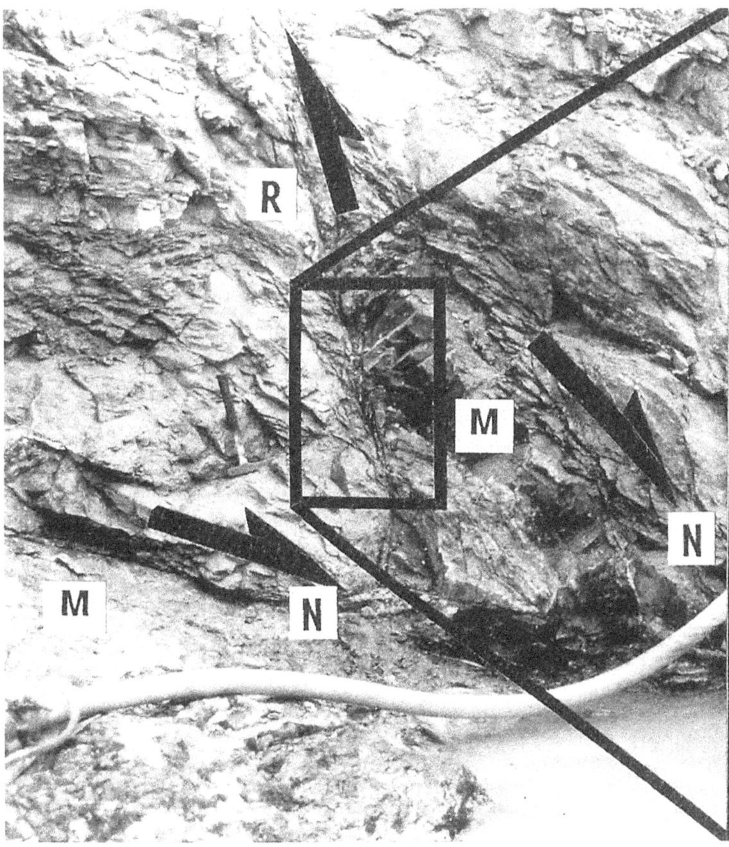


3 C

# 4. KINEMATIC INDICATORS IN BRITTLE FAULT ZONES

MARTIN G. MILLER

Numerous dextral fault zones are developed along west-striking bedding planes in steeply dipping, Jurassic turbidites near Lillooet, British Columbia, Canada. These fault zones display a variety of brittle features which yield reliable and consistent kinematic information. The features shown here include imbricated bedding, imbricated fault rocks, and internal shear fractures, all of which are prominent in many of the zones. In each example, foliation and/or bedding within the zones are oriented obliquely to the edges of the zones. The sense of obliquity is everywhere the same in these fault zones; foliation and/or bedding typically intersect fault-zone edges at an acute angle when measured counterclockwise from the fault-zone edge.

**4A.** Imbricated bedding in dextral fault zone. This photograph shows a strike-slip duplex; each horse consists of fragments of a single sandstone bed (light gray) overlying shale (dark gray). The wide thickness variations and locally intrusive nature of the shale (see arrow) suggest that it behaved ductilely during the deformation. The sigmoidal appearance of individual horses on the left side of the zone suggests that they were deformed during subsequent emplacement of the horses on the right. Therefore imbrication in this zone likely proceeded from left to right.

**4B.** Imbricated fault rocks in dextral fault zone. The asymmetric orientation of these fault rocks with respect to the edge of the fault zone resembles the P foliation of Rutter and others (1986). Foliation in these rocks, however, is not penetrative, but marks sites of slip. Note the conjugate shear fractures which terminate at the zone. These fractures, labeled sf, suggest a component of shortening normal to, but outside of, the fault zone. Local development of pinch-and-swell structures within the fault rock suggests a component of stretching parallel to the foliation *within* the fault zone. A penknife is used for scale.

**4C.** Rotated shear fractures and bedding in dextral fault zone (Figure 4.1). Slip surfaces within this fault zone resemble $R_1$ shears (e.g., Bartlett and others, 1981). The rotated bedding between them, however, implies that the shear fractures rotated from an initially higher angle to the edge of the fault to their present position. Significantly, this rotation of bedding and shear fractures implies a component of shortening normal to the fault zone. Furthermore, the rotated bedding forms a planar fabric which resembles the fabrics illustrated in 4A and 4B. A penknife is used for scale.

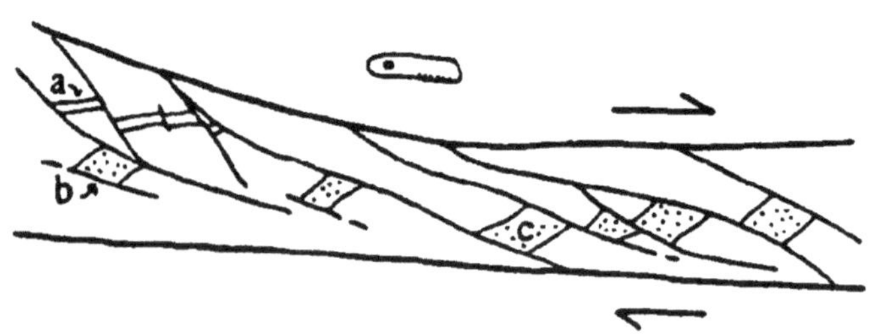


**Figure 4.1.** Line drawing of central part of Plate 4C. Two beds, traceable in the photograph, are cut and offset by shear fractures Letters a, b, and c mark the same features on the drawing and photograph.

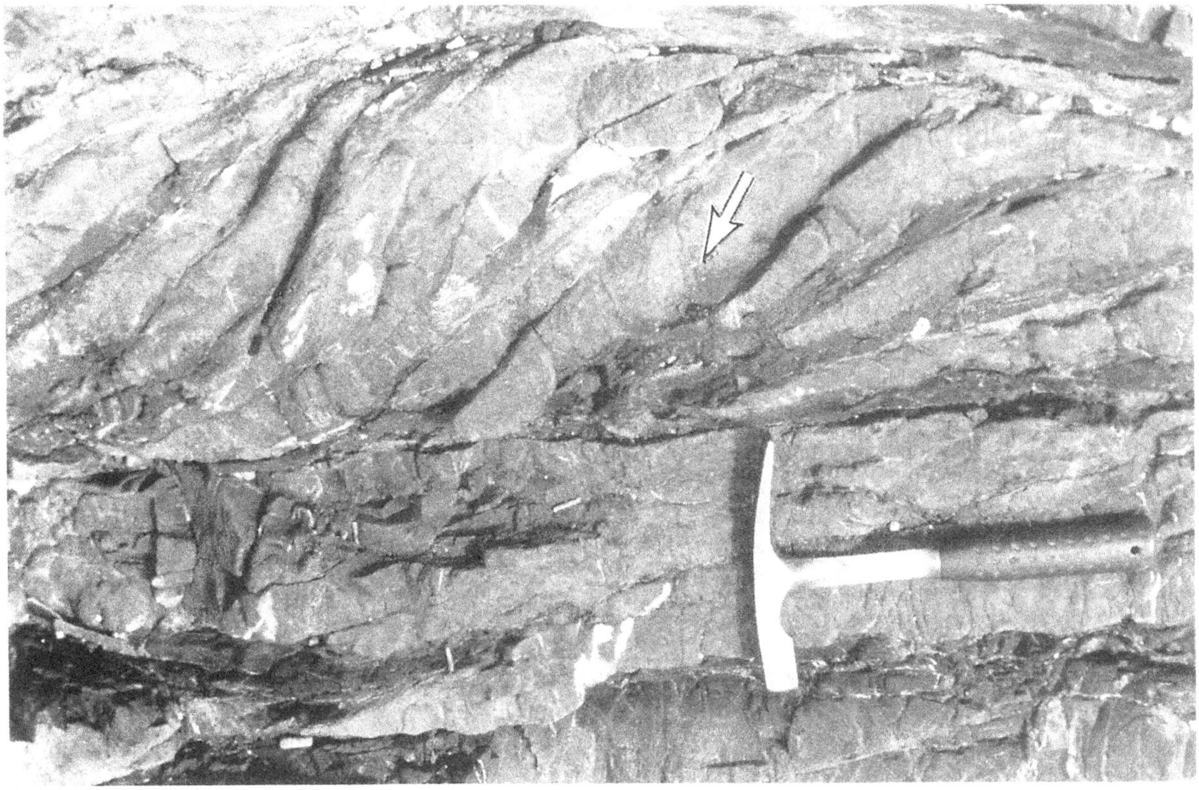

4 A

4 B

4 C

# 5. FAULT-ROCK MICROSTRUCTURES IN BASALT AND DIABASE FROM THE UPPER OCEAN CRUST

SUSAN M. AGAR

Seafloor observations over the last two decades have highlighted the shallow crustal deformation that directly influences the fine-scale relief, magmatism, and fluid circulation around a spreading center (Macdonald and others, 1991). Only a few studies have addressed the mechanisms and distribution of brittle deformation through the ocean crust and timing of deformation relative to magmatic and hydrothermal histories. Hole 504B, drilled during the Deep Sea Drilling Project (DSDP) and Ocean Drilling Program (ODP), currently represents the deepest penetration (2111 m) through the upper ocean crust. The borehole is situated in the eastern equatorial Pacific, midway between the Ecuador and Panama fracture zones, 200 km south of the easternmost segment of the Galapagos Spreading Center–Costa Rica Rift (Figure 5.1). Spreading rates for the southern flank of the Costa Rica Rise are estimated at 38 mm/yr (Lonsdale and Klitgord, 1978). East-west-striking faults in the region around site 504 generate 1–2 km wide, rift-parallel ridges, that separate down-dropped grabens (Langseth and others, 1988). Normal displacements along these faults are estimated to reach a maximum of about 200 m (Figure 5.2). Hole 504B penetrates 5.9 Ma ocean crust (Figure 5.3). The core samples from Hole 504B preserve structures generated during a simple spreading history of orthogonal motion away from the rift zone. Although most of the 504B core samples are oriented for "way-up" only, they still provide valuable constraints on the distribution, conditions, and mechanisms of deformation that complement borehole logging data.

Different structural levels of Hole 504B have been subjected to different alteration histories during successive water-rock interactions (Alt and others, 1993). The extrusive section was affected only by seafloor weathering and low-temperature (<150°C), reducing alteration. The transition zone and sheeted dikes were hydrothermally altered under peak greenschist-facies conditions (250–380°C) (Alt and others, 1986). Anhydrite precipitation, attributed to seawater recharge, postdates the greenschist-facies assemblages in the transition zone. Both the anhydrite and greenschist-facies minerals are overprinted by zeolite alteration (150–200°C).

In the extrusive section a broad spectrum of breccia textures was generated by rapid cooling and synextrusion faulting. Plate 5A (sample location shown in Figure 5.3) shows an example of brittle deformation within the extrusive section interpreted as a minor fault zone that postdated primary brecciation. The polyphase metamorphism lower in the borehole provides critical evidence to distinguish primary and fault-related textures. Geochemical, magnetic, and microstructural data have been used to infer a fault zone that was localized at the boundary between extrusive rocks and sheeted dikes within the upper half of the transition zone, but the exact thickness and sense-of-displacement are unknown (Becker and others, 1988; Agar, 1990, 1991) (Figure 5.3). "Stockwork-like" alteration characterized by quartz- and pyrite-cemented breccias between 910 and 930 mbsf (Honnorez and others, 1983) may be genetically associated with this faulting. Plates 5B–5E show examples of brittle deformation in minor faults and cataclastic zones and "stockwork" textures within the transition zone. There are no absolute age constraints for the deformation.

Orientations of fault planes are given as true dips, but their strike azimuth is unknown. Sample locations are referenced according to Ocean Drilling Program guidelines.

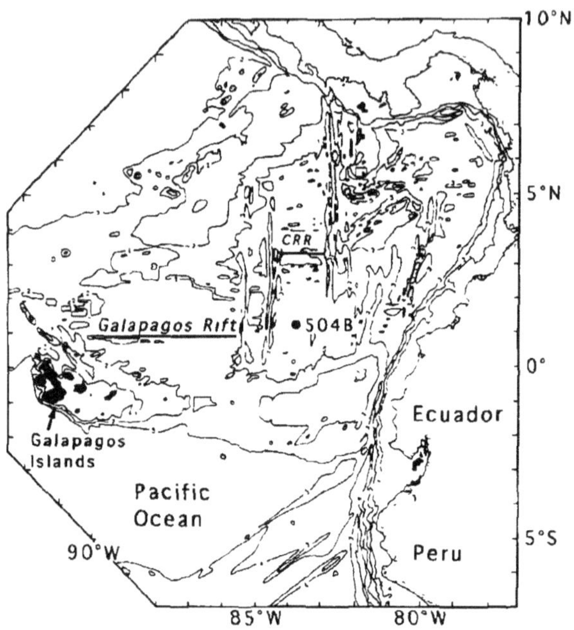


**Figure 5.1.** Location of Hole 504B within the eastern equatorial Pacific. Bathymetric contour map taken from Lonsdale and Klitgord (1978) has contour intervals of 1 km

**5A.** Fault in pillow basalt in the extrusive section of Hole 504B (DSDP Sample 69-504B-49-2, 74–79 cm) (Agar, 1990). A plagioclase phenocryst (p), surrounded by a groundmass of plagioclase and pyroxene crystals, has been broken into angular fragments that are strung out along a narrow fault (arrow) located about 660 mbsf (Figure 5.3). Fracturing during cataclastic deformation was probably assisted by pre-existing thermal cracks in the phenocryst and any hydrous alteration phases precipitated within them prior to faulting. The fault deformation has also promoted water-rock interaction along the fault plane, generating a matrix of clay minerals that is restricted to the immediate surroundings of the phenocryst fragments. As the extrusive section of Hole 504B has not been altered by greenschist-facies mineral assemblages, the timing of deformation relative to brittle structures in the underlying transition and sheeted dike sections is unconstrained. The photo is oriented the same way up as the sampled core piece in the borehole. NX

**5B.** Cataclastic zone in diabase in the transition zone of Hole 504B (DSDP Sample 83-504B-73-2, 30–31 cm). In this

5 A

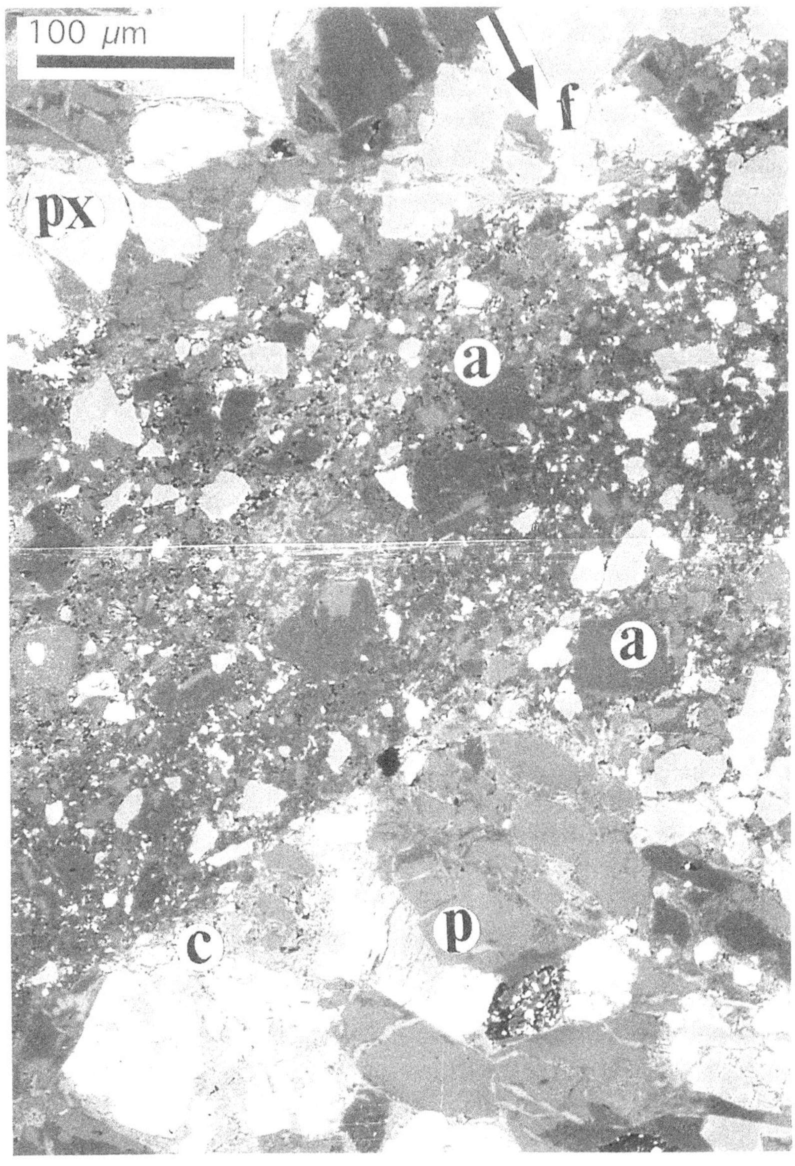


5 B

backscattered electron micrograph, brighter phases have a higher atomic number than darker ones. The image shows the upper and lower margins of a cataclastic zone (860 mbsf, Figure 5.3) in which the host diabase, containing phenocrysts of plagioclase (p) and pyroxene (px), has been pervasively fractured. A steep grain-size gradient across the margins of the cataclastic zone is caused by intense intragranular fracturing within the cataclastic zone. As in 5A, thermal cracking and hydrothermal alteration prior to deformation may have assisted grain-size reduction. Wall-rock fractures are sealed by chlorite (c) and albite (a). Plagioclase fragments within the cataclastic zone appear darker than plagioclase in the wall rock due to the more extensive albite alteration. A transient increase in permeability and an increased reaction surface area caused by cataclastic deformation probably enhanced seawater-rock interaction. A trail of pyrite framboids (f, with arrow pointing to pyrite, the brightest phase in the image) crosses the upper right-hand side of the cataclastic zone and wall rock. The pyrite framboids are undeformed, and the trail is continuous across the contact between the wall rock and cataclastic zone. The pyrite therefore represents a post-kinematic cement. Pyrite precipitation was probably associated with greenschist-facies alteration, indicating that cataclasis occurred near the ridge axis. The sense-of-shear is unknown.

**5C.** Photograph of fault in the transition zone of Hole 504B (DSDP Sample 504B-83-78-1, 105–120 cm) (Agar, 1991). The fault (f) has localized close to the contact between a pillow breccia (p, below) and a lava flow (l, above). Dipping at 30° to the right side of the photograph, the fault is the structurally lowest member in a zone of minor faults preserved within the overlying 40 m of core. The fault rock comprises a chlorite and clay gouge. Chlorite platelets aligned along the dip direction on the upper surface of the fragment of fault rock (small arrow) indicate the synkinematic growth of chlorite, suggesting that the fault was active during the greenschist-facies alteration of this interval. Lighter regions in the wall rock represent hydrothermally altered fragments of the chilled margin of the pillow breccia unit that forms the footwall of the fault. The location of the thin section in Plate 5D is indicated by t. Large arrow at left shows the way up of the core piece.

**5D.** Fault in a porphyritic chilled margin (DSDP Sample 83-504B-78-1R, 110–115 cm) (Agar, 1991). The photograph shows a fault, with a true dip of 70°, located in the hanging wall of the fault in Plate 5C. Note that the thin section orientation is reversed relative to the orientation of the minor fault in Plate 5C. Grain-size reduction of pyroxene (px) and plagioclase (p) phenocrysts along the discrete fault plane has been accomplished by intragranular fracturing along the mechanical contrast between the pillow lava chilled margin and brecciated fragments of the overlying lava flow. The asymmetric, isoclinal folding of a plastically deformed laumontite vein (l) suggests a normal component of motion. As zeolite minerals generally post-date greenschist-facies alteration in the transition zone of Hole 504B, some of the deformation probably postdated or was synchronous with the last phase of low-temperature alteration. NX

**5E.** Breccias sealed by hydrothermal phases in Hole 504B (DSDP Sample 83-504B-90-3R, 120–125 cm and 126–130 cm). These photographs show two adjacent core pieces from the zone of "stockwork-like" alteration at 990 mbsf in the transition zone of Hole 504B (Figure 5.3). The left-hand piece (DSDP Sample 83-504B-90-3R, 126–130 cm) shows a "jig-saw puzzle" texture of quartz (q) veins forming a network around angular basalt (b) fragments. The right-hand piece shows fragments of basalt suspended in a matrix of quartz with minor epidote and pyrite. Both pieces were fractured during greenschist-facies alteration of the transition zone. Laumontite and calcite have sealed small cavities in the veins and matrix during subsequent, low-temperature alteration. The quartz crystallographic fabric within the veins and matrix shows no preferred orientation, nor do the vein orientations. This lack of preferred orientation and the suspension of basalt fragments in the quartz matrix suggest that pore fluid pressures were transiently elevated and may have caused hydraulic fracturing. If the breccias are genetically associated with the inferred fault zone in the upper half of the transition zone (Figure 5.3), they may indicate a seismic pumping mechanism (Sibson, 1987) during faulting (Agar, 1990).

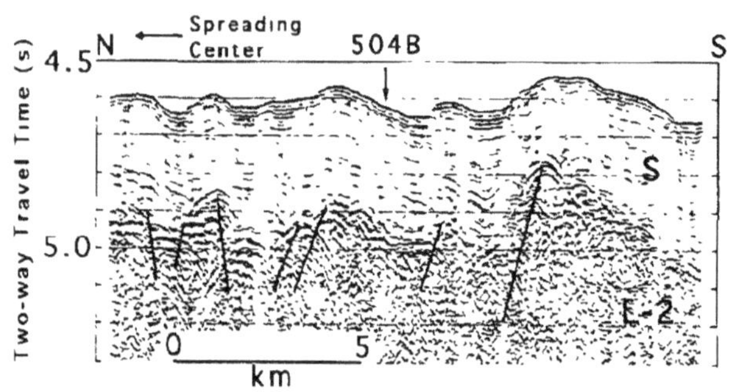


**Figure 5.2.** Single-channel seismic reflection profile across Site 504 showing the approximate location of major normal faults in the upper ocean crust (after Langseth and others, 1988). The location of Hole 504B is projected a few hundred meters east onto the section line which trends 352°, subperpendicular to the Costa Rica Rift axis.

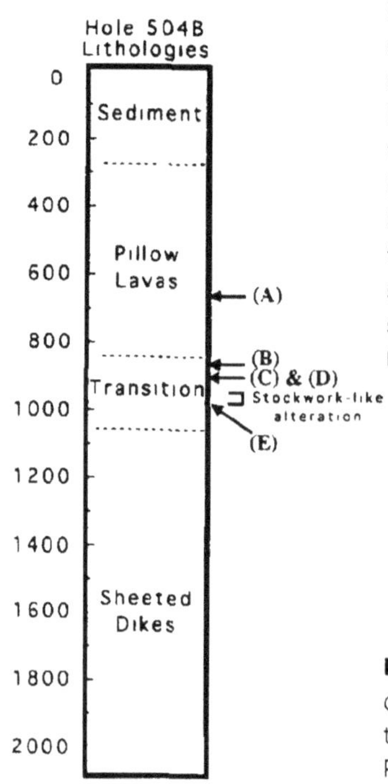


**Figure 5.3.** Lithological divisions of Hole 504B showing the location of samples in Plates 5A–5E

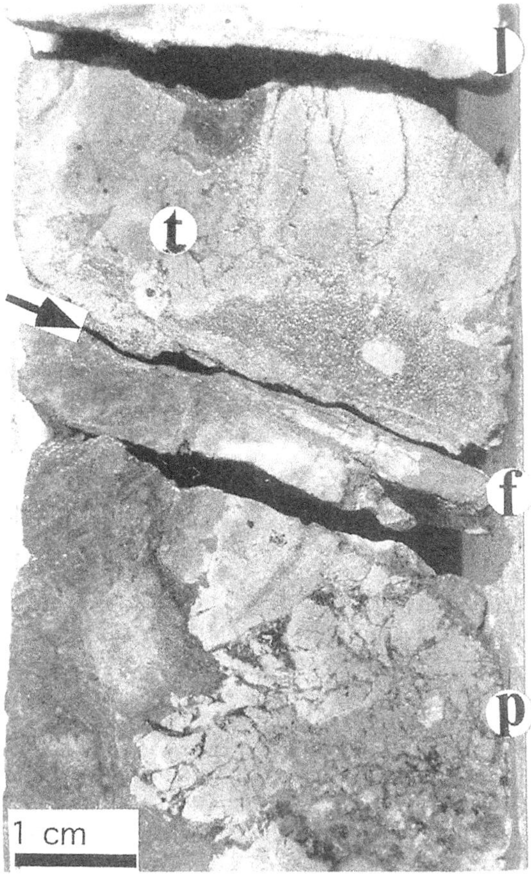


5C

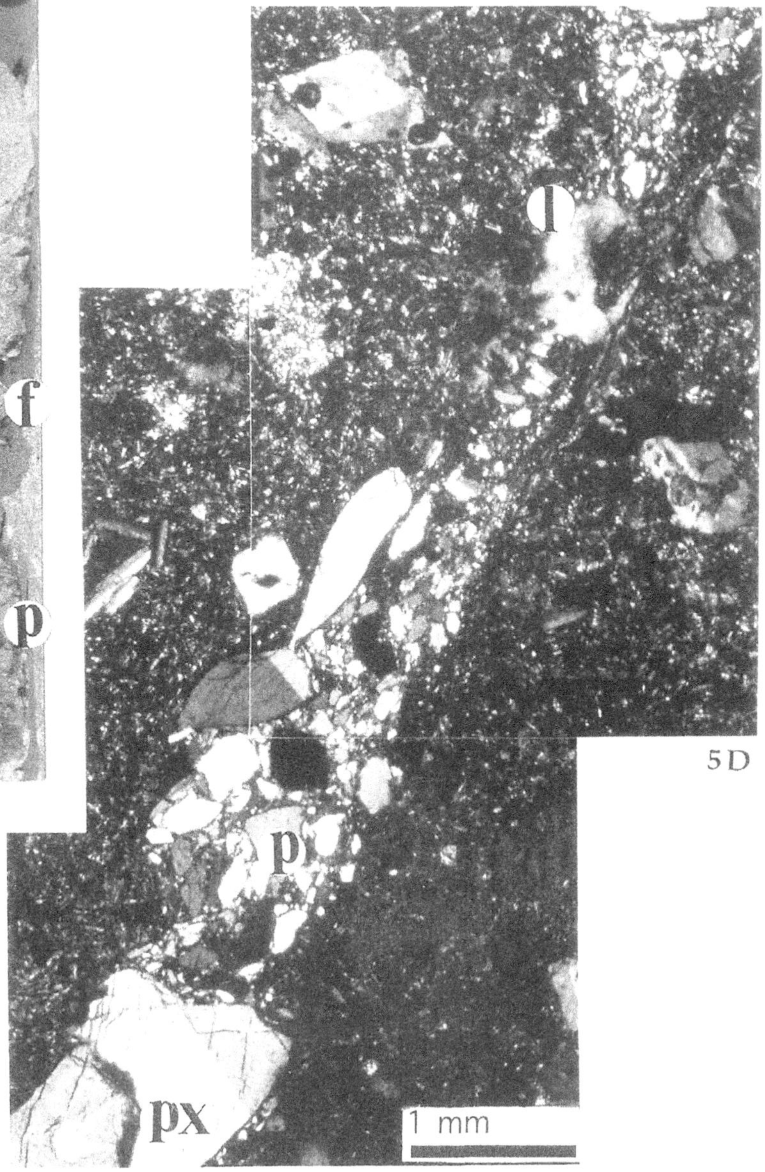


5D

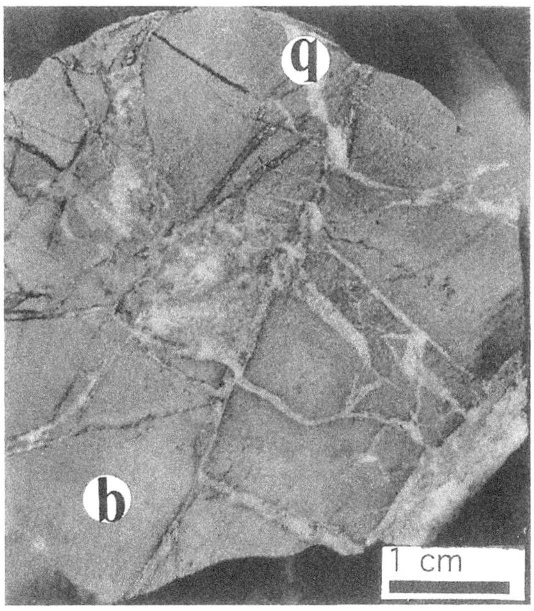


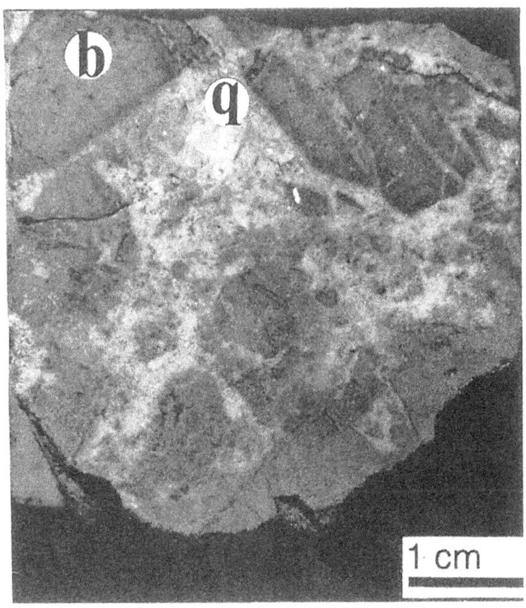


5E

# 6. COHESIVE CRUSH BRECCIAS, MICROBRECCIAS, AND CATACLASITES IN GRANITIC FOOTWALL OF A NORMAL FAULT

G. H. Davis, E. P. Rodriguez, L. J. Cox, R. Ornelas, K. N. Constenius, J. A. Johnson, F. Kong, J. M. Bedford, and M. Pecha

Located just north of Tucson, Arizona, U.S.A. (Figure 6.1), the Pirate fault is a north-striking, steeply west-dipping, ~150 m thick, brittle to semi-brittle shear zone that cuts the 27-Ma Catalina granite (Figure 6.2). The main part of the fault zone contains pervasive, cohesive crush breccias invaded by cataclasite-filled fractures. Near the base of the shear zone, relatively undeformed granite is cut by meter-spaced, cataclasite-filled fractures that dip steeply and strike both parallel and perpendicular to the strike of the Pirate fault. The exposed footwall of Catalina granite resided at a depth of about 4.5 km when faulting initiated at ~15 Ma. By ~5 Ma, the Pirate fault had accommodated at least 3.5 km of total normal-slip displacement.

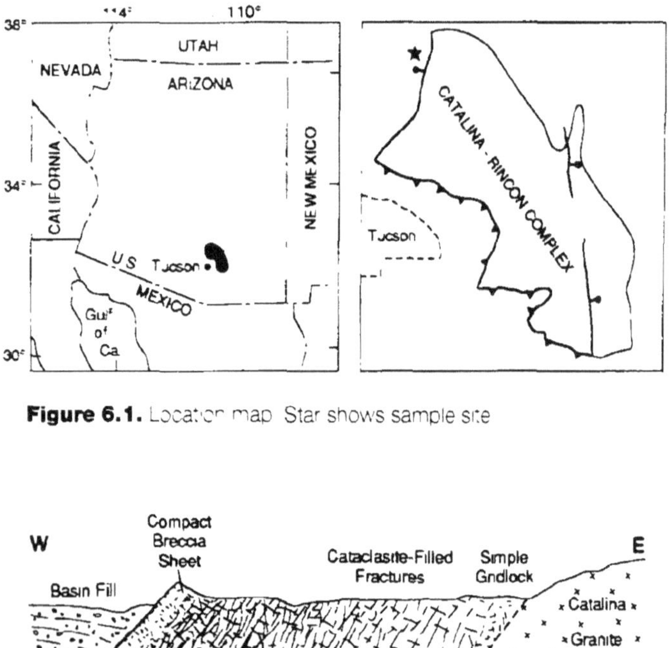


**Figure 6.1.** Location map. Star shows sample site.

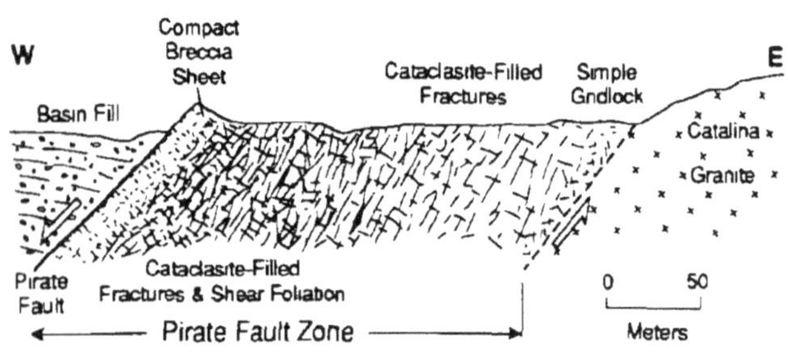


**Figure 6.2.** Schematic structure section through Pirate fault zone.

The cohesive crush breccia in the main part of the Pirate fault zone appear to be the cataclastic equivalent of shallow, incohesive breccia belts. Such belts, according to Stewart and Hancock (1994), form by synfaulting fragmentation, brecciation, and frictional sliding within an original fault-precursor shatter zone that advances upward in front of the tip-line of the fault. The cataclasite-filled fractures near the base of the Pirate fault zone occur in orthogonal arrangements resembling "fracture gridlock" patterns as also described by Stewart and Hancock (1994).

**6A.** Hematitic cataclasite fracture-filling separated by shear surface from microbrecciated and locally cataclastically foliated granite. On the right, cataclasite ($cv_1$) contains angular to subangular fragments of feldspar, quartz, and quartz-feldspar aggregates that generally range from 0.05 to 0.25 mm but may be as much as 2.5 mm across. The dark, fine-grained matrix consists of hematite and quartz. A sharp shear surface separates $cv_1$ from a thin streak of finer-grained ultracataclasite ($ucv_2$). To the left of the shear surface is moderately deformed Catalina granite that is in part microbrecciated ($g_m$) and cataclastically foliated ($g_f$). Some fragmented edges of grains are sharp, whereas others are diffuse due to mixing with the matrix. PPL

**6B.** Cataclasite fracture-fillings in granite. The field of view is dominated by a tension fracture filled with two cataclasite fillings ($cv_1$ and $cv_2$). The right-hand margin of the fracture is in contact with relatively undeformed, coarse-grained Catalina granite (g) containing myrmekitic feldspar slightly altered to sericite. The older cataclasite fracture filling ($cv_1$) consists of comminuted angular fragments of Catalina granite, potassium feldspar, and quartz-feldspar aggregates that are as large as 3.5 mm but are generally <1 mm across. The very fine-grained, dark matrix, is composed mostly of hematite and quartz. The younger, coarser-grained filling ($cv_2$) contains spalled fragments of Catalina granite and angular fragments of feldspar, biotite, and quartz-feldspar aggregates in a matrix of hematite and quartz. Outside of the field of view shown here, rock fragments of cataclasite filling $cv_1$ are surrounded by the younger $cv_2$. PPL

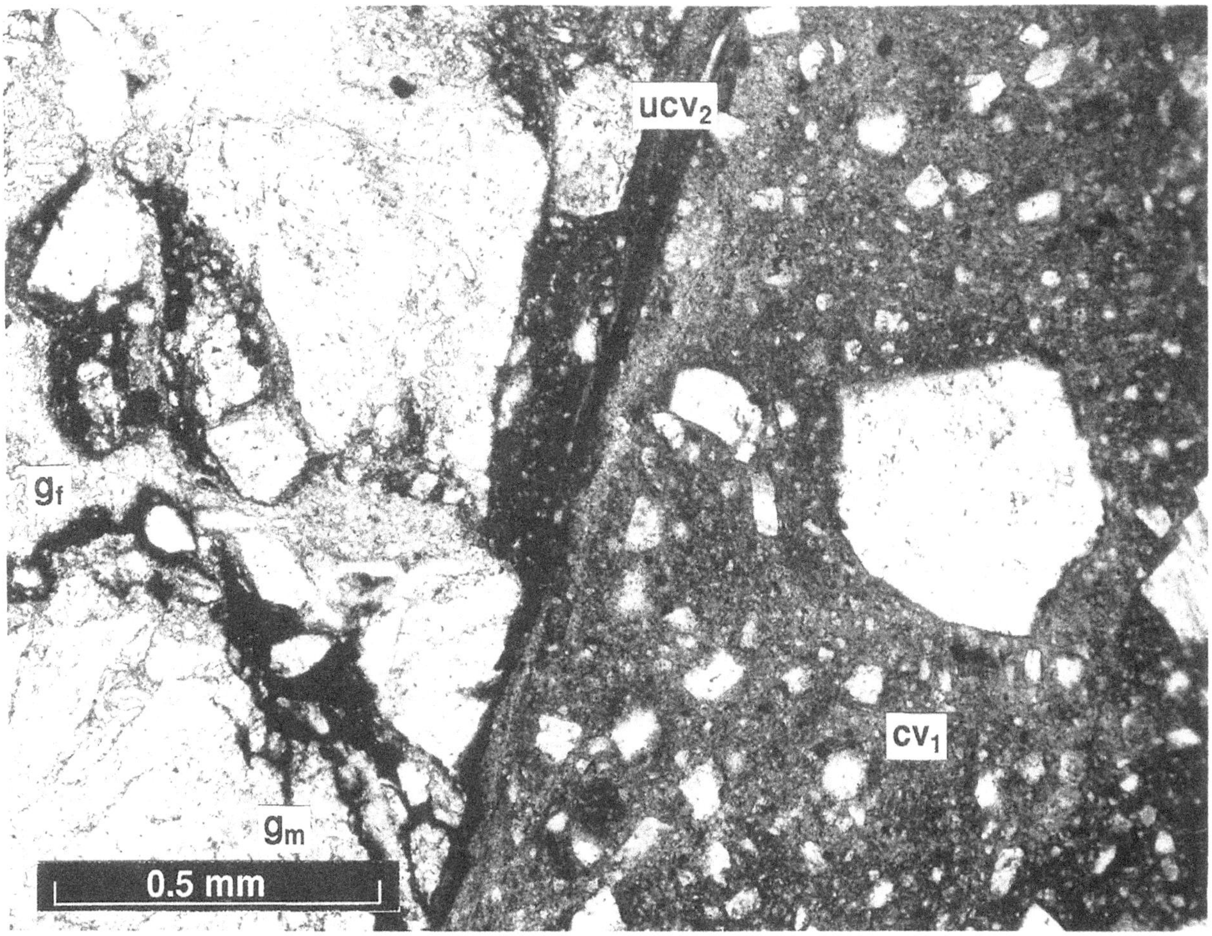


6 A

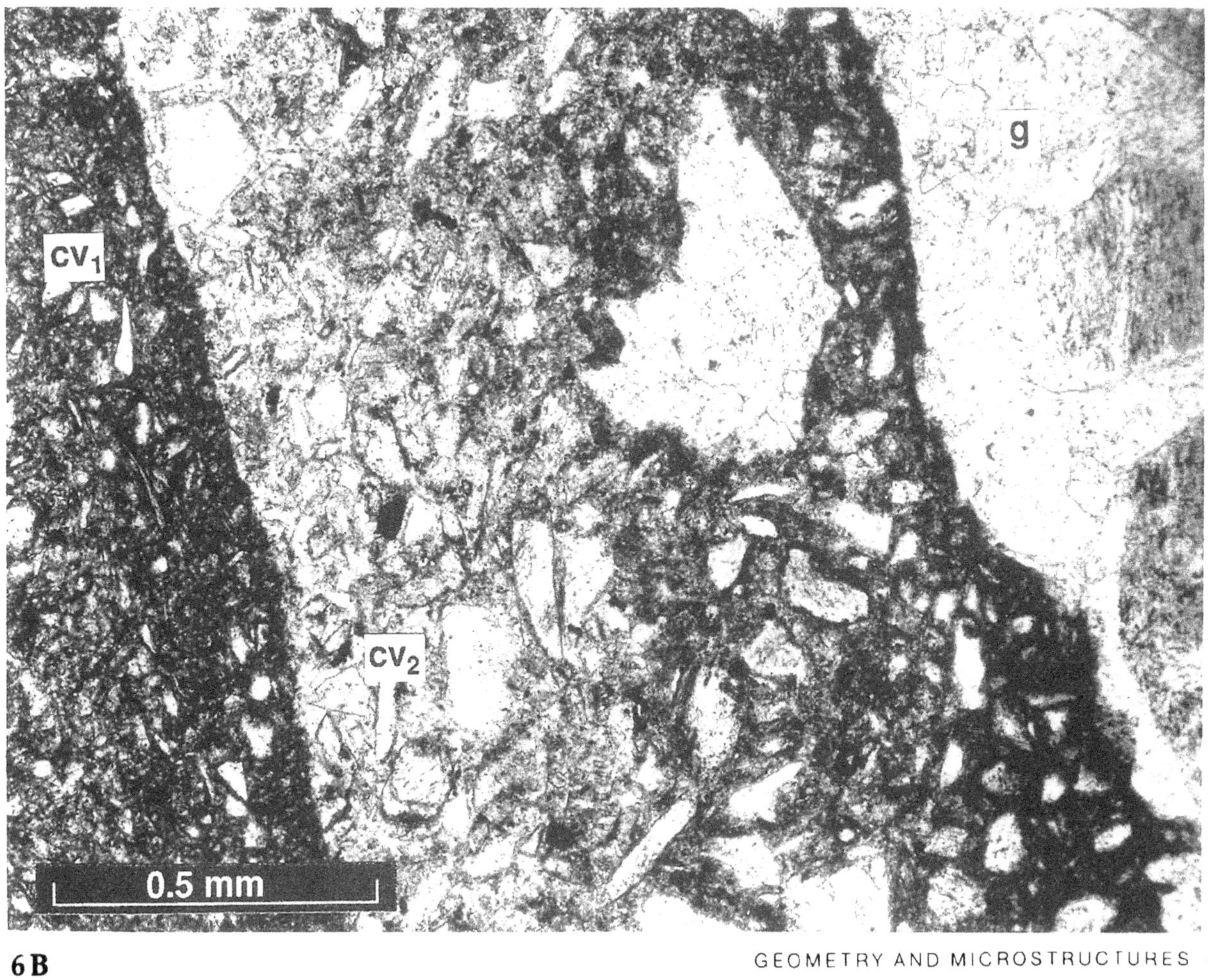


6 B

# 7. DEFORMATION OF GRANITIC ROCKS AT SHALLOW CRUSTAL LEVELS

JAMES P. EVANS

This example of deformation in granitic rocks comes from fault zones developed in granitic gneisses in the Washakie Range, northwestern Wyoming, U.S.A. (Evans, 1988, 1990, 1993). Early Tertiary thrusting of Archean gneisses occurred by slip on narrow faults 1–10 cm thick, and on wide zones 1–7 m thick. Foliation in narrow faults is well expressed in the outcrop, whereas the wider faults consist of irregular pods of fractured and altered protolith enclosed by fine-grained foliated gouge and cataclasite. Deformation occurred at less than 5 km depth.

**7A.** Negative image of an entire thin section from a narrow foliated fault. Foliated cataclasite lies above the two white arrows, and foliation is defined by elongate grains of quartz which lie in a matrix of sericite and feldspar, and by Fe-rich bands at the edges of the zone, here marked by white. Outside of the fault, dark zones of quartz (Q) display few fractures. In contrast, gray feldspar grains (F) contain numerous intragranular fractures (zones labeled I), which commonly follow cleavage planes and transgranular fractures (T) parallel to the fault zone. Sense-of-slip is top to the right. Image is a direct exposure of the thin section in a photo enlarger, so shades are inverted, e.g., dark material is white.

**7B.** Mosaic of photomicrographs from a portion of the fault shown in 7A. Elongate quartz grains (Q) form by intragranular fracture. In some cases, the highly fractured masses of fine-grained quartz resemble dynamically recrystallized quartz, but deformation in this case was dominated by brittle processes. Lack of preferred orientation is indicated when viewed with the first-order plate inserted. Quartz grains lie in a fine-grained sericite-feldspar matrix which deformed by cataclastic flow. Planar fabrics include Riedel fabrics, with the R plane at a shallow angle to the fault-parallel foliation (F). Sense-of-slip is top to the right. NX

**7C.** Photomicrograph of feldspar grain (F) with numerous intragranular fractures, fine-grained sericitic matrix (S), and biotite grain (B) which has deformed by kinking and slip along the basal [0001] plane. NX

**7D.** Photomicrograph showing random-fabric cataclasites from a wide fault. Angular to subrounded feldspar and quartz fragments "float" in a matrix of very fine-grained feldspar grains and syntectonically produced sericite. Deformation is largely dominated by cataclastic flow of the fine-grained particles in the matrix, and the fragments rarely touch other fragments. NX

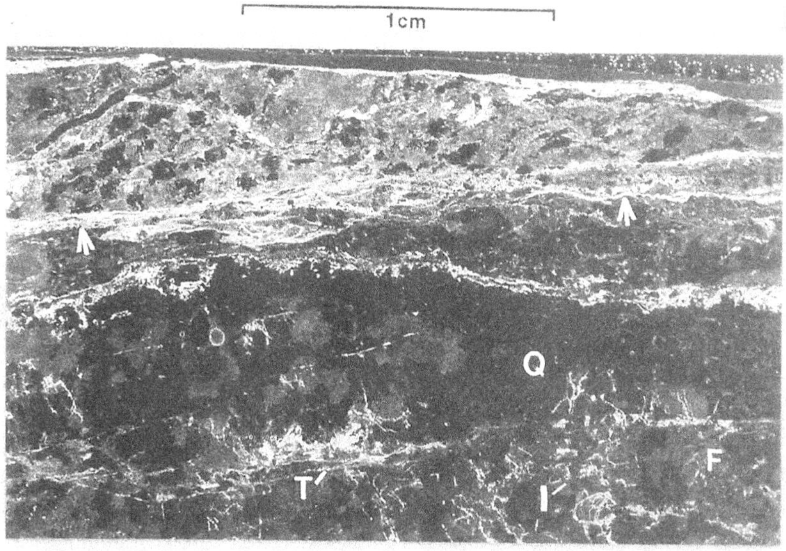


7 A

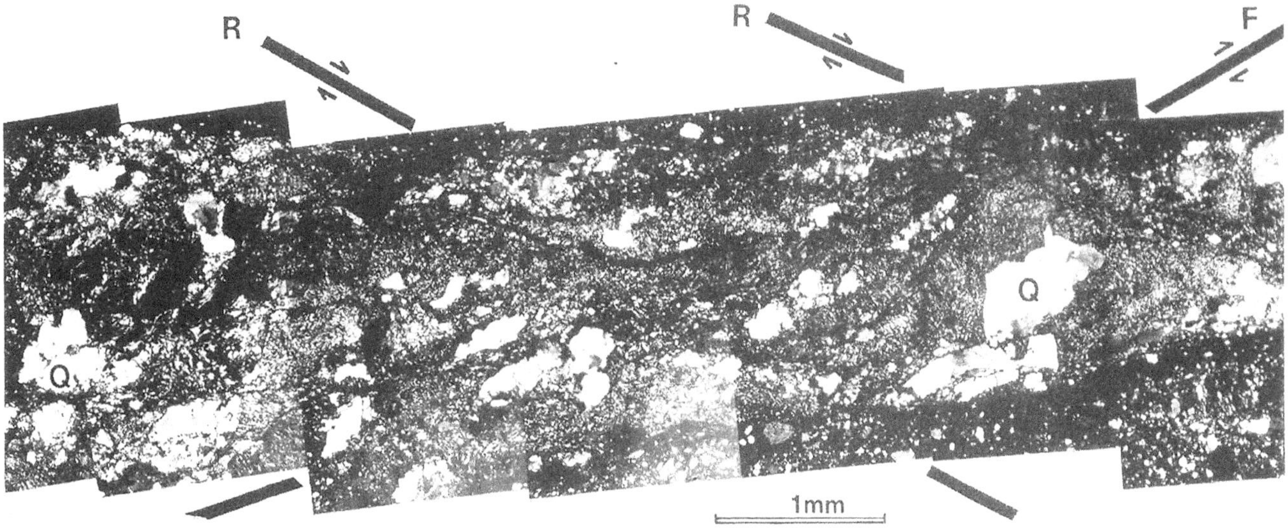


7 B

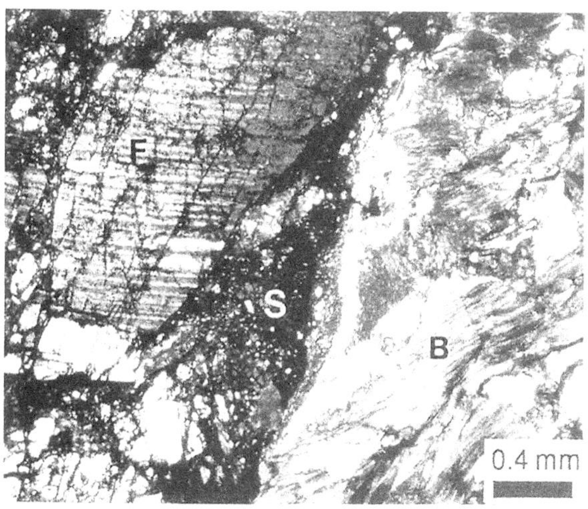


7 C

7 D

# 8. INFLUENCE OF BIOTITE AND FOLIATION ON BRITTLE DEFORMATION OF GNEISS

JAMES P. EVANS

Experimental work of Borg and Handin (1966) and Gottschalk and others (1990) indicates that relatively small quantities of biotite and gneissic foliation may influence the orientation, location, and density of fractures and subsequent faults in brittlely deformed rocks. The examples presented here are from naturally deformed granite gneisses in southwestern Montana, and are analogous to the features described in previous experimental work. Archean gneisses in the footwall and hanging wall of the Scarface thrust were deformed at depths no greater than 4 km during early Tertiary, east-directed thrusting (Schmidt and others, 1993). The protolith consists of well-foliated, medium-grained quartz-feldspar-biotite gneiss. Few fractures exist in undeformed samples of the rock.

**8A.** Slightly deformed gneiss contains numerous intragranular fractures (IF) which are parallel to foliation and connect grains of biotite. Fractures which appear to link biotite grains (L) and incipient zones of cataclasis (C) are evident. PPL

**8B.** Slip along some of the fractures results in narrow fault zones. Faults contain angular fragments of quartz and feldspar with a mantling of fine-grained biotite. The fault zone is continuous with a transgranular fracture out of the field of view, indicating that faults nucleate along transgranular fractures, and that biotite accommodated much of the deformation within the fault. PPL

**8C.** Higher magnification view of a portion of a fault reveals importance of biotite grains in fracture nucleation (F), slip localization (S), development of a wavy fabric (W), and mantling of stiffer grains. PPL

**8D.** In the brittle field, biotite deforms by at least four mechanisms. In addition to kinking and basal slip, biotite which lies at high angles to fault planes is shredded and entrained into the fault zone (arrow). NX

**8E.** A second mechanism of biotite deformation is interpenetration of stiffer grains such as (S) into biotite (B). This mechanism results in biotite mantling stiffer grains, and along with shredding and smearing, results in the distribution of very fine-grained biotite in fault zones isolating stiffer grains. NX

**8F.** Smearing of biotite along its dominant cleavage. This occurs where biotite grains are at shallow angles to the slip direction. Fracture of quartz (Q) and feldspar (F) grains, along with smearing and shredding of biotite, results in the development of a random-fabric cataclasite in lower right part of photo. NX

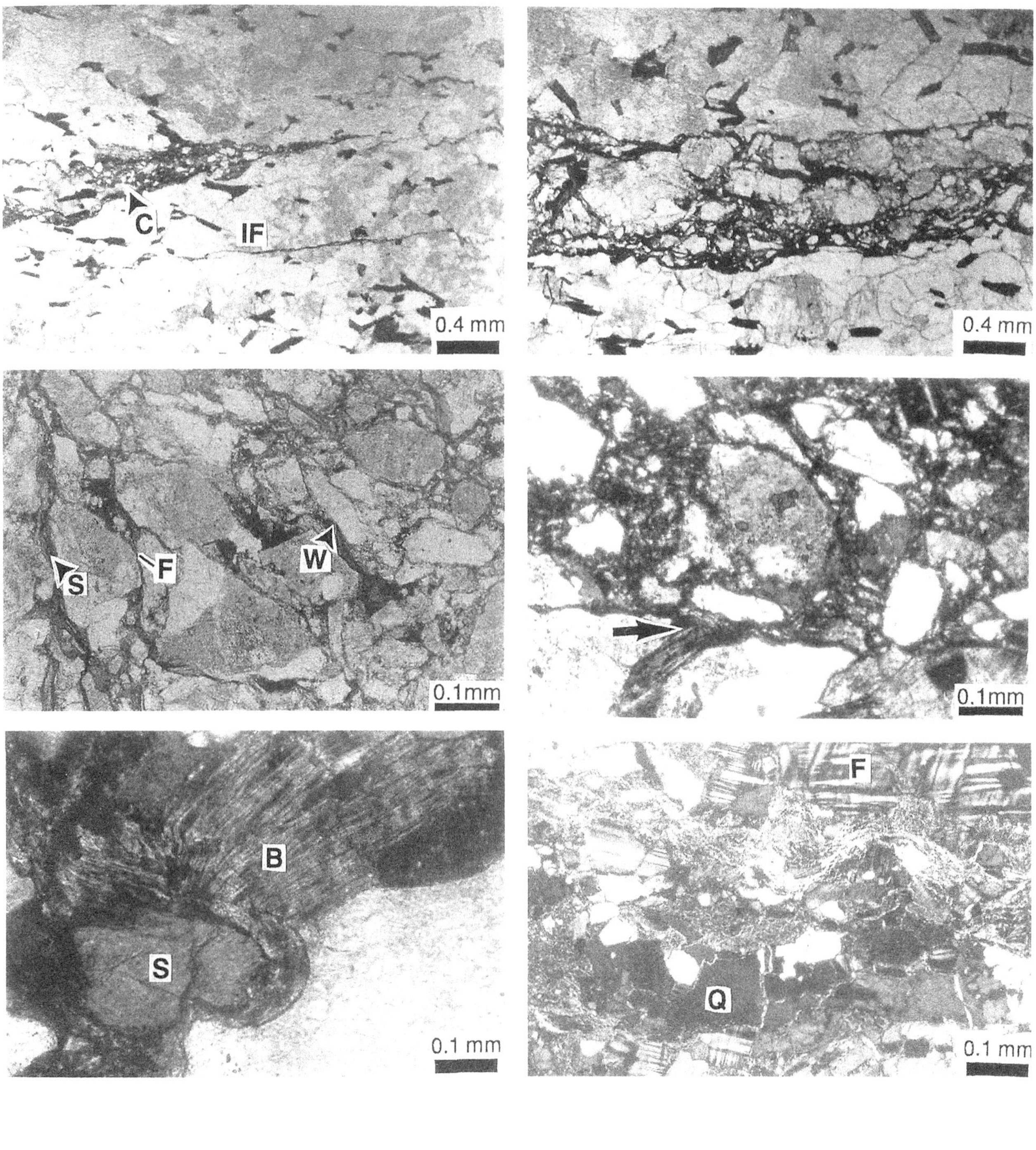


8 A 8 B 8 C 8 D 8 E 8 F

# 9. STRENGTHENING OF FAULT BREECCIA BY K-FELDSPAR CEMENTATION

ROBERT P. WINTSCH

Microbreccias fill fractures that occur along several terrane boundaries in southern New England, especially the Alleghanian Honey Hill fault zone in southeastern Connecticut, U.S.A. (Lundgren and Eblin, 1972). Shown here are microbreccias from the Canterbury granitic gneiss, sampled from drill core (Wintsch and Ambers, 1989) in the hanging wall of the Honey Hill fault zone where it crosses the Connecticut River in Hadlyme, Connecticut. These structures that were formed under lower greenschist-facies conditions overprint higher-grade Alleghanian and Acadian gneissic fabrics (Wintsch and others, 1992). These breccias are significant, because they show at least four cycles of brecciation and recementation. All cements contain an intergrowth of randomly oriented 2–10 μm sized grains of K-feldspar, chlorite, and sphene in varying proportions in successive cycles of cementation.

Alkali feldspars as a major component of these cements have both chemical and mechanical significance. First, precipitation of feldspars requires a strongly alkaline solution, as could be produced in rock-dominated, fluid-poor systems by ion exchange on newly exposed surfaces in the breccia (Wintsch and Knipe, 1983; Wintsch and others, 1995). Thus the ingress of aqueous fluids into this fault zone may have been quite limited, and the fluid pressures may not have been high. Second, K-feldspar-cemented microbreccias will have a strength similar to that of the host feldspathic gneiss. In fact, repeated (seismic?) brecciation of the earlier breccias is evidence that successive generations of breccia were strong, and suggests that cementation occurred on the time scale of the earthquake cycle. Only when the chlorite mode is high do the breccias become foliated. Thus, brecciation in felsic rocks in a rock-dominated fluid system may not lead to significant weakening of fault zones, and experiments using simulated, uncemented fault gouges may greatly underestimate the strengths of microbreccias.

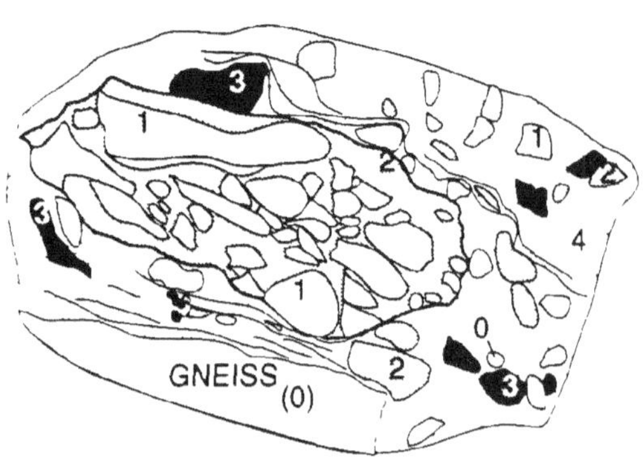


**9A.** A relatively large clast in a microbreccia filling a Mesozoic fracture in K-feldspar-plagioclase-quartz gneiss (PPL), ductilely deformed in the Acadian and Alleghanian orogenies (Wintsch and others, 1993). Most fragments in these clasts (Figure 9.1) are microbreccias with generation 1 cement (albite > K-feldspar > chlorite > sphene; 1, pale gray). These microbreccias are cemented by generation 2 intergrowth of K-feldspar > chlorite > sphene (2, medium gray, Figure 9.1). A third generation of microbreccia (3, dark gray, Figure 9.1) contains clasts of prehnite derived from veins that cut across gneiss and breccia at a high angle. Clasts of generations 1, 2, and 3 are cemented by an intergrowth of chlorite > K-feldspar > sphene (4, dotted, Figure 9.1). A local, weak preferred orientation of chlorite in narrow zones (wavy lines, Figure 9.1) suggests that a small amount of plastic strain was accommodated in generation 4 cements.

**9B.** Composite clast of plagioclase from the gneiss and generation 1 microbreccia (1) floating in a matrix of generation 4 microbreccia (NX). Bright grains are plagioclase (P), quartz (Q), and prehnite (R). The plagioclase of the composite grain is weakly sericitized, albitized, and fractured. Of significance is the fact that this fragment broke across the microbreccia-plagioclase boundary, rather than along this boundary. This confirms that the cements of the microbreccias can be as strong as the plagioclase gneiss host.

**9C.** Backscattered electron image of the boxed portion of 9B. The small pale gray regions in (1) are 3–10 μm grains of K-feldspar among similar grains of albite and scarce chlorite. The microbreccia is cut by two small K-feldspar > chlorite veins (V), that may represent a generation of cements similar to cycle 2 clasts. Clasts of a second cycle contain fragments of first cycle breccia and first cycle or protolith quartz (Q) clasts. Generation 2 cements are K-feldspar > chlorite > albite. Pale patches and bands in clasts of prehnite (R) broken from an adjacent vein are richer in ferric iron. The cement to these clasts (cycle 4 cement) is randomly oriented chlorite > K-feldspar.

**Figure 9.1.** Index map of the large clast in Plate 9A identifying the host gneiss and gneiss fragments (0), and progressively younger generations of cemented breccias 1 through 4

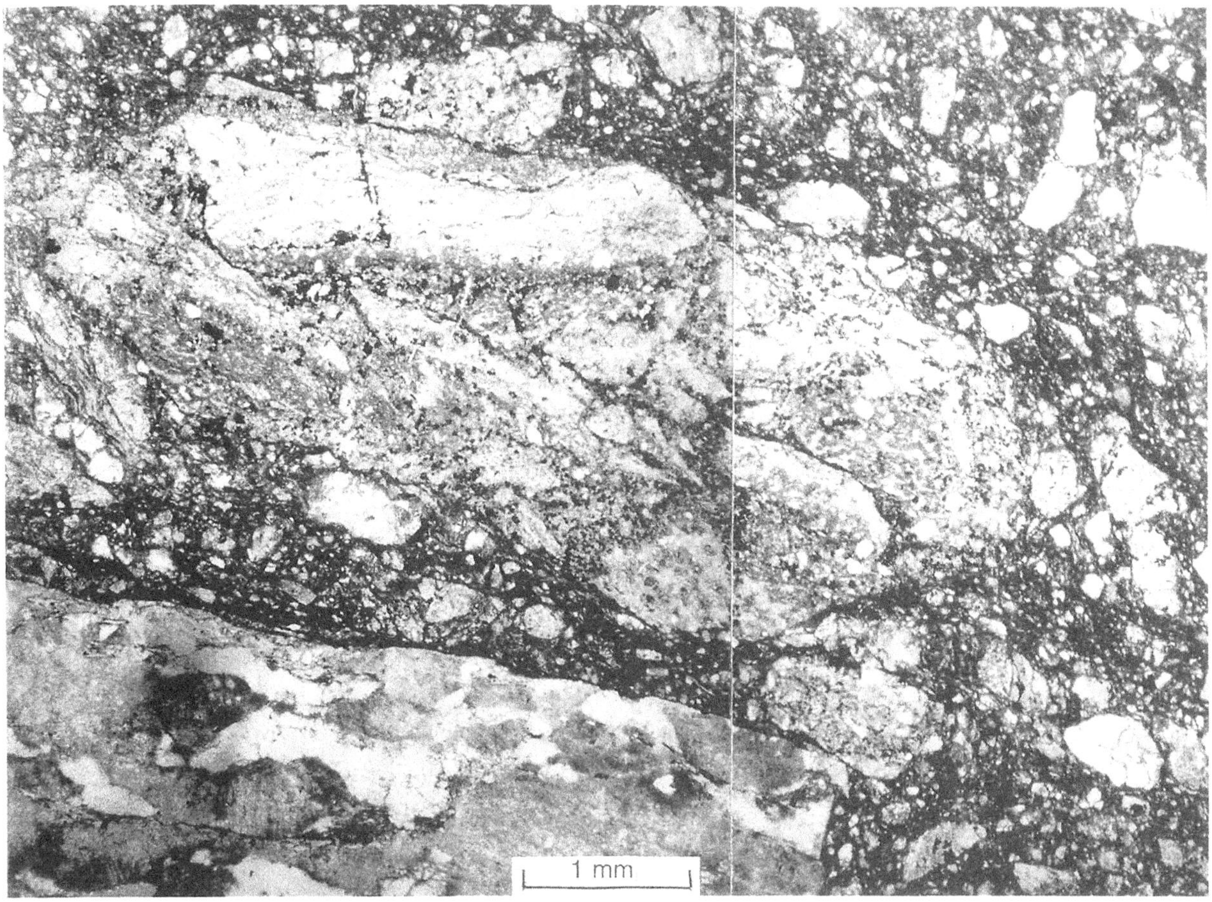


9 A

9 B

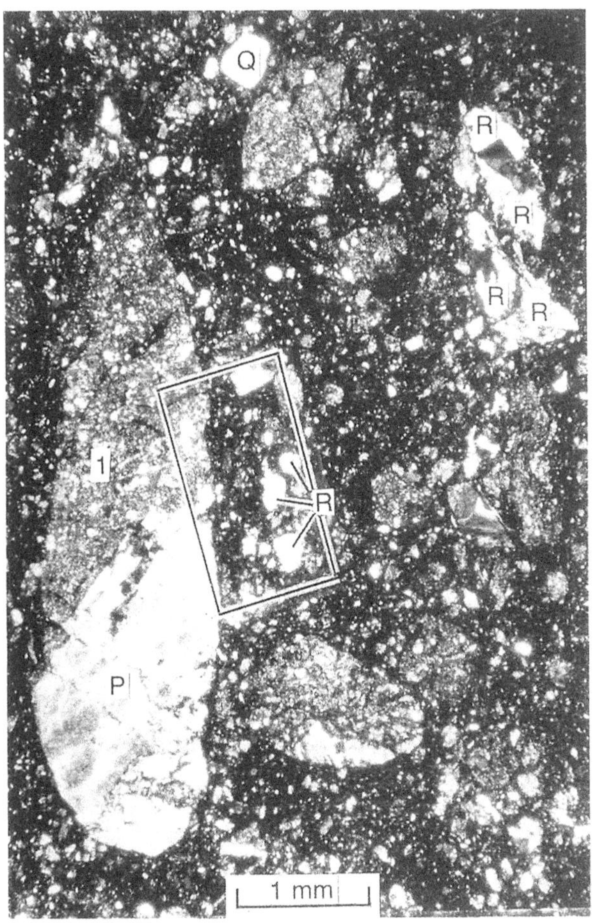


9 C

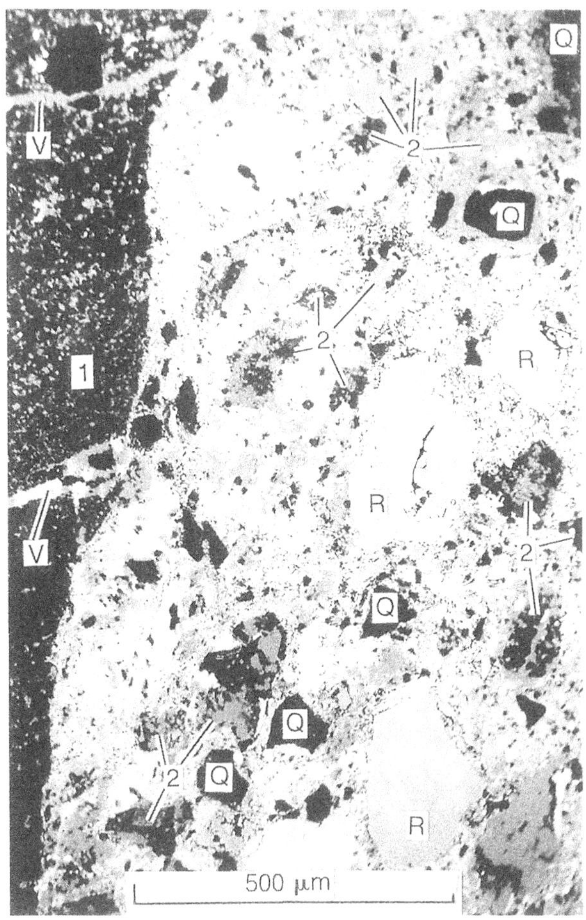

# 10. QUARTZ MICROSTRUCTURES IN A FAULT PROCESS ZONE

T. G. BLENKINSOP AND M. R. DRURY

The microstructures described here come from the Bayas fault in the Ordovician Barrios quartzite, exposed on the north coast of Spain (Blenkinsop and Drury, 1988). The area is part of the Cantabrian zone, a foreland fold-and-thrust belt that is the most external zone of the Variscan orogen in northwest Spain. The deformation can be related to the main phase of Variscan tectonism between Fammenian and Upper Westphalian (Blenkinsop, 1987). The maximum temperatures during deformation, estimated from illite crystallinities, were 148–248°C, and the minimum confining pressure from the stratigraphic depth of burial was 100 MPa. Pore fluid pressures probably fluctuated along the fault zone, but may have been hydrostatic in the wall rock, where the effective confining pressure was 60 MPa, assuming a ratio of pore fluid pressure to lithostatic stress ($\lambda$) of 0.4.

The Barrios quartzite is a fine- to medium-grained quartzite (grain size = 100–200 μm) with a quartz cement. The Bayas fault cuts the hinge zone of a gently inclined, southeast-verging anticline, and the fault surface has lineations plunging northeast at 30°. Bedding in the quartzite has a dextral strike separation of 100 m. The fault zone consists of a layer of cataclasite several tens of millimeters thick, comprising angular fragments of grains in a fine-grained matrix with iron oxides. Both the unsheared wall rock within 3 m of the fault, and fragments of wall rock in the cataclasite, have three important microstructures: microfractures, deformation bands, and deformation lamellae.

**10A.** Microfractures (1), deformation bands (2), and deformation lamellae (3) in wall-rock grains in the cataclasite. Intragranular microfractures are seen as lines of fluid inclusions, which have a strong preferred orientation and extend to grain boundaries but do not offset them. Deformation bands are spindle-shaped bands of extinction contrast, with poorly defined edges. Deformation lamellae are sharply defined, extremely narrow (~1 μm) bands perpendicular to deformation bands. NX

**10B.** TEM view of microfractures. A low density of straight dislocations, and a high density of radiation damage centers in bands parallel to the sharp edges of the microfracture, are diagnostic of precipitated quartz which was deposited as the microfracture healed.

**10C.** TEM view of the banded substructure that defines the deformation lamellae. The boundaries of the substructures consist of curved dislocation walls made up of dense, well-ordered arrays of two or three sets of dislocations.

**10D.** Average dislocation density and high density of fluid inclusions within the sub-basal lamellae. Note occurrence of very low-angle dislocation walls and networks, and the interlinking of dislocations forming a coarse network with fluid inclusions at the nodes. These structures are within the lamellae; the lamellae walls (not seen) are parallel to the vertical direction. TEM

Deformation lamellae form only at high stresses in the "power law breakdown" regime (Drury and Humphreys, 1986), and indicate differential stresses of 100–200 MPa in quartz (Drury, 1993). Minimum differential stresses of 116 MPa and 180 MPa are suggested by partly recovered dislocation densities (McCormick, 1977) and lamella spacing (Koch and Christie, 1981), respectively. These high stresses are consistent with models of the stress concentration in the process zone ahead of the propagating Bayas fault. The modeling shows that remote (tectonic) stresses have geologically reasonable values of less than 100 MPa. The distinction between process zone and remote applied stresses is important in interpreting deformation microstructures.

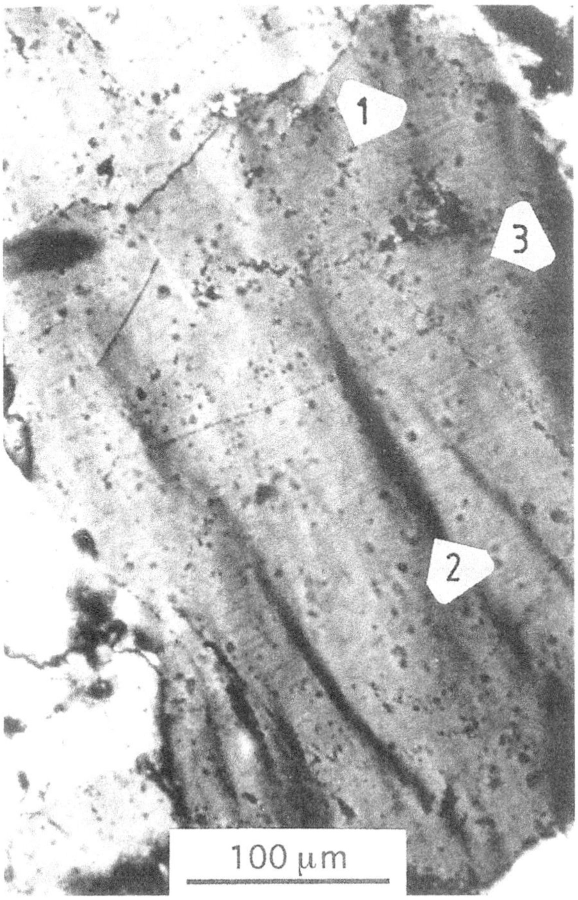


10 A

10 B

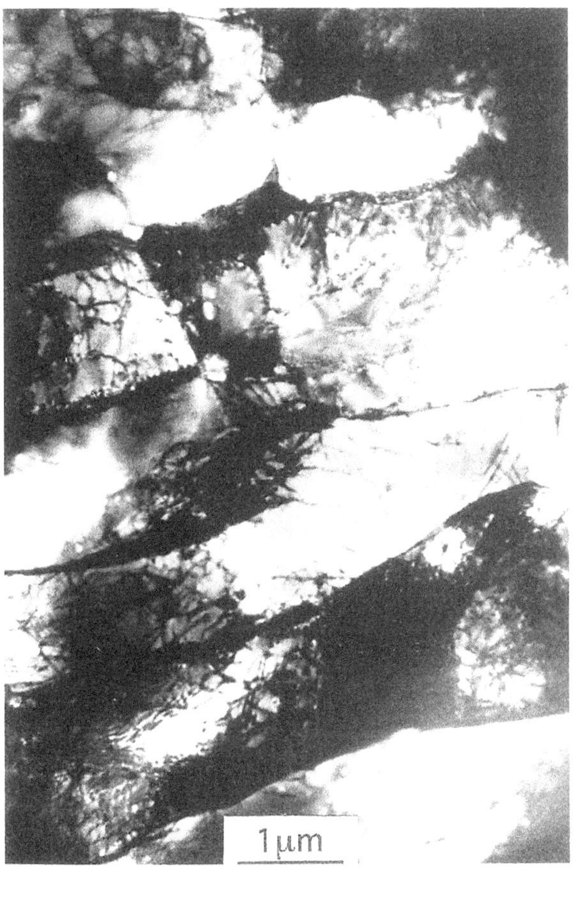


10 C

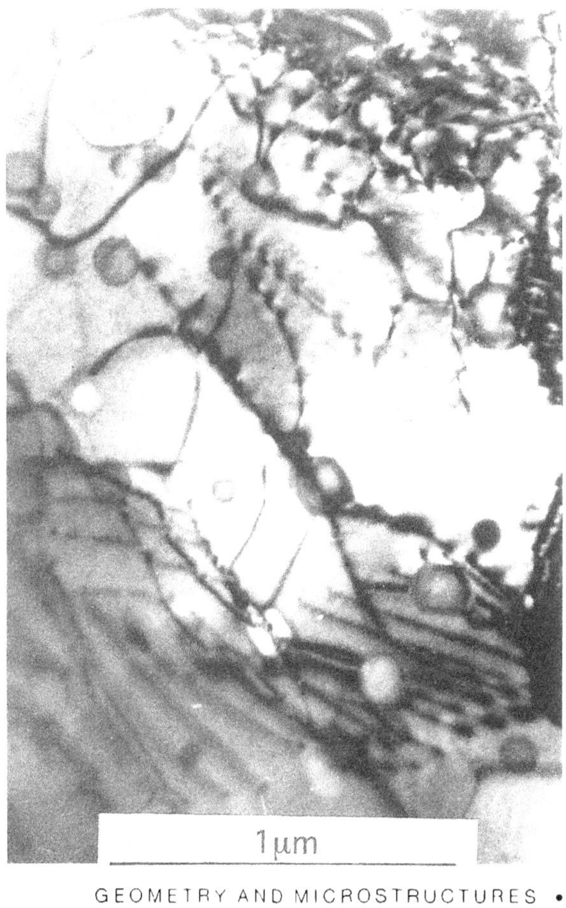


10 D

# 11. CATACLASTIC DEFORMATION OF QUARTZITE REVEALED BY CATHODOLUMINESCENCE

T. G. BLENKINSOP AND E. H. RUTTER

These microstructures come from Cambrian quartzites in the southern part of the Moine thrust zone on the Sleat peninsula, Isle of Skye, Scotland. The Moine thrust zone is here over 10 km wide and consists of a number of thrusts (including the Ord thrust) carrying nappes toward the northwest, emplaced during the Caledonian orogeny. The maximum temperature of deformation was 200°C since there is no evidence of crystal-plasticity in quartz (White, 1976), and illite crystallinity gives a metamorphic grade at the diagenetic-anchizone boundary. This temperature suggests a maximum depth of 7 km during deformation (Blenkinsop and Rutter, 1986).

The quartzite contains angular to subrounded detrital grains with an average diameter of 200 mm. The samples described come from the hinge and limb of a large antiform in the hanging wall of the Ord thrust, where they are found in variably deformed states from intact quartzite through protobreccia (>75% intact quartz) to breccia (25–75% fragments) and ultrabreccia (<25% fragments). In the first stage of deformation, intragranular extension microfractures formed at grain impingement points, implying that the rock was then more porous. The microfractures have a strong preferred orientation at high angles to the shear fractures in all samples except the ultrabreccia. Pressure solution also occurred at grain contacts. Shear fractures formed by the linking of and displacement on microfractures, generating a matrix of comminuted grain fragments, and iron oxide and quartz cement were deposited. Extension microfracture densities increased with deformation from 5 to 36 $mm^{-1}$.

Many of these deformation features are revealed by examining the samples in cathodoluminescence. Polished thin sections were excited in the electron beam of a scanning electron microscope (SEM), and luminescence was detected by a photomultiplier. Different generations of quartz are revealed by contrasts in luminescence: most spectacularly, microfractures are shown as lines of nonluminescing quartz, and the matrix of the shear fractures is seen to consist of cement (nonluminescing) and grain fragments (variable luminescence).

**11A.** Intact quartzite. Detrital grains have variable luminescence, indicating a variety of provenances. Structures seen within grains are inherited. Arrows show a typical pressure solution contact between grains. SEM, cathodoluminescence.

**11B.** Intergranular (1) and intragranular (2) microfractures clearly revealed as lines of nonluminescing quartz. Intragranular microfractures have impingement geometries, commonly radiating from points of contact between grains (e.g., 2), but also have a good overall preferred orientation at high angles to shear fractures. Both pores and microfractures are filled with nonluminescing quartz, implying formation under similar conditions. SEM, cathodoluminescence.

**11C.** High density of microfractures. Shear fracture (between arrows) forms by linkage of and displacement on microfractures. SEM, cathodoluminescence.

**11D.** Shear fracture (1) containing matrix of luminescing grain fragments surrounded by nonluminescing cement. Detachment of grain fragments along microfractures is seen at 2. SEM, cathodoluminescence.

The cataclasis described here has many features in common with samples experimentally deformed in the laboratory, but occurred on a much larger scale. The study shows that cataclasis can accommodate folding under low-grade metamorphic conditions. This cataclastic deformation is typical of the southern end of the Moine thrust zone, and contrasts with the dominantly plastic deformation at the northern end, where there may have been greater uplift (Blenkinsop and Rutter, 1986).

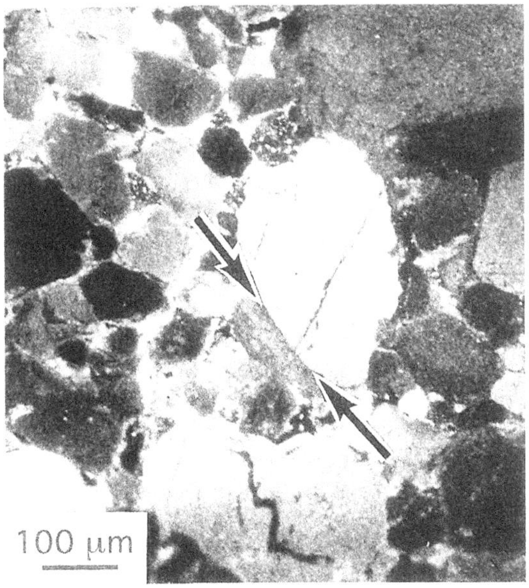


11 A

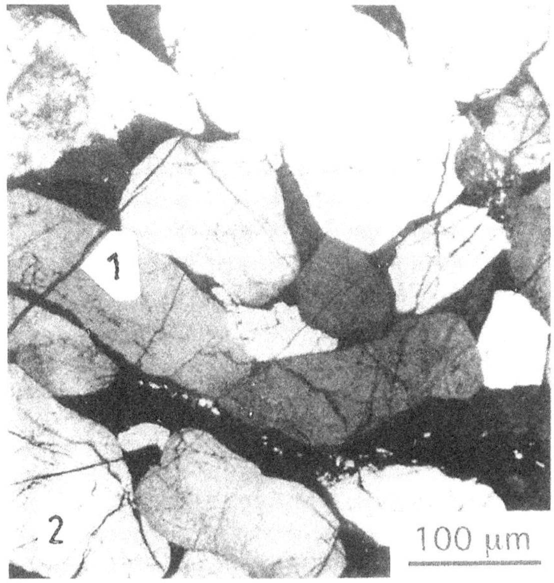


11 B

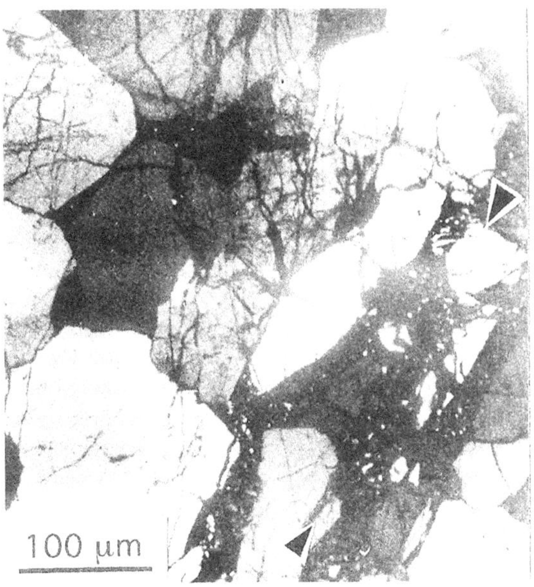


11 C

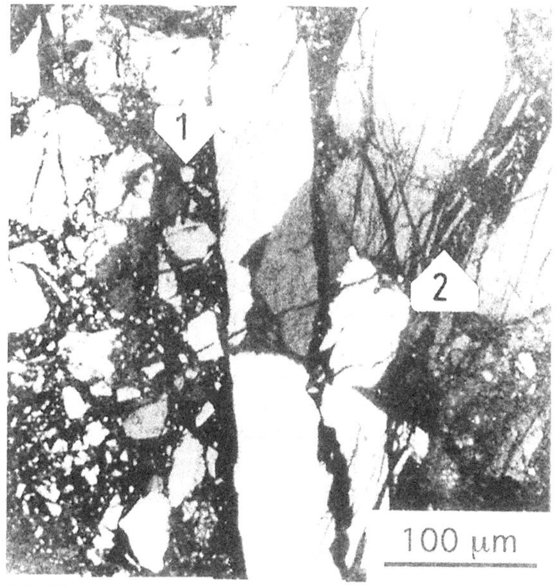


11 D

# **12.** MICROSTRUCTURAL VARIATIONS IN A CATACLASTIC QUARTZ FAULT ROCK

G. E. LLOYD

Faulting under upper crustal conditions, although dominated by fracturing, can also involve other deformation mechanisms and processes, resulting in characteristic but different microstructures. This combination of deformation mechanisms is shown by minor faults (e.g., Skiag Bridge reverse fault) in Cambrian quartzite from the Moine thrust zone at Assynt, northwest Scotland, which developed under conditions of <250°C and ~2–3 MPa (Lloyd and Knipe, 1992; Knipe and Lloyd, 1994). However, it is not always possible to recognize the microstructures indicative of different processes by using a single analytical technique, because different techniques are sensitive to different aspects of the microstructure. Operation of the SEM in the backscattered electron (BSE) electron channeling (EC) orientation contrast (OC) mode but at relatively long working distances (i.e., operating conditions: ~15 mm working distance; solid-state BSE detectors; 20 kV accelerating voltages; high probe currents, 20–50 nA) results not only in images based on changes in specimen crystallographic orientation (see Plate 82) but also in significant increases in other emission signals and improved spatial and atomic number resolutions, because the emission signals originate from more restricted depths in the specimen (Lloyd, 1994). It is possible to gain additional information by combining OC with other imaging modes, such as cathodoluminescence (CL). Such a combined approach is particularly important for microstructural analysis of fault zone microstructures in quartzite, because OC images are most sensitive to microstructures resulting from crystal-plasticity, whereas CL images are most sensitive to microstructures resulting from fracture and diffusive mass transport (including cementation) processes. Indeed, it is likely that SEM CL analysis will provide the best estimate of microcrack density and distribution in quartzose rocks.

Simultaneous SEM OC and CL analyses of the Skiag Bridge fault zone (12A and 12B) reveal the effects of faulting on the original sedimentary rock, and the operation of fracture, crystal-plasticity, and diffusive mass transfer (dmt) deformation mechanisms/processes. A variety of intragranular microstructures can be recognized, although these microstructures vary from grain to grain, presumably due to different crystallographic orientations with respect to the local stress field, and some grains may remain essentially undeformed. The microstructures observed in this sample include (1) subgrain arrays (s), particularly at grain contacts due to local stress concentrations relieved by "slow/soft indentation"; (2) intragranular (extension) microfractures (f), usually healed by authigenic quartz (which has a dark CL contrast) via localized dmt processes, especially originating at grain contacts due to local stress concentrations relieved by "hard/fast indentation"; and (3) deformation lamellae (d), commonly oriented normal to the microfractures and/or the direction of grain indentation at contact points. Intergranular regions ("pores") and grain boundaries commonly show authigenic quartz (darker CL contrasts), precipitated via local dmt processes. Transgranular fracture zones represent the concentration and localization of deformation and displacement, and are characterized by intense grain-size reduction via cataclasis.

**12A.** Orientation contrast image. Note the apparent dominance of crystal-plastic microstructures: intragranular contrast variations in wall-rock grains are due to subgrain formation(s) and/or deformation lamellae (d); whereas the fracture zone has a coherent appearance and seems to consist of much finer grains and/or subgrains.

**12B.** Cathodoluminescence image. Note the dominance of fracture microstructures: healed intragranular extension fractures (f) occur in wall-rock grains (their darker contrast is due to cementation by authigenic quartz), whereas the fracture zone has a fragmented appearance, with parental grain origins of individual fragments revealed by similar contrasts, which also imparts a "foliated" appearance to the fracture microstructure.

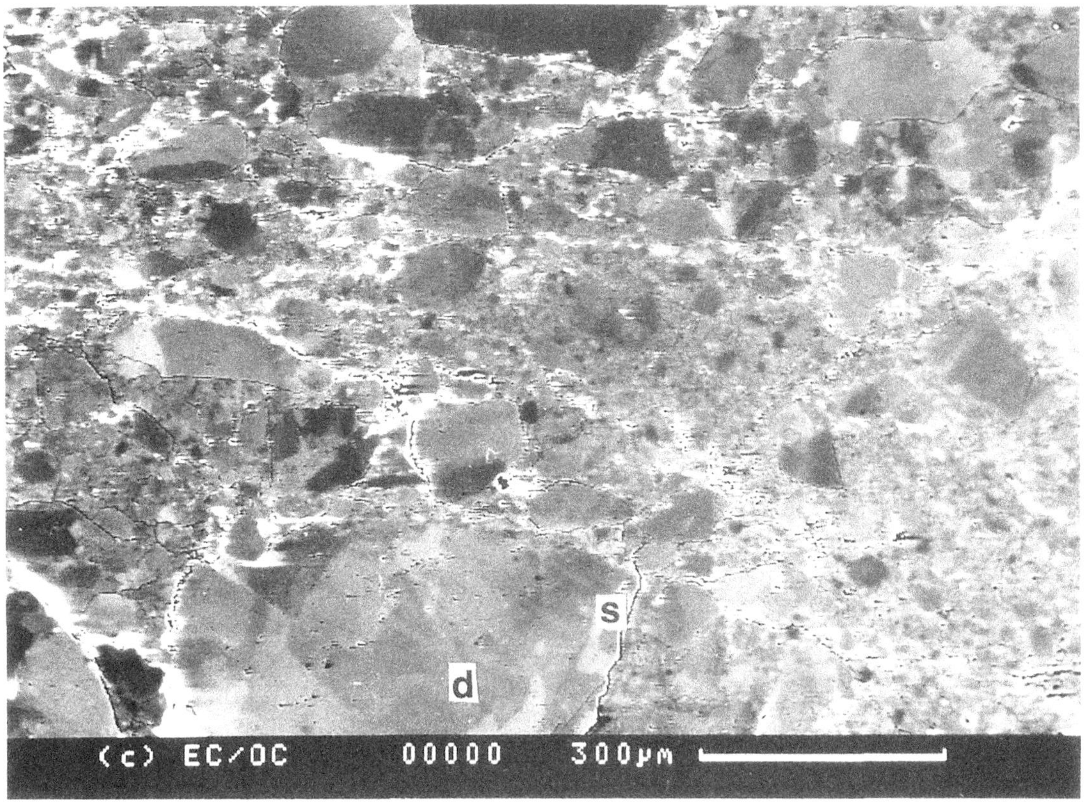


12 A

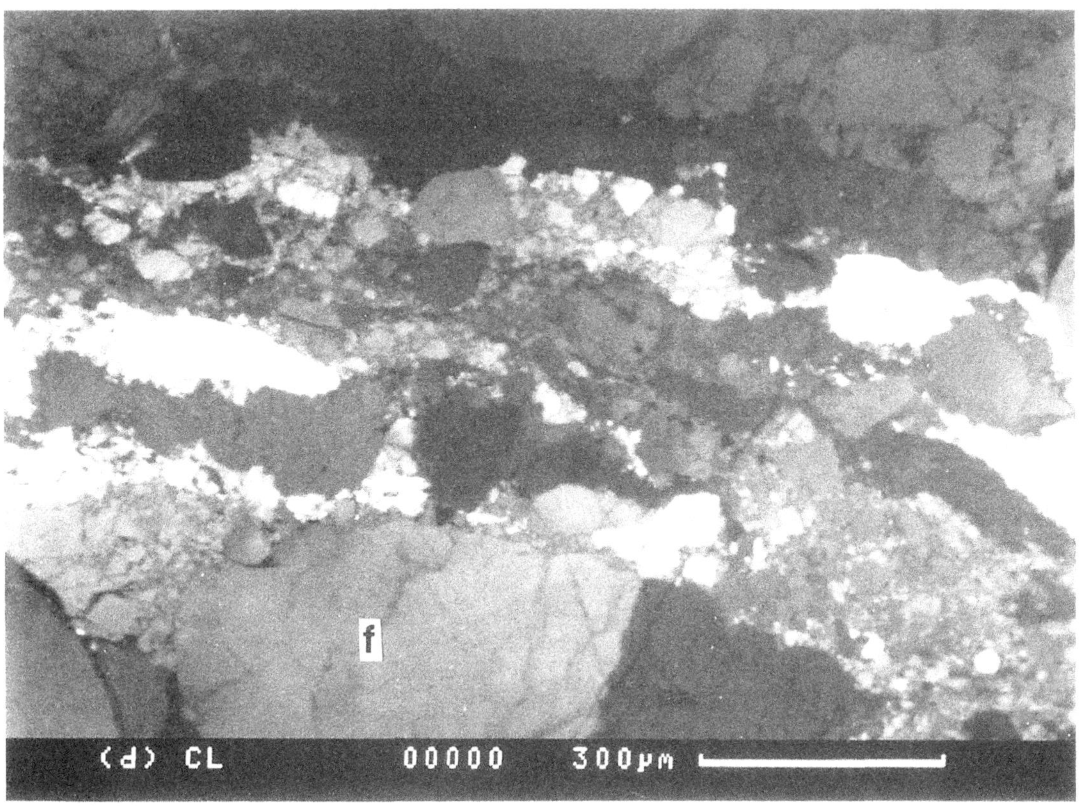


12 B

# BRITTLE BEHAVIOR

## Cataclasis and gouge development



Common abbreviations used in regard to the plates in this section include PPL = plane polarized light and NX = crossed nicols.

Although the basic focus of this set of nine plates overlaps in part with the previous examples from brittle fault zones, they emphasize the processes and nature of gouge and cataclasite from natural and experimental faults. The dominant process in all cases is fracturing on all scales and frictional sliding of the fragments. The ultimate product is extremely fine grained, incohesive fault gouge, or if cementation occurs by precipitation from fluids or other processes, cohesive cataclasite. Many of these contributions illustrate Riedel shears, which develop in both experimental and natural fault gouge, in orientations similar to S, C, or C′ planes of higher temperature mylonites. The formation of foliation in fault gouge is also illustrated in several plates. The macroscopic ductility of fault gouge is emphasized in one plate, whereas another illustrates microstructures possibly indicative of stick-slip behavior.

# 13. PROGRESSIVE DEVELOPMENT OF FOLIATION IN A BRITTLE DEFORMATION ZONE

GAUTAM MITRA

Late Cretaceous–early Tertiary Laramide deformation in Precambrian granitic rocks of the Wind River Mountains, Wyoming, U.S.A., gave rise to brittle deformation zones (BDZs) under zeolite-facies metamorphic conditions (Mitra and Frost, 1981; Mitra, 1993; Yonkee and Mitra, 1993). The zones have a high density of brittle fractures characterized by breccia and/or gouge. The photographs show stages in the development of a foliated coherent gouge along the White Rock fault, a major deformation zone near Green River Lakes on the west side of the northern Wind River Mountains (Mitra, 1984).

**13A.** Brittle fractures form in both feldspars and quartz after only small amounts of plastic deformation within the grains. The photograph shows cleavage cracks of different orientations in one feldspar grain coalescing with one another to form continuous fractures, which then propagate unstably across grain boundaries and adjacent grains. PPL

**13B.** Unstable fractures (arrows) that have propagated across grain boundaries and through both quartz and feldspar grains. Anastomosing fractures give rise to coarse fragments. Sliding on fractures produces wear fragments which are finer grained. Room problems created by sliding on intersecting fractures may produce subsidiary fractures, breaking up the intervening material into small fragments. F = feldspar, Q = quartz. NX

**13C.** Continued fracturing produces a breccia with angular fragments floating in a finer-grained matrix. The coarse fragments may continue to fracture (arrow), but the fractures are stable, dying out at grain boundaries. This reduces the overall grain size, forming a gouge. The finer material may continue to reduce its own grain size by abrasion, but once it is below a certain grain size pressure solution may become important (see 13D). F = feldspar, Q = quartz. PPL

**13D.** Pressure solution in the fine-grained matrix of the fault gouge concentrates insoluble opaque material into thin zones which eventually run together to form thick opaque zones. These zones, together with the preferred dimensional orientation in the matrix (also produced by pressure solution), define a foliation. The foliation swings around coarse brittle fragments of quartz (Q), feldspar (F), and epidote (E). PPL

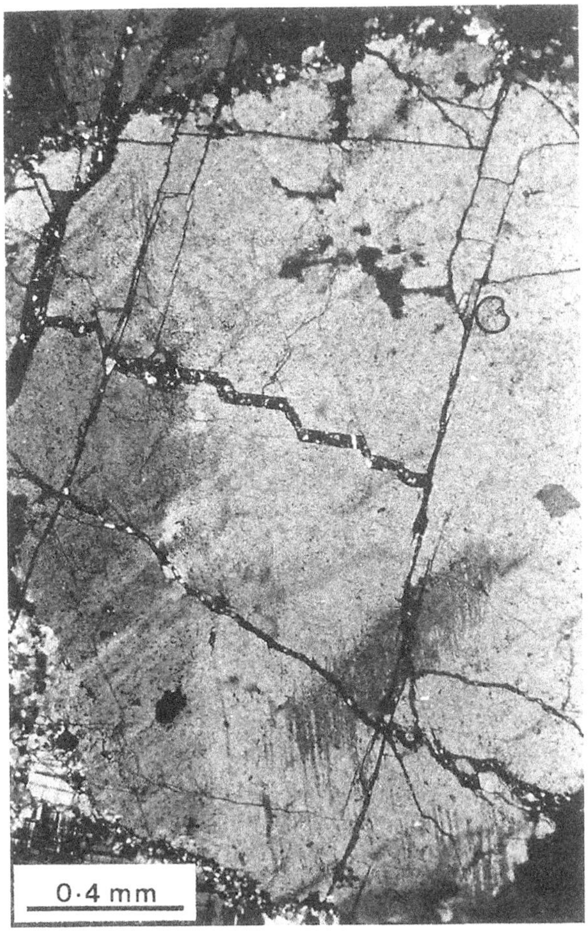

13 A

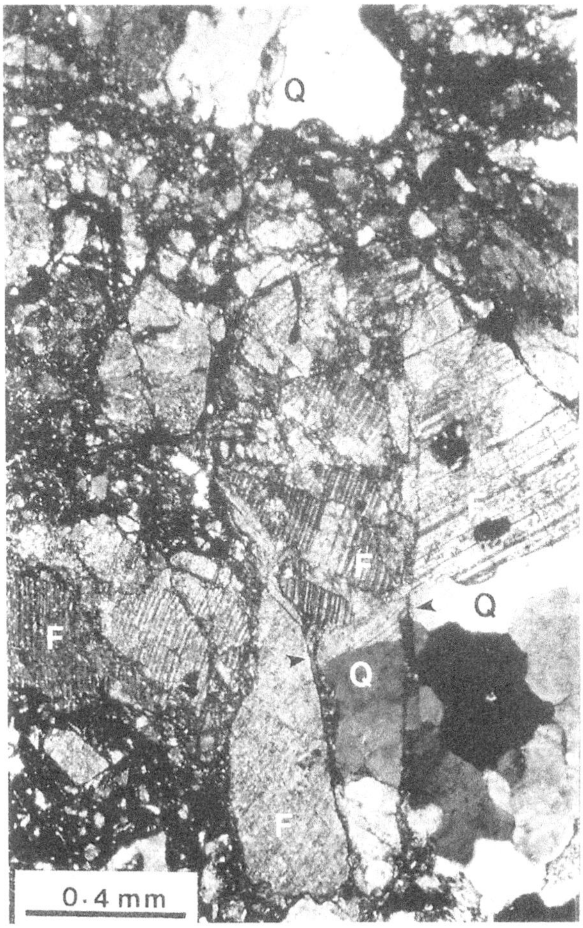

13 B

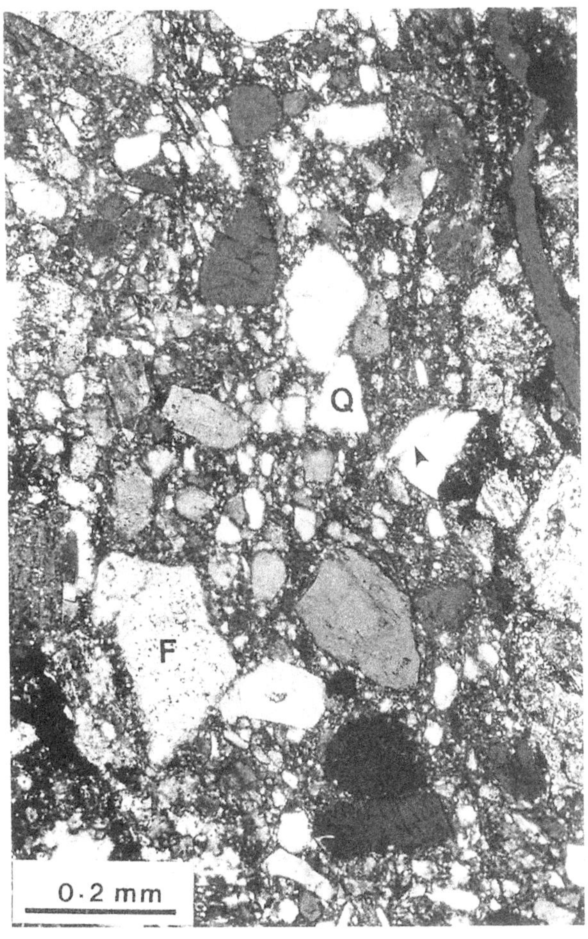

13 C

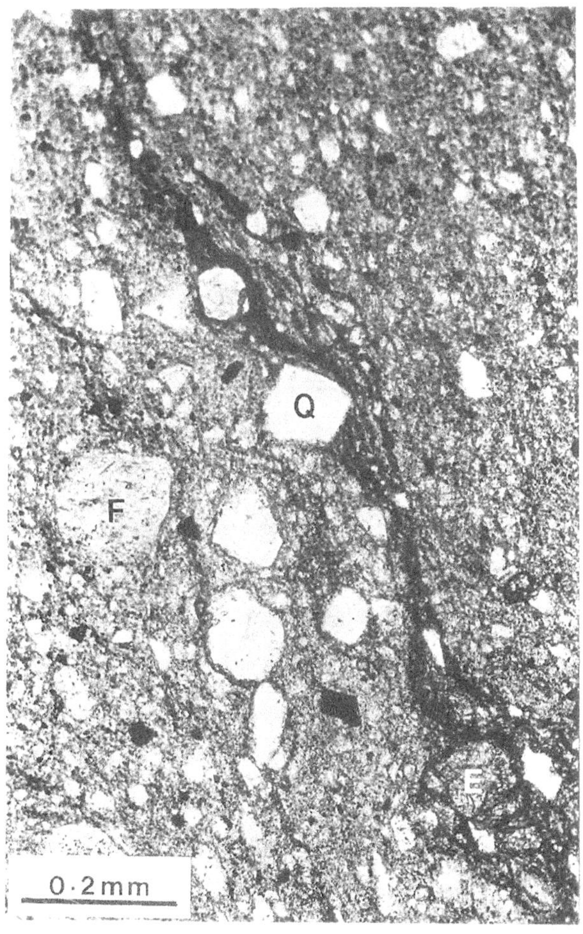

13 D

# 14. MULTIPLE GENERATION CATACLASITES

JÜRGEN E. STREIT

A cataclastically overprinted mylonite belt (40–60 m wide) marks the western contact of a late Proterozoic granite intrusion into metasedimentary rocks along the west coast of King Island, Tasmania, Australia. The south-southeast-striking, greenschist-facies mylonite belt has been active as a strike-slip shear zone and is stepwise offset (sinistral sense) along major east-west-striking faults.

Many steeply dipping (southeast to east strike), narrow fault zones (<5 cm wide) overprint mylonitic fabrics and reveal horizontal displacements of centimeters to several tens of centimeters. The narrow fault zones contain multiple generations of thin cataclasite layers, which usually have sharp contacts with the host rock. Distinct hydrothermal alteration associated with individual cataclasites requires chemical fluid-rock reaction related to individual cataclasite-forming events. The structural and geochemical relationships strongly suggest that the cataclasite layers regained a cohesive strength due to hydrothermal cementation before shear stress accumulation along the faults led to their reactivation. Some cataclasites contain clasts which are locked together along sutured grain boundaries and provide evidence for pressure solution.

**14A.** Multiple generation cataclasite in mylonitic granite, Currie shear zone, King Island, Tasmania. Cataclasite layers contain up to 25% mineral clasts larger than 50 μm, which are largely albite but also include quartz. The clasts are dominantly angular to subangular in shape, are poorly sorted, and are usually embedded in a fine-grained (<50 μm) matrix. Up to 35% of younger cataclasites can consist of fragments from reworked cataclasites including larger mineral clasts (>50 μm). Fragments of older cataclasite layers, larger than 400 μm in diameter, are usually pulled apart and show spalling of sharp tips. Isolated small fragments are generally rounded. Analyzer at 30°.

The matrix bulk composition of successive generations of cataclasites was analyzed by electron microprobe scanning of areas (n = 10) of 50 $\mu m^2$. Average matrix compositions are listed below, as well as a typical X-ray fluorescence analysis of the host rock (mylonitic granite) (Table 14.1).

High losses of MgO and $K_2O$ in cataclasite layers relative to their host rock require mass transfer via migrating fluids, because these element oxides are not contained in any clasts preserved during crushing and do not have increased concentrations in the immediate wall rock. Increased $SiO_2$, slightly increased $Na_2O$, but decreased $Al_2O_3$ concentrations in the matrix of the younger cataclasite layer (c) cannot be due solely to the comminution of large albite and quartz clasts during reworking of older cataclasites (layers a and d). The alteration requires the mobility of several elements and mass transfer via fluid infiltration.

**14B.** Pressure solution in a cataclasite from the Currie shear zone, King Island, Tasmania. Reactivated cataclasite from the same narrow fault as shown in A, but sampled a few meters along strike. Quartz clasts contained in a fine-grained matrix are locked together along sutured grain boundaries and provide evidence for pressure solution. The matrix consists of quartz, albite, epidote, and titanite. Pressure solution is one of the processes active during compaction of gouge that lead to a decrease in porosity and an increase in shear strength. PPL

**Table 14.1.** Matrix analyses of cataclasites and whole-rock analysis of mylonitic granite

| *cataclasite* | *$SiO_2$* | *$TiO_2$* | *$Al_2O_3$* | *FeO** | *CaO* | *$Na_2O$* | *$K_2O$* | *Total* | *n* |
|---|---|---|---|---|---|---|---|---|---|
| a) mean | 62.07 | 1.91 | 18.77 | 2.44 | 11.33 | 4.55 | ≤0.01 | 101.11 | 10 |
| stand. dev. | 3.64 | 0.35 | 1.62 | 0.41 | 1.58 | 0.99 | | 1.08 | |
| c) mean | 80.11 | 0.46 | 12.10 | 0.54 | 2.66 | 5.60 | 0.28 | 101.70 | 10 |
| stand. dev. | 3.88 | 0.12 | 2.15 | 0.17 | 0.63 | 1.17 | 0.17 | 0.76 | |
| d) mean | 47.66 | 2.54 | 23.87 | 5.18 | 17.46 | 3.13 | ≤0.07 | 100.14 | 10 |
| stand. dev. | 4.06 | 0 71 | 1.20 | 0.74 | 2.76 | 1 10 | | 2.21 | |

| *host rock* | *$SiO_2$* | *$TiO_2$* | *$Al_2O_3$* | *FeO** | *CaO* | *$Na_2O$* | *$K_2O$* | *MnO* | *MgO* |
|---|---|---|---|---|---|---|---|---|---|
| myl. granite | 65.84 | 0.72 | 14.78 | 4.47 | 2.65 | 5.97 | 1.34 | 0.08 | 2.18 |
| ± error | 0.33 | 0.02 | 0.07 | 0.04 | 0.03 | 0.29 | 0.01 | 0.0014 | 0.05 |

*Note*: MnO ≤ 0.11, MgO ≤ 1.12
*FeO as total Fe

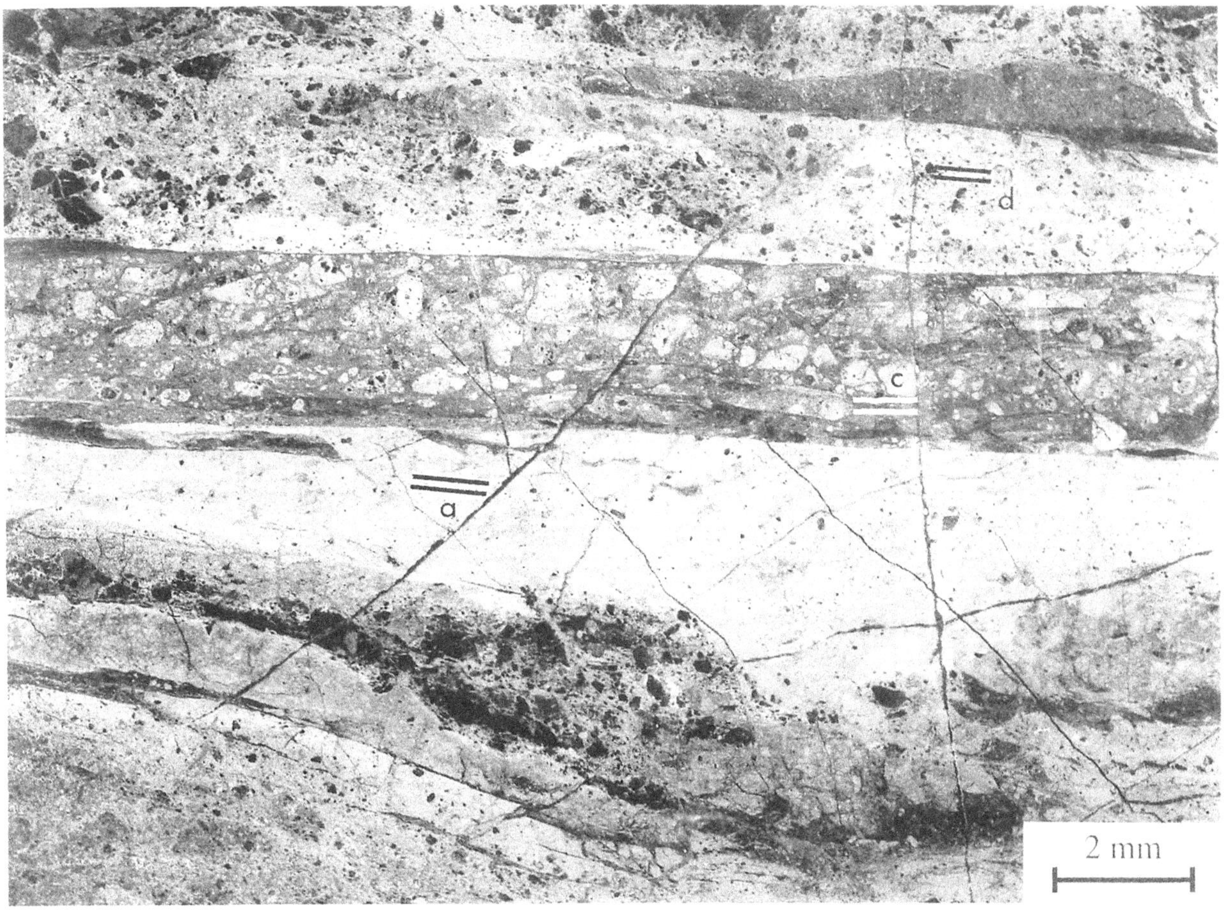


14 A

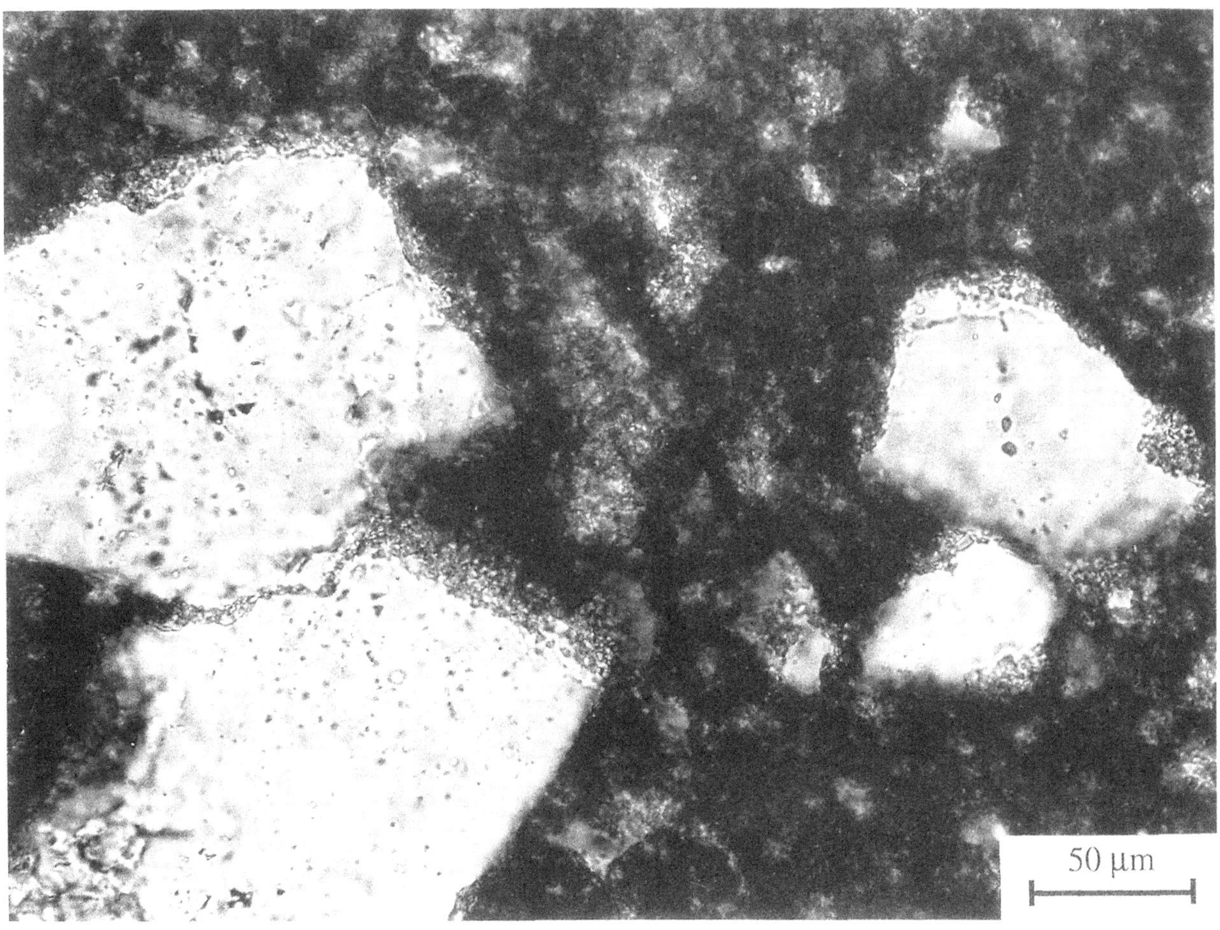


14 B

# 15. BANDED STRUCTURES IN GOUGE

RAMON JULIÀ AND PERE SANTANACH

The northeast-southwest-striking Vallès-Penedès fault, northeastern Iberian Peninsula, Spain, has affected both the Hercynian basement, consisting chiefly of phyllites with quartz veins, and its Alpine cover (Figure 15.1).

During the Paleogene, the fault experienced sinistral displacement with transpressive effects in some areas. The fault had a reverse movement where it strikes eastward, but a dextral-reverse movement in WNW-ESE-striking sections. Later, in the Neogene, the fault acted as a normal fault bounding the Vallès-Penedès graben.

The Paleogene fault movement produced a wide zone of fault gouge (minimum 150 m), 500 m wide where phyllites are present. X-ray powder diffraction analysis from fine gouge shows an iron-rich chlorite (near thuringite), illite, and quartz. This iron-rich chlorite is corroborated by the occurrence of an endothermic peak at 550°C shown by differential thermal analysis.

The gouge shows banded structures that are more or less parallel to the strike of the fault but have variable dips, resulting in an anastomosing pattern. The banded structures consist of layers with different matrix-to-clasts ratios and a planar arrangement of blocks of phyllite and quartz. The banding probably was produced by differential movements along minor faults related to the overall displacement of the Vallès-Penedès fault zone.

**15A.** Banded structures in a segment of the Vallès-Penedès fault that underwent dextral-reverse movement. See Figure 15.2 for additional explanation.

**15B.** Detail of banded structures showing variable degrees of crushing as well as the foliated aspect of the fault zone. The anastomosing character of the foliation is due to the intersection of shear surfaces (C planes) and cleavage surfaces (S planes) defined by preferred orientation of phyllosilicates. See Figure 15.3 for additional explanation.

**Figure 15.2.** Sketch of Plate 15A. Black = quartz blocks, grey = phyllite blocks, and lines = traces of crush planes

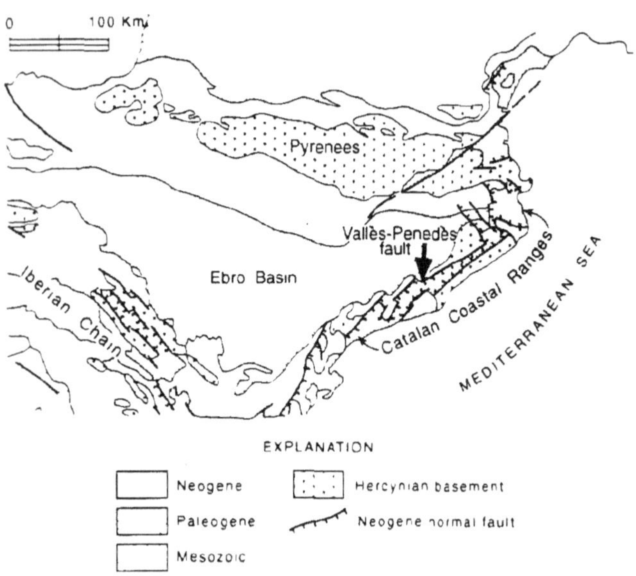


**Figure 15.1.** Geologic sketch map of the Valles-Penedes fault, northeastern Iberian Peninsula

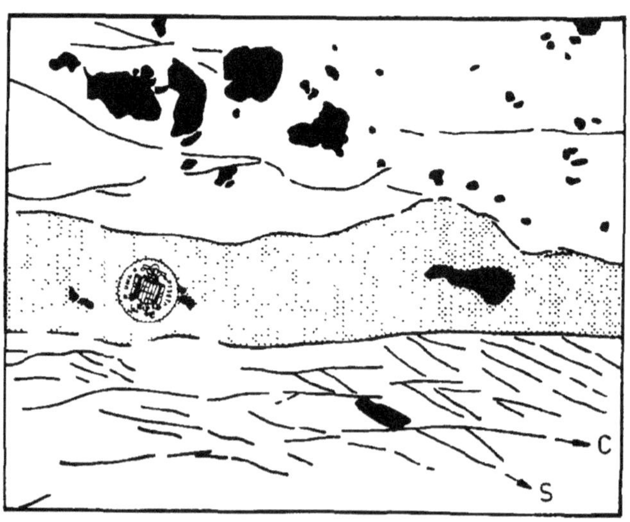


**Figure 15.3.** Sketch of composite gouge fabric exhibited in Plate 15B. Black = quartz clasts, stippled = layer made up essentially of fine-grained gouge. Scale = coin, 2 cm diameter

15 A

15 B

HIDEO TAKAGI

Plates 16A–16D are foliated fault gouges from outcrops of the Median Tectonic Line (MTL) in Shikoku, Chubu, and Kanto regions, Japan (Figure 16.1). The terminology of P foliation, $R_1$ shear, and Y shear is after Rutter and others (1986). P foliation is a preferred alignment of platy minerals that generally develops between 135° and 180° to the shear direction and corresponds to the S surface (*sensu lato*) in mylonitic rocks. $R_1$ shear is a Riedel shear surface in an orientation between 10° and 45° to the shear direction and corresponds to the C′ surface in mylonitic rocks. Conjugate Riedel shears $R_2$ are not common. Y shear is the principal displacement shear surface parallel to the shear direction and corresponds to the C surface in mylonitic rocks.

**16A.** XZ polished slice of P-Y-$R_1$ fault gouge from the MTL (local name: Tobe thrust), Tobe town, Ehime Prefecture, Shikoku (Takagi and others, 1992). The Tobe thrust is a well-known segment of the MTL, and it strikes N 70° E and dips 30° northward. Sandstone and mudstone of the Upper Cretaceous Izumi Group form the hanging wall of the fault, whereas conglomerate beds of the Eocene Kuma Group are on the footwall. A 1 m thick slice of Sambagawa schist (dolomitic schist) occurs between the Kuma and Izumi Group rocks. The P-Y-$R_1$ gouge fabric resembles the S-C-C′ mylonite fabric in Plate 98, and indicates top-to-the-north extensional shear. The Tobe thrust was first active in the mid-Tertiary and later reactivated as a normal fault, probably in the Miocene (14–15 Ma), based on K-Ar ages of micaceous minerals separated from fault gouges (Takagi and others, 1992) and on felsic igneous rocks, inferred to be associated with the mid-Miocene extensional regime, intruded along the MTL (Tazaki and others, 1990).

**16B.** XZ thin section of P-Y-$R_1$ fault gouge from the MTL (local name: Okitano-Iwayama line), Shimonita town, Gunma Prefecture, Kanto region. The MTL strikes E-W and dips 50–70° to the north. The early Miocene Tomioka Group (sandstone) is on the hanging wall (north) of the fault, whereas the Mikabu green rocks of the Sambagawa belt are on the footwall (south). The brecciated zone is about 2 m wide and consists chiefly of foliated fault gouge. The slickenlines on the shear surfaces of the foliated fault gouge plunge NNE to NE at 50°. The attitude of the slickenlines and the composite planar fabric (P-Y-$R_1$) indicates top-to-the-north extensional shear. $R_1$ shear is highly developed in this sample; the Y surface can be defined as an enveloping surface of P and $R_1$. PPL

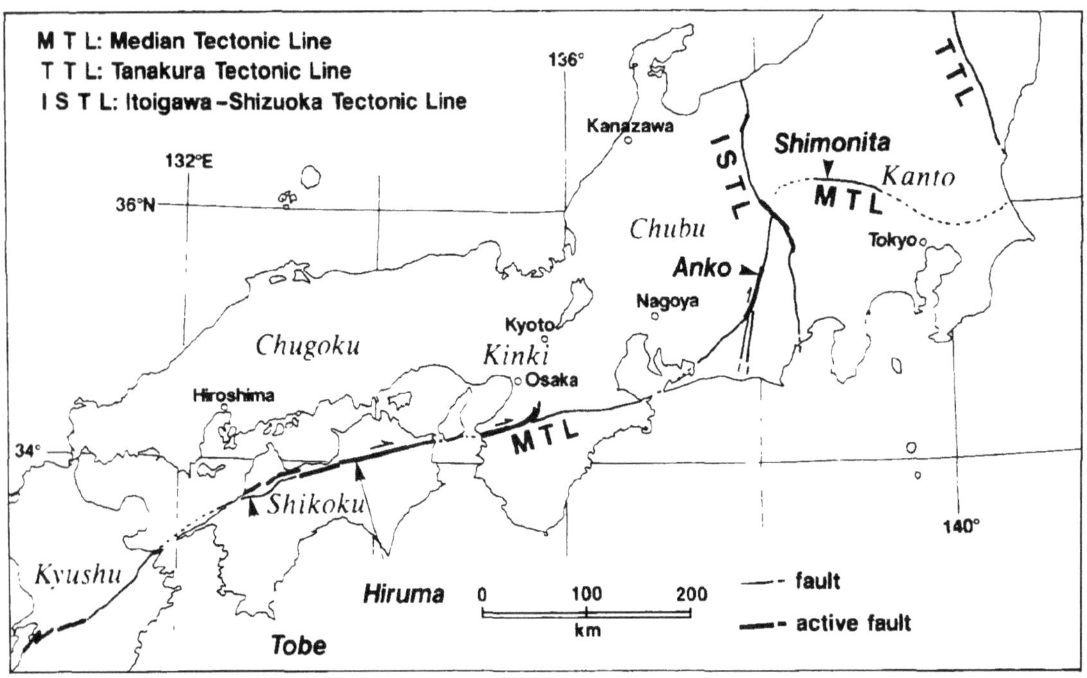


**Figure 16.1.** Sketch map of Southwest Japan, showing the location of three major tectonic lines of Japan. The MTL exists partly as active faults from Kinki to Shikoku (including Hiruma) and partly in Chubu (including Anko). Some MTL active faults extend westward, located north of the nonactive MTL (including the Tobe thrust) west of Shikoku.

16 A

16 B

**16C.** XZ polished slice of foliated and banded fault gouge from the MTL (local name: Ikeda fault), Hiruma, Miyoshi town, Tokushima Prefecture, Shikoku. The sample location is on a part of the active fault of the MTL, where dextral displacement has been reported from topographic data (Okada, 1968). The banding (=Y) of the fault gouge strikes N 84° W and dips 70° to the north. The width of the gouge zone is at least 5 m. Dextral sense-of-shear can be identified from asymmetric tails (AT) around clasts and the P-Y(-$R_1$) fabric in foliated fault gouge. Interstitial water in matrix clay minerals probably plays an important role in promoting "ductile" deformation in the near-surface brittle crustal regime. Conjugate microfaults (MF) formed after shearing. The active fault segments of the MTL in Shikoku and Chubu originated in the Late Cretaceous to Paleogene as sinistral strike-slip faults (shear zones) and were later reactivated as dextral strike-slip faults in the Quaternary.

**16D.** XZ polished slice of foliated fault gouge from the MTL, Anko, Oshika village, Nagano Prefecture, Chubu region. Several fault gouge zones are developed both in the Ryoke and Sambagawa belts. The MTL strikes N 20° E and dips 86° to the northwest; the cataclastic lineation (Tanaka, 1992) trends S 58° W and plunges at 44°. Typical P-Y fabric of the bottom of photograph resembles S-C fabric (*sensu lato*) in mylonites. This P foliation is well defined by the alignment of the long axes of clasts, and a sinistral sense-of-shear is indicated. A dextral sense-of-shear has also been identified from other foliated fault gouges, and several small-scale thrust faults cross-cut the fault gouge zone in the same outcrop. Thus, the MTL has a complex movement history.

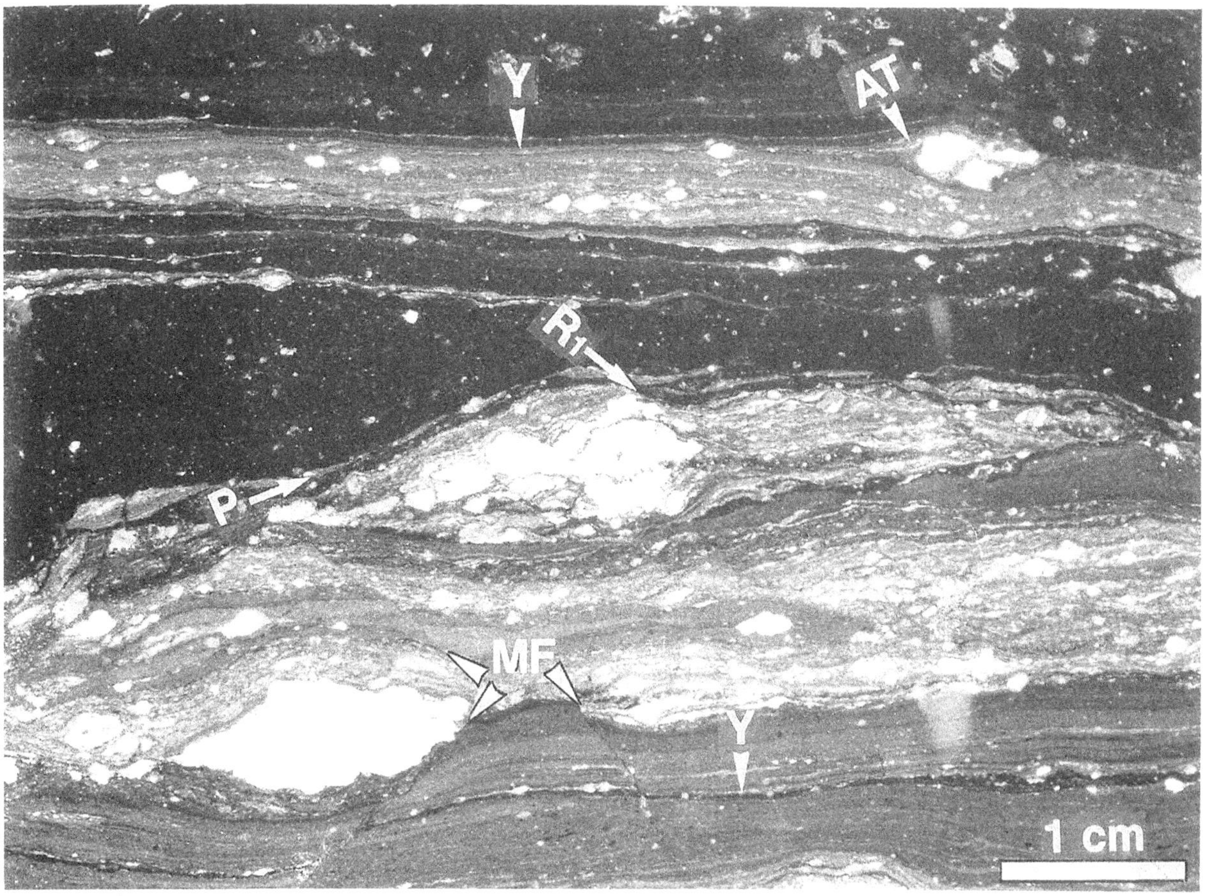


16 C

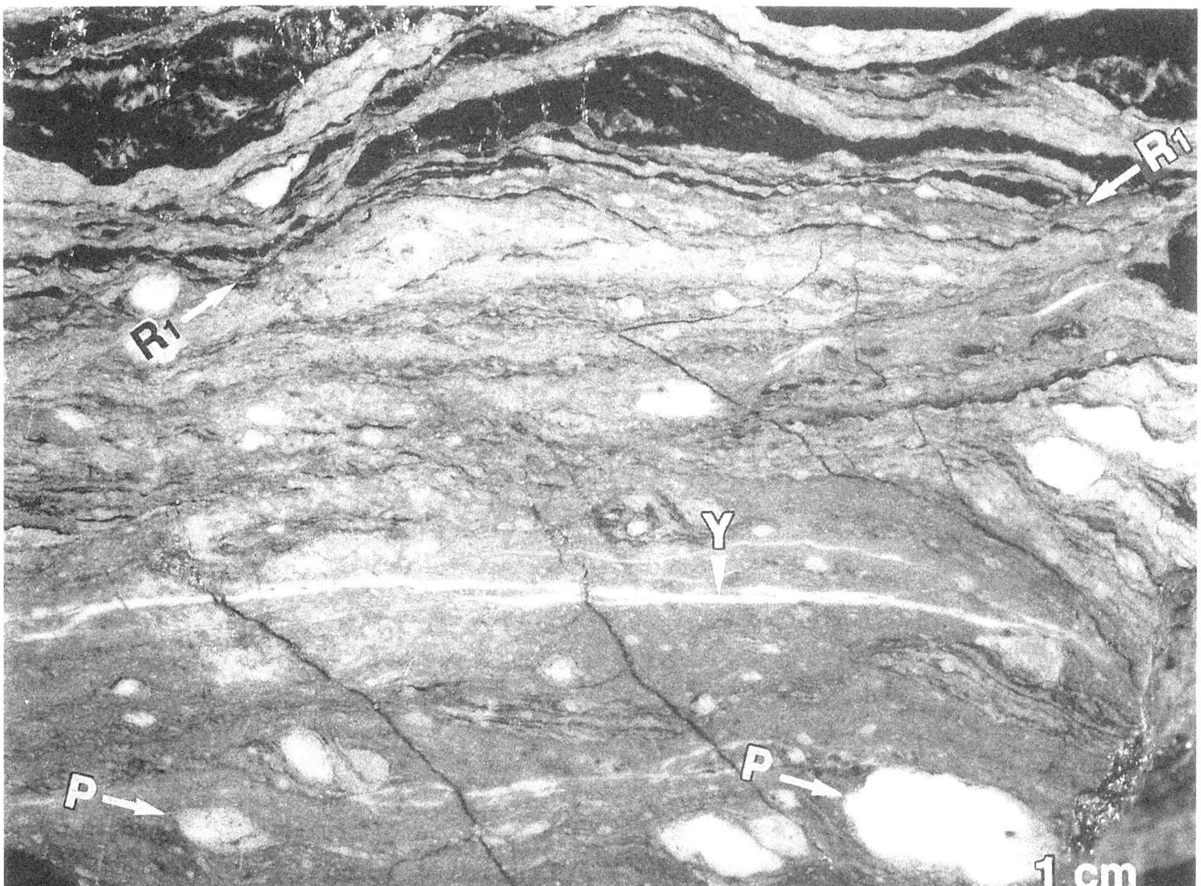


16 D

# 17. MICROSTRUCTURES IN NATURAL SERPENTINITE GOUGES

EILARD H. HOOGERDUIJN STRATING AND REINOUD L. M. VISSERS

The microstructures shown are from natural serpentinite gouges developed in brittle thrusts exposed in the Voltri ophiolite massif (Liguria, northwestern Italy; Hoogerduijn Strating and Vissers, 1994). The faults form part of a foreland-directed imbricate system, developed during Eocene/Oligocene collision between the African/Apulian and European plates. The host rocks of the faults are massive peridotites. The ambient conditions during brittle thrusting were most likely in the range 150–300°C, at pressures not exceeding 200 MPa. These serpentinite gouges represent a class of brittle microfabrics which are characterized by extensive localization of deformation at the scale of the gouge. In contrast to most natural clay-rich gouges, the strain in the serpentinite gouges has been accommodated by movement along a few discrete slip planes separating domains of massive, unfoliated microbreccia.

**17A.** Dextral fault zone with well-developed fabric of synthetic Riedel ($R_1$) shears and boundary parallel (Y) shears bounding lenses of microbreccia. There are no apparent gouge foliations visible in outcrop or hand specimen. A characteristic geometric feature is the asymptotic bending of $R_1$ shears into the Y shears. This curved geometry most likely originated during the initial stages of $R_1$ fracture growth, and was probably controlled by local perturbations of the stress field near heterogeneities on the Y shears. The maximum angle between the $R_1$ and Y shears varies between 25° and 35°, i.e., much larger than the 10–20° angles commonly observed in natural clay-rich gouges. The development of these high-angle relationships can most likely be attributed to an interplay between the apparent coefficient of friction on the Y shears and the internal coefficient of friction of the gouge material (Gu and Wong, 1994).

**17B.** Microbreccia domain dissected by an $R_1$ shear which bends into a Y shear orientation (indicated by arrows). The microbreccia consists of millimeter-scale, angular to subrounded fragments of the wall rock, embedded in a fine-grained serpentinized matrix. There is virtually no alignment nor elongation of fragments leading to any macroscopic foliation in the microbreccia domains. Pervasive, shear-induced gouge foliations are only developed in the immediate vicinity of the slip planes (see 17C and 17D). Fragmentation of the olivine crystals during cataclasis was largely controlled by crack propagation along olivine grain boundaries and along serpentine meshwork in olivine grains. Fragmentation of serpentinized pyroxene grains involved a combination of slip on [100] lattice planes and initiation of transgranular fractures in areas of lattice bending and stress concentration (PPL). The detail outlined in the box is shown in 17D.

17 A

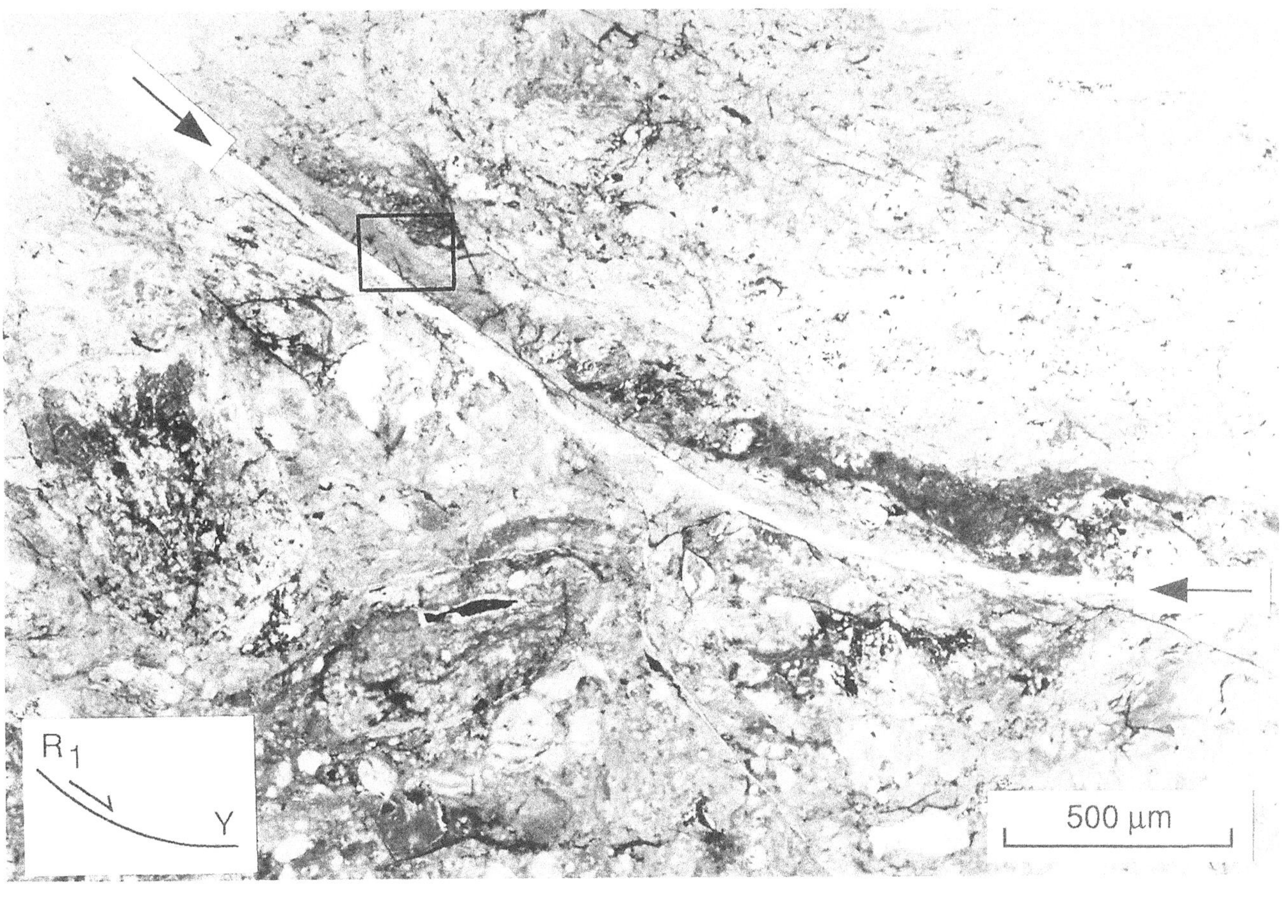


17 B

**17C.** Gradual transition from a synthetic $R_1$ shear (left) to a Y shear (right). The domain between the $R_1$ shears is completely serpentinized and shows a weak shear foliation defined by aligned serpentine and magnetite grains. The upper and lower parts of the micrograph show a completely serpentinized olivine microbreccia devoid of any foliations. The average angle between the shear foliations and the slip planes is about 30°. Assuming nondilatant simple shear, this angle would correspond to a shear strain of about 1. Note, however, that synkinematic transformation of olivine and pyroxene to serpentine minerals results in a significant volume increase (i.e., more than 50%); hence the inferred shear strain represents a minimum estimate only. The slip planes are entirely coated with magnetite, and a hexagonal magnetite neoblast (arrow) nucleated in the foliated domain. Even at the micron scale, backscatter electron microscopy of the magnetite coating and neoblast reveals no deformational features, suggesting essentially post-kinematic growth of the magnetite. PPL

**17D.** Detail of 17B, showing a well-developed gouge foliation adjacent to a magnetite-coated $R_1$ shear plane. High-angle dextral shear bands dissect and offset the foliation as well as the boundary between the foliated domain and the serpentinized microbreccia (top right-hand corner). These shear bands are tentatively interpreted as secondary synthetic Riedel ($R_1'$) shears developed within the primary Riedel structure. A distinct strain gradient in the foliated domain is apparent from the gradual decrease of the angle between the shear foliations and the slip planes, from about 25° away from the slip plane to less than 8° close to the slip plane. From the small angles adjacent to the slip planes it can be inferred that individual Riedel shears may have accommodated shear strains of up to 7 or more. It follows that considerable displacements must have been accommodated by movement along the discrete slip planes in the gouge. Both the distinct localization of the deformation on discrete sliding surfaces and the large angles between the $R_1$ and Y shears have been quoted to be characteristically associated with frictional (stick-slip) fault sliding behavior (Moore and Byerlee, 1991). PPL

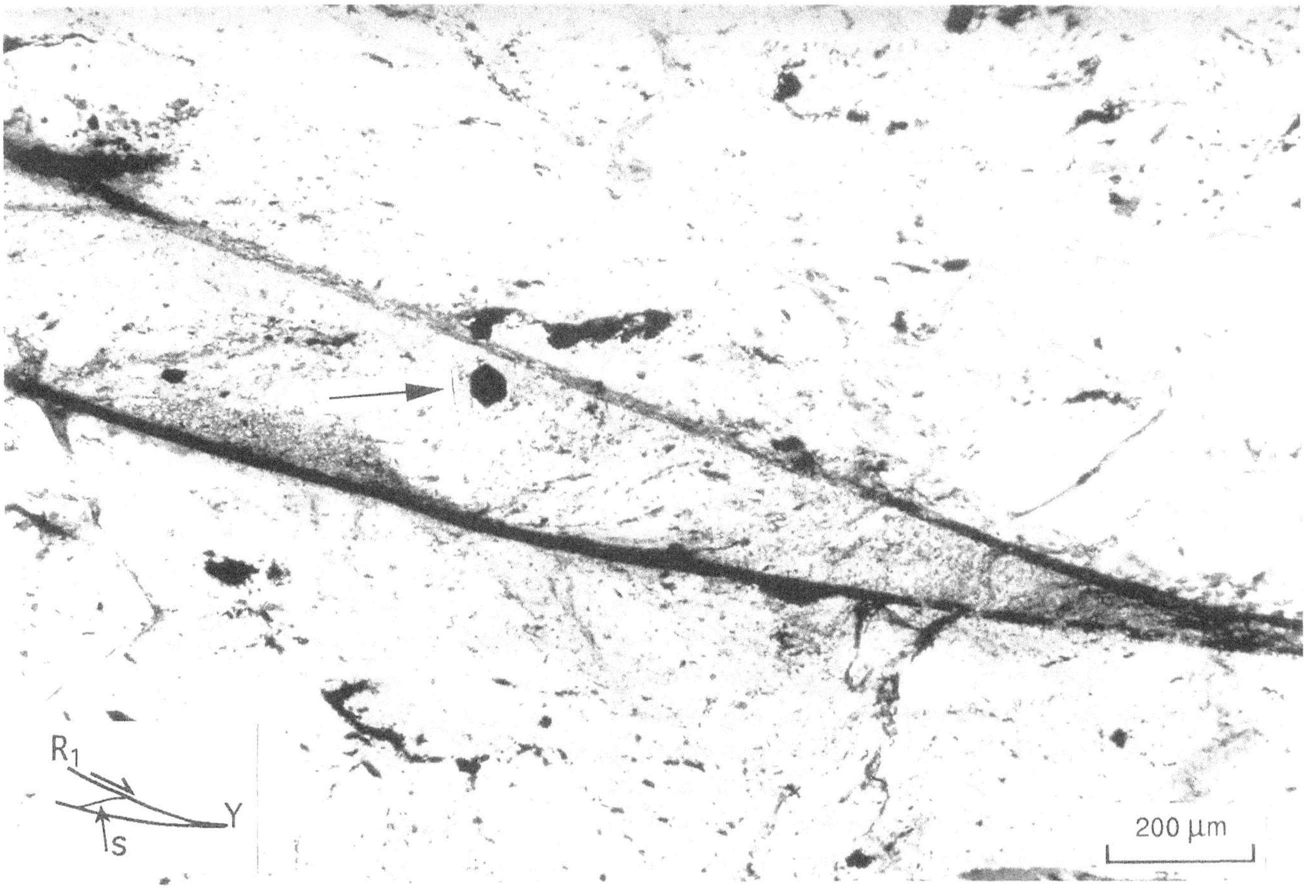


17 C

17 D

# 18. MESOSCOPIC DUCTILITY IN FAULT GOUGES

MARTIN G. MILLER AND DARREL S. COWAN

Fault gouge, the fine-grained by-product of brittle faulting events, typically exhibits mesoscopic ductility even though the actual mechanisms of deformation on a microscale are cataclastic and particulate flow. These photographs illustrate evidence for mesoscopic ductility in fault gouges from three different fault zones. Faulting in each case took place at near-surface conditions because it involved weakly lithified or unlithified materials in the hanging walls. The evidence for ductility includes composite planar fabrics (18A), rootless fold hinges within the gouge (18B), variably sized porphyroclasts surrounded by fine-grained material (18A, 18B), and centimeter- to meter-scale intrusion of gouge along an adjacent fault or into country rock (18C). Structures and fabrics produced by ductile flow in fault gouges may provide reliable kinematic information for the fault zone.

**18A.** Composite planar fabric in fault gouge of range-bounding normal fault, Tucki Mountain, Death Valley, California, U.S.A. The hanging wall consists of unlithified Pleistocene(?) fanglomerates, whereas the footwall consists of Proterozoic metasedimentary rocks. Slip on the fault was top-to-the-right (northeast). This gouge illustrates a composite planar fabric which resembles the C-S fabric in mylonites (Berthé and others, 1979b): discrete slip surfaces (Y shears; e.g., Logan and others, 1979) and separate zones of penetrative inclined foliation (P foliation of Rutter and others, 1986). In this example, surfaces of the P foliation are polished and typically striated.

**18B.** Rootless fold hinges in fault gouge of Badwater Turtleback fault, Death Valley, California, U.S.A. The hanging wall consists of unlithified Pleistocene(?) fanglomerates, whereas the footwall consists of mylonitic Proterozoic basement rocks. Latest slip on the fault surface (arrow) was top-to-the-right (west). Rootless fold hinges in this gouge, displayed by color banding, indicate transposition of the gouge by ductile flow during fault slip. Because the prominent foliation is defined by color banding, rather than a visibly distinct planar preferred orientation of mineral grains, this gouge appears structureless in places where color banding is absent.

**18C.** Intrusion of fault gouge along minor thrust fault, Squaw Peak fault zone, Cajon Pass, California, U.S.A. Rocks involved in this deformation are weakly lithified, coarse clastic rocks of the Miocene Cajon (Punchbowl) Formation. Clay-rich gouge in this photograph intruded from one thrust zone into a smaller, centimeter-scale back-thrust. The gouge exhibits thickness changes along the trace of the smaller thrust, presumably related to intrusion into empty spaces in the broken, sandstone hanging wall. The gouge also displays a planar fabric, defined by polished slip surfaces, which is subparallel to irregularities on the fault surface. Much of the material in this gouge may be derived from minor shale layers within the stratigraphic section. Penny (P) for scale.

18 A

18 B

18 C

# 19. SHEAR BAND FABRIC IN GOUGE

FREDERICK M. CHESTER AND JOHN M. LOGAN

Microscopic shear bands are observed in gouge from the Punchbowl fault, Devil's Punchbowl, Los Angeles County Park, California, U.S.A. (Chester and Logan, 1987). The Punchbowl fault is an ancient, exhumed fault of the San Andreas system. In the Park, the fault juxtaposes a complex of igneous and metamorphic rocks and arkosic sandstones and conglomerates of the Miocene-Pliocene Punchbowl Formation along a continuous layer of brown gouge (Chester and Logan, 1986, 1987). The final stages of faulting in the brown gouge occurred during the Pliocene at less than several kilometers depth at conditions favoring mineralogic alteration to smectite and clinoptilolite (zeolite). The gouge layer contains a microscopic composite planar fabric that consists of an S foliation defined by the preferred orientation of clay minerals, porphyroclasts and compositional lamination, and a planar anisotropy defined by microscopic shear bands. This fabric is typical of foliated cataclasites (Chester and others, 1985).

The geometry of the shear bands in a petrographic thin section oriented perpendicular to the gouge layer and parallel to the inferred slip direction in a right-lateral sense is shown in three photomicrographs taken with different illumination. Preferential orientation of clay in the shear bands and in the intervening gouge matrix may be defined by birefringence under NX illumination. Scale bar is 0.5 mm and applies to all photos. The line drawing (Figure 19.1) generated by tracing the structures identified in the three photomicrographs shows the geometry of the shear bands and offset vein fragments (stippled).

**19A.** NX photomicrograph with polarizers at 45° to the gouge layer illuminates shear bands approximately parallel to the layer.

**19B.** NX photomicrograph with polarizers parallel and perpendicular to the gouge layer illuminates shear bands at approximately 45° to the layer with right-lateral shear offset of vein fragments.

**19C.** PPL photomicrograph of the same structure as shown in 19A and 19B. Vein fragments appear as light areas, and shear bands appear as thin dark seams.

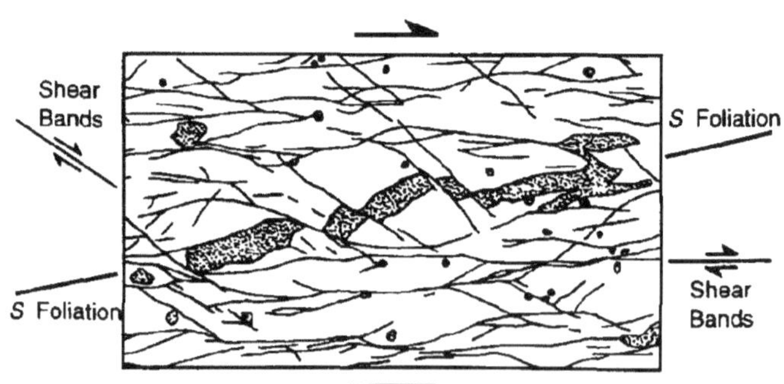


**Figure 19.1.** Composite sketchs of structures visible in Plates 19A, 19B, and 19C

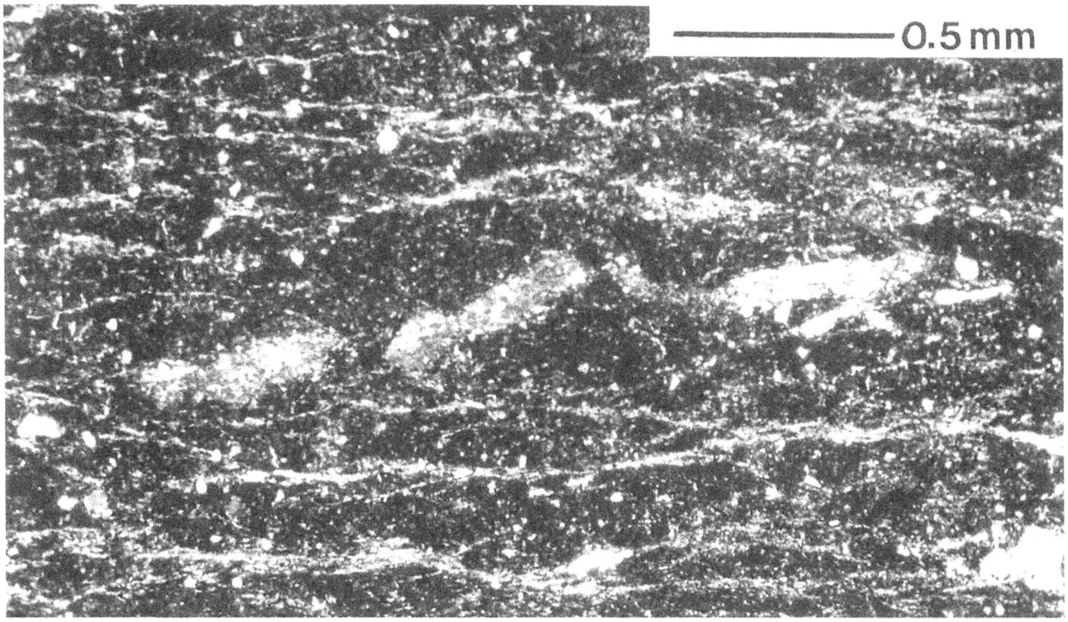


19 A

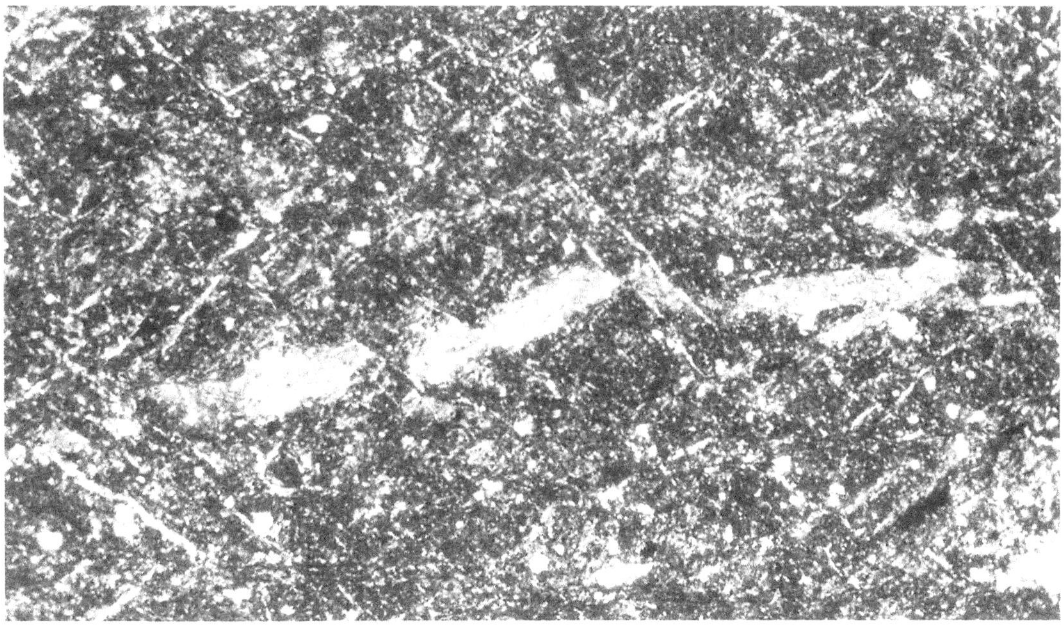

19 B

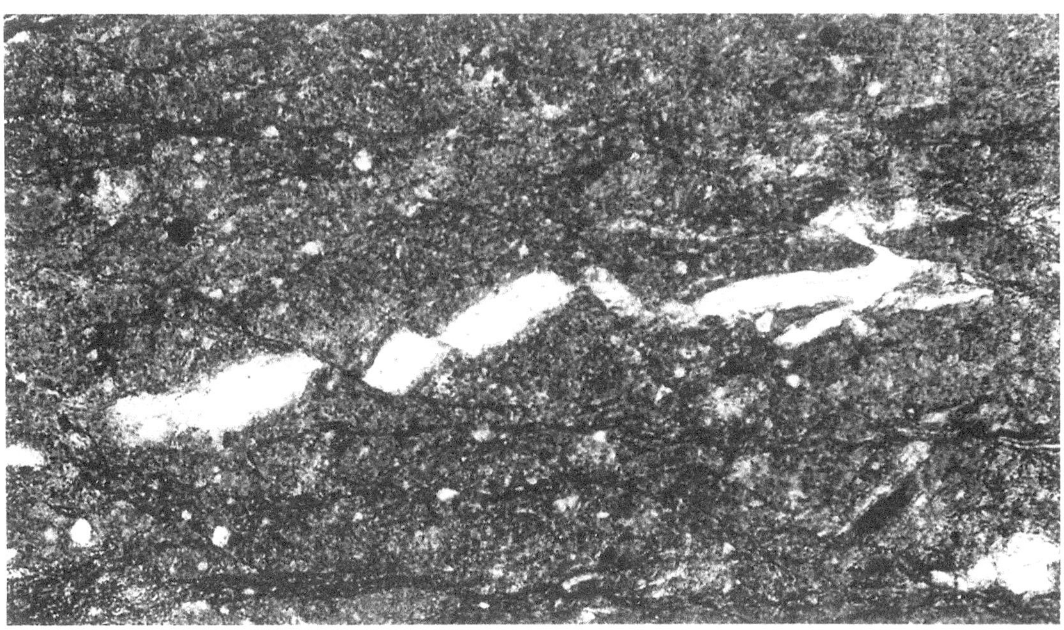

19 C

# 20. CATACLASTIC FABRICS IN EXPERIMENTAL SHEAR ZONES

M. FRIEDMAN, J. M. LOGAN, AND T. SHIMAMOTO

Both natural and experimental shear zones formed in the brittle regime where cataclasis is the dominant mechanism of deformation invariably exhibit the Riedel-fracture geometry (20A) regardless of material composition [e.g., naturally deformed Recent alluvium (Tchalenko and Ambraseys, 1970); experimentally sheared calcite gouge (Friedman and Higgs, 1981); intact limestone (Bartlett and others, 1981); and monomineralic and dimineralic gouges of twelve different minerals (Shimamoto and Logan, 1981)]. Herein, we demonstrate photographically (20B–20E) this penetrative fabric in quartz gouge (Shimamoto and Logan, 1981). The fabric is preserved by impregnation of blue-stained epoxy. Specimens were deformed along 35° precuts in Tennessee Sandstone at 250 MPa confining pressure, at 24°C, and a shortening rate of $2 \times 10^{-4}$ cm/s.

**20A.** Orientation of fracture sets $R_1$, $R_2$, T, X, P, and Y developed within gouge relative to an overall left-lateral sense-of-shear. Assuming $R_1$, $R_2$, and P are "Coulomb shears," $\alpha = \phi/2$, $\lambda = \phi/2$, and $\angle R_1 \wedge R_2 = \beta = 90° - \phi$ where $\phi$ is the angle of internal friction. Tan $\phi$ derived empirically from these angles (2.6) is in good agreement with those calculated from ultimate strength data (Logan and others, 1979).

**20B–20E.** Each photomicrograph is oriented parallel to the line drawing and the arrows point to different fracture sets within the array. Senses-of-shear for $R_1$ and P are indicated by microscopic feather fractures as in 20E (Conrad and Friedman, 1976). Scale line corresponds to 0.6 mm in 20B and 20D, 0.15 mm in 20C, and 0.1 mm in 20E. 20B–20D are in PPL, 20E in NX.

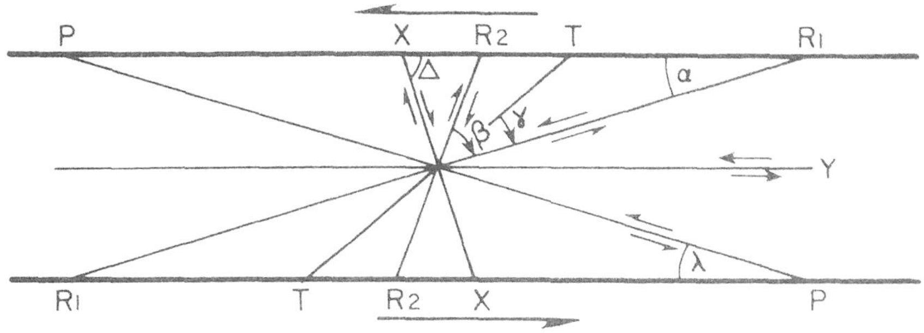


20 A

20 B

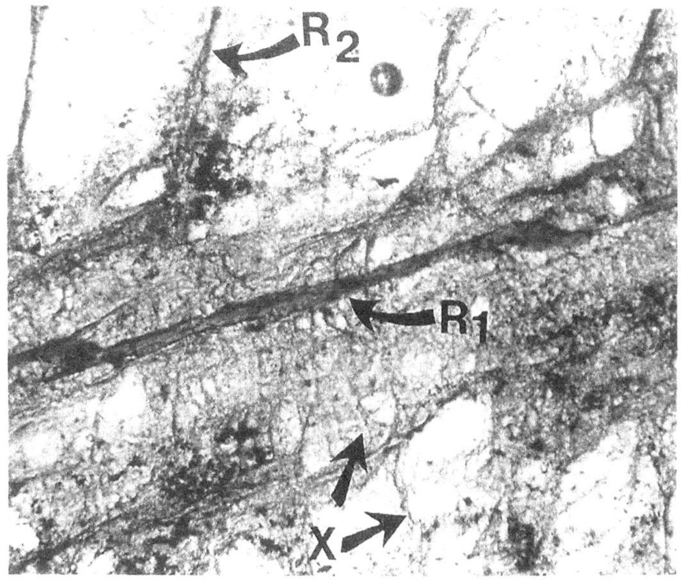


20 C

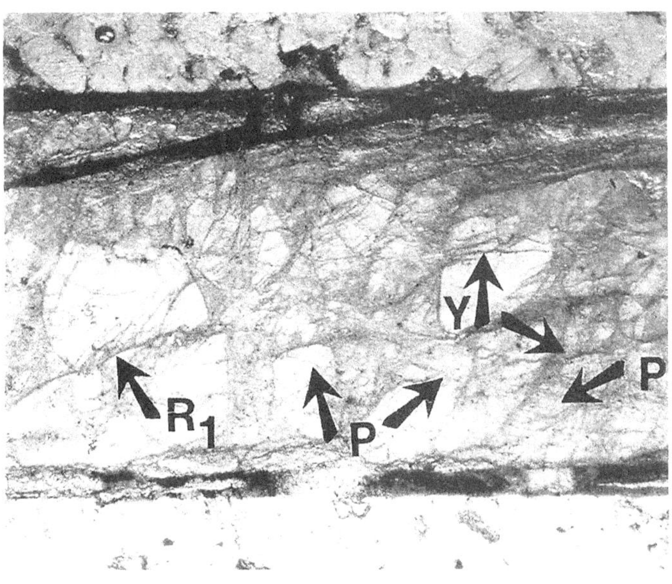


20 D

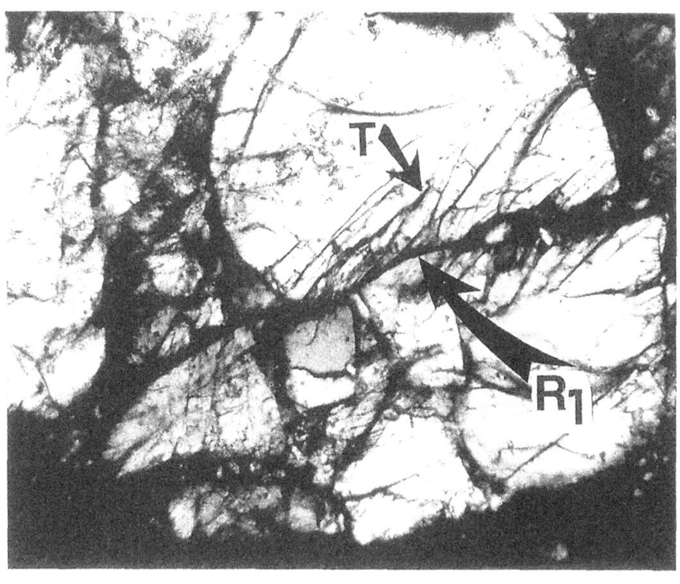


20 E

# 21. EVOLUTION OF GOUGE TEXTURES IN EXPERIMENTAL GRANITE FAULTS

MICHAEL L. BLANPIED AND NICHOLAS M. BEELER

The rotary shear geometry (Figure 21.1) offers the opportunity to study the microstructural development of laboratory faults and shear zones to unlimited shear strain. The photomicrographs shown are from a series of room-temperature, room-humidity, 21 MPa confining pressure, and 25 MPa normal stress sliding experiments on a thin (≈1 mm) layer of simulated granite fault gouge (<90 μm granite powder), sheared between roughened Westerly granite hollow cylinders. The apparatus, methods, and the textural development and mechanical behavior of a larger suite of samples are described in more detail elsewhere (Beeler and others, 1995). Figure 21.1 shows the orientation of photomicrographs; the direction of slip is within 5° to the plane of the photomicrographs. The sense-of-shear in all photomicrographs is left-lateral. Initial shearing is concentrated at or near the gouge-rock interface. With further slip, there is a progression of deformation into the interior of the gouge layer. Intense particle size reduction in the band of localized shear is accompanied by continued comminution throughout the layer. At the TEM scale, the sheared gouge consists of mostly submicron crystalline fragments, plus 5–10 volume percent of a material that appears amorphous and has a mixed-feldspar composition (Yund and others, 1990).

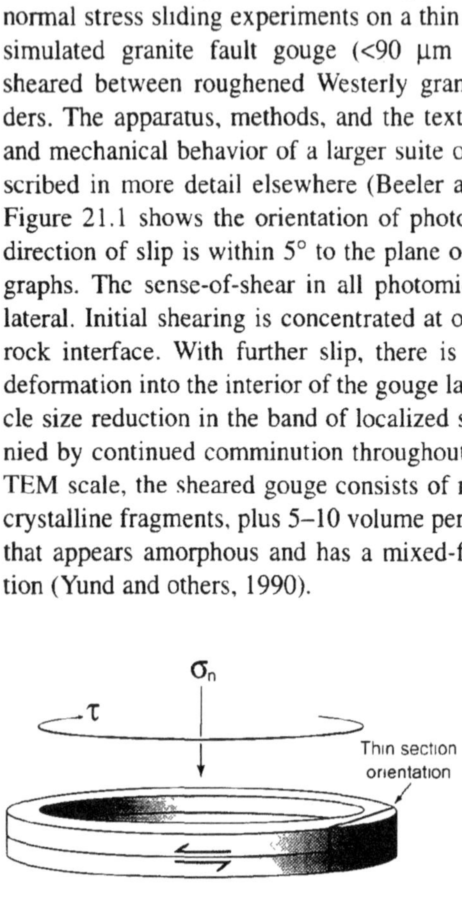


**Figure 21.1.** Rotary shear geometry

**21A.** Experimental fault in Westerly granite slid to 65 mm ($\gamma = 75$). The upper portion of the gouge layer is nearly identical to the undeformed gouge, with a wide distribution of particle sizes. Shear was localized near the lower interface, reflected by a reduction in the number of large particles in the lower 200 μm of the layer. A dark-colored band has developed near the rock interface, appearing to intersect the "country rock" (C) only at high points. This narrow band is extremely fine grained, and appears to possess a foliation defined by many interwoven, semi-parallel surfaces which we identify as Y shears (Y). These Y shears are dark in color and contain no particles visible at the optical scale. Apparently, distributed deformation near the lower rock surface during the first increment of slip allowed the subsequent localization of strain in the narrow band. Width of gouge layer = 850 ± 40 μm. PPL

**21B.** Experimental fault slid to 194 mm ($\gamma = 255$). With increasing slip the band of localized shear expands at the expense of the less deformed gouge. Here the band is about 100 μm thick, and is contained within a layer of fine-grained gouge that is nearly half the thickness of the full gouge layer. In the upper half of the gouge layer the average particle size is also somewhat reduced from the starting material, and evidence for homogeneous deformation can be seen in the flattening of biotite grains. These grains form an inclined foliation (F) suggesting the orientation of a strain ellipse for bulk simple shear. This coarser half of the gouge contains gently inclined $R_1$ shears (R) that are not seen in the lower, finer-grained gouge. Two artifacts of the thin sectioning process are marked, air bubbles in the epoxy (B) and open spaces filled with epoxy (G). Width of gouge layer = 850 ± 40 μm. PPL

**21C.** Experimental fault slid to 407 mm ($\gamma = 530$). The fine-grained, foliated band now occupies the lower 250 μm of the gouge layer. This complex network contains ten or more distinct boundary-parallel Y shears, plus numerous other shears which cross-cut the gouge at low angles. Several short $R_1$ shears can be seen in the central portion of the micrograph, and a longer shear cross-cuts the flattened fabric in the upper half of the gouge layer. It appears that the gouge near the upper boundary of the foliated band becomes finely comminuted by the formation of $R_1$ shears; the band then widens when a Y shear forms in this fine-grained gouge. Few markers exist with which to measure offset across any of the shears. Width of gouge layer = 770 ± 40 μm. PPL

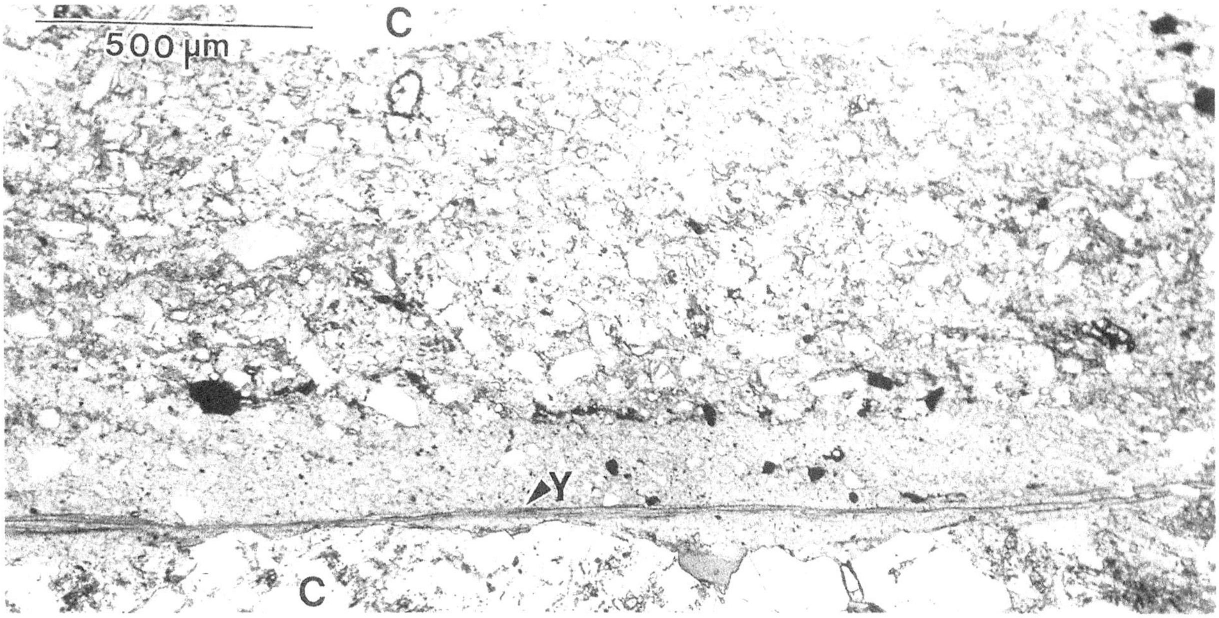


21 A

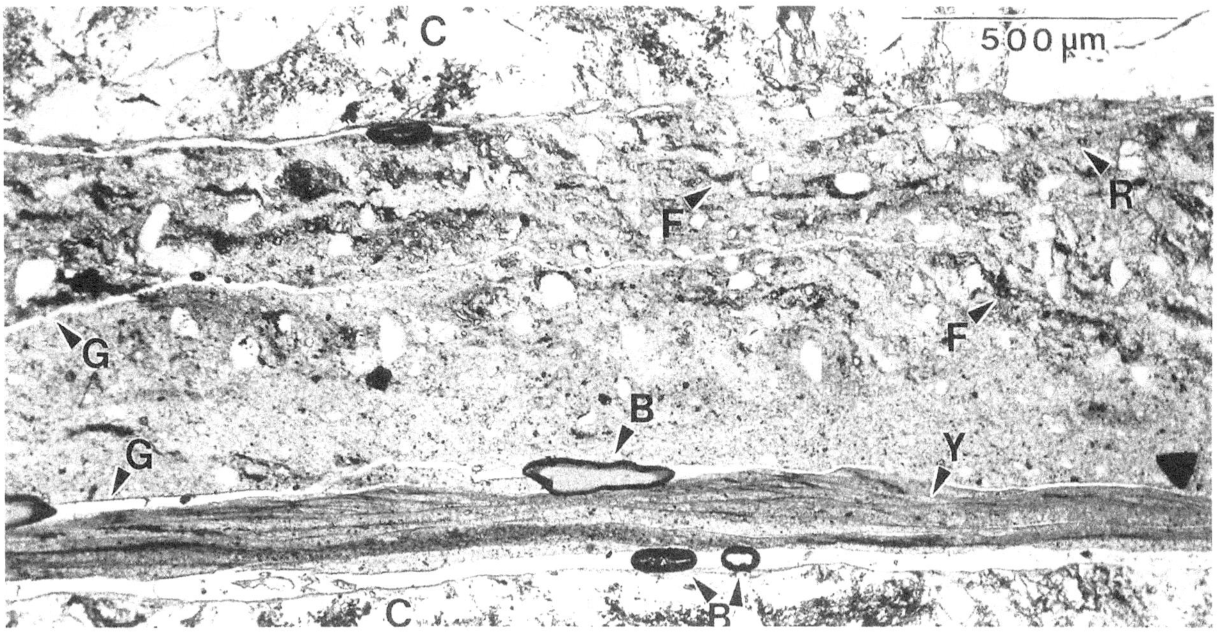


21 B

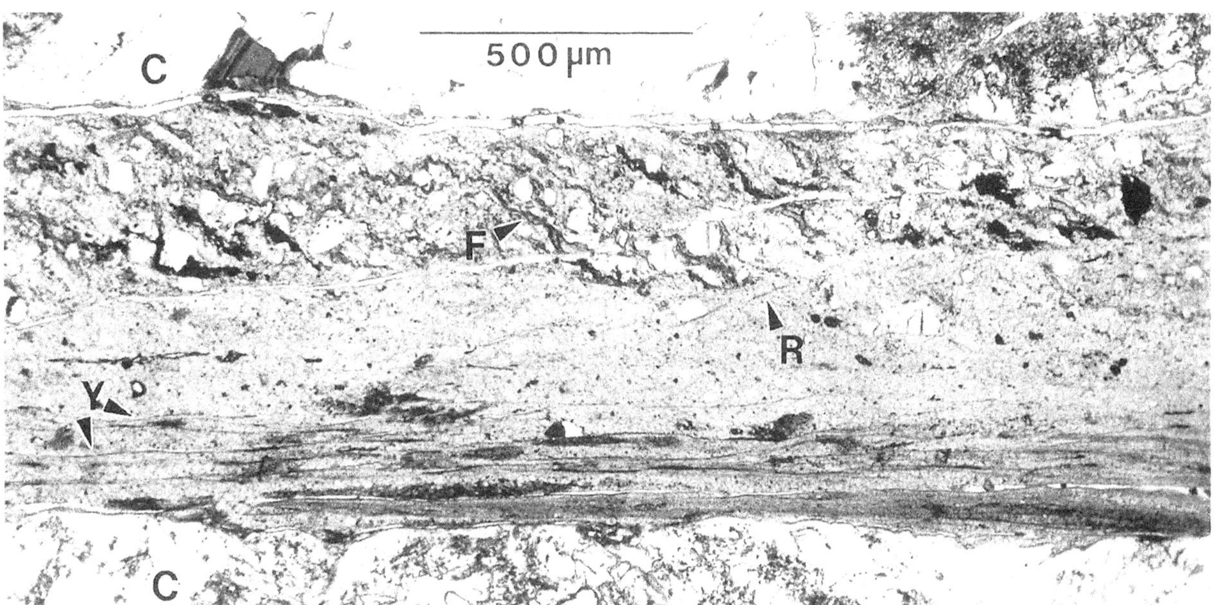


21 C

## Pseudotachylyte



Common abbreviations used in regard to the plates in this section include PPL = plane polarized light, NX = crossed nicols, SEM = scanning electron microscope, and TEM = transmission electron microscope.

Pseudotachylyte is a fault rock of great interest, because frictional melting is the most certain microstructural indicator of seismicity on ancient faults. This set of plates illustrates features of pseudotachylytes on scales from outcrop to thin section to SEM and TEM. The original type locality of pseudotachylyte in the Vredefort Structure, South Africa, is illustrated. Although this feature is now known to be an impact structure, the pseudotachylyte resulted from frictional sliding on faults, just as it does on tectonically induced faults. Recent experimental studies (Spray, 1995) have shown that cataclasis is a necessary precursor to frictional melting, and many of these contributions illustrate the intimate association of pseudotachylyte with mylonites and cataclasites. Although pseudotachylyte is generated along a sliding surface, it produces a volume increase and thus it also injects laterally into the host rock, producing complex geometries. Many of the plates illustrate details about the microstructures within pseudotachylytes, including flow banding patterns that allow one to determine flow direction and multiple events. Although most pseudotachylytes appear to be generated at about 10–15 km depth, one contribution presents evidence for pseudotachylyte generated by deep, high-temperature, but anhydrous plastic instabilities, and another illustrates pseudotachylytes formed at great depth in association with eclogite formation. Although some of the pseudotachylytes illustrated preserve a large amount of glass, others have devitrified, and still others have been subsequently deformed to produce an ultramylonite.

# 22. PSEUDOTACHYLYTE TYPE AREA: THE VREDEFORT STRUCTURE, SOUTH AFRICA

JOHN G. SPRAY

The Vredefort Structure of South Africa is located 120 km south-southwest of Johannesburg in the Witwatersrand Basin (Figure 22.1a). Its origin has been the subject of controversy, with two theories prevailing: (1) an endogenic (terrestrial) origin, resulting from one or more mantle-derived explosion(s) (e.g., Nicolaysen and others, 1963), and (2) an exogenic (extraterrestrial) origin, due to impact by a large meteorite (e.g., Boon and Albritton, 1936; Daly, 1947) The impact theory has gained wider acceptance, particularly since the discovery of shatter cones (Dietz, 1961), which require near-surface stresses of 2–6 GPa (Roddy and Davis, 1977) and, within quartzites of the collar rocks (Martini, 1978), the high-pressure $SiO_2$ polymorphs coesite, formed at <3 GPa (Stöffler, 1971) and stishovite, formed at <7.5 GPa (Sclar and others, 1962).

The Vredefort Structure is the type area for the rock type pseudotachylyte, which was first described and named by Shand (1916). Pseudotachylyte is generated under conditions of very high strain rates, as realized during seismic faulting or shock (e.g., Spray, 1992). Recent research has shown that most pseudotachylytes originate as friction melts that behave as low-viscosity, polyphase suspensions when in the molten state (Spray, 1993). Consequently, they cool to produce a heterogeneous assemblage consisting of a glassy to crystalline matrix containing incompletely digested wall-rock clasts and mineral fragments.

At Vredefort, pseudotachylyte occurs as thin veins (<<1 cm), as thicker dike-like bodies (up to tens of meters thick), and as irregularly shaped "breccia" zones (e.g., Killick and Reimold, 1990; Reimold and Colliston, 1994). It typically possesses dark, glassy or fine- to medium-grained crystalline matrices containing angular to rounded rock and/or mineral clasts (e.g., Bisschoff, 1962; Wilshire, 1971). Martini (1991) refers to the very thin (≤1 mm) pseudotachylyte veins as Type A which, in the appropriate lithology, can be coesite- and stishovite-bearing. Dating of these pseudotachylytes yielded an age of 2.02 Ga, which indicates the age of the impact event (Spray and others, 1995). Type A veins may show no obvious evidence of displacement and have been interpreted as being purely shock induced. The thicker, crystalline veins and dikes are termed Type B and are the variant that Shand described, examples of which are shown here. These postdate Type A, may be fault- (slip-)related and, to date, have not yielded high-pressure polymorphs. The Type B pseudotachylytes are probably related to the modification stage of crater development, whereby the cavity produced by the impact collapses gravitationally via large-displacement fault systems (Lambert, 1981; Thompson and Spray, 1994). In this respect, impact-related Type B pseudotachylytes are produced by the same process that generates endogenic pseudotachylytes. The difference is one of scale. The large displacements that occur during gravity-driven transient cavity collapse result in the production of larger volumes of friction melt.

The pseudotachylyte at Vredefort shows widespread distribution, but appears to be focused in broadly semi-annular zones (Killick and Reimold, 1990; Reimold and Colliston, 1994). An innermost zone occurs between the granulite-facies metagranitoid core and the Outer Granite Gneiss, and a second zone is developed along the core-collar boundary between the Outer Granite Gneiss and Witwatersrand Supergroup (Figure 22.1b). A third pseudotachylyte-rich, semi-annular zone occurs ~80 km from the center of the dome in the northern and northwestern portions of the Witwatersrand Basin. This may equate with an "outermost rim crest" at 150 ± 10 km diameter (Pike, 1985), although it has been suggested that the diameter could be as much as 300 km (Martini, 1992; Therriault and others, 1993). The implied concentric pseudotachylyte zoning indicates that Vredefort is a multiring impact structure (Spudis, 1993; Spray and Thompson, 1995).

**22A.** Pseudotachylyte exposed in the bed of the Vaal River, Parys. Spectacular occurrences several meters in thickness can be seen here, the size of which contrasts with the centimeter-thick pseudotachylytes typical of endogenic fault systems. Shand (1916) provided one of the first detailed descriptions of this rock type and coined the name based on its similarity with tachylyte (glassy basalt). At that time, the best exposures were along the beds and banks of the Vaal River, as shown in this photograph. Here, the entire 2 m wide field of view consists of angular to subangular blocks and smaller fragments of ≥2.8 Ga "Outer Granite Gneiss" set within a black, fine-grained to microcrystalline matrix, both of which, together, constitute the pseudotachylyte.

**22B.** Pseudotachylyte at Otavi Quarry, near Parys. Better (three-dimensional) pseudotachylyte exposures were created in quarries when the Archean granite gneiss was excavated for decorative stone (1920s–present). At Otavi Quarry (now abandoned), ~9 km east-northeast of Parys, quarry walls reveal well-rounded fragments of granite gneiss set in the black matrix. The matrix itself contains smaller rock and mineral fragments derived from the host gneiss. The rounded forms of the rock inclusions are the result of mechanical attrition (wear) and thermal corrosion (melting and partial assimilation) of the initially angular fragments. The development of igneous crystallites and micro-igneous textures within the matrix indicates that the pseudotachylyte originated as a melt-bearing suspension. This is in keeping with an origin by comminution leading to frictional melting during faulting (Spray, 1995). Note hammer and hand lens for scale.

22 A

22 B

**22C.** Pseudotachylyte "dike" at Leeukop Quarry (cross section). An excellent example of the fault-related nature of pseudotachylyte is afforded by Leeukop Quarry, ~4 km west of Parys. A 5 m wide pseudotachylyte "dike" has been freshly exposed (early 1990s) in cross section in Archean granitoid gneiss. Note the angular to subangular nature of the gneiss inclusions, some of which can be fitted back together jigsaw fashion. Smaller, more rounded clasts occupy intervening spaces. The whole is now "bonded" by the once molten, dark matrix. Uwe Reimold (University of Witwatersrand) for scale.

**22D.** Pseudotachylyte "dike" at Leeukop Quarry (view of dip plane). The same "dike" viewed through 90° at the main quarry face. This view is perpendicular to the strike of the "dike" with the succession of ledges neatly revealing the 55–60° dip surface. The same "dike" is also exposed at the bottom of the rectangular pit below the main work level (toward the base of the photograph). Note how the gneiss blocks have sunk within the black matrix, leaving predominantly matrix material occupying the upper portions of the body. This testifies to the low-viscosity condition of the molten matrix during and shortly after pseudotachylyte formation. The Leeukop example shows how large and impressive impact-related (type) pseudotachylytes can be, even though both impact-generated and endogenic examples form by the same process: high-speed slip, comminution, and frictional melting along fault walls.

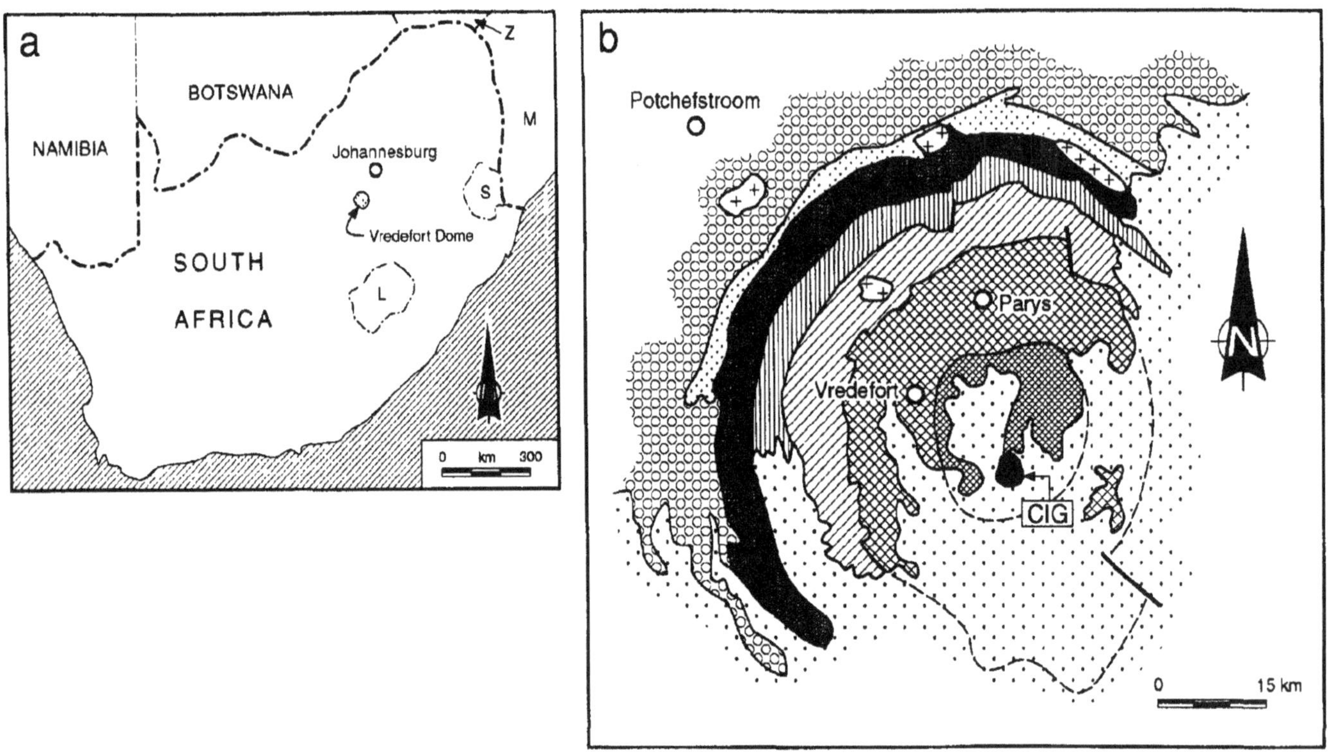


**Figure 22.1** (a) Geological setting of the Vredefort Dome within southern Africa. Z = Zimbabwe, M = Mozambique, S = Swaziland, L = Lesotho; (b) simplified geology of the main central part of the Vredefort Structure. CIG – Central Intrusive Granite

22 C

22 D

# 23. PSEUDOTACHYLYTE FORMED BY FRICTIONAL FUSION

ROBERT H. MADDOCK

Pseudotachylytes are fault rocks generally interpreted to have formed by frictional fusion during palaeoseismic fault movements, although the relative importance of melting and cataclastic processes has, historically, been questioned (Sibson 1975; Wenk, 1978; Maddock, 1983; Magloughlin and Spray 1992). The combination of traumatic, nonequilibrium processes active during pseudotachylyte formation together with the variety of host rock lithologies (i.e., wall rock and protolith) means that these fault rocks show an extraordinary diversity of textures and microstructures. The purpose of the following four plates is not, therefore, to provide a compendium of all possible pseudotachylyte textural types but to illustrate a few well-studied examples from a classic British Isles locality, the Outer Hebrides thrust. All of the photomicrographs presented here provide unequivocal textural evidence for crystallization from a melt phase.

The Outer Hebrides thrust is a low-angle fault that cuts Precambrian gneisses of the Lewisian complex on the Hebridean archipelago, which is situated offshore and to the west of the Moine thrust zone of northwest Scotland. The Outer Hebrides thrust is a crustal-scale structure which may have originated as long ago as 2.5 Ga and was certainly active until the early Mesozoic: the younger movements on the fault were extensional. Fault rocks associated with the thrust include mylonites, phyllonites, cataclasites, and pseudotachylytes. The timing of the pseudotachylyte faulting has been disputed, although formation during Caledonian-age, thrust-sense movements (e.g., Sibson 1977; Lailey and others, 1989) is supported by recent geochronological studies. These dating studies are highlighted in the caption for Plate 23C.

**23A.** Optical photomicrograph (PPL) of pseudotachylyte vein generated in quartzo-feldspathic gneiss, from South Uist, Scotland (OS grid ref. NF709298). The host rocks are composed of plagioclase, quartz, alkali feldspar, biotite, and amphibole. The pseudotachylyte vein occurs as an extensive subhorizontal pavement up to 5 cm thick. The photomicrograph shows poorly sorted, randomly arranged quartz porphyroclasts (Q) and newly crystallized plagioclase microlites (Pl), both set in an ultra-fine-grained groundmass containing an abundance of high-relief phases. The quartz porphyroclasts show spongy internal textures and magmatically corroded margins. The plagioclase microlites ($An_{25-45}$) are skeletal and display acicular and boxwork morphologies. These morphologies together with the An and iron contents (up to 1.5 wt% FeO*) of the plagioclases indicate primary "igneous" crystallization from a rapidly frozen melt (Maddock, 1983).

**23B.** Optical micrograph (PPL) of pseudotachylyte from the same vein as 23A. A large plagioclase porphyroclast (Pc) shows fingerprint texture developed along an internal grain boundary (G), and heterogeneous nucleation of boxwork plagioclase microlites on the clast margin (Pl). Both the fingerprint melting texture and the heterogeneous nucleation microstructure are comparable to the textures of felsic xenocrysts and xenoliths seen in some fine-grained basic igneous rocks.

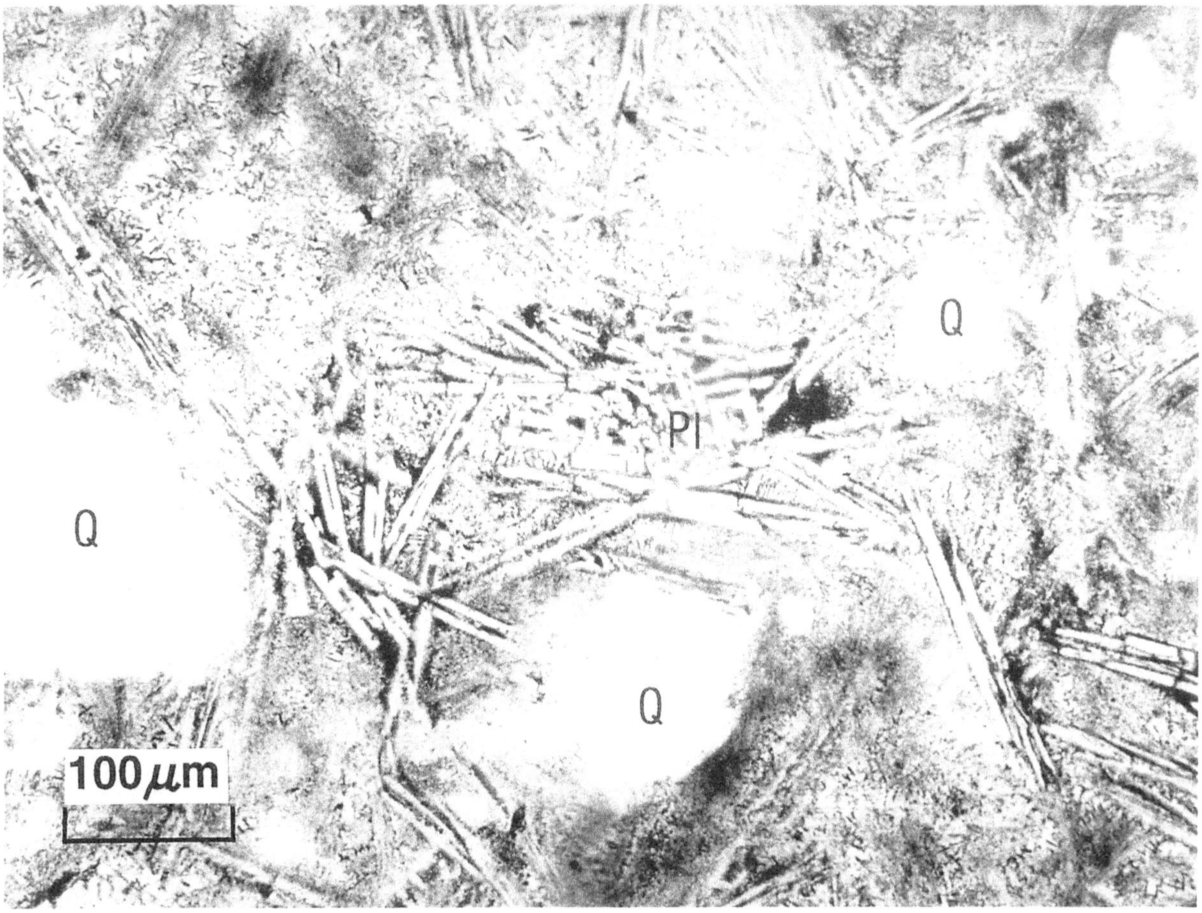


23 A

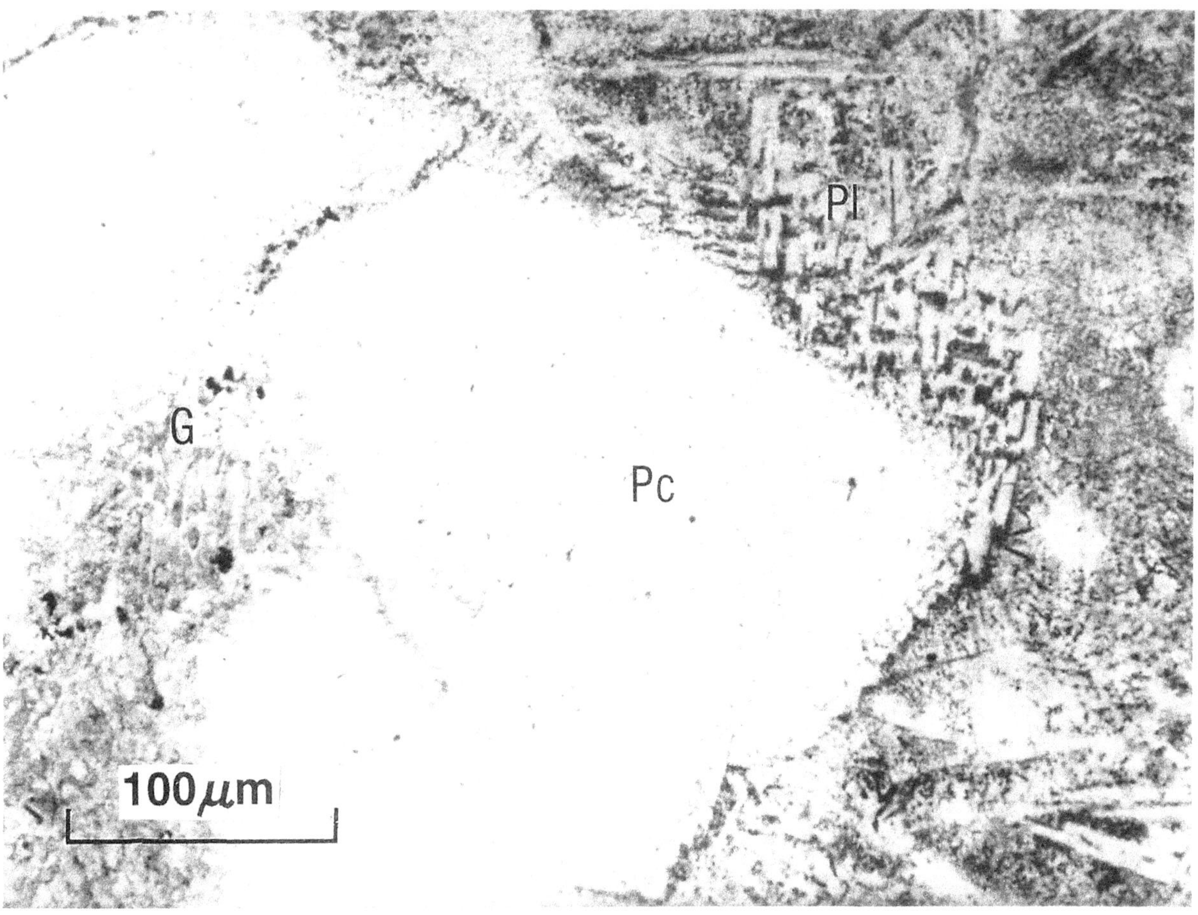


23 B

The two optical photomicrographs (PPL) 23C and 23D are from ultrathin sections (<30 μm thick) from the same sample illustrated in 23A and 23B.

**23C.** Groundmass of pseudotachylyte away from large porphyroclasts showing decussately arranged plagioclase laths (Pl) fringed by small magnetite grains (M) and other high-relief phases. A fan-spherulite (Fs) composed of fine (ca. 1–2 μm diameter) plumose fibers of plagioclase and biotite(?) projects from the blunt termination of one plagioclase lath. This type of spherulitic growth is typical of crystallization during fast cooling of a melt or glass (Lofgren, 1971a). The finest groundmass interstitial to the plagioclase laths and spherulites is composed of ultra-fine-grained (<1 μm) equant quartz, plagioclase, and biotite resolvable only by transmission electron microscopy. This groundmass is believed to have devitrified from a glass.

**23D.** Pseudotachylyte groundmass showing cruciform and dendritic habit magnetite which has grown from the melt phase following crystallization of the plagioclase laths. As well as the dendritic habit and grains fringing the plagioclase microlites (see 23C), patches of magnetite grains also occur as a palimpsest texture clearly derived from the breakdown of ferromagnesian porphyroclasts. The high overall content of magnetite in this pseudotachylyte renders it strongly magnetic.

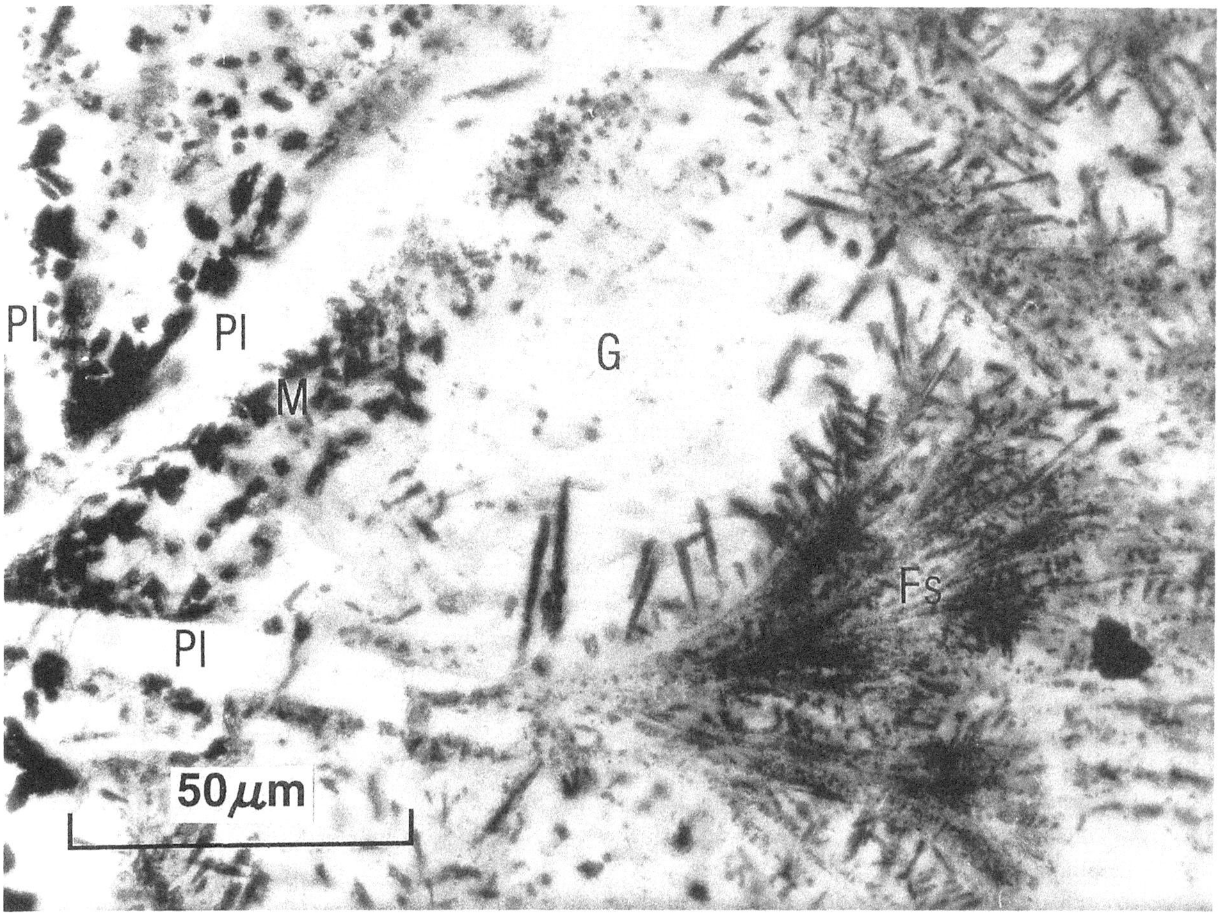


23 C

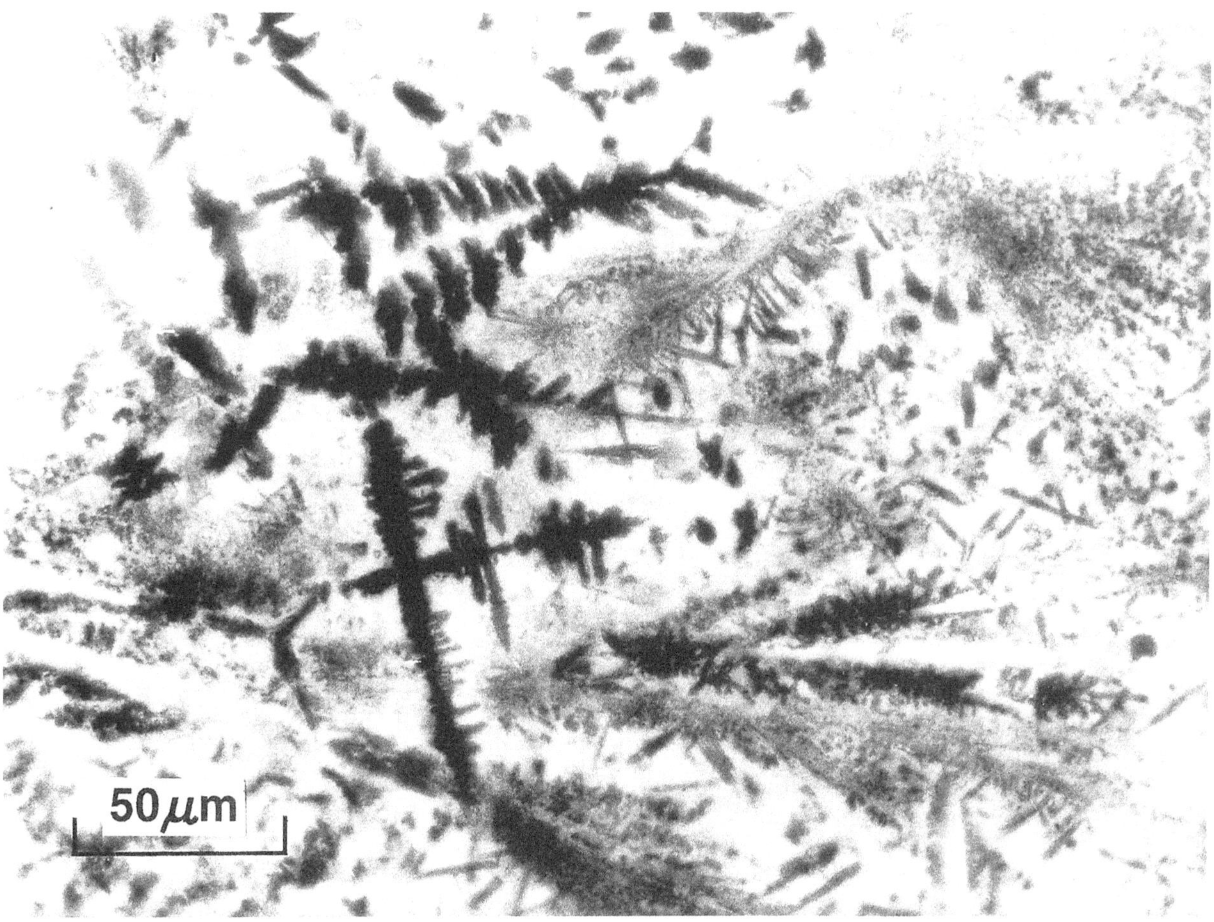


23 D

Plates 23E and 23F show pseudotachylyte generated from Precambrian Lewisian Gneiss, Outer Hebrides thrust, northwest Scotland (island of Grimsay, OS grid ref. NF877547). The host rocks from which the pseudotachylyte was formed consist of moderately well-foliated paragneisses with a modal content of plagioclase (39%), quartz (24%), biotite (28%), and garnet (9%), and an overall tonalitic composition. The pseudotachylyte occurs as foliation-parallel veins up to 8 cm in width and traceable for tens of meters along strike, and as narrower and less continuous veins which cut the foliation. (All of the examples shown in this plate and in 23A–23D are from thick, foliation-parallel veins.)

Whole-rock and electron microprobe analyses of this pseudotachylyte and its host rocks indicate that the pseudotachylyte formed by selective cataclasis and fusion of the various mineral phases that make up the host rock (Maddock, 1992). In particular, chemical and petrographic evidence confirms that biotite preferentially entered the melt phase that formed the pseudotachylyte. The textures of partially melted lithic porphyroclasts within the pseudotachylyte veins demonstrate the importance of dehydration melting of biotite (Maddock, 1986).

The textures and microstructures shown in Plates 23E–23H again confirm the melting and largely primary crystallization origin of this pseudotachylyte. An attempt has been made to date this paleoseismic frictional fusion event by laser-probe $^{40}Ar/^{39}Ar$ analysis of the pseudotachylyte and its host rock (Kelley and others, 1994). A weighted mean age of 430 ± 6 Ma obtained in the center of the vein indicates that pseudotachylyte faulting took place during the Caledonian orogeny, and corresponds closely with movement ages inferred for the associated Moine thrust.

**23E.** Optical micrograph (30 µm thin section) showing globulitic microstructure developed in dark brown, porphyroclast-laden pseudotachylyte. The globulites consist of a porphyroclastic nucleus, generally feldspar or lithic, successively haloed by microlitic feldspar laths, dark-brown matrix, and a narrow selvage of light-brown-colored matrix. The spacing of the globulites varies from discrete (top right) to coalesced (center), probably controlled by an interplay of the rates of crystal nucleation and crystal growth. The quartz porphyroclasts (Q) rarely act as globulite nuclei but often have a narrow rim of light-colored matrix. The pseudotachylyte vein is cut by a narrow quartz + chlorite vein with an associated bleaching halo several times the quartz vein thickness. PPL

**23F.** SEM backscattered electron micrograph showing partially melted lithic porphyroclast in same vein as 23E. The porphyroclast now comprises three phases: quartz with sulfide blebs (Q), plagioclase (P), and melt zones dominated by plagioclase microlites (M). The melt zones are inferred to occupy the former locations of biotite and plagioclase, decomposed by selective dehydration melting. The contact between the melt texture controlled by the porphyroclast and that of the matrix, indicated by a dashed line, corresponds to a globulite margin.

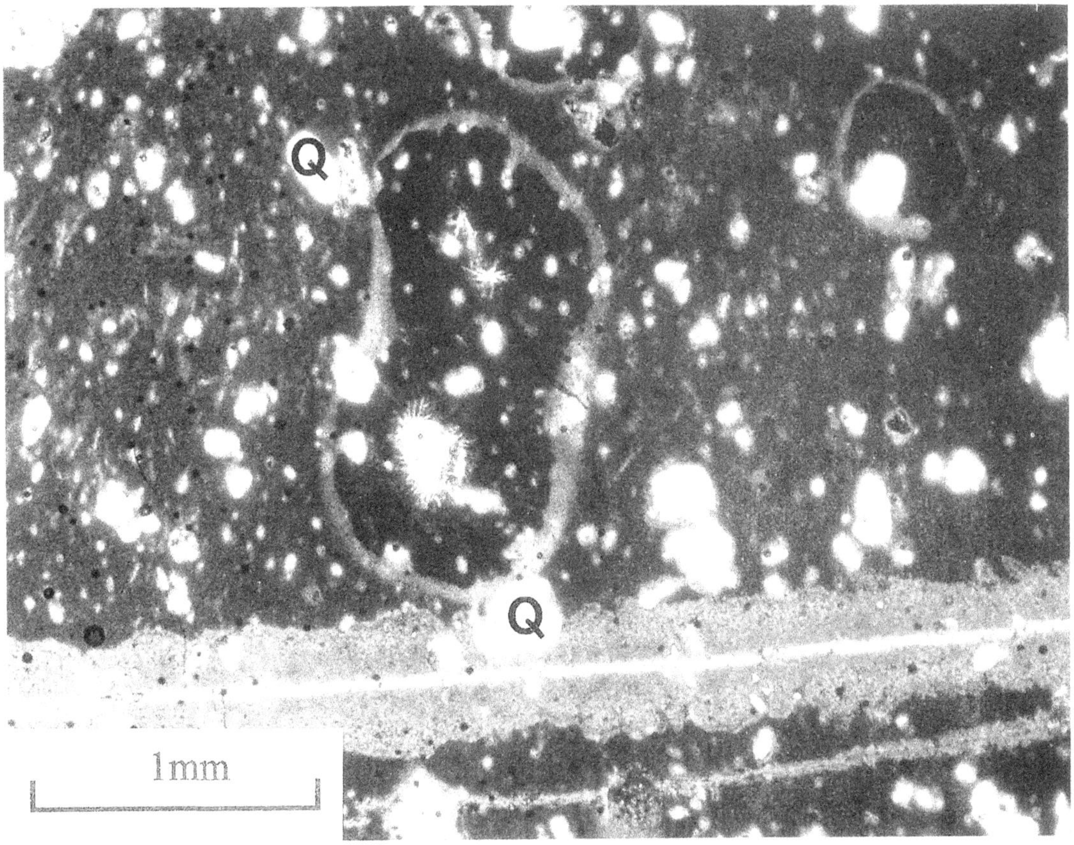


23 E

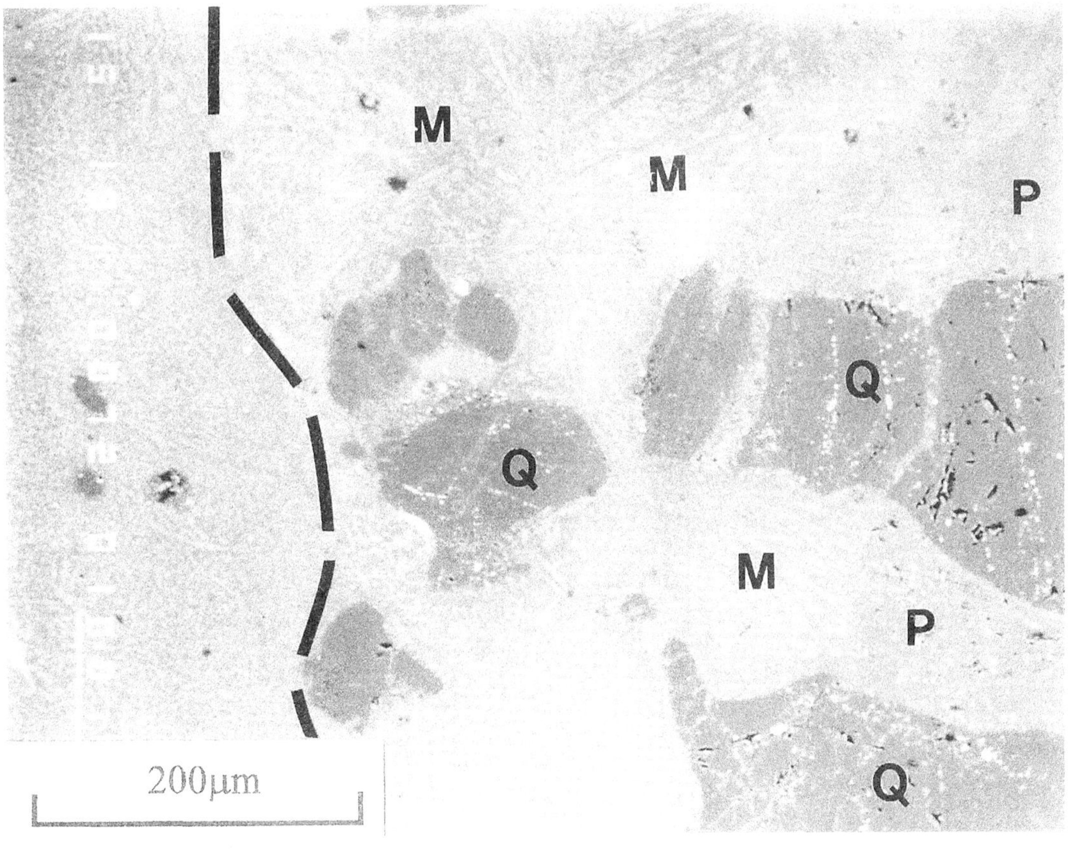


23 F

Plates 23G and 23H are SEM backscattered electron micrographs showing details of the mineralogy and microstructure of the matrix of the pseudotachylyte vein shown in 23E and 23F.

**23G.** Matrix of pseudotachylyte showing random and systematically arranged orthopyroxene dendrites, blebs of sulfide (arrowed) including pyrite, chalcopyrite, and sphalerite, and quartz porphyroclasts (Q). The systematically arranged orthopyroxene dendrites correspond to a globulite structure, whereas the randomly arranged dendrites correspond to the matrix surrounding the globulites (see 23E). The differences in texture therefore must relate to the kinetics of crystal nucleation and growth.

**23H.** High-magnification detail of systematically arranged orthopyroxene dendrites. The dendrites are up to 75 μm long and 5 μm wide, with secondary fibrillar growths or cock's comb texture; this habit is repeated at smaller scales visible only by transmission electron microscopy (TEM). SEM and TEM examination of the optically irresolvable groundmass interstitial to the pyroxene dendrites indicates that it is largely composed of small equant grains of biotite, quartz, plagioclase, and chlorite. This interstitial groundmass is believed to be a product of crystallization or devitrification rather than porphyroclastic debris.

The orthopyroxene (analyzed by microprobe and analytical TEM) has a composition of $Fs_{47}$-$En_{51}$-$Wo_2$ and is unusual in being strongly aluminous (ca. 12 wt% $Al_2O_3$). The conditions that favor crystallization of orthopyroxene in pseudotachylyte are not fully understood, but in the present example the crystal habit and high alumina content clearly indicate disequilibrium freezing of the melt.

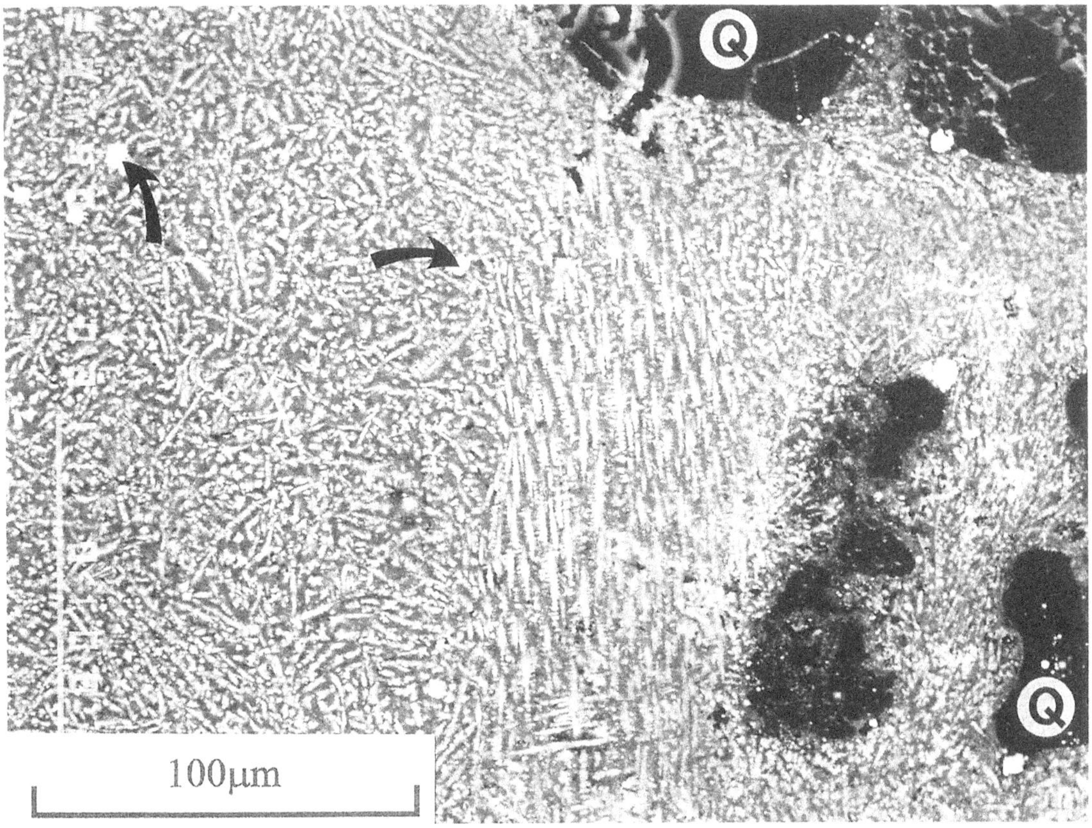


23 G

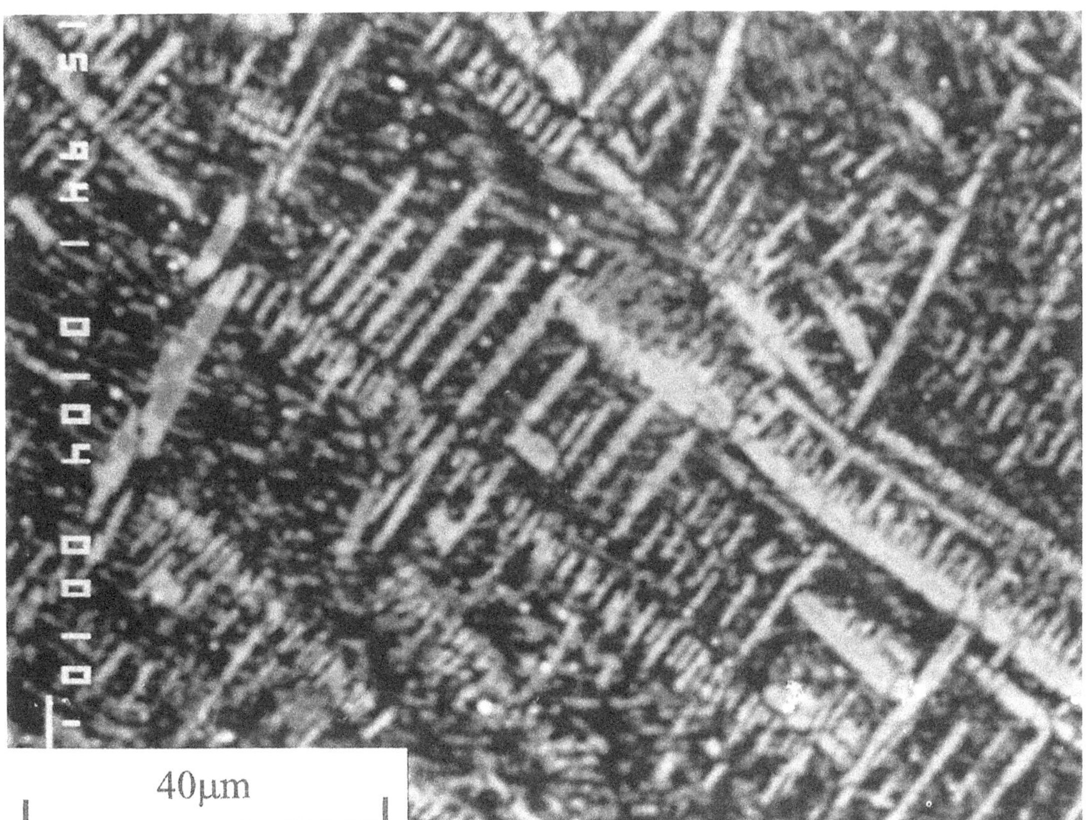


23 H

JERRY F MAGLOUGHLIN

The pseudotachylytes illustrated in the following plates (24–27) are from the Nason terrane (Tabor and others, 1987) in the North Cascade Mountains, Washington State, U.S.A. (Figure 24.1). The terrane underwent amphibolite-facies metamorphism in the middle to late Cretaceous, but the pseudotachylytes and associated fault rocks developed in the very latest Cretaceous or early Tertiary during rapid unroofing of the terrane. Unlike most occurrences, these pseudotachylytes are not associated with a single fault or shear zone, and instead are distributed throughout the northeastern corner of the terrane south of the White River shear zone (Magloughlin and McEwan, 1988). They are most commonly associated with sinistral slip along steeply dipping foliation surfaces within pelitic schist (Magloughlin, 1989), and developed at fairly shallow depths as suggested by the local presence of glass and vesicles (Magloughlin, 1992).

These two photomicrographs are from a single sample displaying several microstructural domains suggesting that pseudotachylyte developed along preexisting ultracataclasite zones. Both samples are from Dirty Face Ridge (see Figure 24.1).

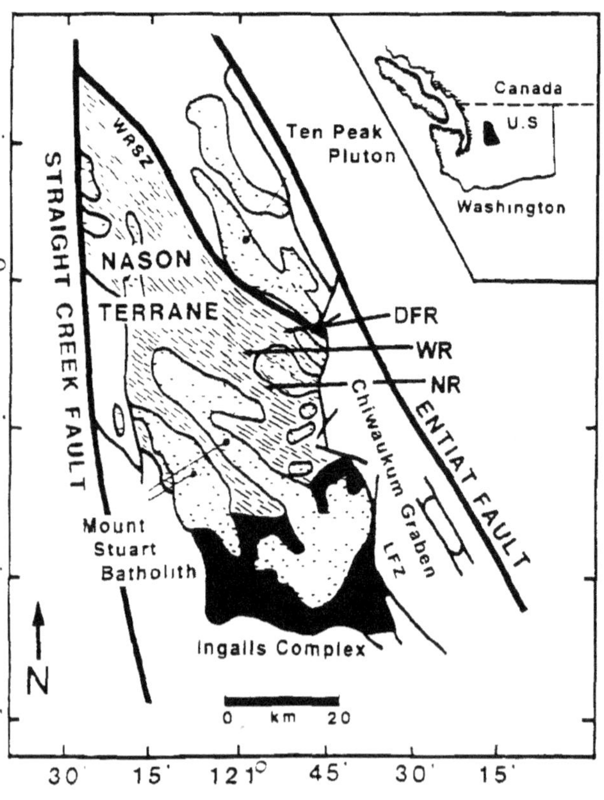


**Figure 24.1.** Geologic sketch map of the Nason terrane, North Cascade Mountains, Washington State, U.S.A. The Nason terrane is bounded by the Straight Creek fault to the west, the Leavenworth fault zone (LFZ) to the east, the Ingalls complex to the south, and the White River shear zone (WRSZ) to the north. Pseudotachylytes described in Plates 24–27 are from Dirty Face Ridge (DFR, Plates 24 and 25), Wenatchee Ridge (WR, Plate 26A), and Nason Ridge (NR, Plate 27).

**24A.** Pseudotachylyte vein margin relationships, North Cascade Mountains, Washington State, U.S.A. (Magloughlin, 1992). The host rock is a quartz biotite schist, and near the pseudotachylyte vein the quartz displays undulose extinction, deformation bands, and fluid inclusion bands (a). Biotite and muscovite are sheared along the vein margin, and the biotite forms long "fish" with a dextral sense-of-shear (b). Separating the host rock from the pseudotachylyte is a band of quartz-rich cataclasite (c) with angular fragments of quartz, plagioclase, and minor mica. Opaque ultracataclasite with only minor quartz borders the quartz-rich cataclasite (d). Above the ultracataclasite, light-gray glassy pseudotachylyte is interfingered with dark pseudotachylyte, and both contain numerous clasts of cataclasite and crystal fragments of quartz. Scale bar: 0.25 mm. PPL

**24B.** Pseudotachylyte vein margin relationships, North Cascade Mountains, Washington State, U.S.A. The margin of this pseudotachylyte vein is marked by a thin band of ultracataclasite (bottom 2 cm of the photograph). The contact with the host rock is parallel to and just below the bottom edge of the photograph. In the ultracataclasite and throughout the vein are strained quartz clasts (white) and fragments of quartz-rich cataclasite (white and dark mottled). Above the ultracataclasite is dark brown pseudotachylyte that displays prominent flow features (b). This pseudotachylyte interfingers with the light-colored glassy pseudotachylyte along the top half of the photograph (c). The black, mostly elongate blebs in the glassy part of this vein (e.g., d) generally contain iron sulfide and plagioclase, and were elongated by the flow of the melt (from right to left as seen in this photograph). Despite their very different appearances, the dark pseudotachylyte and light-colored, glassy pseudotachylyte have very similar compositions (Magloughlin, 1992), although the the light, glassy pseudotachylyte is chemically more homogeneous. Scale bar: 0.15 mm. PPL

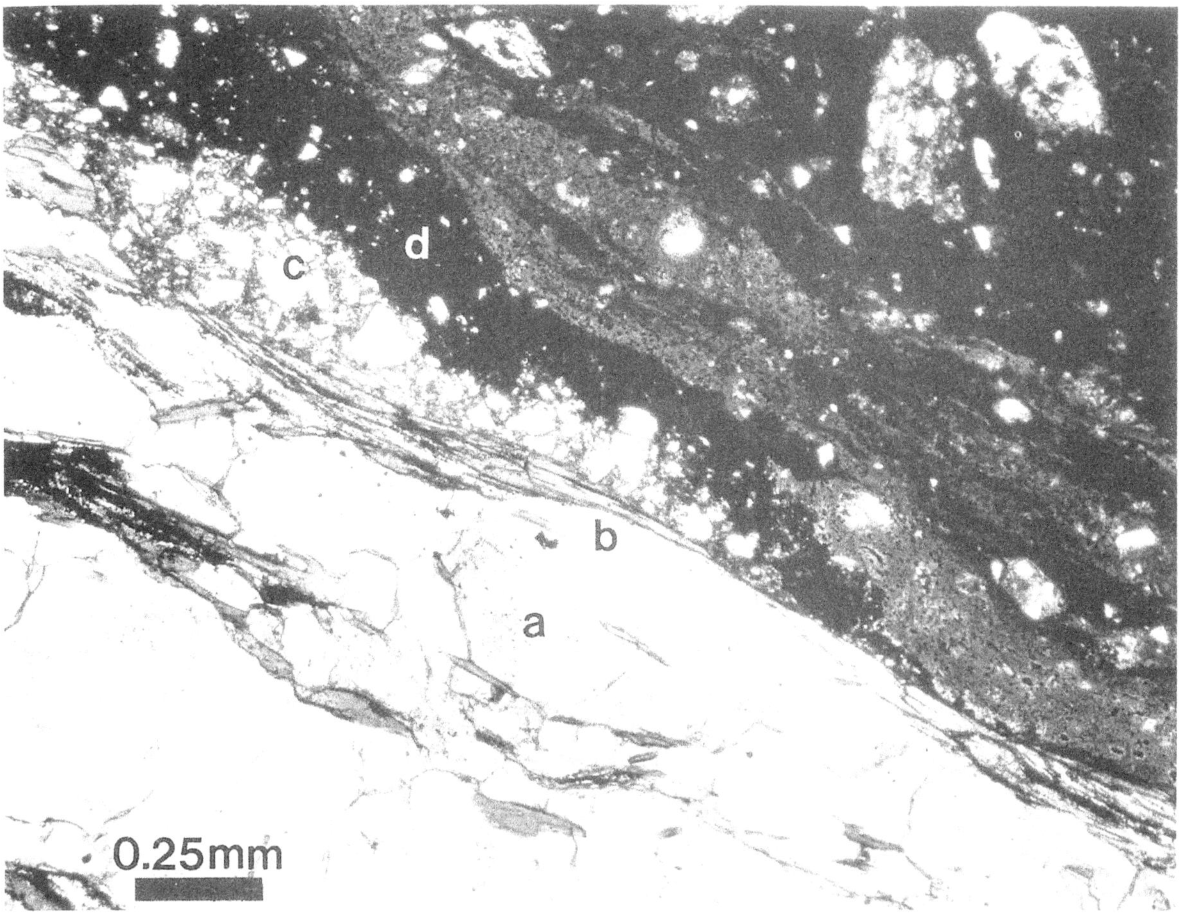


24 A

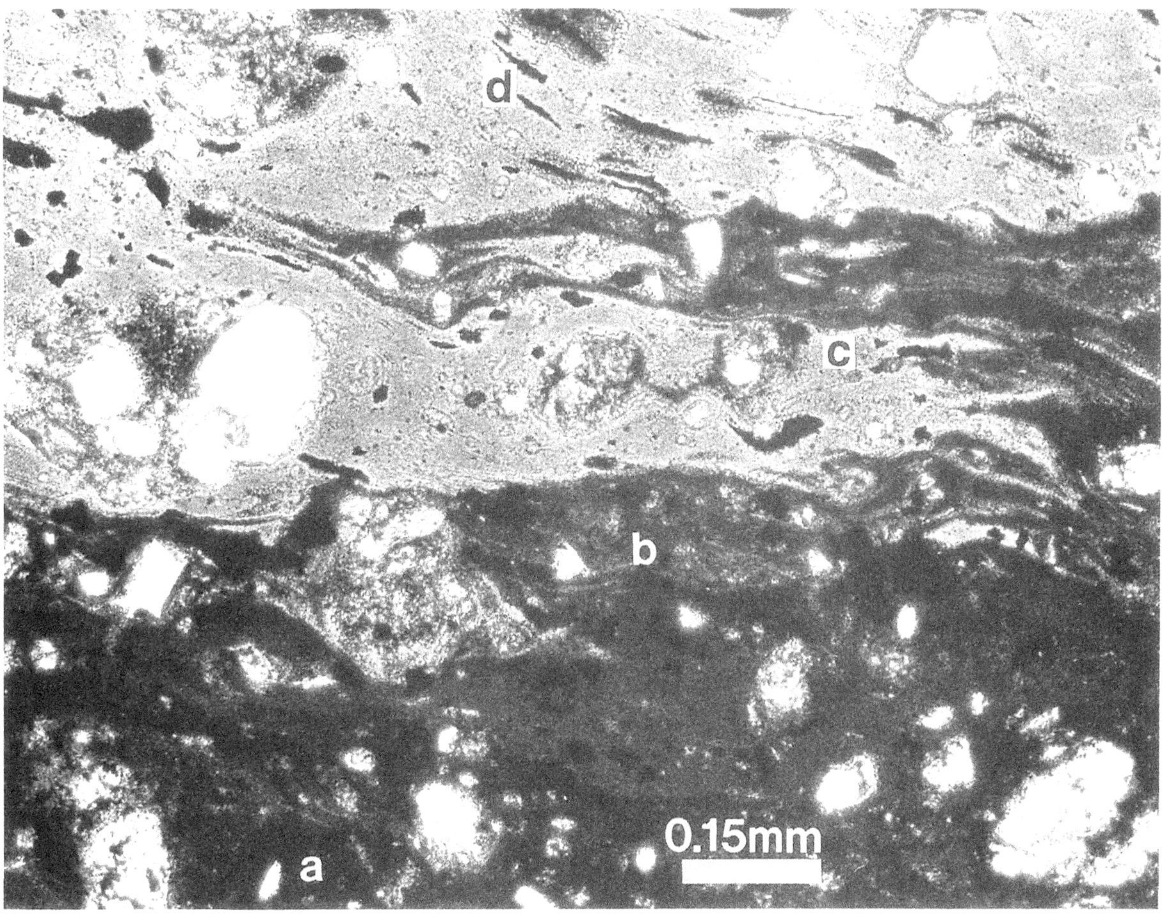


24 B

# 25. MELTING RELATIONS WITHIN LITHIC CLASTS IN PSEUDOTACHYLYTE

JERRY F. MAGLOUGHLIN

As the melt that quenches to form a pseudotachylyte remains liquid for only a short period of time, there is a good potential for recording details of the melt-rock interface. Such interfaces occur at the contact of the vein with the host rock, and at the contact of both lithic clasts and crystal fragments with the enveloping pseudotachylyte. The adjacent photomicrographs are from two samples and illustrate melting within lithic clasts. See Figure 24.1 for the geologic background and location.

**25A.** Melting relations in a plagioclase-quartz-biotite-hornblende lithic clast within pseudotachylyte, North Cascade Mountains, Washington State, U.S.A. (Magloughlin, 1993). The clast is composed dominantly of quartz (light) and light brown biotite (a). Pseudotachylyte (b) is composed of acicular grains of calcic plagioclase, up to about 30 μm long but only a few micrometers wide; smaller aluminous clinopyroxene microlites; and locally crudely dendritic biotite. Melting has proceeded into the clast along biotite (c), and there is a generally narrow transition zone from dark brown pseudotachylyte to light brown biotite (d). Such transition zones can be complex, with biotite and orthopyroxene(?) microlites developed within a matrix having a composition of approximately calcic plagioclase. Microlites up to about 10 μm long occur within the biotite up to about 150 μm from the margin of the brown pseudotachylyte (near e). Melting proceeded into the clast along hornblende-rich zones in a very similar fashion (f). Scale bar: 0.15 mm. PPL

**25B.** Melting relations in a quartz-plagioclase lithic clast and a quartz-rich cataclasite fragment, North Cascade Mountains, Washington State, U.S.A. (Magloughlin, 1992). In the polycrystalline lithic clast at left, melt has formed at three locations (central location at a). As there is no evidence for dilation, melt was generated in situ, especially along quartz-plagioclase grain boundaries. The melt pockets seem to be connected to the pseudotachylyte, perhaps by thin conduits along grain boundaries (b). Fluids may have migrated into the clast from the pseudotachylyte along grain boundaries, inducing melting. The large cataclasite fragment at right is composed dominantly of quartz with minor plagioclase. The grain size within this fragment varies from approximately 200 μm to less than 1 μm. Melting is more pervasive than in the lithic clast, forming networks and pockets (c, d). This is likely due to the highly fractured nature and small size of the grains in the cataclasite, which allow many more pathways along grain boundaries and microcracks for the melt and/or fluids to penetrate the clast and initiate melting. Greatly grain-size-reduced ferromagnesian minerals may also be present within the cataclasite fragment, and this material will have a relatively low melting point, further aiding interior melting of the cataclasite. Scale bar: 0.15 mm. PPL

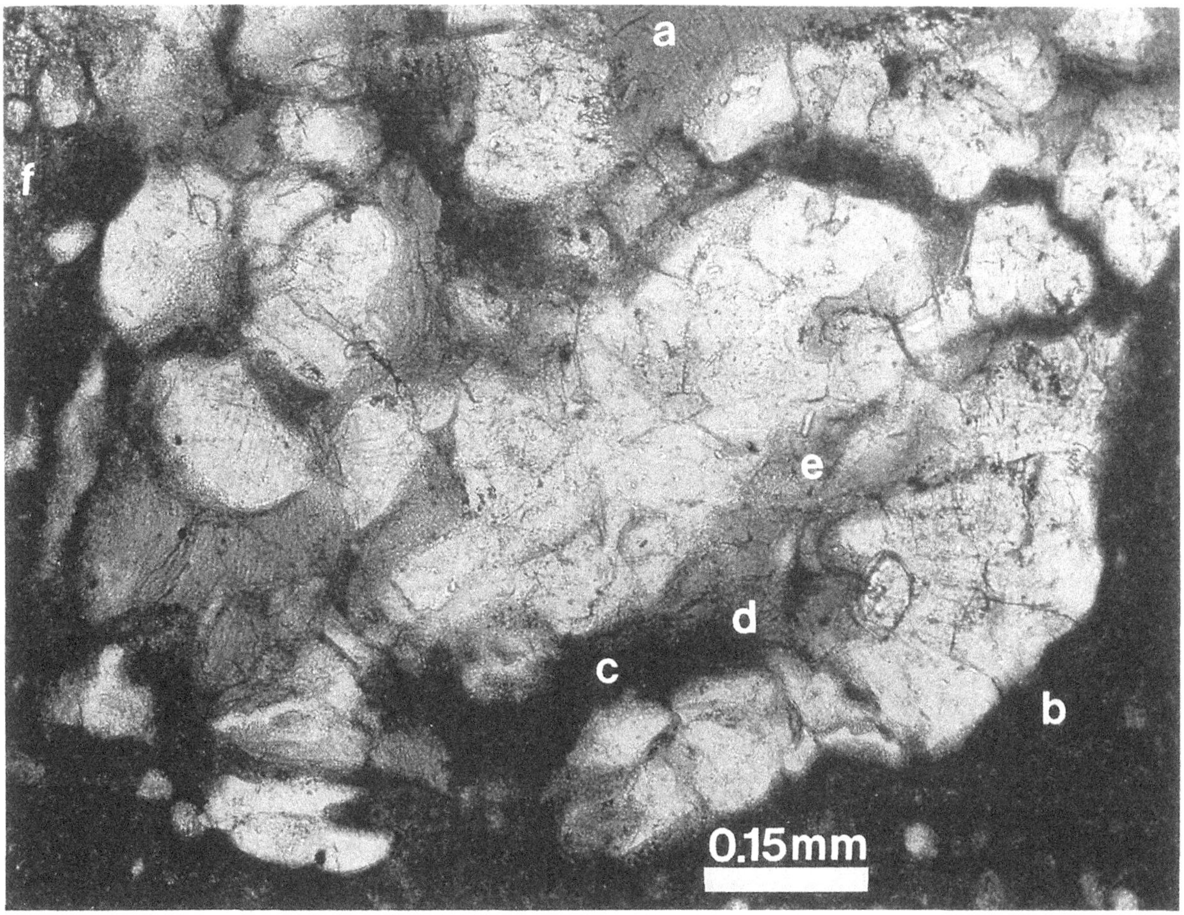

25 A

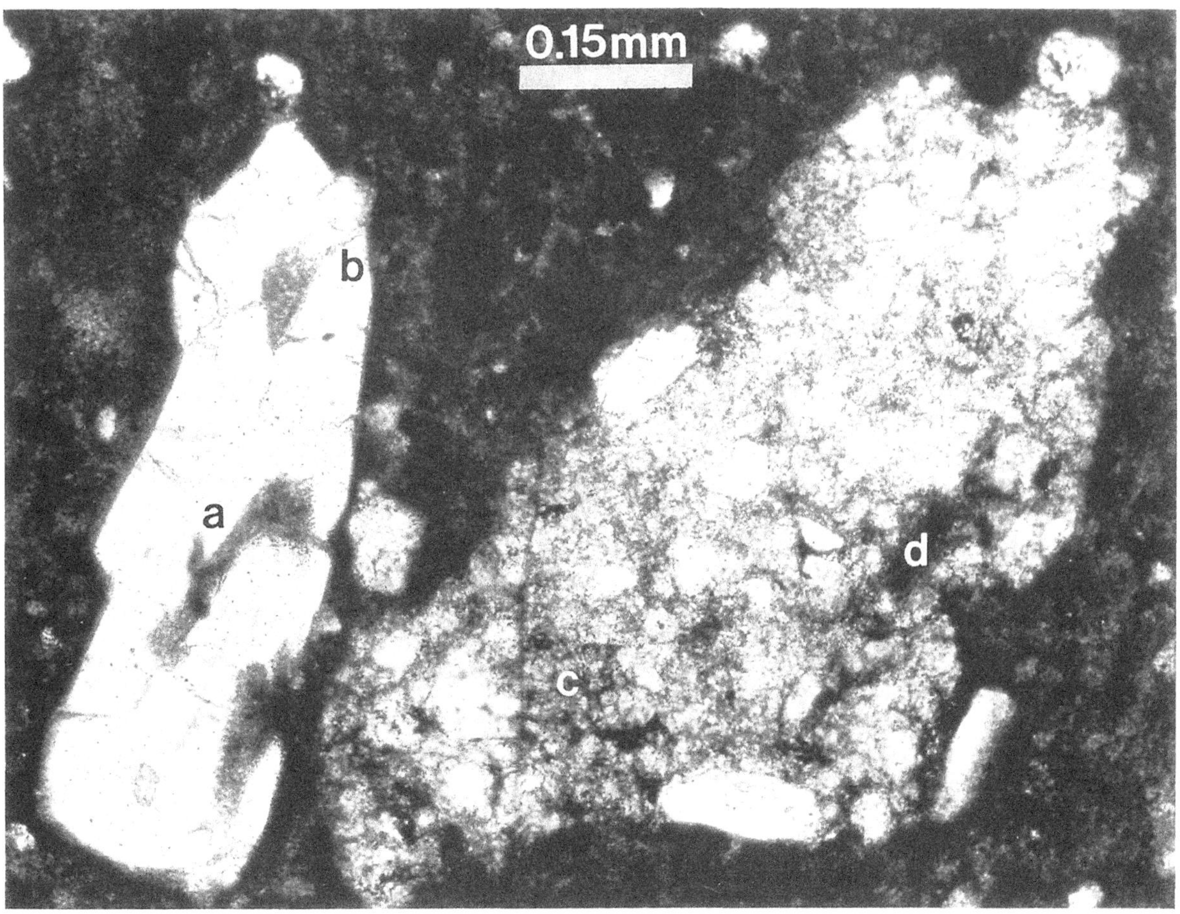

25 B

# 26. AMYGDULES WITHIN PSEUDOTACHYLYTE

JERRY F. MAGLOUGHLIN

The melts that quench to form pseudotachylytes, while short-lived, have similarities to typical crustal melts. Analogous to magmas intruded at shallow crustal levels or extruded onto the surface, pseudotachylyte melts can be saturated in volatiles, and this leads to exsolution of a fluid phase, forming vesicles. Vesicles, and amygdules formed as a result of infilling with minerals precipitated from fluids, seem to be uncommon among pseudotachylyte occurrences but have been described (e.g., Maddock and others, 1987; Magloughlin, 1989; also see Plate 27). In part, this may be due to collapse of the vesicles prior to complete solidification of the pseudotachylyte or during subsequent deformational overprinting. See Figure 24.1 for geologic background and location.

**26A.** Quartz amygdule-rich zone within a pseudotachylyte from the North Cascade Mountains, Washington State, U.S.A. (Magloughlin, 1993). The large amygdule in the upper left is approximately 300 μm across in its longest dimension, while dozens of smaller amygdules, heavily coated with opaque minerals, are also present in this field of view. The amygdule at (a) is mostly within the thin section, with the very top cut through so that the opaque minerals are removed and the quartz is exposed. Within the large amydule, a distinct 10–15 μm wide band of fluid inclusions within the quartz begins approximately 10 μm inside the edge of the amygdule. Apparently the vesicle within the pseudotachylyte had begun filling with quartz inward from the margin, when a pulse of fluids (associated with continuing seismicity?) infiltrated the fault zone and was recorded as the fluid inclusion band when it was overgrown by continued precipitation of quartz. Also visible are acicular, high-relief micron-scale crystals of apatite projecting from the margin of the amygdule into the quartz. Some prominent needles are visible in the smaller of the two large amygdules. Scale bar: 0.1 mm. PPL

**26B.** An "amygdule patch" within a pseudotachylyte from the North Cascade Mountains, Washington State, U.S.A. (Magloughlin, 1993). While some veins display a uniform distribution of amygdule densities across the vein (except at the extreme margin), others only contain amygdules in isolated, commonly light-colored patches up to a few mm across, as shown here. Three prominent quartz-filled amygdules are present at left, each containing a patch of opaque material generally offset slightly to the lower left. Within the amygdule at (a) is a small microbreccia accumulation similar to that described in Plate 27. In cross-polarized light, the quartz in these amygdules consists of roughly a dozen individual crystals, many with a feathery radiating habit. In the dark pseudotachylyte groundmass, small biotite microlites are visible in several places. The large, light-colored patch (boxed b) is silica-rich with regions having a potassium feldspar composition. The microlites (c) within the large, light-colored region are complex, appear to be a silica polymorph, and are slightly fuzzy due to partial replacement. Some attain 100 μm in length and may display a radiating habit as at small (b). This "amygdule patch" may have formed when a hydrous crystal fragment dissolved within the pseudotachylyte melt and liberated water to form the original vesicles. Scale bar: 0.1 mm. PPL

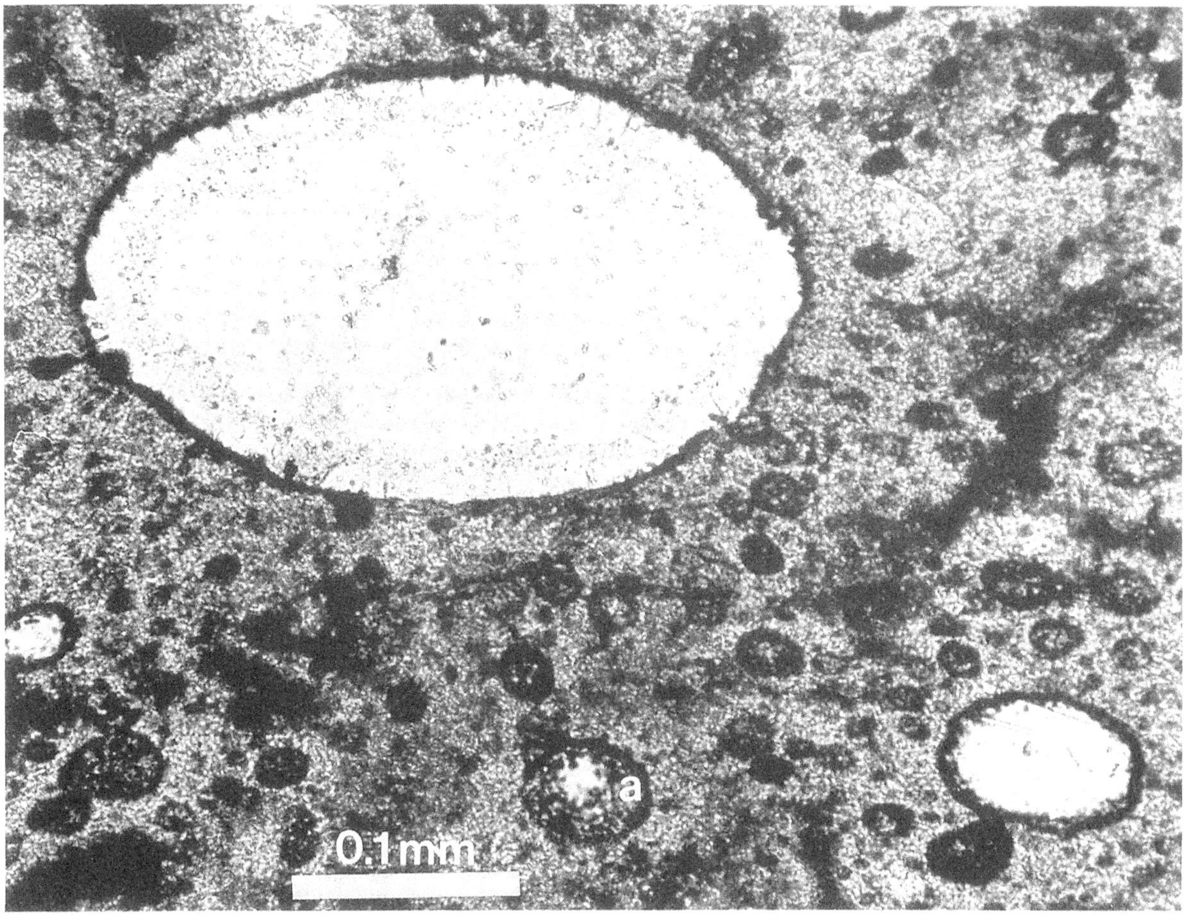


26 A

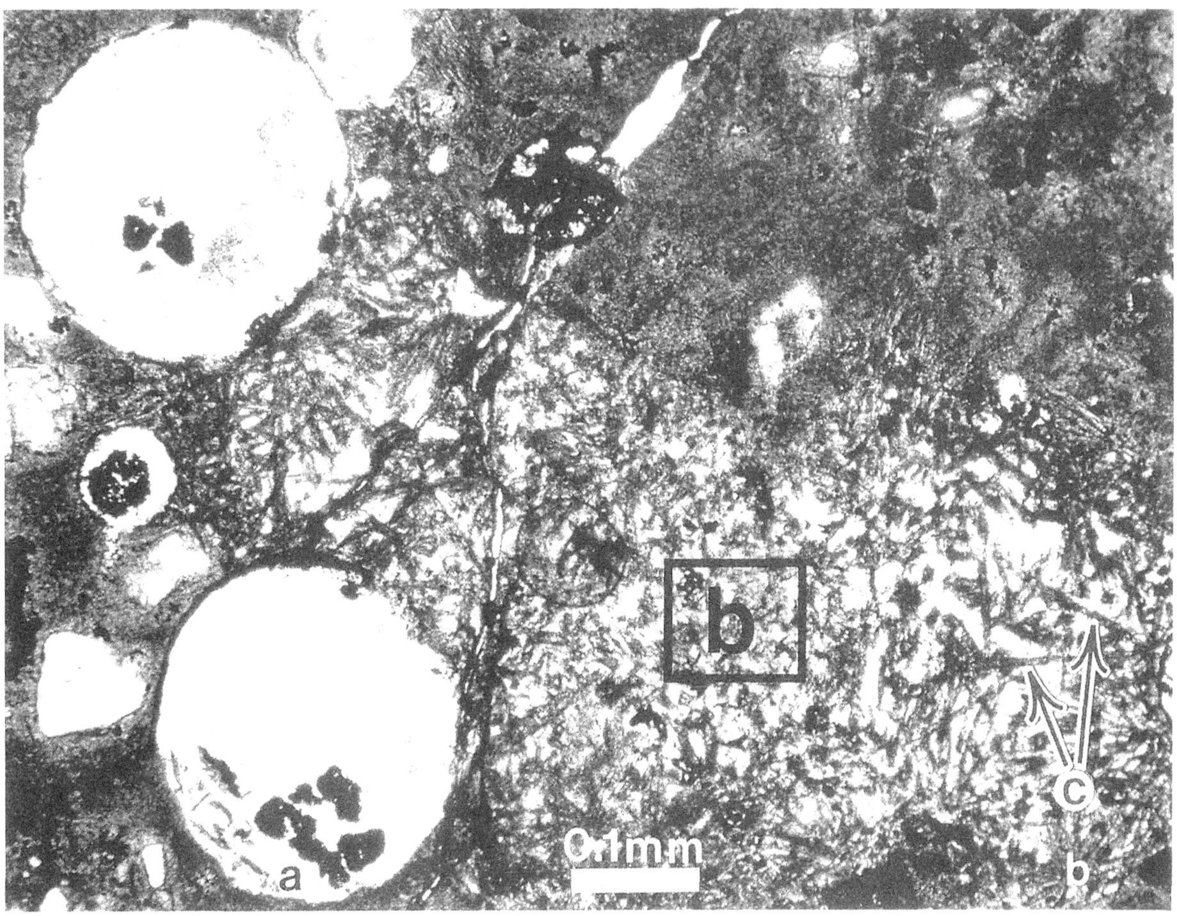


26 B

# 27. AMYGDULES AND MICROBRECCIA COLLAPSE STRUCTURE IN A PSEUDOTACHYLYTE

JERRY F. MAGLOUGHLIN

Vesicles and amygdules are apparently uncommon in pseudotachylytes but have been reported from several localities (e.g., see Plate 26). Once formed, vesicles may collapse prior to complete solidification of the pseudotachylyte, or brittle processes may modify the appearance of the vesicle prior to or during infilling by minerals precipitating from fluid migrating along the fault or shear zone in which the pseudotachylyte developed. The example below is a large vesicle within a pseudotachylyte from the Cascade Mountains in Washington State, U.S.A. (see Figure 24.1 for geologic background and location). It displays evidence for brittle deformation following solidification of the pseudotachylyte, and is partially infilled with quartz.

**27A.** Large quartz amygdule in a pseudotachylyte. North Cascade Mountains, Washington State, U.S.A. (Magloughlin, 1993). The large amygdule at the top left is mostly filled with quartz, and probably had a void at the center (a). It is surrounded by dark pseudotachylyte, whereas a brittlely deformed quartz fragment is at the lower right. In cross-polarized light some of the quartz in the amygdule displays a coarse radiating fibrous habit. The amygdule walls are commonly lined with extremely fine-grained opaque minerals, mostly pyrite, as at (b) and (c). There are two unusual features about this large amygdule. First, near the bottom left, thin white quartz-filled cracks are present parallel to the margin of the amygdule. Apparently the wall of the vesicle fractured and began to spall off during precipitation of quartz in the amygdule. Second, at the presumed bottom of the amygdule (just to the right of the center of the photograph) is a pseudotachylyte microbreccia, composed of angular pseudotachylyte fragments and fine-grained quartz. Most of this microbreccia formed prior to significant quartz infilling, but some larger fragments (d) may have spalled off of the wall of the vesicle after some of the quartz had been precipitated. Such features may serve as paleotops indicators with respect to the time of quartz precipitation. At (e) is a small dark patch of smectite that grew in the void on the quartz at less than 150°C. The area of 27B is outlined. Scale bar: 0.25 mm. PPL

**27B.** Close-up of 27A showing details of the microbreccia. Note the angularity of the clasts, numerous opaque minerals, and the fine-grained quartz "cement." The fragment at (a) displays a concave, opaque lower margin (mostly pyrite) suggesting that it was part of the lining of the vesicle. Scale bar: 0.15 mm. PPL

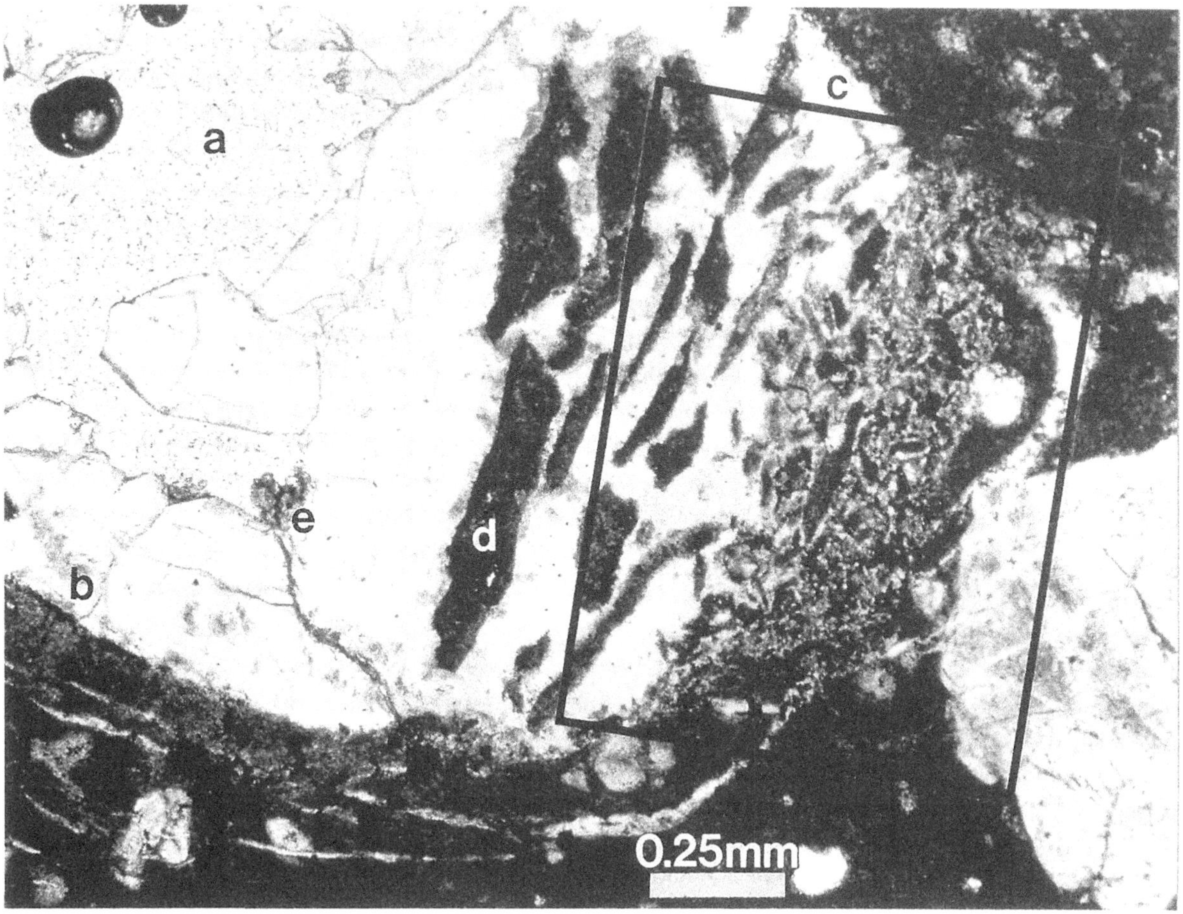


27 A

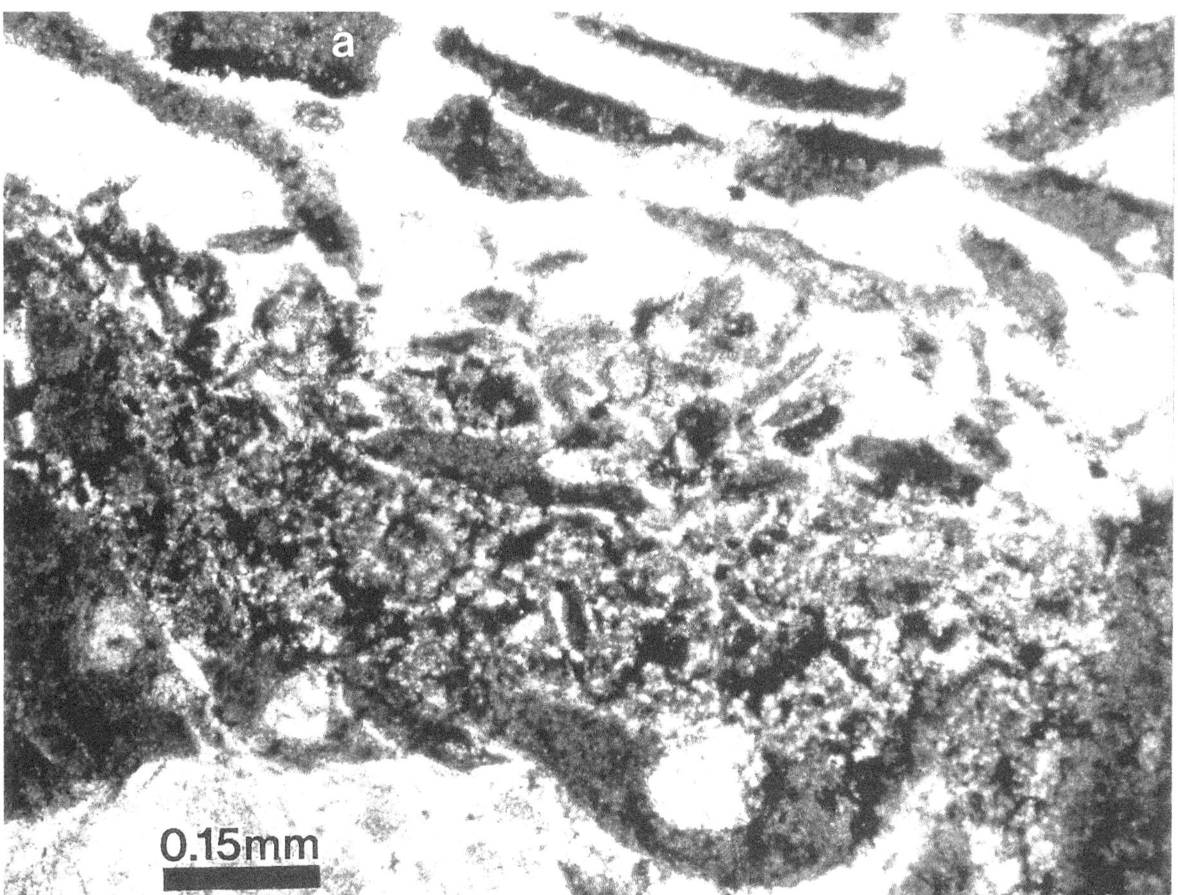


27 B

# 28. MESOSCOPIC AND MICROSCOPIC ASPECTS OF PSEUDOTACHYLYTE

K. S. TECHMER

The Ivrea-Verbano Zone (IVZ), northern Italy, is part of the crystalline basement of the Southern Alps. It chiefly consists of mafic, psammitic, and pelitic rocks with minor intercalated carbonate lenses, which have experienced amphibolite- to granulite-facies metamorphism (Zingg, 1983). Ultramafic rocks are also locally common in the IVZ, especially near the northwestern margin of the zone.

The origin of pseudotachylyte along fault planes in the IVZ is attributed to paleoseismicity along the faults. In the IVZ, pseudotachylytes are formed in the upper crust during the retrogressive, tectonometamorphic evolution of the region (Weber and others, 1989; Techmer and others, 1992). During these natural frictional sliding processes, selective melting occurs within brecciated rocks.

Pseudotachylytes in the IVZ are preferentially generated in metasedimentary and mafic rocks. No pseudotachylyte veins with evidence of frictional melting have yet been found in the ultramafic rocks of the IVZ. Metacarbonates are not source rocks for the formation of pseudotachylytes, although single veins of pseudotachylyte, ranging in thicknesses from centimeters to meters, have been injected into metacarbonate rocks. These pseudotachylyte veins were originally generated along shear planes in mafic source rocks and were subsequently injected into adjacent metacarbonate rocks.

**28A.** Pseudotachylyte veins in the IVZ, northern Italy (Techmer, 1989). Pseudotachylytes in the IVZ are generated along individual shear planes (GP). In this example, the fine-grained matrix was injected into a vein (IV) at a high angle to an older foliation in the host metasedimentary rock. These pseudotachylyte veins are extension fractures, filled with fragments in a fine-grained matrix.

**28B.** Intrusive pseudotachylyte vein, IVZ, northern Italy (Techmer, 1989). Pseudotachylyte formation can result in either a netlike fragmentation of the host rocks or in variably thick pseudotachylyte veins.

Commonly, pseudotachylyte breccias ranging from centimeters to several meters in thickness can be observed. Their appearances are interpreted as representing "pull-apart" extension features. During seismicity, shearing will be variably distributed yielding areas of local extension. During pseudotachylyte generation, the fine-grained pseudotachylyte matrix mixes with coarse fragments. With increased pseudotachylyte generation, the fine-grained matrix and incorporated larger fragments inject into adjacent rocks. For the mixture to become mobile, the critical ratio of pseudotachylyte matrix to fragments is about 30%.

**28C.** Microscopic contact between the neighboring rock (nr, left side) and the pseudotachylyte matrix (pt, right side), IVZ, northern Italy (Techmer, 1989). The contact between the pseudotachylyte vein and the neighboring rock is sharp. The pseudotachylyte matrix (pt) is composed of a fine-grained matrix and larger incorporated fragments of the adjacent neighboring rock, which are chiefly feldspars and pyroxene in this example. Newly formed minerals in the matrix were generated by the rapid cooling of a frictional melt or by secondary alteration. At the contact, a feldspar of the neighboring rock is intensively brecciated and cut by smaller veins, several mm thick. The host rock is a metagabbro. PPL

**28D.** SEM backscattered image of dendrites, generated by rapid cooling of a frictional melt generated within the pseudotachylyte matrix. Newly formed minerals resulting from rapid cooling of a frictional melt can be clearly identified by their skeletal crystal habits, forming dendrites and radially grown crystals. Dendrite occurrences are concentrated in pseudotachylytes within metasedimentary host rocks (Techmer and others, 1992). Dendritic crystal growth occurs after rapid cooling of a frictional melt, generated by the selective melting of K-feldspar, plagioclase, biotite, and garnet. Individual, thermally corroded fragments of plagioclase can be identified by an enrichment in the anorthite component towards their grain boundaries. Feldspar-, quartz-, garnet-, biotite-, and sillimanite-rich host rocks are common source rocks for the formation of frictional melts in the IVZ.

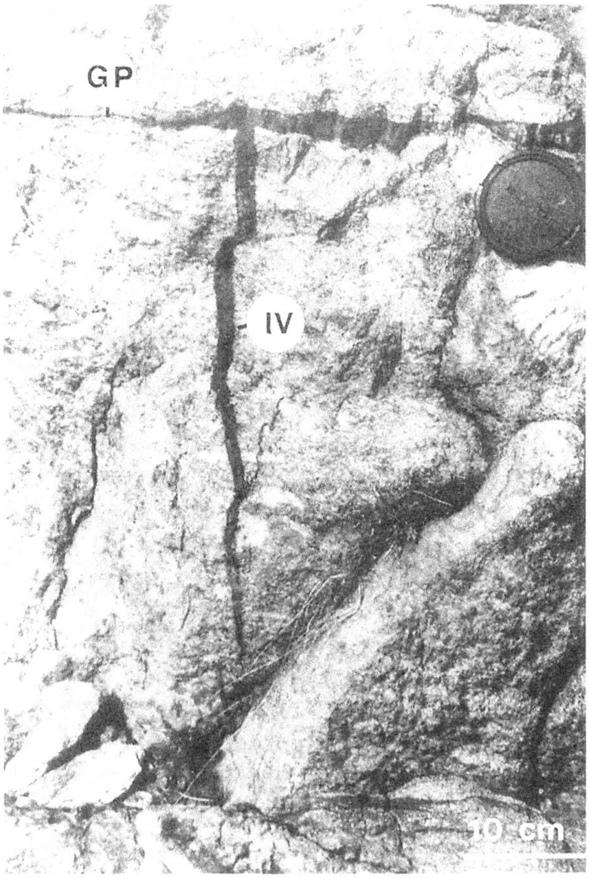


28 A

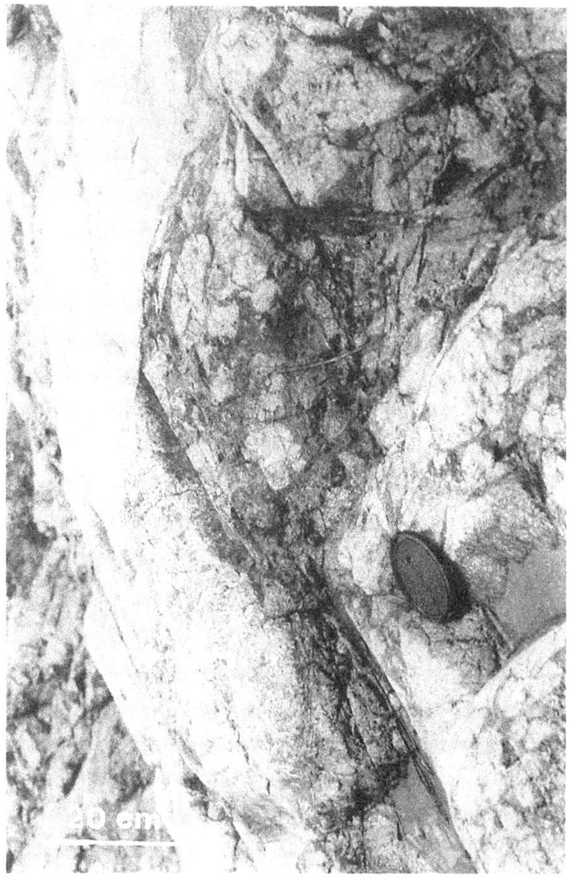

28 B

28 C

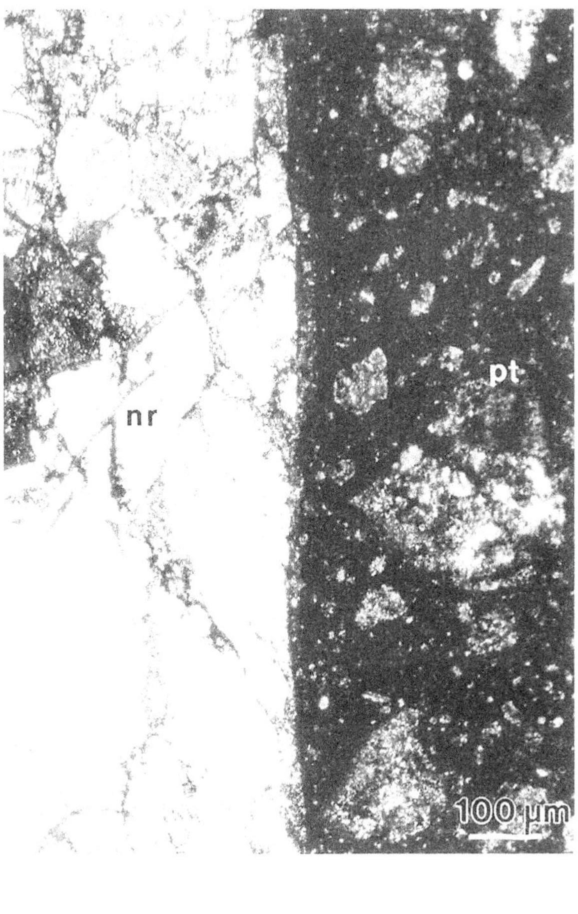


28 D

JOSEPH CLANCY WHITE

The coexistence of large volumes of pseudotachylyte with mylonites within the ductile regime raises the question of how and under what conditions these pseudotachylytes form. The Outer Hebrides thrust zone, Scotland (Figure 29.1), contains examples of pseudotachylyte (Sibson, 1980) that were cyclically injected at amphibolite-grade conditions (≈500°C), concomitant with intense ductile deformation of the host rock. The ubiquity of plastic processes, the absence of evidence for clear-cut brittle processes, and the large volume of pseudotachylyte collectively suggest that pseudotachylyte is produced by repeated generation of plastic instabilities. This response appears to require relatively anhydrous, originally deep-seated rocks with low pore fluid pressures that are exhumed along thrusts throughout which pore fluid pressures remain low. In such an environment, more commonly observed brittle behavior is suppressed in the absence of fluid pressure, enabling high-strength plastic processes to be activated. Paleostress estimates and associated inferred strain rates are compatible with the initiation of plastic instabilities in quartzo-feldspathic crust (Hobbs and Ord, 1988). Ambient shear zone strain rates of $10^{-12}s^{-1}$ at 200 MPa and ≈500°C require only an order of magnitude perturbation in local strain rate in order to move the rock into an instability field. The occurrence of locally heterogeneous flow simply reflects the inherent heterogeneity and non-steady state flow of deforming rock masses.

**29A.** Pseudotachylyte and mylonite, Seaforth Head, Isle of Lewis, Scotland. Amphibolite-grade gneisses deformed within the Outer Hebrides thrust zone contain cyclically injected pseudotachylyte veins that are progressively deformed, folded, and transposed by the ambient mylonitization (Myl). The color of the pseudotachylyte (PST) changes during deformation, becoming lighter with age (i.e., with progressive deformation), allowing different generations of original pseudotachylyte, now ultramylonite, to be identified. PPL

**29B.** Ductile deformation of pseudotachylyte, Seaforth Head, Isle of Lewis, Scotland. A feldspar-quartz clast in pseudotachylyte has been ductilely deformed to produce a tailed porphyroclast. Dark-colored pseudotachylyte (top) is younger than light-colored pseudotachylyte (bottom). The attenuation of pseudotachylyte fragments emphasizes the intensity of subsolidus intracrystalline deformation characteristic of the environment in which the pseudotachylyte formed. PPL

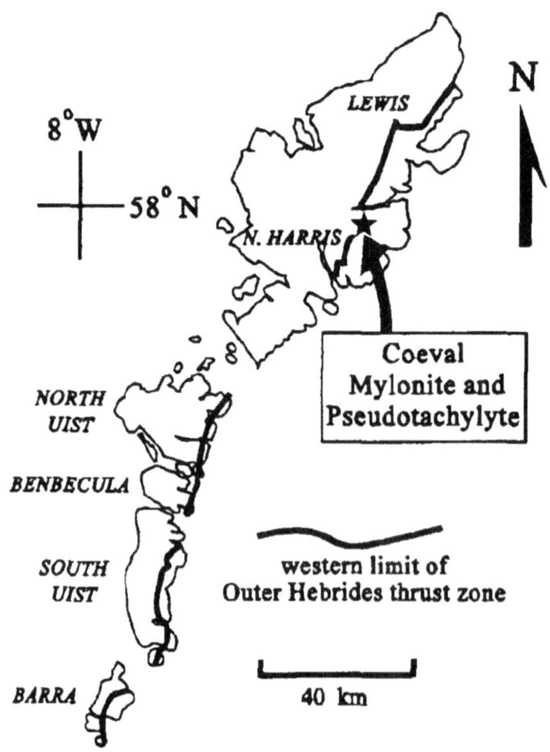


**Figure 29.1.** Geologic sketch map of the Outer Hebrides thrust zone, Scotland. Sample locality of coeval mylonite and pseudotachylyte is noted

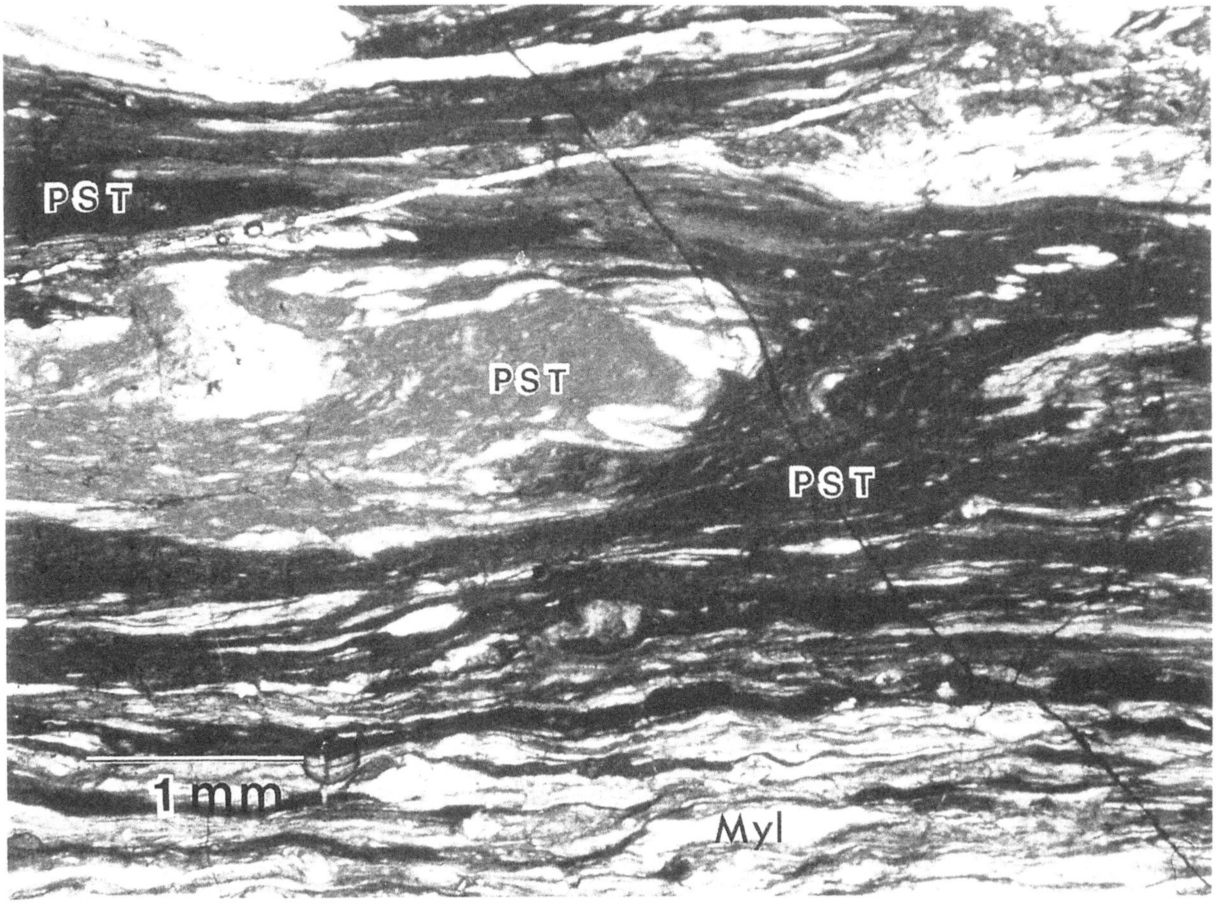


29 A

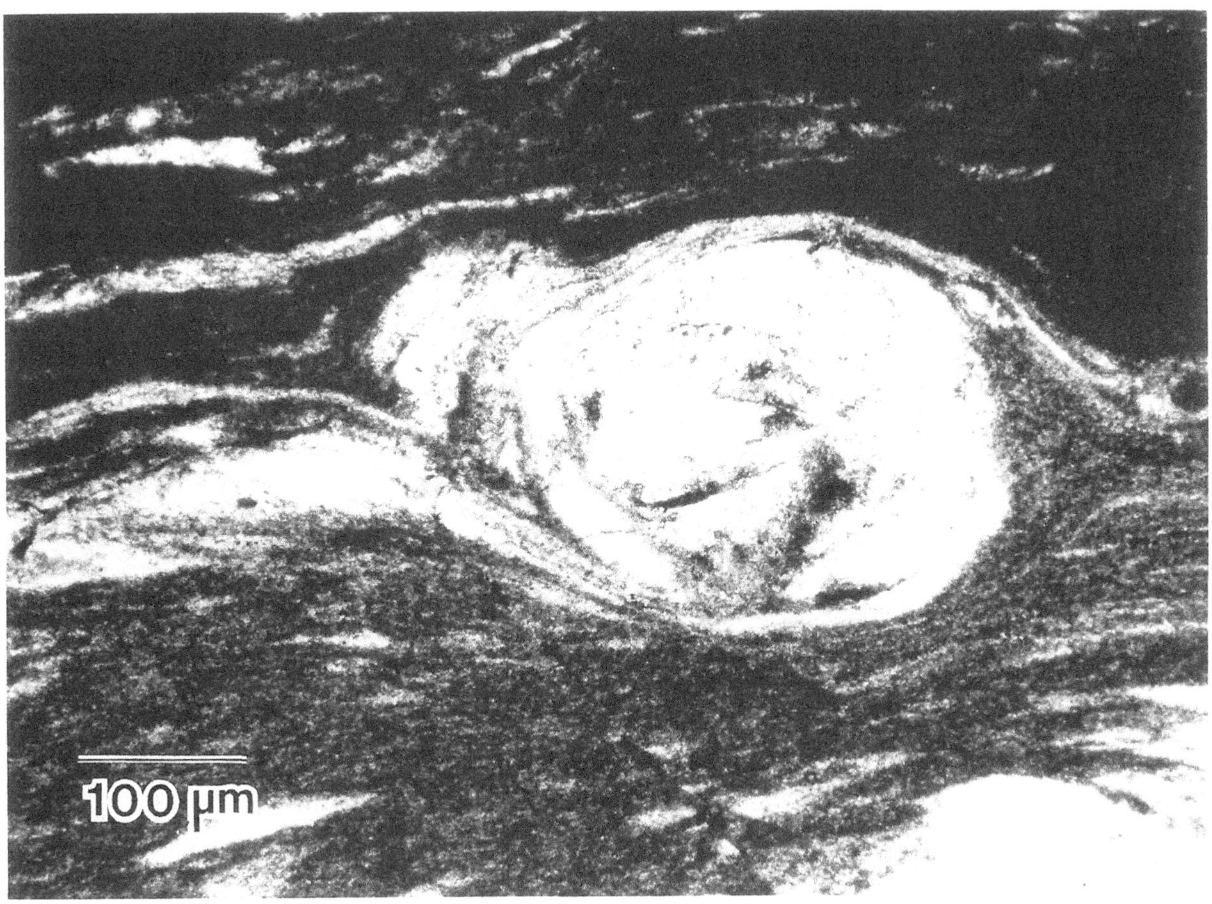


29 B

**29C.** Undeformed pseudotachylyte. A veinlet of pseudotachylyte exhibits distinctive crystallization texture with amphibole grains surrounded by plagioclase that confirms its origin as a melt. Plagioclase clasts are included in a matrix of hornblende (A; light mineral) and plagioclase (P; dark mineral). Matrix grains are commonly elongate and variably oriented with sharp grain boundaries. Tensile fracturing of clasts and injection of pseudotachylyte into a clast is seen at T. SEM backscattered electron image

**29D.** Deformed pseudotachylyte, now ultramylonite, Seaforth Head, Isle of Lewis, Scotland. Upon crystallization, pseudotachylyte is deformed to produce ultramylonites comprising plagioclase (P), hornblende (A), K-feldspar (k), and quartz (q), in proportions that reflect the original melt. Grains are more equant and slightly larger, whereas grain boundaries (arrows) are more curviplanar than in the undeformed pseudotachylyte. Grain size increases with deformation to that generated independently by dynamic recrystallization (see 29F). Dislocations are generally uncommon. The texture resembles multiphase alloys in which the mutual pinning of grain boundaries maintains a very small grain size which in turn favors grain-size-sensitive flow mechanisms. Superplastic flow of the ultramylonites is consistent with the low dislocation densities and large macroscopic strain and complexity of flow in the layers. Notably, no retrograde minerals form during deformation of the pseudotachylyte. TEM energy-dispersive X-ray analysis of the phases confirms the amphibolite grade (≈500°C) origin of the melt and subsequent ductile deformation. The absence of hydration in such fine-grained material emphasizes the low fluid activity in these rocks. TEM bright field (BF)

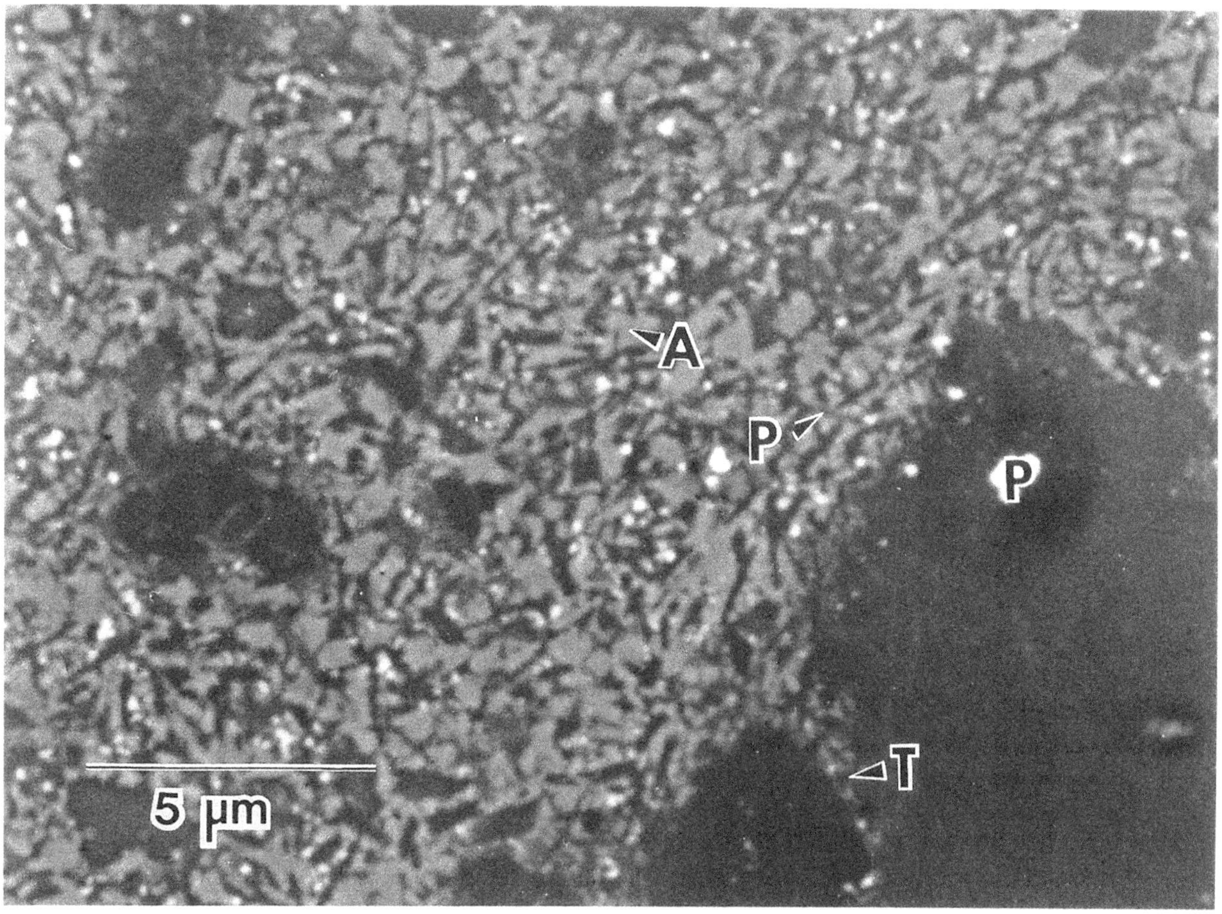


29 C

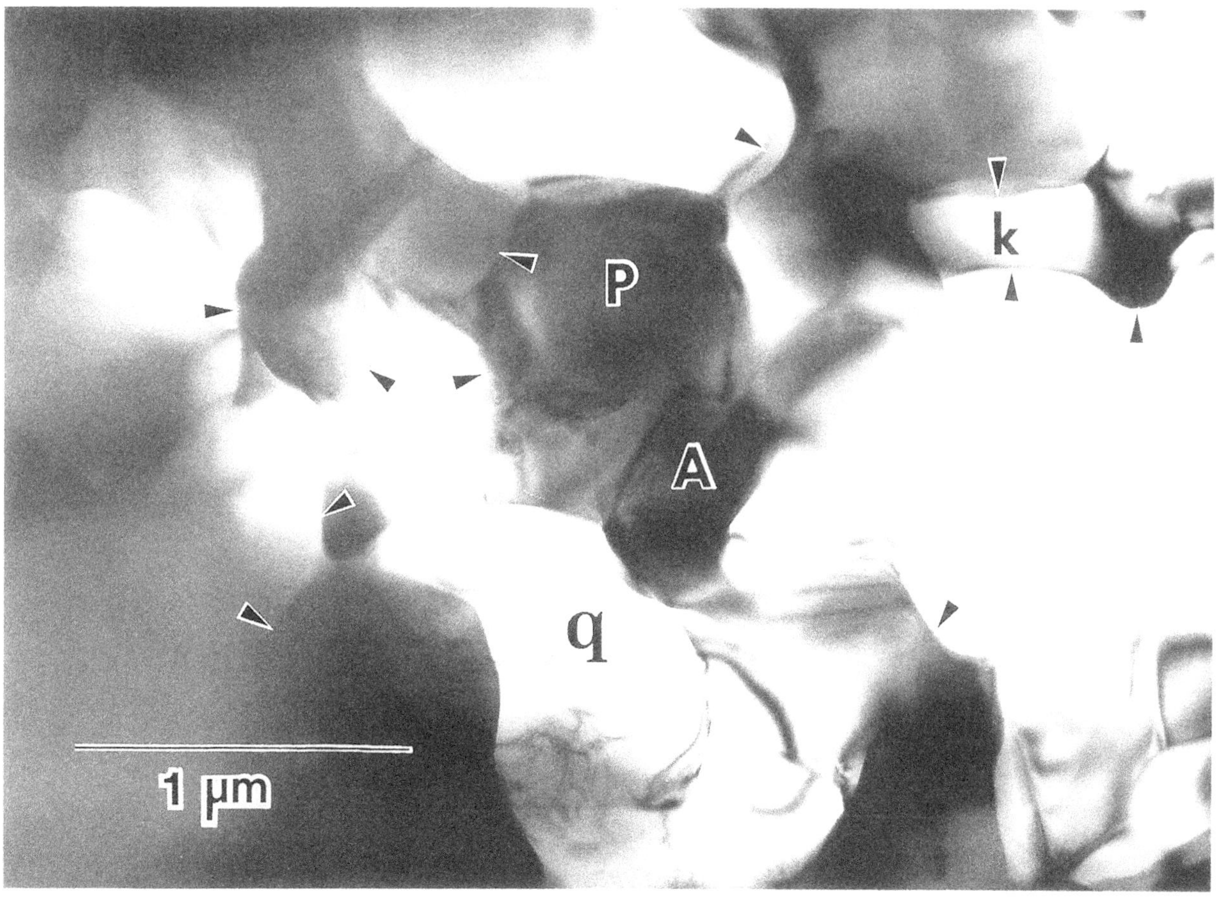


29 D

**29E.** Dislocations in K-feldspar porphyroclast from mylonite that hosts and deforms pseudotachylyte, Seaforth Head, Isle of Lewis, Scotland. K-feldspar, plagioclase, and, to a somewhat lesser degree, quartz exhibit intense dislocation glide with minimal development of recovery substructures or dislocation tangles characteristic of strain-hardening during cold working. Dislocation densities can be very high (K-feldspar = $10^{14}$ $m^{-2}$; plagioclase = $>10^{13} m^{-2}$; quartz = $10^{13}$ $m^{-2}$) and, together with submicrometer recrystallized grain sizes, imply ambient stresses within the mylonites of ≈200 MPa on which are juxtaposed transient differential stresses in excess of 300 MPa. Paleostress estimates (see Plate 126C for pertinent references) suggest a background shear strain rate of $10^{-12}$ $s^{-1}$, on which are imposed transient high-stress/high-strain-rate events consistent with periodic plastic instabilities that may lead to melting. The conditions calculated to be necessary for initiation of local instabilities correspond closely with the predicted onset of athermal dislocation glide, e.g., at differential stresses of about 370 MPa for feldspars. This suggests a link between the activation of extensive dislocation glide, the observed high dislocation densities, and the initiation of melt-generating instabilities. TEM BF

**29F.** Recrystallization of K-feldspar in mylonite that hosts and deforms pseudotachylyte, Seaforth Head, Isle of Lewis, Scotland. High dislocation densities (d) in a mylonite porphyroclast (H) lead to dynamic recrystallization (Rx) by strain-induced boundary migration, with no evidence for brittle deformation. The K-feldspar host grain recrystallizes to modulated, intermediate orthoclase, consistent with amphibolite-grade mylonitization. The absence of brittle deformation features, despite the extremely high dislocation densities along the recrystallization front (arrows), indicates the operation of high-stress/high-temperature glide plus recrystallization at the expense of fracturing. Suppression of fracturing and development of high stresses compatible with plastic deformation require that pore fluid pressures be low. Recrystallized grains have very few, if any, dislocations and are observed to accommodate large macroscopic strains without any evidence for subsequent dislocation activity that is characteristic of cyclic intracrystalline deformation/recrystallization. The absence of dislocations in the presumed weakest material (i.e., no stored strain energy to inhibit glide), the equant grain shape, and absence of a crystallographic preferred orientation argue for extensive intercrystalline deformation comparable to that inferred for the deformed pseudotachylyte/ultramylonite. TEM BF

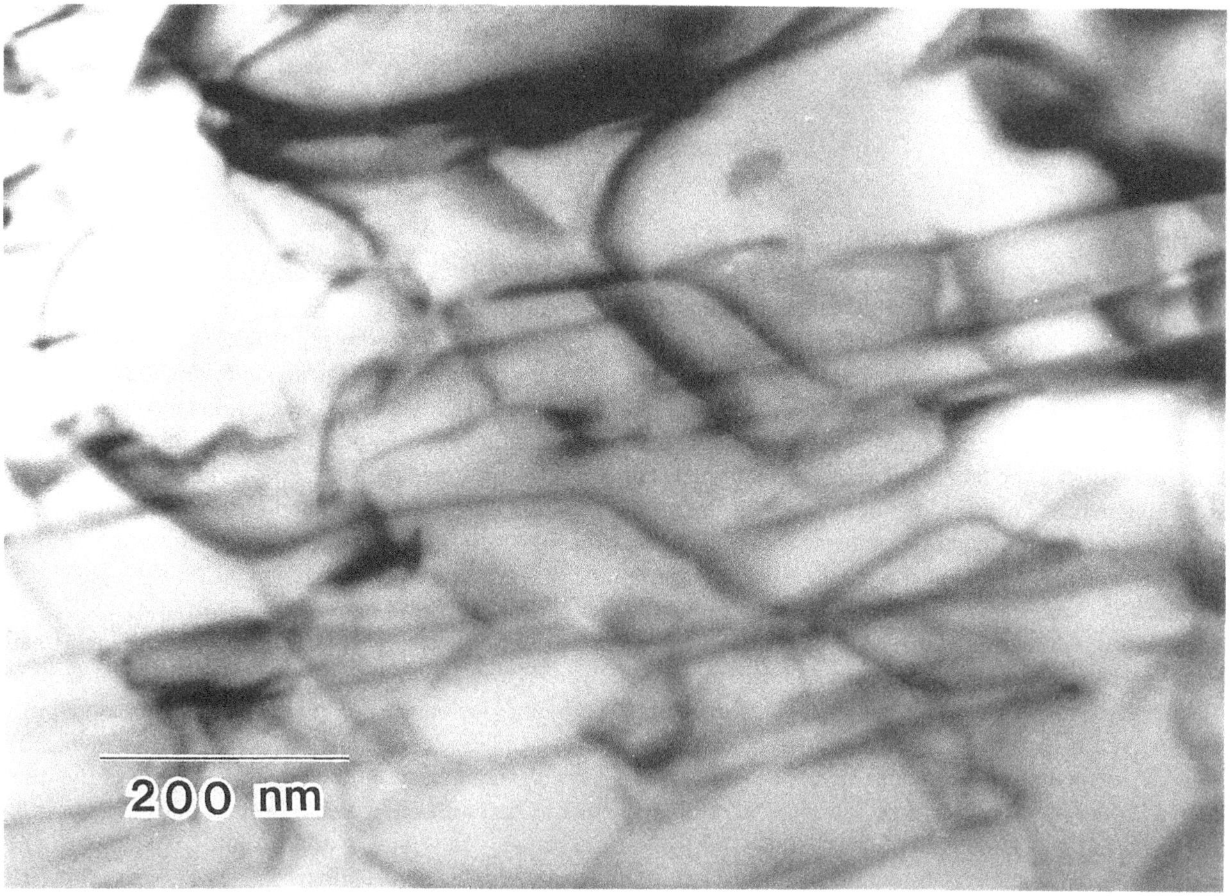


29 E

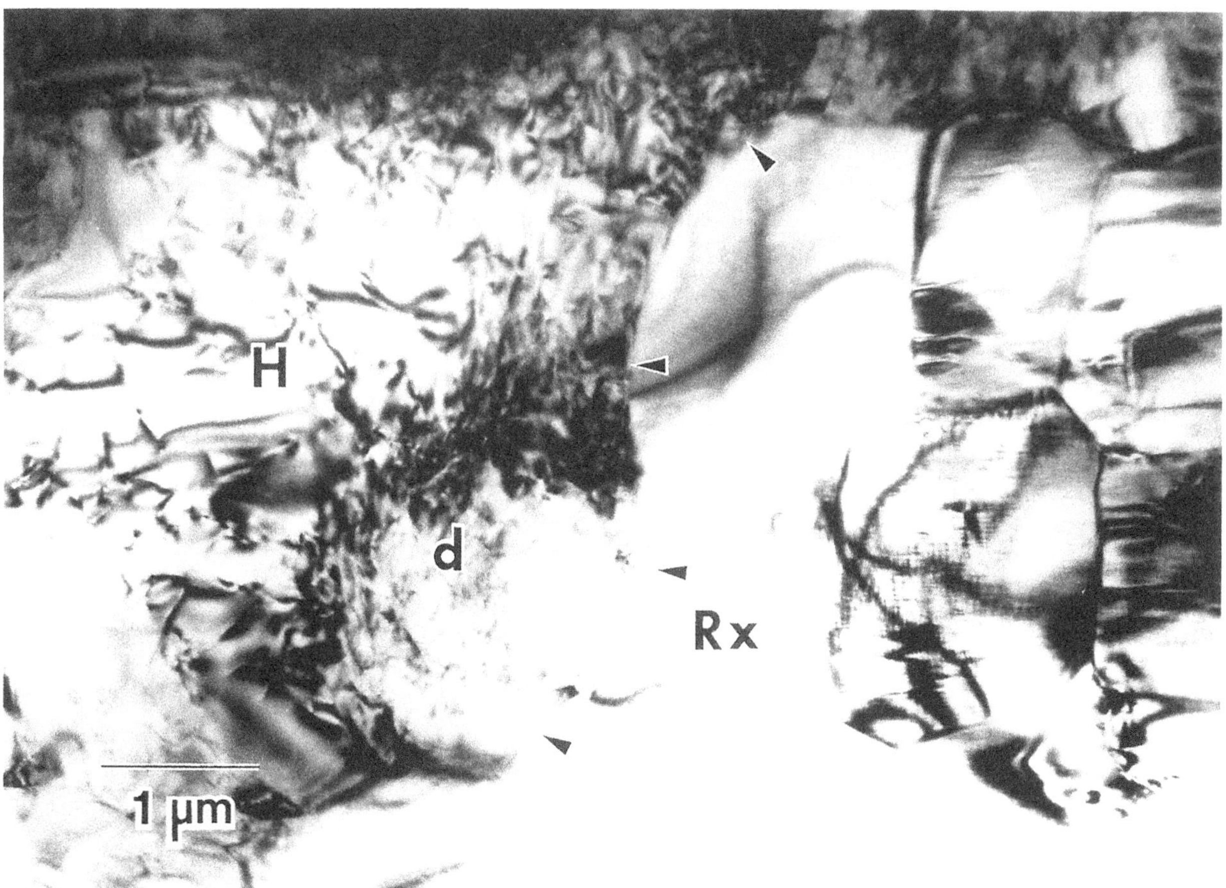


29 F

# 30. PSEUDOTACHYLYTE-BEARING STRIKE-SLIP FAULTS IN MYLONITIC ROCKS

MARK T. SWANSON

The Fort Foster brittle zone was developed within mylonitic rocks of the Rye Formation of southern Maine and New Hampshire, U.S.A. (Swanson, 1988). The Rye Formation consists of pegmatite-injected, high-grade metasedimentary rocks (schist, amphibolite, and marble) of the Merrimack Group which exhibit a complex Paleozoic ductile shear and metamorphic history. The late Paleozoic brittle structures consist of hundreds of layer-parallel, dextral, strike-slip surfaces clustered along the ductile Southern Mylonite/Great Common fault zone within the interior of the Rye Formation outcrop belt. Pseudotachylyte has been abundantly produced by friction melting during rapid, presumably coseismic, slip along layer-parallel fault surfaces. The exposed pattern of repeated brittle faulting with rapid slip and friction melting suggests a preserved history of paleoseismicity. In outcrop, the pseudotachylyte appears as layer-parallel "fault veins" typically only a few millimeters in thickness, discordant orthogonal "injection veins" commonly several centimeters in length, and other dilatant reservoir structures. Displacements are difficult to determine in the absence of cross-strike markers but may be on the centimeters to meters scale. Paired or coupled layer-parallel slip surfaces with internal extensional and contractional minor fracture assemblages have been referred to as "pseudotachylyte generation zones" (Grocott, 1981), or as "strike-slip duplexes" (Swanson, 1988). This paired or duplex geometry may reflect an en echelon relationship between adjacent layer-parallel slip surfaces or the layer-parallel bounding surfaces to slab-shaped sidewall ripout structures (Swanson, 1988, 1989, 1992a). Pseudotachylyte from this locality has been dated using the Rb-Sr technique at 270 ± 31 Ma (Boeckeler, 1994) indicating a late Paleozoic (Alleghanian) dextral strike-slip history. Ductile deformation of some of the pseudotachylyte veins, as seen in outcrop and thin section, suggests that this exhumed fault zone had passed through the brittle/plastic transition at approximately 270 Ma, with the youngest brittle zone development at ~10–12 km depth and ~250–300°C, as approximate conditions for the lower part of the strike-slip seismogenic zone.

**30A**. Orthogonal injection vein (15 cm in length) between two coupled, layer-parallel, dextral strike-slip surfaces in amphibolitic mylonite. Production of pseudotachylyte during layer-parallel slip by friction melting results in volumetric expansion and the concomitant hydrofracturing of the sidewalls under conditions of near-zero layer-parallel shear stress. This deformation presumably occurs during unrestrained slip due to melt lubrication and pressurization *or* under static post-slip conditions. Note the distinct injection flow layering (etched on weathered surfaces) indicating flow from bottom to top of photo, the cryptic nature of the layer-parallel slip surfaces (arrows) that produced the friction melt, and the slight discordance between the adjacent layering along the lower slip surface. Paired slip surfaces as generation zones or duplexes may arise as an offset zone between en echelon slip segments during displacement transfer through this highly anisotropic host rock.

30 A

**30B.** Thin-section view of a paired layer-parallel fault vein system with intervening injection veins developed during dextral slip (with accompanying line diagram; see Figure 30.1). Note the extensional linkage that cascades from the upper to the lower slip surface on the left side of the photo. This stepover zone or duplex is extensional in nature and suggests displacement transfer through a right-stepping, en echelon offset or at the trailing end of an adhesive sidewall ripout (Swanson, 1988, 1989). Note the thin, centimeter-long, orthogonal injection veins that extend from the upper slip surface in the upper part of photo, and the blunt triangular injection zones with rounded corners where modified by melting of fracture walls and assimilation into the melt. The thinning of the upper fault vein from left to right may have resulted from draining of the melt from the fault vein into the intervening dilatant injection vein reservoir (shown in more detail in 30C). PPL

**30C.** A closer view of the system of injection veins that makes up the intervening dilatant reservoir. Melt has been mostly channeled from left to right along the upper fault vein, into the larger dilatant injection vein and out along the lower slip surface, as indicated by the core material within the pseudotachylyte vein. Note the dilatant nature of the larger injection vein due to fault-parallel extension, the modification of edges and corners by melting and assimilation, the linkage between adjacent injection veins along transform-type slip surfaces, and the development of widened asymmetric reservoir zones along the lower slip surface due to plucking during slip. PPL

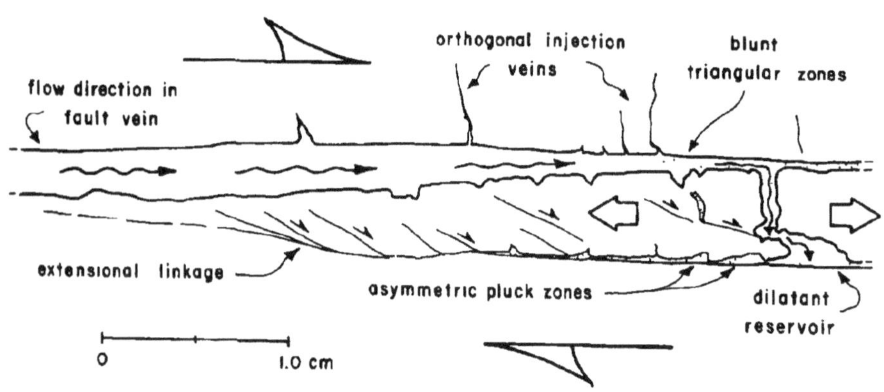


**Figure 30.1.** Line drawing of photomicrograph 30B showing the extensional linkage between major slip surfaces, orthogonal injection veins, dilatant reservoir, modified zones, and interpreted flow direction in the pseudotachylyte.

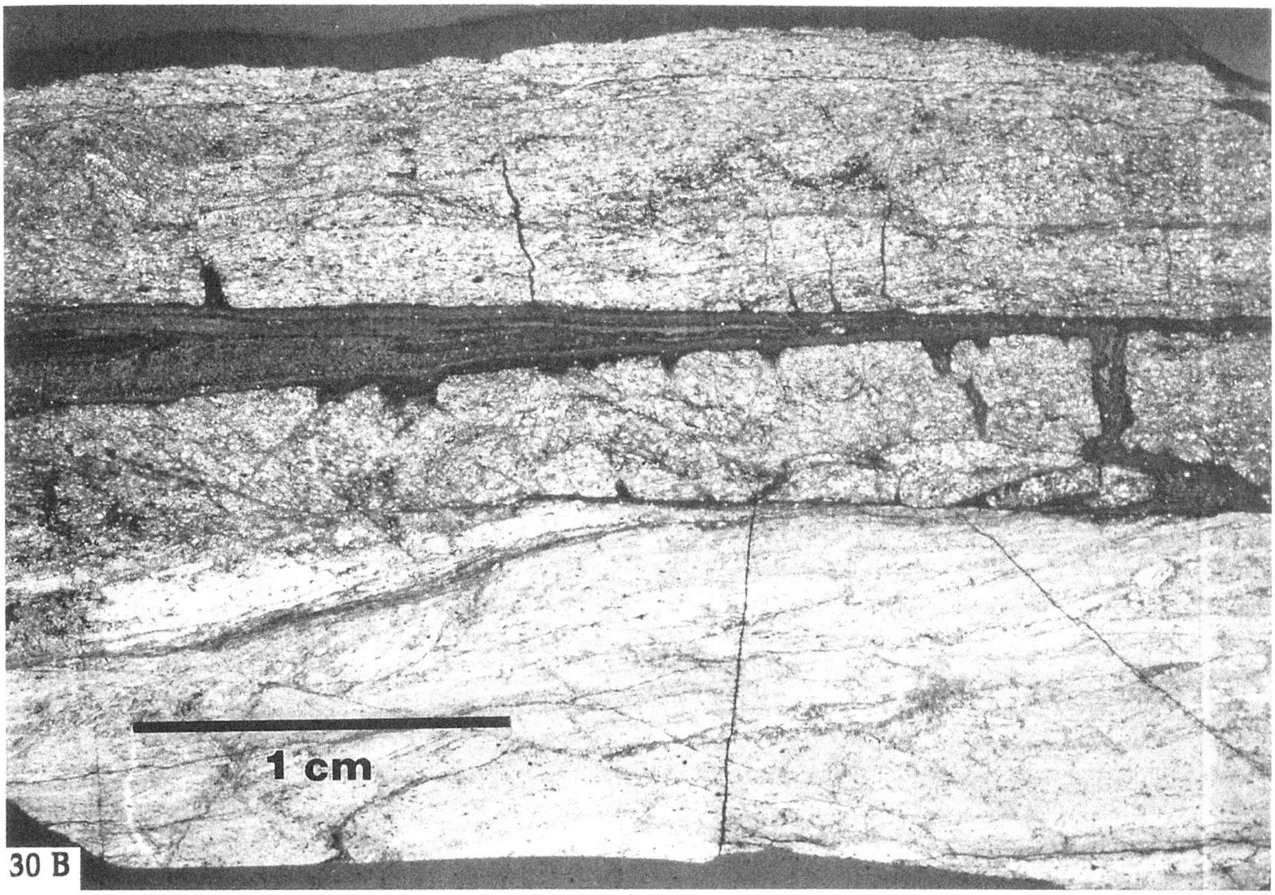


30 B

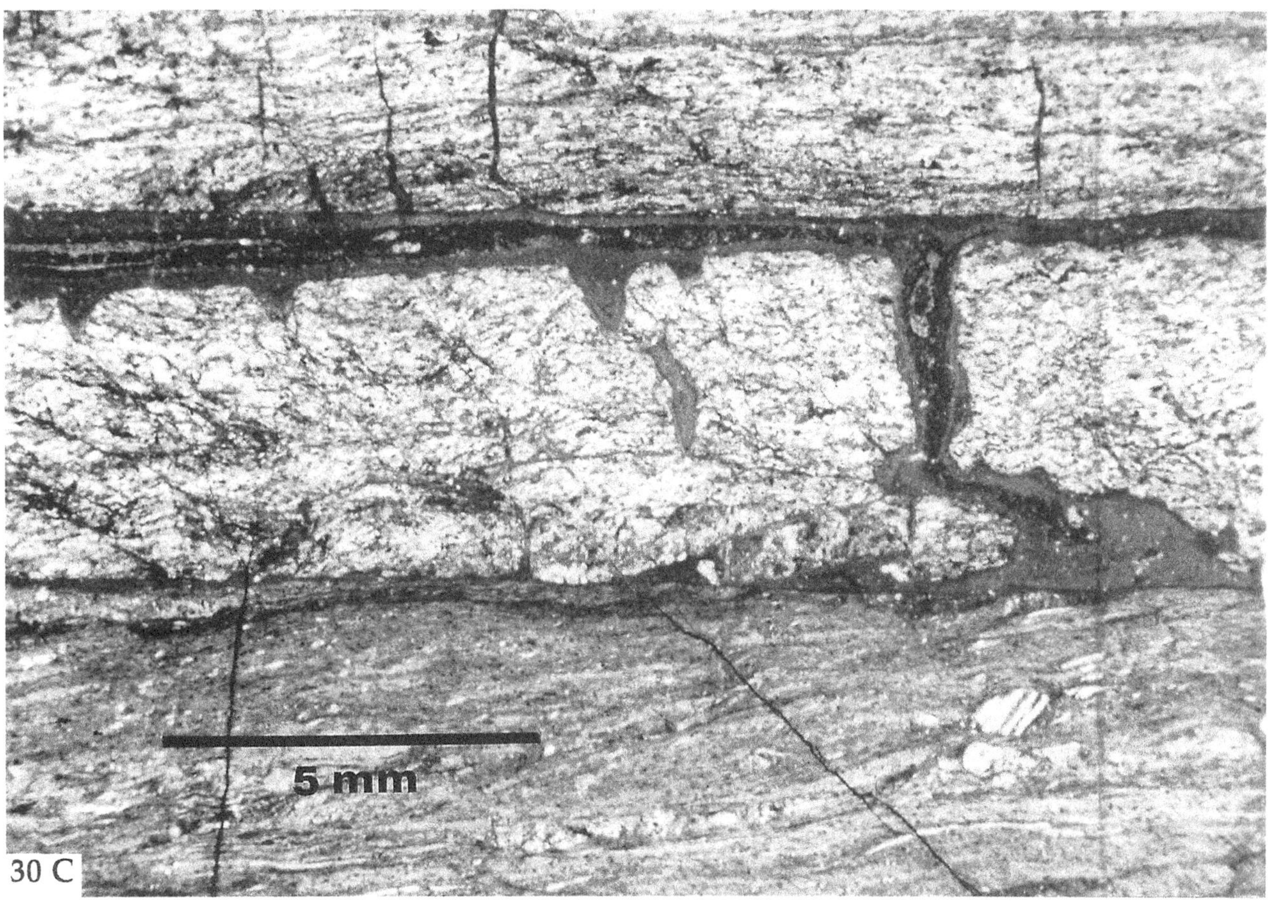


30 C

# 31. PROGRESSIVE FRACTURING AND TWINNING RELATED TO PSEUDOTACHYLYTE GENERATION

JEAN MACAUDIÈRE AND WILLIAM L. BROWN

The Roineabhal meta-anorthosite of South Harris (Scotland) suffered both ductile deformation in Proterozoic shear zones (see Brown and Macaudière, Plate 148, this volume) and cataclastic deformation of Caledonian age associated with, but not directly linked to, the Outer Hebrides thrust zone. This later deformation occurs in irregular domains which are progressively affected by cataclasis on more penetrative shear fractures, sometimes accompanied by pseudotachylyte generation (Macaudière and Brown, 1982). The composition of the rock varies from that of an anorthosite to a gabbro and that of the plagioclase from bytownite to calcic andesine as the mafic index increases. On deformation, the rock is whitened by cataclastic fracturing, but no mineralogical changes occur. There is no evidence for fluid penetration in these deformed rocks, in total contrast to the intense saussuritization near the thrust zone to the southeast (Figure 31.1).

Three stages in the deformation are illustrated on the basis of the development of transcrystalline fracturing and pseudotachylyte generation (31A–31C) and also on the intensity of plagioclase twinning and intracrystalline fracturing (31D–31F).

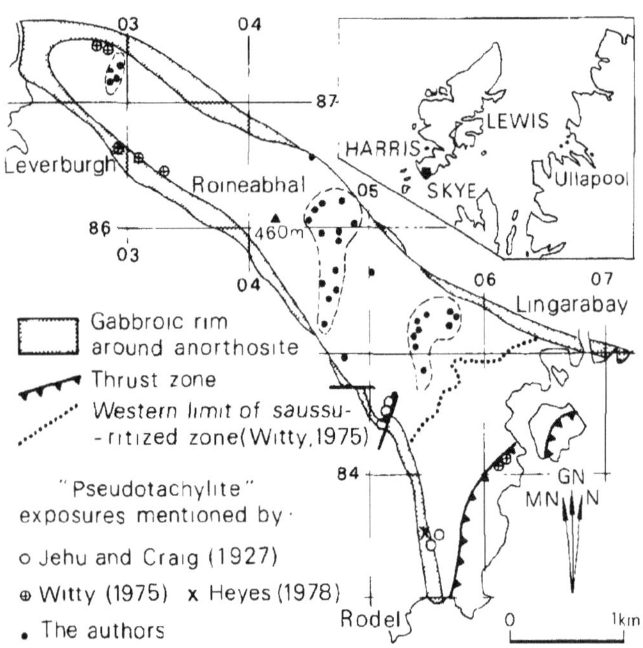


**Figure 31.1.** Map of the Roineabhal meta-anorthosite showing pseudotachylyte localities.

**31A.** Occurrence of isolated fractures. Isolated shear fractures cut the rock, which shows in this case an older Proterozoic mortar texture (see Brown and Macaudière, Plate 148). The rare fractures (a; E-W in the photograph) have produced little or no intracrystalline deformation of plagioclase. A more important shear zone (nearly N-S in the photograph) cuts the former set of fractures and is accompanied by a thin film of pseudotachylyte (b), which is never present in thick veins at this stage. NX

**31B.** Development of a network of fractures. At higher rock strain the transcrystalline shear fractures are much more numerous and intersecting. They are lined by cataclasite and associated with more intense deformation of plagioclase in the intervening rock. Pseudotachylyte is more abundant at this stage and occurs both as thin films and thicker cross-cutting veins (a). NX

**31C.** Penetrative fracturing and deformation. The entire rock is intensely deformed both by numerous trans- and intracrystalline fractures and by plastic deformation of plagioclase (mechanical twinning, lattice bending, etc.). Plagioclase crystals and grain boundaries become indistinct (a). The rock is cut by a thick pseudotachylyte vein with numerous plagioclase clasts in the center and a small dead-end vein to the left (b). NX

31 A

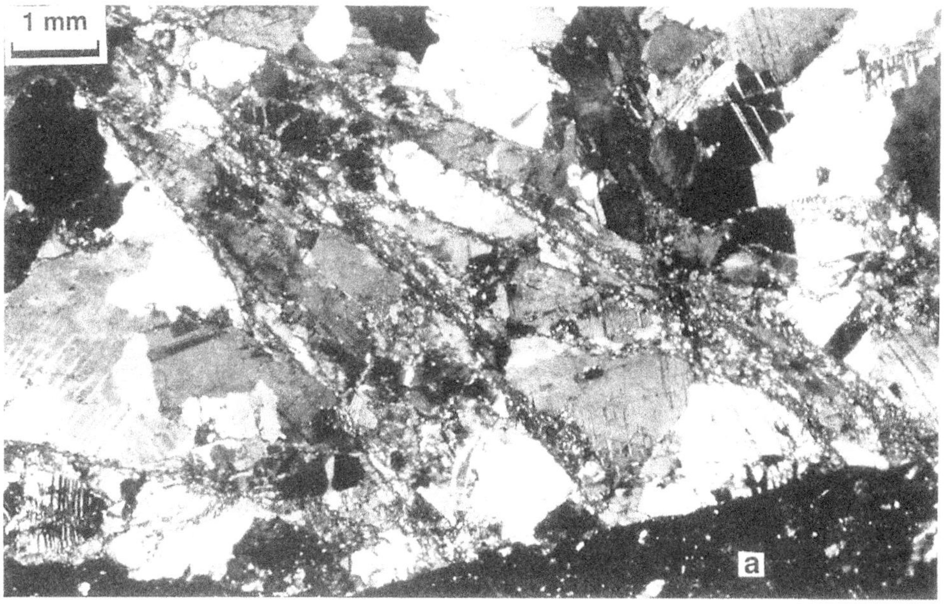


31 B

31 C

Characteristic features of plagioclase deformation depend on the amount of local strain in the rock and particularly on the position of the crystals with respect to the sets of shear fractures (for example, 31B). The deformation includes mechanical twins (Albite and Pericline laws), kink bands, undulose extinction, and intracrystalline fracturing. The photographs in this plate are oriented perpendicular to the a axis to show both twin laws.

**31D.** Weak kinking with mechanical twins. Plagioclase 12 mm from a pseudotachylyte vein and 3–6 mm from shear fractures. Tapering N-S Albite twins develop at or near a kink and on grain boundaries. E-W Pericline twins occur almost only within the Albite deformation twins where they form ladders; they do not intersect the Albite twins and are thus contemporaneous with, or younger than, them. The spacing of the Pericline twins decreases as the Albite twins thin (a). NX

**31E.** Intersecting twins with window texture. Plagioclase from a garnet meta-anorthosite about 6 mm from an isolated shear fracture. In 31D, Pericline twins are restricted (or nearly so) to one of the Albite twin orientations, namely, the one produced by the deformation. In this photograph, E-W Albite and N-S Pericline twins occur. In some areas (a) Pericline twins stop against Albite twins as above; in other areas (b) Albite twins stop against Pericline twins. In still other areas (c and d), the two sets appear to intersect, giving a window or chess-board texture well known in microcline. NX

**31F.** Closely spaced mechanical twinning with intracrystalline fractures. Plagioclase from same rock specimen as 31D but closer to the shear fractures. Variably spaced Albite (N-S) and Pericline (E-W) twins are restricted to large N-S mechanical Albite twins. In many areas the two sets of fine twins appear to intersect. The crystal is cut by NE-SW intracrystalline fractures with a few conjugate NW-SE fractures. The fractures are later than the broad N-S Albite twins, whereas the fine twins stop on them. In only one small area to the right of (a) do fine Pericline twins cross the broad Albite twins and produce a window or chess-board texture as in 31E. NX

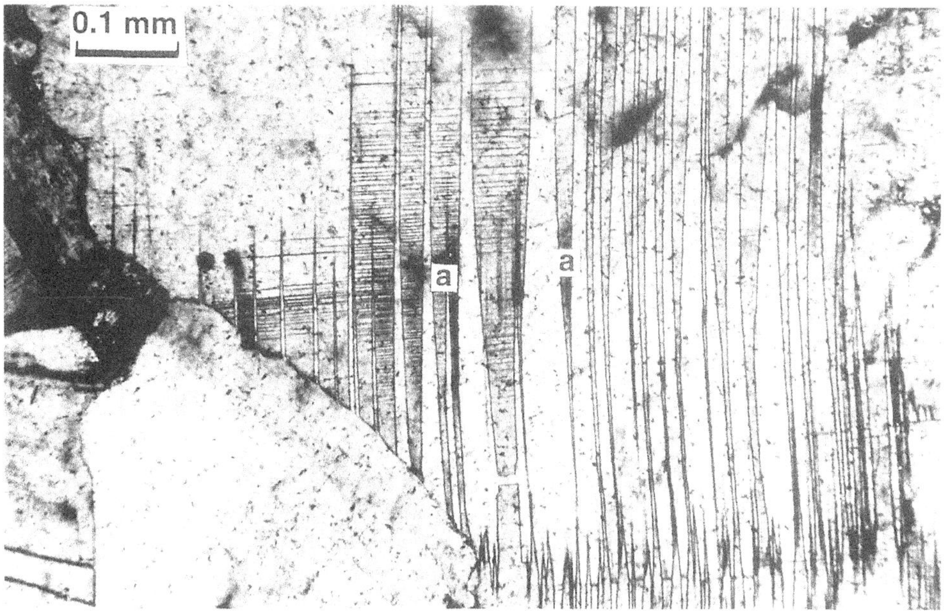


31 D

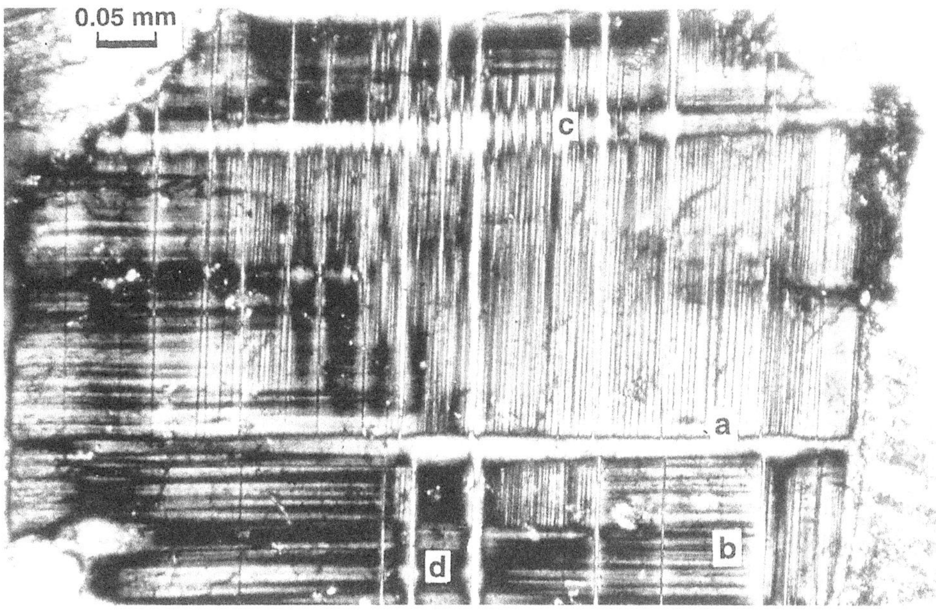


31 E

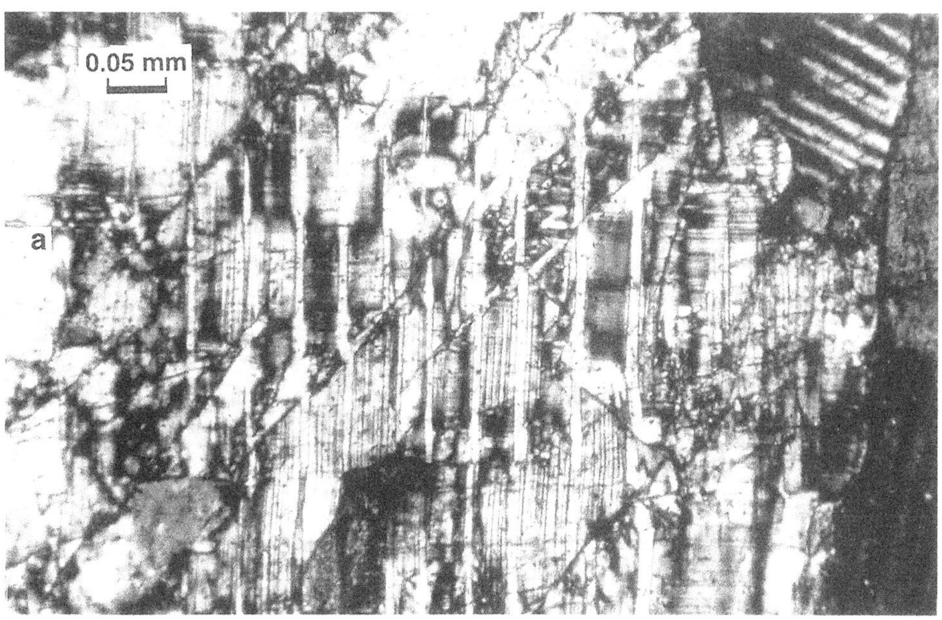


31 F

AIMING LIN

Glassy and microlitic pseudotachylytes occur in granitic mylonite from the Fuyun fault zone, northwest China (Figure 32.1). Microscopic characteristics and powder X-ray diffraction data indicate that the fine-grained matrices in the Fuyun pseudotachylytes vary from almost crystalline to complete glass (a few to 90 wt %). X-ray diffraction patterns also indicate that almost all clasts in the glassy-type pseudotachylyte veins consist of quartz grains. This means that the other minerals such as micas and feldspars contained in the host rocks have been preferentially melted during pseudotachylyte formation (Lin, 1991, 1994a). Locally, there are some transparent glasses with quartz and plagioclase compositions, generally rimming quartz fragments.

Microlites showing quenching or rapid cooling textures such as acicular, skeletal, dendritic, and complex spherulitic shapes are found in the Fuyun pseudotachylytes. There is a gradation in size and shape of microlites from the margins toward the center in the pseudotachylyte vein. In the margins, the microlites are finer in size and simple in shape. Toward the center, the microlites become progressively larger. The microlites are mainly made up of sanidine, anorthoclase, pyrope, plagioclase ($An_{30}$–$An_{70}$), biotite, and hornblende with a high Ti component; these minerals are not present in the host rocks which are upper greenschist-facies rocks. The presence of these high-temperature minerals as microlites indicates that they were the primary crystals formed from a melt (Lin, 1991, 1994b). Vesicles, amygdules, and flow structures are very common in the Fuyun pseudotachylytes.

These textures demonstrate clearly that the fine-grained matrices of the Fuyun pseudotachylyte formed from a melt generated by frictional heating, rather than crushing.

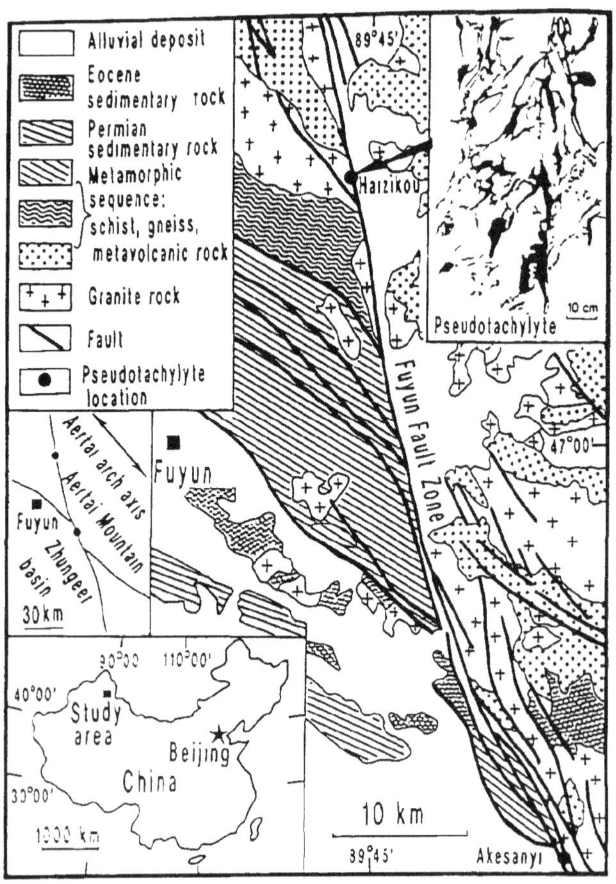


**Figure 32.1.** Geologic map of the Fuyun region, northwest China, simplified from the Regional Surveying Team of the Xingjiang Geological Bureau, 1978.

## Fault topography and field occurrences of the pseudotachylytes

**32A.** Photograph showing fault topography of the Fuyun fault zone at Akesanyi location (Figure 32.1), where the pseudotachylyte outcrop (PO) occurs near an earthquake fault (E.F) developed during a magnitude 8.0 earthquake in 1931. The basement consists of Paleozoic metamorphic rocks and Mesozoic granitic rocks as shown in Figure 32.1. Along the fault zone, granitic rocks have been mylonitized under upper greenschist-facies conditions at an estimated depth of 10–15 km in a zone ranging from a few meters to 1 km wide (Lin and Fan, 1984). A zone of fracturing also occurs along the fault trace, ranging from a few tens of meters to several hundred meters in width, in which schist, gneiss, and mylonite have been cataclastically deformed.

**32B.** Photograph of a complex network of pseudotachylyte (PT) veins in granitic mylonite, which progressively branch, thin, and eventually terminate or end abruptly.

**32C.** Pseudotachylyte vein (PT) injected into granitic mylonite exhibits an irregular shape.

32 A

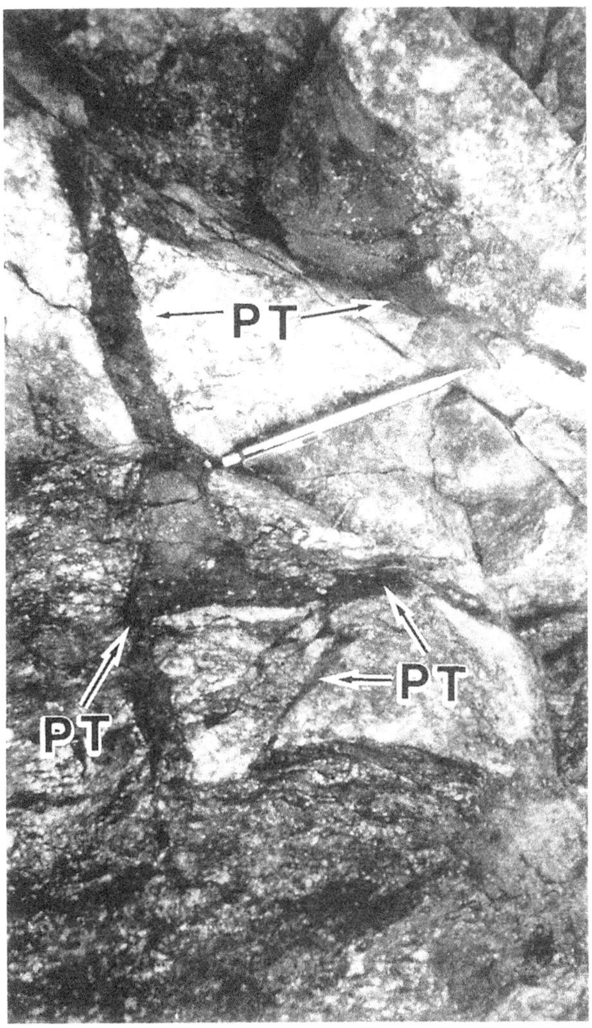


32 B

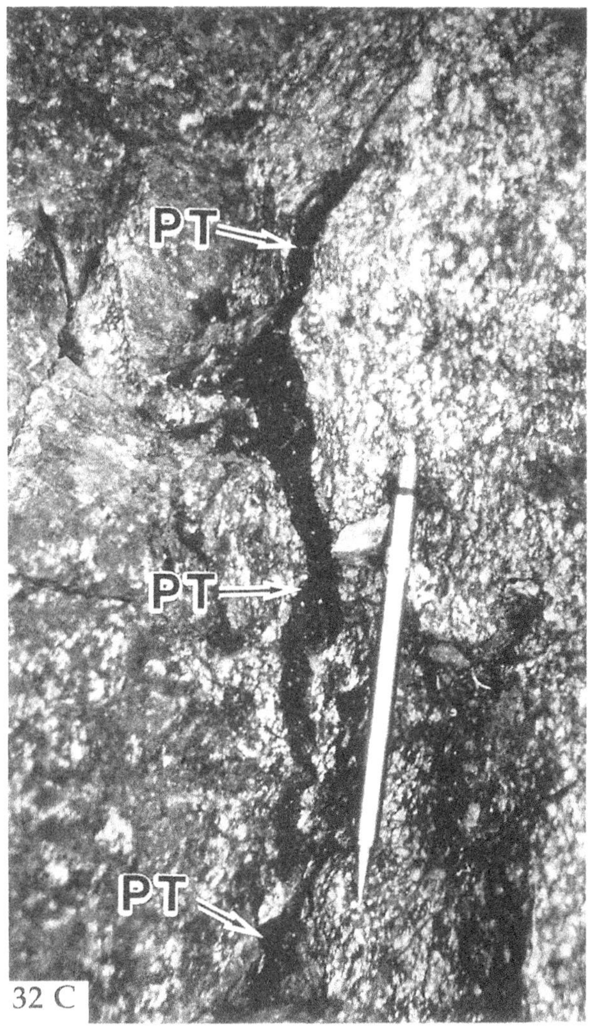


32 C

## Microstructures of pseudotachylyte veins

**32D.** Photomicrograph showing the textures of a complex network of micropseudotachylyte (PT) veins which progressively branch and commonly thin out as also observed in the field (32B). The pseudotachylyte veins consist of a glassy matrix with fragments of various size, from submillimeter to a few millimeters. NX

**32E.** Pseudotachylyte vein (PT) obliquely cutting the mylonite foliation (FO). The boundary (BO) between the pseudotachylyte and mylonite is generally sharp. The flow streaks marked by yellowish-brown and brownish dark-colored layers are parallel to the boundary near the margins of the pseudotachylyte vein, but bend in the center of the vein. PPL

**32F.** New pseudotachylyte vein (NPT) injected into an older pseudotachylyte vein (OPT). The flow structures marked by color banding can be observed clearly along the margins of the new pseudotachylyte veins. Both veins consist of a fine-grained, cryptocrystalline matrix with larger fragments. This reinjection indicates that faulting events generating the pseudotachylyte veins occurred repeatedly at the same location (Lin, 1994a). PPL

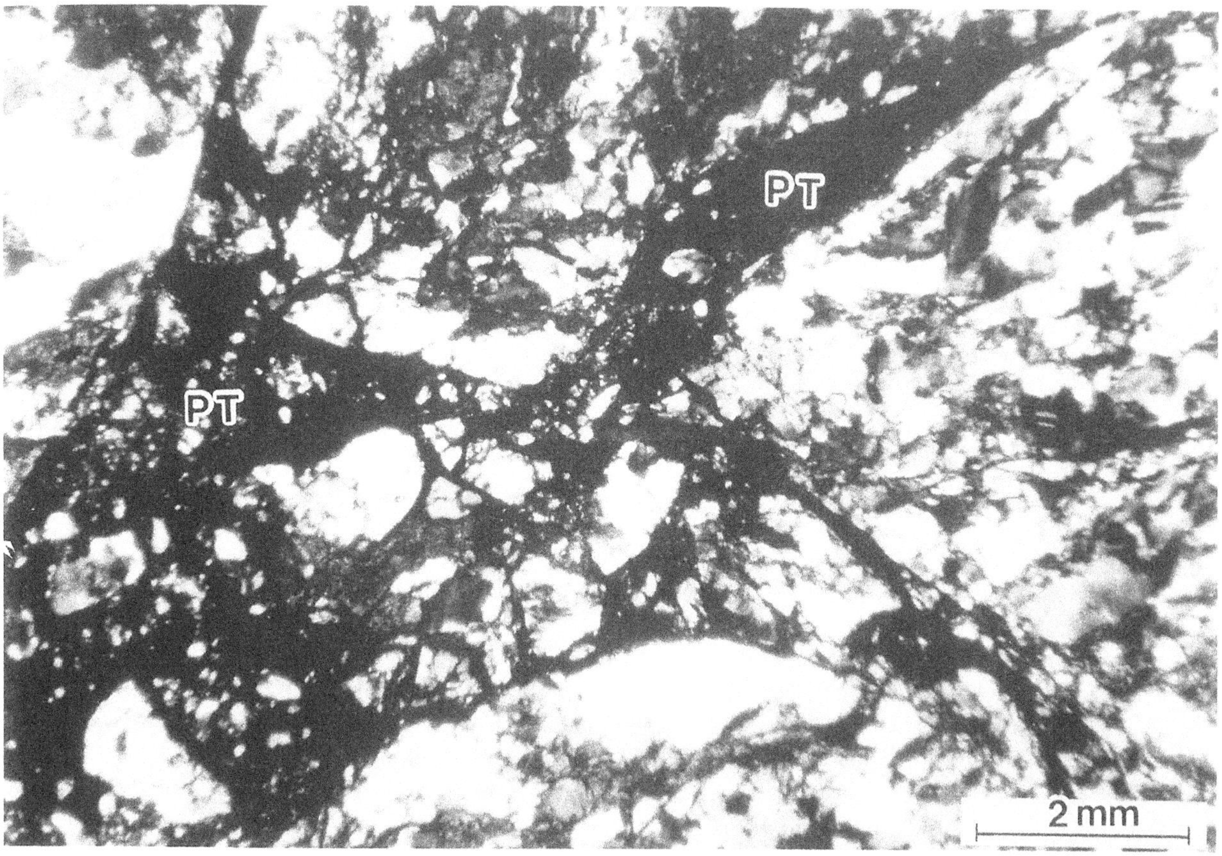


32 D

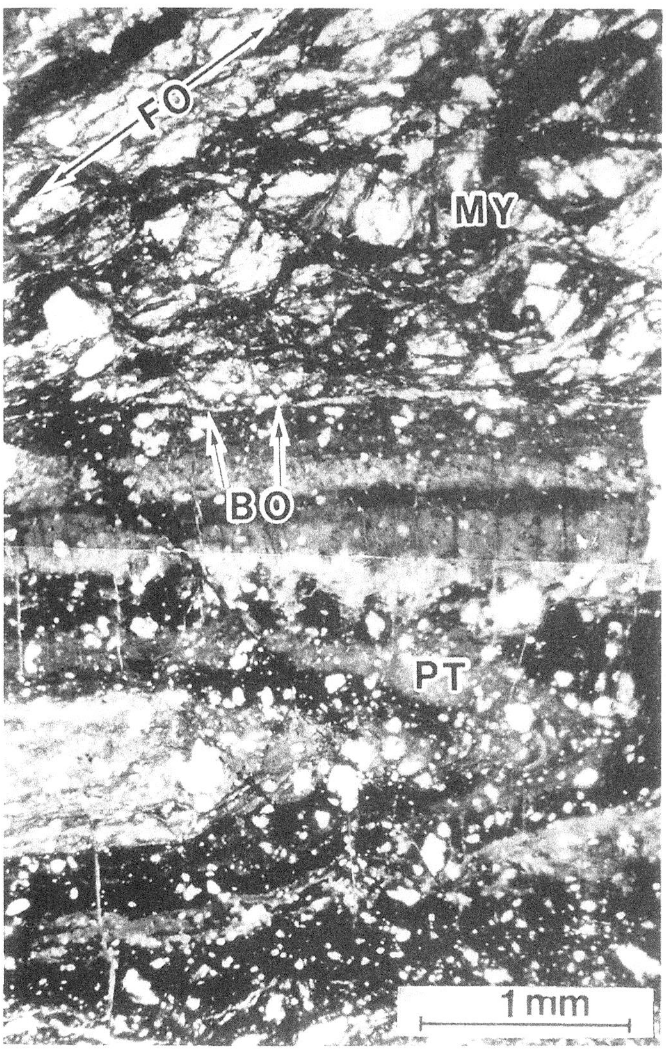


32 E

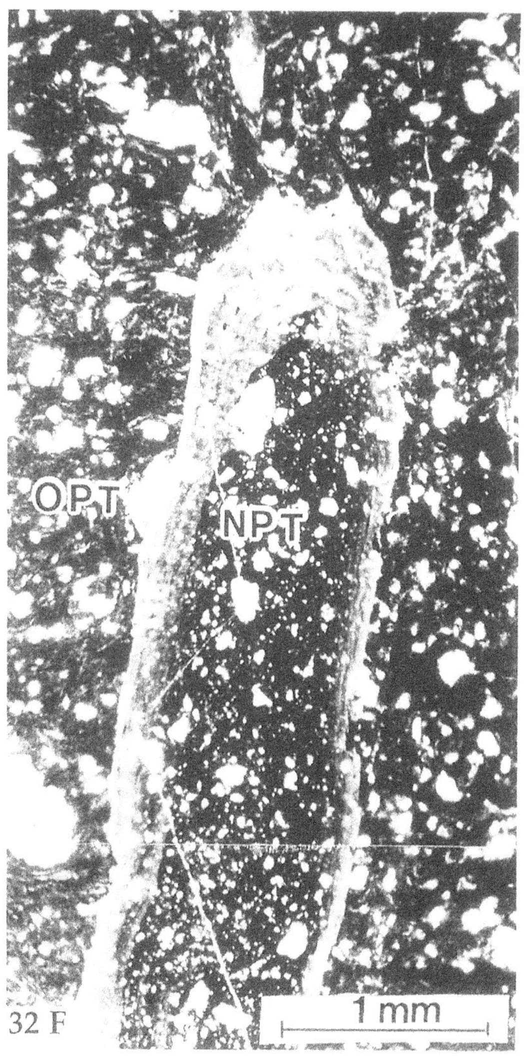


32 F

## Glassy matrices

**32G.** Glassy matrices (GL) of pseudotachylyte are isotropic under the optical microscope and vary in color from brown to yellowish-brown to grayish-yellow. The flow streaks marked by color banding bend around the fragments. Most of the fragments are rounded in shape and aligned parallel to the flow streaks. Powder X-ray diffraction data show that this glassy-type pseudotachylyte vein consists of about 90 wt % glassy or noncrystalline material and 10 wt % quartz fragments (QZ) (Lin, 1991, 1994a). PPL

**32H and 32I.** Transparent glass matrix (TG) occurs as a rim around a quartz fragment (QZ) which has an irregular shape. The yellowish-brown matrix (GL) is also glass. The transparent glass around the quartz fragment (see also Figure 32.2) consists of pure silica. Transparent glass with a pure feldspar composition also occurs locally (Lin, 1991, 1994a). This observation indicates that some quartz and feldspar fragments were melted but did not mix with the other components of the melt. 32H: PPL; 32I: NX

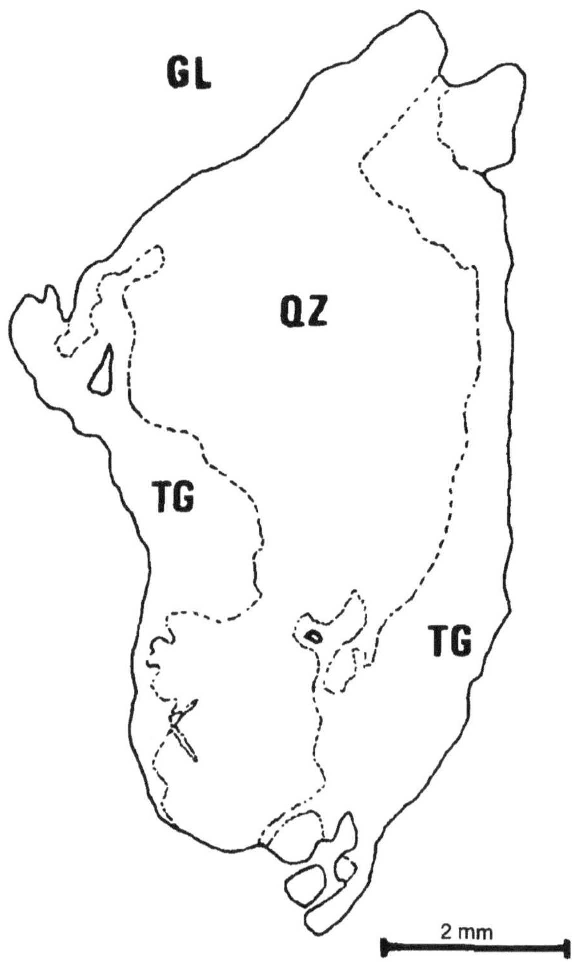


**Figure 32.2.** Sketch of Plates 32H and 32I showing irregular shape of quartz (QZ) partially surrounded by a transparent glass (TG). GL is a yellowish-brown glass, also part of the matrix.

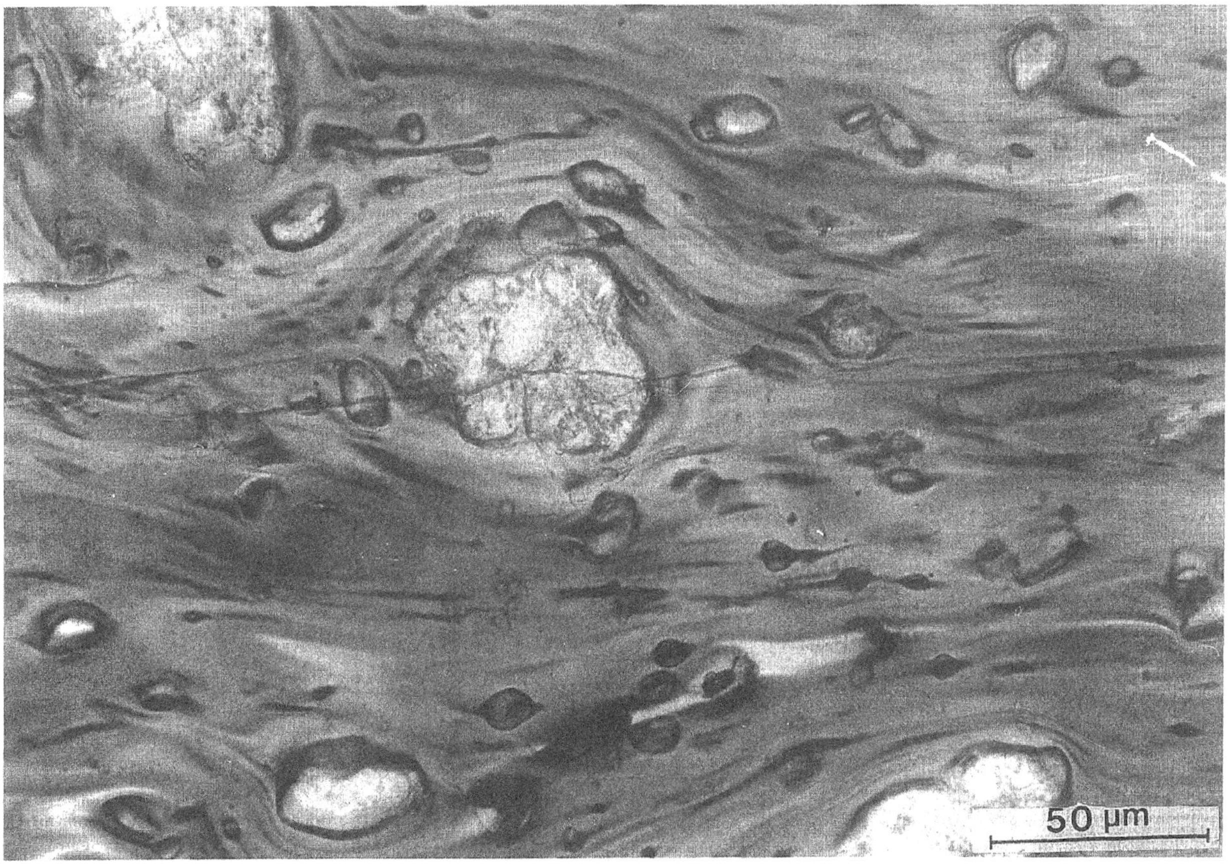


32 G

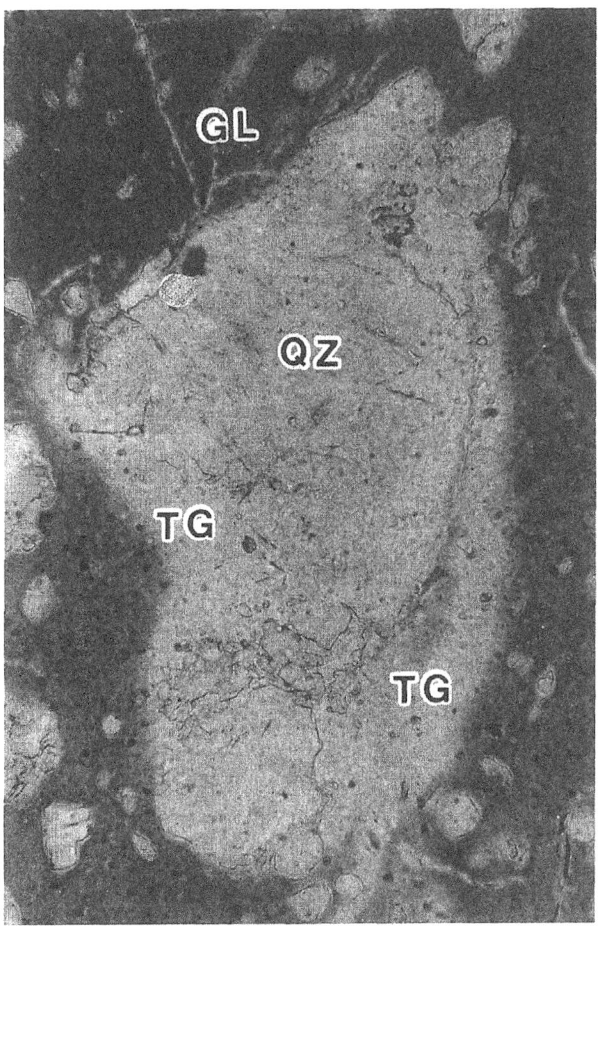


32 H

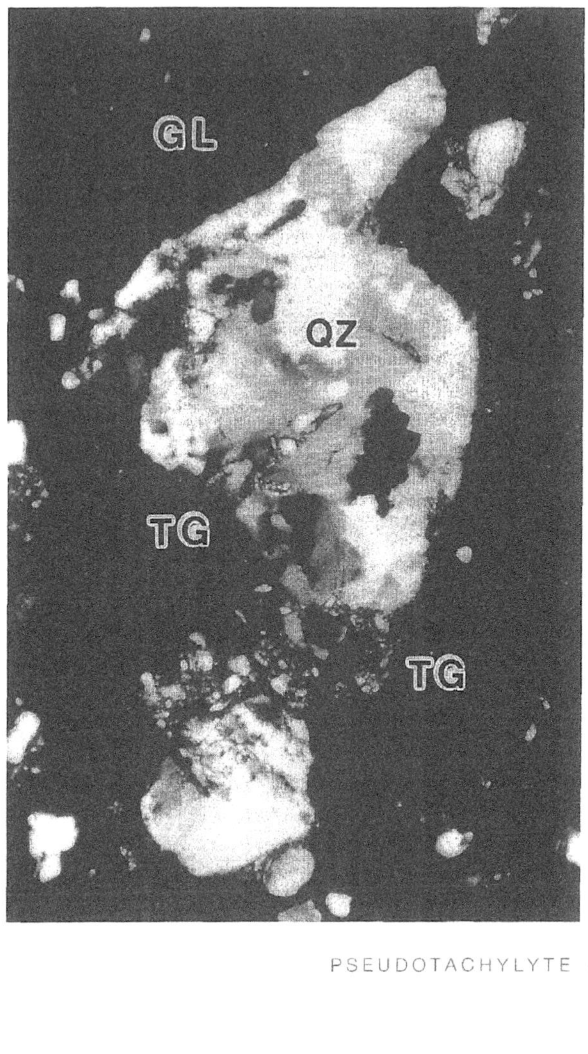


32 I

### Flow structures and amygdules (vesicles)

**32J.** Flow structures deformed into a similar fold. The flow streaks, marked by color banding, bend in the center of the pseudotachylyte vein and become straight and parallel near the margins of the pseudotachylyte vein. The intervals between the flow streaks become progressively larger from the margins toward the center of the pseudotachylyte vein. This interval difference between flow streaks from the margins toward the center is attributed to a difference in flow velocities of the melt between the margins and center of the vein. Generally, the flow velocity of a liquid in the center of a channel is greater than at the margins. The convex direction may indicate the injection direction of the melt during pseudotachylyte formation. PPL

**32K.** Photomicrograph showing a flow structure resembling a sheared porphyroclast in a mylonite. The white band parallel to the boundary (BO) between the pseudotachylyte vein (PT) and the mylonite (MY) is composed of a microcrystalline matrix with a composition close to feldspar and contains rounded quartz fragments (QZ). The flow structures are also characterized by color bands varying from brown to grayish-yellow. It seems that the matrix of the white band formed by melting of feldspar fragments, and this melt never mixed with other adjacent components of the melt. PPL

**32L.** Spherical amygdules (AM) occur in the fine-grained matrix, these vary from a few millimeters to several tens of millimeters in size. Most of the amygdules were filled by carbonate and iron-rich material. PPL

**32M.** Amygdules occur in the glassy matrix; they are almost always filled with carbonate material. Some microfractures (FA) cut the amygdules and were also filled with carbonate material. It is possible that both amygdules and microfractures underwent coeval filling after pseudotachylyte formation. PPL

**32N.** SEM BSEI (backscattered electronic image) photograph showing the filled structure of an amygdule. The circles within the amygdule, defined by color variations, formed by the in-filling of carbonate (Ca) and iron-rich material (Fe). One vesicle (Ve) was not completely filled. SEM BSEI.

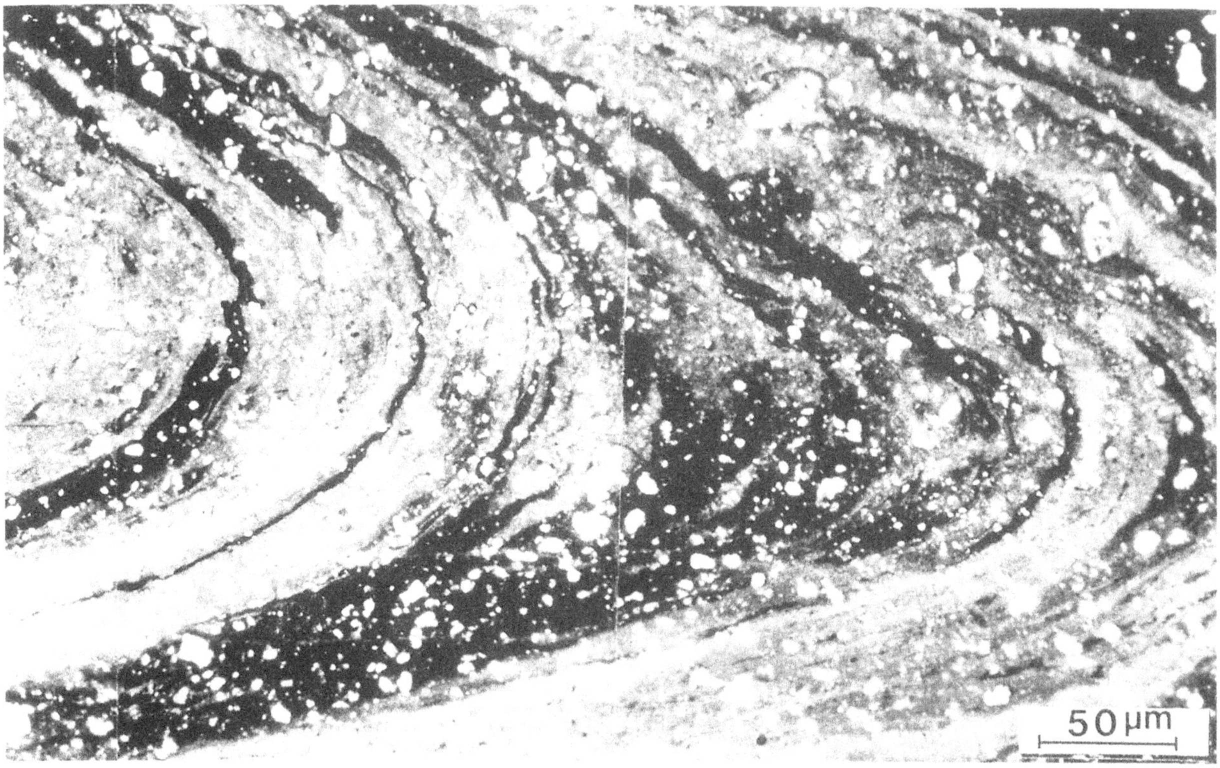


32 J

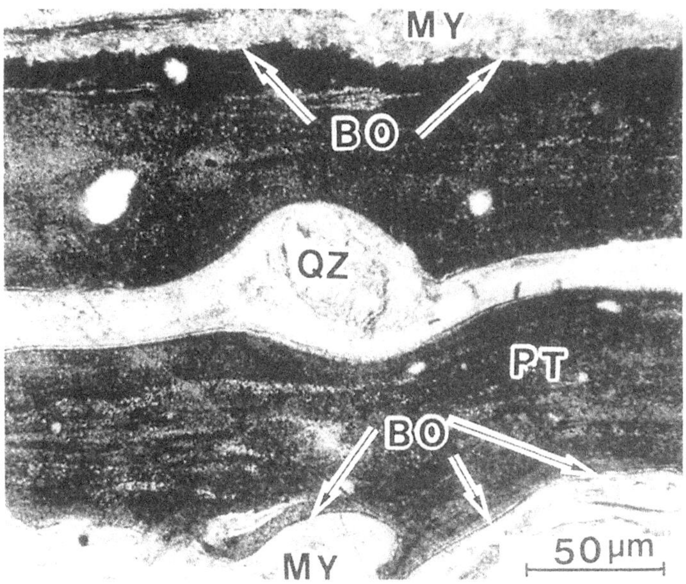


32 K

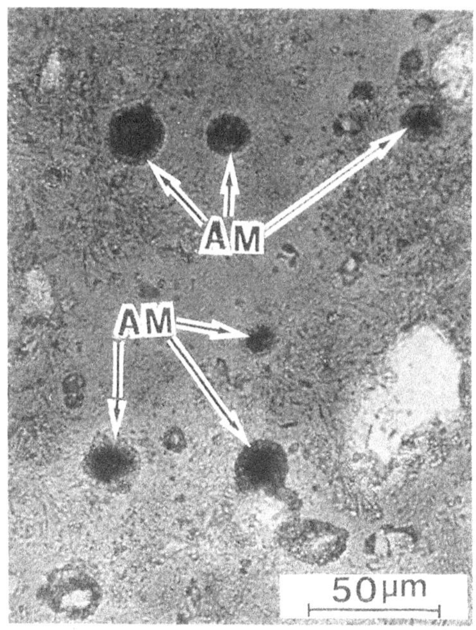


32 L

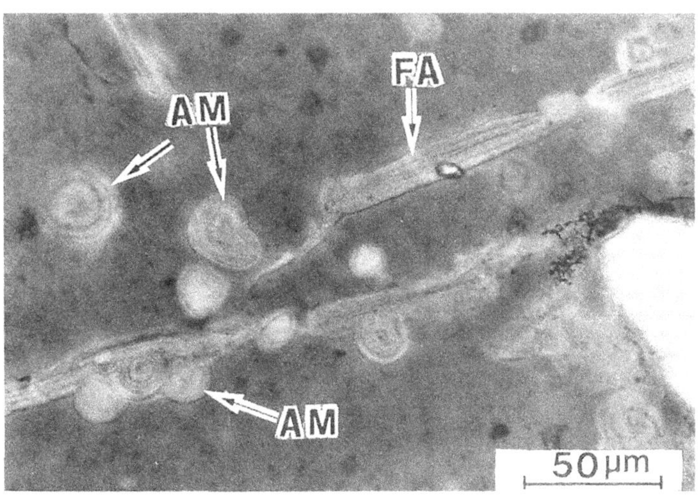


32 M

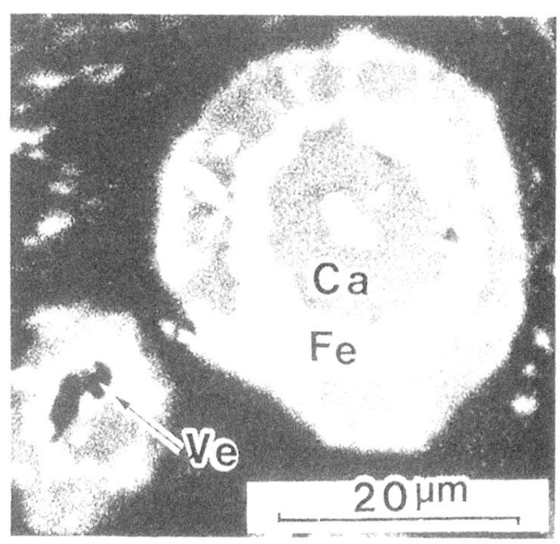


32 N

### Textures and morphology of microlites

The microlites in the pseudotachylyte show a great variety of morphologies ranging from simple to complex. The microlites can be divided into four morphological groups: (1) simple, (2) skeletal, (3) dendritic, and (4) spherulitic. These microlites vary from small to large in size and from simple to complex in shape from the margins to the center of a pseudotachylyte vein (Figure 32.3). Similar textures were also generated in crystallization experiments (Lofgren, 1974, 1980; Doherty, 1980). These textural variations of microlites were explained by the difference in the degree of undercooling between the margins and center of a pseudotachylyte vein (Lin, 1991, 1994b). The fact that the microlites in the margins of the vein are small in size and simple in shape indicates that they formed primarily by supercooling or quenching of a melt. The complex dendritic-spherulitic microlites represent the latest crystallization products.

**32O.** Glassy matrix and simple shape microlites (MI) occurring near the margin of pseudotachylyte vein. The matrix near the boundary between the pseudotachylyte and mylonitic country rock (MY) shows the optical characteristics of glass. The microlites near the boundary (BO) are generally smaller in size than those far from the boundary. The microlites aligning parallel to the boundary are generally smaller than 6–7 mm in size. QZ: quartz fragment. PPL

**32P.** Microlites (MI) occurring near the center of the pseudotachylyte vein show complex spherical and spherulitic shapes. PPL

**32Q.** Photomicrograph showing a complex texture of spherulitic microlites occurring in the center of the pseudotachylyte vein. Acicular and dendritic microlites (MI) overgrowing the quartz fragments (QZ) form the spherulites. PPL

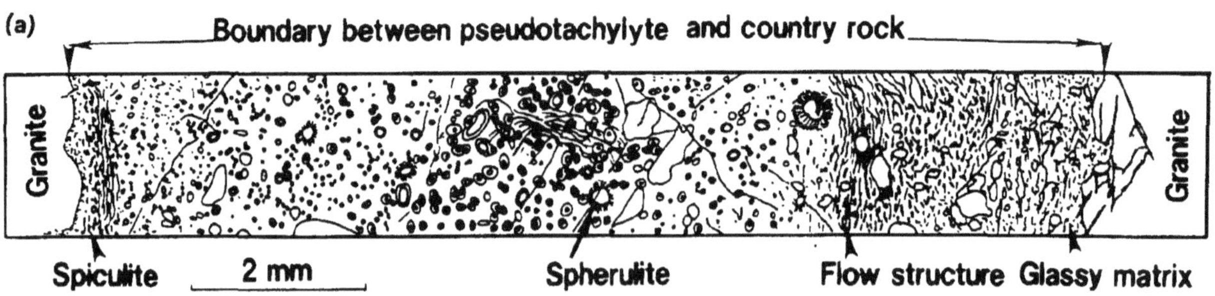


**Figure 32.3.** Sketch across pseudotachylyte vein showing variations in microstructures, especially the character of the microlites.

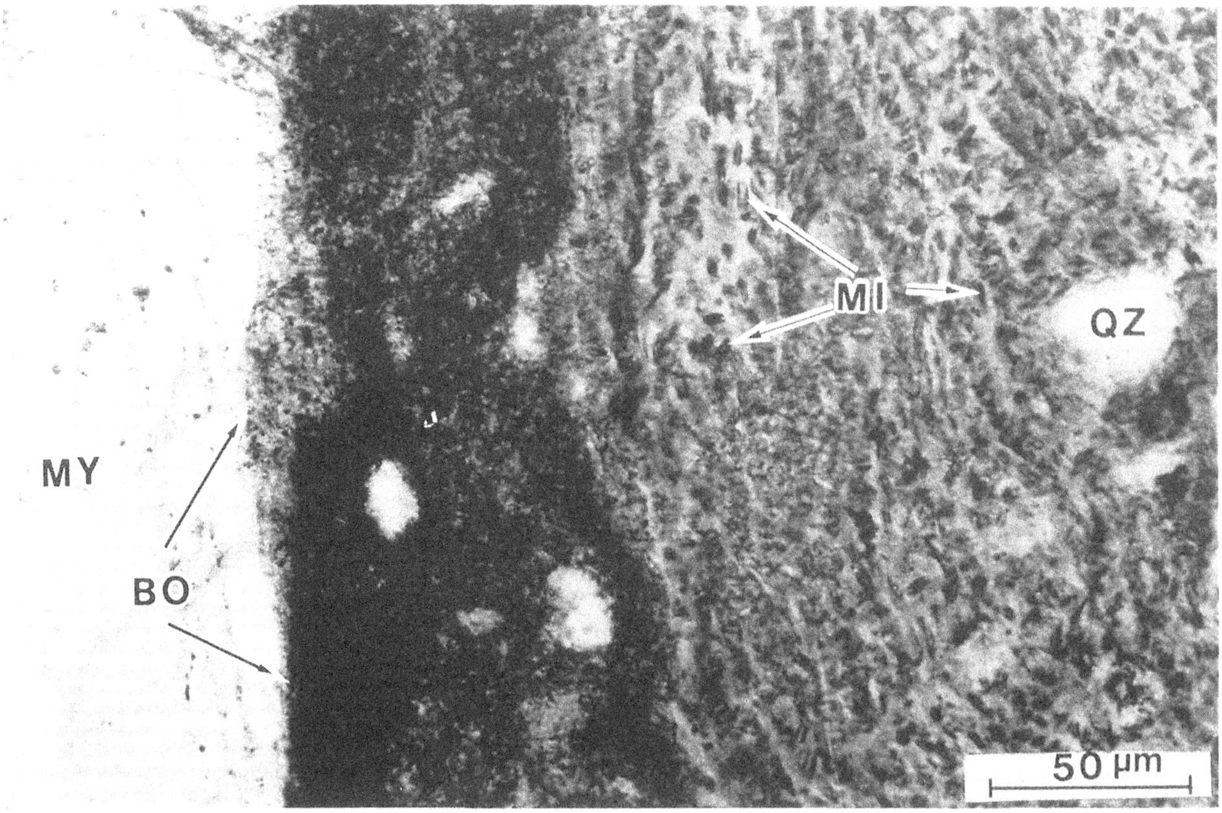


32 O

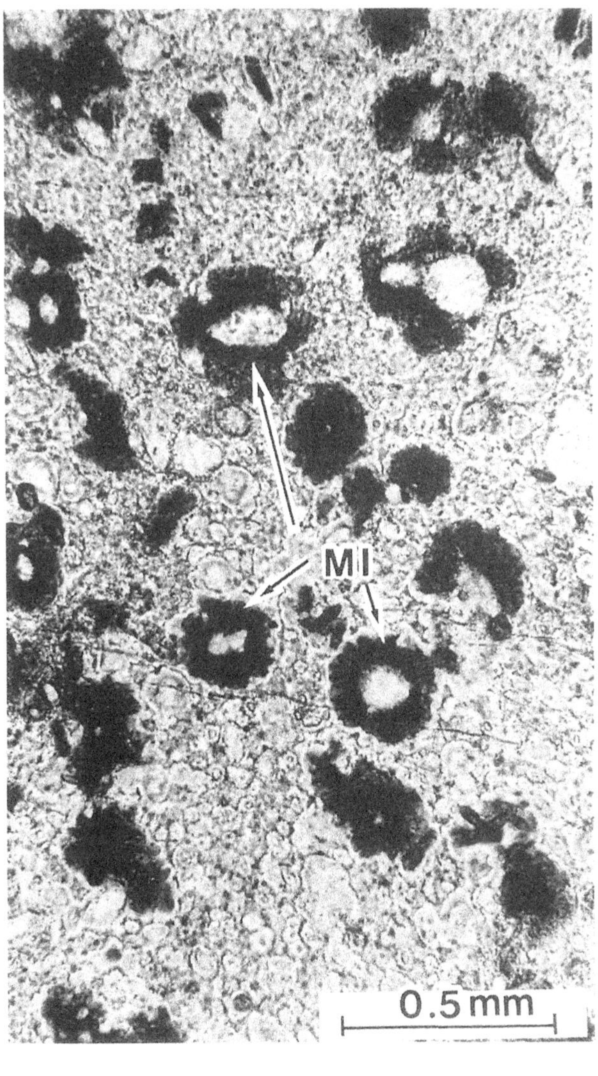


32 P

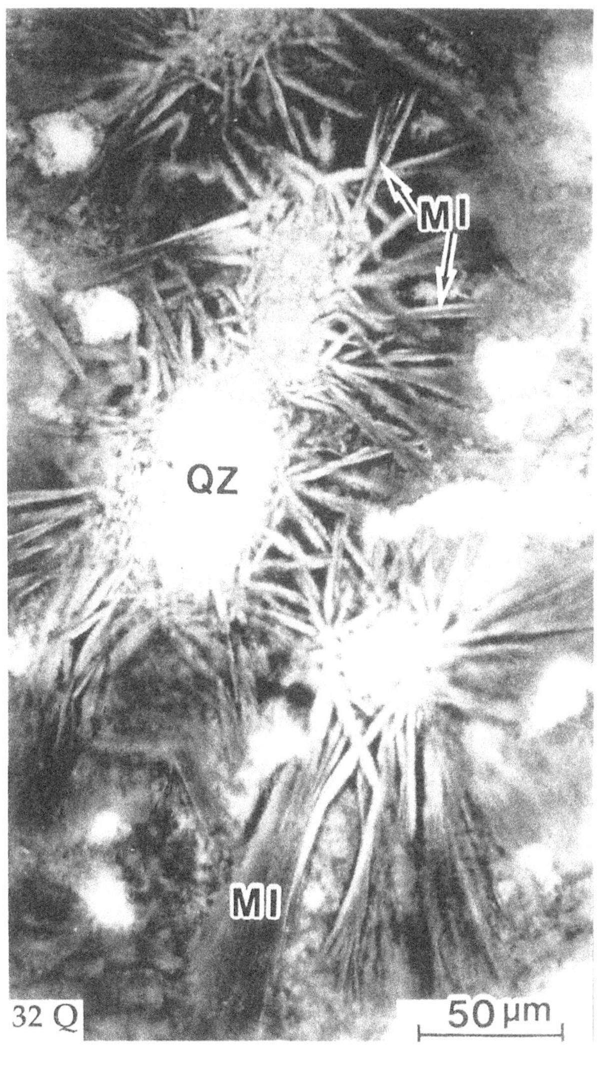

# 33. PSEUDOTACHYLYTE FROM A METAMORPHIC CORE COMPLEX

LAUREL B. GOODWIN, STEPHEN J. REYNOLDS, CHARLES J. FERRANTI,
PHILLIP D. ELLZEY, AND GORDON S. LISTER

The South Mountains, Arizona, U.S.A., typical of Cordilleran-type metamorphic core complexes formed by crustal extension, contain a gently dipping zone of middle Tertiary mylonitic rocks, cataclasite, and breccia formed during the ductile-to-brittle evolution of a low-angle extensional shear zone (Reynolds and Rehrig, 1980; Reynolds, 1985; Reynolds and Lister, 1987, 1990). The mylonitic fabrics formed first and were overprinted by breccia and cataclasite as the rocks were uplifted in the footwall of an extensional detachment fault. Associated with the mylonitic rocks and the cataclasites are millimeter- to centimeter-thick veins of pseudotachylyte. Most pseudotachylyte veins are subparallel to the gently dipping shear zone, cross-cut the mylonitic fabric, and show varying degrees of mylonitic overprint.

The pseudotachylyte veins cut a syntectonic granitic pluton dated at ~22 Ma (Reynolds and others, 1986). Stable isotopic, fluid inclusion, and petrographic data indicate that deformation was ductile at ~600°C, whereas brittle deformation (including pseudotachylyte formation) was dominant at ~400°C (Smith and others, 1991). Dikes and plutons show evidence of shallow to moderate depths of intrusion (Reynolds, 1985). Pseudotachylyte formation occurred after the main magmatic pulse and, therefore, under relatively low confining pressure (less than 300 MPa).

**33A.** Pseudotachylyte vein (pt). Main vein is subparallel to C-surfaces (vertical in photograph, gently dipping in outcrop) in host granodiorite mylonite, a typical orientation. Clasts in vein are dominantly quartz. The dark color of the pseudotachylyte is due to the presence of glass and fine-grained biotite. Note that subsequent ductile deformation (d, to right of vertical line) of the pseudotachylyte vein is localized along the main vein, parallel to C surfaces; the subsidiary branch of the vein is largely undeformed (u, to left of vertical line). PPL

**33B.** Quartz clasts (Q) in pseudotachylyte. Black material is isotropic and is interpreted to be glass. Medium-gray material is biotite. Note fine grain size, concentration of glass along the margins of clasts, and preferred crystallographic alignment of biotite grains (indicated by uniform color). PPL

**33C.** Quartz clasts and biotite crystals in nearly undeformed pseudotachylyte. The dark clasts are quartz (Q), white grains are biotite (B), and the gray matrix is interpreted to be glass. Edges of many quartz clasts are highly irregular, probably as a result of partial resorption of the clasts. The composition of the glass is consistent with a feldspar-rich parent with a minor quartz component. Biotite grains are interpreted to be crystallites that crystallized from the fault-generated melt. Backscattered SEM

**33D.** Glass (G) within pseudotachylyte vein. The presence of glass in pseudotachylyte from the South Mountains is unequivocally indicated by TEM studies. The TEM reveals featureless regions characterized by amorphous rings in diffraction pattern (upper left-hand corner, generated from area indicated by arrow). These regions are common within all four pseudotachylyte veins studied to date. TEM

33 A

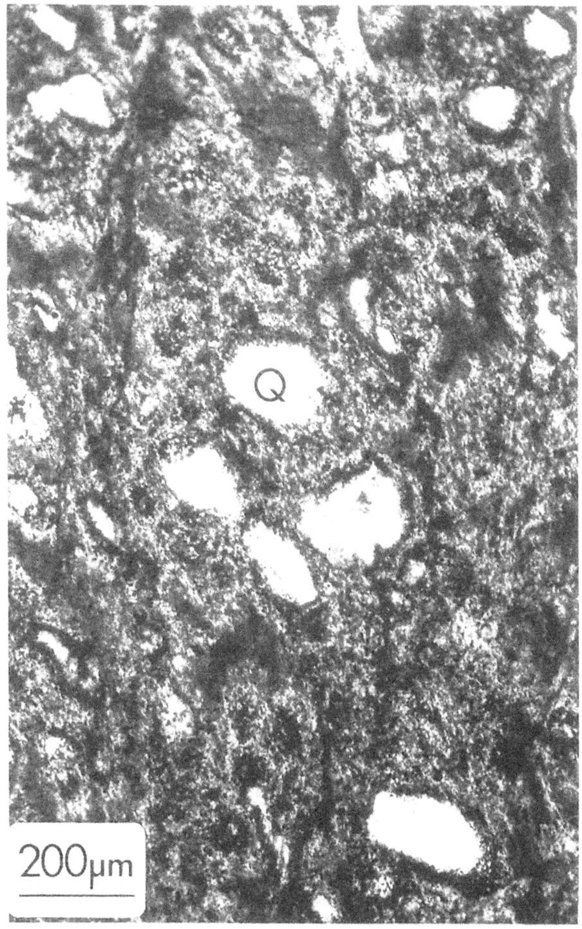

33 B

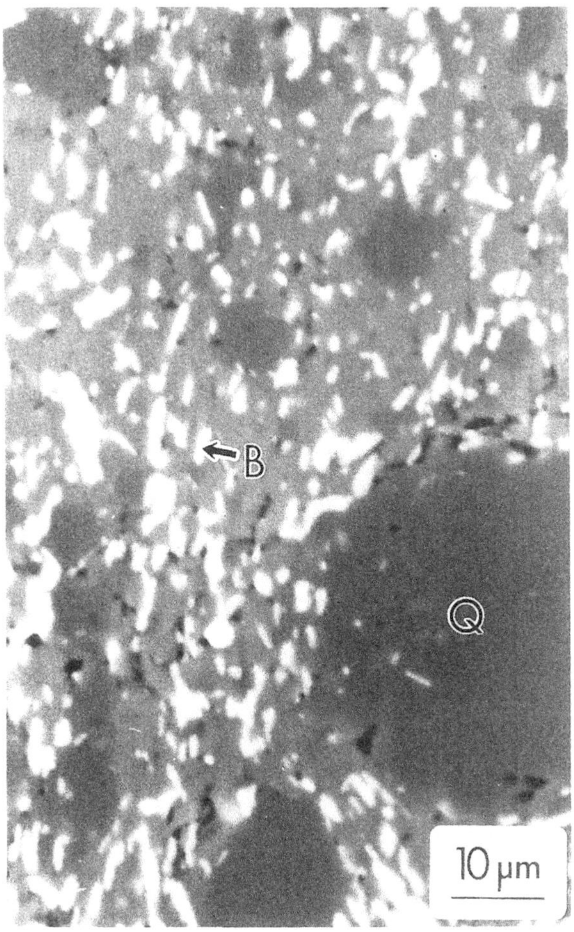

33 C

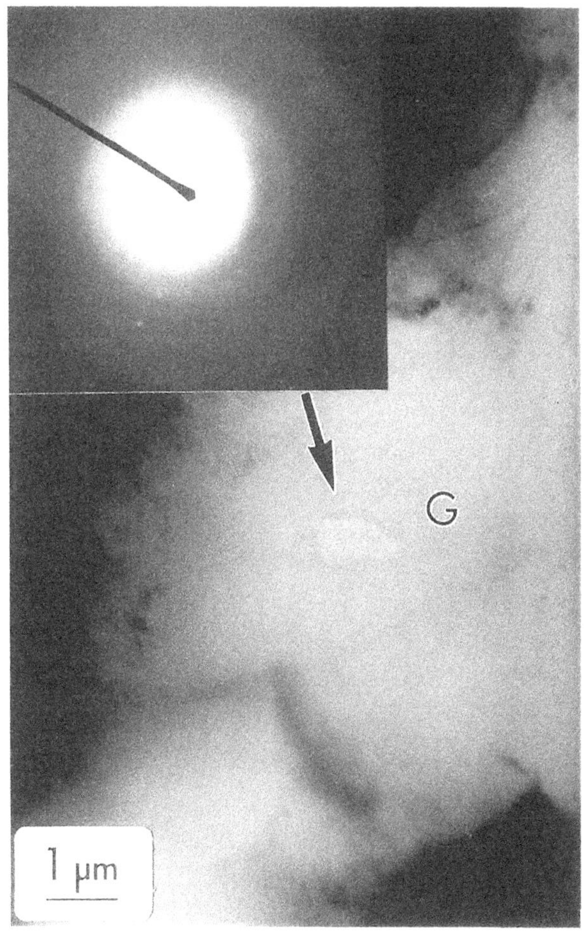

33 D

# 34. DEVELOPMENT OF ULTRAMYLONITE FROM PSEUDOTACHYLYTE IN A METAMORPHIC CORE COMPLEX

Stephen J. Reynolds, Laurel B. Goodwin, Gordon S. Lister, Phillip D. Ellzey, and Charles J. Ferranti

Deformation within the South Mountains metamorphic core complex, Arizona, U.S.A., occurred over a range of temperatures as plutonic rocks in the core crystallized and were exhumed by a low-angle detachment fault (Reynolds, 1985; Reynolds and others, 1986; Reynolds and Lister, 1987, 1990; Smith and others, 1991). The plutonic rocks record this full history, beginning with mylonitization at ~600°C (Smith and others, 1991) and ending with brittle detachment faulting at near-surface temperatures (Fitzgerald and others, 1994). Cross-cutting relationships indicate that pseudotachylyte postdates much of the mylonite formation, but many pseudotachylyte veins have a mylonitic overprint. These relationships suggest that pseudotachylyte formed during the waning stages of mylonitization, probably at ~400°C, when brittle deformation began to dominate (Smith and others, 1991).

Pseudotachylyte veins strongly localized post-pseudotachylyte mylonitization, and many veins were converted into ultramylonite. This localization of mylonitization (see also Sibson, 1980; Passchier, 1982b) probably reflects a decrease in strength of the pseudotachylyte relative to the host mylonitic plutonic rock due to its finer grain size and the presence of glass (see Plate 33). The pseudotachylyte veins formed as the rocks were drawn up through the brittle-ductile transition, but the veins were able to deform ductilely after the adjacent, coarser-grained rocks were within the brittle field.

**34A.** Mylonitic overprint of pseudotachylyte vein cutting Tertiary granodiorite. The main pseudotachylyte vein (M) localized subsequent mylonitization, and has been nearly totally converted into ultramylonite. A thicker, steeply inclined injection vein (I) of pseudotachylyte largely escaped the mylonitic overprint, probably because it was less favorably oriented and was shielded by the coarser-grained wall rocks. PPL

**34B.** Close-up view of an intersection between a gently inclined, mylonitized pseudotachylyte vein (M) and a steep, nonmylonitic pseudotachylyte injection vein (I). Note the shape fabric and preferred orientation of clasts within the mylonitized vein. Banding within the injection vein is interpreted as due mostly to flow and cooling of a fault-related melt, and not to mylonitization. PPL

**34C.** Complexly interfingered pseudotachylyte and ultramylonite. Note variable development of mylonitic fabric. Most layers have an extremely well-developed ultramylonitic fabric derived from pseudotachylyte, clasts, and wall rocks. Pseudotachylyte with less mylonitic fabric is present in protected embayments within the wall rocks and near some larger, less deformed clasts. PPL

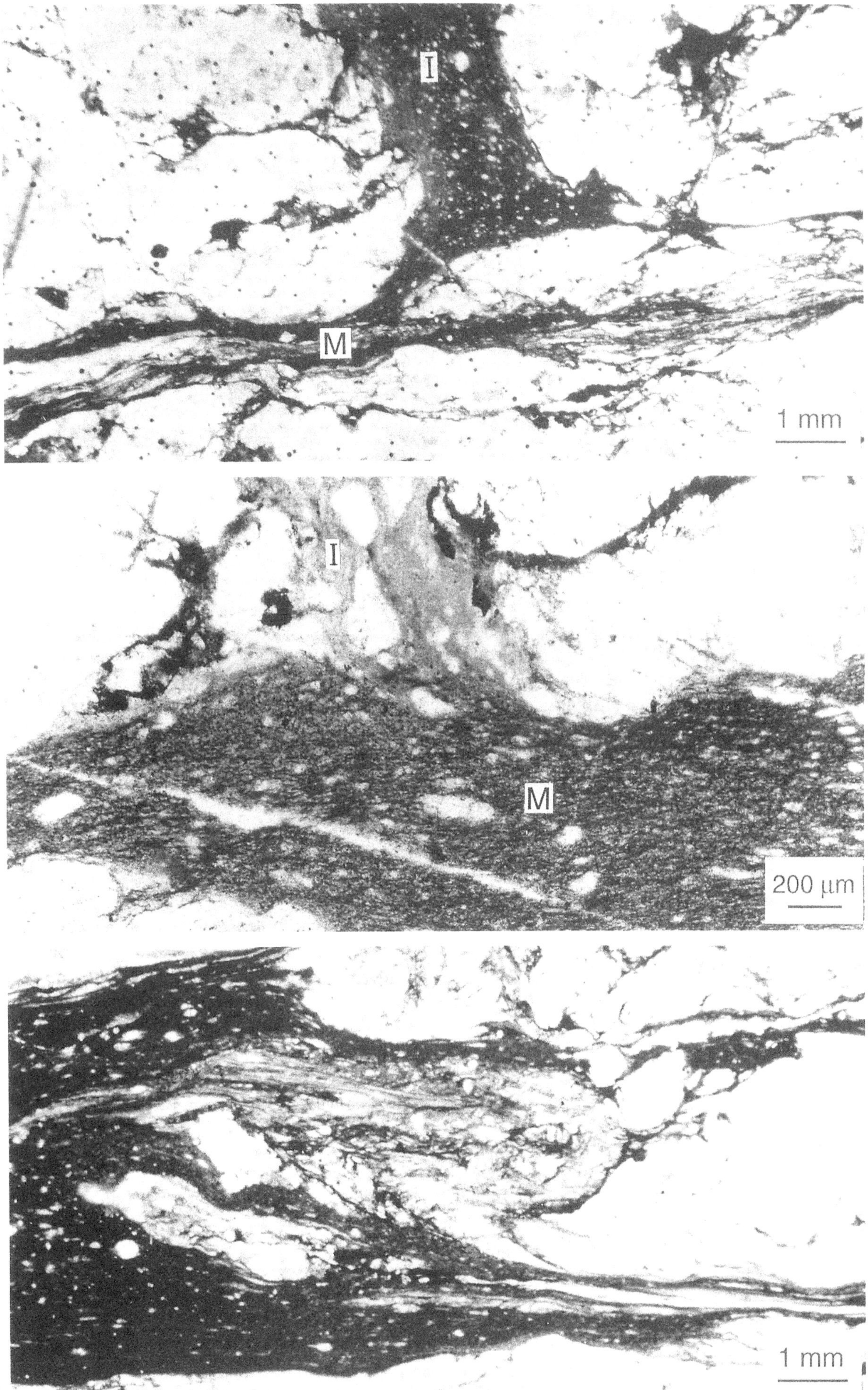


34 A

34 B

34 C

# 35. DEEP CRUSTAL ECLOGITE-FACIES PSEUDOTACHYLYTES

THERESA M. BOUNDY AND HÅKON AUSTRHEIM

Pseudotachylytes are typically interpreted to have formed by frictional melting during coseismic faulting within the upper to middle crust. Recently, however, we discovered pseudotachylytes in an exceptionally deep crustal environment. Eclogite-facies pseudotachylytes from the Bergen Arcs of western Norway are interpreted to bear witness to rapid faulting and seismicity at depths of >60 km in a thickened section of continental crust. We postulate that the pseudotachylytes formed in the roots of the Caledonian orogen in a region that was undergoing fluid-induced eclogitization. The pseudotachylytes may have formed in response to rapid relaxation of stresses caused by metamorphic transformation from granulite to eclogite (Austrheim and Boundy, 1994). The descriptions that follow are based largely on Austrheim and Boundy (1994) and our recent field data.

A slice of the former root zone to the Caledonian continent-continent collisional orogen is exposed in a thrust sheet within the Bergen Arcs of western Norway (Austrheim, 1986/1987). The rocks brought up from this lower part of the lithosphere are Grenville-age, granulite-facies rocks (ca. 850°C, <1000 MPa), some of which were locally transformed to eclogite facies (ca. 650°C, >1500 MPa) during the Caledonian orogeny apparently as a result of burial, deformation, and interaction with fluids (Austrheim and Griffin, 1985; Austrheim, 1986/1987; Jamtveit and others, 1990; Boundy and others, 1992). Eclogite initially formed along fractures within the granulites, and the process culminated in the development of major, eclogite-facies shear zones that are as much as 120 m thick and are laterally continuous over several kilometers (Boundy and others, 1992). The eclogite-facies pseudotachylytes reflect contemporaneous brittle behavior during the development of the eclogites. The pseudotachylyte veins occur within garnet granulites at the border with eclogitized regions. The veins, which are typically about 1 to 5 cm thick, can be traced over >10 m. Offsets of 0.5 m along the pseudotachylyte veins have been observed. At several localities, the veins are discontinuous and form a network completely enclosed by granulite.

**35A.** Hand specimen (Sample HA25/92) from Holsnøy, western Norway, showing two pseudotachylyte veins (dark rocks) in a garnet granulite (Austrheim and Boundy, 1994). The granulite host rocks are anorthositic to gabbroic in composition and consist of garnet, clinopyroxene, and plagioclase locally with pleonaste, orthopyroxene, scapolite, and hornblende. The grain size of the host rock is 1 to 5 mm.

Contacts between the veins and the granulite wall rock are sharp and typically subparallel to granulite foliation. Veins consist of a dark aphanitic matrix with variable (5 to 20%) amounts of wall rock fragments, mainly plagioclase and minor garnet. The smaller vein (left) shows evidence of dextral movement, suggesting ductile deformation prior to pseudotachylyte formation. Several minor injection veins splay off of this vein (as indicated by the black arrow).

**35B.** Photomicrograph showing the sharp contact between a pseudotachylyte vein and garnet granulite wall rock (Sample HA25/92, small vein on left in 35A). The granulite-facies garnet (Grt I) of the wall rock is sharply truncated and is penetrated by numerous fractures. Initial eclogitization is indicated in the wall rock by the formation of omphacite along fractures in the garnets and as selvages (dark rims) between garnet and plagioclase (Plag) grains. Kyanite and (clino)zoisite (the two could not be distinguished because the grain size was too small) formed within the plagioclase grains.

The matrix of the pseudotachylyte veins consists of a fine-grained aggregate of eclogitic microlites including omphacite, garnet (Grt II), kyanite, (clino)zoisite, hornblende, plagioclase, potassium feldspar, quartz, and pyrite. Garnets (Grt II) are euhedral and have a dusty appearance due to minute (typically 1 to 5 μm across) inclusions of all of the other vein minerals except amphibole. The grain size of the vein minerals increases from <10 μm at the margins of the veins toward their centers where microlites of garnet (Grt II) and (clino)-zoisite are as large as 100 μm. The fine grain size at the vein margins suggests chilling against wall rock. PPL

**35C.** Backscattered electron image showing euhedral garnet microlites (Grt) from a pseudotachylyte vein (Sample HA25/92, vein not shown in 35A). Inclusions in the garnets are omphacite, quartz, and kyanite. Garnets are enclosed in a matrix composed of omphacite (Omph) and minor amphibole (Amph) with acicular needles of kyanite (Ky) and plagioclase (Plag). Omphacite in the veins ranges in jadeite (Jd) content from $Jd_{30}$ to $Jd_{50}$. In contrast, the pyroxene in the granulite wall rock is an aluminous diopside. Garnet in the pseudotachylyte shows a wide range in grossular (Gr) content ($Gr_{15-85}$) suggesting a high degree of disequilibrium, whereas the composition of garnet in the granulite host is $Gr_{17}Al_{30}Py_{52}$ (where Al is almandine and Py is pyrope). Plagioclase in the host granulite has a composition of $An_{27}$, whereas that in the vein is $An_{19}$. The textures of the vein minerals, including the small euhedral character of the garnets, the inclusions of omphacite, quartz, and plagioclase within these garnets, and the general increase in grain size toward the centers of the veins, are consistent with crystallization from a melt. Together with the significant compositional differences between the wall rock and vein minerals these textures demonstrate that the minerals in the veins are not inherited fragments, but instead crystallized from a melt under eclogite-facies conditions. SEM

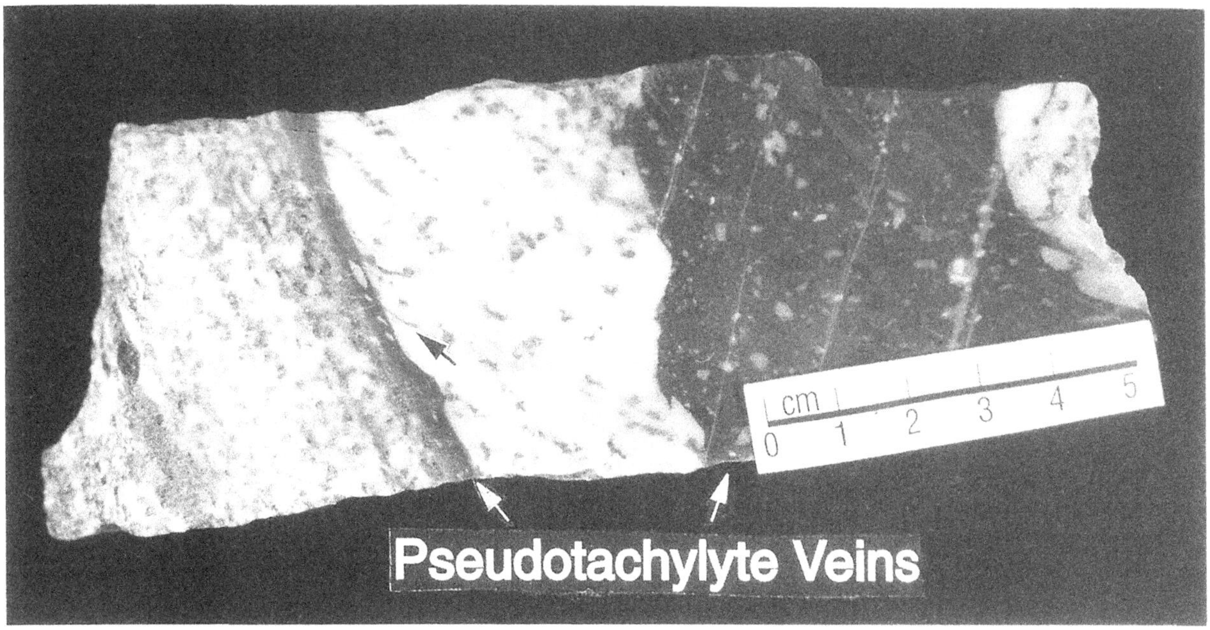
cm
0
1
2
3
4
5
Pseudotachylyte Veins
35 A

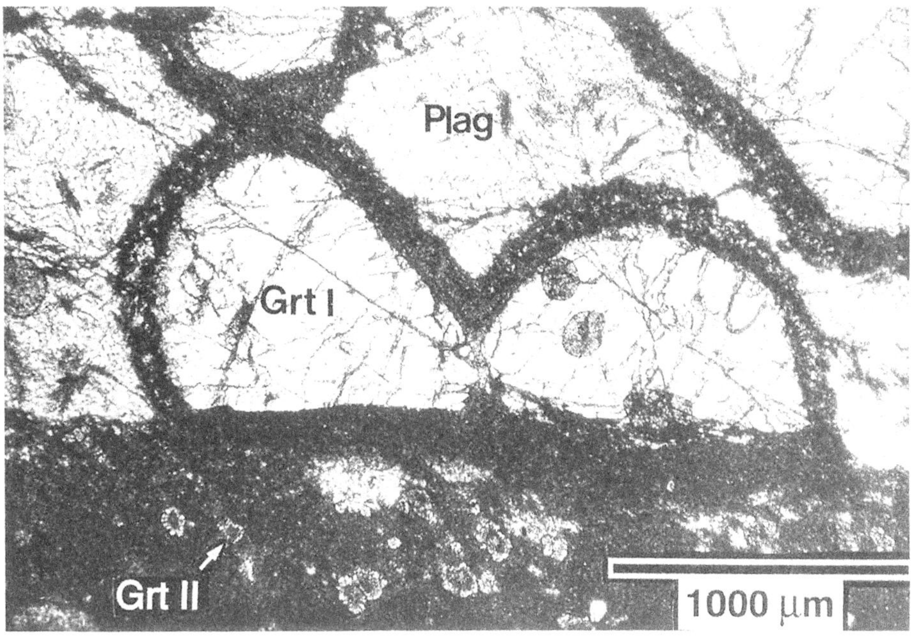
Plag
Grt I
Grt II
1000 μm
35 B

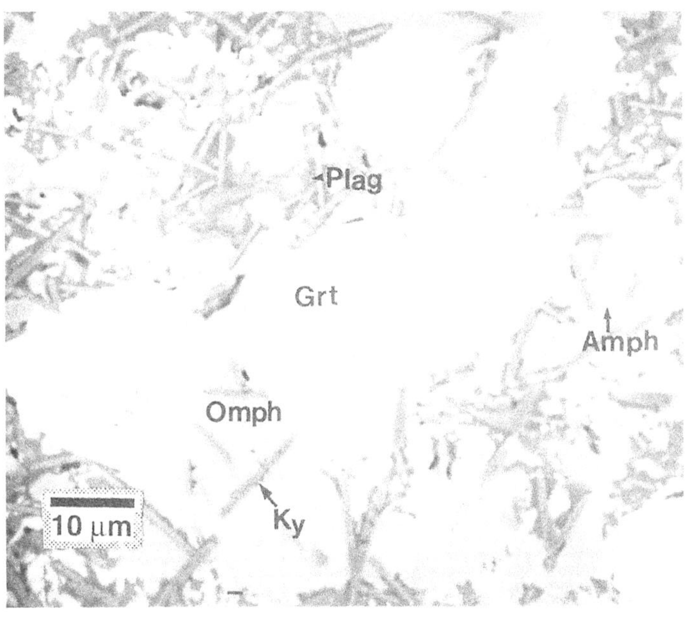
Plag
Grt
Amph
Omph
Ky
10 μm
35 C

## Fluid-related features



Common abbreviations used in regard to the plates in this section include PPL = plane polarized light, NX = crossed nicols, and SEM = scanning electron microscope.

This set of plates, consisting of observations based on both naturally and experimentally deformed samples, illustrates the important role of fluids in faulting. It is now recognized that veins, precipitated from fluids, are a common constituent of many fault zones, indicating that faults tend to focus the upward migration of metamorphic and other fluids. Two of the contributions illustrate portions of fault-related vein systems, and they present clear evidence for cycles of rupture and cataclasis, healing and sealing due to precipitation, and a subsequent increase of fluid pressure until supralithostatic values are reached and a new rupture occurs. Some of the plates show multiple episodes of vein formation, followed by folding and disruption; others show multiple crack-seal events within a single vein. Illustrations of experimental fault surfaces in sandstone slid at high temperature in the presence of water dramatically testify to the fast rates of dissolution and precipitation that occur. These experiments serve as a reminder that at very modest temperatures and in the presence of fluids, fault slip is not accomplished solely by "brittle" processes.

# 36. VEIN DEVELOPMENT IN REVERSE FAULT ZONES

S. F. Cox

Reverse faulting has been associated with the extensive development of quartz vein systems in the Wattle Gully fault zone (WGFZ), near Castlemaine, central Victoria, Australia. This high-angle reverse fault zone (Figure 36.1) has developed at temperatures near 300°C and at a depth of about 10 km late during mid-Devonian regional crustal shortening. The WGFZ forms part of a network of faults that has focused the upward migration of fluids generated by metamorphism and devolatilization at deeper crustal levels (Cox and others, 1991a,b; Cox, 1995). The fault has a length of about 800 m and a maximum displacement of 50 m. Two steeply dipping, bedding-parallel fault segments are localized on the west-dipping limbs of adjacent folds and linked by a less steeply dipping fault segment that is discordant to bedding on the intervening, east-dipping fold limb. The less steeply dipping fault segment forms a dilatant jog which hosts 1.3 million tonnes of vein quartz.

The geometry and internal structure of veins in the WGFZ provide evidence that cyclic changes in fluid pressure and shear stress are associated with repeated slip events during "fault-valve" behavior (Sibson, 1989).

Veins occur as moderately to steeply dipping fault-fill structures, and as areally extensive arrays of subhorizontal extension veins adjacent to numerous slip surfaces in the dilatant jog segment of the WGFZ. Extension veins are present up to 30 m laterally from the core of the fault zone in the jog segment. Extension veins are much less abundant in the steeply dipping, bedding-parallel fault segments. The association of steeply dipping fault-fill veins with subhorizontal extension veins indicates that faulting occurred at relatively low shear stresses and in a transiently supralithostatic fluid pressure regime. The orientation of the WGFZ at 60° to 80° to the far-field maximum principal stress during faulting also indicates that this severely misoriented fault had a low shear strength.

**36A.** Irregularly shaped but generally gently dipping quartz-carbonate extension veins near the margin of the Wattle Gully fault zone, Wattle Gully gold mine. Width of field is 3 m.

**36B.** Quartz extension veins associated with a minor reverse fault in the dilatant jog segment of the WGFZ, Wattle Gully gold mine. Shear sense shown by arrows. Width of field is 1 m.

**36C.** Moderately dipping, fault-fill quartz vein and associated gently dipping, wedge-shaped extension veins in a small reverse fault having a displacement of several meters, Wattle Gully gold mine. Shear sense shown by arrows. Width of field is 3 m.

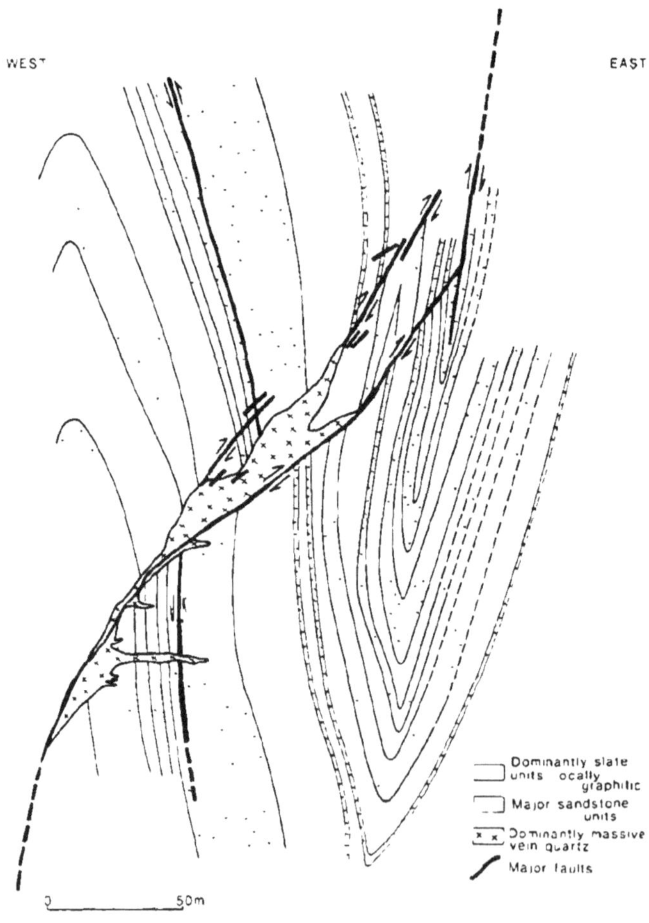


**Figure 36.1.** East-west cross section of the Wattle Gully fault zone, Wattle Gully mine, Chewton, central Victoria, Australia.

36 A

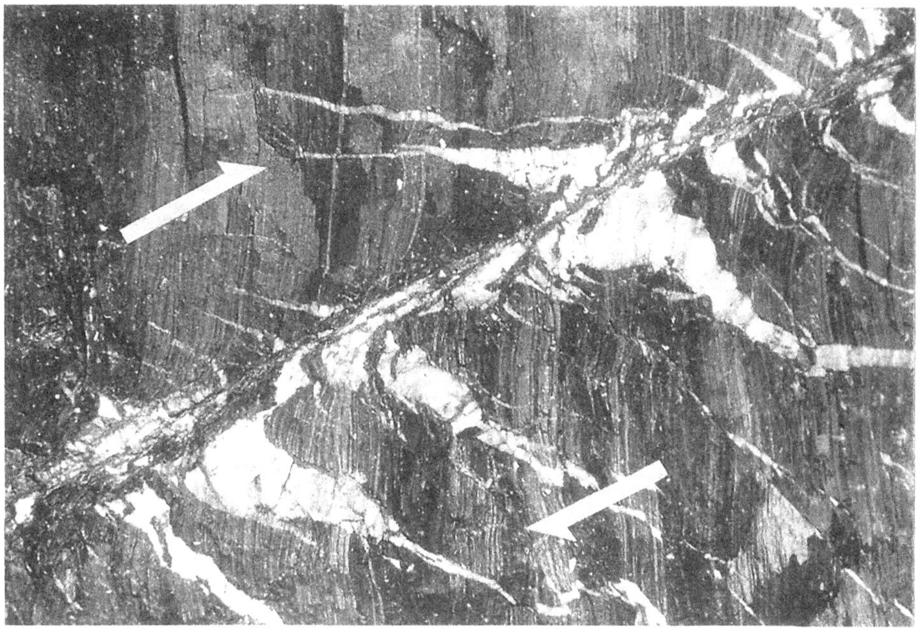
36 B

36 C

*Extension veins*, which are associated with the WGFZ and related faults, commonly have fibrous to massive internal structures; however, crustiform to idiomorphic fabrics also occur. Formation of fibrous to massive extension veins has involved crack-seal growth mechanisms (Ramsay, 1980b; Cox and Etheridge, 1983), with individual growth increments ranging from 20 μm to 100 μm in thickness. Syntaxial and stretched-crystal types of crack-seal microstructures are present in quartz-rich veins in sandstone units, whereas antitaxial microstructures are more usual in quartz veins in slate units. Crustiform, idiomorphic, and some massive vein fabrics are interpreted to have formed by a more simple growth history involving only a single stage of vein opening and sealing.

The delicate, rhythmically repeated structure of successive crack-seal inclusion bands in antitaxial veins indicates that many fault-related extension veins have opened by subcritical crack growth mechanisms (Boullier and Robert, 1992; Cox, 1995).

The repeated opening and sealing of veins indicated by crack-seal microstructures is interpreted to have been controlled by fluctuations in fluid pressures associated with repeated cycles of fault rupture and interseismic fault sealing during fault-valve behavior (Cox and others, 1991a; Cox, 1995).

**36D.** Crack-seal inclusion bands parallel to the vein wall in an antitaxial fibrous quartz extension vein. The inclusion bands are oriented at a high angle to the incremental extension direction, and have developed at each crack-seal increment by (1) overgrowth of fine-grained layer silicates and opaque minerals in the wall rock at the vein margin, and (2) detachment of slivers of wallrock from the vein wall. PPL

**36E.** Stepping of twin boundaries in lamellar twinned albite fiber in a crack-seal quartz-albite-ankerite extension vein. The twin offsets developed during fiber growth due to the twin plane being inclined to the incremental extension direction during repeated intrafiber crack-seal increments. NX

**36F.** Crack-seal inclusion trails and bands in a quartz-rich vein which formed by repeated increments of vein opening at low angles to the vein walls. Trails of ankerite (a) and mica (m) formed during successive crack-seal growth increments by epitaxial overgrowth of carbonate and mica grains at the vein wall. More rapid quartz growth occluded growth of ankerite, leaving ankerite grains in the same trail isolated from each other after each crack-seal increment. The inclusion trails track the displacement path of the vein walls during vein growth and indicate vein opening at an acute angle to the vein margin. In contrast, the long axes of columnar vein quartz grains are oblique to the vein margin and do not track the displacement path of the vein walls (see Cox, 1987). Some inclusion bands (b) have grown on localized steps in the vein wall. NX

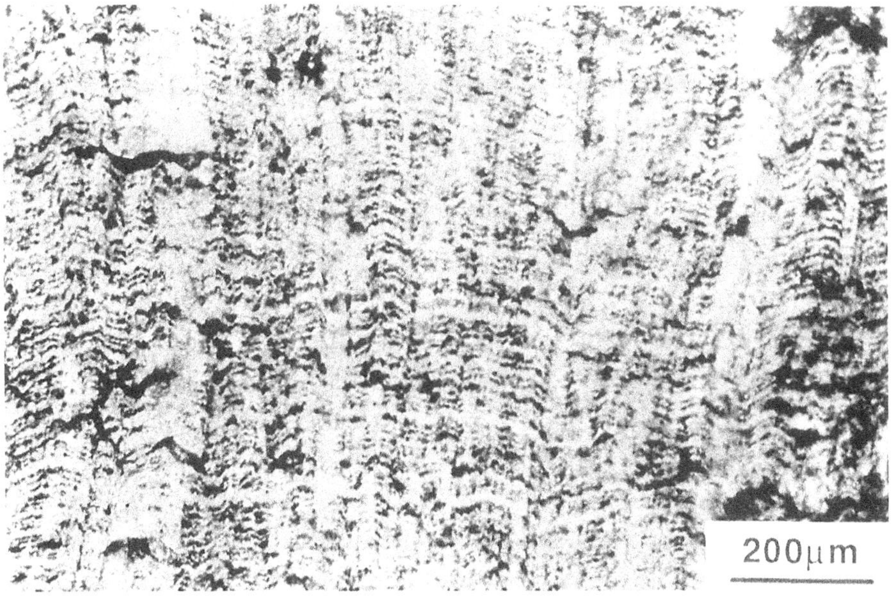

36 D

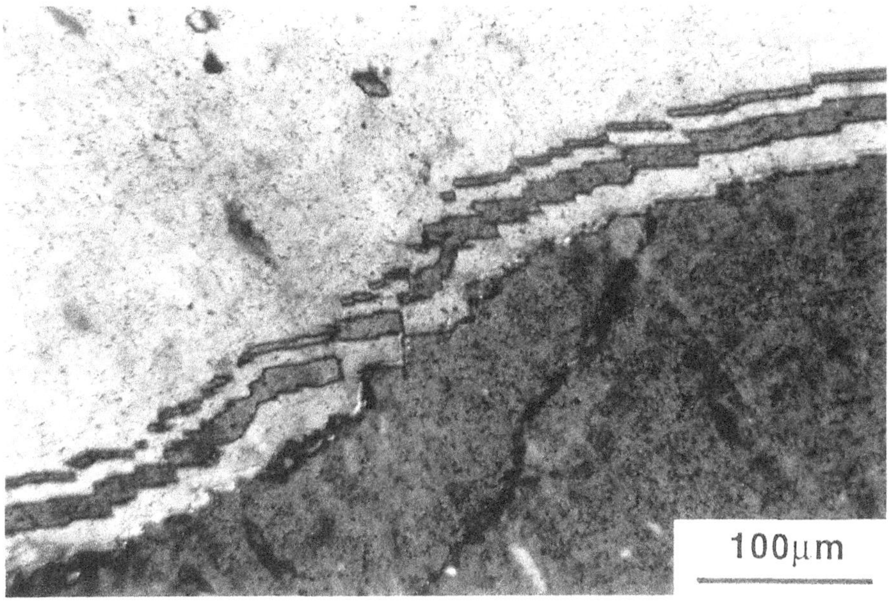

36 E

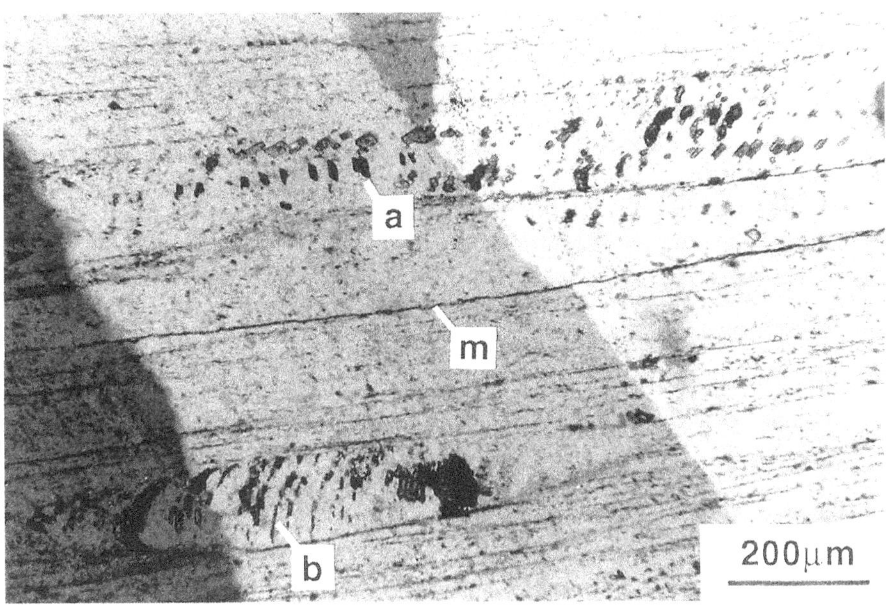

36 F

The internal structures of *fault-fill veins* in the WGFZ provide insights about fluid pressure and shear stress histories associated with repeated fault failure and the operation of fault-valve behavior at near-lithostatic fluid pressures.

In the steeply dipping, bedding-parallel fault segments, fault-fill veins commonly have laminated internal structures (36G). Primary growth textures are preserved in some quartz laminae, whereas others contain fabrics that have been modified during repeated episodes of fault slip involving brittle, plastic, and dissolution-precipitation processes. The geometry of crack-seal inclusion bands and trails indicates that some quartz laminae have formed by numerous crack-seal events involving dilation at high angles to the fault surface, rather than by dilation due to fault-parallel displacement at localized fault bends or jogs. Overprinting relations between undeformed and deformed quartz laminae, as well as cross-cutting microveins, indicate that the development of fault-fill veins has involved repeated fault dilatation, sealing, and shear failure. Fault displacement has also occurred along some of the slaty laminae that separate quartz laminae (36H).

In the dilatant jog segment of the fault zone, fault-fill veins usually have coarse-grained primary growth textures that are also variably overprinted by the effects of plastic, brittle, and dissolution-precipitation deformation processes. Localized brittle shear failure has produced narrow quartz cataclasite bands (36I). Stylolitized margins to slaty septa in fault-fill veins indicate episodic shortening at high angles to the fault zone.

The history of cyclic fault slip, dilatation, and sealing, together with repeated opening of subhorizontal extension fractures adjacent to faults, is interpreted to have been controlled by cyclic variations in shear stress during the seismic cycle, and by fluctuations in fluid pressure which were associated with deformation-induced changes in fault permeability during fault-valve behavior (Figure 36.2). In regimes where faults focus large fluid fluxes, the evolution of fault shear strength, and the nucleation and recurrence of rupture events, are expected to be influenced by rates of recovery of fluid pressure and shear stress after rupture events (see also Cox and others, 1991a; Cox, 1995; Sibson, 1992).

**36G.** Laminated structure in a bedding-parallel fault vein due to interlayering of quartz-rich laminae with thin sheets of slaty wall rock. The laminae are subparallel to the fault orientation. Quartz-rich laminae have formed by vein infill during fault dilatation. The slaty bands have been detached from the wall rock during successive fault dilatation events. Note that some slaty bands have been stylolitized by shortening at high angles to laminae boundaries. PPL

**36H.** Detail of deformed slaty layers and interlaminated vein quartz in a bedding-parallel, fault-fill vein. PPL

**36I.** Fine-grained quartz cataclasite formed within much coarser-grained, fault-fill quartz by localized brittle shear failure and subsequent hydrothermal sealing. NX

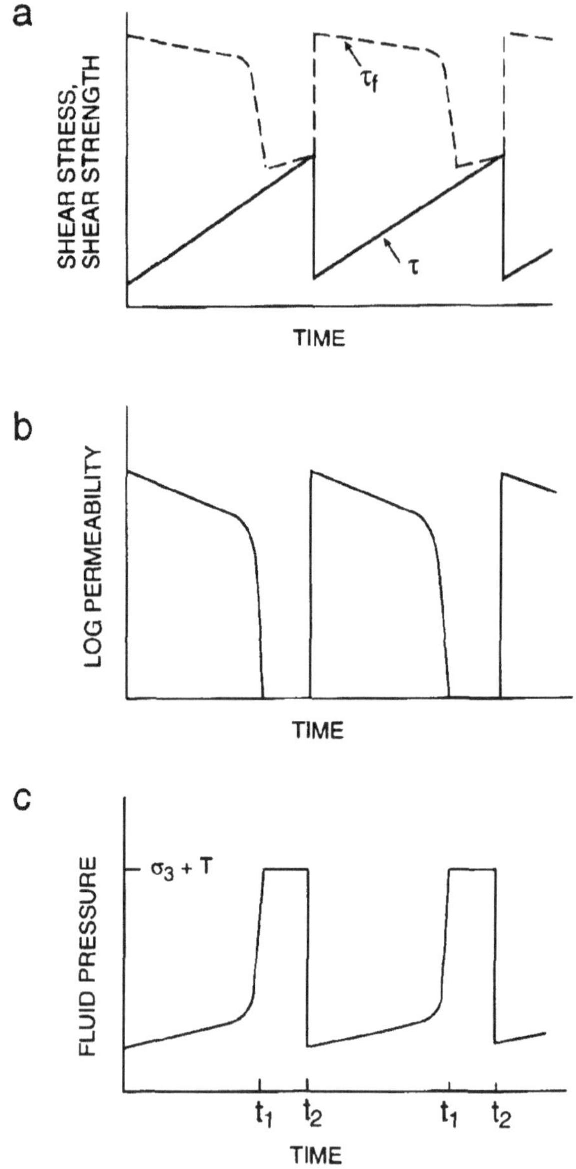


**Figure 36.2.** Schematic illustration of possible variations of (a) shear stress ($\tau$) and fault shear strength ($\tau_f$), (b) permeability of fault seals, and (c) fluid pressure in the fault core, as a function of time during fault-valve behavior in a fault that is a site of focused upward fluid flow. Fluid pressure history is dependent on the evolution of fault permeability. Formation of low-permeability pressure seals before times $t_1$ allows rapid buildup of fluid pressures below seals. The opening of hydraulic extension fractures adjacent to the fault when fluid pressure ($P_f$) reaches the sum of minimum principal stress ($\sigma_3$) and the tensile strength (T) limits the maximum fluid pressure. Shear failure events occur at times $t_2$ when fault shear strength equals the shear stress. Note that fault shear strength is dependent on both fluid pressure and shear stress. Dilatation of steeply dipping, bedding-parallel fault segments (to form laminated fault-fill veins) requires near-total shear stress relief at rupture events and recovery of fluid pressures to near-lithostatic levels before substantial shear stress recovery.

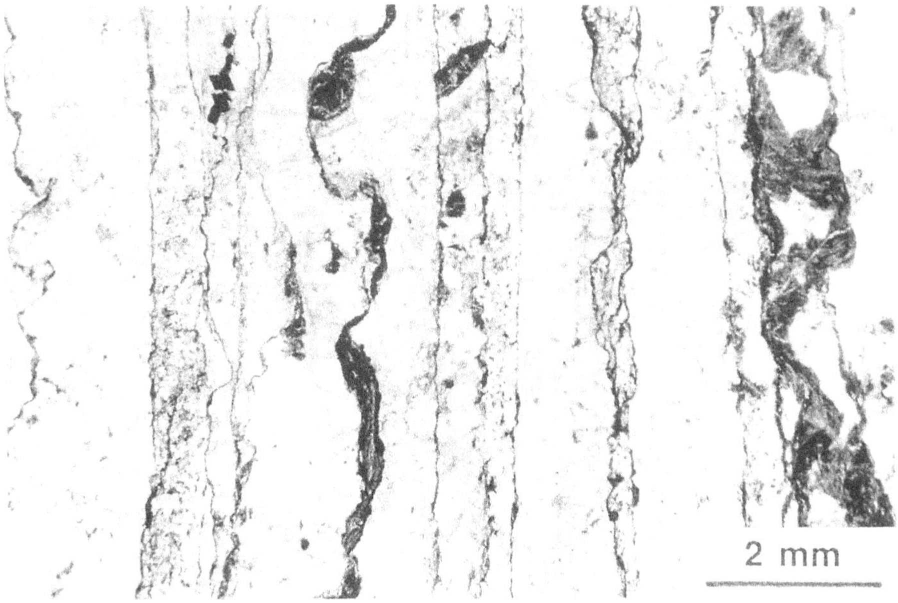


36 G

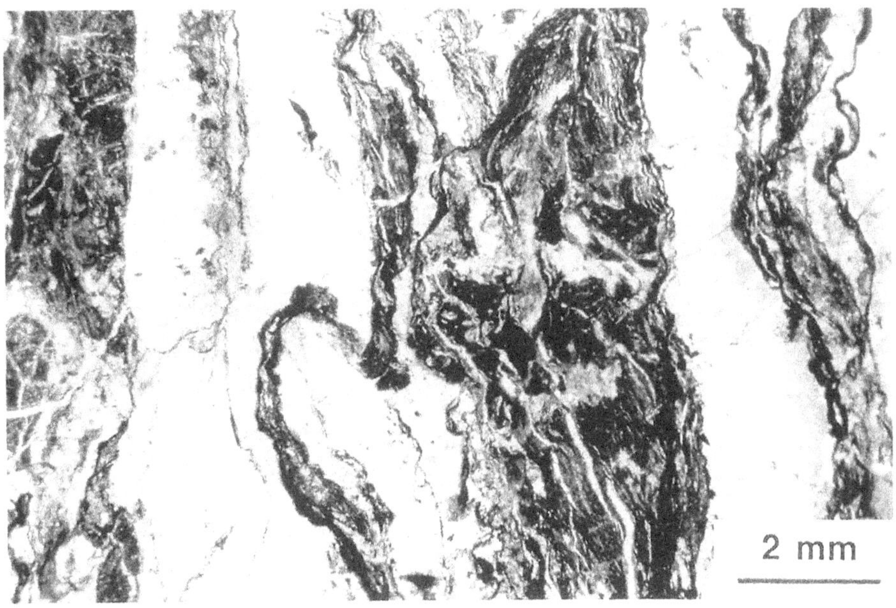


36 H

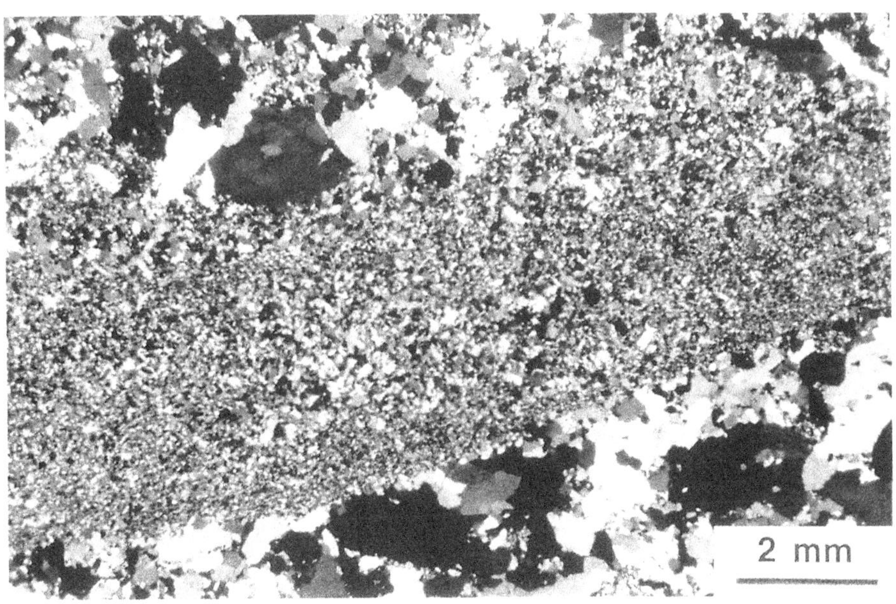


36 I

S. F. Cox

The microstructures produced during experimental sliding on simulated, bare faults at elevated temperatures and pressures illustrate some spectacular differences between processes operating in the presence of reactive pore fluids and those operating in the presence of nonreactive pore fluids.

The experiments have been performed at a confining pressure of 300 MPa, a pore pressure of 200 MPa, and at a temperature of 1200 K in an argon confining medium, triaxial testing apparatus. Tests have been conducted at an axial shortening rate of 0.8 $\mu m\ s^{-1}$ with either argon or water as a pore fluid. At the experimental conditions, water promotes rapid dissolution-precipitation reactions on quartz grain surfaces (Cox and Paterson, 1991), whereas argon is unreactive.

The test specimens consist of two cylindrical forcing blocks that have been fabricated from quartz sandstone (Fontainebleu Sandstone). The sliding interface between the blocks is inclined at 30° to the maximum compression direction (Figure 37.1). The test piece has a diameter of 10.0 mm and a length of 27.5 mm. The sandstone has a porosity of about 12%, and is composed of equant quartz grains with an average diameter of 0.25 mm.

In the presence of argon pore fluid, the simulated fault is characterized by slip hardening behavior and a coefficient of friction of approximately 0.5 (Figure 37.2). However, in the presence of pore water, the simulated fault exhibits a post-yield stress drop and subsequent stable sliding with a substantially reduced coefficient of friction of 0.3.

The differences in mechanical behavior between the "wet" and "dry" experiments demonstrate the potential importance of reactive pore fluids in modifying the resistance to sliding (Blanpied and others, 1992; Cox, unpublished data).

The surface structures that develop during "wet" and "dry" sliding have remarkable differences. The deeply etched wear structures formed during "wet" sliding illustrate the rapidity with which reactive pore fluids can modify the structure of sliding surfaces. The reduced resistance to sliding in the "wet" experiments, relative to the "dry" experiments, is interpreted to be influenced by more rapid asperity removal by processes such as dissolution or stress corrosion cracking in the presence of reactive pore fluid. Additionally, surface forces associated with a reactive water film between sliding surfaces might also have promoted boundary lubrication effects not exhibited by the unreactive argon pore fluid at high temperatures (Parks, 1984).

The micrographs illustrate the importance of reactive pore fluids in controlling the development of structures on sliding interfaces.

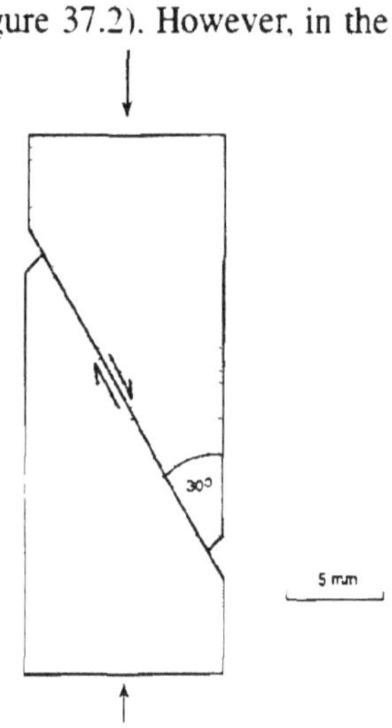


**Figure 37.1.** Specimen configuration during sliding tests. Shortening and sliding directions are indicated.

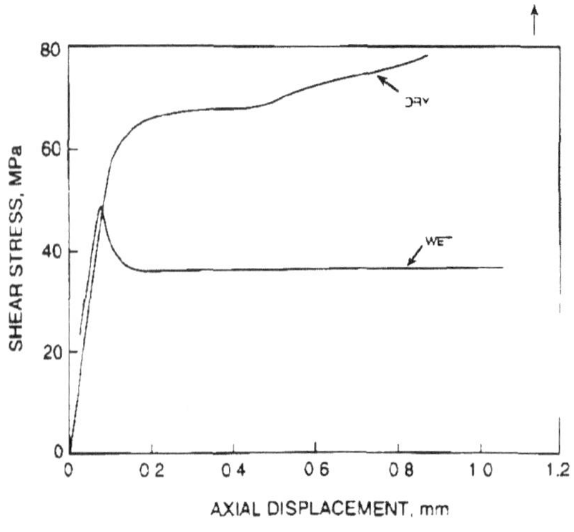


**Figure 37.2.** Shear stress versus axial displacement curves for the "wet" and "dry" sliding experiments.

**37A.** SEM micrograph illustrating the microstructure on the surface of a Fontainebleu Sandstone slider block that has slipped 1 mm in the presence of argon pore fluid. Fine, low-amplitude wear tracks (subhorizontal lines on the micrograph) have formed parallel to the slip direction by asperity ploughing. Many of the small, irregularly shaped pits were formed during specimen preparation and have been preserved during sliding.

**37B.** SEM micrograph illustrating details of the microstructure on the surface of a Fontainebleu Sandstone slider block after just over 1 mm of slip with water as the pore fluid. Despite the total slip being so small, the new microstructure has totally obliterated the surface structure that was present prior to sliding. The slip direction is parallel to the long axis of the micrograph. The well-developed slickenlines are continuous for lengths up to about 150 μm and have transverse amplitudes much greater than those developed during "dry" sliding.

Wear tracks, similar to those illustrated in 37A, are produced mainly during the rapid slip event associated with post-yield stress drop in "wet" experiments. These become deeply etched by the pore fluid during subsequent stable sliding. The microstructure illustrated here developed during a period of about twenty minutes of sliding. Note the cluster of euhedral quartz crystals that has grown during the experiment (near the top right-hand corner of the micrograph).

**37C.** SEM micrograph illustrating slickenline structure over a larger region of the slip surface in the "wet" sliding experiment illustrated in 37B.

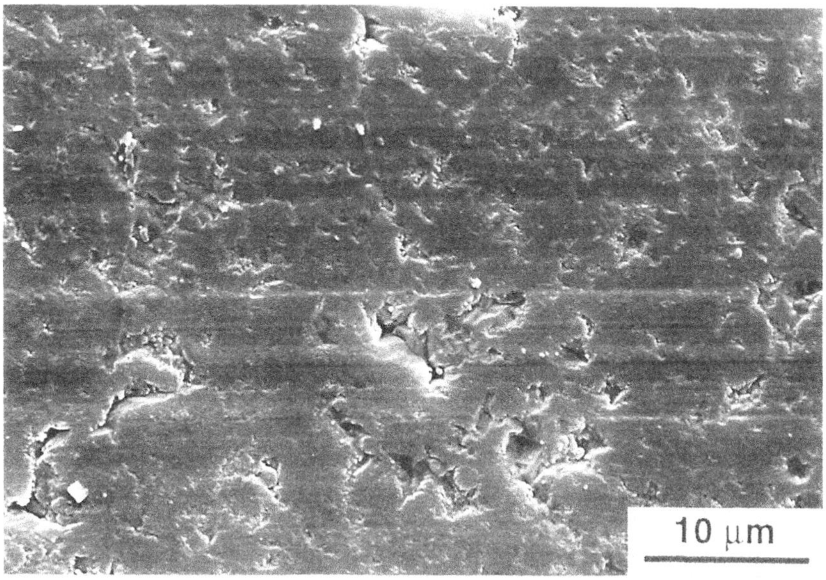


37 A

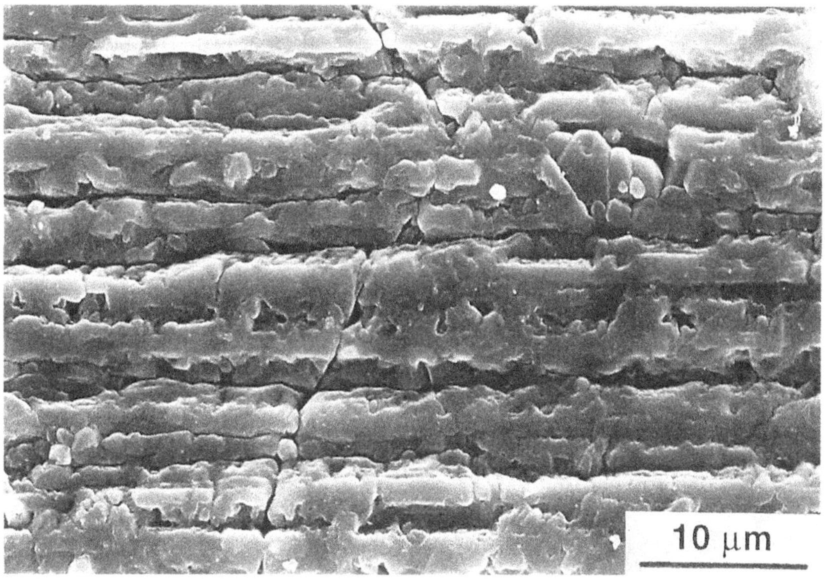


37 B

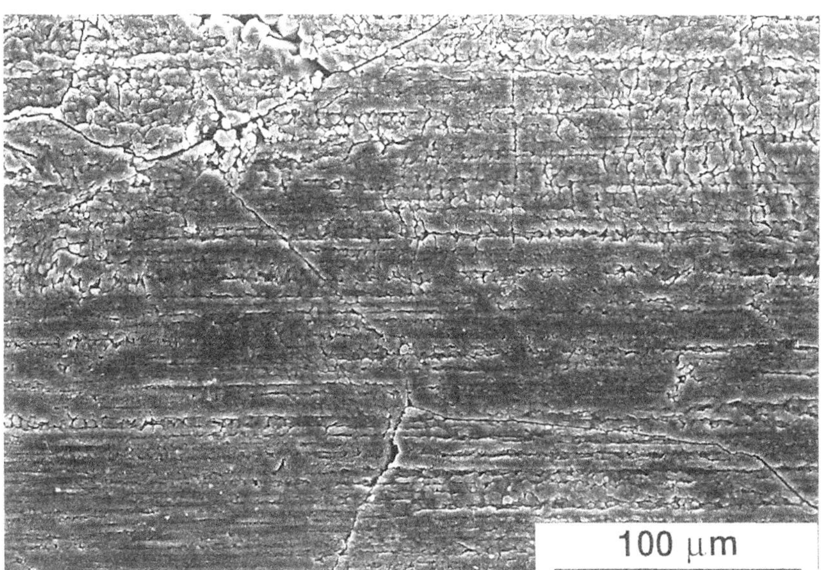


37 C

# 38. MICROSCALE VEIN MINERALIZATION AND PORPHYROCLAST DEVELOPMENT IN LOW-GRADE CATACLASITES

FREDERICK M. CHESTER AND JAMES P. EVANS

Microscopic veins and porphyroclasts derived from veins are observed in the cataclastic rocks from the north branch of the San Gabriel fault, San Gabriel Mountains, California, U.S.A. (Chester and others, 1993). The San Gabriel fault is an ancient, exhumed fault of the San Andreas system. The north branch of the San Gabriel fault cuts a complex of igneous and metamorphic rocks within the San Gabriel terrane and is defined as a zone of fractured rock containing a core of foliated cataclasite and ultracataclasite. Faulting occurred primarily during the Miocene, and exposures record faulting at several kilometers depth and at conditions favoring mineralogic alteration to smectite/illite and laumontite (zeolite).

Although the dominant mechanism of deformation in the fault zone was cataclasis, there is ample evidence of fluid-rock interaction in the form of alteration to clay minerals and neomineralization in veins. Neomineralized material, primarily laumontite, constitutes up to 50% by volume of the cataclastic rock within the fault core at some locations. Textural evidence records many episodes of neomineralization and cataclasis. Most porphyroclasts in the cataclasites are fragments of older veins. The photomicrographs show the progression from relatively intact veins to disrupted veins to isolated porphyroclasts, and some of the textural evidence for multiple episodes of mineralization and cataclasis.

The photomicrographs are of petrographic thin sections oriented perpendicular to the fault and parallel to the inferred slip direction in a right-lateral shear sense. Scale bar is 0.5 mm. PPL

**38A.** Laumontite veins and vein fragments (light regions) within ultracataclasite (gray and black regions) offset on localized shears within the fault core. Although veins occur at all orientations, many are formed parallel to the fault.

**38B.** Vein fragments (light regions) and porphyroclasts of older ultracataclasite (light gray regions) containing vein fragments in ultracataclasite (black regions). Compared to the veins in 38A, these veins have undergone greater cataclasis and disruption. In some cases, the veins are folded and rolled within the ultracataclasite.

**38C.** A large porphyroclast of older veins and ultracataclasite within younger ultracataclasite. The elliptical porphyroclast is outlined by release fractures filled with epoxy. Within the porphyroclast there are older veins within ultracataclasite offset along localized shears that are themselves cut by younger veins. These types of structures within porphyroclasts indicate repeated episodes of vein formation within deforming cataclastic rocks.

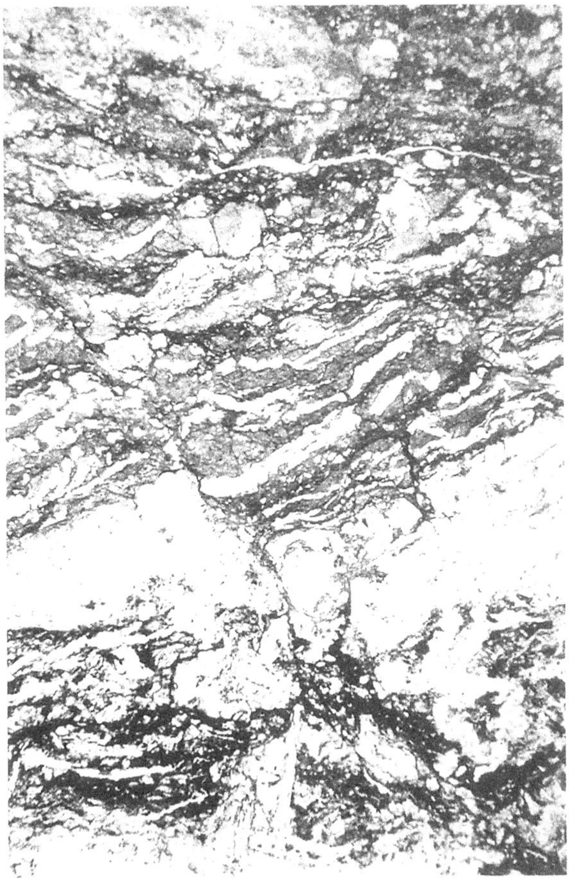

38 A

38 B

38 C

# SEMI-BRITTLE BEHAVIOR

## Geometrical characteristics and microstructures



Common abbreviations used in this section include PPL = plane polarized light, NX = crossed nicols, PNX = partial crossed nicols, and TEM = transmission electron microscope.

The contributions in this section illustrate the complexity of deformation in fault and shear zones at depths spanning the brittle-to-plastic and seismic-aseismic transitions. Most of the contributions illustrate features from naturally deformed quartzo-feldspathic rocks, on outcrop and thin-section scales; other contributions illustrate features from naturally deformed gabbroic rocks, and experimentally deformed halite and feldspar rocks. In all cases, the deformation has occurred partly by crystal-plasticity and partly by fracturing. In some examples, deformation occurred at one set of conditions but different minerals in the rock responded differently; for example, at low- to middle-greenschist facies, it is common for feldspar and amphibole to undergo brittle fracturing, whereas quartz and micas deform by crystal plasticity. In other examples, the deformation conditions have changed with time, and earlier higher-temperature mylonites have been succeeded by later lower-temperature cataclasites. Many of these plates illustrate that fracturing is commonly accompanied by fluid influx, which promotes reaction (especially of feldspars) to fine-grained and weaker phases. This reaction softening may allow exhumed fault-zone rocks to undergo an initial transition from ductile-to-brittle behavior, followed by a second transition back to ductile behavior. The experimental studies illustrate the optical- and TEM-scale microstructures produced by the cataclastic flow of feldspars when neither metamorphic mineral reactions nor crystal-plasticity are involved.

# 39. ANASTOMOSING DEFORMATION ZONES

GAUTAM MITRA

Mylonite zones are commonly regarded as zones with a simple strain history involving progressive simple shear or inhomogeneous simple shear (e.g., Ramsay and Graham, 1970). However, mylonite zones can be produced during either bulk, coaxial, inhomogeneous shortening or bulk, noncoaxial, inhomogeneous shortening (Bell, 1981), producing anastomosing deformation zones in three dimensions. Such networks of deformation zones are observed in rocks of all metamorphic grades ranging from the brittle (Aydin and Reches, 1982) to the ductile (Mitra, 1979), and at all scales (Mitra and others, 1988). The photographs show examples of anastomosing deformation zones formed at different metamorphic grades in granitic rocks. Similar patterns may be observed in other coarse-grained rocks that lack predeformational planar weaknesses.

**39A.** Brittle deformation zone in the Archean Middle Mountain granite along Torrey Creek in the Wind River Mountains near Dubois, Wyoming, U.S.A. The gently dipping zone, which formed during the late Cretaceous–early Tertiary Laramide orogeny, shows anastomosing shear fractures and thin shear zones, each with less than 2 cm of movement. The shear zones surround rhombic blocks (within which the rock is essentially undeformed) whose acute bisectrix plane is subhorizontal Within individual thin shear zones both feldspars and quartz are highly fractured and granulated with little or no chemical alteration (Mitra, 1993). The shear fractures show slickensides formed by frictional sliding (Mitra and Frost, 1981). Hammer is 30 cm.

**39B.** Lower greenschist-facies deformation zones developed during the late Paleozoic Alleghanian orogeny in medium- to fine-grained Precambrian Marshall Granite, Blue Ridge basement, near Marshall, Virginia, U.S.A. The thin zones, which show up as dark lines on a lighter background, show concentrations of fine micaceous minerals which develop from alteration of the feldspars during deformation. The angles between anastomosing zones vary considerably depending on the amount of deformation. The zones initially develop at approximately 90° to one another, and the angle between them decreases as the zones rotate during progressive deformation (Mitra, 1979). Knife is 10 cm.

**39C.** Lower greenschist-facies deformation zones in coarse-grained Precambrian Lovingston quartz monzonite gneiss, Blue Ridge basement, on Old Rag Mountain, Shenandoah National Park, Virginia, U.S.A. The deformation zones, which formed during the late Paleozoic Alleghanian orogeny, divide the rock into rhomb-shaped blocks whose elongation defines the regional cleavage (Mitra, 1979; Mitra and Elliott, 1979). The rhombs show relatively little internal deformation. The deformation zones surrounding them show intense grain-size reduction and alteration of feldspars to quartz and micas. Coarse quartz grains deform by fracturing, whereas finer-grained quartz shows both pressure solution and recrystallization. Pen is 15 cm.

**39D.** Upper amphibolite-facies, Precambrian deformation zones in Horse Ridge granitic gneiss along Glacier trail in the Wind River Mountains, south of Dubois, Wyoming, U.S.A. The zones formed under bulk noncoaxial (simple shear) deformation. The zones show an anastomosing pattern separating lenticular areas that show less deformation. Within individual zones, both feldspar and quartz show grain-size reduction by recrystallization. Pen is 15 cm.

39 A

39 B

39 C

39 D

# 40. DILATIONAL FRACTURES THAT PRECEDED FORMATION OF SHEAR ZONES

Paul Segall

Late Cretaceous granodiorite of Lake Edison, Mount Abbot quadrangle, Sierra Nevada, California, U.S.A.

**40A.** Subvertical dilational fracture (arrows) cross-cutting aplite dike (Ap), and granodiorite (Gd) (Segall and Pollard, 1983). There is no measurable strike separation of the aplite-granodiorite contacts across the fracture. Other fractures in this locality, which are subparallel to this fracture, display significant strike-slip movement (see Plate 41, this volume; Segall and Pollard, 1983). Centimeter divisions on scale.

**40B.** Photomicrograph of dilational fracture in granodiorite (Segall and Pollard, 1983). Intersection of fracture with large biotite grain (Bt) shows relative displacement of biotite in a direction nearly perpendicular to strike of fracture; there is no observable shear component. The jagged, interlocking nature of the fracture surface also indicates a dilational origin of this fracture. The fracture is filled with undeformed hydrothermal vein filling (VF), including zeolites and small fragments of fractured biotite. Epidote, chlorite, and quartz also form vein fillings in these fractures. PPL

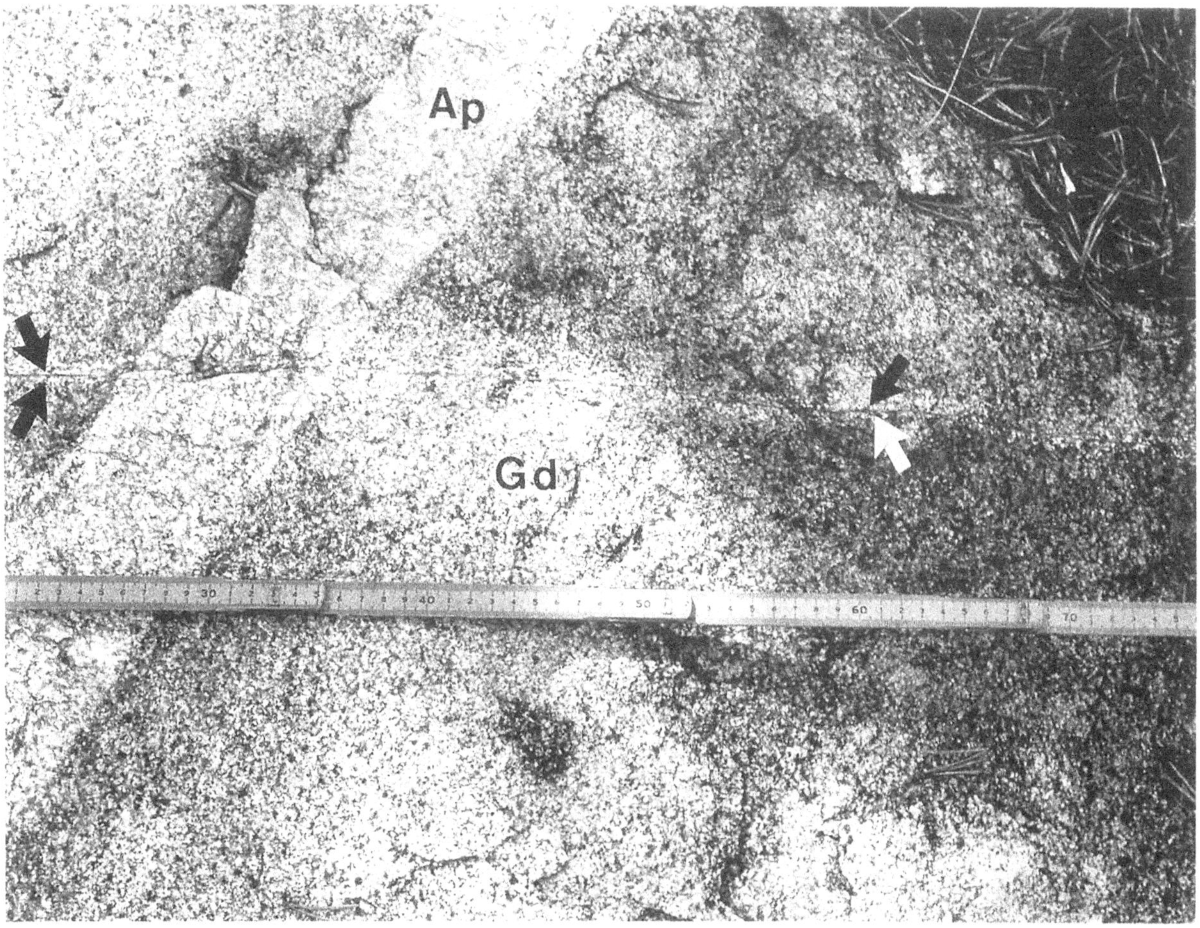


40 A

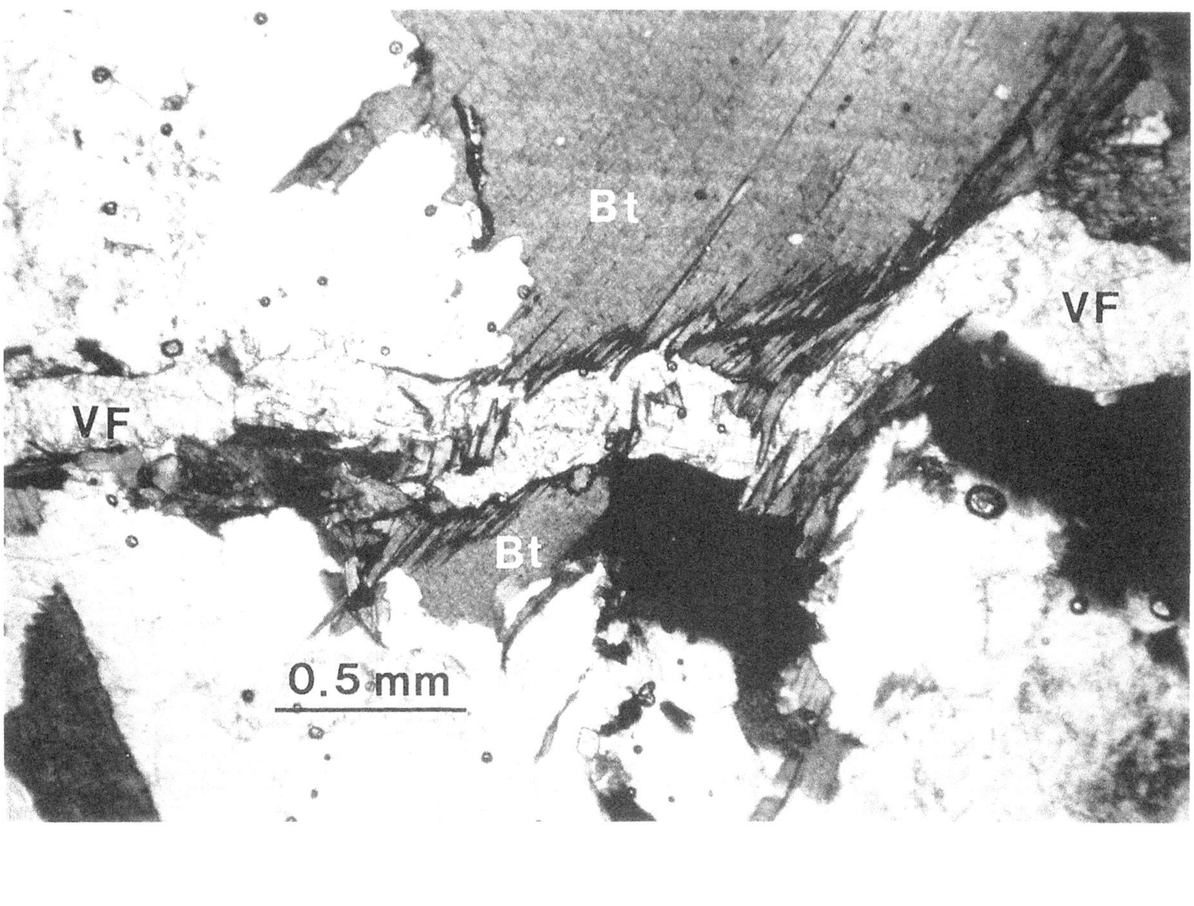


40 B

# 41. STRIKE-SLIP SHEAR ZONE IN GRANODIORITE

PAUL SEGALL

Late Cretaceous granodiorite of Lake Edison, Mount Abbot quadrangle, Sierra Nevada, California, U.S.A.

**41A.** Polished surface of subvertical shear zone and host granodiorite. Left-lateral strike separation across this zone, determined from offset aplite dike, is 0.39 m (see Figure 41.1). Width of shear zone is 2 mm to 4 mm. The average shear strain within the zone is thus 100 to 200. Note that despite the high shear strain within the zone, there is very little deformation of the adjacent wall rock. This is also shown by the absence of deformation of the offset aplite dike (Figure 41.1). Samples of these shear zones tend to break along one margin of the shear zone; the thin dark line at the margin of the zone (E) is a layer of epoxy.

**41B.** Photomicrograph of shear zone illustrated in 41A (Segall and Pollard, 1983). Photograph covers entire width of shear zone which is 2 mm in this section. Dark band (E) is epoxy layer. Elongate aggregates of quartz, as well as bands of epidote and chlorite define a foliation striking 10 to 20° from the shear zone boundaries (also visible in 41A). Hydrothermally precipitated quartz is deformed and dynamically recrystallized (Q). Rounded feldspar porphyroclasts (P) are thought to have been mechanically derived from the wall rock. This texture contrasts with the undeformed textures of minerals filling nearby fractures with no strike separation (see Plate 40). The fractures illustrated in Plate 40 are subparallel to this shear zone and less than 20 m away. These field and petrographic observations indicate that the shear zone nucleated on a preexisting dilational fracture (Segall and Pollard, 1983). NX

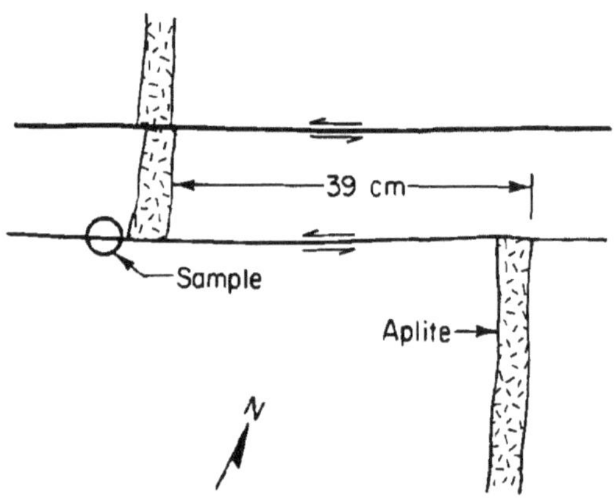


Figure 41.1. Sketch of offset aplite dike showing position of sample illustrated in Plate 41A.

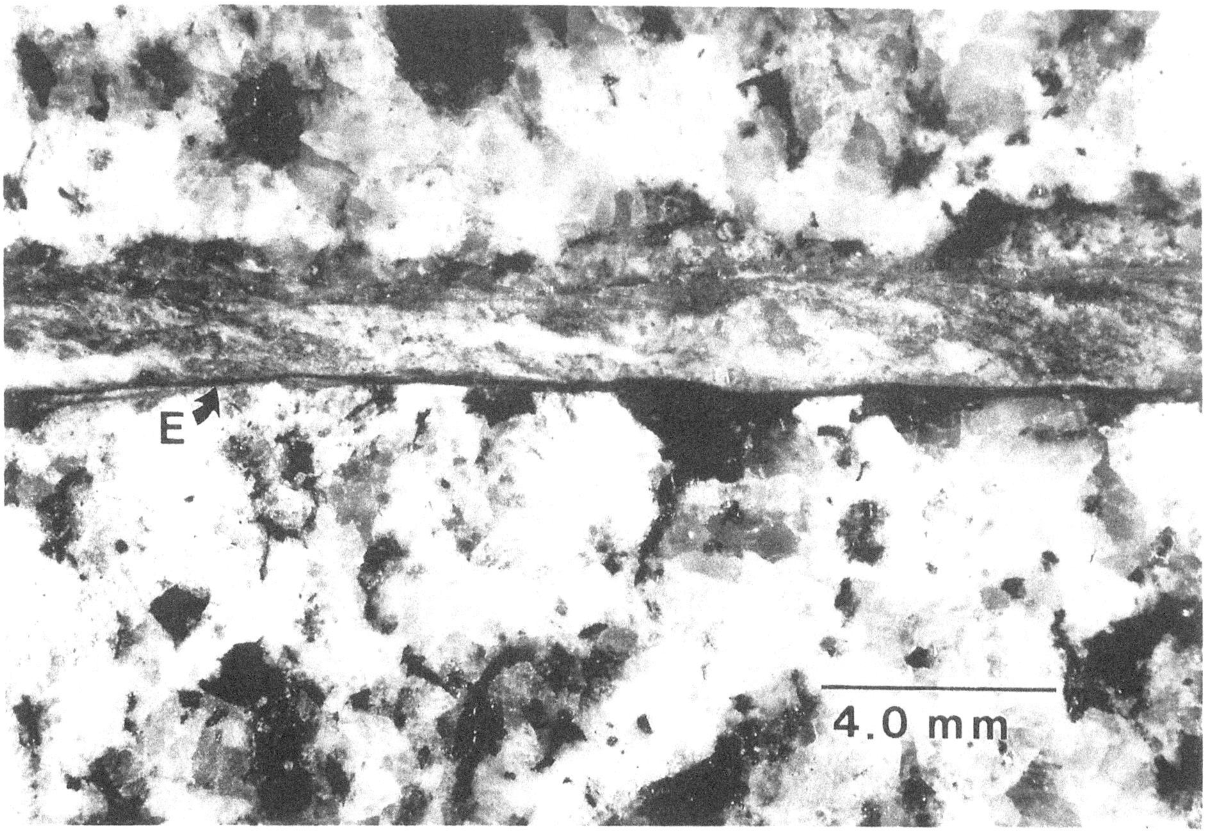


41 A

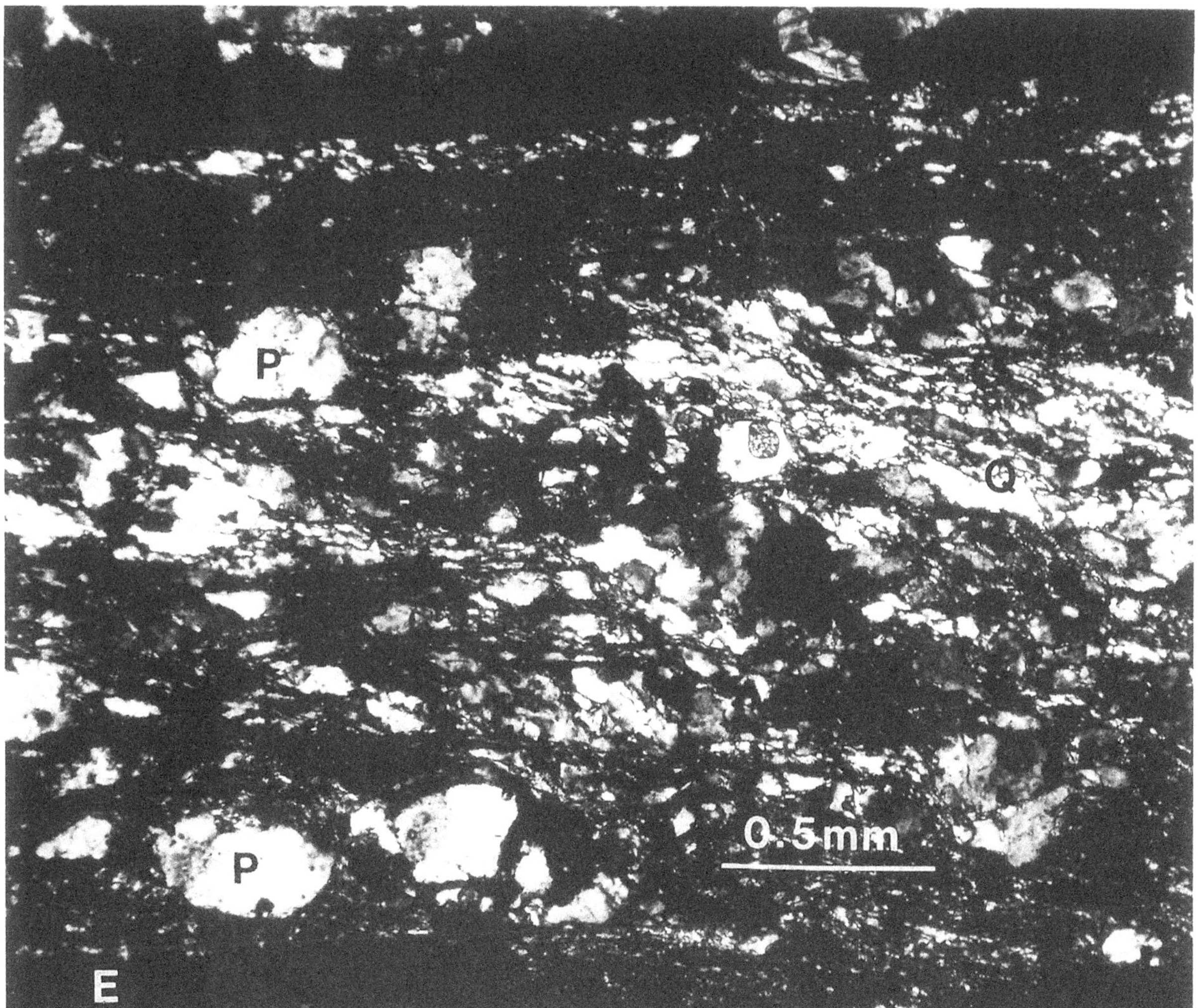


41 B

**41C.** Subvertical shear zone in glacially polished granodiorite (Segall and Simpson, 1986). Slickensides (not visible in photograph) plunge 25° to the east (left). There is 1.3 m of left-lateral strike separation across this shear zone approximately 10 m west of the site at which the photograph was taken. The width of the shear zone ranges from 4 mm to 14 mm. These data indicate that the average shear strain within the zone is approximately 90 to 325. A prominent foliation within the zone, defined by elongate quartz aggregates and chlorite-epidote-rich bands, strikes 10 to 20° from the overall strike of the zone. Knife is 9 cm long.

**41D.** Photomicrograph of central portion of shear zone illustrated in 41C. Shear zone exhibits mylonitic texture, with feldspar porphyroclasts (P) enclosed in a fine-grained, quartz-chlorite-epidote-rich matrix. Highly deformed quartz aggregates (Q) contain dynamically recrystallized grains on the order of 0.05 mm in average dimension. Elongate quartz aggregates (Q) and bands of epidote and chlorite define the macroscopic foliation (see 41C). In this photomicrograph, the foliation is oriented 20° (clockwise) with respect to the trace of the shear-zone boundaries (SZB). The long dimensions of the recrystallized quartz grains within the aggregates are oriented 45° (clockwise) with respect to the shear-zone boundaries.

41 C

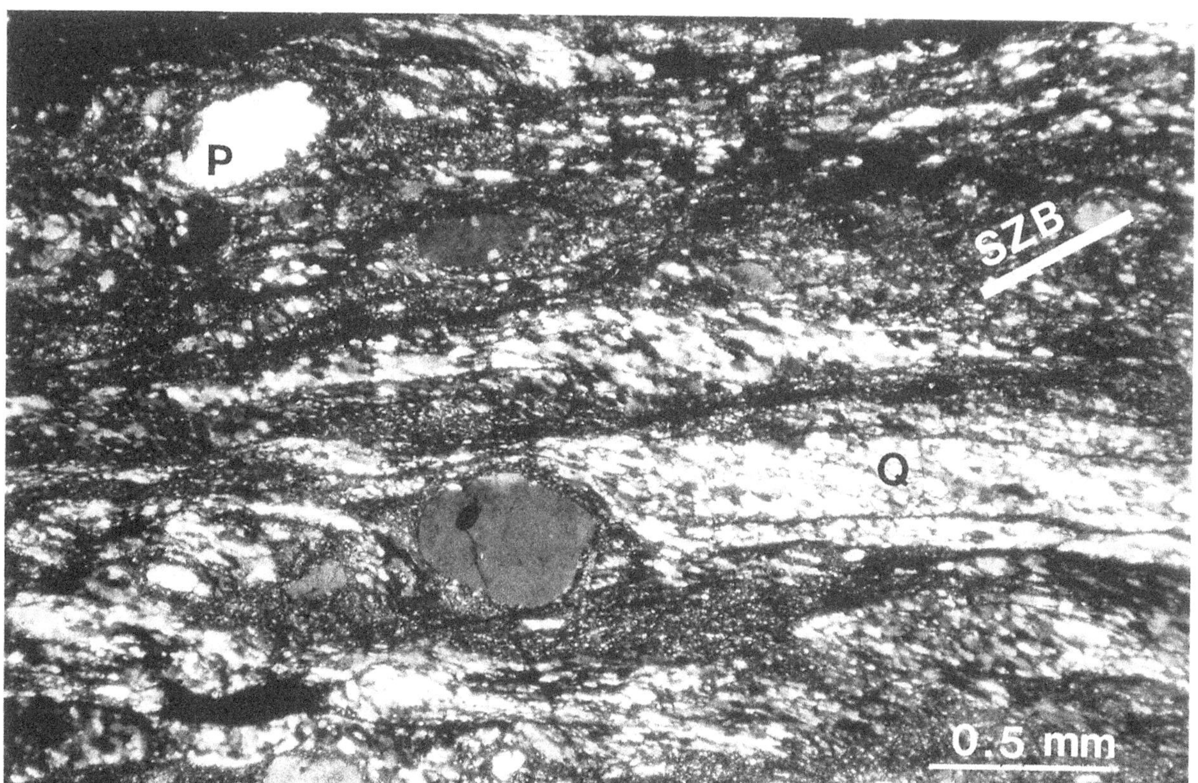


41 D

**41E.** Photomosaic and accompanying map illustrate several aplite veins that are offset and deformed by an east-west-striking shear zone. The rectangular box on the map indicates the area shown in the photomosaic. The shear zone dips 70° to the south, and slickensides on the walls of the shear zone (not shown) plunge 4° to the east. The net left-lateral displacement across the shear zone is nearly 2 m; note that a, b, and c indicate matching aplite veins on either side of the shear zone. Aplite a is strongly deformed near the shear zone, particularly on the south side of the zone. Also on the south side of the zone, aplite b is less deformed than a, and aplite c is nearly undeformed. Aplite d, on the north side of the shear zone, also shows no evidence of deformation.

A quartz vein within the central portion of the shear zone is visible near the eastern edge of the photograph (QVF: quartz vein filling). The quartz within the vein is highly deformed, demonstrating that the vein formation preceded the deformation. This evidence, as well as evidence presented in Plate 40, indicates that the shear zone nucleated on the preexisting, quartz-filled fracture or vein.

A prominent foliation is locally developed in the granodiorite adjacent to the shear zone. Away from the shear zone, the foliation is weakly developed and oriented approximately 45° from the strike of the zone. Approaching the zone, the foliation tends to be oriented more nearly parallel to the shear zone. The foliation also tends to become more strongly developed near the zone; however, the degree of foliation development varies spatially, both along the strike of the shear zone and across the zone. This is particularly vivid near aplite d, where there is strong foliation developed on the south side of the zone, and little or no foliation on the north side. The degree of foliation development correlates well with the magnitude of the strain in the rock as evidenced by the deformation of the aplite veins. The reason for the marked spatial variation in strain and intensity of the foliation is not presently known. The heavy lines in the photograph are fractures which crosscut the foliation, and must therefore be younger than the shear zone.

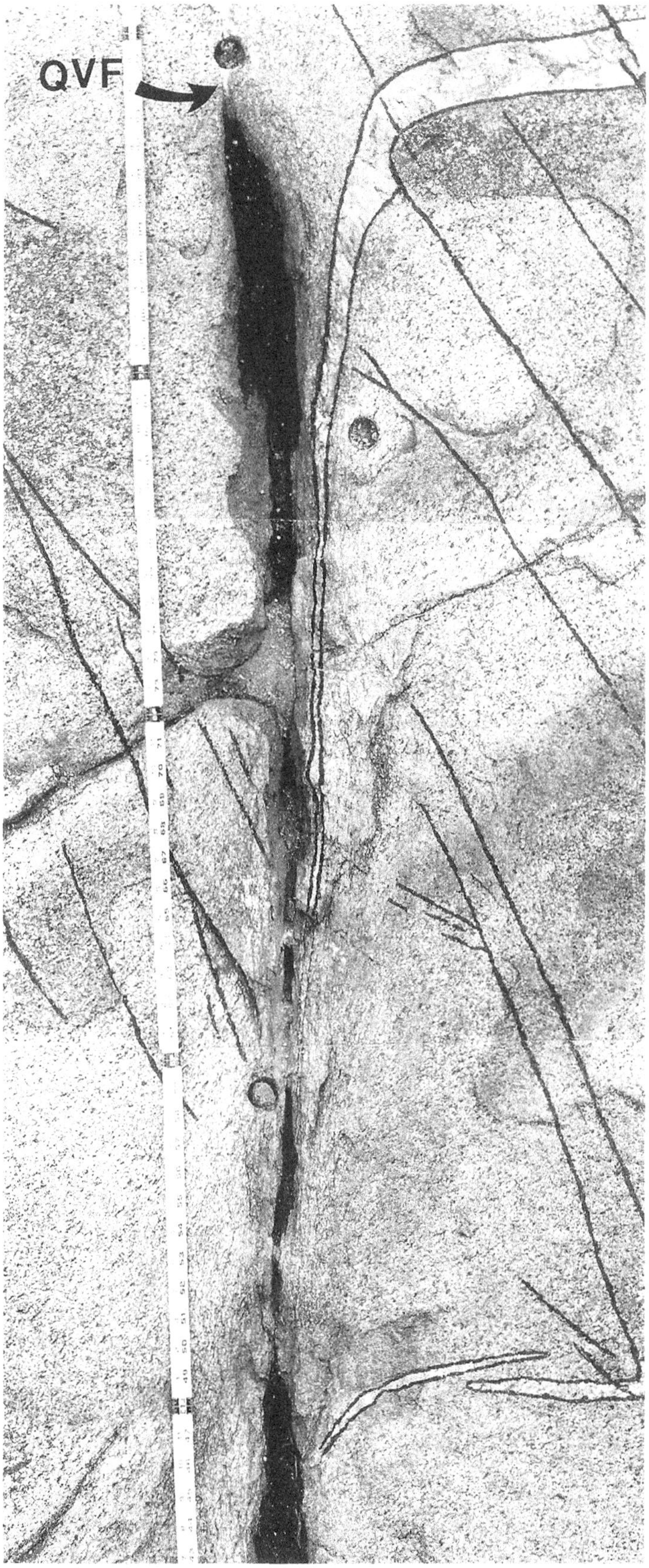


41 E

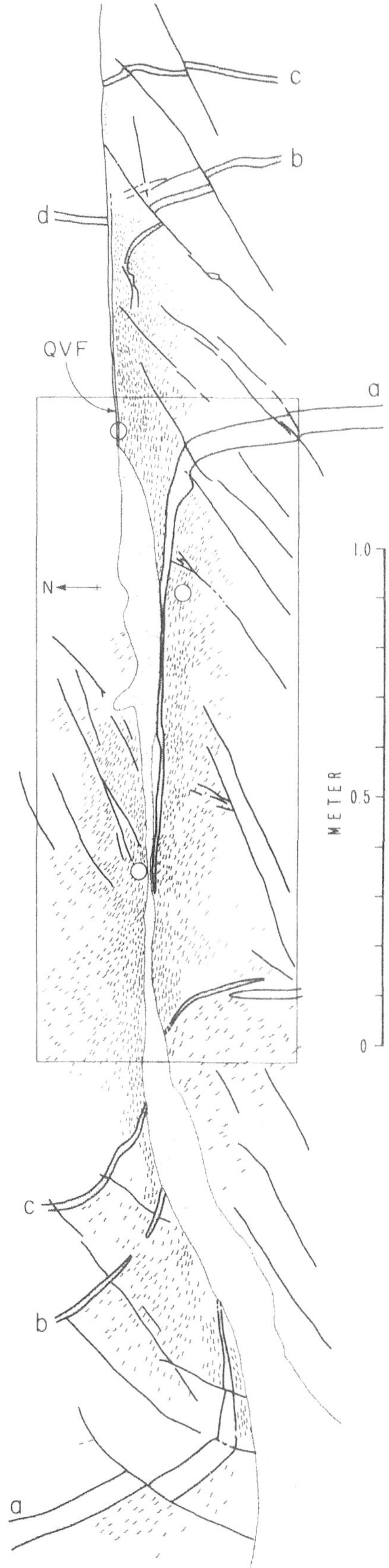


EXPLANATION

Foliated Granodiorite | Aplite | Core sample

Unfoliated Granodiorite | Covered | Fracture

# 42. CATACLASITES AND MYLONITES

S. H. White

## Geologic map

The structural geology of the Outer Hebrides fault zone in North Uist, Scotland is shown in Figure 42.1. The fault zone was a late Caledonian thrust structure that has subsequently been reactivated as a normal fault. It is marked by a wide belt of crushed Lewisian gneiss that has been described as crush melange by Sibson (1977). The melange consists of blocks of Lewisian gneiss, ranging in size from centimeters to meters, separated by veins of fine-grained cataclasite. Individual fault planes within the crush melange are marked by narrow anastomosing zones of cataclasites and ultracataclasites in the west and by phyllosilicate-rich mylonites (phyllonites) in the east. Although lithological variations on either side of the zone indicate a thrust fault (Sibson, 1977), the rotated clasts in the cataclasites, together with the orientation of intrafolial folds and shear bands in the mylonites, indicate normal-sense movement; that is, across these zones the thrust was reactivated as a normal fault in post-Caledonian times. The mylonites and cataclasites described in 42A–42F are all associated with the late normal movement. The determination of the movement sense within fault zones from macroscopic and microstructural features has been discussed by White (1982) and Simpson and Schmid (1983) and summarized in Hanmer and Passchier (1991). The sequence of micrographs is one of increasing deformation from 42A to 42F. Detailed microstructural features are shown in 42G–42J.

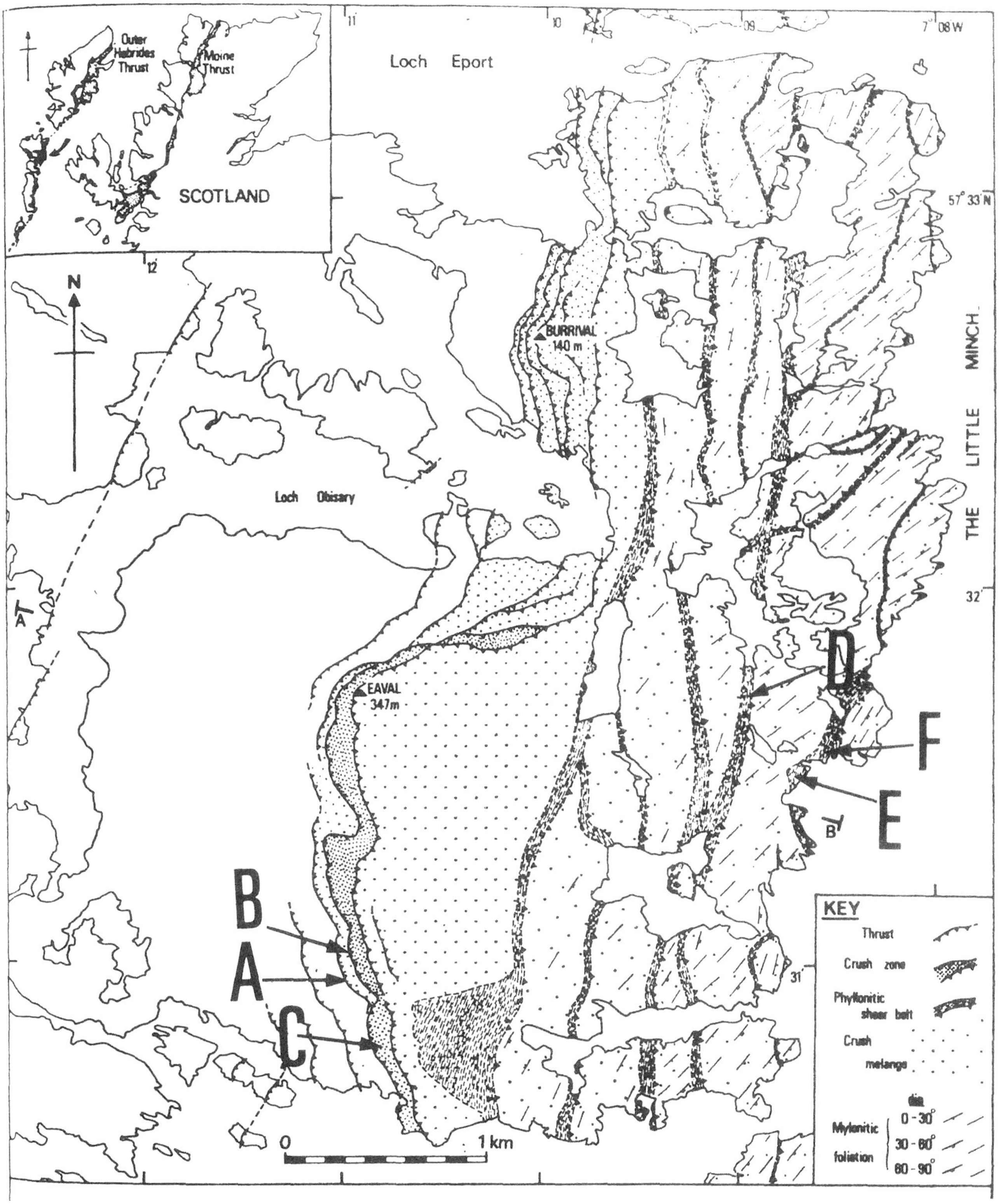


**Figure 42.1.** Map of the Outer Hebrides fault in the Eaval area of the Isle of North Uist. The locations of the specimens from which the micrographs 42A to 42F were taken are marked by the same letters on the map. [Map reproduced from Sibson (1977) with permission from the Geological Society of London.]

## Optical and backscattered electron micrographs

Optical and backscattered scanning electron micrographs showing the microstructures of the cataclasites and mylonites.

**42A.** An optical micrograph of a fine-grained vein of cataclasite in the crush melange. The cataclasite consists of fragments of quartz (q) and of altered alkali-feldspar (f). NX

**42B.** An optical micrograph (PPL) of a cataclasite at the base of the crush zone at Eaval. The cataclasite consists of fragments of quartz in a matrix of mica and chlorite. The matrix is cut by bands of dark, semi-opaque, ultrafine-grained material which cannot be resolved with an optical microscope.

**42C.** A backscattered scanning electron micrograph from an ultracataclasite zone in the crush zone at Eaval. Optical micrographs from the ultracataclasite had the same appearance as the dark bands in 42B. Nothing could be resolved. The backscattered micrograph shows finely comminuted fragments of retrogressed granitic gneiss.

42 A

42 B

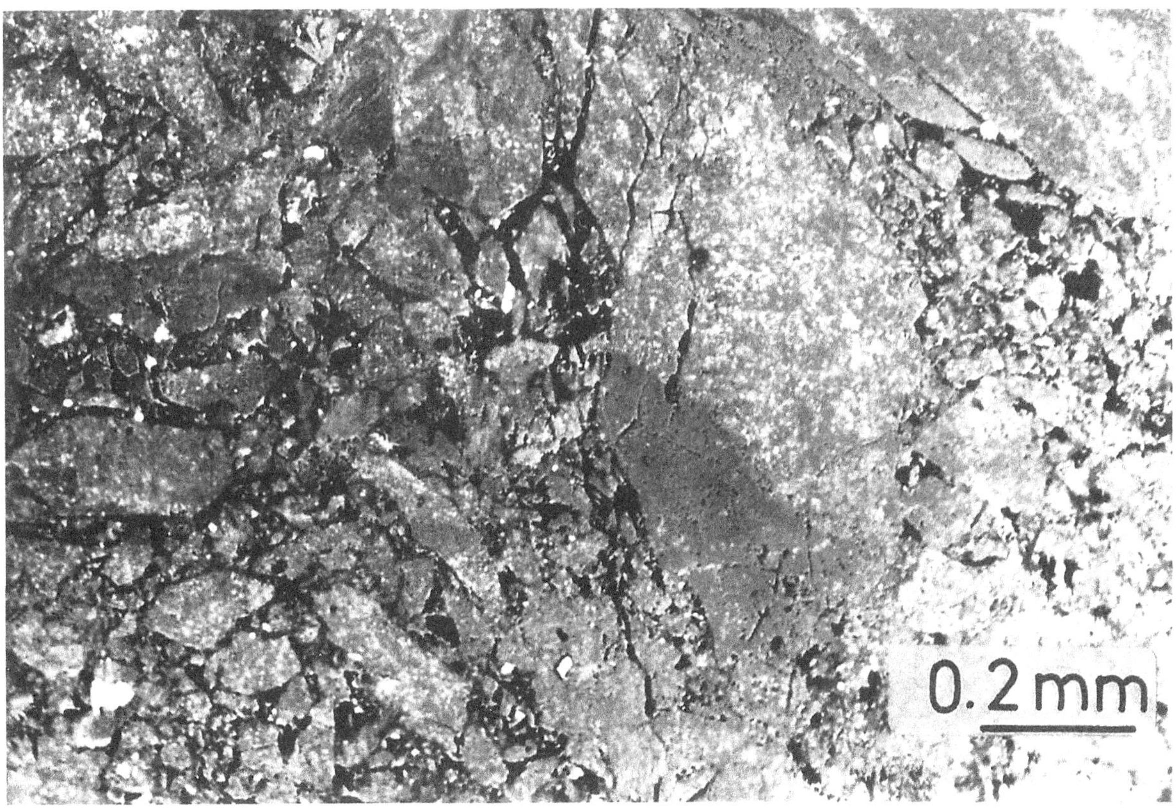


42 C

## Optical micrographs

**42D.** Optical micrograph (PPL) of a weakly foliated mylonite. The microstructure includes domains of quartz clasts which lie in a matrix of sericite and chlorite and which are enclosed by bands of semi-opaque ultrafine-grained material.

**42E.** Optical micrograph (PPL) of a strongly foliated mylonite showing alternating bands of streaked-out quartz and mica clasts in a banded matrix of ultrafine-grained material.

**42F.** Optical micrograph (PPL) from a mylonite zone on the west coast of North Uist. Shear bands (Reidel shears) are well developed. They form a second foliation ($S_s$) which rotates the mylonite foliation ($S_m$) and result from late, normal movements within the mylonite zone. $S_s$ and $S_m$ form a C-C′ fabric.

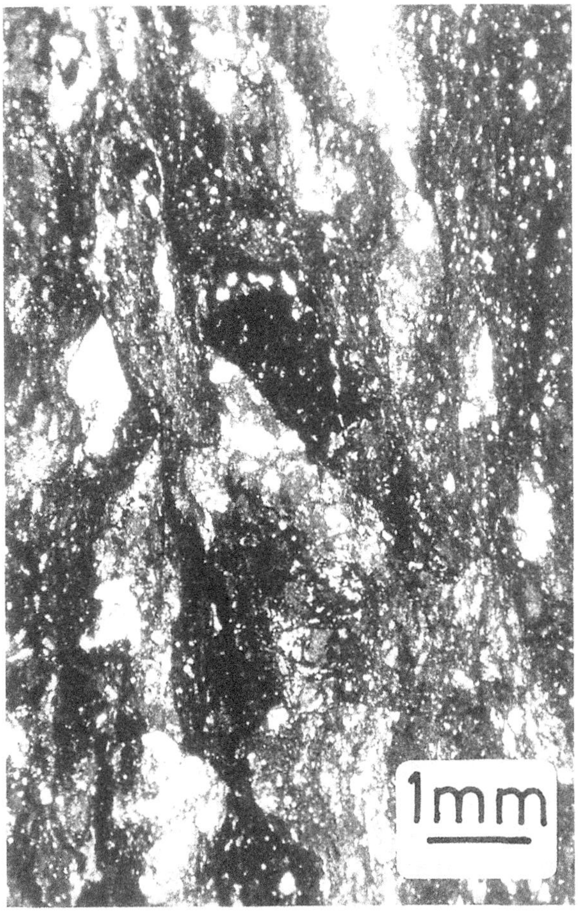


42 D

42 E

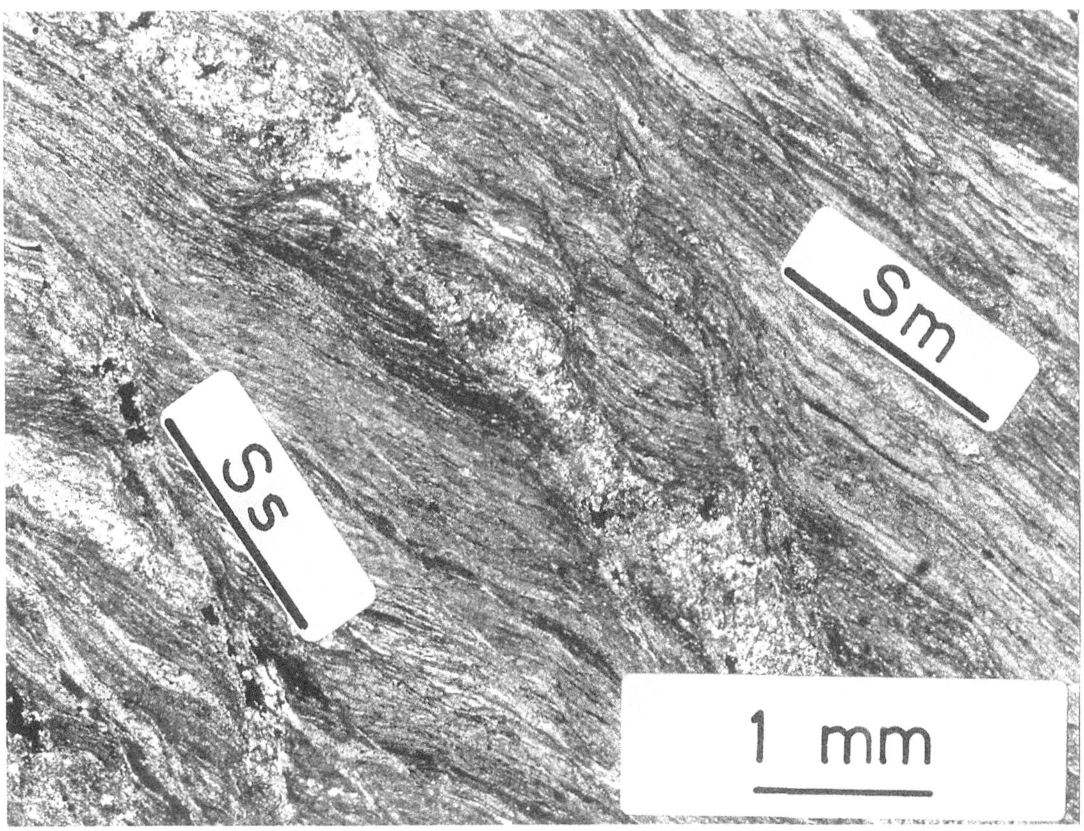


42 F

### Backscattered electron micrographs

Backscattered electron micrographs showing details of the sericite and chlorite matrix that surrounds the quartz fragments in the cataclasite and mylonite and of the ultrafine-grained dark matrix in both.

**42G.** Sericite-chlorite matrix in a cataclasite. The matrix consists largely of quartz and albite (dark) with small sericite and chlorite grains (intermediate tone) and contains patches of epidote (lightest tone). Although most grains have a random orientation, preferentially oriented bands are present.

**42H.** Ultrafine-grained dark matrix in cataclasites and ultracataclasites. The microstructure is dominated by clusters of small epidote grains (lightest tone) with small chlorite and sericite grains (intermediate tone) in quartz matrix (dark areas).

**42I.** Sericite-chlorite matrix in a weakly foliated mylonite. The microstructure consists of alternating bands of oriented sericite and chlorite (one is marked with arrows) and of quartz with randomly oriented sericite and chlorite. Small clusters of epidote and grains of calcite (c) occur in both bands.

**42J.** Sericite-chlorite matrix in a strongly foliated mylonite. The matrix has a well-developed foliation marked by oriented sericite and chlorite grains and elongated quartz (q) grains. Epidote clasts (lightest tone) are distributed throughout the matrix.

The random orientation of the sericite and chlorite grains in the cataclasites and in some bands of the weakly foliated mylonites indicates that post-tectonic grain growth followed cataclasis. However, this grain growth was not present in the strongly foliated mylonites, and this difference is interpreted as indicating that the mylonite zones represent zones of concentrated deformation after an initial widespread cataclastic deformation had produced retrogression of the crushed granitic gneiss. The cataclastic deformation appears to be associated with the reactivation of the zone. None of the fault rocks described above are the direct products of the thrust movements.

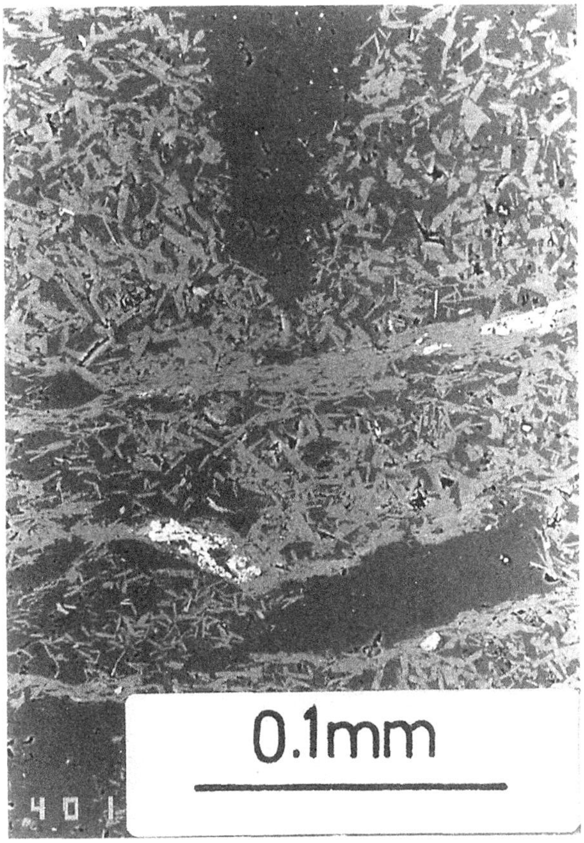


42 G

42 H

42 I

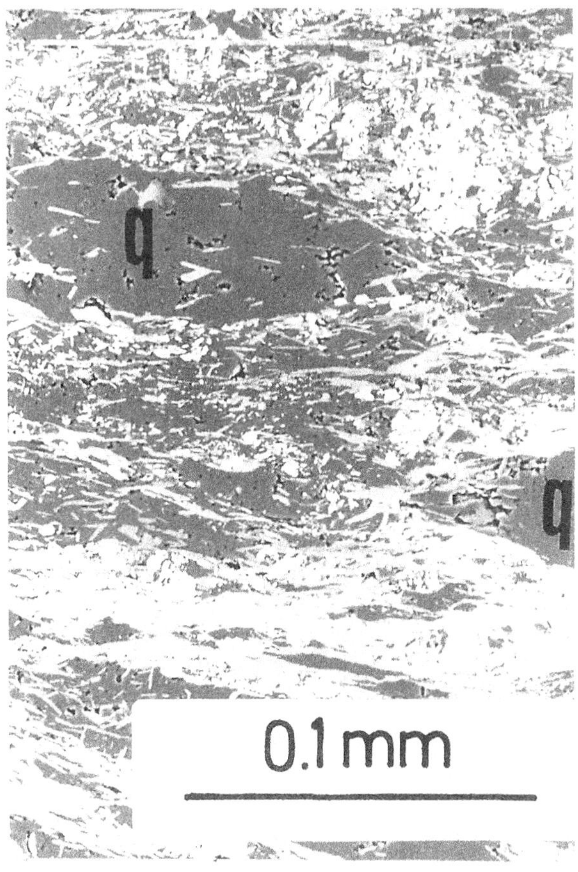


42 J

# 43. PROTOMYLONITE, AUGEN MYLONITE, PROTOCATACLASITE, AND CATACLASITE IN GRANITIC FOOTWALL OF A DETACHMENT FAULT

G. H. DAVIS, L. J. COX, R. ORNELAS, K. N. CONSTENIUS, J. A. JOHNSON, F. KONG, J. M. BEDFORD, M. PECHA, AND E. P. RODRIGUEZ

The Catalina detachment fault is a brittle-ductile shear zone ~3 km thick, located just north and east of Tucson, Arizona, U.S.A. (Figure 43.1), defining the southwestern margin of the Catalina-Rincon metamorphic core complex. At the Martinez Ranch study site, the Catalina detachment fault dips gently (~17–25°) and separates upper-plate nonmylonitic Precambrian-Paleozoic-Miocene rocks from lower-plate cataclastic and mylonitic rocks derived mainly from Precambrian and Cenozoic quartz monzonite protoliths (Figure 43.2).

The quartz monzonite lower plate, from which samples 43A–43D were collected, resided at a depth of about 12 km when ductile shearing was initiated at about 25 Ma. S-C relations in mylonites reveal normal-slip sense-of-shear. By about 18 Ma, the Catalina detachment fault had accommodated at least 15 km of total top-to-the-southwest displacement. Mylonites were created during ductile shearing, and as deformation proceeded they were raised incrementally to higher and higher structural levels, whereupon brittle shearing took over.

In the Martinez Ranch area, mylonites are transformed upward into protocataclasite and cataclasite. The most extreme cataclasis is found along the detachment fault itself. Where the cataclastic overprint is weak to moderate, original mylonitic fabrics are recognized both in outcrop and in thin section. However, where the cataclastic overprint is intense, the fault rock is extremely fine grained and the relict mylonitic fabrics may be unrecognizable in thin section.

**43A.** Protomylonite derived from quartz monzonite protolith (section cut perpendicular to foliation, parallel to lineation). Porphyroclasts of relatively rigid orthoclase (f) form augen in a finer-grained, foliated matrix composed of quartz (q) and biotite (b) ribbons, plagioclase, and muscovite. The orthoclase porphyroclasts range in diameter from about 0.2 mm to 10 mm. S planes are defined by the orientations of feldspar, mica fish (mf), and quartz and biotite ribbons, all of which are deflected into higher-strain C planes parallel to horizontal compositional layering. The feldspar porphyroclasts have wedge-shaped σ-type tails (t) (see Passchier and Simpson, 1986). The "step-up" from the lower left tail to the upper right tail reveals right-lateral shear within the field of the photomicrograph and corresponds to normal slip. PPL

**43B.** Augen mylonite derived from protomylonite (section cut perpendicular to foliation and parallel to lineation). Orthoclase porphyroclasts (f) are set in a fine-grained matrix composed of quartz (q) and biotite (b) ribbons, muscovite, sericite, and opaque minerals. The feldspar porphyroclasts range in diameter from about 0.04 mm to 5 mm. Although this sample is from the same shear zone as 43A, it has undergone more intense deformation. Local fold perturbations such as those illustrated at the top of 43A occur as asymmetrical microfolds that are best expressed by quartz ribbons. Some feldspar porphyroclasts display two sets of tails which are representative of a complex σ-δ porphyroclast system (see Passchier and Simpson, 1986). The thin, elongated "first generation" tails (δt) resemble normal δ symmetry in that embayments of matrix material occur adjacent to the porphyroclast. The shorter "second generation" tails (σt) resemble normal σ-type tails that, in this case, as in 43A, step up from left to right. The tails reveal a sense-of-shear that is consistent with right-lateral shear across the field of view and corresponds to normal slip. PPL

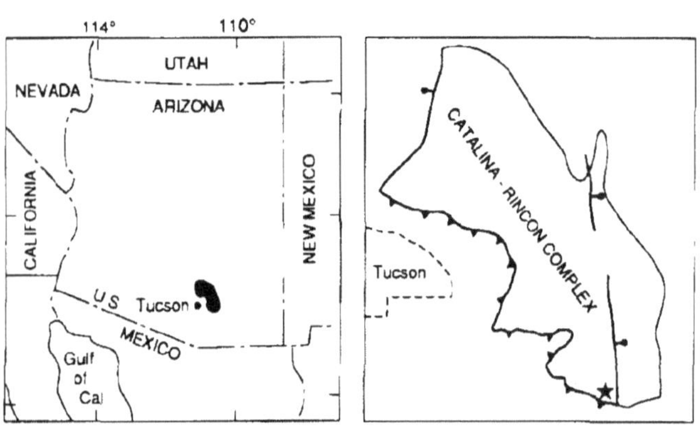


**Figure 43.1.** Location map. Star shows sample site

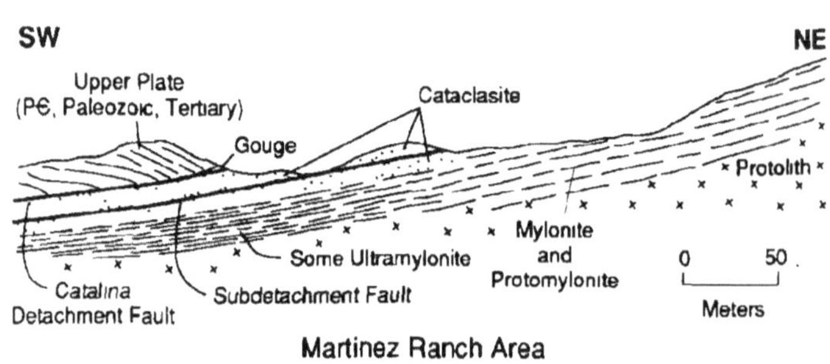


**Figure 43.2.** Schematic structure section through Catalina detachment fault

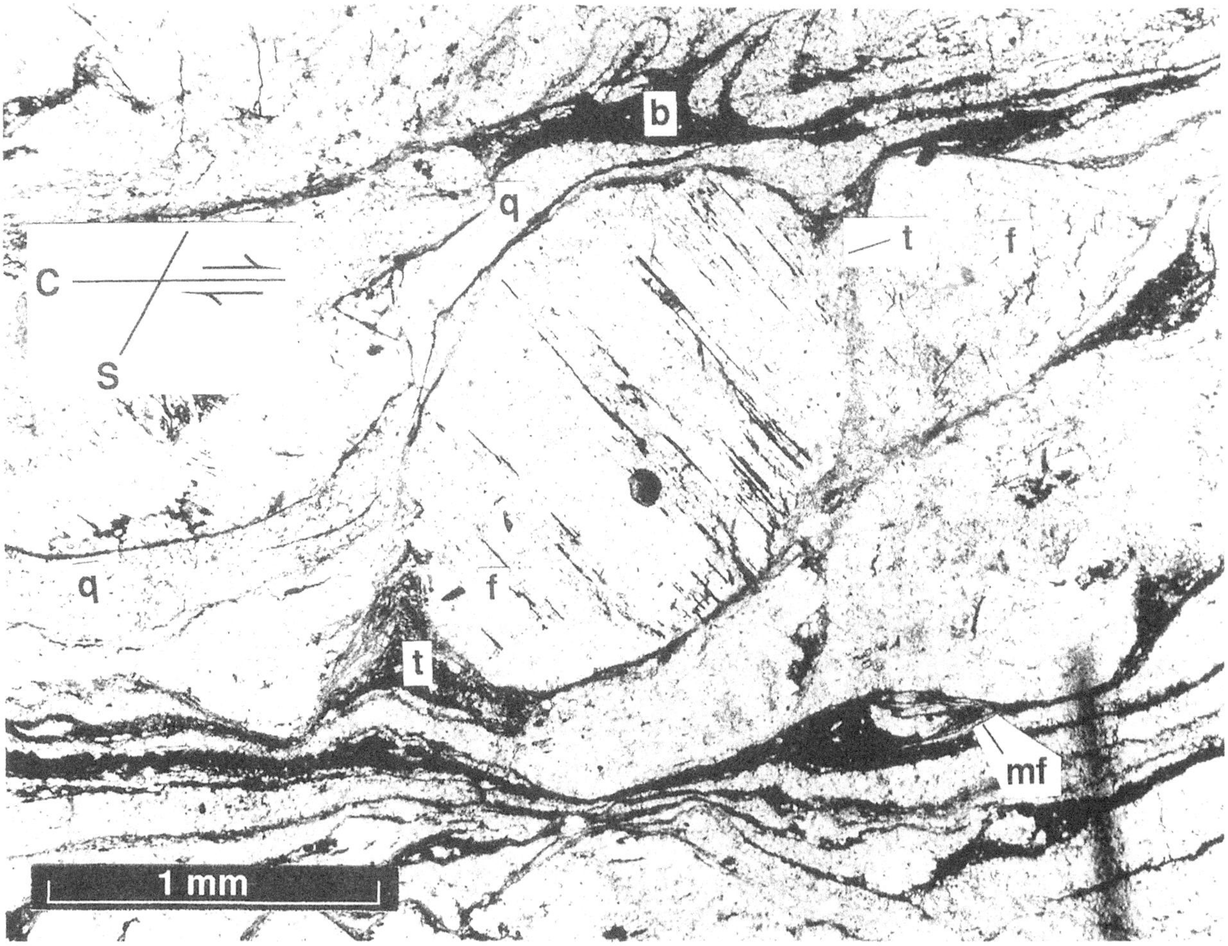


43 A

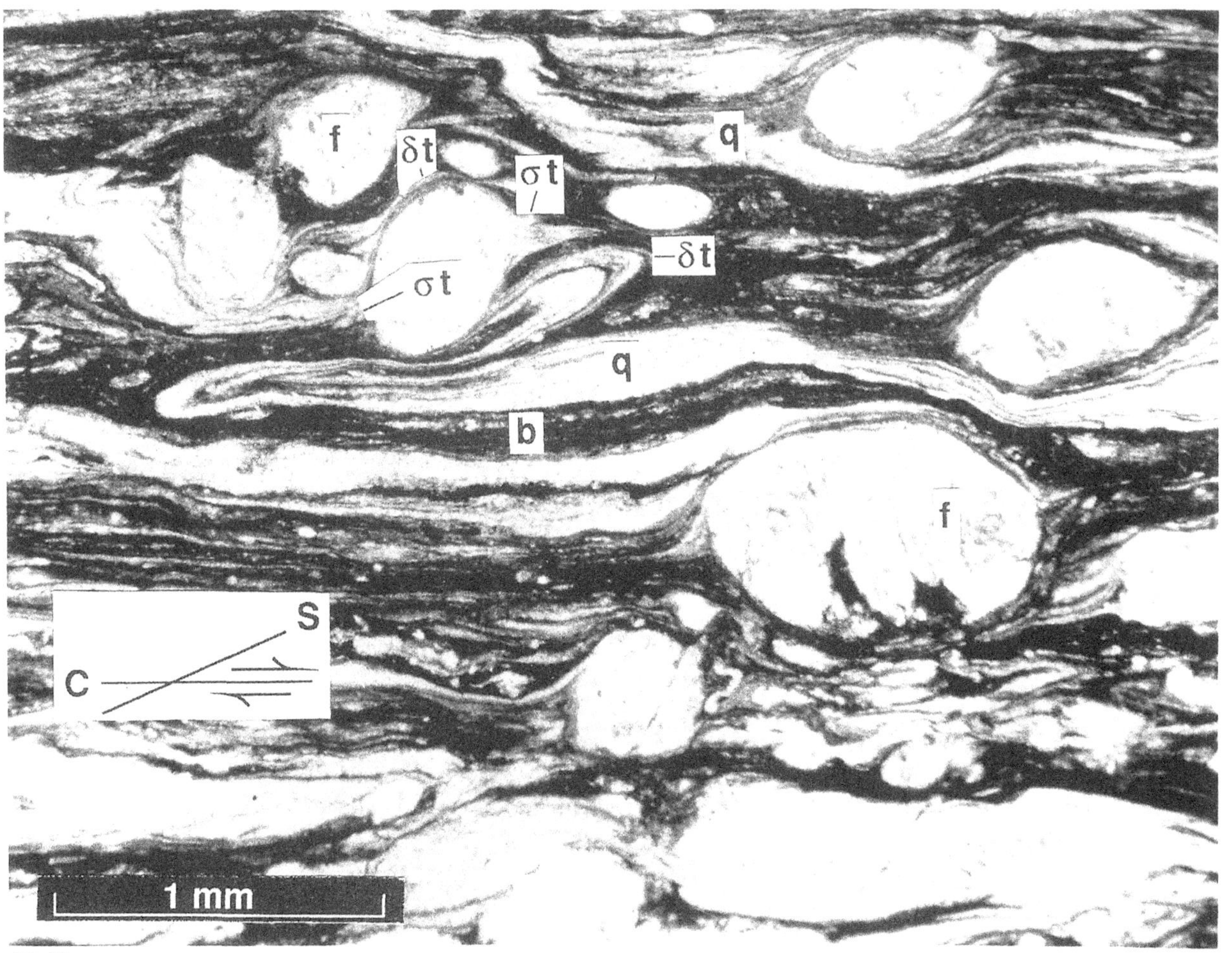


43 B

**43C.** Protocataclasite derived from protomylonite and augen mylonite. Broken relict feldspar porphyroclasts ($f_r$), fragments of ribbon quartz (q), and fragments of mylonite lie within bands of finer-grained quartz, feldspar, muscovite, sericite, and hematite. Clast sizes range from about 0.04 mm to >2 mm in diameter. Cataclasis of mylonite has obliterated earlier quartz and biotite ribbons and severely degraded the porphyroclasts. Some surviving feldspar porphyroclasts show strained extinction ($f_e$) under crossed nicols; offset along cleavage planes is common. Both the sharp-edged feldspar clasts and the bands of finer crushed material define a new shear band foliation ($C_n$) that is oblique to the relict mylonitic shear plane ($C_o$). PPL

**43D.** Cataclasite derived from protocataclasite. Ghosts of feldspar porphyroclasts ($f_p$) defined by minute, relict orthoclase fragments float in a matrix-dominated rock of quartz, sericite, and feldspar. The rock is at most 90% matrix. Clasts constituting 10% or more of the rock range from about 0.04 mm to 0.6 mm in diameter. Porphyroclast ghosts are larger (>1 mm) than actual clasts but are only half the size of the largest clasts in the protocataclasite 43C above. Intense cataclasis has resulted in a nearly random rock fabric due to almost total destruction of the earlier formed $C_n$ and $C_o$ foliations. Euhedral pyrite (p), jarosite (j), and adularia (a) occur in narrow, randomly oriented later-formed veins (v). The vein-filling minerals are larger than most of the clastic fragments. PPL

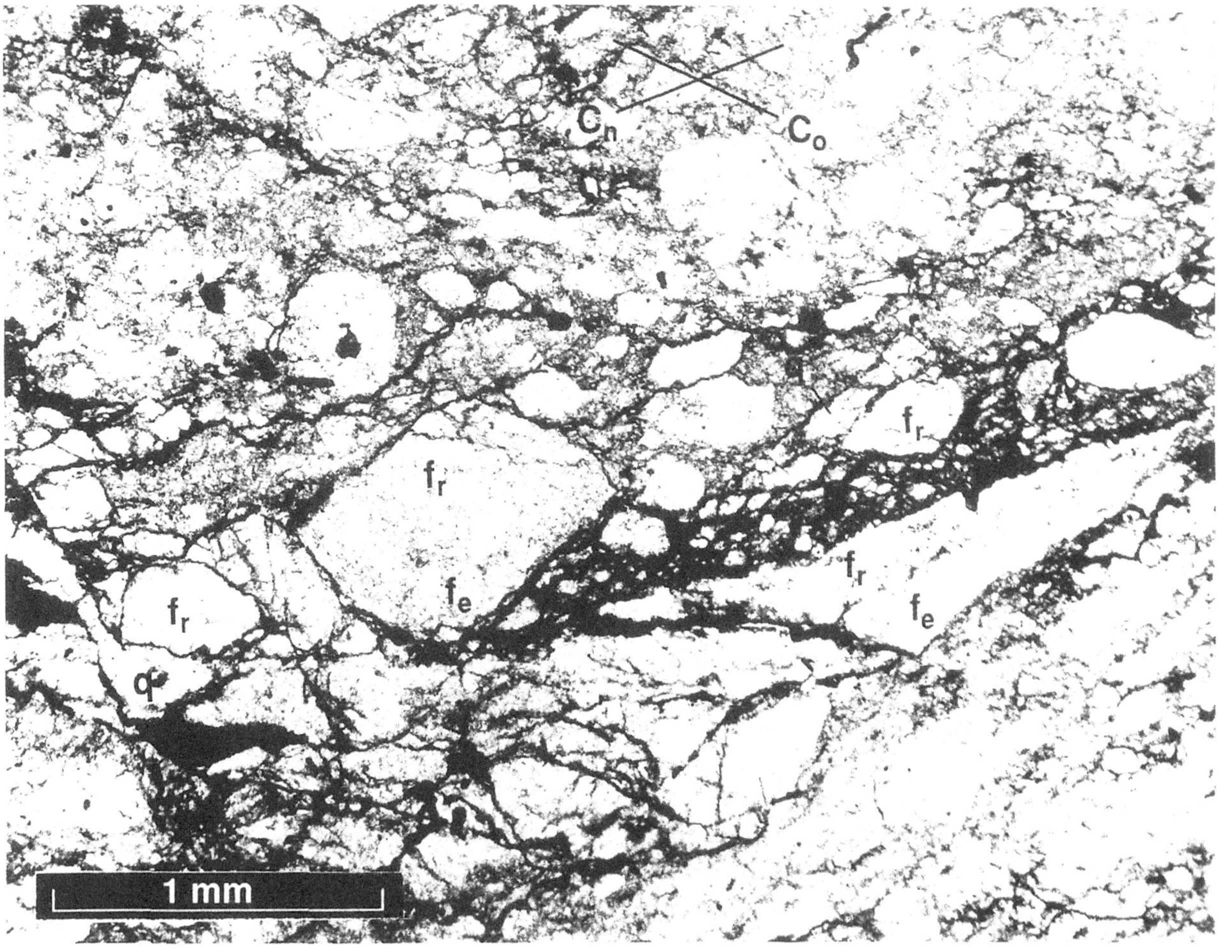


43 C

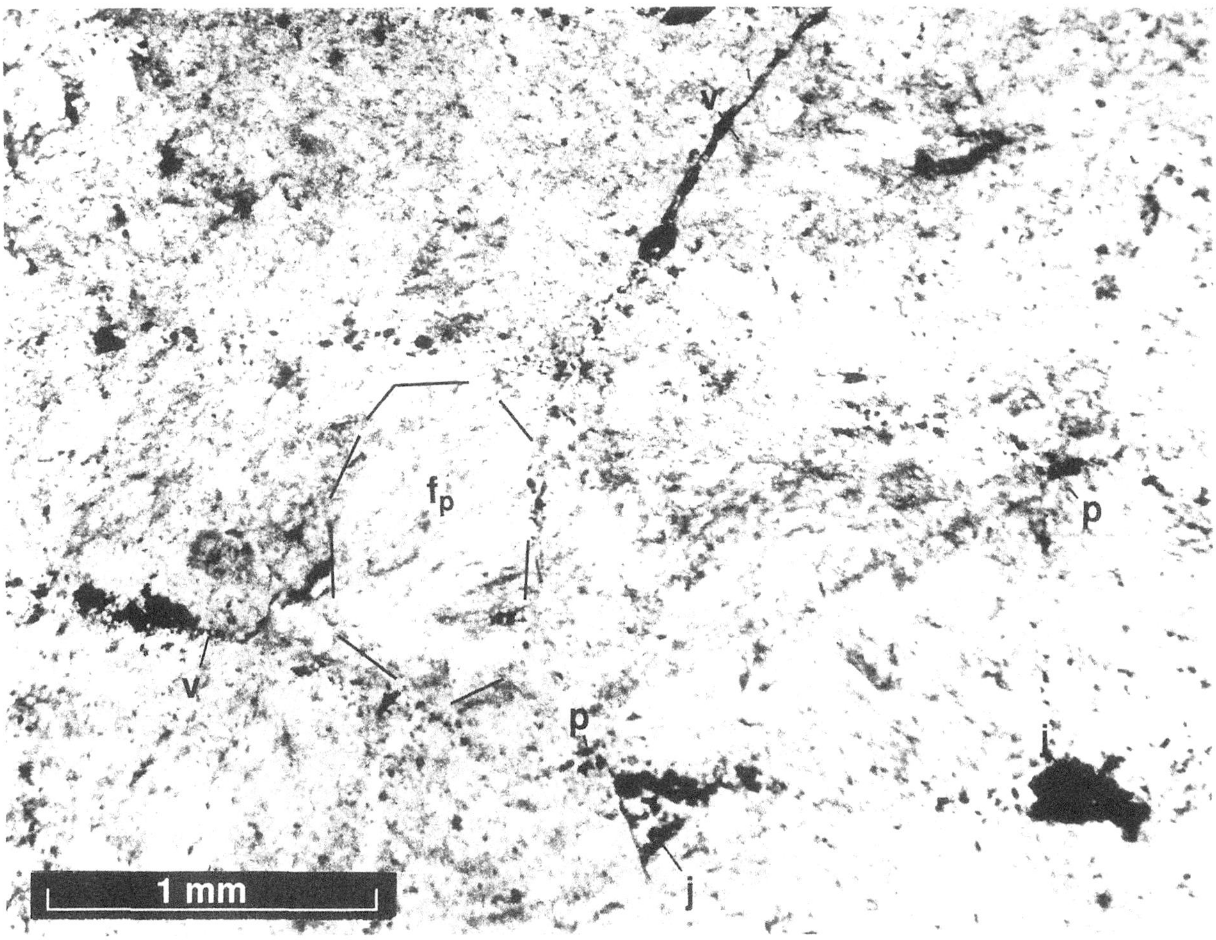


43 D

Mark R. Handy

The samples pictured in Plates 44A–44F come from the Pogallo ductile fault zone (PDFZ), a low-angle normal fault that exhumed lower continental crust during early Mesozoic time. The PDFZ presently forms part of the subvertical contact between the Ivrea and Strona-Ceneri units in the westernmost part of the southern Alps, northern Italy (Figure 44.1). Together, these units make up an eroded cross section of intermediate to lower continental crust. This crustal section underwent repeated uplift during Paleozoic and early Mesozoic time. However, its emplacement near the surface in a subvertical orientation resulted from late Cretaceous to Tertiary thrusting and folding during the Alpine orogeny (Zingg and others, 1990; Handy and Zingg, 1991).

Erosion has exposed the PDFZ in a profile nearly perpendicular to its mylonitic foliation and slightly oblique to its stretching lineation. This affords a rare glimpse of the anatomy of a low-angle normal fault. In map view, the PDFZ is about 30 km long and at least 2 to 3 km wide. Syntectonic metamorphism and deformational style vary both across and along strike of the PDFZ, suggesting that prior to subvertical emplacement the PDFZ was a gently to moderately dipping, intracrustal extensional shear zone. At the originally shallower southwest end of the PDFZ, cataclasites grade into greenschist-facies mylonites, which in turn splay into a zone of diffuse, predominantly crystal-plastic, amphibolite-facies deformation at the originally deeper northeast end of the PDFZ. The steep metamorphic gradient from northwest to southeast across the PDFZ reflects rapid, asymmetrical cooling of the shear zone as the hot, uplifting Ivrea footwall was juxtaposed against cooler parts of the Strona-Ceneri hanging wall. There, cataclasites and greenschist-facies mylonites overprint amphibolite-facies mylonites, as depicted on the plates in the following spreads.

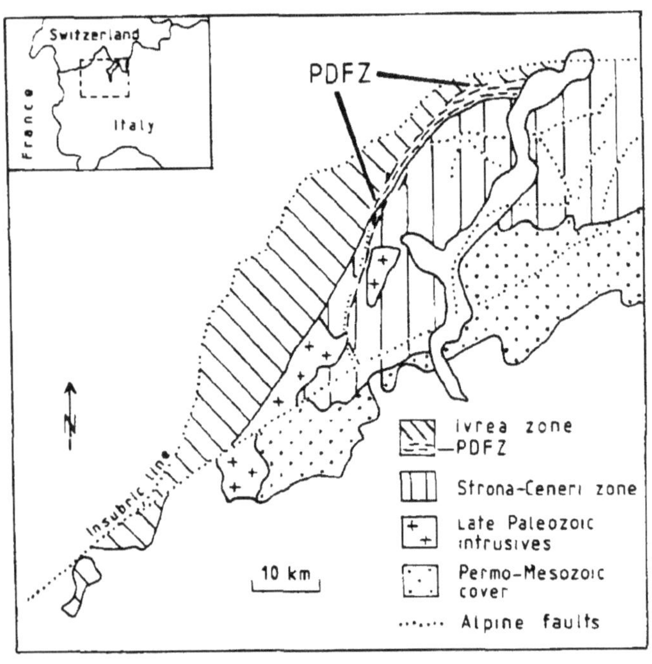


**Figure 44.1.** Map of the Ivrea crustal cross section showing location of the Pogallo ductile fault zone (PDFZ; see text).

### The transition from mylonitic creep to cataclastic flow

**44A.** Cracks at intersection of conjugate mylonitic shear zones. Anastomosing, greenschist-facies mylonitic shear zones truncate the amphibolite-facies mylonitic foliation of granitoid gneiss. The outcrop surface is perpendicular to foliation and subparallel to the stretching lineation. Sinistral, greenschist-facies shear zones are synthetic with respect to the overall sinistral sense-of-shear within the Pogallo ductile fault zone. Where conjugate shear zones meet, the dextral shear zone becomes narrower and bends into concordance with its sinistral partner. Arcuate extensional cracks (labeled e) emanate from the top of the antithetic shear zone. Conjugate high-angle extensional cracks (indicated with pen cap) form mutually cross-cutting relationships with the mylonitic shear zones, suggesting that brittle deformation was contemporaneous with the final stages of greenschist-facies mylonitization. Pen cap for scale is 6 cm long. Photo from figure 1c in Handy (1994b).

**44B.** Gradation from mylonite to cataclasite. Retrograde, greenschist-facies mylonite (top) grades to a brecciated cataclasite (bottom) within the dextral shear zone in the lower right-hand part of 44A. Dark bands within, and especially between, the mylonitic and brecciated domains are weakly foliated. They contain small clasts of mylonite and very fine-grained trains of opaque minerals, epidote, and occasionally quartz. The dominant deformation mechanism in such bands could not be identified, but may have involved some kind of (viscous or frictional?) grain-boundary sliding. Note that both mylonitic and cataclastic domains accommodated ductile (i.e., distributed) flow on the scale of this micrograph. NX, section parallel to the XZ fabric plane. Micrograph from figure 10b in Handy (1987).

44 A

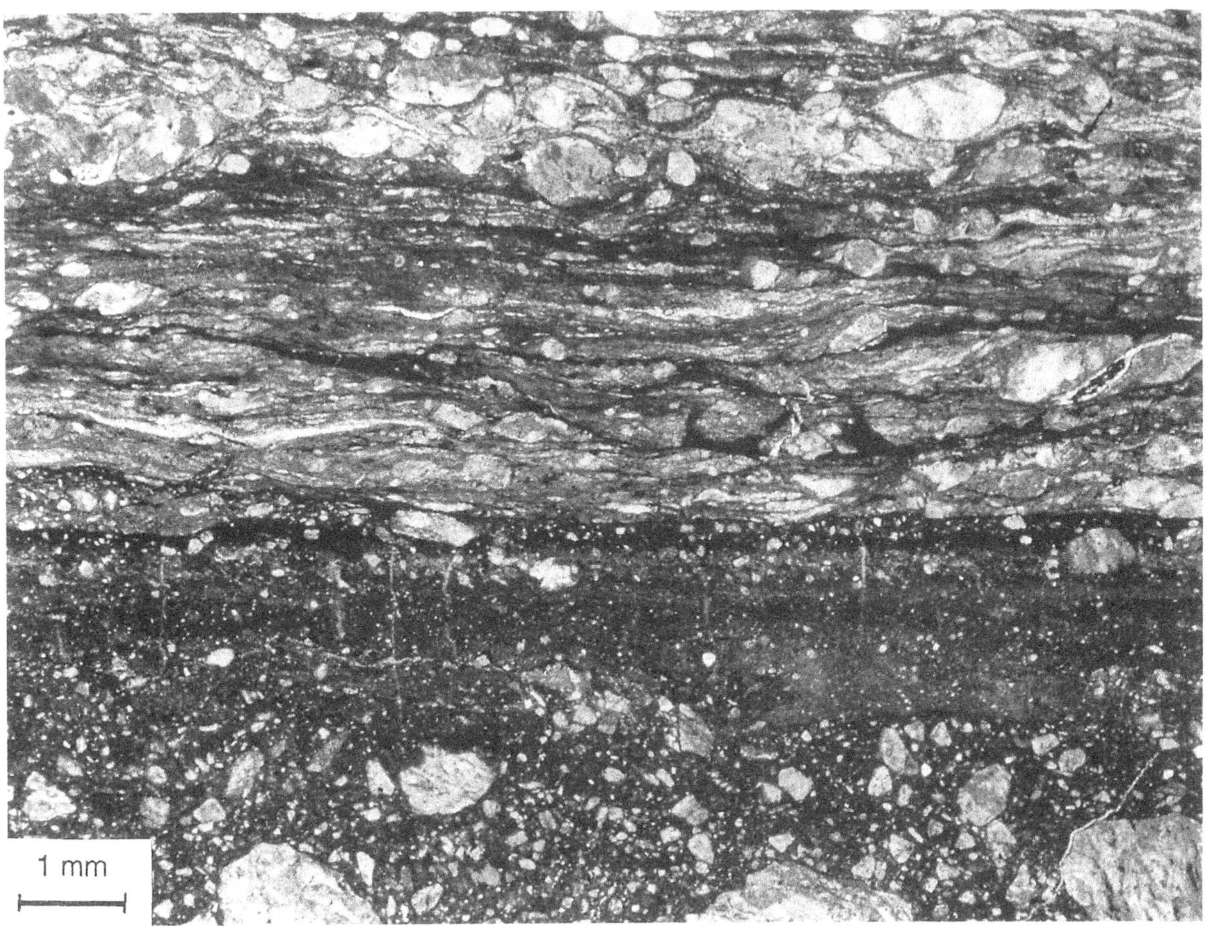


44 B

## Reaction-enhanced changes in microstructure and deformation mechanism

**44C.** Strain localization. Thin bands of retrograde, greenschist-facies mylonite (top, parallel to print length) discordantly overprint amphibolite-facies mylonitic foliation (middle and bottom) at the border of a sinistral shear zone along the cool margin of the PDFZ. Note the strain-dependent change in mineralogy from a quartz-plagioclase-white mica-biotite mylonite below to a fine-grained chlorite-white mica-clinozoisite-albite-quartz mylonite above. NX, section parallel to the XZ fabric plane. Photo from figure 10a in Handy (1987).

**44D.** Syntectonic breakdown of plagioclase. Close-up of plagioclase clast within a weak matrix of dynamically recrystallized quartz in the lower mylonitic domain of 43C. The clast is undergoing incipient reaction to fine-grained white mica, clinozoisite, albite, and quartz. Note the nucleation of these reaction products along the boundaries and fractures of the clast. Strain compatibility between clast and matrix is maintained by the nucleation and growth of reaction products in the strain shadows of the clasts. The rheology of the rock is inferred to be governed by dislocation creep of the quartz matrix. NX, section parallel to the XZ fabric plane. Photo from figure 5a in Handy (1990).

**44E.** Viscous creep in fine-grained micaceous mylonite. Close-up of fine-grained, micaceous foliation in upper mylonitic domain of 44C. Micaceous reaction products comprise a weak, fine-grained matrix that surrounds elongate boudins of fine, dynamically recrystallized quartz. The initially weaker quartz in 44D is now inferred to be slightly stronger than the micaceous reaction products. The dominant deformation mechanism in this mylonitic domain is inferred to be dislocation- and/or diffusion-accommodated, grain-boundary sliding in the fine-grained, micaceous matrix. NX, section parallel to the XZ fabric plane. Photo from figure 5b in Handy (1990).

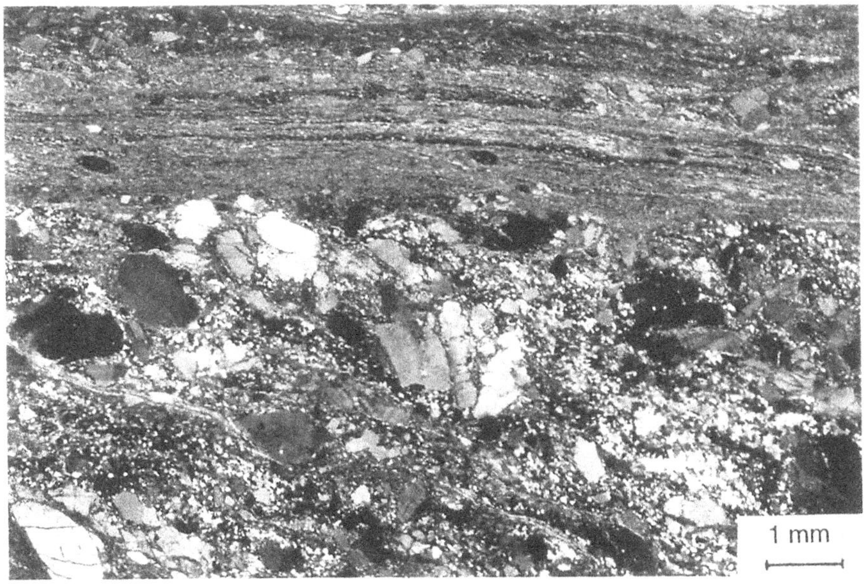


44 C

44 D

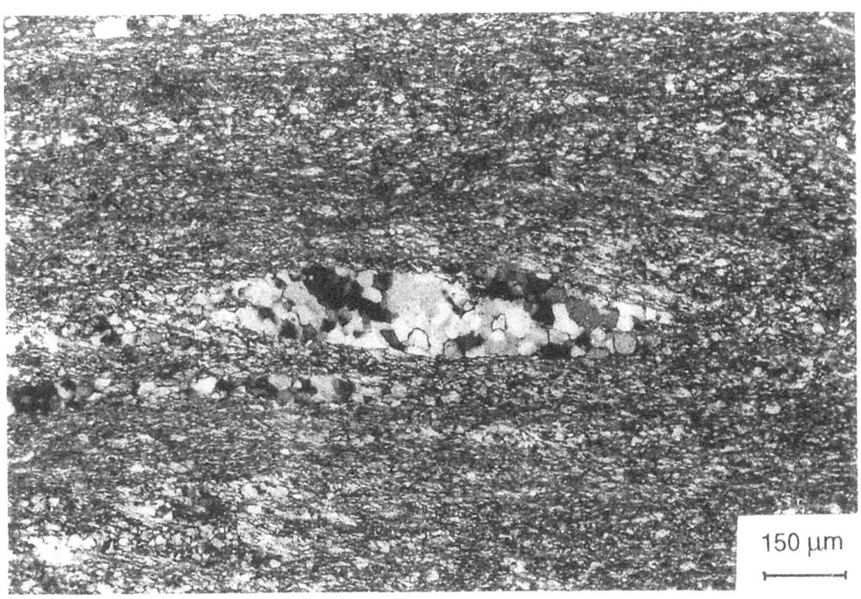


44 E

### Contrasting deformation of minerals in a mylonite

**44F.** Viscous flow of quartz between feldspar clasts. Dynamically recrystallized quartz between two feldspar grains (labeled f) in greenschist-facies, granitoid mylonite. Note that directly between the impinging feldspar clasts, quartz grains are both smaller and more elongate than where the feldspars are further apart. Dynamically recrystallized grain size is inversely and nonlinearly proportional to creep stress (e.g., Etheridge and Wilkie, 1981). Therefore, the gradients in grain size and shape within the quartz matrix confirm previous inferences that stress, strain, and strain rate were locally much higher between the feldspar grains than elsewhere (Lister and Price, 1978). Numerical models of two-phase aggregates predict a similar concentration of stress and strain in a weak, viscous matrix flowing around rigid bodies (Masuda and Ando, 1988). The preservation in quartz of arcuate grain boundaries and grain triple junctions with unequal angles suggests that recrystallization was dynamic. Evidently, decreases in differential stress and temperature during extensional exhumation were fast enough to freeze in stress-dependent grain size on a very small scale (Handy, 1990; Prior and others, 1990). NX, section parallel to the XZ fabric plane. Micrograph published as figure 3 in Handy (1994a).

44 F 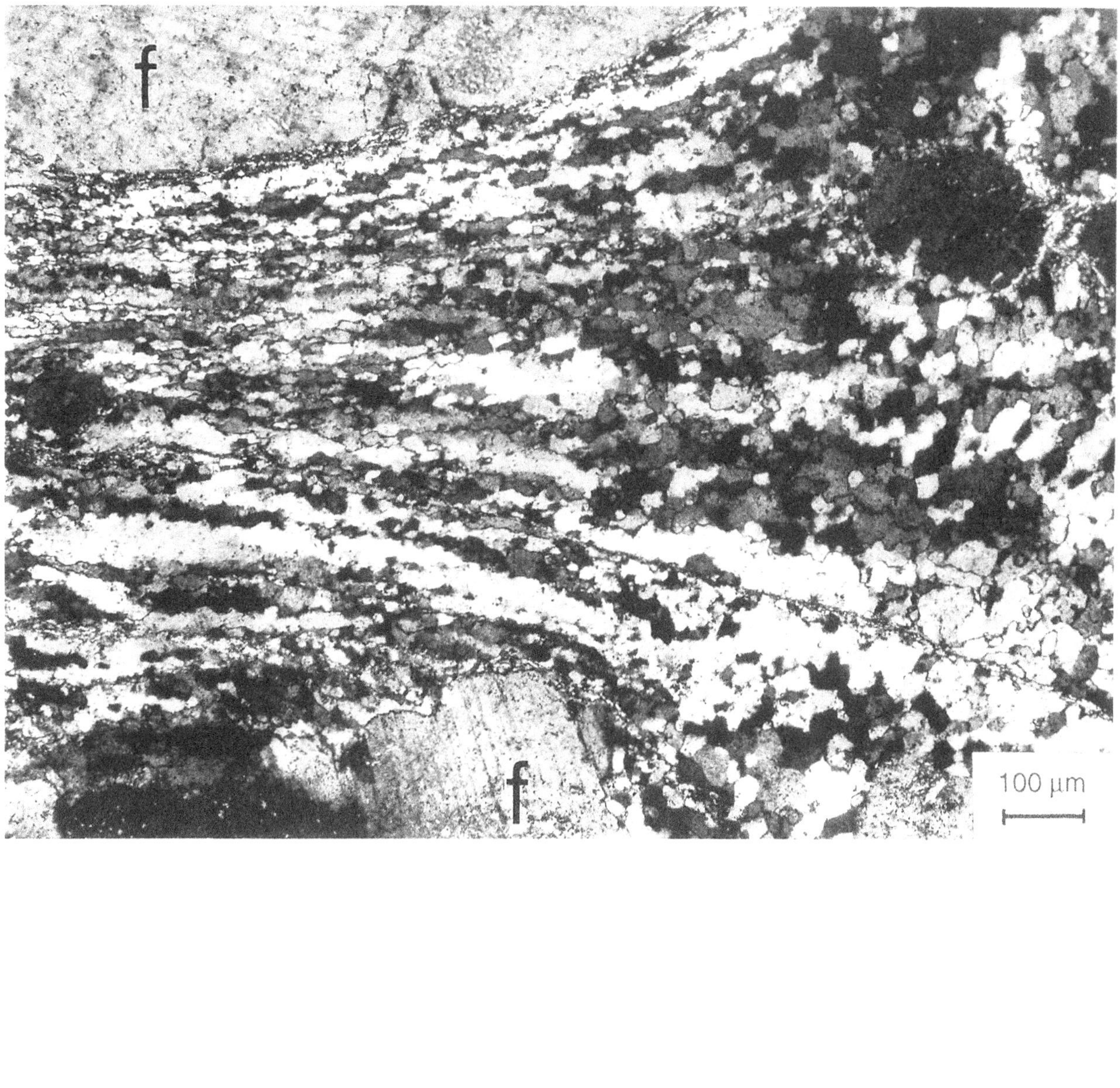

# 45. SEMI-BRITTLE DEFORMATION IN QUARTZO-FELDSPATHIC MYLONITES

CAROL SIMPSON

Photomicrographs of semi-brittle deformation in lower greenschist-facies, quartzo-feldspathic mylonites from tonalite of the Borrego Springs section of the eastern Peninsular Ranges mylonite zone, southern California, U.S.A. (Simpson, 1985); pegmatite vein in Roses granodiorite, northeast Spain (Simpson, 1981; Simpson and others, 1982); and biotite granitoid from the Rockfish Valley fault zone, Charlottesville, Virginia, U.S.A. (Bailey and Simpson, 1993; Bailey and others, 1994).

In each of these mylonites, feldspar constitutes a significant proportion of the protolith rock and is stronger than all other phases except accessory minerals. All examples fall within the "clast-matrix field" of Handy (1990), as the strong minerals neither form a supporting framework, nor do they show significant ductility.

**45A**. Deformed tonalite from the Borrego Springs mylonite zone shows striking contrast in behavior of relatively strong feldspar grains and weaker quartz grains. Feldspar (F) deformed entirely in the brittle field, to produce smaller angular fragments that rotated as they were drawn away from the host clasts. Quartz grains (Q) are ribbonlike and show undulatory extinction; they deformed predominantly by dislocation glide, consistent with deformation in dislocation creep Regime 2 of Hirth and Tullis (1992), with minor grain-boundary recrystallization along deformation band boundaries. Fine-grained mica, quartz, and feldspar grains in the matrix between larger quartz and feldspar grains most likely accommodated the strain by grain-boundary sliding. NX

**45B**. Mylonitized pegmatite dike from outcrops of deformed Roses granodiorite, northeast Spain, shows reduction in grain size of feldspars (F) mainly by brittle fracture (arrowed A and B) and reaction-softening to fine-grained white mica, clinozoisite, and quartz. Some feldspar porphyroclasts contain strain-induced twins (arrowed A and C). The reaction-softening assisted brittle failure in producing the subrounded porphyroclasts. Quartz-rich zones (Q), such as across the center of the photograph, show well-developed, oblique grain shape, preferred orientation and evidence for grain-boundary migration recrystallization. NX

**45C**. In this mylonitic biotite-granitoid of the Rockfish Valley fault zone west of Charlottesville, Virginia, U.S.A. all deformation is related to Alleghanian thrusting (Bailey and Simpson, 1993). Feldspar grains were reduced in size by a combination of brittle fracturing and reaction to form white mica and quartz, which also comprise the bulk of the matrix. Stable cracks in feldspar grains (F) contain face-controlled quartz fibers, 50–100 µm wide by up to 500 µm long, with minor white mica, biotite, chlorite, and stilpnomelane (Bailey and Simpson, 1993). Fractured and boudinaged feldspar grains are extended along the foliation plane and are connected by transverse veins (TV) within which all quartz has rotation-recrystallized to 20–30 µm sized new grains (Bailey and others, 1994). However, ribbonlike single grains of quartz (Q), 200–300 µm wide by up to 5 mm long, which began as large, irregular blebs in undeformed rock (Bailey and others, 1994), show undulatory extinction, indicative of deformation by dislocation glide. The lack of recrystallization of the original igneous quartz, versus the complete recrystallization of quartz precipitated between boudinaged feldspar fragments, may be due in part to differences in both original grain size and shape. NX

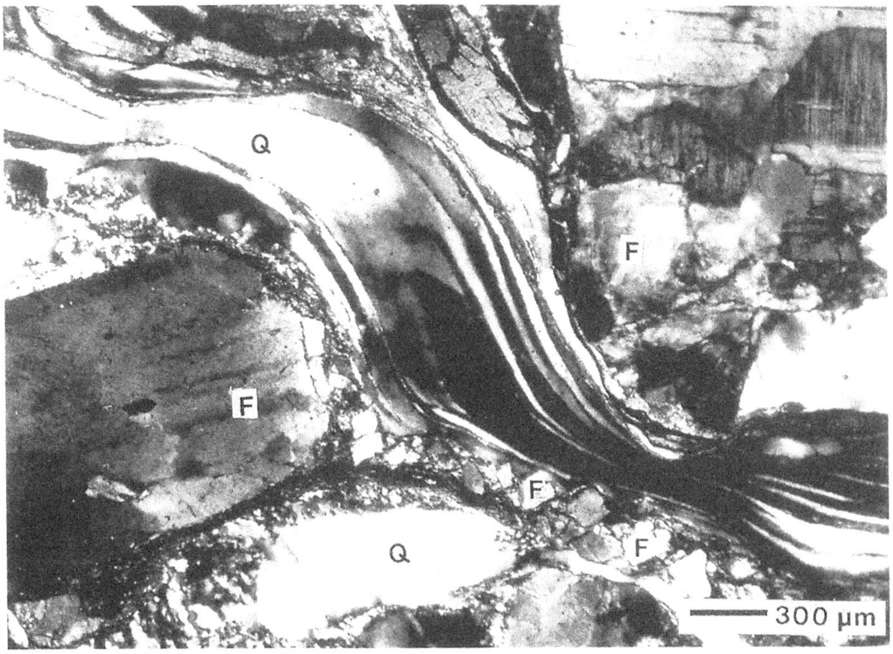

45 A

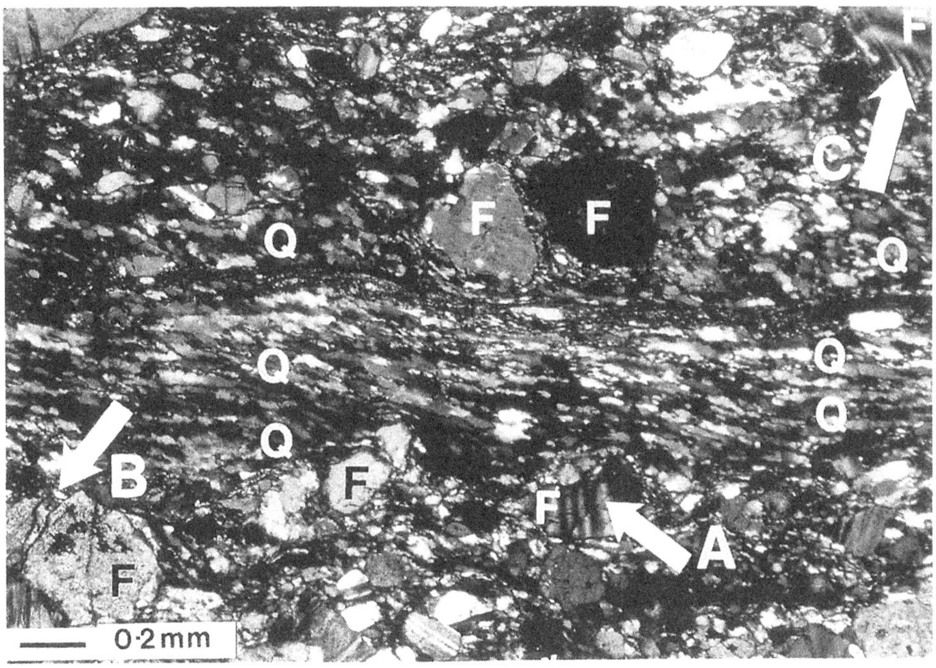

45 B

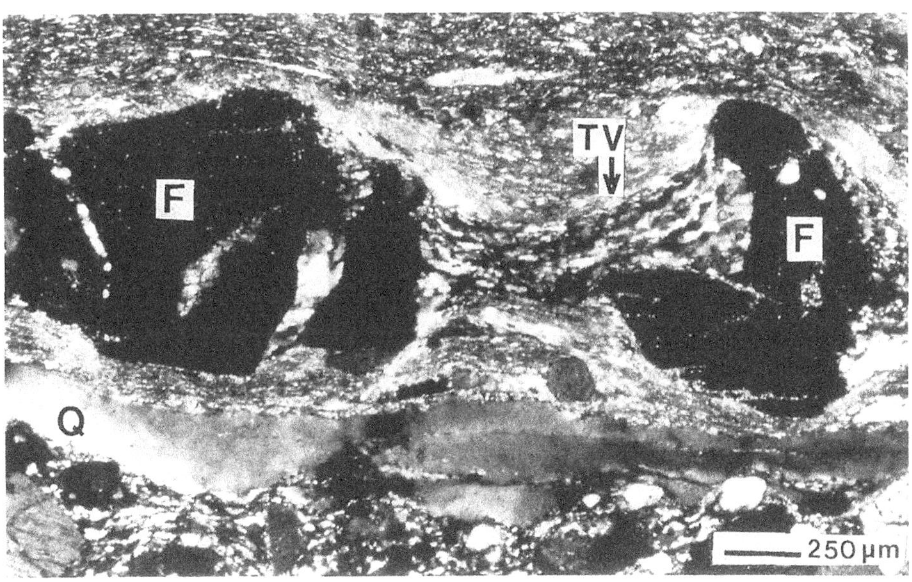

45 C

## 46. BRITTLE-DUCTILE DEFORMATION OF FELDSPAR

KEITH S. POLLMAN AND SHARON MOSHER

Progressive mylonitization of a granite gneiss from within the Garlock fault zone of southern California, U.S.A., reveals that deformation in albite and microcline feldspar can be a continuous process from brittle failure to recrystallization along fracture planes. Feldspar deformed in a semi-brittle manner, whereas quartz was completely ductile.

The suite of rocks examined was collected from a ductilely deformed granite gneiss of unknown age that crops out in the Quail Mountains of southern California within the 2 km wide Garlock fault zone. The granite gneiss (approximate mineralogy: 35% quartz, 30% microcline, 25% albite, 10% muscovite, biotite, and opaque minerals) varies from undeformed to progressively more ductilely deformed and mylonitized approaching the fault traces, culminating in total recrystallization of all constituent mineral phases. Cataclastic deformation associated with the brittle faults comprising the Garlock fault zone overprints the ductile deformation. The foliation within the granite gneiss is vertical and parallels the fault traces. Although an apparent spatial relationship between the mylonites and fault traces exists, no genetic relationship has been established. Elevations at sample locations vary, but no direct correlation between elevation and extent of deformation can be inferred. Also, no attempt has been made to establish physical parameters during ductile deformation, as little is known about confining pressure, temperature, strain rate, or fluid contents in the area.

Although the whole-rock response to strain has been ductile, the initial reaction of albite and microcline is brittle. Fracturing occurs in noncrystallographic orientations in both minerals, along (010) twin planes in albite, and along the (001) and (010) cleavage planes in microcline. With continued strain, development of new grains occurs along these fractures, resulting in thin zones of cryptocrystalline material derived from the host feldspar. The zones of cryptocrystalline grains become more pervasive as strain progresses, until the original host grain has been completely recrystallized.

**46A.** Undeformed granite gneiss from the Quail Mountains of southern California. Sample locality is within the Garlock fault zone, but this outcrop was relatively unaffected by either cataclastic or ductile deformation observed along nearby fault traces. Texture is typical of a coarse-grained granite: massive and relatively equigranular with no trace of a foliation. NX

**46B.** Albite porphyroclasts (A) from a moderately deformed granite gneiss outcrop from the Quail Mountains. The original granitic texture of this rock has been markedly altered. Igneous quartz grains are elongate, partially to totally recrystallized (B), and appear to bend around the more rigid feldspars, thus defining an incipient foliation. Albites are oriented with the (010) twin planes nearly parallel to the local shortening direction, resulting in parting, offset, and separation along the twin planes after some bending of the grains. The parting planes are straight and uniform, and the gaps formed during separation have been filled with quartz (C). NX

46 A

46 B

**46C.** Fractured albite porphyroclast from a moderately deformed granite gneiss from the Quail Mountains. The extent of whole-rock deformation is similar to that in the previous example. Elongate and partially recrystallized quartz grains are shown (A). This albite grain is oriented with (010) twin planes almost perpendicular to the local shortening direction, and fracture planes do not appear to coincide with crystallographic planes. Fractures are conjugate and produce marked offset of twin planes. Close inspection of the fracture planes in this albite indicates some are straight (B) and the margins of others are irregular (C). The latter have become sites for new grain growth, and microprobe analysis confirms that cryptocrystalline grains along the fractures are recrystallized albite (D). This is an example of the transition from brittle to ductile behavior in albite. During brittle failure, slip along the fracture planes may have induced lattice dislocations and subsequent tangling. The resultant difference in stored strain energy on either side of the fracture plane, in turn, may cause the boundaries to migrate and form new recrystallized grains. NX

**46D.** Perthitic microcline in deformed granite gneiss from Quail Mountains. Whole-rock deformation is more extensive than in the previous examples; rock is foliated, feldspar porphyroclasts are pervasively fractured and have been flattened; original quartz grains are completely recrystallized. This perthite has fractured along the (010) and (001) cleavage planes, and, as in the previously described albite, slip along the fracture planes may have induced dislocation tangling and subsequent boundary migration. The cryptocrystalline grains along the fracture planes (C) are albite and potassium feldspar (confirmed by microprobe analysis) that have nucleated on the margins of the fractures. The entire transition is shown; note straight (A) and irregular fractures (B) and zones of recrystallized grains parallel to fractures (C). NX

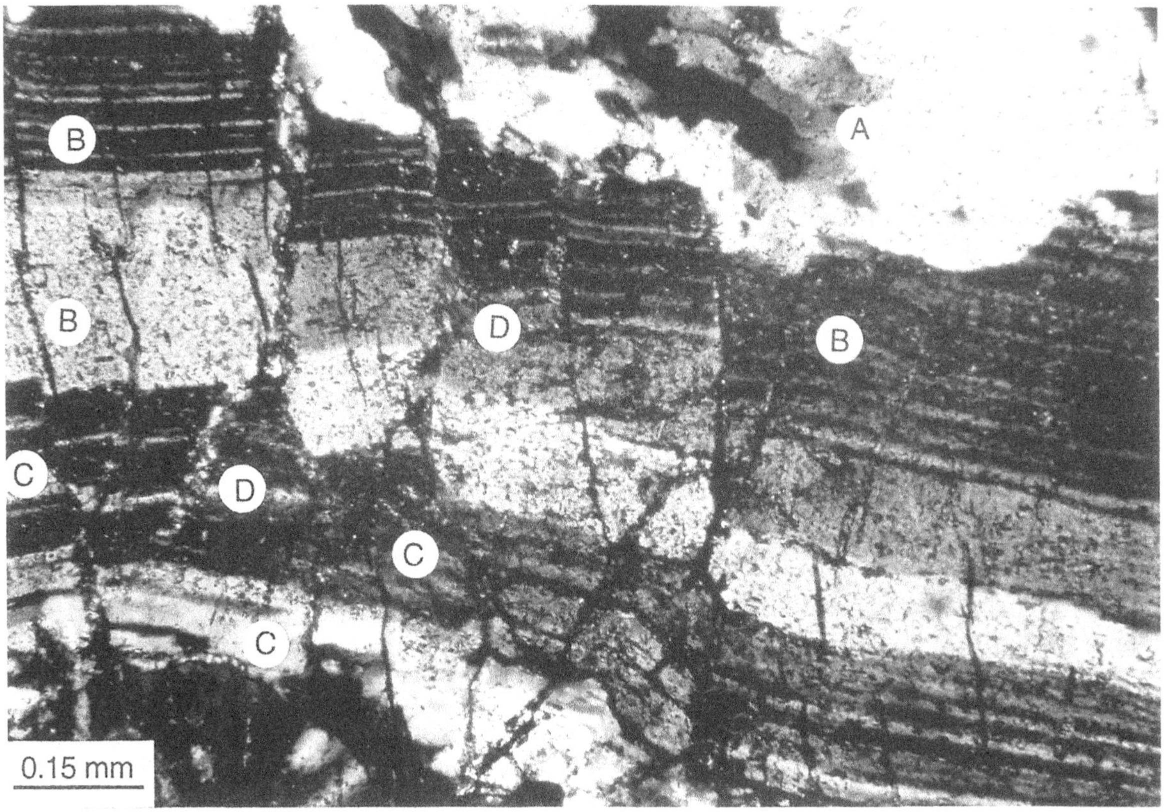


46 C

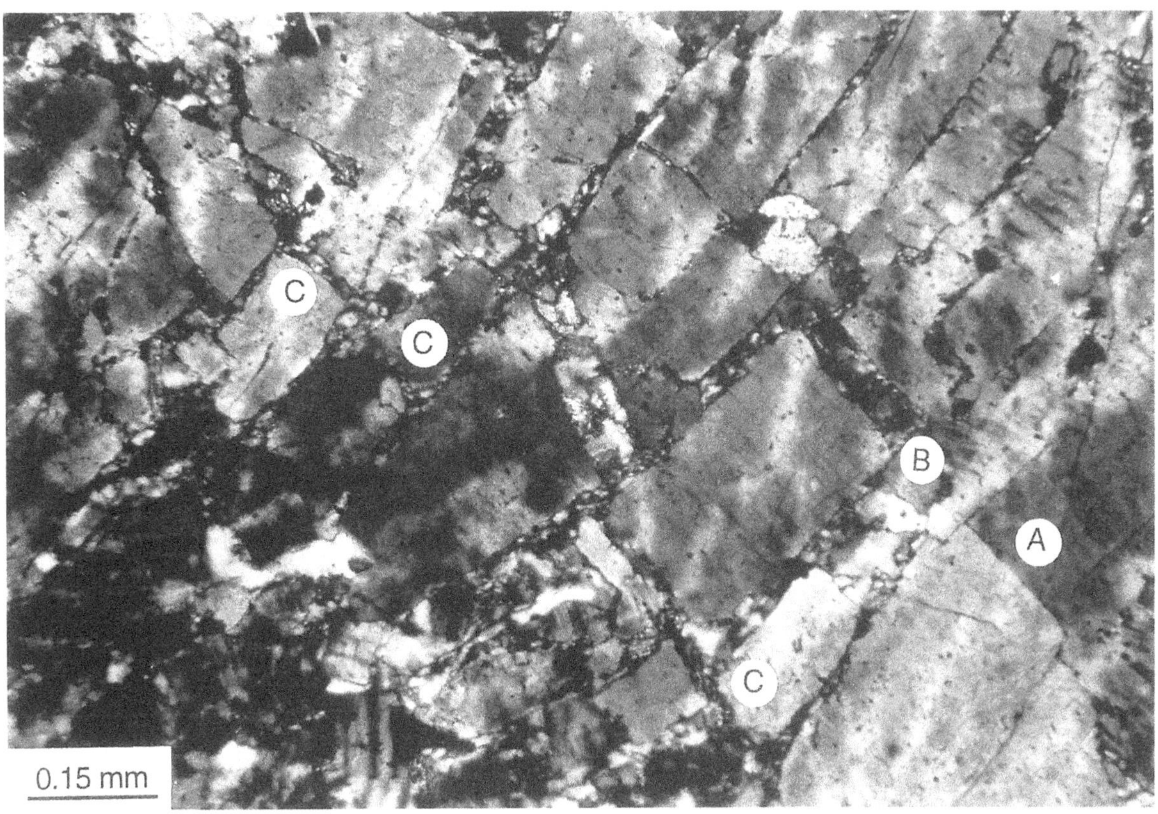


46 D

# 47. PLASTIC-TO-BRITTLE DEFORMATION OF MICROCLINE DURING DEFORMATION AND COOLING OF A GRANITIC PLUTON

Ernest M. Duebendorfer and Clint H. Christensen

The 1425 Ma Beer Bottle Pass pluton, Lucy Gray Range, southern Nevada, U.S.A. (Figure 47.1), is part of a transcontinental belt of ca. 1400 Ma granitoids that conventionally have been interpreted as anorogenic. Recent field, microstructural, and geochronologic data from the pluton indicate that major northeast-striking, northwest-dipping mylonite zones recording dextral-contractile strains were probably active during pluton emplacement (Duebendorfer and Christensen, 1995). Four photomicrographs show microstructures in microcline megacrysts from the pluton that record deformation at temperatures possibly ranging from greater than 550°C to less than 400°C. Deformation mechanisms in microcline megacrysts vary from dynamic recrystallization in the samples deformed at higher temperatures to cataclasis in samples deformed at lower temperatures.

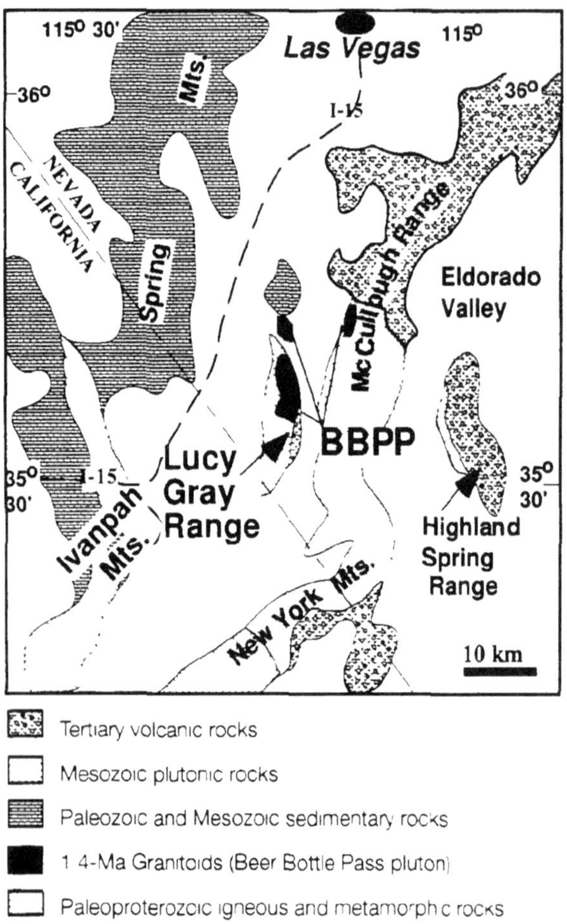


**Figure 47.1.** Generalized geologic map showing location of Lucy Gray Range and surrounding area. BBPP = Beer Bottle Pass pluton. I-15 = U.S. Interstate Highway 15.

**47A.** Dynamically recrystallized microcline ribbon (center) from an 80 m wide mylonite zone that cuts the Beer Bottle Pass pluton, Lucy Gray Range, Nevada. The elongate, fine-grained "tail" is composed exclusively of dynamically recrystallized, new microcline grains, suggesting that deformation took place dominantly by recrystallization-accommodated dislocation creep (Tullis and Yund, 1991; Yund and Tullis, 1991). By analogy with other studies in naturally deformed rocks (e.g., Vidal and others, 1980; Olsen and Kohlstedt, 1985; Pryer, 1993), these feldspar microstructures suggest deformation at temperatures greater than 550°C. This interpretation is corroborated by the presence of highly necked, dynamically recrystallized, but stable hornblende within mylonitic amphibolites from the mylonite zone. Note the completely recrystallized quartz ribbon directly above the microcline ribbon. NX

**47B.** Incipient dynamic recrystallization of microcline porphyroclast (nearly the entire field of view) from an 80 m wide mylonite zone within the Beer Bottle Pass pluton. The very fine-grained aggregates within the porphyroclast (arrow) are composed of newly recrystallized microcline grains formed at the expense of the host grain. Incipient recrystallization may have been localized along internal deformation bands within, and at the margins of, the large microcline grain. Preferential dynamic recrystallization along internal deformation bands has been documented by Tullis and Yund (1985). NX

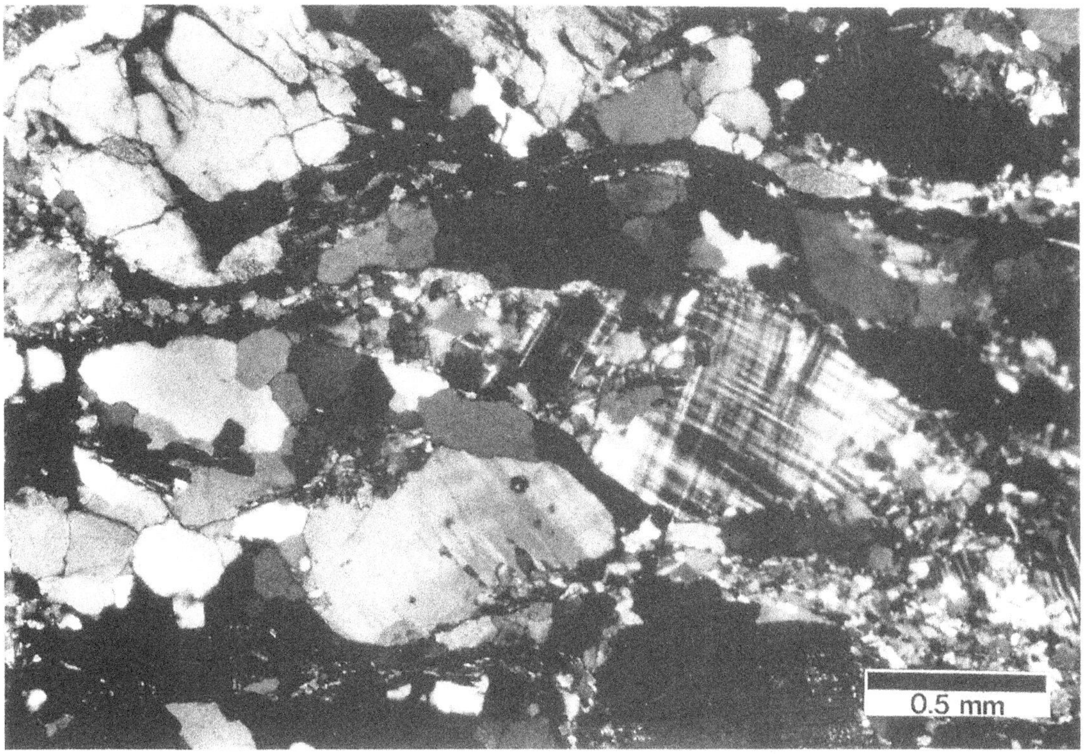


47 A

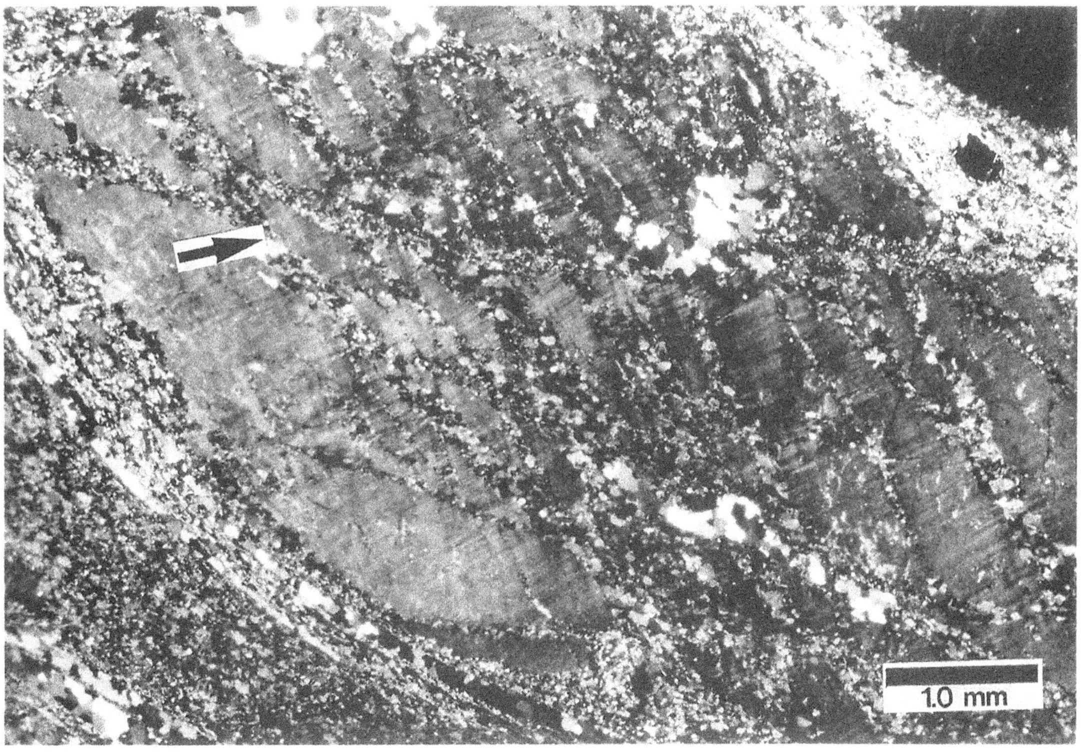


47 B

**47C.** Antithetic microfracture in microcline porphyroclast (center) from retrograde mylonite zone that cuts the Beer Bottle Pass pluton. Note the contrast between cataclastic deformation in microcline and plastic deformation in elongate quartz ribbons (arrow). Quartz ribbons show no optical evidence of subgrains or recrystallized grains. A small, δ-type, microcline porphyroclast is present at the upper right. Sense-of-shear is dextral in this view and corresponds to dextral, northwest-side-up in the field. Retrogression of hornblende to synkinematic biotite and chlorite in this sample indicates deformation at lower greenschist-facies conditions. PPL

**47D.** Multiple antithetic microfractures in microcline porphyroclast from retrograde mylonite zone that cuts the Beer Bottle Pass pluton. A quartz ribbon with no optically visible subgrains or recrystallized grains is draped across individual microfractures in the microcline grain (arrow). As in the above sample, the presence of synkinematic biotite and chlorite indicates deformation at lower greenschist-facies conditions (less than 400°C). NX

47 C

47 D

# 48. SEMI-BRITTLE DEFORMATION AND CATACLASTIC FLOW OF HORNBLENDE-QUARTZ ROCK IN A DUCTILE SHEAR ZONE

HASSAN A. BABAIE AND TIMOTHY E. LA TOUR

The 4000 km long Grenville Front of the Canadian shield is one of the world's great structural discontinuities, marking the boundary of Archean and early Proterozoic terranes with the middle Proterozoic Grenville terrane. Near Coniston, Ontario, Canada, a part of the Grenville Front known as the Coniston shear zone separates the high-grade Grenville province from the low-grade Southern province of early Proterozoic age. Within the Coniston shear zone, metamorphic grade varies from lower to upper amphibolite facies on the Grenville province side to greenschist facies on the Southern province side. Garnet-biotite geothermometry on metapelites (La Tour, 1981), corroborated by oxygen isotope studies on a variety of rocks in the zone (Kerrich and others, 1984), suggests a temperature of about 540°C for the hornblende-quartz rocks discussed here.

**48A.** Mesoscopic fabric. Cluster of mesoscopic "dominoes" in the ductile Coniston shear zone, near Coniston, Ontario. The hornblende-quartz rock formed by pervasive quartz microveining of amphibolite which formed under lower amphibolite-facies conditions. The quartz microveining and semi-brittle deformation occurred under retrograde, greenschist-facies conditions (La Tour, 1981; Babaie and La Tour, 1994).

Mesodomino clusters constitute pods up to 40 cm long and are enveloped by anastomosing shear zones, which define the dominant C foliation. The premylonitic schistosity of the mesodominoes was fractured, sheared, and transposed into the C foliation except in pods where the original schistosity is preserved in mesodominoes. The shearing has straightened the otherwise steplike profile of the mesodominoes, which are separated from each other by thin quartz-rich shear zones.

The space problem, associated with domino kinematics, was alleviated on the mesoscopic scale by smoothing corners of the translated segments of individual mesodominoes (see Figure 48.1). The overall sense-of-shear in the rock is given by the arrow couplet.

**48B.** Microscopic fabric. Microscopic dominoes of hornblende crystals (hb) in the hornblende-quartz rock. The hornblende cleavage (110) is at high angles (60°–90°) to the interdomino microfaults. Small apatite crystals are boudinaged (1) in quartz (q) layers, and boudinaged and bent at the corners of the microdominoes (2). Like quartz, Fe-Ti oxides (o) deformed plastically and commonly fill voids between hornblende fragments. The long axis of quartz (q) ribbons parallels the mylonitic C foliation. PPL

On the microscopic scale, the space problem associated with the rotation of dominoes was obviated by plastic deformation of quartz layers that envelope the hornblende microdominoes.

The slip and rotation of small hornblende dominoes and boudinaged apatite crystals along the C foliation indicate that boundary conditions allowed extension within the bulk shear plane. The overall sense-of-shear in the rock is given by the arrow couplet.

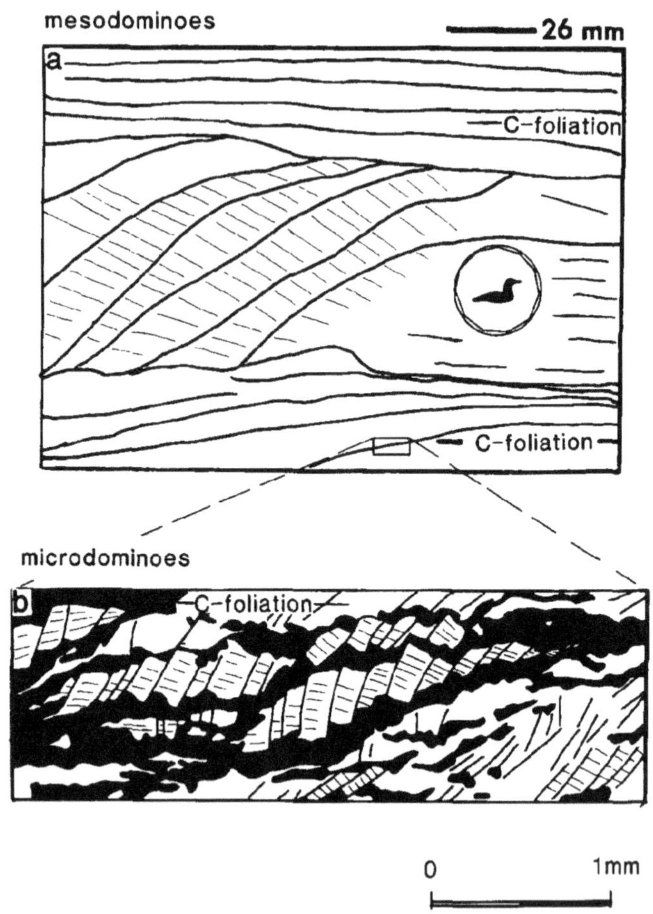


**Figure 48.1.** (a) Sketch of photograph in Plate 48A, showing relationship of different structural elements Mesodominoes are separated from each other by thin, quartz-rich shear zones. Early foliation within each mesodomino is at high angle to the mesodomino boundaries. The mesodomino cluster is enveloped by the dominant C foliation The imbricate mesodomino fabric is transposed into the C foliation to the right of the coin (b) Representative zone of microdomino deformation, developed in the zone of transposed fabric in sketch (a). Note the similarity in style on the different scales of deformation.

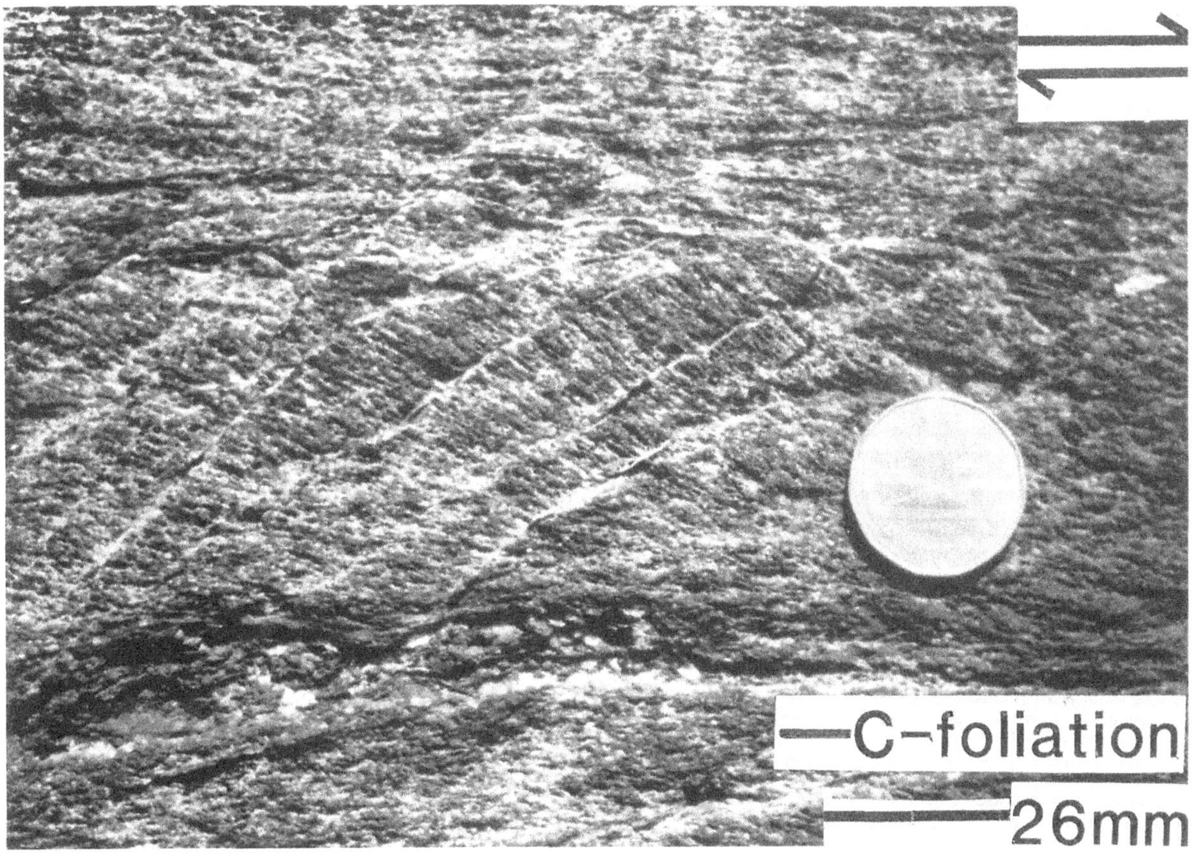


48 A

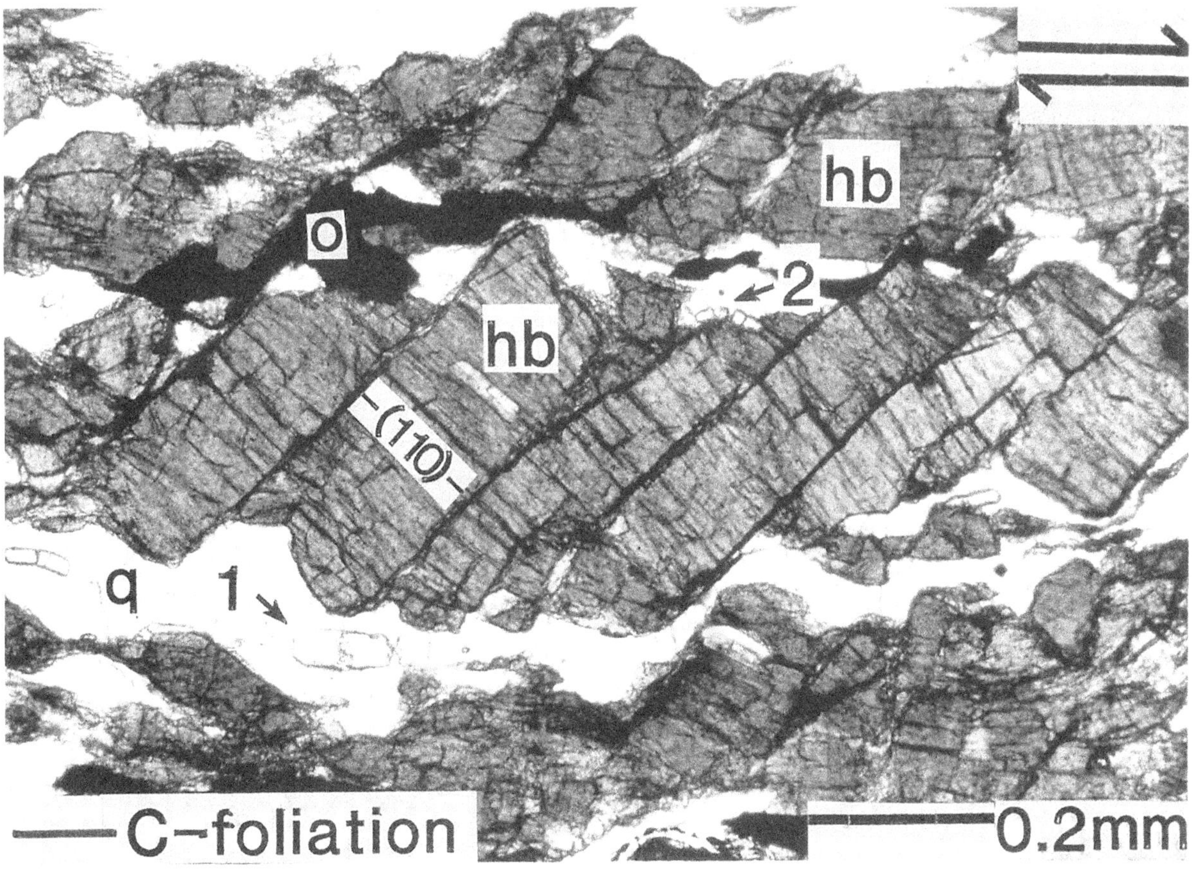


48 B

**48C.** Riedel shear bands cutting across hornblende-rich layer. Coniston shear zone, Coniston, Ontario. Hornblende growth and deformation which formed the original schistosity occurred under lower amphibolite-facies metamorphic conditions. The hornblende fractures and Riedel shears formed during later semi-brittle deformation under retrograde, greenschist-facies conditions (La Tour, 1981). During the retrograde event, characterized by reduction in grain size, hornblende behaved brittlely and quartz behaved plastically. Chlorite fibers (ch) grew locally in gaps formed by separating fragments of hornblende. PPL

Numerous $R_1$-type Riedel shears (Babaie and La Tour, 1994) divide hornblende-dominated layers into wedge-shaped domains that contain brittlely deformed hornblende (hb). The Riedel shears ($R_1$) are oblique to the dominant rock layering, and extend completely across hornblende-dominated layers where they merge asymptotically into the C foliation in the flanking quartz layers (q). The sense-of-shear on the Riedel shears is shown by the small arrow couplets, and the overall sense-of-shear in the rock is shown by the large arrow couplet. The quartz layers consist of quartz mylonite.

All hornblende grains in the rock are broken along regularly spaced fractures subnormal to the c crystallographic direction (Allison and La Tour, 1977). The Riedel shears were superimposed locally on this preexisting fracture fabric. Deformation in the Riedel shears was by cataclastic flow of hornblende fragments, forming ultracataclasite. On the margins of the shears, milder cataclastic deformation produced rhombic fragments of hornblende that formed by breakage along the two amphibole cleavages.

**48D.** Riedel shear band magnified from 48C. Cataclastic flow of hornblende has produced a band of ultracataclasite with indistinct boundaries. Near the band, hornblende crystals (hb) are fragmented into rhombic grains (1), with faces parallel to the two amphibole cleavages. The fragments of hornblende get smaller with proximity to the shear band, eventually becoming gouge in the band itself. Opaque Fe-Ti oxides (o) have flowed plastically into and along the gouge of the shear band. PPL

Microfaulting along (110) in hornblende occurred near the shear bands (2 and 3) but rarely elsewhere in the rock. Apparently, microfaulting on (110) was favored where there was severe grain-size reduction, whereas microfaulting on the fractures subnormal to the c axis was dominant at lower levels of strain.

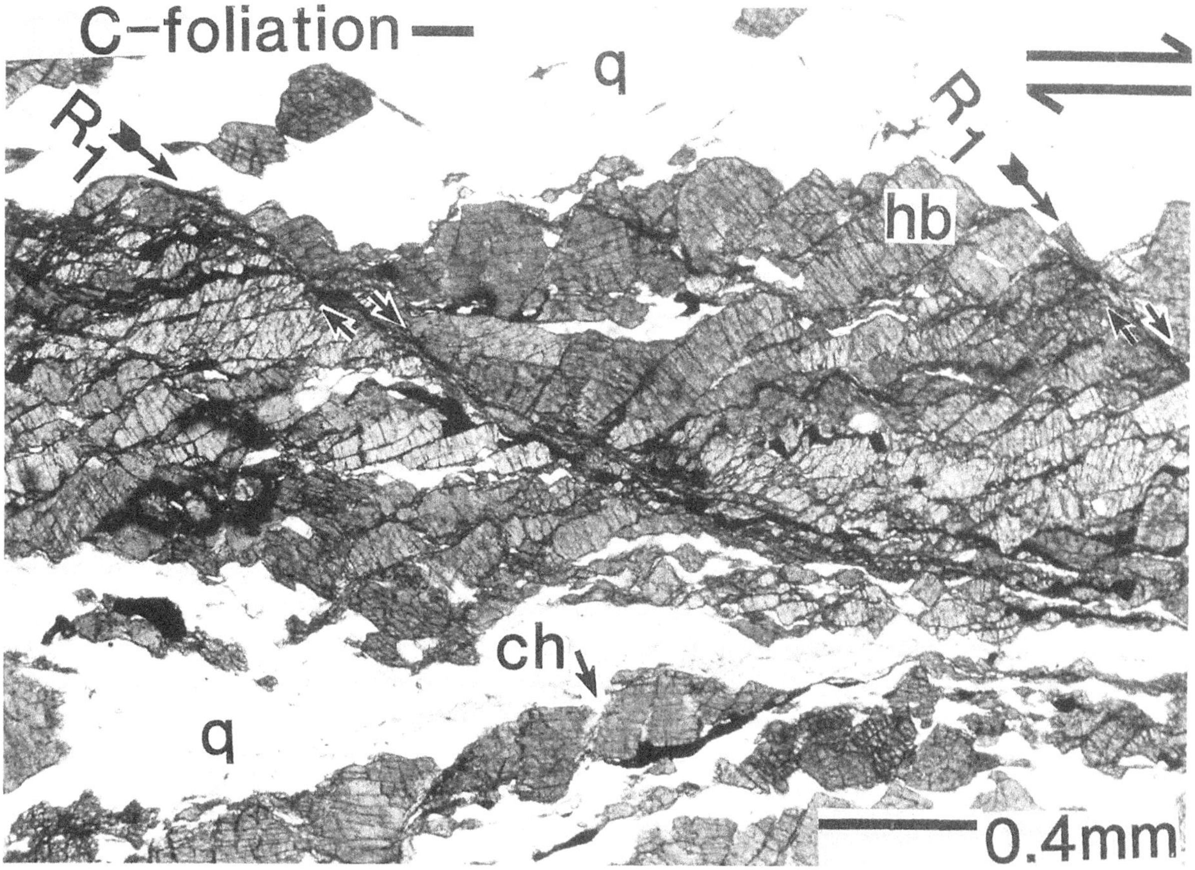

48 C

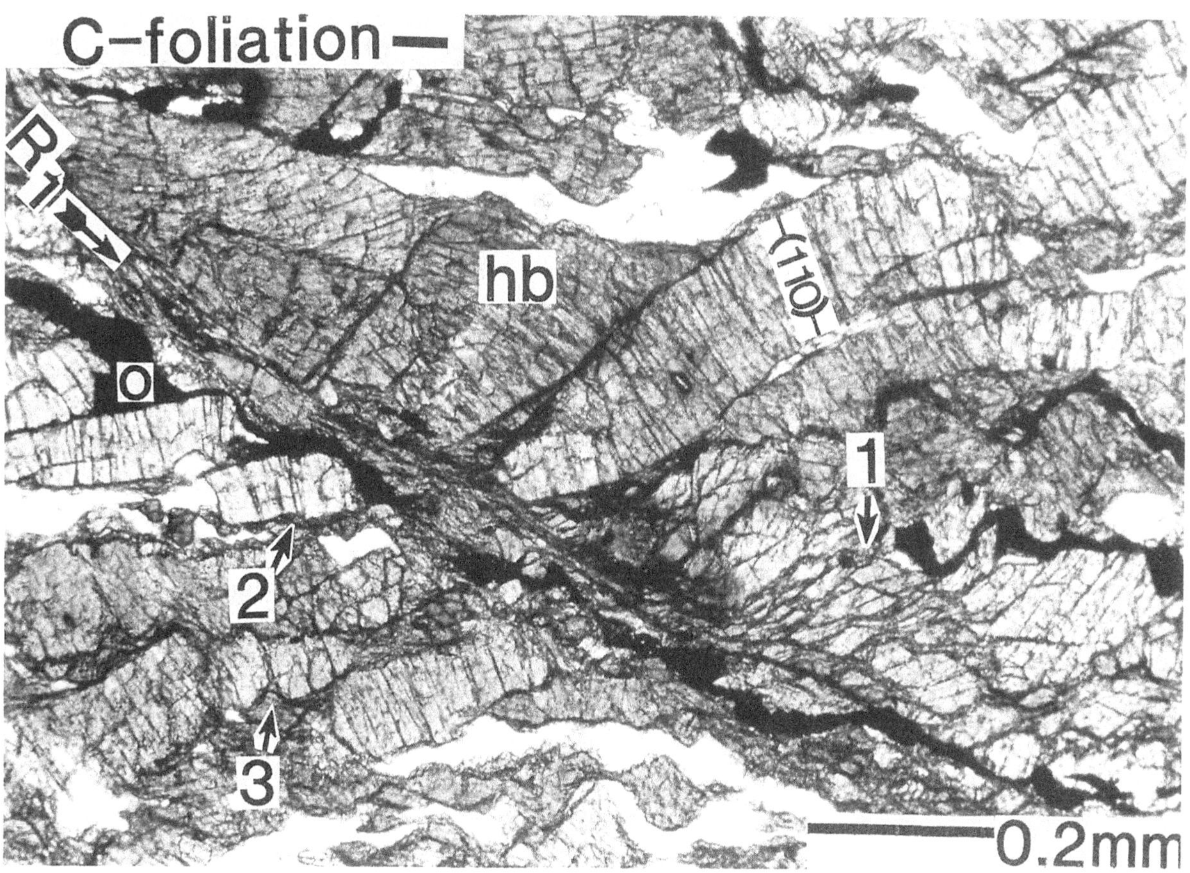

48 D

# 49. REACTION WEAKENING IN PHYLLONITES DERIVED FROM GRANITE

JAMES P. EVANS

These examples of microstructures and textures associated with reaction weakening come from the Gessaman fault in southern Arizona, U.S.A., and were first described in Janecke and Evans (1988). Phyllonitic textures are concentrated in a 2–3 m wide zone formed in Proterozoic Oracle Granite. The protolith granite consists of 10–25 mm diameter quartz grains, 5–12 mm diameter orthoclase grains, 5–10 mm plagioclase grains, and minor amounts of biotite and accessory minerals. Deformation occurred during the early Tertiary, probably at temperatures of 250–400°C.

**49A.** Negative image of an entire thin section from highly foliated quartz-muscovite-biotite phyllonite. This image shows the end result of the fracture, comminution, and syntectonic alteration of feldspars to mica. Feldspar grains form a weaker matrix, and alteration to muscovite results in a very well-defined foliation parallel to the slip plane. Quartz grains (here shown as black grains labeled as Q) are not as fractured, and form rounded porphryoclasts in a white muscovite-biotite-fine-grained quartz matrix, labeled as MBQ. Sense-of-slip is top to the left. Image is a direct exposure of the thin section in a photo enlarger, so shades are inverted, e.g., dark material is white.

**49B.** Photomicrograph of an intermediate stage of feldspar (F) fracture and alteration. Fine-grained muscovite-quartz reaction products (MQ) fill fractures and begin to define foliation. Brittlely fractured quartz (Q) lies to the right of the feldspar grains. Sense-of-slip is left side down. NX

**49C.** Quartz porphryoclasts may develop by intragranular fractures which open parallel to the slip direction and are filled with fine-grained reaction products from the reaction of feldspars to muscovite + quartz. Sense-of-slip is top to the left. NX

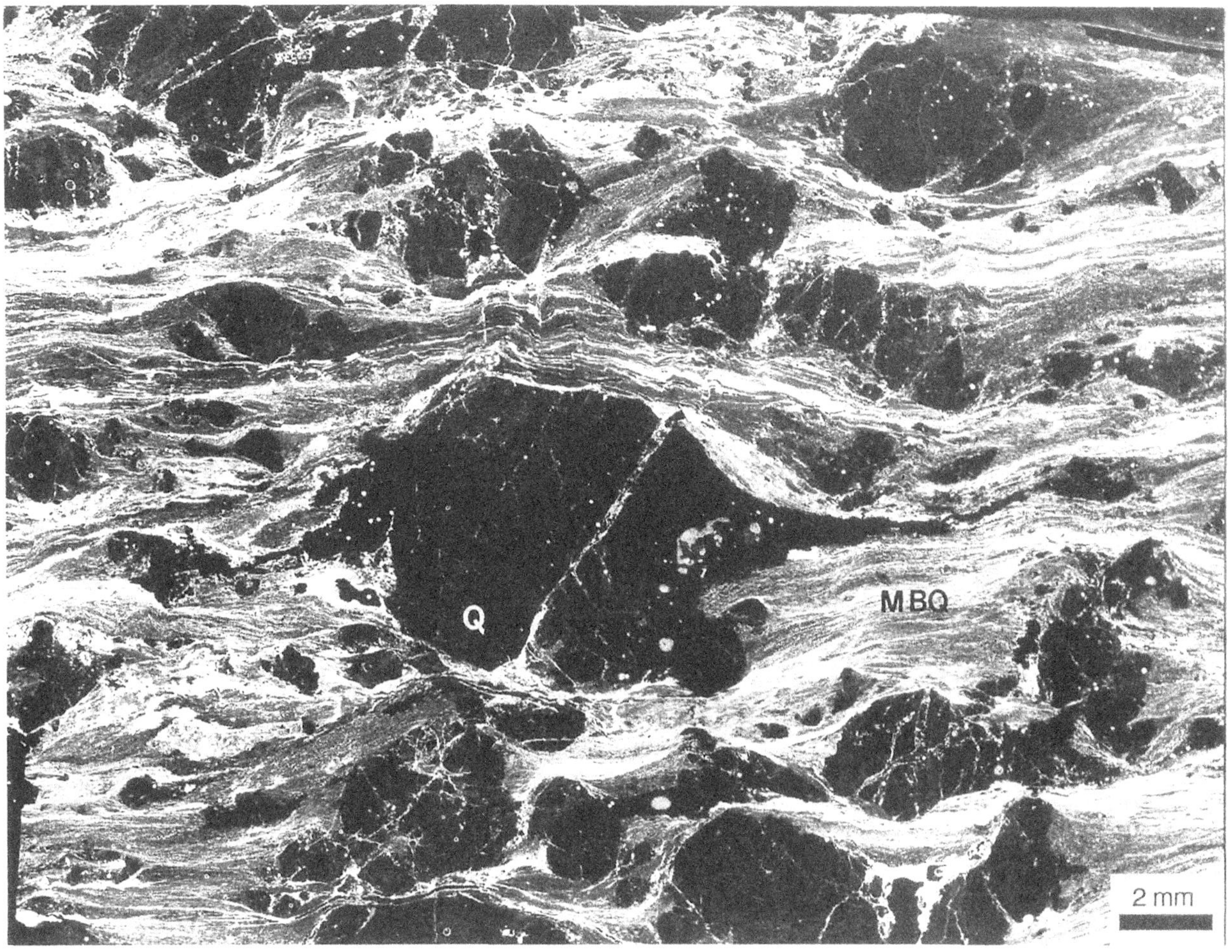


49 A

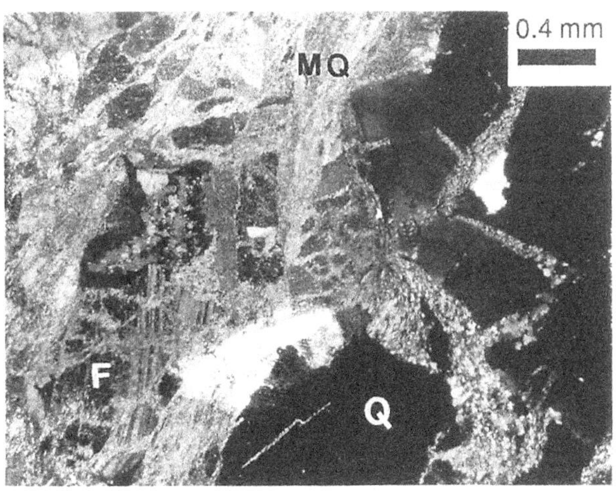


49 B

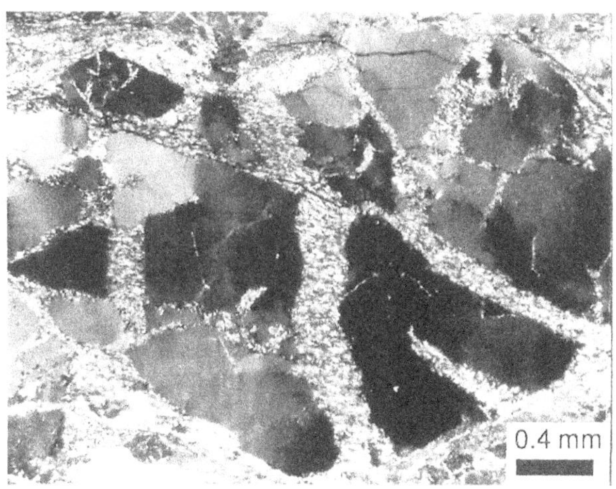


49 C

# 50. BRITTLE-DUCTILE FAULT ROCKS IN A BASEMENT-CORED THRUST SHEET

W. Adolph Yonkee

## Tectonic setting

Fault and shear zones formed by a mixture of cataclastic, crystal-plastic, and fluid-rock processes within a basement-cored anticlinorium in northern Utah, U.S.A. These zones, including the Ogden thrust and Weber Canyon shear zone (Figure 50.1), developed during regional thrusting between 140 and 110 Ma and are characterized by retrograde, greenschist-facies alteration and concentrated strain (Yonkee, 1992). Both the Ogden thrust and shear zones were later passively rotated and uplifted as the anticlinorium grew during movement on structurally lower thrusts between 90 and 50 Ma.

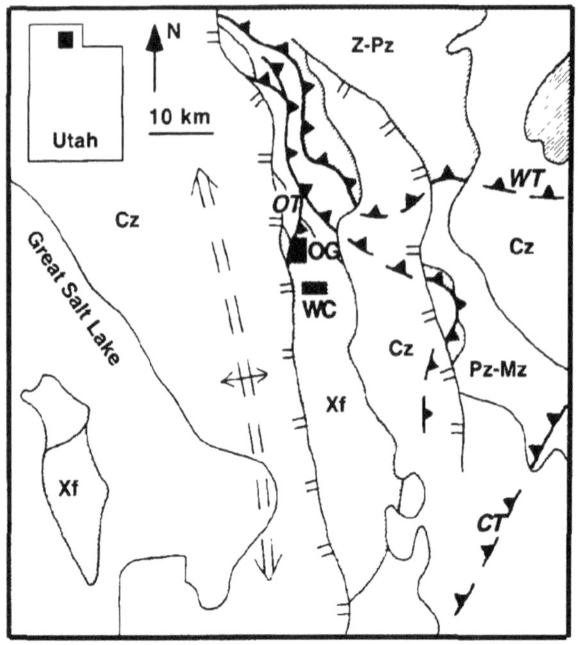


**Figure 50.1.** Index map of northern Utah. Study areas for the Ogden thrust (OG) and Weber Canyon shear zone (WC) are located on the eastern limb of a basement-cored anticlinorium (crest indicated by double-dash fold symbol). Thrust faults are *CT*—Crawford; *OT*—Ogden, and *WT*—Willard thrusts. Units are Z-Pz = Proterozoic and Paleozoic rocks of Willard thrust sheet (oblique lines), Xf = Precambrian basement rocks (close stipple); Pz-Mz = Paleozoic and Mesozoic sedimentary cover rocks (light stipple), and Cz = post-thrusting Cenozoic deposits (unpatterned)

## Ogden thrust: Deformational styles

The Ogden thrust places Precambrian basement rocks over Cambrian sedimentary rocks and has an estimated displacement of 10 to 15 km. Deformation occurred at temperatures of 300 to 350°C, depths of 10 to 12 km, and at elevated fluid pressures (Yonkee and others, 1989). A traverse from the hanging wall downward toward the fault reveals three broad zones: (i) wallrock consisting of relatively unaltered granitic gneiss; (ii) a 50 to 200 m thick, low-strain transition zone of variably fractured and altered gneiss; and (iii) a 20 to 50 m thick, high-strain fault zone of pervasively deformed and altered cataclasite and phyllonite (Yonkee and Mitra, 1993). Although cataclasite and phyllonite are mixed in many outcrops, phyllonite increases in overall abundance to the south as the Ogden thrust ramps obliquely down section about 2 km. Distinct changes in microtextures observed toward the fault zone include greatly increased intergranular fracturing and microcracking of feldspar; greater alteration of feldspar to mica; increased crystal-plastic deformation of quartz and mica; and pronounced reduction in grain size. Progressive deformation and fluid-rock interaction converted relatively strong, coarse-grained, quartzo-feldspathic gneiss into relatively weak, fine-grained, micaceous phyllonite and cataclasite.

**50A.** Granitic gneiss protolith. Quartz (Q), K-feldspar (K), plagioclase (P), and hornblende (H) grains have a granoblastic texture. Quartz has widely spaced fluid-inclusion trains and limited undulatory extinction. Intergranular fractures and intragranular microcracks (c) are widely spaced, and feldspar has undergone limited alteration to fine-grained mica (m). Scale bar = 500 μm. NX

**50B.** Fractured gneiss from transition zone. Quartz (Q) displays smooth to patchy undulatory extinction with scarce subgrains (s), reflecting limited dislocation glide and recovery. Plagioclase (P) and K-feldspar grains are cut by many microcracks, including cracks along cleavage planes (c). Intergranular fractures (f and arrow) and quartz-mica-filled veins (v) are widespread. Feldspars and mafic minerals are partly altered to a fine-grained matrix (m) of chlorite, muscovite, and quartz, especially along cracks and grain margins. Scale bar = 500 μm. NX

**50C.** Cataclasite from northern part of fault zone. Randomly oriented clasts sit in a matrix of fine-grained chlorite, muscovite, quartz, and feldspar fragments. Polycrystalline quartz clasts (Q) are cut by cracks and display undulatory extinction with localized recrystallization, recording periods of fracturing and crystal-plastic deformation. Plagioclase (P) and K-feldspar clasts are partly altered to very fine-grained mica. Clasts of cemented cataclasite (CC) record overlapping periods of cataclasis, fluid influx, and cementation. Layering (L) and parallel concentrations of oxides in the matrix, which wrap around some clasts, may record cataclastic flow aided by diffusive mass transfer processes. Scale bar = 500 μm. NX

**50D.** Phyllonite from southern part of fault zone. Quartz (Q) forms elongate ribbons with subgrains (s) and recrystallized grains (r), including some recrystallized grains with bulges along deformation bands (arrow), reflecting extensive dis-

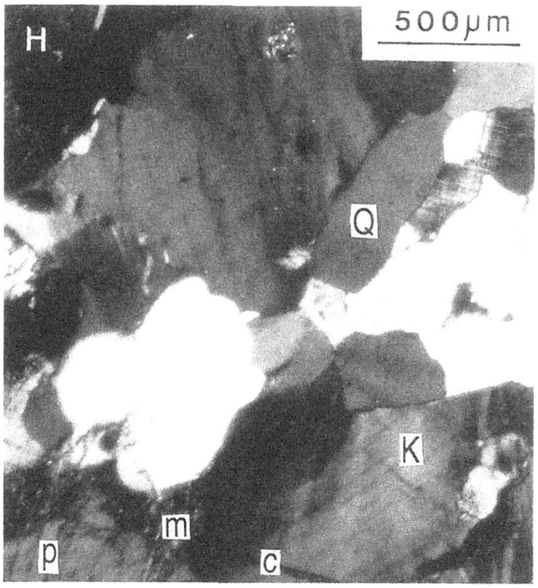


50 A

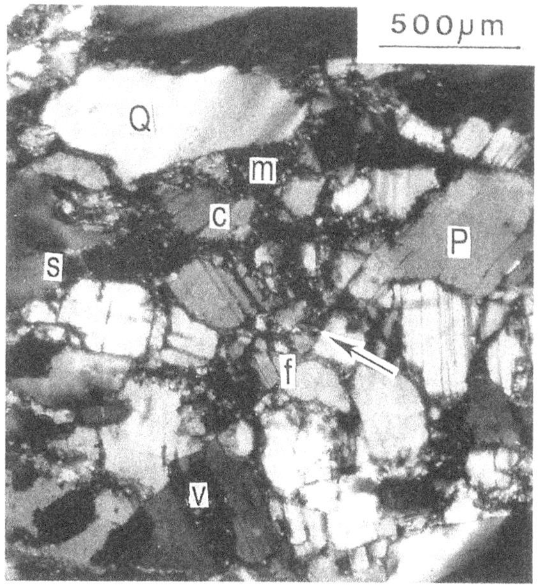


50 B

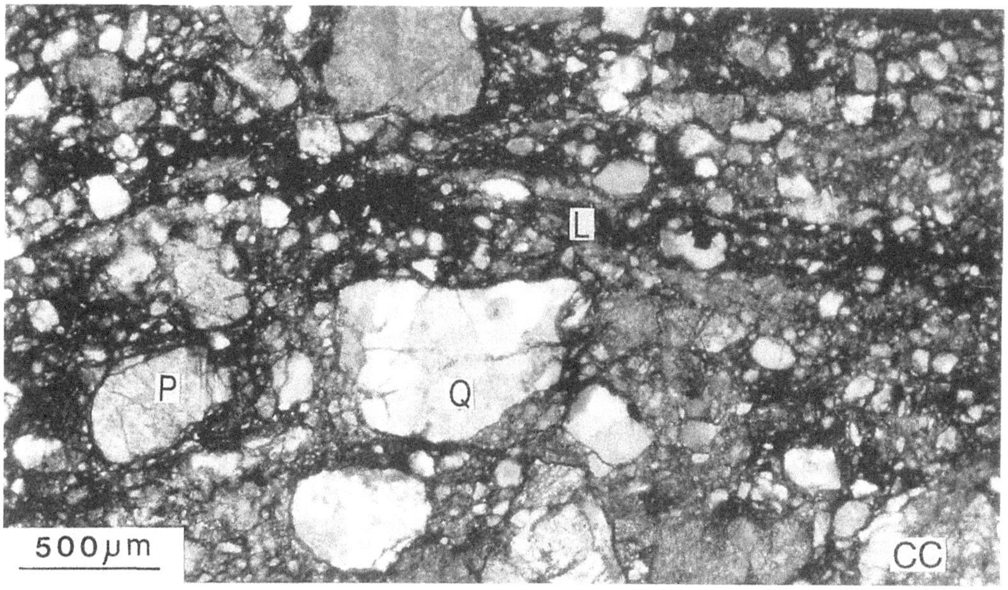


50 C

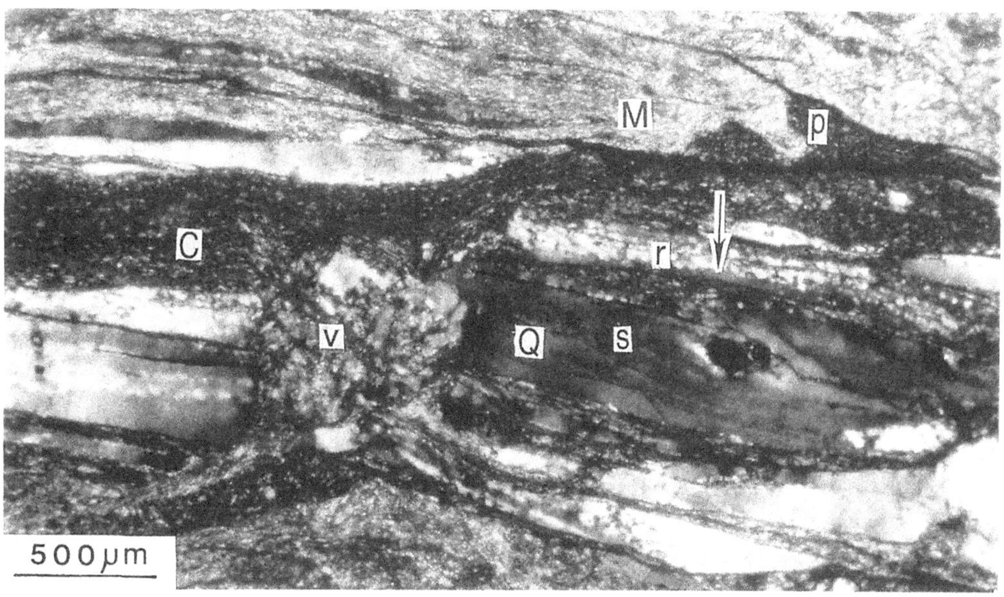


50 D

location creep. Quartz-mica-filled veins (v) cut quartz ribbons and are deformed into "bow-tie" shapes, recording overlapping fracturing, sealing, and plastic deformation. Foliated aggregates of fine-grained muscovite (M) parallel to quartz ribbons also deformed plastically. Very fine-grained material with abundant chlorite and quartz (C) occurs along shear bands and locally fills podlike regions (p). Scale bar = 500 μm. NX

## Ogden thrust: Fluid-rock interaction

Spatially and temporally overlapping periods of fracturing, fluid influx, alteration, sealing, and crystal-plastic deformation produced a wide range of fault rock textures along the Ogden thrust. Recemented cataclasite clasts, complex fracture patterns, and multiple sets of deformed veins record a complex history, in which periods of semi-brittle microcracking and brittle intergranular fracturing that opened fluid pathways alternated with periods of sealing and cementation that removed fluid pathways. Episodic influx of reactive fluids resulted in progressive alteration of feldspar to mica, dissolution of material along seams and some grain contacts, and precipitation of mica and quartz in dilatant regions. Concentrated plastic deformation of quartz that overlapped with cracking and fluid influx may record hydrolytic weakening and deformation of initially small, strain-free grains that precipitated in dilatant sites.

The progression from granitic gneiss to cataclasite and phyllonite reflects a complex interplay between deformation and fluid-rock interaction, and was accompanied by distinct changes in mineral abundances and matrix fraction (Figure 50.2). The granitic gneiss protolith consists of 30 to 40 vol% quartz, 45 to 60% total feldspar, and lesser amounts of mafic minerals and mica (Figure 50.2). Fractured gneiss from the transition zone contains slightly less feldspar, more mica, and a greater fraction of matrix (defined as newly crystallized grains and fragments with sizes less than 50 μm) produced mostly by fracturing, frictional wear, and alteration of feldspar and mafic minerals to mica. Cataclasite from the fault zone contains >20% mica and >40% matrix produced by continued fracturing and alteration. Phyllonite contains >30% mica, increased quartz contents, and >60% matrix produced by pervasive alteration of feldspar to mica and dynamic recrystallization of quartz. Distinct changes in whole-rock chemistry, marked by depletion of Ca, Na, and K and enrichment of Mg and $H_2O$ in fault-zone rocks, also accompanied deformation and influx of reactive fluids. Similar average abundances of Si, Al, and Ti for gneiss and fault-zone rocks indicate that only limited large-scale volume change accompanied deformation, although variable Si contents in fault-zone samples record local mobility of quartz.

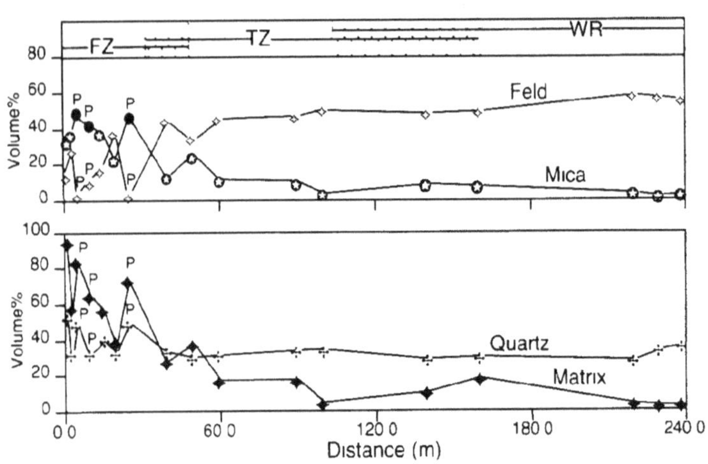


**Figure 50.2.** Mineral abundances and matrix fraction on composite traverse from wall rock (WR), through transition zone (TZ), into Ogden fault zone (FZ). Feld = total feldspar. Matrix = fraction of newly crystallized grains and fragments with sizes less than 50 μm. Phyllonite samples indicated by P. Distance measured upward from the fault contact with footwall sedimentary rocks. Results for two along-strike traverses are combined in the plot. Gradational boundaries between WR, TZ, and FZ are indicated by stippled patterns and vary in position along strike.

**50E.** Cataclasite with widespread alteration from fault zone. Variably altered and cemented clasts sit within a weakly foliated, chlorite-rich matrix that contains mica- and oxide-rich seams (s) and corroded quartz fragments. Some larger clasts of recemented cataclasite (CC) contain smaller fragments of cataclasite that had been cemented previously (cc), recording repeated episodes of cataclasis, fluid influx, and cementation. Clasts of quartz aggregates (Q) with undulatory extinction and localized recrystallization record earlier plastic deformation. Quartz-filled veins (v) along clast margins and dissolution of some clasts at contacts with other clasts (arrow) record diffusive mass transfer processes. Scale bar = 500 μm. PNX

**50F.** Fractured phyllonite from fault zone. A fracture (f) offsets quartz ribbons (Q) and very fine-grained material rich in chlorite (C). Neocrystallized quartz grains (r2) along dilatant steps in the fracture may be related to fluid influx and localized strain, and are larger than subgrains and recrystallized grains (r1) along margins of quartz ribbons. Fluid inclusion trains (arrow) at high angles to ribbon boundaries probably represent healed microcracks. Foliated aggregates of muscovite (M) are cut by a kink band and thin quartz-mica vein (v). Scale bar = 500 μm. NX

**50G.** Folded quartz vein from fault zone. Quartz within a tightly folded vein (V) displays widespread undulatory extinction, subgrains (s), and recrystallized grains (r). This vein cuts strongly foliated, fine-grained material with abundant chlorite and quartz (C), and in turn is cut by smaller quartz-mica-filled veins (arrows) that show less extensive plastic deformation. Multiple sets of variably deformed veins record repeated fracturing, fluid influx, sealing, and crystal-plastic deformation. Scale bar = 500 μm. PNX

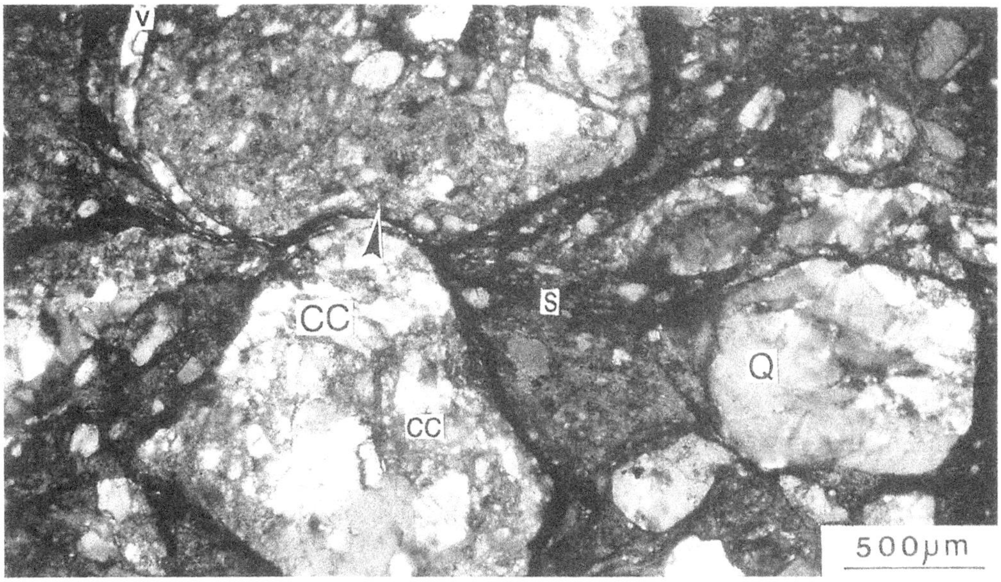


50 E

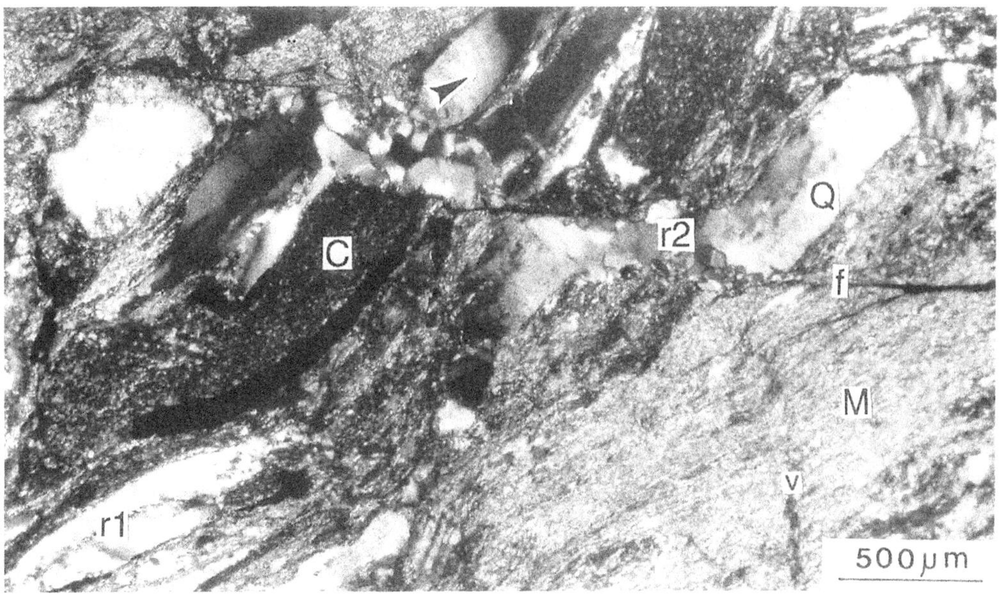


50 F

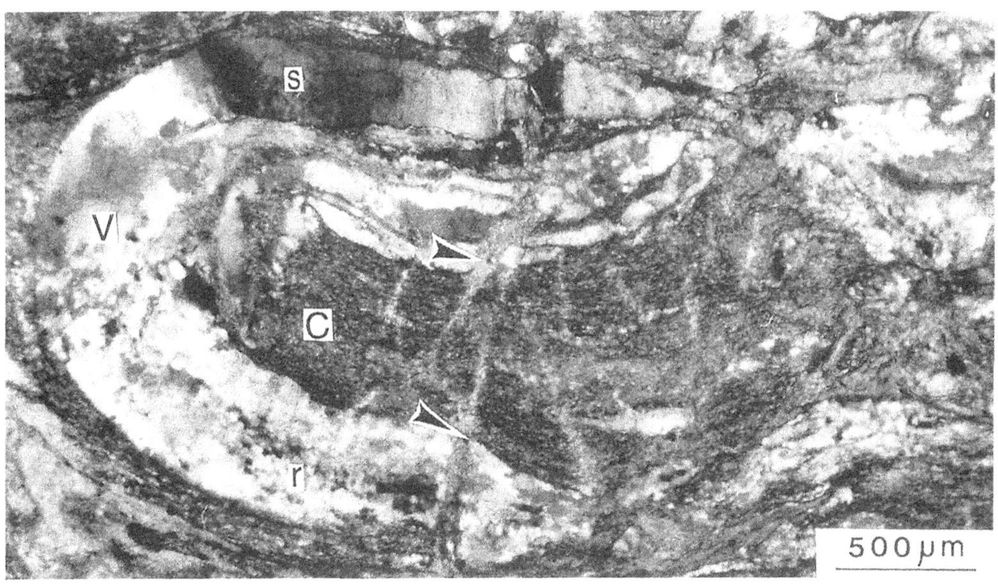


50 G

### Weber Canyon shear zone: Deformation styles and fluid-rock interaction

A gently dipping shear zone, with an estimated displacement of about 50 to 100 m, is well exposed along the steep slopes of Weber Canyon. This shear zone is part of a regional network of zones that accommodated internal deformation of basement during regional thrusting (Yonkee, 1992). Strongly foliated phyllonite, closely spaced C surfaces and shear bands, sheath folds, and complex fibrous pressure shadows indicate high strain within these zones. Deformation here occurred at slightly higher temperatures (≈ 350–400°C) and deeper levels compared to the Ogden thrust (Yonkee, 1990). A traverse across the Weber Canyon shear zone reveals three rock types: (i) wall rock away from the shear zone that consists of relatively unaltered quartzo-feldspathic migmatitic gneiss; (ii) a 3 to 10 m thick transition zone along both the upper and lower shear-zone margins that consists of variably altered, fractured gneiss and protomylonite; and (iii) a 10 to 15 m thick central shear zone of pervasively altered phyllonite with lenses of less deformed protophyllonite (Yonkee and Mitra, 1993). Distinct changes in microtextures observed toward the shear zone include greatly increased crystal-plastic deformation of quartz and mica; widespread alteration of feldspar to mica; pronounced grain-size reduction; and an overall increase in intergranular fracturing and microcracking of feldspar within the transition zone. Compared to the Ogden thrust, crystal-plastic deformation of quartz and alteration of feldspar to mica are more pervasive, and intergranular fracturing is less widespread.

Overlapping deformation and fluid-rock interaction produced distinct changes in mineral abundances and matrix fraction as migmatitic gneiss was converted into phyllonite. Migmatitic gneiss contains 30 to 40 vol% quartz, 40 to 55% total feldspar, and lesser amounts of mica (Figure 50.3). Fractured gneiss and protomylonite of the transition zone contain less feldspar, more mica, and increased matrix fractions produced by alteration of feldspar to mica and neocrystallized albite, local plastic recrystallization of quartz, and microcracking. Phyllonite contains >40% mica, slightly increased and more variable quartz contents, and >80% matrix produced by widespread alteration of feldspar to mica and dynamic recrystallization of quartz. Distinct changes in whole-rock chemistry, marked by depletion of Ca, Na, and K and enrichment of Mg and $H_2O$ in shear-zone phyllonite, also accompanied alteration (Yonkee, 1990). Similar average abundances of Si, Al, and Ti for gneiss and phyllonite record <10% large-scale volume change during deformation, although more variable Si contents in phyllonite record local mobility of quartz.

**50H.** Migmatitic gneiss protolith. Quartz (Q), K-feldspar (K), plagioclase (P), and biotite (B) grains have a granoblastic texture. Quartz has undulatory extinction and widely spaced fluid-inclusion trains. Feldspar has undergone very limited alteration to fine-grained mica and albite along grain margins (arrow). Intragranular and intergranular fractures are scarce. Scale bar = 500 μm. NX

**50I.** Fractured gneiss from transition zone. Quartz (Q) displays undulatory extinction with local subgrains and recrystallized grains (r), reflecting limited dislocation creep. Plagioclase and K-feldspar (K) grains are cut by many microcracks (c), and fine-grained mica-quartz veins (v) separate feldspar fragments. Feldspar is partly altered to muscovite and albite (a) along cracks and grain margins. Scale bar = 500 μm. NX

**50J.** Protophyllonite from central shear zone. Quartz grains (Q) in a less deformed region display undulatory extinction with local recrystallization along grain margins, whereas in more deformed regions quartz is pervasively recrystallized into small grains (r) that sit in a matrix of foliated muscovite and chlorite (M). A large K-feldspar clast (K) displays undulatory extinction, mica-filled cracks (arrows), and neocrystallized albite (a) along its margin. An incipient shear band contains very fine-grained material with abundant chlorite and quartz (C). Scale bar = 500 μm. NX

**50K.** Phyllonite from central shear zone. Quartz (Q) forms ribbons with abundant subgrains (s) surrounded by recrystallized grains (r), recording extensive dislocation creep. Foliated aggregates of muscovite (M) contain highly elongate bands of chlorite and minor sphene (S), recording pervasive plastic deformation of micas. Shear bands contain very fine-grained material with abundant chlorite and quartz (C). A chlorite-quartz-filled vein (V) cuts foliated muscovite, and in turn is plastically deformed. Scale bar = 500 μm. NX

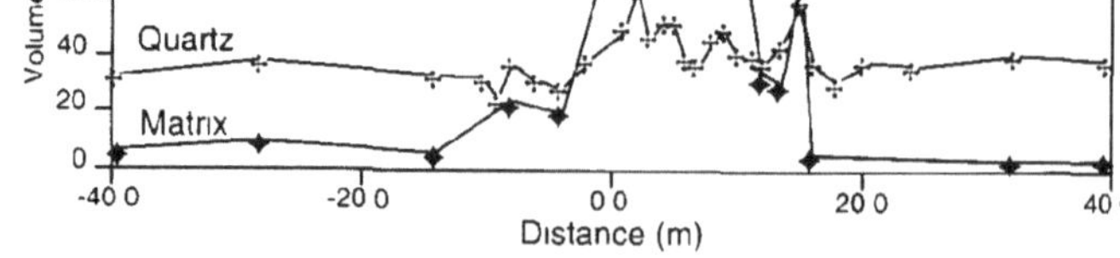


**Figure 50.3.** Mineral abundances and matrix fraction on traverse across Weber Canyon shear zone. Feld = total feldspar. Matrix = fraction of newly crystallized grains and fragments less than 50 μm in size. Distance measured from base of shear zone. General locations of wall rock (WR), transition zone (TZ), and central shear zone (SZ) indicated. Modified from Yonkee and Mitra (1993).

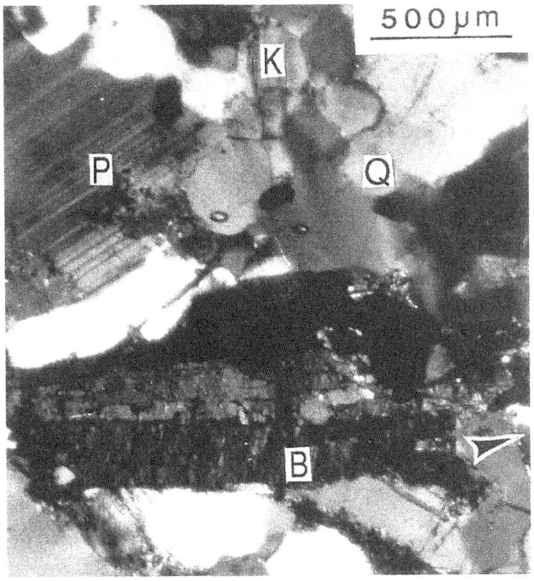

50 H

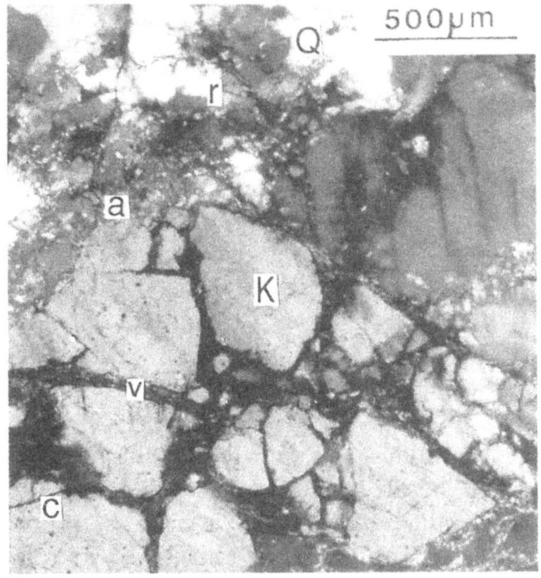

50 I

50 J

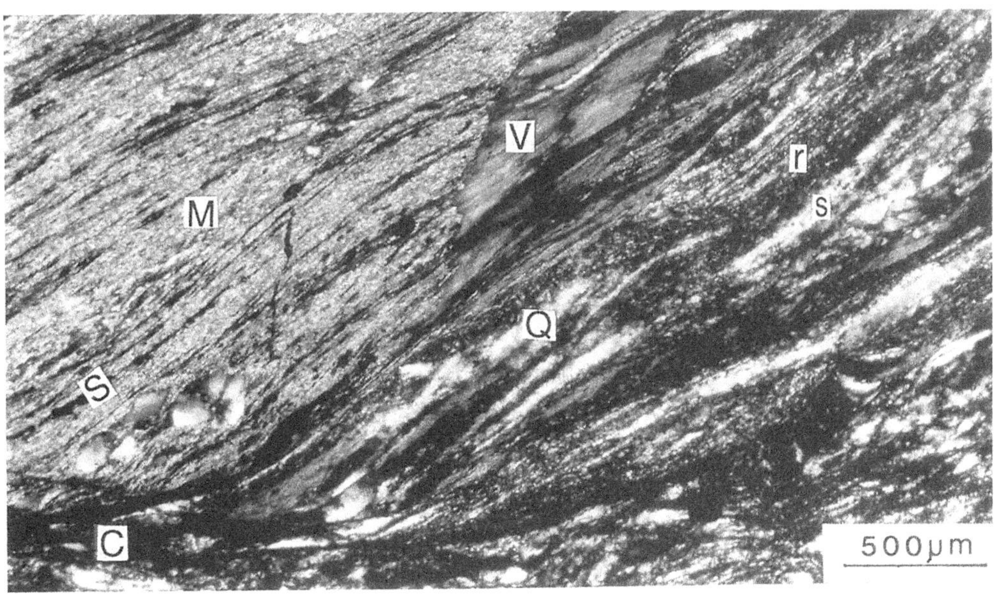

50 K

## Minor shear zones: Deformation styles and processes of shear zone growth

Minor shear zones, with thicknesses from <1 cm to ≈1 m, are developed locally within migmatitic and fractured gneiss along the margins of the Weber Canyon shear zone. The central parts of these minor zones display variations in microtextures, with thicker zones having overall stronger foliation and more pervasive alteration. Most minor shear zones are also bounded by thin transition zones of microcracked and altered gneiss and protomylonite. Thicknesses of minor shear zones are correlated positively with shear displacements, indicating that many zones probably grew in thickness with increasing displacement and time, although there is much scatter (Yonkee and Mitra, 1993). Thus, spatial variations in microtextures on traverses across individual zones and variations in microtextures between zones of varying thickness provide a crude record of deformation styles and processes of shear-zone growth during progressive deformation.

Thinner, smaller-displacement shear zones display overall weakly developed foliation and many appear to have nucleated along intergranular fractures and thin cataclastic zones (50L). Zones of fracturing appear to have provided pathways for initial fluid influx that produced partial alteration of feldspar to mica and localized plastic deformation of hydrolytically weakened(?) quartz. Many thicker zones display moderately developed protomylonitic to mylonitic foliation and increased matrix fractions produced by continued alteration of feldspar and dynamic recrystallization of quartz (50M). However, porphyroclasts of quartz and feldspar remain and cataclastic zones continued to form, especially along shear-zone margins. Some thicker shear zones have larger displacements and display strong mylonitic to phyllonitic foliation (50N). The centers of these zones consist almost entirely of very fine-grained matrix within which strain, dynamic recrystallization, and pervasive alteration were concentrated. These zones are also bounded by fractured gneiss in which matrix continued to be produced by alteration and recrystallization concentrated along fractures. These varying textures record an overall brittle-to-ductile transition during progressive deformation with repeated episodes of fracturing, fluid influx, alteration of feldspar to mica, and dynamic recrystallization of quartz producing progressively greater amounts of ductile matrix. Reaction softening and possible hydrolytic weakening resulted in significant strain softening within the shear zones, but zones also continued to grow in thickness along their margins by fracturing and fluid influx.

**50L.** Cataclasite and protophyllonite in 2 cm thick shear zone (arrows mark edges of shear zone). The zone has limited shear displacement concentrated within matrix near its margins and is in an early stage of development. The shear zone appears to have nucleated on a cataclastic zone composed of fractured fragments of quartz aggregates (Q), K-feldspar (K), and plagioclase (P), all of which sit in a matrix (m) that is partly altered to mica. Some quartz grains are little deformed, whereas others display numerous subgrains and local recrystallized grains (r). Feldspar grains are cut by cracks (c) and partly altered to mica. Scale bar = 500 μm. NX

**50M.** Protomylonite and cataclasite near edge of 8 cm thick shear zone. A fracture (f) marks the lower margin of the shear zone. Within the shear zone recrystallized quartz grains (Q) are elongate parallel to foliation. K-feldspar fragments (K) display undulatory extinction and microcracks, and are separated by quartz-mica fillings deformed into "bow-tie" shapes (arrows). A plagioclase porphyroclast (white P) is partly altered to fine-grained muscovite and albite. Shear bands (SB) and mica-filled veins (V) are locally developed. Cemented cataclasite (CC) along the shear-zone margin consists of angular fragments sitting in a chlorite-rich matrix. Below the margin of the zone, quartz (q) and plagioclase (black p) grains are locally fractured, but overall less deformed and altered. Scale bar = 500 μm. NX

**50N.** Mylonite along edge of 5 cm thick shear zone (SZ) bounded by fractured transition zone. The shear zone consists of finely recrystallized quartz, muscovite, chlorite, actinolite, and albite that define a strong mylonitic foliation. Gneiss along the margin of the shear zone is cut by intergranular fractures (large arrows) that are subparallel to the shear zone and become more widely spaced away from the zone. Quartz (Q) has widespread undulatory extinction with recrystallized grains (r) concentrated along fractures and fluid inclusion trains (small arrow) that were probably pathways for fluid influx. K-feldspar (K) has undulatory extinction and is altered to muscovite (m) along fractures. Mica fibers at moderate angles to the fracture boundaries record minor extensional and shear displacement having the same sense as the mylonitic shear zone. Scale bar = 500 μm. NX

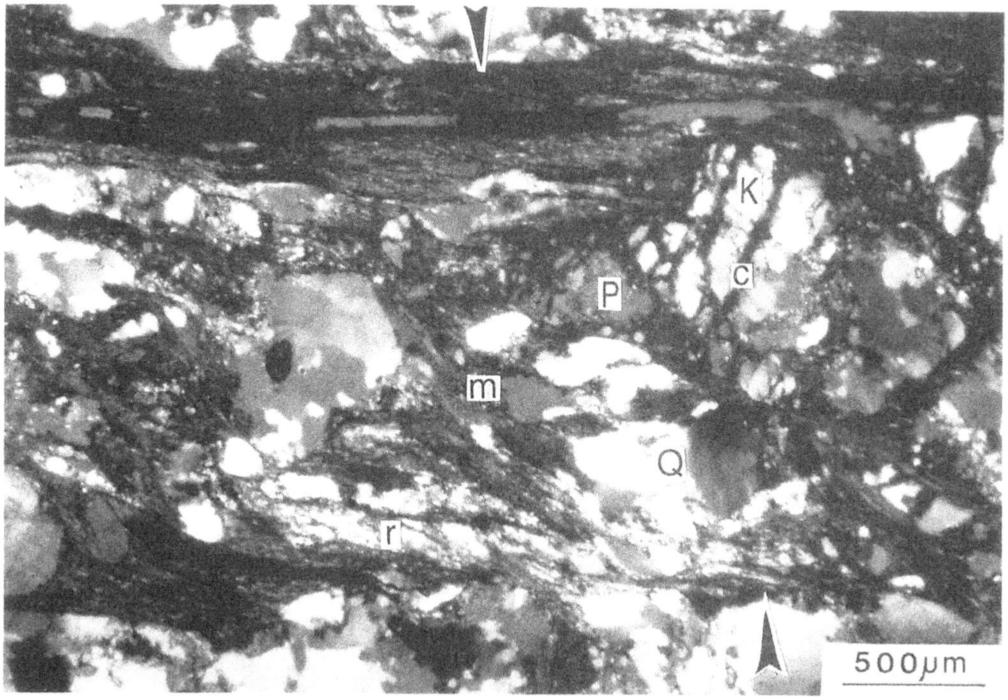

50 L

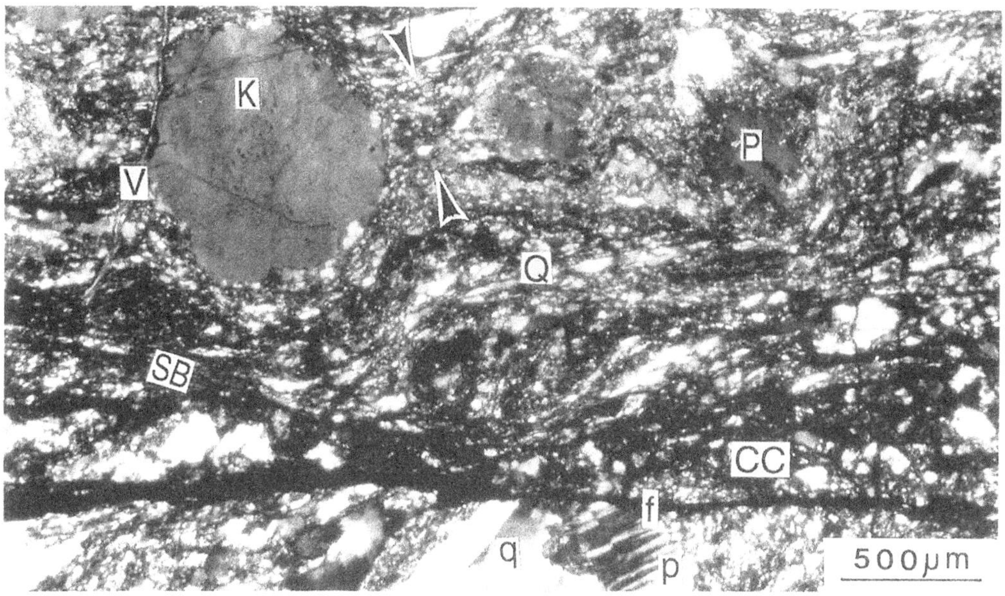

50 M

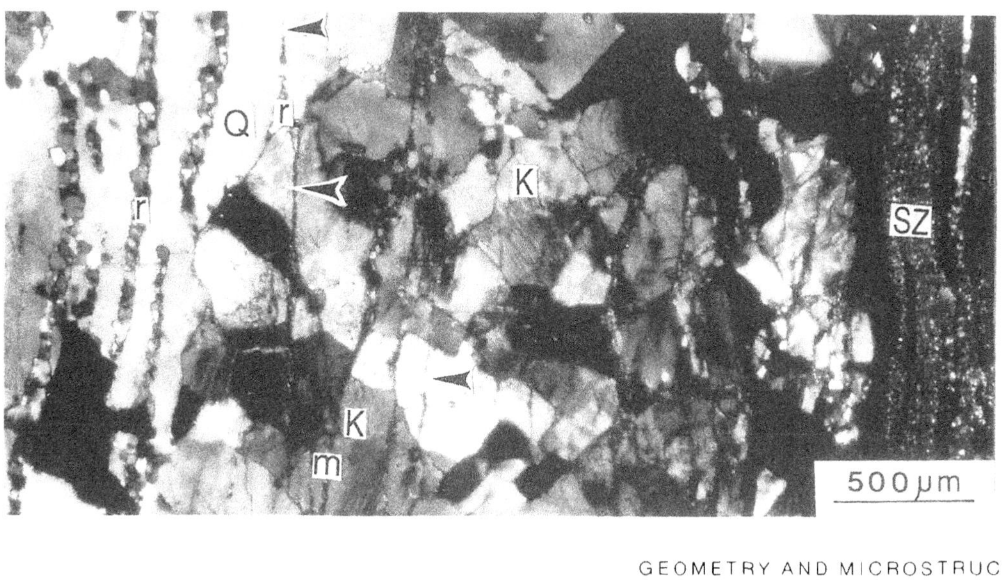

50 N

J. A. KARSON

## Field relations

The axis of the Mid-Atlantic Ridge (MAR) at 24°N appears to be typical of many slow-spreading (20 mm/yr) ridge segments where mechanical extension of the lithosphere dominates over magmatic construction. The rift valley has an asymmetrical cross section with a relatively steep, high, western median valley wall (MVW), a broad median valley floor (MVF), and a relatively gently sloping, lower, eastern median valley wall (Figure 51.1). Manned submersibles and deeply towed camera sleds have been used to map the geology of this area (Karson and Dick, 1983; Karson and others, 1987; Mével and others, 1991). The MVF is a rugged volcanic terrane, less than $5 \times 10^5$ yr old, dominated by basaltic pillow lavas cut by numerous small (tens of meters high) normal faults and fissures. The eastern MVW consists exclusively of block-faulted basaltic rocks. In contrast, the western MVW has extensive exposures of variably deformed (mylonitic to cataclastic) and metamorphosed (amphibolite to subgreenschist facies) gabbroic rocks and is interpreted as the exhumed footwall of a major detachment fault in a highly extended oceanic core complex. Low-angle (~30°), east-dipping shear zones and fault surfaces essentially parallel the slope of the wall. These are interrupted by active or very young, steeply dipping (~60°) faults with vertical relief of up to 200 m. The gabbroic rocks are approximately 1 Ma and are believed to have crystallized at crustal depths of 2 to 5 km before being uplifted and exhumed. Major low-angle detachment faults may be common features of crust produced at slow-spreading ridges (Mutter and Karson, 1992).

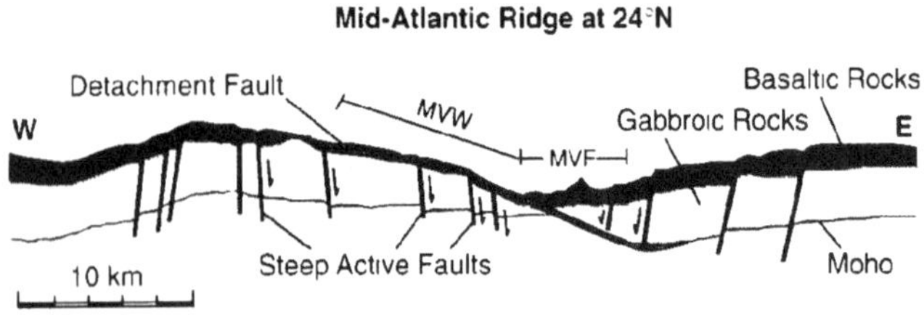


**Figure 51.1.** Schematic cross section across the Mid-Atlantic Ridge at 24°N

**51A.** Vertical fissure in pillow basalts of the median valley floor of the MAR. The fissure is parallel to the MAR axis. Bulbous basaltic pillows and angular talus are partially blanketed by smooth, light-colored carbonate ooze. Vertical incident photo from camera sled.

**51B.** Moderately east-dipping normal fault cutting massive, jointed gabbroic rock on the western median valley wall of the MAR. Shear zones and faults such as this one combine to form a major detachment fault. This fault surface is truncated by an active, steeply dipping normal fault just outside the field of view. Photograph from side viewport of the manned submersible *Alvin*.

51 A

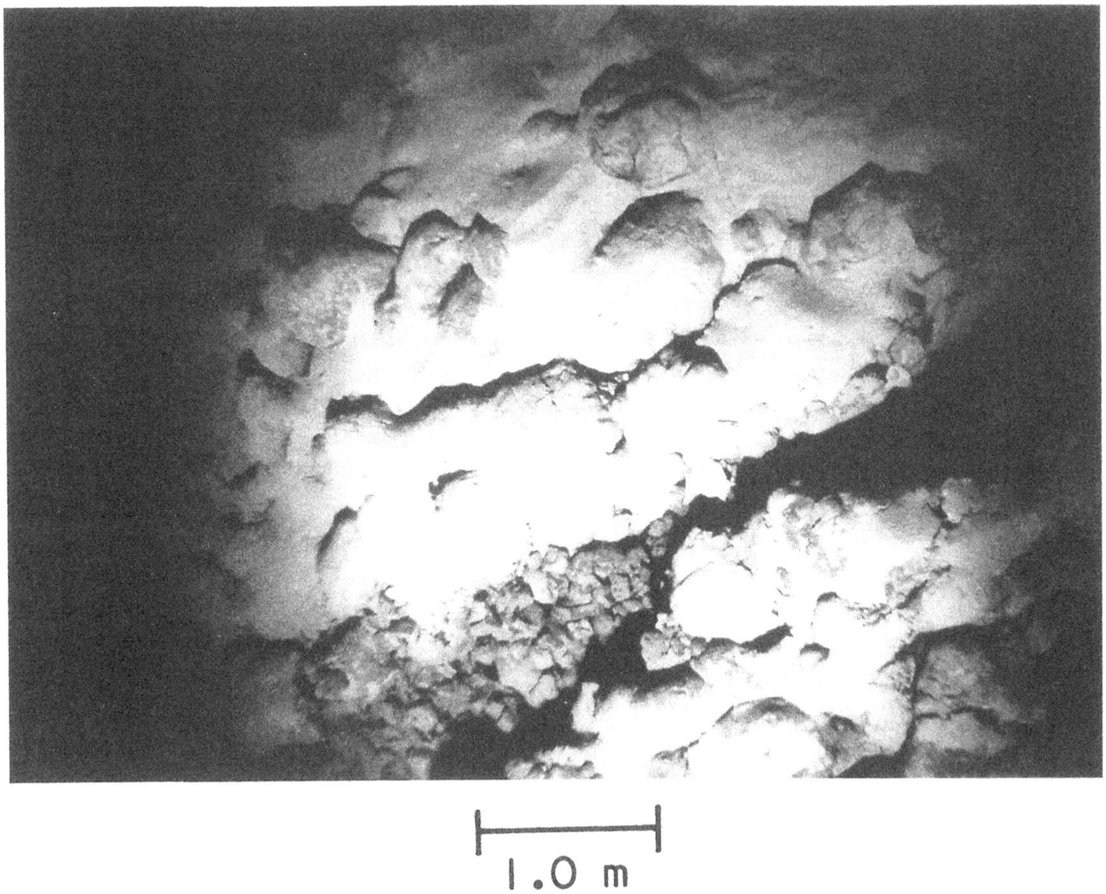


51 B

## Petrography

Gabbroic rocks and their deformed and metamorphosed derivatives crop out extensively on the western median valley wall of the Mid-Atlantic Ridge (MAR) at about 24°N, just south of the Kane Transform. Extensive collections have been made with the manned submersibles *Alvin* and *Nautile* (Karson and Dick, 1983; Mével and others, 1991) and by drilling on Ocean Drilling Program Leg 153 (Cannat and others, 1995). Field and petrographic relations of these rocks demonstrate that deformation, metamorphism, and hydrothermal processes have been confined to discrete fractures, faults, and shear zones that cut unaltered gabbroic rocks. Greenstone microbreccias, cataclasites, chlorite schists, and mylonites are the most common fault rocks. These are considered to be derived from progressively uplifted and retrograded midcrustal rock units that are approximately 1 Ma old. The rocks are exposed in the footwall of a major detachment fault in a highly extended oceanic core complex (Karson and others, 1987; Karson, 1990).

**51C.** Coarse-grained cumulate gabbro cut by a shear fracture filled with very fine-grained chlorite, albite, epidote, and quartz. Note very fine fractures in clouded plagioclase and pyroxene grains near the fracture. NX

**51D.** Greenstone microbreccia composed of porphyroclasts of plagioclase, quartz, and greenstone in a finely comminuted matrix of chlorite, albite, and quartz. The large prominent porphyroclast (P) has fine mylonitic layering ($S_m$) cut by a hydrothermal quartz veinlet (q). Multiple generations of cross-cutting, chlorite-rich shear fractures (f) and open extension fractures (white) cut the matrix. PPL

**51E.** Chlorite schist with spaced, curviplanar shear fractures ($S_c$) cutting an earlier pervasive anastomosing fabric $S_a$. $S_a$ is defined by finely laminated mylonite enclosing numerous small angular porphyroclasts (plagioclase and quartz) and larger phacoidal domains (greenstone). Hydrothermal quartz veinlets (q) cut an earlier generation of mylonitic lamination ($S_l$) in some of the greenstone domains. Both the veinlets and $S_l$ are truncated by $S_a$. PPL

51 C

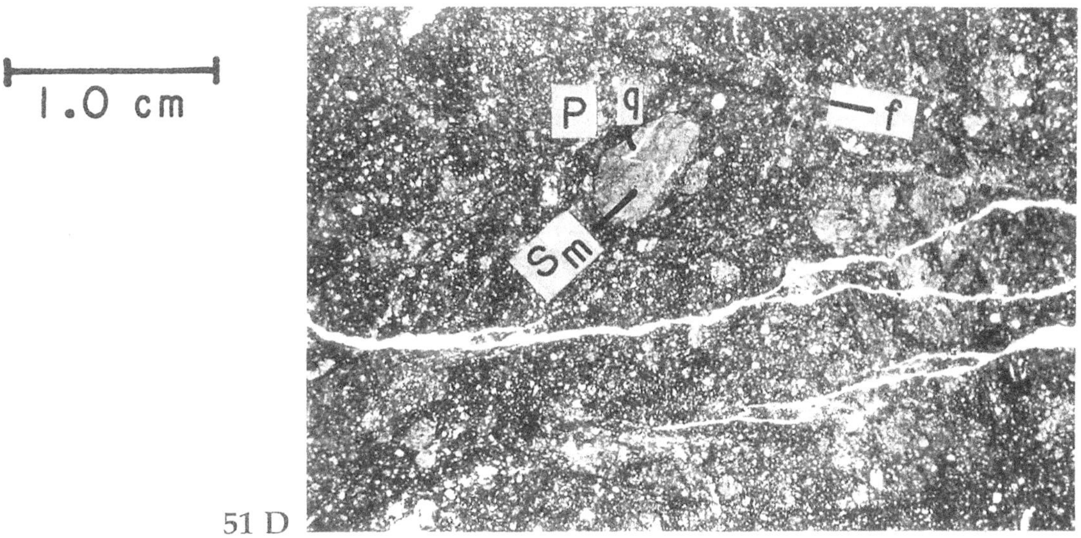


51 D

51 E

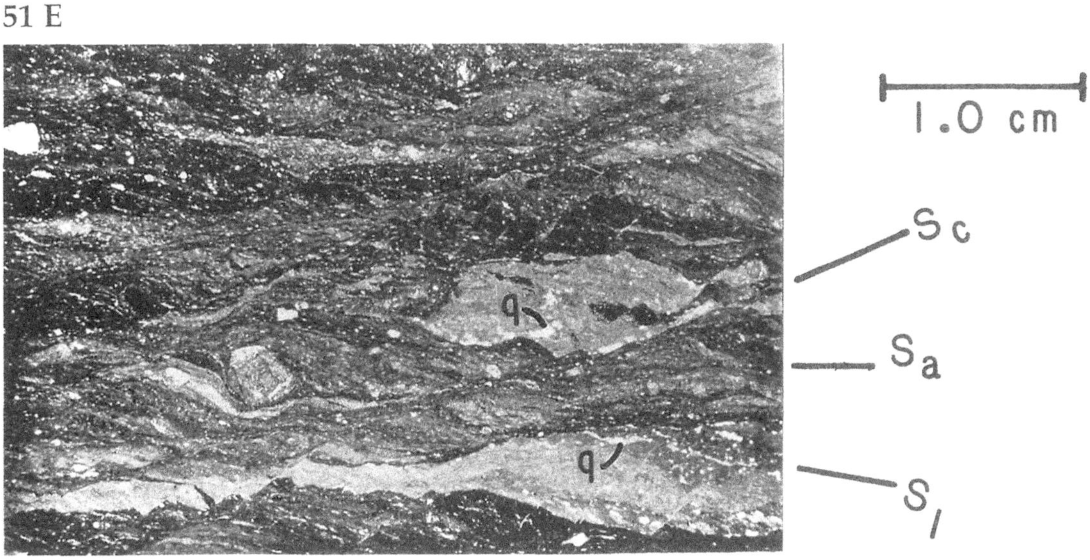

# 52. CATACLASTIC TO CRYSTAL-PLASTIC TRANSITION IN HALITE SHEAR ZONES

FREDERICK M. CHESTER

The microstructures of halite layers developed during simple shear deformation under controlled conditions of temperature, pressure, and shear displacement rate have been studied (Chester, 1988). The halite layers are approximately 0.5 mm thick and were sheared approximately 4 mm in a triaxial rock deformation apparatus. The shear deformation occurred at conditions of normal stress (across the layer) from 20 to 70 MPa, temperatures from 22°C to 250°C, and shear displacement rates from 0.030 μm $s^{-1}$ to 300 μm $s^{-1}$ (Chester, 1988, 1989; Chester and Logan, 1990). An increase in normal stress, an increase in temperature, or a decrease in shear rate tends to promote a transition in the dominant deformation mechanism from cataclasis to intracrystalline plasticity. Mechanical behavior and microstructures were used to define the conditions over which different deformation mechanisms were dominant. The photographs illustrate the microstructure of three important deformation mechanism fields identified at room temperature: (52A) localized slip, (52B) cataclastic flow, and (52C) semi-brittle flow.

The photomicrographs are of petrographic thin sections oriented perpendicular to the halite layer and parallel to the slip direction in a right-lateral shear sense. Scale bar is 0.5 mm; PPL. Small arrows at the sides of the photos locate the boundary of the halite layer in contact with sandstone blocks used to shear the layers. Separation of the halite layer occurs during depressurization after shearing to form a gap that is filled with epoxy during thin-section preparation; the location and extent of the epoxy is indicated with a black bar and e.

**52A.** Localized slip is the dominant mechanism at low normal stress and high shear displacement rates. A layer sheared at 20 MPa normal stress, 22°C, and 3.0 μm $s^{-1}$ shear displacement rate is shown. Shear displacement occurs almost entirely along a slip surface near one boundary of the halite layer (large open arrow). The slip surface is more planar than the boundary of the sandstone block; halite has filled in the sandstone irregularities and occurs on both sides of the slip surface. Extension fractures are oriented at approximately 45° to the layer and are preferentially located along the sliding surface (small open arrow). Grain boundaries also are evident due to impurities (solid arrows).

**52B.** Cataclastic flow is the dominant mechanism at low normal stress and low shear displacement rates. A layer sheared at 30 MPa normal stress, 22°C, and 0.03 μm $s^{-1}$ shear displacement rate is shown. Shear displacement occurs via fracturing and cataclasis throughout the layer. Localized (Riedel) shears are apparent within the layer (open arrow). A crude S foliation is developed locally along the boundary of the layer by reorientation of grain boundaries (solid arrow).

**52C.** Semi-brittle flow is the dominant mechanism at all shear displacement rates at high normal stress. A layer sheared at 45 MPa normal stress, 22°C, and 3.0 μm $s^{-1}$ shear displacement rate is shown. Shear displacement is distributed throughout the layer and occurs by intracrystalline plasticity and some fracturing. The halite undergoes dynamic recrystallization, but semi-brittle behavior is implied by the frictional (pressure-dependent) shear strength. Localized (Riedel) shears are apparent within the layer (open arrow). A uniform and penetrative S foliation is developed by reorientation of grain boundaries (solid arrows). The orientation of the S foliation is consistent with homogeneous simple shear in the halite.

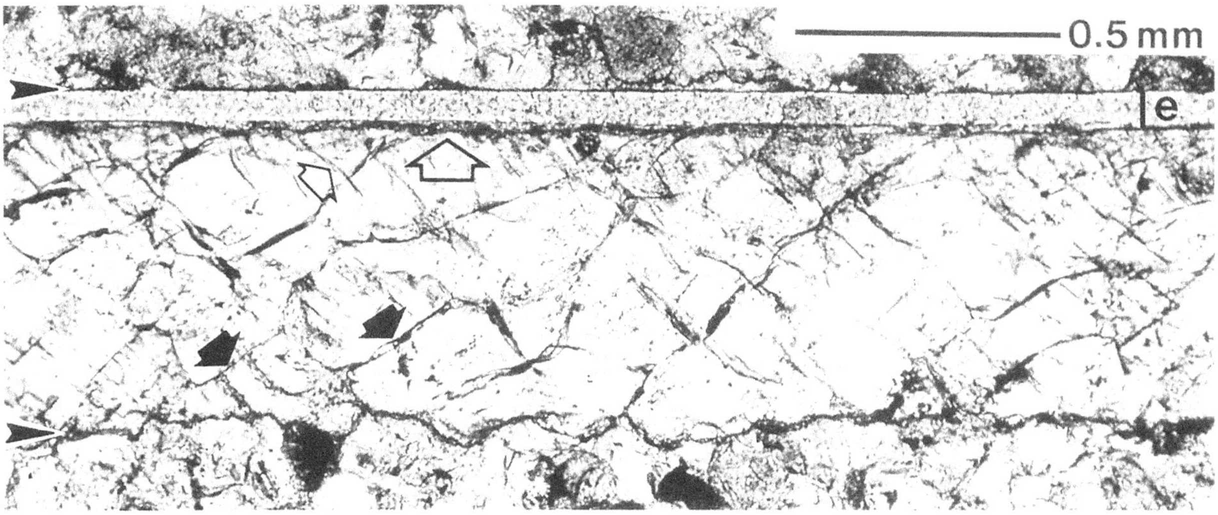

52 A

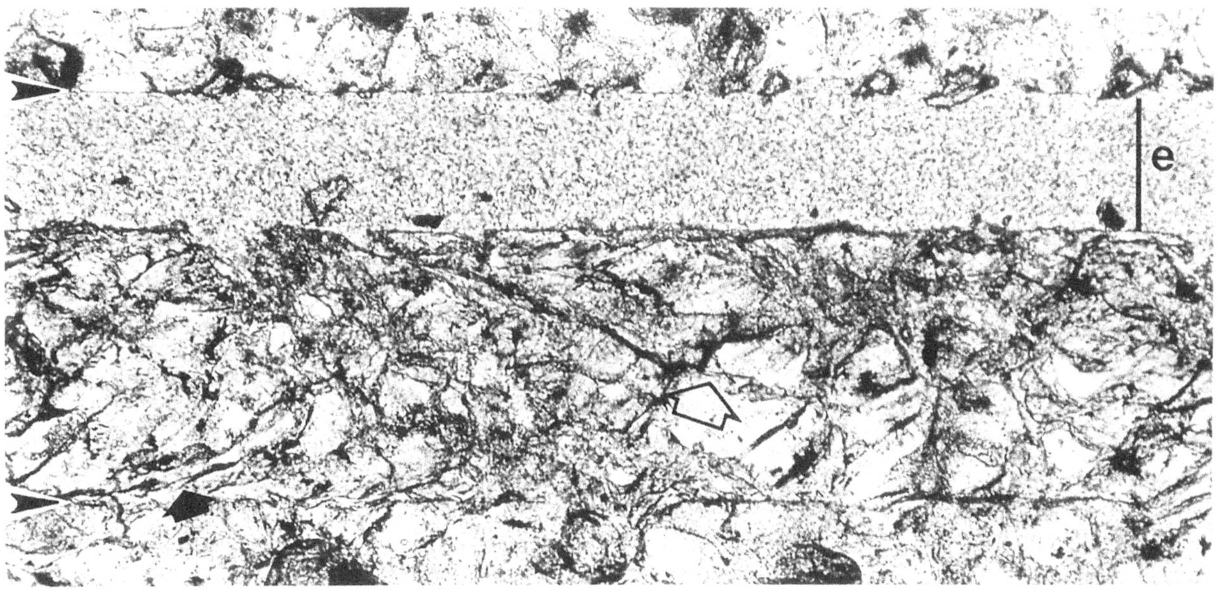

52 B

52 C

# 53. CATACLASTIC FLOW OF FELDSPAR AGGREGATES

JAN TULLIS AND RICHARD A. YUND

Feldspar aggregates experimentally deformed at intermediate temperatures and pressures deform by macroscopically ductile cataclastic flow (Figure 53.1), a process involving distributed cracking with no contribution from dislocation processes (e.g., Tullis and Yund, 1987, 1991). The presence of two cleavages in feldspars is important in promoting this deformation mechanism; in contrast, quartz aggregates progress directly from faulting to semi-brittle flow with increasing temperature and pressure (Hirth and Tullis, 1994).

All of the photos are from samples of fine-grained (d ~ 200–400 μm) Bushveld anorthosite ($An_{75}$) axially compressed at a confining pressure of 1000 MPa, a constant strain rate of $10^{-5/sec}$, and a temperature of 300°C. In all of the optical micrographs the compression direction is vertical.

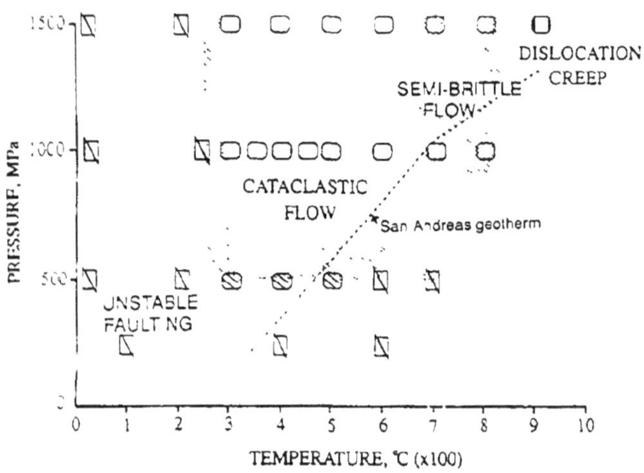


**Figure 53.1.** Plot of confining pressure vs. temperature, with symbols showing deformation experiments on anorthosite at a strain rate of $10^{-5}$ sec. There is transition with increasing pressure and temperature from faulting to cataclastic flow (involving distributed cracks with no dislocations) to semi-brittle flow (involving a combination of cracking and dislocation glide) to dislocation creep. At natural strain rates the temperatures of these transitions would be lower. *Note:* This figure is a slightly modified version of Figure 1 from Tullis and Yund (1992).

**53A.** Optical micrograph of sample shortened 10%. Note multiple grain-scale faults (NW-SE), especially evident where they offset preexisting twins. These grain-scale faults occur primarily along cleavage planes and rarely, if ever, continue across grain boundaries into adjacent grains. Patchy undulatory extinction tends to be developed throughout the grains: although it may appear similar to subgrains, TEM shows it is due to distributed submicroscopic cracks. NX

**53B.** Optical micrograph of sample shortened 30%. Note grain-scale faults (NW-SE, again evident because they offset preexisting twins). The faults tend to be more closely spaced than they are at lower strain. NX

**53C.** Optical micrograph of sample shortened 30%: This grain has developed conjugate grain-scale shear zones of discrete width (mostly NW-SE in this micrograph), with strong patchy undulatory extinction. These optical-scale microstructures are the reflection of submicroscopic distributed cracks and crush zones; no dislocations are involved. NX

**53D.** Optical micrograph of sample shortened 50%. The field of view includes portions of several original grains, but it is almost impossible to distinguish them due to the intense crushing. Conjugate shear zones are wider than at lower strain, and consist of finely crushed material, with no component of dynamic recrystallization. NX

**53E.** TEM micrograph (bright field) from sample shortened 10%. Multiple subparallel cracks terminate in a thin (~1 μm wide) microcrush zone (SE third of micrograph), within which the grain size is <0.1 μm. Streaking of spots in the selected area diffraction (SAD) pattern from the cracked region (upper left corner) indicates some shear on the cracks; extensive streaking in the SAD from the crush zone (lower right corner) reflects rotation of the fragments.

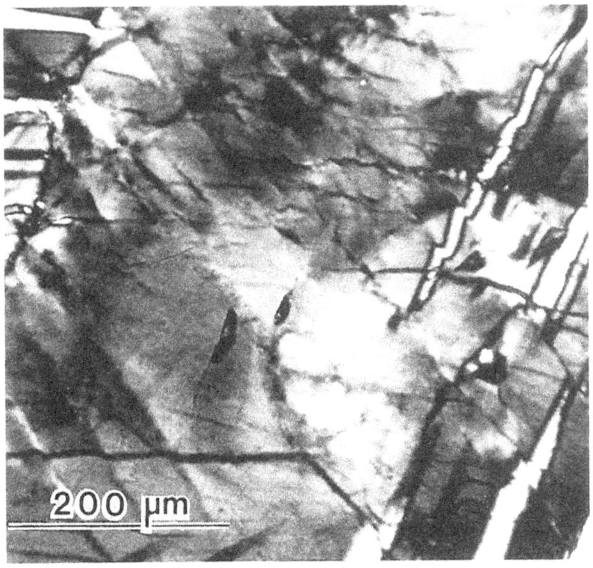


53 A

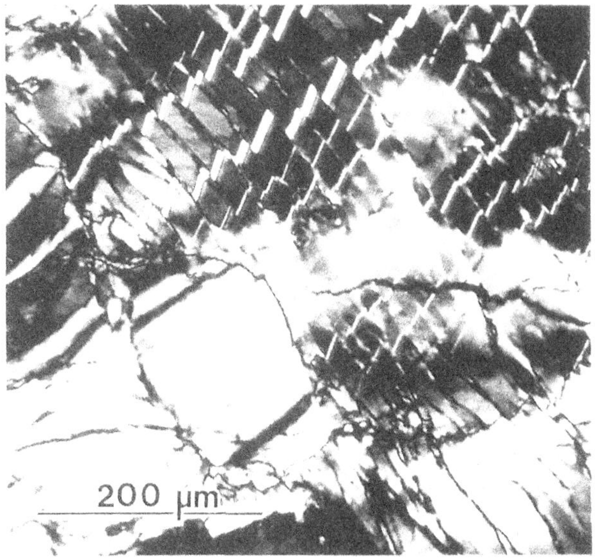


53 B

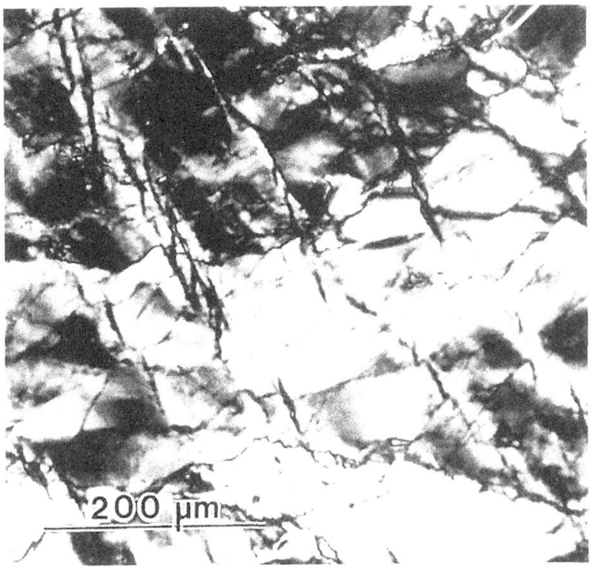


53 C

53 D

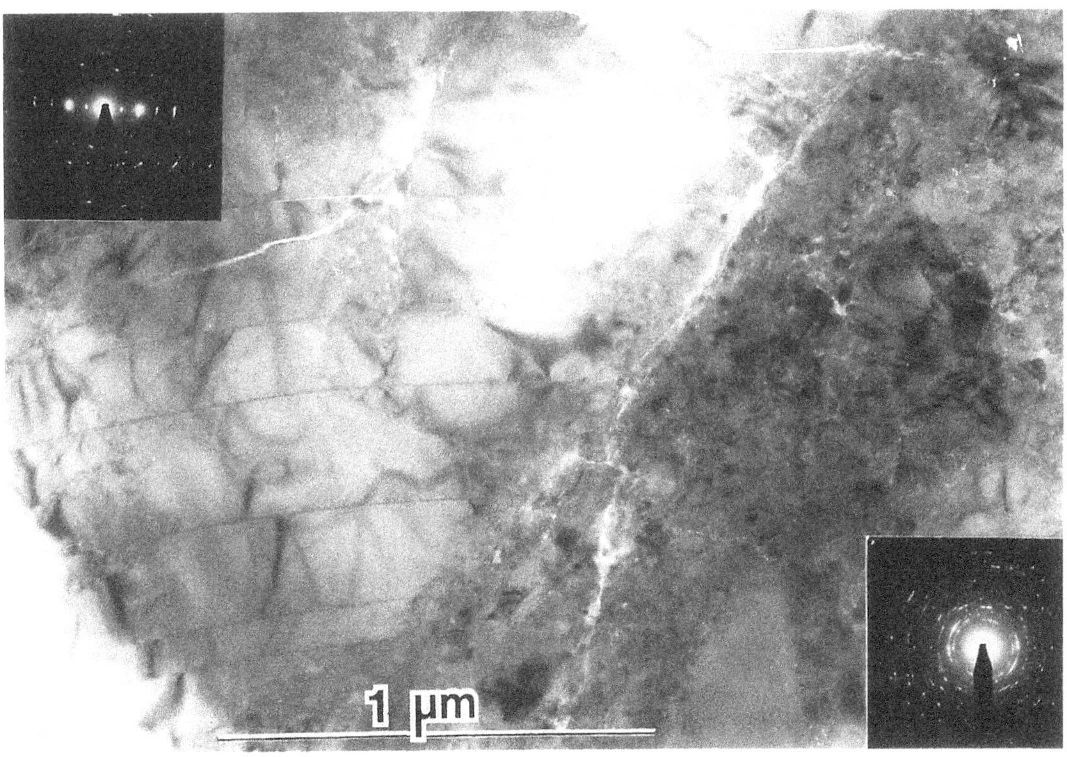


53 E

# SEMI-BRITTLE BEHAVIOR

## Fluid-related features



Common abbreviations used in regard to the plates in this section include PPL = plane polarized light, NX = crossed nicols, and SEM = scanning electron microscope.

This set of plates illustrates a further complexity of rocks deformed at depths spanning the brittle-to-plastic and seismic-aseismic transitions, namely, the influence of fluids. These plates document that even though the dominant deformation in low-grade shear zones may be crystal-plasticity, the permeability in these shear zones is higher than in the undeformed wall rocks. The evidence for the influence of fluids ranges from outcrop features to thin-section and SEM-scale microstructures, as well as oxygen isotope systematics and radiometric dating. The plates illustrate microstructures such as chemical zoning patterns and microscale pores, which indicate that very different flow regimes can exist in adjacent rock layers within a shear zone. The enhanced permeability may be transient and highly channelized in fracture networks in some rock layers, whereas in other rocks it may be relatively pervasive, perhaps aided by grain-boundary sliding. Several of the contributions add to the growing evidence that interphase boundaries are more permeable than same-phase boundaries. Another important aspect of fluids that is illustrated in the plates is its role in promoting retrograde reactions; these mineralogical changes typically decrease grain size, which in turn may allow a switch to grain-size-sensitive creep.

# 54. COUPLED MASS TRANSFER AND MICROFRACTURING IN GABBROIC MYLONITES

Bruno Lafrance and Ron H. Vernon

Gabbroic rocks of the Late Jurassic Guadalupe Igneous Complex (GIC), Sierra Nevada, California, U.S.A., were deformed in the Bear Mountain fault zone shortly after emplacement of the complex. Deformation produced a strongly differentiated mylonitic foliation, consisting of well-defined layers rich in amphiboles alternating with layers of recrystallized plagioclase. The plagioclase-rich layers are the products of plastic deformation by dislocation creep and dynamic recrystallization by grain-boundary migration. The compositions of recrystallized and neocrystallized grains suggest amphibolite-facies conditions during deformation.

Although the plagioclase-rich layers formed essentially in the solid state, the amphibole-rich layers were strongly affected by fluid flow, microfracturing, and mass transfer. Variations in pore fluid pressures, strain rates, and/or stresses caused microfracturing, opening of dilatant sites in "pressure-shadows" adjacent to pyroxene and amphibole porphyroclasts, and the crystallization in these sites of new amphibole grains to form "tails" of smaller, elongate amphibole grains. Coalescence of the "pressure-shadows" and further crystallization of new amphibole grains in gaps formed during grain-boundary sliding produced the amphibole-rich layers.

Spectacular monomineralic veins of scapolite, quartz, and sphene are parallel to the mylonitic foliation. These veins are also products of coupled mass transfer and microfracturing. The sphene veins indicate mobility of titanium under these conditions, as do planar concentrations of ilmenite occurring as layers in recrystallized plagioclase and scarce transgressive veins. All photographs are from Lafrance and Vernon (1993).

**54A.** Differentiated mylonitic foliation in a gabbroic mylonite. Amphibole-rich layers (Amp-layer) alternate with layers of recrystallized plagioclase (Pl-layer). Plagioclase porphyroclasts (PL) in the recrystallized layers are labradorite, recrystallized grains contiguous to the porphyroclasts are andesine, and all other recrystallized grains are oligoclase. Dynamic recrystallization by grain-boundary migration in the presence of a Na-rich fluid homogenized the composition of the recrystallized grains in the plagioclase layers. In the amphibole-rich layers, amphiboles crystallized in "pressure-shadows" adjacent to clinopyroxene porphyroclasts (PX) and in gaps formed during grain-boundary sliding. Veins of scapolite (Scp) and elongate grains of sphene (Spn) appear to have crystallized in foliation-parallel fractures. PPL

**54B.** Quartz plate (Q) parallel to the mylonitic foliation. Quartz plates are veins with sharp, planar boundaries, and are composed of a few large grains that generally span the width of the veins. Quartz presumably formed as a product of the breakdown of pyroxene to amphibole. The plagioclase porphyroclast (PL) has patchy extinction, suggesting strain hardening of the grain during deformation, and is cut by fractures parallel to polysynthetic twins. NX

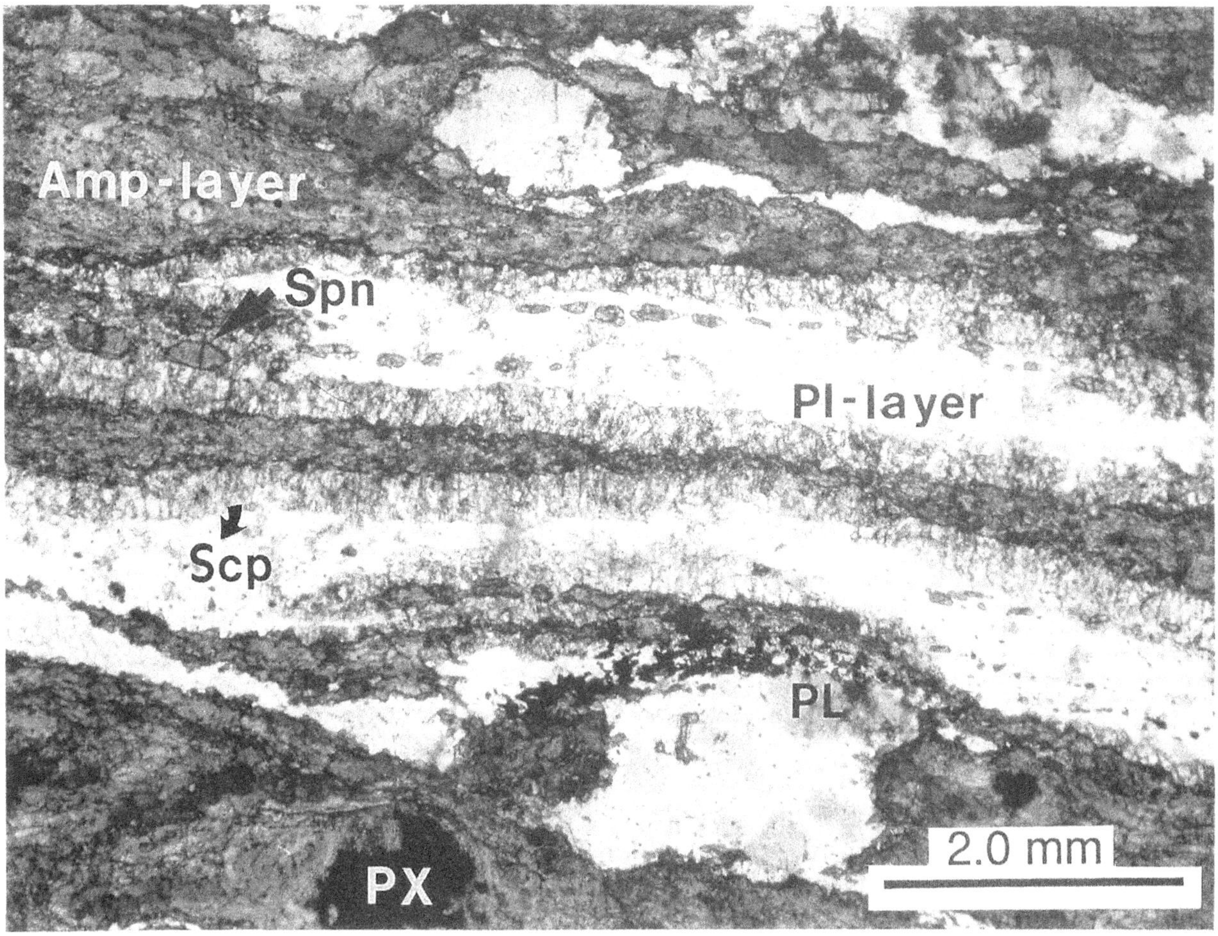


54 A

54 B

**54C.** Two-stage replacement of clinopyroxene by amphiboles. The clinopyroxene (PX) is replaced by an optically continuous actinolite pseudomorph (Act), which is in turn replaced by magnesiohornblende (Hbl). The magnesiohornblende crystallized in the "pressure shadow" adjacent to the actinolite and then grew into the actinolite. Amphibole porphyroclasts in these rocks are produced by the complete replacement of pyroxene porphyroclasts by actinolite pseudomorphs. The arrow shows the area where clinopyroxene was being replaced by actinolite. NX

**54D.** Actinolite porphyroclast (Act) with asymmetrical "tails" of smaller magnesiohornblende grains (Hbl). The actinolite porphyroclast behaved as a rigid object in a more ductile recrystallized plagioclase matrix. Because the porphyroclast was more competent than the matrix, microfractures opened at the interface between the actinolite porphyroclast and the matrix. Fluids flowed to the microfractures and magnesiohornblende crystallized as overgrowths on the actinolite porphyroclast. The new amphibole grains then grew into the porphyroclast, presumably parallel to low-index rational crystallographic planes. NX

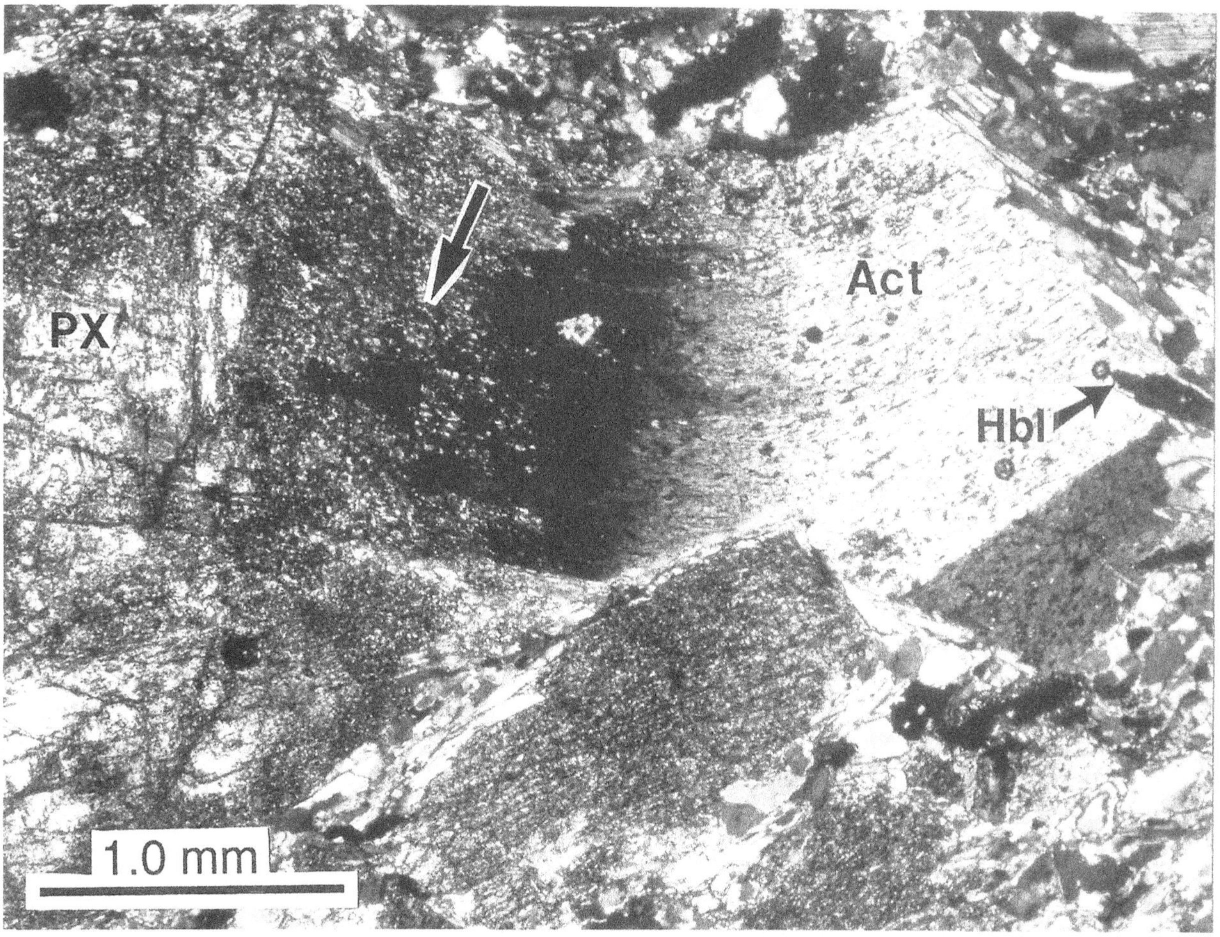


54 C

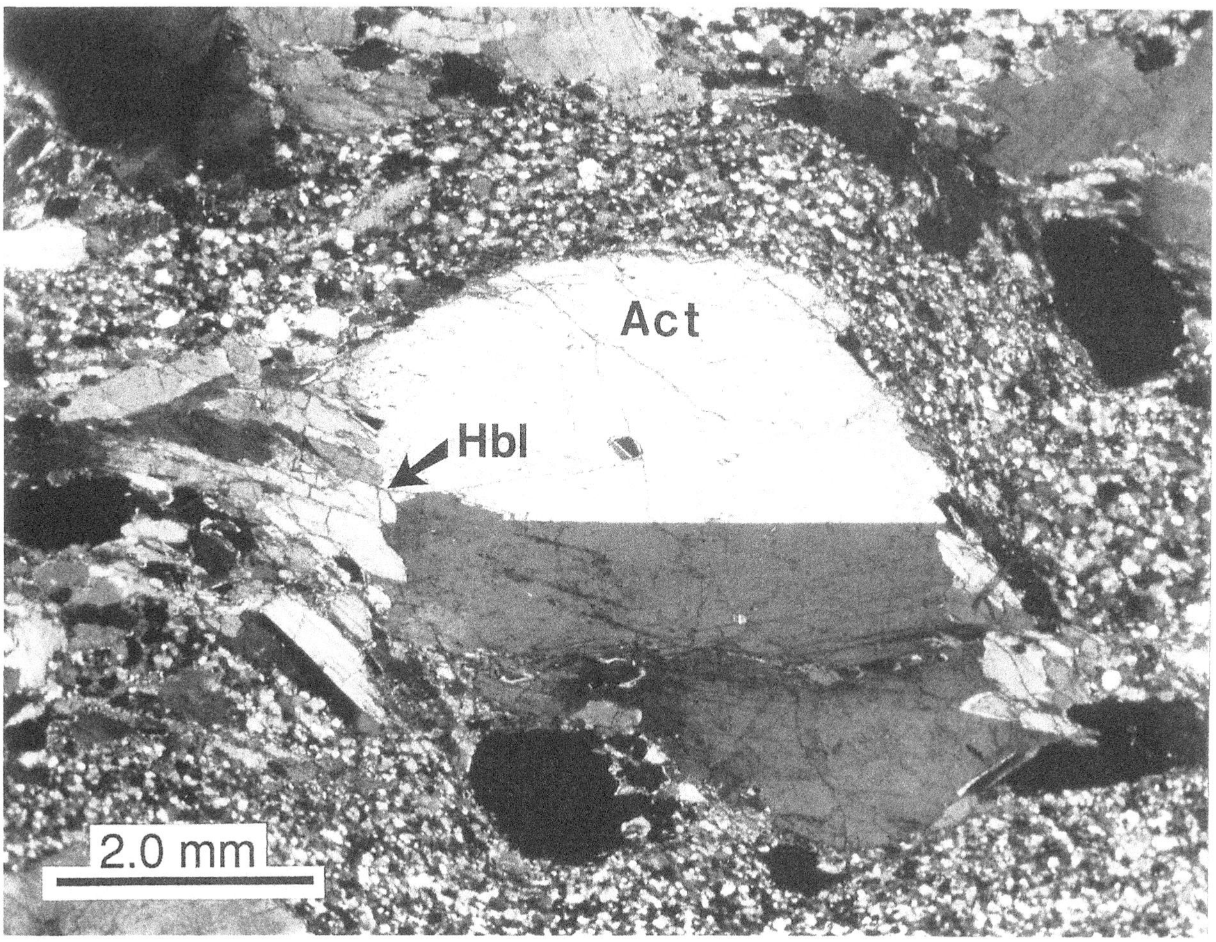


54 D

# 55. FLUID FLOW MECHANISMS IN MYLONITES: EVIDENCE FROM COMPOSITIONAL ZONING PATTERNS

ANDREW MCCAIG

Geochemical evidence shows that mylonitic shear zones are commonly important channelways for fluid flow, and that average permeabilities in mylonites may be considerably higher than those in undeformed wall rocks (McCaig and others, 1990, 1995). This plate and the following two illustrate how microstructural and microchemical data can be combined to infer the permeability structure of mylonites and the extent to which fluid flow was pervasive or channelized at the hand specimen scale, and whether flow was transient or continuous (see McCaig and Knipe, 1990; Knipe and McCaig, 1994).

**55A.** Mylonite from a retrogressive, greenschist-facies shear zone cutting gneisses in the Aston massif, central Pyrenees, Spain. Large volumes of fluid moved through these shear zones during deformation, causing reductions in $\delta^{18}O$ ratios and metasomatic alteration (McCaig and others, 1990). Layers of various compositions including granite (Y) and mixed diorite/metapelite (X) have been inherited from the gneisses, and have been folded and transposed into an ultramylonite band at the top of the photo. PPL

**55B.** Part of a backscattered SEM montage of a dioritic/pelitic layer in the mylonite (at location X in 55A). Labeled phases are epidote (ep), plagioclase (plag), quartz (qz), phengite (ph), and tschermakitic amphibole (amph). Biotite, chlorite, graphite, and apatite are also present, and garnet and sphene occur in nearby parts of this layer. Plagioclases up to 150 μm in size are optically strain-free and considerably smaller than the grain size of the gneissic protolith; they are inferred to be porphyroblasts which grew in the mylonite during deformation. Adjacent plagioclase porphyroblasts (plag 1 and plag 2) contain contrasting zoning patterns (Figure 55.1). Plag 1 is reverse zoned from $An_{26}$ to $An_{36}$, whereas plag 2 is normally zoned from $An_{27}$ to $An_{20}$, with sharp steps in composition at one end. In both cases, the grains are interpreted to have grown lengthwise into their own pressure shadows, with the sharp steps in plag 2 representing fracture events. Closed-system metamorphic reactions driven by P-T change or pressure solution would produce similar directions of zoning in all grains within the same layer. McCaig and Knipe (1990) inferred instead that the zoning was driven by access of fluids with different Na/Ca ratios to different plagioclase growth sites within the mylonite (channelized flow), and that the fluid composition reaching plag 2 changed rapidly with time, probably due to switching of flow paths between different compositional layers (transient flow).

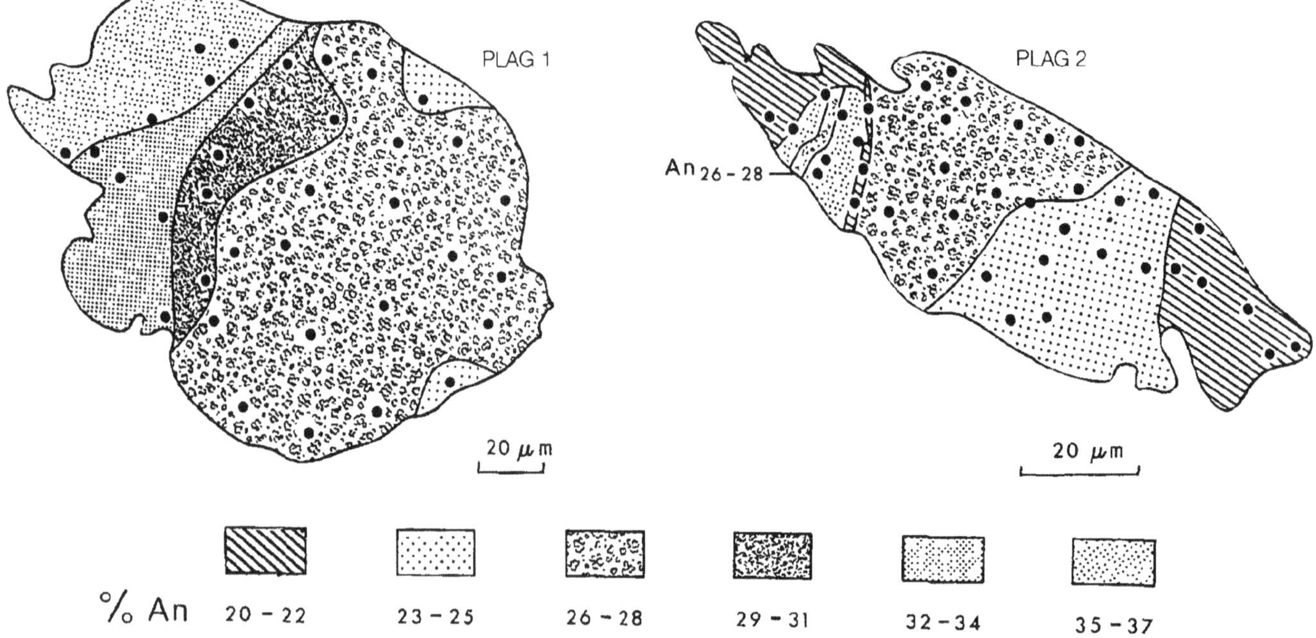


**Figure 55.1.** Zoning patterns in plagioclase porphyroblasts identified in 55B Dots are electron probe analysis sites

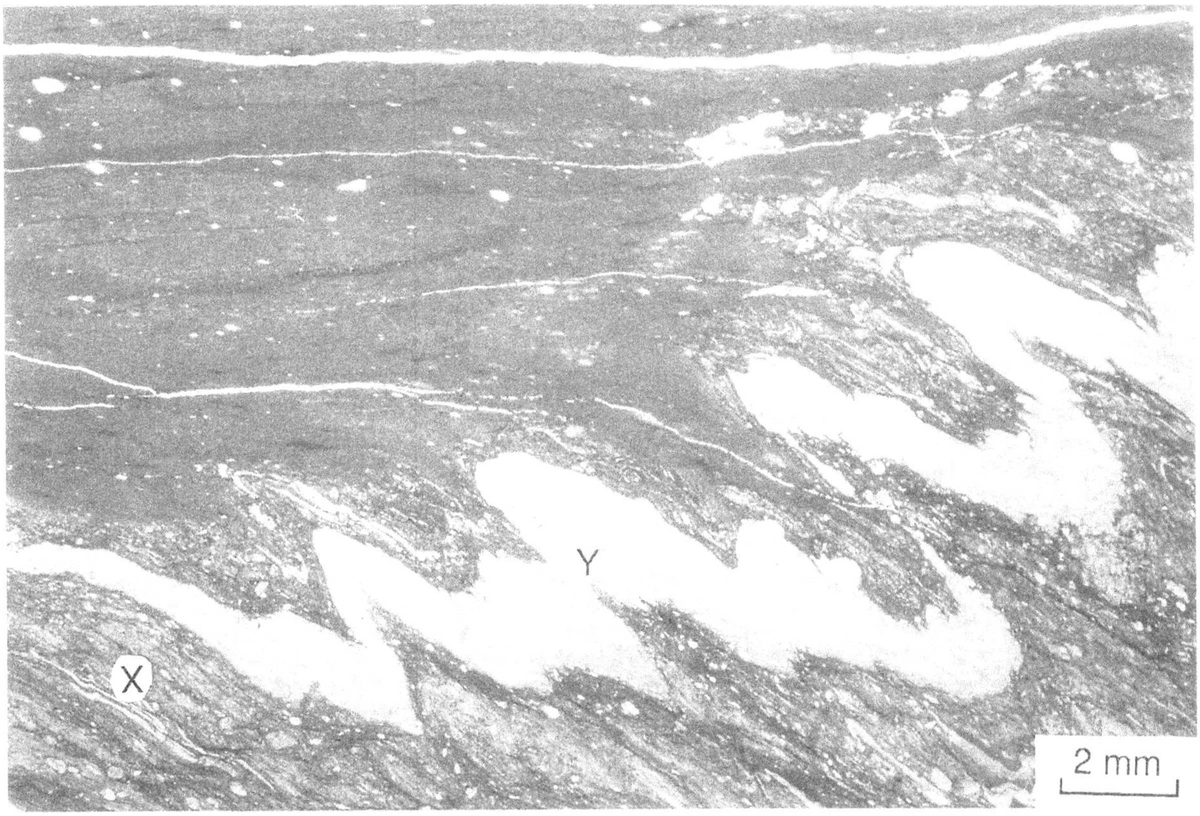


55 A

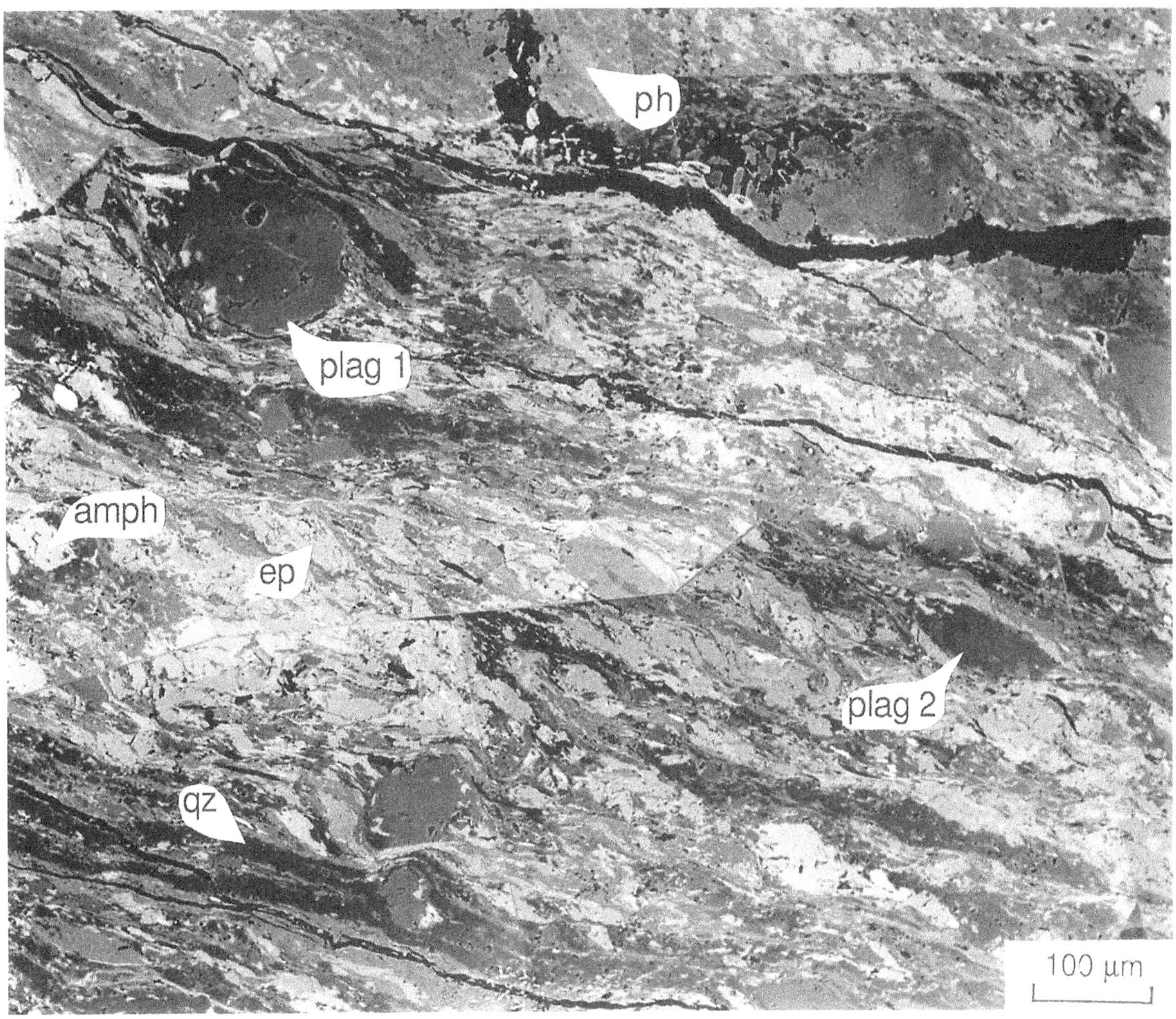


55 B

# 56. FLUID FLOW MECHANISMS IN MYLONITES: EVIDENCE FROM MICROCRACKS

ANDREW MCCAIG

This plate (backscattered SEM montage) illustrates direct evidence for fluid flow through microcrack networks in a greenschist-facies mylonite from the central Pyrenees, Spain (see McCaig and Knipe, 1990; Knipe and McCaig, 1994). Evidence from plagioclase zoning patterns in the same specimen is presented in Plate 55. The microcracks are preserved as veins cutting a large plagioclase grain which comes from a position analogous to Y in Plate 55A, the hinge region of a fold in a granitic layer (this plate illustrates a different cut of the specimen, thus the porphyroclast is not present in Plate 55A). The granitic layer consists of fine-grained (10–20 μm) quartz, phengite, and plagioclase showing a mylonitic fabric, together with larger plagioclase grains up to 400 μm in size, such as that illustrated here. The size of this grain and its relationship to the mylonitic fabric suggest that it is a porphyroclast inherited from the granitic protolith, although it could have grown in an early stage of mylonitization. Electron probe analyses show that in common with most feldspars in the granitic layer, the porphyroclast composition is $An_{15-17}$, and it is cut by numerous veins ranging in composition from $An_{23}$ to $An_{35}$. An approximate sequence of alteration and veining events affecting the porphyroclast is listed below and numbered on the photograph:

1) patchy areas of more albitic feldspar, possibly related to exsolution;

2) thin conjugate veins of $An_{30-35}$ trending 340° and 010° (azimuths relative to the upward direction on the photo);

3) patchy areas of alteration and veins trending 340° of $An_{20-26}$ ± quartz;

4) composite veins of $An_{30-35}$ trending 030° and showing crack-seal textures;

5) a few further veins trending 340° of composition $An_{25-28}$; and

6) marginal growths on the right and bottom left of the grain consisting of $An_{23}$ ± quartz and phengite.

Some of these veining events, e.g., (3) and (4), appear to overlap in time. Most of the more calcic feldspar appears to have grown in veins, but where quartz is present in the veins, partial replacement of the original feldspar is inferred according to the reaction $2NaAlSi_3O_8 + Ca^{2+} = CaAl_2Si2O_8 + 4SiO_2 + 2Na^+$.

The importance of this plate is that it shows evidence for cyclic microcracking and veining over a protracted period of time in a rock deforming predominantly by crystal-plastic processes. The veins are not present in the mylonitic matrix, but a string of fine-grained recrystallized feldspars with compositions in the range $An_{20-30}$ leads away from the porphyroclast through the core of the fold in the granitic layer (Plate 55A). These are inferred to be the deformed and recrystallized remnants of veins (McCaig and Knipe, 1990). The variable compositions in the veins suggest different local fluid sources for different veining events; fluids with a range of Ca/Na ratios would be present in different layers in the mylonite. It is noticeable that the plagioclase porphyroblasts growing in the dioritic/pelitic layers (Plate 55) show a similar range of compositions to the veins in the porhyroclast shown here, implying a similar range of fluid Ca/Na ratios.

To summarize, both the compositional zoning patterns in the preceding plate (Plate 55) and the microveins in this one are evidence for channelized, transient fluid flow. The evidence is preserved only because of the chemically heterogeneous nature of the rock; these processes are probably general in mylonites deformed under similar conditions, but may not always be easy to recognize.

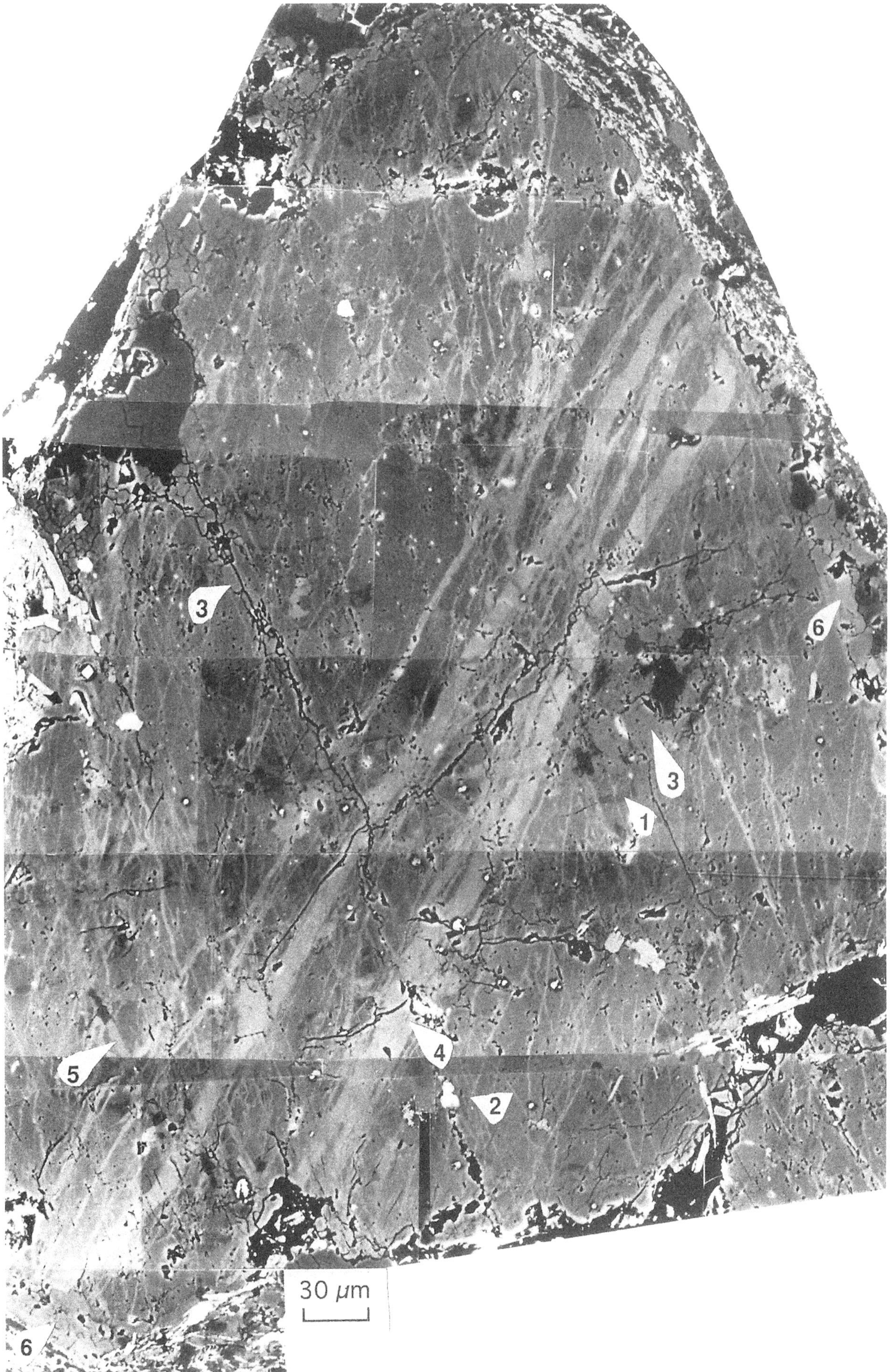
3
6
3
1
4
5
2
6
30 μm

# 57. FLUID FLOW MECHANISMS IN MYLONITES: PERVASIVE AND CHANNELIZED FLOW IN CARBONATES

Andrew McCaig and Liang-Yan Gong

Deformation mechanisms exert a profound influence on mechanisms of fluid flow and on chemical interactions between fluid and rock during deformation (Knipe and McCaig, 1994). In this plate, contrasting fluid flow regimes and deformation mechanisms are seen in adjacent dolomite and calcite layers within mylonitized, Cretaceous carbonates beneath the Gavarnie thrust at the Plan de Larri, 8 km northwest of Bielsa, in the central Spanish Pyrenees. Isotopic and fluid inclusion evidence (McCaig and others, 1995) shows that significant volumes of hypersaline brine flowed southwards along the mylonite zone during south-directed thrusting at temperatures of 250–300°C.

It is important that radically different flow regimes can coexist in adjacent layers. Channelized flow occurred as a result of brittle fracture in the dolomite and during veining events in the calcite mylonite, and fluid may have been stored in dilatant sites within the dolomite. Pervasive flow occurred during quasi-plastic deformation in calcite (probably including a component of grain-boundary sliding), but with variations in fluid source leading to zoning patterns. Alternations of veining and pervasive fluid flow in the calcite mylonite can clearly be linked to variations in deformation mechanism due to stress cycling (Knipe and McCaig, 1994).

**57A.** Part of a backscattered SEM montage from a boudinaged, brittlely deformed 0.5 m thick dolomite layer within the mylonite. Phases present are dolomite (dol), ferroan dolomite with variable Fe contents (Fe-dol), quartz (qz), and calcite (cc). Numbered features are (1) patchy replacement of dolomite by more Fe-rich carbonates, partially controlled by cleavage; (2) crack-seal veins with repeated generations of carbonate of different compositions; (3) irregular quartz masses apparently truncating veins, but concentrated in veined areas; (4) open cavities lined with zoned Fe-dolomite rhombs, and filled with calcite. Fluid flow, veining, and alteration in the dolomite were clearly related primarily to brittle fracture, probably at high fluid pressures.

**57B.** SEM orientation contrast image from a deformed, pure calcite vein within calcite-rich mylonites, collected 1 m below the sample shown in 57A. The veins themselves attest to cyclic fracture events, but the microstructure indicates that once emplaced, the veins were affected by dynamic recrystallization and quasi-plastic deformation. The main source of contrast is lattice orientation variations (Lloyd, 1987). The darker area, crossing the center of the image horizontally, is a recrystallized ribbon grain inherited from the original vein. Truncated zoning patterns (Z) within many of the grains must be due to compositional variations, probably in Fe content, and almost certainly resulted from variations in externally derived pore fluid chemistry, with fluid supplied from dolomitic layers (57A). A similar sequence of dark and light zones is seen within all the grains, consistent with pervasive introduction of fluid into many calcite growth sites simultaneously. Channelized flow would result in less consistent directions of chemical change (see Plate 55). Small (~1 μm) grain-boundary pores (Y) are concentrated on low-stress grain boundaries in the simple shear flow regime, and may have played an important role in fluid flow (Knipe and McCaig, 1994). Experimental data (Walker and others, 1990) suggest that a component of grain-boundary sliding contributes to calcite deformation over a wide range of conditions; this may be important in opening grain-boundary porosity (Ree, 1994).

57 A

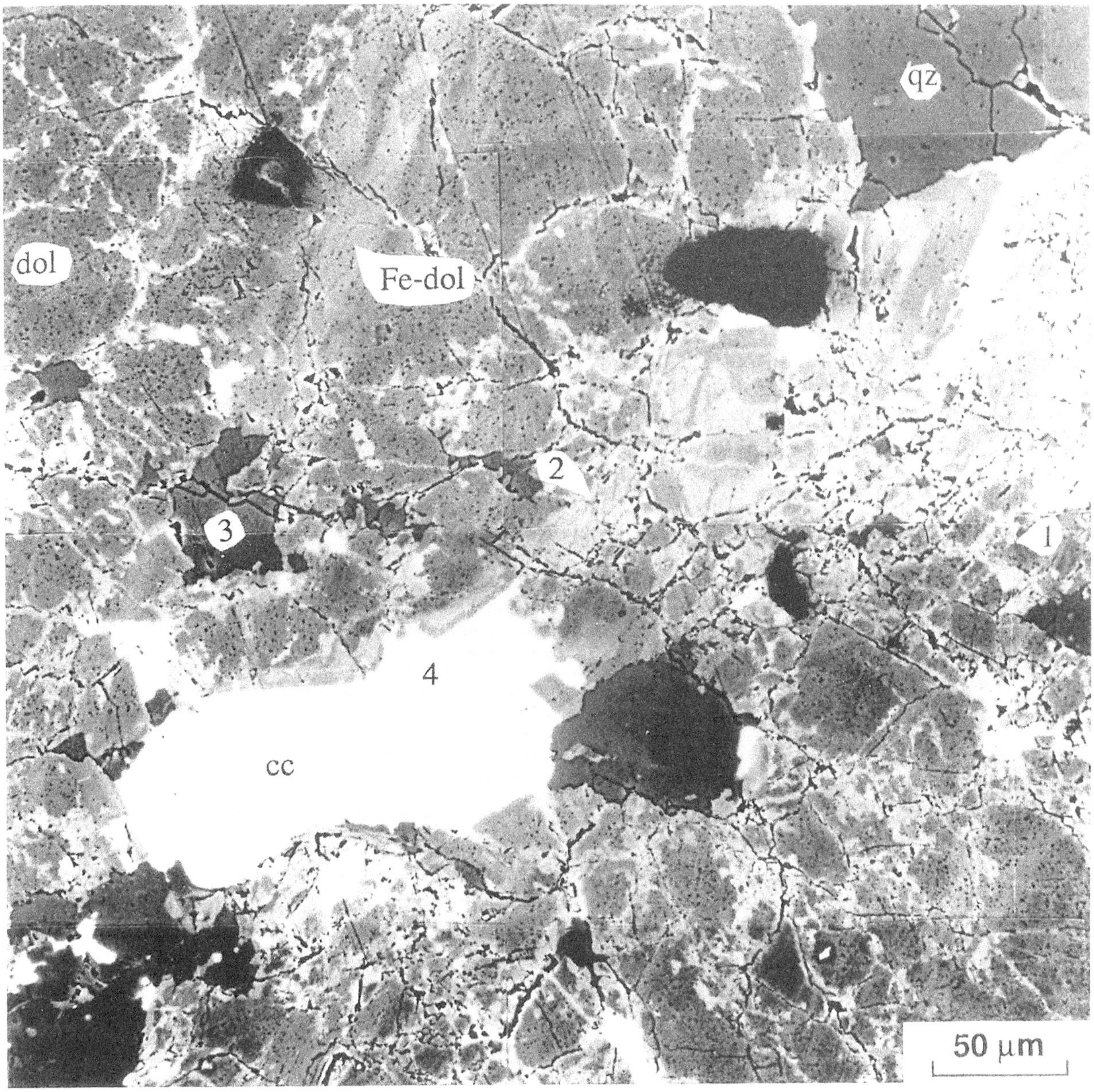


57 B

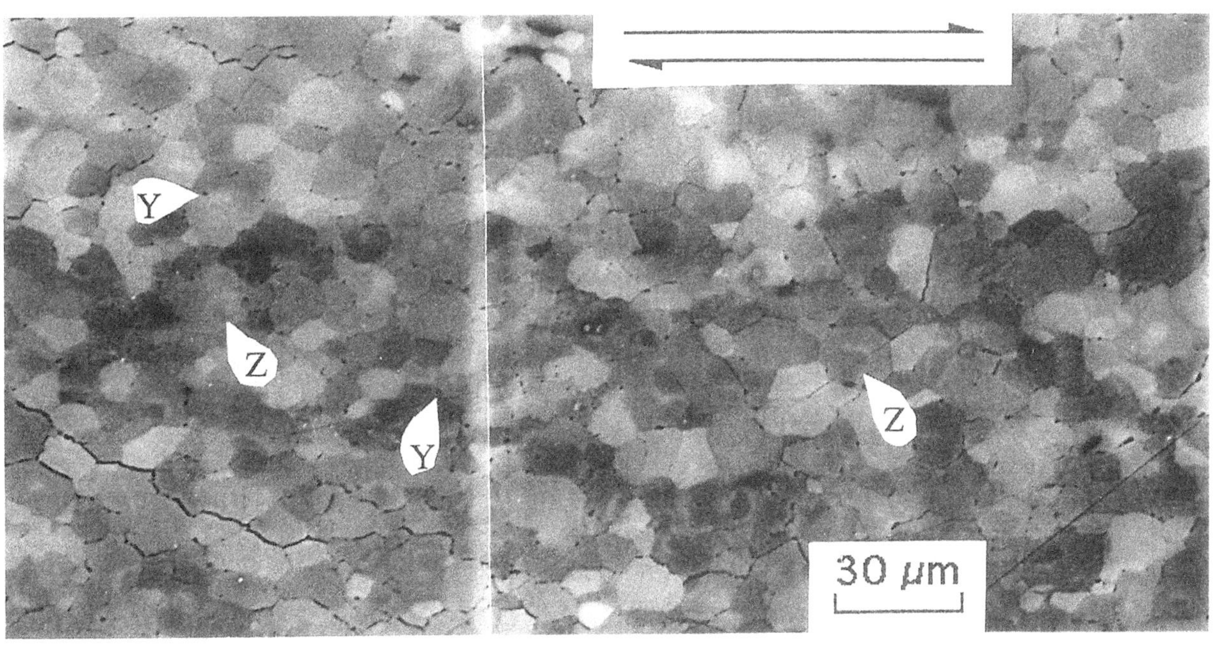

# 58. Rb-Sr STUDY OF AN ULTRAMYLONITE ZONE IN GRANITIC ORTHOGNEISS

ARTHUR W. SNOKE AND PAUL D. FULLAGAR

A well-exposed, meter-scale zone of ultramylonite is superimposed on a regional, mid-Tertiary mylonitic foliation on the northwest flank of the Ruby Mountains, Nevada, U.S.A. (Figure 58.1). The map view of the zone coupled with measured structural data suggest a slight curviplanar form with a northeast strike and a northwest dip. The ultramylonite zone is transitional in both time and deformational style between mid-Tertiary, kilometer-scale, mylonitic shear-zone deformation (Snoke and others, 1990a; Wright and Snoke, 1993; see Plate 111) and younger, brittle Basin and Range normal faulting (listric to planar).

The rocks of the ultramylonite zone display extreme grain-size reduction (58A) as well as retrogression of the granitic gneiss protolith (58B). Field, petrographic, and chemical data suggest that the biotite-muscovite granitic orthogneiss of Secret Peak (Wright and Snoke, 1993) was the protolith of the cataclastic gneiss and flinty rocks of the ultramylonite zone. Chemical data indicate that the ultramylonite is chemically similar to the orthogneiss of Secret Peak, but that the ultramylonite is somewhat enriched in $SiO_2$ and depleted in $Al_2O_3$, $K_2O$, and Rb. U-Pb zircon geochronological data (Wright and Snoke, 1993) indicate that the orthogneiss of Secret Peak is at least 32 Ma and may be significantly older (e.g., Cretaceous). Whole rock-muscovite Rb-Sr analysis of the orthogneiss of Secret Peak yielded a date of 23 ± 1 Ma, whereas a nine-point whole-rock Rb-Sr isochron yielded a date of 19 ± 4 Ma and an initial $^{87}Sr/^{86}Sr$ ratio of 0.7402 ± 0.0006 (Fullagar, 1992; Fullagar and Snoke, unpublished data, Figure 58.2).

A suite of samples was collected from the ultramylonite zone to study the effects of intense retrograde mylonitization on Rb-Sr systematics. Of the seven whole-rock samples analyzed, two samples were cataclastic gneiss considered to be partially reworked granitic orthogneiss and five samples were very fine-grained ultramylonite (similar to 58A). The analytical data are plotted on Figure 58.2, on which the 19 ± 4 Ma isochron derived from samples of biotite-muscovite granitic orthogneiss of Secret Peak has been plotted for comparison. Note that the two cataclastic gneiss samples plot near the 19 ± 4 Ma reference isochron, whereas the five ultramylonite samples lie significantly below the isochron. These five samples define a crude trend toward a hypothetical component "A," which is characterized by low $^{87}Rb/^{86}Sr$ and $^{87}Sr/^{86}Sr$ ratios. Clearly, the trend of the ultramylonitic samples is a mixing line between component "A" and the orthogneiss of Secret Peak. A geologically reasonable age is *not* defined by these data; instead, they indicate an open system involving the interaction of external fluids during cataclasis and mylonitization of the granitic rocks. Such fluids apparently have the ability to "reset" the Rb-Sr system during fault-rock development and to yield "mixing" lines of no geochronologic significance. The fact that the Rb-Sr whole-rock date for the orthogneisses from Secret Peak is significantly younger than the U-Pb age obtained from zircon samples suggests that all of the Secret Peak orthogneiss samples may have been altered with respect to Rb-Sr. Perhaps this alteration also was due to migrating fluids.

**58A.** Close-up view of an outcrop of ultramylonite characterized by aphanitic texture, pervasive foliation, and conchoidal fracture. The strike of the ultramylonitic foliation is subparallel to the length of the photograph, and the dip is moderately steep such that the plane of the photograph is approximately perpendicular to the foliation plane.

**58B.** Outcrop view of the terminus of the ultramylonite zone (black, fine-grained rock with hackly fractured surface) and adjacent cataclastic granitic gneiss. um = ultramylonite; cggn = cataclastic granitic gneiss.

58 A

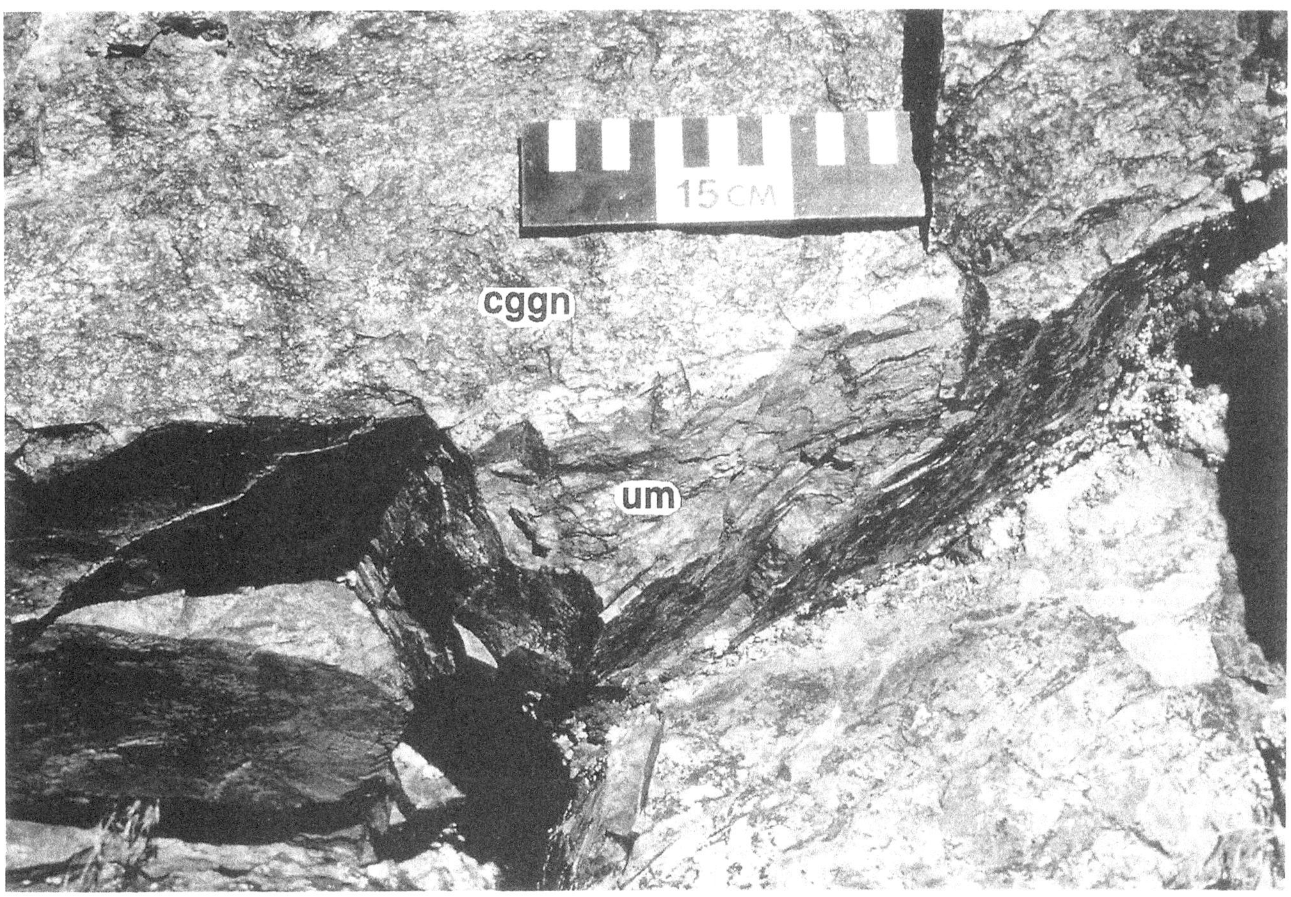


58 B

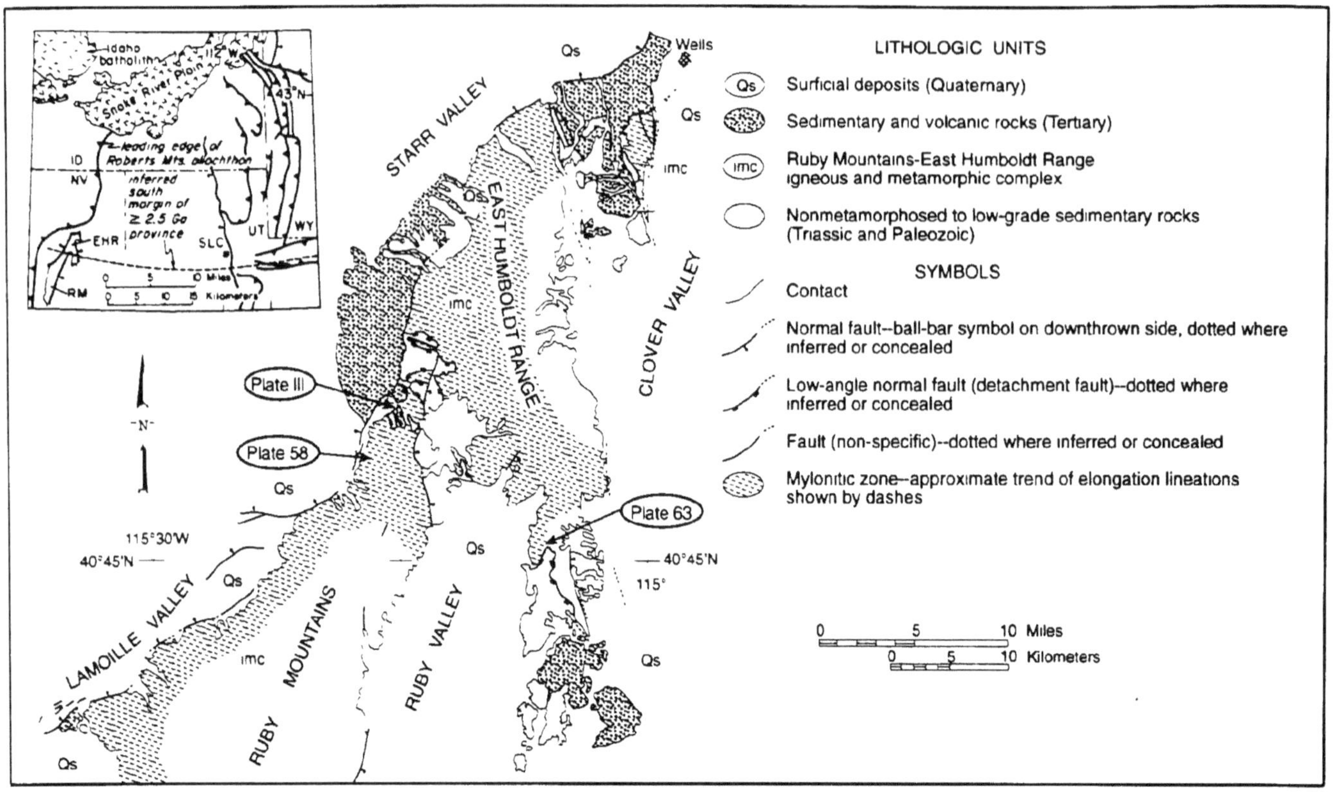


**Figure 58.1.** Geologic sketch map showing the location of the Ruby Mountains and East Humboldt Range, Nevada, and the approximate localities of samples shown in Plates 58, 63, and 111.

**58C.** The cataclastic granitic gneiss is characterized by a network of shear fractures defined by zones of extreme grain-size reduction. Coarse fragments of quartz (q) and scarce feldspar (f) occur in a fine-grained matrix of quartz and phyllosilicate minerals. The occurrence of green chlorite (after biotite?) in some samples of cataclastic gneiss indicates lower greenschist-facies metamorphic conditions during the deformation. NX

**58D.** The ultramylonitic rocks are so fine-grained that precise mineral identification in thin section is difficult, but the mineral assemblage appears to be similar to that in the cataclastic granitic gneiss. A crystallographic preferred orientation of fine-grained recrystallized quartz is apparent in thin section with the aid of the first-order red plate. Anastomosing microbanding, related to variations in phyllosilicate content, also defines a mylonitic fluxion structure. Scattered, lenticular microporphyroclasts composed of undulatory quartz are similar to quartz porphyroclasts found in some quartz-mica phyllonites derived from granite during cataclasis under hydrated conditions (Janecke and Evans, 1988; see Plate 49A, this volume). In such rocks, the comminution of feldspar and its alteration to phyllosilicate are important processes that led to the development of phyllonite. In the ultramylonite of the northern Ruby Mountains, cataclasis and alteration of feldspar appear nearly complete, and the remaining quartz and phyllosilicate minerals deformed chiefly by crystal-plastic mechanisms resulting in a flinty-appearing mylonitic rock. The ultramylonite thus evolved from the granitic orthogneiss of Secret Peak through a combination of cataclasis and crystal-plastic flow. However, rock-fluid interactions, involving the migration and redistribution of various chemical elements, also played a key role in the textural development of these ultramylonitic rocks. NX

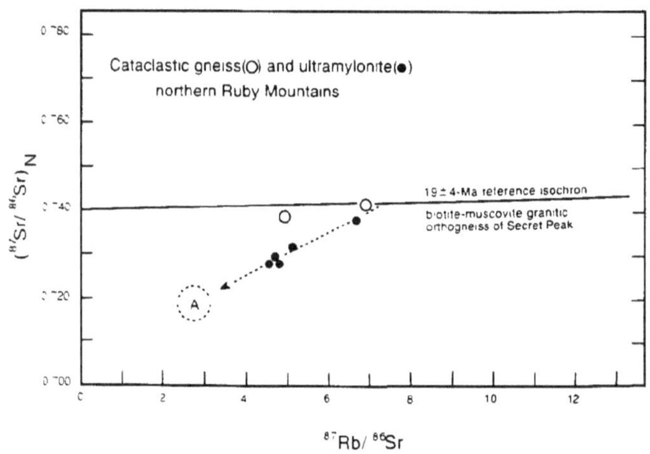


**Figure 58.2.** Rb-Sr diagram for cataclastic granitic gneiss and ultramylonite samples showing the reference isochron for the biotite-muscovite granitic orthogneiss of Secret Peak.

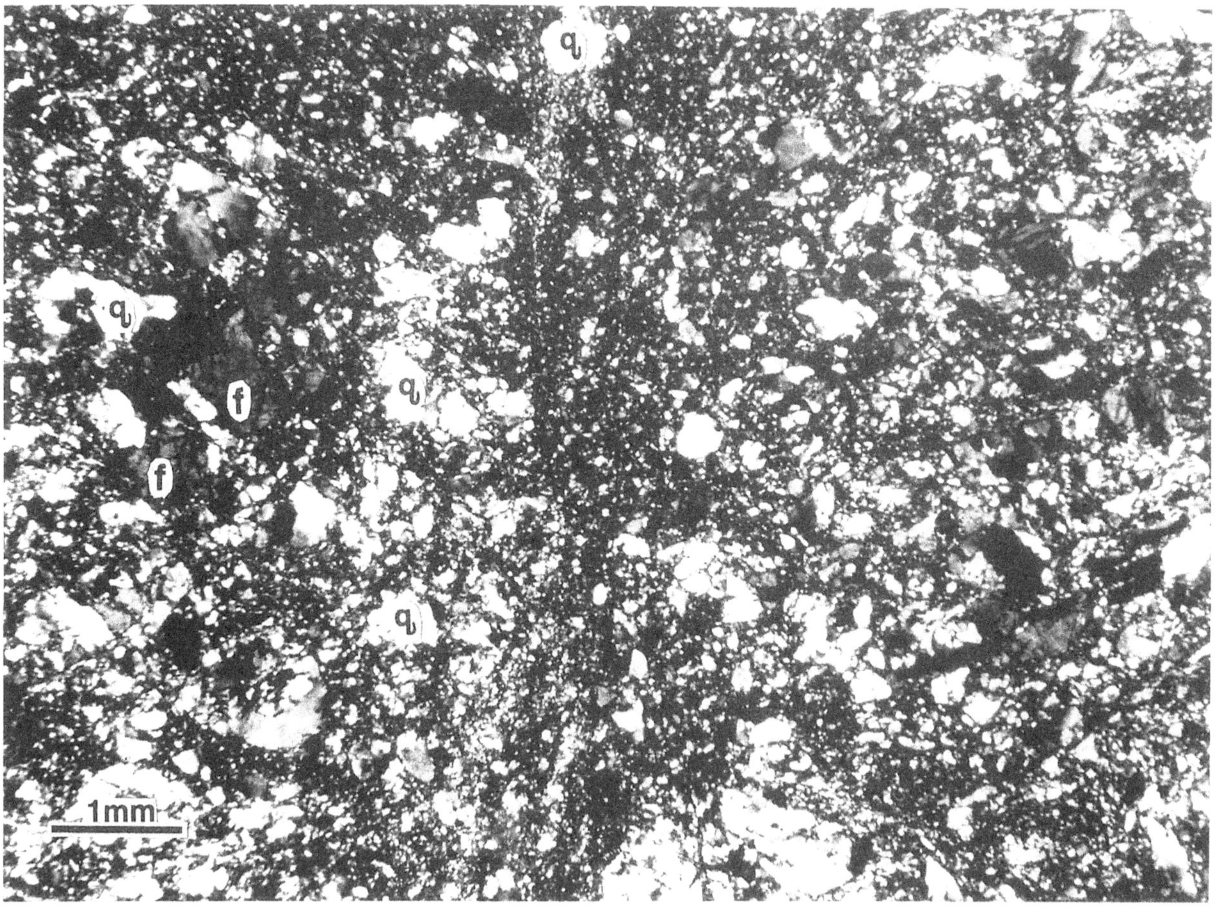


58 C

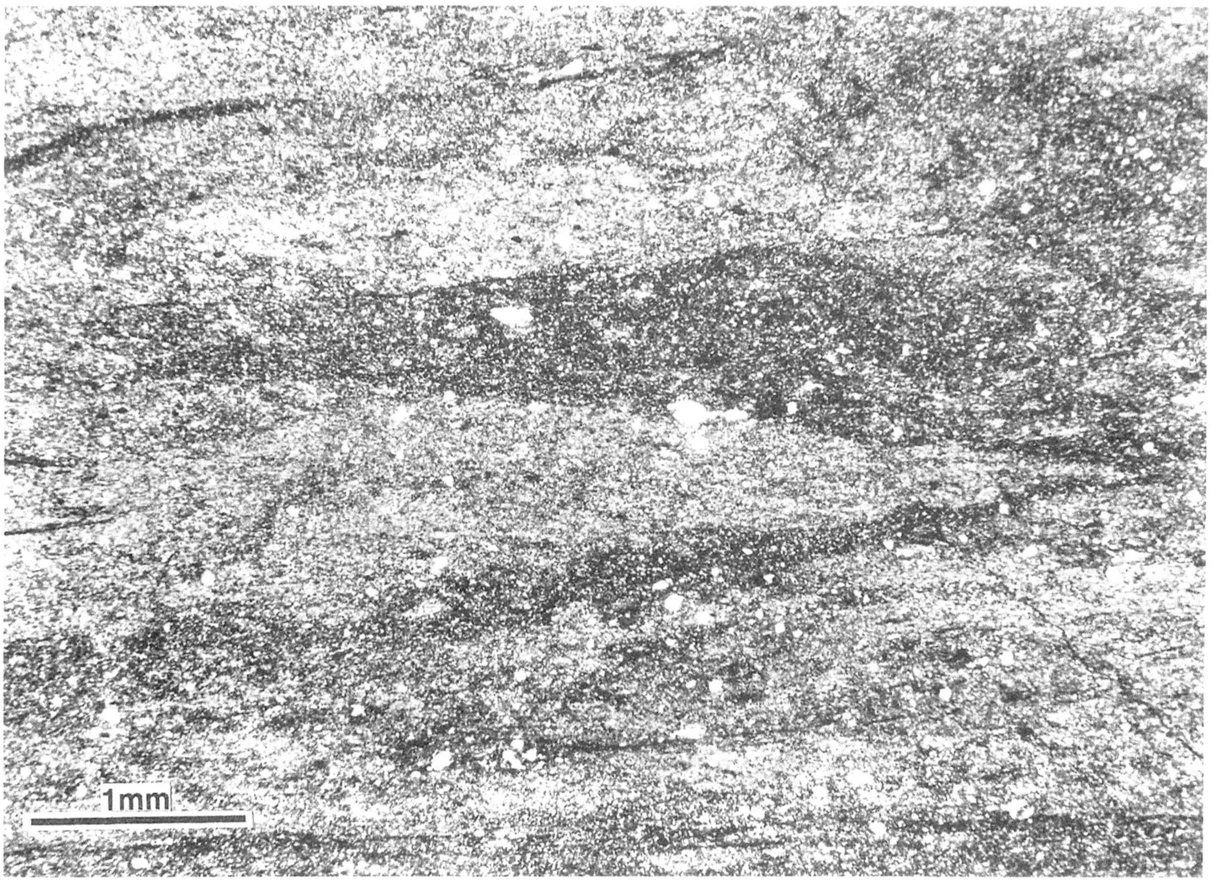


58 D

# 59. DUCTILE-BRITTLE AND REACTION-ASSISTED DEFORMATION IN SHEARED GRANITOID

J. F. HIPPERTT

Granitoids deformed in low metamorphic grade shear zones ($T < 500°C$) offer ideal conditions for the simultaneous operation of different deformation mechanisms that interact and enhance each other. This example, from the Moeda-Bonfim shear zone (MBSZ), Quadrilátero Ferrífero, southeast Brazil, shows the XZ view of the microstructure developed at intermediate levels of strain, i.e., between the undeformed wall rock (a nonfoliated, coarse-grained granite with K-feldspar megacrysts) and the fine-grained quartz-mica mylonite occurring in the highest-strain domain of the shear zone. The MBSZ is a dip-slip, medium greenschist-facies shear zone ($T \sim 400$–$450°C$, $P \sim 4$ kb) that occurs along the contact between Archean gneissic granitic rocks of the Bonfim complex and Proterozoic quartzites of the Minas Supergroup (Moeda Quartzite) and thus affects both lithologies (Hippertt, 1993). Kinematic and geochronological data indicate that this shear zone was formed by the uplift of the gneissic granitic rocks during a Trans-Amazonian (~1.7 Ga) extensional tectonic event in the region (Marshak and Alkmin, 1989). The outcrop where this sample was collected is along the road to Santa Efigênia, 2.5 km from the BR-040 highway.

The deformed granitoid is a typical quartzo-feldspathic mylonite with well-developed, mica-rich S and C foliations forming angles between 15° and 45°. In microscale, the original K-feldspar megacrysts appear as groups of broken fragments intimately associated with secondary muscovite (Kf-mus) which was generated by the breakdown of feldspar via softening reactions involving water. Muscovite developed principally by alteration of the irregular and discontinuous perthitic plagioclase veinlets ($An_{5-18}$) that always cut the K-feldspar fragments at low angles to the overall shear plane, i.e., independent of the crystal orientation of K-feldspar (arrows). This relationship indicates that the perthitic veinlets formed by a kinematically controlled process of K-feldspar replacement by plagioclase (replacement perthite). Sericitization of the perthitic veinlets favors fracturing and displacement of the K-feldspar fragments. In turn, fracturing enhances the access of water and progression of the replacement and softening reactions in feldspar. This is a situation where reaction-assisted and brittle deformation proceed together, impelling each other. The process leads to grain-size reduction, elimination of feldspar, and mica enrichment.

The polygonal aggregates of recrystallized quartz show strong concentration of c axes close to the Y axis of finite strain (Hippertt and Borba, 1992), i.e., perpendicular to the plane of the photograph. This is indicated by the high proportion of dark quartz grains in the quartzose layer. Quartz c axis maxima around Y are normally produced by prism ⟨a⟩ slip during amphibolite/granulite-facies metamorphic conditions. Its activation at lower temperatures as in this shear zone may reflect hydrolytic weakening of quartz due to influx of water during deformation (Paterson, 1989). The overall shear plane (top right corner) as well as synthetic and antithetic micro-shears (couplets of opposing small arrows) are indicated. The thin section is perpendicular to foliation and parallel to the stretching lineation. Key: K-feldspar (Kf), muscovite (mus), quartz (qz), albite (ab). NX

59

# 60. FLUID-FILLED PORES IN MYLONITIC ROCKS

J. F. HIPPERTT

Scanning electron microscopy of fresh fracture surfaces permits three-dimensional observation of intergranular pores (originally fluid filled) commonly present in mylonites formed under greenschist/amphibolite-facies conditions. This plate illustrates the contrast of porosity between a quartz mylonite deformed in relatively "dry" conditions and a phyllonite subjected to pervasive fluid influx that enabled quartz dissolution and mica enrichment. The features seen in these photographs are typical of many quartz mylonites and phyllonites. In the two examples, the pore distribution is kinematically controlled, reflecting syndeformational migration of fluids to grain boundaries oriented at high angles to the maximum extension direction. The pores are not connected and are interpreted as representing a dynamically stable, crystal-fluid microstructure formed during deformation. The views in the photographs are approximately parallel to the XZ plane of finite strain.

**60A.** Fluid pockets in quartz mylonite, Moeda Quartzite, Ouro Preto, southeast Brazil (Hippertt, 1994a,b). Outcrop adjacent to Viaduto do Funil bridge along BR-262 highway. This mylonite was formed under lower amphibolite/upper greenschist-facies conditions in a zone of mylonitic rocks that surrounds the migmatitic Bação complex of the Quadrilátero Ferrífero region. This sample, which was collected in a 1 cm wide, mica-rich shear band (sericite 15%), is totally recrystallized (dynamically) showing a mosaic of equigranular polygonal quartz grains (grain size = 70 μm) pinned by sericite crystals. A polygonal quartz grain has been plucked off showing some isolated fluid pockets (FP) along a grain boundary oriented at a high angle to the shear plane (couplet of opposing arrows). The grain boundaries in other orientations are nearly pore-free at this scale. A typical ellipsoidal pore occurs at the join of grain corners (arrow). The total area of the grain boundaries in contact with pore space (fluid) was estimated at less than 1%. SEM

**60B.** Pores in phyllonite (sericite 75%, quartz 25%), Moeda-Bonfim shear zone, Quadrilátero Ferrífero, southeast Brazil (Hippertt, 1994c). The photograph shows a profusion of intergranular pores occupying up to 70% of the grain boundary area. Even with this very high proportion of crystal-fluid interfaces, a connected porosity was not observed. This phyllonite occurs in the highest-strain domain of the shear zone (see Plate 59) and was interpreted as a product of quartz dissolution and mica enrichment in a precursor micaceous quartzite during deformation at low greenschist-facies conditions, and in the presence of a relatively high volumetric (>3%) aqueous fluid phase. Curved quartz grains, which typically form in the presence of relatively large volumes of fluid (Den Brok, 1992), are pinned by sericite crystals (ser). Pores occur along quartz boundaries oriented at high angles to the shear plane, which is roughly defined in the photograph by the alignment of the sericite crystals. The pore geometry is crystallographically controlled: hexagonal or approximately circular pores form on basal planes of quartz (B); triangular pores with two sides straight and one curved form on prismatic planes (P). A sericite crystal has been plucked off (arrow), revealing a typical flat, pore-free interface between quartz and mica. Interestingly, in mylonitic quartzites pores do not occur along quartz boundaries that are in contact with basal planes of mica. SEM

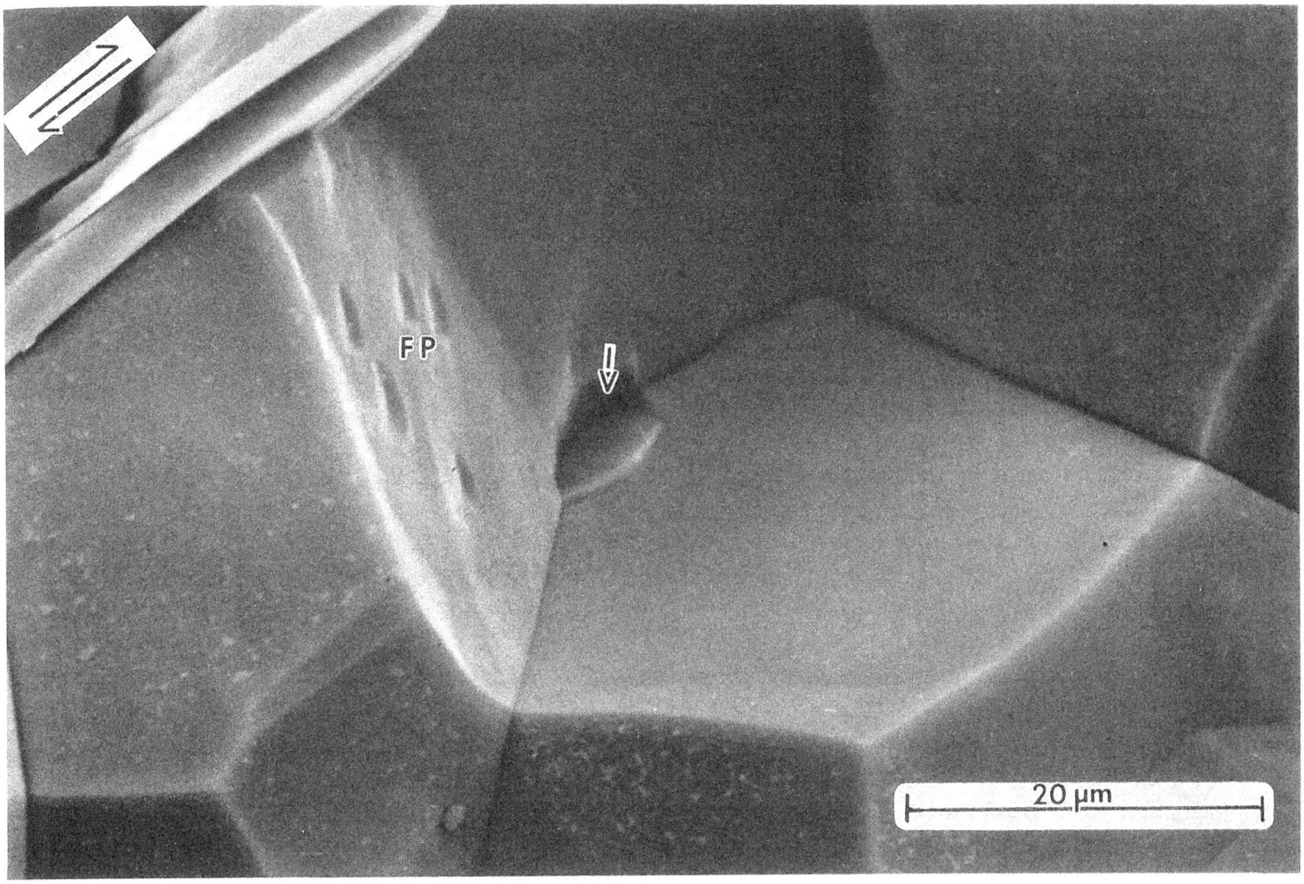


0 A

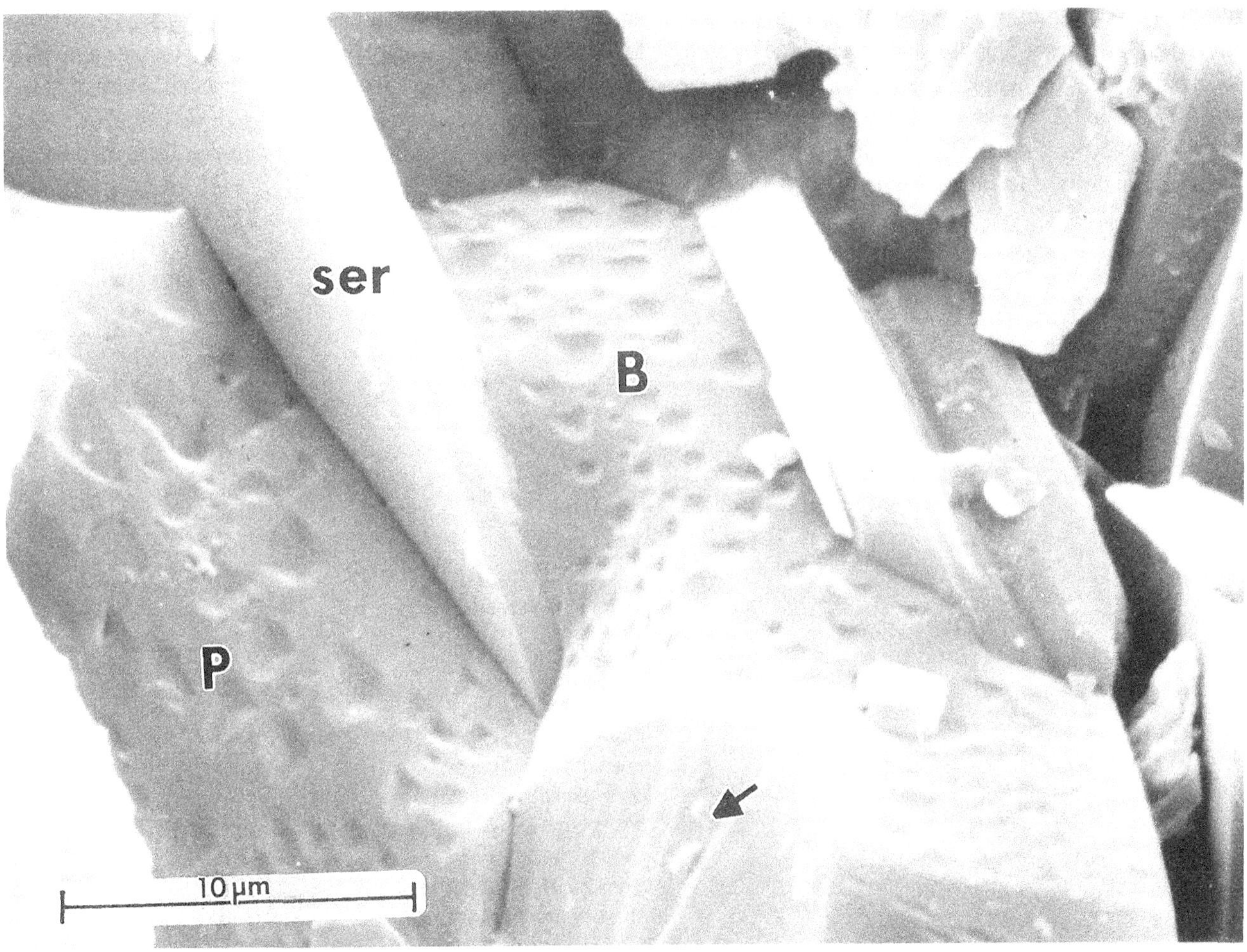


60 B

# 61. "OLD" AND "NEW" GRAINS IN QUARTZ MYLONITE

J. F. HIPPERTT

Shear zones developed under low-grade metamorphic conditions can be pathways for fluid channeling in the upper crust, and their mylonites commonly show strain partitioning between crystal-plastic and fluid-assisted deformation processes (solution transfer). In such mylonites, polygonal (recrystallized) grains and subgrains are normally attributed to dislocation creep, whereas microstructures such as dissolution cleavages, overgrowths, pressure fringes, and sealed cracks are the usual products of solution reprecipitation. This plate, however, illustrates a case where solution reprecipitation accounts for a pervasive formation of "new" grains in a quartz mylonite. The rock is the sericitic Moeda Quartzite deformed under middle greenschist-facies conditions (T ~ 400–450°C, P ~ 4 kb) in the Moeda-Bonfim shear zone, Quadrilátero Ferrífero, southeast Brazil (see Plate 59). Strain determinations by the Fry (1979) method in quartz porphyroclasts indicate a deformation regime close to plane strain, simple shear (maximum $\varepsilon_1 = 0.84$, $\varphi = 65°$). Evidence for solution transfer of $SiO_2$ is given by the noticeable development of phyllonites, presence of fibrous quartz in cracks, truncation of porphyroclast edges (by dissolution), and preferential orientation of quartz c axes at low angles to the stretching lineation (Hippertt, 1994d).

Under the optical microscope, this mylonite shows anastomosing mica-rich folia surrounding groups of polygonal quartz (grain size = 10–80 μm) with variable amounts of relict, deformed quartz porphyroclasts (5–15%). SEM observation revealed that many small polygonal grains adjacent to the older, deformed quartz porphyroclasts are in fact idiomorphic crystals when observed in three dimensions. This observation is not consistent with the ordinary mechanisms of recrystallization, such as subgrain rotation and grain-boundary migration, and suggests that a process of grain growth in a free medium (fluid) accounts, at least in part, for the "new," polygonal grains in this mylonite (Figure 61.1).

**61A.** The photograph shows a small, "new" quartz grain with well-developed prismatic and rhombohedral crystal faces (arrow 1), adjacent to a large, deformed quartz porphyroclast (por). Note that this "new" grain occurs in a concavity of the porphyroclast surface. The geometric cavities in the porphyroclast (arrows 2) correspond to "new" grains that were plucked off during the sample preparation. SEM

**61B.** Detail of a "new" quartz grain with incipient crystal faces (N), adjacent to an older quartz grain in which a number of individual fluid pockets are seen (P). Note that the "new" grain is surrounded by an aureole of pore space (arrow). This photograph was taken in a quartzose domain surrounded by sericite-rich folia (ser). SEM

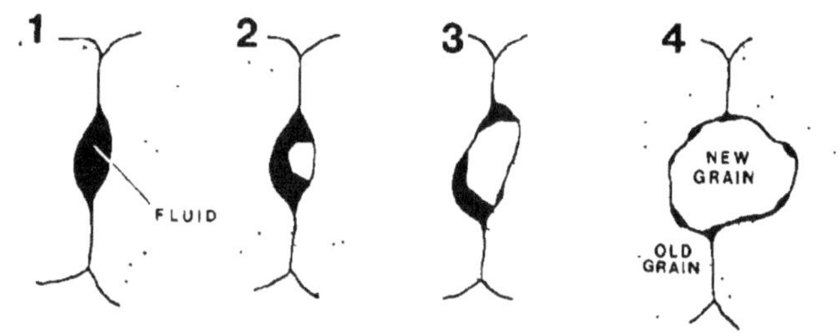


**Figure 61.1.** Suggested mechanism of solution redeposition in situ where the "old," deformed grains are dissolved in parallel with the nucleation of "new" grains in intergranular, fluid-filled pores. The "new" grains grow with their boundaries largely in contact with fluid, enabling development of crystal faces, at least during the initial stages. This mechanism should be favored in the presence of a high volumetric fluid phase, which causes a larger portion of the grain to be in contact with fluid (see also Plate 60)

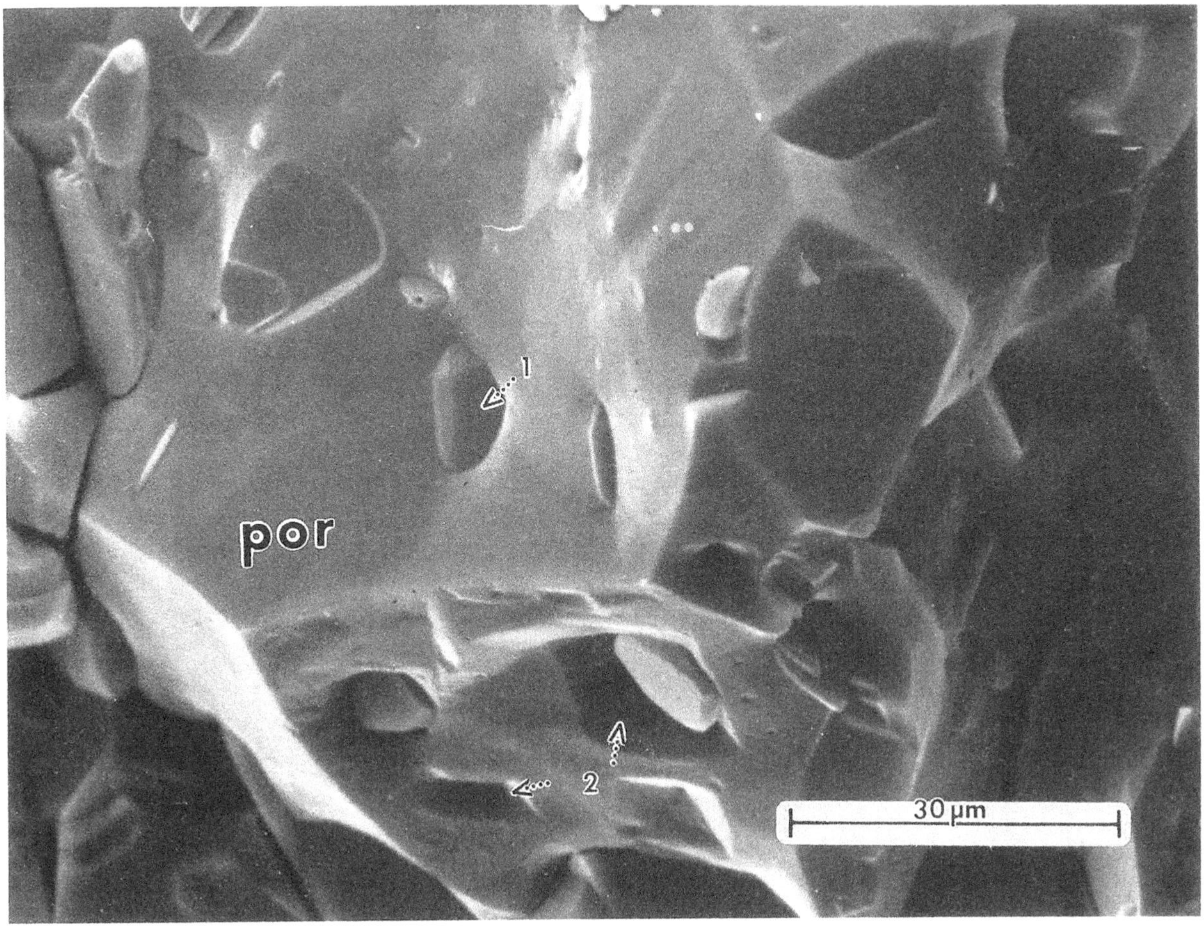

61 A

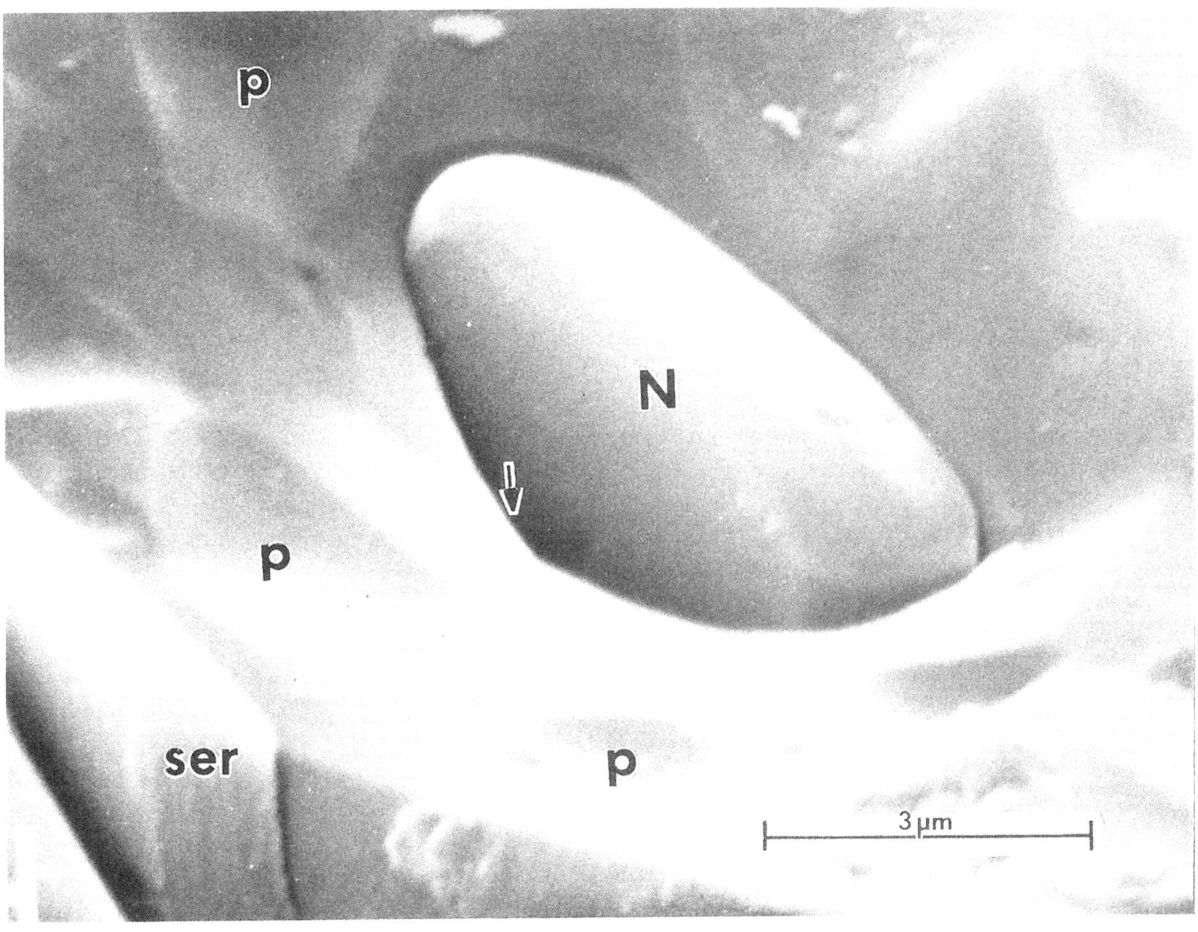

61 B

# DUCTILE BEHAVIOR

## Strain partitioning and other geometric characteristics



Common abbreviations used in regard to the plates in this section include NX = crossed nicols and TEM = transmission electron microscope.

This set of plates illustrates the partitioning of strain on mesoscopic and microscopic scales, from rocks deformed at greenschist through granulite facies, and from rocks of the oceanic as well as the continental crust. Strain partitioning results from rheological heterogeneities; the plates in this section primarily illustrate strain partitioning resulting from compositional heterogeneities. Some of the microstructures that characterize Lapworth's original mylonite are due to grain-scale strain partitioning in a granitic rock, deformed at physical conditions where feldspar is strong and quartz is weak. On a larger scale, strain partitioning in ductile deformation zones may produce folds, boudins, or localized shear zones. Some of the plates illustrate strain partitioning that results from variations in grain size, due to the fact that fine grain sizes may allow a switch to grain-size-sensitive creep. Other plates illustrate how strain partitioning may change with time, as the bulk composition changes due to metamorphic reactions in part related to fluid loss or addition, or the grain size changes due to cataclasis, dynamic recrystallization, or neomineralization.

# 62. HETEROGENEOUS SHEAR ZONES IN GRANODIORITE

CAROL SIMPSON

Outcrops and hand specimens of mylonitic shear zones formed under lower greenschist-facies conditions in Roses granodiorite, northeast Spain (Simpson and others, 1982; Segall and Simpson, 1986).

**62A.** Discrete shear zones in late Hercynian granodiorite that contain basic xenoliths and are intruded by an aplite dike. The shear zones are part of the late Hercynian mylonite belt that deforms the crystalline rocks of the Axial Zone of the eastern Pyrenees (Simpson and others, 1982). In the Roses outcrop, the narrow left-lateral shear zones are all northwest-striking and steeply dipping. Although several of the narrow (<3 mm thick) shear zones in the center of the photograph terminate in brittle Mode I fractures (Segall and Simpson, 1986), deformation along most of their trace is plastic, and the aplite dike can be traced continuously through each zone. Coin is 2.5 cm in diameter.

**62B.** Outcrop of deformed granodiorite (approximately 80 m to the west of outcrop in 62A) contains a 3 cm thick mylonite zone (D in Figure 62.1) that records the maximum finite strain. Rock in area A is essentially undeformed. In area C, a weak foliation is oriented at a low angle to the mylonite zone boundary. Within areas B and E, a weak penetrative foliation is deflected into a series of left-lateral shear bands that do not transect the mylonite or area C. Close spatial association of the shear bands with the mylonite zone, their consistent angle, and their minor offset suggest that they represent a late stage of the mylonite-forming event and were not significantly rotated after formation. They allowed the rock on either side of the mylonite zone to extend parallel to foliation, probably in response to a compressional component of pure shear across the zone.

**62C.** Close-up of mylonite zone (D) illustrated in Plate 62B. Polished hand specimen illustrates heterogeneous deformation of granodiorite into a 3 cm thick mylonite that offsets an aplite dike (not illustrated) with a left-lateral shear sense. Relict feldspar porphyroclasts (arrowed F) and small lenses of less highly strained rock (arrowed L) occur within the more highly strained rock. In the high-strain zone, biotite is retrogressed to chlorite, and feldspar is altered to fine-grained white mica, clinozoisite, and quartz. Development of foliation becomes gradually less intense on either side of the mylonite zone, passing into relatively undeformed rock in the lower third of the photograph.

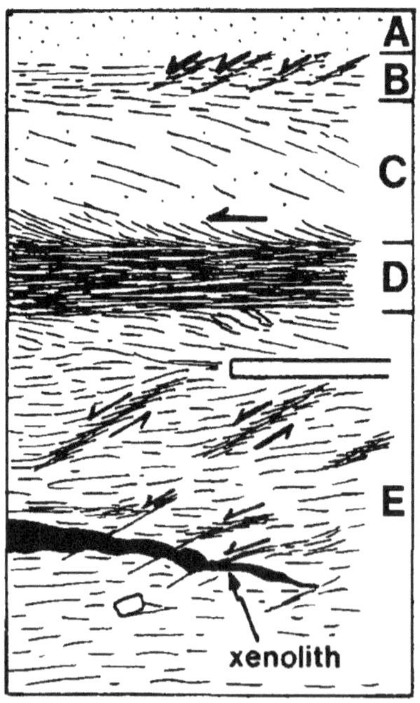


**Figure 62.1.** Line drawing of main features visible in Plate 62B, so as to highlight the different areas of strain intensity in this outcrop. Area A is almost undeformed, area B has a weak foliation and shear bands, area C is very weakly foliated, area D is a high-strain mylonite zone; area E has similar structures to area B. Ruler shows scale in centimeters.

62 A

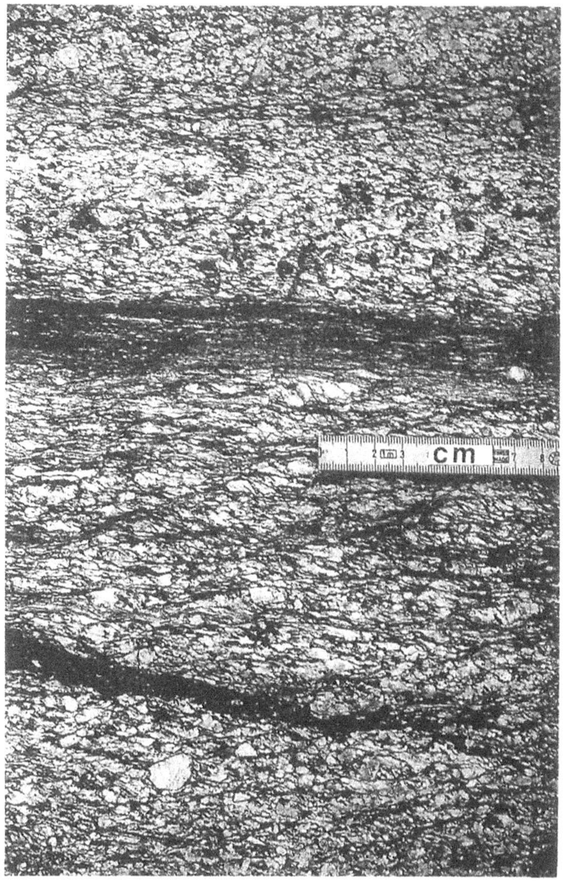

62 B

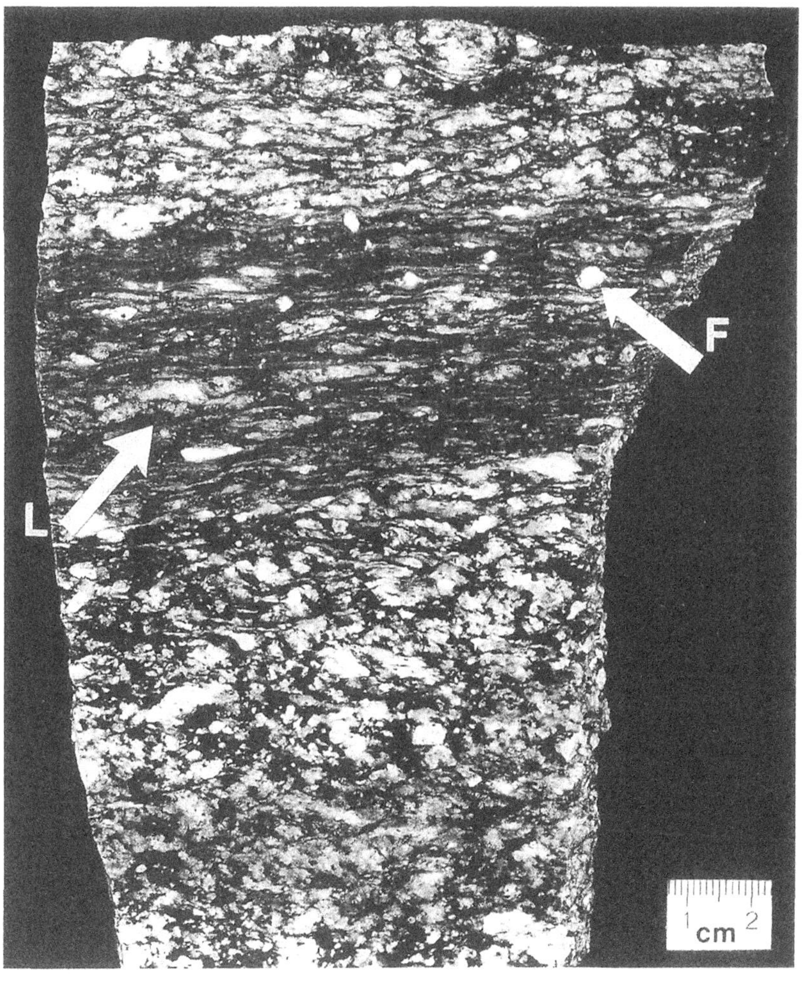

62 C

# 63. STRAIN PARTITIONING ASSOCIATED WITH A MYLONITIC SHEAR ZONE–DETACHMENT FAULT SYSTEM

ARTHUR W. SNOKE

In the southeast East Humboldt Range, Nevada, U.S.A. (see Figure 58.1), a hectometer-scale mylonitic shear zone of the Ruby–East Humboldt metamorphic core complex defines the top of a kilometer-scale lense of metasedimentary rocks intruded by sheet-like, late Cretaceous pegmatitic muscovite leucogranite bodies (Figure 63.1). Structurally deeper and defining the base of the lense of metasedimentary rocks is a kilometer-scale, Oligocene mylonitic shear zone (Snoke and others, 1990b; Wright and Snoke, 1993; also see Plates 111 and 143, this volume). The age relationship between the two shear zones is uncertain. Field relationships only constrain the age of deformation in the upper mylonitic shear zone as post–late Cretaceous, whereas Oligocene and Eocene granitic rocks are involved in the mylonitic deformation associated with the structurally lower shear zone (Wright and Snoke, 1993). A conspicuous elongation lineation, related to the mylonitic deformation, exhibits an overall similar azimuth in both shear zones. Superposed on the upper mylonitic shear zone is a chiefly brittle detachment fault system (southeast East Humboldt Range system) that has partitioned upper and middle Paleozoic sedimentary and low-grade metasedimentary rocks into a sequence of tectonic slices (horses in an extensional duplex) (Figure 63.2). This set of photographs demonstrates the geometry of strain partitioning in the upper mylonitic shear zone at macroscopic and mesoscopic scales in a sequence of lower amphibolite-facies metacarbonate rocks intruded by pegmatitic muscovite leucogranite.

**63A.** Mylonitic shear zone–detachment fault system exposed in the southeast East Humboldt Range, Nevada, U.S.A (see accompanying explanatory sketch, Figure 63.1).

**63B.** Strongly foliated impure calcite marble of the upper mylonitic shear zone with relict, but deformed and recrystallized, lenses of metachert. Mylonitic impure calcite marble (O€m) occurs within a highly deformed band, bounded above by a low-angle normal fault of the detachment fault system and below by a deformed intrusive contact along the top of a sheetlike body of leucogranite (Figure 63.2).

**Figure 63.1.** Sketch of 63A showing the upper mylonitic shear zone, deformed intrusive contact, and detachment fault. The label "mylonitic shear zone" indicates the approximate base of the shear zone as developed in the upper part of an extensive, sheetlike intrusion of late Cretaceous pegmatitic muscovite leucogranite. The overlying O€m unit consists of pervasively mylonitic metacarbonate rocks. Dg = greenschist-facies metalimestone and metadolomite of the Upper Devonian Guilmette Formation. O€m = mylonitic impure calcite marble (Ordovician and Cambrian protolith); and Klg = late Cretaceous pegmatitic muscovite leucogranite

63 A

63 B

**63C.** Undeformed pegmatitic muscovite leucogranite (Late Cretaceous). Note the heterogeneous texture due in large part to scattered, coarse-grained feldspar.

**63D.** Mylonitic leucogranite showing heterogeneous strain manifested by variable grain size and foliation development. Note the anastomosing character of the mylonitic foliation, which has partitioned the rock body into discrete zones that have experienced variable grain-size reduction. Excellent S-C fabrics are locally well developed in these mylonitic leucogranites.

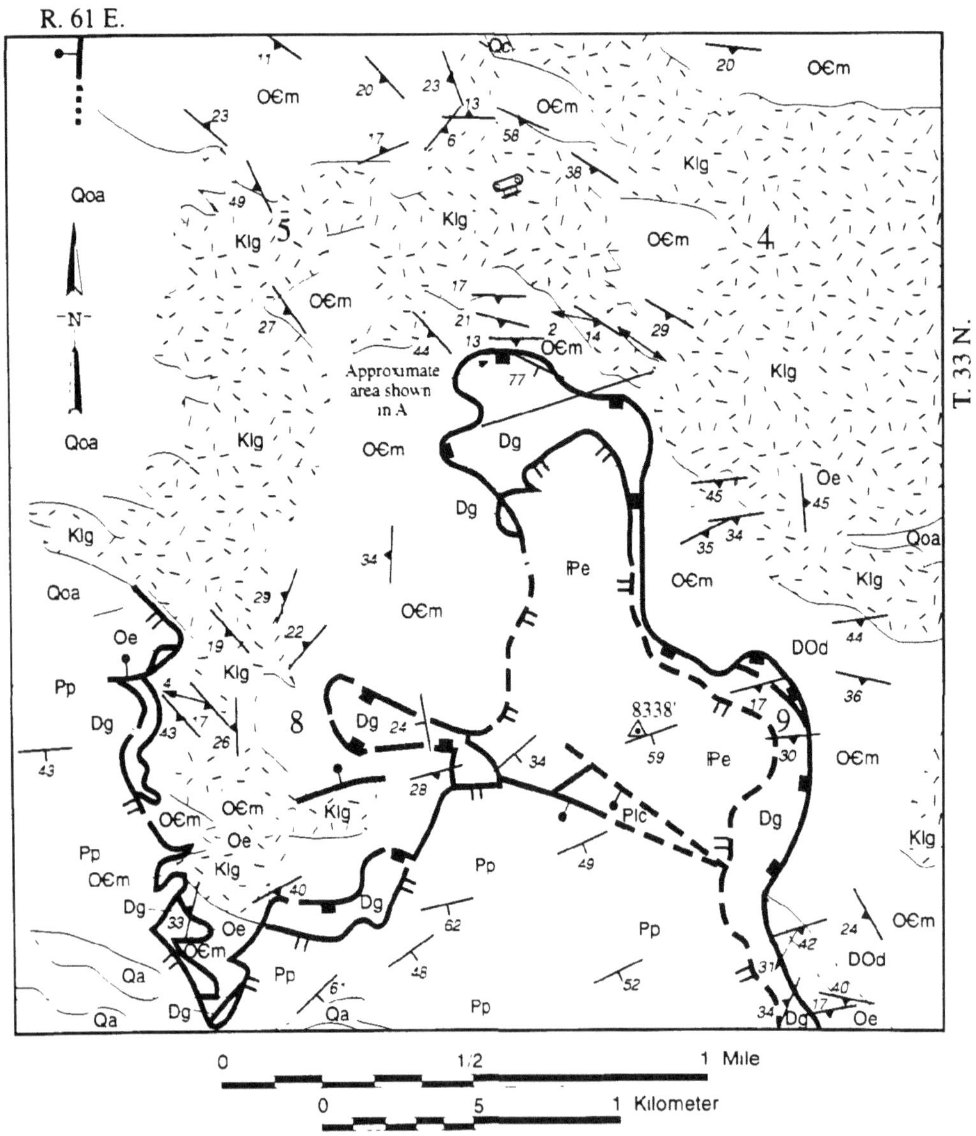


**Figure 63.2.** Geologic map of the southeast East Humboldt Range, Nevada, U.S.A. Qa = alluvium (Quaternary), Qc = colluvium (Quaternary); Qoa = older alluvium (Quaternary); Klg = Pegmatitic leucogranite (Cretaceous); Pp = Pequop Formation (Permian); Plc = limestone and conglomerate (Permian); Dg = Guilmette Formation (Devonian); DOd = metadolomite (Devonian-Ordovician), Oe = Eureka metaquartzite (Ordovician); O€m = impure calcite marble and calc-silicate paragneiss (Ordovician and Cambrian). Standard symbols for attitudes of bedding and foliation, trend and plunge of lineation, geologic contacts, and normal faults [except double tick marks on upper plate of brittle low-angle normal fault (detachment fault) and filled squares on upper plate of plastic-to-brittle low-angle normal fault (detachment fault)]. Faults are dashed where approximate and dotted where concealed.

63 C

63 D

# 64. HETEROGENEOUS DEFORMATION OF TONALITE AND PEGMATITE VEIN

CAROL SIMPSON

Photograph of hand specimen of mylonite and ultramylonite from Borrego Springs section of the eastern Peninsular Ranges mylonite zone, southern California, U.S.A. (Simpson, 1985), part of a major late Cretaceous to early Tertiary, west-directed thrust zone affecting tonalites and granodiorites of the southern California batholith (Simpson, 1984). In this specimen, a quartz-feldspar-biotite pegmatite vein is also deformed in the mylonite zone. All deformation was at middle greenschist facies.

The highly strained pegmatite in the center of the photograph has a relatively large grain size (>50 μm) of recrystallized quartz (light gray). Semi-brittle plagioclase (white, F) and biotite grains (black) are concentrated into narrow bands. The <1 cm thick band of ultramylonite (um) near the top of the photo contains small white porphyroclasts of feldspar that in thin section indicate dextral shear. Larger porphyroclasts and shear bands in the lower-strain mylonitic granodiorite both above and below the ultramylonite band also indicate dextral shear. Despite obvious differences in finite strain state, there is no detectable difference in mineralogy from ultramylonite to mylonitic granodiorite, and all deformation in this hand specimen occurred under the same metamorphic conditions. The grain-size differences may be a consequence of strain-rate variations.

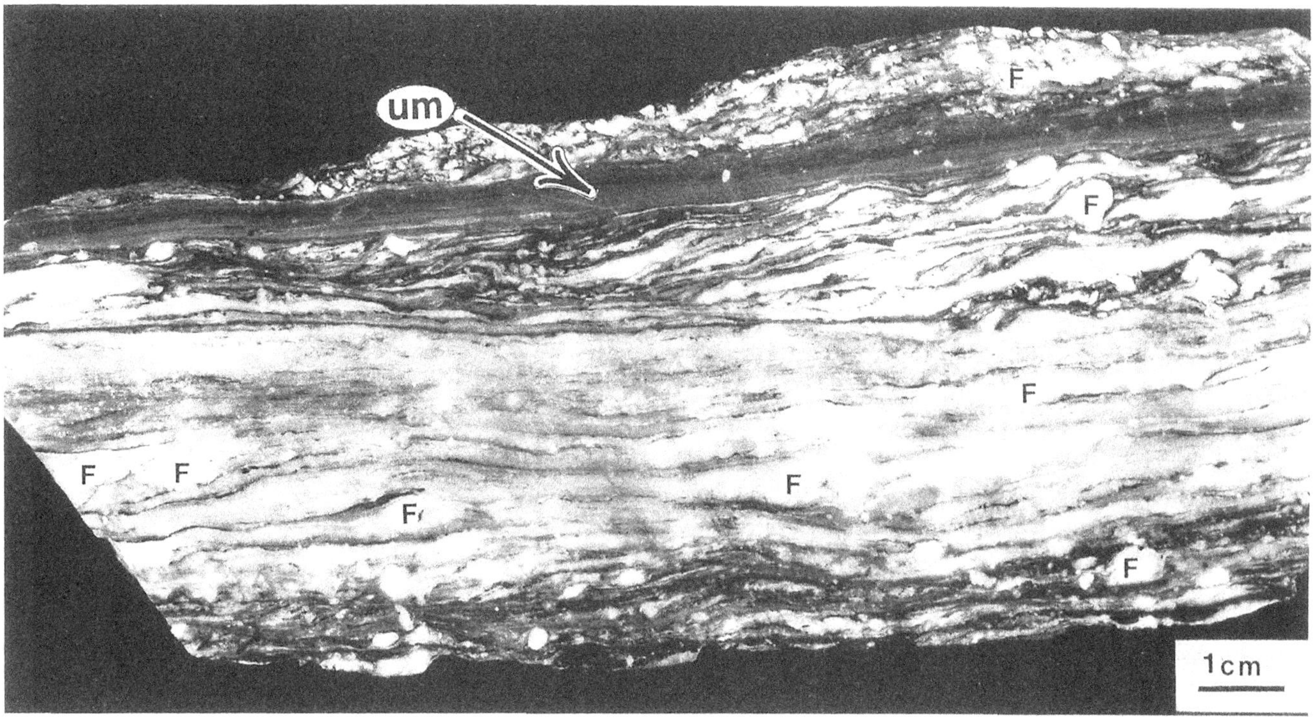


64

# 65. SOME MESOSCOPIC ASPECTS OF MYLONITIC SHEAR ZONES

J. Carreras, M. Losantos, and J. M. Casas

In the Cap de Creus peninsula (easternmost Pyrenees, Spain), two northwest-southeast-striking Hercynian mylonite belts exist. The northern one (Carreras and Casas, 1987) cuts across medium- to high-grade schists, the southern one (Carreras and Losantos, 1982, Simpson and others, 1982) traverses Hercynian granodiorites emplaced into low-grade metasedimentary rocks. Both belts are separated by complexly folded, low-grade metasedimentary rocks (Figure 65.1). Similar to other mylonite belts in the Pyrenean domain, these mylonites develop under greenschist-facies conditions in ductile shear zones. The shear zones have millimeter to hectometer thicknesses. The relationship between displacement and shear-zone width ranges usually between one and two orders of magnitude. A stretching lineation is commonly well marked, and the kinematic history of the shear zones is indicated by the bulk shear-zone geometry and various kinematic indicators (e.g., shear bands, asymmetric *c*-axis quartz fabric girdles). Strike-slip movement is dominant in both mylonite belts. The northern and southern belts are considered to be contemporaneous on the basis of similar relative timing with regard to the structural and metamorphic evolution.

**65A.** Shear zone in Cap de Creus mylonite zone showing progressive bending of the regional metamorphic foliation ($S_r$) into parallelism with the strike of the shear zone. A new mylonitic foliation ($S_m$) has transposed the earlier regional fabric and associated folds.

**65B.** Close-up of outcrop showing multiple shear zones in granodiorite. Two shear zones (1 to 5 cm thick) are developed in foliated Roses granodiorite of Hercynian age.

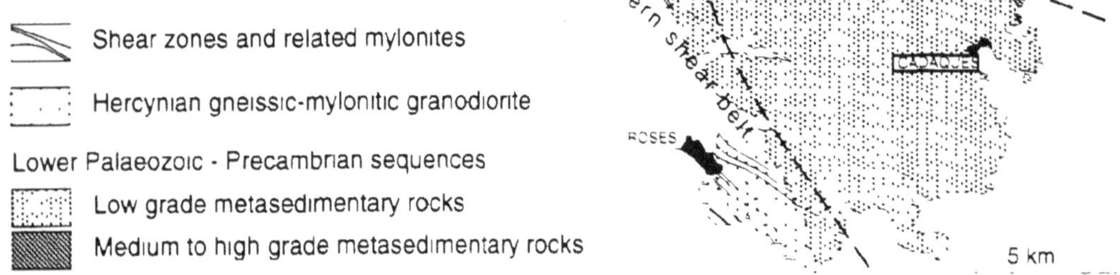


**Figure 65.1.** Geologic sketch map of the Cap de Creus peninsula (easternmost Pyrenees, Spain) showing the location of the mylonite belts

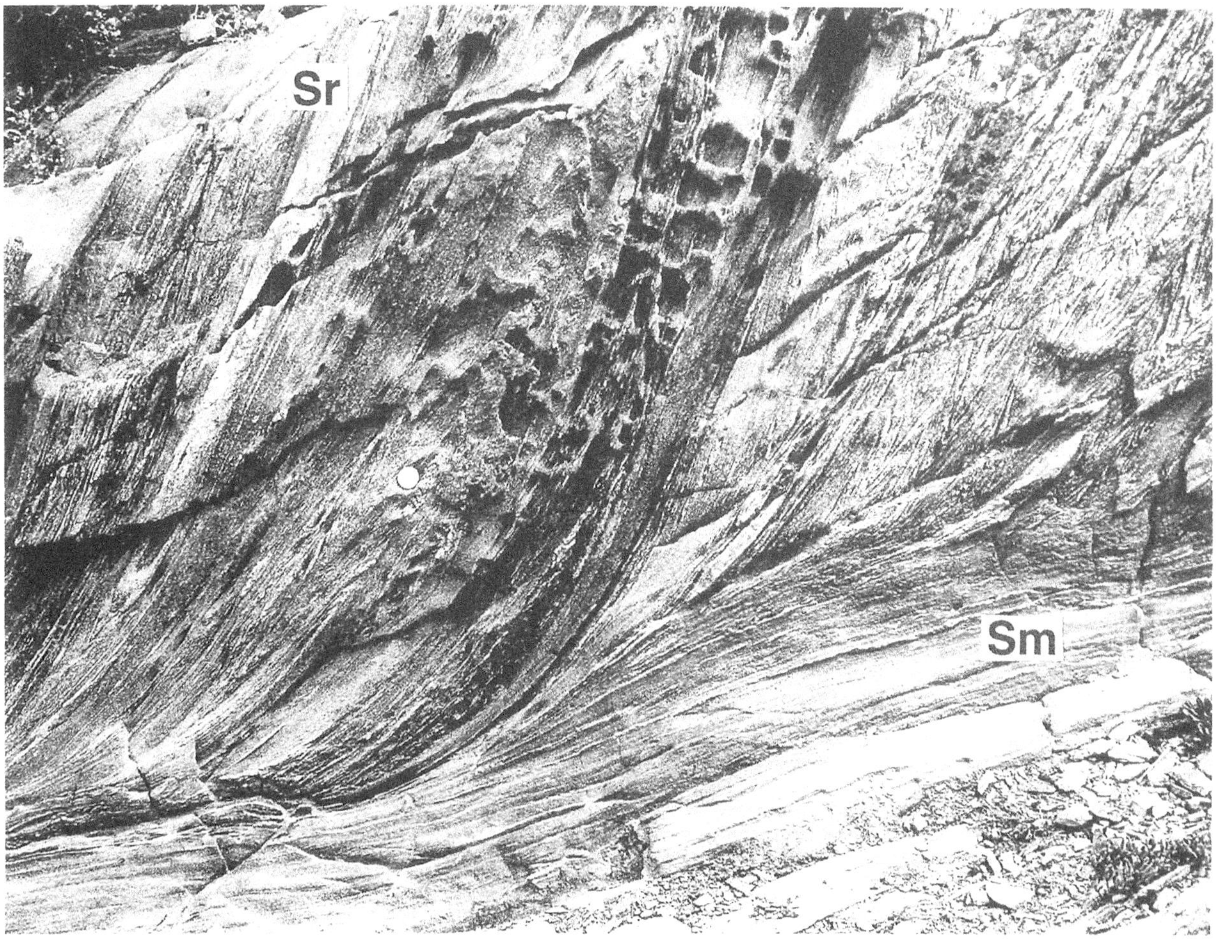

65 A

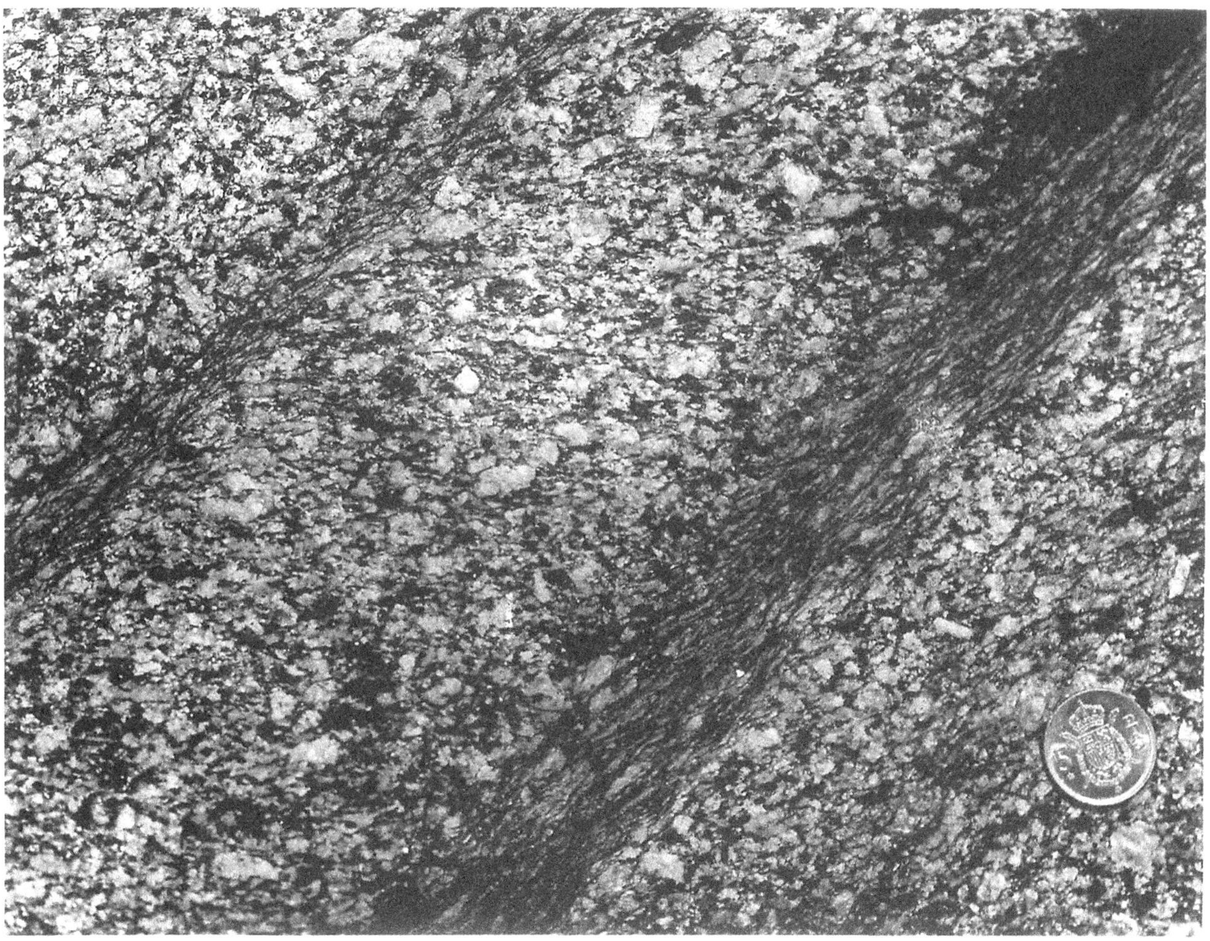
65 B

# 66. ANASTOMOSING CHARACTER OF MYLONITIC FOLIATION

T. H. Bell

Mylonitic foliations commonly anastomose on a large variety of scales from that of a thin section upward. This characteristic is commonly a function of inhomogeneous strain due to primary competency variation in the rock on the grain scale or larger. However, it is also a necessary consequence of deformation histories with a bulk shortening component (e.g., Bell, 1981) and could locally be independent of primary competency variation at a mesoscopic scale.

**66A and 66B.** These photographs show two views (taken at right angles to one another) of a "flying-saucer-shaped" pod of mylonitic schistosity. Adjacent fractured outcrops of similar pods revealed that the mylonitic schistosity anastomoses around ellipsoidal pods of less deformed country rock (Bell, 1978a). This outcrop almost totally exposes in three dimensions the ellipsoidal pod with its carapace of mylonitic foliation (the hammer in both photographs is in the same location in the outcrop). These mylonites formed under upper amphibolite-facies conditions in the Woodroffe thrust, central Australia, from original granulite-facies felsic gneisses.

Note that the lineation seen on the mylonitic foliation in 66B is gently curved. This lineation, the result of the intersection of an earlier layered schistosity and the mylonitic foliation ($S_m$), lies at a low angle to the stretching lineation. It is slightly curved due to inhomogeneous strain within $S_m$. Stretching lineations are also commonly slightly curved on such anastomosing surfaces, due to strain variation around an ellipsoidal pod. This characteristic is important as it provides an explanation for variation in orientation of the stretching lineation often seen on successive parallel planes of the mylonitic foliation. Careful examination of outcrops showing this latter feature commonly reveals the presence of two stretching lineations on some of the mylonitic foliation planes. The younger of these is generally considerably weaker in appearance but correlates with the average stretching lineation in the mylonite zone. The variation in the earlier formed stretching lineation can be attributed to the anastomosing character of earlier generated mylonitic foliation as shown above (Bell, 1978a).

66 A

66 B

JOHN P. SIMS AND CHRIS J. L. WILSON

Granulite-facies, noncoaxial deformation in the Rauer Group, East Antarctica, which developed due to a crustal-scale, sinistral, transpressive event at 1000 Ma (Sims and others, 1995), partitioned into zones of high and low strain (Figure 67.1, Plate 67A). The high-strain zones, which may be several kilometers in width, are marked by mesoscopic layering defined by transposed and extremely attenuated mafic, felsic, and pelitic gneiss (67A). These zones are characterized by a millimeter- to centimeter-scale gneissic layering that evolved through transposition of earlier fabrics and synkinematic metamorphic equilibration (67B; see also Plate 68). The low-strain zones, in contrast, preserve either pretectonic relationships (67A), or synkinematic foliations that developed earlier in the deformation. Hinge lines in complex fold interference patterns in low-strain zones parallel the mineral and elongation lineation in high-strain zones. Progressive deformation was accompanied by both repartitioning of strain and localization of high-strain deformation (Figure 67.1, 67B).

**67A.** Multiple suites of Proterozoic mafic dikes, which cross-cut Archean felsic gneiss on Lunnyy Island, act as markers for incremental and finite strain distribution. Two generations of synkinematic high-strain zones ($HSZ_{1\&2}$) are identified by the isoclinally folded, transposed, and attenuated mafic dikes, which partly define a mesoscopic layering. A continuous low-strain zone (LSZ) is apparent due to the preservation of cross-cuts between the pretectonic dikes (e.g., arrowed) and between the dikes and older gneissic foliations (Sims and others, 1995). An incremental low-strain zone coincident in time with $HSZ_2$ is apparent due to the reorientation of $HSZ_1$.

**67B.** Narrow zone of high strain on Macey Peninsula with a millimeter- to centimeter-scale gneissic layering ($GL_2$), into which a gneiss with centimeter- to decimeter-scale layering ($GL_1$) is transposed. The transposed layered gneiss ($GL_1$) represents an earlier, synkinematic increment of high strain. The mineral assemblage in the latest layering ($GL_2$), however, is not simply an attenuated version of the earlier layering but is an equilibrium assemblage for the bulk chemistry of the transposed, composite rock. The hinge lines developed in the earlier layered gneiss are parallel to the elongation lineation within the late zone. Note that the early layering is boudinaged (arrowed) prior to development of the narrow high-strain zone.

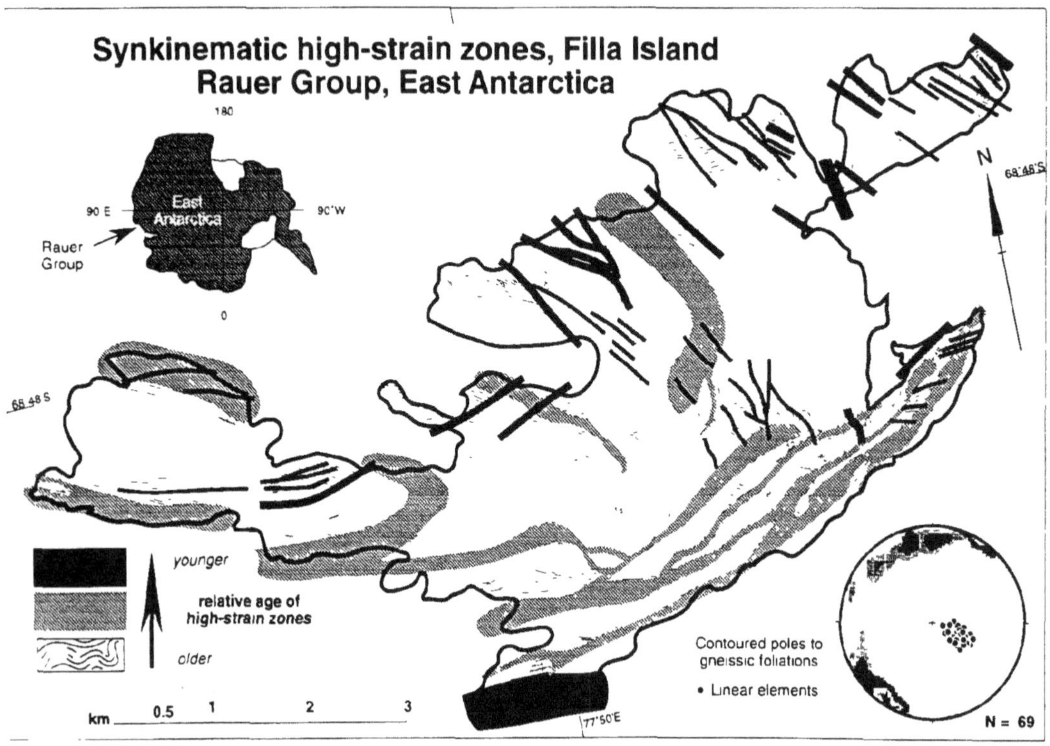


**Figure 67.1.** The spatial distribution and relative age of progressively developed zones of high strain in a portion of a major, crustal-scale shear zone. The foliations in the shear zone are near vertical, and all linear features are coaxial and steeply dipping. Pervasive, early formed foliations were reoriented, as strain was repartitioned into more discrete high-strain zones separating progressively larger, low-strain domains. In general, while the oldest high-strain zones display the most complex foliation geometries, the greatest finite strains are recorded in the latest formed high-strain zones, which form subplanar and anastomosing networks of apparently simple planar foliations

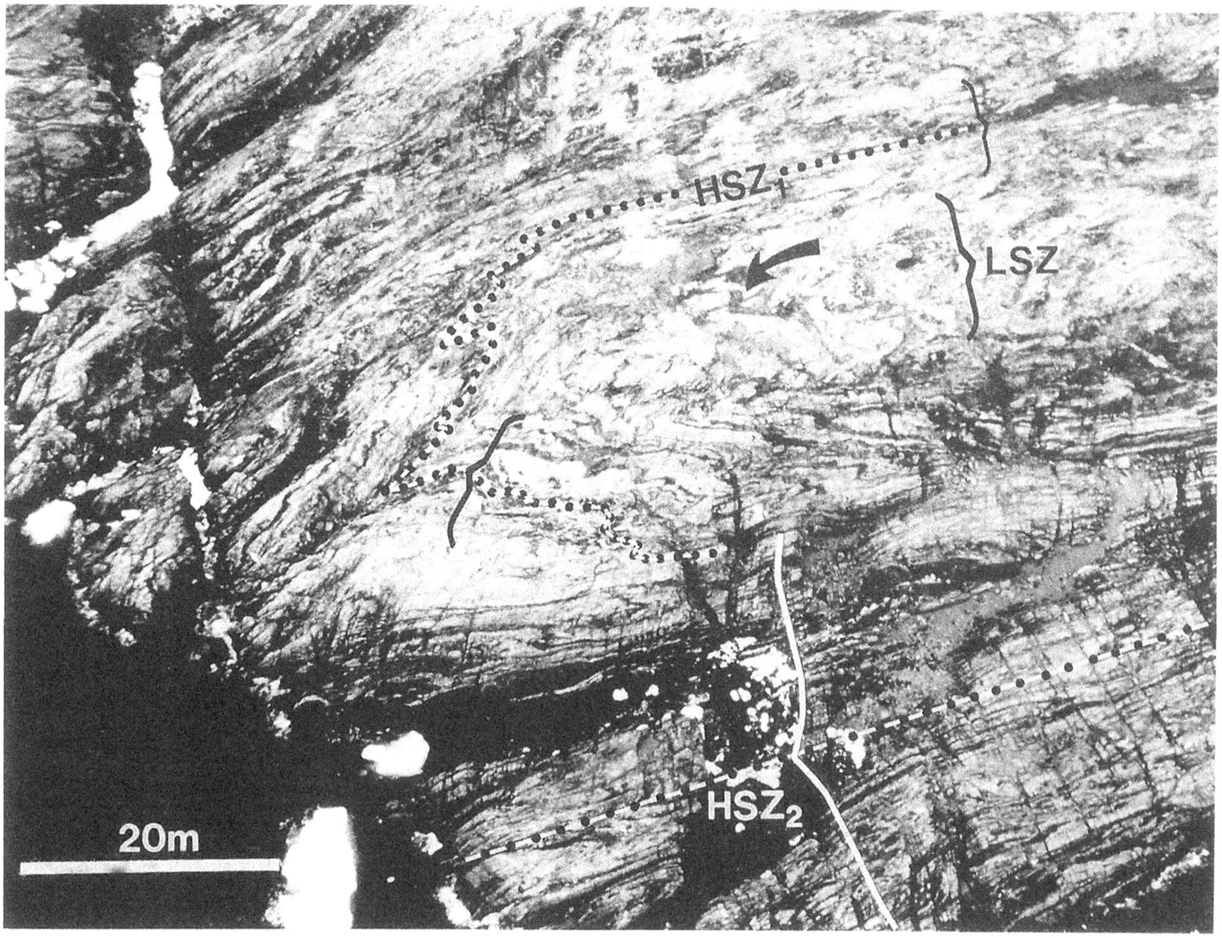


67 A

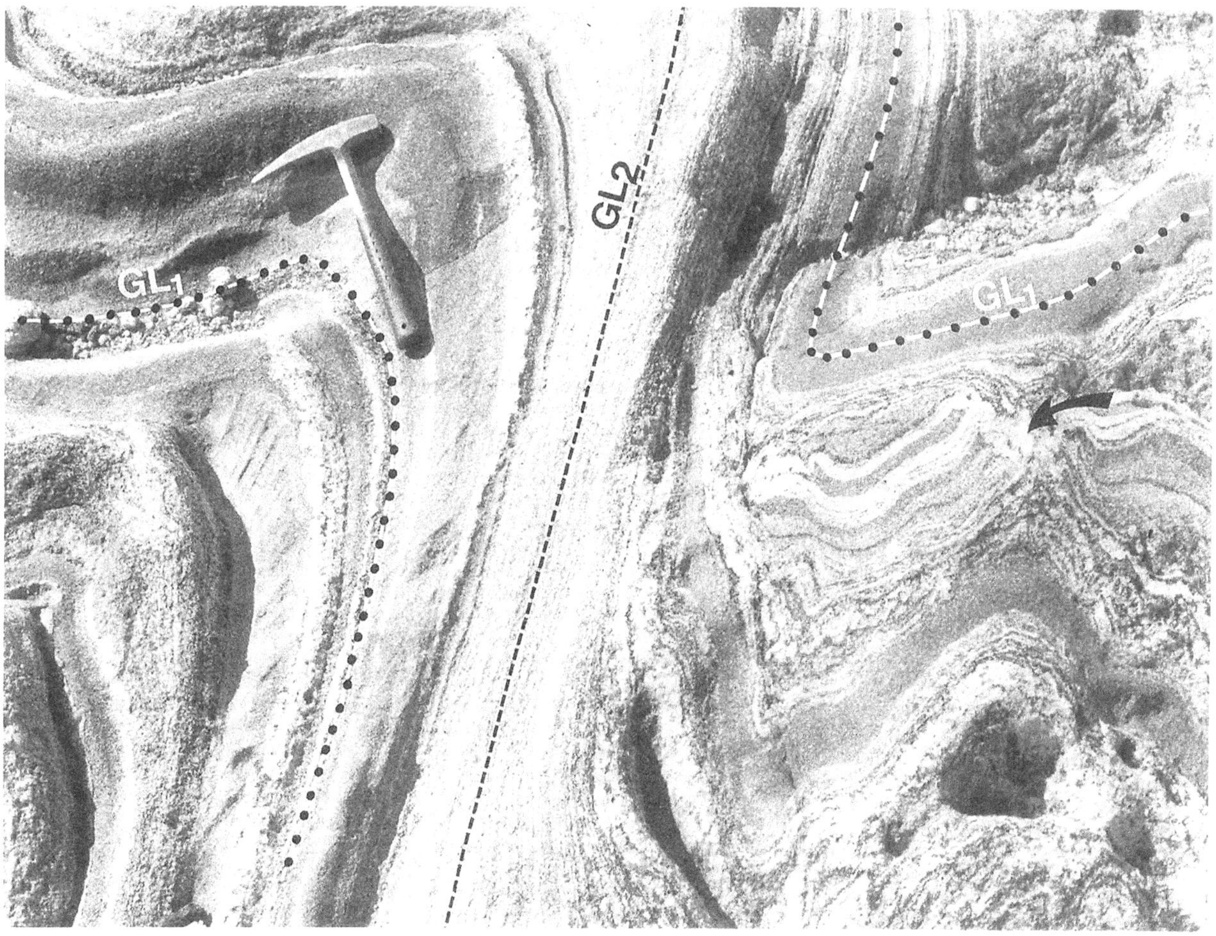


67 B

# 68. STRAIN PARTITIONING IN A GRANULITE-FACIES SHEAR ZONE: II. MICROSCOPIC FEATURES

JOHN P. SIMS AND CHRIS J. L. WILSON

Granulite-facies, metamorphic mineral assemblages and reaction textures that developed in a crustal-scale shear zone, exposed in the Rauer Group, East Antarctica, indicate that following peak conditions of 1000–1200 MPa and >1000°C, the rocks experienced at least 600 MPa of decompression while temperature remained above ~800°C (Harley and Fitzsimons, 1991). Microstructures and mineral reaction textures are contrasted between zones of high and low noncoaxial strain that developed during a single episode of progressive deformation coincident with the decompression. The progressive deformation was accompanied by repartitioning of strain distribution and gradual localization of high-strain deformation. In zones of high strain, mineral assemblages formed with a shape orientation consistent with the extension direction of the deformation (68A), and the equilibrium mineral-assemblage paragenesis indicates a continuous deformation path in P-T space (e.g., Figure 68.1). In zones of low strain, in contrast, decompression produced unoriented corona textures around peak assemblages that reflect grain-scale disequilibrium (68B). Unlike the high-strain assemblage paragenesis, the low-strain peak and corona assemblages represent increments on the continuous P-T path.

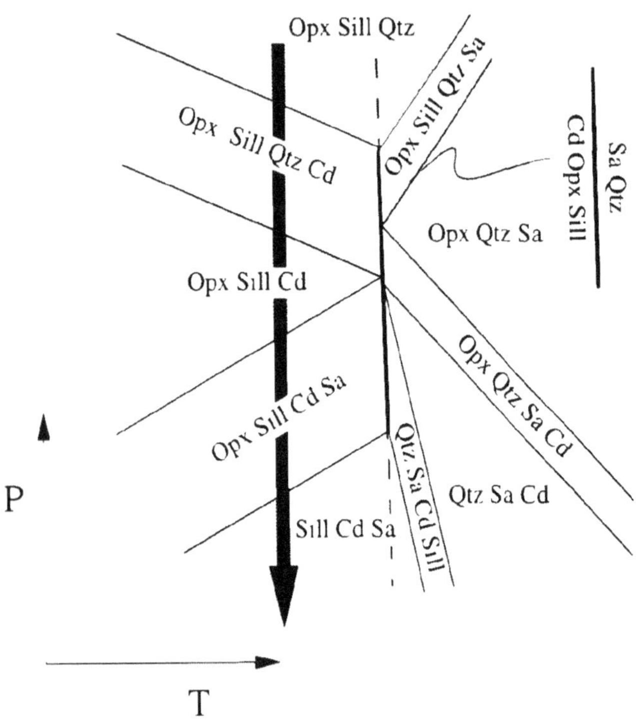


**68A.** Microstructures and mineral reaction textures in a high-Mg metapelitic rock, indicating decompression accompanying progressive deformation in a zone of high strain, Mather Peninsula (XZ section). There is a well-developed shape fabric associated with the peak assemblage (orthopyroxene + sillimanite + quartz) that is collinear with the shape fabric of the final (high-temperature) decompression assemblage (sapphirine + cordierite + sillimanite; Figure 68.1). The cordierite displays intracrystalline deformation in the form of twinning and recrystallization, suggesting continued accommodation of strain. The oriented biotite is unlikely to be stable at the P-T conditions of these decompression reactions and is considered to be a late crystallizing phase. NX

**68B.** Microstructures and mineral reaction textures in a mafic metasedimentary rock, indicating decompression (Harley, 1988) in a zone of low strain, Mather Peninsula. The peak assemblage [garnet (Gt) + clinopyroxene (Cpx)] is overprinted by an unoriented decompression assemblage [plagioclase (Plag) + orthopyroxene (Opx)], suggesting that the texture developed under low (local) differential stress. Ilmenite (ilm) may have been a peak phase occurring on the grain boundary between garnet and pyroxene, whereas biotite (Bi) is a late formed phase. Note that the plagioclase in the symplectite is in optical continuity, and there are only minor intracrystalline deformation features. This peak metamorphic assemblage occurs in atoll-like segregations overprinting a gneissic foliation defined by millimeter-scale domains dominated by hornblende and plagioclase, respectively, suggesting that these rocks were subjected to high strains prior to the metamorphic peak. NX

**Figure 68.1.** A schematic FeO-MgO-$Al_2O_3$-$SiO_2$ system P-T pseudosection for the assemblage orthopyroxene (Opx), sillimanite (Sill), quartz (Qtz), cordierite (Cd), and sapphirine (Sa), based on the topology of figure 2.5a in Hensen and Harley (1990; drawn along the garnet-absent univariant reaction emanating toward higher pressure from the spinel-absent invariant point). The progression of stable mineral assemblages observed in 68A that formed concurrently with the deformation is indicated with the arrow.

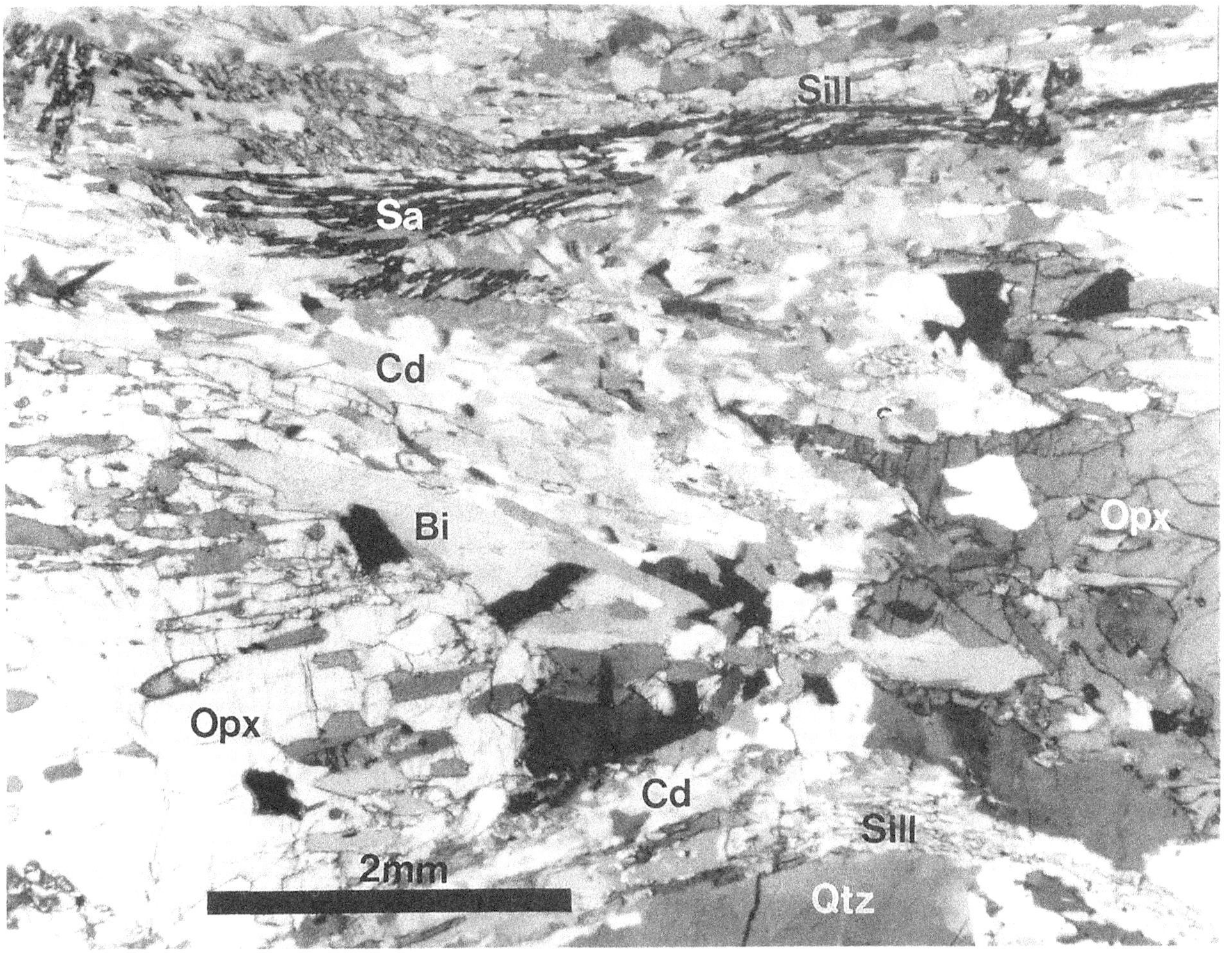
Sill
Sa
Cd
Bi
Opx
Opx
Cd
Sill
2mm
Qtz

68 A

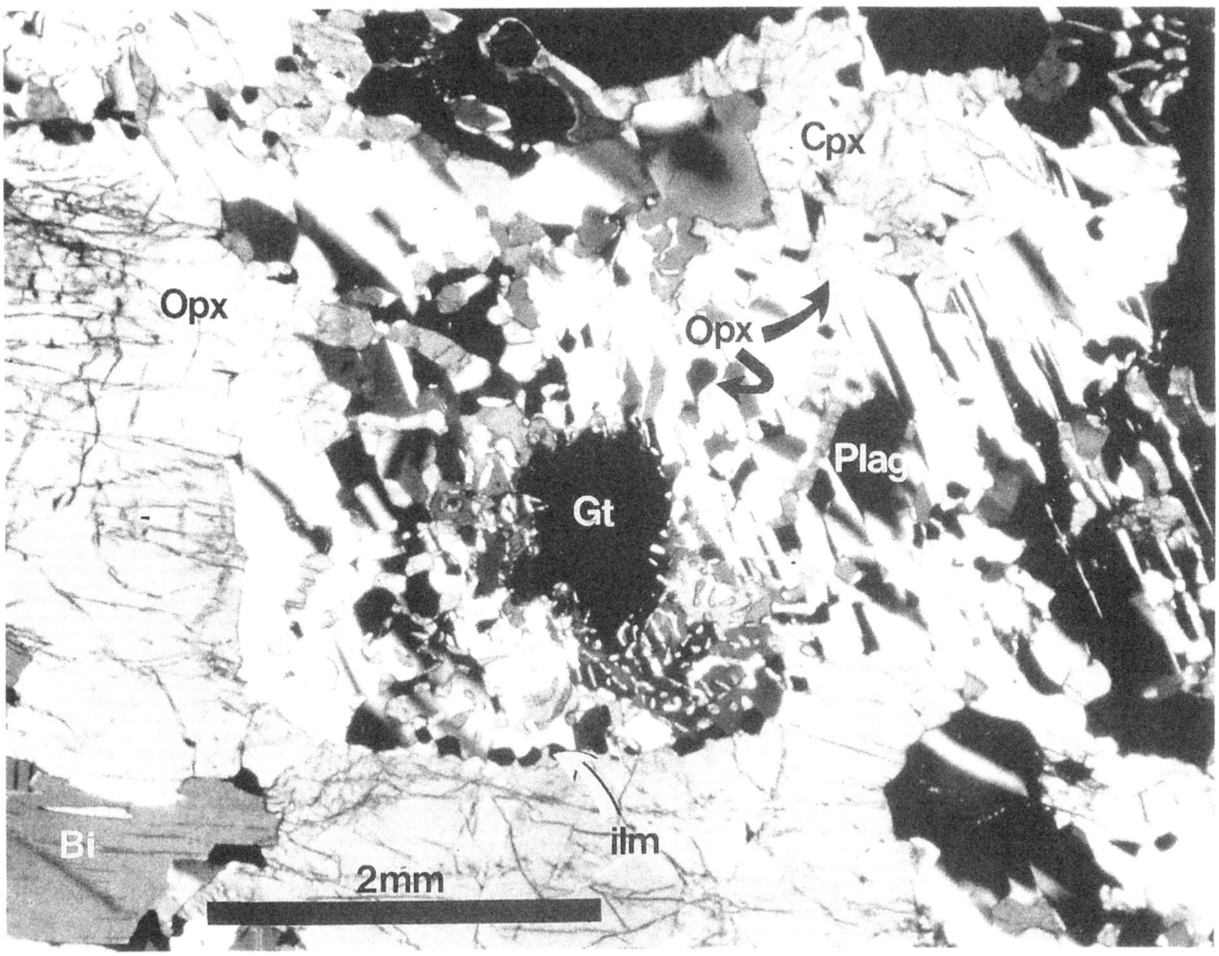
Cpx
Opx
Opx
Plag
Gt
ilm
Bi
2mm

68 B

Laurel B. Goodwin

Deformation of granodiorite within the Santa Rosa mylonite zone, southeastern California, U.S.A., was initiated during thrusting at amphibolite-facies conditions (Simpson, 1985) at about 80 Ma (Goodwin and Renne, 1991). The types and patterns of structures in the mylonite zone reflect, at all scales, variations in deformation mechanisms over space and time.

**69A.** Strain is partitioned during mylonitization between narrow zones of ultramylonite and phyllonite and broad regions of mylonite (Goodwin and Wenk, 1995). The outcrop face is perpendicular to foliation and parallel to lineation. Note the sharp boundaries between mylonite and thin, laterally discontinuous (at the outcrop scale) ultramylonite (u). The boundaries are parallel to C surfaces. Shear bands (sb) record sinistral (top-to-the-southwest) displacement. S-C structures and shear bands can be seen in the mylonite, whereas only C surfaces are macroscopically visible in the ultramylonite and phyllonite. Chisel is 15 cm long.

**69B.** Cataclasite with a partial ductile overprint. Initial deformation in zones of ultramylonite and phyllonite was accomplished through whole-rock cataclasis (Goodwin and Wenk, 1995). Subsequent, macroscopically ductile flow produced ultramylonite or phyllonite from cataclasite, and mylonite from the host rock. Strain was therefore heterogeneous over both space and time within the Santa Rosa mylonite zone. The boundary (B) between cataclasite and macroscopically undeformed granodiorite is knife-sharp; sharp boundaries are maintained during ductile flow (see 69A). Within the cataclasite, fragments are angular and biotite grains (b) are locally strongly aligned. Quartz (q) exhibits undulatory extinction and subgrains both within and adjacent to the cataclasite, and monocrystalline ribbons in the adjacent rock. Both undulatory extinction and fractures are common in feldspars (p). NX

**69C.** Intracrystalline folding in biotite in mylonite (Goodwin and Wenk, 1990). Grain-size reduction in mylonite, where whole-rock cataclasis did not take place, occurred in part through intracrystalline folding and cataclasis in biotite. Strain therefore was partitioned at the scale of microns. Micrograph viewed looking down the basal planes of the biotite. Folding of laminae parallel to (001) within coherent crystal. TEM

**69D.** Lattice image of a biotite grain in phyllonite (Goodwin and Wenk, 1995). Subsequent to cataclasis, deformation of biotite in phyllonite and ultramylonite took place by dynamic recrystallization in concert with intracrystalline slip and mechanical rotation. Intracrystalline slip is recorded in part by irregularly spaced stacking faults in biotite (e.g., st), which cause streaking in the diffraction pattern (s). Spacing between lattice planes is generally 1 nm, typical of the most common, 1M, polytype of biotite. Mechanisms accommodating deformation in these fine-grained regions are therefore different from mechanisms accommodating grain-size reduction in mylonite (e.g., 69C). C* is normal to both the a and b crystallographic axes. TEM

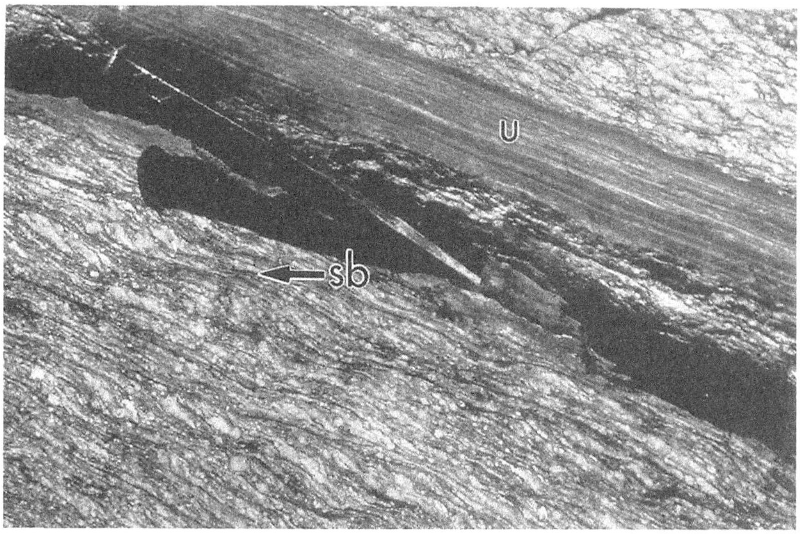


69 A

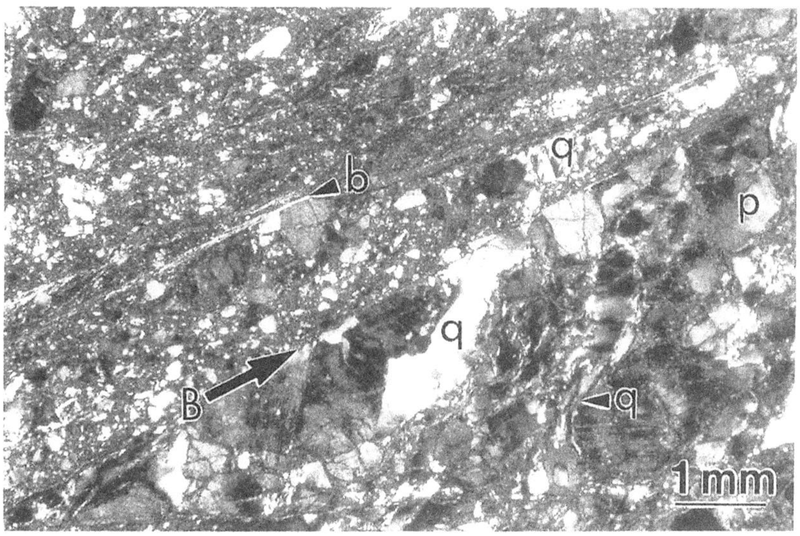


69 B

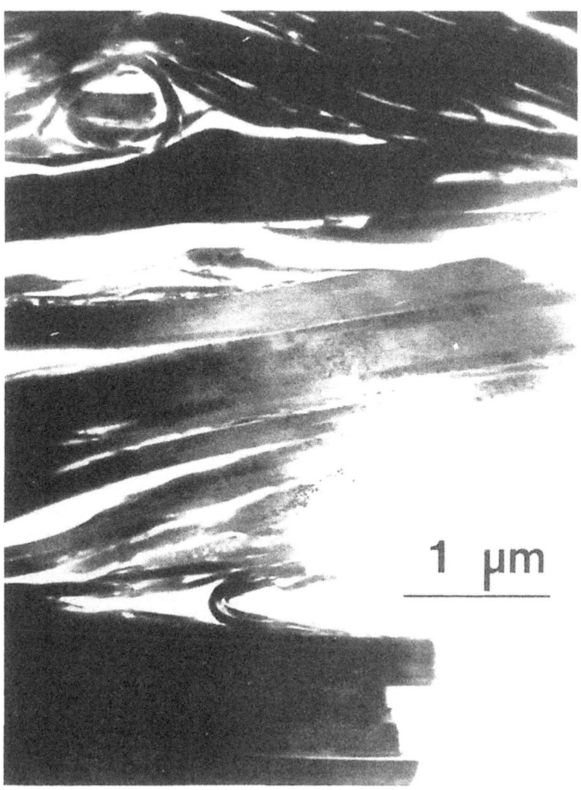


69 C

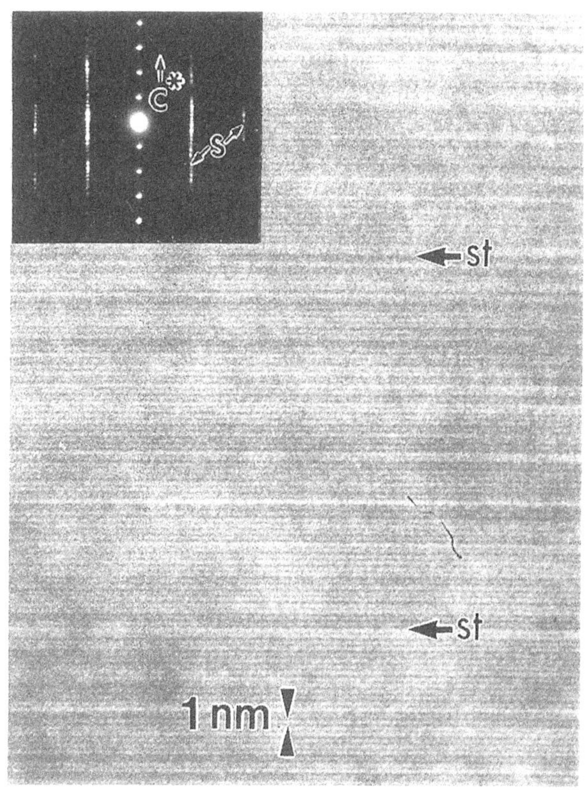


69 D

# 70. OLIVINE AND QUARTZ TECTONITES RELATED TO INTRA-OCEANIC THRUSTING

JEAN LUC BOUCHEZ, GEORGES CEULENEER, AND ADOLPHE NICOLAS

In many ophiolite complexes the base of the peridotite sequence overlies a sole of metamorphic rocks grading downwards from garnet-amphibolite facies to lower metamorphic grades. These rocks, derived from materials of the upper oceanic crust (basalts, sediments), were sheared during the thrusting of the oceanic lithosphere (Nicolas and Le Pichon, 1980) in geodynamic situations that were analyzed in Boudier and others (1982). Up to a few tens to hundreds of meters above the contact plane, the peridotite exhibits mylonitic microstructures (70A) indicative of large strains under high-stress conditions (~100 MPa). This stress level is deduced from the average recrystallized grain size of a few tens of microns, according to the experimental calibrations of Karato (1984). The temperature of deformation of these peridotites, estimated at 700° to 900°C (in equilibrium with the immediately underlying metamorphic rocks), is in the "low-temperature" domain for plastic deformation of olivine, contrasting with the peridotites deformed at "high temperature" (see Borradaile and others, 1982, plate 163, as well as Plate 175, this volume) which form the overlying ultramafic sequence and originate from asthenospheric flow. The sole of metamorphic rocks, made of alternating layers of amphibolite, marble, and quartzite, has undergone the same shearing episode as the peridotites. Quartzites located close to the peridotites have been deformed in the "high-temperature" domain for the natural plastic deformation of quartz (70B). However, due to the lower temperature threshold for plastic flow of quartz, secondary, lower-temperature tectonic events may affect the quartzites, leaving the peridotites undeformed by these late events. In 70A and 70B: the foliation is normal to the plate and N-S; lineation is parallel to the plate. Half-crossed nicols.

**70A.** Low-temperature peridotite mylonite (Oman Mountains, Sultanate of Oman). Twenty to 150 μm grains of recrystallized olivine, orthopyroxene, and spinel (sp in boxed area) surround a few porphyroclasts of orthopyroxene with either a globular (cross) or a ribbon-shaped (arrow) habit. The lattice preferred orientation (LPO) of olivine reveals crystal-plasticity on multiple low-temperature slip systems (Boudier and Coleman, 1981). In the pyroxenes, the obliquity of the unique (100) slip plane (clearly visible cleavage) with respect to foliation (N-S) reveals the sense of shear (dextral), which is related to the sense of the oceanic thrusting itself. Late serpentinization is reponsible for the network of fractures.

**70B.** High-temperature quartzite mylonite (Oman Mountains, Sultanate of Oman). A few meters below the contact plane, the quartzite consists of 1 to 5 mm thick, completely recrystallized quartz layers that alternate with layers of fine-grained amphibole-garnet-chlorite-epidote-calcite. Numerous dark inclusions within quartz grains, mainly garnet and chlorite, demonstrate grain growth of the quartz. The oblique preferred orientation of the *c*-axes, marked here by northwest-striking prismatic sub-boundaries, indicates the same dextral sense of shear as inferred from the peridotites (70A). The sub-boundaries themselves are deformation features that appeared during a late strain pulse, probably at lower temperature, which did not substantially alter the original grain arrangement and LPO (Bouchez and Pêcher, 1981). Note the numerous nearly basal (0001) deformation lamellae (faint bright striae), normal to the sub-boundaries, also marked by trails of minute fluid inclusions, and the calcite-filled fracture.

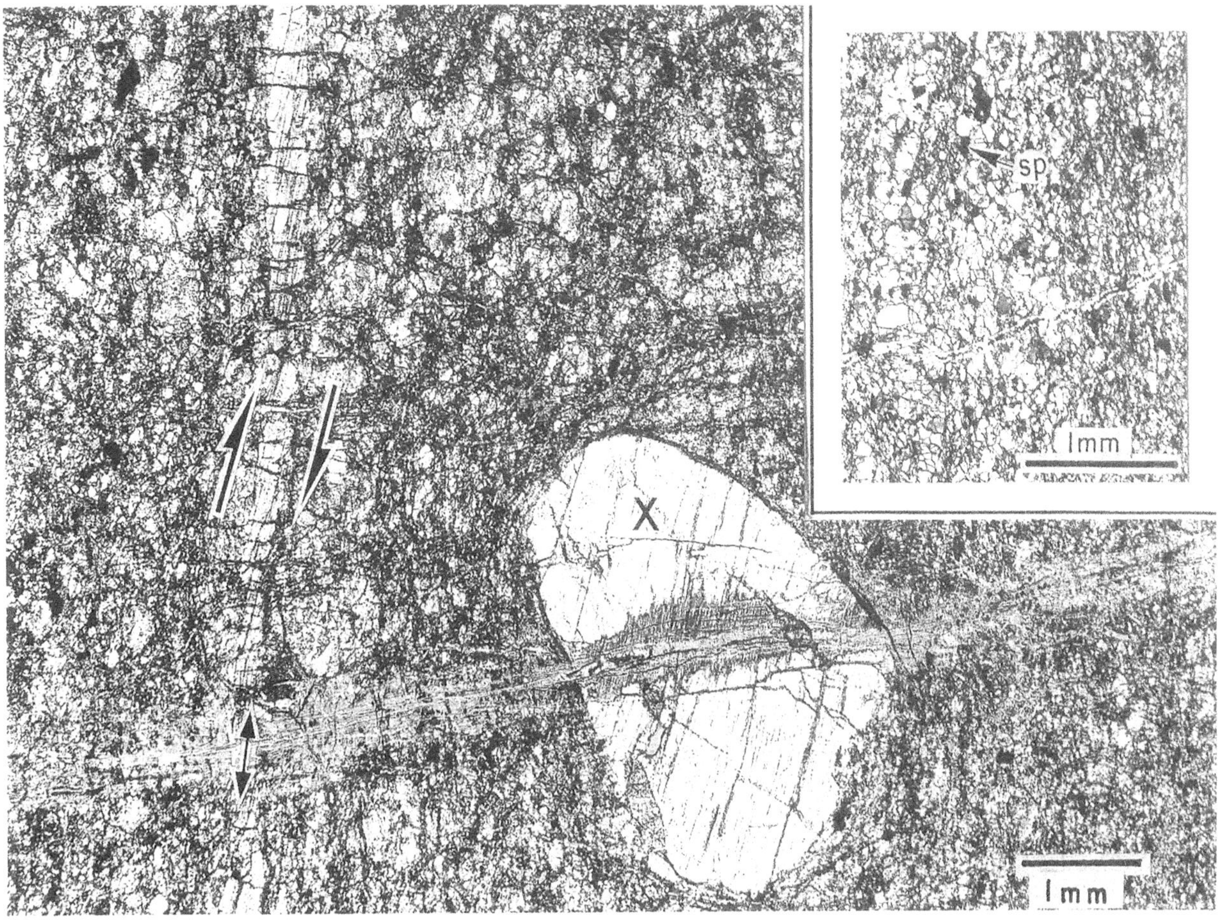

70 A

70 B

# 71. ASYMMETRIC BOUDINAGE IN MYLONITIC GNEISSES

MARK T. SWANSON

Steeply dipping mylonitic gneisses within the Cushing Formation (Precambrian to Ordovician?) of the Casco Bay area, coastal Maine, U.S.A., have developed asymmetric boudinage due to elongation and clockwise rotation during dextral strike-slip shear. This regional shearing is related to transpression along the Flying Point restraining bend segment of the Norumbega fault zone during late Devonian to Permian deformation. Competent-layer and foliation boudin partings (commonly expressed as quartz veins and pegmatite intrusions) were initiated orthogonal to the mylonitic shear layering. Dextral shear has resulted in the clockwise rotation of these veins and partings to progressively lower angles ($\beta$) relative to this shear layering. Partings and veins show sinistral slip during clockwise rotation along with increasing amounts of vein-parallel elongation. Initially orthogonal quartz veins and pegmatite intrusions have typically evolved into strings of boudin pods at low angles to shear layering upon rotation and elongation during this dextral shear (Swanson, 1992b). Measured $\beta$ angles can be used to approximate minimum shear strains in these mylonitic rocks assuming a dominantly simple-shear deformation (Swanson, 1994).

The photographs are of asymmetric boudin features within rocks at Simpson's Point, upper Casco Bay, approximately 2 km southeast of the high-strain core of the Flying Point fault zone. A range of orientations for these features suggests the repeated initiation and modification of orthogonal boudin partings during progressive dextral shear. $\beta$ angles of only a few degrees in these lithologies indicate minimum shear strain ($\gamma$) values of ~10 to 20.

**71A.** Slight clockwise rotation of intervening quartz-filled parting surface with modification of the ends of competent-layer boudins. Note sinistral slip along the parting surface during rotation. $\beta$ angles of ~85 to 50° correspond to minimum $\gamma$ value of only ~0.2–0.8 within this more competent layer.

**71B.** Additional clockwise rotation and sinistral slip along a similar parting surface. $\beta$ angle of ~40° corresponds to a minimum $\gamma$ value of ~1.2.

**71C.** Increased clockwise rotation to much lower $\beta$ angle relative to the mylonitic shear layering. $\beta$ angle of ~25° corresponds to a minimum $\gamma$ value of ~2.1.

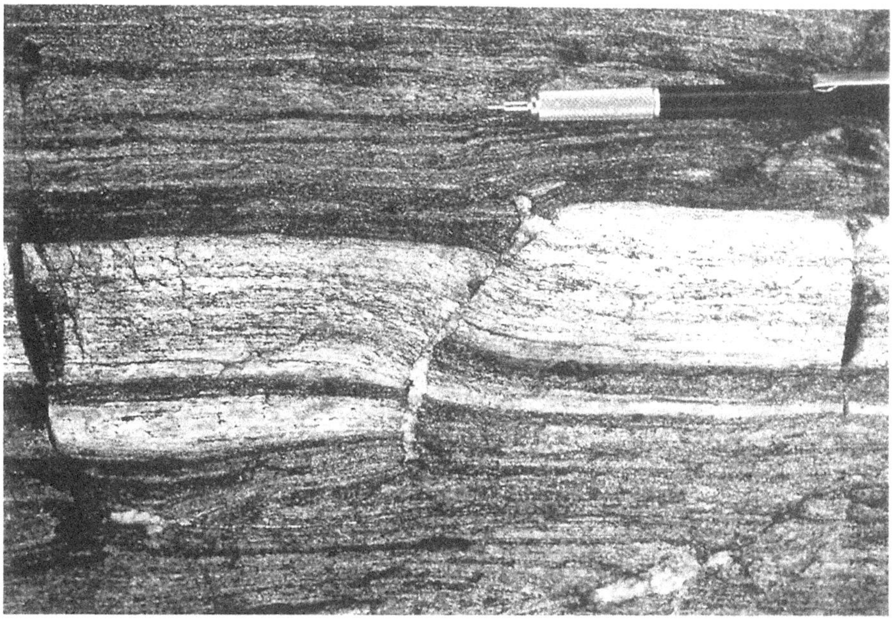

71 A

71 B

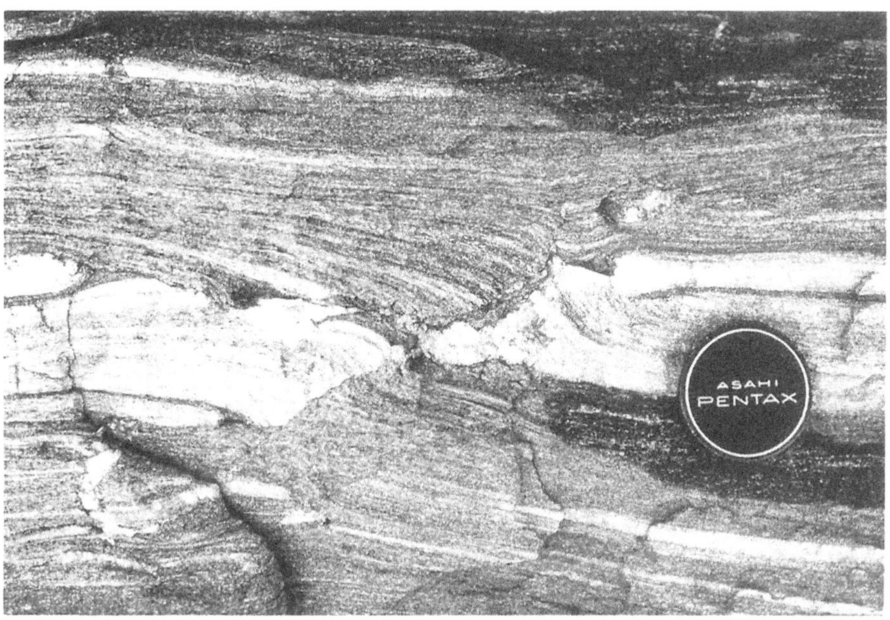


71 C

# 72. TENSILE FRACTURE WITHIN ELASTIC-BRITTLE INCLUSIONS OF TOURMALINE

SHAOCHENG JI

Microboudins of tourmaline occur in a mylonitized migmatite collected from the Ximeng metamorphic complex (22°40′N, 99°30′E), Yunnan Province, China. This metamorphic complex is located in the Mesozoic Lancang-River collision-tectonic zone between the Baoshan and Lancang blocks (Li, 1988). The collision resulted in the formation of a united Indochina block (Tapponnier and others, 1986, 1990) or Qiangtang block (Dewey and others, 1988). Both foliation and stretching lineation in the tourmaline-bearing mylonite dip steeply (65–80° on average) at the scale of the outcrop.

The tourmaline crystals grew during migmatization (prograde), and then fractured in lower amphibolite-facies conditions during mylonitization of the migmatite (retrograde stage) (Ji and Zhao, 1993). All fractures show rectangular forms and are plane breaks normal to the lineation. This indicates that the fracture of tourmaline crystals was controlled by elastic-brittle behavior rather than elastic-plastic fracture, because the latter would have led to "pinch and swell" boudin shapes. The above conclusion indicates also that the fractures were probably formed in a coaxial strain regime, since noncoaxial strain would have caused the microboudins to be arranged obliquely to the lineation. Gaps between the microboudins have been filled with deformed and dynamically recrystallized quartz, suggesting that the fracturing and subsequent pull-apart event was syntectonic. The gap lengths are not uniform, reflecting the temporal sequence of fracturing. Large gaps are associated with the sites of earlier fractures and smaller gaps with later fractures.

Breakage of elastic-brittle inclusions in the ductile flowing matrix can be qualitatively explained according to fiber-loading theory (Cox, 1952). Fracture occurs at the position of maximum internal tensile stress ($s_{tm}$) in a fiber. Once the $s_{tm}$ within the fiber exceeds its tensile-fracture strength, fiber breakage occurs and results in two segments. These segments are in turn newly loaded during further creep because the surrounding matrix continues straining. The magnitude of $s_{tm}$ in the fiber decreases when its aspect ratio (length/width) decreases. Sequential fracturing will decrease aspect ratio and in consequence will reduce $s_{tm}$ in the fiber. The breakage process will continue until a fiber is reduced to segments, all of which are shorter than some critical length for which the $s_{tm}$ transferred from the matrix to the fiber segments is less than, or equal to, the tensile fracture strength of the fiber. It is also important to point out that the distribution of $s_{tm}$ and thus the location of tensile fracture in the fiber depend also on its shape (Ji and Zhao, 1993).

**72A.** Successive midpoint fracturing occurs within a tourmaline crystal of constant thickness. XZ section. NX

**72B.** A tourmaline crystal of nonuniform thickness is sequentially fragmented into two segments of unequal length, the longer of which has the greater thickness. XZ section. NX

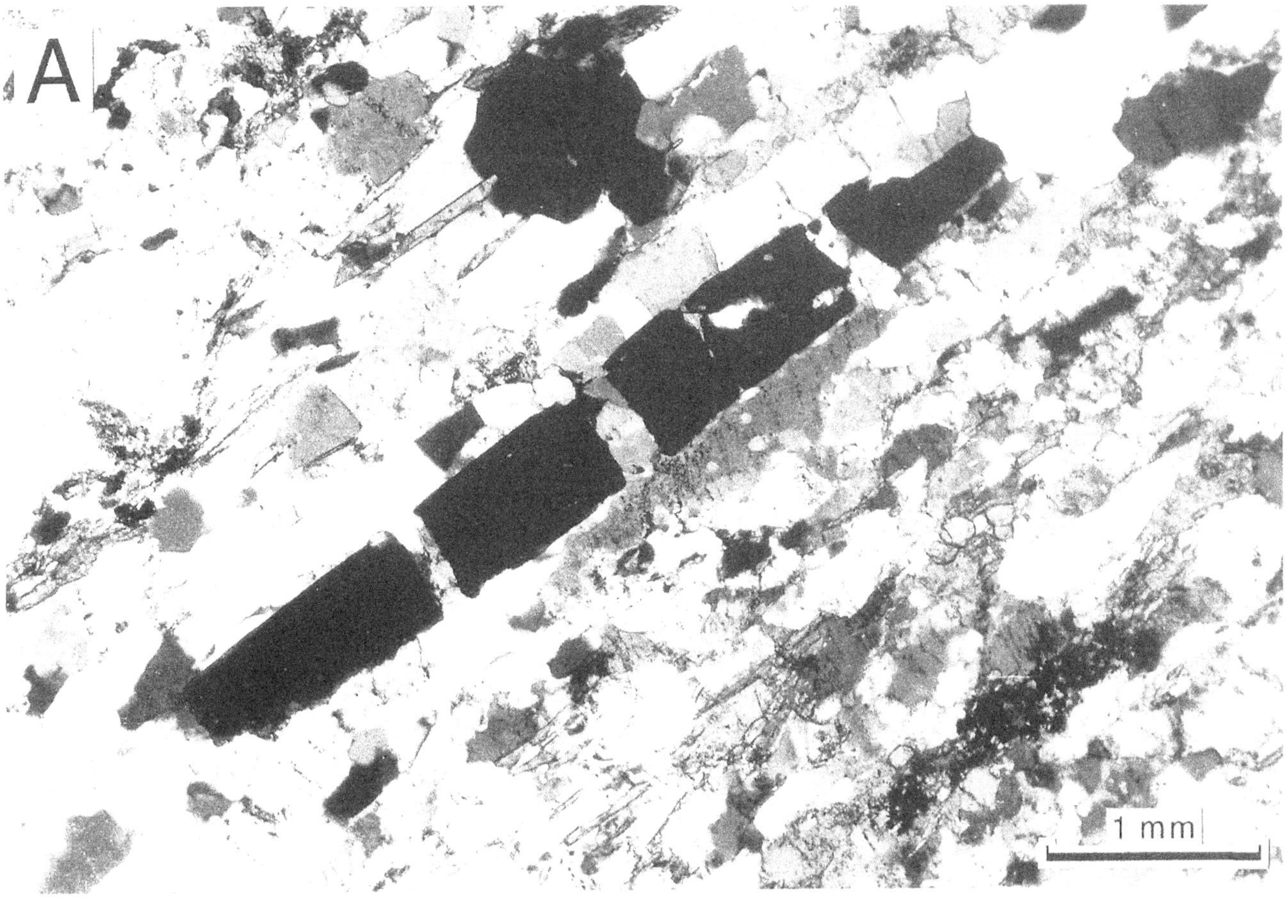


72 A

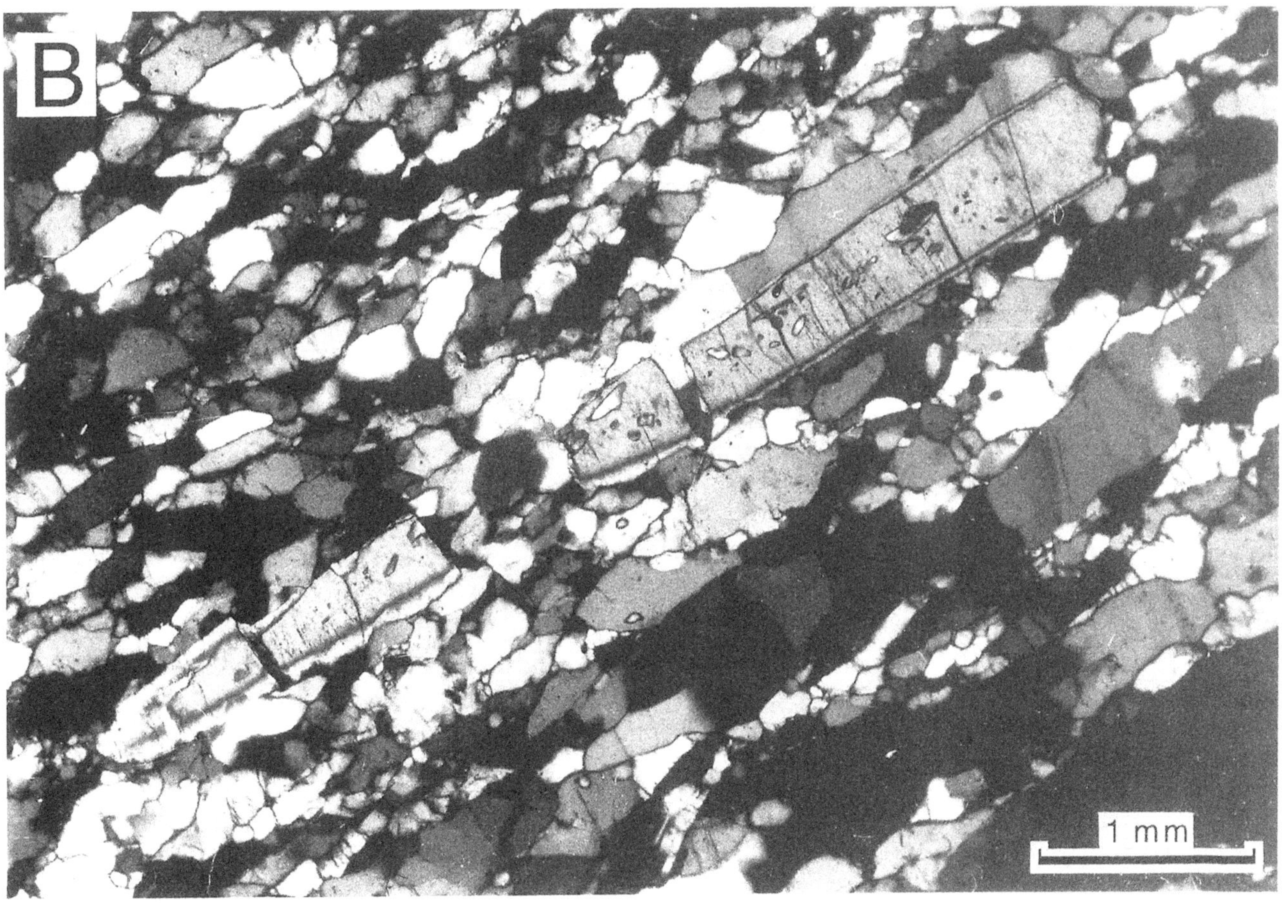


72 B

# 73. MICROSTRUCTURAL DEVELOPMENT IN APLITE WITH PROGRESSIVE STRAIN

LISA N. DELL'ANGELO AND JAN TULLIS

These photographs illustrate the strain partitioning that occurs during progressive axial compression of a fine-grained granite (aplite) at conditions where the stronger feldspars deform by recrystallization-accommodated dislocation creep (Regime 1 of Hirth and Tullis, 1992; see Plate 157, this volume) and the weaker quartz deforms by climb-accommodated dislocation creep (Regime 2 of Hirth and Tullis, 1992; see Plate 146, this volume). Similar microstructures are exhibited by granitic aggregates naturally deformed at lower greenschist-facies conditions. In Regime 1 dislocation creep, the difficulty of dislocation climb causes original grains to become work-hardened; grain-boundary migration recrystallization occurs along grain or deformation band boundaries, and strain is partitioned into these initially strain-free recrystallized grains. In Regime 2, easy dislocation climb allows homogeneous strain of original grains, and formation of coarser recrystallized grains by subgrain rotation. As the initially dispersed quartz grains become highly flattened and more interconnected, strain is partitioned into the quartz-rich regions.

These samples were deformed at a confining pressure of 1500 MPa, a constant strain rate of $10^{-6}$/s, and a temperature of 900°C; all samples showed progressive strain weakening. Part of the weakening is due to the progressive Regime 1 recrystallization of the feldspar and part of it is due to a reversal in the connectedness of the weaker quartz (originally the dispersed phase) and stronger feldspars (originally the interconnected phase) (Figure 73.1). In all photographs the compression direction is horizontal, and the scale is the same as for 73A, where the bar is 200 μm. These photographs demonstrate that mylonitic textures, lacking asymmetric features, can be induced by high axial strains; compare these photographs with those from quartz-feldspar aggregates subjected to simple shear at similar conditions (see Plate 101, this volume).

**73A.** Starting material is aplite from a dike cutting the Black Mountain granite in Enfield, Vermont, U.S.A. (Naylor, 1971). Average grain size is 150–200 mm, and the composition is 30% quartz (Q), 35% microcline, 33% oligoclase, and 2% biotite and iron oxides. Because the deformation behavior of the feldspars is almost identical, the aplite can be idealized as a two-phase aggregate consisting of isolated quartz grains within a continuous stress-supporting framework of feldspar grains (see Figure 73.1). NX

**73B.** Sample shortened 40% (E-W). Deformation lamellae are present in some quartz grains (Q) and minor recrystallization occurs along quartz-quartz grain boundaries (arrows). TEM shows that the quartz grains contain subgrains, whereas the feldspar grains (F) have a very high density of tangled dislocations. NX

**73C.** Sample shortened 65% (E-W). Feldspar grains (F) have deformed inhomogeneously, undergoing microboudinage and separating into less deformed lozenges with thin, more deformed necks or tails that are recrystallized on the TEM scale. Quartz grains (Q) have flattened much more homogeneously and have started to become interconnected. NX

**73D.** Sample shortened 85% (E-W). Feldspar grains (F) have separated into less deformed augen separated by thin, more deformed tails of very fine (~1–5 mm) recrystallized grains, formed by grain-boundary migration. Quartz grains (Q) are highly flattened and partially recrystallized (average size ~10 μm) by progressive subgrain rotation. NX

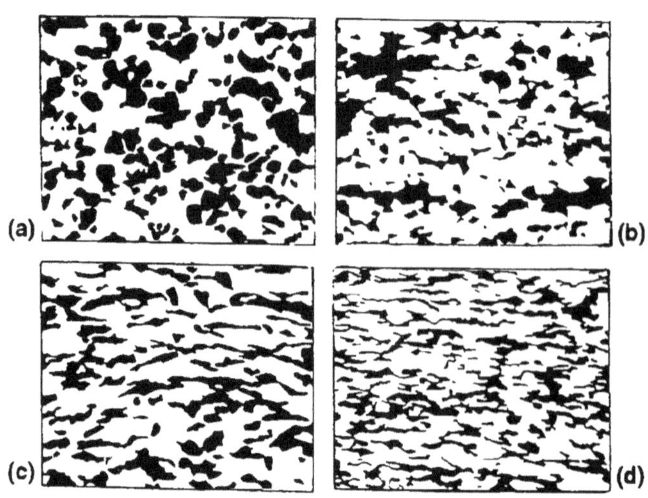


**Figure 73.1.** Tracings of optical micrographs of aplite illustrating changes in arrangement of quartz and feldspar phases with progressive strain at 900°C, 1500 MPa, $10^{-6}$/s; black is quartz, white is feldspars (both oligoclase and microcline). (a) Undeformed starting material, quartz is dispersed, feldspars form the interconnected matrix. (b) Sample with 24% shortening (NS). (c) Sample with 55% shortening (NS). (d) Sample with 85% shortening (NS), quartz has become much more interconnected. Width of one box = 2 mm.

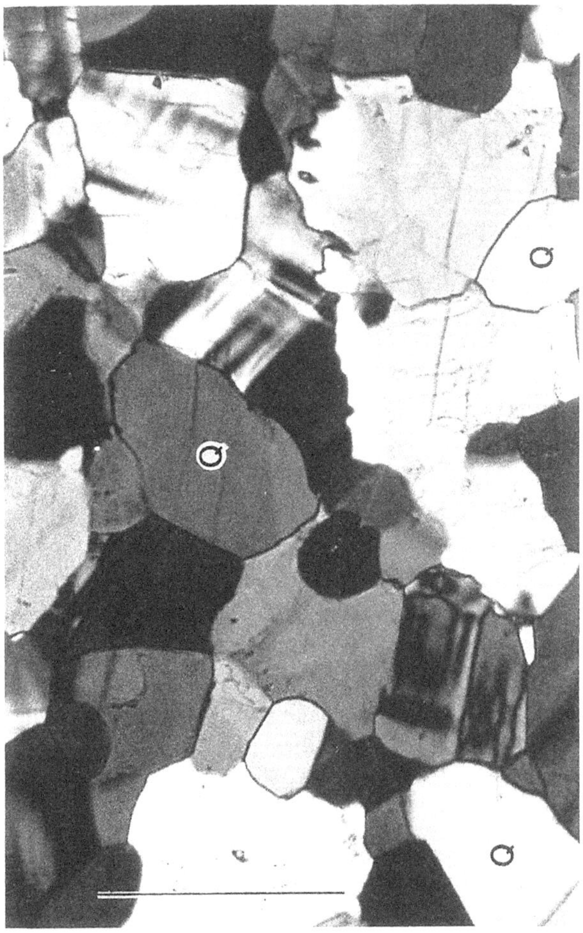


73 A

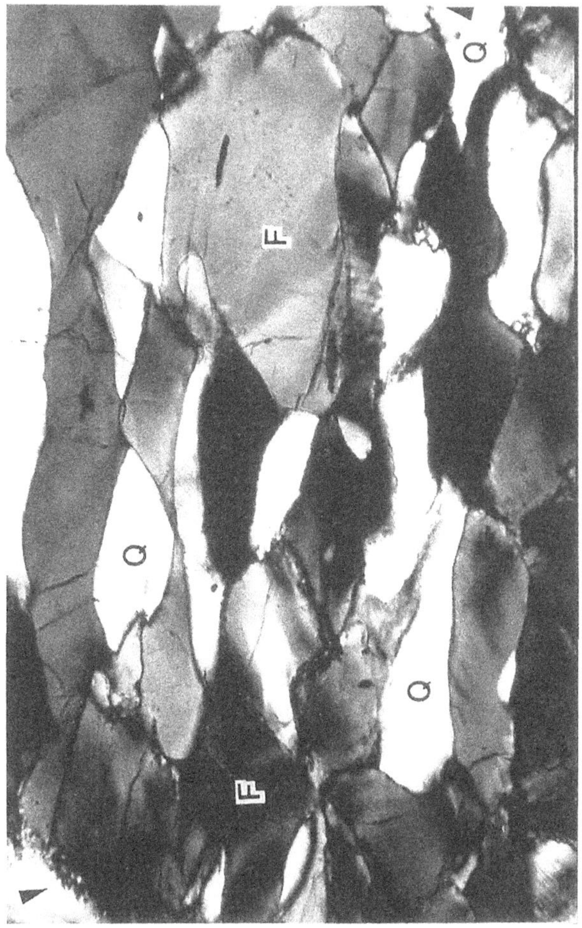


73 B

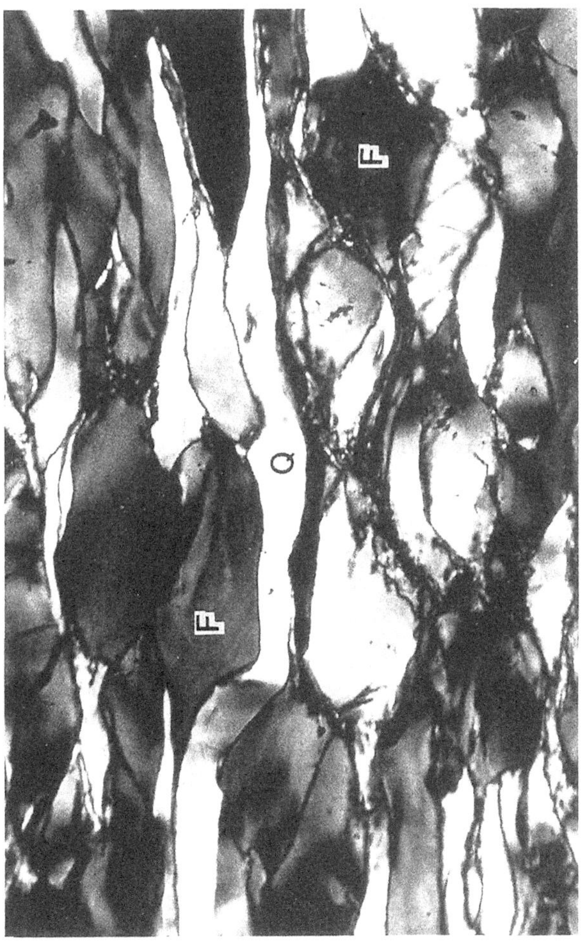


73 C

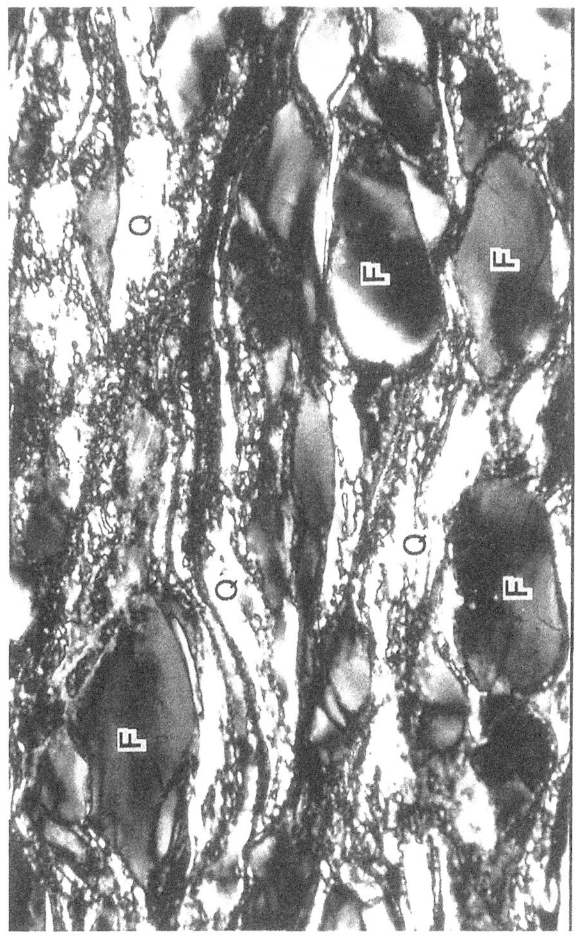


73 D

# 74. FORMATION OF DUCTILE SHEAR ZONES ALONG PREEXISTING FAULTS IN FELDSPATHIC ROCKS

JAN TULLIS, RICHARD A. YUND, AND LISA N. DELL'ANGELO

These photographs illustrate how the process of low-temperature grain-boundary migration recrystallization (Regime 1 dislocation creep of Hirth and Tullis, 1992) in feldspars can produce strain partitioning leading to ductile shear zones (Tullis and others, 1990). In this regime, dislocation climb is very limited, and thus gliding dislocations are soon tangled, producing work hardening. The recovery mechanism is grain-boundary migration recrystallization. Grain-boundary bulging occurs due to local differences in dislocation density, and the bulges pinch off to form very small (~1 μm) recrystallized grains. These grains are initially dislocation-free and thus can undergo a small amount of easy glide before becoming work-hardened themselves and replaced during later episodes of grain-boundary migration. Because the very fine, recrystallized grains are weaker than coarser original grains, strain will be partitioned into them.

Coarse-grained feldspar aggregates deformed within this dislocation creep regime undergo macroscopically homogeneous shortening (microstructures are shown in Plate 157, this volume), although strain is partitioned into the recrystallized regions. In aggregates containing preexisting zones of finer grain size, deformation within this regime results in extreme strain partitioning, and ductile shear zones form along the original fine-grained zone. This occurs because the finer grains can more rapidly recover by grain-boundary migration, allowing increments of easy glide, whereas the larger grains remain work-hardened.

**74A.** Tanco albite rock from the border zone of the Tanco pegmatite, Ontario, Canada; grains are randomly oriented laths with an average width of 40 μm and length of 150 μm. This sample was deformed (compression direction N-S) at low pressure and temperature and a fast strain rate to produce a (seismic) fault (NW-SE-trending irregular zone of fine-grained material). The sample was then unloaded and taken to higher temperature and pressure (900°C, 1500 MPa) for 10 hr, during which time the fault gouge underwent compaction and minor grain growth. NX

**74B.** Sample of Tanco albite rock treated identically as the sample shown in 74A but then deformed at a constant strain rate of $10^{-6}$/s (at 900°C, 1500 MPa) to 60% sample shortening. Almost all of the strain has occurred by ductile shearing along the preexisting fault zone. TEM shows that this shearing has occurred by Regime 1 dislocation creep: the grains within the shear zone are all the same size (~1 μm), are polygonal in shape with no pores, and have dislocation densities ranging from very high to very low. It might seem that this macroscopically "superplastic" behavior has occurred by grain-boundary sliding, but the local presence of high dislocation densities as well as the extremely strong crystallographic preferred orientation (note the parallel extinction within the zone) indicate that the deformation mechanism was low-temperature dislocation creep. Scale same as 74A. NX

**74C.** Sample of Enfield aplite (composition ~30% quartz, 35% microcline, 33% oligoclase, and 2% biotite and oxides) that was faulted at the same conditions as the sample shown in 74A, and then deformed at the same conditions as the sample shown in 74B. Again the sample strain has been highly partitioned into a ductile shear zone along the trace of the preexisting fault, because of the initially finer grain size of the feldspar within that zone. It is important to note that a prefaulted quartzite sample deformed under the same conditions does not show strain partitioning because it deforms by climb-accommodated dislocation creep, a mechanism in which the strength does not depend upon grain size. Because the aplite consists of roughly ⅔ feldspar and ⅓ quartz, the behavior of the feldspar (recrystallization-accommodated dislocation creep) controls the deformation and causes strong strain partitioning. Note the significantly larger size of the recrystallized grains of quartz (arrow; white regions along margins of shear zone) as compared to feldspar (optically unresolvable; darker center of shear zone). The recrystallized quartz grains have formed by a different mechanism (subgrain rotation).

The implication of these experiments is that naturally deformed feldspathic rocks will tend to show strain partitioning into ductile shear zones under metamorphic conditions where recrystallization-accommodated dislocation creep dominates, but will deform more homogeneously at higher temperatures (upper amphibolite- and granulite-facies conditions) where climb-accommodated dislocation creep becomes dominant.

74 A

74 B

74 C

# DUCTILE BEHAVIOR

## Microstructures



Common abbreviations used in regard to the plates in this section include: PPL = plane polarized light, NX = crossed nicols, PNX = partial crossed nicols, and SEM = scanning electron microscope.

This set of plates illustrates some of the grain-scale microstructures characteristic of highly strained rocks or mylonites. Most examples are from quartz or feldspar, although there are several illustrations of other phases, including micas, garnet, calcite, and analog materials. Many of these plates illustrate aspects of dynamic recrystallization, which is one of the hallmarks of a ductilely deformed mylonite, and a process that is very important for reducing the grain size. One especially important aspect illustrated in several plates is oblique, recrystallized grain-shape fabric. This is a common textural aspect of monomineralic layers in mylonitic rocks; when coupled with compositional layering or a mica foliation, it can be used as a sense-of-shear indicator in some mylonitic rocks. These plates are useful for illustrating the difference in character of grain-size reduction resulting from dynamic recrystallization as compared to cataclasis (see numerous examples in the section "Brittle behavior"). They are also useful for illustrating the differences in character of recrystallized grains formed by different mechanisms of recrystallization, which in turn can serve as a rough guide to temperature during deformation. One contribution demonstrates the importance of microstructural analysis for direct dating of deformation by radiometric techniques. Many additional examples of microstructures are illustrated in the later section "Microstructural development of mylonitic rocks of specific composition" (Plates 141–177).

# 75. VARIATION IN RELATIVE RECRYSTALLIZED GRAIN SIZES OF QUARTZ AND FELDSPAR WITH DEFORMATION TEMPERATURE

CAROL SIMPSON

Photomicrographs of quartzo-feldspathic mylonites deformed under a range of conditions demonstrate that the recrystallized grain size of quartz always exceeds that of feldspar. Samples are of low and middle greenschist-grade, mylonitic tonalite of the Borrego Springs section of the eastern Peninsular Ranges mylonite zone, southern California, U.S.A. (Simpson, 1985); lower amphibolite-grade, mylonitic granodiorite of the Santa Rosa section of the eastern Peninsular Ranges mylonite zone, southern California, U.S.A. (Simpson and Wintsch, 1989); and upper amphibolite-grade, quartzo-feldspathic mylonitic gneiss, Parry Sound, Ontario, Canada (Gower and Simpson, 1992).

**75A.** Fractured microcline and ribbon quartz grains from lower greenschist-grade mylonitic tonalite of Borrego Springs. Microcline (F) has undergone reaction along fractures and grain boundaries to form very fine-grained (<1 μm) white mica and quartz. Quartz ribbons, 50–100 μm wide and defined by foliation-parallel deformation bands and new grain boundaries, show minor grain-boundary migration recrystallization (r) to 3–5 μm sized new grains along their margins. NX

**75B.** Recrystallized K-feldspar and quartz grains from middle greenschist-grade mylonitic tonalite of Borrego Springs section. Original orthoclase grain relics (O) have irregular outlines and are surrounded by a matrix of 3–5 μm sized new K-feldspar grains with a locally strong, sinistral, grain-shape preferred orientation. Without TEM work it is not possible to determine for certain if the newly recrystallized K-feldspars are the product of rotation recrystallization (Regime 2, climb-accommodated dislocation creep) or grain-boundary migration recrystallization (Regime 1) (Hirth and Tullis, 1992). However, the lack of flattening of relict K-feldspar grains and elongate shapes of the recrystallized new grains suggest that Regime 1, grain-boundary migration recrystallization, was predominant. White mica has concentrated on the upper right and lower left sides of the orthoclase porphyroclast in extinction (white O on dark background), consistent with sinistral movement. Quartz ribbon grains (Q) contain 20 μm sized subgrains and rotation-recrystallized new grains that also show a sinistral grain-shape preferred orientation. NX

**75C.** Recrystallized feldspar, mica, and quartz from middle to upper greenschist-grade mylonitic tonalite of Borrego Springs section. Relict feldspar porphyroclasts (F) have irregular edges and are surrounded by 8–10 μm sized elongate new grains in feldspathic bands. The elongate shapes of the new feldspar grains and lack of flattening of relict porphyroclasts suggest that Regime 1, low-temperature grain-boundary migration recrystallization, was active in the feldspars. Muscovite grains (mu) are oblique to main foliation in a sinistral sense. Quartz ribbons contain 50 μm sized, equant, strain-free grains (Q) with straight boundaries that meet in 120° triple points, indicating Regime 3, high-temperature, grain-boundary migration recrystallization. Ribbons are separated by trails of recrystallized mica and feldspar/mica bands. NX

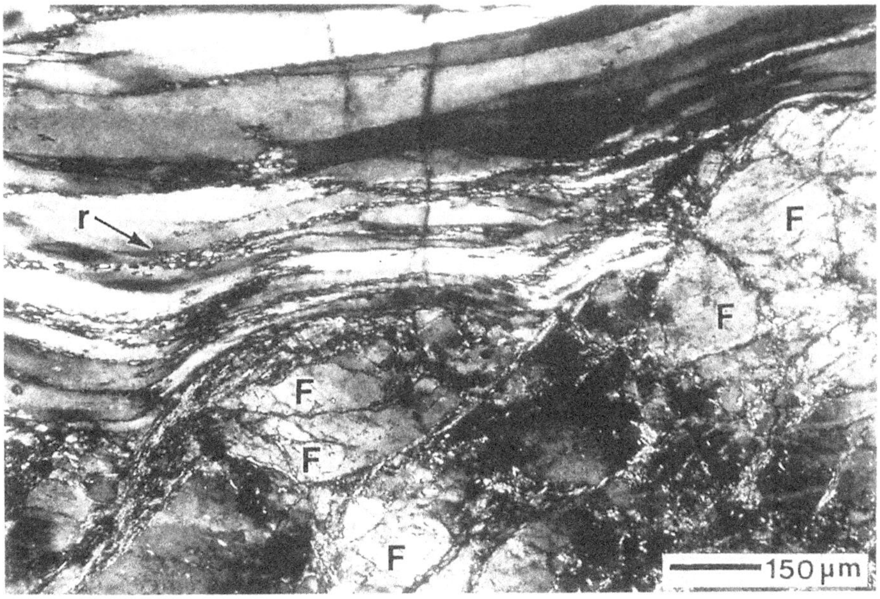


75 A

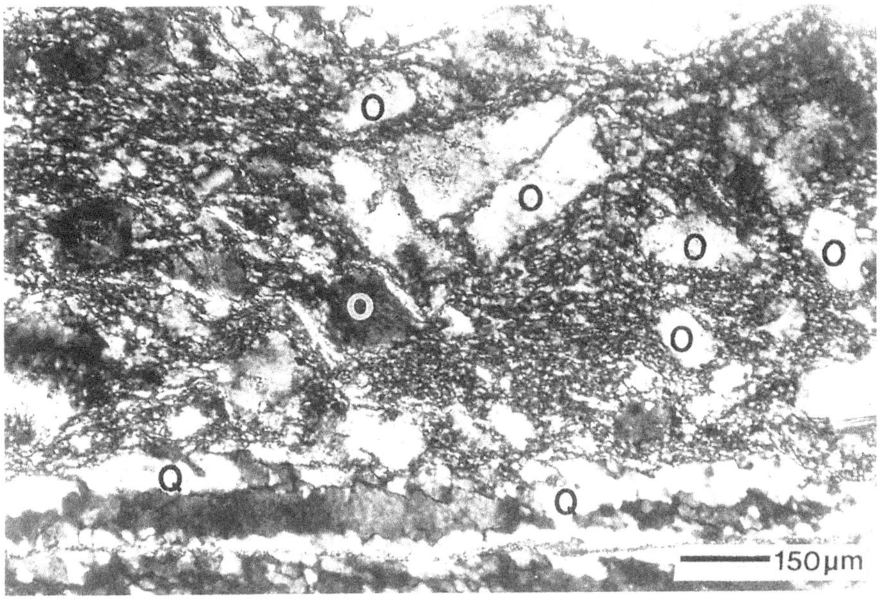


75 B

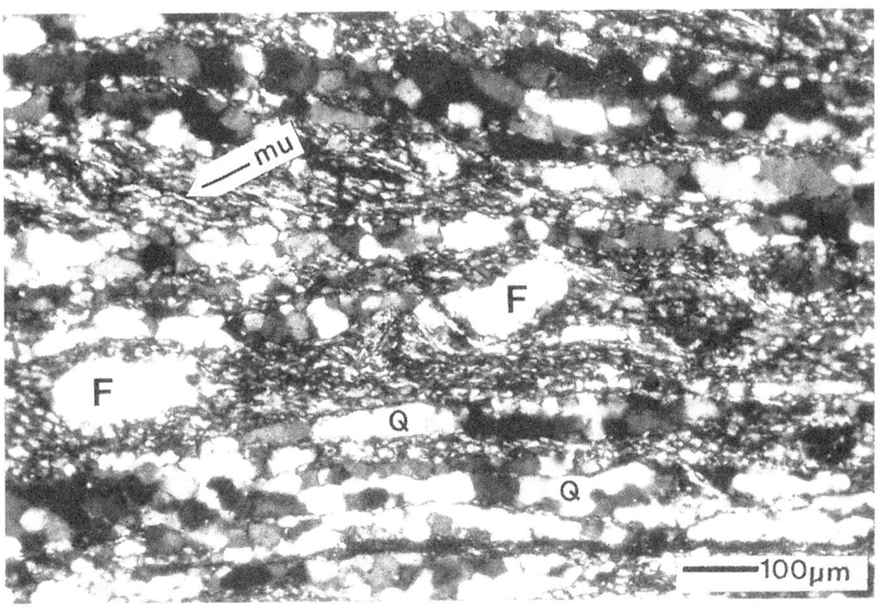


75 C

**75D.** Recrystallized feldspar, mica, and quartz from lower amphibolite-grade, mylonitic granodiorite of Santa Rosa section. Relict K-feldspar porphyroclasts (F) have irregular edges due to reaction of K-feldspar to form myrmekitic oligoclase and quartz (Simpson and Wintsch, 1989). They are surrounded by 60 µm sized equant new feldspar and mica. Recrystallized biotite has concentrated in a shear band across center of photograph (sense-of-shear dextral). Quartz ribbons contain 150–200 µm sized strain-free grains (Q) that have straight boundaries which meet in 120° triple points or are at a high angle to ribbon boundaries. Some of the quartz grains have coalesced by grain-boundary migration into larger, irregularly shaped grains. Quartz ribbons are separated by recrystallized mica bands. NX

**75E.** Recrystallized feldspar and quartz in quartzo-feldspathic mylonitic gneisses of Parry Sound, Ontario. Deformation temperatures were >650–750°C (Gower and Simpson, 1992). Plagioclase porphyroclasts (P) show extreme lattice bending and 100 µm sized subgrains (sg), and are surrounded by 100–200 µm sized rotation-recrystallized grains (rr). The recrystallized plagioclase grains have straight grain boundaries that meet in 120° triple points and contain straight twins. Quartz occurs in irregular ribbons and amoeboid shapes, from 200 µm to 2 mm wide, and is optically strain-free. Quartz grain boundaries are smoothly curved or cuspate, and most of the boundaries between quartz and feldspar are optically clean. NX

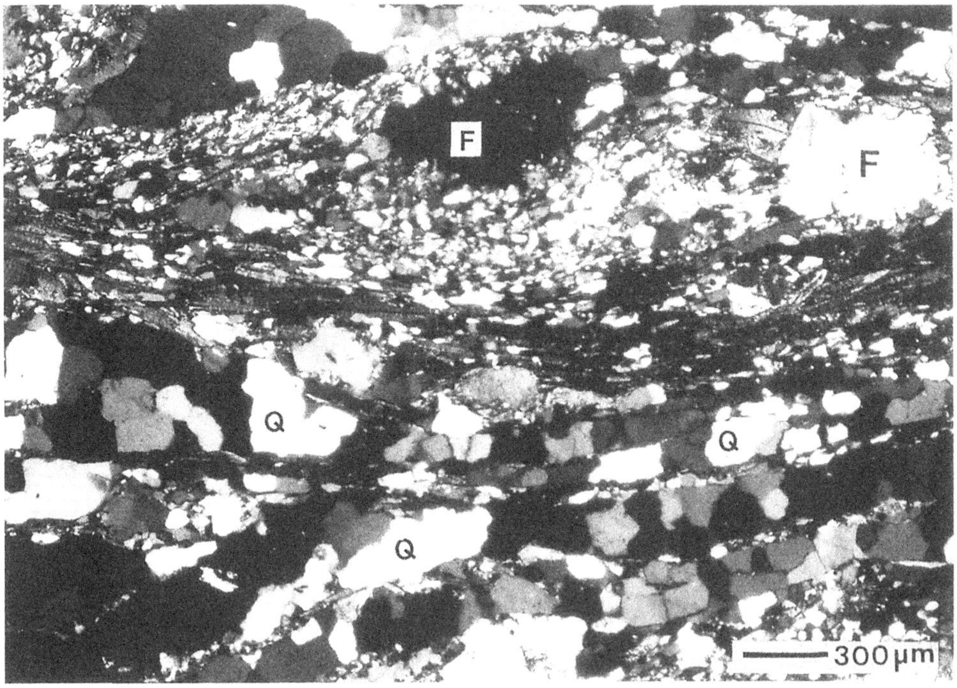

75 D

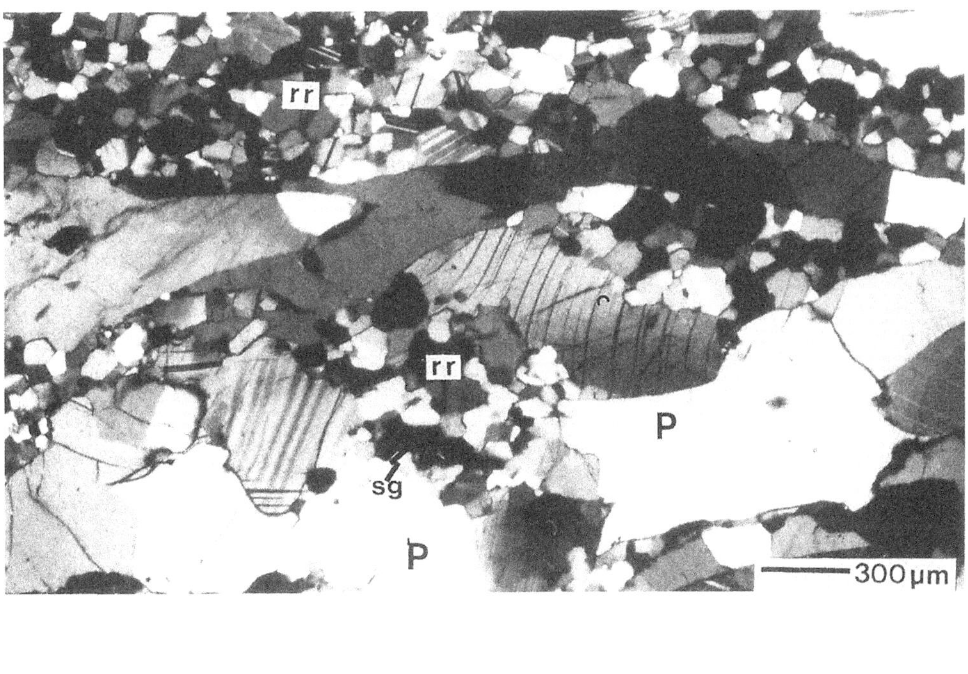

75 E

# 76. OBLIQUE GRAIN-SHAPE PREFERRED ORIENTATION IN QUARTZO-FELDSPATHIC AND CALCITIC MYLONITES

CAROL SIMPSON

Photomicrographs of grain-shape preferred orientation (GSPO) from a variety of naturally deformed rocks illustrate the common obliquity of recrystallized grain aspect ratios with respect to the bulk foliation. Samples are taken from the Borrego Springs section of the eastern Peninsular Ranges mylonite zone, southern California, U.S.A. (Simpson, 1985; Passchier and Simpson, 1986); quartz-rich Orocopia schist, lower plate of the late Cretaceous to early Tertiary Chocolate Mountains fault system, Gavilan Hills region, southeastern California, U.S.A. (Simpson, 1990); and Catalina core complex mylonite zone, near Tucson, Arizona, U.S.A. (sample courtesy of Anne Bykerk-Kaufmann).

The illustrations show how GSPO in quartzo-feldspathic and calcitic mylonites may develop initially as a consequence of rotation recrystallization, but becomes enhanced by subsequent grain-boundary migration recrystallization. Once formed, the elongate new grains rotate toward the maximum finite elongation direction as straining continues (Simpson and De Paor, 1993). However, these are neither rigid grains nor passive strain markers, and grain-boundary migration occurs continuously to elongate the grains in a direction close to the maximum incremental elongation direction (Urai and others, 1986). This combination of rotation of the grains' long axes toward the finite maximum elongation direction, and recrystallization to reorient their long axes toward the incremental maximum elongation direction, maintains the grain aspect ratios in a close to "steady state" orientation that is somewhere between the two.

**76A.** GSPO developed at 32° to the main foliation in a low to middle greenschist-facies, quartz-rich mylonite from the Borrego Springs section of the eastern Peninsular Ranges mylonite zone (Simpson, 1985; Passchier and Simpson, 1986). The sense-of-shear is sinistral. Feldspar porphyroclasts (F) have reacted to form micaceous tails. Most of the quartz in this view is in 10 × 30 μm sized subgrains (sg), with rare rotation-recrystallized new grains (rr), consistent with deformation in dislocation creep Regime 2 of Hirth and Tullis (1992). Grain-boundary migration recrystallization (gbm) has commenced at top and right of photo. NX

**76B.** GSPO in quartz-rich Orocopia schist, Gavilan Hills region, southeastern California (Simpson, 1990), here deformed at low to middle greenschist facies. The sense-of-shear is sinistral. Elongate subgrains (sg) are visible in the larger quartz grains and are the same size and aspect ratio as adjacent rotation-recrystallized new grains (rr), consistent with deformation in dislocation creep Regime 2 of Hirth and Tullis (1992). Grain-boundary migration recrystallization (gbm) has increased the axial ratio of arrowed new grains. NX

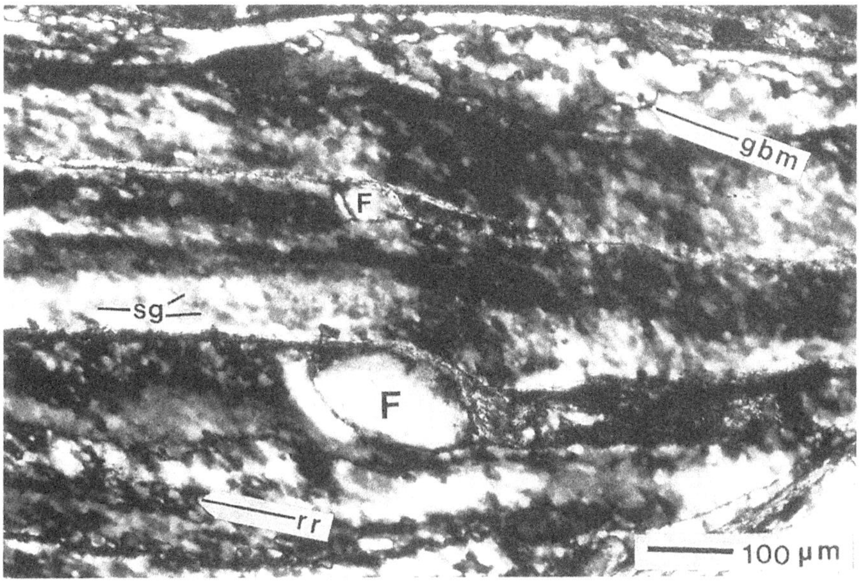


76 A

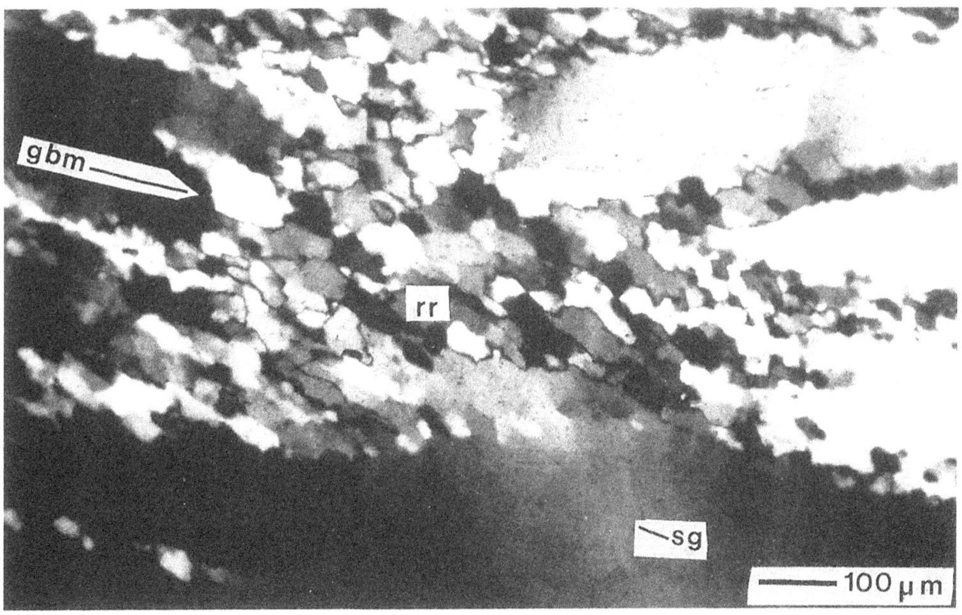


76 B

**76C.** Quartzo-feldspathic mylonite from the Borrego Springs mylonite zone (see 76A), here deformed at middle greenschist facies. Subgrains (sg) of quartz are still visible in patches, but most quartz grains are fully recrystallized with aspect ratios of between 2:1 and 3:1. Grain-boundary migration recrystallization has caused some coalescence into even longer grains (arrowed), consistent with deformation in dislocation creep Regime 3 of Hirth and Tullis (1992). Feldspar porphyroclasts (F) occur as relics within the narrow central zone of dynamically recrystallized feldspar, which is mostly comprised of elongate new feldspar grains of aspect ratios between 4:1 and 7:1, formed by grain-boundary migration recrystallization (Regime 1 of Hirth and Tullis, 1992). Although the GSPO in quartz makes an angle of about 53° to the main foliation, GSPO in the feldspar makes an angle of only 35°, probably reflecting the predominant influence of the rotational strain component over grain-boundary migration on these stiff and elongate feldspar grains (Simpson and De Paor, 1993). NX

**76D.** Calcitic mylonite deformed at greenschist facies from the Catalina core complex mylonite zone, near Tucson, Arizona, shows a strong GSPO at 25° to the main foliation (foliation is parallel to long axis of photograph but not visible in this view). Most grains have an aspect ratio of about 3:1. Evidence for grain-boundary migration recrystallization is provided by indentations and suturing of grain boundaries (arrowed). NX

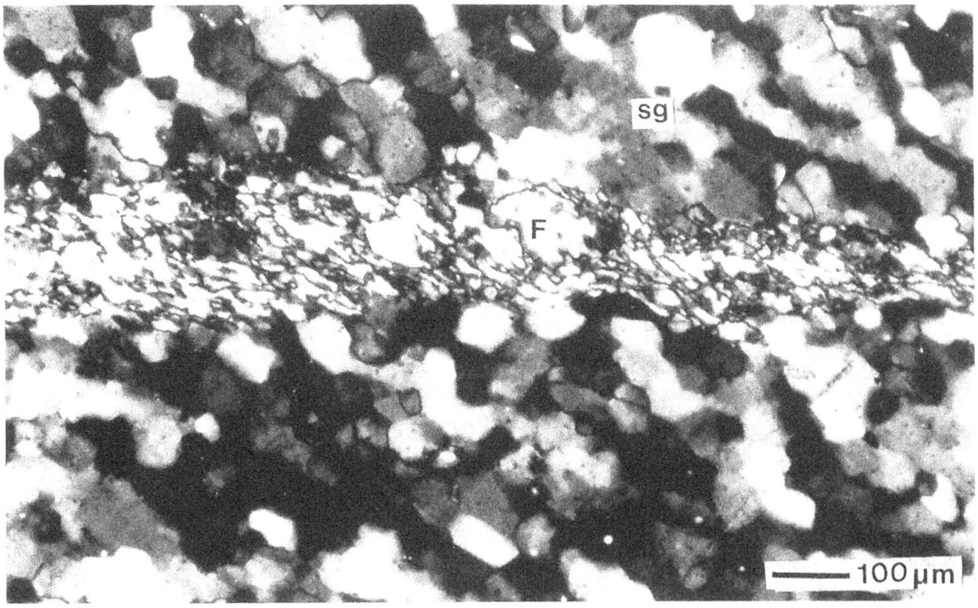


76 C

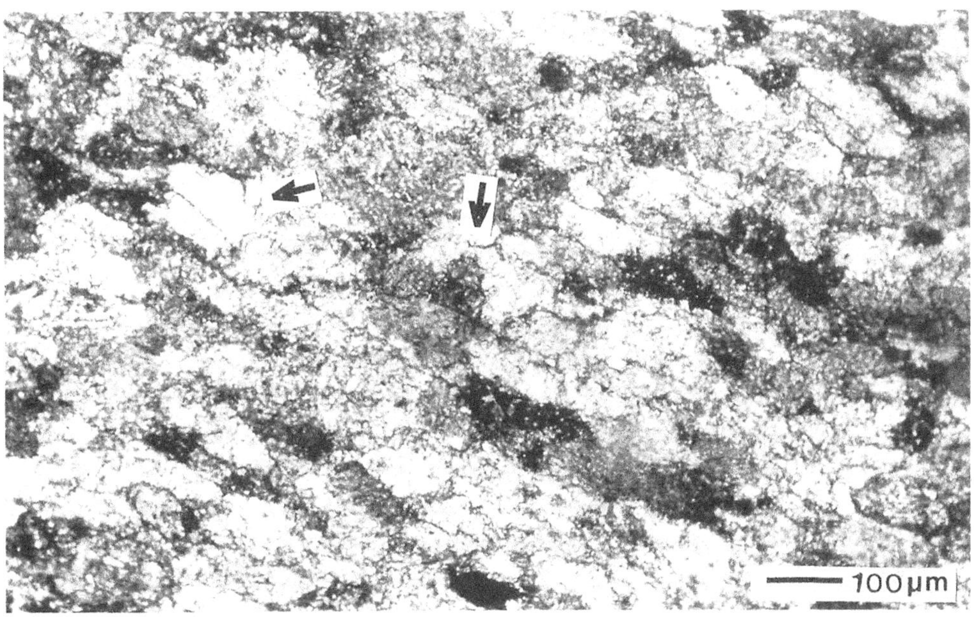


76 D

# 77. OBLIQUE GRAIN-SHAPE FABRICS IN A MYLONITIC QUARTZ VEIN

R. D. LAW

The sample illustrated is from a 30 cm wide, deformed, planar quartz vein. The vein, which lies along a joint plane in Proterozoic gneiss, was deformed by crystal-plastic processes when the rigid joint blocks moved in opposite directions parallel to the vein margins. In thin section, the deformed quartz vein is a classic Type II S-C mylonite in the sense of Lister and Snoke (1984). Microstructures and crystallographic fabrics from this vein have previously been described by Law and others (1990) and Lloyd and others (1992). Deformation conditions are thought to have closely approximated to bulk simple shear. Deformation temperatures are poorly constrained. Feldspar porphyroclasts in the vein generally exhibit brittle deformation features, although Lloyd and others (1992) have recorded intragranular microstructures (e.g., subgrains) and growth of metamorphic minerals (e.g., zoisite) associated with feldspar deformation, which may indicate at least upper greenschist-facies conditions of metamorphism.

**77A.** Polished surface of sample from the center of the quartz vein cut perpendicular to mylonitic foliation and parallel to the grain-shape lineation. Mylonitic foliation ($S_A$) is picked out by planar domains of translucent quartz 1 mm–1 cm wide, separated by pink-colored domains (white in photograph) composed of extremely fine-grained feldspar and quartz that occasionally form tails on large feldspar porphyroclasts.

**77B and 77C.** Optical micrographs of thin section cut perpendicular to mylonitic foliation and parallel to grain-shape lineation. Two microstructural domains defining the mylonitic foliation seen in hand sample are distinguished: Type A domains (dark zones in micrograph 77B) are composed of fine-grained feldspar and quartz; the domains range from 20 to 600 µm in width. Type B domains are composed of much larger, dynamically recrystallized quartz grains and range from 40 to 700 µm in width. Individual quartz grains in Type B domains are commonly elongate (50 µm by 150 µm), with their long axes oriented at 18° to 36° to the mylonitic foliation. The obliquity between foliation ($S_A$) and elongate grain-shape alignment ($S_B$) indicates a dextral shear sense. This shear sense is confirmed by both the pattern of mylonitic foliation trajectories traced across the quartz vein and the asymmetry between quartz crystallographic fabrics and finite strain features (foliation and lineation) (Figure 77.1).

Two types of foliation-parallel compositional layering are observed in Type A domains: layers composed of equant to slightly elongate (5–20 µm) feldspar grains with interstitial quartz, and monomineralic layers composed of equant-elongate (5 µm by 15 µm) dynamically recrystallized quartz grains. Elongate quartz grains in the fine-grained Type A domains also display a preferred alignment oblique to $S_A$, indicating a dextral shear sense, although the angle of obliquity (<5°) is much smaller than in the coarser-grained Type B domains. Flat stage optical microscope examination of the elongate quartz grains in the Type A domains indicates that they have a similar, although weaker, crystallographic fabric than quartz in the Type B domains. Using the flat stage, feldspar grains appear to exhibit a random crystal orientation.

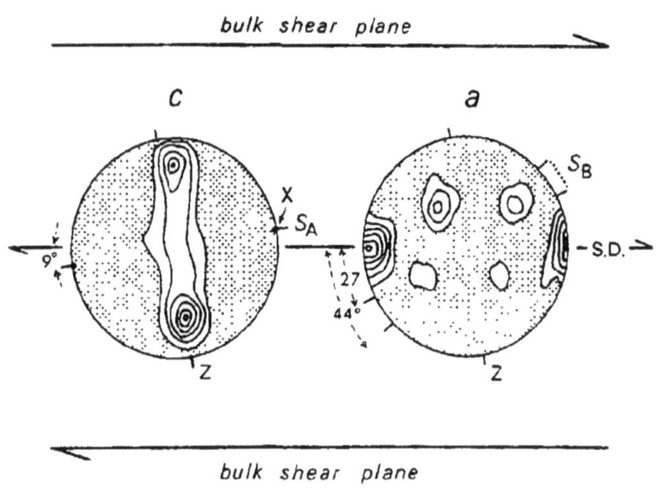


**Figure 77.1.** Angular relationships between mylonitic foliation ($S_A$), pole to mylonitic foliation (Z), lineation (X), variation in oblique recrystallized quartz grain-shape alignment ($S_B$), quartz *c*- and *a*-axis fabrics, and simple shear kinematic framework (bulk shear plane and shear direction S.D.). Kinematic framework is inferred from macroscopic shear zone measurements, e.g., bulk shear plane is parallel to margins of quartz vein. Diagram after Law and others (1990).

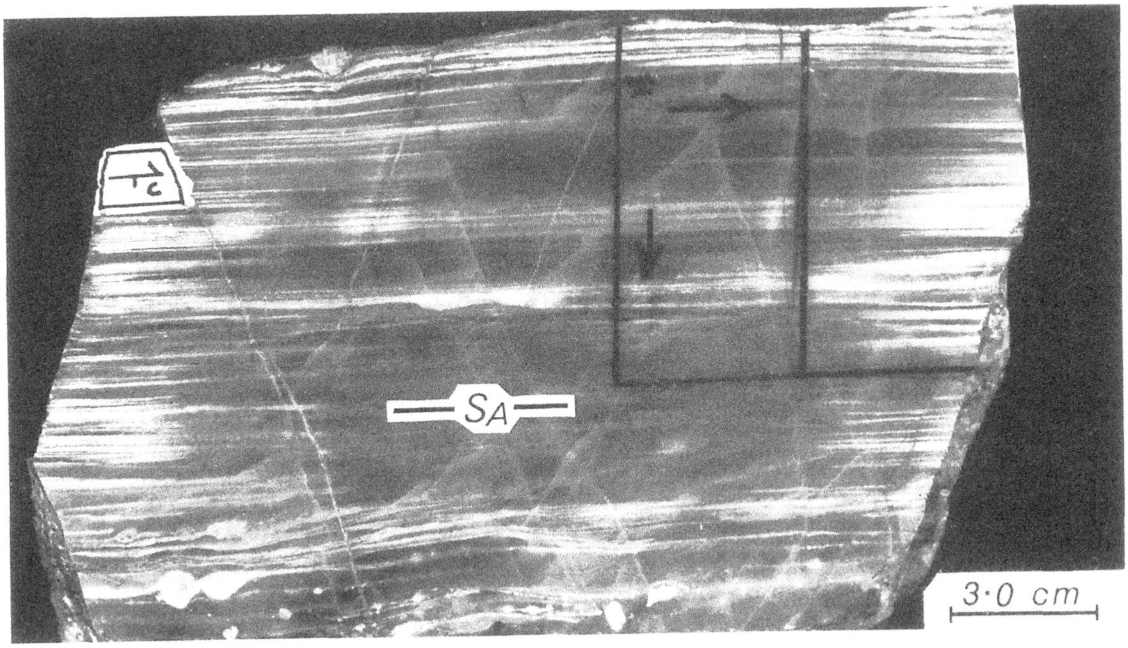


77 A

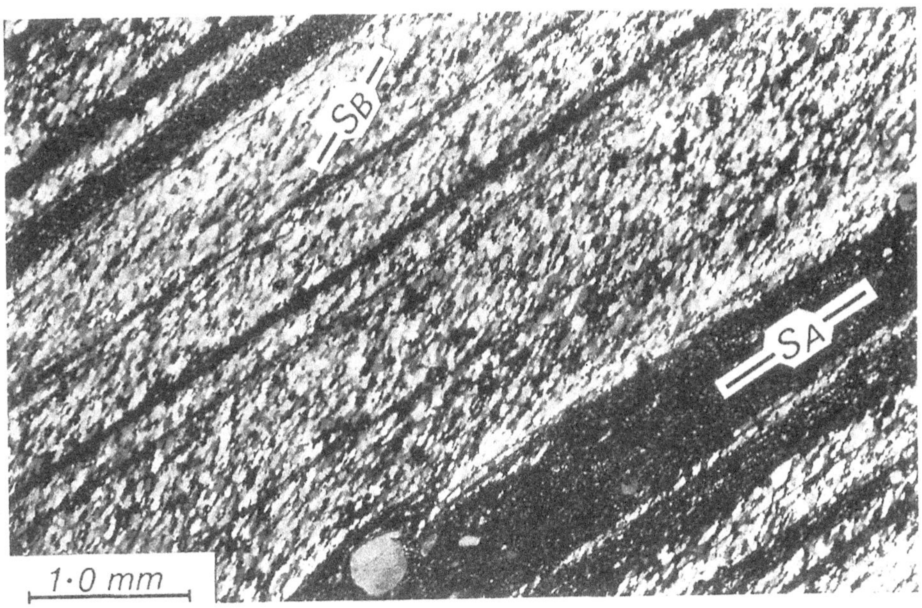


77 B

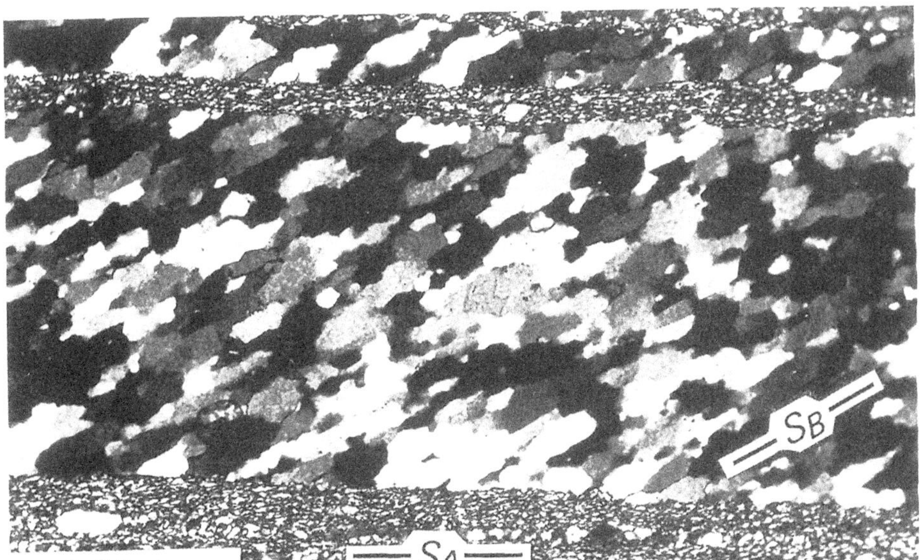


77 C

# 78. NATURAL EXAMPLES OF LOW-TEMPERATURE, GRAIN-BOUNDARY MIGRATION RECRYSTALLIZATION

CAROL SIMPSON

Photomicrographs of arrested grain-boundary migration recrystallization from naturally deformed quartz, Borrego Springs section of the eastern Peninsular Ranges mylonite zone, southern California, U.S.A. (Simpson, 1985), and feldspar, Port Deposit tonalite, Maryland, U.S.A.

Syntectonic grain-boundary migration is thought to be a response to differences in lattice strain across grain boundaries (Guillope and Poirier, 1979; Urai and others, 1986). Hirth and Tullis (1992), in their experimental work, recognized grain-boundary migration recrystallization as the *predominant* recovery mechanism of their dislocation creep Regimes 1 and 3. Regime 1 is low-temperature, grain-boundary migration recrystallization, where the grain-boundary mobility is low but the driving force is very high. In Regime 2, climb-accommodated dislocation creep, the predominant recovery mechanism is that of dislocation climb, although grain-boundary migration is also active. Regime 3 is high-temperature, grain-boundary migration recrystallization where the driving force is low but the grain-boundary mobility is high; under these conditions, recrystallized grains tend to be large and, as dislocation climb is also very easy, they also have low internal stored strain. Similar features in naturally deformed minerals may provide clues about deformation conditions. In these examples, grain-boundary migration has occurred in conjunction with different degrees of strain recovery.

**78A.** Grain-boundary migration recrystallization along the margins of quartz ribbon grains, from Borrego Springs mylonitic tonalite deformed under middle greenschist conditions. Individual quartz ribbons show undulatory extinction and minor subgrain formation, indicative of dislocation creep Regime 2. Grain-boundary migration-recrystallized new grains along ribbon boundaries are approximately one-eighth the size of subgrains. NX

**78B.** Enlargement of one of the grain boundaries between two quartz ribbon grains shown in 78A shows size difference between subgrains (sg) and grain-boundary migration lobes (gbl) and their associated grain-boundary migration-recrystallized new grains (ng). The grain-boundary lobes and new grains may represent the component of grain-boundary migration that is always present in Regime 2 before subgrains are rotated enough to form new grains, or they may represent a late stage, colder, Regime 1 deformation. NX

**78C.** Preserved grain-boundary migration recrystallization in plagioclase of Port Deposit tonalite, a Late Cambrian to Early Ordovician pluton in the Appalachian Piedmont, Maryland. Mylonitic deformation was at middle to upper amphibolite facies. A single auge of plagioclase contains a mosaic of >100 μm sized, roughly polygonal, recrystallized grains that are mostly strain-free and likely formed initially by rotation recrystallization in dislocation creep Regime 2. No subgrains are visible. However, many of the polygonal grains show undulatory extinction (ux), and all grain boundaries are highly sutured, indicative of a late-stage deformation event that activated Regime 1 grain-boundary migration recrystallization. Recrystallized new grains (arrowed) of 10–15 μm size occur along the sutured grain boundaries. NX

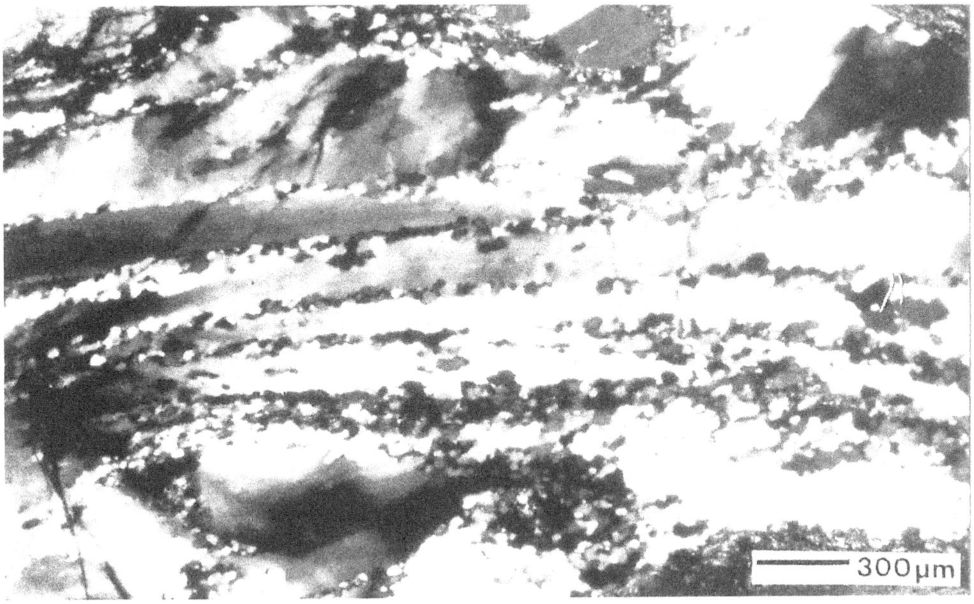


78 A

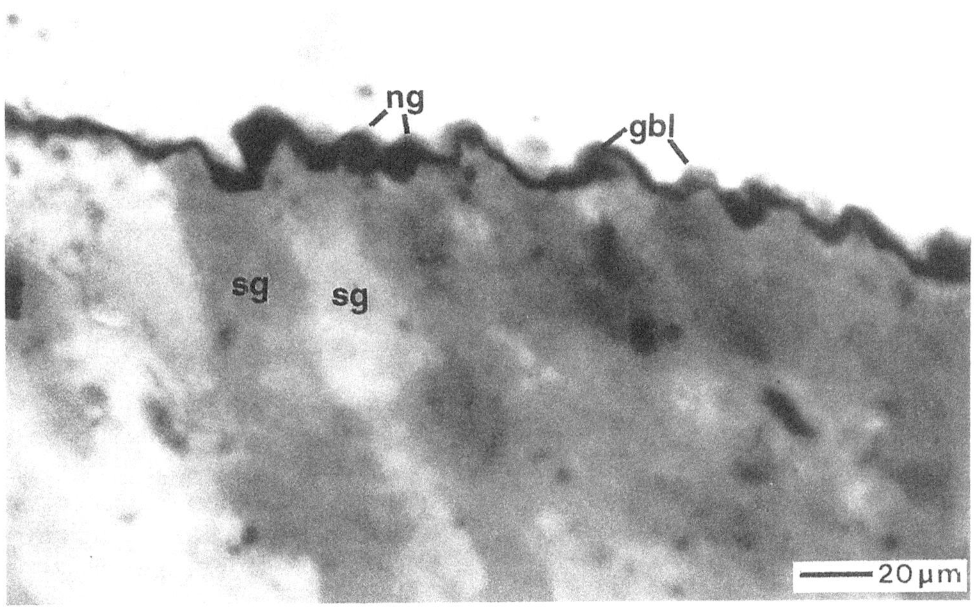


78 B

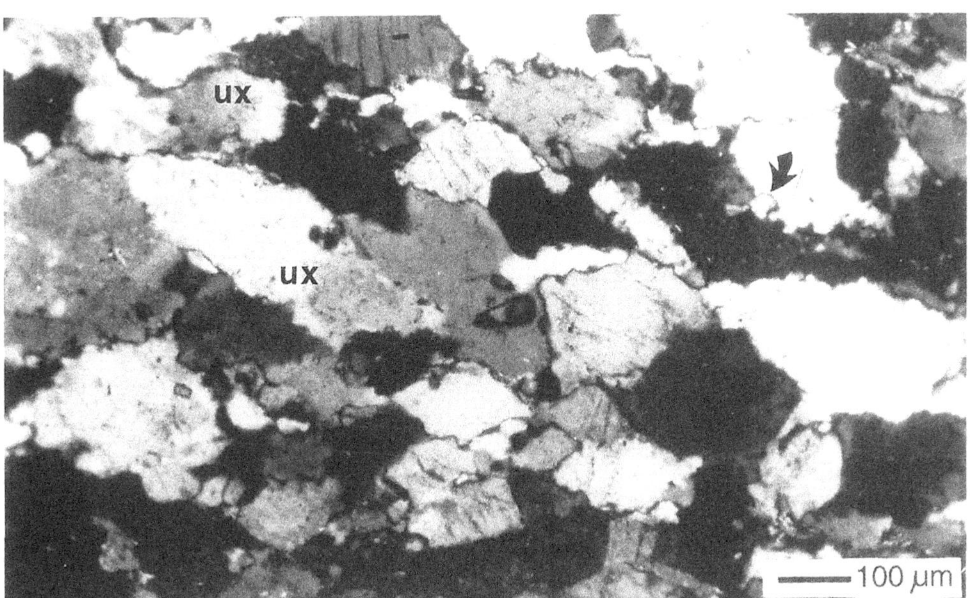


78 C

# 79. NATURAL EXAMPLES OF ROTATION RECRYSTALLIZATION

CAROL SIMPSON

Photomicrographs of rotation recrystallization in naturally deformed quartz from the Borrego Springs section of the eastern Peninsular Ranges mylonite zone, southern California, U.S.A. (Simpson, 1985), and in feldspar from the Parry Sound domain gneiss, Ontario, Canada (Gower and Simpson, 1992).

Formation and progressive rotation of subgrains result in rotation recrystallization (White, 1976; Guillope and Poirier, 1979; Urai and others, 1986). Subgrains are visible in polarized light as regular patches of differing extinction from adjacent parts of the crystal. A subgrain boundary is visible in polarized light as a sharp boundary beween areas of differing extinction, but it is not generally visible in plane light. Rotation of newly formed subgrains produces independent, rotation-recrystallized new grains with clearly visible grain boundaries in plane light. This mechanism of recrystallization predominates in dislocation creep Regime 2 of Hirth and Tullis (1992).

Both examples show classic "core and mantle texture" (White, 1976), in which the central core of the grain is relatively undeformed, grading into a region of equidimensional subgrains and then a mantle of rotation-recrystallized new grains of the same size as the subgrains.

**79A.** Rotation recrystallization in a quartz ribbon grain from the Borrego Springs mylonite zone, where tonalite is deformed at middle greenschist facies. Feldspar has fractured and reacted to form white mica and quartz, which with biotite line the quartz ribbon boundaries. The relatively undeformed core of the quartz grain (Qc) has a mantle of subgrains (sg) surrounded by an outer mantle of rotation-recrystallized new grains (rr). Minor grain-boundary migration (gbm) recrystallization has modified the outline of some of the larger new grains. NX

**79B.** Core and mantle texture in plagioclase from a pegmatite vein deformed at temperatures of 660–680°C and pressures of about 1000 MPa in the Parry Sound domain, Ontario (Gower and Simpson, 1992). Slight undulatory extinction and lattice bending is visible in the core of the grain (Pc). A zone of subgrains (arrowed sg) in the center of the photo passes out into rotation-recrystallized new grains (rr) of similar size. Subsequent grain-boundary migration may have enlarged the rotation-recrystallized grain size to produce the strain-free grains with straight boundaries that meet in 120° triple points in the upper part of the photo. NX

**79C.** Rotation recrystallization in quartz from same locality as 79A. Subgrains (sg) are visible in the mantle of the quartz grain at upper left. Progressive misorientation of subgrains produced rotation-recrystallized new grains (rr) at lower right. New grains have similar sizes to subgrains, but show a wider range of extinction angles. NX

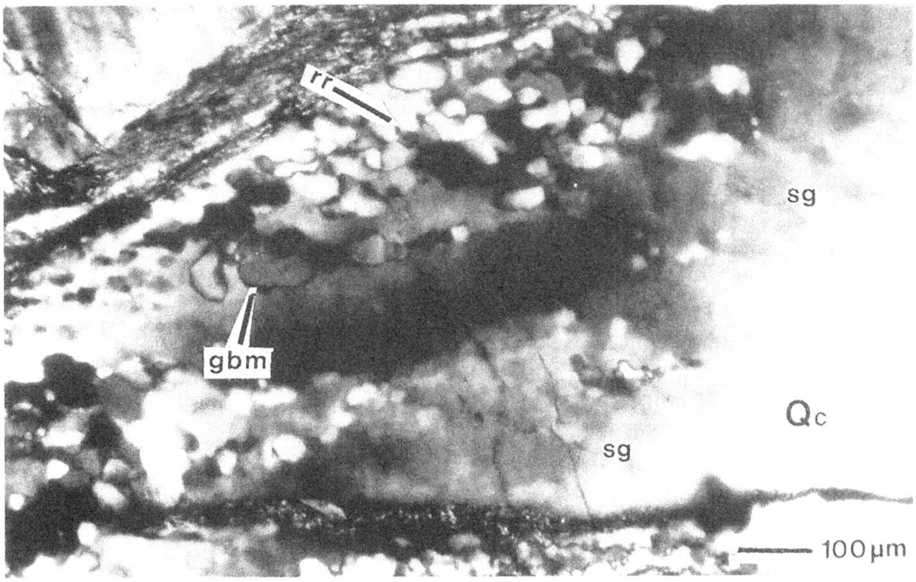


79 A

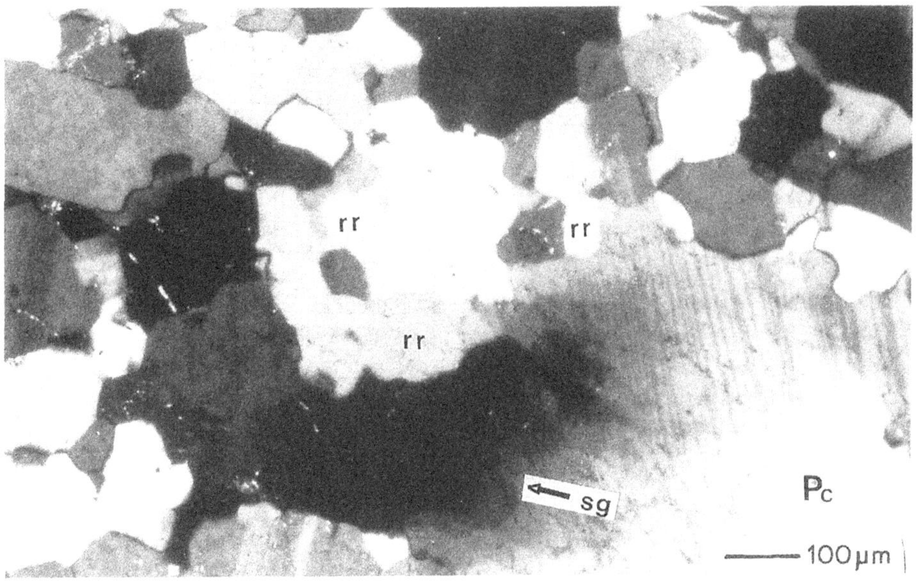


79 B

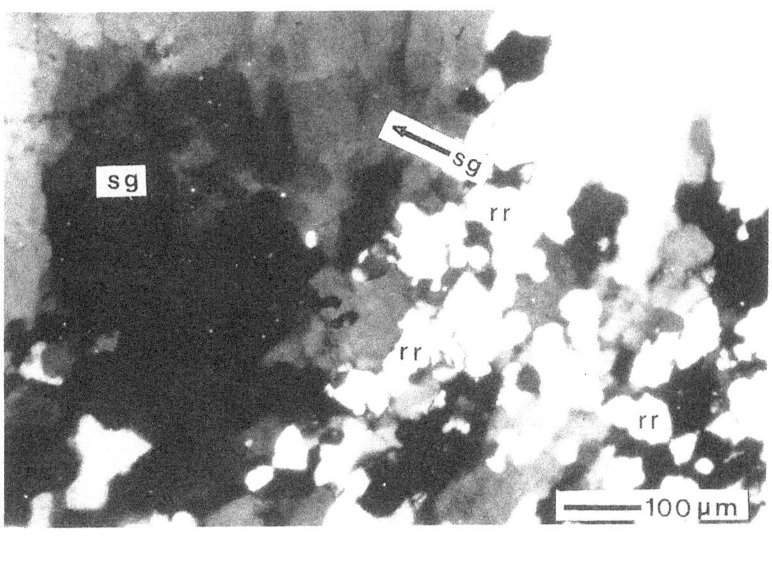


79 C

# 80. HIGH-TEMPERATURE GRAIN-BOUNDARY AND PHASE-BOUNDARY MIGRATION

CAROL SIMPSON

Photomicrographs of grain-boundary migration-recrystallized quartz and phase-boundary migration between quartz and feldspar, Parry Sound domain gneiss, Ontario, Canada (Gower and Simpson, 1992).

**80A.** Isolated lensoid plagioclase grains surrounded by large quartz grains from quartzo-feldspathic mylonitic gneisses of Parry Sound, Ontario. Deformation temperatures were >650–750°C (Gower and Simpson, 1992). High-angle grain boundaries between two large, optically strain-free quartz grains, one in extinction and the other labeled Q, indicate Regime 3 (Hirth and Tullis, 1992) grain-boundary migration recrystallization. Many of the quartz grain boundaries in this gneiss are "pinned" by feldspar grains, such as in the lower third of the photo. Plagioclase grains (F) are elongate parallel to bulk foliation, and are also strain-free even at their narrow tips, with smoothly curved grain boundaries that truncate twins and cleavage planes. Although most deformation in the feldspars probably involved Regime 2 dislocation creep mechanisms, these preserved features suggest that quartz-feldspar phase boundaries were extremely mobile during the high-temperature deformation, and their motion was accommodated in part by feldspar dissolution along foliation-parallel, quartz-feldspar grain boundaries. Other cuspate phase-boundary structures, not illustrated here, indicate solid-state, grain-boundary-accommodated diffusional creep (Gower and Simpson, 1992). NX

**80B.** Isolated lensoid grains of plagioclase (P) in quartz (Q) matrix from same locality as 80A. Gently curved and cuspate quartz-quartz boundaries abut feldspar grains at a high angle. Quartz grains are almost strain-free but contain a few deformation bands; their relatively straight, high-angle boundaries indicate predominantly Regime 3 grain-boundary migration recrystallization. Feldspar long axes are parallel to the bulk foliation, their grain boundaries are smooth, and there is no evidence for internal strain. As in 80A, the feldspar is thought to have undergone dissolution along highly mobile quartz-feldspar phase boundaries during high-temperature deformation. NX

**80C.** Backscattered SEM photomicrograph of lensoid plagioclase (P) grain isolated in quartz (Q) matrix, from locality near 80A. Pointed cusps of the plagioclase grain contain K-feldspar (K). Long axis of grain is parallel to lineation. Although there is no optical evidence for internal strain in either the quartz matrix or the plagioclase grain, the presence of stretched rutile needles in the quartz matrix and flattening of the plagioclase grain parallel to foliation are consistent with phase-boundary migration-assisted, dislocation creep processes (Gower and Simpson, 1992). The size and lenticular shape of the plagioclase grains suggest that phase-boundary migration and diffusive mass transfer were active over length scales greater than 200 μm.

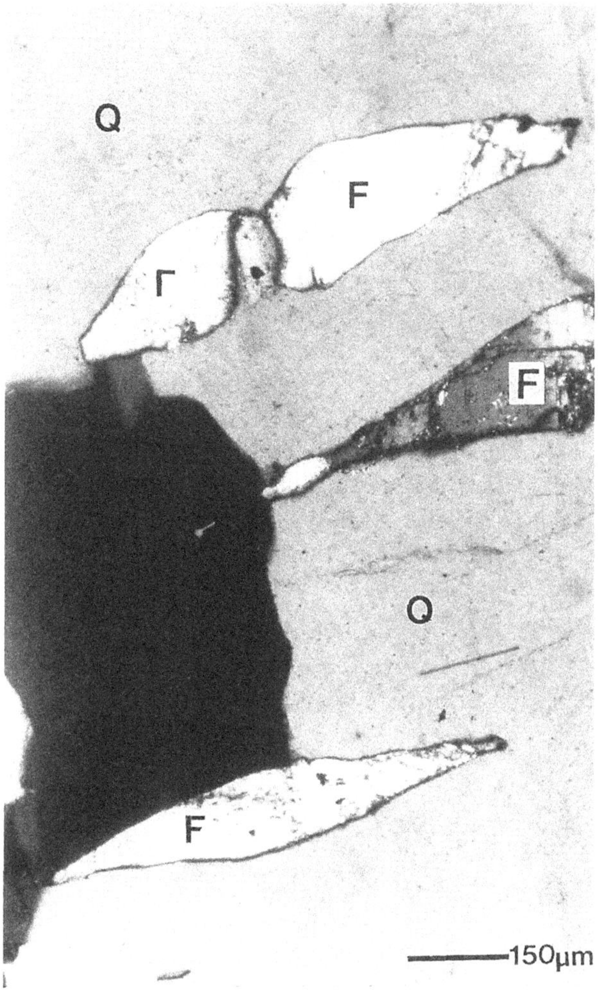

80 A

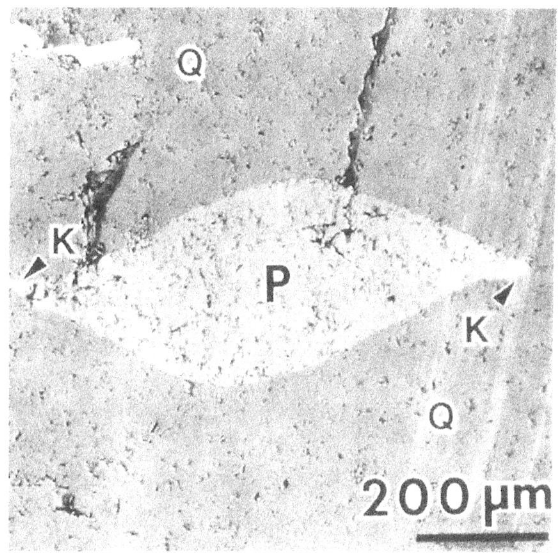

80 C

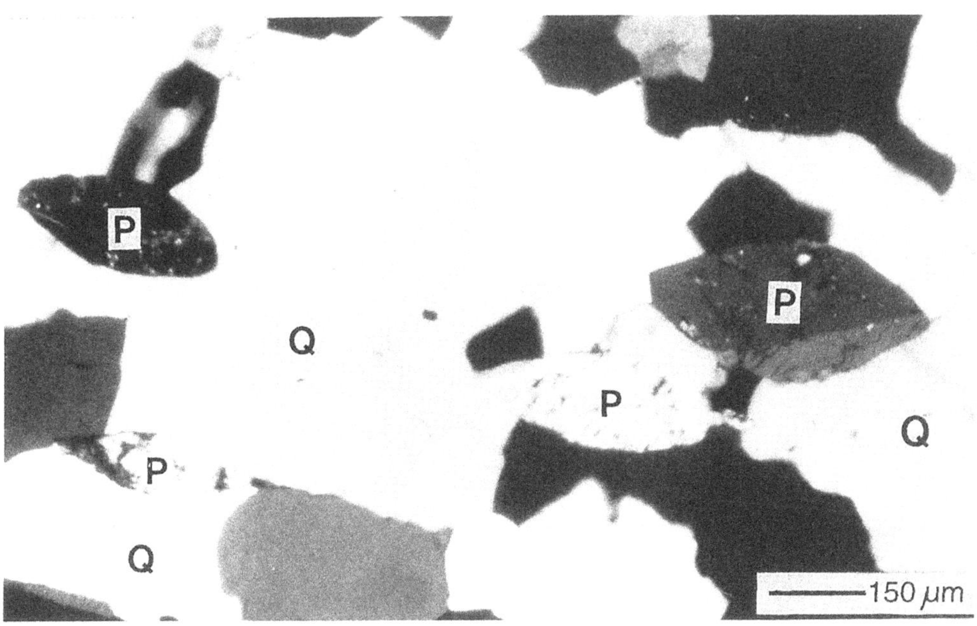

80 B

# 81. RECRYSTALLIZATION OF BIOTITE BY SUBGRAIN ROTATION

T. H. Bell

Mica generally recrystallizes by mechanisms other than subgrain rotation (Vernon, 1977; Bell, 1978b). However, in amphibolite-facies, mylonitic felsic gneisses, originally derived from granulite-facies rocks of the hanging wall of the Woodroffe thrust, central Australia, biotite recrystallized by a subgrain rotation mechanism in a fashion similar to quartz and feldspar (Bell, 1979). This microphotograph of an ultra-thin section shows host (original) biotite with (001) oriented just west of N-S, partially recrystallized in zones of high strain via a subgrain rotation mechanism. The new biotites show little or no change in mineral chemistry. The black grain in the center is apatite, and the kink-banded grain at the bottom right-hand corner is clinopyroxene. Subgrains and new grains have formed in the biotite around both of these grains; these locations were apparently zones of high strain. Subgrains and new grains have also formed in a central, approximately N-S, linear zone of high strain and a short E-W zone, both right of the center of the photograph. A sharp, narrow kink band also passes N-S through the center of the photograph. Some kink-band boundaries bulge around subgrains, and new grains can be seen in the center of the photograph beneath the apatite. The width of the photograph is 170 μm. NX

81

# 82. MICROSTRUCTURAL DEVELOPMENT IN A QUARTZO-FELDSPATHIC MYLONITIC SHEAR ZONE

G. E. LLOYD

Backscattered electron (BSE) SEM electron channeling (EC) orientation contrast (OC) photomicrographs show the progressive microstructural evolution in an amphibolite-facies, dextral shear zone developed in a quartzo-feldspathic vein in Lewisian gneiss, Torridon, northwestern Scotland (see Lloyd and others, 1992). The operating conditions involved short working distances, 5–15 mm; solid-state BSE detectors; 20–30 kV accelerating voltages; and high probe currents, 20–50 nA. This approach reveals specimen crystallographic microstructure (e.g., grains, subgrains, etc.) in terms of image contrast variation (Lloyd, 1985, 1987, 1994; Lloyd and others, 1991). Although quartz undergoes grain-size reduction (from a few millimeters to <30 μm) with increasing shear strain from the edge to the center of the shear zone, the progression is not simple and significant microstructural variations and strain perturbations occur. These are especially pronounced along the shear-zone boundary (see 82A), where the potential for mechanical incompatibilities between adjacent grains exists because of differential strain accommodation and/or shear stress distribution due to different grain crystallographic orientations. However, most original grains undergo grain-size reduction by the localization of intragranular deformation and the cyclical formation of arrays of progressively smaller subgrains, inclined initially at relatively high angles to the (horizontal) shear plane (see 82B). XZ section: X parallel to lineation, Z normal to mylonitic foliation.

**82A.** The formation of different intragranular microstructures in adjacent quartz grains due to mechanical incompatibility caused by their different crystallographic orientations (determined via SEM electron channeling pattern analysis). In the lower grain, the c axis lies in the horizontal shear (XY) plane, an r pole is subparallel to X and an m pole is subparallel to Z; whereas in the upper grain, the c axis is slightly inclined to Z within the YZ plane and an a axis is subparallel to X (Lloyd and others, 1991). The basal planes of the two grains are therefore almost perpendicular to each other. The lower grain has a core region (c) of relatively large equant subgrains, surrounded by a mantle (m) of smaller subgrains (i.e., conventional "core-and-mantle" microstructure). The upper grain exhibits alternating equant (eq) and elongate (el) subgrains. These differences in behavior between the two grains are accommodated by localization of deformation adjacent to their common boundary and the formation of distinctive grain-boundary microstructures. In the lower grain, a surface region (r) has begun to detach ("delaminate") along a grain-boundary-parallel shear zone (d-d). In contrast, in the upper grain (s), intense grain-size reduction (<10 μm) has occurred.

**82B.** Linear features (L1) that rotate in the dextral shear sense are typical intragranular deformation microstructures of quartz grains at the edge (low-strain region?) of the shear zone and are associated with a rapid grain-size reduction (i.e., from a few millimeters to ~100 μm in ~1 mm). There is a subtle crystallographic orientation change and (sinistral) displacement across these linear features, but both die away in the direction of decreasing shear strain (i.e., downward). This suggests that the linear features accommodate displacement and/or rotation between adjacent intragranular regions. A second set (L2) of much shorter, but otherwise similar, linear features occurs subnormal to the first set. Both sets are interpreted as subgrain boundaries: the former is responsible for the formation of relatively large elongate subgrains which progressively overturn into the shear direction; the latter divide the elongate subgrains into smaller equant subgrains, which may eventually develop further sets of boundaries subparallel to the first pair. This cyclical microstructural development results in a progressive reduction of grain size, and a continual change in grain-subgrain shape, which accommodates the increasing displacements associated with increasing simple shear strain. As such, this behavior of intragranular boundaries is analogous to the accommodation of intragranular deformation by the operation of several crystal-slip systems: the boundaries represent narrow zones of concentrated deformation (i.e., "slip planes") to which specific crystallographic indices can perhaps be applied (i.e., "slip plane and direction"). Finally, the upper part of the image shows the abrupt decrease in grain size that marks the transition from relatively low to significantly higher shear strains; the shear strain profile is not continuous.

1mm

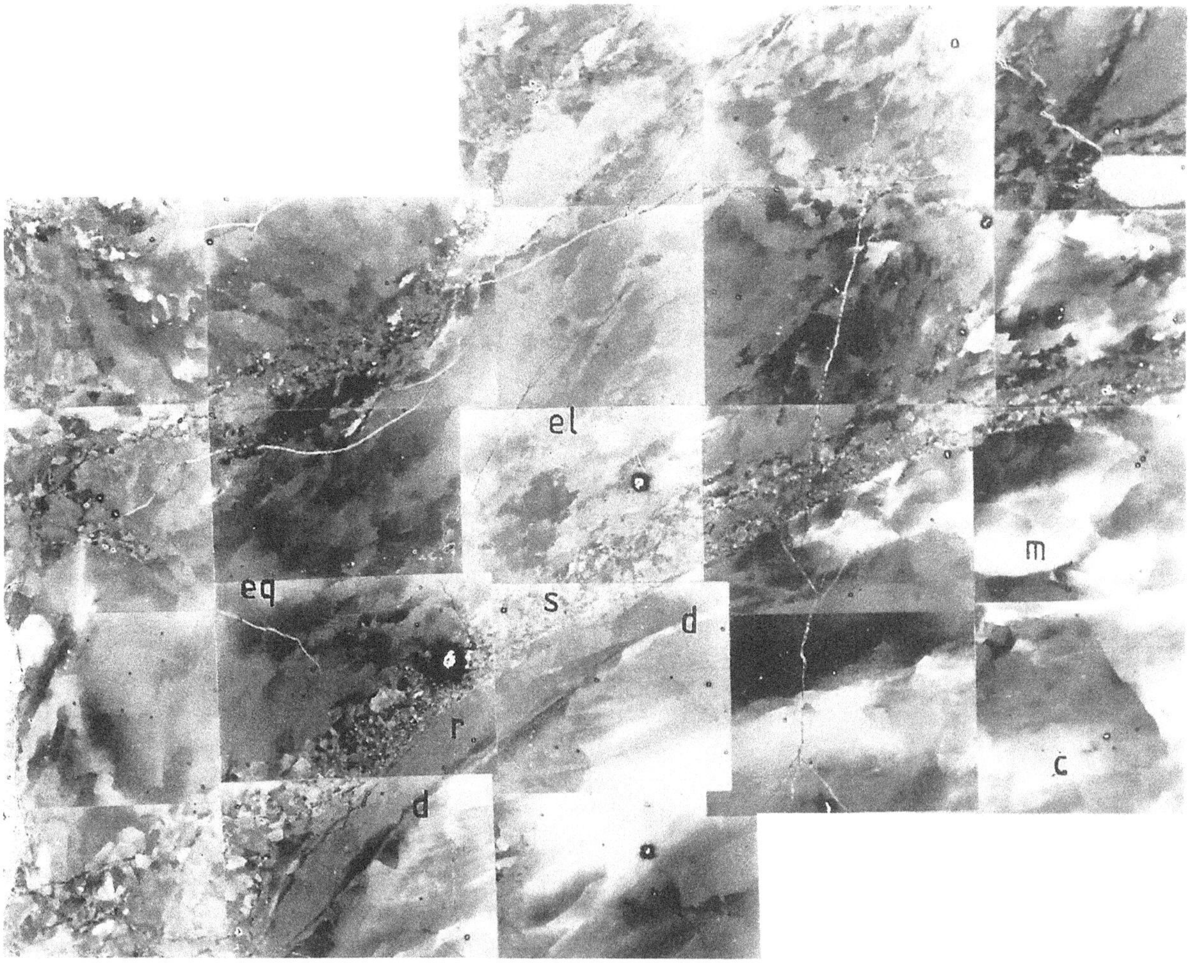


82 A

82 B

1mm

# 83. DEFORMATION MICROSTRUCTURE OF NATURAL PLAGIOCLASE

Shaocheng Ji

Naturally deformed plagioclase ($An_{45\pm5}$) in an anorthosite mylonite from a shear zone in the Jotun nappe complex, Jotunheimen, Norway (Ji and Mainprice, 1988, 1990). The metamorphic temperature during deformation has been estimated at about 600°C (Boullier and Guéguen, 1975). cpx = clinopyroxene, gt = garnet, pl = plagioclase.

**83A.** S-C anorthosite mylonite. The S foliation is defined by the elongate plagioclase and clinopyroxene porphyroclasts, whereas the C plane is defined by a set of parallel microshear zones composed of fine-grained, recrystallized neoblasts. The angular relationship between the S foliation and C plane indicates a dextral shear sense (top to the right). Petrofabric measurements show that the (010) plane of plagioclase porphyroclasts is preferentially aligned subparallel to the foliation, and the [001] direction is oriented within the foliation plane with a strong concentration subparallel to the lineation. XZ section. NX

**83B.** Porphyroclast-matrix structure in the anorthosite mylonite. Ribbon porphyroclasts of plagioclase are aligned parallel to the foliation and stretching lineation. These porphyroclasts display abundant optical evidence of plastic deformation, such as tapered deformation twins, undulatory extinction, and kink bands. Dynamic recrystallization produced fine neoblasts in the intragranular microshear zones or kink band boundaries, and along the margins of the porphyroclasts. At the same time, clinopyroxene neocrystallized into extremely fine-grained amphibole. This indicates that deformation took place during retrograde metamorphism, from granulite to amphibolite facies. XZ section. NX

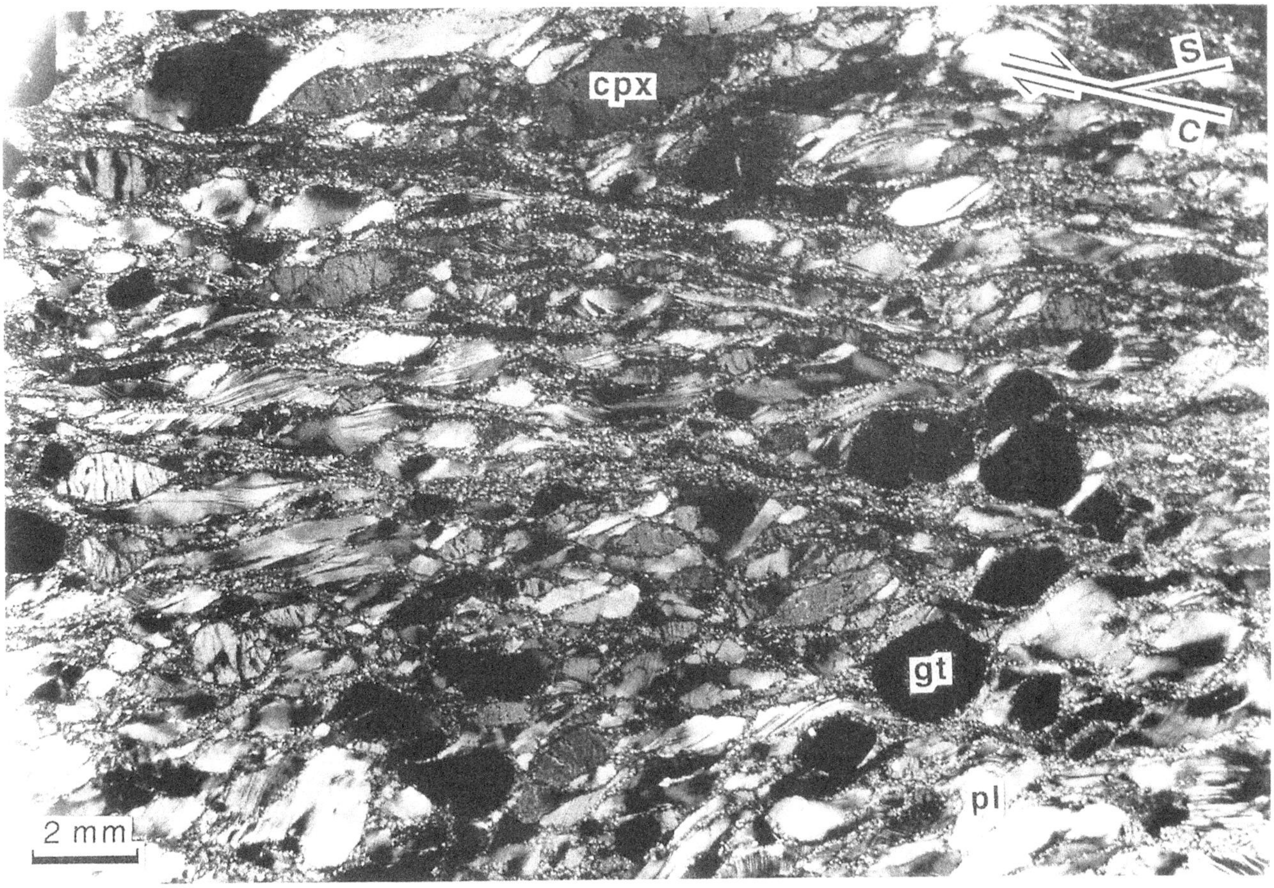


83 A

83 B

# 84. KINK BANDS AND RECRYSTALLIZATION IN PLAGIOCLASE

Shaocheng Ji

Kink bands of plagioclase ($An_{45\pm5}$) in an anorthosite mylonite from a shear zone in the Jotun nappe complex, Jotunheimen, Norway. The metamorphic temperature during deformation has been estimated at about 600°C (Boullier and Guéguen, 1975). Kink bands in plagioclase were produced by (010) [001] slip (Ji and Mainprice, 1990). Kink band boundaries (indicated by the arrow) are parallel to the foliation (S) and stretching lineation. Fine-grained neoblasts formed by dynamic recrystallization occur preferentially along the kink band boundaries where relatively large strain has accumulated. Recrystallization has led to the complete replacement of narrow kink bands by small, new grains. Consequently, the host grain is divided into several ribbon grains by the completely recrystallized kink bands. The (010) planes (albite twin planes) of the neoblasts along the kink band boundaries are aligned preferentially in a position intermediate between those of their two host individuals. Kinking recrystallization is an important softening mechanism for ductile deformation of plagioclase. NX

(010)
S
1 mm

# 85. DUCTILITY OF GARNET INDICATING HIGH-TEMPERATURE DEFORMATION

SHAOCHENG JI

Disk-shaped garnets ($Alm_{63}$ $Prp_{33}$ $Grs_3$ $Sps_1$) occur in the foliation plane of lineated, quartzo-feldspathic mylonites of the Morin shear zone located along the eastern boundary of the Morin anorthosite complex, Grenville tectonic province, Québec, Canada. TEM observations indicate dislocation slip and recovery as the main mechanisms for the deformation of garnet (Ji and Martignole, 1994). In the mylonites, the flattened shapes of the garnets indicate a coaxial strain ($X = Y >> Z$) at very high temperature (≥900°C), whereas the shapes and microstructures of quartz and feldspars indicate a rotational shear strain ($X > Y > Z$) under middle-temperature conditions (750–550°C). The coaxial ductile strain recorded by the garnets could be produced by an earlier local, high-temperature deformation at the roof of the rising anorthositic diapir (~1155 Ma). Subsequent, middle-temperature rotational deformation is related to strike-slip movement along the Morin shear zone (~1020 Ma) (Ji and Martignole, 1994).

Feldspar and particularly quartz, which are easily deformed under amphibolite- and greenschist-facies conditions, generally do not preserve a record of earlier high-temperature deformations. Their microstructures only record the last increments of strain. In contrast, deformed garnets may be among the few minerals able to preserve information about an early, extremely high-temperature deformation, because they become rigid below a temperature where quartz and feldspar continue to deform plastically.

**85A.** Lens-shaped garnet grains in a quartz-rich layer of the quartzo-feldspathic mylonite. XZ section. gt = garnet, fd = feldspar, qtz = quartz. NX

**85B.** A single crystal of garnet showing a typical pinch-and-swell structure. PPL

**85C.** Deformed garnet single crystals in the quartzo-feldspathic mylonite. Tensile fractures within the garnet grains, which are perpendicular to the lineation, were developed during the last stage of deformation. XZ section. Reflected polarized light micrograph.

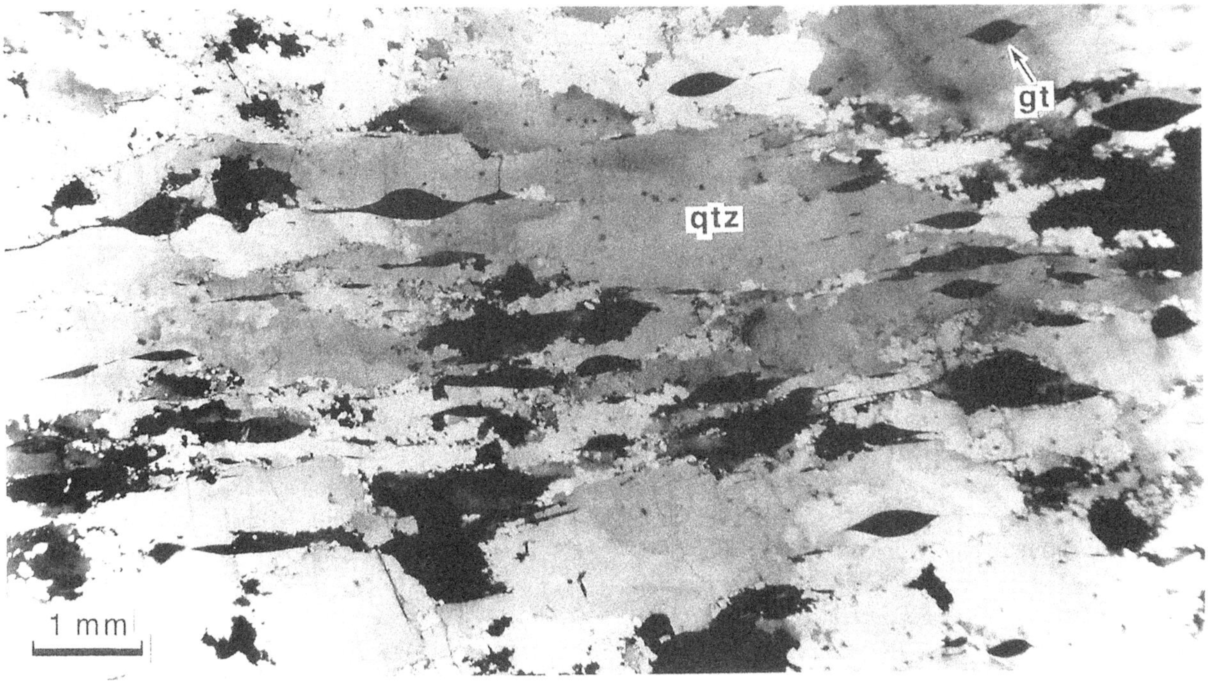


85 A

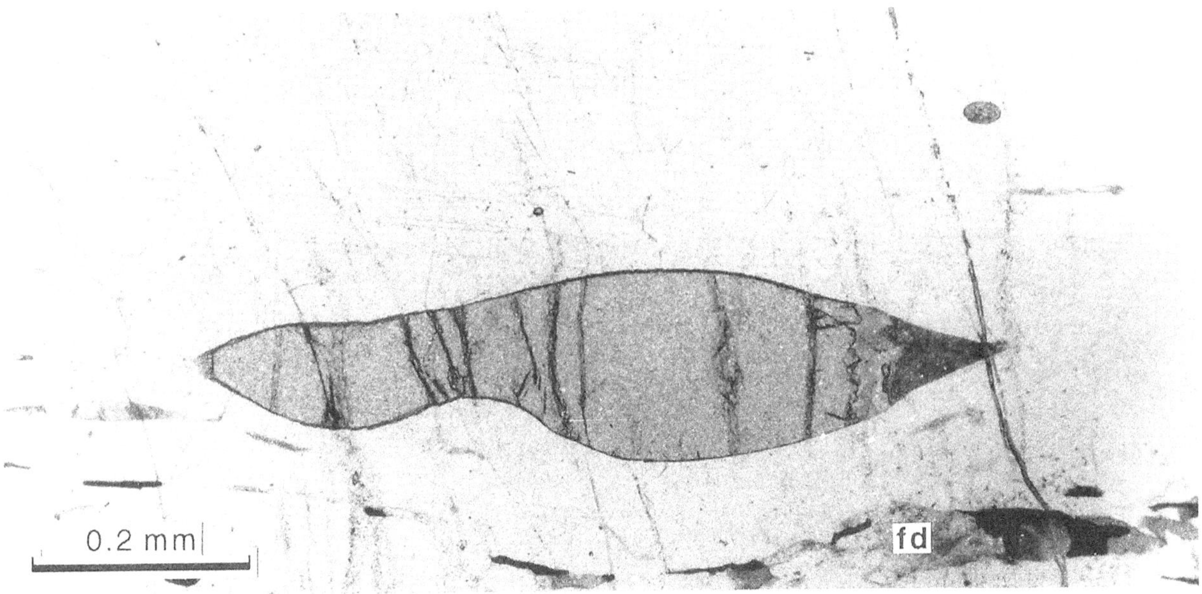


85 B

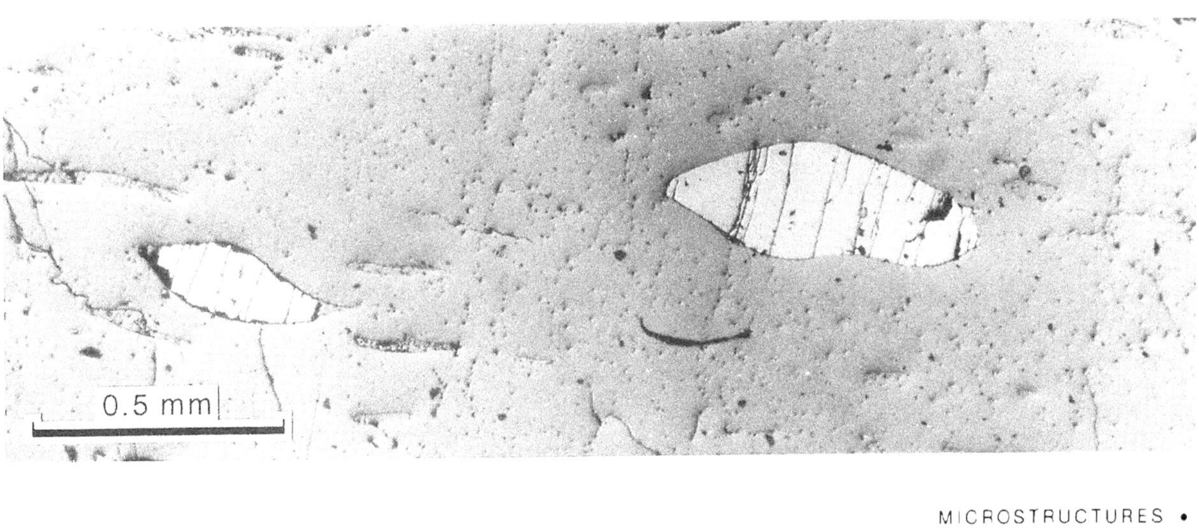


85 C

# 86. DATING DEFORMATION IN MYLONITES

W. James Dunlap and Christian Teyssier

This contribution illustrates the microstructure of micaceous mylonites used to date deformation in the Ruby Gap duplex of central Australia (Figure 86.1). White micas and K-feldspars were dated by K-Ar and $^{40}$Ar/$^{39}$Ar methods in order to constrain the thermal and kinematic history of the duplex (Dunlap and others, 1991, 1995). A remarkable correlation exists between the apparent ages of the white micas and their respective microstructures. Cleavage-forming (C) white micas that grew during duplex formation give apparent ages that are about one billion years younger than white-mica porphyroclasts, which are relict grains. Plateau-like ages of the C white micas decrease systematically toward the roof thrust, suggesting some type of structural control. Analysis of the isotopic data shows that subsequent to crystallization the micas did not lose significant argon by diffusion, thus the ages may record when cleavage formation ceased.

**86A.** Least deformed Heavitree Quartzite of sheet 1. Detrital quartz grains ($Q_c$) were unaffected by the deformation, except for development of deformation bands (DB) and dissolution features (D). An alternative interpretation is that the DB formed in the source rock of the quartzite. Detrital white mica (M) yields a K-Ar apparent age of 1569 ± 16 Ma. Quartz grains are coated with sericitic white mica (C) that grew during deformation. A sericite separate containing a few percent of the older detrital white mica gave a mixed K-Ar age of 376 ± 4 Ma. NX

**86B.** Mica-rich shear zone in sheet 1 quartzite. Detrital white mica (M) is kinked and split apart along cleavage. The geometry and location of fine-grained C white micas and quartz between a quartz pull-apart ($Q_c$) indicate that the grains grew during deformation. NX

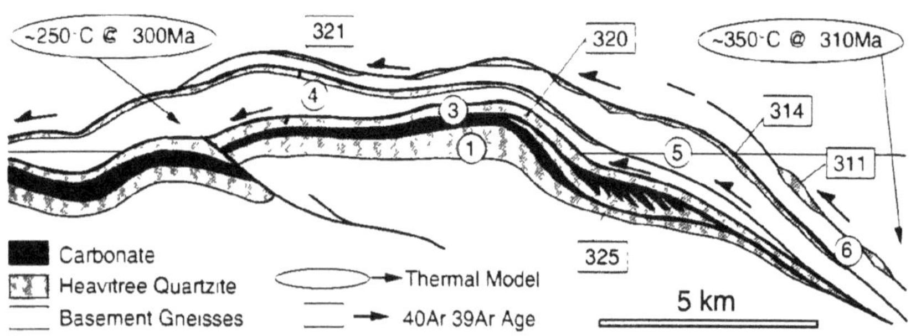


**Figure 86.1.** Cross section of Ruby Gap duplex, central Australia

86 A

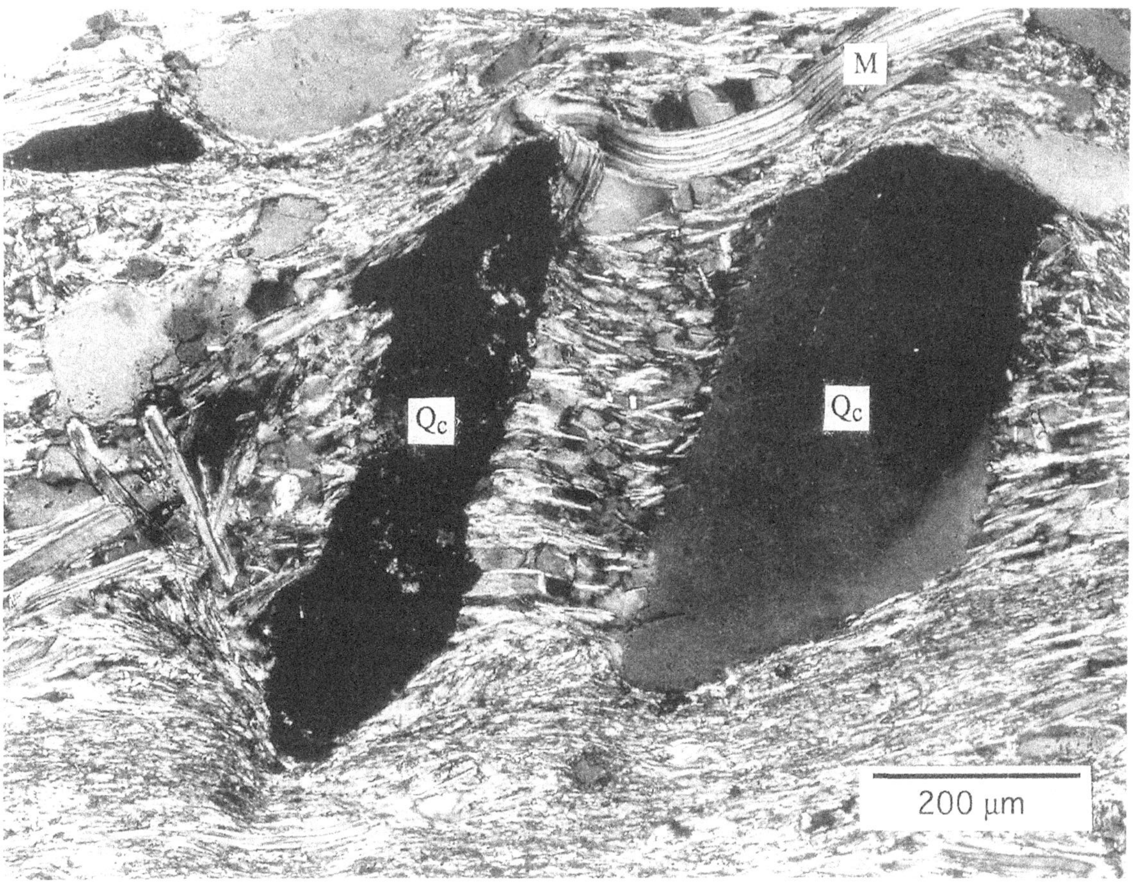


86 B

The Ruby Gap duplex (Figure 86.1) is an antiformal stack of six thrust sheets that were deformed and superposed under greenschist-facies conditions during the 400–300 Ma Alice Springs orogeny (Dunlap and others, 1995). Each thrust sheet is capped by mylonitized Heavitree Quartzite, which was undeformed prior to thrusting. The microstructure of the quartzite varies from nearly undeformed in the lowest sheet to coarsely recrystallized in the upper sheets, with recrystallized grain size of both quartz and C white mica increasing toward the upper sheets (cf. Plate 140 and Plate 147, this volume).

Age spectra for C white micas from quartzite mylonites (Figure 86.2a) contrast sharply with those for white-mica porphyroclasts from granitic basement (Figure 86.2b). The results suggest that the white-mica porphyroclasts were not above their closure temperature (~350°C, depending on kinetic parameters) for any significant period in the Paleozoic. Plateau-like segments of age spectra for C white micas decrease toward the hinterland, which may indicate that shearing and cleavage formation ceased first in the foreland and last in the hinterland (Figure 86.1).

**86C.** Quartzite from northern exposure of sheet 1 (Figure 86.1). Quartz is mostly recrystallized and few porphyroclasts of detrital quartz ($Q_c$) remain. Neocrystallized C white mica is undeformed. Approximately 1% of the white mica may be remnant detrital grains. The discordant age spectrum yields a plateau-like segment in the saddle at 325 ± 1 Ma. Peak estimated temperature ($T_p$) during deformation: 280 ± 20°C (see models, i.e., Figure 86.3b). NX

**86D.** Entirely recrystallized quartzite from northern part of sheet 3 (Figure 86.1). The aspect ratio of the C white micas reaches up to 20:1. Although the white micas may be only a few microns thick, they are undeformed, suggesting that they crystallized late in the deformation history. Plateau age for the white micas is 320 ± 1 Ma. $T_p$=280 ± 20°C. NX

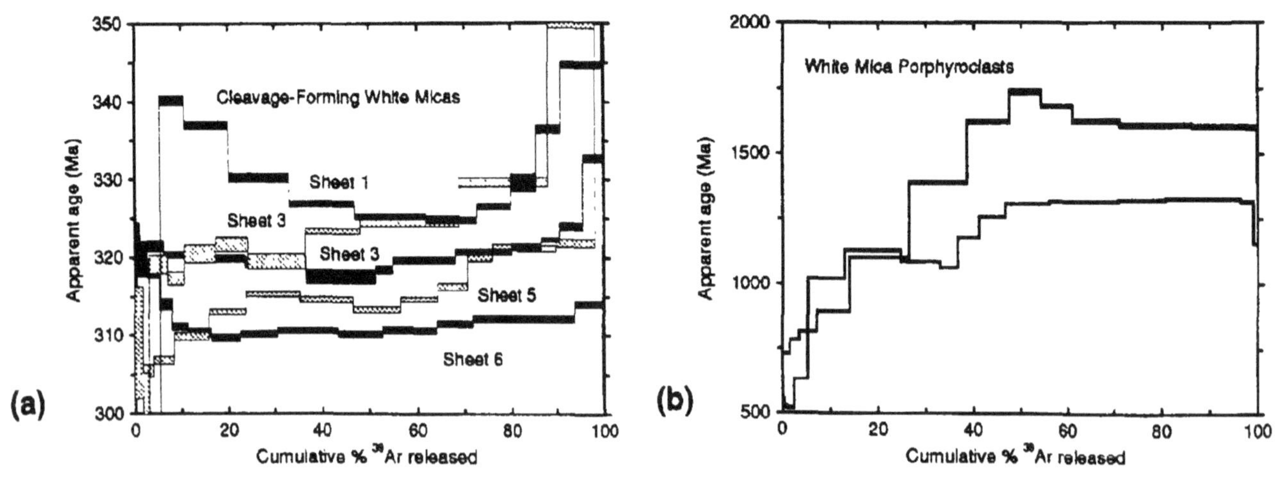


**Figure 86.2.** Age spectra for cleavage-forming (a) and porphyroclast (b) white mica

86 C

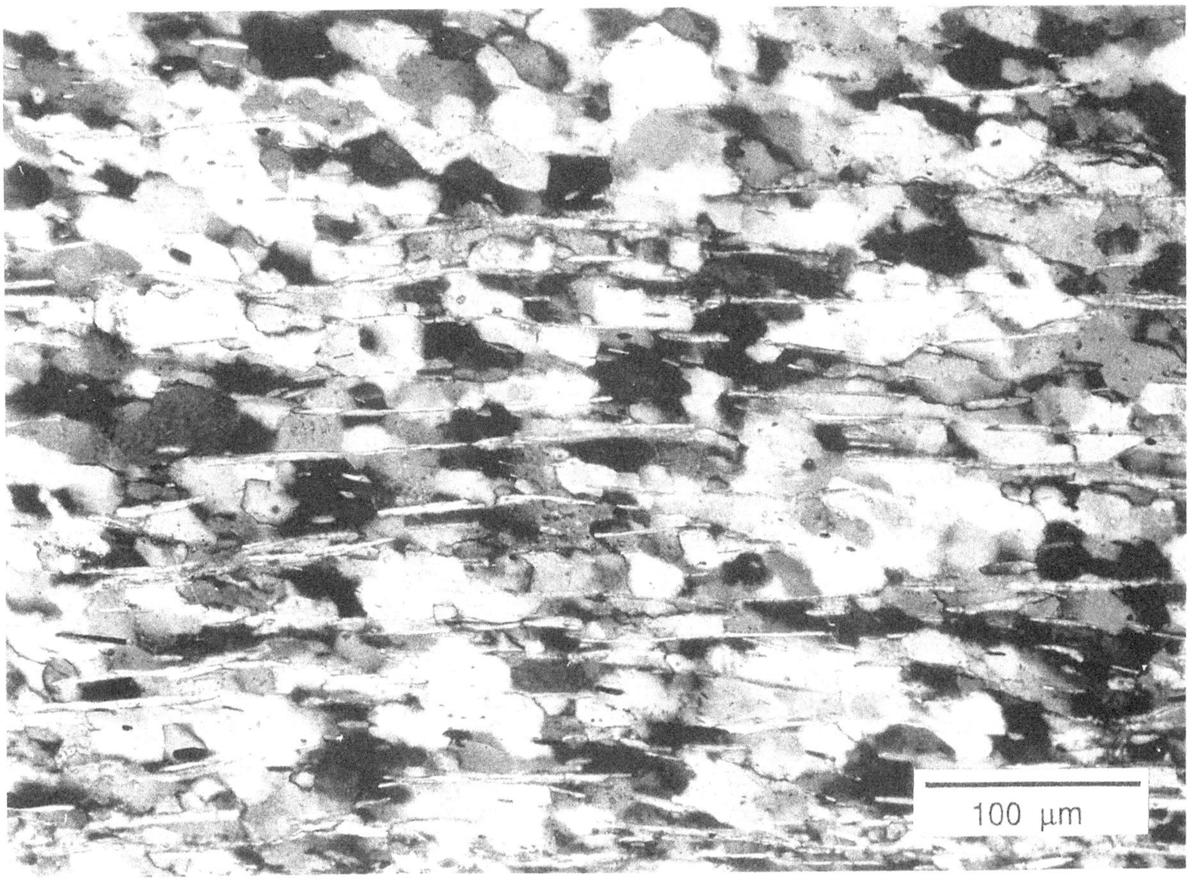


86 D

Thermal modeling of K-feldspar $^{40}Ar/^{39}Ar$ age spectra (Lovera and others, 1989) indicates that the duplex experienced maximum temperatures between ~250 and 320 ± 20° C during formation of the C white micas at ~325–310 Ma (Figure 86.3). According to diffusion theory (and assuming $E_a$=40 Kcal/mol, dT/dt=2°C/Ma, $D_0$=1.38 × $10^{-4}$ $cm^2/s$; Robbins, 1972), grains larger than about 125 µm would not be susceptible to significant outgassing during the thermal history in Figure 86.3 (Ruby Gap duplex). This is only true if the effective diffusion length scale for argon is the mean grain size. The analyzed white micas all have mean grain sizes parallel to cleavage of >125 µm (except the sample from the southern part of sheet 3, ~40 µm). We conclude that the $^{40}Ar/^{39}Ar$ plateau-like segments of age spectra for the C white micas record the progressive cessation of ductile deformation toward the north with time.

**86E.** Recrystallized quartzite from sheet 6, the roof thrust zone of the duplex (Figure 86.1). The coarse-grained microstructure contrasts sharply with that in the lowest sheets of the duplex. Grain-boundary migration in quartz has apparently isolated small grains of C white mica entirely within recrystallized quartz. White-mica plateau age 311 ± 1 Ma. $T_p$=~320 ± 20°C. NX

**86F.** Mylonitized granitic basement, sheet 4. Quartz is recrystallized ($Q_r$), K-feldspar is cataclastically deformed (K). Note inter- and intragrain truncation of cleavage in white-mica porphyroclast (M). M has undergone dissolution, whereas C white micas (C) have neocrystallized to form the cleavage. White-mica porphyroclasts give $^{40}Ar/^{39}Ar$ age steps as old as 1738 ± 7 Ma (Figure 86.2), whereas C white micas yield a mixed K-Ar age of 414 ± 4 Ma. $T_p$=~300 ± 20 °C. NX

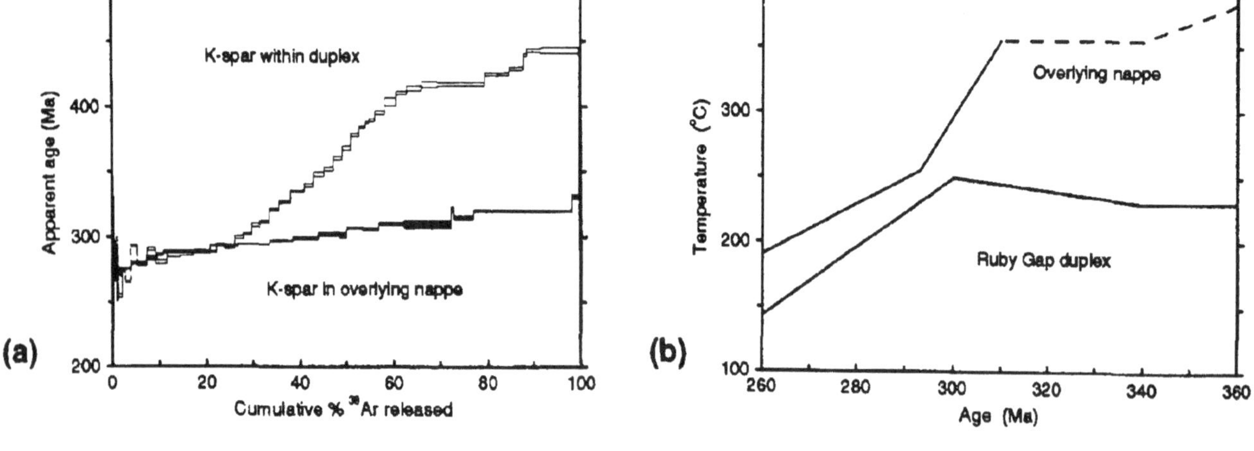


**Figure 86.3.** Age spectra for K-feldspars (a) and associated thermal models (b)

86 E

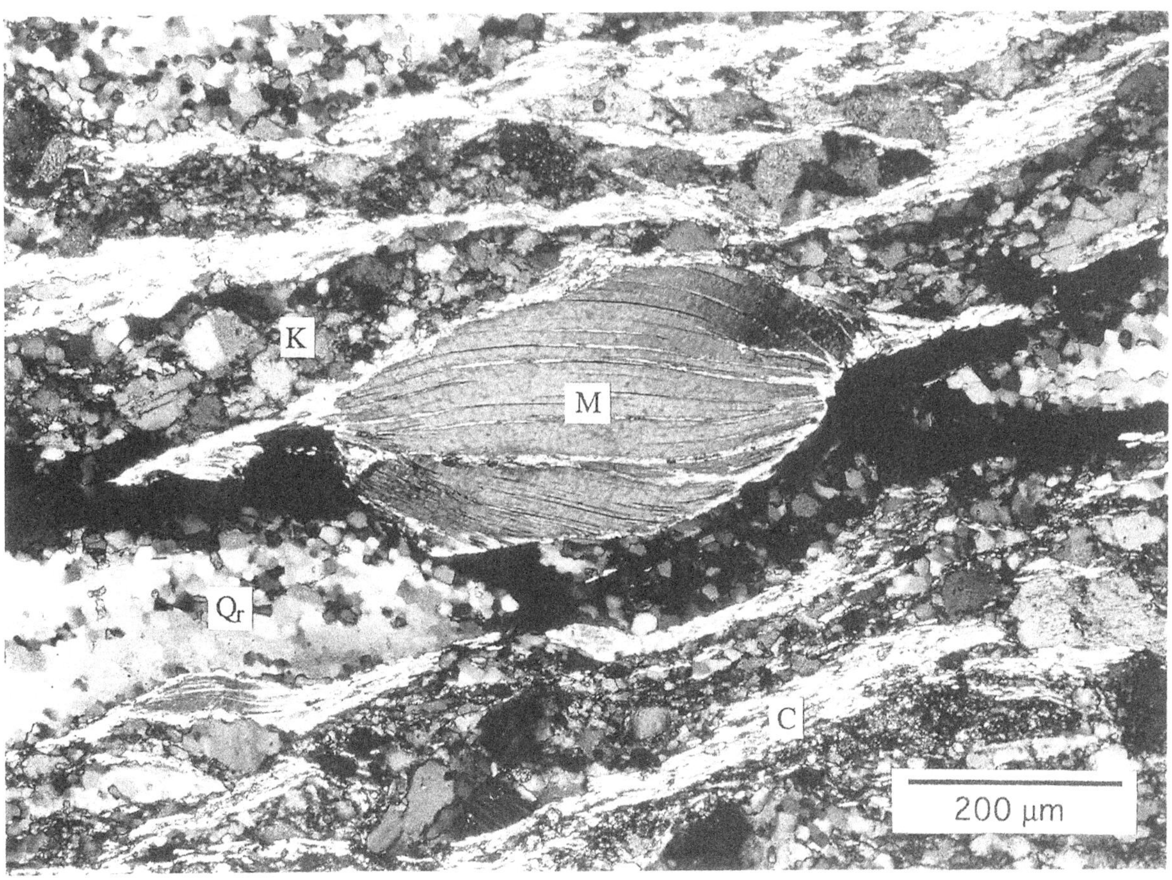


86 F

# 87. GRAIN-SCALE MICROSTRUCTURES IN AN EXPERIMENTAL MYLONITE

PAUL D. BONS AND MARK W. JESSELL

Mylonites are zones of very high shear strain that formed at elevated pressure and temperature. These conditions are difficult to accomplish in experiments on natural rocks. One can, however, make experimental "mylonites" with some soft, organic crystalline rock analogs, such as octachloropropane (OCP) and camphor. These materials are analogs to plastically deforming minerals in the sense that their rheological and microstructural behavior is similar (Means, 1989). The most popular use for these materials is for the transparent deformation cell, because of their relative softness. A thin specimen is sandwiched and deformed between two glass plates. The specimen thus resembles a thin section and can be observed with an optical microscope throughout deformation.

Two photomicrographs are shown here of OCP [hexagonal, moderate birefringence; see Jessell (1986)] with a few faintly visible camphor bands [rhombohedral, low birefringence; see Urai and others (1980)] at a finite shear strain of approximately 200. Deformation is in a transparent deformation cell with a circular shear zone (Jessell and Lister, 1990) at a dextral shear strain rate of approximately $2 \times 10^{-4}s^{-1}$ at room temperature (Figure 87.1).

**87A.** Crystallographic preferred orientation. The deformation causes most c axes to align perpendicular to the horizontal shear plane (light-colored grains, Figure 87.2). OCP with this c axis orientation almost forms a single ribbon grain, divided into a mosaic of subgrains. A second and minor point maximum (black grains) is formed by grains with c axes in the shear plane and perpendicular to the shear direction. These grains occur mainly near the edge of the shear zone, which is at the bottom of the micrograph, where shear stress is highest due to the curvature of the shear zone. PNX (45°)

**87B.** Oblique grain-shape preferred orientation (SPO) and kinks. Interaction between shearing of grains and grain-boundary migration causes an oblique grain SPO. The angle of the SPO with the shear plane decreases with increasing shear intensity toward the edge of the shear zone, which is at the bottom of the micrograph. Folding of the stronger camphor layers causes kinking in the adjacent OCP ribbons (Figure 87.2). PNX (45°)

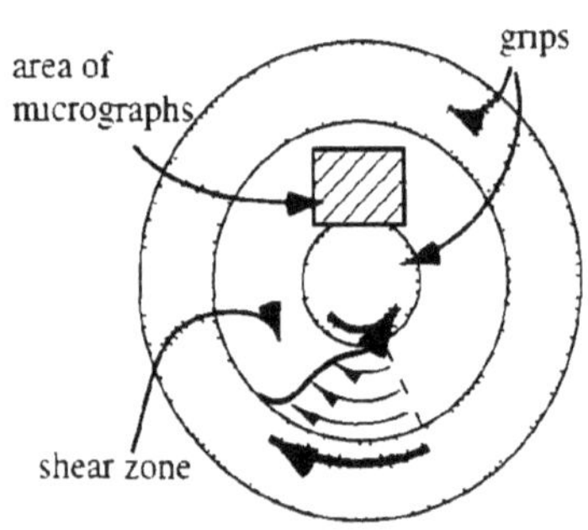


**Figure 87.1.** Circular shear zone in transparent deformation cell

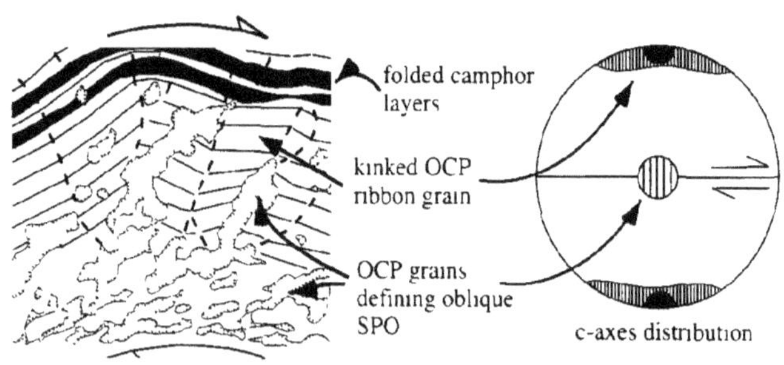


**Figure 87.2.** Sketch of micrograph 87B: distribution of two crystallographic orientations and kinks in OCP grains

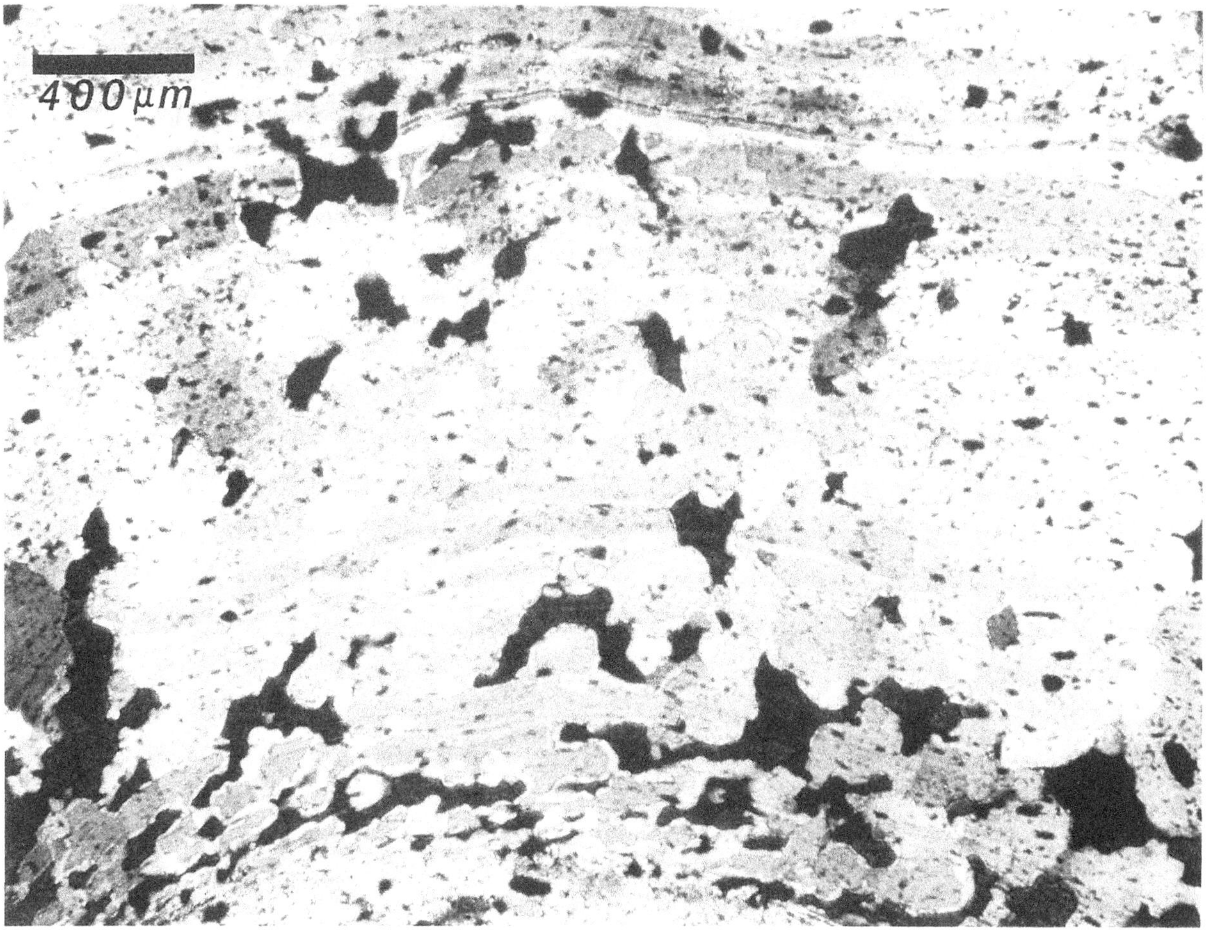


87 A

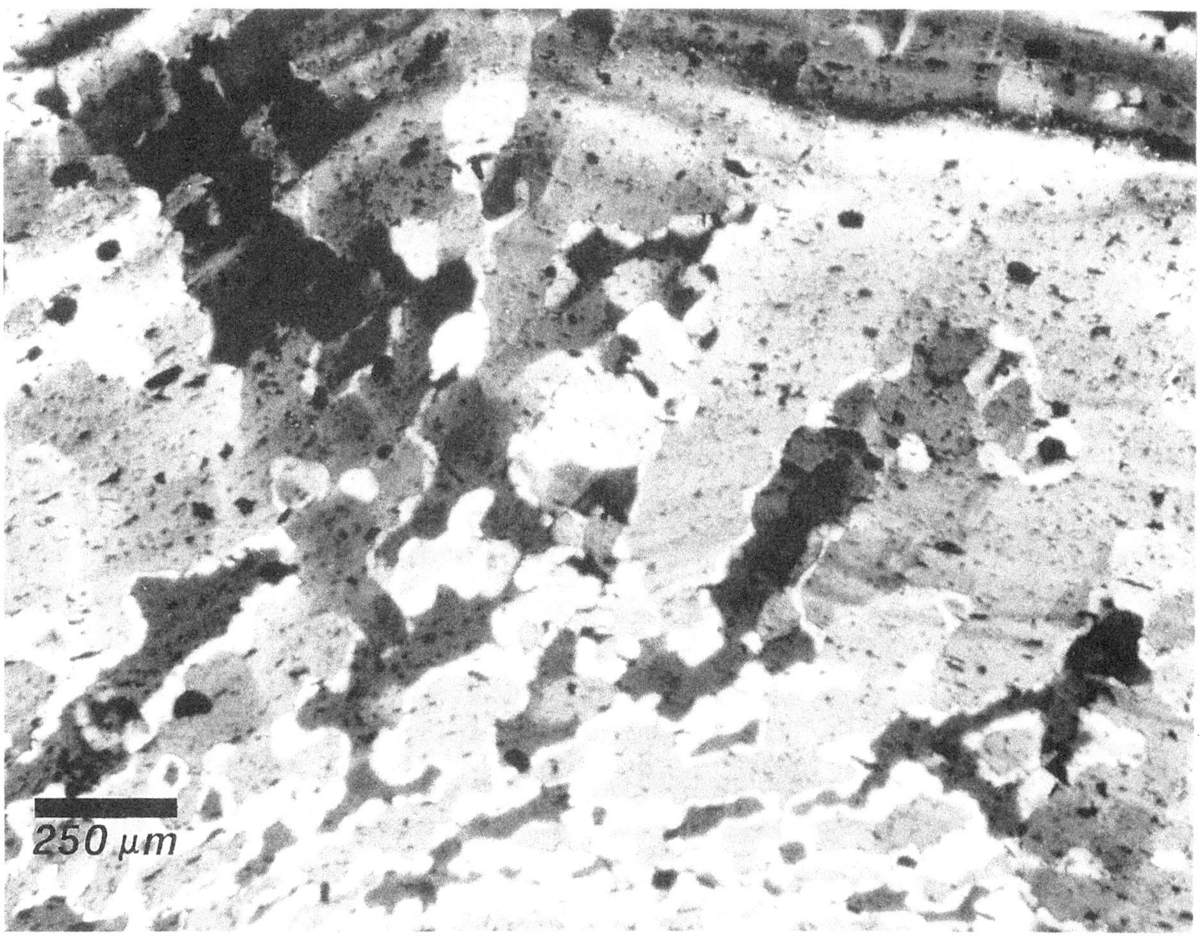


87 B

# DUCTILE BEHAVIOR

## Foliation development



Common abbreviations used in regard to the plates in this section include PPL = plane polarized light and NX = crossed nicols.

In addition to grain-size reduction, another characteristic of mylonites in Lapworth's original definition was fluxion structure or foliation. This set of plates illustrates the development of foliation in mylonites by several processes, including the high-strain attentuation of original compositional layers; the flattening or shearing of originally equant grains or regions; and the synkinematic reaction to micaceous phases. The plates illustrate that foliation may be fairly uniform and planar, but once formed it constitutes a heterogeneity and may become folded with progressive deformation (also shown in several other natural examples in Plate 111 and in experimental analog material in Plate 113).

# 88. DEVELOPMENT OF FOLIATION IN SHEARED ADAMELLITE

CAROL SIMPSON

Hand specimens and photomicrographs of a small-scale, ductile shear zone of epidote-amphibolite grade in granitic gneiss, Valle di Maggia, Ticino, Switzerland (Simpson, 1983). Relatively unfoliated Hercynian granite, adamellite, and diorite occur in the core of the Maggia nappe, a Lower Pennine crystalline nappe of the Lepontine Alps. A penetrative lineation, locally accompanied by a foliation in ductile shear zones, formed during the first phase of Alpine deformation in this region. The photographs illustrate how quartz grain aggregates change shape with increasing strain.

**88A.** Polished hand specimen illustrates the change in shape fabric of quartz (Q) and biotite (black) grain aggregates in feldspar (F) matrix across a narrow, right-lateral shear zone. Grain aggregates in the unsheared rock are roughly equidimensional and individual grains are strain-free. The shear-zone boundary (SZB) is the position of the first visible foliation. With increasing shear strain toward the center of the zone, the degree of planar anisotropy increases and the angle between foliation and SZB decreases.

**88B.** Photomicrograph illustrates part of teardrop-shaped aggregate of quartz grains (Q) from the lower shear zone boundary in 88A. Fine-grained biotite, epidote, muscovite, and feldspar grains decorate discrete planes (arrowed) that parallel the long edges of the quartz aggregate and have an average spacing of 0.3 mm. Biotite, epidote, and muscovite grains in the matrix around the quartz aggregates are aligned parallel to the planar surfaces. Quartz grains contain deformation bands, subgrains (sg), and grain-boundary migration-recrystallized grains (rg) (Regime 3 of Hirth and Tullis, 1992). There is no preferred orientation of quartz *c*-axes in the unsheared rock, but *c*-axes of quartz grains in teardrop-shaped aggregates show a weak single-girdle distribution approximately perpendicular to the SZB, indicating deformation by intracrystalline glide (Simpson, 1983). Across the SZB, there is a progressive reduction in grain size of plagioclase and K-feldspar from 0.2 to 0.08 mm. These feldspars are always strain-free, with straight grain boundaries that meet in 120° triple junctions, indicating that grain-boundary migration recrystallization (Regime 1 of Hirth and Tullis, 1992) and recovery processes were active. NX

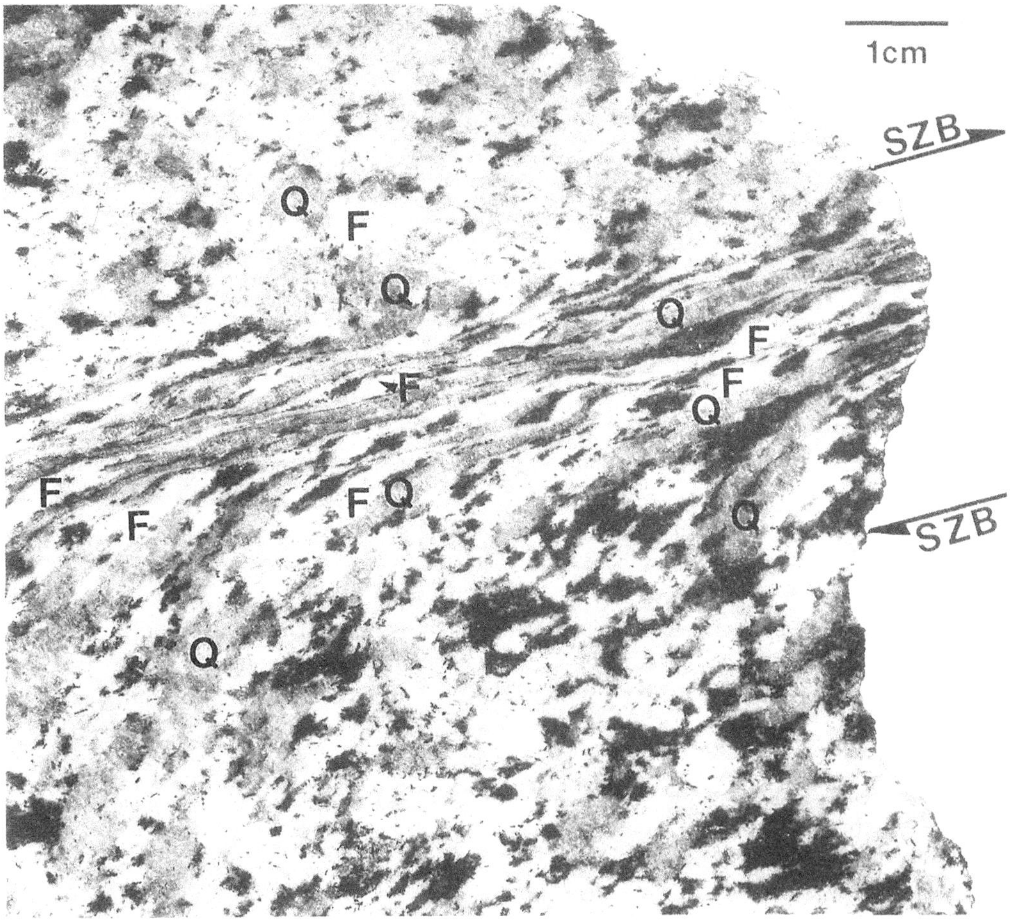


88 A

88 B

**88C.** Enlargement of part of 88B shows two planar surfaces (arrowed) in teardrop-shaped quartz grain aggregate. The planar surfaces are defined by quartz grain boundaries and 0.05 mm epidote and mica grains. The lower planar surface terminates within a quartz grain in a network of healed microcracks (star-arrow); undulatory extinction and patchy subgrains, which may indicate partial healing of a crush zone, accompany this termination. The presence of microcracks suggests that the planar surfaces may have acted as surfaces for intergranular slip, a process that would have been aided by the presence of the mica grains. NX

**88D.** Enlargement of the upper planar surface shown in 88C shows obliquity of long axes of epidote grains with respect to quartz ribbon boundaries. The consistent orientation of the epidote grains, regardless of their aspect ratio, suggests that they grew in this orientation. NX

**88E.** Quartz (Q) in ribbonlike aggregates within the central part of the shear zone shown in 88A. Ribbon width averages 0.3 mm, the same as the average distance between planar surfaces illustrated in 88B. Mica and epidote grains are aligned along quartz ribbon edges. Plagioclase and K-feldspar grains (F) between quartz ribbons are equigranular, strain-free, and have 120° triple junction grain boundaries, indicating complete recrystallization. Deformation bands (db) and elongate recrystallized grains (rg) in quartz are subperpendicular to ribbon edges. A well-developed girdle of quartz *c*-axes, oriented perpendicular to the SZB, has been recorded from the quartz ribbons and indicates deformation by intracrystalline glide (Simpson, 1983). NX

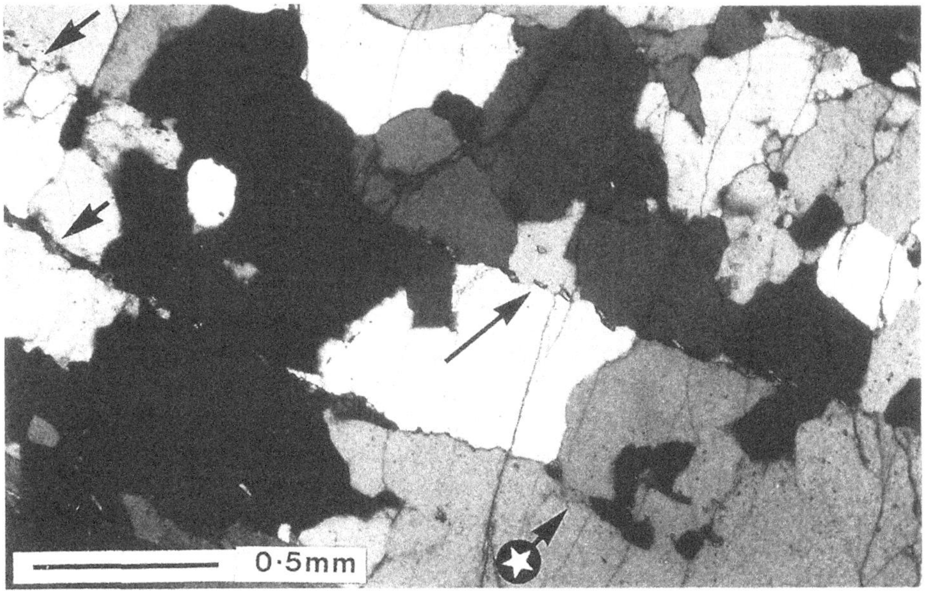

88 C

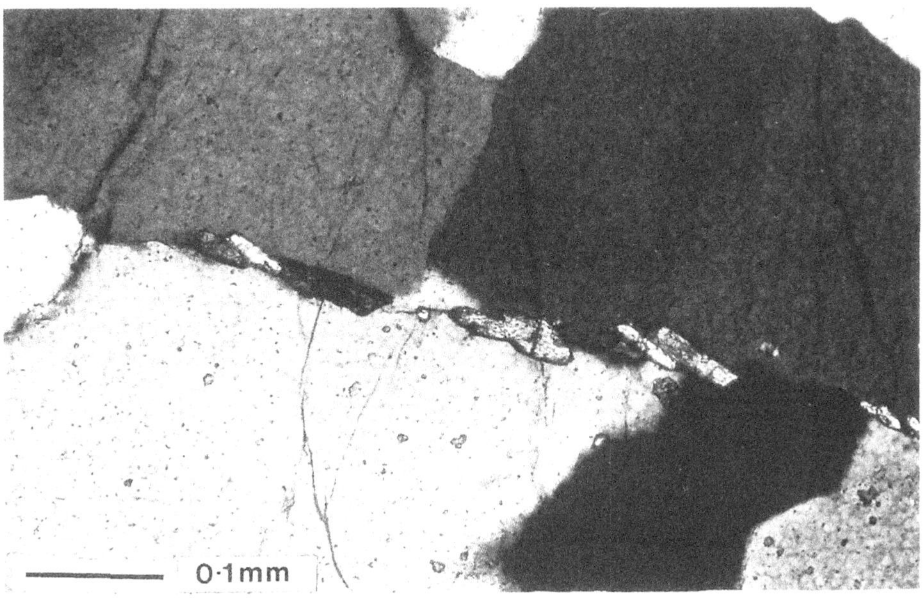

88 D

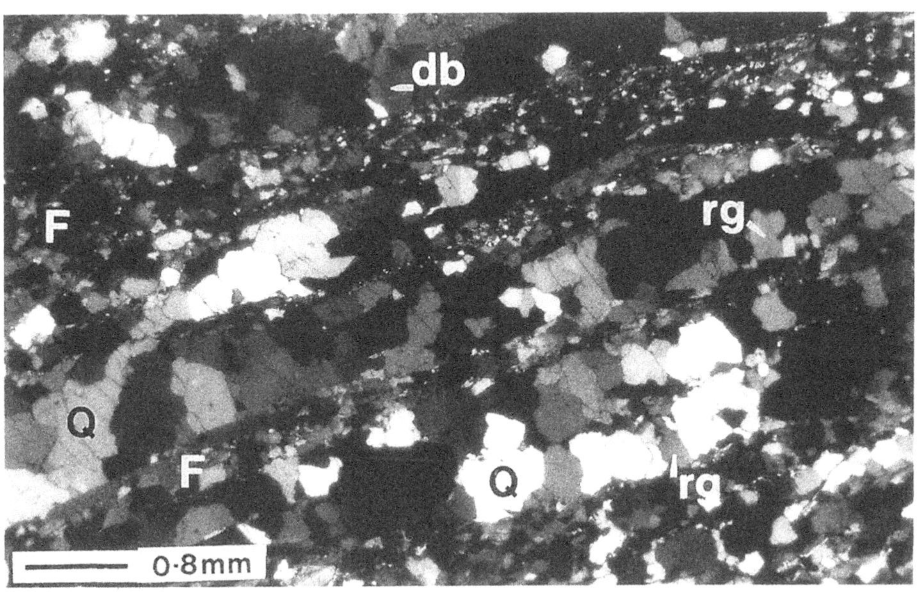

88 E

# 89. BANDING AND FOLIATION IN MYLONITES

J. CARRERAS, J. M. CASAS, AND M. LOSANTOS

In the Cap de Creus peninsula (northeast Spain), medium-grade schists derived from Cambrian or older sedimentary rocks are cut by an anastomosing network of shear zones forming a broadly northwest-southeast-striking belt with prevalent dextral strike-slip displacements (Figure 89.1; see also Plate 65). The shear zones cut across cordierite-andalusite or sillimanite schists which underwent a previous polyphase deformational history and a prograde low pressure/high temperature metamorphism. The metamorphic peak, prior to late shear-zone development, reached upper amphibolite-facies conditions with localized, syntectonic migmatite development, and was followed by widespread intrusion of pegmatitic granite (Carreras and Druguet, 1994). The width of individual shear zones ranges from a few centimeters to a few tens of meters. Shearing occurred during retrograde, greenschist-facies conditions, and partial reequilibration of mineral phases is commonly observed in the mylonites with new growth of biotite, muscovite, albite, epidote, and chlorite, as well as extensive recrystallization of quartz. Mylonitic equivalents of schists and related intrusive rocks are penetratively foliated and commonly banded. The strong penetrative mylonitic foliation formed as the result of intensive shearing with shear strains greater than 10, which has caused the complete obliteration of previous structures in the schists. Sheared, originally coarse-grained pegmatites develop a strong L-S tectonite fabric, and extreme grain-size reduction and porphyroclastic structures are typical. The quartzo-feldspathic mylonites derived from sheared pegmatites and granitoids exhibit a marked stretching lineation commonly defined by mineral elongation (e.g., tourmalines in pegmatites). The banded character of these rocks reflects the extreme stretching of previous compositional heterogeneities of variable origin and shape (e.g., transposed layering in the metasedimentary rocks, segregation quartz veins and nodules, pegmatite dikes, migmatites).

**89A.** Banded mylonite consisting of quartz ribbons and layers of phyllonite containing plagioclase porphyroclasts. The banding resulted from high shear strain ($\gamma > 10$) of a coarse-grained schist that contained segregated quartz veins. PPL

**89B.** Mylonitic granitic pegmatite. The mylonitic foliation is defined by narrow ribbons of fine-grained, granoblastic quartz and by a few elongated quartz grains that in this view are near extinction. Quartz ribbons wrap around feldspar porphyroclasts. A band consisting mostly of small, broken feldspar clasts is denoted by arrows. NX

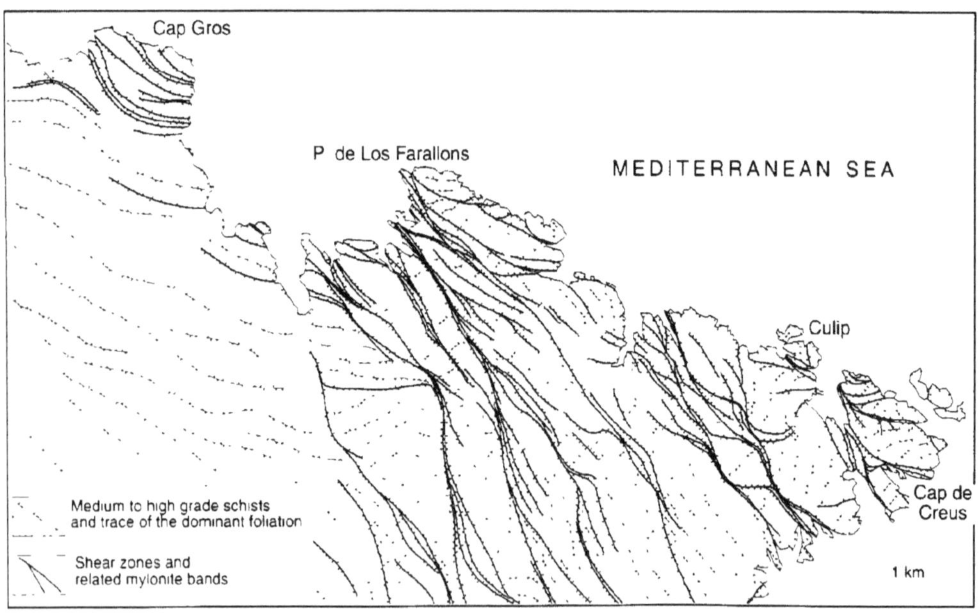


**Figure 89.1.** Structural map showing general orientation of pre–shear zone metamorphic foliation and the distribution and orientation of late, retrograde shear zones in the Cap de Creus peninsula area, northeastern Spain

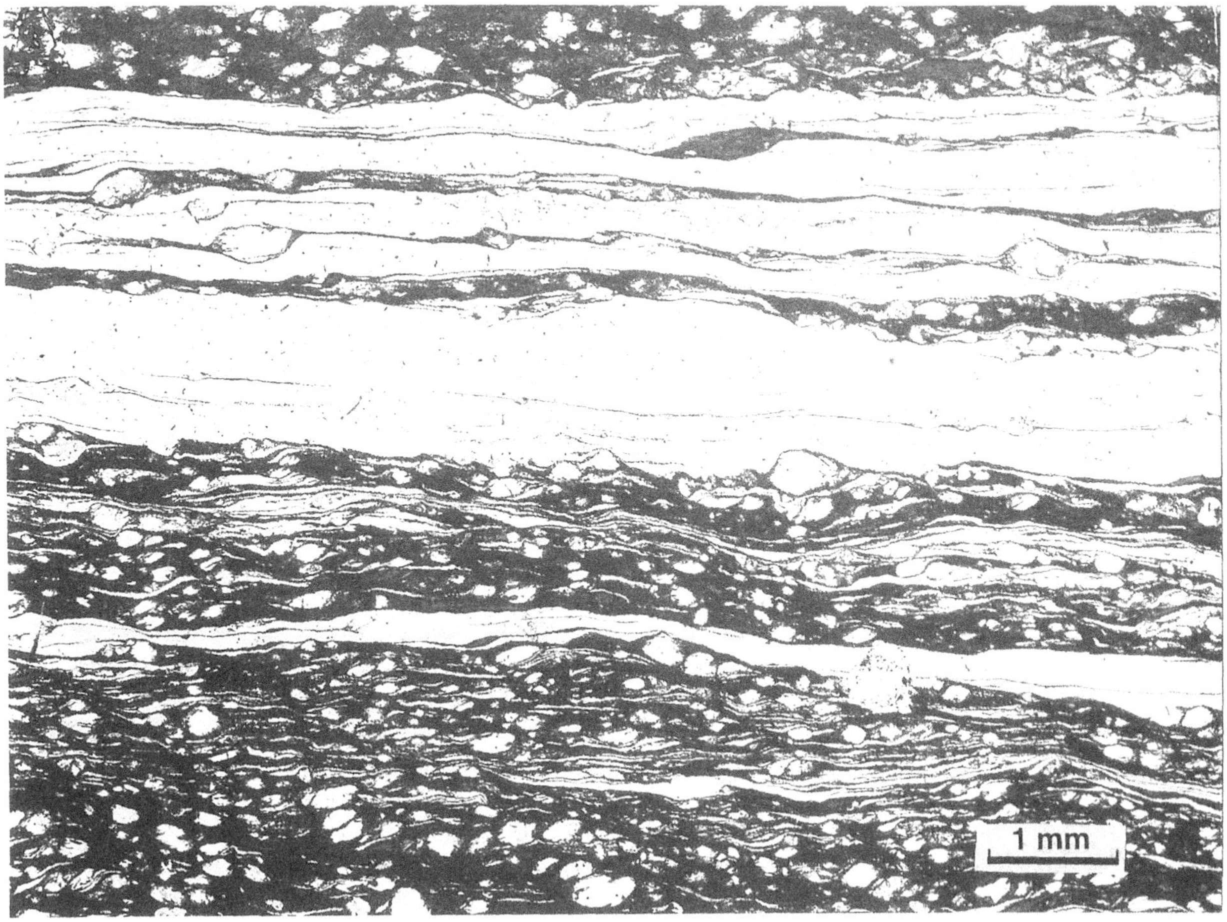

89 A

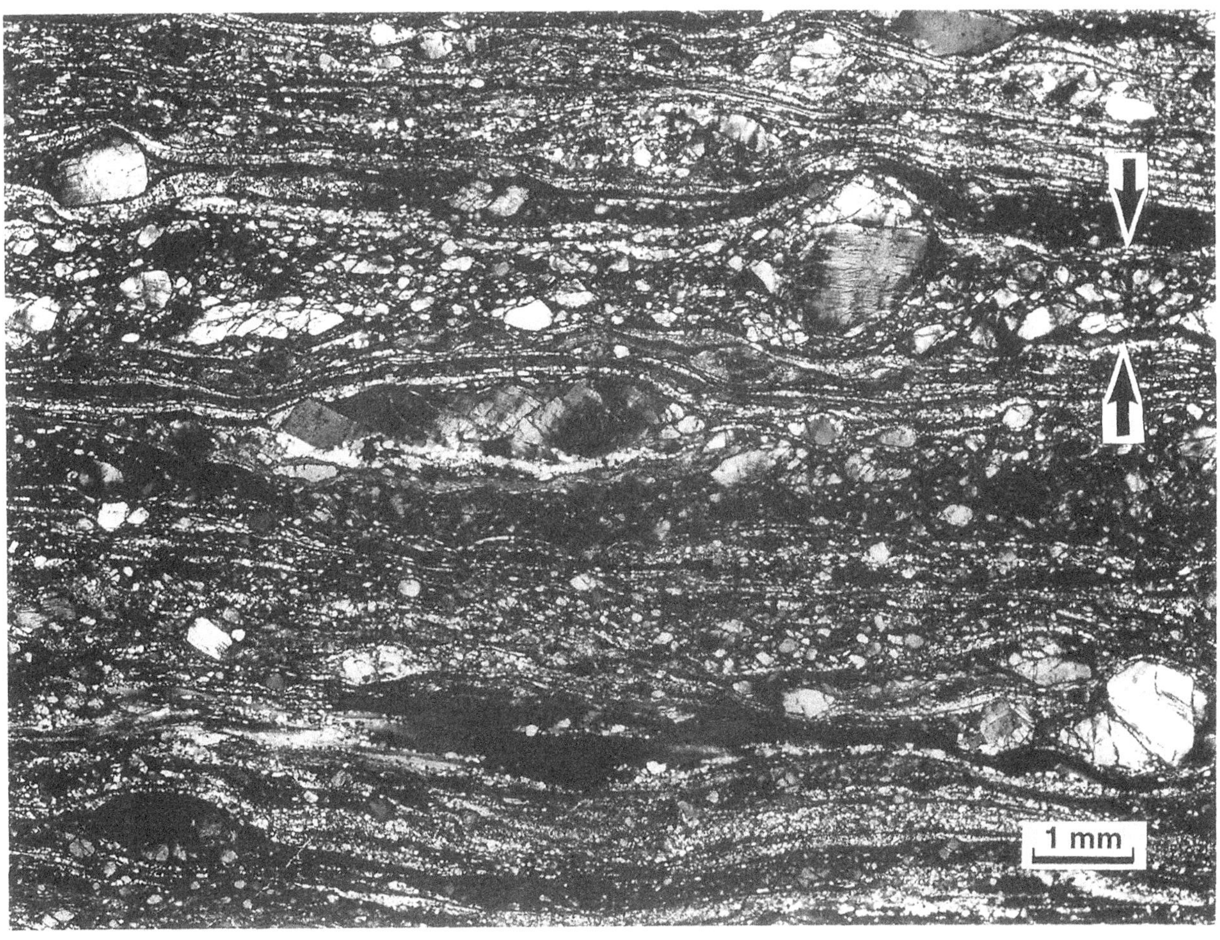

89 B

# 90. PROGRESSIVE DEVELOPMENT OF FOLIATION IN A LOWER GREENSCHIST-FACIES DUCTILE DEFORMATION ZONE

GAUTAM MITRA

Late Paleozoic (Alleghanian) deformation in Precambrian granitic basement of the Blue Ridge anticline (in northern Virginia, U.S.A.) is characterized by the development of ductile deformation zones (DDZs) along which much of the deformation is concentrated (Mitra, 1978). These zones are characterized by grain-size reduction of feldspars and quartz, and chemical alteration of feldspars to quartz and mica (Mitra, 1978, 1992). The zones tend to widen with progressive deformation, as adjacent wall rock is softened by chemical alteration. The photographs show progressive development of structures within a 20 m wide DDZ on Locust Shade Farm, about two miles south of Ben Venue, Virginia.

**90A.** Coarse grains of potassium feldspar, plagioclase, and quartz are initially fractured. The photograph shows fracturing within a single, large K-feldspar grain. The fractures form at high angles to glide planes, perhaps due to stress concentrations produced where dislocation glide is obstructed after some amount of plastic deformation has taken place. The fractures are normally stable and do not cut across grain boundaries. As the fragments pull apart, fibrous quartz is precipitated in the cracks (indicating pressure solution of quartz elsewhere in the rock); the fibers indicate the extension direction (approximately E-W in this photograph). Fracturing of coarse grains continues during progressive deformation; the later formed fractures (left edge of photograph) may not be quartz-filled. F = feldspar, Q + M = quartz-mica fibers. NX

**90B.** Fractures formed in more than one orientation in K-feldspar allow rhombic cleavage fragments to form, which may then pull apart. Spaces between fragments are filled by quartz and mica, at least some of which may be formed from alteration of feldspar. The orientation of the fibers indicates the extension direction (approximately N-S in this photograph). B = biotite, F = feldspar, Q + M = quartz-mica fibers. NX

**90C.** With progressive deformation the feldspar fragments pull further apart, and large parts of the rock are made up of oriented fine quartz and mica, which begin to define a foliation (approximately E-W in this photograph). The feldspar and coarse quartz fragments continue to break up by stable fracturing, forming smaller fragments and rounding off corners on larger fragments. F = feldspar, Q = quartz, Q + M = quartz-mica matrix. NX

**90D.** In the final stages of deformation, fine-grained quartz and mica define a strong foliation in the rock (approximately N-S in this photograph). They also fill pressure shadows behind coarse feldspar grains, with the main foliation swinging around these fragments. The feldspar and coarse quartz grains continue to undergo stable fracturing which further reduces grain size. Finer-grained quartz deforms in a more ductile fashion, by either pressure solution or dislocation-controlled processes; thus, the dominant deformation mechanisms are largely controlled by the grain size. Chemical alteration of feldspars also plays an important role; alteration is typically concentrated along fractures, which allow fluids to penetrate into the grains, thus providing the water necessary for the alteration of feldspars to quartz and mica. F = feldspar, Q = quartz, Q + M = quartz-mica matrix. NX

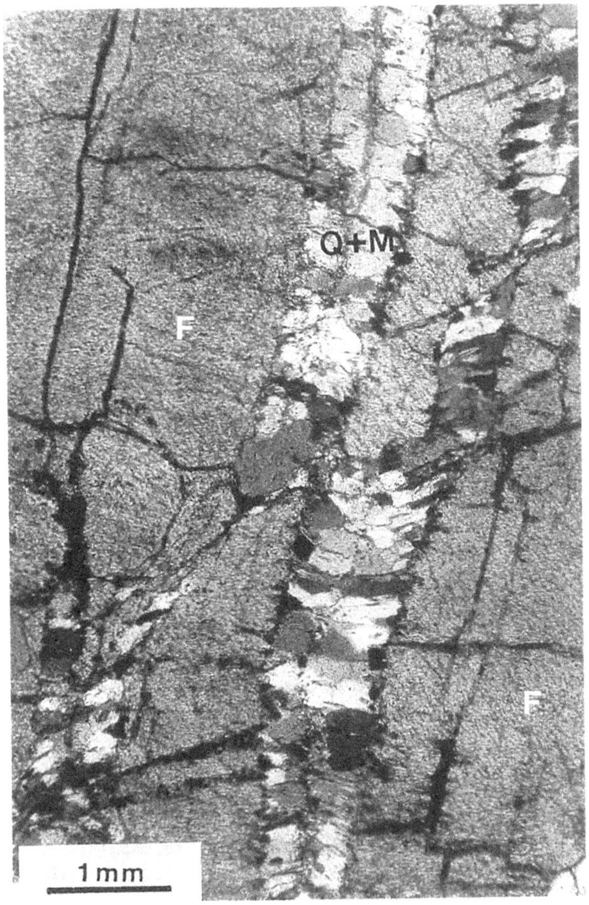


90 A

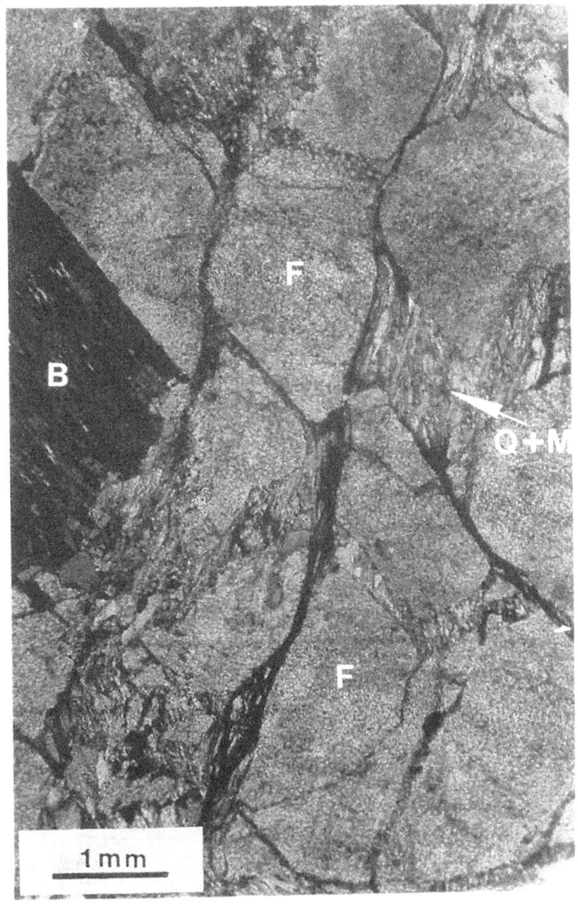


90 B

90 C

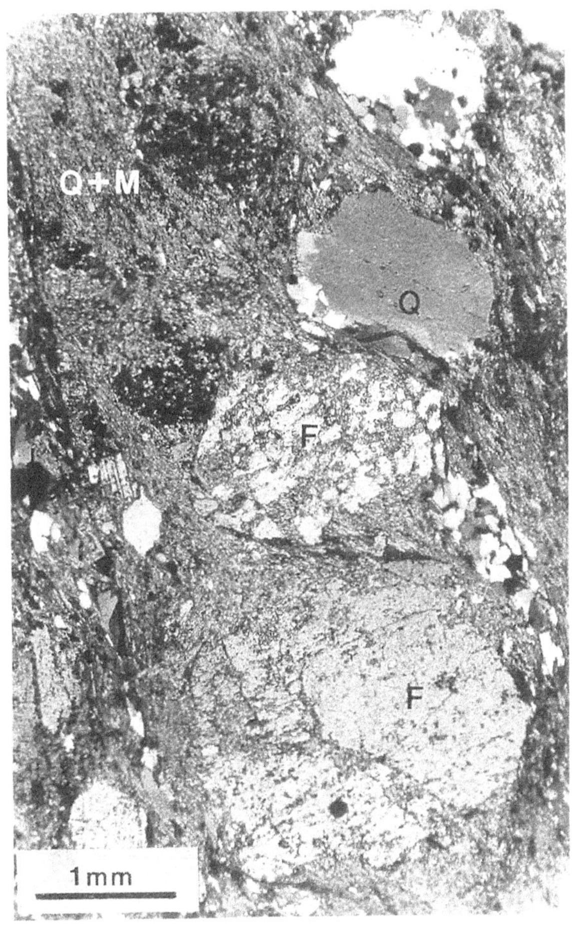


90 D

# 91. DEVELOPMENT OF FOLIATION IN MYLONITES

Andy R. Bobyarchick

The Dover Creek mylonite zone, Virginia, U.S.A., is part of the Alleghanian Eastern Piedmont fault system, which lies in the Piedmont province of the southern Appalachians (Hatcher and others, 1977; Bobyarchick, 1981). The Dover Creek zone is about 7 km long but is only about 10 m thick. Protoliths for the mylonite were coarse-grained, garnet-hornblende-biotite-plagioclase-quartz gneiss or schist and biotite-quartz amphibolite (Bobyarchick, 1976; Bobyarchick and Glover, 1979) that are Grenville age (Glover and others, 1978) and were metamorphosed under amphibolite-facies conditions during the late Paleozoic. Late-stage, retrograde dynamothermal metamorphism during Alleghanian mylonitization produced greenschist-facies minerals.

**91A.** Garnet porphyroclasts in a quartz-rich layer. During deformation, garnet grains were disarticulated into inequant fragments that were then displaced from the host parallel to the developing mylonite foliation, which is E-W in the photomicrograph. Evidence for disarticulation includes the presence of incipiently displaced fragments (1) and numerous garnet porphyroclasts with sharp, faceted edges (2). Most garnets in the protolith are equant and rounded. Streaks of very fine-grained brown chlorite and opaque minerals in the mylonitic foliation (3) formed during low-grade metamorphism and alteration of garnet grains. The porphyroclasts and chlorite films are set in a matrix of recrystallized quartz ribbons (4) that also define the mylonite foliation. Partial NX

**91B.** Micaceous mylonite with plagioclase porphyroclasts. This photomicrograph shows some detail in the very fine-grained brown chlorite and opaque matrix, which is commonly dark even in plane polarized light. Twinned and untwinned plagioclase porphyroclasts (1) are enclosed by a dark chloritic matrix that also contains epidote, pyrite, and scarce biotite porphyroclasts. The protolith of this mylonite was probably a biotite amphibolite. Plagioclase grains range in shape from elliptical to nearly equant and have rounded margins. The mylonite foliation is defined by the preferred orientation of recrystallized quartz lenses (2), trains of fine-grained opaque dust in the matrix (3), and long axes of plagioclase porphyroclasts. NX

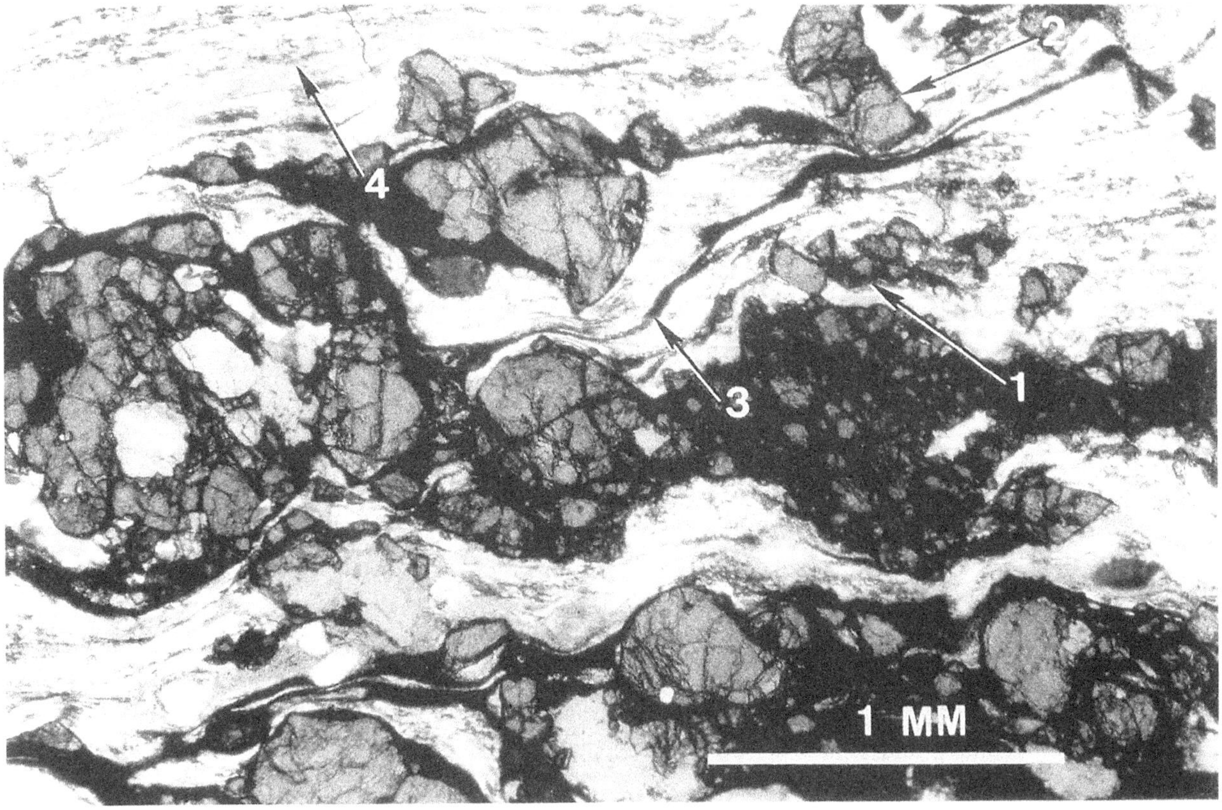


91 A

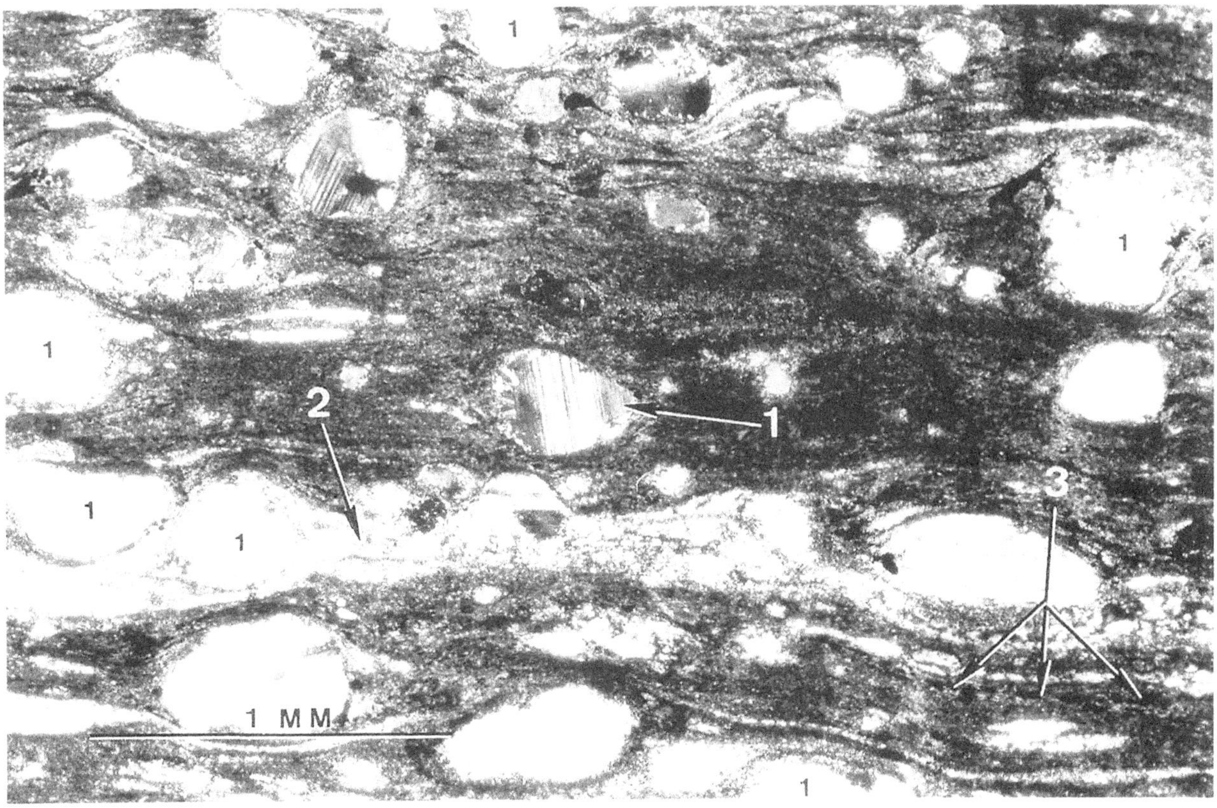


91 B

# 92. FOLIATION HETEROGENEITIES IN MYLONITES

ANDY R. BOBYARCHICK

Mylonites in the Dover Creek zone, Virginia, U.S.A., contain numerous intrafolial folds as a result of flow in a ductile matrix containing rigid porphyroclasts. These folds, which are best defined by isoclinally folded quartz-rich layers, have axial surfaces that, in hand specimen, lie approximately parallel to the mylonite foliation. At the scale of a thin section, the axial surfaces are variably oriented.

**92A.** Porphyroclasts and folds in a chlorite-quartz-plagioclase mylonite. A large potassium feldspar porphyroclast (1) has a folded tail made of recrystallized quartz (2) that lies roughly in the mylonite foliation (E-W in photomicrograph). The quartz tail is folded as a result of interference by an adjacent feldspar porphyroclast. Note that a biotite grain above the quartz tail has deformed by basal slip (3). Interlayered mica films and recrystallized quartz ribbons in the mylonite foliation (4) are deviated by a feldspar porphyroclast (5) located at the boundary between a quartz-rich layer (above) and the dark matrix (below). NX

**92B.** Round to elliptical plagioclase porphyroclasts with rounded margins are enclosed by a matrix that consists of very fine-grained chlorite and opaque minerals. The isoclinally folded layers are lenses of recrystallized quartz (1). PPL

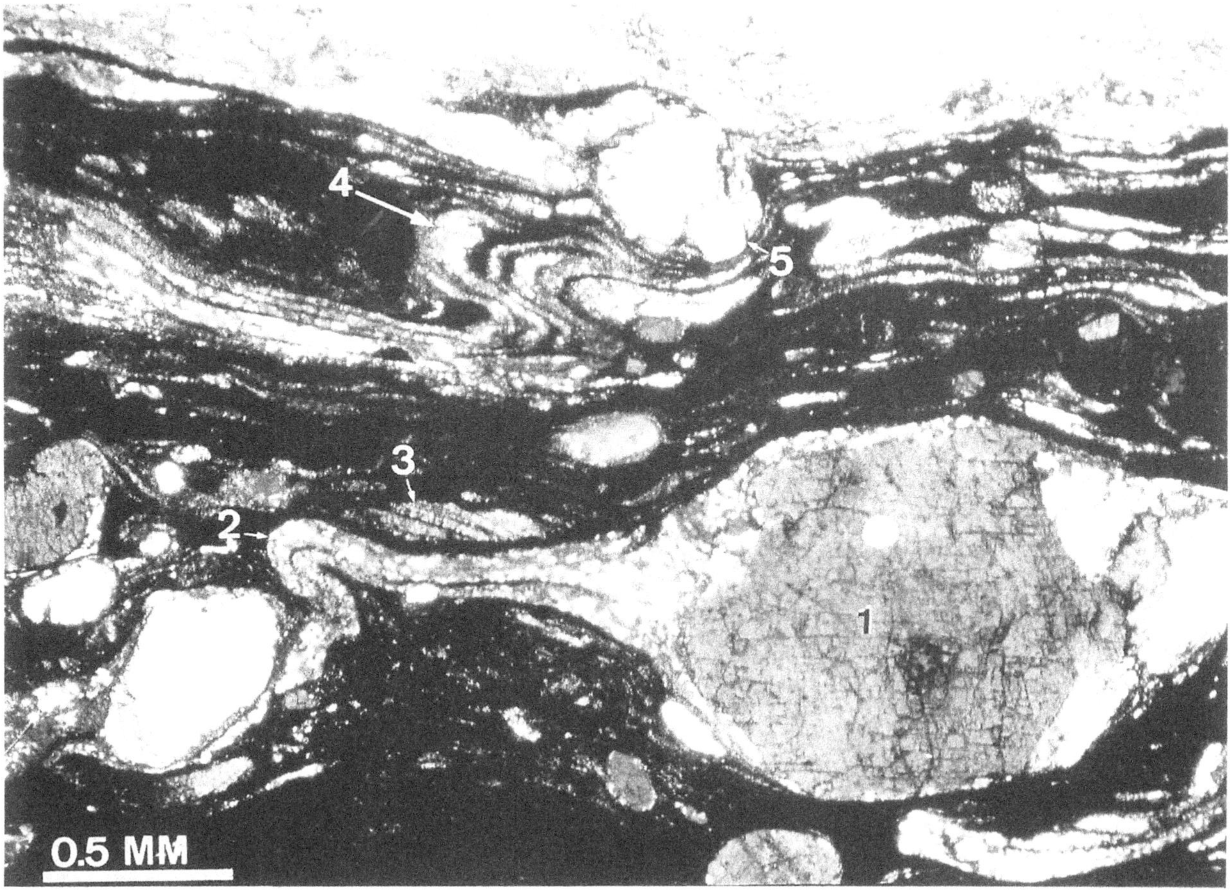

92 A

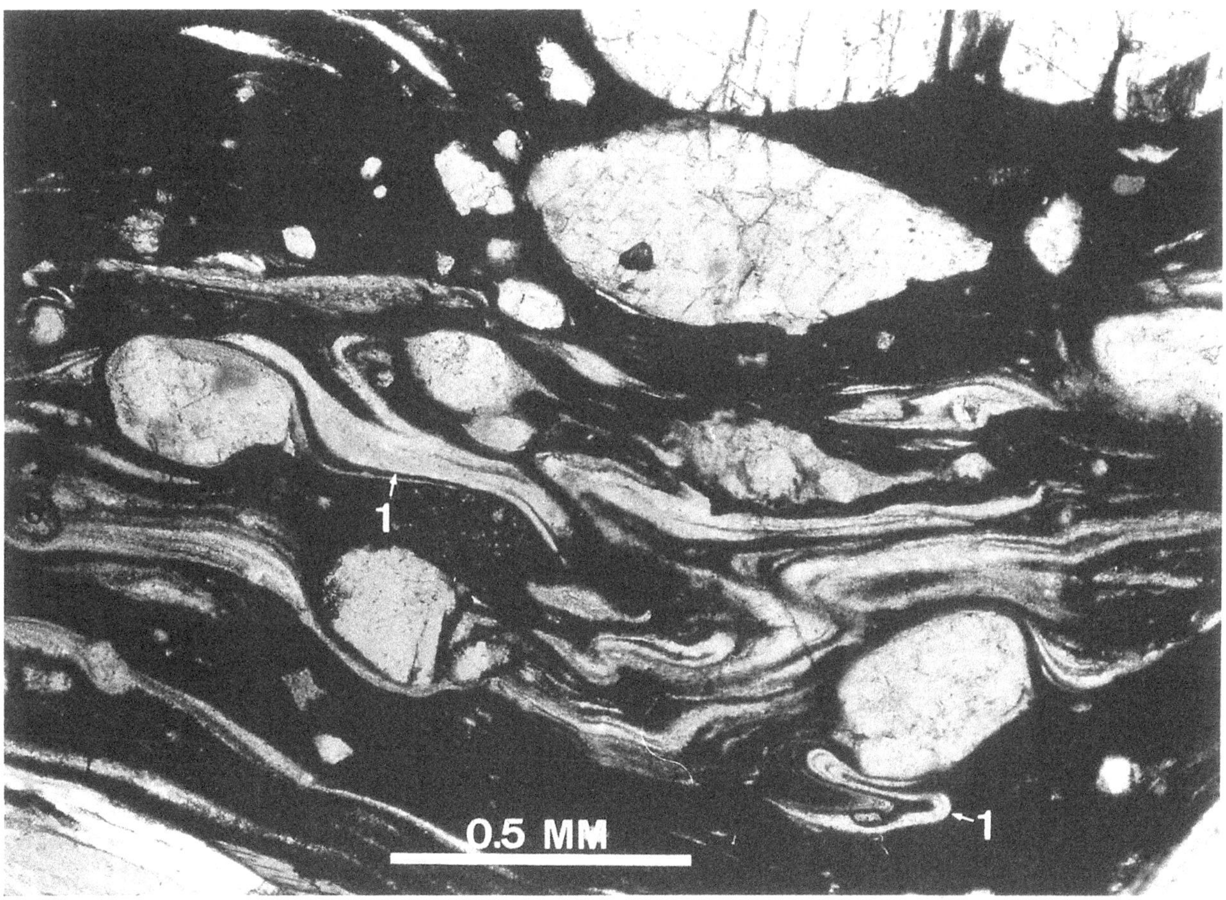

92 B

# DUCTILE BEHAVIOR

## Composite foliations in mylonites



Common abbreviations used in regard to the plates in this section include PPL = plane polarized light and NX = crossed nicols.

An important microstructural feature of many mylonitic rocks is composite foliations. In a widely cited paper on mylonitic granitic rocks, Berthé and others (1979a) introduced the terminology S (for *schistosité* or plane of flattening) and C (for *cisaillement* or plane of shearing). The classic S-C fabric, as originally described by Berthé and others (1979a), was interpreted as a shear-strain-sensitive fabric in that the angle between S and C became smaller with increased strain. One contribution within this set of plates documents this strain-sensitive relationship in S-C granites from the South Armorican shear zone (Brittany, France), the locality originally discussed by Berthé and others (1979a). Other composite fabrics are not strain sensitive and have been described in impure quartzites by Lister and Snoke (1984); such composite fabrics also have been reproduced in laboratory shearing experiments. A variety of names exists in the literature for secondary foliations in highly strained rocks, including C′ foliation, shear band foliation, normal- and reverse-slip crenulation cleavage, and extensional crenulation cleavage. These structures differ from C surfaces as defined by Berthé and others (1979a) because they form oblique to the boundaries of the shear zone. The secondary foliations typically form late in the deformational history of the shear zone, and can occur with S-C texture, thereby resulting in an S-C-C′ fabric as illustrated in several plates. Both S-C fabric and the secondary foliations typically provide consistent sense-of-shear data.

# 93. S-C FABRIC IN GRANITES

D. GAPAIS AND P. JÉGOUZO

The South Armorican shear zone (Brittany, France) is a major dextral transcurrent shear zone of Hercynian age. Syntectonic granites were mylonitized during their emplacement along the shear zone (Figure 93.1) and during their subsequent cooling from upper amphibolite-facies to greenschist-facies metamorphic conditions (around 500 to 300 MPa and 550°C–450°C) (Jégouzo, 1980). The resulting general strain pattern (see diagram on the adjacent photo plate) is well represented by a model of bulk heterogeneous simple shear (Berthé and others, 1979a,b), involving the development of a foliation S (preferred orientation of flattened grains interpreted as principal strain plane) and of narrow (millimeter to centimeter wide) dextral shear bands (C surfaces) which lie close to the bulk shearing plane. Both structures are always found together, are locally seen to develop from isotropic granite, and form during a single deformational event.

**93A–93E.** All sections shown here are cut perpendicular to both foliation and shear bands and parallel to the stretching lineation (preferred orientation of elongate grains). The polished slabs 93A through 93E show the same leucogranite protolith in undeformed state (93A) and four increasingly sheared states [orthogneiss (93B, 93C) to ultramylonite (93E)] as one approaches a fault (a, on Figure 93.1). The mineral phases are quartz, K-feldspar, plagioclase ($An_{10-15}$), brown biotite, muscovite, and accessory minerals. Increasing strain is marked by a gradual decrease in mean angle between foliations (S) and (C) (from about 45° to 0°), by a decrease in spacing between shear bands (C surfaces), and by a marked reduction in grain size. Relict porphyroclasts of K-feldspar (KF) are observed in the ultramylonite.

The specimens and data which are the focus of this contribution are from the combined research of D. Berthé, P. Jégouzo, P. Choukroune, D. Gapais, and J. P. Cogné.

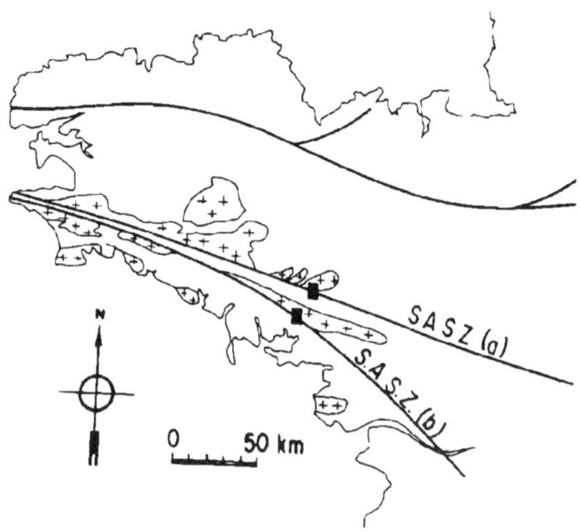


**Figure 93.1.** Sketch map of the South Armorican shear zone (S.A.S.Z.) in Brittany, France, black rectangles, sampling areas; crosses leucogranites, deformation zone across fault (a) and fault (b) is between 1.5 and 2 km wide

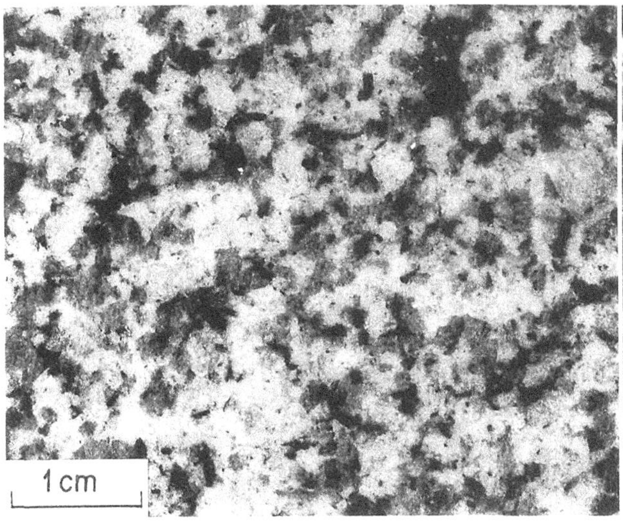


93 A

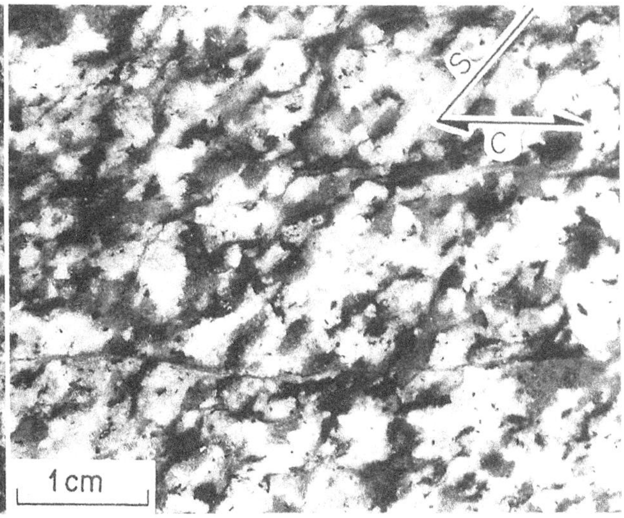


93 B

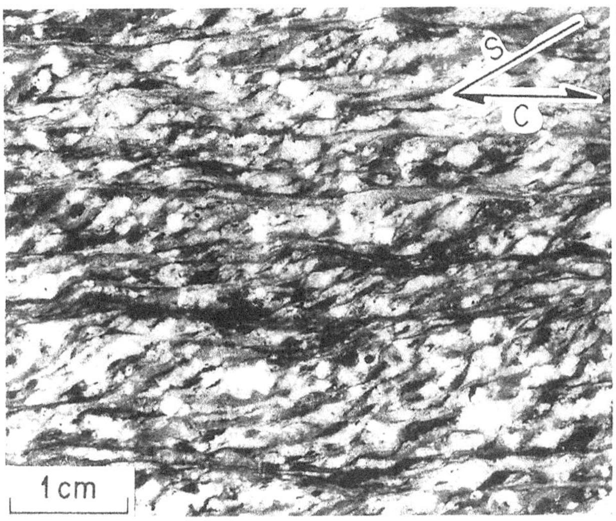


93 C

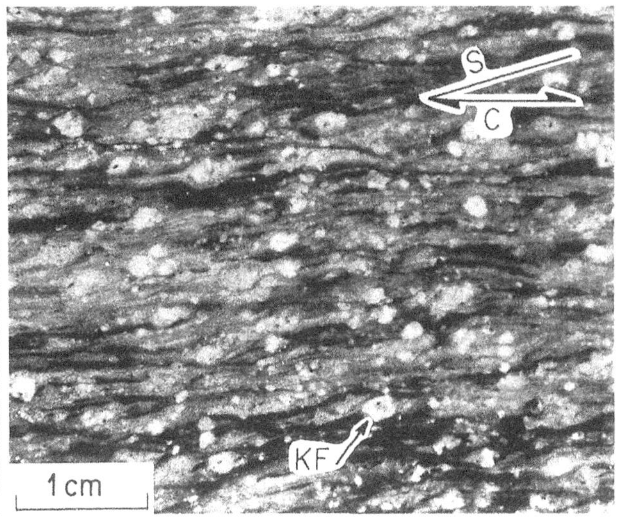


93 D

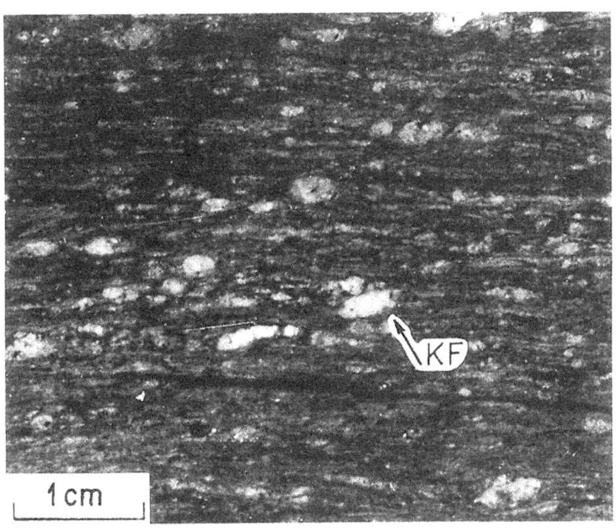


93 E

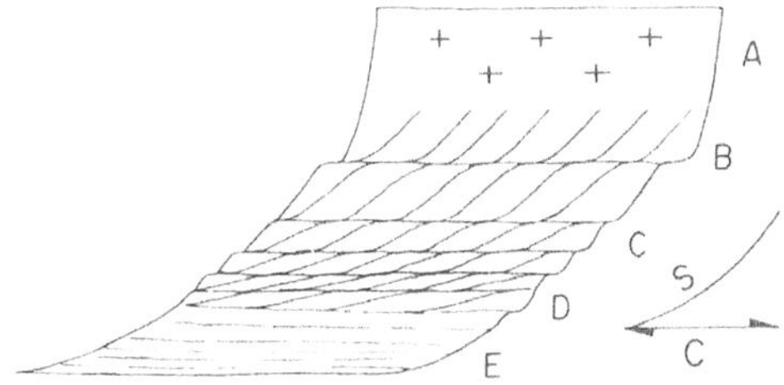

Mesoscopic structures associated with ductile shear deformation within South Armorican granites. Polished slabs cut as in 93A–93E.

**93F.** Small-scale strain gradient from isotropic granite (top of photograph) to ultramylonite (bottom of photograph). The gradual change in the relative attitude between foliation (S) and shearing planes (C) with increasing strain is consistent with a model of progressive shearing where the structures develop together at low strain.

**93G.** Ultramylonite formed in two-mica granite (fault b on Figure 93.1). The starting material is a homogeneous granitic rock (quartz, K-feldspar, plagioclase, biotite, and muscovite). A compositional banding (quartzo-feldspathic layers and mica-rich layers) formed during mylonitization and further shearing developed passive asymmetric folds. Feldspar porphyroclasts (KF) are preserved locally within the ultrafine matrix.

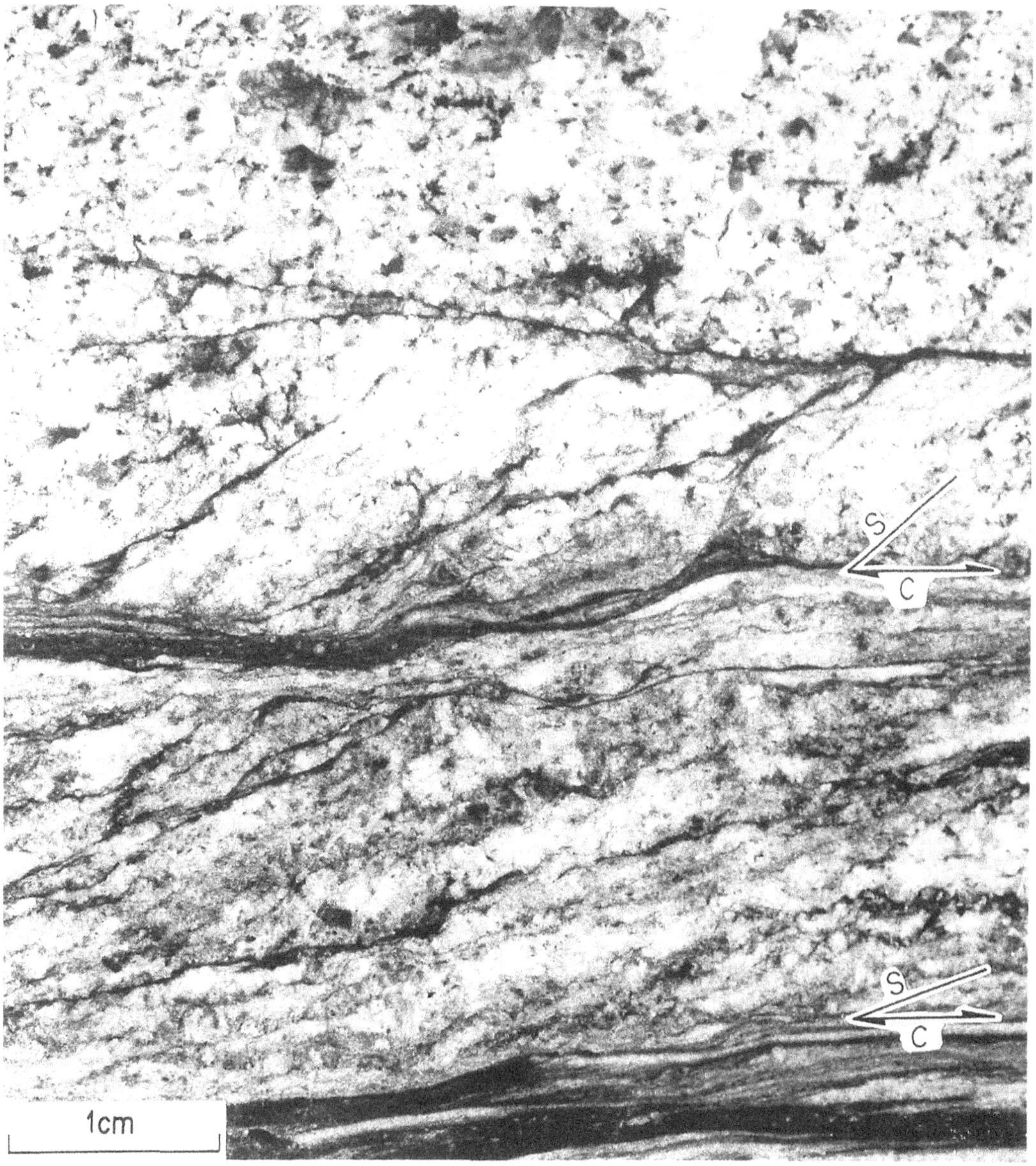

93 F

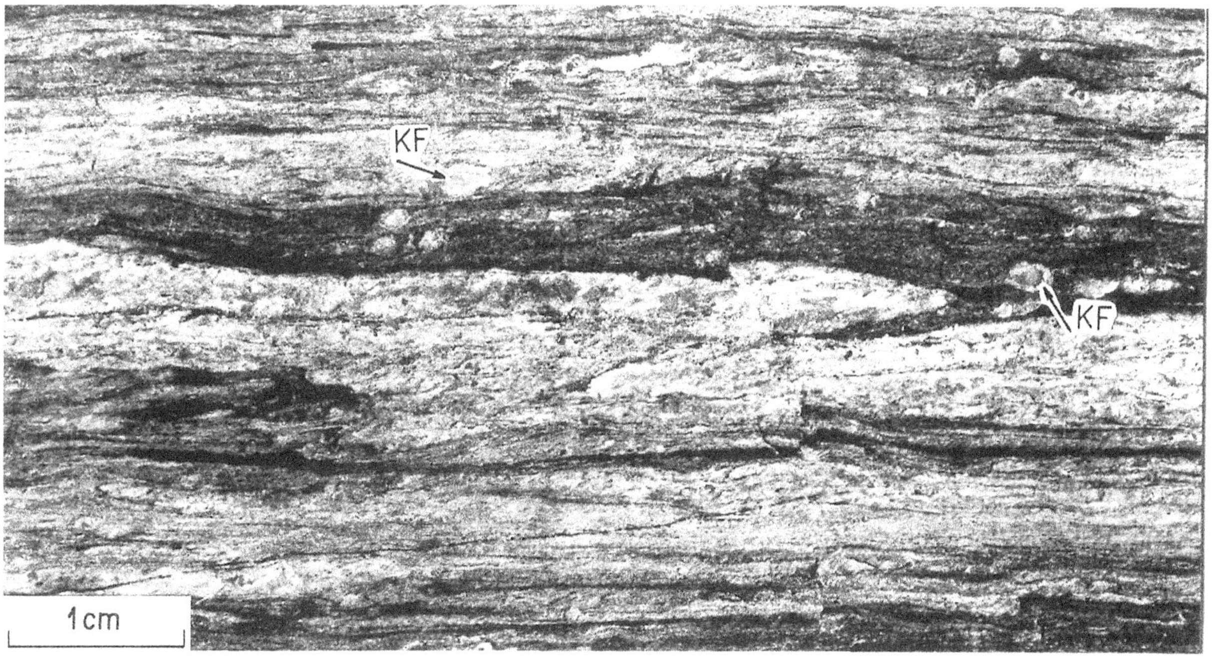

93 G

Microscopic aspects of ductile shear within South Armorican leucogranites. Photomicrographs of thin sections cut as in 93A–93E.

**93H.** Moderately deformed two-mica granite (Plate 93B). Narrow C surfaces are marked by phyllosilicates (biotite and muscovite) and show incipient reduction in grain size. The foliation outside the C surfaces is defined by the grain-shape fabric of the feldspars (F) and the trend of cleavage planes of micas (Ms). PPL

**93I.** More advanced stage of deformation (Plate 93C). C surfaces are wider than in 93H and show considerable reduction in grain size. C surfaces are mainly composed of small, dynamically recrystallized quartz grains and phyllosilicates (biotite and muscovite) aligned parallel to the shearing plane. Relict porphyroclasts of feldspars (F) and phyllosilicates (Bi) are observed outside the C surfaces. Polycrystalline quartz ribbons (Qz) have developed from original quartz grains. Partial NX

93 H

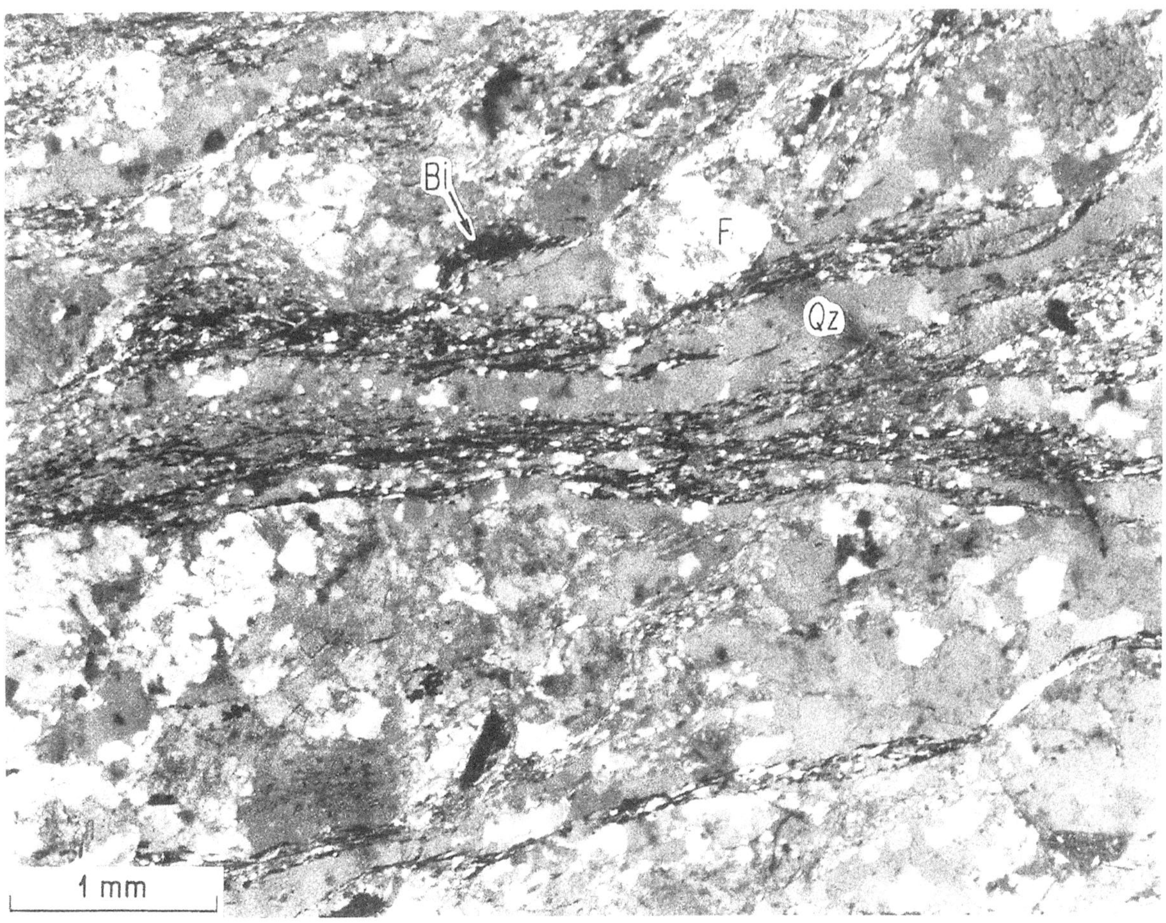

93 I

# 94. COMPOSITE PLANAR FABRICS AND ASYMMETRIC DEVELOPMENT OF MYRMEKITE IN ORTHOGNEISS

CAROL SIMPSON

Hand specimen photograph and photomicrographs of mylonitic gneiss from Palm Canyon, California, U.S.A. (Simpson and Schmid, 1983; Simpson, 1984, 1985; Simpson and Wintsch, 1989).

**94A.** Hand specimen from the Santa Rosa mylonite zone, part of a major, late Cretaceous to early Tertiary, west-directed thrust zone affecting tonalites and granodiorites of the southern California batholith (Simpson, 1984, 1985). Deformation occurred at temperatures of 495–540°C (Simpson and Wintsch, 1989) and produced two foliation surfaces: S (*schistosité*) and C (*cisaillement*, shear) planes (Berthé and others, 1979a; Simpson and Schmid, 1983; Lister and Snoke, 1984). S planes are defined by aligned biotite (black), quartz ribbons (gray), and feldspar (white) grain aggregates. They are thought to represent relict XY planes of a relatively low bulk shear strain. The C planes are discrete, millimeter-thick zones of higher strain that deflect the S planes with a curvature consistent with the shear sense of the overall shear zone. C planes, also known as "extensional crenulation cleavage" (Platt and Vissers, 1980), allow the rocks to extend parallel to foliation and imply a general shear state; i.e., simple shear plus a component of pure shear acting perpendicular to the foliation (Simpson and De Paor, 1993).

**94B.** Photomicrograph of hand specimen shown in 94A. The S planes are defined by ribbonlike quartz aggregates (Q), aligned biotite flakes (B), and minor muscovite. Both recrystallized and relict feldspar porphyroclasts (F) occur between S planes. There is a progressive decrease in grain size of all minerals by syntectonic recrystallization processes where the S planes curve asymptotically into the left-lateral C planes. High-angle and polygonal grain boundaries in quartz ribbon aggregates suggest that grain-boundary migration recrystallization (Regime 3 of Hirth and Tullis, 1992) permitted a complete recovery of lattice strain, which is consistent with the known deformation conditions. NX

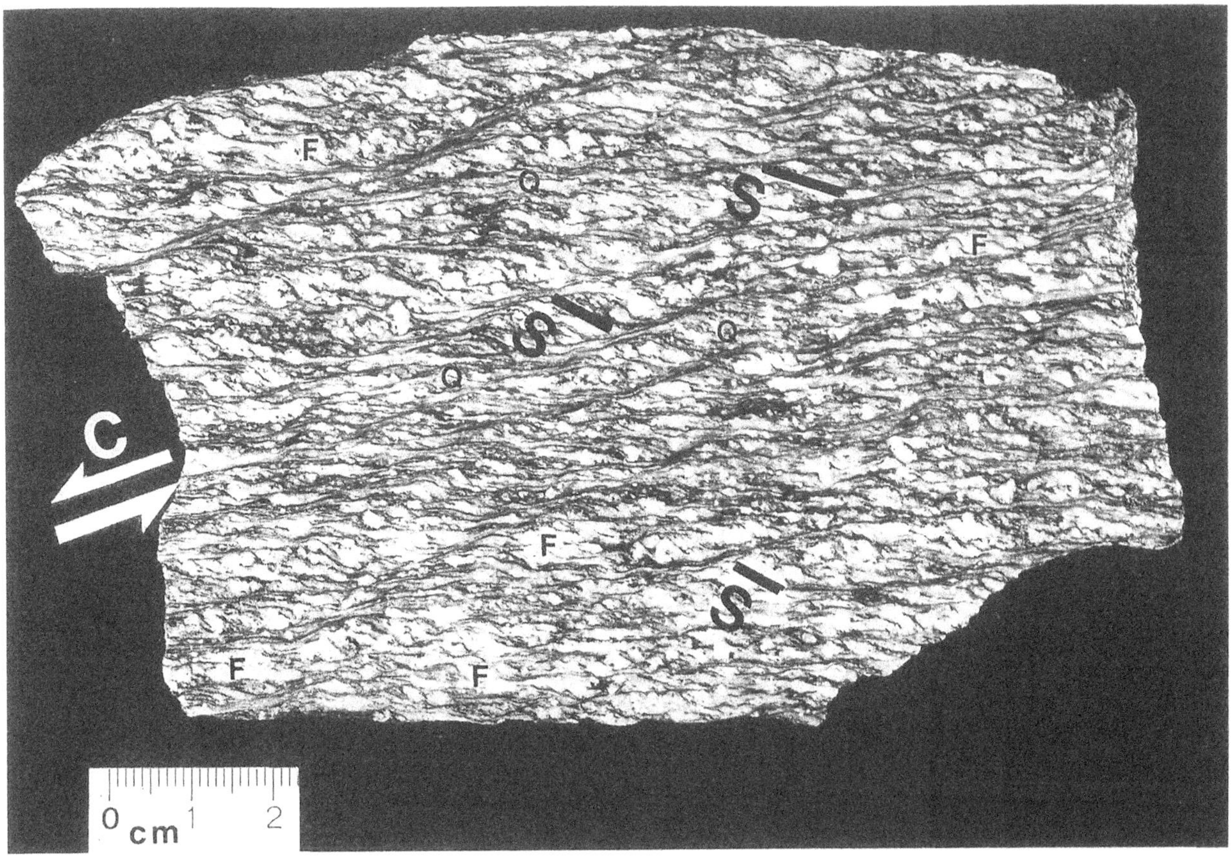
F
Q
S
S
F
Q
Q
C
F
S
F
F
0 cm 1 2

94 A

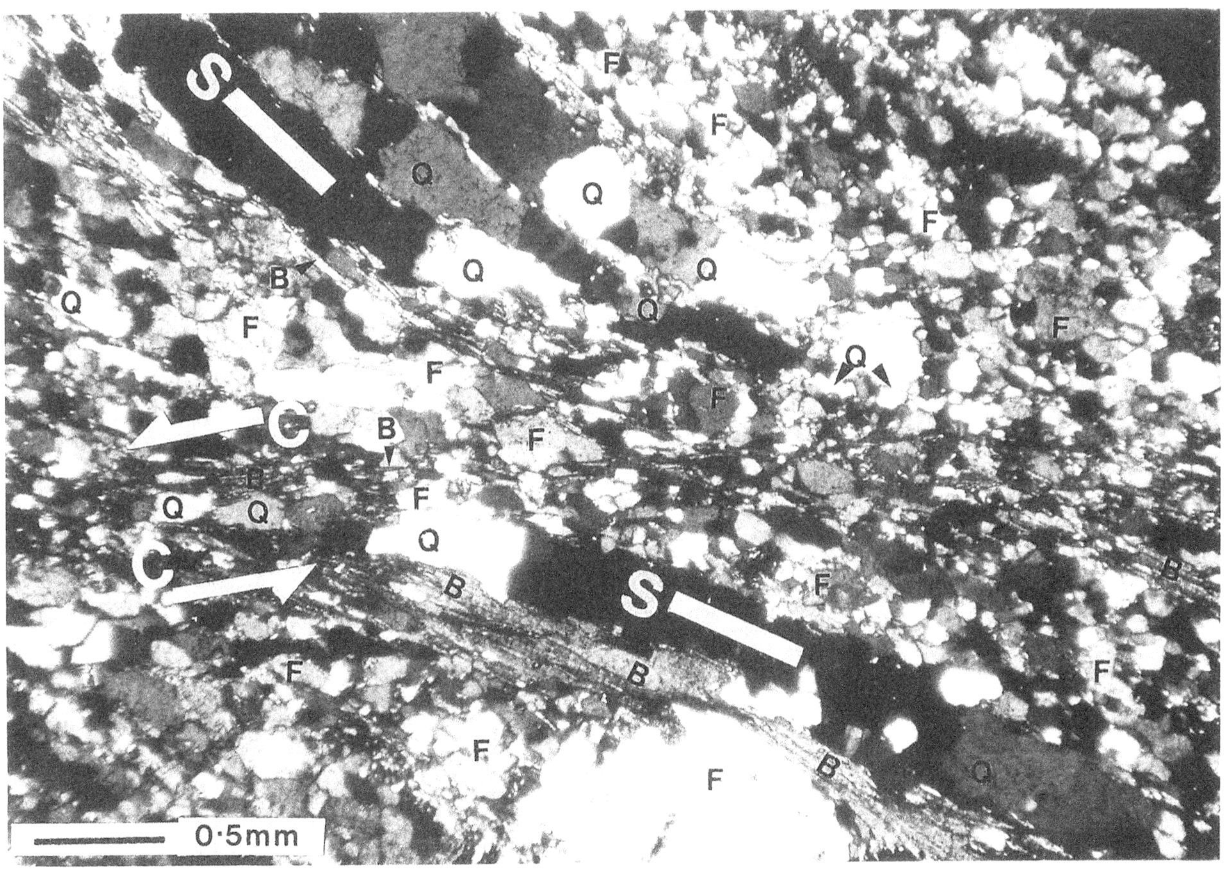
S
F
F
Q
Q
F
B
Q
Q
Q
F
Q
F
F
C
B
F
F
Q
Q
F
C
Q
B
S
F
B
F
B
F
F
F
B
Q
0·5mm

94 B

**94C.** A potassium feldspar porphyroclast (F) preserved between the S planes of 94A shows myrmekitic intergrowths (my) of oligoclase and quartz on the sides parallel to S. Myrmekite is not developed on the ends of the porphyroclast that face the inferred incremental extension direction; in this position, there are equant, polygonal, optically strain-free K-feldspar grains that probably formed by rotation recrystallization. NX

**94D.** Myrmekitic intergrowths (my) on the margins of K-feldspar porphyroclasts (F) curve to the left along C planes in the matrix of 94A, consistent with the known sense-of-shear in this sample, and indicate a syntectonic reaction of K-feldspar to oligoclase and quartz (Simpson and Wintsch, 1989). This reaction produces a fine-grained, reaction-softened matrix of very small, new, initially strain-free quartz and feldspar grains adjacent to the porphyroclasts. NX

**94E.** A bimodal distribution of grain sizes is visible in the fine-grained matrix between K-feldspar porphyroclasts (F) and quartz ribbons in 94A. Many of the 60 μm sized recrystallized grains of feldspar (rf) in the matrix themselves contain myrmekitic intergrowths of oligoclase and quartz (my). Where this reaction has gone to completion, the grain size is reduced to <5 μm. The rock strength is reduced in these regions of small, relatively strain-free new grains because (a) the small new grains are weaker than the original high-dislocation-density porphyroclasts, and (b) their small size facilitates both dislocation creep and grain-boundary sliding deformation mechanisms. Grain-boundary migration in the small quartz grains has locally produced new polygonal grain shapes (arrowed). NX

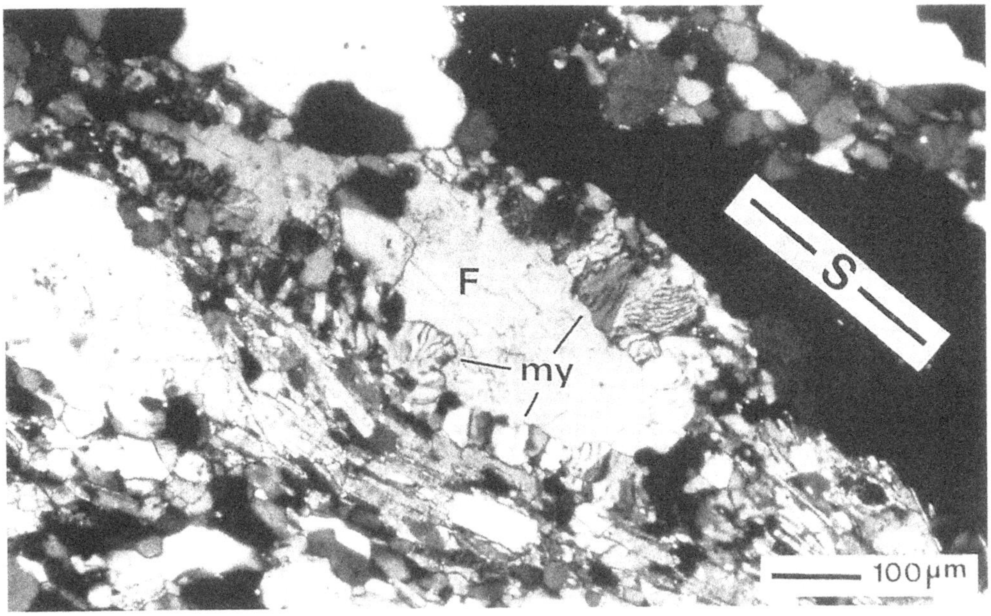

94 C

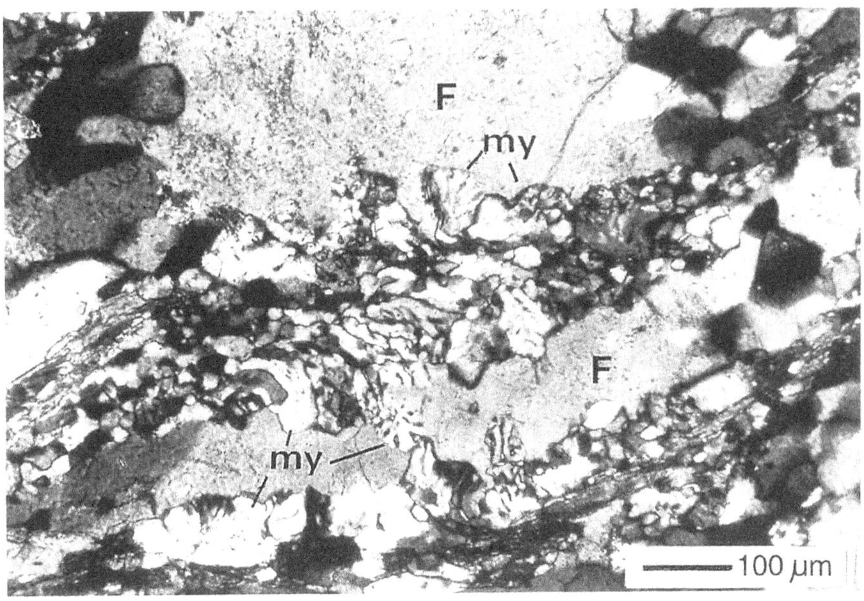

94 D

94 E

# 95. S-C FABRICS FORMED IN SYNTECTONIC PLUTONS DURING THE TRANSITION FROM MAGMATIC TO HIGH-TEMPERATURE SUBSOLIDUS CONDITIONS

ROBERT B. MILLER AND SCOTT R. PATERSON

The margins of plutons intruded into active fault zones at mid-crustal or deeper levels should ideally record deformation through the transition from magmatic to near-solidus to high-temperature subsolidus conditions (e.g., Miller and Paterson, 1994). Microstructural evidence for this deformational history includes a strong magmatic foliation variably overprinted by high-temperature, solid-state S-C fabrics. Such a history is well displayed by the 93 to 96 Ma Mount Stuart batholith of the North Cascades, Washington State, U.S.A. (MSB on Figure 95.1). The northeast margin of this batholith is strongly deformed within a 200 m wide belt of the Tumwater Mountain shear zone. This steep, reverse-slip shear zone places wall rocks of the Late Jurassic, ophiolitic Ingalls complex over the northeast margin of the batholith. Protomylonitic hornblende-biotite tonalites within the shear zone display a strong magmatic foliation overprinted by amphibolite-facies assemblages (hornblende-biotite-plagioclase-quartz) that define solid-state S-C fabrics. The lack of retrogression together with geochronological data indicate that deformation was synchronous with the emplacement of the batholith (Miller and Paterson, 1992, 1994).

The photomicrographs on the facing page illustrate the evidence for strong magmatic, near-solidus, and high-temperature subsolidus deformation in tonalites of the Mount Stuart batholith in the Tumwater Mountain shear zone.

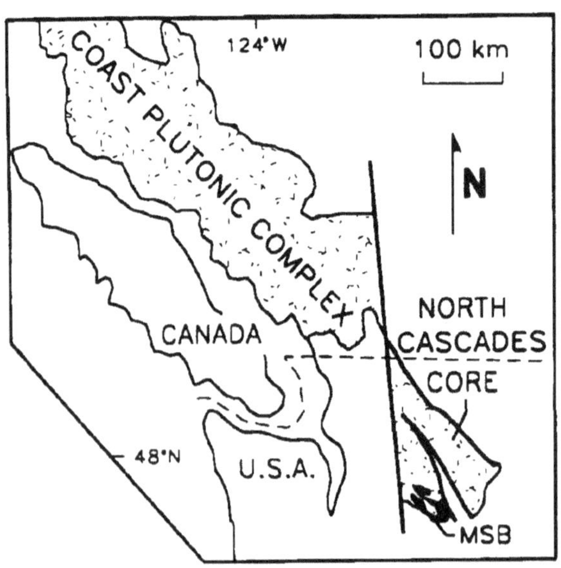


**Figure 95.1.** Simplified map showing location of the Mount Stuart batholith (MSB) at the southern margin of the Late Cretaceous and Paleogene crystalline core of the North Cascades, which is the southeast extension of the ~1800 km long Coast Plutonic complex.

**95A.** Strong magmatic foliation and incipient S-C fabric in tonalite, Mount Stuart batholith, North Cascades, Washington State, U.S.A. (Miller and Paterson, 1992, 1994). Magmatic foliation is well displayed by strongly aligned, euhedral, tabular plagioclase. Igneous biotite and hornblende also lie in the S plane. Solid-state deformation is marked by elongate aggregates of recrystallized quartz, biotite, and hornblende (difficult to distinguish at this scale). Incipient C surfaces are defined by biotite, hornblende, and finely recrystallized quartz and plagioclase. NX

**95B.** S-C protomylonitic tonalite showing orientation of relict plagioclase and evidence for near-solidus (submagmatic) deformation in the Mount Stuart batholith. This sample is extensively recrystallized, but relict tabular, euhedral, zoned igneous crystals of plagioclase preserve magmatic foliation (S) that has been overprinted by a parallel, solid-state foliation. Similar plagioclase grains with twin planes oriented parallel to crystal boundaries are also aligned in C surfaces. A fine-grained mosaic of recrystallized plagioclase, which is nearly identical in composition to rims of the igneous plagioclase ($\sim An_{27-28}$), has variably replaced bent igneous crystals in both orientations. Recrystallization was probably by grain-boundary migration (Miller and Paterson, 1994). We infer that the large plagioclase crystals rotated into C orientations late in the crystallization history of the tonalite, but while sufficient melt was still present to facilitate rotation. Continued shear as the system "locked up" due to further crystallization resulted in the concentration of shear strain into C surfaces. This partitioning of strain continued to varying degrees during subsolidus crystal-plastic flow. NX

**95C.** Plagioclase fabric in S-C protomylonitic tonalite of the Mount Stuart batholith. This sample shows more extensive recrystallization than that in 95B, but strongly bent igneous plagioclase marks a relict magmatic foliation and several igneous grains lie in C surfaces. The plagioclase is interpreted to have rotated into the C surfaces under near-solidus conditions (analogous to the case of 95B). NX

95 A

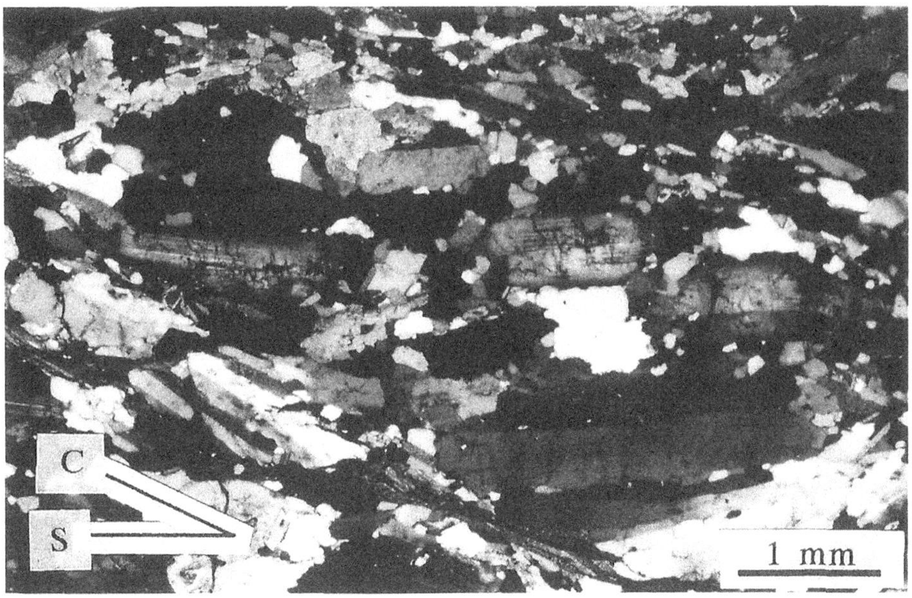


95 B

95 C

# 96. MYLONITIC DEFORMATION OF A GRANITOID ROCK

R. H. VERNON

The predominantly peraluminous Wologorong batholith is one of several elongate, deformed batholiths of Silurian-Devonian age in the Lachlan fold belt of eastern New South Wales, Australia. The batholith has undergone heterogeneous tectonic deformation, mainly at upper greenschist-facies conditions, and in strongly deformed zones has two foliations. One is a gneissic foliation ($S_1$) consisting of relatively discontinuous, fine-grained, micaceous and feldspathic folia anastomosing between feldspar relics and elongate aggregates of recrystallized quartz. The other ($S_2$) is more continuous, the folia being fine-grained and commonly laminated (mylonitic). The foliations could have formed simultaneously or sequentially during a single deformation, in which case they could be labeled S and C instead of $S_1$ and $S_2$. $S_2$ formed obliquely (up to 30°) to $S_1$, apparently in order to accommodate a shear component of the strain in strongly deformed parts of the batholith (Vernon and others, 1983). $S_2$ folia appear to have been initiated on suitably oriented parts of anastomosing $S_1$ folia. Continued deformation locally reduced the angle between $S_1$ and $S_2$ and resulted in the development of mylonite zones. Grain-size reduction by recrystallization and/or neocrystallization involved all minerals during the deformation, but quartz attained a steady-state grain size of about 0.15 mm, which was not reduced further during later, more intense deformation. Inferred deformation mechanisms include microfracturing of feldspar; intragranular slip in internally deformed, elongate quartz and mica grains; and solution transfer in the formation of lenticular mica grains and possibly micaeous folia truncating zoned plagioclase.

**96A.** Typical outcrop of deformed granitoid with two foliations in the Abercrombie River area. A gneissic foliation ($S_1$) is apparently displaced, with a dextral sense of rotation, by a more continuous, finely laminated, mylonitic foliation ($S_2$). The average angle between the two foliations is about 30°. In places, $S_1$ is also finely laminated, suggesting local reactivation during the production of $S_2$ folia. Australian 20 cent coin (28 mm diameter) for scale. From Vernon and others (1983, figure 4B).

**96B.** Deformed granitoid from the same area as the rock illustrated in 96A, showing locally increased strain that produced a laminated mylonite zone (center). In the upper part of the photograph, $S_1$ and $S_2$ folia are generally distinct, but in the lower part the strain is greater, resulting in more continuous, laminated $S_1$ folia (averaging 15° to 20°) between $S_1$ and $S_2$. Australian 20 cent coin (28 mm diameter) for scale. From Vernon and others (1983, figure 4C).

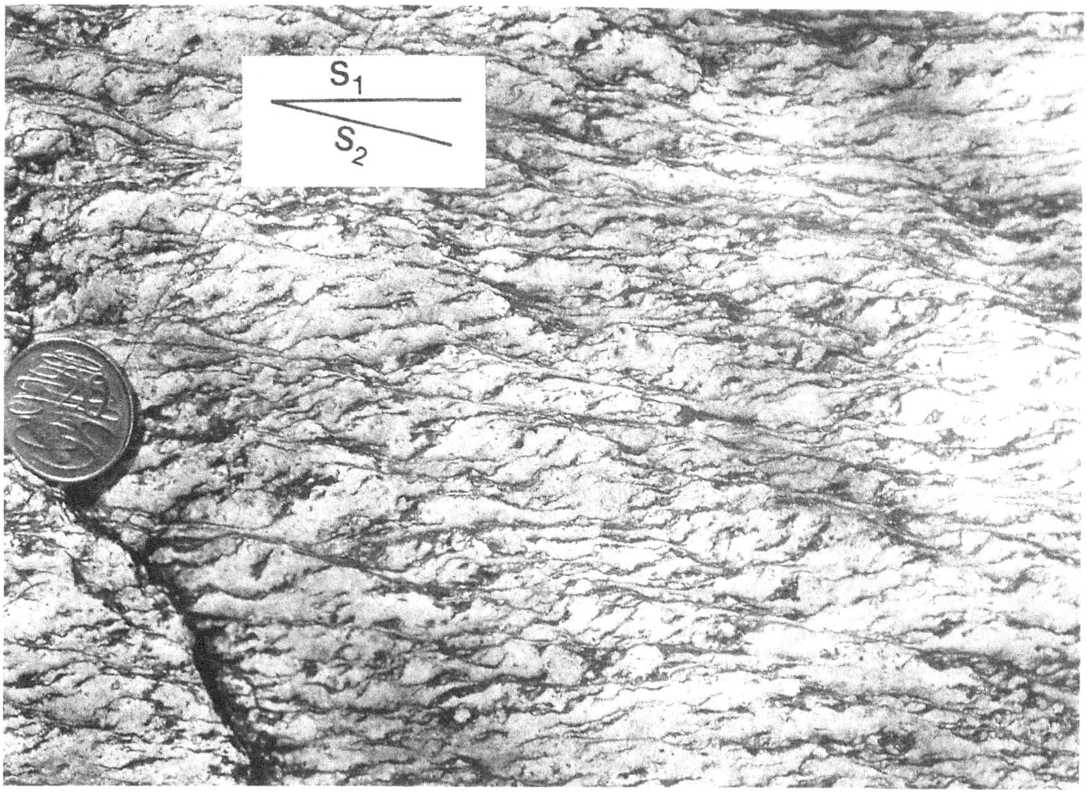


96 A

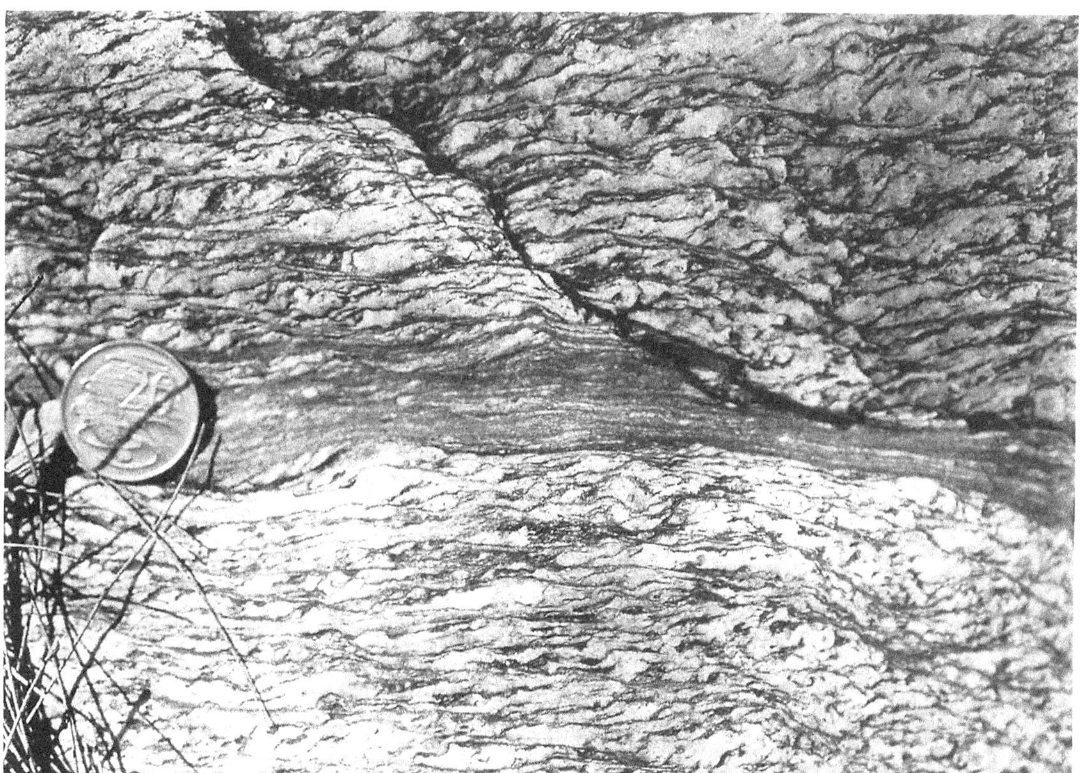

96 B

**96C.** Photomicrograph of rock similar to that shown in 96A, showing relatively continuous, fine-grained, mylonitic $S_2$ folia contrasting with more discontinuous, partly coarser-grained $S_1$ folia. The ends of two $S_2$ folia are linked by an $S_1$ folium. Lenticular relics of deformed biotite, partly neocrystallized to new biotite, muscovite, and sphene, occur in the $S_1$ folia (above center, bottom right). $S_2$ folia consist mainly of neocrystallized aggregates of biotite, quartz, muscovite, and feldspar. PPL. From Vernon and others (1983, figure 7C).

**96D.** Mylonite (similar to that shown in 96B) consisting of quartz, feldspar, and mica, interspersed with resistant relics of clouded plagioclase (P), patches and folia of coarser-grained recrystallized quartz (Q), and micaceous folia (M). Vestiges of $S_1$ oblique to $S_2$ are present. Plagioclase relics are progressively recrystallized internally and at their margins in the most strongly deformed rocks, but small relics generally remain. NX. From Vernon and others (1983, figure 7D).

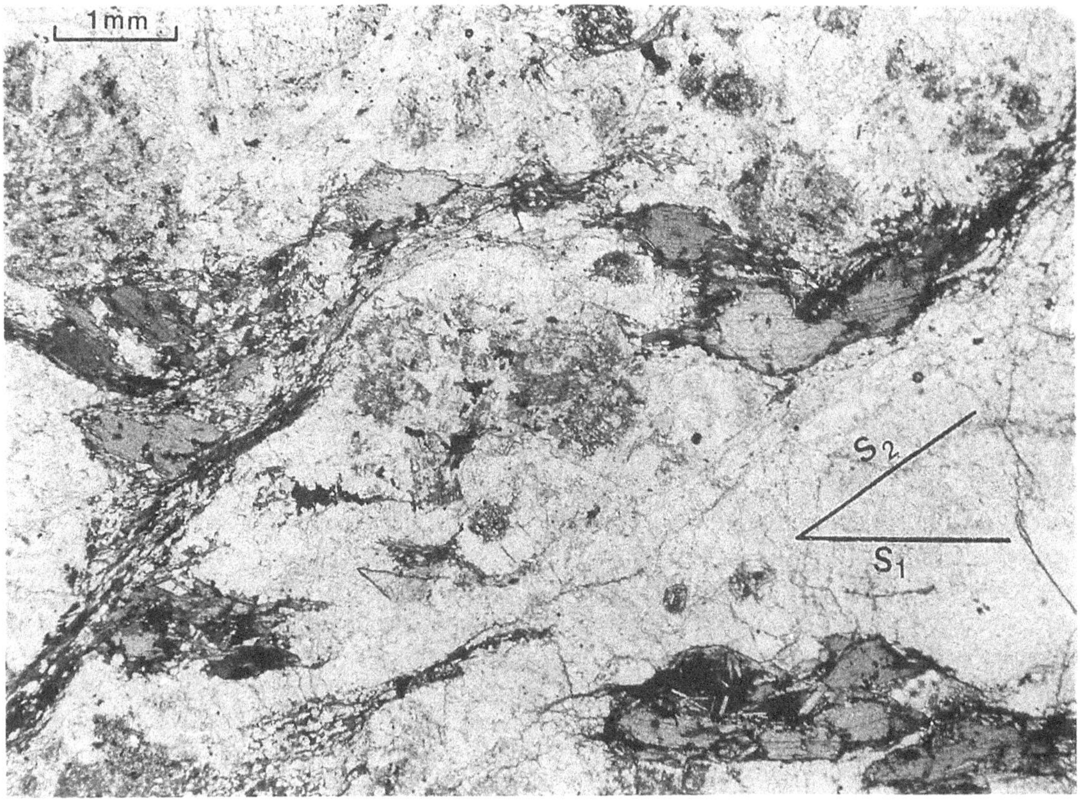


96 C

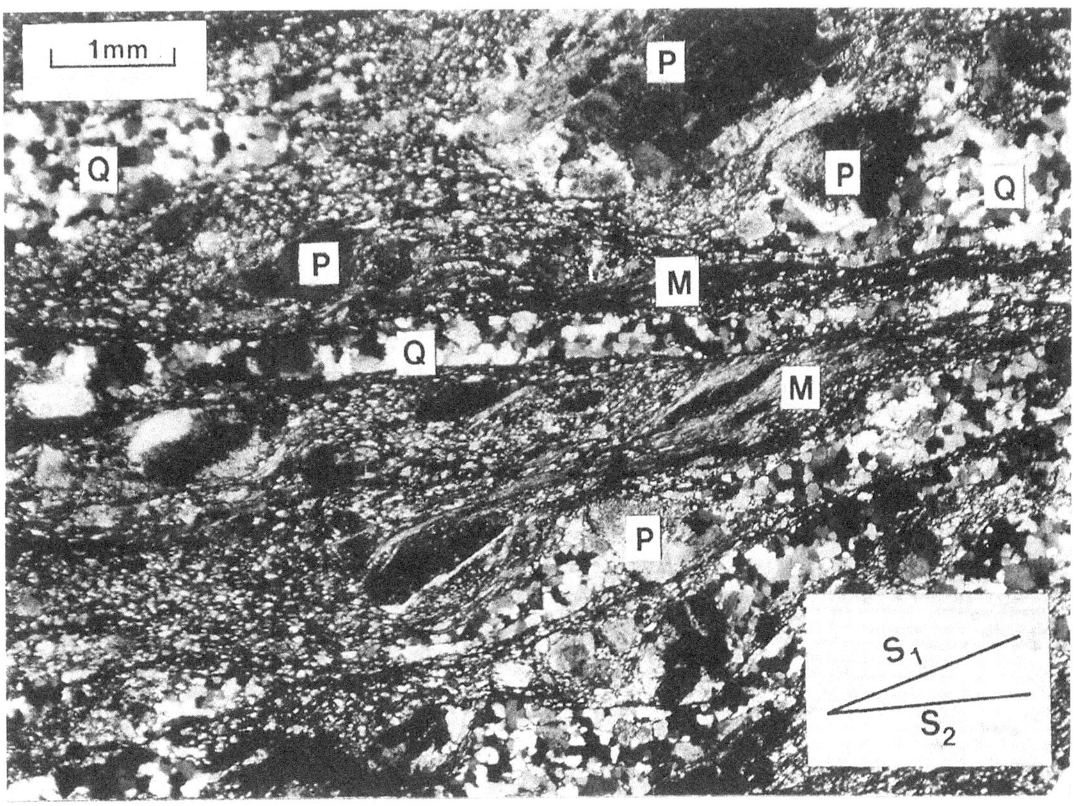


96 D

# 97. LOW-ANGLE CRENULATION CLEAVAGE IN MICACEOUS MYLONITES

ANDY R. BOBYARCHICK

These photographs illustrate low-angle crenulation cleavage in micaceous mylonites from the Brevard zone near Lenoir, North Carolina, U.S.A. The rock is tourmaline-garnet-chlorite-quartz-muscovite schist, where much of the chlorite was derived from alteration of garnet and possibly biotite. Both cleavage and mylonitic foliation formed under conditions of greenschist-facies metamorphism and through the retrogression of amphibolite-facies mineral assemblages. The two foliations are interpreted to be Alleghanian (late Carboniferous and Permian) structures. The relative orientations of the mylonitic foliation (056, 72 SE) and low-angle crenulation cleavage (072, 72 SE) indicate dextral shear in this part of the Brevard zone (Bobyarchick, 1983).

**97A.** Polished slab with low-angle crenulation cleavage. The surface of this slab is perpendicular to both foliations. The mylonitic foliation (parallel to long dimension of paper clip) is an axial surface foliation defined by intrafolial isoclinal folding of a previous foliation. Compositional layering is discontinuous and lenticular because of the isoclinal folds (fold closure at 1) and because of low-angle crenulation cleavage superimposed on the layering and mylonitic foliation (trace at 2). Weathered lenticles of polycrystalline white mica produce the "button" or "fish-scale" texture for which these rocks are known (Roper, 1972). Dark spots are chloritized poikiloblastic garnet porphyroclasts (3). Isolated fold closures are commonly shown by muscovite lenticles. At the scale of the slab, crenulation cleavage is sharply defined by flexures in the mylonitic foliation adjacent to a cleavage trace. The mylonitic foliation between cleavage traces is asymmetrically crenulated. Short, steep limbs of crenulations are parallel to cleavage traces. The maximum angle between a cleavage trace and the mylonitic foliation is about 45° in this segment of the Brevard zone; the statistical average for this angle is 31° (Bobyarchick, 1983). Crenulations in the mylonitic foliation do not fold sharply defined cleavage traces, and no cleavage trace obviously disrupts a crenulation. The crenulations are thus interpreted to be structures that accommodated cleavage formation. At their tips, cleavage traces lose definition by merging with the mylonitic foliation. A low-angle crenulation cleavage plane is not a brittle discontinuity (microfault) nor is it the site of significant volume transfer. The plane is a microscopic ductile shear zone, as can be demonstrated in some Brevard zone mylonites by continuity of deformed quartz veins and muscovite lenticles across cleavage traces. The sense-of-shear along every cleavage trace in the slab is sinistral.

**97B.** Photomicrograph of low-angle crenulation cleavage. The plane of the photomicrograph is perpendicular to cleavage and mylonitic foliation. The mineralogy of the rock is as follows: quartz—clear gray or white grains; muscovite—dark to light gray tabular grains with mottled texture; chlorite—clear dark gray tabular grains; garnet—large poikiloblastic grain; pyrite—large opaque grains; graphite(?)—very small, tabular opaque grains. A single, narrow, sharply defined low-angle crenulation cleavage trace crosses the mylonitic foliation (horizontal) at an angle of 22°. Quartz, white mica, and chlorite are the main minerals within the cleavage zone. Dimensions of quartz grains in deformed veins in the mylonite range between 1.0 and 0.2 mm, but where a vein intersects a cleavage zone, grain size ranges between 0.08 and 0.03 mm (Bobyarchick, 1983). Equivalent scales of grain-size reduction occur in other minerals adjacent to the cleavage zone. Bending of the mylonitic foliation adjacent to the cleavage trace is obvious. Darker gray patches of muscovite grains near the trace reflect a change in optical orientation because of bending and shear parallel to the cleavage. Deformation of the chlorite rim on the garnet porphyroclast is consistent with sinistral shear along the cleavage trace. Note that the garnet grain contains an internal foliation that passes into the chlorite rim without deflection but that the internal foliation is deformed within the cleavage zone. Partial NX

The sequence of events leading to formation of low-angle crenulation cleavage in the Brevard zone is (1) deformation and metamorphism under amphibolite-facies conditions; (2) formation of mylonitic foliation, under greenschist-facies conditions, by isoclinal folding and shear parallel to foliation; (3) formation of cleavage under greenschist-facies conditions. The cleavage is spaced at intervals of between 1 and 6 cm even though the mylonitic foliation, which formed at the same metamorphic grade, is penetrative. The low-angle cleavage is a result of shear instabilities and localized softening processes that operated in the late stages of ductile deformation in the Brevard zone.

Low-angle crenulation cleavage occurs in many ductile shear zones and is known by several names. Recent names include C or C′ (Berthé and others, 1979a,b), extensional crenulation cleavage (Platt and Vissers, 1980), shear bands (White and others, 1980), oblique crenulation cleavage (Bobyarchick, 1983), and normal-slip crenulation (Dennis and Secor, 1987).

97 A

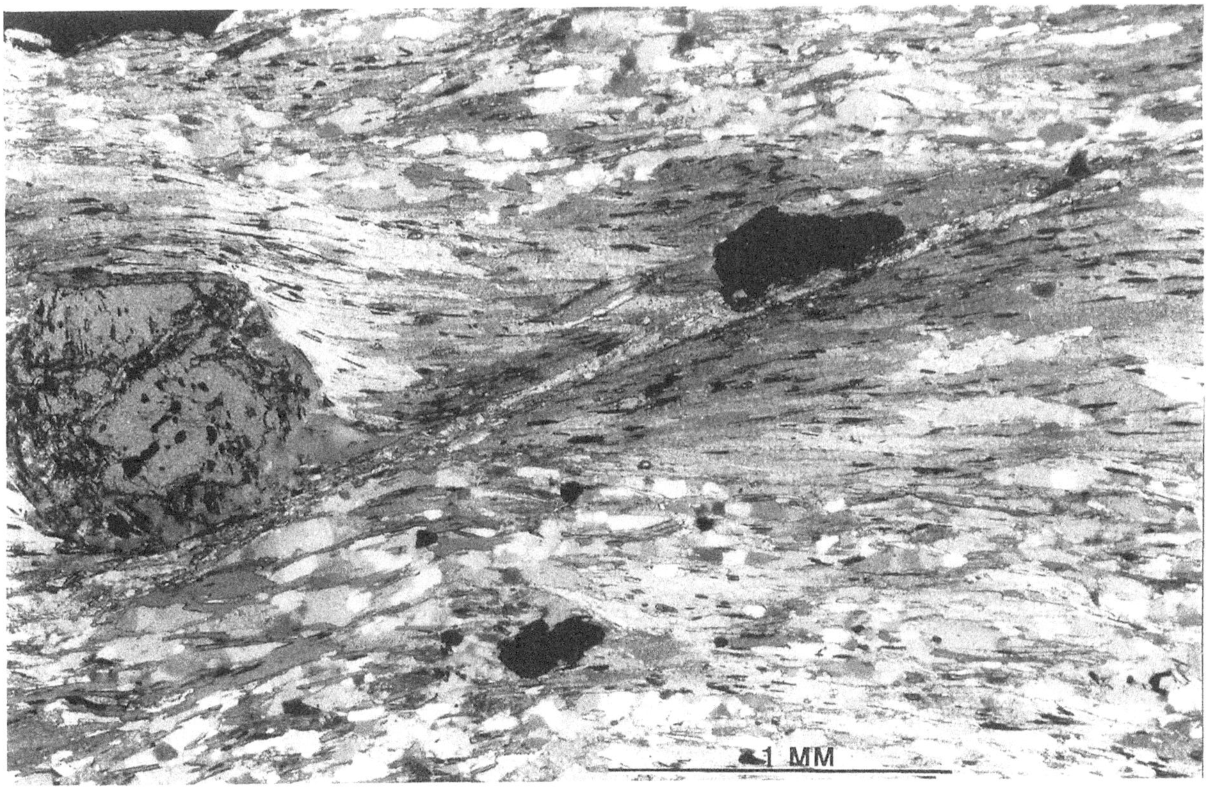


97 B

# 98. COMPOSITE PLANAR FABRIC IN MYLONITES

HIDEO TAKAGI

Mylonites along the Median Tectonic Line, Japan, (MTL, see explanatory comments for Plate 108) commonly contain shear bands comprising S(-C)-C′ fabrics. In this case, the terminology of S and C follows the initial definition by Berthé and others (1979a), where the S surface (*sensu stricto*) was considered as corresponding to the XY (flattening) plane of the finite strain ellipsoid during the initial stages of deformation, and the C surface was defined as a shear surface parallel to the shear-zone boundary. There is no significant change of angle between C′ and S (average 21°–24°) toward the center of the shear zone, and this relationship is thus different from the typical S-C relationship with increased shear strain (Takagi, 1992). The generation of the composite planar fabric can be classified into four different strain-partitioning models as shown in Figure 98.1. The formation of a composite planar fabric in the mylonites along the MTL is probably initiated according to model A, followed possibly by model C in conditions of increasing strain, and then model D (slip-compensation model proposed by Dennis and Secor, 1987) when the angle between S and the shear zone boundary (C surface) becomes small enough to produce S-parallel slip.

**98A.** S(-C)-C′ mylonite derived from granodiorite in the Kashio area. Shear bands (C′) in this rock type are several millimeters long, 20–200 µm thick, and irregularly spaced (200 µm–4 mm). They are composed chiefly of biotite with subordinate amounts of fine-grained, recrystallized quartz. Three-dimensional observations show that the C′ surfaces are commonly curved and surround porphyroclasts (Takagi, 1992). Myrmekite intergrowths (MYR) form along plagioclase/K-feldspar boundaries. The intergrowths are elongate and form fluxion banding with quartz aggregates (Q). The feldspars are stained. XZ thin section. PPL

**98B.** S(-C)-C′ phyllonite derived from pelitic gneiss or schist intercalated with granitic mylonite in the Ichinosé area. Shear bands (C′) in this rock type range from several hundred micrometers in length and 10–30 µm thick, and are irregularly spaced (20–200 µm). They are composed mainly of fine-grained muscovite (sericite) and biotite. The angle between C′ and S tends to be large (20°–30°, maximum 40°) compared with those in quartzose mylonites (e.g., see Plate 99). Some shear bands are not parallel but anastomose, and this feature is also common in the foliated fault gouges (see Plate 16). The S(-C) surface is parallel to the scale bar. XZ thin section. NX

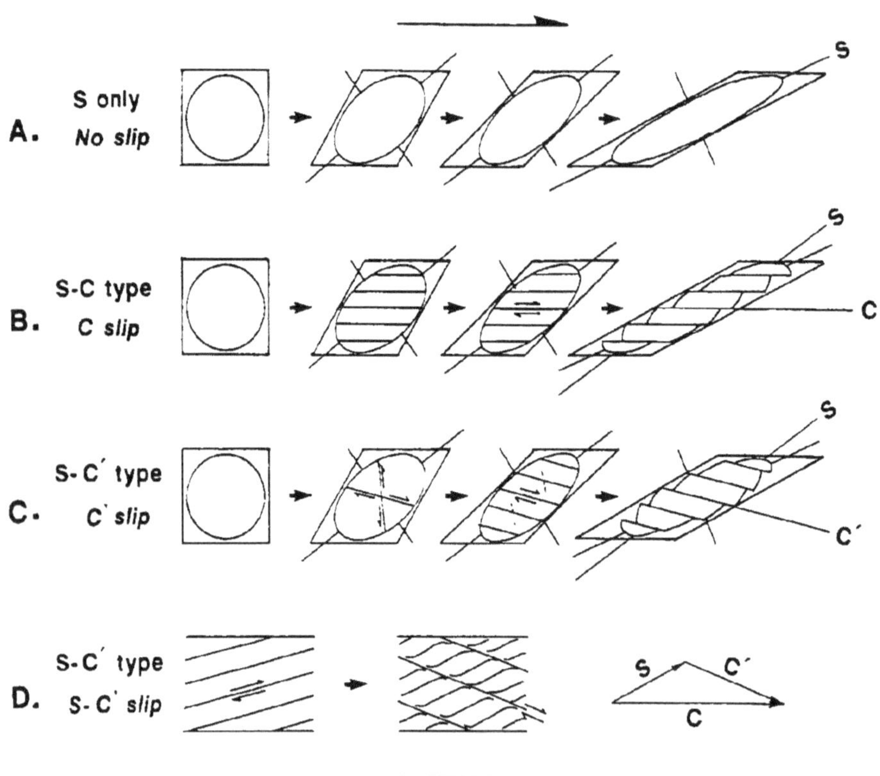


**Figure 98.1.** Possible models for the development of composite planar fabrics during progressive simple shear in the case of dextral shear after Takagi (1992).

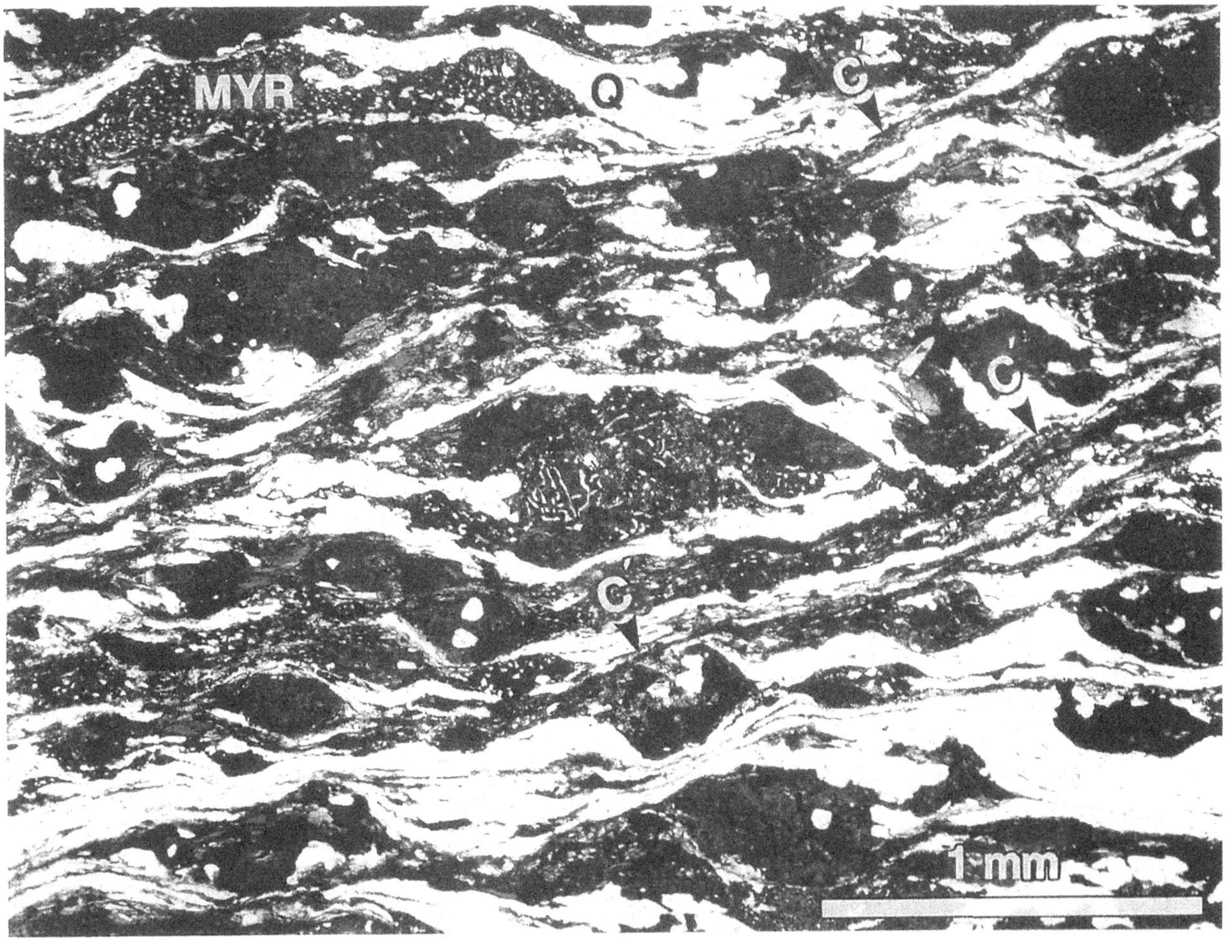


98 A

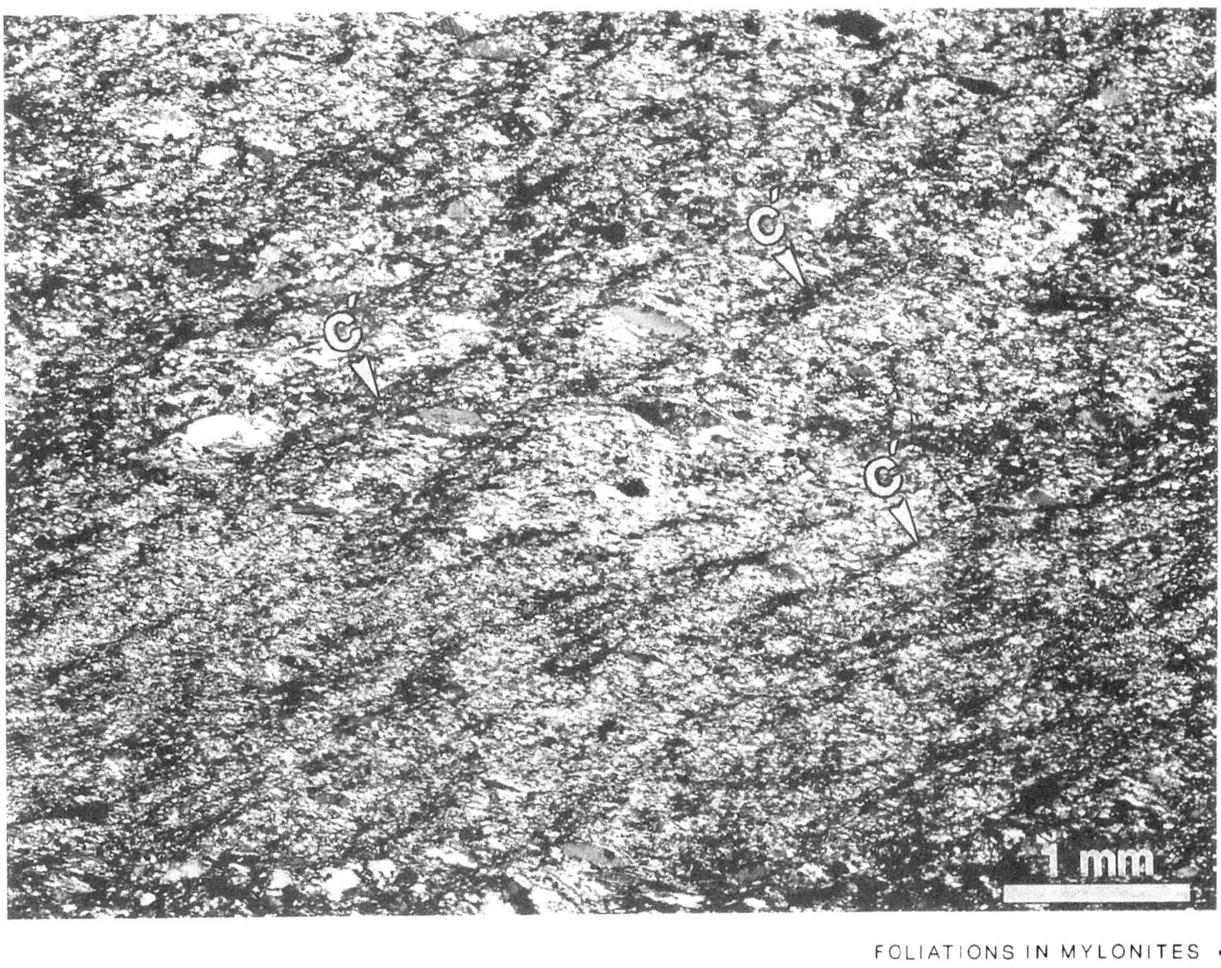


98 B

# 99. SHEAR BANDS IN QUARTZOSE PHYLLONITE

HIDEO TAKAGI

Mylonites along the Median Tectonic Line, Japan (MTL) are derived mainly from granitic rocks (see Plates 98A, 108, and 109), but some intercalations of quartzose and pelitic metasedimentary equivalents to Ryoke metamorphic rocks occur as lenticular bodies in the Ichinosé and Kashio areas. These quartzose and pelitic mylonites are called phyllonite (Takagi, 1984), because of the phyllitic appearance and fissility. An example of a pelitic phyllonite is shown in Plate 98B.

The adjacent photographs show shear bands (C′) in quartzose phyllonites derived from metachert from the Ichinosé area (Takagi, 1992). Mica fish are common, and these phyllonites are examples of type II S-C mylonites described by Lister and Snoke (1984). The shear bands in this rock type are several hundred micrometers long (maximum 700 μm), 10–20 μm thick, and irregularly spaced (>50 μm). The S(-C) surface is parallel to the scale bar.

**99A.** Mica fish in quartzose phyllonite in the Ichinosé area. Lack of fine-grained recrystallized muscovite around the fish and spindle-shaped outline clearly cross-cutting the (001) cleavage suggest that dissolution at the margin of the porphyroclasts occurred. The tails constitute shear bands, and connect fish porphyroclasts in a stair-stepping manner. This suggests that the fine-grained muscovite tails that extend from the fish are precipitated from dissolved porphyroclasts. These shear bands form only around mica fish. The angle between C′ and S tends to be small (10–20°). The shape fabric of recrystallized quartz oblique to the mylonitic foliation suggests a sinistral sense-of-shear. XZ thin section. NX

**99B.** Garnet porphyroclasts with shear bands mainly composed of chlorite. These shear bands (C′) occur only around the porphyroclasts and immediately terminate in the quartzose matrix. The surface between competent porphyroclasts (garnet) and the ductile matrix (quartz) probably served as an effective slip surface during ductile shearing, whereas retrograde chlorite grains define shear bands formed under greenschist-facies conditions. XZ thin section. PPL

99 A

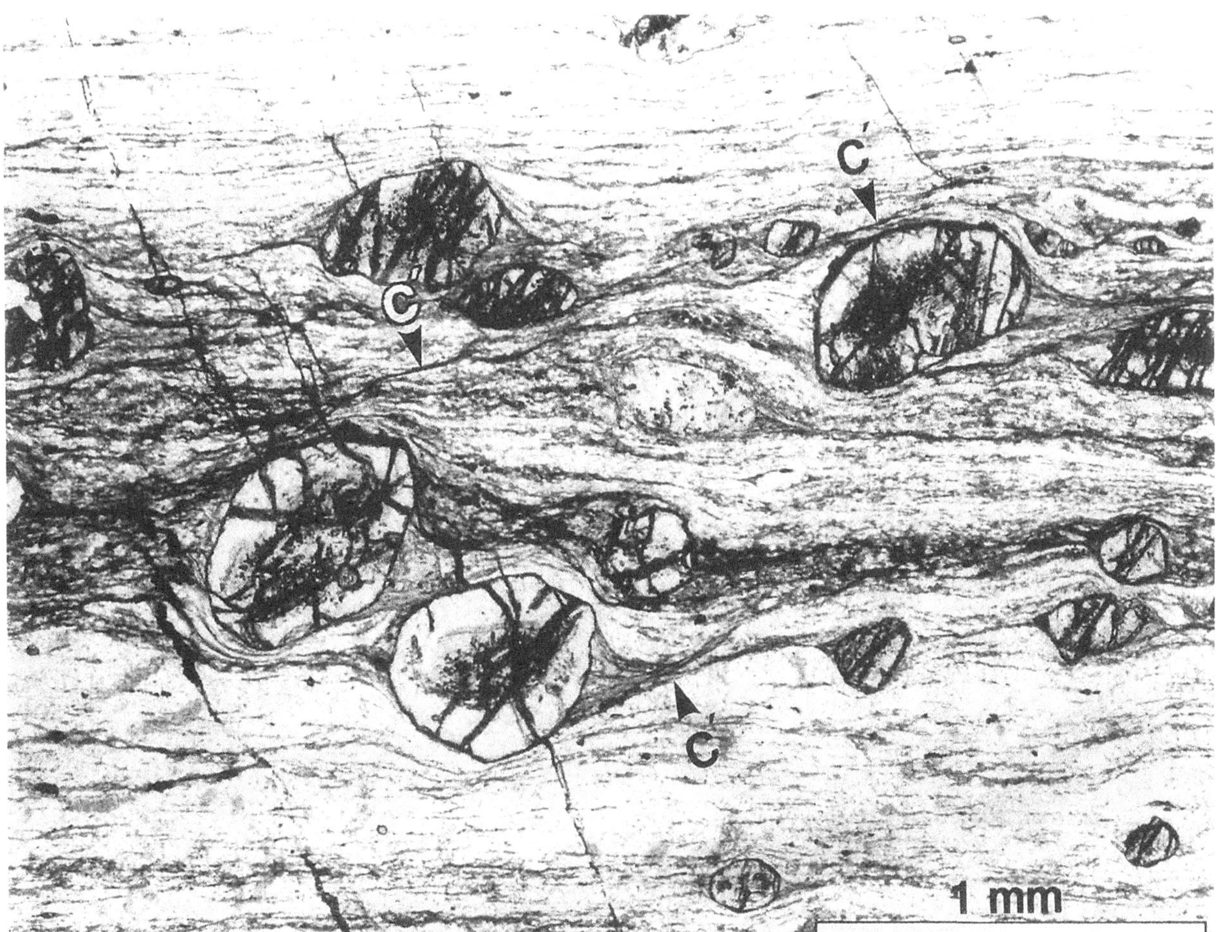

99 B

# 100. NORMAL-SENSE MYLONITES IN GRANITIC ROCKS

GARY H. GIRTY, CELESTE N. THOMSON, DAVID L. CARMICHAEL,
STEVEN L. NETTO, AND RENE WAGEMAKERS

The Cuyamaca-Laguna Mountains shear zone (CLMSZ) is a polygenetic zone of high-temperature (~500°C) ductile strain, which extends for over 50 km along the backbone of the Peninsular Ranges, California, U.S.A. (Todd and others, 1988; Thomson and Girty, 1994). During $D_1$ contractile deformation (about 115 to 118 Ma), a pervasive, northwest-striking, mylonitic to gneissic fabric was produced within the shear zone. Subsequent extension during $D_2$ produced a normal-sense shear zone which we term the Scove Canyon segment (SCs). The SCs cuts obliquely across the older $D_1$ fabric, transecting rocks as young as the ~105 Ma Las Bancas tonalite, but is deflected and truncated by the ~94 Ma La Posta pluton (see Figure 100.1). Evidence for normal-sense movement includes a down-dip stretching lineation and steeply eastward-dipping shear bands like those displayed in 100A, 100B, and 100C. The SCs may be the result of the collapse of an overthickened magmatic welt produced by the subduction of the Farallon plate beneath western North America during the late Cretaceous (Gastil and others, 1992; Thomson and Girty, 1994).

**100A.** Normal-sense shear bands (dextral sense-of-shear) in sawed slab from hand sample of Pine Valley granodiorite. View is looking northwest at the XZ fabric plane. Q = quartz; F = feldspar; and B = biotite.

**100B.** Normal-sense shear band (dextral sense-of-shear) in the Harper Creek gneiss. View is looking northwest at XZ fabric plane. M = white mica. Other symbols are the same as in 100A. NX

**100C.** Normal-sense shear bands (dextral sense-of-shear) in the Harper Creek gneiss. View is looking northwest at XZ fabric plane. Symbols are the same as in 100A and 100B. NX

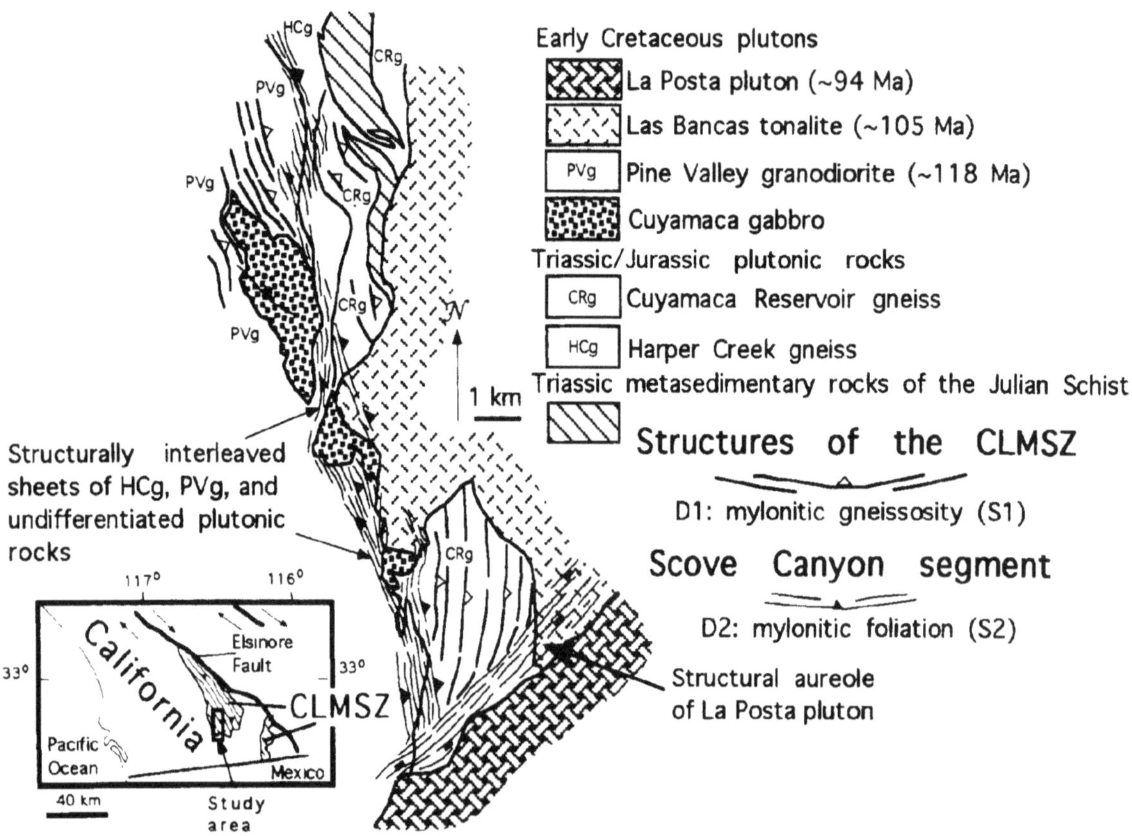


**Figure 100.1.** Geologic map showing structural relationships between the CLMSZ, the Scove Canyon segment, and various Mesozoic plutonic rocks

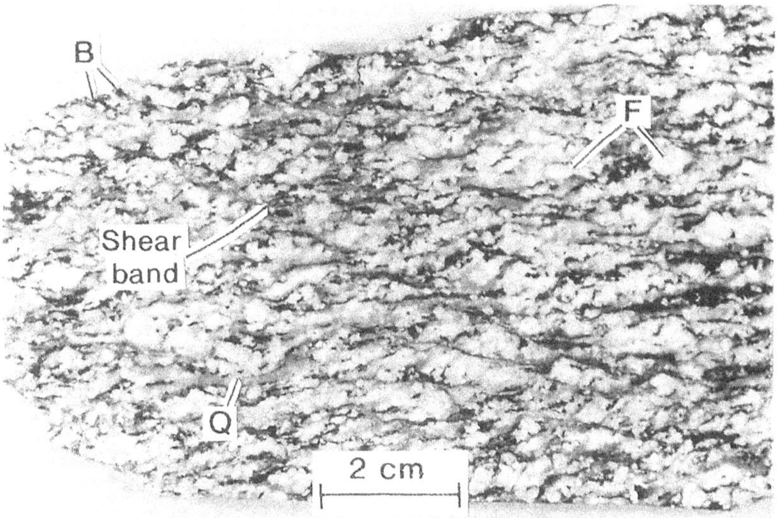

100 A

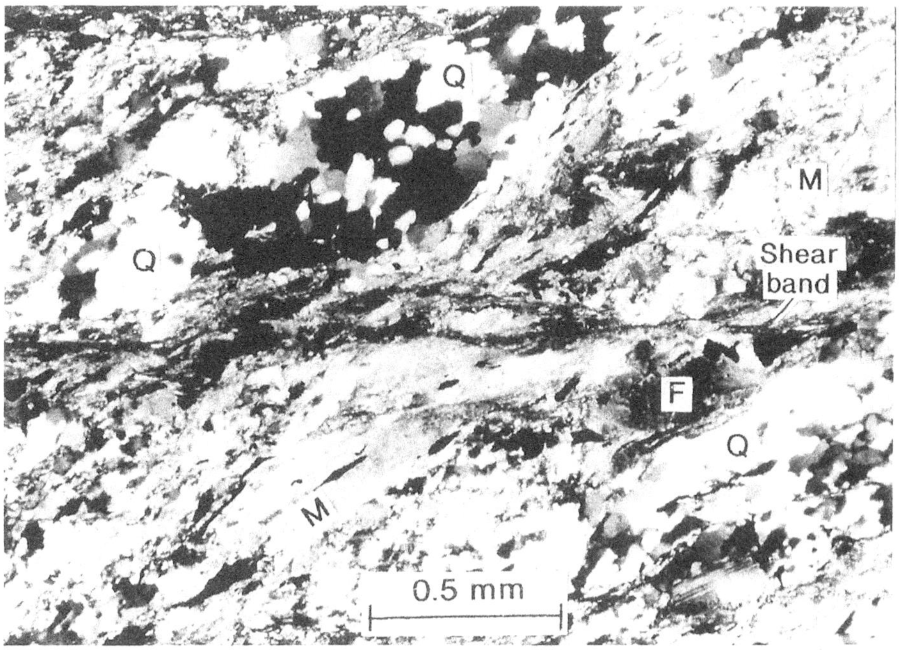

100 B

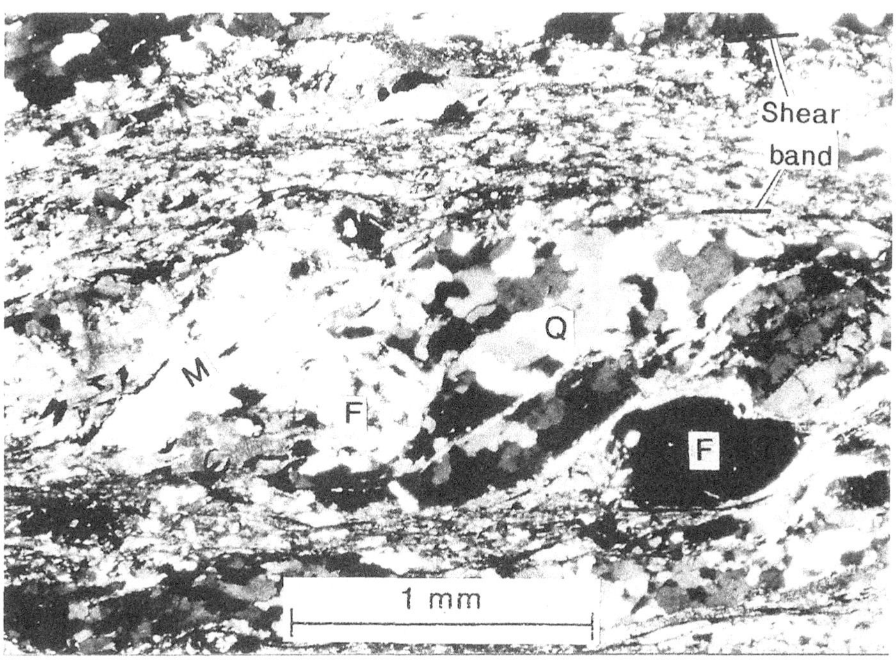

100 C

# 101. EXPERIMENTAL SHEARING OF QUARTZITE AND QUARTZ-FELDSPAR AGGREGATE

LISA N. DELL'ANGELO AND JAN TULLIS

Experimental deformation studies have been remarkably successful in inducing the same deformation mechanisms that are observed in naturally deformed quartzo-feldspathic aggregates, but one major limitation has been the fact that most studies have been restricted to axial compression. These photos illustrate microstructures that were developed in a quartzite and in a quartz-feldspar aggregate under a combination of simple shear and axial compression (Dell'Angelo and Tullis, 1989). Figure 101.1 shows the experimental configuration; the sample slab is inclined at 45° to $\sigma_1$. Experiments were performed at 800°C, 1500 MPa, and $10^{-6}$/s, in the dislocation creep regime.

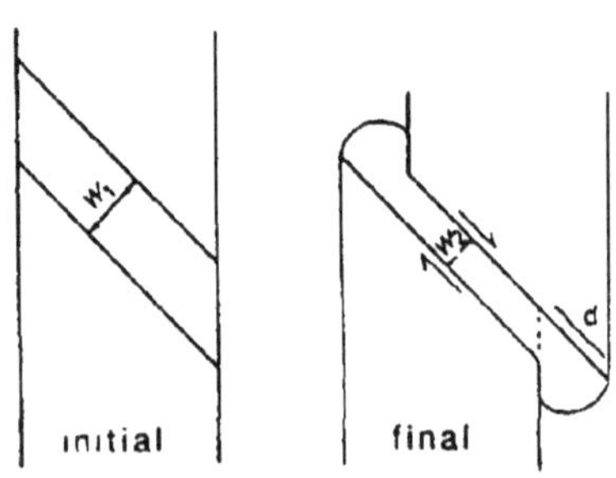


**Figure 101.1.** Schematic drawing of a cross section of the sample and pistons before and after deformation, to illustrate the combined axial shortening and shear strain of the sample. Initial sample thickness $W_1$, final sample thickness $W_2$, shear displacement d. Diameter of pistons 0.25". This figure is a part of figure 3 from Dell'Angelo and Tullis (1989).

**101A.** Sample consisting of a mechanical mixture of 33% quartz (grains from Black Hills quartzite) and 67% albite (grains from Hale albite) with an axial shortening of 40% and an average right-lateral shear strain of 2.4. Shear-zone boundaries are E-W. At these conditions, the quartz undergoes climb-accommodated dislocation creep (Regime 2 of Hirth and Tullis, 1992), with homogeneous flattening of original grains and recrystallization occurring by progressive subgrain rotation, whereas albite undergoes recrystallization-accommodated dislocation creep (Regime 1 of Hirth and Tullis, 1992), with little internal deformation of original grains and recrystallization occurring by grain-boundary migration. Note the rotated feldspar (F) porphyroclasts with very fine-grained recrystallized tails. NX

**101B.** Sample of Black Hills quartzite with an axial shortening of 50% and an average left-lateral shear strain of 2.1. Shear-zone boundaries are E-W. Note the oblique recrystallized grains (NW-SE), which have formed by progressive subgrain rotation. They appear to maintain their oblique orientation by minor grain-boundary migration, because their inclination does not increase with increasing shear strain. NX

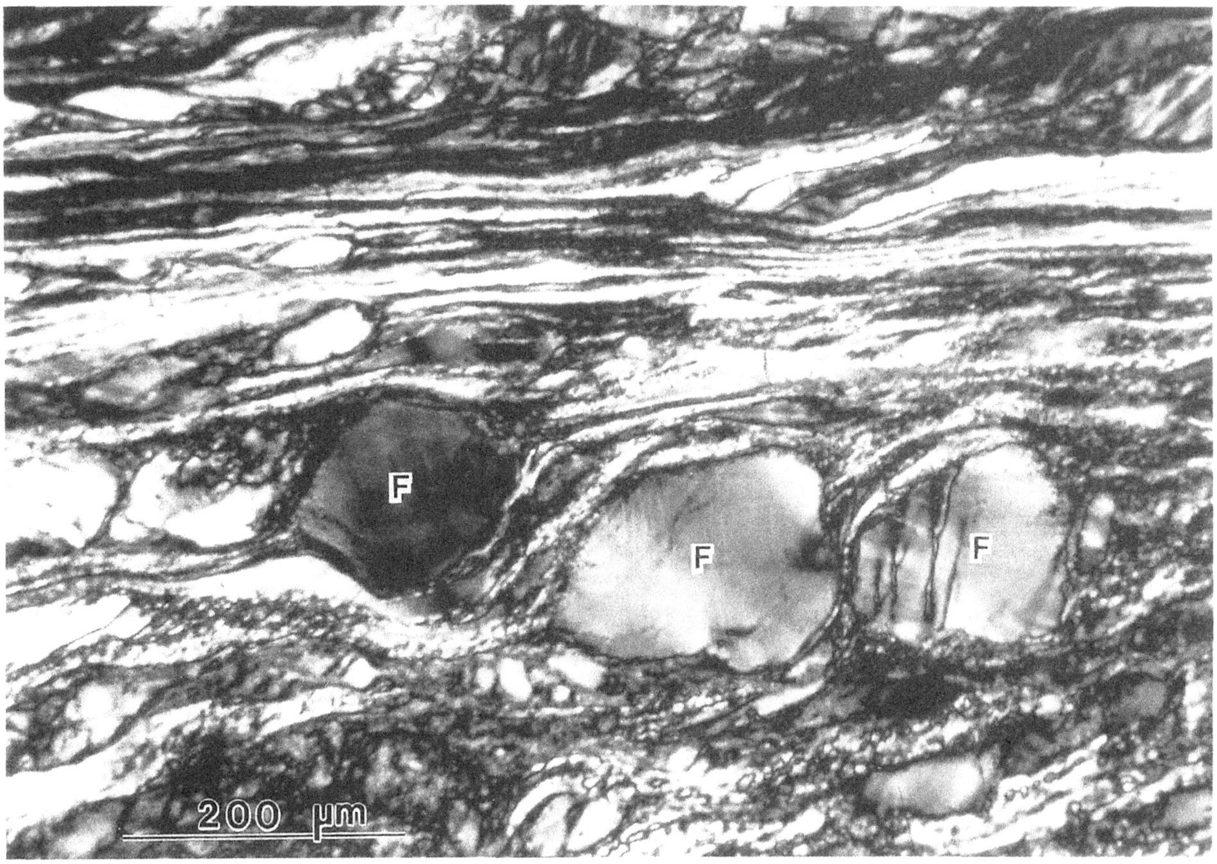


101 A

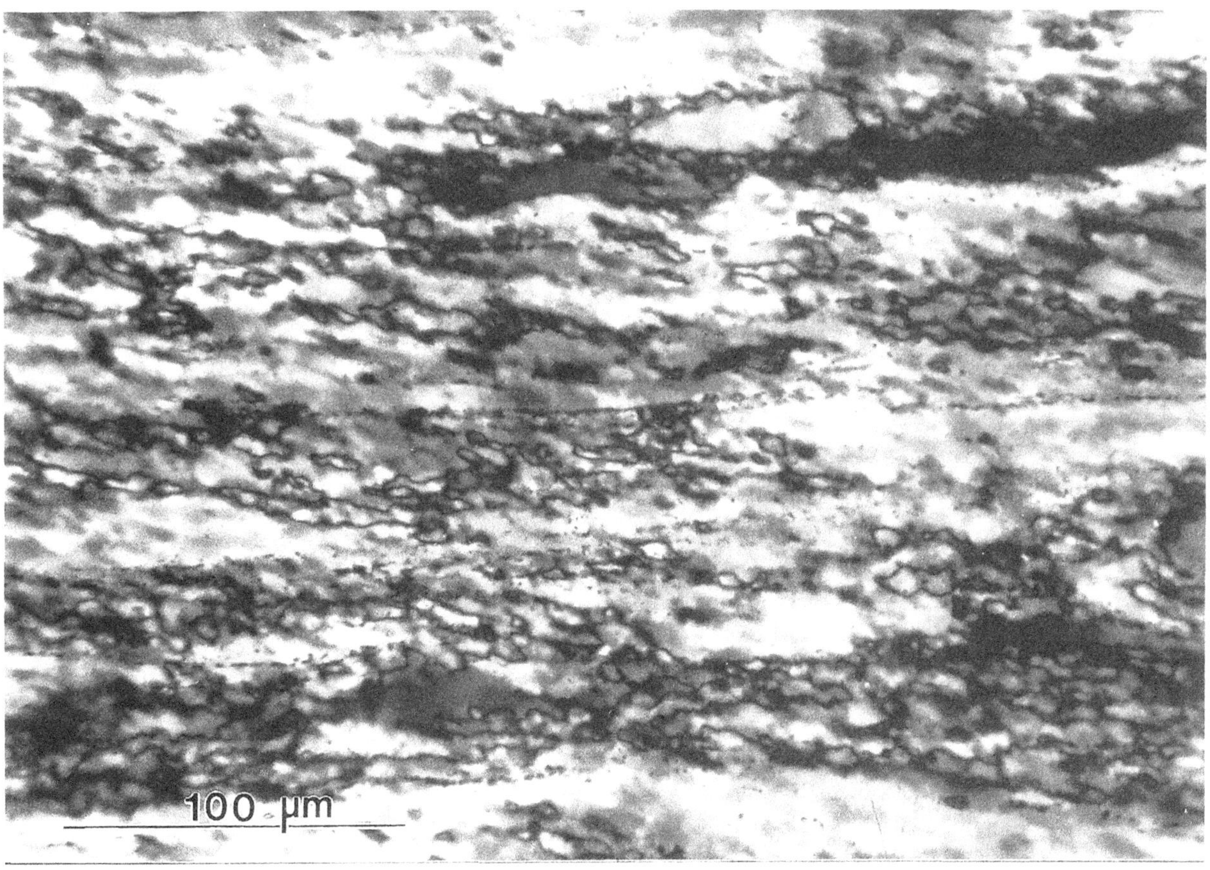


101 B

# DUCTILE BEHAVIOR

## Porphyroclasts in mylonites



Common abbreviations used in regard to the plates in this section include PL = plane light. PPL = plane polarized light, and NX = crossed nicols.

This set of plates illustrates an important microstructural indicator of shear sense in mylonites, namely asymmetric porphyroclasts. A great deal of research in the past fifteen years has been devoted to experimental and theoretical investigations of porphyroclast systems, determining the effect of rheologic contrast, initial shapes, and rate of shearing versus rate of dynamic recrystallization. These plates illustrate natural porphyroclasts of feldspar, mica, amphibole, as well as porphyroclasts in experimental systems where the variables have been carefully controlled.

# 102. ROTATED PORPHYROCLAST SYSTEMS IN QUARTZO-FELDSPATHIC MYLONITES

CAROL SIMPSON

Photomicrographs of rotated porphyroclast systems from the Santa Rosa and Borrego Springs sections of the eastern Peninsular Ranges mylonite zone and the Saint Barthélémy mylonite, French Pyrenees (Passchier and Simpson, 1986); the Pinaleño Mountains mylonite zone in central Arizona; and the Sarv thrust zone in the Swedish Caledonides. Outcrop photograph of ultramylonitic gneiss from the Parry Sound shear zone, Ontario, Canada. All examples are oriented so that sense-of-shear is sinistral.

Porphyroclasts of relatively strong minerals in mylonites commonly have recrystallized tails of weaker, finer-grained minerals that "step-up" (Lister and Snoke, 1984) in the direction of shear. Passchier and Simpson (1986) subdivided porphyroclasts and their tails into sigma (σ) and delta (δ) porphyroclast systems on the basis of their morphology and symmetry (Figure 102.1). Their analysis assumed a steady-state, simple shear flow regime and a single deformation event. Development of σ-type grains was shown to be favored where there was a relatively high recrystallization rate and a relatively low shear-strain rate, whereas δ-type grains were shown to form at low recrystallization rates with respect to shear-strain rates.

Simpson and De Paor (1993) took this analysis further and showed that certain σ-type systems, with tails that extend from the broad sides of the grains, indicate rotation of the grain in a direction *opposite* to that of the bulk flow regime. These back-rotated grains occur in general shear zones, in which there is a component of both pure and simple shear. In any zone that has a component of simple shear, most stiff grains rotate forward, with the same sense as the simple shear component. However, some ellipsoidal stiff grains that are initially oriented with their axes in the acute angle between the eigenvectors of flow (i.e., between the two directions that do not rotate) will attempt to rotate forward with the simple shear component but are pushed backward in response to the pure shear component (Figure 102.2). Where the pure shear component overwhelms the simple shear component, the grain rotates backward. These grains can be used to determine the kinematic vorticity number of the mylonite by plotting their axial ratios and long axis orientations on a hyperbolic net (Simpson and De Paor, 1993).

The plates illustrate the three principal types of porphyroclast system: forward-rotated σ types, backward-rotated σ types, and forward-rotated δ types. An example of a complex σ-δ grain is included to illustrate the morphology of grains that develop in nonsteady flow.

**102A.** Forward-rotated σ-type potassium feldspar porphyroclast in a quartz-mica matrix from Santa Rosa section of the eastern Peninsular Ranges mylonite zone (Passchier and Simpson, 1986). Deformation was at epidote-amphibolite facies and tails contain 10 × 20 μm sized recrystallized K-feldspar grains as well as white mica. Compare with Figure 102.1. NX

**102B.** Forward-rotated σ-type feldspars in quartz-mica ultramylonite band, deformed at middle greenschist facies, from Borrego Springs section of the eastern Peninsular Ranges mylonite zone (Passchier and Simpson, 1986). Fine-grained tails are formed from the breakdown reaction of feldspar to white mica and quartz. Compare with Figure 102.1. PL

**102C.** Forward-rotated σ-type feldspars in quartz-rich, Borrego Springs mylonite, deformed at middle greenschist facies. Although not visible in plane light, most of the quartz is in 10 × 30 μm sized subgrains, with rare rotation-recrystallized new grains. Tails are composed of fine-grained biotite and white mica. Compare with Figure 102.1. PL

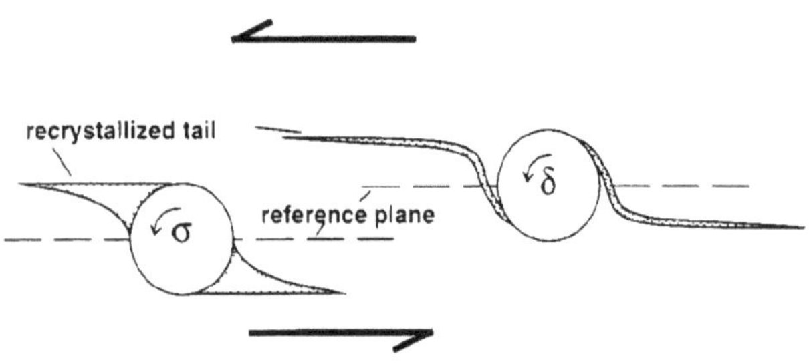


**Figure 102.1.** Sigma (σ) and delta (δ) porphyroclasts in a left-lateral shear zone. Median line of σ tails does not cross reference plane (parallel to foliation) but that of δ grain does. In both cases tails step-up to the left (after Passchier and Simpson, 1986).

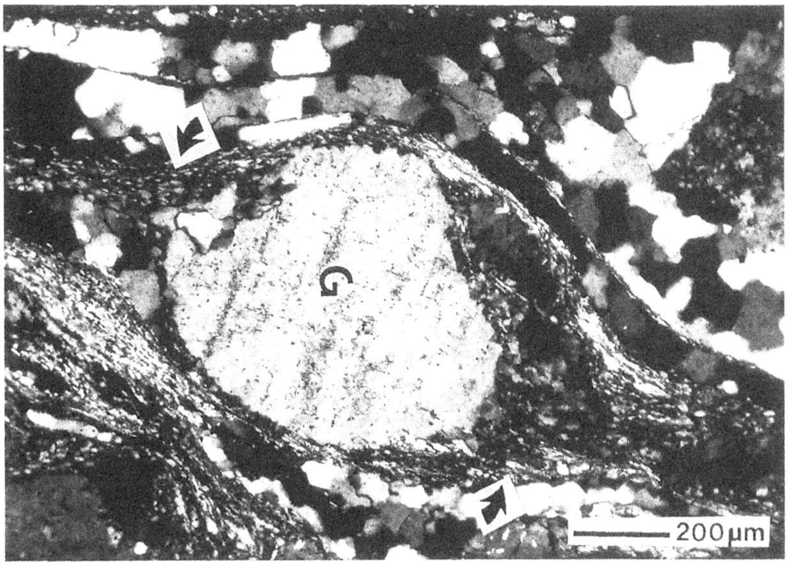


102 A

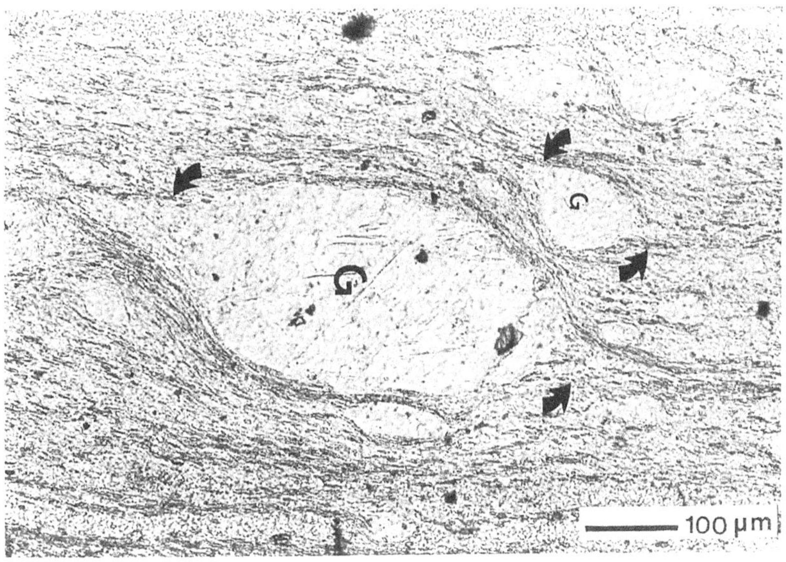


102 B

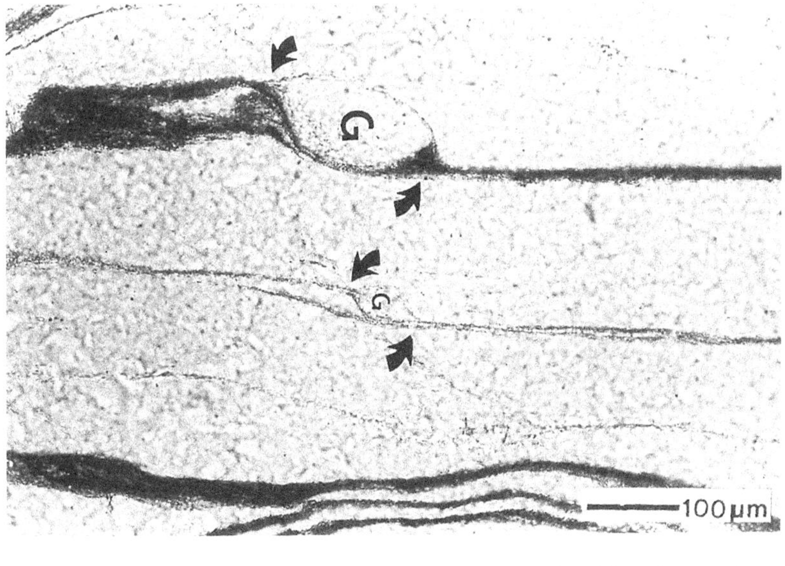


102 C

**102D.** Backward-rotated σ-type feldspar in mylonitic Upper Proterozoic arkoses of the Tossåsfjället Group, Särv thrust zone, Hällberget, Sweden. Deformation occurred during the Caledonian orogeny, at temperatures of 450–500°C and depths of 15–20 km (Gilotti, 1989). Tails "step-up" to the left, indicating sinistral shear. Recrystallized feldspar and mica tails have grown from the broad side of the grain, rather than from the narrow end as shown in 102A–102C, indicating that the grain itself has rotated *against* the bulk shear direction. This back-rotation, together with the steep inclination of the long axis of the elliptical grain with respect to the foliation, indicates that there was a significant component of pure shear across the zone (compare with Figure 102.2). PL

**102E.** As 102D, but with NX.

**102F.** Complex σ-δ feldspar grain in quartz-mica matrix from Pinaleño Mountains ultramylonite. Deformation was at middle greenschist facies and tails formed from fine-grained recrystallized feldspar and mica. Following the model of Passchier and Simpson (1986), the first set of tails to form were the left-stepping δ tails, in a regime of high strain rate and relatively low recrystallization rate. The left-stepping σ tails would have developed later, as the recrystallization rate slowed with respect to strain rate. This microstructure can be interpreted as the result of an increase in deformation temperature with time; a decrease in the bulk strain rate with time; or a decrease in the local, grain-scale strain rate as the porphyroclast approached a stable position in the flow (Simpson and De Paor, 1993). PL

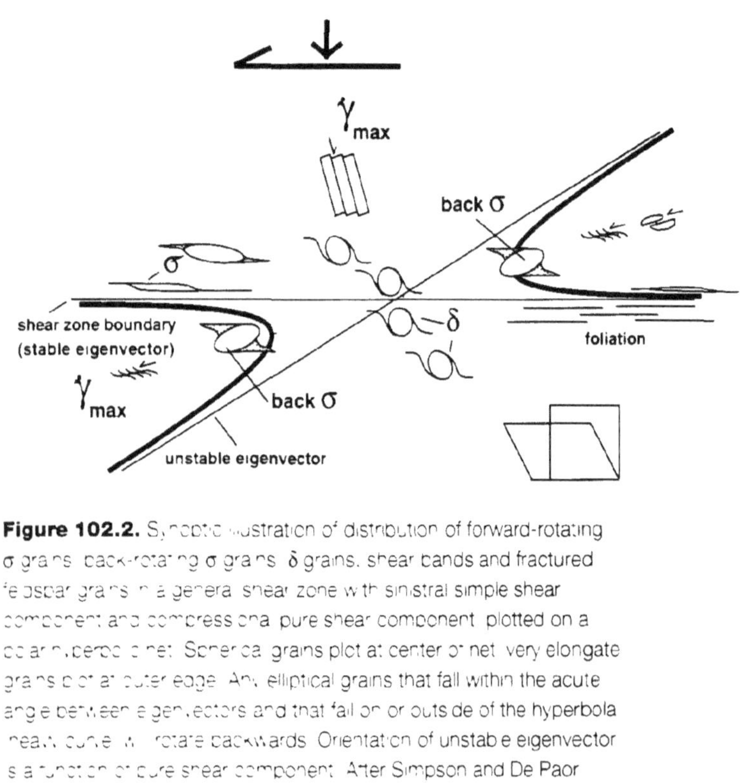


**Figure 102.2.** Synoptic illustration of distribution of forward-rotating σ grains, back-rotating σ grains, δ grains, shear bands and fractured feldspar grains in a general shear zone with sinistral simple shear component and compressional pure shear component plotted on a polar hyperbolic net. Spherical grains plot at center of net; very elongate grains plot at outer edge. Any elliptical grains that fall within the acute angle between eigenvectors and that fall on or outside of the hyperbola (heavy curve) will rotate backwards. Orientation of unstable eigenvector is a function of pure shear component. After Simpson and De Paor (1993).

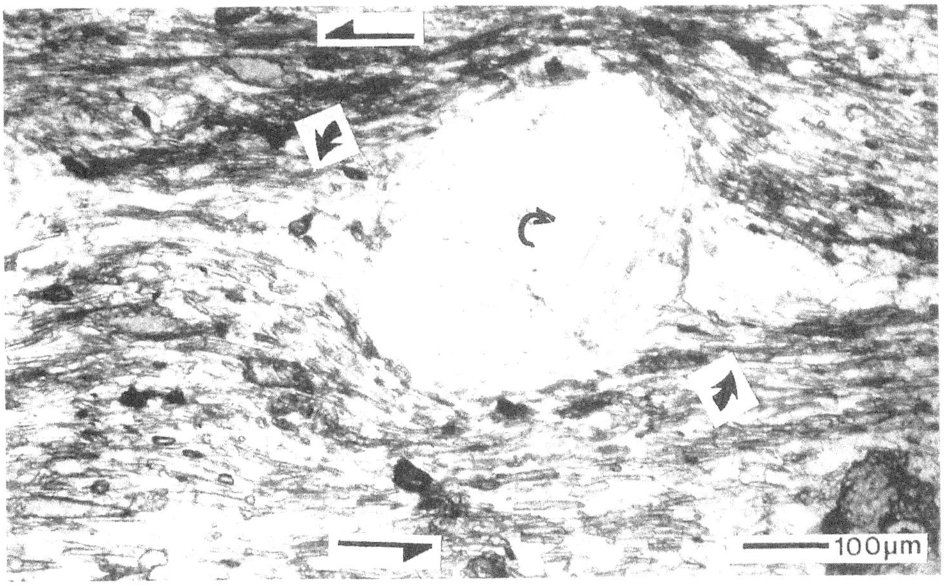


102 D

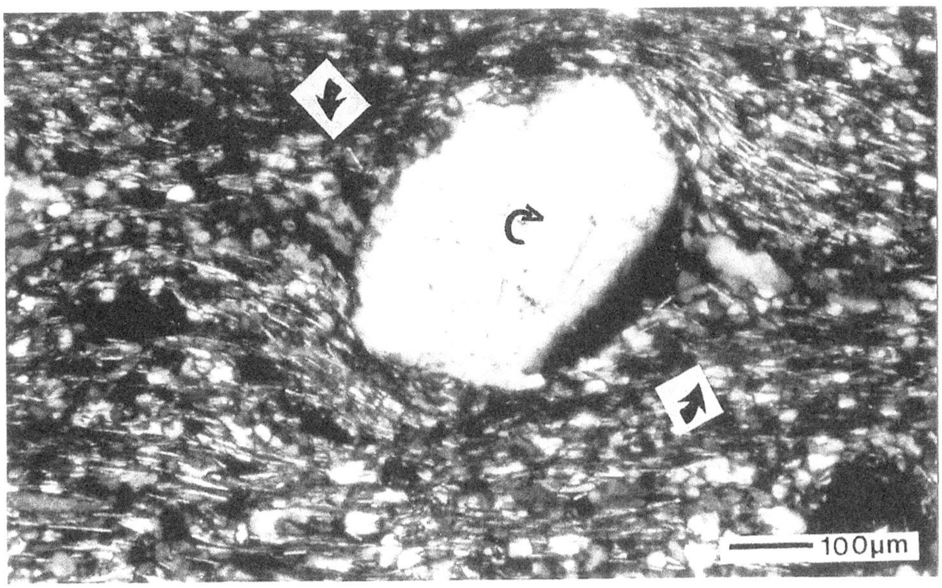


102 E

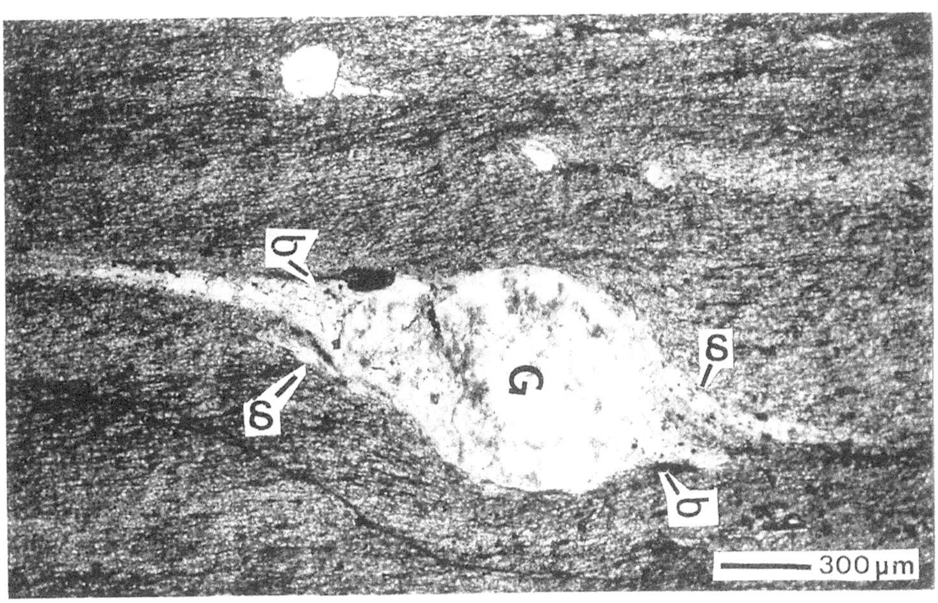


102 F

**102G.** Forward-rotated δ-type feldspar porphyroclast system from ultramylonites of Saint Barthélémy, central French Pyrenees (Passchier and Simpson, 1986). Deformation occurred at 450–550°C and 200–300 MPa (Passchier, 1982a). Matrix grains are 2–5 μm sized biotite, plagioclase, quartz, and white mica. Compare with Figure 102.1. PL

**102H.** Forward-rotated δ-type feldspar porphyroclast system from ultramylonites of Saint Barthélémy (Passchier and Simpson, 1986). Deformation conditions same as for 102G. Compare with Figure 102.1. PL

**102I.** Outcrop photograph of forward-rotated δ-type feldspar porphyroclast system; this is all that remains of a sheared pegmatite vein in upper amphibolite-grade ultramylonitic gneisses, Parry Sound shear zone, Ontario, Canada (Gower and Simpson, 1992; photo taken by R. J. W. Gower). Recrystallized feldspar tails wrap around a single large feldspar crystal that has now fractured into two pieces. Matrix is composed of equigranular feldspar, quartz, and amphibole. Compare with Figure 102.1.

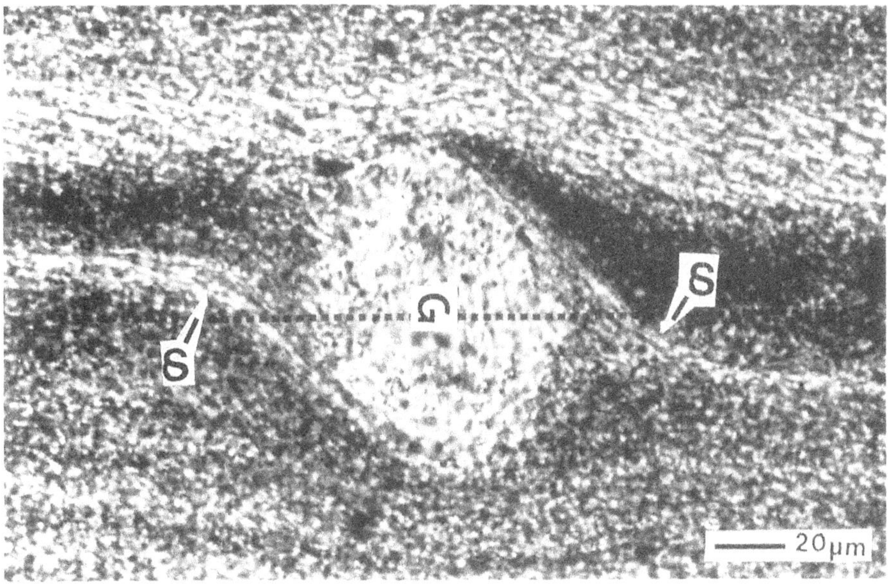


102 G

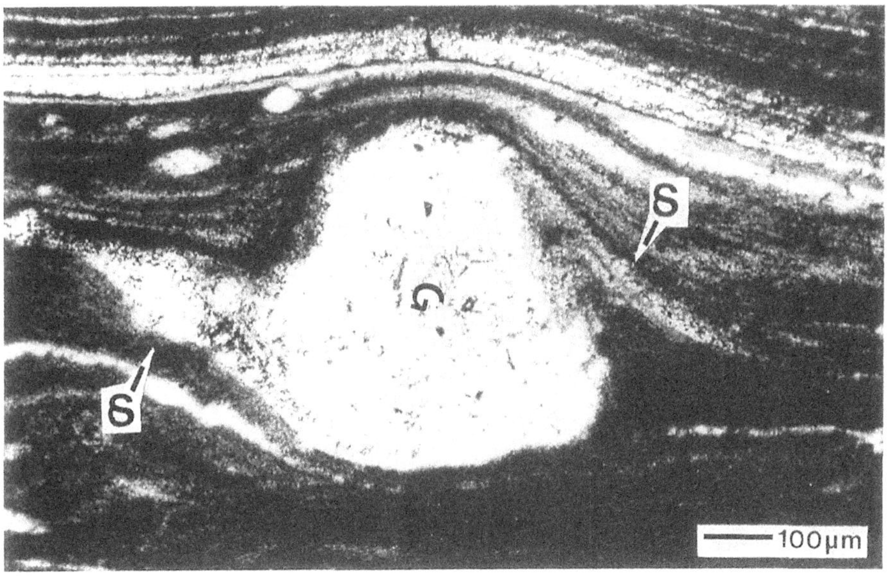


102 H

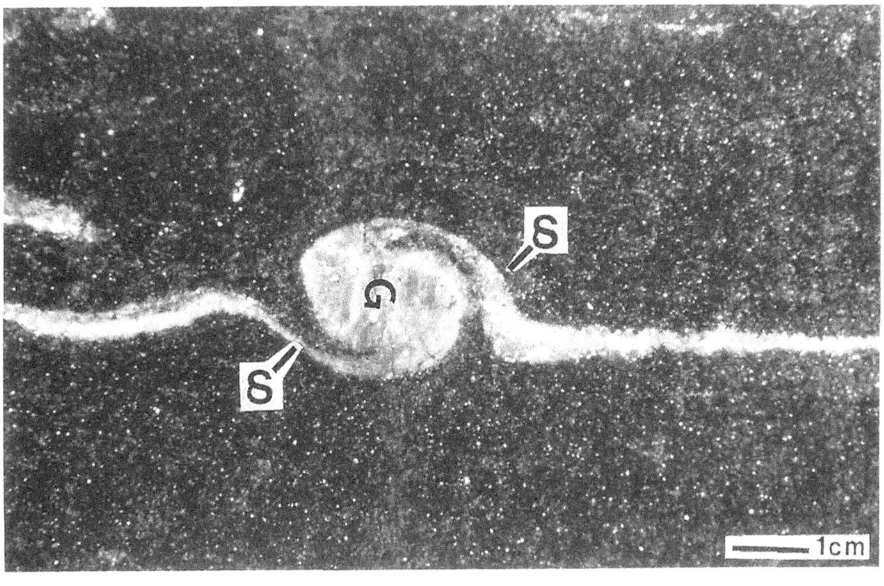


102 I

# 103. MICA FISH IN QUARTZITE MYLONITES

LARRY S. LANE

The Monashee décollement (MD, Figure 103.1) is a shallowly eastward-dipping ductile mylonite zone initiated during the Jurassic-Cretaceous collapse of the western continental margin of North America and reactivated during early Eocene extension in the Shuswap core complex (Brown and others, 1986, 1992; Lane and others, 1989). A variety of meso- and microstructures document west-over-east sense-of-shear, including rotation of minor fold hinges into the extension direction, asymmetric quartz *c*-axis fabrics, σ-type porphyroclast rotation, asymmetric strain shadows, C-S fabrics, and shear band (C′) fabrics. In many cases, several types of microstructures are evident within a single thin section (Lane and others, 1989).

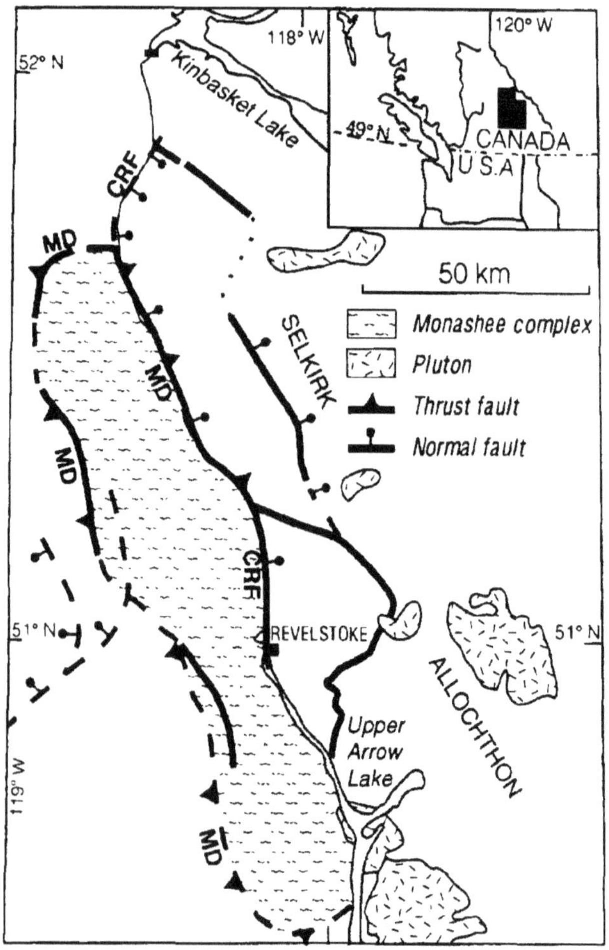


**Figure 103.1.** Geologic index map showing major tectonic elements near Revelstoke, British Columbia, Canada. The Monashee complex (horizontal dash pattern) forms the footwall of the Monashee décollement (MD). CRF = Columbia River fault

**103A.** Mica fish in quartzite mylonite (see Lane and Brown 1987, figure 3d; Lane and others, 1989, figure 6c; Hanmer and Passchier, 1991, figure 58a). Mylonitic foliation (C plane) is marked by synkinematic pressure-solution seam at top. Mica fish are back-rotated, with thin "tails" in C′ fabric orientation (see inset). Scale bar is 0.5 mm long. PPL

**103B.** Mica fish in muscovite-bearing, quartzite mylonite (see Hanmer and Passchier, 1991, figure 58b). If the two muscovite porphyroclasts near the center of the photo had been joined originally, a fracture orientation oblique to the basal cleavage plane of the mica would be required. Such a fracture orientation is undocumented for muscovite (cf. Plate 102C). Right-hand porphyroclast shows significant distortion, indicated by curved basal cleavage laminae. Scale bar is 0.5 mm long. PPL with ¼ λ plate inserted.

**103C.** Elongate muscovite porphyroclast showing asymmetrical necking, back-rotation, and incipient C′ plane development (same thin section as 103B). Planar micaceous pressure-solution seams define the local shear plane (C plane). Asymmetry of quartz *c*-axis girdle (Figure 103.2) is characteristic of advanced stage of mylonitic microstructural development, and is consistent with top-to-the-right shear sense in all three photographs (Lane and others, 1989, figures 6a and 6b). Scale bar is 0.5 mm long. PPL with ¼ λ plate inserted.

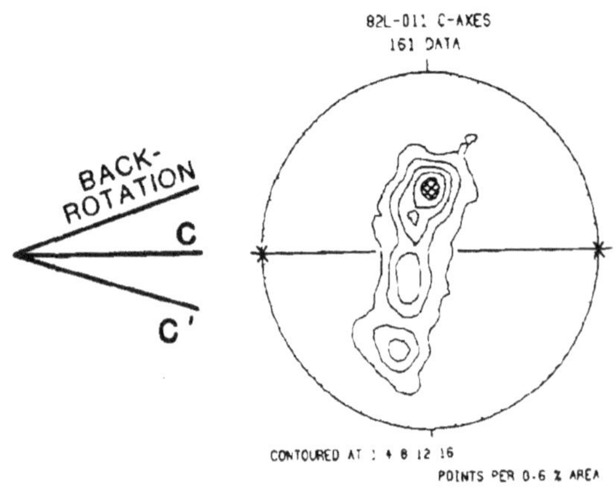


**Figure 103.2.** Quartz *c*-axis fabric for sample shown in Plates 103B and 103C (GSC contribution no. 37394)

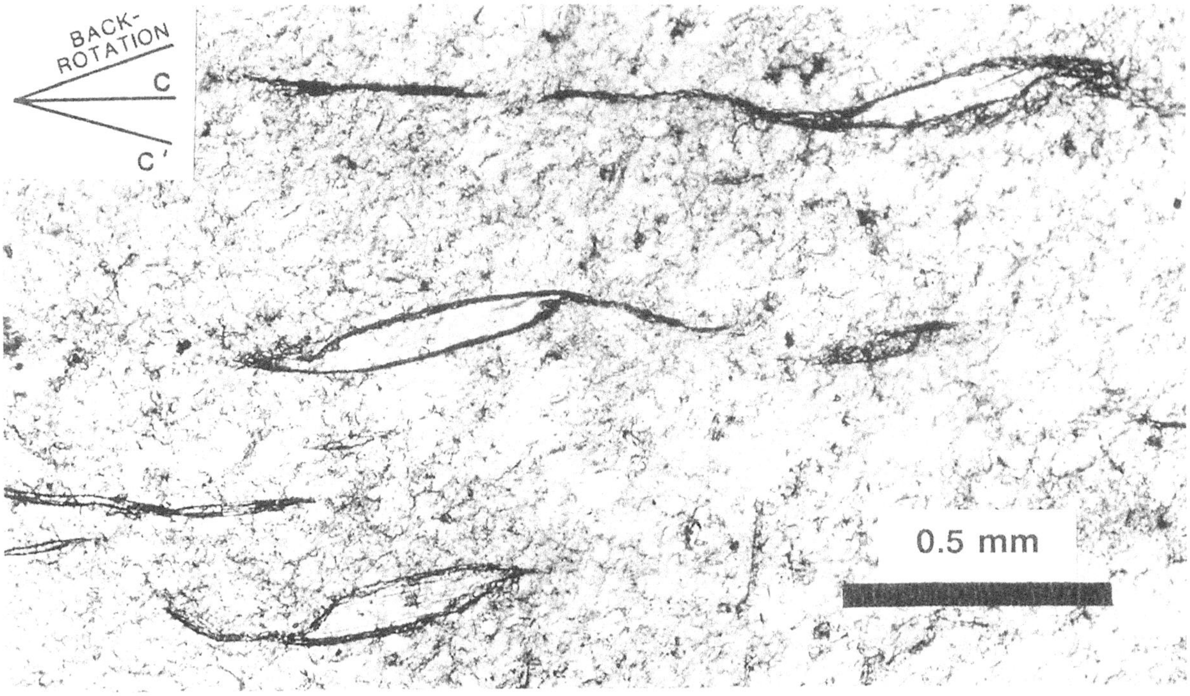


103 A

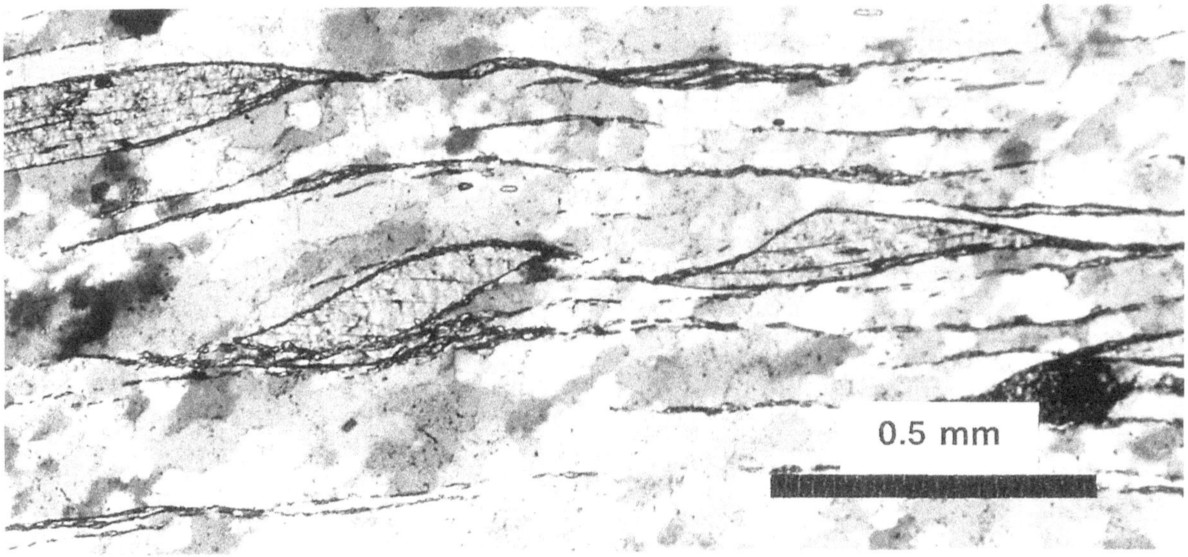


103 B

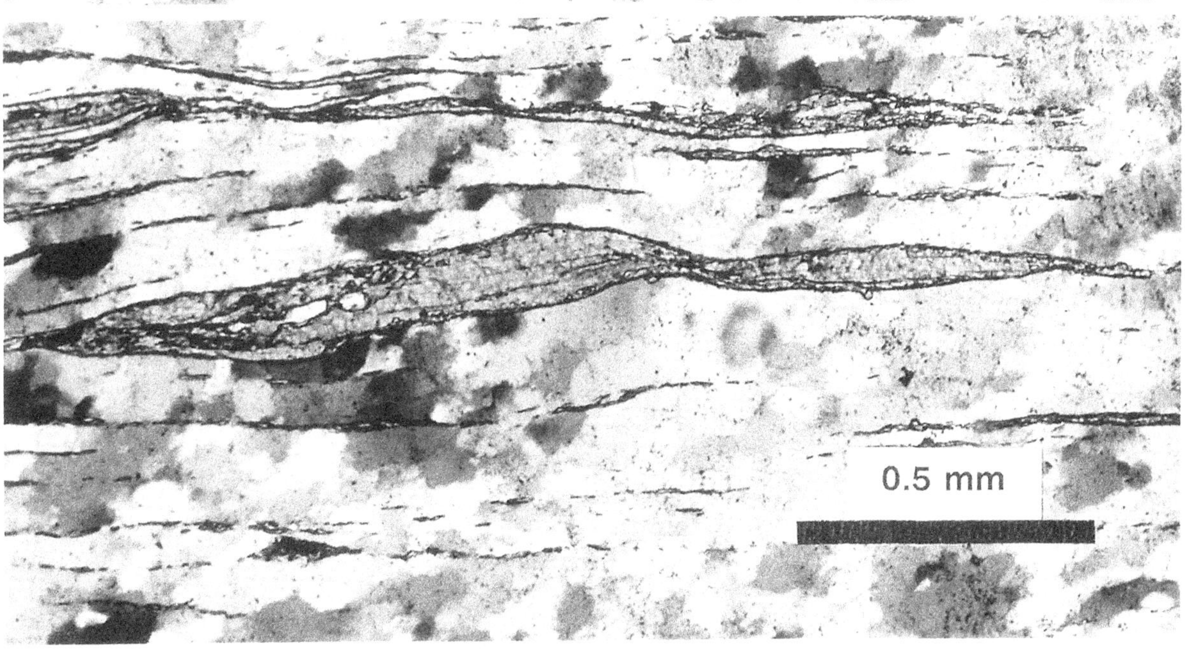


103 C

# 104. SHEAR BANDS AND ASYMMETRIC PORPHYROCLASTS IN A MYLONITIC AMPHIBOLITE

ARTHUR W. SNOKE, DAVID W. ROWE, AND J. DOUGLAS YULE

Amphibolite-facies metabasitic rocks form a thin (<250 m structural thickness), nearly continuous, ENE-WSW-striking belt between the North Coast Schist and an intrusive plutonic complex (ultramafic to tonalitic) on Tobago, West Indies (Figure 104.1). Where exposed and unfaulted, the southern boundary of the belt is a southward-dipping, commonly sharp, intrusive contact between massive ultramafic rocks of the plutonic complex and strongly foliated and lineated amphibolite. Metamorphic grade within the belt decreases with structural depth, exhibiting an inverted metamorphic gradient. The amphibolite-facies belt is interpreted to be a dynamothermal metamorphic aureole along an intrusive plutonic contact. Field and petrographic data suggest that the mafic, volcanogenic Parlatuvier Formation of the North Coast Schist was the protolith of the amphibolitic rocks of the aureole.

The inner, structurally higher part of the amphibolite-facies aureole consists of black to gray, locally gneissic, intensely foliated and lineated amphibolite. Toward the outer or structurally lower part of the aureole, the amphibolitic rocks grade into gray-green to green epidote-rich amphibolite and the intensity of the penetrative fabric decreases. The structurally lower boundary of the amphibolite-facies aureole is marked by a brittle fault that dips moderately to steeply southward, generally subparallel to foliation in the amphibolite-facies aureole.

Samples of mylonitic hornblende schist (fine-grained amphibolite) from the Louis D'Or River area contain asymmetric hornblende and leucoxene/sphene porphyroclasts as well as shear bands (104A–104C). These kinematic indicators yield a consistent sense-of-shear that is down-the-plunge of the elongation lineation (i.e., normal sense in present orientation). The thin section used for the photomicrographs (104A–104C) was cut perpendicular to foliation and parallel to the elongation lineation of the mylonitic hornblende schist.

**104A.** Mylonitic hornblende schist from amphibolite-facies aureole, Tobago, West Indies exhibiting well-developed shear bands (sb) and hornblende porphyroclasts (hb). The shear bands cut across the main foliation of the hornblende schist and are defined by fine-grained, fibrous hornblende with similar optical properties to those of the porphyroclasts. Sense-of-shear in this view is dextral. NX

**104B.** Mylonitic hornblende schist from amphibolite-facies aureole, Tobago, West Indies exhibiting shear bands (sb) and retort-shaped leucoxene/sphene porphyroclast (l/sp). Hornblende porphyroclast (hb) has a tail that merges into a shear band. The shear bands, as in 104A, are defined chiefly by fine-grained, fibrous hornblende. Sense-of-shear in this view is dextral. The area of 104C is outlined. NX

**104C.** Close-up of a part of the view in 104B showing shear bands (sb) defined by fibrous hornblende and a retort-shaped, leucoxene/sphene porphyroclast (l/sp). A hornblende porphyroclast (hb) exhibits variations in pleochroism (light brown in the core and pale green in the rim), indicating compositional zoning. PPL

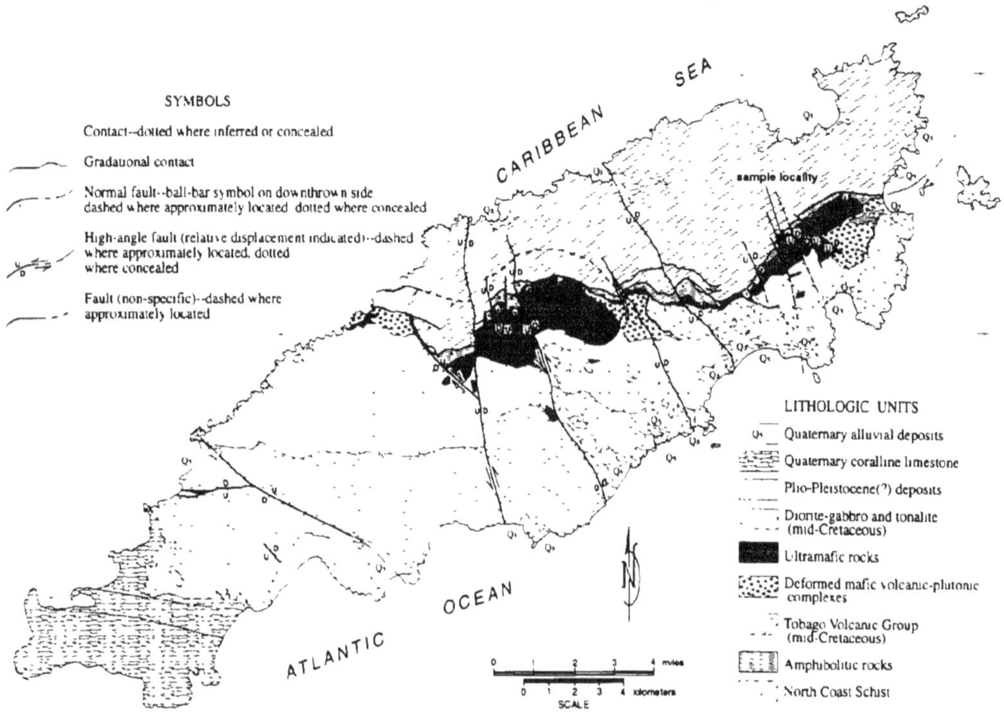


**Figure 104.1.** Generalized geologic map of Tobago, West Indies, showing sample locality of the mylonitic amphibolite sample illustrated in photomicrographs 104A–104C. Mapped by A. W. Snoke, D. W. Rowe, J. D. Yule, and G. Wadge (1985–88)

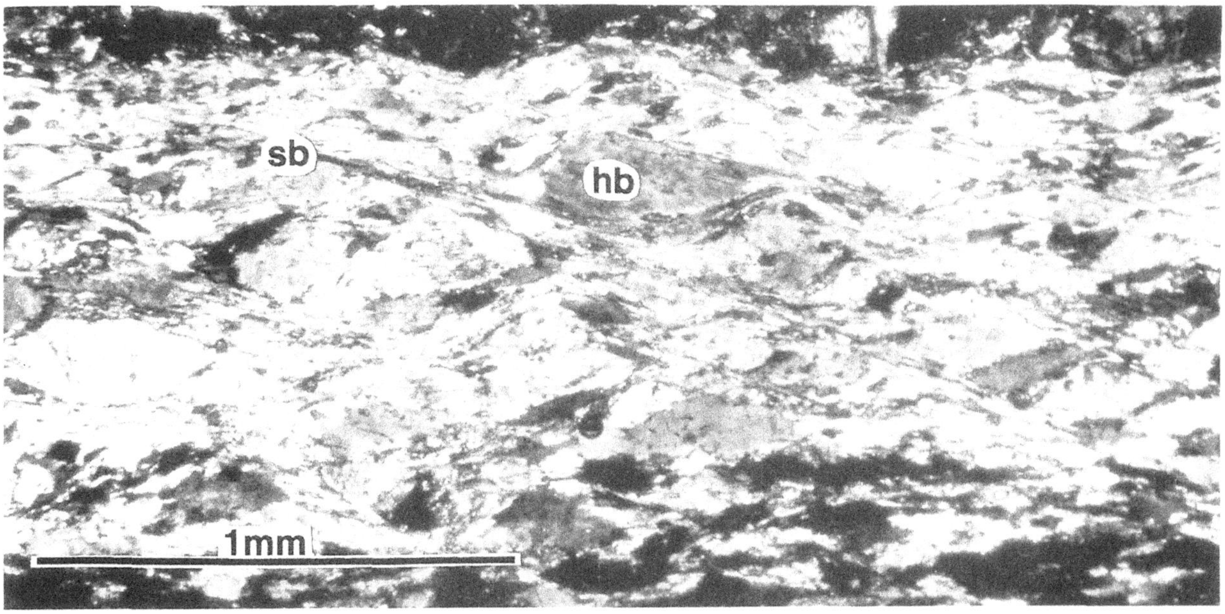

104 A

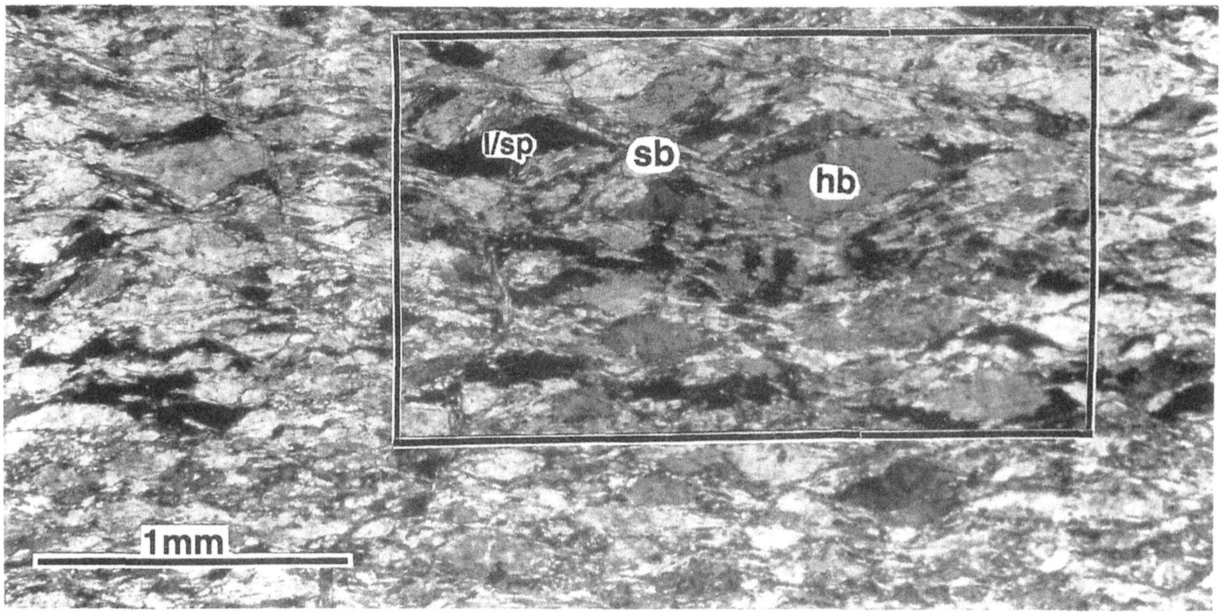

104 B

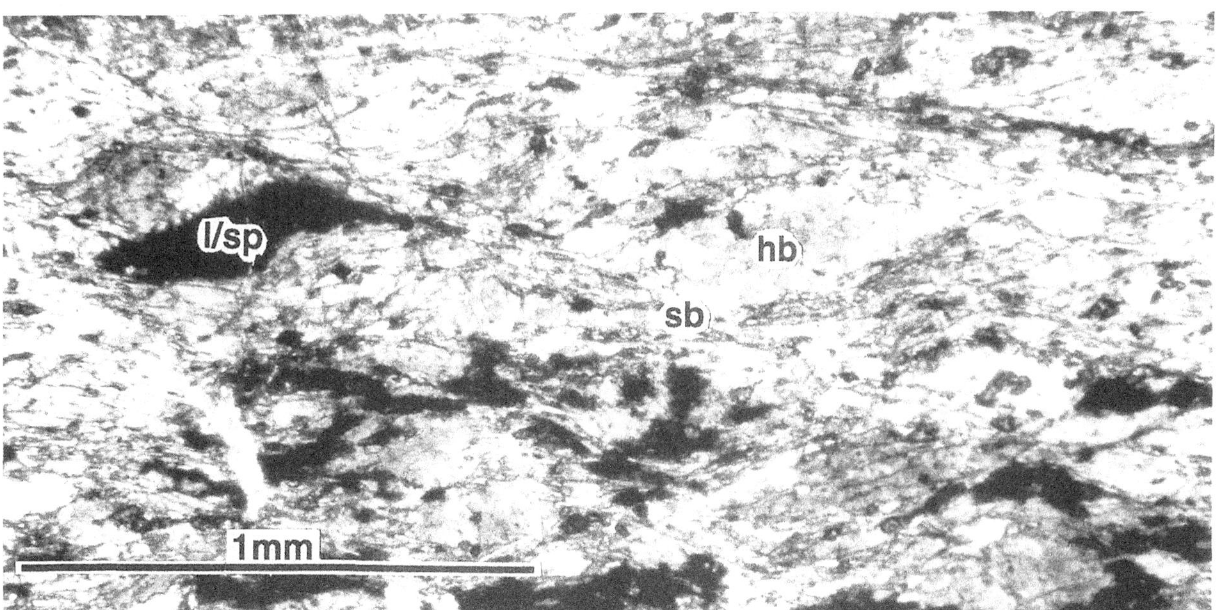

104 C

# 105. WINGED PORPHYROCLASTS: SIMPLE SHEAR/GENERAL NONCOAXIAL SHEAR

Mel R. Stauffer

The fabric of some winged porphyroclasts can be used to identify a component of flattening across a ductile shear zone. This example is from the Needle Falls shear zone (NFSZ), an Early Proterozoic shear zone in the Trans-Hudson orogen of northern Saskatchewan, Canada (Figure 105.1) (Stauffer and Lewry, 1993). The NFSZ is steeply west-dipping, has a width of about 1.5 km, and records an estimated 40 km of chiefly dextral strike-slip displacement. Delta-winged porphyroclasts (Passchier and Simpson, 1986) are well-developed in mid-amphibolite facies mylonites derived from a megacrystic granitoid body, the Wathaman batholith (Figure 105.1, Plate 105A). In many outcrops, the porphyroclast fabrics appear consistent with deformation having been by simple shear only, and with shear-induced rotation of some porphyroclasts (105B). However, such fabrics are not unequivocal, and a considerable amount of flattening could have taken place across the shear zone without modifying the fabric enough to be recognizable. In a few outcrops, evidence of flattening comes from the draping of well-developed "wings" over adjacent porphyroclasts (105C).

**105A.** Protolith of NFSZ: a megacrystic granitoid of the Wathaman batholith (ca. 1860 Ma), which locally displays an igneous flow fabric of aligned K-feldspar phenocrysts. The black patches are lichen.

**105B.** Delta-winged porphyroclasts from the NFSZ (XZ plane), indicating dextral shear. Although apparently consistent with simple-shear-induced rotation, this fabric is not unequivocal and considerable general noncoaxial strain could have taken place. Apparent shear strain is approximately 11.

**105C.** Winged porphyroclasts indicating a component of flattening across the NFSZ (XZ plane). The extremely long wing is deflected by an adjacent porphyroclast. Apparent shear strain is about 16, based on the ratio of the wing length to porphyroclast radius.

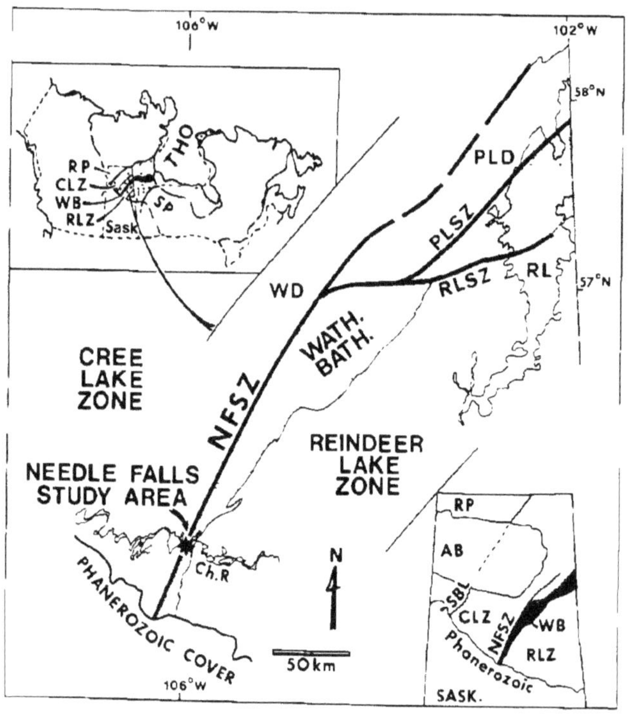


**Figure 105.1.** Location of the NFSZ. Trans-Hudson orogen (THO) includes Cree Lake zone (CLZ), Reindeer Lake zone (RLZ), and Wathaman batholith (WB). This orogen is bounded by the Rae (RP) and Superior (SP) provinces. Branch faults include the Parker Lake (PLSZ) and Reilly Lake (RLSZ) shear zones. Other features are the Wollaston (WD) and Peter Lake (PLD) domains, the Snowbird line (SBL), Athabasca Basin (AB), and the Churchill River (Ch R)

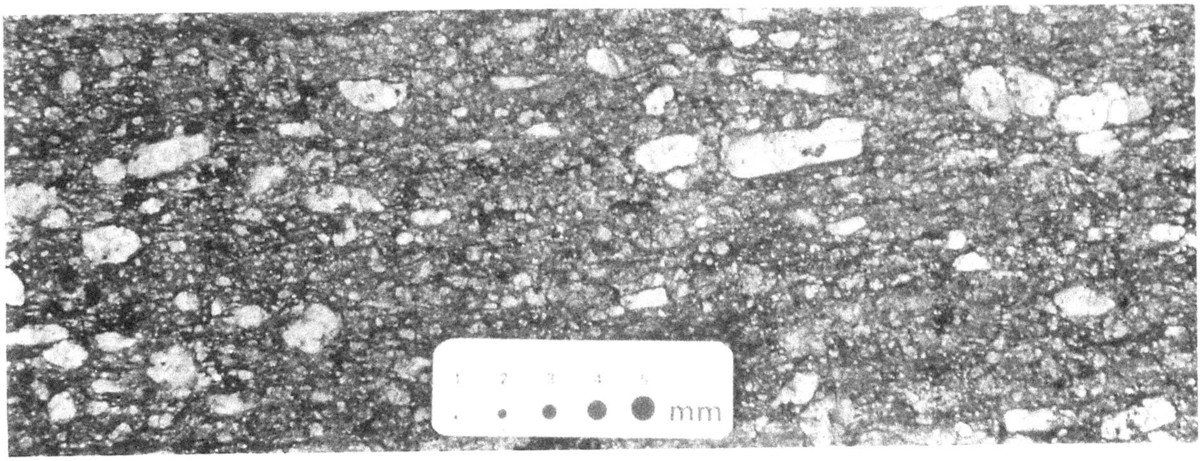


105 A

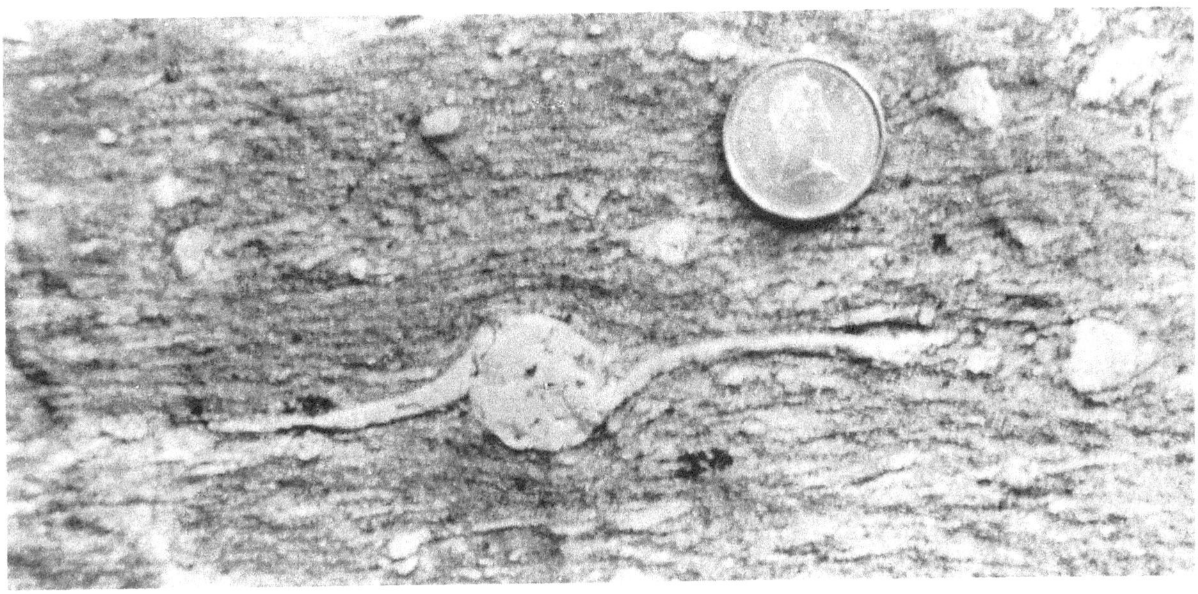

105 B

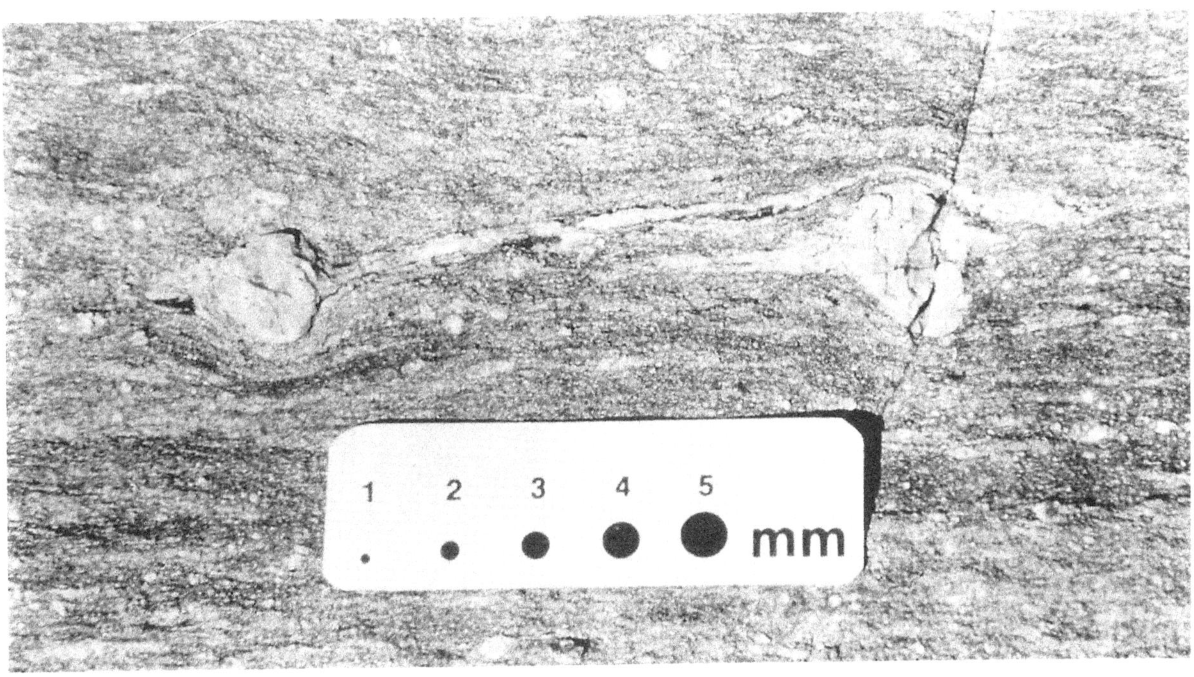


105 C

# 106. STRING-BEAD FABRIC: HOMOGENEITY OF NONCOAXIAL STRAIN

Mel R. Stauffer

The shapes of porphyroclasts in mylonites depend partly on the ductility contrast between porphyroclasts and matrix. If the ductility contrast was low during noncoaxial shear, grains will be stretched into long strings (ribbon mylonites), commonly with little size reduction. Where the ductility contrast was high, however, porphyroclasts will have undergone rotation and size reduction, producing a "beaded" fabric (106A). Intermediate ductility contrast results in a combination of strings and beads (string-bead fabric; Stauffer and Lewry, 1993) (106B), or in winged porphyroclasts. Interlayered rocks, each with different ductility contrasts between porphyroclasts and matrix, result in a layered mylonite in which each layer has developed its own porphyroclast fabric: some ribboned, some beaded, and some winged (106C).

This example is from the Needle Falls shear zone (NFSZ) Saskatchewan, Canada (see Plate 105 for index map and examples of winged porphyroclasts). The protolith (granitic rocks of the Wathaman batholith) consists of various layers, including primary igneous layers with different proportions of mafic minerals (mainly biotite), slabby mafic xenoliths, and felsic veins. Both primary layers and xenoliths contain K-feldspar megacrysts of about the same size. K-feldspar porphyroclastic fabrics within the shear zone vary dramatically with the proportion of biotite in the matrix of each layer. Mainly beaded porphyroclasts develop where the biotite content is greater than 20%, as in the mafic xenoliths (106C) and in the most mafic primary layers (106A). String-bead fabric and winged porphyroclasts develop in primary layers where the biotite is between 10 and 20% (106B and 106C). Ribbon mylonite forms from primary layers with less than 10% biotite and in the felsic veins (106C). Most of the ribbons are K-feldspar, but in the felsic veins some are quartz.

**106A.** Beaded mylonite in NFSZ from granitic rock of the Wathaman batholith containing just over 20% biotite.

**106B.** String-bead mylonite in NFSZ from granitic rock of the Wathaman batholith containing 15% biotite. Same outcrop as 106A. These two lithologies are interlayered on the decimeter scale. Shear strain is estimated to be about 10–15.

**106C.** Layered mylonite in NFSZ from Wathaman batholith: a mafic xenolith (M) with beaded porphyroclasts; biotite-poor primary layers (R) of ribbon mylonite; thin felsic veins (V) of ribbon mylonite; and primary layers with 15% biotite (S) of string-bead mylonite. The shear strain is estimated to be greater than 50, based on the shapes of the K-feldspar ribbons in R layers.

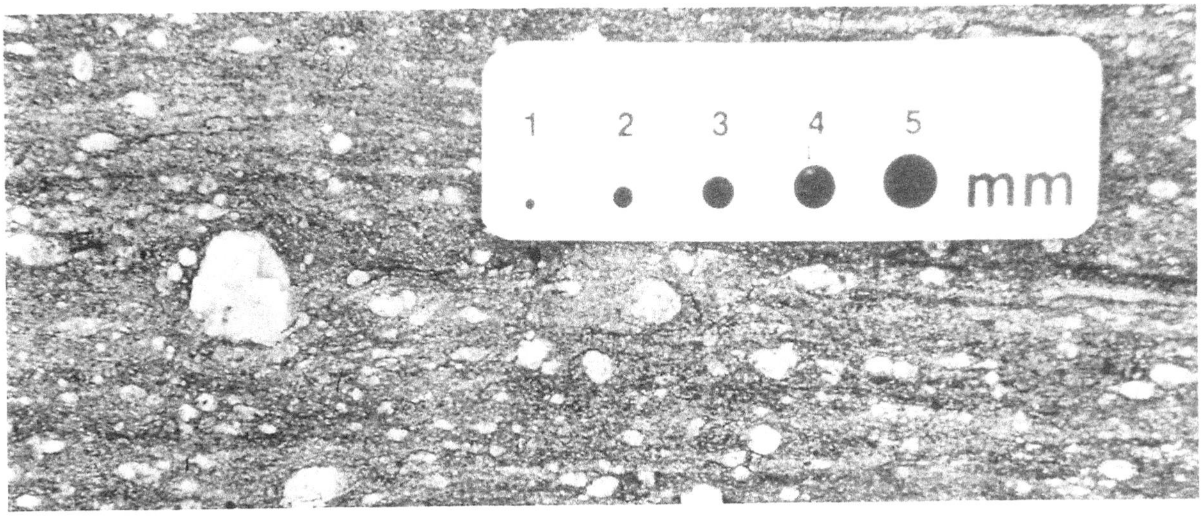


106 A

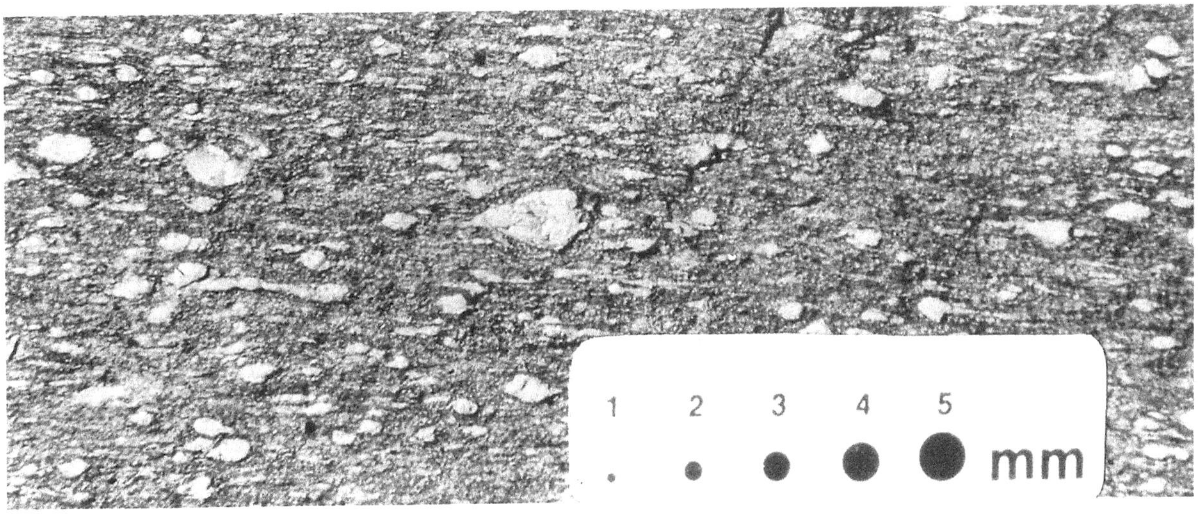


106 B

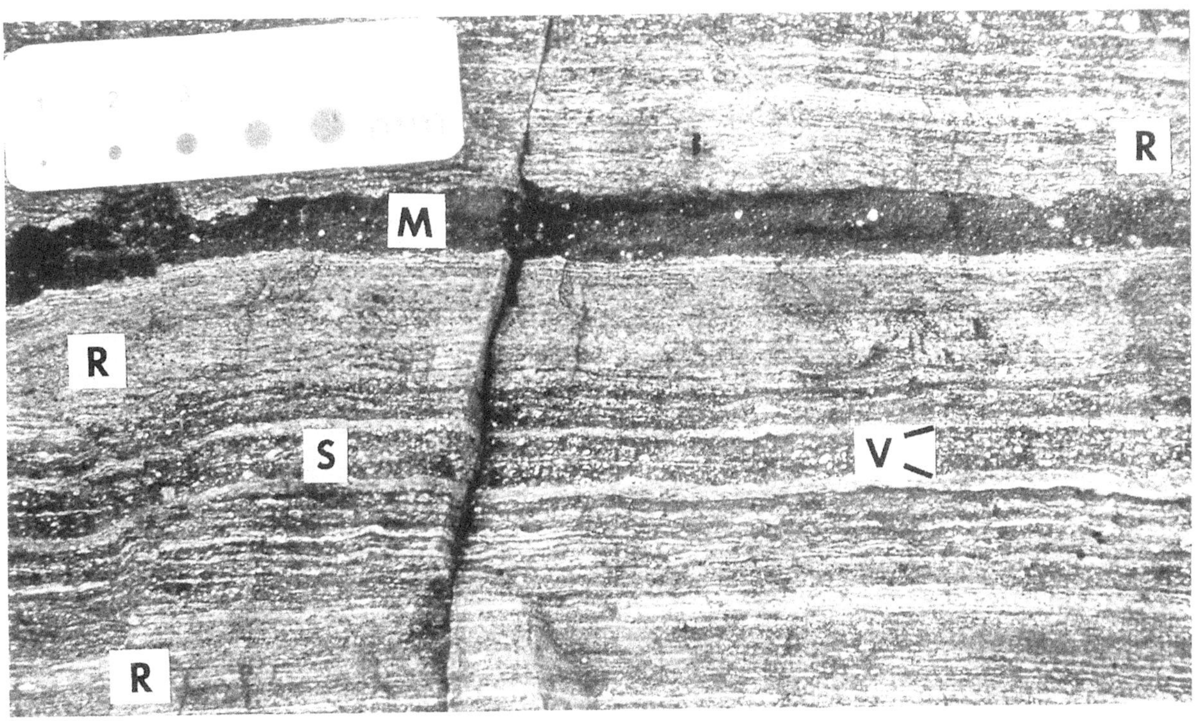


106 C

# 107. MANTLED PORPHYROCLASTS

CEES W. PASSCHIER AND COEN E. TEN BRINK

Porphyroclasts in a mylonite, surrounded by a fine-grained aggregate of the same composition, are known as mantled porphyroclasts or mantled objects (107A). They are interpreted to form by recrystallization in the rim of a porphyroclast in response to crystal-plastic flow in the matrix. The fine-grained mantle can be deformed into wings (or tails) that extend on both sides of the object parallel to the shape fabric in the mylonite (Passchier and Simpson, 1986). Wings are thought to stretch and change shape, while the porphyroclast continues to recrystallize in the rim and shrink in size. Wing shape can be used as a shear-sense indicator when observed in sections parallel to the stretching lineation and normal to the foliation in a mylonite (107B). Common types of mantled objects are shown in 107B (Hanmer, 1984; Passchier and Simpson, 1986; van den Driessche and Brun, 1987; Hooper and Hatcher, 1988b). θ-type objects lack wings, ϕ-type objects have a mantle with orthorhombic symmetry (Passchier, 1994). σ-type, δ-type, and complex objects have monoclinic shape symmetry. σ-type objects have wide mantles near the porphyroclast and wings that are parallel but lie at different elevations on both sides. This difference in elevation is known as stair-stepping (Lister and Snoke, 1984) and can be used as an independent shear-sense indicator (107B). δ-type objects have narrow wings and characteristic bends in the wings adjacent to the porphyroclast. Complex objects have more than one set of wings.

The shape of mantled porphyroclasts is mainly determined by the ratio of synkinematic recrystallization rate of the porphyroclast to strain rate in the matrix (RR/SR), the shape of the porphyroclast, and the rheology of the matrix (Passchier and Simpson 1986; Passchier and Sokoutis, 1993). Presently available data indicate that σ- and ϕ-type objects develop at high RR/SR, δ-type objects at low RR/SR, and θ-type objects at very low RR/SR (Passchier, 1994). Stair-stepping of δ- and σ-type objects may be due to a non-Newtonian matrix (Passchier and others, 1993). Complex objects form around elongate porphyroclasts or during flow with a complex deformation history.

**107A.** Isolated mantled plagioclase porphyroclasts in upper greenschist-facies quartz mylonite, St Barthélémy massif, France. The host rock is a quartz vein in granite, deformed during the Variscan orogeny. A central δ-type object is mantled by recrystallized plagioclase, and small σ-type objects occur in the left-hand wing of the δ-type object and in other bands of recrystallized feldspar. All mantled objects show stair-stepping. The quartz matrix shows a shape fabric defined by elongate recrystallized grains, running from lower left to top right. Dextral shear sense. Scale bar is 1 mm. PPL

**107B.** Schematic drawing of types of mantled porphyroclasts that are commonly observed in mylonites. Dextral shear sense.

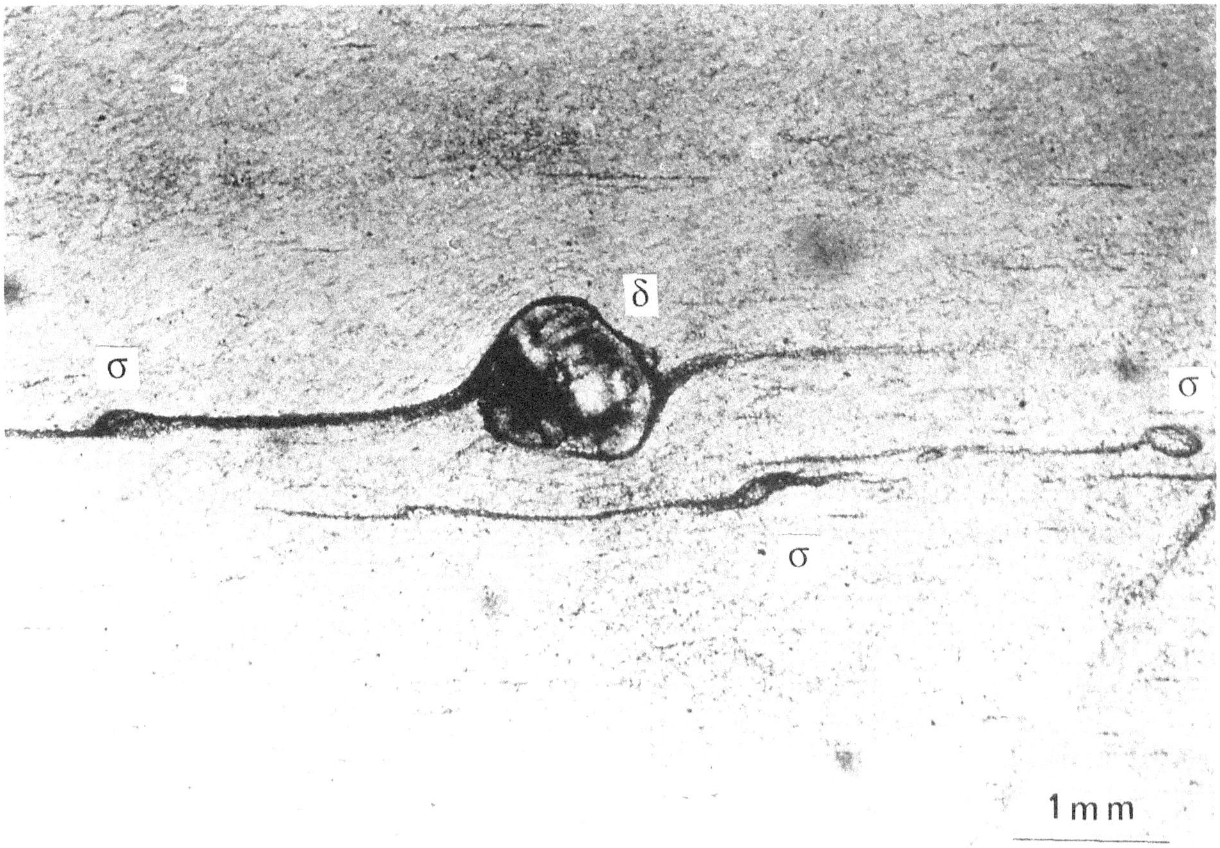
δ
σ
σ
σ
1 mm

107 A

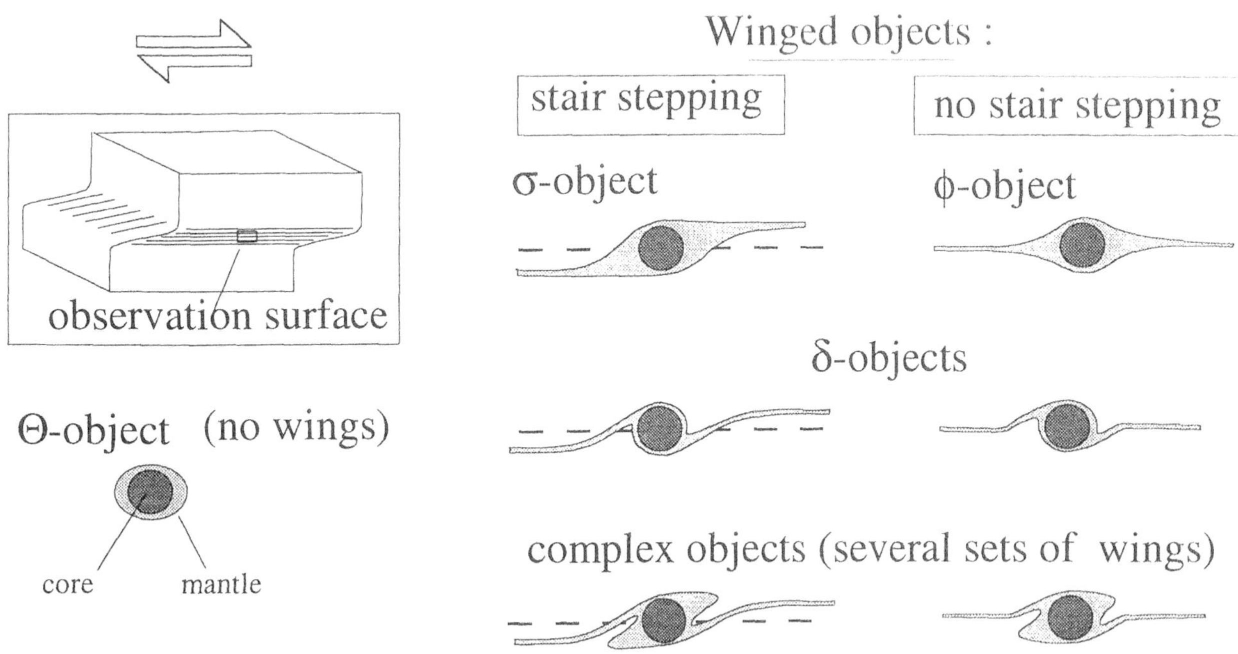
observation surface
Θ-object (no wings)
core
mantle
Winged objects :
stair stepping
no stair stepping
σ-object
ϕ-object
δ-objects
complex objects (several sets of wings)

107 B

# 108. SENSE-OF-SHEAR INDICATORS (IN THREE DIMENSIONS) IN MYLONITIC GRANITIC ROCKS

HIDEO TAKAGI

The Median Tectonic Line (MTL) is a major tectonic boundary dividing southwest Japan into Inner (northern) and Outer (southern) zones. It has a strike length of nearly 1000 km and extends from the Kanto region through Chubu, Kinki, and Shikoku to the Kyushu region. It forms the boundary between paired metamorphic belts: one is a low-pressure/high-temperature belt of cordierite-sillimanite rocks closely associated with granitic rocks in the north (Ryoke belt), and the other is a high-pressure/low-temperature, glaucophane-bearing belt of crystalline schists in the south (Sambagawa belt) (Figure 108.1). In some areas, however, the Ryoke belt is in contact with the Chichibu, Kurosegawa, or even Shimanto terranes: the MTL thus can be defined as the southern marginal fault of the Ryoke belt. A several hundred meter thick mylonite zone has developed along the MTL in the Ryoke belt (so-called Kashio mylonite) in the Chubu and Kinki regions and partly in the Kanto region (Figure 108.1).

The adjacent photographs show polished rock slabs of typical mylonitic granitic rocks along the MTL, Kashio area, Nagano Prefecture, Chubu region. The protolith is banded gneissose tonalite and granodiorite (Hiji tonalite, which is one of the oldest of the Ryoke granites). The recrystallized mineral assemblage suggests that the mylonitization occurred under epidote-amphibolite- to upper greenschist-facies metamorphic conditions. K-Ar age data for deformed and nondeformed granitic rocks suggest that the mylonitization took place during the Late Cretaceous (90–70 Ma; Shibata and Takagi, 1988).

The mylonitic foliation strikes nearly parallel to the MTL and dips west vertically to steeply. Stretching lineations trend subparallel to the MTL; the plunge is nearly horizontal (Takagi, 1986).

**108A.** Polished rock slab of strongly deformed mylonite (XZ plane of the finite strain ellipsoid) composed of fine-grained matrix minerals (quartz, biotite, plagioclase, K-feldspar, and various accessory phases) and porphyroclasts of plagioclase and K-feldspar with asymmetric pressure shadows that indicate a sinistral sense-of-shear. Original, centimeter-scale compositional banding is preserved, whereas newly developed, millimeter-scale fluxion banding composed of stretched, quartzo-feldspathic tails or pressure shadows is also present.

**108B.** Polished rock slab of strongly deformed mylonite (XZ plane) derived from gneissose granodiorite. Fluxion banding is highly developed. Kinematic indicators such as asymmetric pressure shadows, displaced broken grains (DBG), and fine-grained, fish-shaped hornblende porphyroclasts (H) indicate a sinistral sense of shear.

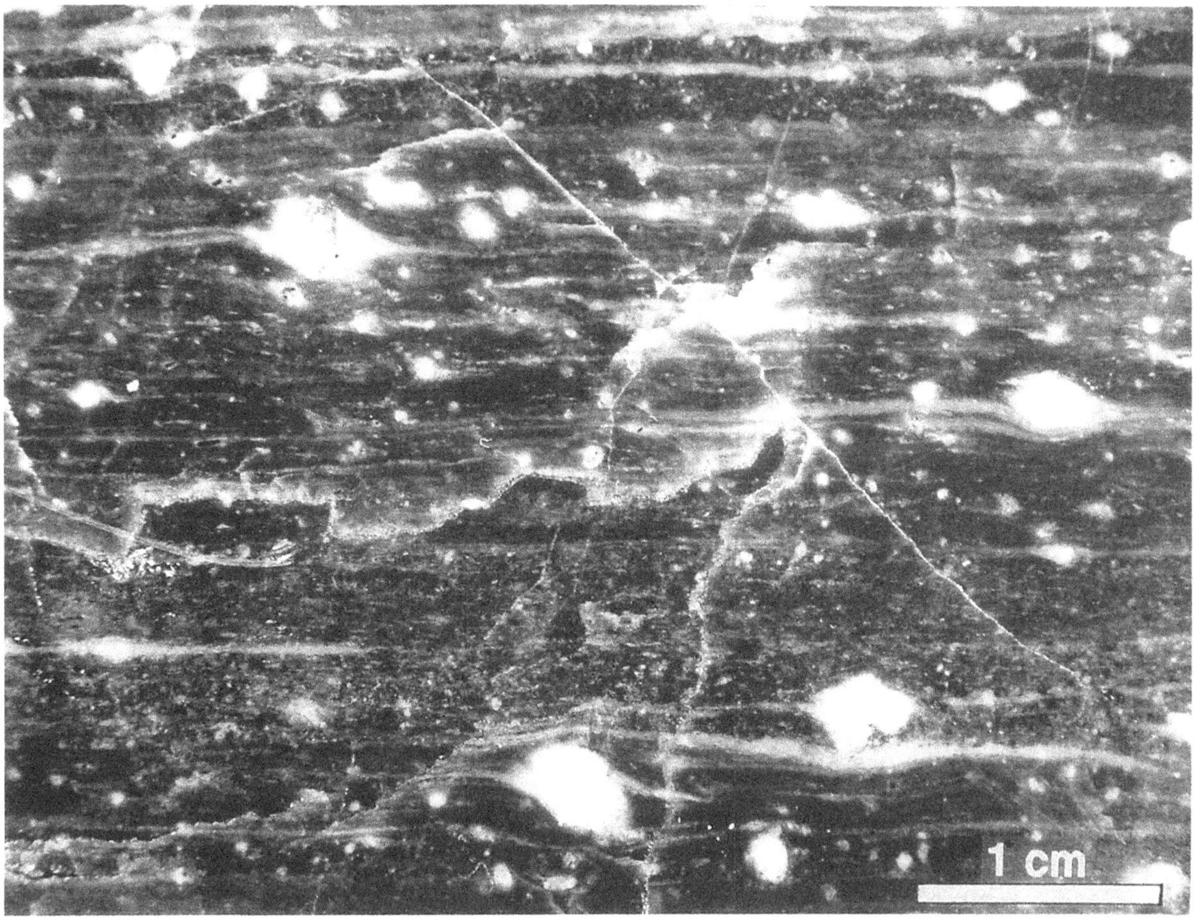


108 A

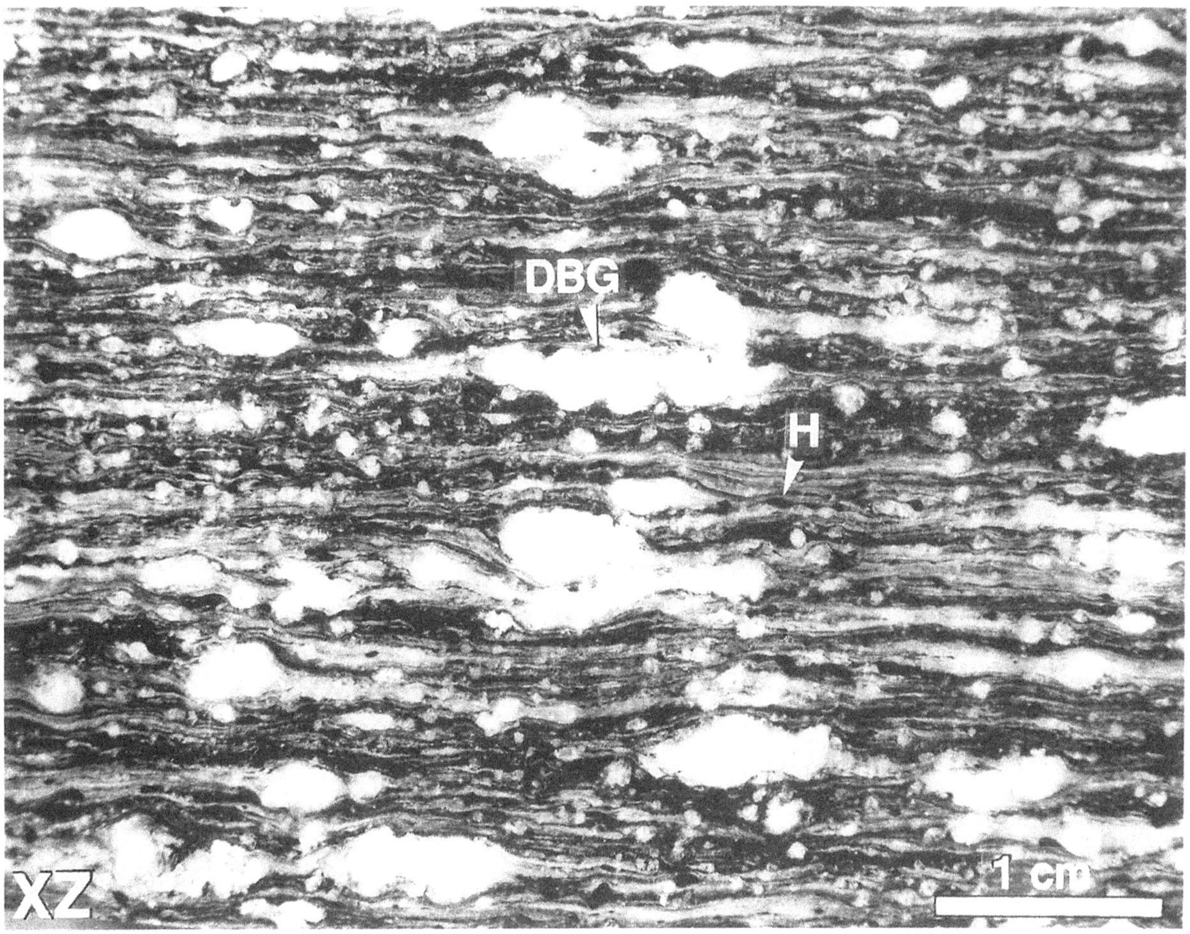


108 B

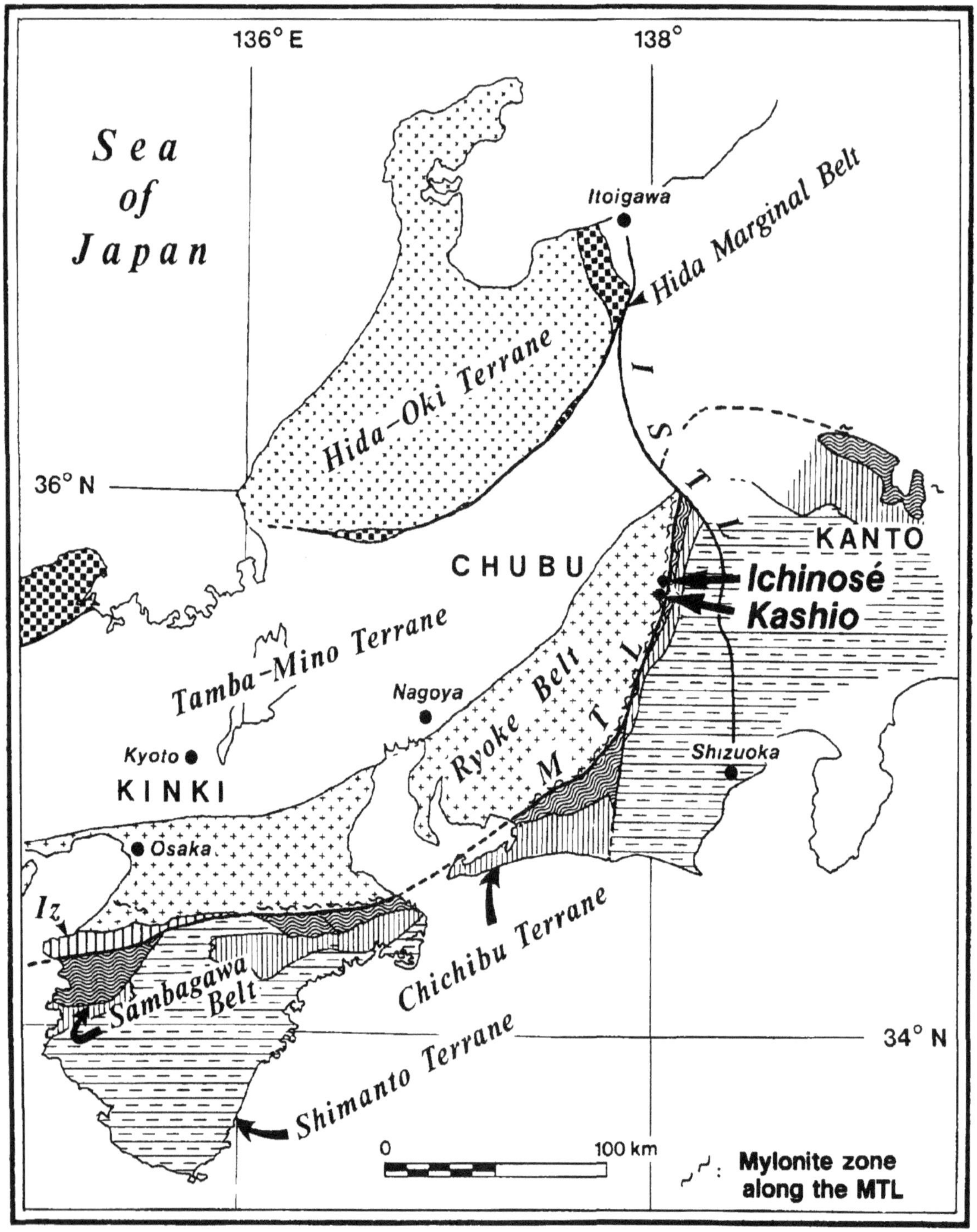


**Figure 108.1.** Geologic sketch map of the Chubu region, southwest Japan. The sample locations (Kashio and Ichinose areas) of Plates 108A–108D are indicated, as well as those featured in Plates 98, 99, and 109. MTL = Median Tectonic Line, ISTL = Itoigawa-Shizuoka Tectonic Line.

**108C.** Polished rock slab of the same rock as in 108B, but here cut parallel to the XY plane. A stretching lineation defined by stretched mineral aggregates and stretched pressure shadows is parallel to the scale bar. Note beardlike pressure shadow domains (arrows).

**108D.** Polished rock slab of the same rock as in 108B, but here cut parallel to the YZ plane (i.e., perpendicular to the stretching lineation). No asymmetric microstructure can be observed.

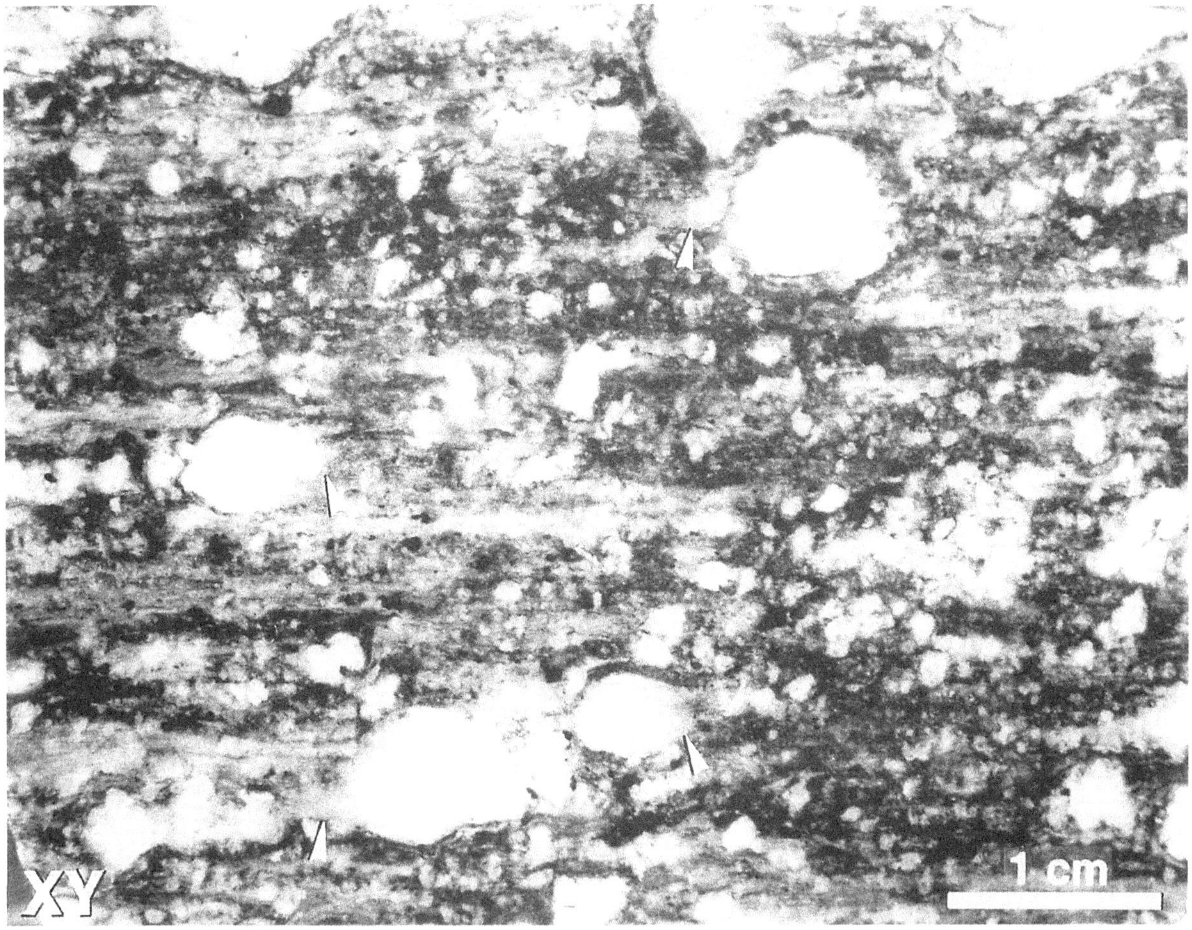


108 C

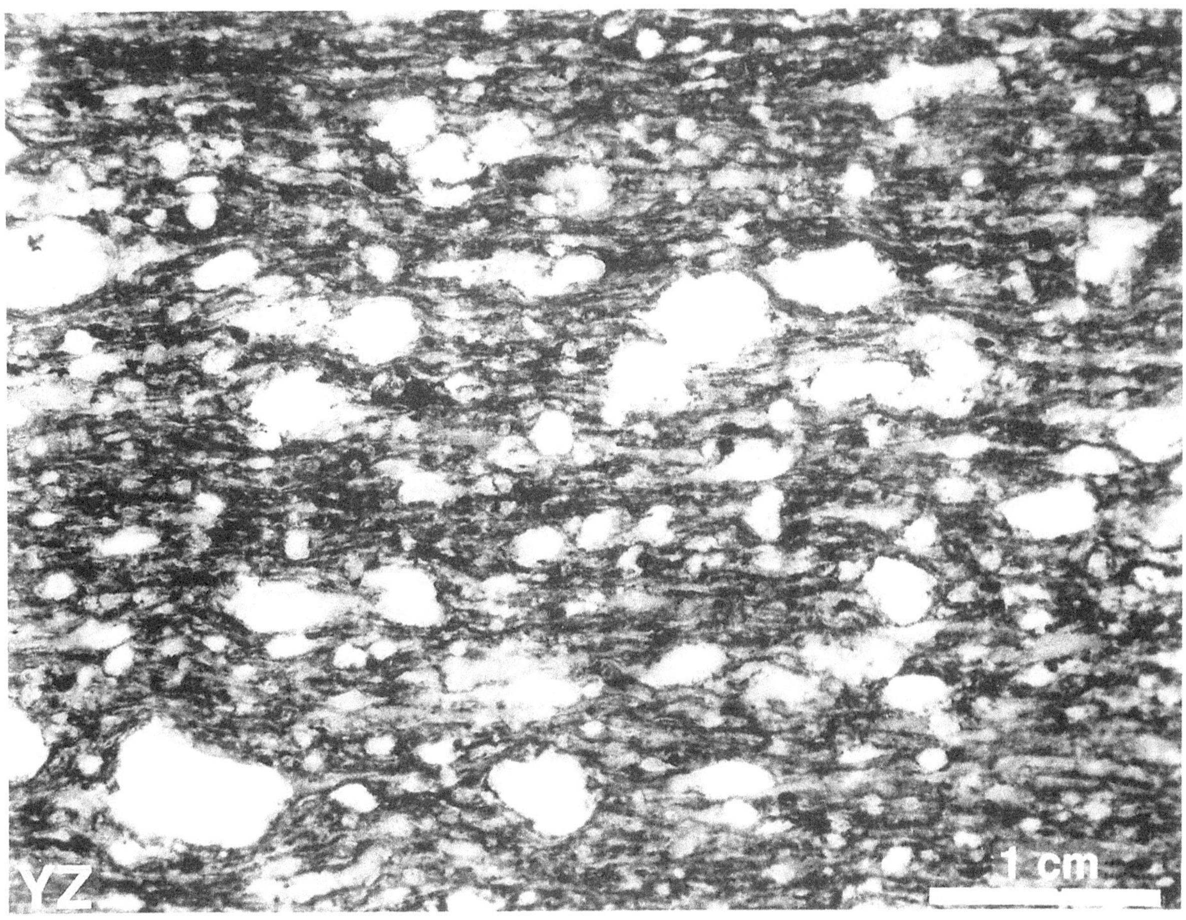


108 D

# 109. ROTATED PORPHYROCLASTS IN MYLONITES

HIDEO TAKAGI

Asymmetric pressure shadows on both sides of rotated porphyroclasts are a useful kinematic indicator in mylonitic granitic rocks. The drag angle (see Figure 109.1) between the boundary of the shadow domain (s) and the matrix and mylonitic foliation ($S_m$) is larger in the upper right-/lower left-hand sides (βi, βiii) than in the upper left-/lower right-hand sides (βii, βiv) in the case of mylonites showing a sinistral sense of shear, as illustrated in the adjacent photographs. This "drag angles method" allows the determination of the sense of shear without any ambiguity, whether the microstructure is a clast-pressure shadow system or clast-tail system, or whether the system is of σ type or δ type (Takagi and Ito, 1988).

**109A.** σ-type asymmetric pressure shadows on both sides of K-feldspar porphyroclasts. The shadows are composed of an aggregate of quartz and K-feldspar. This sample is the same one as used in Plate 108A. The thin section is cut parallel to the XZ plane of the finite strain ellipsoid. NX

**109B.** δ-type asymmetric pressure shadows on both sides of a plagioclase porphyroclast. The shadows are composed of an aggregate of K-feldspar and quartz with minor muscovite. The stretched shadow on the left-hand side connects on top of the other porphyroclast in a stair-stepping manner (outside the photograph), which is also an indicator of the sense of shear and rotation of porphyroclasts during ductile shear. The thin section was cut parallel to the XZ plane of the finite strain ellipsoid. NX

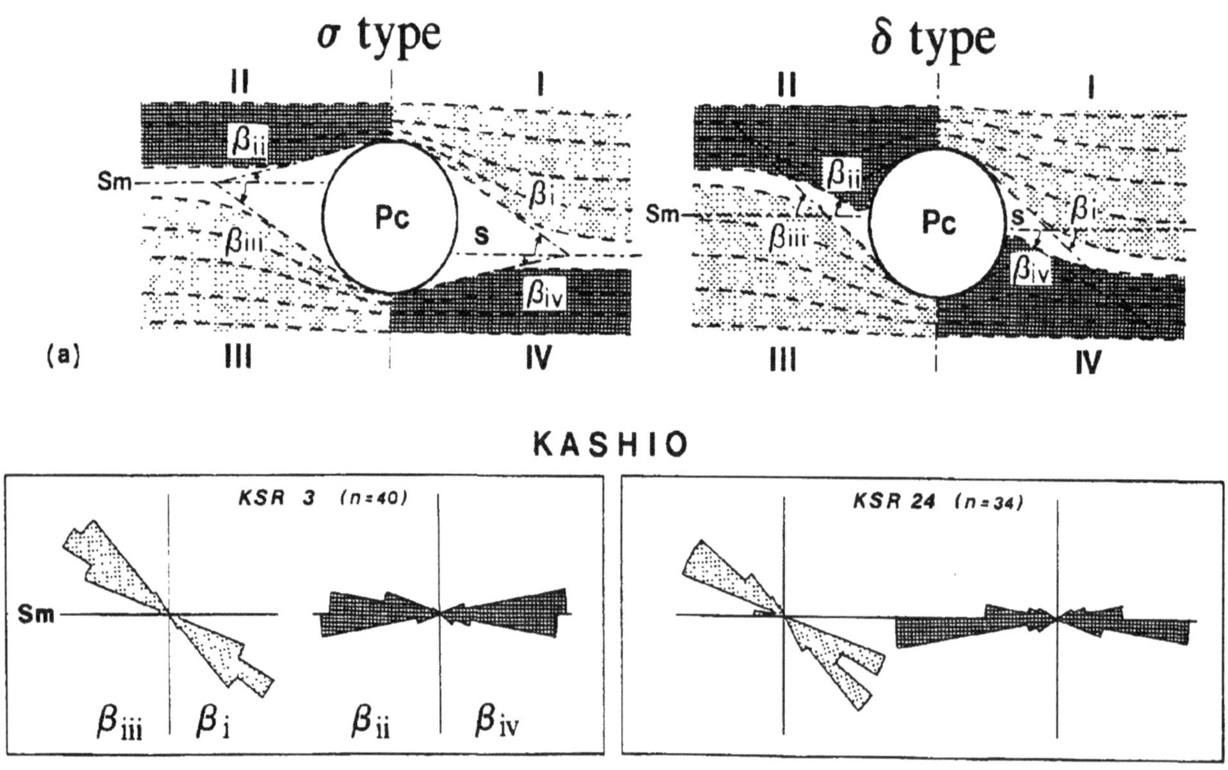


**Figure 109.1.** (a) Definition of drag angles (βi, βii, βiii, βiv) in quartered domains around a porphyroclast (Pc). s: shadow domain. (b) Rose diagrams showing the frequency distributions of the drag angles in each sample from the Kashio area. After figure 12 of Takagi and Ito (1988). The sample KSR 3 is shown in Plate 109A.

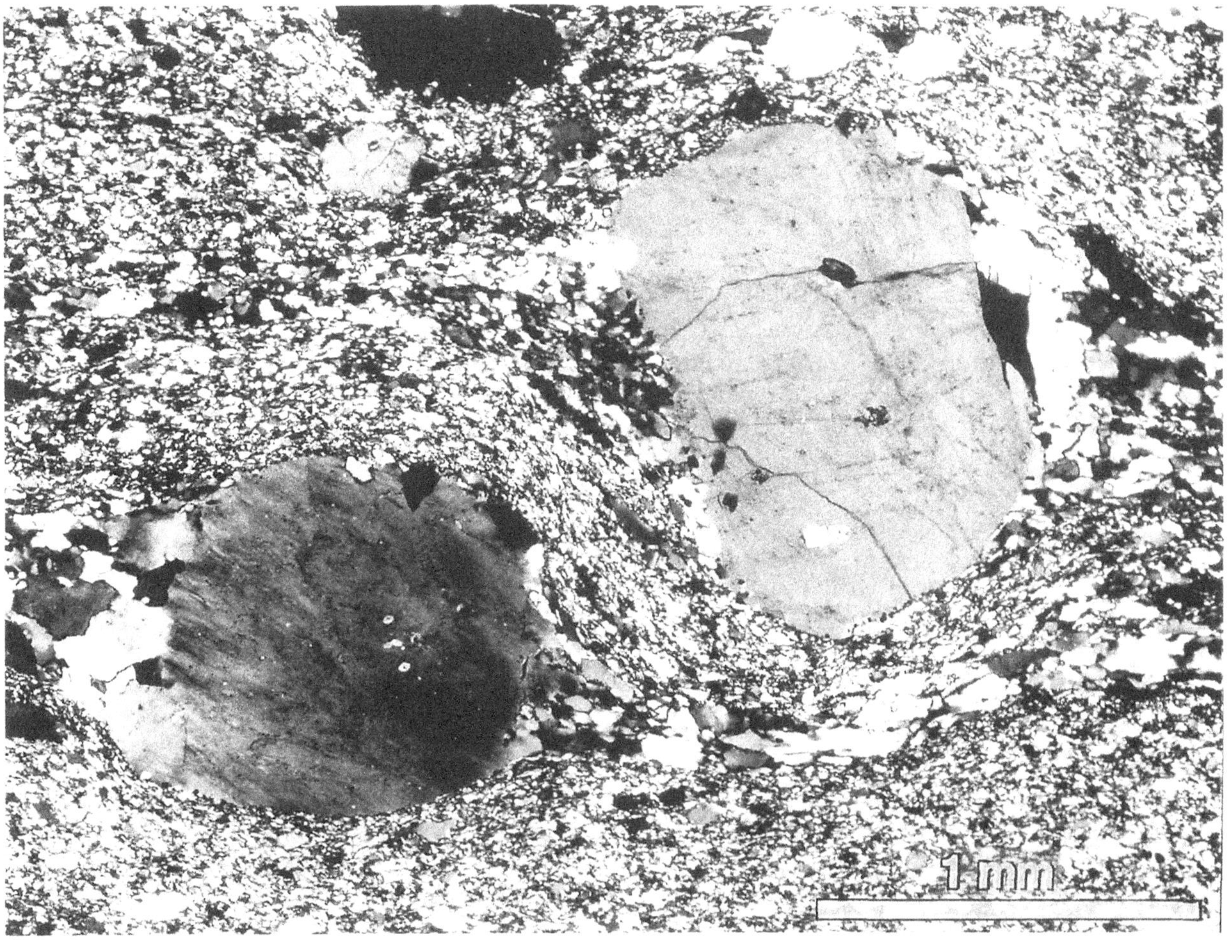

109 A

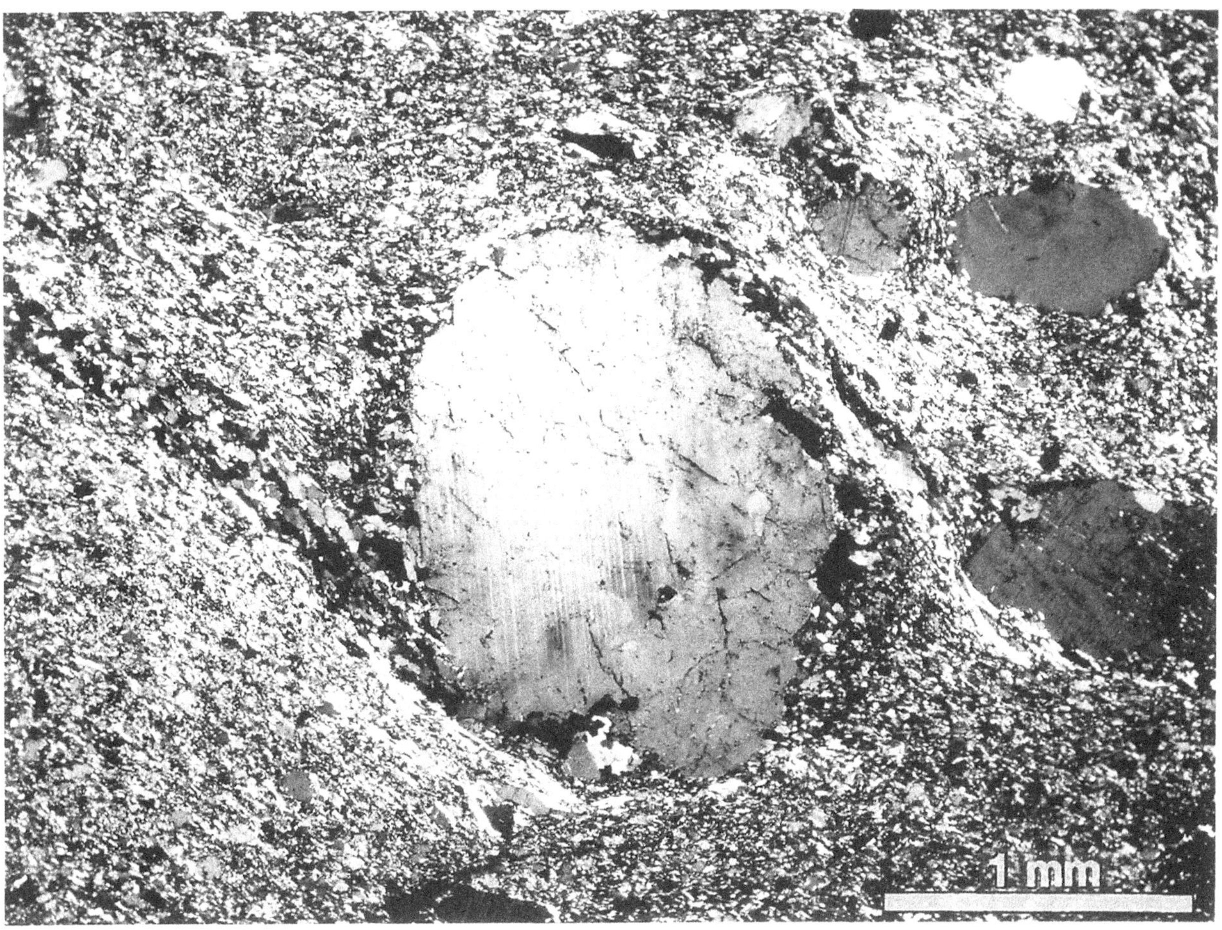

109 B

# 110. δ CLAST IN AN EXPERIMENTAL MYLONITE

Coen E. ten Brink and Cees W. Passchier

Example of the development of a δ-type porphyroclast in an experimental mylonite in organic crystalline rock analogs (ten Brink and Passchier, 1995). Relatively competent clast consists of camphor (Urai and others, 1980; Bons, 1993); octachloropropane (OCP) is matrix material (Means, 1989; Jessell, 1986; Ree, 1991). In order to achieve high strains, a torsional transparent deformation cell with a circular shear zone was used. Shear strain rate was $2.6 \times 10^{-3}s^{-1}$ at a temperature of 60°C (approximately 0.75 homologous temperature).

**110A–110F.** Evolution of a camphor object during deformation. The initial rectangular object (110A) started to shed fine-grained material from its rim, probably by dynamic recrystallization. Up to a shear strain (γ) of 19 (110B), this fine-grained material stretched into the surrounding OCP matrix at the two opposite corners of the rectangular object that were in the extensional quadrants (top right and bottom left) of the flow. With continuing deformation the other two corners rotated into a similar position and also started to shed material (110C, γ = 37). The resulting "wings" of fine-grained material did not detach from the camphor core and with increasing strain they stretched, thinned, and even "wrapped around" the central object (110D, γ = 56; 110E, γ = 75; 110F, γ = 94). Embayments of OCP formed between the smoothly curved wings and caused the typical δ shape. After initial development of the wings, the camphor objects did not shed any new fine-grained material but continued to deform internally. The wings did not stretch along a single "flow plane" through the center of the object, but show stair-stepping; they step up over the object in the shear direction. Darker area below broken line in 110F is edge of inner grip (see ten Brink and Passchier, 1995, for apparatus).

Dextral shear, PPL, center of rotation is downward, black dots are inert marker particles that were used to trace the flow pattern (Figure 110.1).

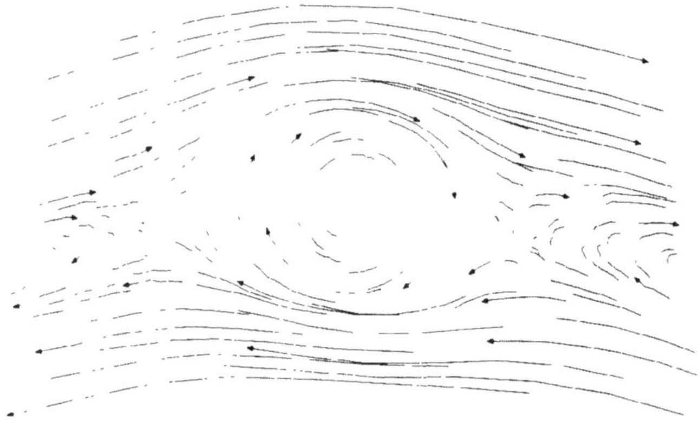

**Figure 110.1.** Flow pattern with center of clast as origin for a shear interval that included Plate 110F Stair-stepping is the result of the "bow-tie" flow pattern, which is thought to occur only in materials that show a power-law strain-rate dependency on stress (Passchier and others, 1993)

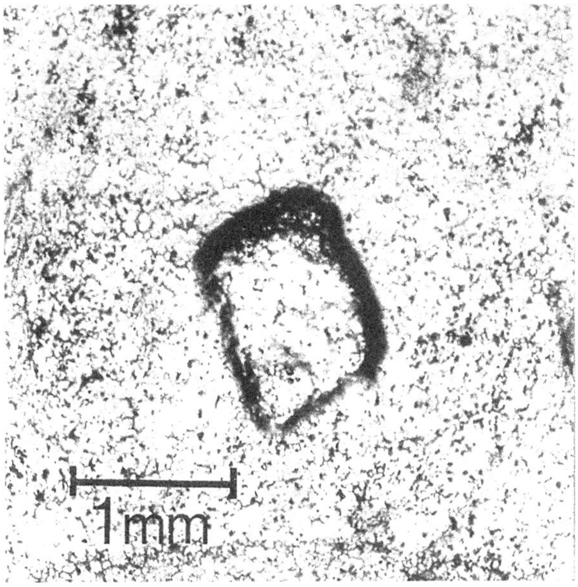


110 A

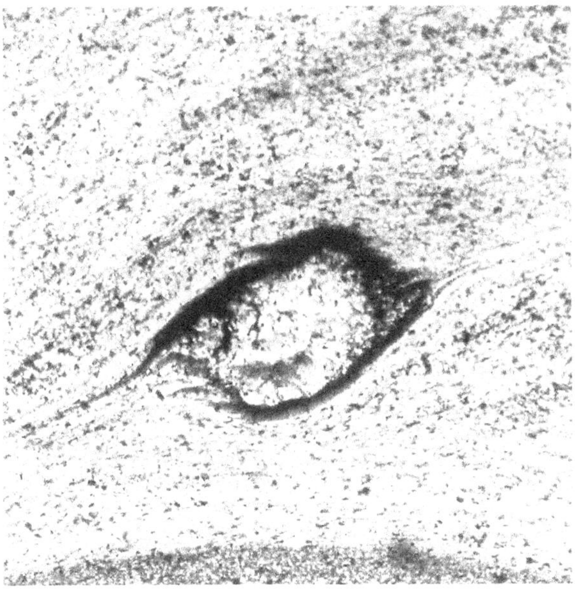

110 B

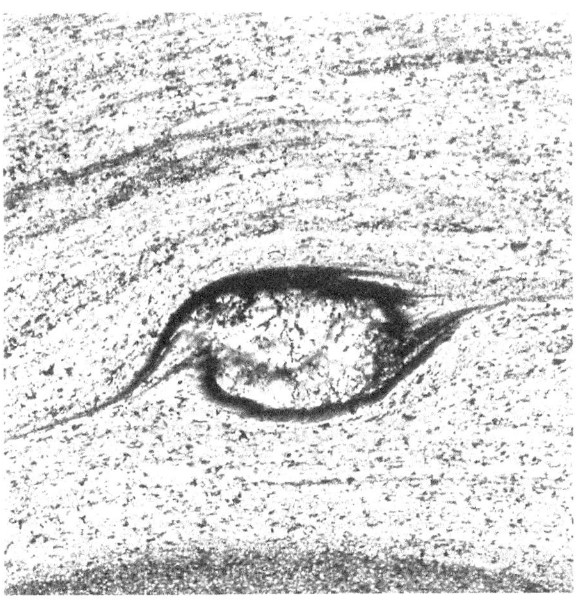

110 C

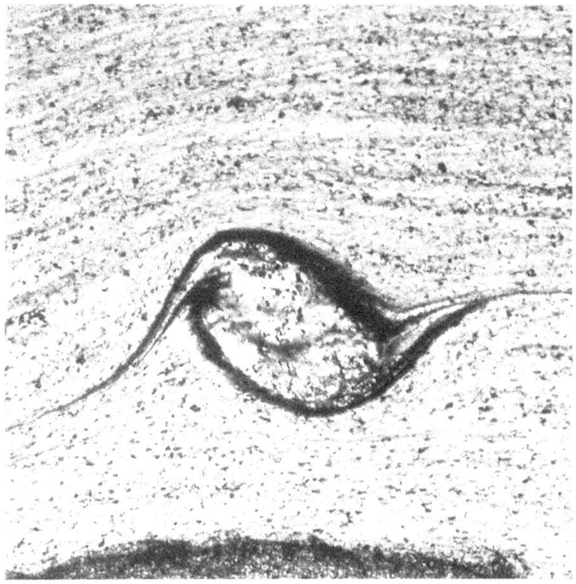

110 D

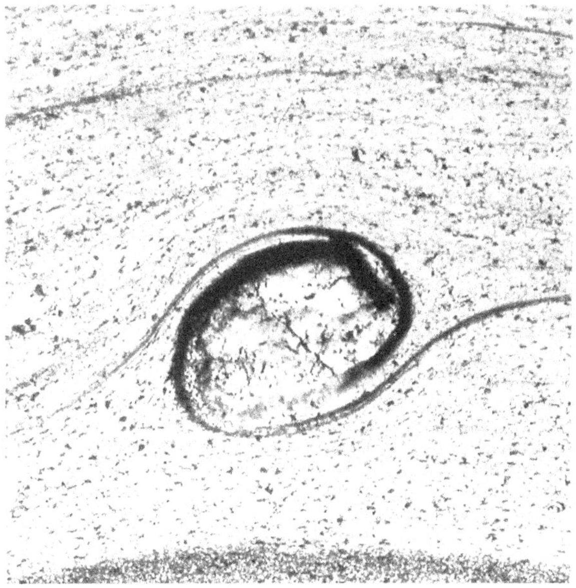

110 E

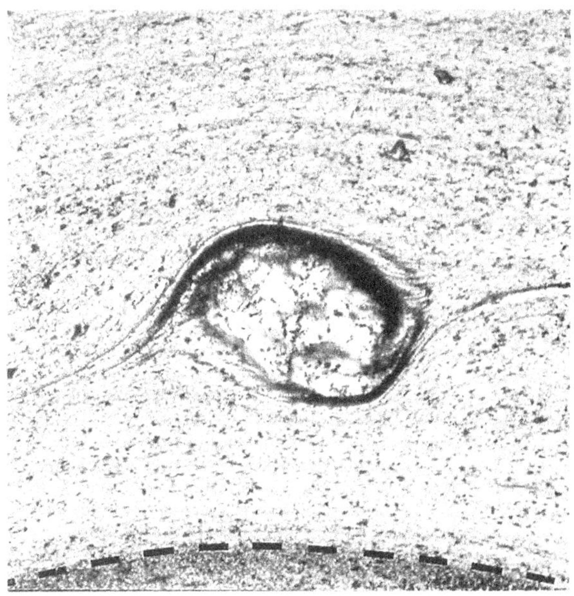

110 F

# DUCTILE BEHAVIOR

## Folds in mylonitic zones



Abbreviation used in regard to the plates include PPL = plane polarized light.

Progressive general shear tends to amplify initial heterogeneities and to produce folds on various scales. After high strain, the limbs will be parallel and the hinge line may be curvilinear, as in a sheath fold. Folds in shear zones are commonly difficult to recognize; but if hinges can be identified and a consistent vergence established, the folds can serve as sense-of-shear indicators, especially if the folding and high strain were synchronous. However, folds in mylonitic shear zones can be confusing for kinematic analysis, because they may represent earlier features that were subsequently flattened and stretched during the shear-zone deformation. The most useful and consistent folds are those that formed late in the mylonitic deformation. This set of plates illustrates folds of various scales associated with mylonite zones; in some cases the mylonitic foliation itself has become folded during progressive strain. One plate illustrates folds developed in two-phase, organic analog material in an experimental mylonite zone.

# 111. FOLDS AND OTHER MESOSCOPIC STRUCTURAL FEATURES ASSOCIATED WITH AN EXTENSIONAL MYLONITIC SHEAR ZONE

ARTHUR W. SNOKE

The most fundamental structural features in the Ruby–East Humboldt metamorphic core complex, northeast Nevada, U.S.A., are a kilometer-scale, mid-Tertiary extensional mylonitic shear zone and brittle detachment fault system (Wright and Snoke, 1993) (see Figure 58.1). The shear zone is characterized by top-to-the-west-northwest and normal-sense displacement (Snoke and Lush, 1984). The physical conditions of the mylonitization are locally well constrained (e.g., Hacker and others, 1990; Hurlow and others, 1991), but include a broad range of P-T conditions, because different structural levels in the shear zone are exposed along strike for >100 km (Snoke and others, 1990a). Furthermore, as the shear zone evolved, the ambient P-T conditions changed in response to the tectonic denudation of the cover rocks. Consequently, the character of the deformation changed from crystal-plastic to brittle (see Plate 58) and low-angle, semi-brittle normal faults were localized in the mylonitic shear zone. The mylonitic shear zone and detachment fault system are exposed spectacularly in the Secret Creek gorge area, northern Ruby Mountains (see Figure 111.1). The mesoscopic structural features shown here are all (with the exception of 111B) exposed in this area.

A variety of folds and other mesoscopic structural features is associated with the mylonitic shear zone. Some folds predated the shear-zone deformation, whereas others developed early or late during mylonitization. The older folds have been stretched and rotated during shear-zone deformation, whereas the later folds deform the mylonitic foliation but share a similar slip line with the mylonitic zone (Howard, 1968; Snoke and Lush, 1984).

**111A.** Rotated clast of competent paragneiss (left of knife handle) in mylonitic calcite marble. Note that a mylonitic fluxion structure is deflected around the clast, which itself contains a mylonitic foliation. Both foliations developed during a continuous, noncoaxial extensional deformation.

**111B.** Mylonitic foliation surface with strong WNW-trending elongation lineation in granitic orthogneiss. The white part of the foliation surface is a mylonitic quartz vein within the orthogneiss. Note conjugate joint system associated with the elongation lineation. The granitic orthogneiss is part of the Horse Creek assemblage (Snoke and others, 1990b) which is exposed in the southern East Humboldt Range, Nevada, U.S.A.

**111C.** Boudinaged folds defined by competent paragneiss layers in mylonitic calcite marble. Folding and boudinage are parts of a continuum of noncoaxial deformation that is characteristic of the mylonitic shear zone.

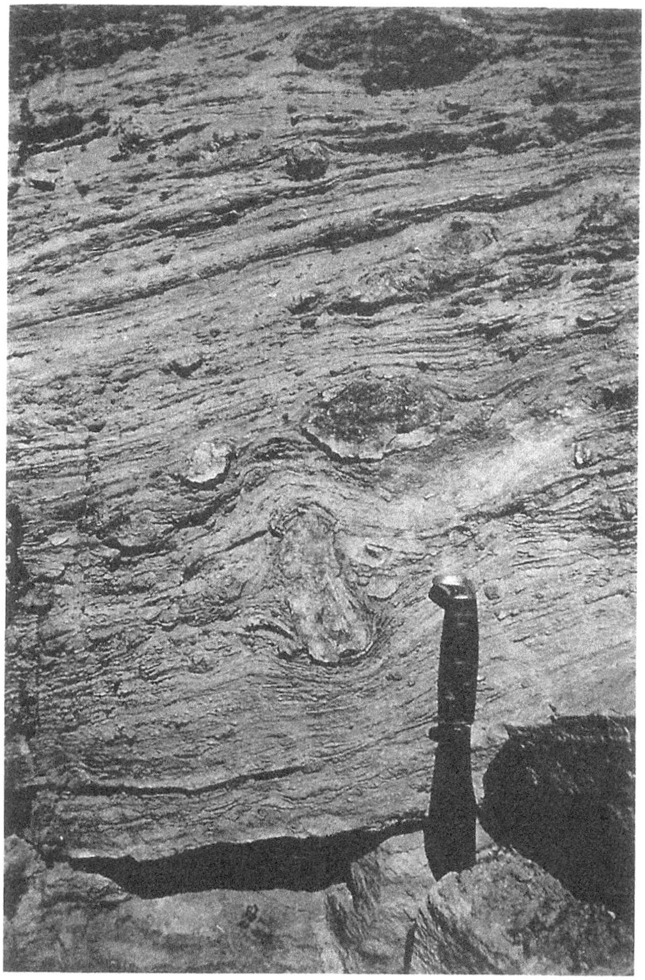

111 A

111 B

111 C

**111D.** Late, post-mylonitic folds in the mylonitic shear zone. The main rock types are m. impure calcite marble with pelitic interlayers; gn. biotite granodiorite orthogneiss; and qtz. flaggy white quartzite. This exposure is part of a cascade of asymmetric, west-verging folds within the mylonitic shear zone. These folds illustrate the noncoaxial character of deformation in the shear zone, wherein layers were folded in the shortening field despite an overall extensional deformation. Note the disharmonic nature of the folding, controlled by contrasting layer thickness and competency.

**111E.** Fold in mylonitic orthogneiss with thickened hinge zone and attenuated limbs. Note that the mylonitic foliation (i.e., gneissic layering) of the shear zone is deformed by this fold.

**111F.** Tightly folded interlayered pelitic schist and mylonitic leucogranite. Note that the mylonitic foliation in the leucogranite is folded. Foliation in the pelitic schist in the hinge zones of the folds has been rotated subparallel to the axial surfaces of the folded leucogranite layers. The vergence of the folds is westward.

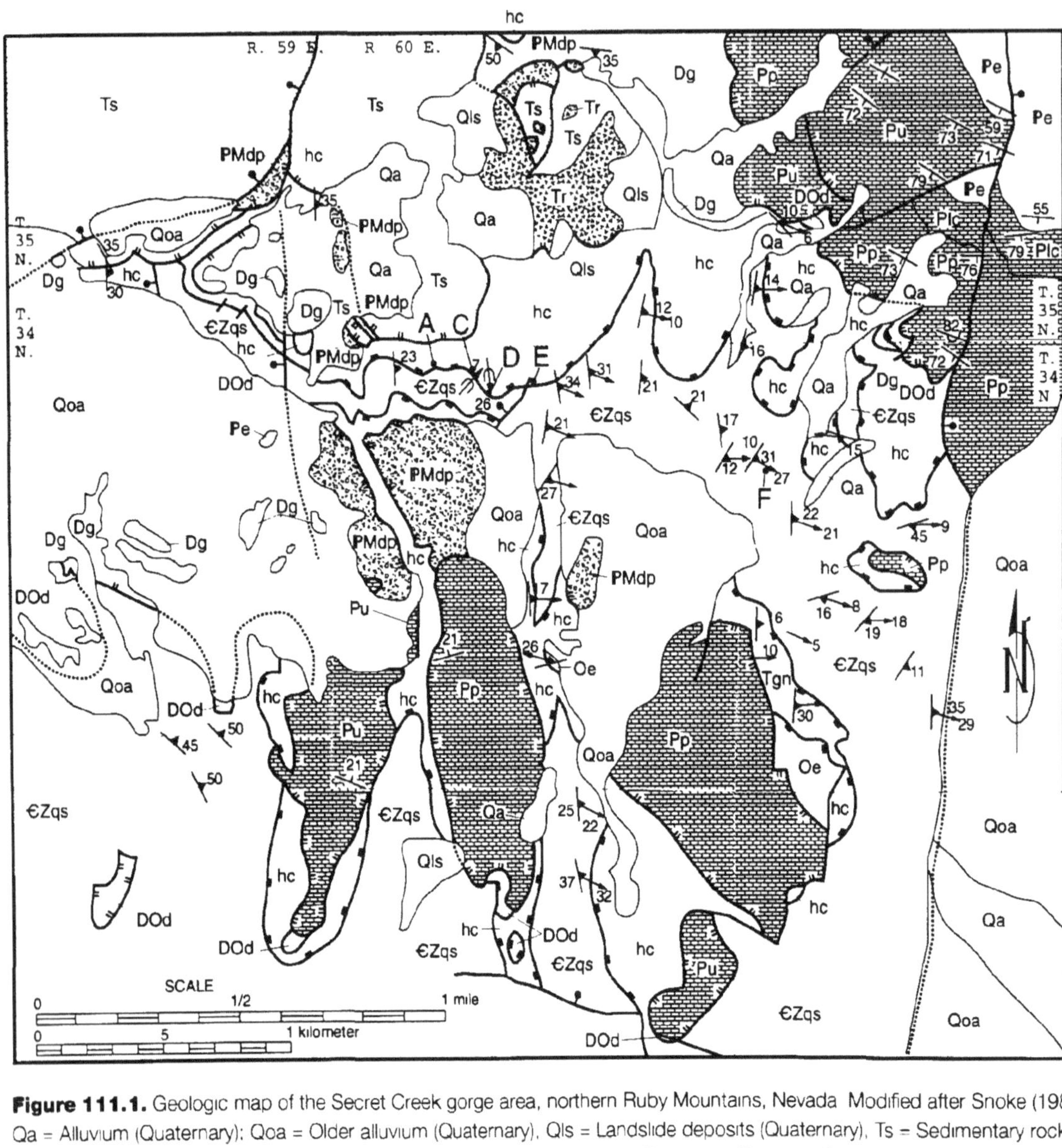


**Figure 111.1.** Geologic map of the Secret Creek gorge area, northern Ruby Mountains, Nevada. Modified after Snoke (1980, figure 5). Qa = Alluvium (Quaternary); Qoa = Older alluvium (Quaternary), Qls = Landslide deposits (Quaternary), Ts = Sedimentary rocks (Miocene); Tr = Rhyolite (Miocene), Tgn = Granitic orthogneiss (Tertiary); Pp = Pequop Formation (Permian); Plc = Limestone and conglomerate (Permian); Pu = Permian undivided; Pe = Ely Limestone (Pennsylvanian), PMdp = Diamond Peak Formation (Pennsylvanian-Mississippian); Dg = Guilmette Formation (Devonian), DOd = Metadolomite (Devonian-Ordovician), Oe = Eureka metaquartzite (Ordovician), hc = Horse Creek assemblage (metasedimentary and granitic rocks), €Zqs = Impure metaquartzite and schist (Cambrian and Neoproterozoic). Standard symbols for attitude of bedding or foliation, trend and plunge of lineation, hinge line of mesoscopic fold, geologic contacts, and normal faults [except for double tick marks on upper plate of brittle low-angle normal fault (detachment fault) and filled squares on upper plate of plastic-to-brittle low-angle normal fault (detachment fault)]. Faults are dotted where concealed. The localities of Plates 111A, and 111C–111F are indicated on the map; Plate 111B was taken in the southern East Humboldt Range, approximately 12.5 km to the east-southeast of the center of the map area.

111 D

111 E

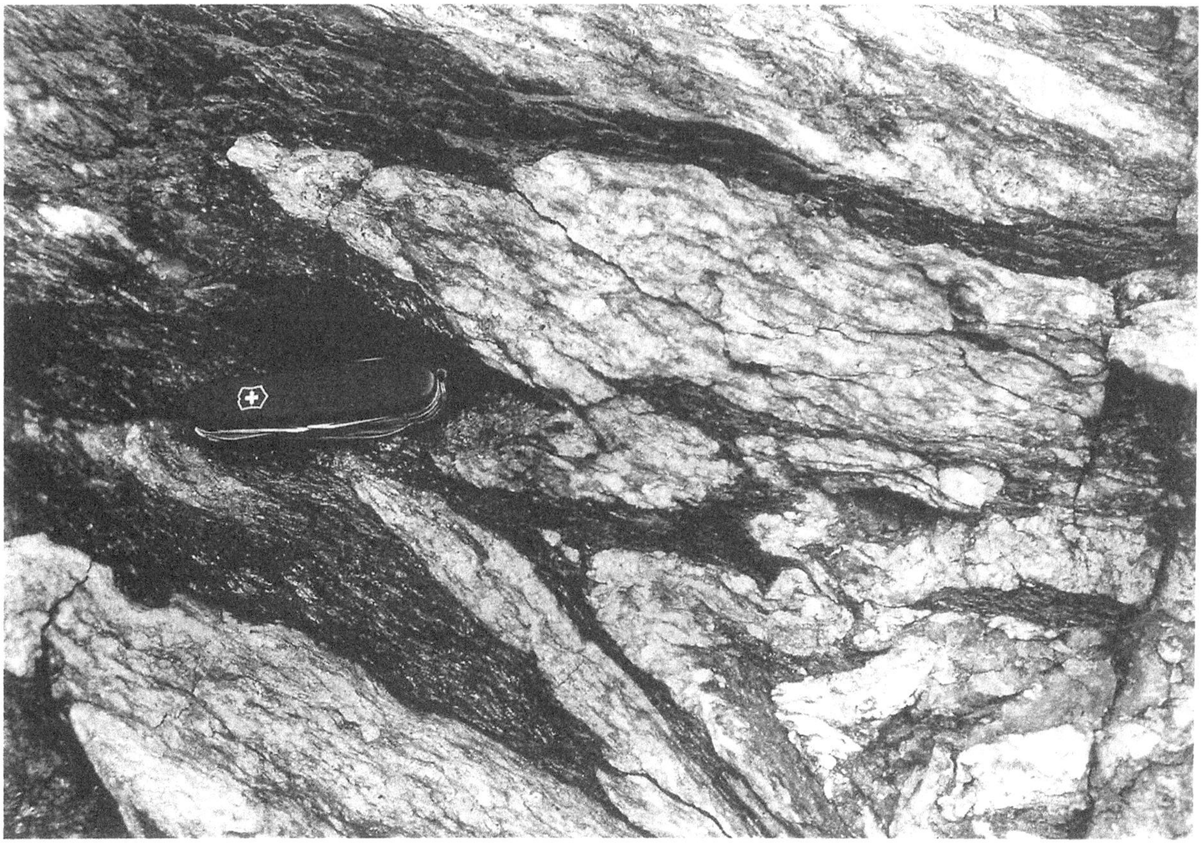
111 F

# 112. CURVATURE OF HINGE LINES ABOUT A STRETCHING LINEATION

T. H. BELL

Plates 112A and 112B show curved hinge lines in mylonitic foliation ($S_m$) from the Woodroffe thrust mylonite zone, central Australia. Curvature of hinge lines about a stretching lineation can occur in three ways (Bell, 1978a): (1) rotation of hinge lines toward the stretching lineation due to inhomogeneous strain; (2) apparent rotation of hinge lines due to folding of an originally anastomosing foliation; or (3) some combination of these two.

**112A.** This photograph shows apparent rotation of the hinge line due to folding of an originally anastomosing portion of $S_m$. Mylonitic foliations commonly anastomose on a large variety of scales around less deformed ellipsoidal pods (see Plate 66). Such variable $S_m$ surfaces are commonly refolded during progressive mylonitization of the ellipsoidal core (Bell, 1978a). The axial surfaces of these later folds intersect the anastomosing $S_m$ such that associated hinge lines and intersection lineations are *initially* strongly curved. As a consequence, these hinge lines can form subparallel to stretching lineations on the earlier anastomosing foliation, although the folds can be quite open. Such folds are common in mylonite zones.

**112B.** This photograph shows a rock that has fractured parallel to a mylonitic foliation ($S_m$) on two separate levels 10 mm apart such that a curved lineation is exposed on one surface and a straight lineation on the other. Both lineations are due to the intersection of an early formed mylonitic foliation by a later one. A weak stretching lineation can also be observed on the surface containing the curved intersection lineation, and it lies exactly parallel to the straight intersection lineation on the other surface. The curvature of the intersection lineation can be attributed to mechanisms (1), (2), or (3) above. The straight intersection lineation has been rotated to be, and/or formed close to, parallel to the stretching lineation as described above in 112A. Small-scale folds of the earlier $S_m$ are preserved on the edges of this specimen at right angles to the surface photographed, and their hinge lines are parallel to the intersection lineations.

112 A

112 B

# 113. FOLDING IN EXPERIMENTAL MYLONITES

Paul D. Bons and Mark W. Jessell

Examples are given here of folds that developed in experimental "mylonites" composed of the organic crystalline rock analogs octachloropropane (OCP) and camphor (Means, 1989). OCP is softer than camphor (Bons and Urai, 1994).

**113A–113C.** Microstructural development in a deforming aggregate of relatively soft OCP (68 vol. %, dark) and more competent camphor (32 vol. %, light) deformed in a hollow-cylinder shear apparatus (Bons, 1993; Bons and Cox, 1994) (Figure 113.1).

**113A.** Starting material of camphor inclusions surrounded by OCP.

**113B.** $\gamma = 16$. Camphor inclusions have deformed into lenses and layers.

**113C.** $\gamma = 122$. The camphor inclusions were sheared into layers that subsequently folded. Folding is disharmonic and caused by instabilities arising from the interaction of the camphor layers and lenses. Folding starts with buckling with the axial plane perpendicular to the shear plane, followed by rotation of the axial plane and tightening of the folds (Figure 113.2). Dextral shear-strain rate is $5 \times 10^{-5}\ s^{-1}$ at a temperature of 30°C.

**113D–113G.** Fold developing in a camphor layer surrounded by OCP in a transparent deformation cell [see Jessell and Lister (1990) for apparatus]. The sequence commences (113D) after the initial phase of buckling. This is followed (113E and 113F) by tightening of the fold, by a combination of hinge rollover, rotation, and stretching of the overturned forelimbs until final transposition where all limbs are approximately parallel (113G). Sinistral shear-strain rate is $1 \times 10^{-3}\ s^{-1}$ at room temperature. PPL

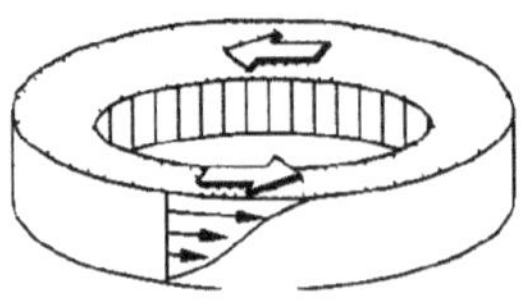

**Figure 113.1.** Hollow-cylinder deformation geometry

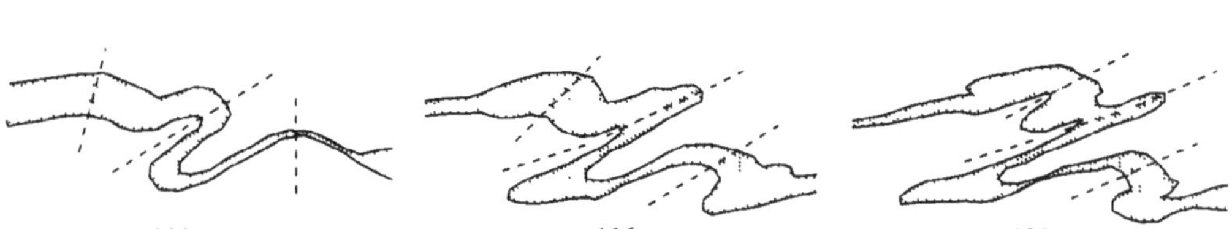


**Figure 113.2.** Drawing of the development of one fold train visible in Plate 113C

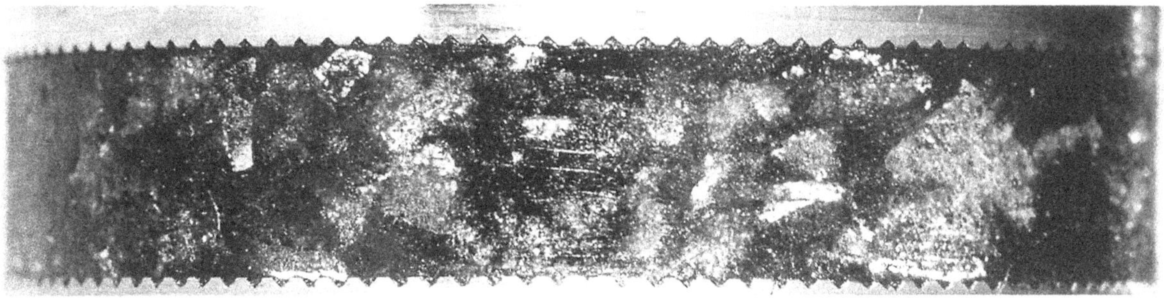

113 A

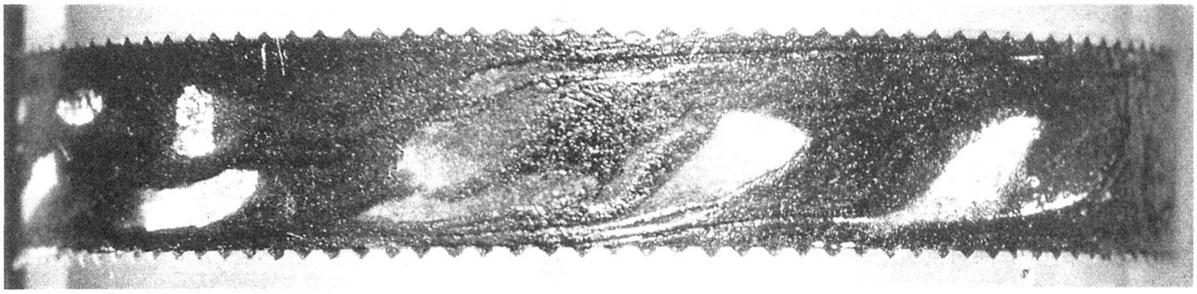

113 B

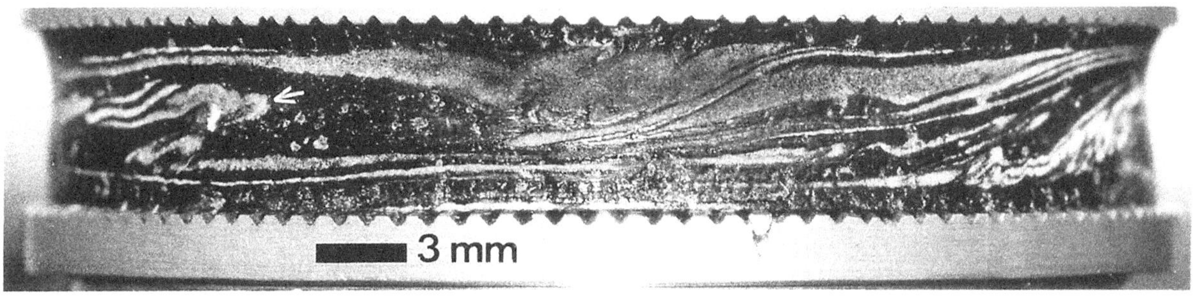


113 C

113 D

113 E

113 F

113 G

# DUCTILE BEHAVIOR

## Evidence of coaxial deformational histories



Common abbreviations used in regard to the plates in this section include PL = plane light, NX = crossed nicols, and TEM = transmission electron microscope.

Most mylonites are thought to result from simple or general shear, and the abundance of asymmetric structures such as composite foliations, rotated porphyroclasts, and asymmetric crystallographic preferred orientations is consistent with that interpretation. However, highly localized strain can sometimes result from coaxial deformation. For example, mylonitic textures have been produced in high-strain axial compression experiments. The plates in this section illustrate some examples of natural mylonites interpreted to have resulted from coaxial deformation. The evidence for coaxial deformation includes the presence of relatively undeformed globular grains poorly oriented for slip; the pancake shapes of grains and xenoliths; and the presence of crystallographic preferred orientations that are symmetric to the foliation.

# 114. COAXIAL DEFORMATION IN QUARTZITE MYLONITE

C. K. Mawer

A mylonite zone (the Fish Hole shear zone) that is 3–10 m thick and in excess of 10 km long is developed at the margins of the Chewings Range Quartzite, a pure orthoquartzite consisting of about 98% quartz and 2% sillimanite, the latter retrogressed to muscovite during mylonitization. The Chewings Range Quartzite is part of the Middle Proterozoic Arunta Complex of central Australia (see Figure 114.1 for location information). The quartzite was mylonitized during the Carboniferous Alice Springs orogeny (340 ± 18 Ma) at a temperature of approximately 375°C and under a differential stress of approximately 120–160 MPa (determined from the measurement of dynamically recrystallized quartz grain size). Mylonite formed in narrow, steeply dipping zones at or near the boundaries of the Chewings Range Quartzite with the surrounding schists and gneisses, as a result of deformation concentration at these lithological interfaces. The finite strain determined from measurements of deformed quartz grains is an extreme flattening strain (Mawer, 1983), but is quite inhomogeneous on both mesoscopic and microscopic scales. The mylonite deformed principally by dislocation creep. The deformation in the mylonite zone leading to the formation of these mylonites involved essentially coaxial straining, superimposed on an earlier annealing event which was itself superimposed on an even earlier, strong noncoaxial straining history (Mawer and Fitz Gerald, 1993).

All four photographs are taken from thin sections oriented *perpendicular* to both lineation and foliation; the mean trace of foliation is horizontal in all photographs.

**114A.** Sample 1. Note extreme aspect ratios of quartz grains, strong development of deformation bands, and minor dynamic recrystallization at grain and deformation band boundaries. NX

**114B.** Sample 2. "Augen" quartz grain with strong, irregular, undulose extinction and ribbon grains wrapped around it. Parts of ribbons above and below augen grain are more strongly deformed than the ribbon material to the sides. Compare with plate 149B by J.-L. Bouchez in Borradaile and others (1982); such augen grains are also sometimes referred to as *globular* grains. The augen grain has its ⟨*c*⟩ axis perpendicular to the mean trend of the foliation (flattening plane), whereas the ribbon grains have ⟨*c*⟩ axes at low angles (~20–30°) to the mean foliation trend. Thus, the augen grain was in an orientation relatively resistant to deformation during the relatively coaxial straining conditions recorded in these mylonites. NX

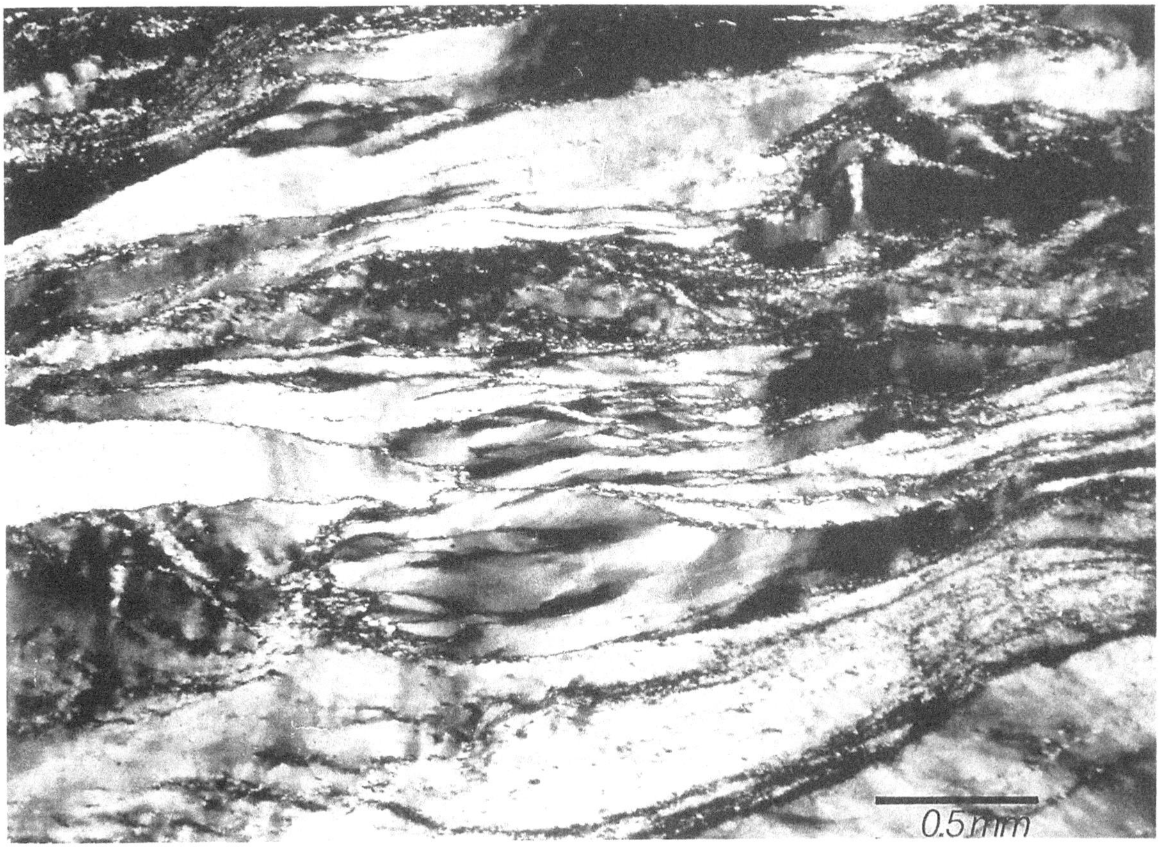


114 A

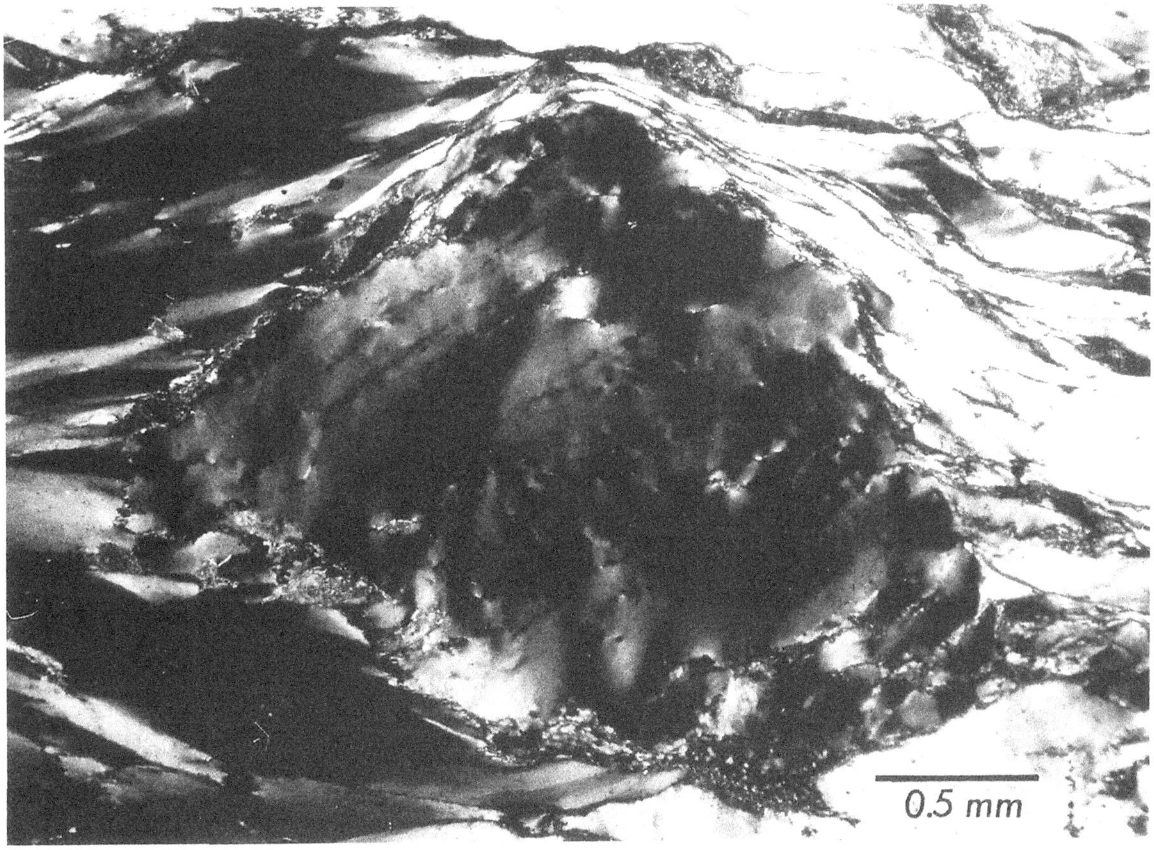


114 B

**114C.** Sample 2. Pronounced development of deformation bands in a single-ribbon quartz grain. Note lack of recrystallization. NX

**114D.** Sample 2. Grain-boundary bulging and migration at the end of a ribbon quartz grain. NX

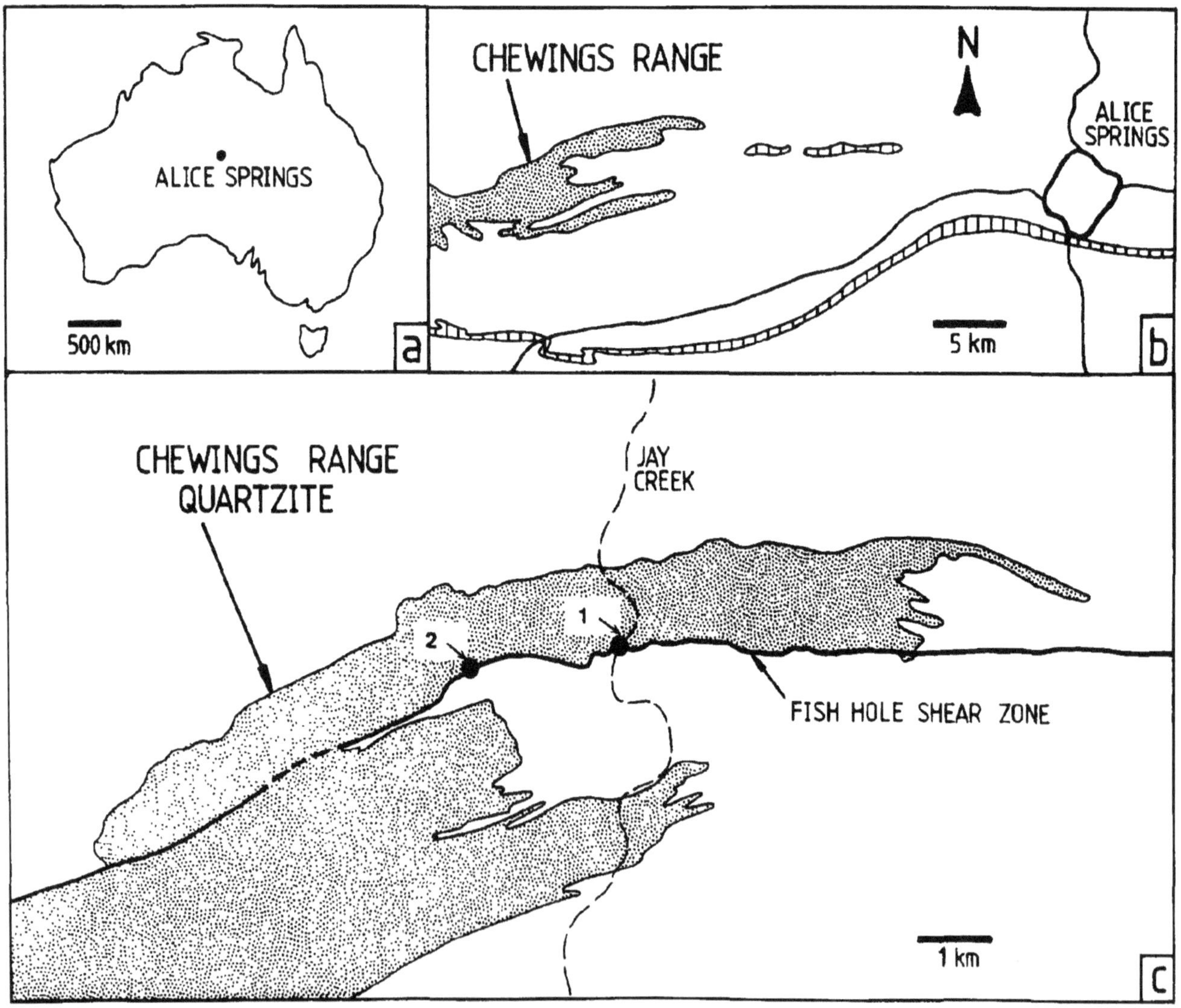


**Figure 114.1.** Location map (a) location of Alice Springs, central Australia; (b) location of the Chewings Range, west of Alice Springs; and (c) the Fish Hole shear zone, Chewings Range, showing the locations of samples 1 and 2

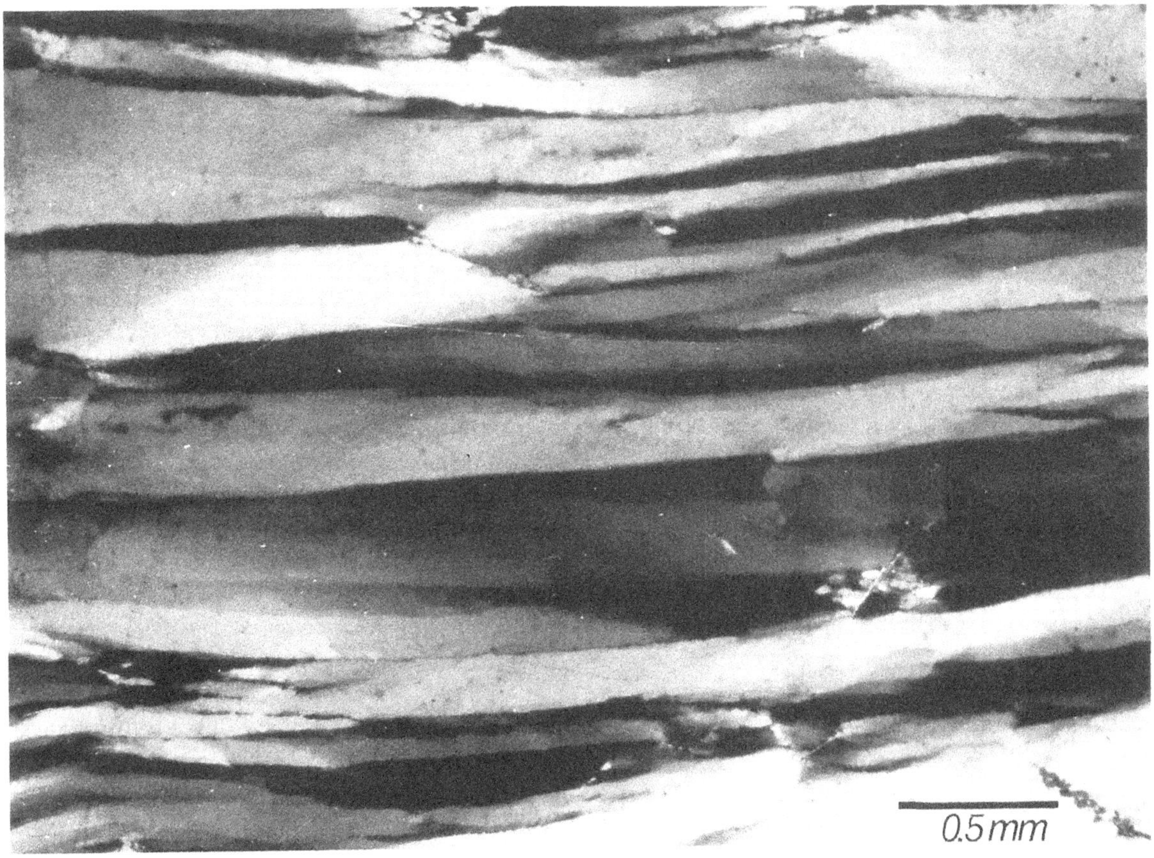


114 C

114 D

# 115. DEFORMATIONAL HISTORY INDICATORS?

T. H. Bell

Mylonites are commonly assumed to be the products of inhomogeneous, simple shear deformational histories. However, there is evidence within some mylonitic rocks which suggests that this is not always the case, including quartz *c*-axis fabric diagrams symmetrical about the foliation plane and changes in sense of shear along mylonitic foliation planes (Garcia-Celma, 1983).

A number of unique fold and foliation geometries exist which enable us to recognize the deformational history a rock has undergone (Bell, 1981). Other geometries, as shown in the adjacent three photographs, are not so clearcut, and I argue that they are difficult to explain if the deformational history had simply involved inhomogeneous simple shear.

**115A.** Transition from amphibolite-facies country rock into mylonite at the base of the Woodroffe thrust mylonite zone, central Australia (Bell, 1978a). The amphibolite-facies country rock foliation $S_2$, oriented parallel to the layers striking N-S up the center of the photograph, is folded progressively into tighter folds from the bottom to the top of the photograph. Above the hammer, the strain has intensified to a degree where the mylonitic foliation ($S_m$), which has developed parallel to the axial surfaces of the folds, dominates the weathering pattern of the outcrop. Within this zone, folds of the country rock foliation ($S_2$) are isoclinal. The orthogonal relationship of $S_2$ to $S_m$ across the zone of progressive increase in strain is more readily explained by a deformational history involving bulk, inhomogeneous shortening (Bell, 1981) than inhomogeneous simple shear.

**115B.** Extremely disharmonic folds in mylonites from the Mann fault, central Australia. Although the mineral elongation lineation in this outcrop is perpendicular to the photograph and the change in geometry along the axial surface could possibly be explained in terms of rotation of hinge lines through the horizontal, the extreme disharmony of the folds is not indicative of a deformational history involving inhomogeneous simple shear. Production of this fold geometry by general bulk, inhomogeneous shortening seems most likely.

**115C.** Thin-section photograph of the hinge of an $F_m$ fold (that is, a fold of mylonitic foliation generated during the same deformation that produced the foliation itself; see Bell, 1978a). The presence of the feldspar porphyroclasts has caused extreme anastomosing of the axial surfaces of some of the small microfolds. The resulting geometry can be more readily explained by a deformational history involving bulk, inhomogeneous shortening than inhomogeneous simple shear. Vertical width of photograph is 20 mm. PL

115 A

115 B

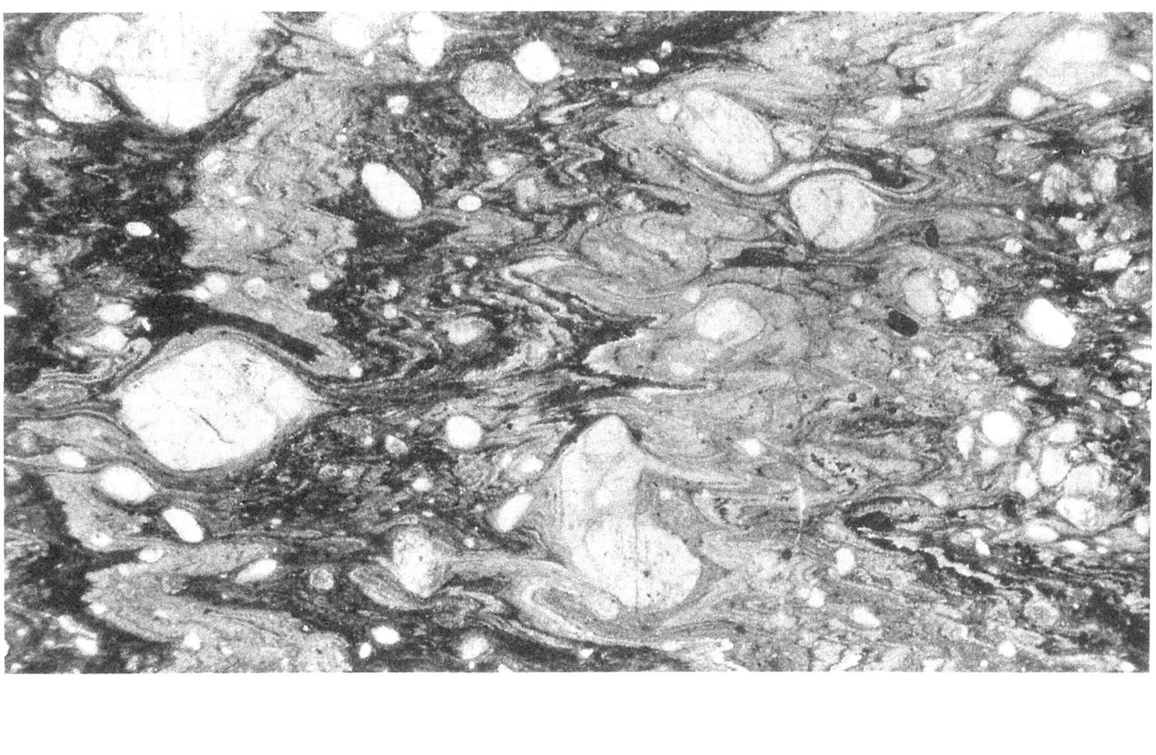

115 C

Bradley G. Erskine and H.-R. Wenk

The Santa Rosa mylonite zone (SRMZ) is a major deformational belt along the eastern margin of the northern Peninsular Ranges batholith (NPRB) that extends about 80 km southward from Palm Springs, California, U.S.A. (Simpson, 1984). Highly deformed plutonic rocks of the border facies of the Peninsular Ranges batholith are structurally overlain by deformed amphibolite-facies metamorphic rocks (Palm Canyon Formation) and the fairly undeformed Asbestos Mountain granodiorite (Figure 116.1). Geochronological studies (Goodwin and Renne, 1991) indicate that deformation occurred in the late Cretaceous during cooling of the NPRB, initially in a ductile fashion at amphibolite-facies conditions and extending into a semi-brittle stage at greenschist-facies conditions. Spectacular deformation features are observed in the mylonitized protolith with changes in microstructures, grain size, and preferred orientation at constant chemical and mineralogical composition (O'Brien and others, 1987; Goodwin and Wenk, 1995). The transition from the relatively undeformed granodiorite to mylonite and phyllonite is gradational to abrupt. Fine-grained phyllonites with a slaty habit and massive, flinty, black ultramylonites are the most highly deformed rocks; they are concentrated in zones of 5 cm to 10 m width and large lateral extent. The absence of a preferred orientation of quartz in these phyllonites suggests that deformation may have occurred by grain-boundary sliding, whereas in the mylonites deformation occurred by intracrystalline slip followed by recrystallization (Wenk and Pannetier, 1990). Photographs 116A–116C illustrate the transition from relatively undeformed plutonic rocks (116A), to mylonites with rounded feldspar porphyroclasts and bands of fine-grained recrystallized quartz (116B), and ultimately to very fine-grained phyllonite (116C). Although S-C structures document some component of simple shear, the overall deformation was by flattening as indicated by deformed xenoliths which have a pancake shape and display extreme strains of a factor of 30–100 (Figure 116.2).

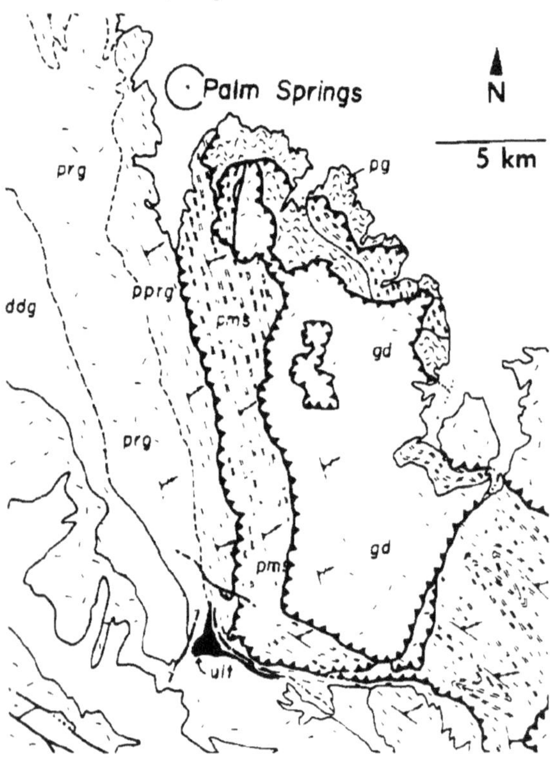


**Figure 116.1.** Geologic sketch map of the northern Santa Rosa mylonite zone. Barbs on faults point to the upper plate *ddg*, prebatholithic Desert Divide metasedimentary screen, *prg*, relatively undeformed undifferentiated Peninsular Ranges granitic rocks; *pprg*, porphyroclastic (mylonitic) granitoids, *pms*, mylonitic Palm Canyon metasedimentary rocks, *pg*, porphyroclastic orthogneiss; *gd*, relatively undeformed upper plate granodiorites

**116A.** Hand specimen of relatively undeformed NRPB granodiorite sample in upper Palm Canyon near the mylonitic front of the SRMZ. Optical photomicrograph of relatively undeformed NRPB granodiorite. Incipient ductile deformation is expressed by undulatory extinction in quartz. NX

**116B.** Hand specimen of mylonite from the SRMZ showing typical well-developed foliation and down-dip elongation lineation. Sampled in upper Palm Canyon (locality = *ult*, Figure 116.1). Optical photomicrograph of mylonite. Cataclastic deformation of feldspar has resulted in abundant well-rounded porphyroclasts. Thin bands of dynamically recrystallized quartz produce the fluxion structure characteristic of mylonites. NX

**116C.** Hand specimen of ultramylonite from upper Palm Canyon showing the typical slaty fabric of phyllonites. Note elongation lineation on foliation surface. Optical photomicrograph of ultramylonite. Few rounded feldspar porphyroclasts and quartz bands remain in these extremely fine-grained rocks. NX

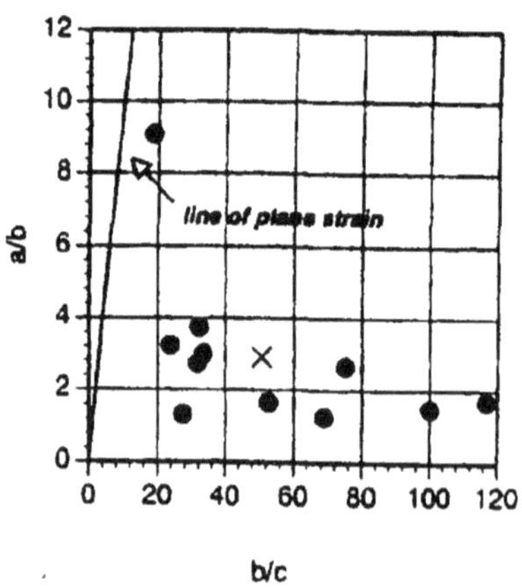


**Figure 116.2.** Aspect ratios of xenoliths in deformed protolith in Palm Canyon The data document that deformation was predominantly by flattening and that total strains were very large (a > b > c, c is the length perpendicular to the foliation, a parallel to the lineation) Cross is an average of 100 xenoliths on which only a/b or b/c could be measured.

116 A

116 B

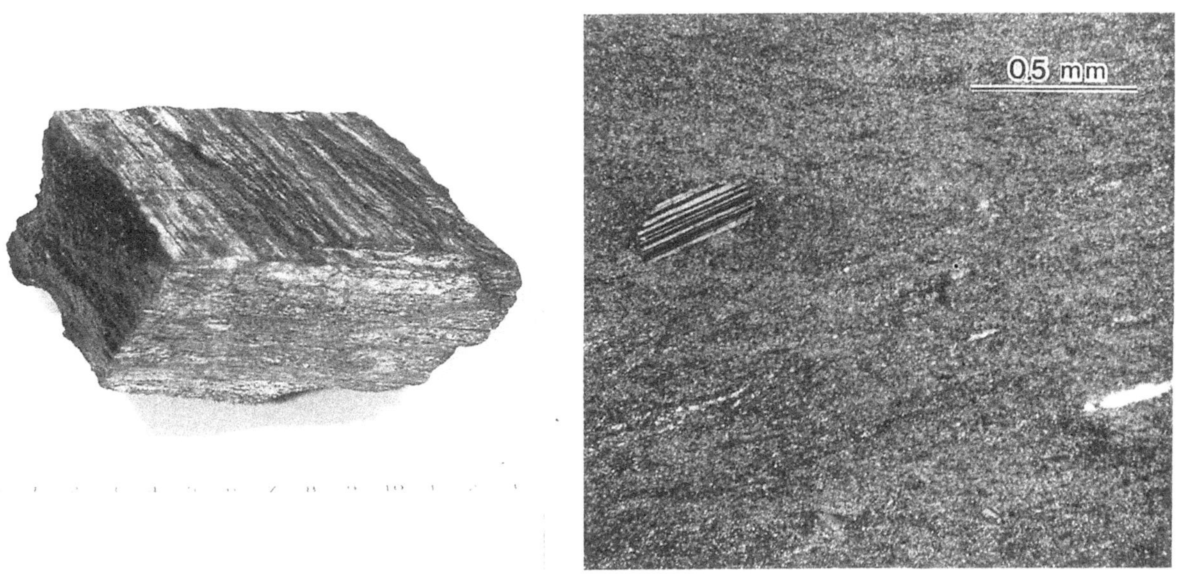


116 C

# 117. PREFERRED ORIENTATION AND MICROSTRUCTURE OF MYLONITIC CARBONATE ROCKS

Bradley G. Erskine and H.-R. Wenk

Deformation of carbonate rocks in Palm Canyon produced strong crystallographic preferred orientation, which yields information on the deformational history of the Santa Rosa mylonite zone (SRMZ, see Plate 116). [11$\bar{2}$0] pole figures of calcite from the SRMZ are characterized by a girdle in the plane of foliation with two maxima symmetrically disposed 60° about the lineation (Figure 117). These orthorhombic pole figures are remarkably similar to experimentally produced (Kern and Wenk, 1983; Wenk and others, 1986) and theoretically predicted (Wenk and others, 1987) textures for deformation by pure shear. This suggests that deformation within the SRMZ was primarily coaxial with a relatively minor noncoaxial component, and that these mylonites are the result of crustal thinning rather than deep-seated thrusting (Erskine and Wenk, 1985). Similar observations were made in mylonitic marbles of the core complexes in the Cordillera of the western United States (Erskine and others, 1993).

The transition from a coarse-grained (>1.5 mm) marble tectonite to a fine-grained mylonitic or ultramylonitic marble is the result of a complex interaction of many mechanisms. In the early stages, mechanical twinning and kinking are of great importance as illustrated in Plate 117A. Recrystallization along grain boundaries transforms the large, highly strained crystals into a fine-grained (25 μm) aggregate (117B). TEM studies reveal similar microstructures in both coarse and fine grains indicating that the recrystallized matrix is also deformed. Plates 117C and 117D show mechanical *e*-twins and tangles of dislocations with little evidence of recovery. The deformation mechanisms changed from glide and twinning to dislocation creep accompanied by dynamic recrystallization, and perhaps ultimately to grain-boundary sliding in the most fine-grained samples. Although recrystallization and grain-boundary sliding are clearly important in the microstructural development, dislocation glide and twinning were responsible for the development of the lattice preferred orientation, and thus were important processes throughout the deformational history.

**117A.** Optical photomicrograph of a coarse-grained marble showing a complexly twinned and kinked calcite crystal. Note recrystallization at grain boundaries. NX

**117B.** Optical photomicrograph of a fine-grained marble with relict coarse grains. All but a few grains have recrystallized to form a fine-grained aggregate. NX

**117C.** Bright field TEM photomicrograph of a highly twinned crystal from a coarse-grained marble (see 117A).

**117D.** Bright field TEM photomicrograph of a crystal from a fine-grained marble. Note that twinning is still present.

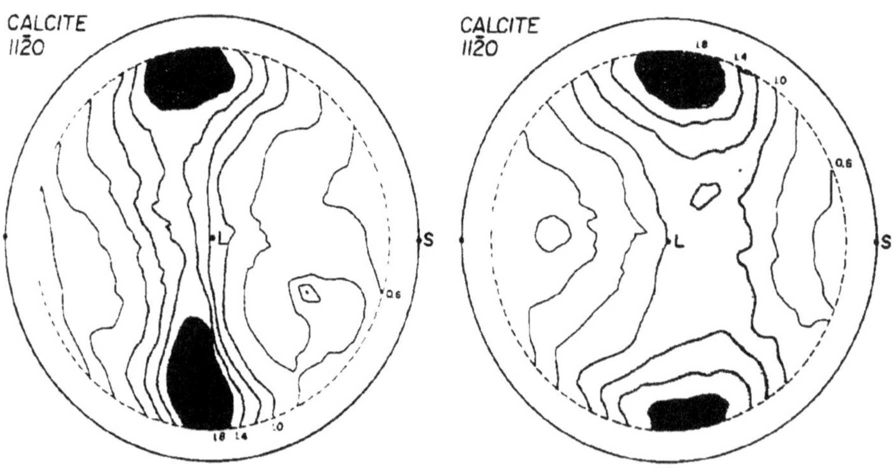


**Figure 117.1.** Pole figures of deformed carbonates from the SRMZ measured by X-ray diffraction with a pole figure goniometer. Contours in multiples of a random distribution. L lineation, S pole to foliation. (Left) Pole figure of a fine-grained marble. Same sample as shown in Plate 117B. Right: Pole figure of a coarse-grained marble with a significant fine fraction.

117 A

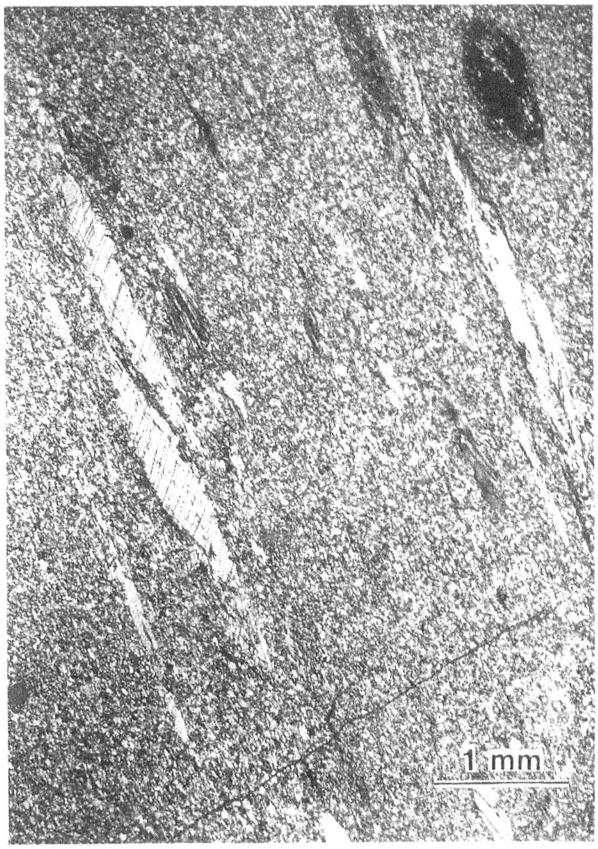


117 B

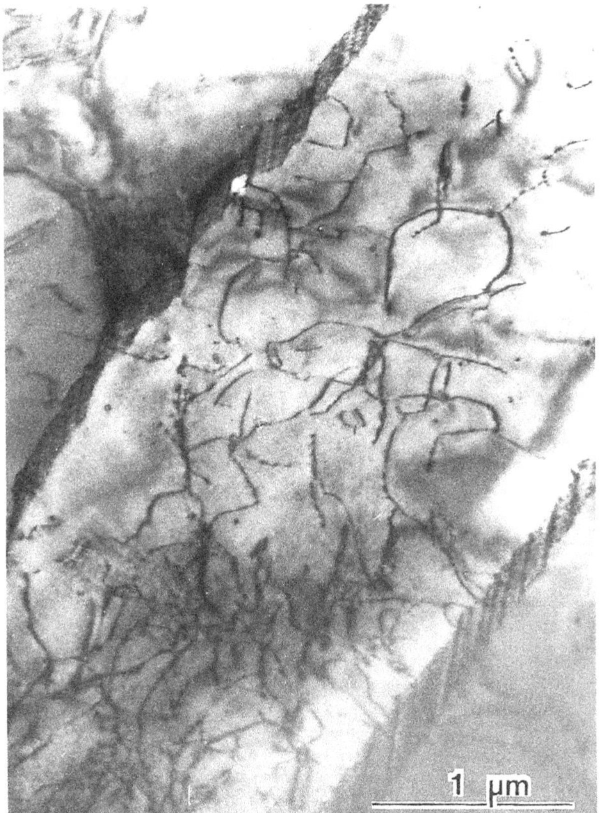


117 C

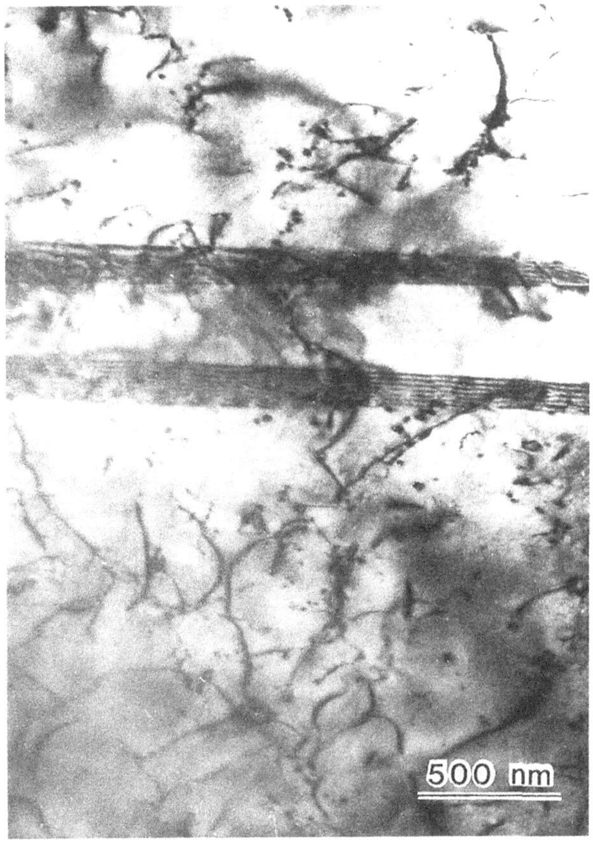


117 D

# DUCTILE BEHAVIOR

## Metamorphic grade variations

### Low- and medium-grade shear zones



Common abbreviations used in regard to the plates in this section include PPL = plane polarized light, NX = crossed nicols, PNX = partial crossed nocols, SEM = scanning electron microscope, and TEM = transmission electron microscope.

This set of plates includes a contribution from the type locality of mylonite at Ben Arnaboll Hill within the Moine thrust zone at Eriboll, northwest Scotland. The rocks illustrated from this classic locality, as well as those illustrated in several other plates, are quartzo-feldspathic. This group of plates also illustrates low- and medium-grade mylonites developed in other rock types, including mafic and ultramafic rocks and chalcopyrite ore. Many of the plates illustrate how features have developed as a function of increasing strain. Several examples illustrate a sequence developed during decreasing temperature associated with exhumation; these show lower-temperature and more localized deformation superimposed on earlier higher-temperature and more homogeneous deformation, producing a sequence from blastomylonite to mylonite to ultramylonite. Several plates illustrate mylonites developed in polyphase aggregates in which one or more phases has deformed by dislocation creep, whereas other phases have deformed by cataclasis, illustrating that "low-" or "high-temperature" deformation microstructures depend on the specific minerals in the rock. Several plates illustrate the importance of three possibly concurrent processes of grain-size reduction (dynamic recrystallization, cataclasis, and neomineralization) in causing strain localization due to a switch to grain-size-sensitive creep.

# 118. FAULT ROCKS FROM BEN ARNABOLL, MOINE THRUST ZONE, NORTHWEST SCOTLAND

S. H. White

Lapworth (1885) based much of his original description of mylonites on samples selected from Ben Arnaboll Hill within the Moine thrust zone at Eriboll, northwest Scotland. Teall (1918) gave an accurate location for Lapworth's type samples. These were situated adjacent to the Arnaboll thrust (Figure 118.1), which brings a sliver of Precambrian Lewisian acid gneiss over Cambrian quartzite. The samples described below, which were taken adjacent to the Arnaboll thrust, illustrate the progressive development of mylonitic microstructures in the Lewisian acid gneiss with increasing shear strain. Detailed descriptions are published elsewhere (White and others, 1982). The Lewisian quartzo-feldspathic gneiss has a massive appearance to within about 1.5 m of the Arnaboll thrust plane (118A). At this stage, a weak mylonitic foliation develops parallel to the thrust plane and increases in strength until at 1 m above the plane a strongly foliated quartzo-feldspathic mylonite is present. This subsequently changes into a quartz/phyllosilicate ultramylonite adjacent to the thrust plane. The retrogression of the feldspars is strain dependent and is in contrast to the cataclasite to mylonite sequence from the Outer Hebrides thrust zone (see Plate 42).

Optical photomicrographs shown in 118B–118E illustrate the sequential development of cataclasite, mylonite, and ultramylonite. Fine details in the cataclasites and ultramylonites are highlighted in electron micrographs in 118F–118M.

**Figure 118.1** *(opposite)* A map of the structural geology of the Moine thrust zone in the Eriboll area of northwest Scotland (from Coward and McClay, 1979) The locality from which the specimens were taken is marked with an arrow

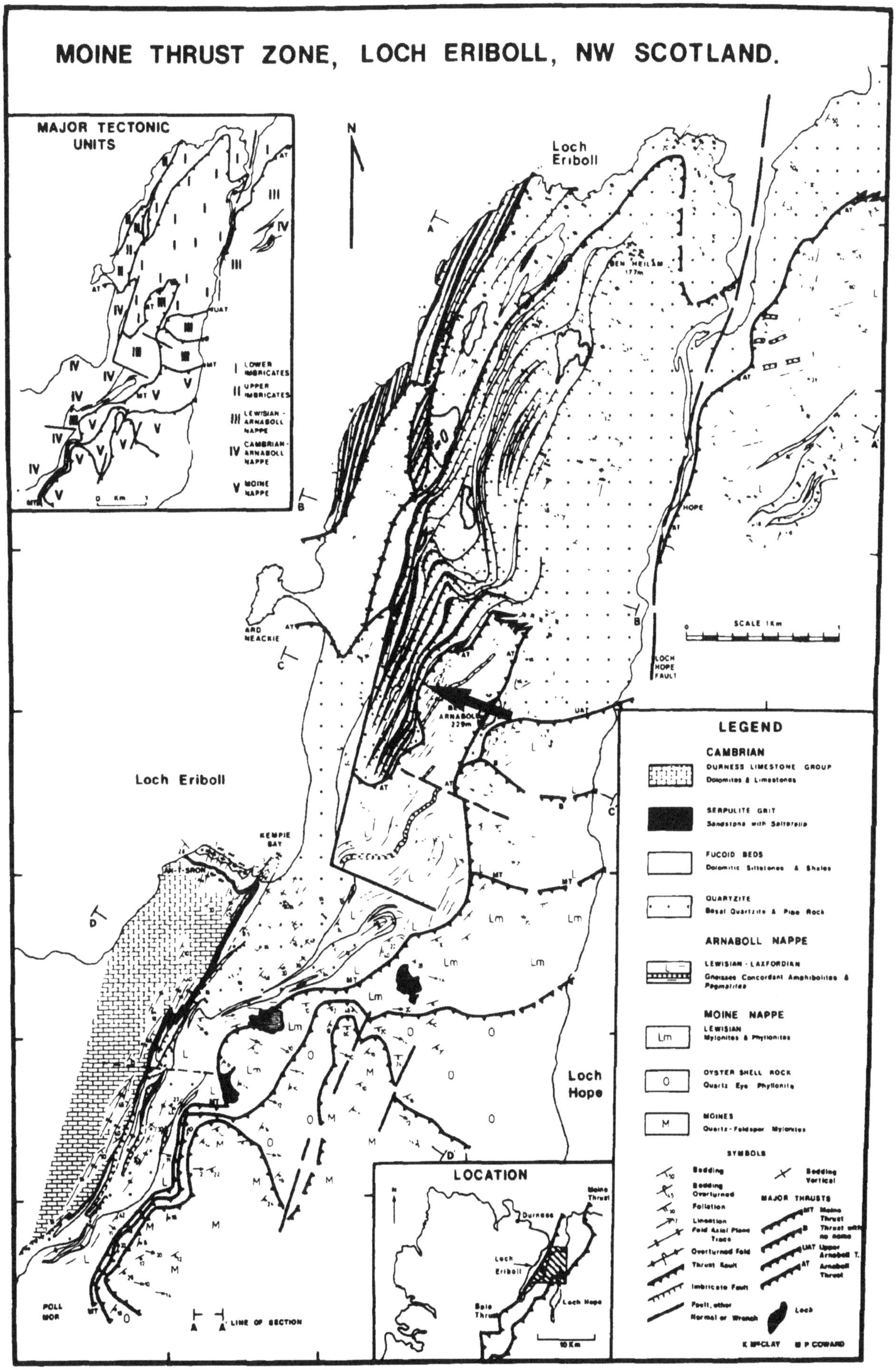
MOINE THRUST ZONE, LOCH ERIBOLL, NW SCOTLAND.
MAJOR TECTONIC UNITS
I LOWER IMBRICATES
II UPPER IMBRICATES
III LEWISIAN - ARNABOLL NAPPE
IV CAMBRIAN - ARNABOLL NAPPE
V MOINE NAPPE
Loch Eriboll
Loch Eriboll
Loch Hope
ARD NEACKIE
KEMPIE BAY
AN-T-SRON
POLL MOR
LOCH HOPE FAULT
HOPE
SCALE 1Km
LEGEND
CAMBRIAN
DURNESS LIMESTONE GROUP
Dolomites & Limestones
SERPULITE GRIT
Sandstone with Salterella
FUCOID BEDS
Dolomitic Siltstones & Shales
QUARTZITE
Basal Quartzite & Pipe Rock
ARNABOLL NAPPE
LEWISIAN - LAXFORDIAN
Gneisses Concordant Amphibolites & Pegmatites
MOINE NAPPE
LEWISIAN
Mylonites & Phyllonites
OYSTER SHELL ROCK
Quartz Eye Phyllonite
MOINES
Quartz-Feldspar Mylonites
SYMBOLS
Bedding
Bedding Vertical
Bedding Overturned
Foliation
Lineation
Fold Axial Plane Trace
Overturned Fold
Thrust Fault
Imbricate Fault
Fault, other Normal or Wrench
MAJOR THRUSTS
MT Moine Thrust
B Thrust with no name
UAT Upper Arnaboll T.
AT Arnaboll Thrust
Loch
LINE OF SECTION
LOCATION
Moine Thrust
Durness
Loch Eriboll
Loch Hope
10 Km
K McCLAY
M P COWARD

## Setting

**118A.** A field photograph showing the foliated Lewisian gneiss above the Arnaboll thrust plane. The hammer rests on the contact between mylonitized gneiss and quartzite that defines the thrust plane. Note the increase in the strength of the mylonite foliation as the Arnaboll thrust plane is approached. The sites from which samples 118B–118E were taken are marked by similar letters.

118 A

## Optical photomicrographs

**118B.** Cataclasite. The massive Lewisian gneiss above the Arnaboll thrust has developed a cataclastic texture. The major mineral constituents, alkali feldspar and plagioclase, are extensively fractured with slight alteration to albite, sericite, chlorite, biotite, and epidote along the original grain boundaries and along fractures. Some plagioclase clasts have bent twin lamellae attesting to slight intracrystalline plasticity. The quartz has deformed plastically, recrystallized, and shows no evidence of fracturing. NX

**118C.** Cataclasite with an incipient mylonitic foliation. The initial foliation that has formed within the cataclasite is marked by zones of intense deformation parallel to the Arnaboll thrust plane. Along the bands (one is marked by arrows) granulation of the feldspars is severe, and quartz has recrystallized completely. These zones of intense deformation are C planes (see Berthé and others, 1979b). NX

**118D.** Mylonite. The mylonite which forms from the cataclasite at higher strains consists of well-developed, alternating zones of recrystallized quartz and finely granulated feldspars together with their alteration products. There is a progressive decrease in clast size and increase in the amount of alteration products with increasing shear strain. NX

**118E.** Ultramylonite. The ultramylonite consists of alternating zones of fine-grained quartz and of ultrafine-grained quartz-mica-chlorite. Zones of darker material that cannot be resolved with an optical microscope occur within the phyllosilicate-rich zones. Discordant quartz veins, marked by arrows, are common. They become parallel to the mylonitic foliation and are completely recrystallized within the phyllosilicate-rich zones. These mylonitic veins indicate that shear strains are greater in the phyllosilicate-rich zones than within the quartz-rich zones. Shear bands (bottom left-hand corner) form a C-C′ fabric. NX

118 B

118 C

118 D

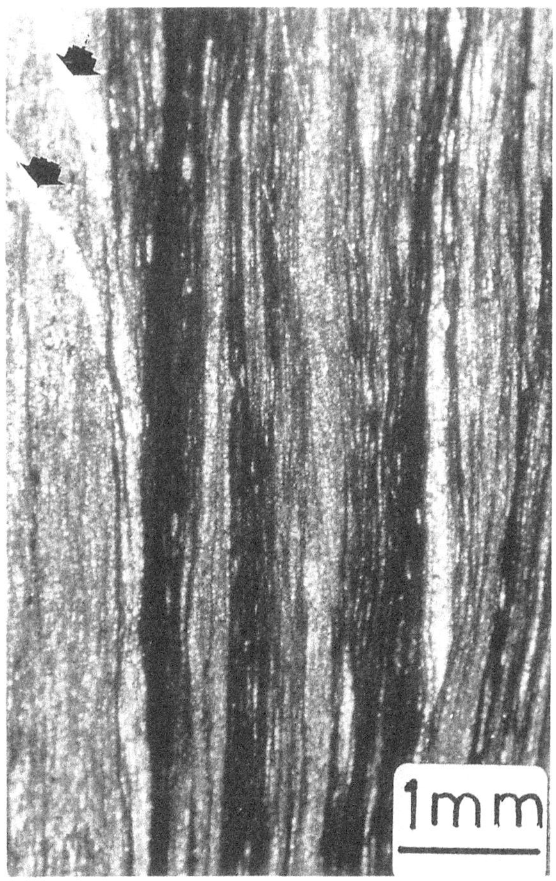


118 E

## Electron micrographs

There is little evidence in optical micrographs for the ductile deformation of either plagioclase or alkali feldspar during cataclasis. Transmission electron micrographs from these two minerals reveal high densities of dislocations and also narrow microtwins. Furthermore, the micrographs from quartz are characterized by well-developed dislocation substructures, which support the optical observations that the quartz deformed plastically.

**118F.** Twinning, tt, in an alkali-feldspar clast in cataclasite. The dark areas at the top and at the bottom left of the micrograph are areas of high dislocation density. Details of these areas are shown in 118G and 118H. TEM

**118G.** Small subgrains, marked by walls of dislocations in the above alkali-feldspar clast. The subgrains have high internal dislocation densities and are shown in 118H. TEM

**118H.** Dislocation structures within the subgrains shown in 118G. They form a three-dimensional tangle. A twin boundary, marked tt, is cut by the dislocations. TEM

**118I.** A well-developed subgrain structure in a quartz grain in a cataclasite. Internal dislocations are present within the subgrains. They are less tangled and have lower densities than those in the alkali-feldspar clast. TEM

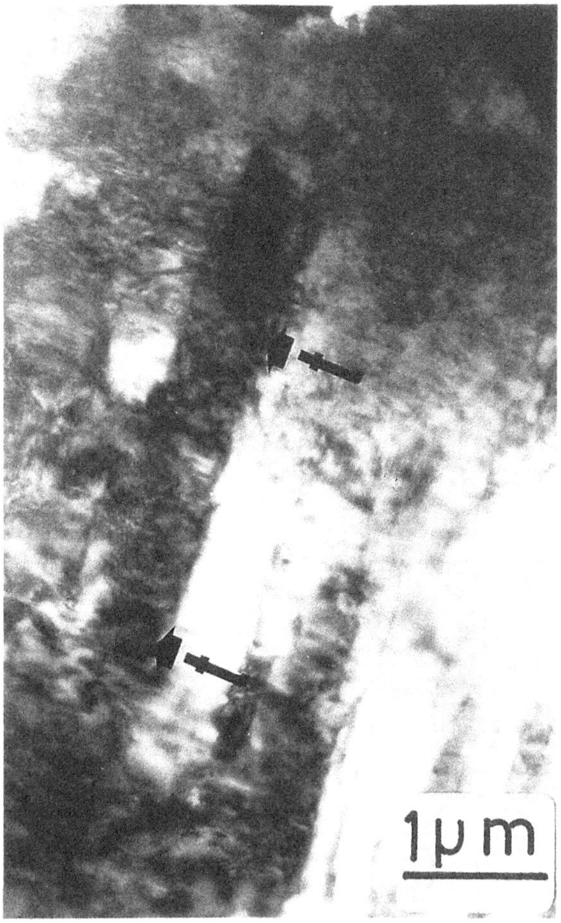


118 F

118 G

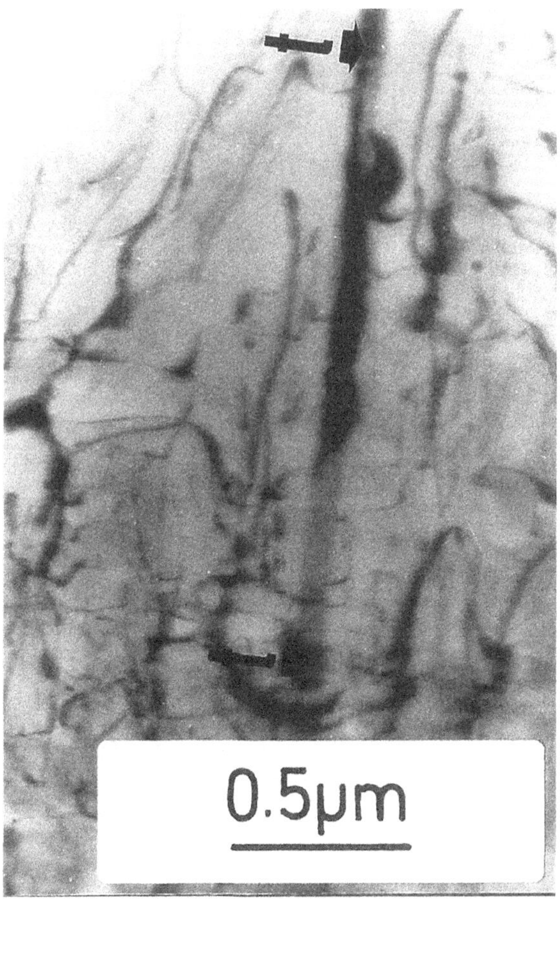


118 H

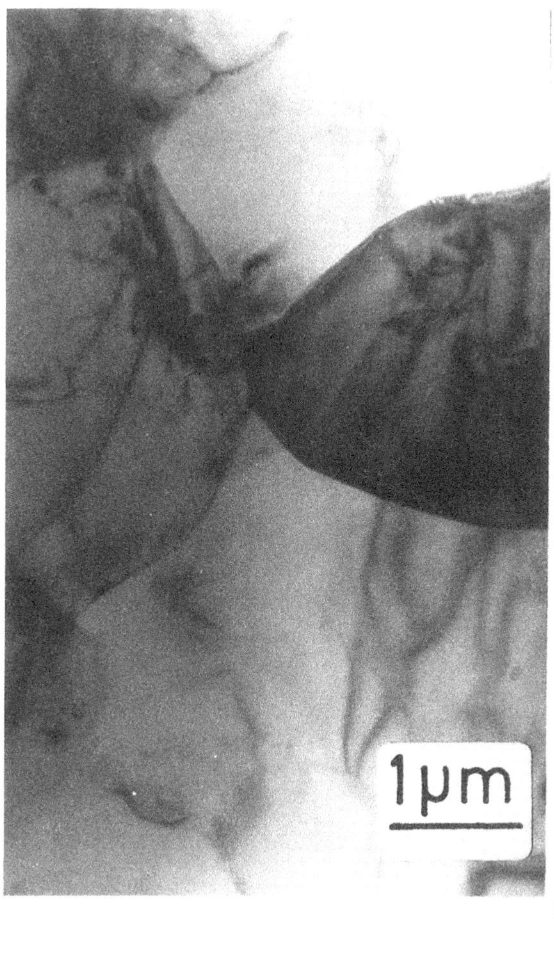


118 I

### Electron micrographs

**118J.** A micrograph from a quartz-rich band that contains mica and chlorite grains. Grain boundaries are traced out on the micrograph, and the grains are numbered 1 to 6. Note that the grains do not have internal subgrains and that there are low densities of dislocations within the grains. Small sericite grains are marked by arrows. TEM

**118K.** Coarse quartz grains in zones free from chlorite and mica grains. Grain boundaries are marked, and the individual grains are numbered 1–3. In contrast to quartz grains shown in 118J, the coarse grains have a higher dislocation density and may contain subgrains. TEM

**118L.** The microstructure typical of an ultrafine-grained quartz-mica-chlorite band. The main difference between this view and 118J is the presence of sericite and chlorite, not only along grain boundaries but also within the quartz grains. The chlorite grains, marked by an arrow, are much smaller than the sericites (see 118J) and may pin dislocations within the grains. TEM

**118M.** A backscattered scanning electron micrograph of a dark lamella in a quartz-mica band. The lamella contains small albite grains (black areas) in a matrix of sericite and chlorite (intermediate tone) that contains zones rich in epidote (lightest tones). The dark lamellae are the remnants of the feldspar bands seen at the mylonite stage. SEM

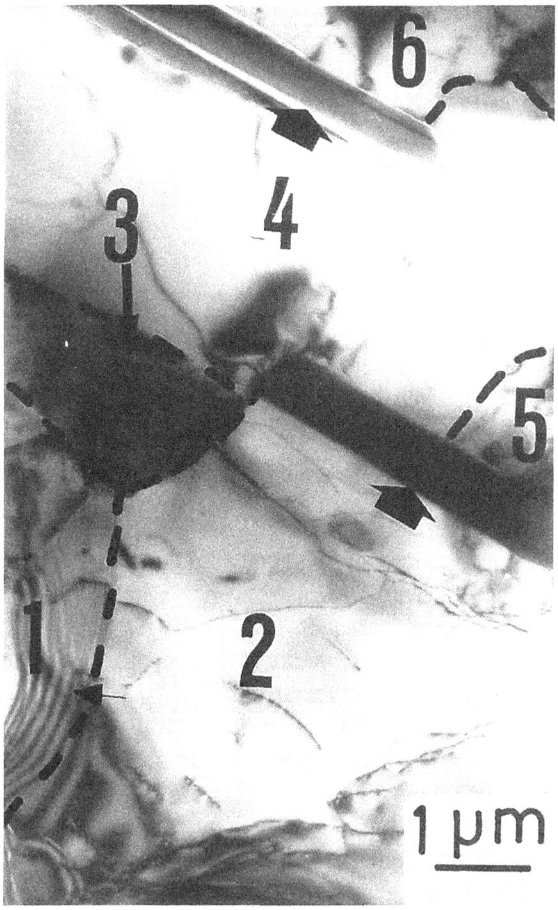

118 J

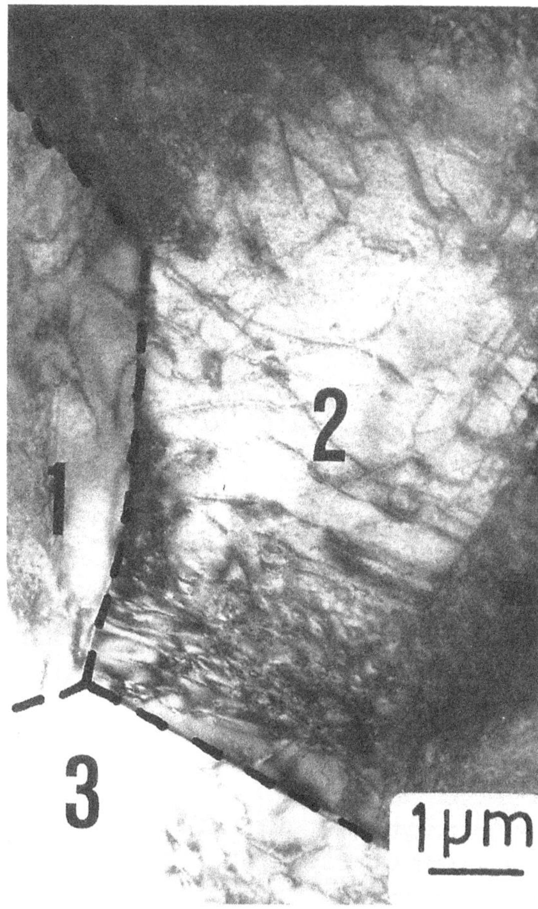

118 K

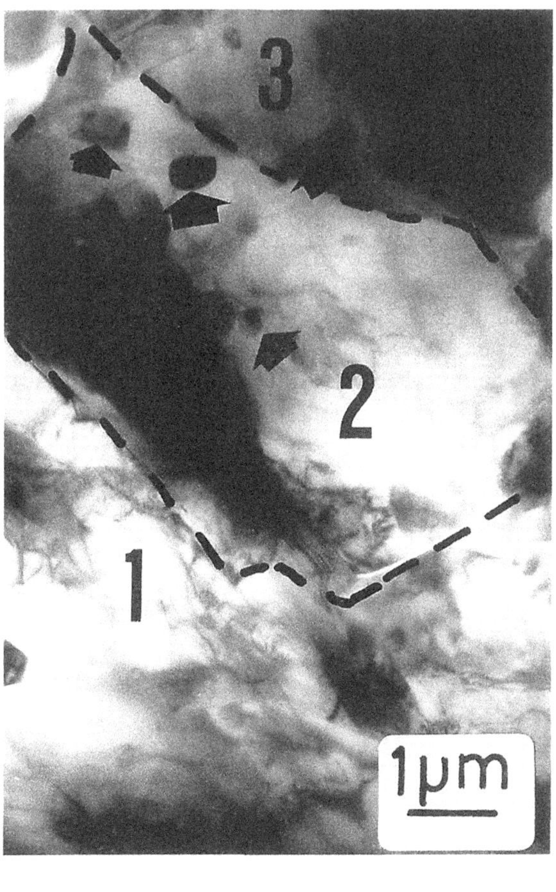

118 L

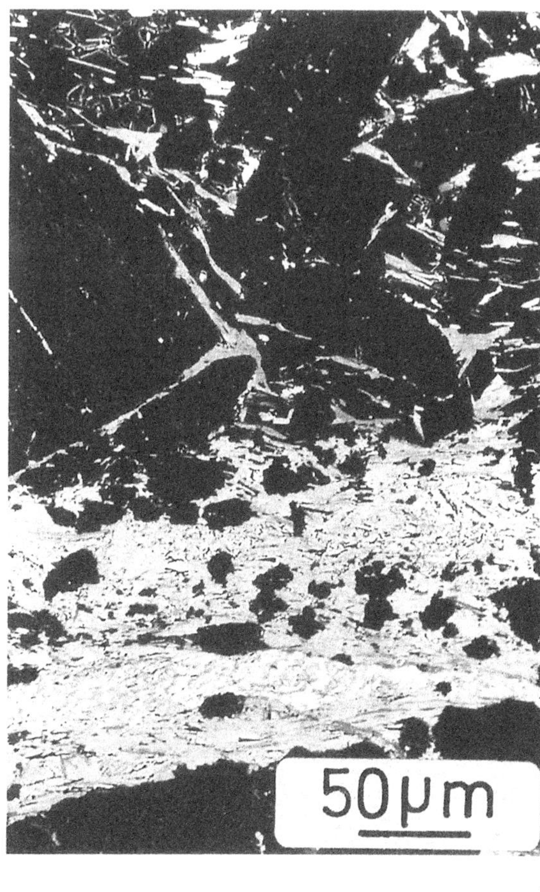

118 M

# 119. GREENSCHIST-FACIES MYLONITE ZONE IN QUARTZO-FELDSPATHIC GNEISS

PATRICIA B. KAYGI

The Fries fault zone is a major tectonic discontinuity in the Blue Ridge geologic province of the southern Appalachians, southeastern U.S.A. The Fries fault is the easternmost of a series of low-angle faults that bound the southern Blue Ridge on its northwest side. Rb-Sr mineral and whole-rock analyses performed on mylonitized samples from the Fries fault zone in the study area (Dietrich and others, 1969) gave an age of 330 ± 15 Ma for the timing of displacement along the fault. The Fries fault has traditionally been interpreted as a thrust fault related to regional contractile deformation (e.g., Rankin and others, 1973; Bartholomew, 1981). However, Simpson and Kalaghan (1989) argued that microstructural data indicate a normal-sense displacement for part of the Fries fault zone and related the mylonitization to late Precambrian crustal extension.

In southwest Virginia, the fault has juxtaposed basement rocks of the Little River Gneiss (1128 ± 25 Ma; Truman, 1976) on the southeast against Precambrian basement rocks of the Pilot Gneiss and overlying Lower Cambrian metasedimentary rocks on the northwest. The fault is marked by a 2.5 km wide zone of mylonitic rocks that exhibit progressive stages of continuous ductile deformation. These stages have been identified as protomylonite, mylonite, and ultramylonite subzones (after Sibson, 1977), each characterized by a fabric and microstructures that reflect increasing intensity of deformation toward the fault.

### Protomylonite subzone of Fries fault

The protomylonite subzone is 1.2 km wide, and the rocks herein exhibit incipient mylonitization that was accompanied by retrogressive metamorphism to greenschist facies. This change is rather subtle in the Little River Gneiss, but much more apparent in overlying metasedimentary rocks and mafic intrusive rocks. It is based on the greenschist-facies mineral assemblage in metasedimentary rocks, the disappearance of microcline, the formation of albite, the change from large muscovite crystals to fine-grained phengite and sericite, and from large, brown biotite crystals to fine-grained, pale brown and green biotite that is oriented parallel to and helps to define the mylonitic foliation. Mineralogic reactions constrain a maximum temperature range of 375 to 450°C during faulting. A geothermal gradient of 30°C/km would indicate a depth of about 12–15 km. This is consistent with Sibson's (1977) fault model, wherein the transition from brittle failure to ductile faulting occurs between 10 and 15 km depth and at 250 to 350°C. This suggests that temperatures may have been lower than those indicated above, and that the faulting occurred in the ductile zone, just below brittle-ductile transition.

**119A.** Typical outcrop of protomylonite shows a weakly developed mylonitic foliation ($S_m$) that is defined by slightly flattened feldspar augen (arrow) and quartz grains and by oriented micas. $S_m$ is parallel to subparallel to $S_1$, the original foliation, suggesting that $S_1$ controlled the orientation of the mylonitic fabric.

**119B.** Quartz in the protomylonite has undergone ductile flattening with aspect ratios from 2:1 to 3:1 (arrow), extensive subgrain development, and 8 to 10% recrystallization. Alkali-feldspar grains are fractured and partially disaggregated (arrow). NX

**119C.** Sericitic and quartzose stringers are deflected around large alkali-feldspar augen (arrow) in which fractures are oriented at a high angle to $S_m$. The recrystallization of muscovite to sericite and the highly strained quartz within these stringers reflect higher strain states adjacent to the more resistant augen. NX

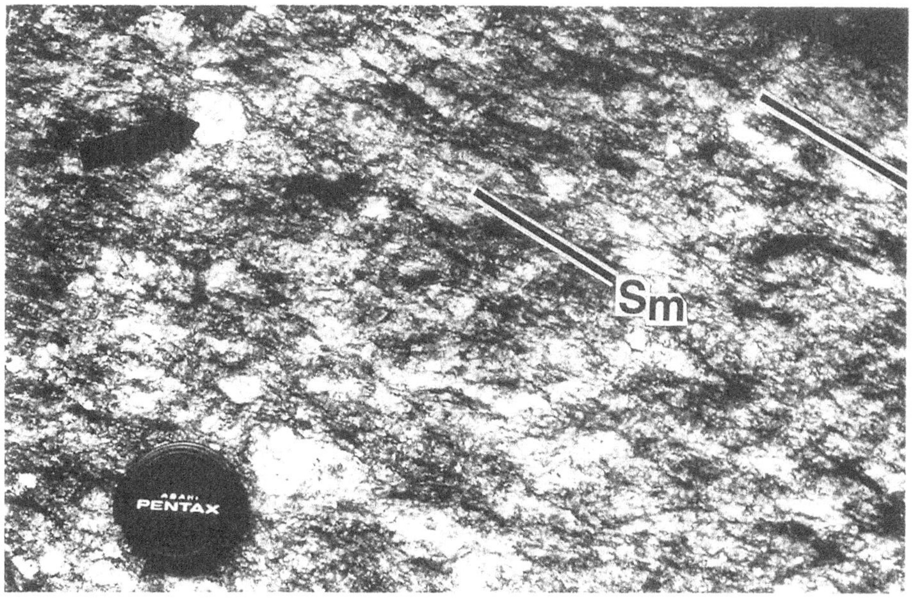

119 A

119 B

119 C

### Mylonite subzone of Fries fault

The mylonite subzone is up to 1.0 km wide and is characterized by a well-developed mylonitic foliation and lineation, the latter defined by elongate quartz grains and feldspar augen oriented parallel to $S_m$. Grain deformation includes ductile flattening, recovery, and recrystallization in quartz, ductile and brittle yielding in plagioclase, and brittle fracturing in alkali feldspar.

**119D.** Outcrop of Little River Gneiss with well-developed mylonitic foliation marked by large, elongate alkali-feldspar augen (arrow).

**119E.** Large alkali-feldspar augen set in a finer-grained matrix that consists of quartz, feldspar, and micas, fractured systematically at a high angle to $S_m$ (arrow). Orientation of these fractures with respect to $S_m$ is consistent throughout the fault zone. Elongation of feldspar augen in the $S_m$ plane is a function of dextral displacement along these fractures, associated with minor counterclockwise rotation of segments. The matrix adjacent to the large augen consists of fine, recrystallized quartz and sericite; the extensive recrystallization in these zones is the result of higher, localized strain in these areas. NX

**119F.** Mylonitic matrix away from the influence of the larger (up to 5 cm) augen. $S_m$ is defined by alternating micaceous and quartzo-feldspathic layers. Only a trace of relict brown biotite remains; most of the original biotite has been absorbed by reactions producing alkali feldspar, chlorite, and stilpnomelene, or replaced by fine-grained, green biotite. Deformation of quartz grains was accomplished by ductile flattening (aspect ratios up to 12:1), development of subgrains, and 40–50% recrystallization (arrow). Plagioclase is extensively altered to sericite and fine-grained epidote; albite is present in minor amounts (arrow). NX

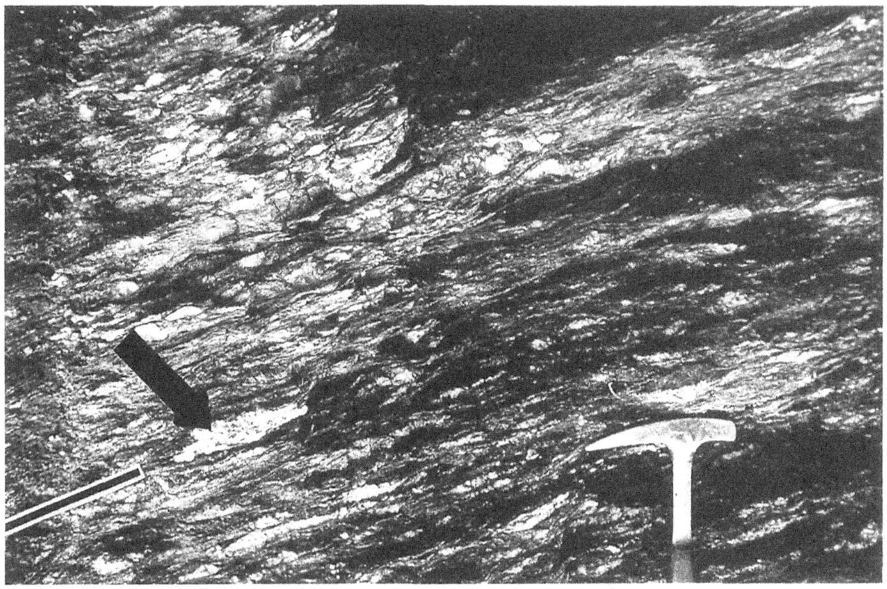

119 D

119 E

119 F

### Ultramylonite subzone of Fries fault

Narrow zones of ultramylonites (greater than 95% matrix), centimeters to tens of meters in width, are discontinuously interspersed throughout the mylonites, but are most common near and along the fault trace. These zones do not appear to anastomose or enclose less deformed mylonitic zones as has been observed in other fault zones (e.g., Subimal, 1977; Mitra, 1979), but are parallel to the mylonitic zones. Apparently, the planar fabric ($S_1$) of the Little River Gneiss controlled the orientation of the mylonitic foliation, indicating that the degree of anastomosing is, in part, dependent on the initial fabric of the host rock.

**119G.** Typical outcrop of an ultramylonite. The rock is very fine-grained, phyllitic, and very friable.

**119H.** The mylonitic foliation is defined in thin section by alternating fine-grained, recrystallized quartz and plagioclase layers with micaceous zones. A few relict plagioclase and quartz grains are present locally (arrow). Quartz has undergone 80–90% recrystallization with a grain size in the range of 20–30 μm, and plagioclase has been almost completely altered to fine-grained albite, epidote, and sericite. Relict alkali-feldspar grains and porphyroblasts are virtually absent, having been altered to muscovite and quartz. The overall grain size has thus been significantly reduced by recrystallization, brittle fracturing, and chemical changes. NX

119 G

119 H

# 120. PROGRESSIVE MYLONITIZATION OF ARKOSE ALONG A MAJOR THRUST FAULT

JANE A. GILOTTI

Arkosic sandstones dominate the 4.5–6 km thick sequence of Upper Proterozoic clastic sedimentary rocks that, together with an Iapetan diabase dike swarm, comprise the Särv thrust sheet (Kumpulainen, 1980). The arkoses are rarely cleaved and preserve primary sedimentary structures to within a few meters of the mylonite zone that marks the base of the Caledonian Särv thrust. The arkosic mylonites are L-S tectonites; the thin sections used for the accompanying photographs are cut perpendicular to the foliation and contain the lineation. Grain size was reduced during mylonitization, whereas modal mineralogy remained nearly fixed, with up to 28% feldspar present in some ultramylonites. The production of fine-grained aggregates deforming via grain-size-sensitive mechanisms is far more important in localizing and promoting strain in the arkoses than growth of intrinsically weak phases (Gilotti, 1992).

The continuity of deformation exhibited by the extensive, alternating planar layers of quartzo-feldspathic and metadiabase mylonites suggests that the arkoses attain the same extremely high strains as the dikes (see Plate 121) (Gilotti and Hull, 1990). Although the dominant deformation mechanisms are quite different in the diabases and the arkoses during progressive mylonitization, both lithologies probably accumulate strain via grain-size-sensitive flow once a stable microstructure is established.

**120A.** Weakly foliated arkose protolith (U-22) located 1 m above the upper shear-zone boundary of the Särv thrust at Ulvberget, Jämtland County, Sweden. Detrital myrmekite grain in center. Modal composition is 36% quartz, 33% feldspar, and 28% muscovite with accessory epidote, sphene, and magnetite. NX

**120B.** Quartz-rich augen mylonite (U-30) located 3.5 m below the upper shear-zone boundary of the thrust at Ulvberget. Elongate ribbons of quartz, a strong planar foliation defined by the alignment of phyllosilicates parallel to ribbon boundaries, feldspar augen, and 50–90% recrystallized fine-grained groundmass characterize the augen mylonites. High strains are indicated by strongly attenuated quartz ribbons, some only one grain thick, with axial ratios of 100:1 and a mean grain size of 15 μm. NX

**120C.** Ultramylonite (Al-29) located 1.5 m below the upper shear-zone boundary of the Särv thrust at Hällberget, Jämtland County, Sweden. A very fine-grained, recrystallized matrix of quartz, feldspar, and muscovite comprises 100% of the mylonite. Extremely high strains are indicated by submillimeter-scale grain size and compositional banding. The coarse bands are quartz rich. The disappearance of ribbons in favor of a mixture of grains is consistent with a change in the dominant deformation mechanism from dislocation creep to grain-boundary sliding. NX

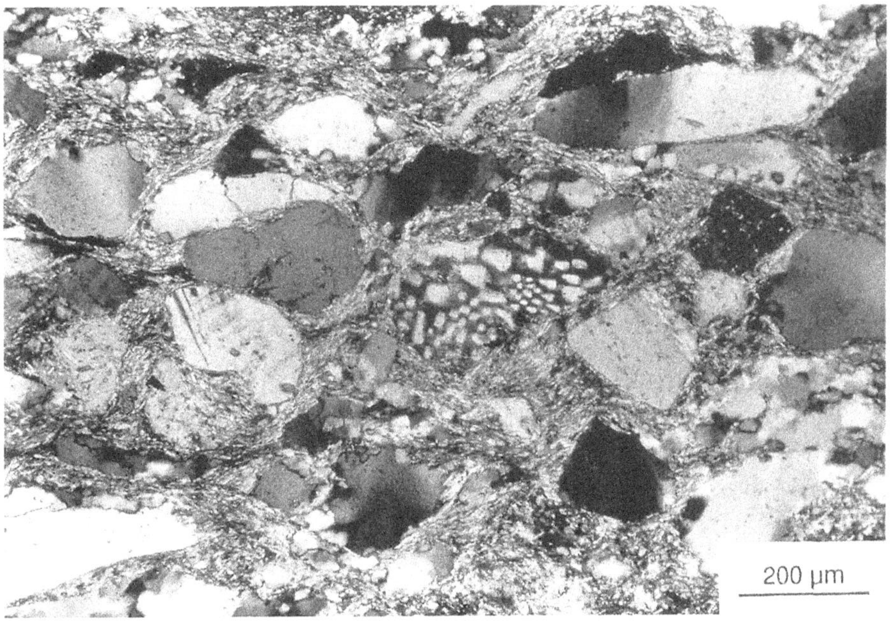


120 A

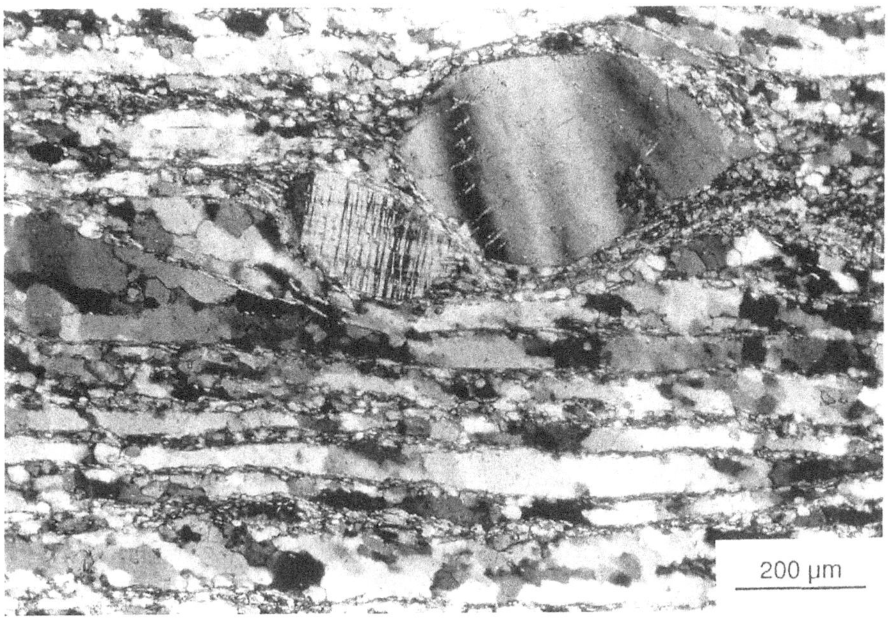


120 B

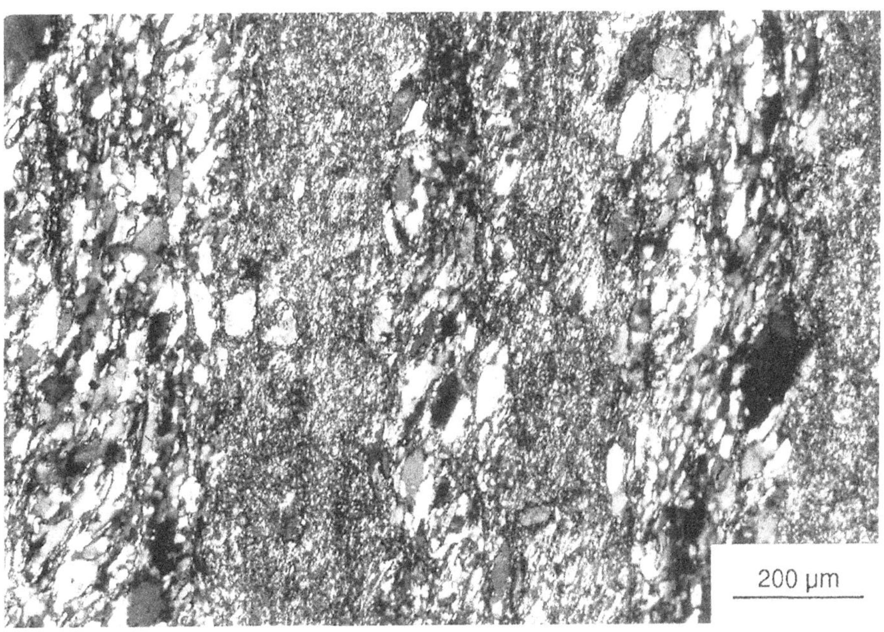


120 C

# 121. PROGRESSIVE MYLONITIZATION OF DIABASE DIKES ALONG A MAJOR THRUST FAULT

JANE A. GILOTTI

The Särv thrust in the Swedish Caledonides displaced a thick sequence of Upper Proterozoic sandstones intruded by diabase dikes a minimum of 80 km east-southeast onto Baltoscandian granitic basement (Gilotti and Kumpulainen, 1986). Rocks within the Särv thrust sheet are relatively undeformed, and deformation is concentrated in a thin (10–100 m) mylonite zone at the base of the thrust sheet. Dikes record extremely high strains ($\gamma \approx 1000$) and underwent syntectonic, greenschist-facies metamorphism as they rotated into the fault zone (Gilotti, 1989; Gilotti and Hull, 1990) (see Figure 121.1). The breakdown of amphibole to weaker phyllosilicates was the important strain softening mechanism below $\gamma \approx 100$. Neither the microstructure nor the metamorphic assemblage changes at higher strains, suggesting that the dikes attain a steady-state microstructure (Means, 1981) above this critical strain. The persistence of this stable microstructure could be due to dynamic recrystallization or grain-size-sensitive mechanisms.

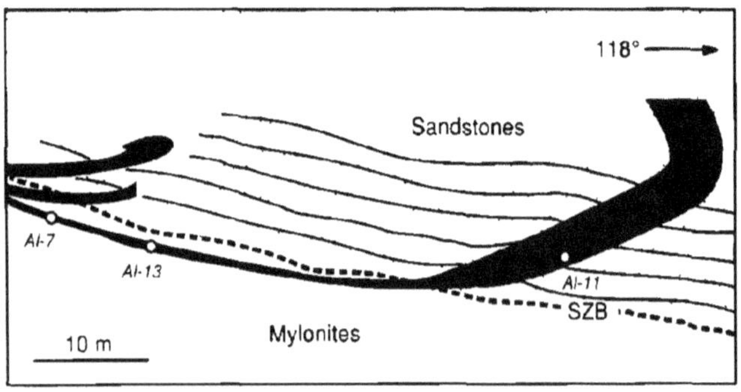


**Figure 121.1.** Strain plane projection of the Sarv thrust zone at Hallberget, Sweden, showing a dike that records a simple shear strain of $\gamma = 103$ at sample locality Al-7. The upper shear-zone boundary (SZB) marks an abrupt change from unfoliated rocks above to strongly foliated and lineated mylonites beneath. The projection plane is perpendicular to the foliation and contains the transport direction.

**121A.** Undeformed, but altered, diabase (Al-11) taken 3 m above the upper SZB of the Särv thrust at Hällberget, Jämtland County, Sweden (see Figure 121.1 for sample location). Relict igneous clinopyroxene (cpx) is armored by amphibole (amp); plagioclase phenocrysts (not shown) are saussuritized. The unfoliated groundmass consists of amphibole + epidote + sphene + ilmenite. PPL

**121B.** Amphibole-bearing metadiabase mylonite (Al-13) from the same dike at Hällberget (see Figure 121.1 for sample location). The assemblage amphibole + epidote + plagioclase + biotite + chlorite + sphene + quartz is typical of the intermediate strain state (i.e., $\gamma < 40$). Clinopyroxene is completely replaced by amphibole, which now comprises 20–40% of the rock. Metamorphism is isochemical (Gilotti, 1989). A strong foliation is defined by the preferred orientation of the phyllosilicates, as well as by a weak shape fabric in amphibole and epidote. The label abbreviations are ep = epidote, sp = sphene, pl = plagioclase, amp = amphibole, and bi = biotite. PPL

**121C.** High-strain, amphibole-free metadiabase mylonite (Al-7) from the same dike at Hallberget (see Figure 121.1 for sample location). The stable assemblage is biotite + chlorite + albite + epidote + sphene + quartz; saussuritized plagioclase phenocrysts (not shown) still persist as augen. K metasomatism was required to produce the excess biotite found in the high-strain assemblages. Grains are optically strain-free with uniform shapes and sizes. Same label abbreviations as in 121B. PPL

121 A

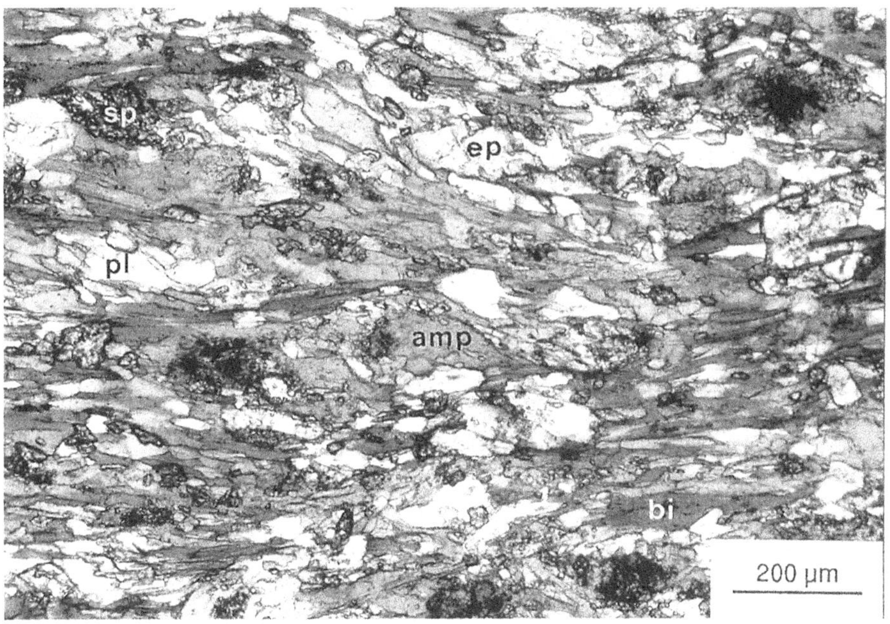

121 B

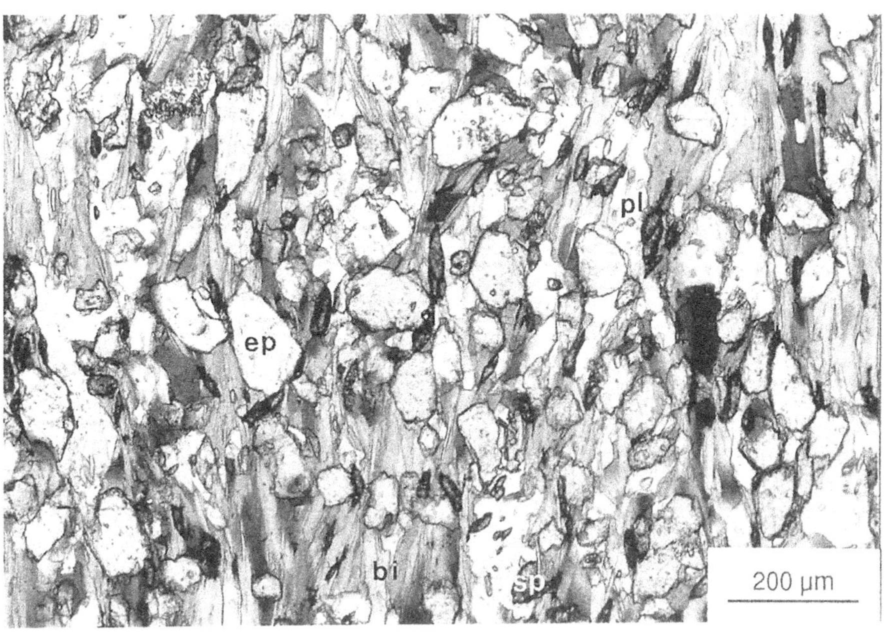

121 C

# 122. RETROGRESSIVE MYLONITIC METABASIC FAULT ROCKS

K. H. Brodie

The Kamila shear zone, a deep to mid-crustal structure developed within the Kohistan island-arc complex of northern Pakistan (Figures 122.1 and 122.2), contains a variety of fault rocks developed in metabasic protoliths (Treloar and others, 1990). Initial metamorphism of gabbroic rocks under granulite-facies conditions was followed by amphibolite-facies static hydration with pyroxenes replaced by pargasitic hornblende. Evidence is preserved of early deformation under upper amphibolite-facies conditions. A subsequent history of exhumation during decreasing temperature is documented by a progressive sequence of down-temperature retrogression and deformation in superimposed shear zones. These predominantly affected coarse-grained rocks unaffected by the earlier high-temperature deformation. The deformation textures document a range of processes from high-temperature, crystal-plastic deformation through a sequence of variably brittle and plastic deformation processes depending on the extent of retrogression (i.e., the change in mineralogy), to late-stage high-level, low-temperature deformation that included the development of fault gouge.

**122A** (PPL) **and 122B** (NX). Sample from granulite-facies metabasic shear zone showing alternations of plagioclase- and amphibole-rich layers. Porphyroclasts of brown amphibole and orthopyroxene are present in a matrix of plagioclase ($An_{55-60}$), amphibole (tshermakitic to ferroan pargasitic hornblende), ilmenite, and rutile, with some quartz layers. Little evidence of crystal-plasticity is preserved, and the plagioclase and amphibole show a recrystallized granoblastic microstructure inferred to have resulted from high-temperature equilibration. However, textural coarsening may occur in a material that is being slowly strained, and it can be difficult to distinguish dynamic textures from those due to static recovery and grain growth. Segregation into amphibole-rich and plagioclase-rich layers is a common feature in deformed amphibolite-facies metabasic rocks. The formation of the layers as a result of deformation of initially larger grains would require extremely high strains through either (a) cyclic dynamic recrystallization, or (b) diffusion-accommodated grain-boundary sliding, if grain size was sufficiently reduced. See Table 122.1 for representative mineral analyses.

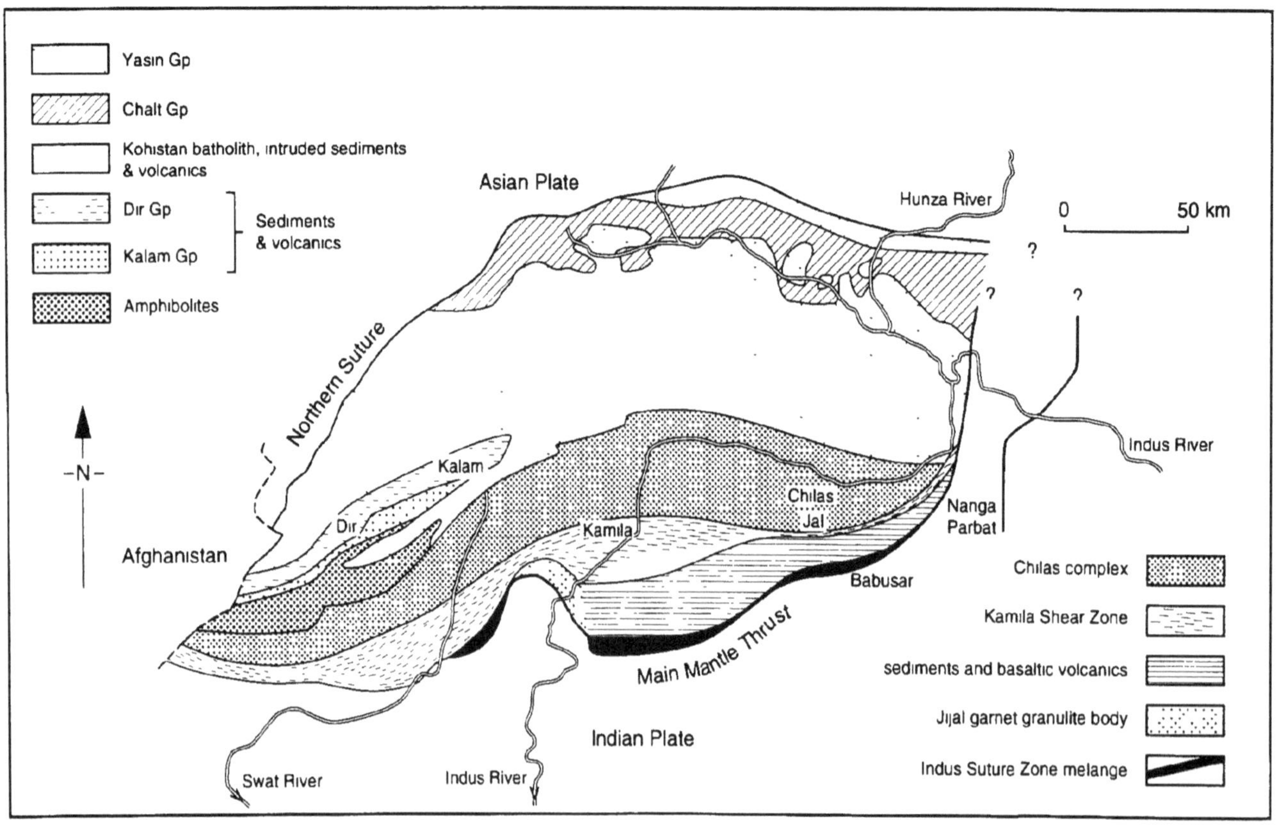


**Figure 122.1.** Geological map showing the location of the Kamila shear zone. The samples illustrated here come from the Indus River section (from Treloar and others, 1990)

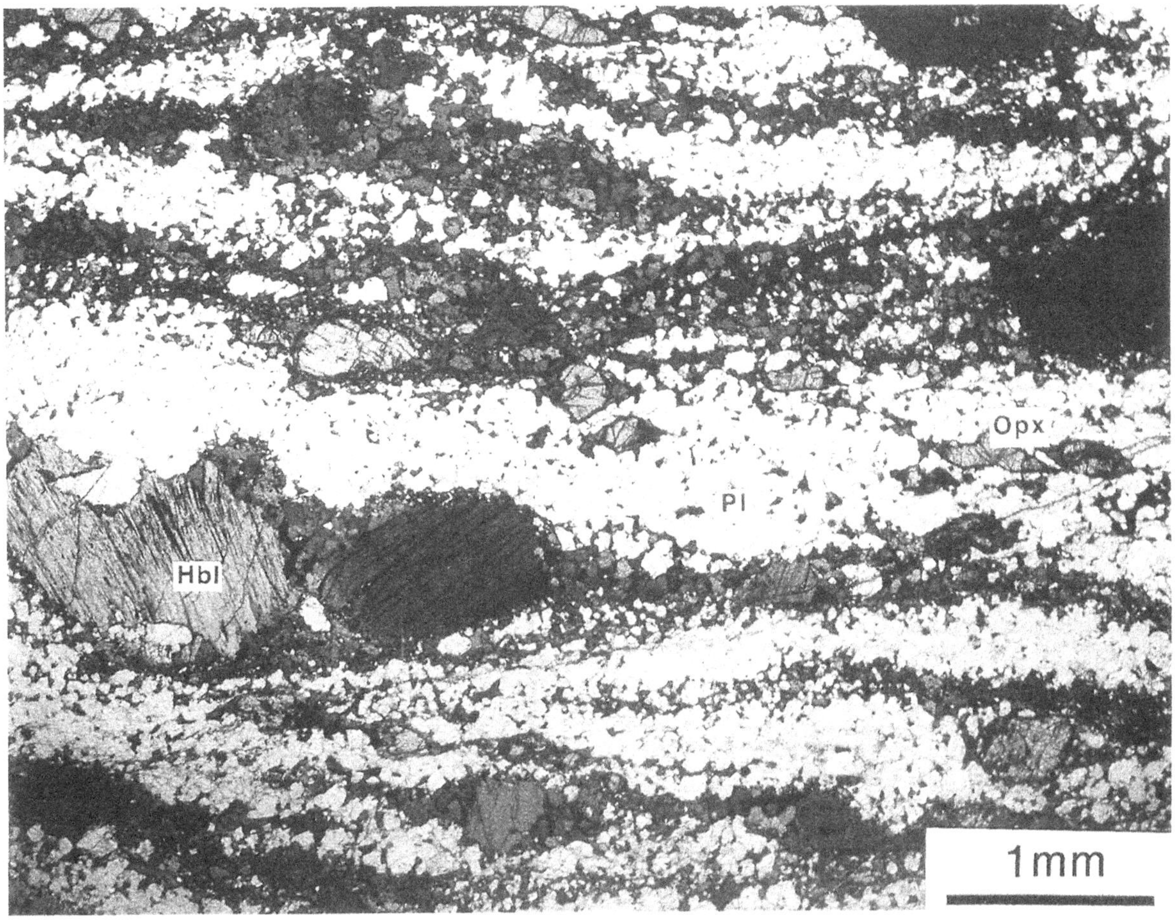


122 A

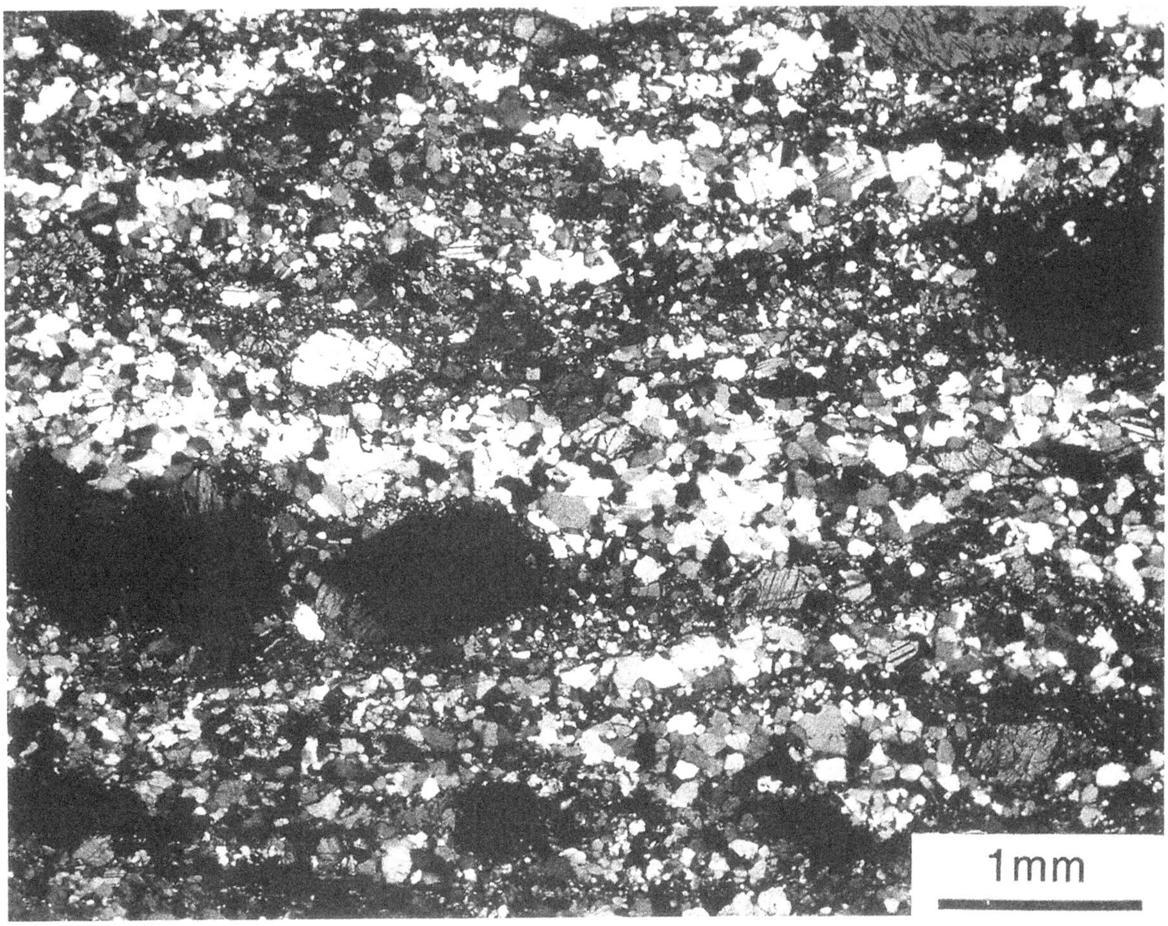


122 B

**122C** (PPL) **and 122D** (NX). Sample from medium- to low-grade, cataclastic metabasic shear zone in which brittle fragmentation (cataclasis) of plagioclase (~$An_{65}$) and static recrystallization of magnesio-hornblende occurs. The margins of the plagioclase fragments are altered to a more sodic composition (~$An_{40}$), and fine-grained epidote laths grow in the cracks (see inset SEM backscattered electron image in 122D). The amphibole has recrystallized to random aggregates of lath-shaped grains, commonly rimmed by fine-grained clinozoisite aggregates and chlorite. The deformation was predominantly cataclastic, with static recrystallization of random amphibole aggregates, and growth of new epidote along fractures in the plagioclase. Following this brittle deformation and associated retrogression, continuing deformation produced localized highly foliated zones cutting the rock. One such zone is present at the bottom of the field of view, where a narrow, highly foliated mylonite zone cuts the rock producing further grain-size reduction and retrogression, and rounding of the relict plagioclase and amphibole grains and clasts. The presence of chlorite in these narrow zones is thought to be responsible for the change to more ductile deformation. See Table 122.1 for representative mineral analyses.

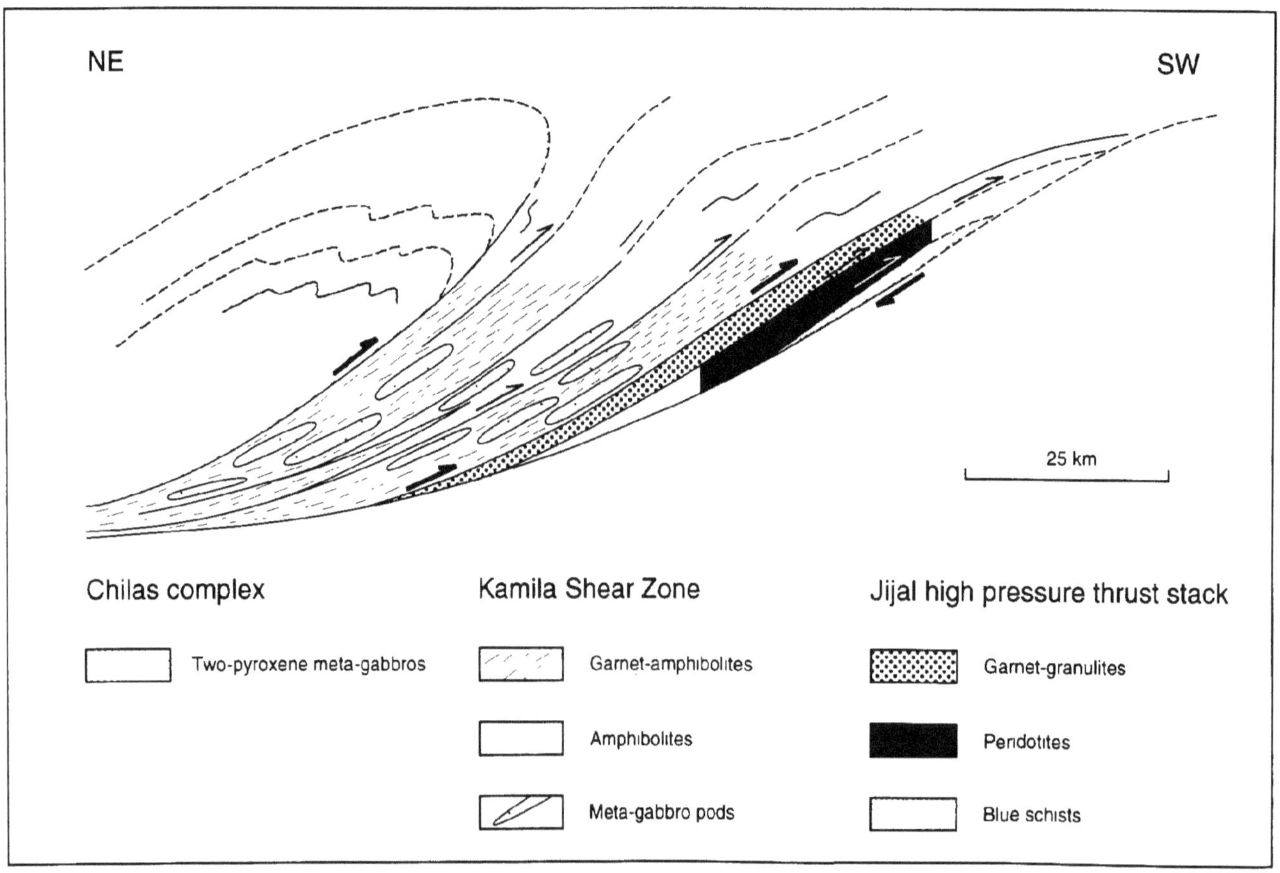


**Figure 122.2.** Cross section through the Kamila shear zone (from Treloar and others, 1990).

122 C

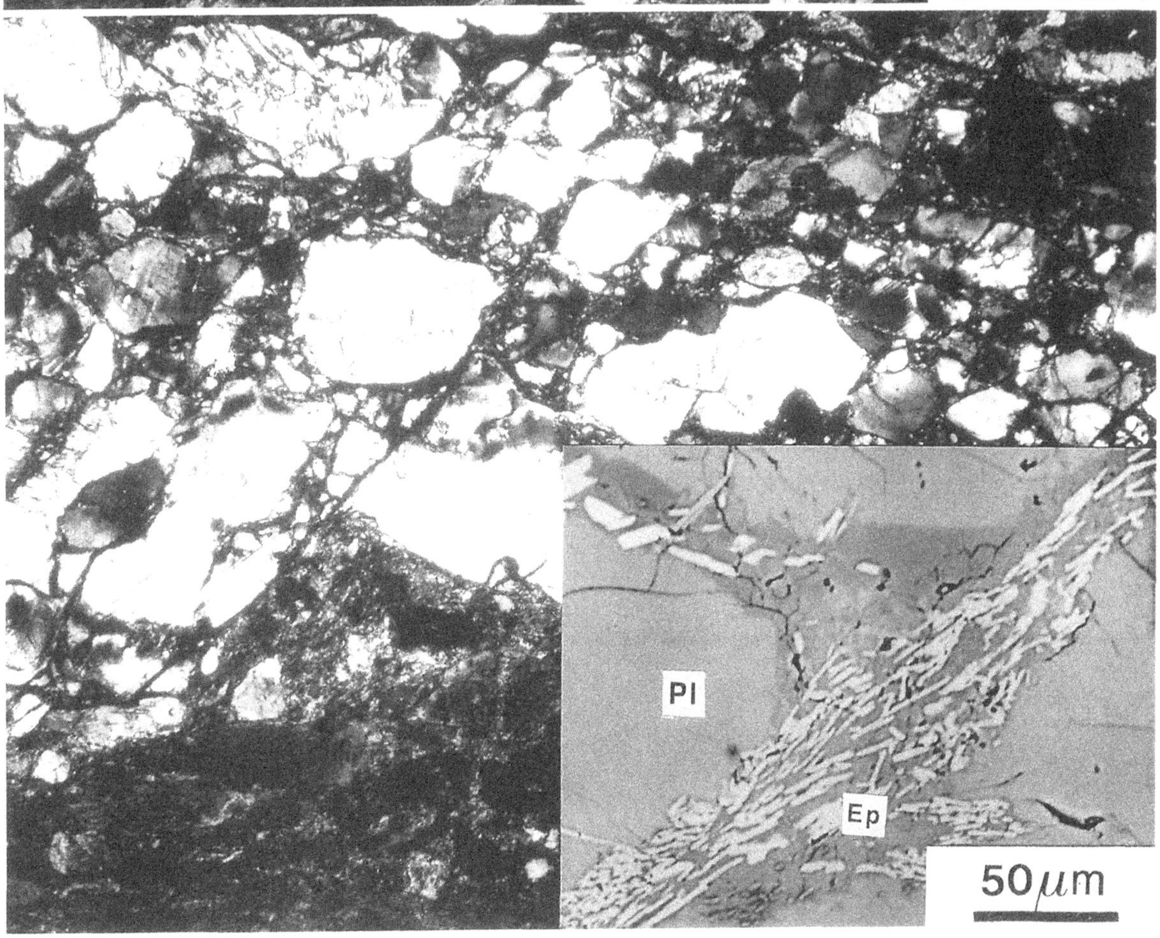


122 D

**122E.** Highly foliated, low-grade (greenschist-facies) mylonite containing porphyroclasts of green magnesio-hornblende and calcic plagioclase ($An_{65}$) in a very-fine grained matrix of actinolite-clinozoisite-biotite-chlorite-plagioclase-quartz. The porphyroclasts are rounded and the plagioclase shows marginal alteration in pressure shadow areas to a more sodic composition (122G). These foliated zones vary from narrow anastomosing networks that cut the cataclastic metabasic rocks (e.g., 122C and 122D), to discrete mylonite zones up to 50 cm thick. These zones are characterized in part by the development of fine-grained fibrous chlorite-actinolite-plagioclase-quartz aggregates that define S-C fabrics. Much of the strain within these rocks seems to be accommodated by deformation of chlorite which underwent slip on (001), as commonly observed in phyllosilicates at temperatures as low as 250–350°C (Bons, 1988), together with limited plastic deformation of quartz. This kind of deformation tends to be localized within rocks that have undergone earlier cataclastic deformation, which would have enhanced permeability and influx of fluids to produce the retrogression. The box indicates the location of 122F. PPL

**122F.** Backscattered electron micrograph (SEM) of part of the foliated greenschist-facies mylonite shown in 122E. The plagioclase porphyroclasts show marginal alteration to a more sodic composition, and the fine-grained matrix contains chlorite, epidote, and sodic plagioclase.

**122G.** Higher-magnification backscattered electron micrograph (SEM) showing the microstructure of the very fine-grained matrix. The more sodic plagioclase has a darker contrast than the calcic plagioclase. The light laths are epidote.

**Table 122.1.** Representative mineral analyses from high-, medium-, and low-grade shear zones within the Kamila shear zone.

| | *High-grade mylonite* | | | | *Medium- to low-grade cataclastic mylonite* | | | | *Low-grade mylonite* | | | |
|---|---|---|---|---|---|---|---|---|---|---|---|---|
| | *Amph* | *Amph p/c* | *Opx* | *Pl* | *Amph* | *Pl p/c* | *Pl edge* | *Chlorite* | *Amph p/c* | *Pl p/c* | *Pl matrix* | *Epidote* |
| $SiO_2$ | 43.93 | 44.36 | 52.10 | 52.96 | 46.12 | 53.14 | 54.21 | 27.29 | 44.35 | 51.91 | 60.92 | 39.03 |
| $TiO_2$ | 1.69 | 1.90 | 0.00 | 0.00 | 0.35 | 0.00 | 0.00 | 0.00 | 2.00 | 0.00 | 0.00 | 0.00 |
| $Al_2O_3$ | 11.30 | 10.85 | 1.60 | 29.06 | 13.23 | 30.39 | 31.79 | 22.62 | 11.00 | 30.63 | 24.92 | 26.20 |
| FeO | 12.64 | 12.63 | 21.79 | 0.00 | 11.36 | 0.00 | 0.00 | 13.72 | 14.47 | 0.00 | 0.00 | 8.17 |
| MnO | 0.23 | 0.33 | 0.82 | 0.00 | 0.23 | 0.00 | 0.00 | 0.24 | 0.00 | 0.00 | 0.00 | 0.00 |
| MgO | 12.67 | 13.45 | 22.76 | 0.00 | 13.40 | 0.00 | 0.00 | 24.00 | 12.27 | 0.00 | 0.00 | 0.00 |
| CaO | 11.63 | 11.49 | 0.51 | 12.38 | 11.78 | 13.24 | 7.30 | 0.00 | 11.60 | 15.16 | 7.85 | 25.06 |
| $Na_2O$ | 1.91 | 2.10 | 0.00 | 5.45 | 1.30 | 4.09 | 5.74 | 0.34 | 1.81 | 3.24 | 6.86 | 0.00 |
| $K_2O$ | 0.19 | 0.07 | 0.00 | 0.00 | 0.16 | 0.00 | 0.00 | 0.00 | 0.57 | 0.05 | 0.00 | 0.00 |
| Total | 96.19 | 97.18 | 99.58 | 99.85 | 97.93 | 100.86 | 99.04 | 88.21 | 98.07 | 100.99 | 100.55 | 98.46 |
| | | | | | | | | | | | | |
| Si | 6.54 | 6.54 | 1.95 | 2.41 | 6.64 | 2.39 | 2.44 | 5.37 | 6.54 | 2.34 | 2.69 | 6.17 |
| Al | 1.98 | 1.89 | 0.07 | 1.56 | 2.24 | 1.61 | 1.68 | 5.24 | 1.91 | 1.63 | 1.30 | 4.88 |
| $Fe^{2+}$ | 1.57 | 1.56 | 0.68 | 0.00 | 1.37 | 0.00 | 0.00 | 2.26 | 1.78 | 0.00 | 0.00 | 1.08 |
| Mg | 2.81 | 2.96 | 1.27 | 0.00 | 2.88 | 0.00 | 0.00 | 7.04 | 2.70 | 0.00 | 0.00 | 0.00 |
| Ca | 1.86 | 1.82 | 0.02 | 0.60 | 1.82 | 0.64 | 0.35 | 0.00 | 1.83 | 0.73 | 0.37 | 4.25 |
| Na | 0.55 | 0.60 | 0.00 | 0.48 | 0.36 | 0.36 | 0.50 | 0.13 | 0.52 | 0.28 | 0.59 | 0.00 |
| K | 0.04 | 0.01 | 0.00 | 0.00 | 0.03 | 0.00 | 0.00 | 0.00 | 0.11 | 0.00 | 0.00 | 0.00 |
| Ti | 0.19 | 0.21 | 0.00 | 0.00 | 0.04 | 0.00 | 0.00 | 0.00 | 0.22 | 0.00 | 0.00 | 0.00 |
| Mn | 0.03 | 0.04 | 0.03 | 0.00 | 0.03 | 0.00 | 0.00 | 0.04 | 0.00 | 0.00 | 0.00 | 0.00 |
| No oxygens | 23.00 | 23.00 | 6.00 | 8.00 | 23.00 | 8.00 | 8.00 | 28.00 | 23.00 | 8.00 | 8.00 | 25.00 |
| | | | | | | | | | | | | |
| Albite | | | | 44.34 | | 35.86 | 58.73 | | | 27.81 | 61.26 | |
| Anorthite | | | | 55.66 | | 64.14 | 41.27 | | | 71.91 | 38.74 | |
| Enstatite | | | 64.38 | | | | | | | | | |
| Ferrosilite | | | 34.58 | | | | | | | | | |
| Wollastonite | | | 1.04 | | | | | | | | | |

122 E
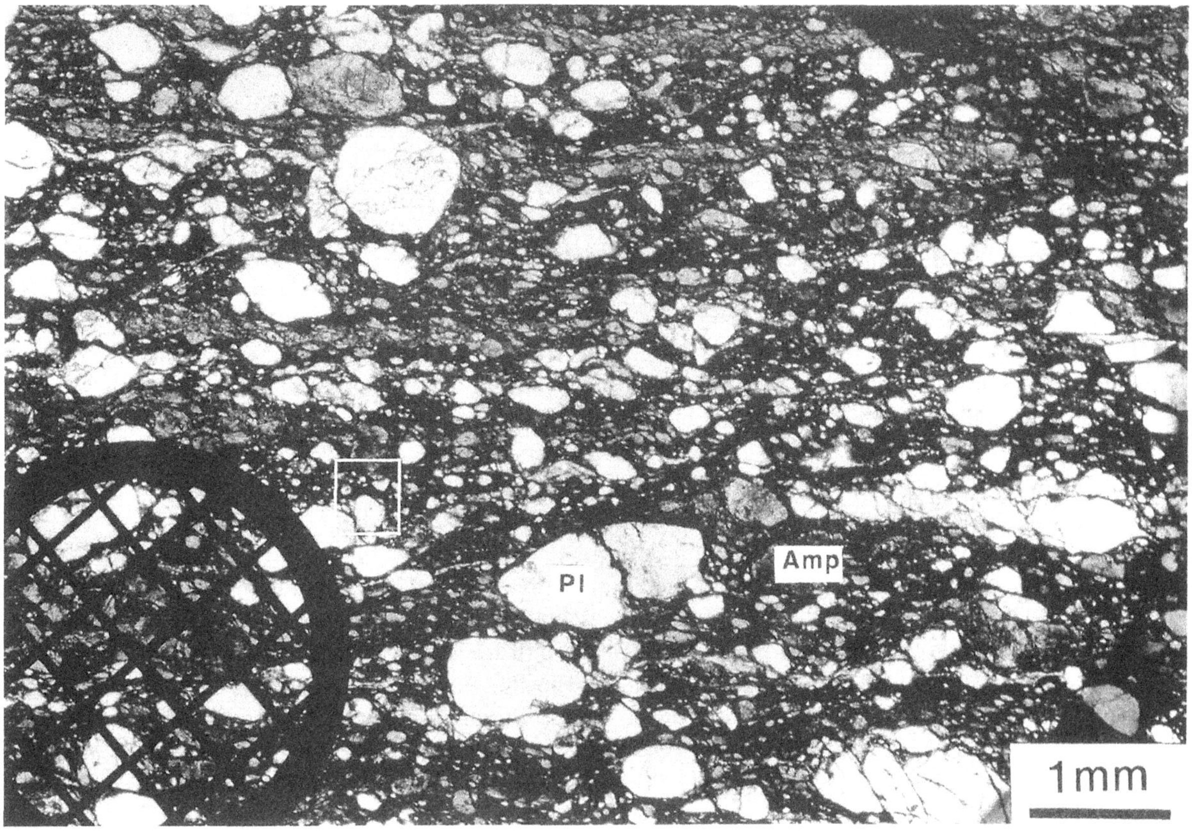


122 F
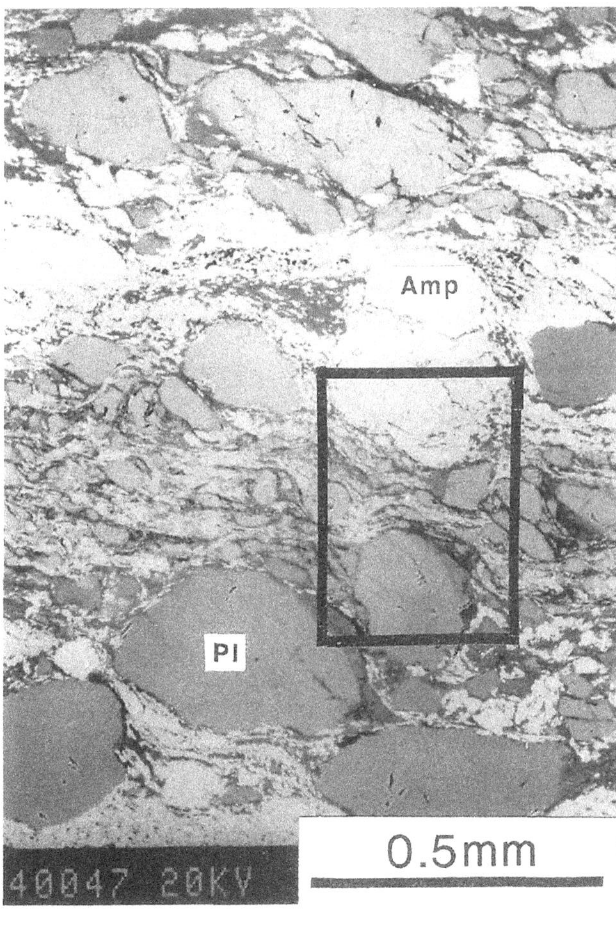


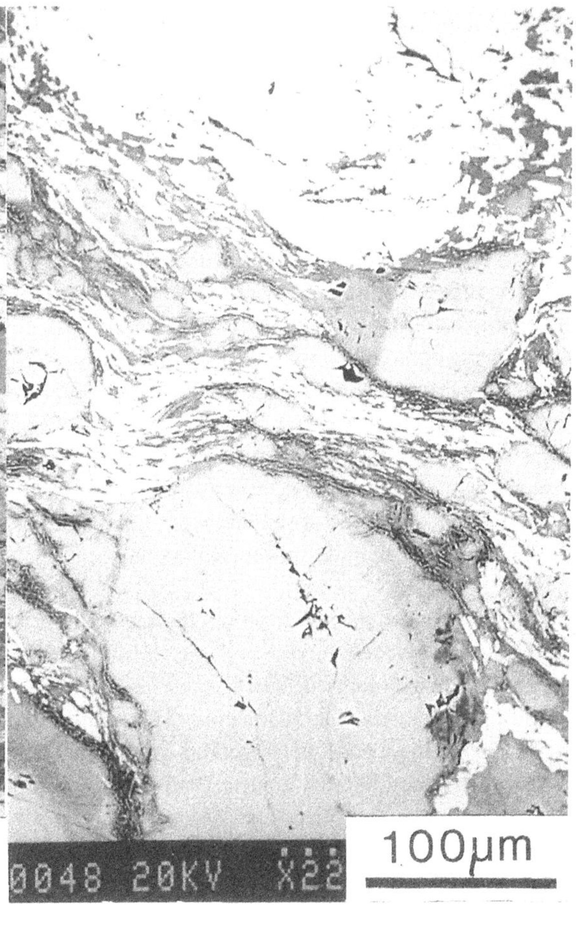

122G

# 123. SERPENTINITE MYLONITES FROM AN OPHIOLITE

ANGELA J. COULTON

Serpentinite mylonites are found within several serpentinized fault zones in the peridotite section of the Josephine ophiolite, northwestern California, U.S.A., shown schematically in Figure 123.1. They occur along the basal sole of the ophiolite (Grady, 1990; Harper and others, 1990) in a regionally extensive mylonite zone, located approximately 1 km below the base of the cumulate sequence (Norrell and Harper, 1988, 1990; Norrell, 1989; Coulton and others, 1995) and within a fault zone at the base of the cumulate sequence. All examples shown are from the mylonite zone 1 km below the base of the cumulates, indicated by an asterisk in Figure 123.1.

The mineralogy and textures of the serpentinite mylonites provide some constraints on the conditions of their formation. The temperature of formation of the mylonite has been estimated from oxygen isotope geothermometry on serpentine-magnetite pairs as ~500°C (Norrell, 1989), close to the upper limit of antigorite stability. Deformed amphibolitic dikes within the shear zone also indicate temperatures of 500–600°C.

Some constraints are available on the relative timing of serpentinization and mylonitization. Unrodingitized dikes, geochemically related to late dikes in the sheeted dike sequence, cross-cut low-temperature, serpentinized fault zones. They indicate that serpentinization along certain fault zones took place during the oceanic stage, prior to ophiolite emplacement (Coulton and others, 1995). The mylonite along the basal sole developed during ophiolite emplacement (Harper and others, 1990), and the other mylonites developed either before or during ophiolite emplacement based on cross-cutting relationships of mafic dikes (Coulton and others, 1995). The serpentinite mylonites therefore may represent oceanic fault zones reactivated during ophiolite emplacement.

Serpentinite mylonites contain many of the features typical of more commonly described quartzo-feldspathic mylonites, including strongly developed foliations and lineations, asymmetric rotated porphyroclasts, and shear bands. Plates 123A–123E show the main features of serpentinite mylonites from the Josephine ophiolite in hand specimen and thin section.

**123A.** Foliation, rotated magnetite porphyroclasts, and shear bands in hand sample of serpentinite mylonite, Red Mountain shear zone (Norrell and others, 1989). The cut face is parallel to lineation and perpendicular to foliation. The foliation is defined by alternating bands of serpentine-group minerals and magnetite that are offset by a shear band (arrow) in the upper left corner of the specimen. The shear band indicates a sinistral sense of shear. The abundant porphyroclasts are composed of orthopyroxene bastites and altered chrome spinel. The porphyroclasts are commonly rimmed by magnetite, which is chromiferous where it rims spinel. Asymmetric neomineralization trails, also composed of magnetite, commonly provide sense-of-shear criteria. Scale bar is 2 cm.

**123B.** Serpentinite mylonite of 123A, face cut perpendicular to foliation and lineation. Comparison of 123A and 123B, which were cut from the same specimen, shows the variation in strain parallel and perpendicular to lineation. The foliation appears to be planar in 123A, but is seen to have an anastomosing character in the third dimension, wrapping around the porphyroclasts. Scale bar is 2 cm.

**123C.** Asymmetric magnetite porphyroclast within fine-grained antigorite mylonite matrix, Toll Road shear zone (Norrell and others, 1989). Asymmetry of magnetite neomineralization trails on altered chrome spinel indicates a dextral sense of shear. Antigorite blades are nearly parallel to the foliation defined by magnetite trails, making identification of individual blades difficult. However, insertion of the first-order plate shows antigorite to have a strong crystallographic preferred orientation, thus indicating high strain in this part of the shear zone. Scale bar is 0.5 mm. PNX.

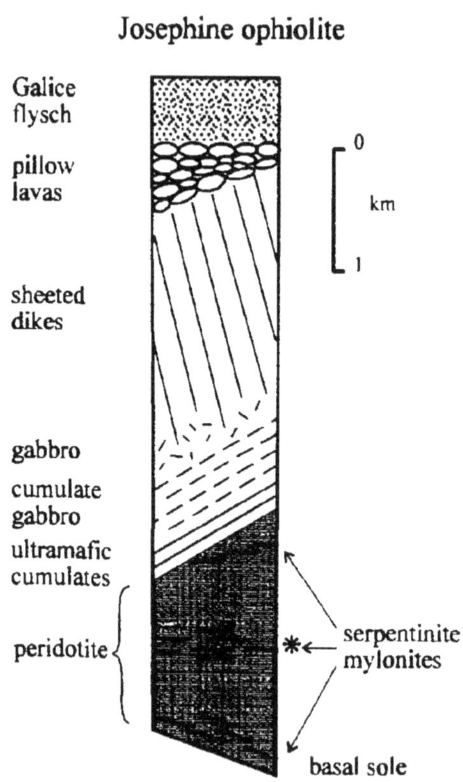


**Figure 123.1.** Pseudostratigraphic section of the Josephine ophiolite, Klamath Mountains, northwestern California, U S A

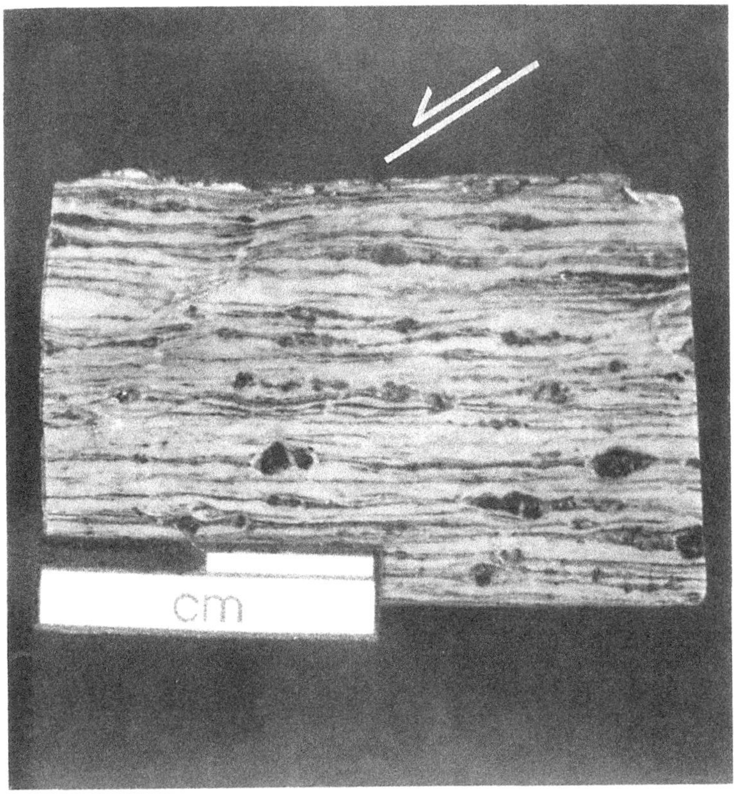


123 A

123 B

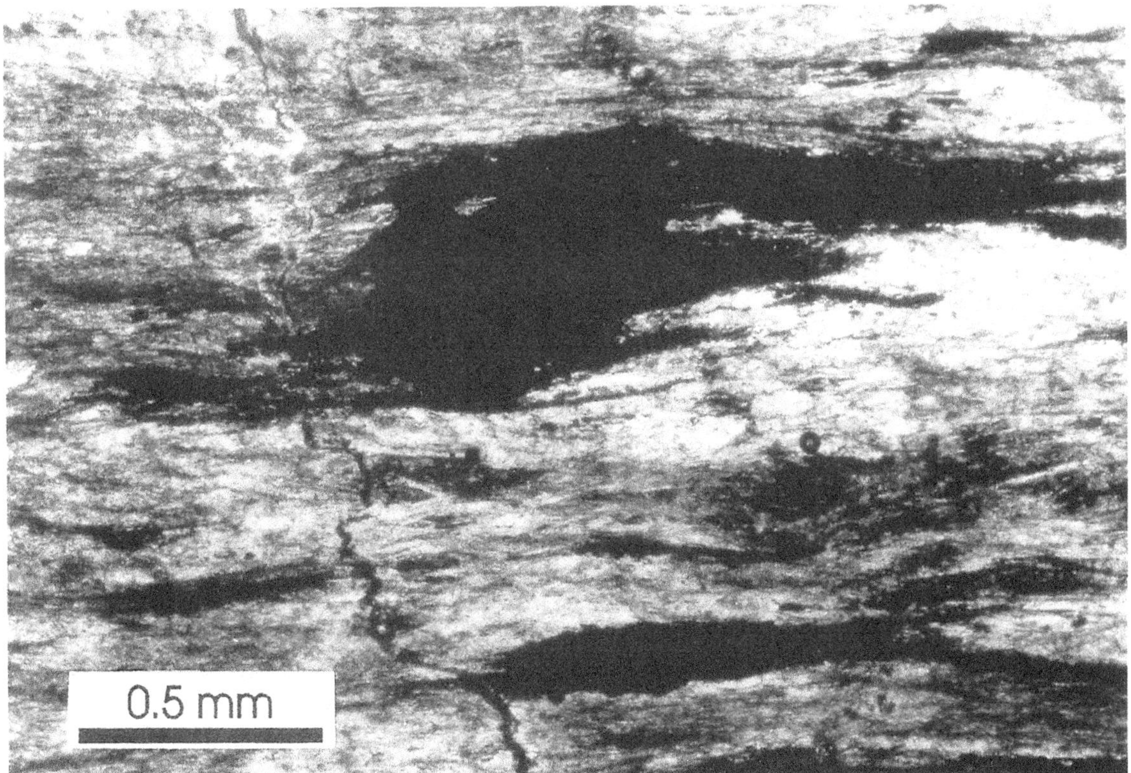


123 C

The serpentine-group minerals are antigorite, lizardite, and chrysotile. All have been found within serpentinite mylonites in the Josephine ophiolite, but antigorite is the dominant mineral and clearly formed during mylonitization. Lesser amounts of lizardite, usually with magnetite, and scarce chrysotile are also found. Antigorite is generally more stable at higher temperatures (~250–500°C) than lizardite and chrysotile (<250°C) (O'Hanley and others, 1989).

The serpentinite mylonites, with the exception of the mylonite at the base of the ultramafic cumulates, are located within massive, partially serpentinized peridotite. Virtually fresh peridotite occurs in some areas below the shear zone marked with an asterisk in Figure 123.1. The mylonites have previously been interpreted as retrogressive shear zones in which antigorite formed directly from olivine (Norrell and others, 1989). However, 123D and 123E show textural relationships which indicate that formation of lizardite predated antigorite. The presence of early lizardite suggests that the shear zone had a low-temperature serpentinization history and was subsequently heated and reactivated to form the antigorite mylonites. Antigorite mylonites therefore formed along preexisting, lizardite-bearing fault zones, and not directly from olivine. Lizardite-bearing fault zones probably provided the necessary zones of weakness, within otherwise massive, largely unserpentinized peridotite, to cause localization of fluids and noncoaxial deformation during formation of the antigorite mylonites.

Antigorite mylonites formed by neomineralization and deformation of an earlier, lower-temperature serpentinite, not from serpentinization of a high-temperature peridotite mylonite. The antigorite blades are coarser-grained than the lizardite that they replaced. Therefore, during the development of these serpentinite mylonites grain-size reduction did not occur. In this respect, these serpentinite mylonites depart from the strict definition of a mylonite (Tullis and others, 1982).

**123D.** S-C fabric in antigorite mylonite, Toll Road shear zone (Norrell and others, 1989). An S-C fabric is defined by coarse-grained antigorite blades (Figure 123.2). The C foliation is defined by bands containing bundles of antigorite blades alternating with domains of lizardite. Lizardite has a very fine-grained mesh texture, pseudomorphous after olivine. The S foliation is defined by antigorite blades, which cross-cut lizardite domains at angles of 10–45° to the C foliation (compare with the higher degree of strain in 123C). The domains of lizardite, cross-cut by antigorite, indicate that a low-temperature serpentinite fault zone developed prior to the formation of the antigorite mylonite. Scale bar is 0.5 mm. PNX

**123E.** Antigorite shear band with late chrysotile vein (Figure 123.3), Toll Road shear zone (Coulton and others, 1995). Close-up view of a dextral shear band within an antigorite mylonite. This photomicrograph shows the relative timing of formation of all three serpentine-group minerals with respect to formation of the mylonite. Fine-grained lizardite + magnetite matrix is cross-cut by antigorite blades having a strong shape and crystallographic preferred orientation. The chrysotile + magnetite extension vein, parallel to the antigorite foliation, is undeformed and therefore postdates formation of antigorite mylonite. The serpentinite mylonites throughout the Josephine ophiolite consistently exhibit this sequence of events in their serpentinization history. Scale bar is 0.05 mm. PNX

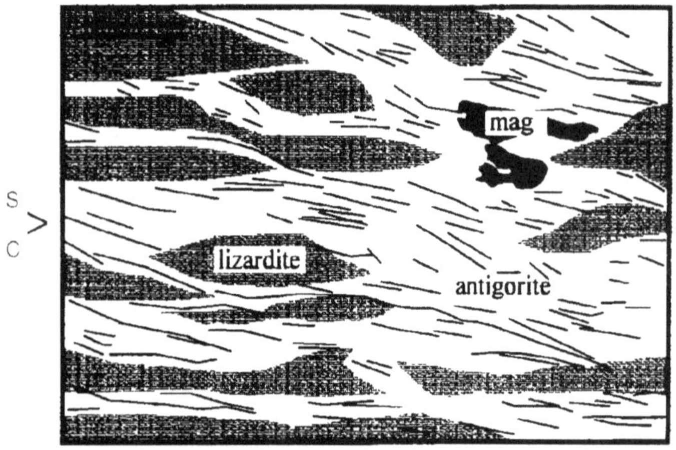


**Figure 123.2.** Sketch of Plate 123D

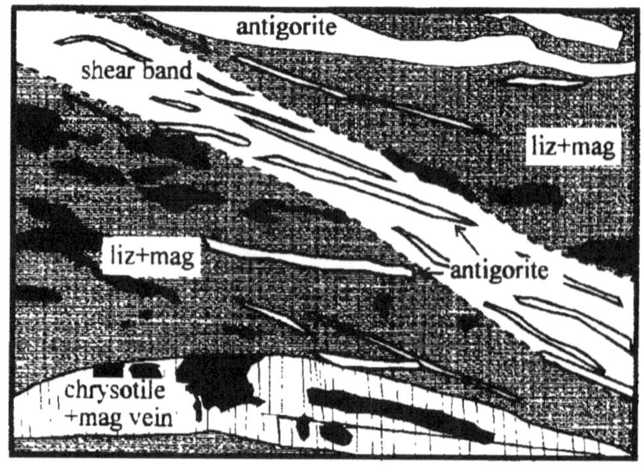


**Figure 123.3.** Sketch of Plate 123E

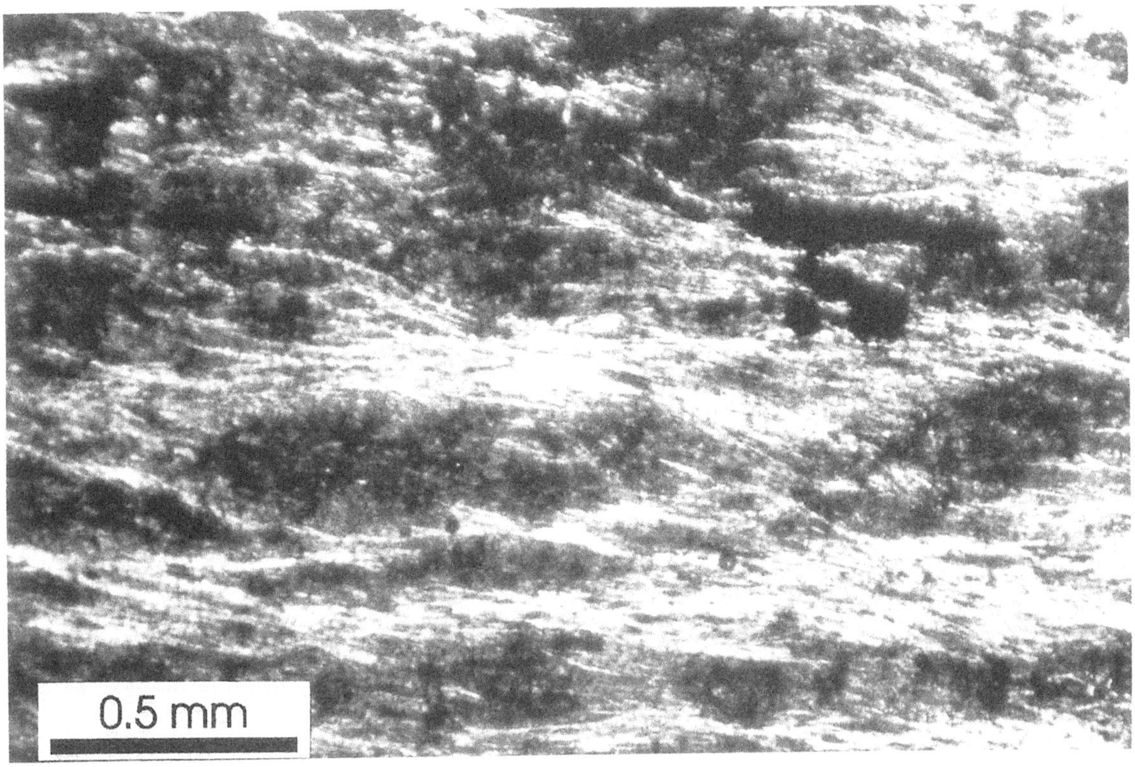


123 D

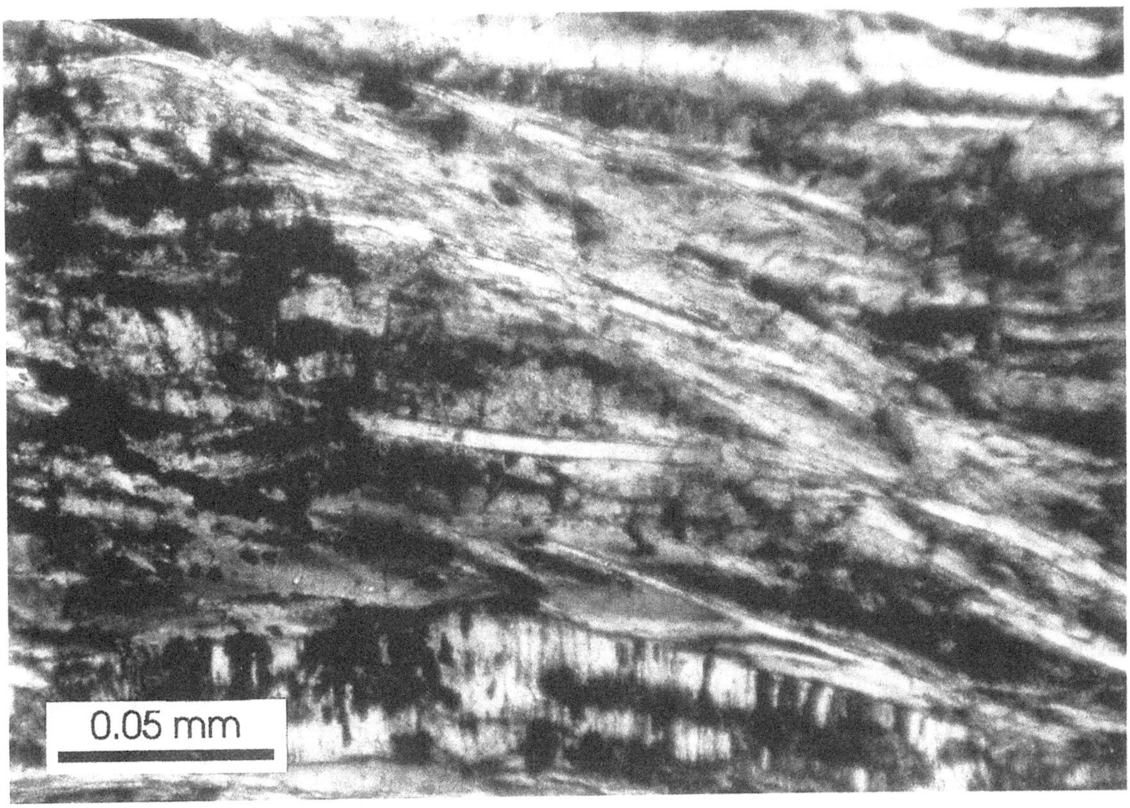


123 E

S. F. Cox and M. A. Etheridge

The accompanying micrographs illustrate the progressive microstructural transition developed in a small (approximately 10 mm wide) shear zone within a massive chalcopyrite lens from the Mt. Lyell volcanic-hosted, base metal deposits in Tasmania, Australia (Cox, 1981; Cox and Etheridge, 1984). Development of the shear zones occurred during regional, Middle Devonian deformation in a low-grade metamorphic regime. Deformation of the chalcopyrite has occurred by dislocation creep mechanisms. Elongate grains and well-developed crystallographic preferred orientations have formed dominantly by {112}⟨110⟩ and ⟨201⟩ dislocation glide. The presence of recovered dislocation substructures indicates that dislocation climb was relatively easy.

With increasing strain there is a microstructural transition involving the development of elongate original grains, subgrain development within the original grains, grain-boundary migration, and ultimately grain-size refinement by recrystallization. As the grain elongation fabric becomes more strongly developed, the resulting foliation curves toward parallelism with the shear plane. Figure 124.1A illustrates the orientation ($\theta$) of the foliation (relative to the shear plane orientation) as a function of distance from the center of the shear zone. In Figure 124.1C, strain expressed as $X/Z$ and calculated from the foliation geometry, assuming plane strain and constant volume deformation within the shear zone, is illustrated. In Figure 124.1B, it may be seen that the degree of recrystallization increases rapidly with increasing strain, so that at the center of the shear zone the aggregate is almost totally recrystallized.

Chalcopyrite ($CuFeS_2$) is an opaque, tetragonal phase and thus isotropic in incident light. The microstructure has been revealed by etching specimens with a $NH_4OH/H_2O_2$ solution. In each of the micrographs, the shear-zone boundary is parallel to the long axis of the micrograph and the sense of shear is dextral

**124A.** Generally subequant, undeformed to weakly strained chalcopyrite at the margin of the shear zone. Plane-polarized incident light (PPIL).

**124B.** Deformed elongate chalcopyrite grains showing evidence of lattice bending, subgrain development, and grain-boundary bulging. Grain-boundary recrystallization has commenced in a few areas. Shear strain is about 0.7. PPIL

**124C.** At higher strains, the deformed grains become more elongate and develop strong deformation bands. Dynamic recrystallization along grain boundaries is consuming the old grains. Shear strain is about 2. PPIL

**124D.** Near the center of the shear zone, a few relict deformed grains are present. Dynamic recrystallization has resulted in the development of a fine-grained aggregate of polygonal and equant to weakly elongate grains. Shear strain is about 20. PPIL

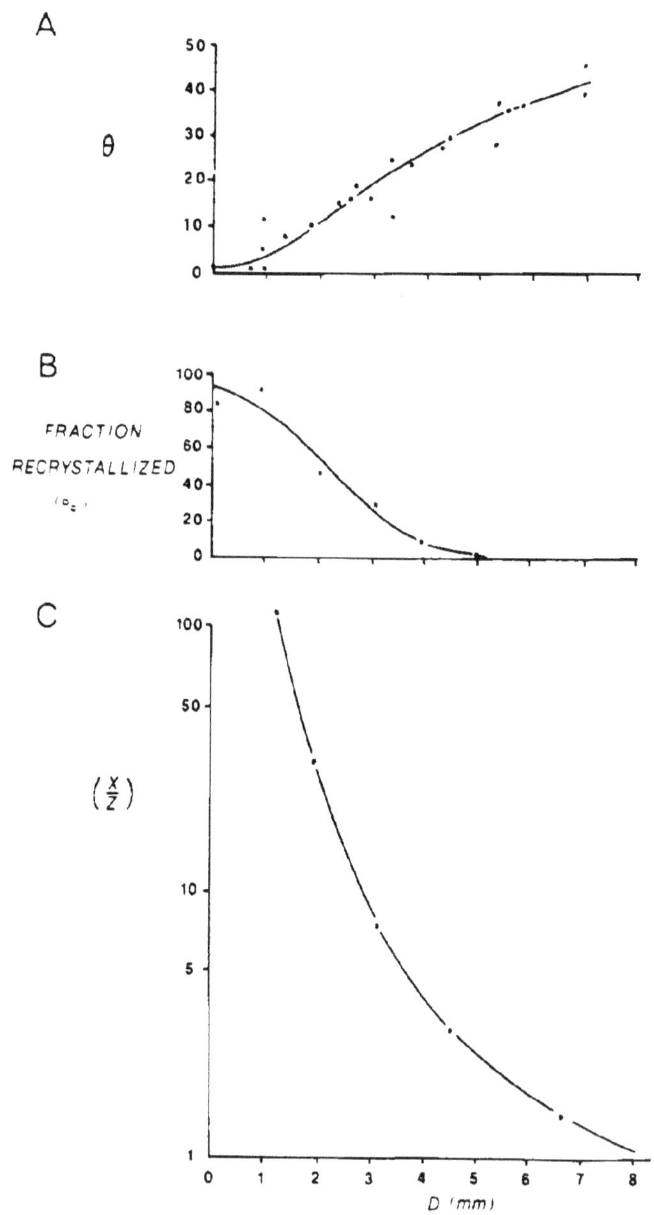


**Figure 124.1.** Variation of θ (explained in text), percent of recrystallization, and strain as a function of distance from the center of the shear zone

124 A

124 B

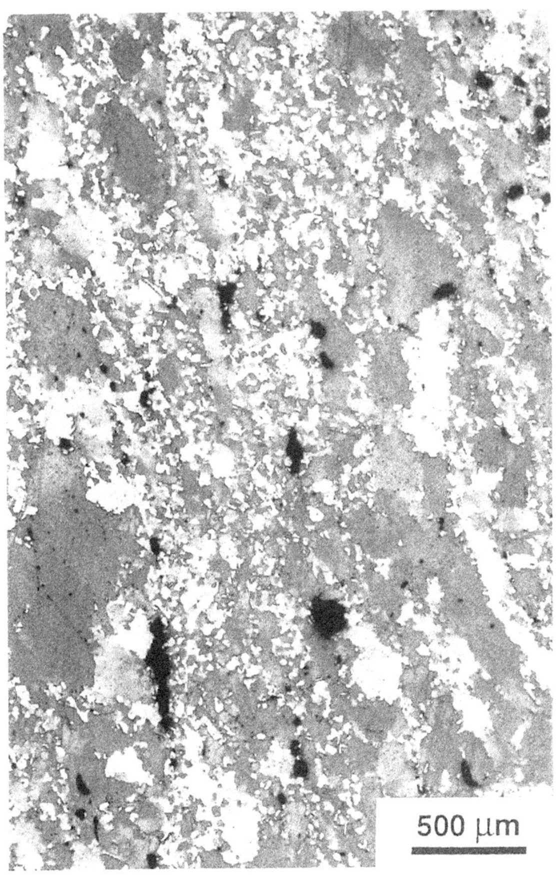


124 C

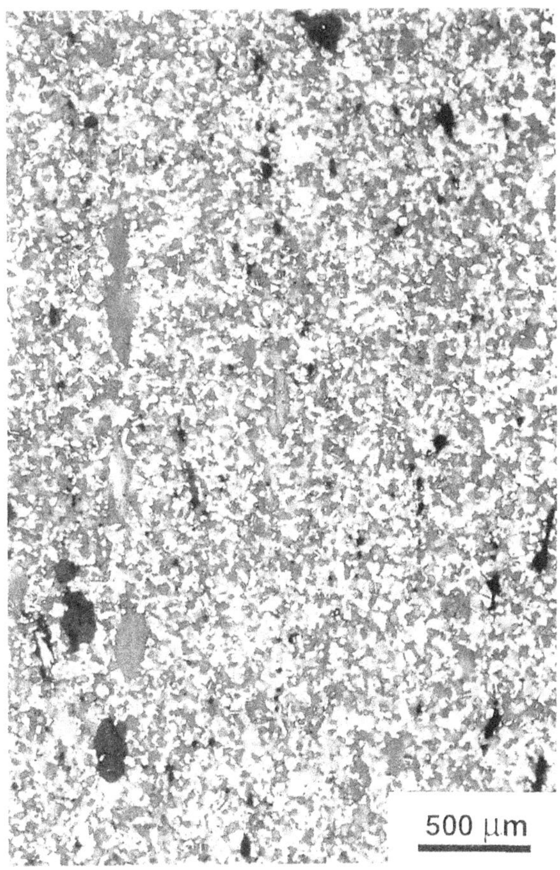


124 D

The microstructures within chalcopyrite mylonites from the Mt. Lyell area in Tasmania (Australia) have many similarities to those developed in some silicates, ionic materials, and metals deformed to high strains by dislocation creep mechanisms. Apart from illustrating the types of microstructures developed in chalcopyrite during dislocation creep, the accompanying micrographs also illustrate the use of different etching techniques to highlight various aspects of the microstructure.

**124E.** Deformed original grains, elongate grain structure, subgrain development, and deformation bands in strongly deformed, coarse-grained chalcopyrite as revealed by the $NH_4OH/H_2O_2$ etch. PPIL

**124F.** Electrolytic etching of chalcopyrite in concentrated acetic acid clearly illustrates the geometry of subgrain boundaries within this deformed single chalcopyrite grain. Subgrains typically have highly oriented boundaries and blocky shapes. PPIL

**124G.** Partly recrystallized chalcopyrite mylonite containing elongate deformed grains and smaller recrystallized grains. The latter are typically equant to weakly elongate in shape, and have straight to gently curved boundaries. Some broad, parallel-sided growth twins are also present in the recrystallized grains.

At the initial stages of recrystallization, new grains develop dominantly along the grain boundaries of the large deformed grains, and are spatially associated with grain-boundary bulges having dimensions similar to those of the recrystallized grains. This suggests that, at least in the initial stages of recrystallization, grain-boundary bulging mechanisms have been important. However, subgrain rotation and subgrain coalescence mechanisms of recrystallization appear to have been important following initial dynamic recrystallization of the original grain-boundary regions of host grains (Cox and Etheridge, 1984). In some cases, significant grain growth by twin coalescence has followed new grain nucleation. $NH_4OH/H_2O_2$ etch. PPIL

**124H.** Subgrain structure within a recrystallized chalcopyrite mylonite. Some recrystallized grains are elongate and have well-developed subgrains, whereas others are more equant and have poorly developed subgrains. Acetic acid electrolytic etch. PPIL

124 E

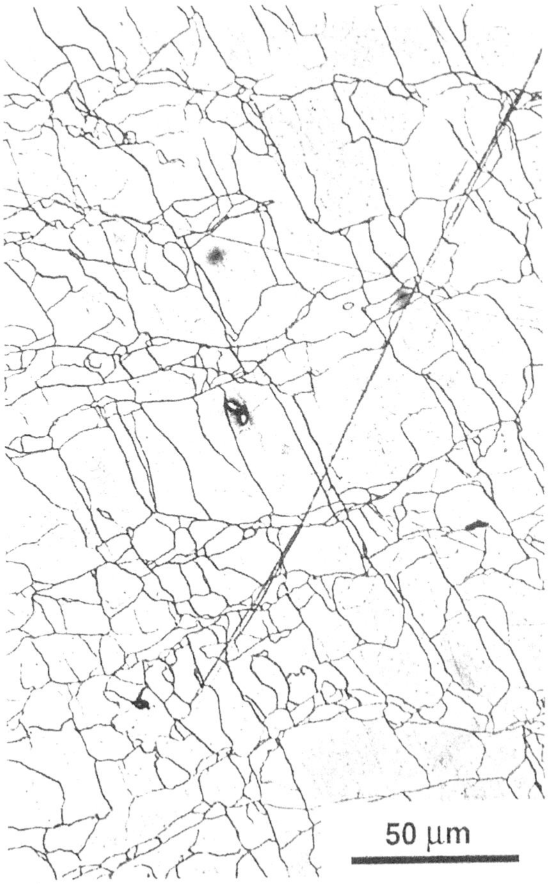

124 F

124 G

124 H

# DUCTILE BEHAVIOR

## Metamorphic grade variations

### High-grade shear zones



Common abbreviations used in regard to the plates in this section include PPL = plane polarized light, NX = crossed nicols, SEM = scanning electron microscope, and TEM = transmission electron microscope.

This set of plates illustrates features developed in shear zones formed at upper-amphibolite to granulite facies, and during the transition from granulite to eclogite facies. High temperature promotes crystal plasticity in plagioclase, amphibole, and pyroxene which are strong at low to moderate grades. However, high-temperature microstructures are only well preserved if the rocks remain dry during exhumation. Several plates illustrate the interplay of deformation and metamorphism; others emphasize the role of water, e.g., in controlling the reaction of granulite-facies rocks to eclogites in shear zones. Several plates demonstrate that plagioclase requires much higher temperatures than quartz for equivalent microstructures, e.g., it requires amphibolite-facies conditions for recrystallization-accommodated dislocation creep and granulite-facies conditions for climb-accommodated dislocation creep (depending somewhat on water availability). Pyroxene and garnet are even stronger. Several contributions emphasize the importance of polymineralic aggregates which undergo dynamic recrystallization to a much smaller grain size where grain boundaries are pinned; this may promote a switch to grain-size-sensitive creep. Sense-of-shear indicators, although not as common in high-grade shear zones as in low- to medium-grade shear zones, are illustrated in several of the plates.

JOSEPH CLANCY WHITE

Deep-crustal mylonites can exhibit intense single-crystal distortion of "low-ductility" minerals. Synkinematic metamorphic relationships indicate that these mylonites from the Parry Sound shear zone, Ontario, Canada (Figure 125.1), formed at temperatures of 700–800°C and pressures on the order of 800 MPa, related to probable Himalayan-type collisional tectonics and crustal thickening during the Proterozoic Grenville event.

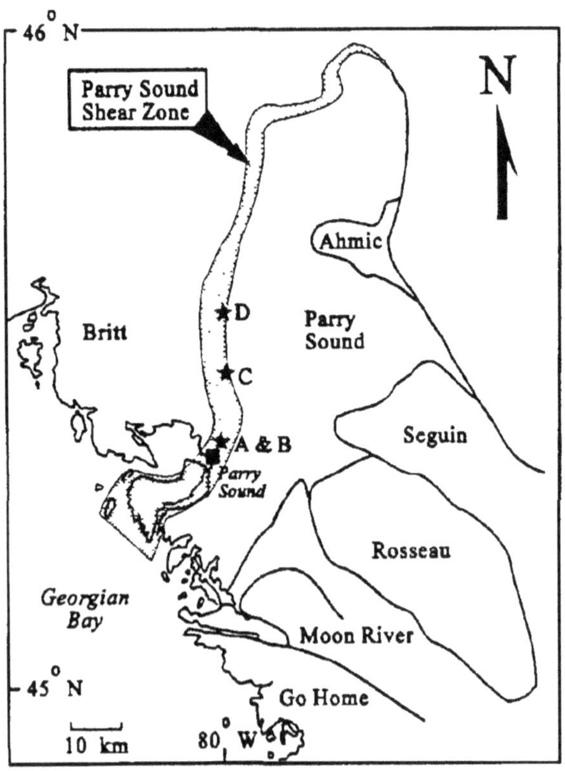


**Figure 125.1.** Geologic sketch map of the Parry Sound shear zone, Ontario, Canada. Approximate localities of samples used for the photographs of Plate 125 are noted. After Davidson and others (1982) and Culshaw and others (1989).

**125A.** Folded perthite porphyroclast (K), Parry Sound shear zone, Parry Sound, Ontario, Canada (White and Mawer, 1986). The mylonite matrix is predominantly dynamically recrystallized K-feldspar and plagioclase derived from the host perthite. In contrast to the distortions observed in the feldspar, hornblende porphyroclasts (H) record much smaller single-crystal strains prior to dynamic recrystallization. Porphyroclasts are derived from synkinematic pegmatites. The orientation of the stretching lineation, L, varies within the foliation. The perthite porphyroclast may indicate a reversal in noncoaxiality (as opposed to shear sense). This texture is representative of the distribution and partitioning of deformation typical of heterogeneous, non-steady-state flow in rocks (Jiang, 1994) and in this case reflects a local kinematic vorticity number $W_k < 1$. Polished surface.

**125B.** Strain gradient in a perthite porphyroclast, Parry Sound shear zone, Parry Sound, Ontario, Canada (White and Mawer, 1986). Detailed, analytical TEM has demonstrated (White and Mawer, 1988) that plagioclase exsolution domains (p) in the perthite evolve through deformation-enhanced diffusion along incoherent boundaries, growth and coalescence in a K-feldspar host (k), and eventually produce discrete grains. Exsolution domains become extremely elongate near the porphyroclast-matrix boundary in response to increased intracrystalline strain within the porphyroclast. The long axes of these domains are consistent with slip on (010), parallel to the observed albite-law deformation twins (T). Progressive misorientation of exsolution domains during deformation leads to recrystallization of the perthite into discrete K-feldspar and plagioclase grains, forming a two-phase matrix favorable to the operation of intercrystalline deformation mechanisms. The latter is indicated by the dramatic changes in matrix grain orientations as one moves away from the recrystallization front, and is consistent with accumulation of the observed large bulk strains (attenuation, transposition, folding) in the mylonite that do not involve continued extensive intracrystalline deformation. NX

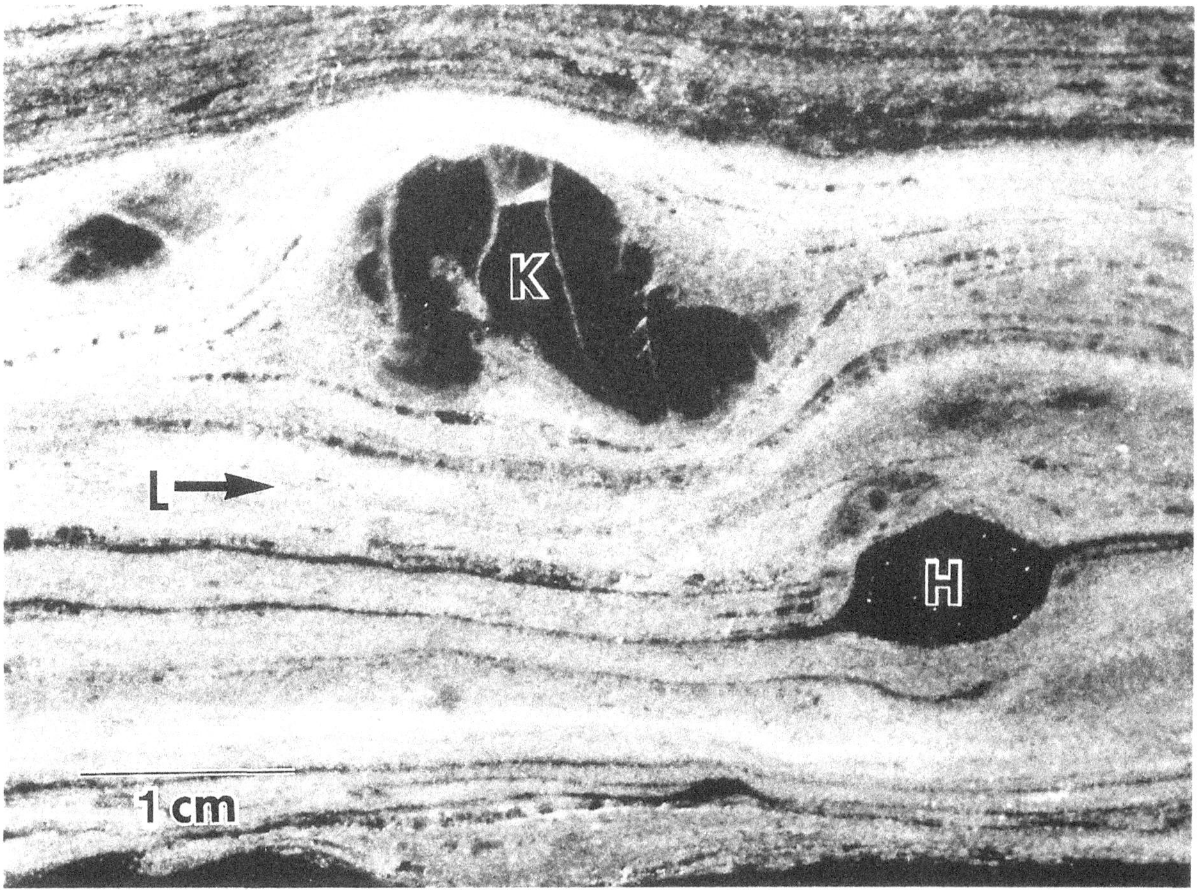

125 A

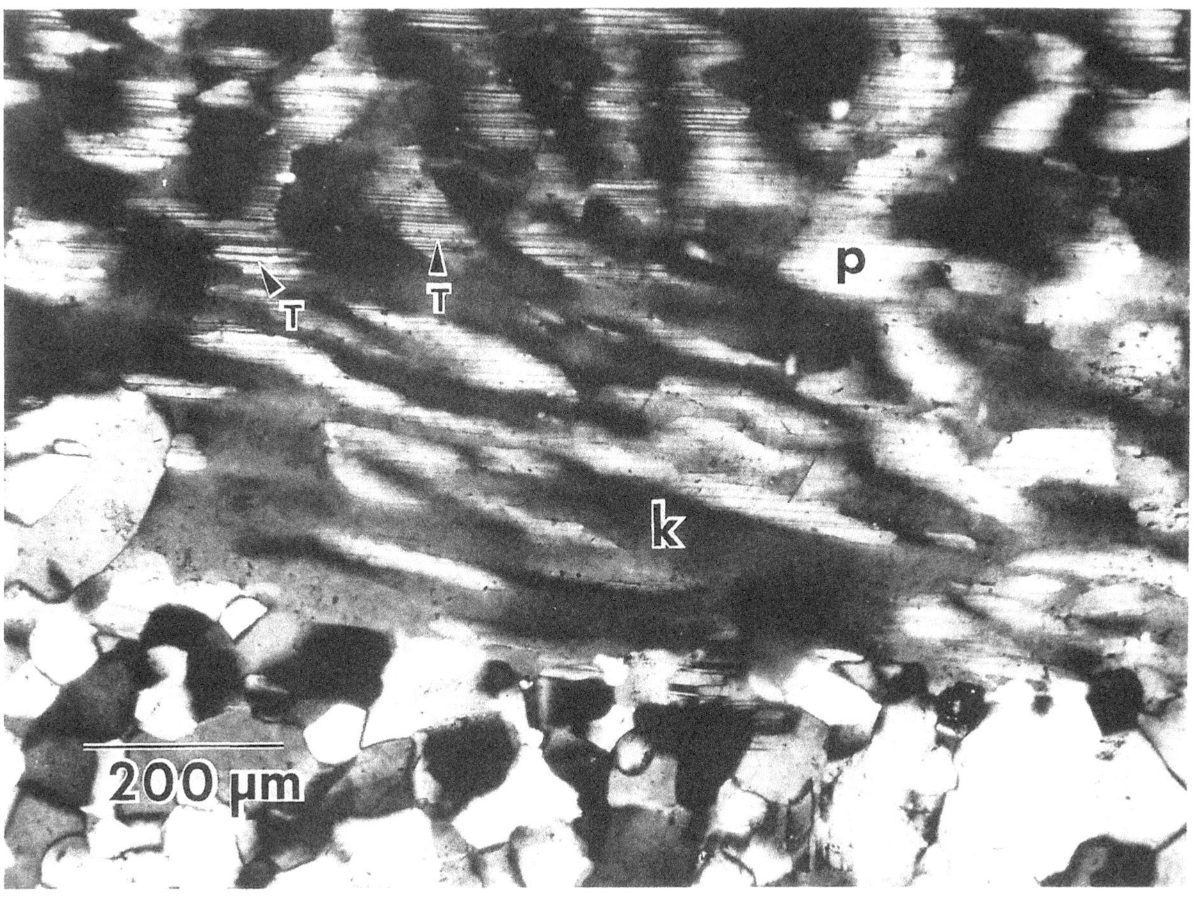

125 B

**125C.** Elongate orthopyroxene crystal, Parry Sound shear zone, Waubamik, Ontario (White and Mawer, 1992). Granulite-grade metasedimentary units contain sillimanite-orthopyroxene-plagioclase-garnet subassemblages. Highly elongate orthopyroxene grains (Opx) are surrounded by plagioclase grains (Plag), obviating the possibility that the high aspect ratio reflects only preferential recrystallization. The elongate orthopyroxene grains are oriented in the foliation to allow easy slip on the dominant (100) slip plane, traces of which are indicated by arrows. NX

**125D.** Folded augitic clinopyroxene, Parry Sound shear zone, Dunchurch, Ontario (White and Mawer, 1992). Sweats or veins comprising garnet, clinopyroxene, and plagioclase formed in intermediate to mafic granulites and were subsequently deformed at hornblende-granulite conditions. Hydration has produced incipient amphibole layers (A) in the clinopyroxene (Cpx) that serve as markers for the intense intracrystalline distortion that develops at the clinopyroxene grain boundary, adjacent to deformed and recrystallized plagioclase (Plag). Under these deformation/metamorphism conditions, the apparent rheological contrasts among clinopyroxene, orthopyroxene, and plagioclase can be minimal, although the general sequence of increasing ductility remains clinopyroxene → orthopyroxene → plagioclase. Clinopyroxene expresses a stronger tendency to recrystallize with minimal grain elongation, whereas both orthopyroxene and plagioclase grains can achieve high aspect ratios before dynamically recrystallizing. PPL

125 C

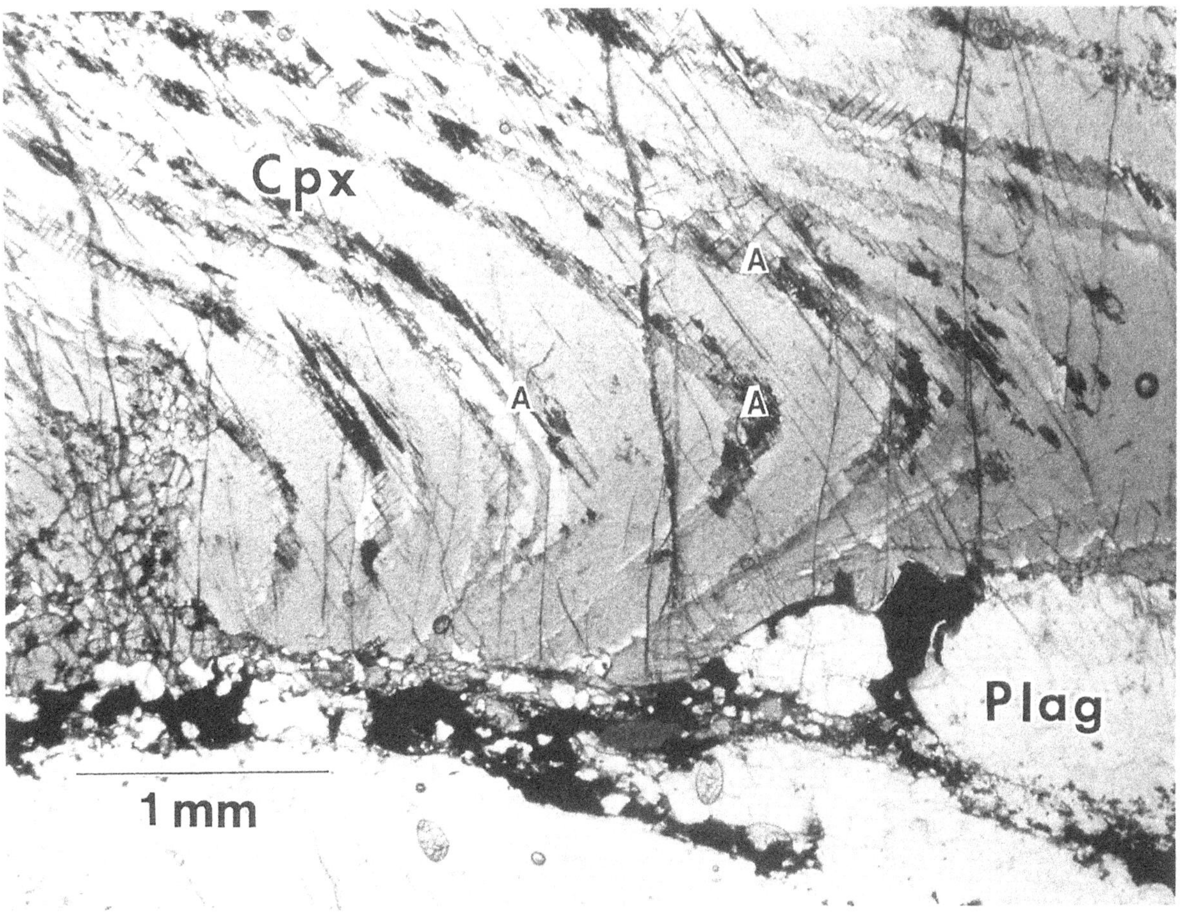

125 D

# 126. HIGH-TEMPERATURE MYLONITES: II

JOSEPH CLANCY WHITE, DAZHI JIANG, AND SIMON HANMER

The preservation of macroscopic and microscopic deformation features from deep-crustal regimes is particularly favored in anhydrous rocks in which fluid-enhanced overprints are suppressed. Archean granulite mylonites from the Snowbird tectonic zone (Hanmer and others, 1994), northwest Saskatchewan, Canada (Figure 126.1), preserve a complete sequence of deformation, recovery, and dynamic recrystallization for minimum synkinematic P-T estimates of 800–900 MPa and 780–800°C.

**126A.** Granulite mylonite, East Athabasca mylonite zone, Saskatchewan, Canada. Gabbroic anorthosites have been mylonitized to produce orthopyroxene porphyroclasts in a matrix of recrystallized and deformed plagioclase, garnet, orthopyroxene, and clinopyroxene grains. Orthopyroxene (Opx) commonly forms high-aspect-ratio grains (up to 120:1) that are oriented for easy slip. Plagioclase grains (Plag) exhibit intense optical preferred crystallographic orientations indicative of crystal-plasticity. Recrystallized grains of pyroxene and plagioclase are commonly on the order of 10 μm or less in diameter. NX

**126B.** Highly deformed orthopyroxene grain, East Athabasca mylonite zone, Saskatchewan, Canada. The shear-induced orthoenstatite → clinoenstatite transformation is temperature-stress-strain rate dependent (Raleigh and others, 1971) and can be used to confirm the formation of these mylonites at deep-crustal conditions. Clinoenstatite (CLEN) is observed in "shard-like," high-aspect-ratio orthoenstatite (OREN) grains and exists as bands of 0.9 nm fringes within a host of 1.8 nm fringes. The proportion of transformed material is highly variable. A twin boundary serves as a locus for dislocations (D). For the minimum temperature estimates of 780–800°C, differential stresses on the order of 75–120 MPa are required for the transformation, corresponding to strain rates in the range of $10^{-11}$ $s^{-1}$–$10^{-10}$ $s^{-1}$ (Raleigh and others, 1971). The highest stress estimates (≈200 MPa) in the mylonite exceed these limits, with the range of inferred deformation conditions within the mylonites spanning the boundary for the OREN → CLEN transformation. The rheological response corresponding to the transformation requires the occurrence of localized high strain rates in strong lower crust. TEM bright field (BF)

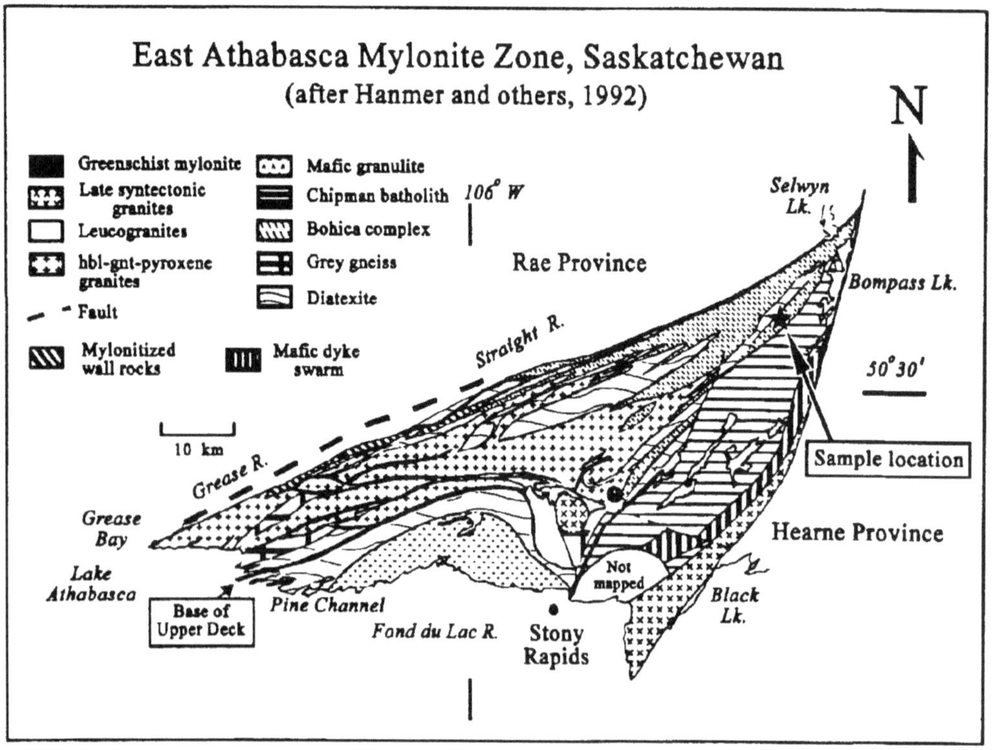


**Figure 126.1.** Geologic sketch map of the East Athabasca mylonite zone, Saskatchewan, Canada

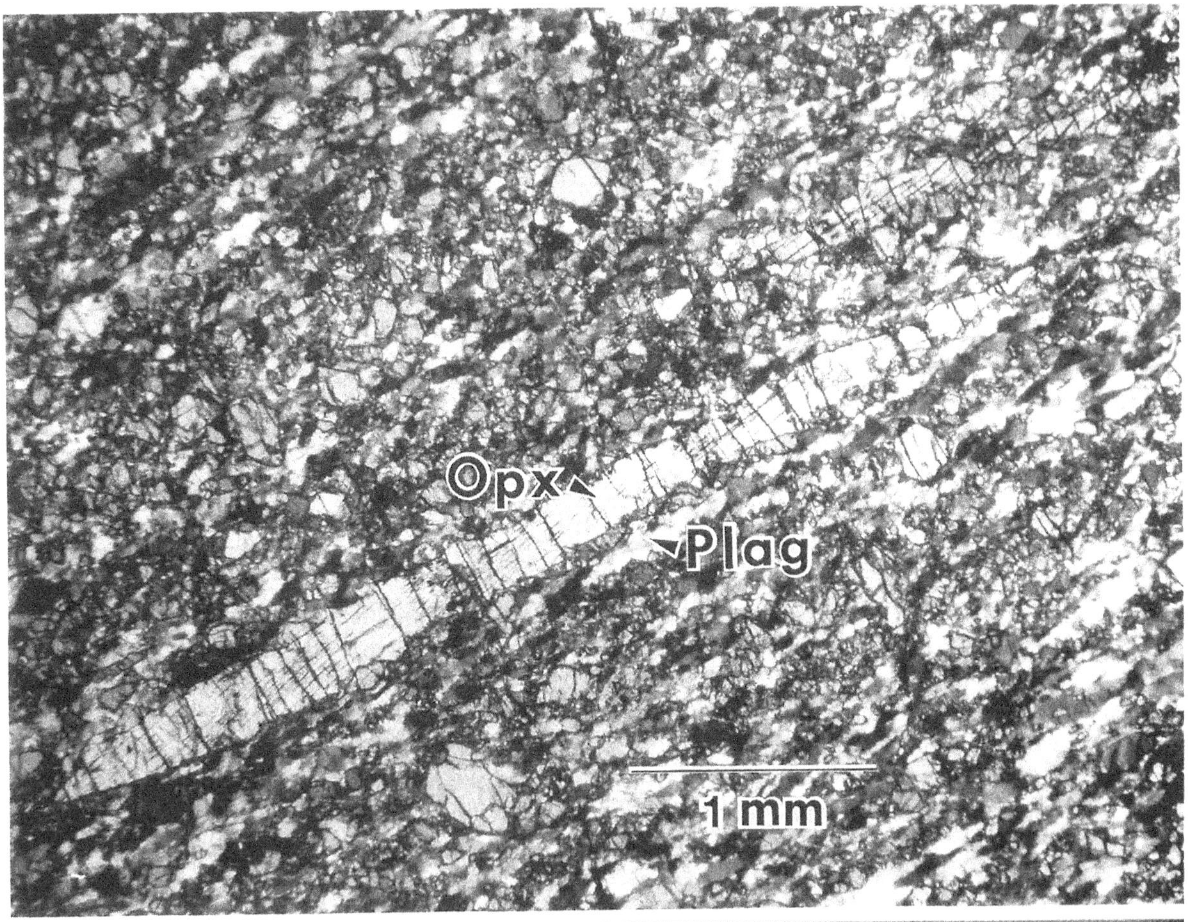


126 A

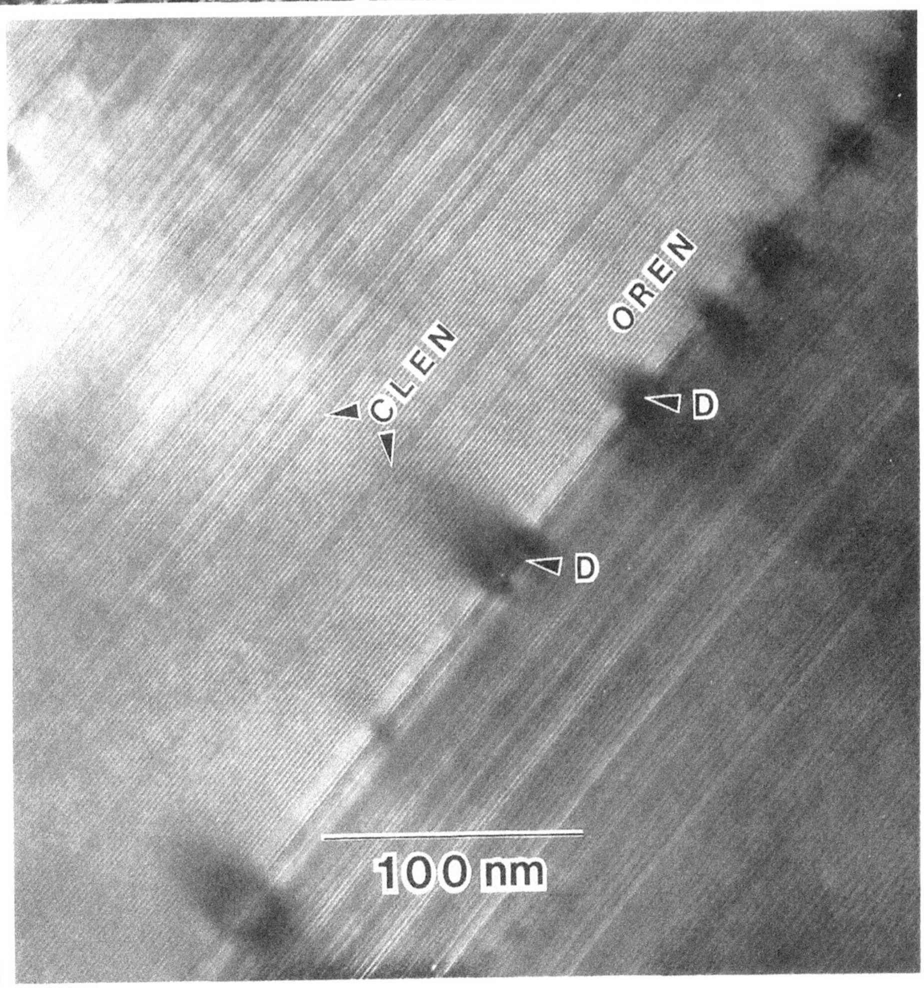


126 B

**126C.** Defect substructures in andesine, East Athabasca mylonite zone, Saskatchewan, Canada. Dislocation glide (D) on several slip systems is common in plagioclase within the mylonites, with only limited recovery to form subgrain walls (W) prior to dynamic recrystallization. Paleopiezometers for plagioclase have been adapted from Weathers and others (1979) and White (1982), for dislocation density, and constructed from experimental data (Tullis and Yund, 1985; Tullis and others, 1990) for recrystallized grain size. These are clearly first-order estimates, but are consistent with estimates made using quartz, from both experiments and rocks that contain both minerals. Dislocation densities range from $10^{11}$ $m^{-2}$ (26 MPa) in "coarse" mylonite (30 μm grains, 20 MPa) to $10^{13}$ $m^{-2}$ (260 MPa) in the finest-grained mylonites (1 μm, 220 MPa, e.g., see Plate 126D). The latter stresses are consistent with ambient mylonite zone deformation at $10^{-12}$ $s^{-1}$–$10^{-11}$ $s^{-1}$, within which occur localized higher-strain-rate transients. Evidence for late-stage, low-temperature overprints is absent. TEM BF

**126D.** Dynamically recrystallized andesine, East Athabasca mylonite zone, Saskatchewan, Canada. Strain-free neoblast developed during recrystallization by strain-induced boundary migration; such neoblasts commonly experience subsequent intracrystalline deformation. This microstructure indicates that at higher strain rates, diffusion-controlled recovery is not able to accommodate dislocation glide, and dynamic recrystallization is the process by which stored strain energy can be dissipated. Adjacent orthopyroxene (with clinopyroxene lamellae; Opx) and plagioclase (Plag) grains have significant dislocation densities ($\approx 1 \times 10^{12}$ $m^{-2}$ and $5 \times 10^{12}$ $m^{-2}$, respectively). The variation in dislocation density observed in the rocks reflects the cyclic nature of the deformation-recovery-recrystallization process, and argues against extensive late-stage overprints. Subgrain rotation recrystallization is largely limited to rocks interpreted to reflect the lower stress/strain-rate background environment ($\approx$10 MPa, $10^{-13}$ $s^{-1}$) in which the mylonites are found. TEM BF

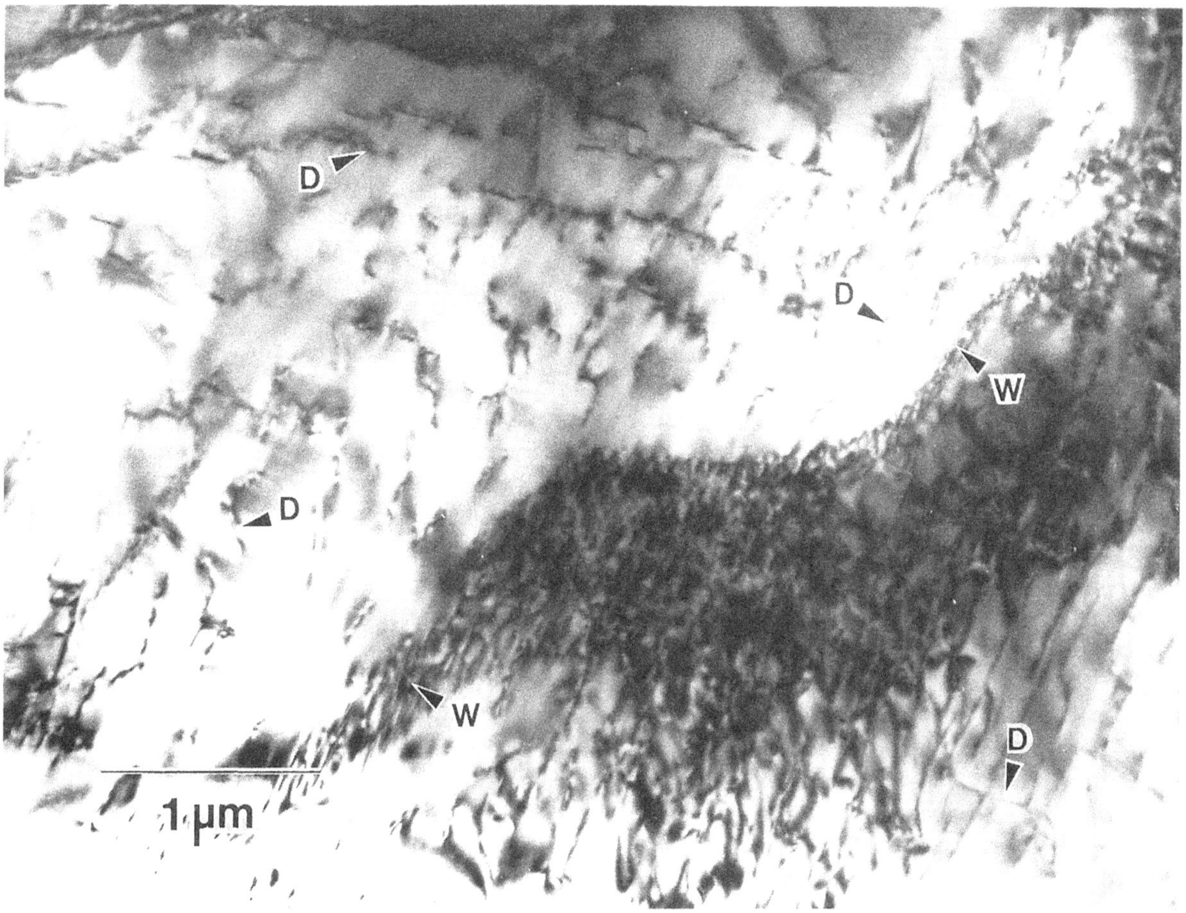


126 C

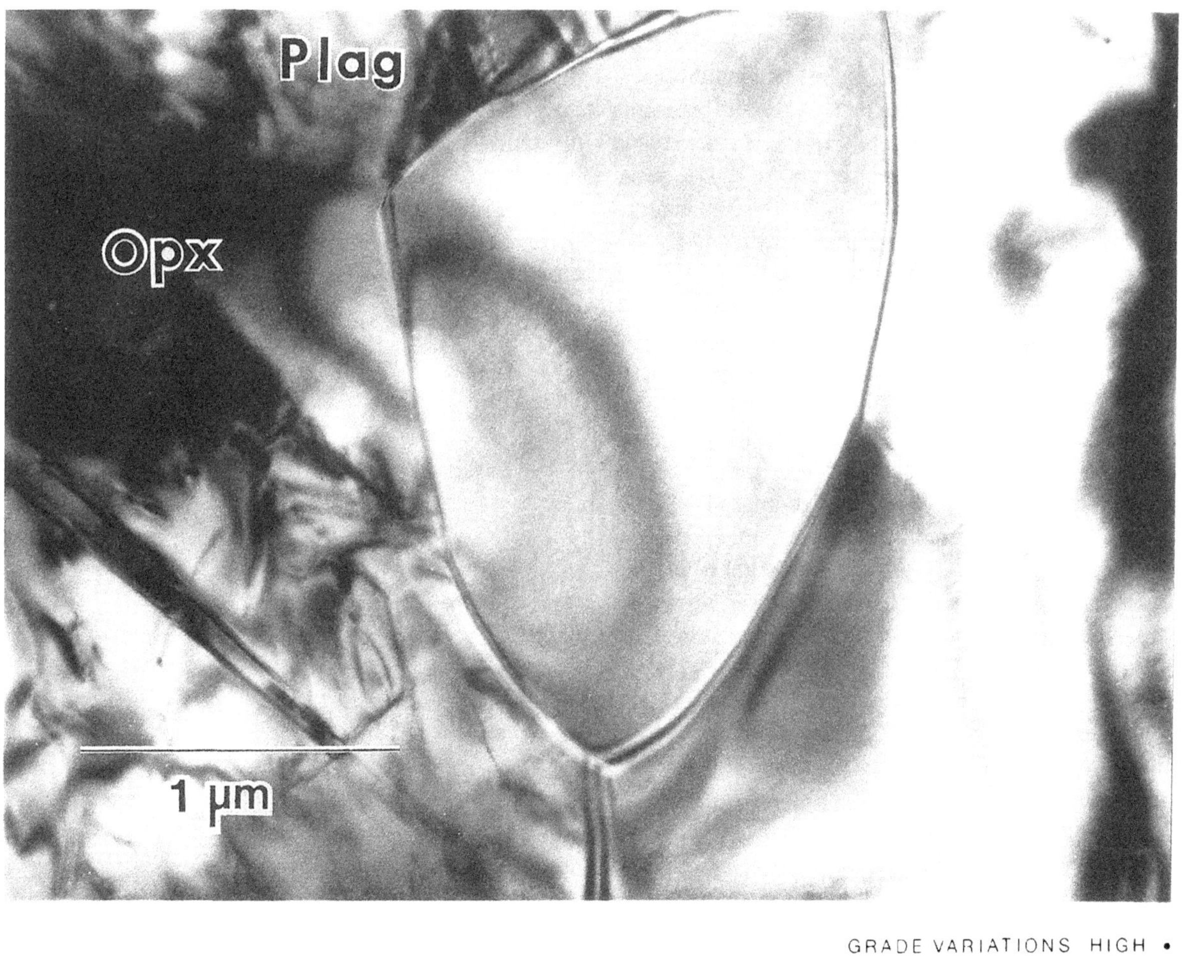


126 D

TSUYOSHI TOYOSHIMA

Large lenticular bodies of gabbro mylonite (GM in Figure 127.1) occur in the basal zone of the northern and southern parts of the Hidaka metamorphic belt (HMB), Hokkaido (H), northern Japan. Some of the gabbro mylonite is also found as smaller sheets or lenses in the basal zone of the south-central parts of the belt. The gabbro mylonite consists of orthopyroxene, plagioclase, small amounts of clinopyroxene, and opaque minerals. Mylonitic foliation and mineral lineation are subparallel to the general north-south strike of the Hidaka metamorphic belt. The gabbro mylonite developed in lower-crustal rocks of the Hidaka metamorphic belt (Figure 127.1), along an intracrustal basal décollement under granulite-facies conditions (about 600 MPa and 800°C; Komatsu and others, 1994) during the early Paleogene (Toyoshima and others, 1994). The gabbro mylonites display asymmetrical microstructures and textures such as S-C fabric (Berthé and others, 1979a; Lister and Snoke, 1984) as well as porphyroclasts with asymmetrical tails of recrystallized fine-grained minerals, all of which indicate a dextral sense-of-shear. Orthopyroxene and plagioclase show strong crystallographic preferred orientation with monoclinic symmetry in the gabbro mylonite. Metamorphic and structural studies indicate that the mylonitic foliation and mineral lineation were formed subhorizontally, and that the mylonitization was caused by top-to-the-south, subhorizontal displacement of detached, lower-crustal rocks of the Hidaka metamorphic belt along a subhorizontal detachment surface linked to a subvertical strike-slip fault, i.e., a wrench tectonic regime (Toyoshima and others, 1994). During exhumation of the Hidaka crustal rocks after the mylonitization of the gabbro, the Hidaka rocks were rotated around horizontal north-south axes, and the foliation, which originally formed subhorizontally along the detachment surface, was tilted steeply to the east.

**127A.** Asymmetrical microstructures in gabbro mylonite. Asymmetrical tails (Ato) of orthopyroxene porphyroclasts (Opxpc) are composed of recrystallized fine-grained orthopyroxene with or without small amounts of clinopyroxene. Asymmetrical tails (Atc) of clinopyroxene porphyroclasts (Cpxpc) are formed by recrystallized, fine-grained clinopyroxene with or without small amounts of orthopyroxene. Orthopyroxene porphyroclasts are sometimes linked by thin tails of recrystallized, fine-grained orthopyroxenes, showing a stair-stepping from one orthopyroxene porphyroclast to the next (Opxstep).

Porphyroclasts of orthopyroxene, clinopyroxene, and plagioclase are replaced at their rims by fine-grained orthopyroxene, clinopyroxene, and plagioclase, respectively, showing that their grain-size reduction and dynamic recrystallization occurred during the mylonitization. Most of the porphyroclasts in the gabbro mylonite are orthopyroxene, but some are clinopyroxene. Plagioclase porphyroclasts (Plpc) are scarce. Most of the plagioclase occurs as a constituent of the recrystallized, fine-grained matrix. Cpxpc is less elongated than Opxpc. These observations suggest that clinopyroxene was the strongest, whereas plagioclase was the weakest phase during mylonitization. PPL

**127B.** Gabbro mylonite containing strongly elongated orthopyroxene porphyroclasts (Lopx) as well as weakly elongated porphyroclasts (Sopx). Both Lopx and Sopx porphyroclasts contain deformation lamellae (DL) parallel to the (100) plane, but the orientation of the lamellae with respect to the mylonitic foliation (S) differs for the two porphyroclast types. The deformation lamellae of Lopx lie in a favorable orientation for easy slip on the (100) plane, whereas the deformation lamellae of Sopx lie in an unfavorable orientation. Lopx porphyroclasts were elongated by slip on (100) planes with glide direction parallel to the [001] axis. Bending of the (100) plane and deformation lamellae (Bop) occurred during the elongation of the orthopyroxene porphyroclast. The degree of elongation of orthopyroxene porphyroclasts apparently depended upon the initial orientation of potential glide planes relative to the kinematic axes of shear deformation during mylonitization (Toyoshima, 1991).

The gabbro mylonite has a recrystallized, fine-grained matrix in which orthopyroxene, clinopyroxene, and plagioclase are commonly mixed on the grain scale. However, the gabbro mylonite also usually exhibits a banded structure, which is formed by the alternation of distinctive unmixed layers [recrystallized, fine-grained orthopyroxene layers (Rop); recrystallized, fine-grained clinopyroxene layers (Rcp); and recrystallized, fine-grained plagioclase layers (Rpl)], orthopyroxene-clinopyroxene-mixed layers, and orthopyroxene-clinopyroxene-plagioclase-mixed layers (Rm). PPL

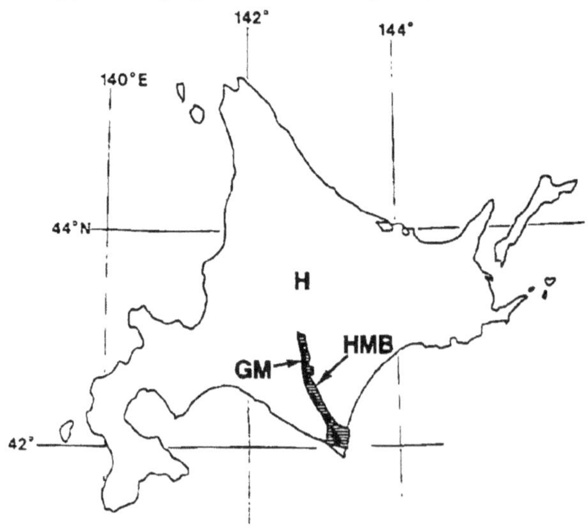


**Figure 127.1.** Sketch map of Hokkaido, Japan, showing the location of the Hidaka metamorphic belt (HMB) and the distribution of gabbro mylonites (GM) in the belt.

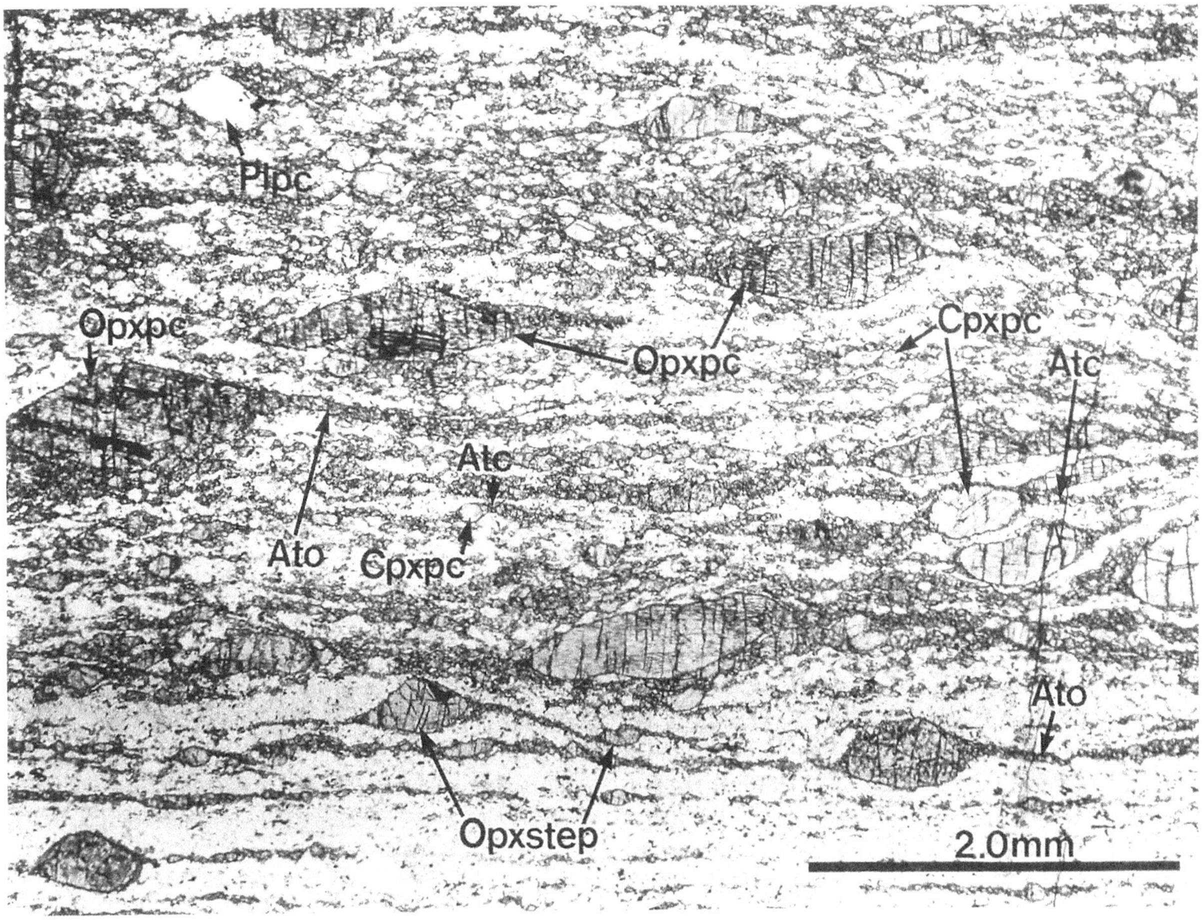
Plpc
Opxpc
Opxpc
Cpxpc
Atc
Atc
Ato
Cpxpc
Ato
Opxstep
2.0mm

127 A

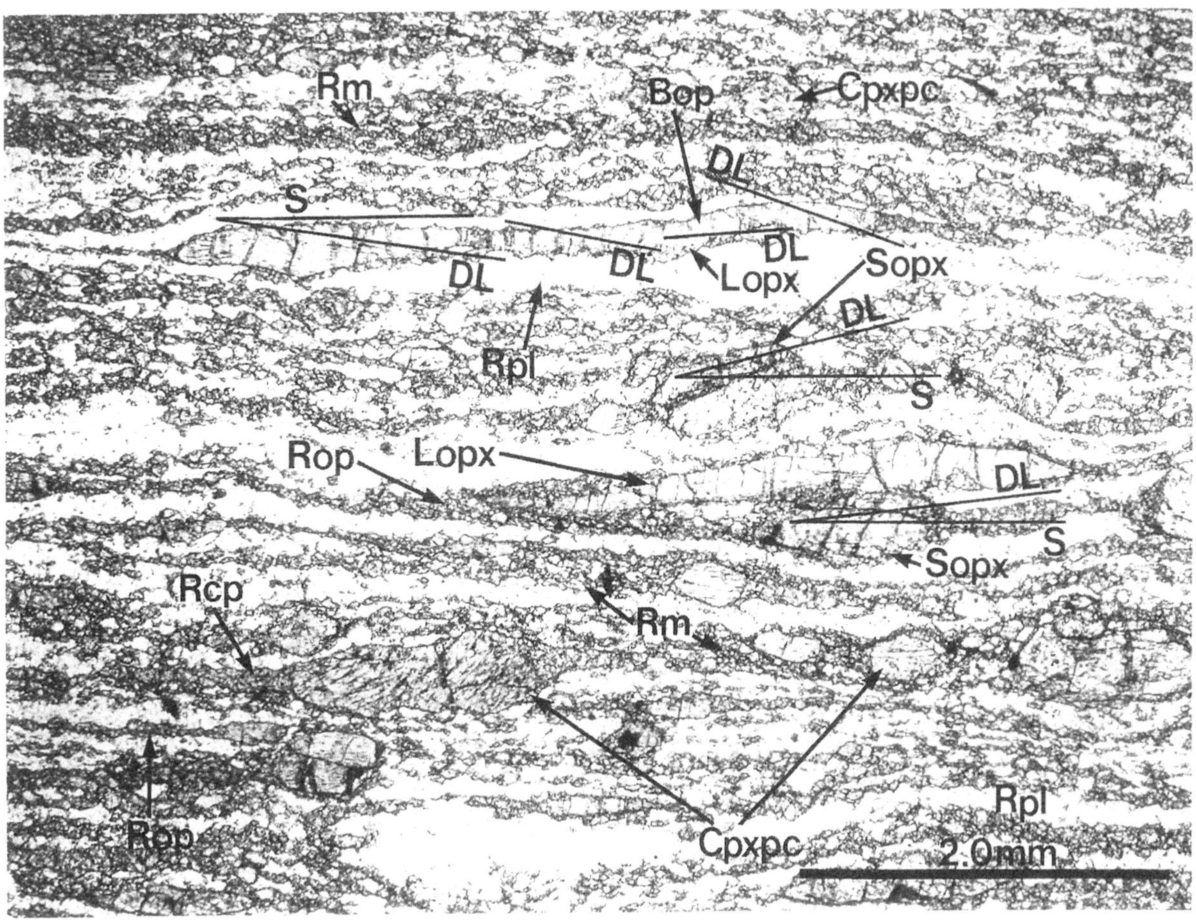
Rm
Bop
Cpxpc
DL
S
DL
DL
DL
Lopx
Sopx
DL
Rpl
S
Rop
Lopx
DL
S
Sopx
Rcp
Rm
Rop
Cpxpc
Rpl
2.0mm

127 B

# 128. HIGH-TEMPERATURE MYLONITES: I. METABASIC MYLONITES

K. H. Brodie

The Ivrea-Verbano zone in northern Italy (Figure 128.1) represents a section of post-Hercynian lower continental crust that has been uplifted and emplaced into its present position during the Alpine collision. The microstructures of the rocks in most of the high-grade parts of the zone were developed under conditions of low to zero deviatoric stress, following deformation and high-grade regional metamorphism (Rutter and others, 1993). Superimposed on these peak metamorphic equilibrated microstructures, but still under high-temperature (upper amphibolite- to granulite-facies) conditions are localized zones of high plastic strain. A network of such high-temperature shear zones occurs within the high-grade part of the zone, cutting metabasic, ultrabasic, and metasedimentary rocks (Figure 128.2; Brodie and Rutter, 1987; Rutter and others, 1993). Zones of more distributed, post-peak metamorphic plastic deformation may be found, bounding these localized high-temperature shear or mylonite zones. The mylonites in the shear zones are characterized by extreme grain-size reduction by dynamic recrystallization, and progressive deformation from protomylonite to ultramylonite is observed (see the review of lower-crustal rheology by Rutter and Brodie, 1992).

Deformation of the metabasic rocks involves grain-size reduction by crystal-plastic deformation of plagioclase, and slip on ⟨001⟩ in orthopyroxene producing elongate grains. Garnet and clinopyroxene, the other main phases present in the rocks, appear to have undergone grain-size reduction by more brittle granulation, but in many of the zones are no longer stable due to the reaction: garnet + clinopyroxene = orthopyroxene + plagioclase, indicative of decreasing pressure conditions but within the granulite facies (Brodie, 1995).

**128A** (PPL) **and 128B** (NX). Optical micrographs showing the microstructures of a protomylonite at the edge of a high-strain zone (Rio del Ponte, Valle d'Ossola). The original coarse-grained garnet-pyroxene-plagioclase granofels has undergone crystal-plastic deformation of the plagioclase, producing recrystallization along grain boundaries. In addition, in this sample, wormlike symplectitic intergrowths of plagioclase and orthopyroxene are developed around the garnet grains. These are well oriented and define the stretching direction within the mylonite, i.e., they are growing in the direction of greatest finite extension and are developed only on potentially dilatant interfaces. They have developed synkinematically with the shear deformation, confirming that the deformation occurred under high-temperature, granulite-facies conditions (the area outlined is the field of view shown in 128G). Representative mineral analyses are shown in Table 128.1.

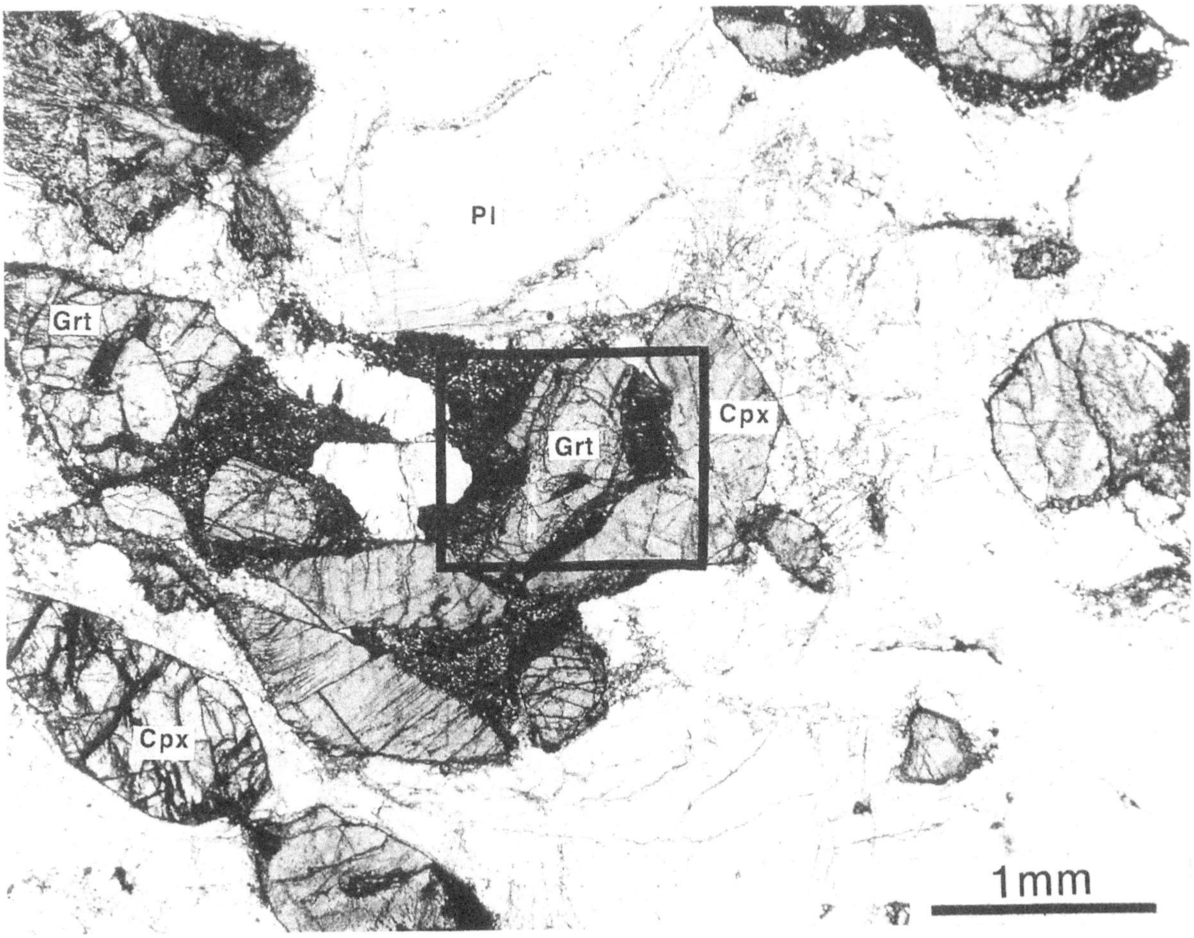

128 A

128 B

**128C** (PPL) **and 128D** (NX). Optical micrographs of a metabasic mylonite developed with increasing strain into the shear zone. The plagioclase is recrystallized to an ultrafine grain size (~10–15 μm), although some larger rounded porphyroclasts remain. The orthopyroxene grains are commonly elongated by slip. Scarce rounded porphyroclasts of garnet and clinopyroxene are also present.

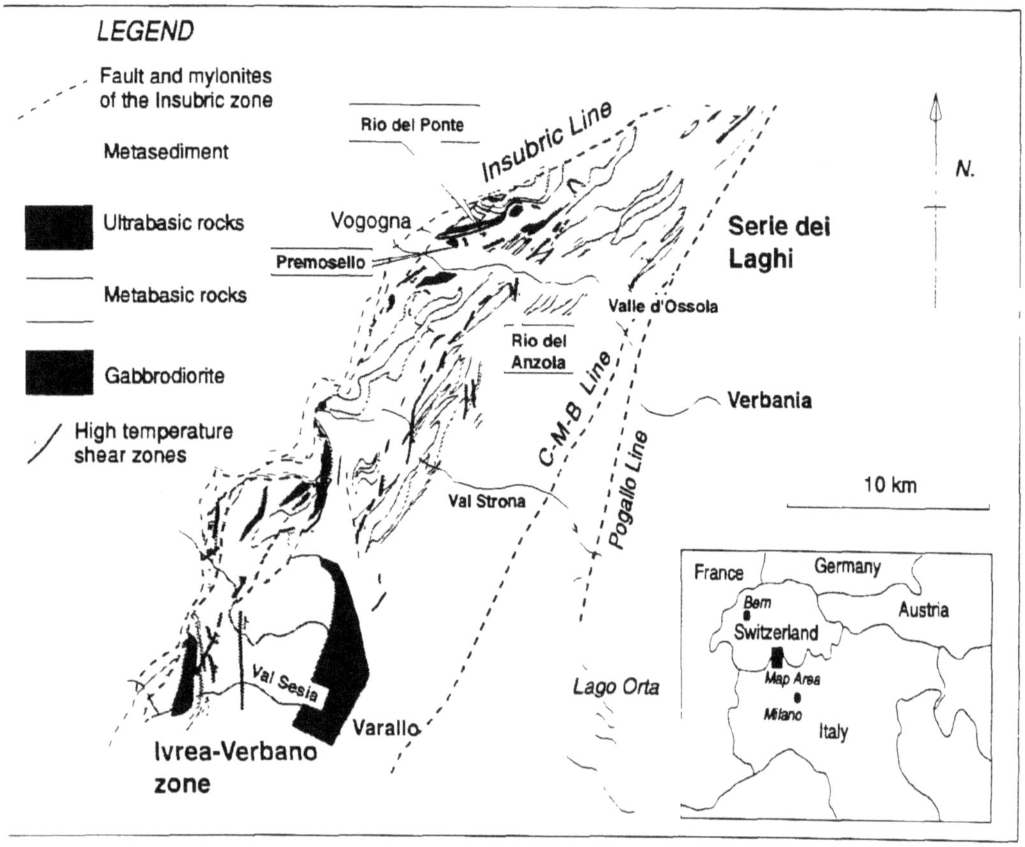


**Figure 128.1.** Map of the Ivrea-Verbano zone showing the location of the samples illustrated

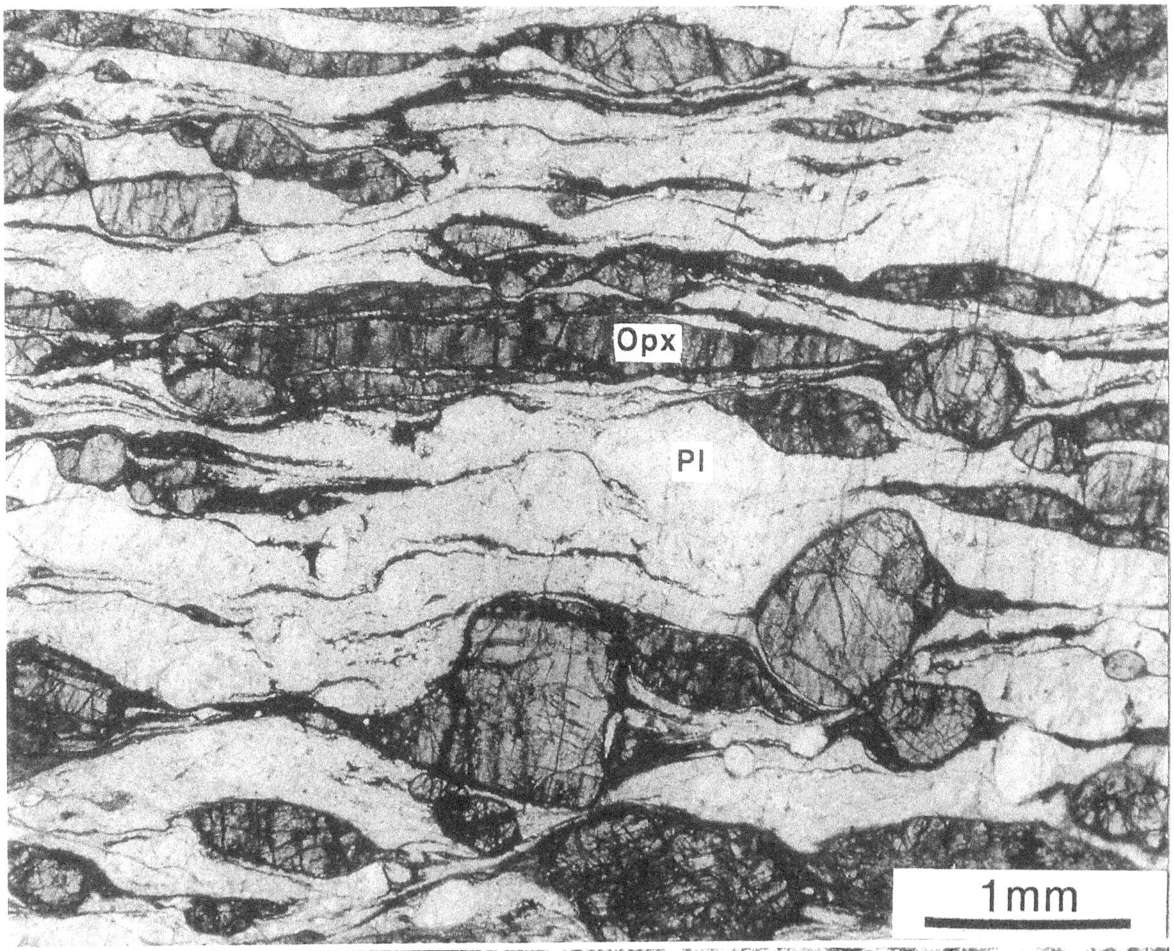


128 C

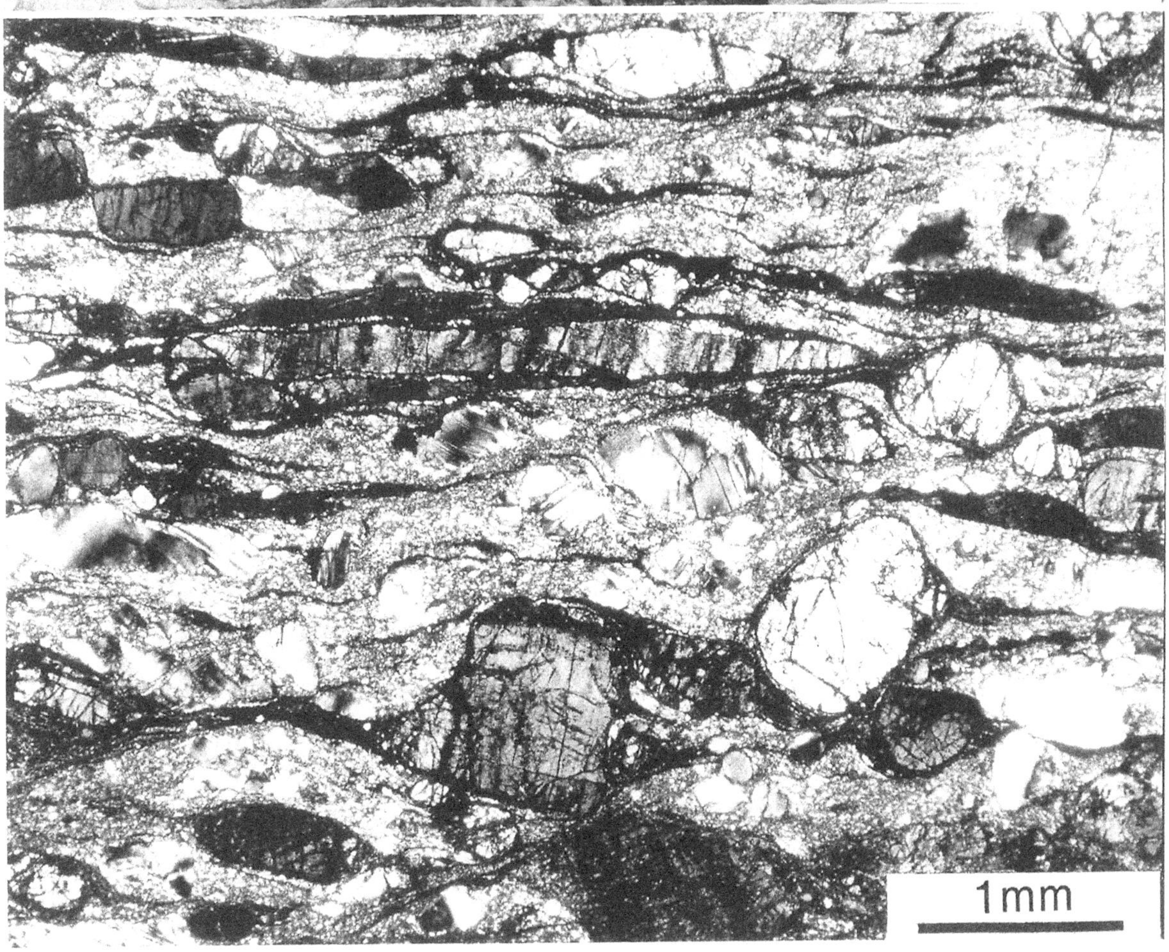


128 D

**128E** (PPL) **and 128F** (NX) Optical micrographs of the ultrafine-grained ultramylonite (~5 μm) formed in the center of the metabasic shear zones. The rock is strongly foliated and few porphyroclasts remain. The matrix (SEM backscattered micrograph inset in 128F) is composed of intergrown plagioclase (dark contrast) (~75%), orthopyroxene (light contrast) (~12–15%), an opaque phase (very bright contrast) (~10–12%), and minor quartz. Sense-of-shear criteria are uncommon in these ultramylonites, although rotated porphyroclasts can sometimes be found (see inset). Garnet and clinopyroxene are no longer stable in the ultramylonites, although occasionally rounded porphyroclasts are preserved.

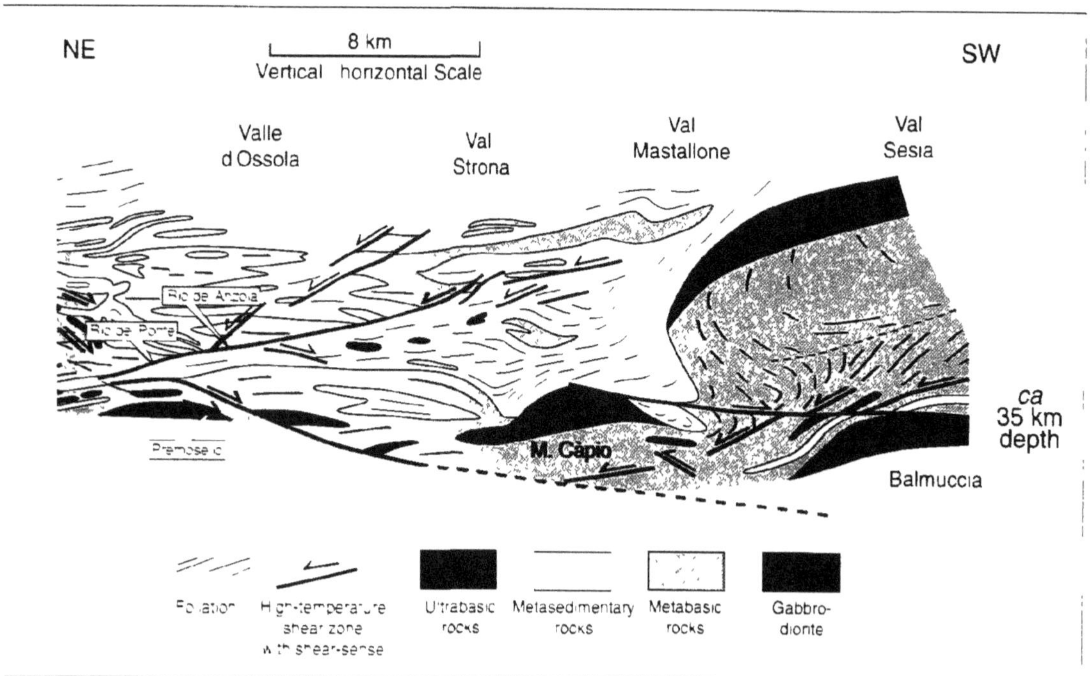


**Figure 128.2.** Restored cross section through the structurally lower part of the Ivrea-Verbano zone between Valle d'Ossola and Val Strona after the removal of late low-temperature faulting and folding. The section is drawn parallel to the slip direction of the shear zones. The locations of the samples illustrated are indicated (after Brodie and others 1992; Evans 1995).

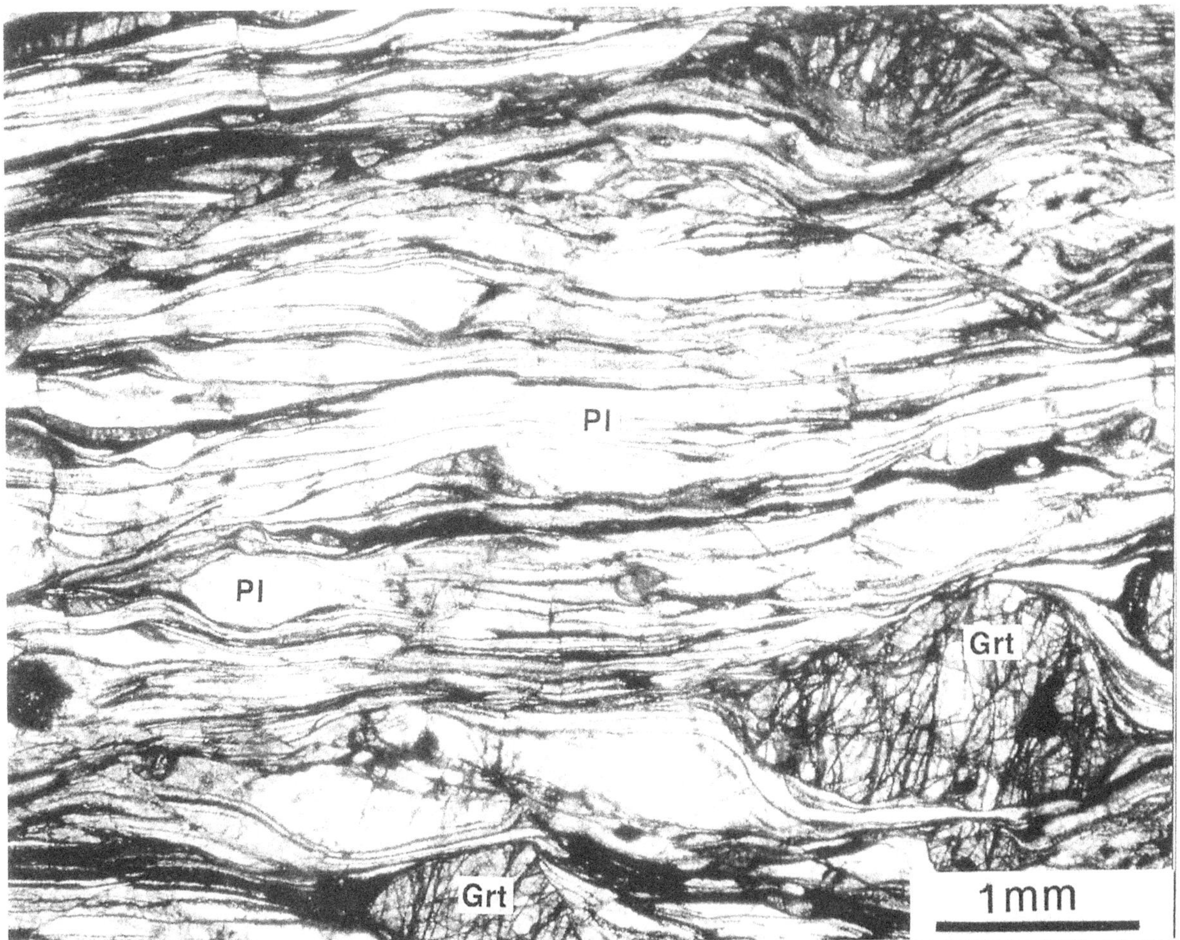


128 E

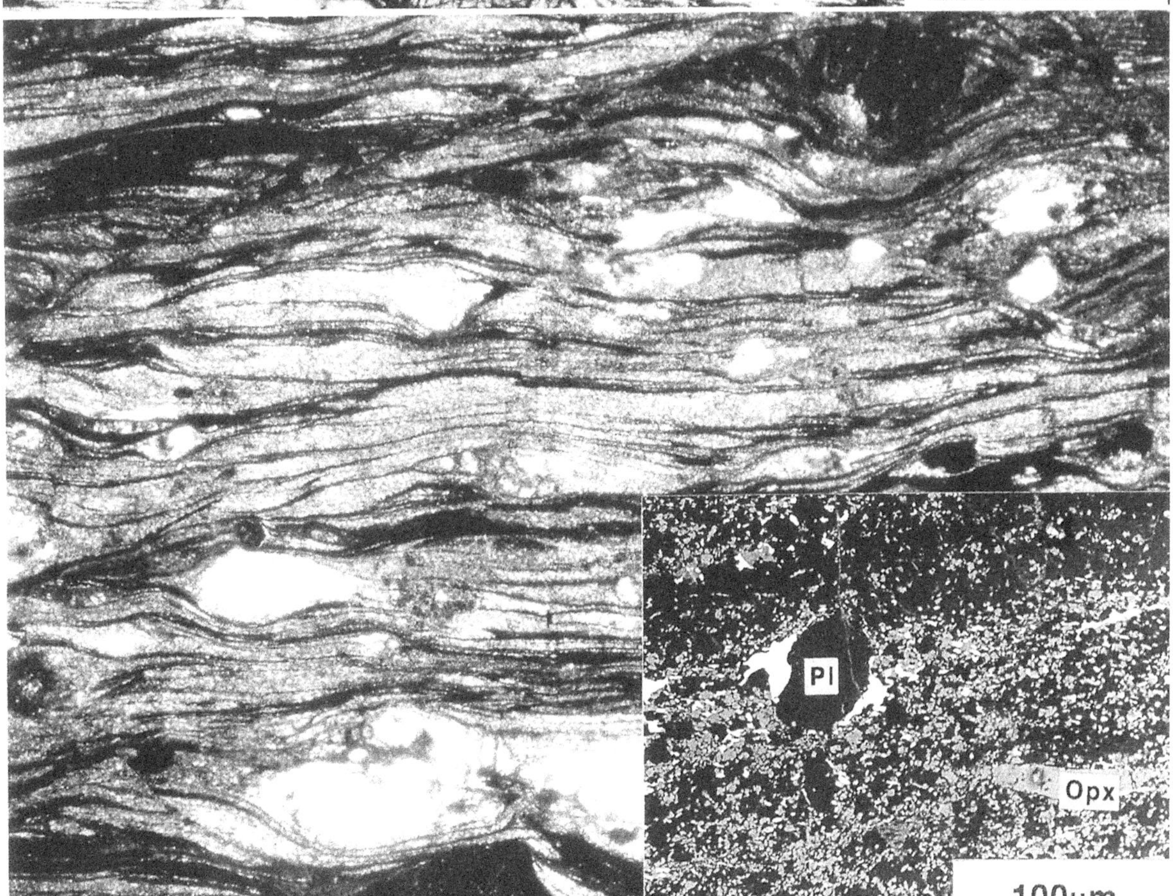


128 F

**128G.** Backscattered electron micrograph (SEM) of the symplectite developed around a garnet grain in the protomylonite, from the same general area as illustrated in 128A. The symplectites are developed on interfaces originally in contact with plagioclase (left side) and clinopyroxene (right side). Note that the symplectite is embayed into the garnet, indicating that the reaction is proceeding at the expense of the garnet. The edges of the garnet show a slightly lighter contrast than the center, and this represents an increase in the Fe/Mg value at the margins of the garnet (Figure 128.3), due to diffusional exchange during cooling.

**128H.** Higher magnification micrograph of the area outlined in 128G showing the intergrowth of orthopyroxene and plagioclase. Bright areas represent spinel, and the light phase intergrown with the orthopyroxene is an iron oxide which may have exsolved during cooling. SEM

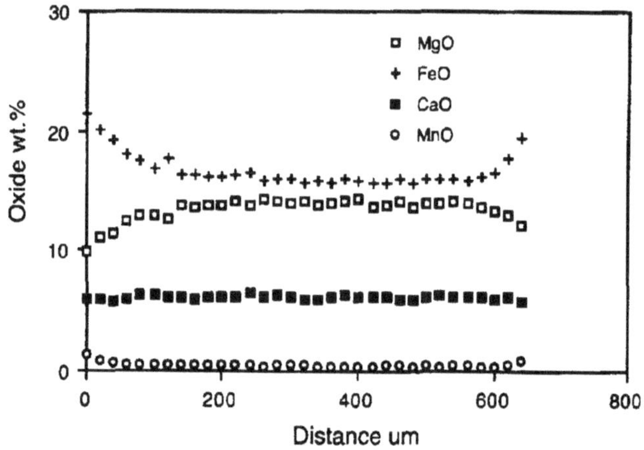


**Figure 128.3.** Graph showing variation in composition in a traverse across the garnet grain shown in Plate 128G, which has symplectites developed at both margins.

**Table 128.1.** Representative mineral analyses from the Rio del Ponte protomylonite.

| *Mineral* | *Grt core* | *Grt edge* | *Pl* | *Pl in symplec.* | *Cpx* | *Opx inclusn.* | *Opx in symplec.* | *Spl in symplec.* |
|---|---|---|---|---|---|---|---|---|
| $SiO_2$ | 40.23 | 39.56 | 45.89 | 44.33 | 49.00 | 52.58 | 52.57 | 0 |
| $TiO_2$ | 0 | 0 | 0 | 0 | 1.01 | 0.08 | 0 | 0 |
| $Al_2O_3$ | 22.82 | 21.91 | 34.81 | 35.34 | 6.39 | 4.94 | 3.47 | 60.35 |
| $Fe_2O_3$ | 2.15 | 1.11 | 0 | 0 | 1.78 | 1.35 | 2.74 | 4.35 |
| FeO | 14.64 | 19.17 | 0 | 0 | 3.73 | 12.12 | 14.82 | 24.13 |
| MnO | 0.42 | 1.01 | 0 | 0 | 0.05 | 0.05 | 0.33 | 0.12 |
| MgO | 14.09 | 10.88 | 0 | 0 | 13.79 | 28.21 | 26.77 | 11.36 |
| CaO | 6.19 | 6.03 | 18.02 | 18.75 | 23.44 | 0.39 | 0 | 0 |
| $Na_2O$ | 0 | 0 | 1.53 | 1.05 | 0.75 | 0 | 0 | 0 |
| Total | 100.54 | 99.67 | 100.25 | 99.47 | 99.94 | 99.72 | 100.7 | 100.31 |
| Si | 2.95 | 2.99 | 2.11 | 2.06 | 1.81 | 1.88 | 1.89 | 0.00 |
| Al | 1.97 | 1.95 | 1.89 | 1.94 | 0.28 | 0.21 | 0.15 | 15.29 |
| $Fe_3$ | 0.12 | 0.06 | 0.00 | 0.00 | 0.05 | 0.04 | 0.07 | 0.70 |
| $Fe_2$ | 0.90 | 1.21 | 0.00 | 0.00 | 0.12 | 0.36 | 0.45 | 4.34 |
| Mg | 1.54 | 1.23 | 0.00 | 0.00 | 0.76 | 1.50 | 1.43 | 3.64 |
| Ca | 0.49 | 0.49 | 0.89 | 0.93 | 0.93 | 0.02 | 0.00 | 0.00 |
| Na | 0.00 | 0.00 | 0.14 | 0.10 | 0.05 | 0.00 | 0.00 | 0.00 |
| Ti | 0.00 | 0.00 | 0.00 | 0.00 | 0.03 | 0.00 | 0.00 | 0.00 |
| Mn | 0.03 | 0.07 | 0.00 | 0.00 | 0.00 | 0.00 | 0.01 | 0.02 |
| No. oxygens | 12 | 12 | 8 | 8 | 6 | 6 | 6 | 32 |

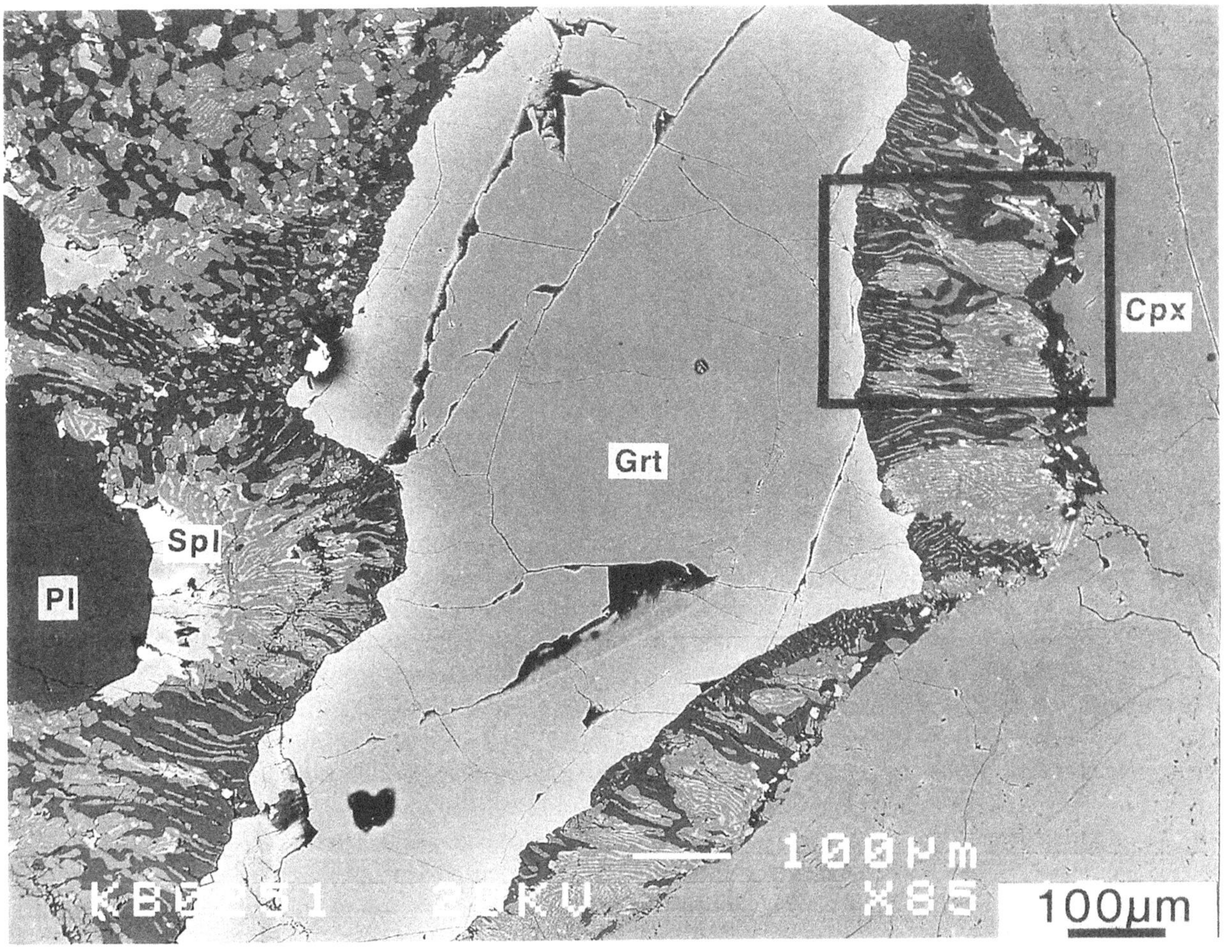


128 G

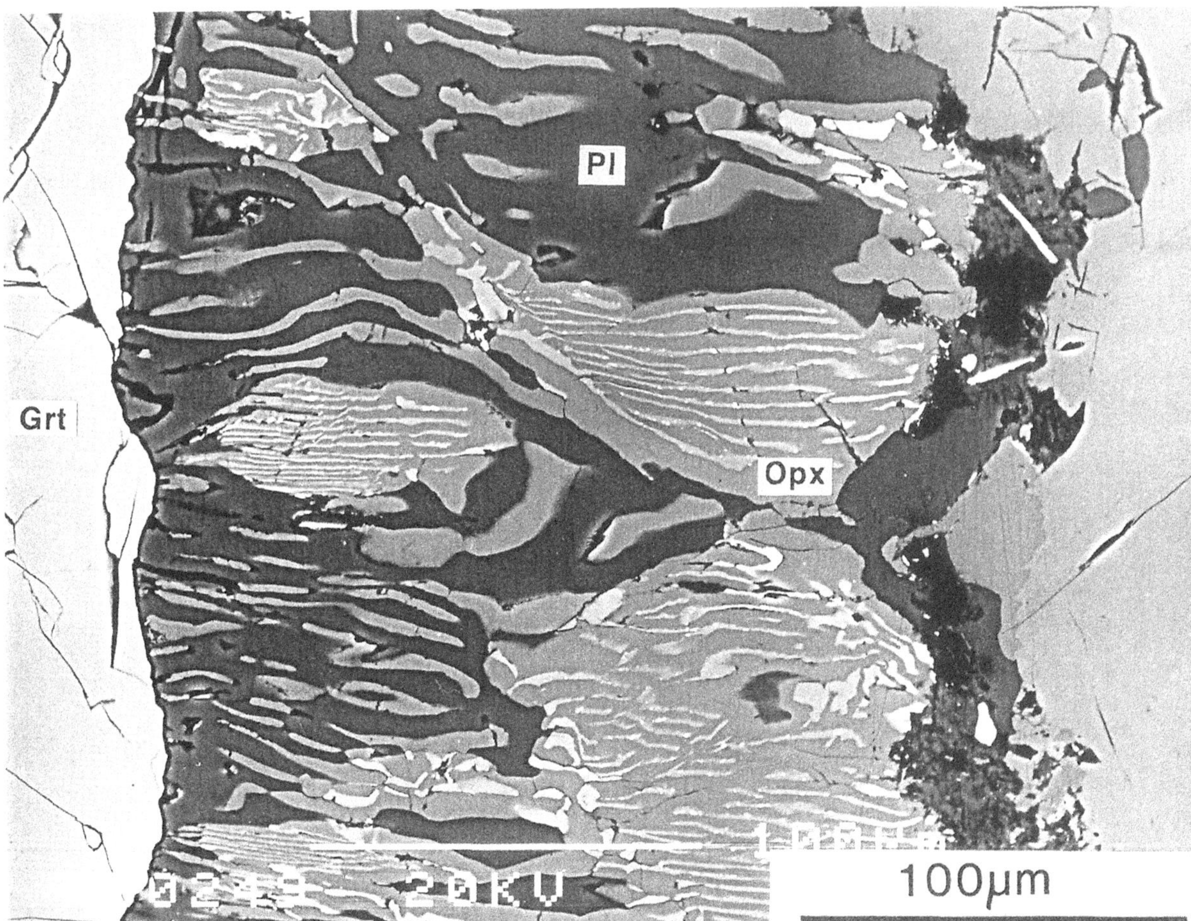


128 H

# 129. HIGH-TEMPERATURE MYLONITES: II. ULTRABASIC MYLONITES

K. H. BRODIE

Ultrabasic rocks (peridotites, dunites, and pyroxenites) are present in the high-grade parts of the Ivrea-Verbano zone, northern Italy. They generally show little evidence of strain, preserving relatively coarse-grained, granoblastic polygonal microstructures. Localized zones of high strain are found, however, in both the olivine- and the pyroxene-rich rocks. One example of an olivine mylonite is found in the structurally lowest part of the Ivrea-Verbano zone in the Valle d'Ossola region, near Premosello (Figures 128.1 and 128.2). In hand specimen the rock is difficult to recognize as a mylonite, the only obvious foliation being produced by streaked-out spinel grains. However, in thin section it is clear that intense grain-size reduction of olivine has occurred, despite its high strength relative to many other minerals.

**129A.** Optical micrograph of the relatively undeformed olivine-rich peridotite (dunite) at the margins of a high-temperature high-strain zone. Deformation bands are common in the olivine grains, but the rock is coarse-grained, and little recrystallization is evident. Premosello, Valle d'Ossola. NX

**129B.** Optical micrograph of the protomylonitic peridotite showing the initial stages of recrystallization of the olivine. Deformation of the peridotite involves grain-size reduction by crystal-plastic deformation of olivine. Recrystallization of the olivine along grain boundaries is present, and the olivine grains are elongated producing a foliation in the rock. Optical strain features in the form of subgrain walls are present in the relict olivine grains. NX

129 A

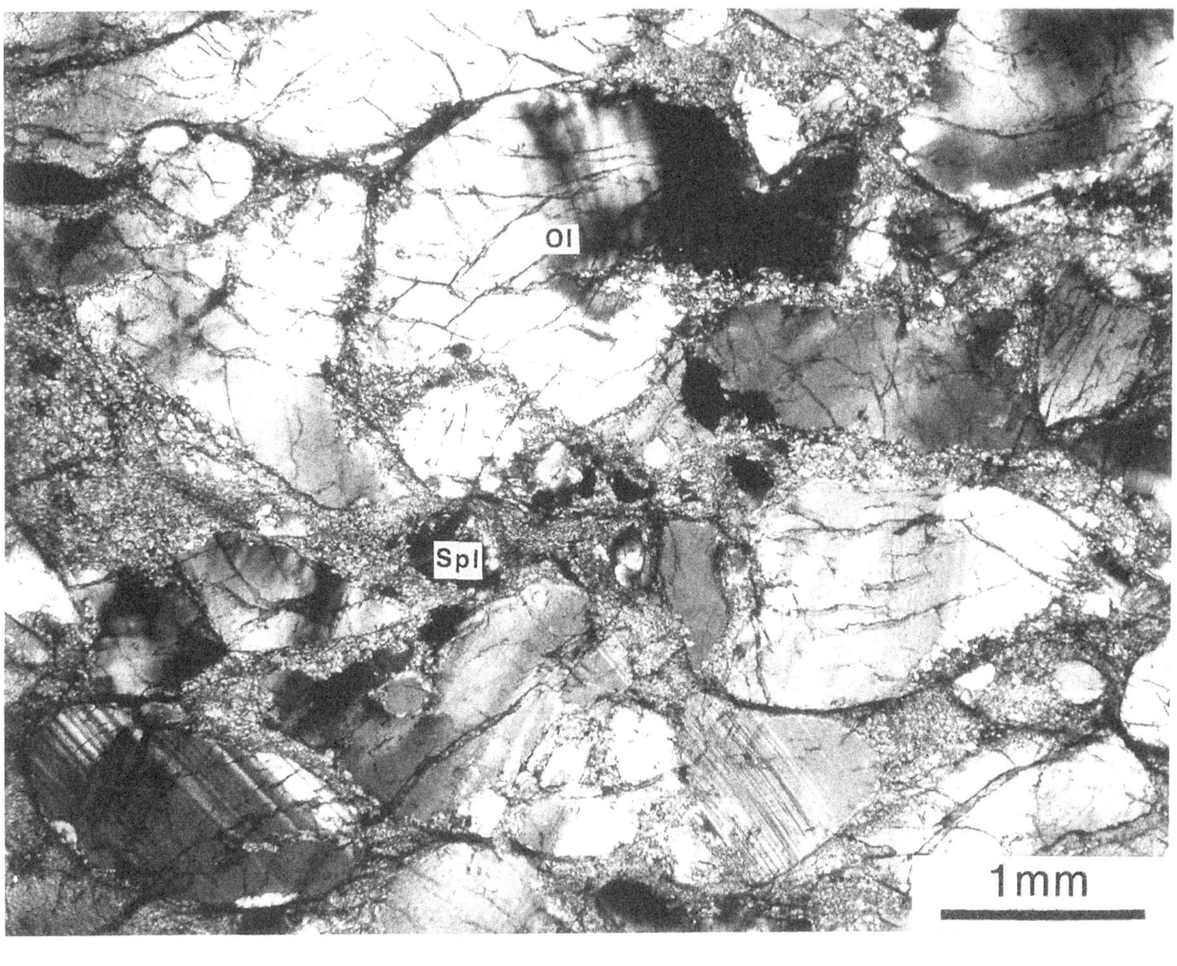


129 B

**129C.** Optical micrograph of the peridotite mylonite, which contains porphyroclasts of olivine, and lesser pyroxene, in a fine- to ultrafine-grained matrix of recrystallized olivine and clinopyroxene. Where the two are mixed on a grain scale the grain size tends to be finer. Elongate trails of fine-grained spinel are also present. Premosello, Valle d'Ossola. NX

**129D.** TEM micrograph showing the dislocation structure of an olivine porphyroclast from the peridotite mylonite. Bound dislocations form a subgrain wall (causing the optical subgrain features), and free dislocations are also relatively abundant. Such structures are indicative of high-temperature plastic deformation.

**129E.** TEM micrograph of the ultrafine-grained matrix of the peridotite mylonite showing intergrown clinopyroxene (containing planar stacking faults) and olivine of grain size 1–2 μm. From the very fine-grained nature of these rocks it is tempting to infer that grain-size-sensitive flow processes have been involved in their deformation, but this has not yet been proven to be the case (Rutter and Brodie, 1988).

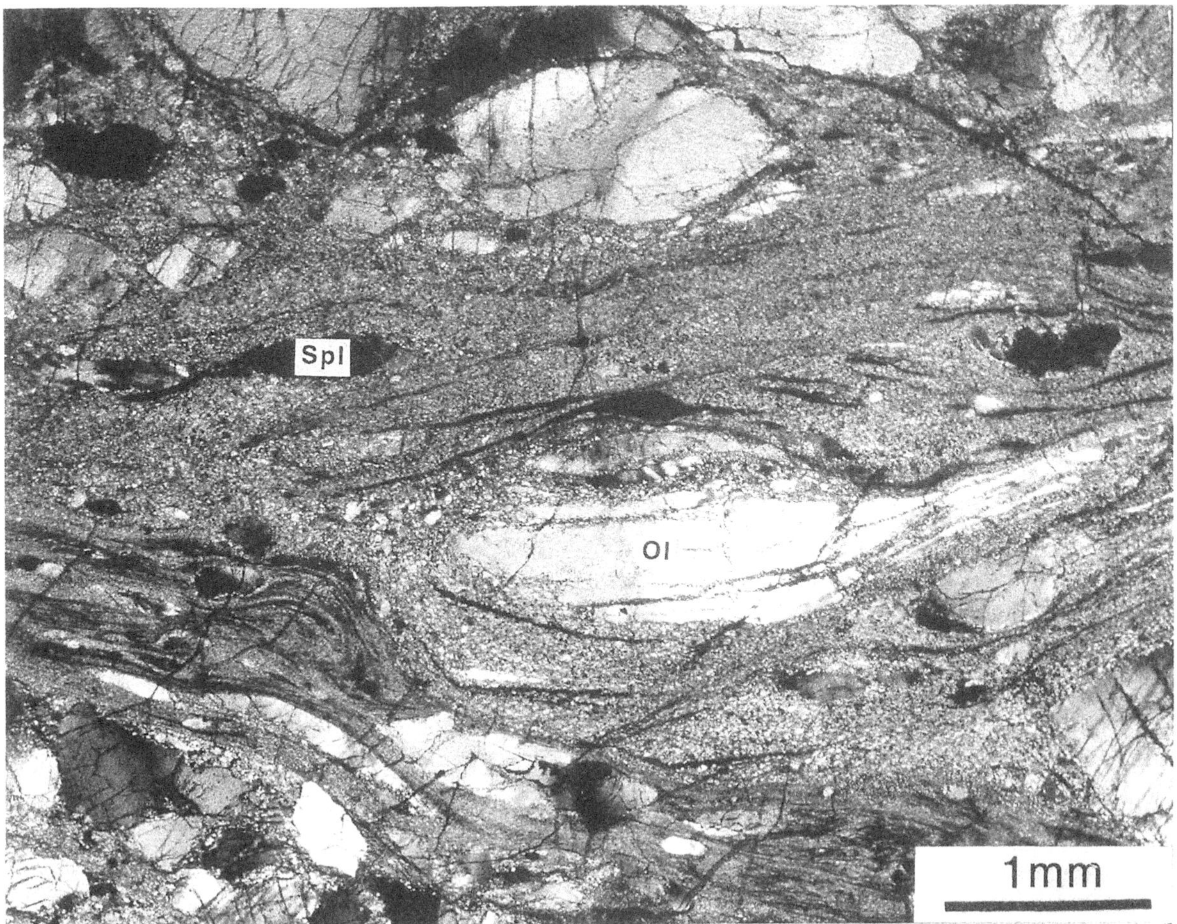


129 C

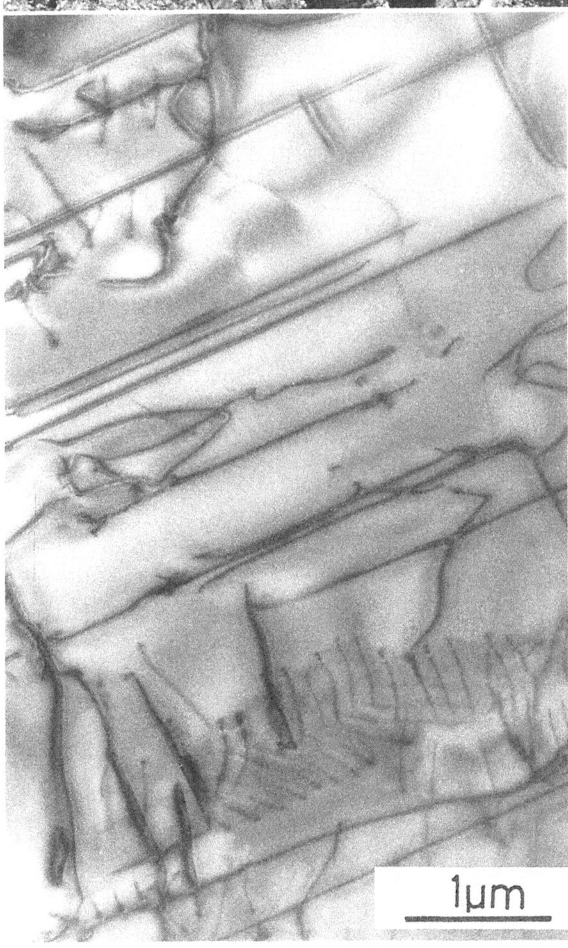


129 D

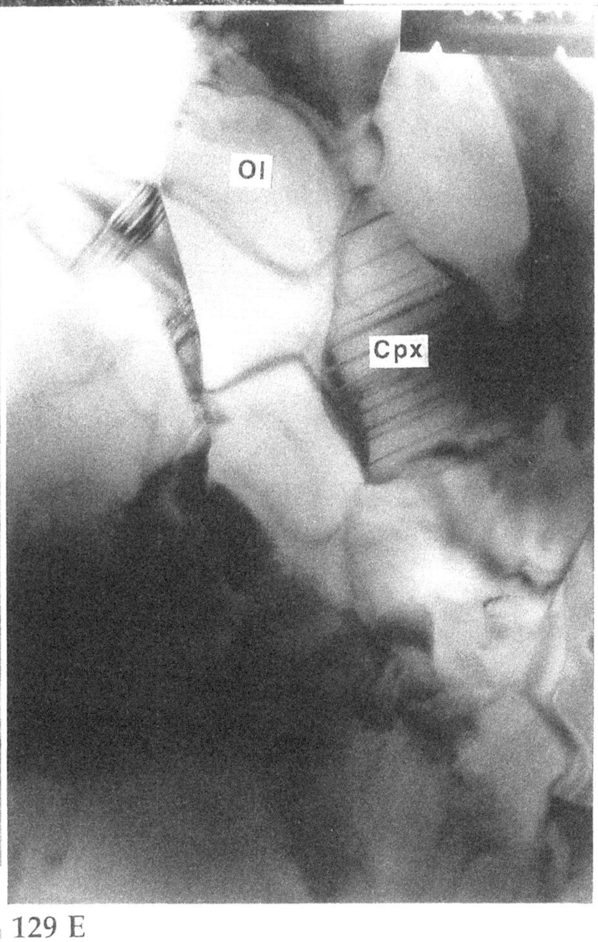


129 E

# **130.** HIGH-TEMPERATURE MYLONITES: III. METASEDIMENTARY MYLONITES

K. H. BRODIE

The high-grade metasedimentary gneisses of the Ivrea-Verbano zone, northern Italy, generally show evidence of pervasive strain, but localized high-strain zones are also present. In lower-grade parts of the zone, where the metasedimentary rocks contain more quartz and mica, the deformation is distributed, and localized zones of high strain have not been observed.

**130A** (PPL) **and 130B** (NX). Optical micrographs of a mylonite formed in an upper amphibolite-facies pelitic paragneiss. Intense grain-size reduction by dynamic recrystallization of plagioclase, and quartz, if present, produces a strongly foliated fine-grained mylonite. Sillimanite and garnet undergo brittle fracture, and elongate grains of sillimanite are commonly pulled apart in the stretching direction. Garnet porphyroclasts are rounded due to marginal granulation. Rio del Anzola, Valle d'Ossola.

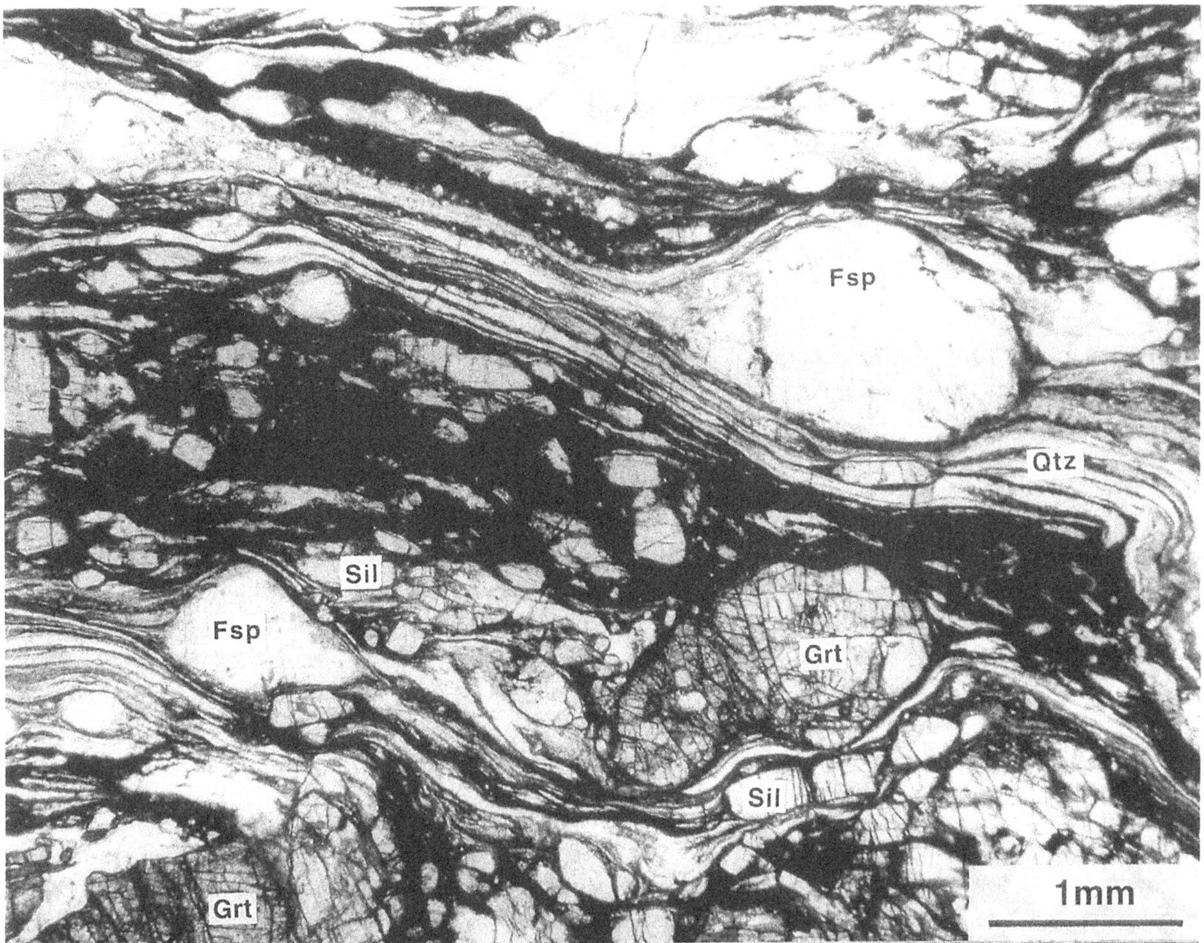


130 A

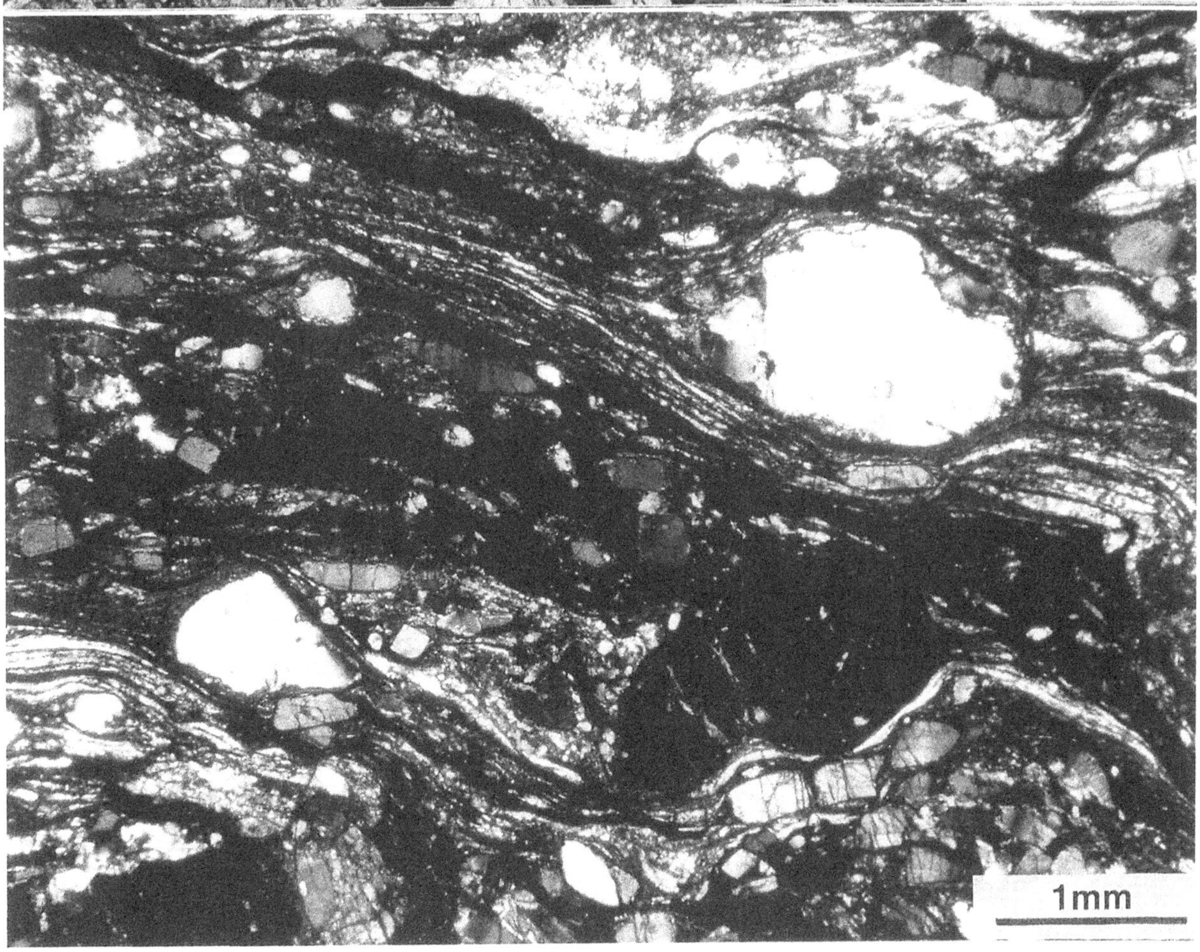


130 B

# 131. MESOSCOPIC KINEMATIC INDICATORS IN A MELT-RICH, HIGH-GRADE SHEAR ZONE

Arthur W. Snoke and Thomas J. Kalakay

The Ivrea-Verbano zone (IVZ), northern Italy, exposes lower continental crust that was upthrusted and exhumed during the Alpine orogeny (Zingg and others, 1990). The IVZ is a subvertical cross section of Southern Alpine basement (see Figure 128.1, this volume) that formed during a complex, late Paleozoic orogenic history involving early contraction, mafic to felsic magmatism, and late extension (or transtension) (Handy and Zingg, 1991; Schmid, 1993). Metabasic and metasedimentary rocks (the "Kinzigite Formation") were intruded at lower crustal levels in the late Paleozoic by a heterogeneous mafic to intermediate igneous complex (Quick and others, 1994). Granulite-facies metamorphism and widespread anatexis accompanied the emplacement of multiple batches of mantle-derived magma. Detailed geologic mapping of the intrusive contact between the igneous complex and its wall rocks revealed that a steeply dipping, kilometer-scale zone of intense plastic deformation is localized in the wall rocks (Kalakay and Snoke, 1995). This high-temperature, migmatitic shear zone, which can be traced for at least 20 km, is characterized by the transposition of earlier foliation(s), isoclinal folding, boudinage of competent layers, moderate north-plunging lineation, local mylonitic fabrics, and melt-filled extensional fractures (131A). Sense-of-shear data (131A and 131B), although not ubiquitous, indicate an oblique-sinistral, noncoaxial deformation in the present steep orientation of the shear zone. It is inferred that the abundance of anatectic granitic melt in this zone has affected its overall rheological properties and has led to a zone of distributed shear strain.

**131A.** Adjacent exposures that exhibit domino-style blocks of calc-silicate paragneiss with the extended blocks outlined by leucocratic granitic melt (see Figure 131.1). The sense-of-shear is sinistral. In the photograph on the right, note that several small-scale, asymmetric folds with sinistral vergence, as viewed down-plunge, occur at the base of the domino-style blocks. Leucocratic granite is involved in this folding. The locality is along the west bank of the Sesia River ~1 km south of Varallo, Italy.

**131B.** Rotated mafic clast in highly deformed migmatitic pelitic paragneiss, about 250 m east of the intrusive contact with the igneous complex in the Strona di Postua. The mafic clast shows δ-type structure that indicates a sinistral sense-of-shear. In this outcrop, boudinage and dismemberment of competent layers within the paragneiss resulted in a chaotic-appearing deformation style, but foliation, lineation, and sense-of-shear are surprisingly uniform along the strike of the kilometer-scale shear zone that parallels the intrusive contact.

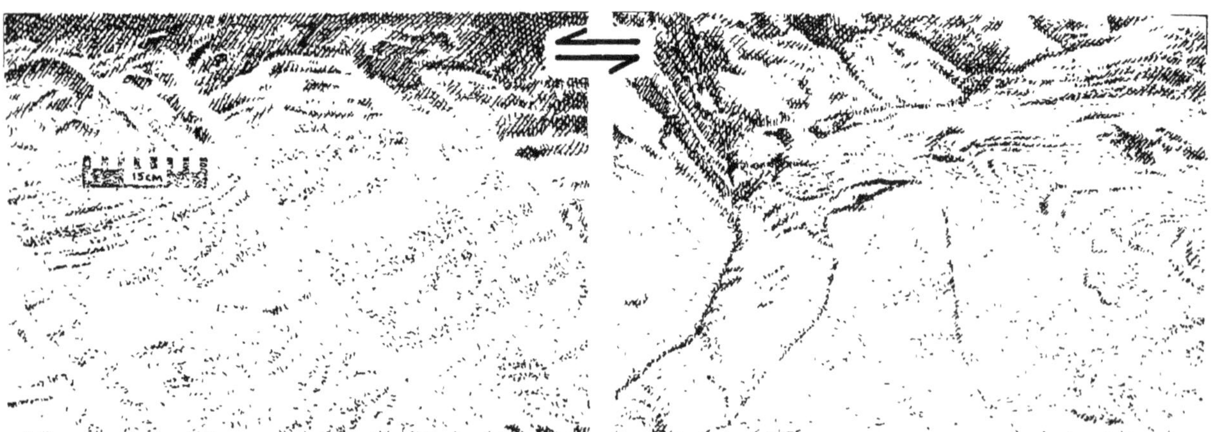


**Figure 131.1.** Sketch of photographs in Plate 131A

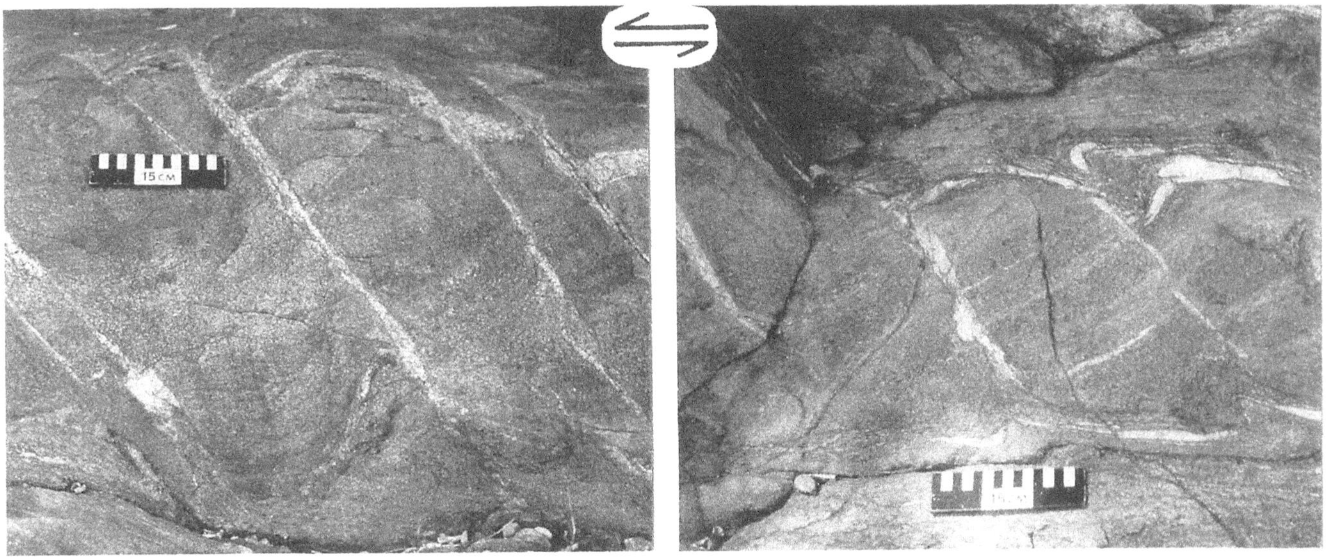


131 A

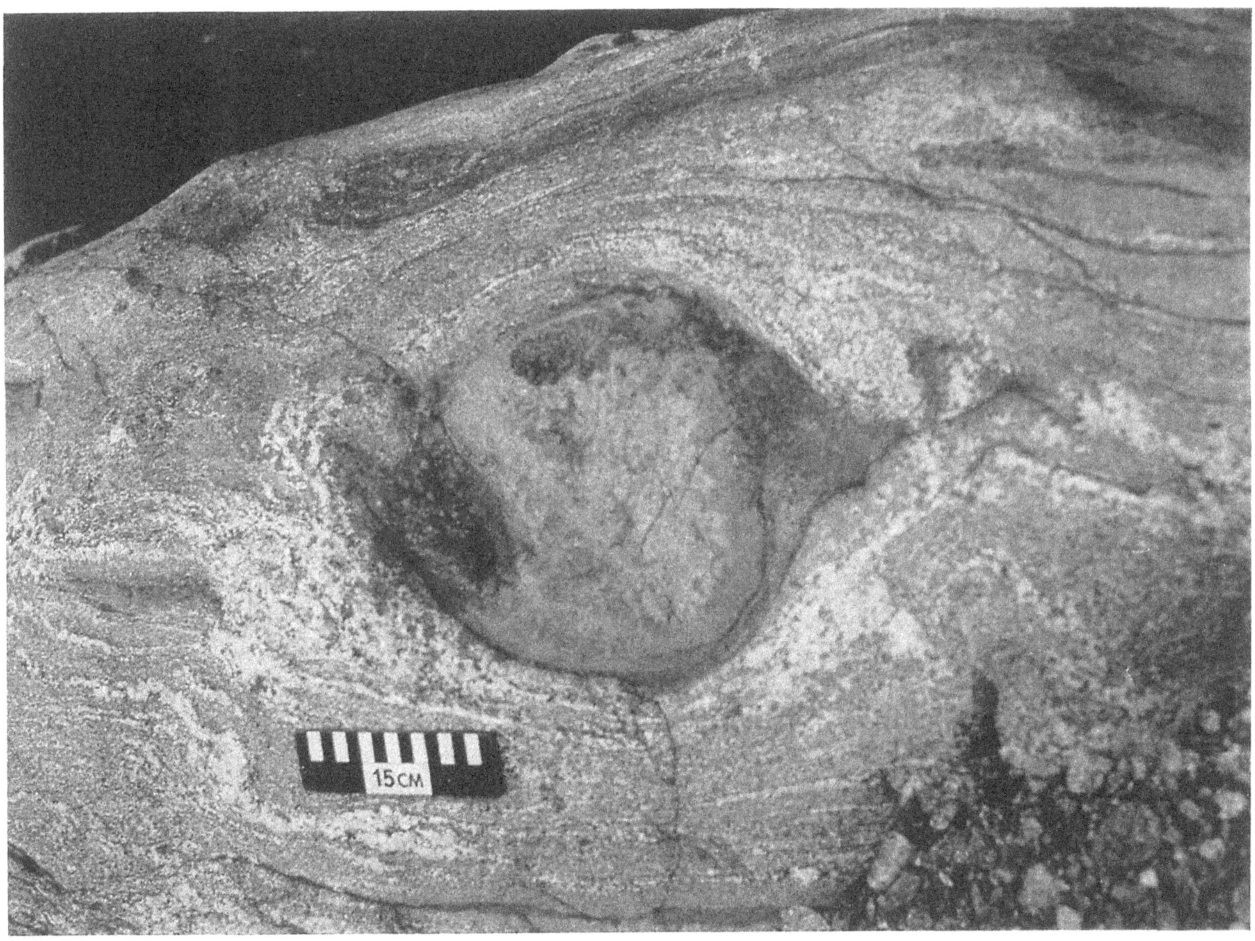


131 B

# 132. KINEMATIC INDICATORS IN ECLOGITE-FACIES SHEAR ZONES

PATRICE REY, DAVID M. FOUNTAIN, AND THERESA M. BOUNDY

Eclogite-facies shear zones, although scarce, occur in terranes that experienced deep tectonic burial and rapid exhumation. These shear zones provide important clues about the deformation processes within the deeper portions of contractional orogenic belts. Here we demonstrate, using an example from the Bergen Arcs, that the fundamental principles of kinematic analysis developed for lower-grade shear zones can be successfully applied to the study of eclogite-facies shear zones, and that, in this particular example, the eclogite-facies shear zones are characterized by noncoaxial strain.

Caledonian eclogite-facies shear zones developed within a Grenville-age, granulite-facies anorthosite-gabbro-jotunite-mangerite complex on Holsnøy in the Bergen Arcs, Norway (Austrheim and Griffin, 1985). Metamorphic conditions of the eclogite-facies event were approximately 700°C and between 1600 and 2100 MPa. The conversion of the granulite-facies rocks into eclogites was facilitated by fluid infiltration, initially along thin veins (Austrheim, 1986/87). Strongly deformed eclogite occurs in anastomosing, subparallel, 30–150 m thick shear zones that are laterally continuous along strike over distances up to several kilometers (Figure 132.1).

**132A.** View of the eclogite-facies Hundskjeften shear zone (HSZ). The cliff is about 25 m high, and the view is generally toward the northeast. The shear zone is surrounded by a breccia (the generally white area in the upper part of the photograph) that consists of granulite- and eclogite-facies rocks in about equal proportions. The breccia is composed of zones of foliated eclogite that enclose and wrap around blocks of granulite-facies rocks, which occur as angular, lensoidal blocks typically less than 5 m across. These zones are flanked above and below by granulite-facies anorthosites and gabbroic anorthosites. The eclogite-facies shear zone (dark horizontal band) exhibits a pronounced millimeter-scale layering defined by alternating omphacite/garnet- and kyanite/zoisite-rich layers and a strong shape-fabric foliation defined by aligned omphacite, kyanite, zoisite, and phengite. The foliation of the shear zones generally have a north to northeast dip of about 10° to 30° (Figure 132.1).

**132B.** Photograph of a sigmoid-shaped mineral cluster in an eclogite-facies shear zone. The view is to the north-northwest, and the pencil is approximately parallel to the east-northeast-plunging lineation. The dark areas, including the sigmoid, are relict granulite-facies mineral aggregates (primarily pyroxene) and the light areas are composed of sheared eclogite-facies rocks. The structure indicates a top-to-the-right (northeast) sense-of-shear.

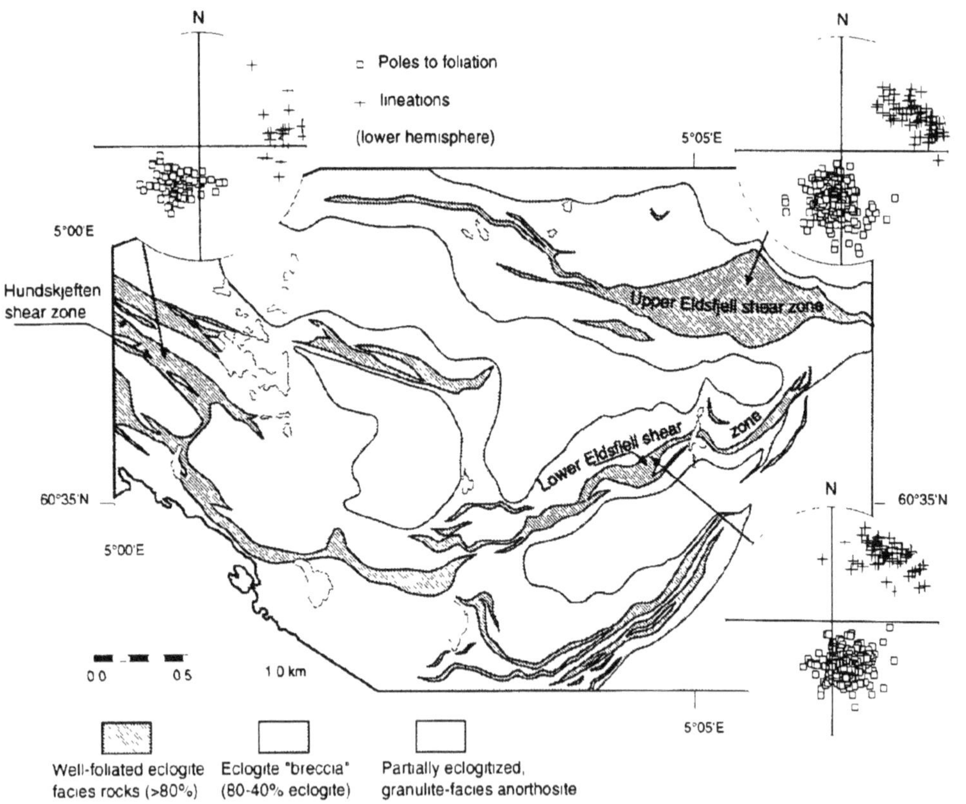


**Figure 132.1.** Geologic map of a portion of Holsnøy (from Boundy and others, 1992, Fountain and others, 1994) showing eclogite-facies shear zones and structural data (squares = poles to foliation; + = lineations). Outcrop widths are not indicative of true thickness because topography is not shown

132 A

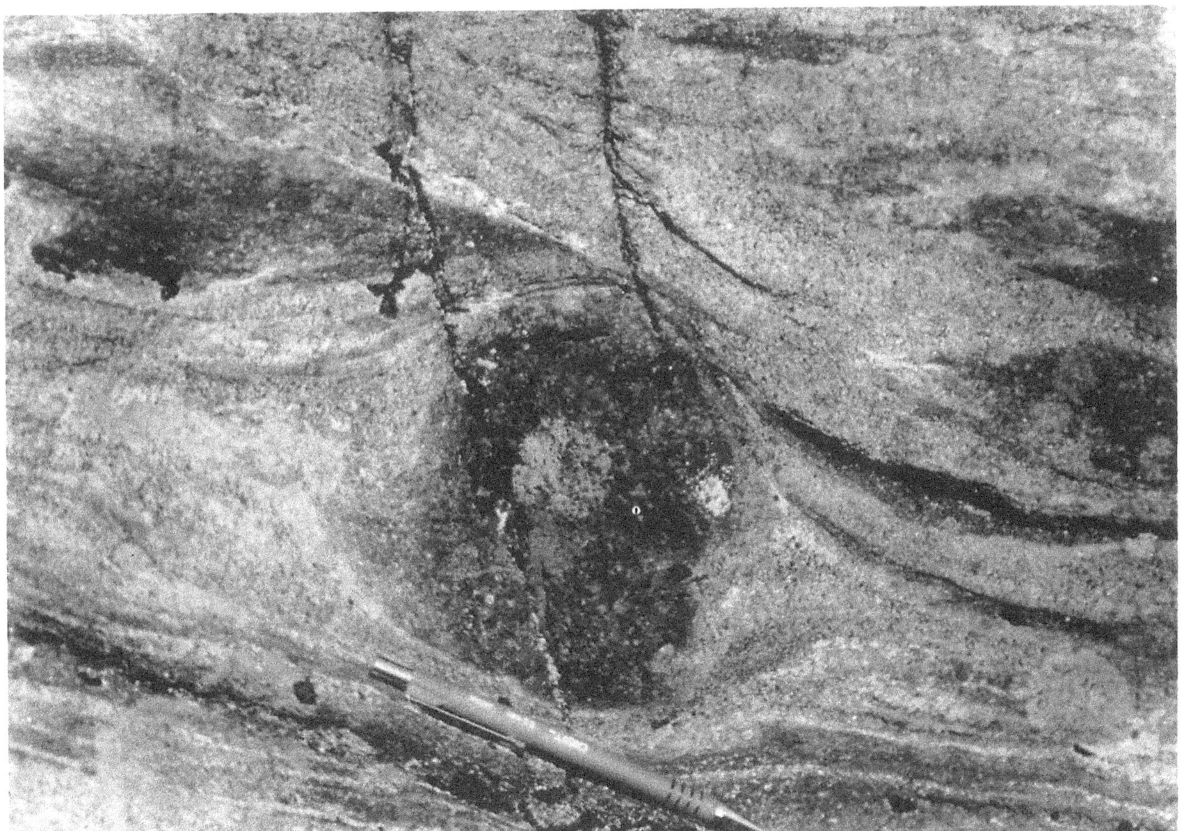

132 B

**132C.** Photomicrograph of thin section cut parallel to foliation (XY plane) of deformed eclogite-facies gabbroic anorthosite from the HSZ showing a strong shape fabric of zoisite, omphacite, kyanite, and phengite. This mineral lineation, indicated by arrows, is parallel to lineations that are defined, in thin section, by a shape fabric of pressure shadows around garnet and, in outcrop, by rod-shaped mineral aggregates (mainly garnet and omphacite) and elongate relict corona structures. The eclogite-facies assemblage in the shear zones is typically omphacite ($Jd_{44-49}$), garnet [including two generations: relict granulite-facies ($Pyr_{48}Alm_{31}Gross_{21}$) and eclogite-facies ($Pyr_{38}Alm_{40}Gross_{21}$)], zoisite, and kyanite, with minor Na-rich phengitic muscovite, rutile, quartz, and amphibole. Scale bar is 2 mm. NX

**132D.** Photomicrograph of thin section cut perpendicular to foliation and parallel to lineation (XZ plane) of the sample shown in 132C. The shape fabric of large zoisite, phengite, and omphacite grains defines the foliation plane (S plane, dashed arrow). Shear bands (C plane, solid arrow), defined by smaller zoisite and phengite grains, form an acute angle of about 30° with the S plane. C planes appear darker than the surrounding material in the photomicrograph because the constituent minerals have different crystallographic orientations than the surrounding phases. The similar mineralogy of the S and C planes suggests they developed simultaneously and during the eclogite-facies metamorphism. Asymmetric omphacite- and phengite-bearing pressure shadows developed around the garnets (Figure 132.2). Tiling microstructures occur between zoisite grains or between garnet grains (Figure 132.3). The geometrical relationships between the S and C planes, the asymmetry of the crystallization tails around garnet, and the right-stepping overlaps of garnet and zoisite grains indicate a dextral sense-of-shear in the photomicrograph that is consistent with the top to the east-northeast sense-of-shear for the HSZ inferred from macroscopic features such as sigmoid-type mineral clusters (132B). Although these features indicate that the shear zones are normal faults in their present orientation, shear-zone orientation during metamorphism and deformation was likely different. The rectangle outlines the area detailed in Figure 132.2. Scale bar is 2 mm. NX

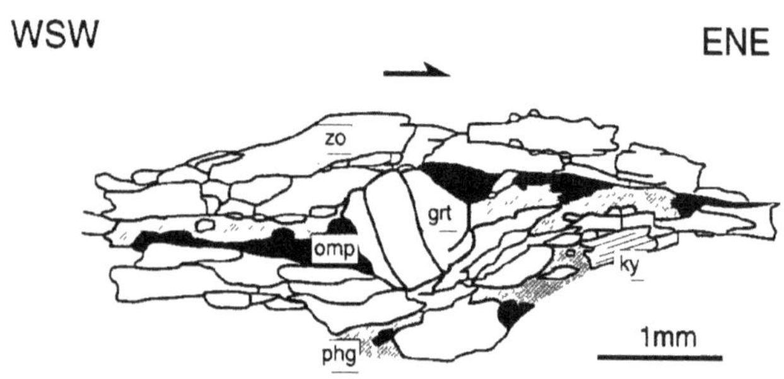


**Figure 132.2.** Sketch of crystallization tails of omphacite (omp) in the pressure shadow of garnet (grt) (from Plate 132D) Also shown are zoisite (zo), kyanite (ky), and phengite (phg)

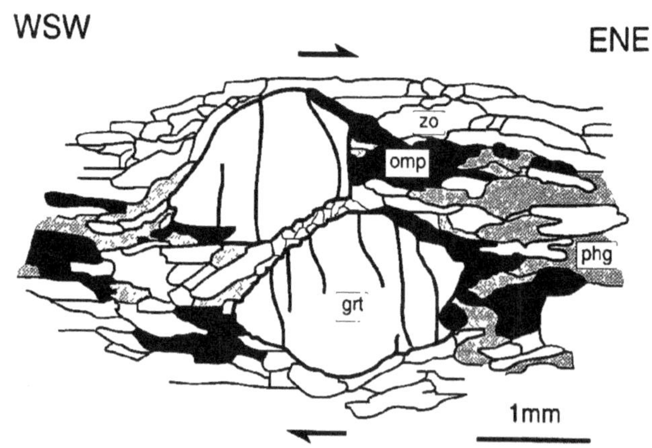


**Figure 132.3.** Tiling microstructures in eclogite-facies shear zone. Garnet grains are blocked at right-stepping overlaps. The contact between the two garnet grains is a pressure zone in which zoisite grains recrystallized A pressure shadow is developed around the two garnets. Abbreviations are the same as in Figure 132.2

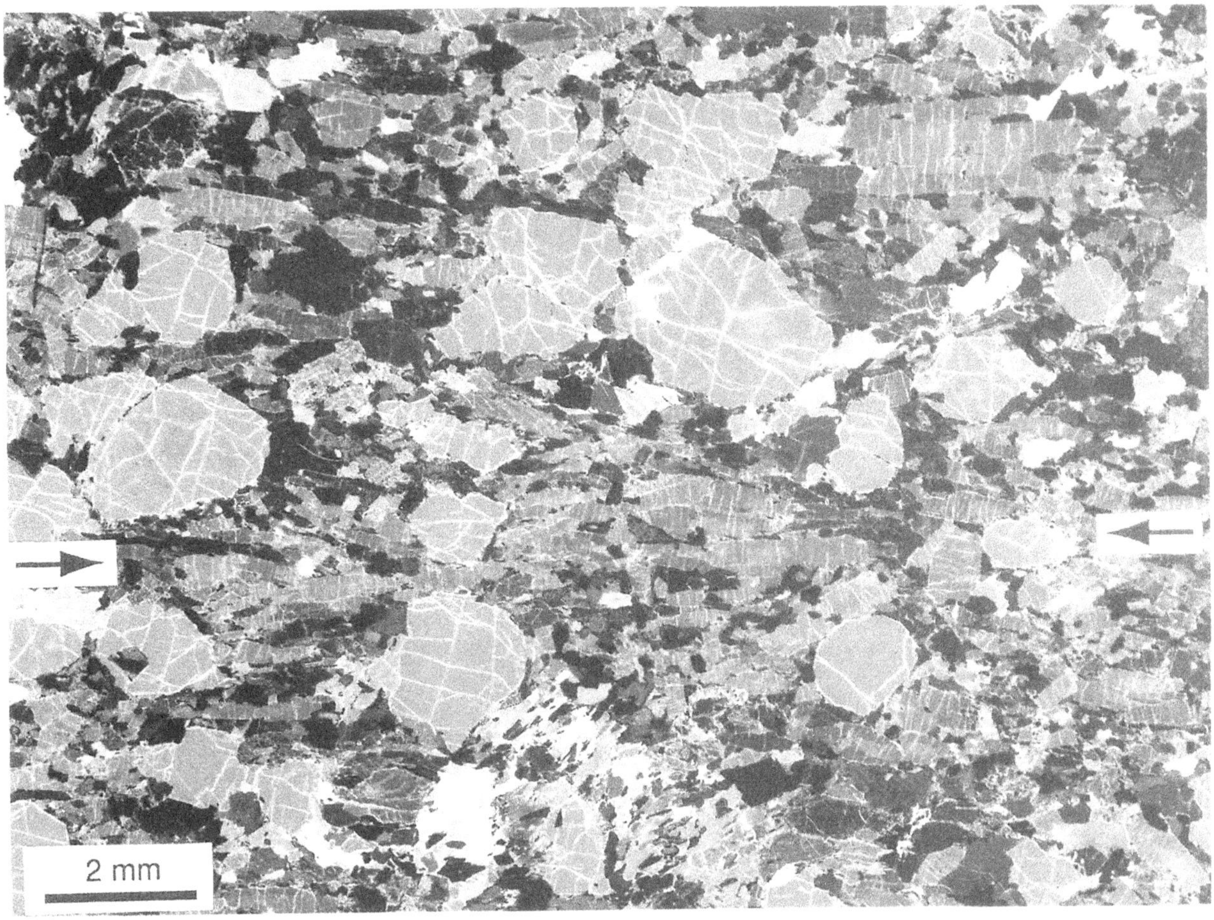


132 C

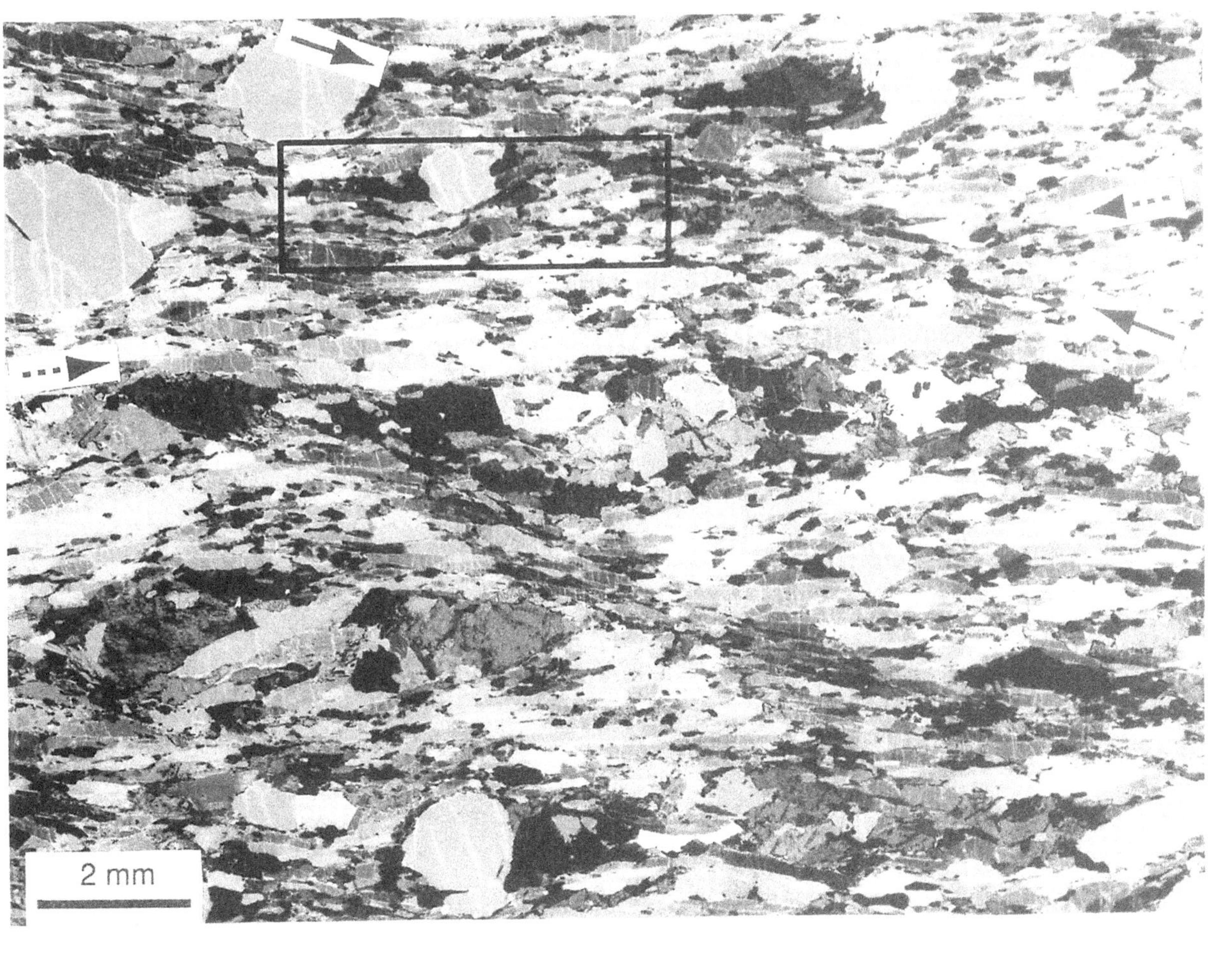


132 D

# DUCTILE BEHAVIOR

## Metamorphic grade variations

### Annealed fault-rock textures



Common abbreviations used in regard to the plates in this section include: PPL = plane polarized light, NX = crossed nicols, and TEM = transmission electron microscope.

High-temperature mylonites tend to be completely recrystallized to an aggregate of polygonal grains that appear internally unstrained. Such rocks may appear to have been statically annealed following deformation, but at high temperatures the rates of dislocation climb and dynamic recrystallization keep pace with the strain rate, and dynamic recrystallization can produce microstructures that are almost almost identical to those of annealing recrystallization. This set of plates includes the results of two experimental studies of static annealing, one on the analog material octachloropropane ($C_3Cl_8$) and the other on quartzite and fine-grained granite. Both studies show that although annealing destroys any preexisting grain-shape foliation, the deformation crystallographic preferred orientation is preserved exactly, indicating that the final recrystallized grains grew from preexisting dynamically recrystallized grains. One contribution dealing with naturally deformed rocks illustrates the difference in meso- and microstructures between retrograde or low-temperature mylonites, in which the strain rate exceeded the recovery and recrystallization rate, and high-temperature prograde mylonites, in which recovery and recrystallization kept pace with deformation and may even have outlasted it. Another plate presents possible evidence for high-temperature healing and annealing of the products of transient brittle deformation at amphibolite-facies conditions; such rocks must exist, but preservation of diagnostic microstructures is probably scarce in nature.

# 133. POST-DEFORMATIONAL RECOVERY OF MICROSTRUCTURE IN SHEARED OCTACHLOROPROPANE ($C_3Cl_8$)

W. D. MEANS AND H. G. DONG

The photographs show the effects of shearing octachloropropane (OCP) at $10^{-3}$ $s^{-1}$ (133A–133C) and then holding it without further deformation at the temperature of shearing (133D–133F). The specimen is 40 μm thick and was sheared at room temperature (0.8 $T_M$) on the stage of a petrographic microscope using the frosted grip technique described in Means (1989). Labels (a) to (e) identify parts of the specimen that can be correlated from one photograph to the next. Their changes in configuration were used to calculate the strains. OCP is a uniaxial positive, hexagonal or trigonal substance of currently unknown crystal structure. Its behavior in pure shearing has been described by Means (1983) and in simple shearing by Jessell (1986) and Ree (1991). All photographs taken with NX with planes of polarization at 45° to the (horizontal) direction of dextral shearing.

**133A.** Undeformed starting material prepared by hot-pressing at 0.9 $T_M$. Grains are mostly equi-axed and define a fairly good, low-energy "foam structure." Pairs of dashed lines represent orientations of material lines that will become $\lambda_1$ directions of strain in 133B (less steeply inclined line) and 133C (more steeply inclined line), in triangles abc, cbd, and cde.

**133B.** Same material after a bulk shear strain of about 0.38. White lines near the centers of triangles abc, cbd, and cde show orientations of the $\lambda_1$ directions, with corresponding principal shortenings of 21%, 20%, and 13%, respectively. Most of the original grains can still be recognized. Many show undulose extinction, and some contain wide deformation bands (e.g., the grain at the center of the bottom of the photograph). A grain-shape foliation has begun to develop.

**133C.** Same material after a bulk shear strain of about 0.65. Percent shortening normal to $\lambda_1$ direction in abc, cbd, and cde is 35%, 35%, and 21%. The grain-shape foliation is now better defined and lies approximately parallel to the $\lambda_1$ direction. Some small, new, recrystallized(?) grains have appeared (e.g., about 1 cm below d). The grain with a wide deformation band in 133B has become subdivided into three parts separated by boundaries that are themselves becoming serrate and probably mobile.

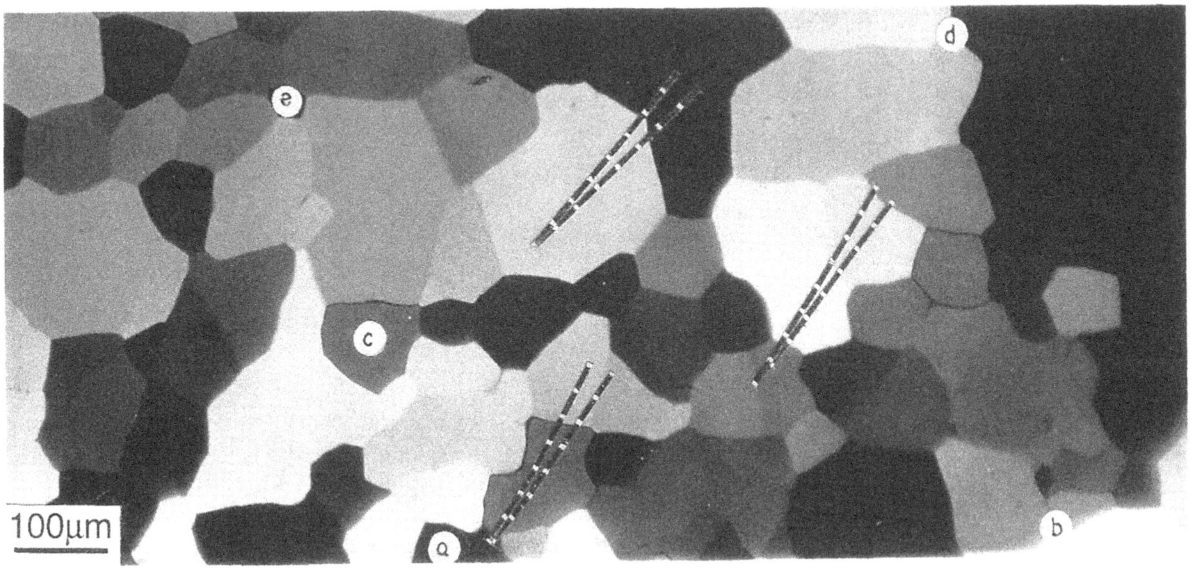


133 A

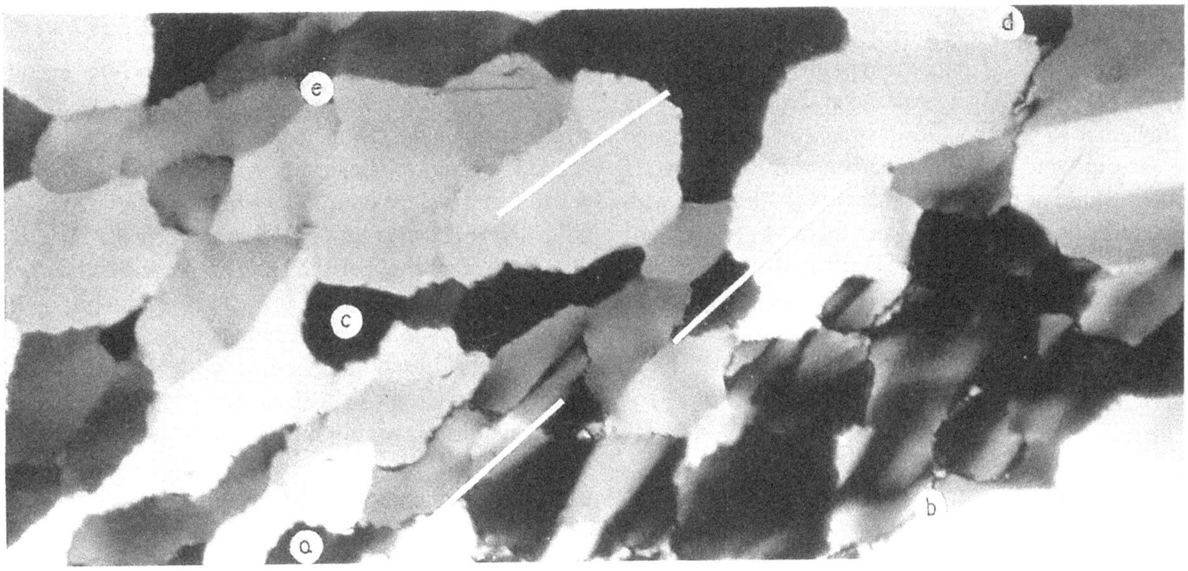


133 B

133 C

Plates 133D–133F show microstructural changes that occur in the material of 133C when it is held at room temperature (0.8 $T_M$) without further deformation. They illustrate how a deformed material can, when held "post-tectonically" at high temperature, lose the microstructural evidence of the deformation to which it was previously subjected.

**133D.** Material of 133C, 2 minutes after end of deformation. The scale of serration of many grain boundaries has become coarser, and some new grains have appeared in grain-boundary regions by an unknown mechanism. Examples are the small gray grain 16 mm "southeast" of marker (c), which can also be seen in Plate 133C, and the somewhat larger white grain about a third of the way from (e) to (d). The grain with a wide deformation band pointed out in 133B is now composed of three parts that are no longer recognizable as portions of a single original grain. The grain-shape foliation remains prominent. NX

**133E.** Same material 9 minutes after end of deformation. The grain-shape foliation is now much weakened, particularly in the lower half of the photograph where the strain was greatest and the foliation was best defined at the end of the deformation. The two new grains pointed out in 133D have both grown larger, as has the grain with the prominent, light-gray–dark-gray sub-boundary about 15 mm south-southeast of (d). NX

**133F.** Same material 15 hours, 38 minutes after end of deformation. The material has recovered a low-energy foam structure not unlike the microstructure of the undeformed material. The grains present are, perhaps surprisingly, not all new grains that developed during or after the deformation. Some are old grains that were present in the original undeformed material. An example of such a grain is the small dark grain 25 mm north-northwest of c that can be traced back through the photographs to a grain of similar shape, size, and extinction color that lies 32 mm west of (c) in 133A. Examples of new grains are the medium-sized gray grain 13 mm southwest of (e), and the previously mentioned white grain, now about 40 mm east of (e). The strain-induced grain-shape foliation that was so prominent earlier in the history of the specimen is now entirely gone. One wonders whether there is any memory of the deformation left in the lattice orientations of the grains. NX

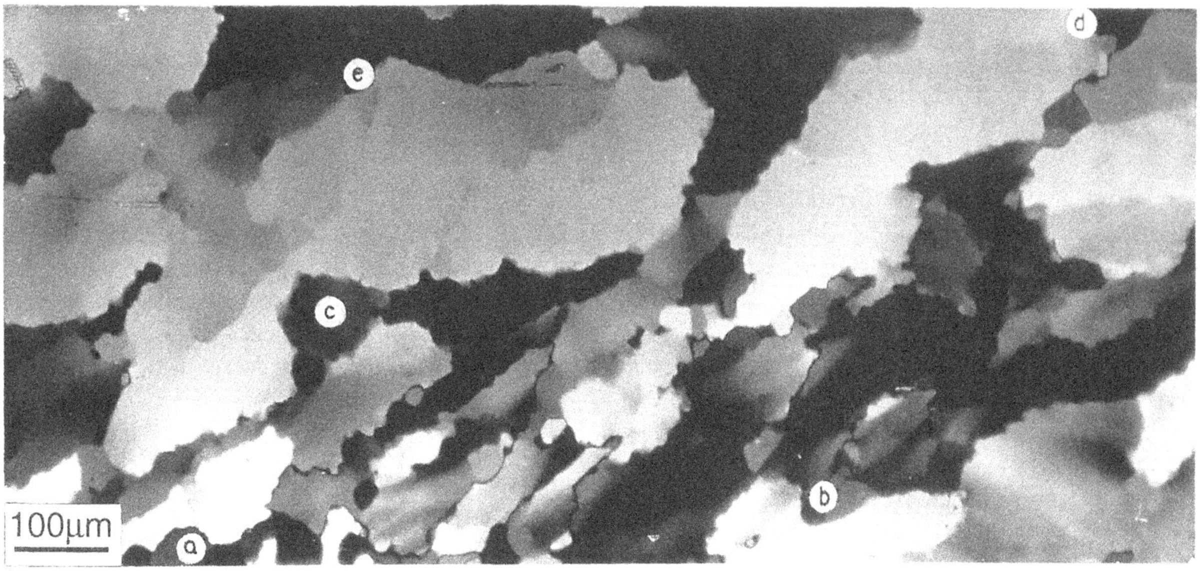


133 D

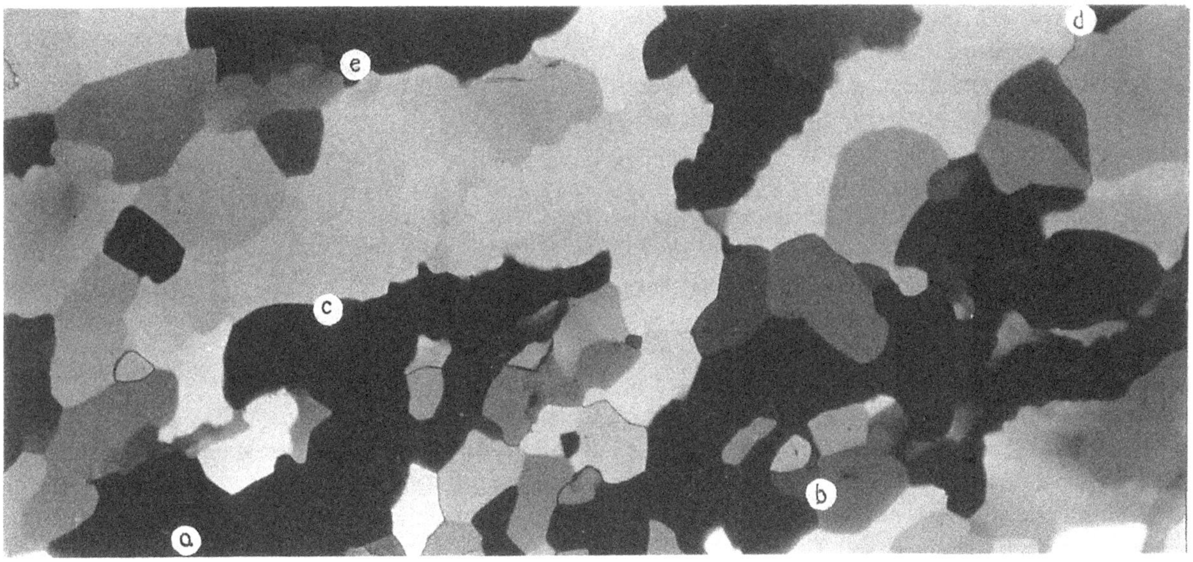


133 E

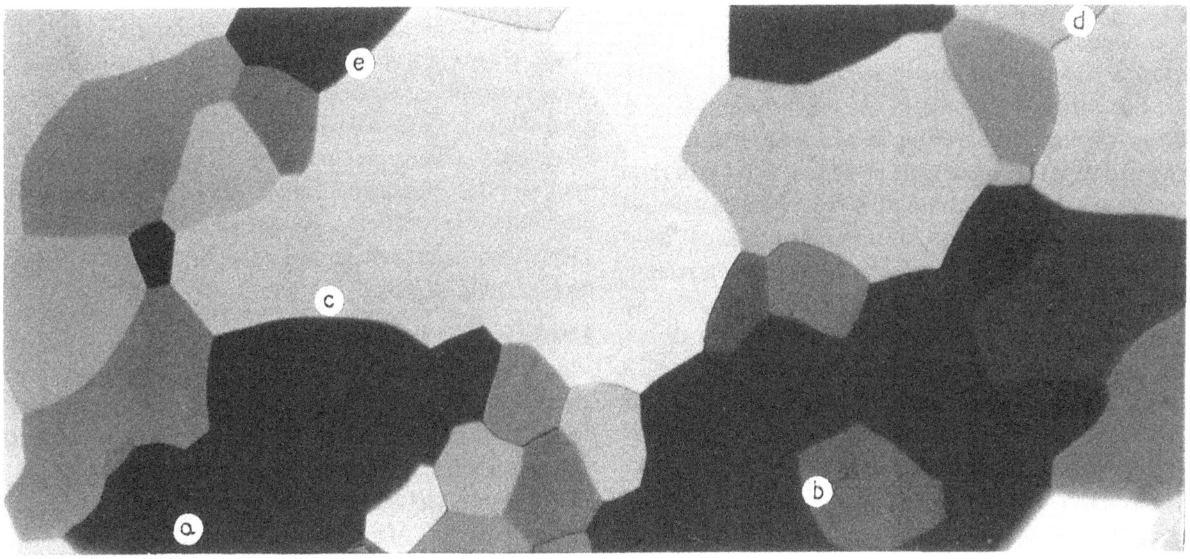


133 F

# 134. *BLASTOCATACLASITE*: A POSSIBLE ROCK TYPE FROM THE BOTTOM OF THE SEISMIC ZONE

ROBERT P. WINTSCH

Earthquakes are believed to produce breccias and pseudotachylytes in the upper crust (e.g., Sibson, 1975, 1980; Hobbs and others, 1986), but few such rocks have been reported from the middle and lower crust (Clarke and Norman, 1993; Austrheim and Boundy, 1994). The apparent scarcity of lower-crustal, high-strain-rate fault rocks probably stems not only from the lower frequency of occurrence of these earthquakes (Chen and Molnar, 1983), but also from poor preservation. The temperatures of the amphibolite and granulite facies of such deeper brittle fault rocks are high enough to allow significant grain growth and/or replacement that could destroy fault-rock textures and structures during normal exhumation and slow cooling (<10–12°C/m.y.). Preservation of pseudotachylytes and breccias requires at least very dry conditions (Clarke and Norman, 1993), and probably also relatively rapid cooling through rapid exhumation. Nevertheless, cataclastic textures from slowly exhumed, amphibolite-facies rocks can be preserved in the abrupt chemical differences between broken grains and cementing and overgrowth grains of amphibole and plagioclase that produce a core-mantle-like structure (Nyman and others, 1992; Nyman and Tracy, 1993).

This contribution describes textures that may be related to those of Nyman and Tracy (1993). These rocks were produced under amphibolite-facies metamorphic conditions in the hanging wall of the Honey Hill fault zone during Acadian–early Alleghanian thrust motion, which deformed and juxtaposed several lithotectonic terranes in southern New England. The fault zone was reactivated during slow Permian-Triassic exhumation (Wintsch and others, 1992) by retrograde, brittle (see Plate 9, this volume), and fabric-forming deformation. The textures described here occur in layers that cut the Hebron Formation near Hadlyme, Connecticut, U.S.A., in the immediate hanging wall of the Honey Hill fault zone. Ubiquitous to the Hebron Formation over tens of kilometers are alternating layers of violet biotite schist and green hornblende schist (locally 0.5 to 1.0 cm thick). However, some outcrops are cut by 5–30 cm thick layers of porphyroblastic granofels that lack the mineralogical layering and strong foliation of the protolith. The question arises: what process could have destroyed the preexisting compositional layering and foliation? The suggestion offered here is that an early Alleghanian, high-strain-rate, cataclasite-producing event randomized the fabric, bent biotite flakes, destroyed compositional layering, and produced cross-cutting layers of granofels. However, the inferred cataclastic texture is strongly overprinted by the growth of biotite and growth and overgrowth of plagioclase porphyroblasts driven by high ambient amphibolite-facies temperatures to produce a "blastocataclasite." The term blastocataclasite parallels Sibson's (1977) term "blastomylonite," a rock in which grain growth has followed and modified an earlier deformation texture. Plates 134A–134D show images documenting the random orientation of biotites (some even bent), both within the matrix and in porphyroblasts, as well as important core-overgrowth structures (boundaries arrowed). Because these rocks have come from the bottom of the seismic zone, the overprinting of some of the textures by subsequent late, lower amphibolite-facies Alleghanian muscovite + biotite–bearing ductile shear zones (134A, 134B) is not unexpected.

**134A.** Plagioclase granofels in a biotite schist (NX). Large plagioclase porphyroblast shows patchy zoning and twinning and includes quartz and calcite grains and randomly oriented biotite flakes (Figure 134.1). These textures suggest that the plagioclase porphyroblast replaced earlier plagioclase and quartz, while including biotite in its preexisting random orientation. Biotite grains of an adjacent schist (S) are part of a ductile shear zone that cuts across plagioclase grains of the granofels, indicating that the schist resulted from a younger, ductile, fabric-forming, (late Alleghanian?) event (see also 134B).

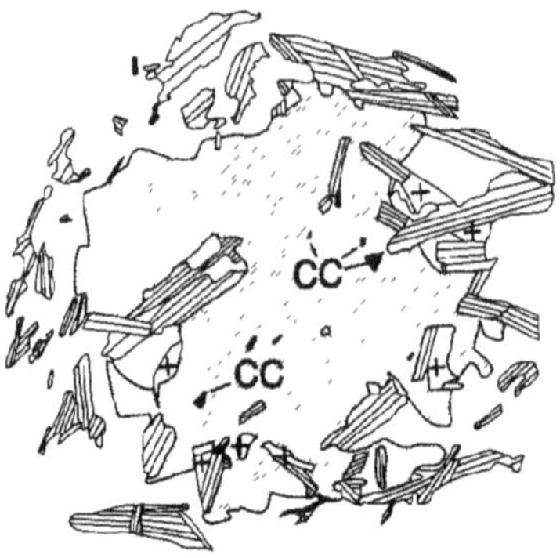


**Figure 134.1.** Sketch of the plagioclase porphyroblast of Plate 134A, showing inclusions of calcite (CC) and biotite Trace of biotite cleavage is indicated with ruled lines, and + indicates biotite c-axis is perpendicular to plane of section

**134B.** A 4 mm thick plagioclase granofels with biotite flakes in random orientation (NX). Abrupt chemical zoning in most plagioclase grains (cores $An_{40-48}$, rims $An_{28-40}$), indicated by different extinction angles, suggests recrystallization above the peristerite solvus (~450°C), thus reflecting amphibolite-facies conditions

**134C.** An enlargement of the lower portion of 134B (NX) showing several zoned plagioclase grains that indicate an abrupt plagioclase growth event producing overgrowths on many randomly oriented plagioclase grains. One plagioclase grain shows biotites in the core (C) oriented at a high angle to a biotite flake (B) in the rim (R). The latter biotite is bent, but the host plagioclase is optically undeformed, suggesting that the biotite was bent before it was included in the plagioclase, and that the plagioclase grew over the biotite before recrystallization could straighten it.

**134D.** A backscattered electron image of the plagioclase grain of 134C, showing that the optical discontinuities coincide with chemical discontinuities between the paler gray, calcic cores (C) and the darker gray rim (overgrowths) (R). Quartz (black) and biotite (white) are interstitial to and included in plagioclase, and both help define the later, cross-cutting foliation surface (S) of 134B. SEM

134 A

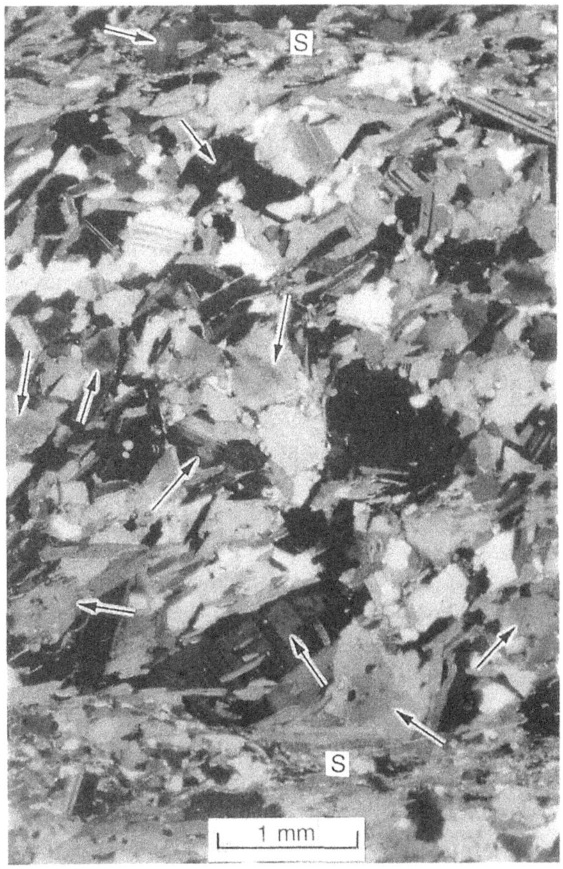


134 B

134 C

134 D

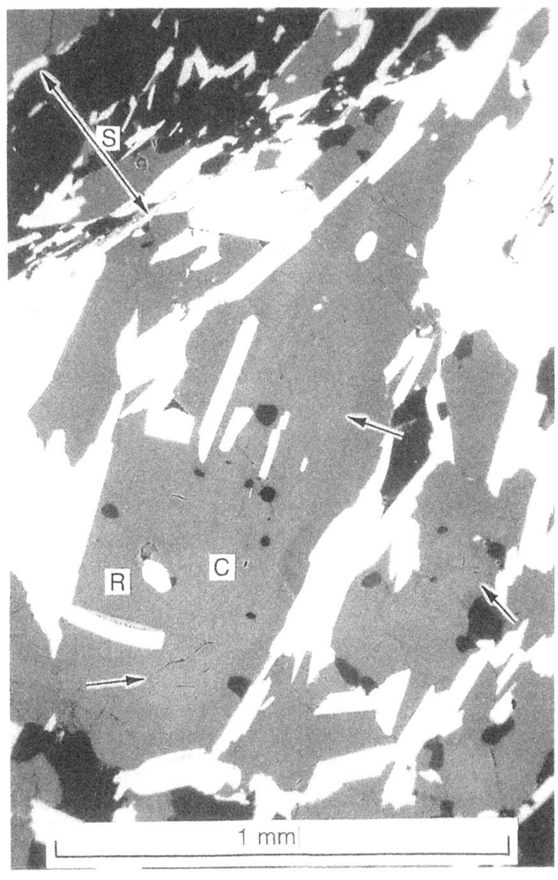

# 135. RETROGRADE AND ANNEALED MYLONITES

Robert D. Hatcher, Jr. and Robert J. Hooper

The idea that different types of mylonites may be influenced by competition between strain rate and recovery/recrystallization rate, which itself depends on temperature, was formulated at least as early as the late 1960s (e.g., Theodore and Christie, 1969; Hatcher, 1978; Hatcher and Hooper, 1981). The importance of this relationship was emphasized by Wise and others (1984) in their classification of fault-related rocks, and by Passchier and Simpson (1986), who suggested that the form of rotated porphyroclasts depends on the ratio of recrystallization rate to shear strain rate. Deformation at lower temperature and/or faster strain rates involves a lower ratio of recovery rate to strain rate, and tends to produce mylonites with finer grain size and higher density of intracrystalline defects. These mylonites may be produced during low- (metamorphic) grade deformation, or they may overprint earlier fabrics during the retrograde portion of a higher-grade deformation (Figure 135.1). Deformation at higher temperatures and/or slower strain rates involves a higher ratio of recovery rate of intracrystalline defects to strain rate, depending on the mineral assemblage in the rock. Dynamically recrystallized high-temperature mylonites may thus be difficult to differentiate microscopically from mylonites that were statically annealed following deformation.

The complex Pine Mountain window in central Georgia and Alabama exposes 1.1 Ga old North American basement and metasedimentary cover rocks, and is framed by an array of syn- and postmetamorphic faults (Figure 135.2). Two of these, the Box Ankle and Towaliga faults, contain syn- (near thermal peak) and postmetamorphic (retrograde) mylonites that formed under sillimanite- and garnet-grade conditions, respectively (Hooper and Hatcher, 1988a, b) (135A–135D). Student and Sinha (1992) reported a 303 Ma U-Pb age for the Box Ankle fault. The Towaliga fault records dextral motion and truncates the Box Ankle fault, whereas a variety of shear-sense indicators in the Box Ankle fault indicate that displacement was northwest-directed, and that the fault is a thrust. The protolith of the Towaliga mylonite (135A) was sillimanite-grade, biotite ortho- and paragneiss. Although the quartz, muscovite, and biotite are dynamically recrystallized, the feldspars behaved brittlely.

**135A.** Low-temperature, retrograde mylonite from the Towaliga fault zone, central Georgia, southeastern U.S.A. The light layer contains dominantly quartz and feldspar, whereas the dark layer additionally contains biotite. The cuspate-lobate structure along the light-dark layer boundary reflects a ductility contrast, with the biotite-rich darker layer being weaker.

**135B.** Towaliga mylonite showing differences in texture on either side of the light-dark layer boundary in 135A. The mineral assemblage present is quartz, plagioclase, biotite + microcline, epidote, and muscovite. Note strong orientation of quartz ribbons at the light-dark boundary. Scale bar is 1 mm. NX plus gypsum plate.

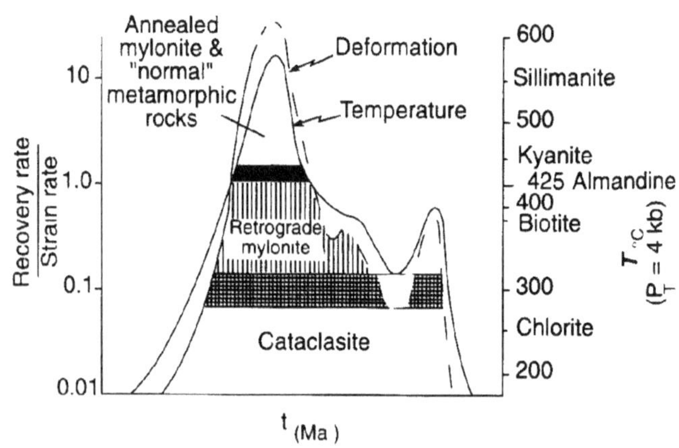


**Figure 135.1.** Temperature (P ~ 400 MPa) and recovery rate/strain rate (a unitless quantity) vs. time, showing the relationships between cataclasite, retrograde mylonite, and annealed mylonite. The ratios indicated along the left vertical axis and the gradational boundaries between fields were derived from observations of mylonitic textures and mineral assemblages in thin sections from many fault zones that formed over a wide range of P-T conditions. Two curves represent deformation intensity and temperature, with both peaks nearly superposed in time. The ratio of recovery rate to strain rate is higher at high temperatures and thus may produce annealed mylonite (assuming deformation ceases at high temperature), but lower at low temperature where it may produce retrograde mylonite or cataclasite.

135 A

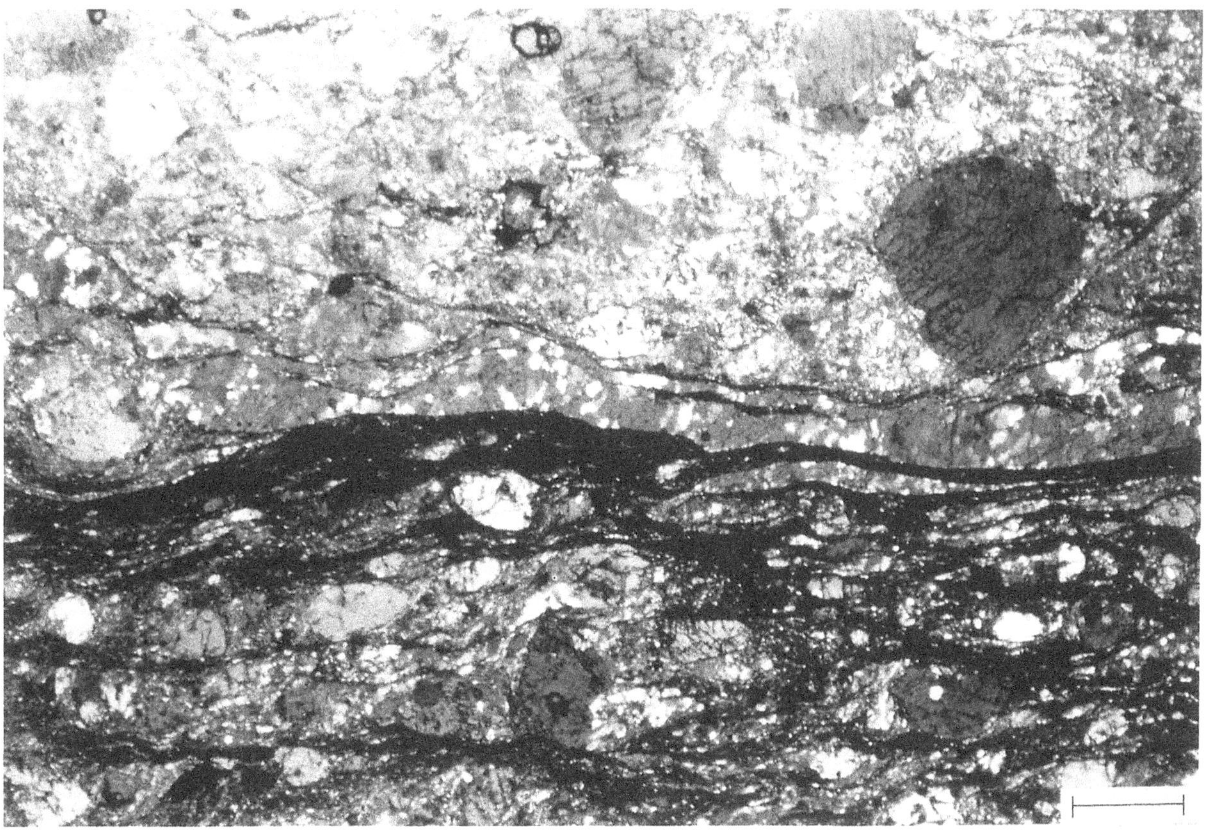

135 B

The protolith of the Box Ankle mylonite (135C) was quartz-biotite-plagioclase paragneiss (metagraywacke) that formed under sillimanite-grade conditions. The mylonite contains unrecrystallized feldspar porphyroclasts with tails of recrystallized quartz and mechanically derived small feldspar grains, and quartz ribbons that are mostly recrystallized. The groundmass is likewise recrystallized and consists of randomly oriented grains. Field and petrographic data both within and outside the fault zone indicate a uniform high metamorphic grade with no fault-related retrogression. This, together with the microfabric, suggests that the fault rocks were either statically recrystallized after motion on the Box Ankle fault ceased, or recrystallized at an exceedingly slow strain rate, producing an annealed microfabric but preserving the megascopic mylonitic texture in ribbons and tailed porphyroclasts.

Both the Towaliga and Box Ankle mylonites formed under conditions well above the threshold of the Hirth and Tullis (1992) Regime 3—at high enough temperatures to promote rapid dislocation climb and recovery facilitated by grain-boundary migration, thus involving steady-state flow.

**135C.** Annealed high-temperature mylonite from east end of the Box Ankle fault zone, central Georgia, U.S.A. Megascopic, salt-and-pepper texture (roughly equant randomly oriented quartz and feldspar, light, and biotite, dark) comprises the recrystallized groundmass. Large quartz ribbons and plagioclase porphyroclasts with tails of both quartz and feldspar preserve the mylonitic texture.

**135D.** Annealed high-temperature mylonite from east end of the Box Ankle fault zone, central Georgia, U.S.A. Strongly oriented, but recrystallized, ribbons and groundmass do not preserve evidence of microscopic strain. Unrecrystallized plagioclase porphyroclasts and flanking annealed quartz ribbons indicate a sinistral (northwest-directed) shear sense. Note the recrystallized groundmass that consists chiefly of randomly oriented biotite and quartz with subordinate feldspar. Scale bar is 1 mm. PPL

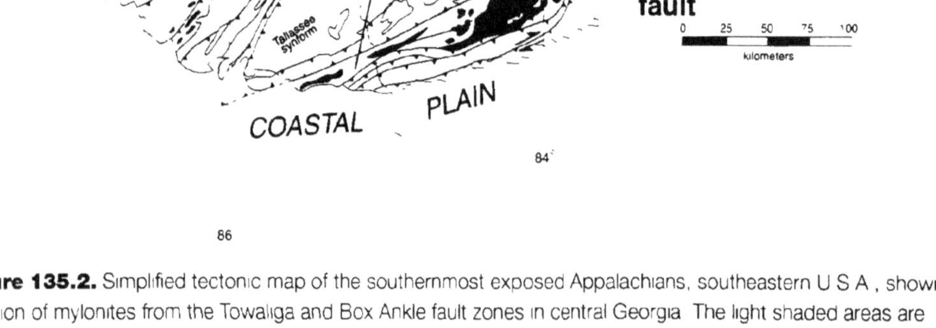


**Figure 135.2.** Simplified tectonic map of the southernmost exposed Appalachians, southeastern U S A , showing the location of mylonites from the Towaliga and Box Ankle fault zones in central Georgia The light shaded areas are North American rocks of the Blue Ridge and Piedmont The darker shaded area represents North American (~1 1 Ga) basement in the Pine Mountain window Mafic bodies are shown in black Carolina terrane rocks southeast of the central Piedmont suture, larger granitoids and the Valley and Ridge (foreland fold-and-thrust belt) rocks to the northwest are not patterned

135 C

135 D

The Jotun thrust sheet, Norway (Figure 135.3) consists of a far-traveled mass of felsic, intermediate, and mafic crust that was emplaced, along with several lower thrust sheets, onto the Baltic shield during the Caledonian orogeny (435–400 Ma) (Milnes and Koestler, 1985). Part of the sheet west of the study area was dismembered by extension during the Early Devonian collapse of the orogen (Fossen and Holst, 1995). During thrusting horses of quartzite, orthogneiss, and carbonate were detached from the underlying platform, transported with the thrust sheet, and later exhumed at the present erosion level. Mineral assemblages within the fault zone, horses, and the underlying metsedimentary rocks indicate that emplacement occurred under prograde chlorite- to (lower-)biotite-zone conditions. All of the quartzite in these horses contains a strong mineral lineation oriented parallel to the transport direction (WSW-ENE; see Plate 135E).

The microfabric of metamorphosed orthoquartzite from these two horses is mylonitic and characterized by only partial recovery and recrystallization. The Bessheim sample (135E and 135F) is partially recrystallized, whereas the sample from northwest of Randsverk (135G) is unrecrystallized, but is strongly deformed. Originally(?) fine-grained layers of quartz sand in the Bessheim sample were likely interbedded with coarse quartz sand grains in the protolith, yielding quartz mylonite layers dominated by large ribbons alternating with layers containing markedly smaller ribbons with roughly the same aspect ratios (10 to 25) as the larger ribbons. Small, newly recrystallized grains have formed in the fine-grained layers, as well as along the boundaries of larger grains (135F). This mylonite is probably a product of deformation near the thermal peak, but, because of its relatively low metamorphic grade, insufficient thermal or mechanical energy may have been present to permit complete recrystallization of the rock. The sample and microstructures in 135E and 135F suggest that this prograde mylonite may have formed under conditions for Regime 2 for quartz of Hirth and Tullis (1992)—extreme flattening of original grains accompanied by development of subgrains. Dynamic recrystallization occurred by progressive rotation of subgrains and formation of new unstrained quartz crystals at grain boundaries, best displayed in 135F. The Randsverk sample (135G) may have been deformed by the Hirth and Tullis Regime 1 process, but recrystallization may have been inhibited by lack either of thermal energy or of fluid to promote recrystallization at grain boundaries.

**135E.** Strongly foliated and lineated quartzite mylonite from a horse located near Bessheim (Figure 135.3). The strong mineral lineation, shear-sense indicators in both the mesofabric of the hanging-wall gneiss and footwall schist, and the quartzite microfabric indicate eastward transport. Pencil point indicates scale; arrow on top of sample is oriented down plunge (west).

**135F.** Sample cut parallel to mineral lineation; transport was top to the left (east). A few small feldspar grains are present (e.g., lower left) and produced local pinning of quartz grain boundaries. Scale bar equals 1 mm. NX

**135G.** Sample collected from a different horse than 135E and 135F, northwest of Randsverk, was cut perpendicular to the mineral lineation. The quartz is highly deformed, but is only minimally recrystallized, possibly because of deformation at lower temperature, but still at chlorite grade. Scale bar equals 1 mm. NX

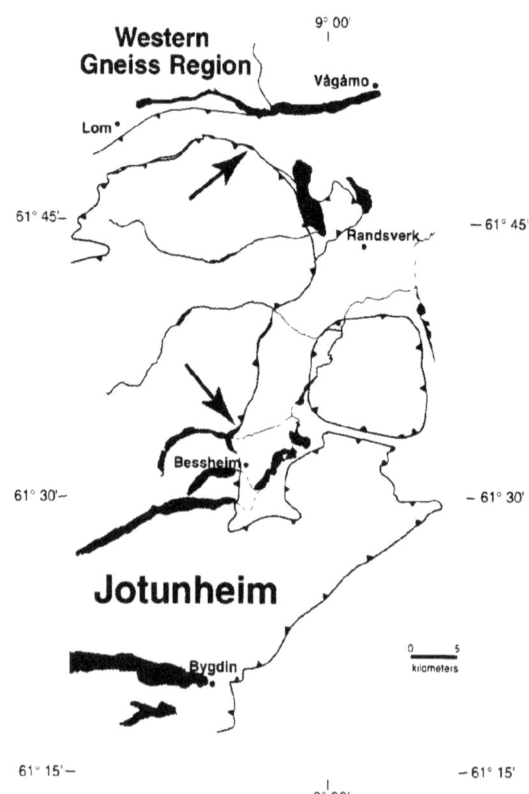


**Figure 135.3.** Location map of part of the Jotun thrust sheet (lightly shaded area). Arrows indicate the horses from which the quartzite mylonite samples were collected Geology by R D H., 1990–91. Darker shaded areas are lakes and streams

135 E

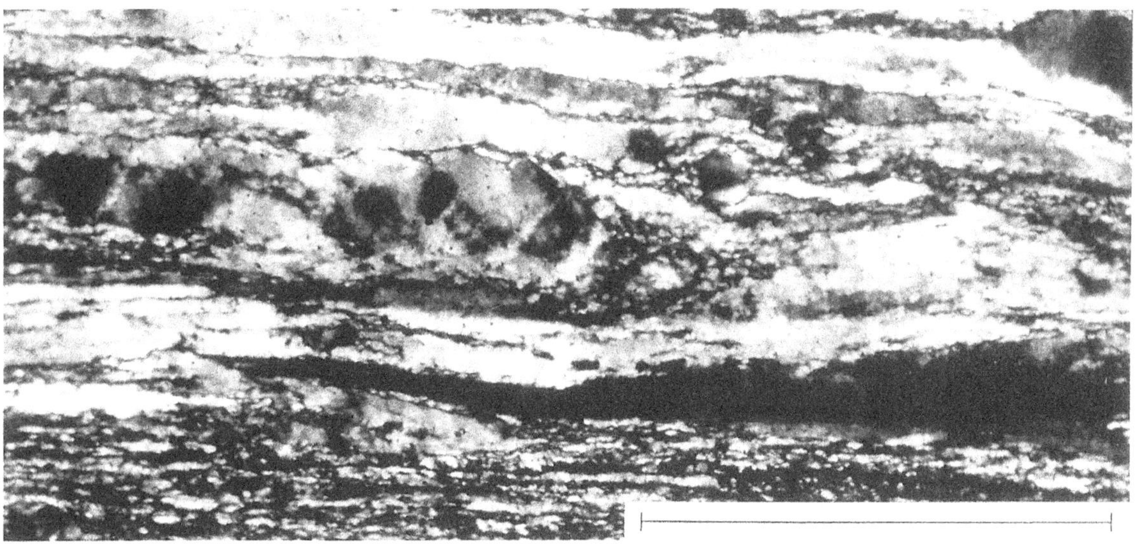

135 F

135 G

# 136. EFFECTS OF STATIC ANNEALING ON EXPERIMENTALLY DEFORMED QUARTZITE AND APLITE

JAN TULLIS AND GAYLE GLEASON

These photographs illustrate how microstructures produced by dislocation creep of quartzite and aplite are modified by later static annealing. These samples were shortened 60% in axial compression at a confining pressure of 1500 MPa, a constant strain rate of $10^{-6}$/s, and a temperature of 800°C. Two samples were subsequently annealed (hydrostatically) at 900°C for 120 hours. It is difficult, if not impossible, to distinguish on the basis of microstructures a sample that has been statically annealed from one that has been deformed at very high temperature and slow strain rate, where the rate of recovery can keep pace with the rate of deformation. The scale for all photographs is the same as that in 136A, where the bar is 200 μm.

**136A.** Sample of Black Hills quartzite (initial grain size of ~100 μm) shortened 60% (E-W) at a confining pressure of 1500 MPa, a constant strain rate of $10^{-6}$/s, and a temperature of 800°C. Deformation occurred by climb-accommodated dislocation creep (Regime 2 of Hirth and Tullis, 1992), producing homogeneously flattened original grains containing well developed subgrains (on the TEM scale) and recrystallized grains (mostly along original grain boundaries) which have formed by progressive subgrain rotation. NX

**136B.** Sample of Black Hills quartzite deformed at the same conditions as the sample illustrated in 136A, but after 60% shortening the motor was backed off, leaving the sample at hydrostatic pressure, and the temperature was raised to 900°C for 120 hours. The recrystallized grains have grown to consume all of the deformed original grains, producing an "equilibrium" microstructure of strain-free polygonal grains. However, this sample retains a crystallographic preferred orientation of the same pattern and strength as that of the deformed sample. Furthermore, a very similar texture of polygonal recrystallized grains is produced by dynamic recrystallization accompanying high-temperature deformation (see Plate 146C, this volume). NX

**136C.** Sample of Enfield aplite (initial grain size ~150 μm; composition ~30% quartz, 35% microcline, 33% oligoclase, and 2% biotite and oxides) shortened 60% (E-W) at a confining pressure of 1500 MPa, a constant strain rate of $10^{-6}$/s, and a temperature of 800°C. Relatively homogeneous flattening of the quartz grains (Q) occurred by climb-accommodated dislocation creep; most grains contain less deformed cores and mantles with subgrains and recrystallized grains produced by subgrain rotation. In contrast, inhomogeneous microboudinage of the feldspars (F) occurred by recrystallization-accommodated dislocation creep (Regime 1 of Hirth and Tullis, 1992). NX

**136D.** Sample of Enfield aplite deformed at the same conditions as the sample shown in 136C, but then annealed hydrostatically at 1500 MPa and 900°C for 120 hours. Note that the greater strain present in the mantles of the quartz grains (Q) has resulted in a necklace of relatively large recrystallized grains surrounding each core. In the feldspars (F), the size of the recrystallized grains remains so small (~2 μm) that they are only resolvable in TEM. NX

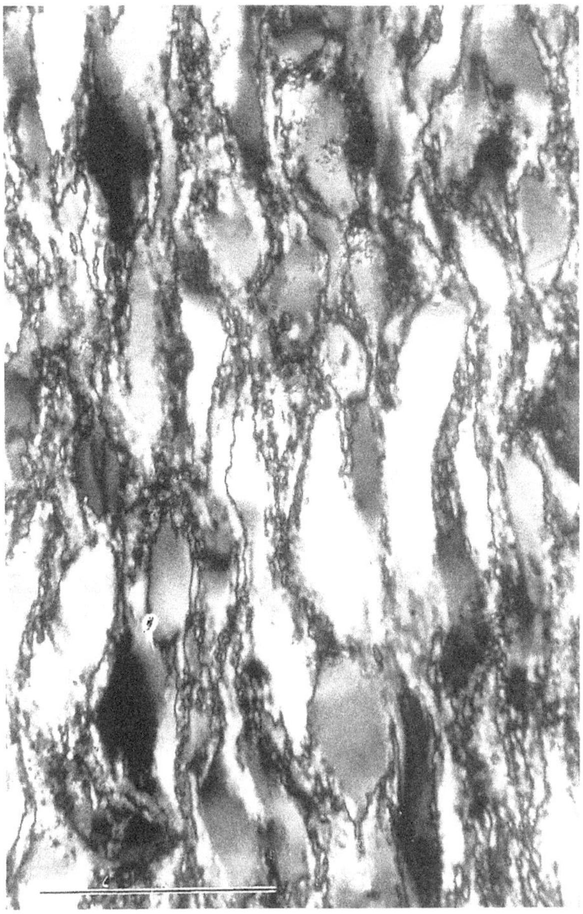

136 A

136 B

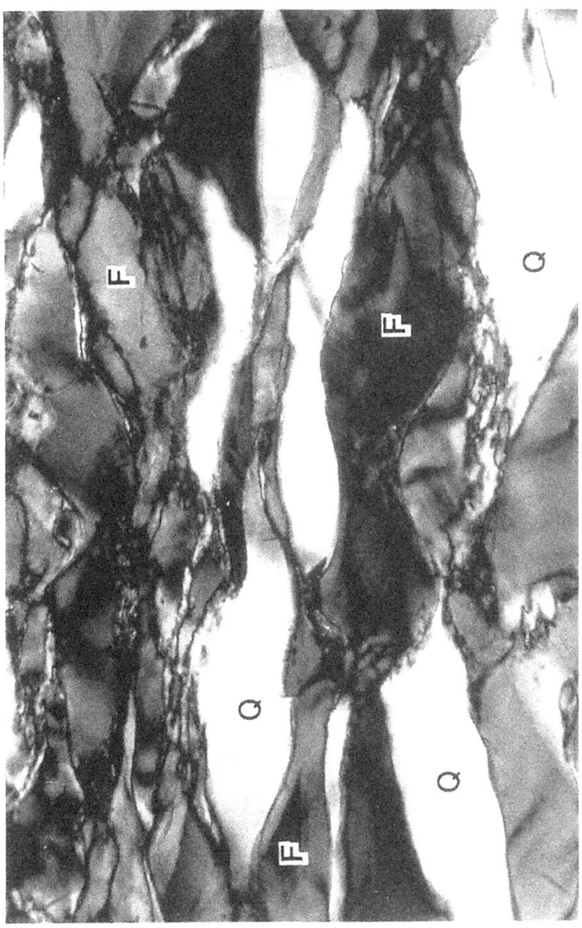


136 C

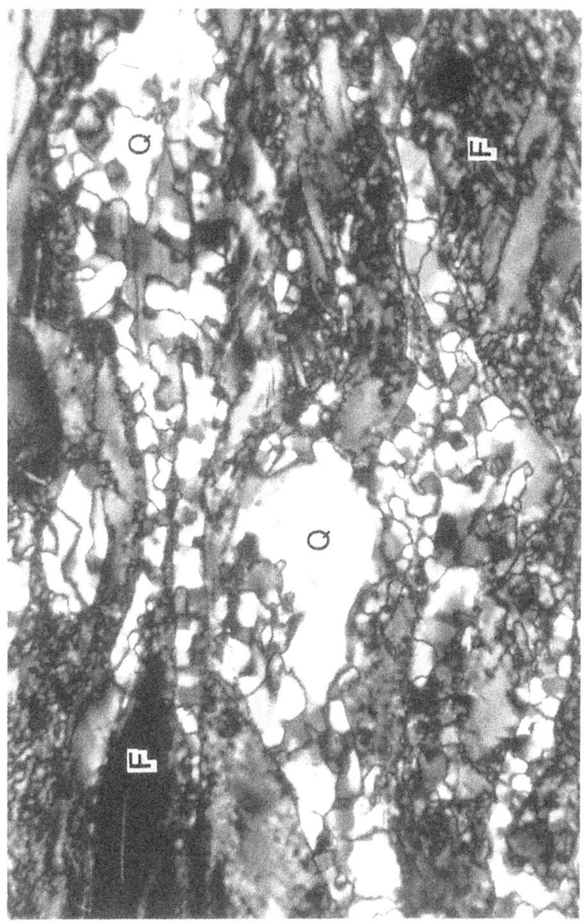


136 D

# DUCTILE BEHAVIOR

## Fluid-related features



Common abbreviations used in regard to the plates in this section include PPL = plane polarized light, NX = crossed nicols, SEM = scanning electron microscope, and TEM = transmission electron microscope.

It is known from experimental studies that aqueous fluids strongly affect all deformation mechanisms; this conclusion is supported by the microstructures of many naturally deformed rocks, and some of these are illustrated in this set of plates. One plate illustrates a possible example of stress corrosion, the process whereby fluids enhance the rate of stable crack propagation. Fluids also enhance the rates of pressure solution, and two plates present microstructural evidence for the operation of this process in greenschist-facies quartzo-feldspathic mylonites. Fluids are commonly assumed to enhance the rate of grain-boundary migration, and thus affect both static and dynamic recrystallization; one contribution demonstrates that isotopic equilibration between quartzite and coexisting fluids is strongly enhanced by grain-boundary migration recrystallization. Two important effects of fluids are not directly illustrated in this section. First, in the dislocation creep regime, the presence of fluid may accomplish hydrolytic weakening; this has been documented in the natural shear zone illustrated in Plate 41 (this volume) by Kronenberg and others (1990). Second, the availability of fluids along a shear zone during retrograde deformation will allow reaction to weaker hydrous phases; this is illustrated for mafic mylonitic rocks in Plate 122, this volume.

# 137. EVIDENCE FOR CHEMICAL PROCESSES IN THE EVOLUTION OF MYLONITES

R. P. WINTSCH AND R. J. KNIPE

Mylonites derived from Lewisian granitic gneiss occurring at the north end of the Moine thrust at Loch Eriboll, Scotland, contain many chemically zoned feldspar grains that suggest pressure solution creep was an important deformation mechanism. The mylonites contain primarily quartz, plagioclase ($An_{0-8}$), muscovite, microcline ($Or_{93-100}$), and chlorite that form domains rich in phyllosilicate (P), quartz (Q), and quartz-plagioclase-phyllosilicate (QFP) (Knipe and Wintsch, 1985), reflecting lower greenschist-facies conditions of deformation. The mylonite contains K-feldspar–quartz veins, and K-feldspar > quartz > plagioclase porphyroclasts (Knipe and Wintsch, 1985) probably derived from these veins. The rocks also contain plagioclase porphyroblasts in a matrix of quartz, albite, and muscovite grains 20–80 μm long, with aspect ratios of up to 1:5 (137A, 137B, 137C, optical NX, and 137D, backscattered SEM). Sections are cut perpendicular to a foliation defined by the preferred orientation of muscovite, and parallel to a lineation of muscovite streaks and quartz rods.

Chemical fluctuations are present in most plagioclase porphyroblasts and in all (49) matrix grains analyzed. Evidence that the grains grew in situ comes from overgrowths defined by inclusion trails (137B) and by abrupt compositional discontinuities (137A, 137B). Their very albitic compositions ($<An_{05}$) further reflect growth at lower greenschist-facies conditions. Structures defined by chemical gradients within single crystals suggest grain growth or overgrowth parallel to the extension direction. Overgrowths and adjacent quartz "beards" of 137A and 137B may also reflect precipitation in "pull-apart" fractures opened during stick-slip motion, where the porphyroblast was large enough to cause stress concentrations.

Surfaces facing these relatively high normal stresses (parallel to foliation) are defined by muscovite flakes that truncate compositional zoning in porphyroblasts (Figures 137.1 and 137.2). These truncations could reflect pressure dissolution and replacement by muscovite of portions of these large grains. In the matrix, however, the small plagioclase grains show only continuous compositional fluctuations (Figure 137.3), suggesting that the matrix underwent pressure solution creep, where chemical fluctuations reflect changes in the rate of extension. The growth of mica (P domains) and feldspar overgrowths may be related through a reversible reaction: musc + qz = $2(K, Na)^+$ = feldspar + $2H^+$. The reaction shows that fluid composition (alkalinity) defines the stability of mica and feldspars even at constant pressure and temperature (e.g., Wintsch and Knipe, 1983). If balancing this reaction by Al conservation is appropriate (Carmichael, 1969), then feldspar produced in regions of low mean stress (e.g., overgrowths) occurs with 10% volume increase, whereas muscovite + quartz replacement of feldspars on grain boundaries facing high mean stress occurs with 10% volume decrease. These volume relationships are those predicted by the contrasting mean stresses at the two reaction sites. The petrological significance is that the deformation mechanism of pressure solution creep is also a reaction mechanism of dissolution and precipitation through the above reaction.

**137A.** A chemically zoned (Figure 137.1) albite porphyroblast (P) in a QFP domain. Muscovite grains, some replacing albite grains, define foliation-producing P domains (M). NX

**137B.** A zoned (Figure 137.2) plagioclase porphyroblast (A) with quartz in a pressure shadow (Q) in a QFP domain. Muscovite in P domains (M) defines foliation and locally replaces albite porphyroblasts. NX

**137C.** Plagioclase (P) and microcline (K) porphyroblasts in a QFP domain. Most larger feldspar grains are outlined by muscovite in P domains (M). NX

**137D.** A backscattered electron image of matrix plagioclase grains (P) near the microcline grain (K, pale gray) in 137C in a matrix of quartz (dark gray) and muscovite (M, pale gray). Microprobe analysis shows that all matrix albite grains are chemically zoned (e.g., Figure 137.3). SEM

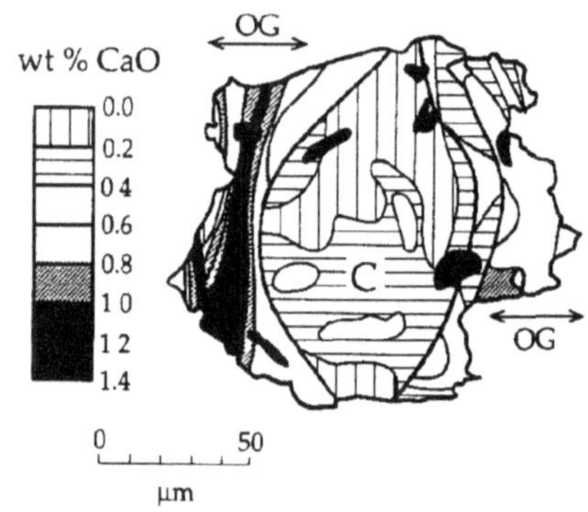


**Figure 137.1.** Structure of porphyroblast in Plate 137A as defined by chemical zoning. Bold lines mark compositional discontinuities between a sodic ($An_{01-02}$) core (C) with quartz inclusions and overgrowths (OG) facing the extension direction defined by compositional discontinuities (Wintsch and Knipe, 1983).

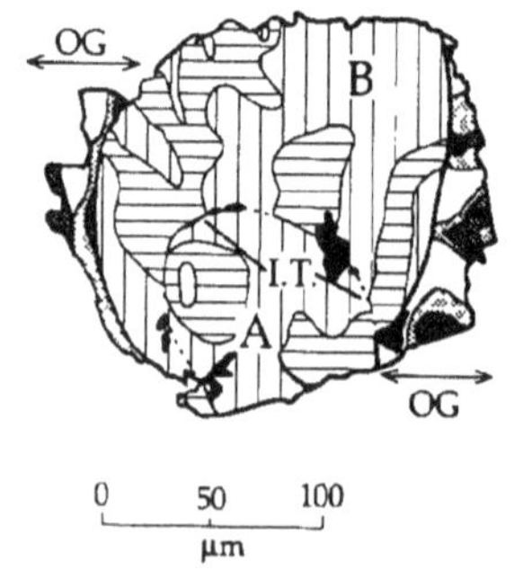


**Figure 137.2.** Structure of chemically zoned porphyroblast in Plate 137B. Overgrowths in this optically continuous porphyroblast are defined by inclusion trails (I.T., dotted line) in very sodic ($An_{01}$) inner (A) and outer (B) cores, and by chemical discontinuities (bold lines) that outline overgrowths (OG) facing the extension direction (after Knipe and Wintsch, 1985).

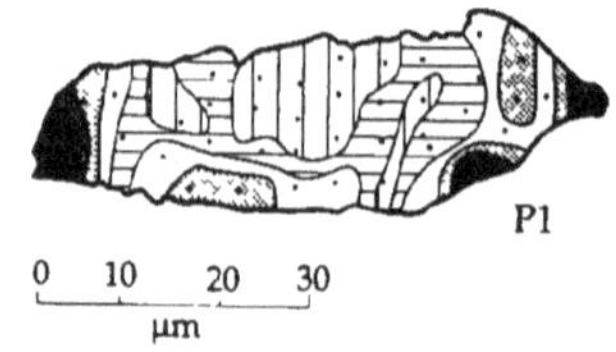


**Figure 137.3.** Internal chemical structure of albite grain (P1) of Plate 137D showing zoning from 0.1 to 1.0 wt% CaO ($An_{1-5}$). Dots indicate locations of probe analyses.

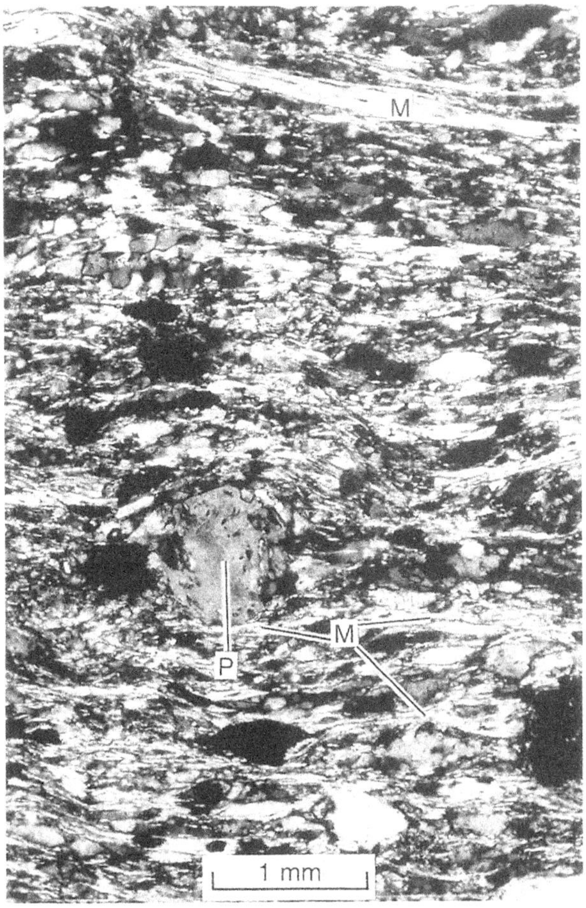


137 A

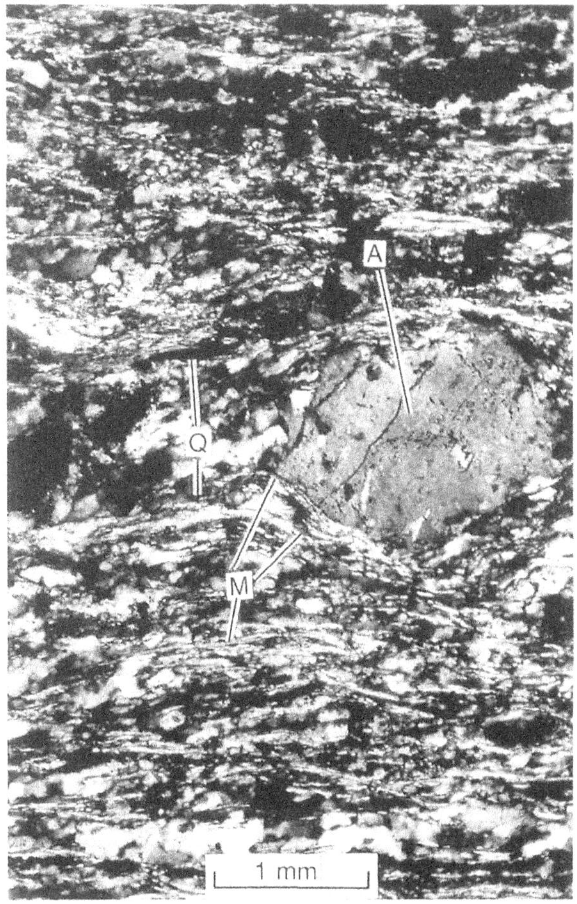


137 B

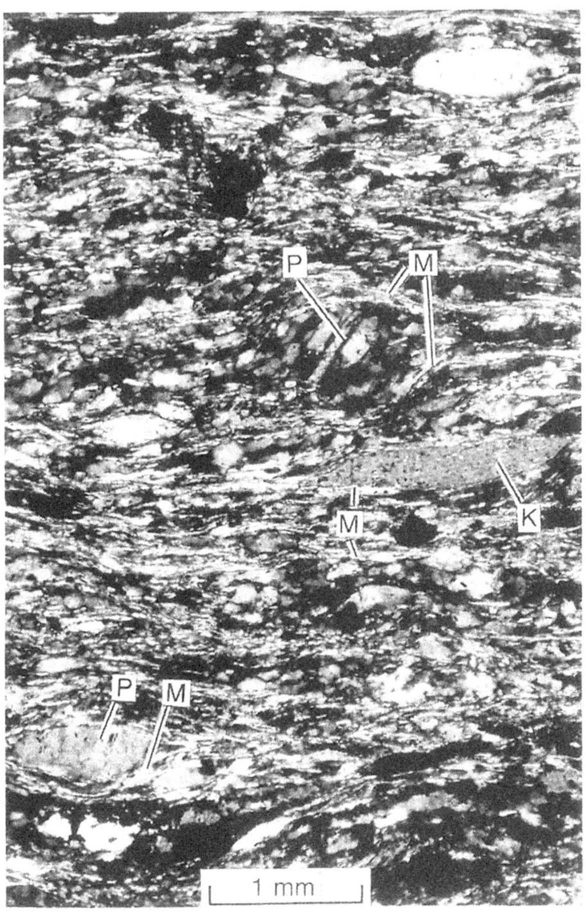


137 C

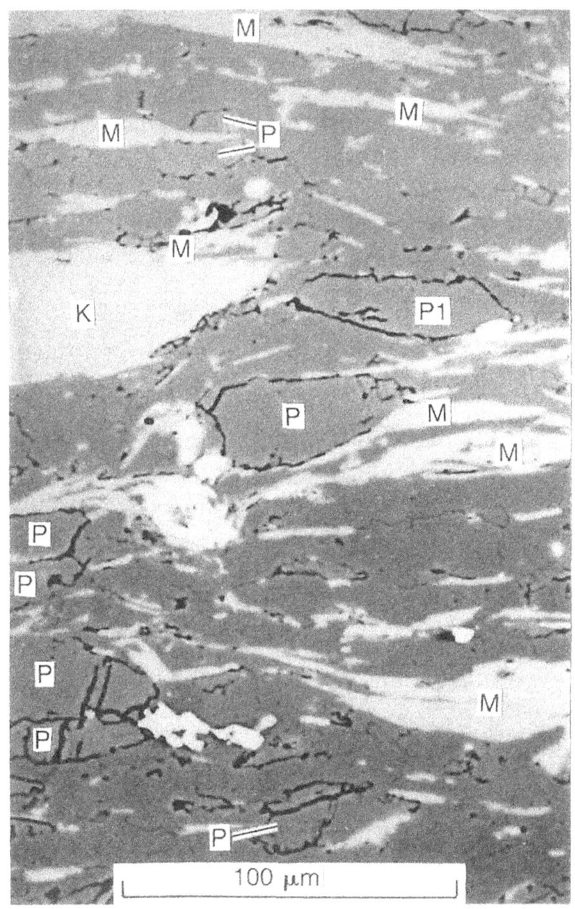


137 D

# 138. MINERAL REACTIONS PARTICIPATING IN INTRAGRANULAR FRACTURE PROPAGATION: SUBCRITICAL CRACK GROWTH?

R. W. KERRICH, T. E. LA TOUR, AND R. L. BARNETT

In mylonite zones, the initial stages of tectonic grain-size reduction typically involve transgranular fracturing of some minerals, specifically feldspars. The photomicrographs opposite illustrate a variety of such fractured grains, where reaction of the parent grain has occurred at crack tips, raising the question as to whether chemical and mechanical factors have interacted in subcritical crack propagation. Subcritical crack propagation involves slow stable crack growth by a "stress corrosion reaction," which proceeds above a threshold stress for catalyzing the reaction, but below the critical stress for fast irreversible brittle rupture characteristic of Griffith failure.

**138A.** Intragranular fractures in biotite, granodiorite mylonite, Miéville, Switzerland (Kerrich and others, 1980). NX. Ambient temperature of deformation 280–320°C. Grains of biotite have fractures that appear not to involve significant displacements as evidenced by the registry of trails of solid inclusions across fractures (a). Fracture openings are occupied by fine-grained secondary ilmenite (b), low-Ti biotite, and high-Ti muscovite. The fractures are bounded by an envelope of pale brown birefringence (c) corresponding to a decrease of Ti and increase of Fe + Mg relative to the composition of grains in areas unaffected by cracking. Randomly oriented ilmenite grains (~10 µm length) are present up to 200 µm ahead of the leading edges of cracks (d).

**138B.** Intragranular fractures in orthoclase, granodiorite mylonite, Miéville, Switzerland (Barnett and Kerrich, 1980). NX. Porphyroclasts of orthoclase (and plagioclase) feldspar are transected by irregular microfractures (a), which may terminate within crystals, are relatively wide (20 µm to 500 µm), and have little discernible evidence of displacement. Fractures are occupied by secondary reaction products of the parent grain, typically secondary feldspar of approximately the parent grain composition. However, oligoclase and orthoclase porphyroclasts react to form more sodic and potassic compositions, respectively, within fractures.

**138C.** Intragranular fracturing of hornblende, deformed amphibolite, Grenville front, Ontario, Canada (Kerrich and others, 1981). NX. Ambient temperature of cracking ~500°C. This photomicrograph illustrates a grain of hornblende (a) with an intragranular fracture (b) occupied by hornblende in close optical continuity with the parent grain. Hornblende in fractures is depleted in Ti, Al, and K, and enriched in Mg relative to parent grains. Fracture intensity within hornblende increases with progressive deformation until fractures coalesce and grains lose cohesion.

A generalized reaction during cracking of the hornblende is as follows:

*average host amphibole*
$(Ca_{1\ 815}Na_{0\ 160}K_{0\ 160})(Mg_{1\ 825}Fe_{2\ 490}Fe_{0\ 043}Mn_{0\ 045}Ti_{0\ 113}Al_{0\ 613})$
$(Si_{6\ 680}Al_{1\ 320})O_{22}(OH)_2 + 0.103\ CaO + 0.446\ MgO +$
$0.0725\ Fe_2O_3 + 0.546\ SiO_2 \rightarrow$

*average crack amphibole*
$(Ca_{1\ 918}Na_{0\ 202}K_{0\ 049})(Mg_{2\ 271}Fe_{2\ 255}Fe_{0\ 188}Mn_{0\ 040}Ti_{0\ 016}Al_{0\ 309})$
$(Si\text{-}_{226}Al_{0\ 774})O_{22}(0H)_2 + 0.425\ Al_2O_3 + 0.097\ TiO_2 +$
$0.235\ FeO + 0.005\ MnO + 0.094\ Na_2O + 0.0165\ K_2O.$

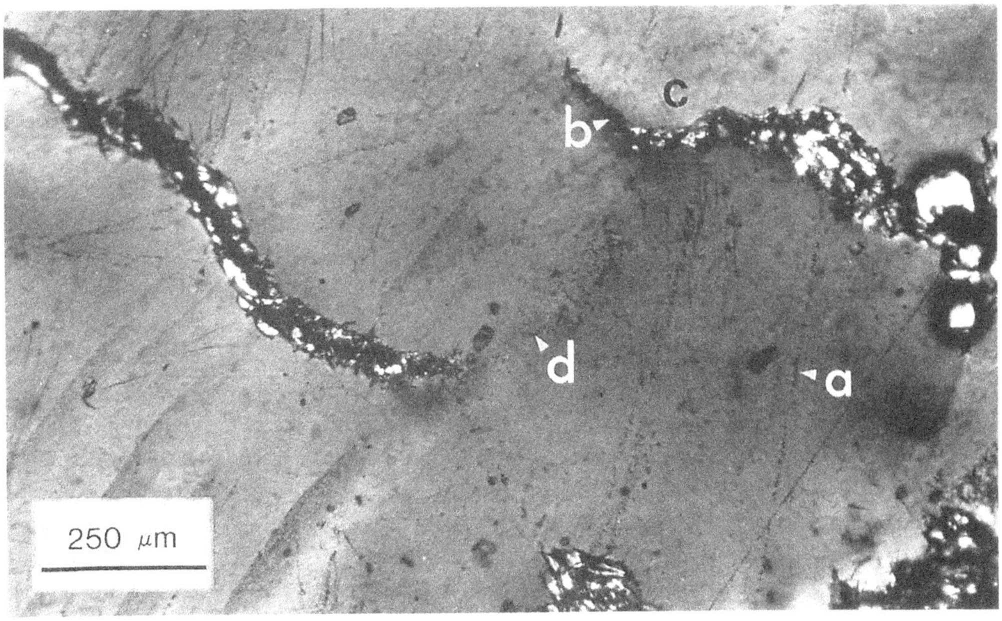


138 A

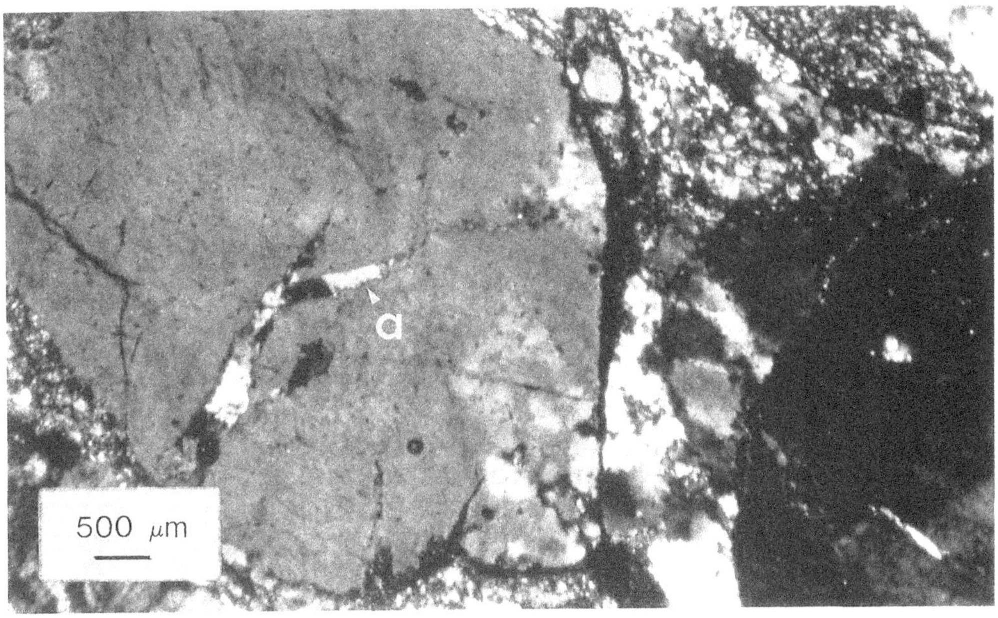


138 B

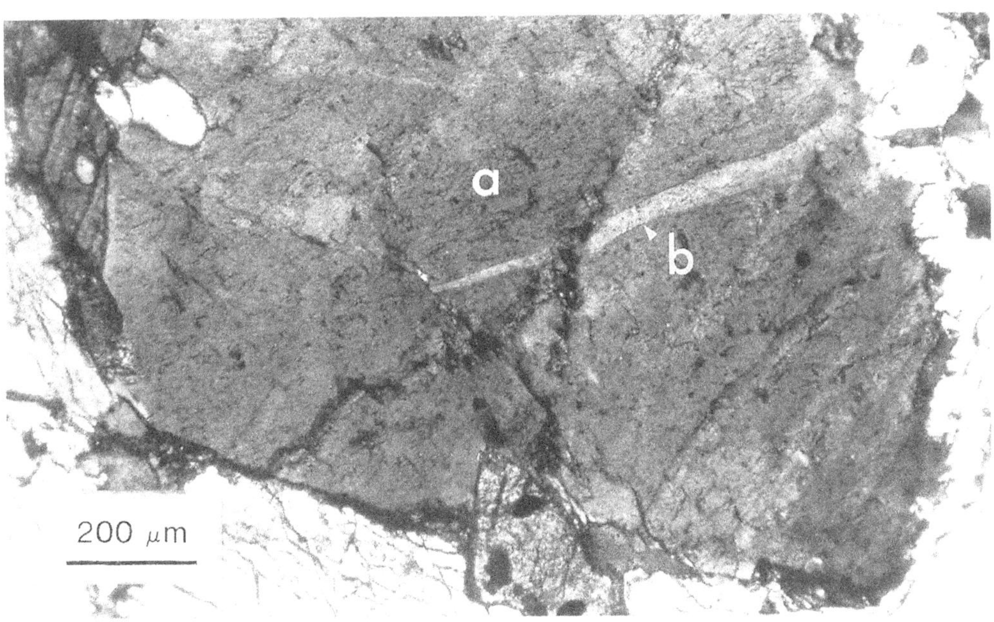


138 C

## 139. INTERACTION OF CHEMICAL AND MECHANICAL FACTORS IN BIOTITE DEFORMATION

R. W. KERRICH, I. ALLISON, AND R. L. BARNETT

This photomicrograph illustrates the deformation of a single biotite grain that recrystallized at its periphery to an aggregate of fine-grained products. The rock that contains this biotite grain is a deformed granodiorite from the Miéville shear zone, Switzerland. At Miéville, located in the Aiguilles Rouges massif, basement rocks of Hercynian age are transected by a major subvertical shear zone, which varies in width from 10 to 500 m, and can be traced laterally for ~15 km. A protracted tectonic history for the shear zone is indicated, with initiation in the Permian, followed by varied intensities of Alpine deformation. Undeformed granodiorite consists of orthoclase and oligoclase phenocrysts, quartz, biotite, and muscovite. The metamorphic mineral assemblage of deformed counterparts (quartz + albite + epidote + chlorite) corresponds to lower-greenschist facies. Ambient temperatures of deformation are estimated at 280–320°C. Overall, deformation of the granodiorite was isochemical (Kerrich and others, 1980).

Biotite has responded to deformation principally by dislocation glide and recrystallization to fine-grained, new pale biotite + ilmenite as a mantle (a) around the relict parent grains (b). Igneous biotite grains have a consistent primary zoning with respect to titanium, from 4% $TiO_2$ in the core to 2% $TiO_2$ at the margin (analyses 1–5; Table 139.1). Aluminum is distributed antithetically to titanium such that the margins of biotite grains are more aluminous ($Al^{VI}$) than the cores. At low states of strain the biotites are distorted and have developed a peripheral domain with pale brown or green birefringence, which is due to a decrease of Ti and increase of Al compared to pristine zoned biotite at either core or rim. With increasing strain, ultrafine-grained (10–50 μm), recrystallized biotite developes in mantles around the parent grains, the secondary biotite being consistently lower in Ti and higher in Fe than the adjacent relict parent (analyses 2, 3; Table 139.1). As the mantles were deformed into elongate trails, ferrimuscovites formed in response to K release accompanying the reaction of secondary biotite to chlorite. In rocks at the highest state of strain, ferrimuscovites progressively reacted to a variety lower in Fe and MgO (0.9% FeO) and higher in $Al^{VI}$ (analyses 10, 11; Table 139.1). At this stage the secondary, low-Ti pale biotite in mantles was largely consumed, with chlorite, muscovite, plus ilmenite as products. NX

**Table 139.1.** Analyses of biotite grain and secondary fine-grained products in the deformed mantle

| | *1* | *2* | *3* | *4* | *5* | *6* | *7* | *8* | *9* | *10* | *11* | *12* |
|---|---|---|---|---|---|---|---|---|---|---|---|---|
| $SiO_2$ | 35.95 | 36.40 | 36.17 | 34.59 | 35.56 | 34.88 | 34.92 | 34.58 | 35.40 | 49.45 | 48.98 | 27.52 |
| $Al_2O_3$ | 17.83 | 18.30 | 19.25 | 18.29 | 20.18 | 19.90 | 19.97 | 20.14 | 19.93 | 32.21 | 35.02 | 20.02 |
| $TiO_2$ | 4.05 | 3.95 | 3.30 | 2.18 | 1.59 | 1.30 | 0.52 | 0.23 | 0.05 | bd | 0.02 | 0.02 |
| FeO | 21.75 | 21.29 | 21.89 | 23.71 | 21.92 | 23.10 | 23.60 | 22.80 | 23.64 | 2.47 | 1.30 | 31.88 |
| MgO | 6.45 | 6.25 | 5.71 | 6.65 | 6.64 | 7.16 | 7.23 | 7.36 | 7.63 | 1.62 | 0.72 | 9.56 |
| MnO | 0.30 | 0.32 | 0.33 | 0.38 | 0.32 | 0.37 | 0.34 | 0.39 | 0.40 | 0.02 | bd | 0.59 |
| CaO | 0.04 | 0.02 | 0.02 | 0.05 | 0.03 | 0.02 | 0.03 | 0.03 | 0.04 | bd | bd | bd |
| $Na_2O$ | 0.08 | 0.13 | 0.12 | 0.04 | 0.03 | 0.13 | bd | 0.05 | bd | 0.19 | 0.22 | bd |
| $K_2O$ | 9.82 | 9.63 | 9.95 | 9.71 | 10.04 | 8.92 | 9.17 | 9.25 | 8.45 | 10.94 | 10.48 | bd |
| Total | 96.27 | 96.29 | 96.74 | 95.60 | 96.31 | 95.82 | 95.78 | 94.83 | 95.54 | 96.90 | 96.74 | 89.59 |
| Si | 5.491 | 5.528 | 5.487 | 5.389 | 5.424 | 5.363 | 5.386 | 5.377 | 5.443 | 6.494 | 6.377 | 5.843 |
| Al IV | 2.509 | 2.472 | 2.513 | 2.611 | 2.576 | 2.637 | 2.614 | 2.623 | 2.557 | 1.506 | 1.623 | 2.157 |
| Sum Z | 8.000 | 8.000 | 8.000 | 8.000 | 8.000 | 8.000 | 8.000 | 8.000 | 8.000 | 8.000 | 8.000 | 8.000 |
| Al VI | 0.700 | 0.803 | 0.928 | 0.748 | 1.052 | 0.968 | 1.015 | 1.067 | 1.054 | 3.479 | 3.750 | 2.852 |
| Ti | 0.465 | 0.451 | 0.376 | 0.255 | 0.182 | 0.150 | 0.060 | 0.027 | 0.006 | nc | 0.002 | 0.003 |
| Fe | 2.778 | 2.704 | 2.777 | 3.089 | 2.796 | 2.970 | 3.044 | 2.965 | 3.040 | 0.271 | 0.142 | 5.661 |
| Mg | 1.468 | 1.415 | 1.291 | 1.544 | 1.510 | 1.641 | 1.662 | 1.706 | 1.749 | 0.317 | 0.140 | 3.025 |
| Mn | 0.039 | 0.041 | 0.042 | 0.050 | 0.041 | 0.048 | 0.044 | 0.051 | 0.052 | 0.002 | nc | 0.106 |
| Sum Y | 5.450 | 5.414 | 5.414 | 5.687 | 5.581 | 5.782 | 5.827 | 5.816 | 5.901 | 4.069 | 4.034 | 11.647 |
| Ca | 0.007 | 0.003 | 0.003 | 0.008 | 0.005 | 0.003 | 0.005 | 0.005 | 0.007 | nc | nc | nc |
| Na | 0.024 | 0.038 | 0.035 | 0.012 | 0.009 | 0.039 | nc | 0.015 | nc | 0.048 | 0.056 | nc |
| K | 1.913 | 1.865 | 1.925 | 1.930 | 1.953 | 1.749 | 1.804 | 1.834 | 1.657 | 1.833 | 1.740 | nc |
| Sum X | 1.944 | 1.906 | 1.963 | 1.950 | 1.967 | 1.791 | 1.809 | 1.854 | 1.664 | 1.881 | 1.796 | nc |
| Fe/Mg | 1.918 | 1.940 | 2.184 | 2.033 | 1.880 | 1.840 | 1.858 | 1.768 | 1.768 | 0.862 | 1.013 | 1.906 |

1 (center) to 5 (margin) of a zoned biotite grain. 6 = pale, lower Ti domain where biotite distorted. 7 to 9 = green biotite zoned with respect to Ti, where parent grain stretched out at end. 10 to 11 = progression from ferrimuscovite to lower Fe muscovite with progressive deformation of mantles into trails. 12 = chlorite coexisting with recrystallized biotite and new muscovite in deformed trails. bd = below detection. nc = not calculated.

139

# 140. EFFECTS OF DEFORMATION ON ISOTOPIC EXCHANGE IN QUARTZITE

DAVID L. KIRSCHNER, CHRISTIAN TEYSSIER, ROBERT T. GREGORY, AND ZACHARY D. SHARP

Deformation is one of the most important parameters affecting stable and radiogenic isotopic exchange in silicates and carbonates at low metamorphic grades. The extent of isotopic exchange is dependent on the deformation mechanism(s) and on the presence of a mobile fluid phase. The examples shown in these photographs are of the Heavitree Quartzite (central Australia), which was deformed during one prograde, greenschist-grade metamorphism ($T_{max}$ < 350 to 400°C) in a mid-crustal ($depth_{max} \approx$ 10 to 15 km) ductile duplex (see Dunlap and Teyssier, Plate 86, this volume). Microstructural and oxygen-isotope analyses indicate that isotopic exchange in deformed Heavitree Quartzite occurred through the activation of dissolution-precipitation mass transfer, microfracturing/veining, dislocation-dominated creep, and dynamic recrystallization. Dynamic recrystallization by grain-boundary migration was the most important mechanism for isotopic exchange during deformation of the Heavitree Quartzite (Kirschner and others, 1995).

**140A.** Undeformed quartzite protolith. The well-cemented, medium- to coarse-grained Heavitree Quartzite contains variable amounts of igneous and metamorphic quartz clasts. Laser-based, oxygen-isotope analyses have documented a range in $\delta^{18}O$ values from 11 to 14.5‰ among individual grains measured from the same hand sample. This isotopic heterogeneity has been modified by deformation and interaction with minor amounts of fluids. PPL

**140B.** Pressure solution transfer in quartz protomylonite. Tectonically induced, planar dissolution contact (dc) between two detrital quartz grains parallels mica-defined foliation. Mass transfer involves dissolution and transfer of Si-O species in thin (nanometer wide), continuous fluid films or discontinuous fluid-filled tubules followed by subsequent precipitation. Isotopic exchange occurs because the newly precipitated quartz reflects oxygen isotope equilibrium with the fluid. Pressure solution transfer (at <10% of the rock volume) did not result, however, in large-scale isotopic exchange in the Heavitree Quartzite. NX

**140C.** Microfractures and vein in quartz protomylonite. Quartz vein (v), which is localized in a zone of intense microfracturing, cross-cuts two detrital quartz clasts ($c_1$, $c_2$) and recrystallized quartz matrix. The apparent stretching of detrital grains resulted from nonsystematic crack-seal growth of the vein (cf. the "stretched crystal" model of Durney and Ramsay, 1973). Quartz precipitation in microfractures (microveins) effectively mixed newly precipitated quartz having an oxygen isotope value of $\delta^{18}O \approx 3‰$ with detrital quartz having an isotopic value of $\delta^{18}O \approx 12‰$ (Kirschner and others, 1993). In the absence of quartz precipitation, isotopic exchange between fluid in microfractures and adjacent fracture walls can occur by diffusion-dominated exchange processes, but only at moderate to high temperatures among closely spaced fractures. Diffusional exchange was not an important mechanism for isotopic exchange in the greenschist-grade Heavitree Quartzite. NX

140 A

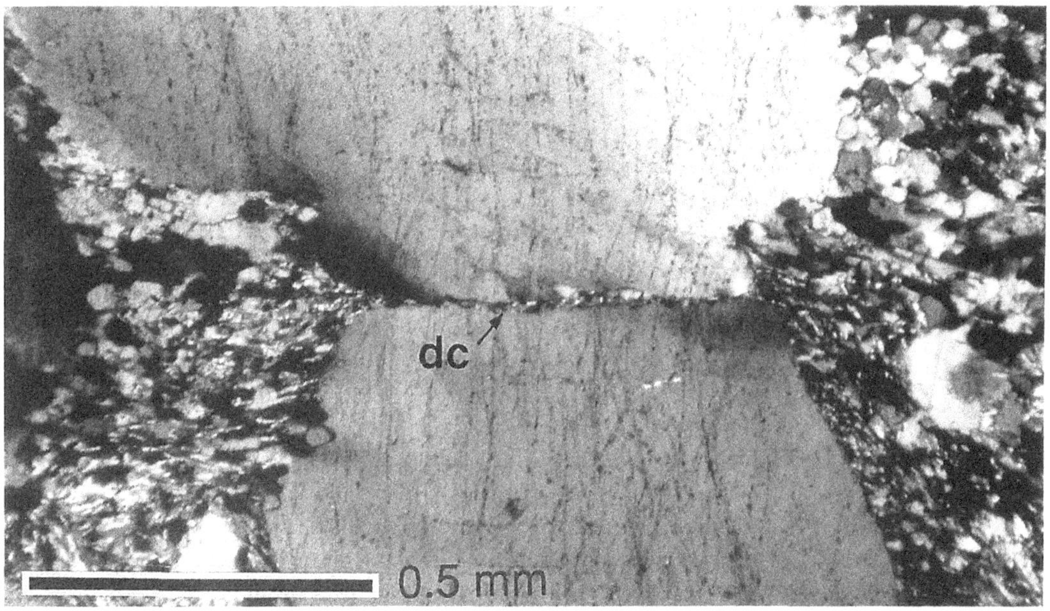


140 B

140 C

**140D.** Grain-shape change and the development of crystallographic preferred orientation (CPO texture) of ribbon grains in quartz protomylonite. Shape change of detrital (sub)spherical grains to ellipsoidal grains increases surface area (up to $\sim$5 times for ribbon grains) and decreases effective grain size for volume diffusion (up to $\sim$10 times for ribbon grains). The decrease in effective grain size is commonly accompanied by the alignment of the quartz *c*-axis with the short or intermediate axis of the ellipsoidal grain (cf. figure 1 in Kirschner and Teyssier, 1992a; Price, 1985). This alignment augments isotopic exchange between grain and grain-boundary fluid, because the *c*-axis is the fastest diffusion direction for oxygen-bearing species by several orders of magnitude in quartz (Giletti and Yund, 1984). These changes (e.g., increased surface area, decreased grain size, alignment of *c*-axis with short axis of ellipsoid grains) did not result in significant isotopic exchange in the Heavitree Quartzite, because the diffusivity of oxygen in quartz was too low during the greenschist-grade deformation.

**140E.** Grain-size reduction due to dynamic recrystallization in quartz mylonite. Dynamic recrystallization by both subgrain rotation and grain-boundary migration during progressive deformation reduced the medium- to coarse-grained Heavitree Quartzite protolith to a fine-grained mylonite with grain size on the order of 30 to 100 μm (Dunlap, 1992). Increased surface area ($\sim$100 times) and decreased grain size ($\sim$10 times) accompanying recrystallization enhance isotopic exchange between quartz and grain-boundary fluid at moderate to high temperatures. This grain-size reduction was unimportant, however, in effecting isotopic exchange in the Heavitree Quartzite, because the calculated diffusivity of oxygen in quartz was too low during the greenschist-grade deformation, and the time scale too short. NX

**140F.** Grain-boundary migration in quartz mylonite. White-mica grains contained in recrystallized quartz grains (this photograph) and grain-shape fabrics in dynamically recrystallized mylonites (140E) are consistent with extensive grain-boundary migration. Material points in the quartz were probably transferred across migrating grain boundaries many times during the protracted deformation. Isotopic exchange between quartz and grain-boundary fluid films or tubules would occur during material transfer across the migrating boundary. Even if the fluid was isolated in an interconnected tubule network along three-grain junctions, fast diffusion of O-bearing species along the grain boundaries (grain boundary diffusion is $\sim 10^5$ times faster than volume diffusion in quartz; Farver and Yund, 1991) would maintain isotopic communication between the fluid and dry sections of the migrating boundaries. Grain-boundary migration during recrystallization was the most important mechanism for facilitating isotopic exchange in the Heavitree Quartzite. NX

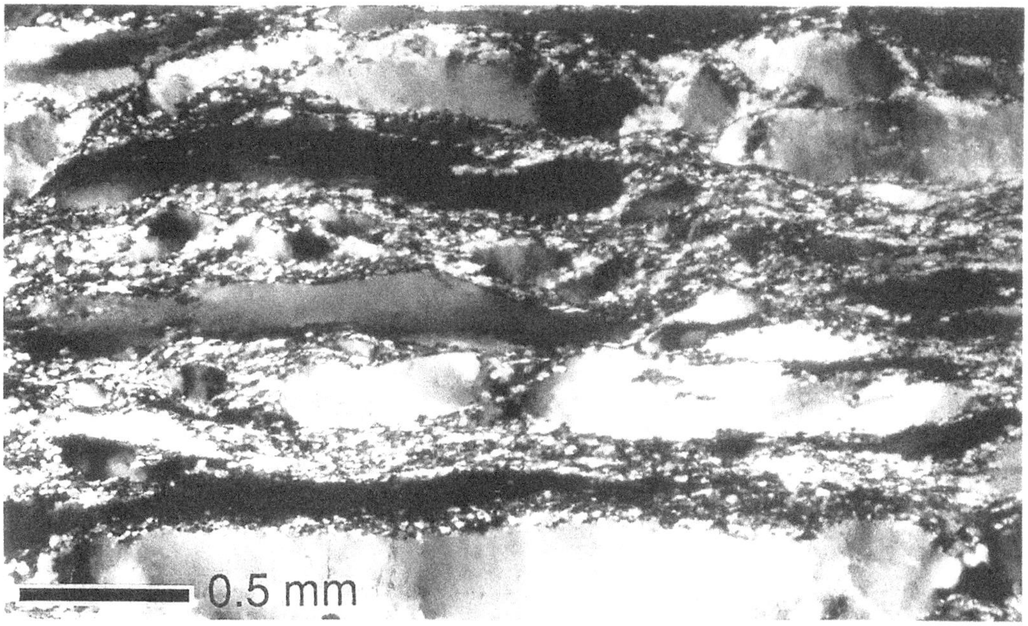

140 D

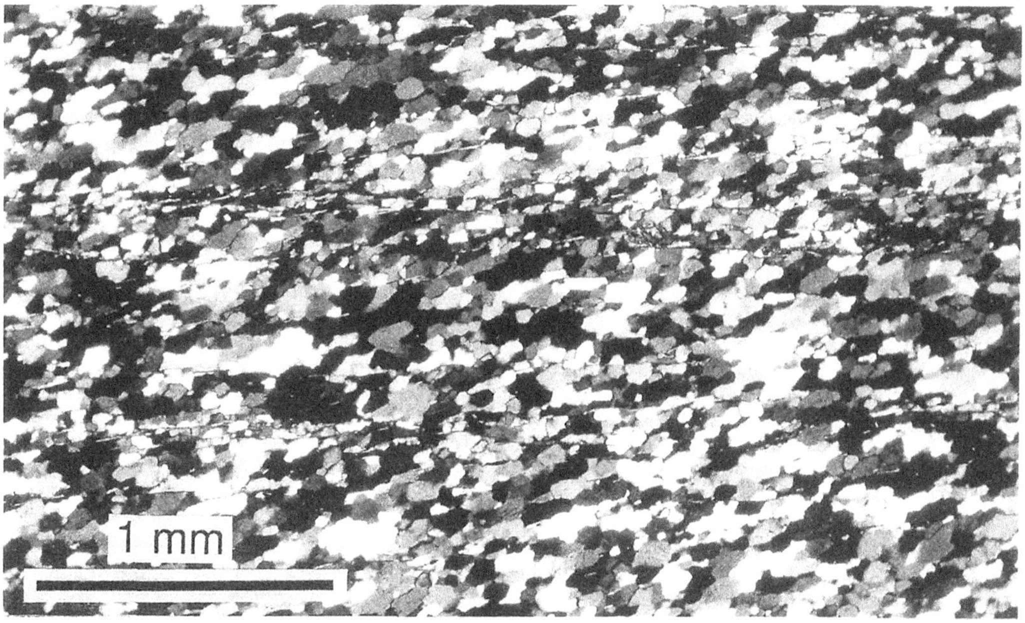

140 E

140 F

# DUCTILE BEHAVIOR

## Microstructural development of mylonitic rocks of specific composition

### Quartzose rocks



Common abbreviations used in regard to the plates in this section include PPL = plane polarized light, NX = crossed nicols, SEM = scanning electron microscope, and TEM = transmission electron microscope.

Quartz is one of the most ductile as well as one of the most studied crustal silicates. It undergoes significant crystal-plasticity under greenschist-facies conditions and develops strong crystallographic preferred orientations (CPOs) that are relatively easy to measure. These plates illustrate microstructures developed in naturally deformed quartzites from lower greenschist through upper amphibolite facies, including some from classic Moine thrust localities in the Scottish Highlands, as well as almost identical microstructures developed in experimentally deformed quartzites. Illustrations include the microstructures developed as a function of increasing strain within a shear zone, as well as the progression in microstructures and CPOs from noncoaxial simple shearing in rocks near discrete thrust zones to coaxial deformation in hotter and deeper rocks. Recent work shows that there are three distinct dislocation creep regimes, based on three different recrystallization mechanisms which operate at different temperatures. The optical and TEM microstructures characterizing these regimes are illustrated for both experimentally and naturally deformed quartzites; these microstructures can be useful qualitative indicators of deformation conditions.

# **141.** PROGRESSIVE DEVELOPMENT OF A QUARTZ MYLONITE

J. CARRERAS, J. M. CASAS, M. LOSANTOS, AND S. H. WHITE

Quartz mylonites at Cap de Creus, Spain, are the result of dynamic recrystallization during mylonitization of quartz veins, originally segregated from pelitic to psammitic schist during prograde Hercynian metamorphism. The mylonitization developed during a late shear-zone deformation associated with the Hercynian orogeny. The deformation occurred under greenschist-facies metamorphic conditions, thereby inducing partial retrogression of earlier, medium-grade mineral assemblages. As shown in Plate 65 and Plate 89, the mylonitization is related to predominantly strike-slip ductile shear zones. Mylonitization of quartz segregation veins induced a strong crystallographic preferred orientation, with *c*-axis patterns evolving with increased shear strain from crossed girdles to an asymmetric single girdle (Carreras and others, 1977; Carreras and Garcia-Celma, 1982).

**141A.** Texture of a nodular quartz vein located 1 m from the boundary of the shear zone. Note the large subgrains within the deformed quartz grains and serrated grain boundaries. NX

**141B.** Texture of quartz vein at the margin of the shear zone. Shear strain ($\gamma$) is less than 1. Incipient recrystallization, manifested by new, small, equant quartz neoblasts, has occurred along the grain boundaries. NX

**141C.** Quartz mylonite from the margin of the shear zone. $1 < \gamma < 5$. Recrystallization is more advanced than in 141B; small, new granoblastic grains coexist with elongated old grains containing numerous subgrains. The thin section is cut parallel to a stretching lineation. NX

**141D.** Quartz mylonite from the center of the shear zone. $\gamma > 10$. This rock consists almost entirely of new grains. A banding parallel to the mylonitic foliation is recognizable by variation in optical orientation of the quartz grains. NX

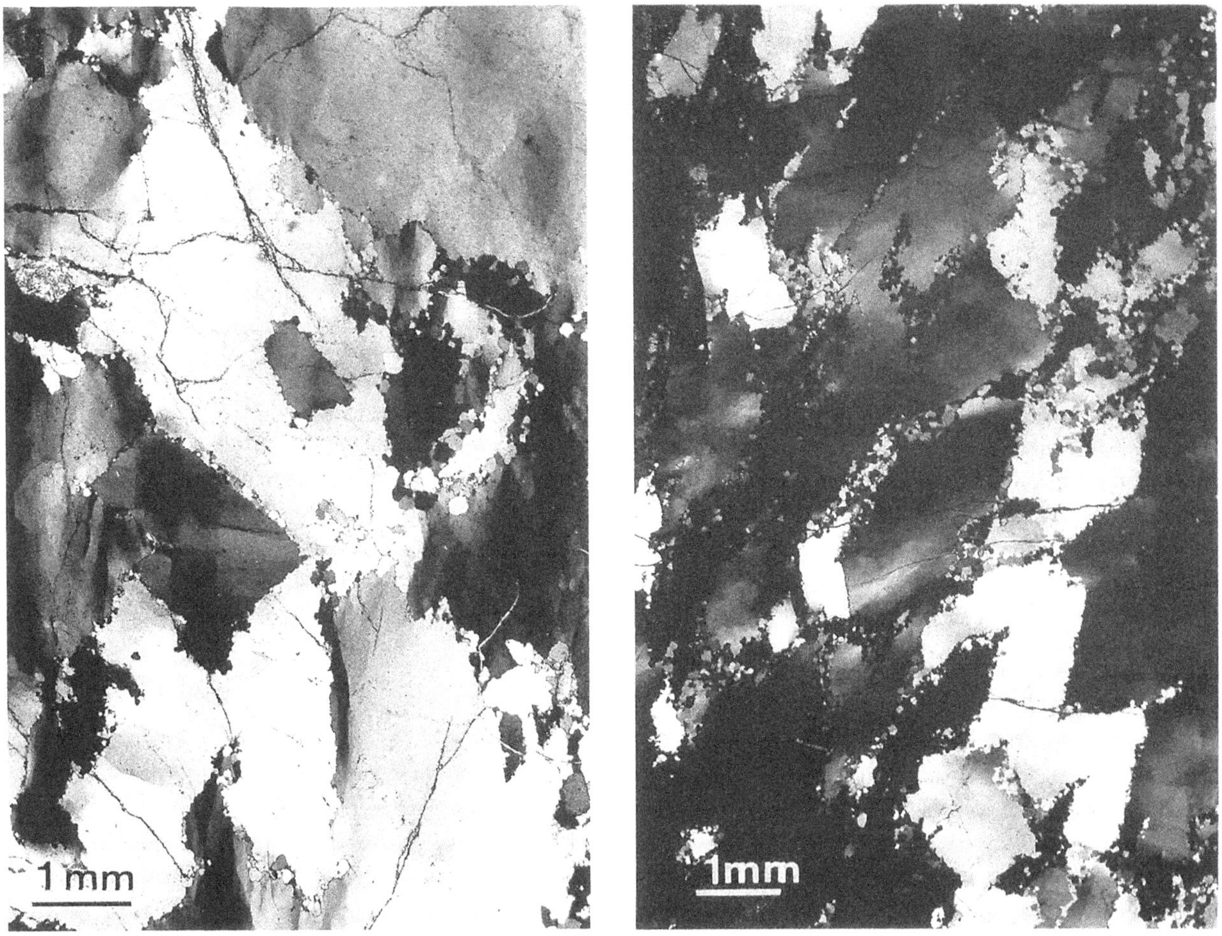


141 A,B

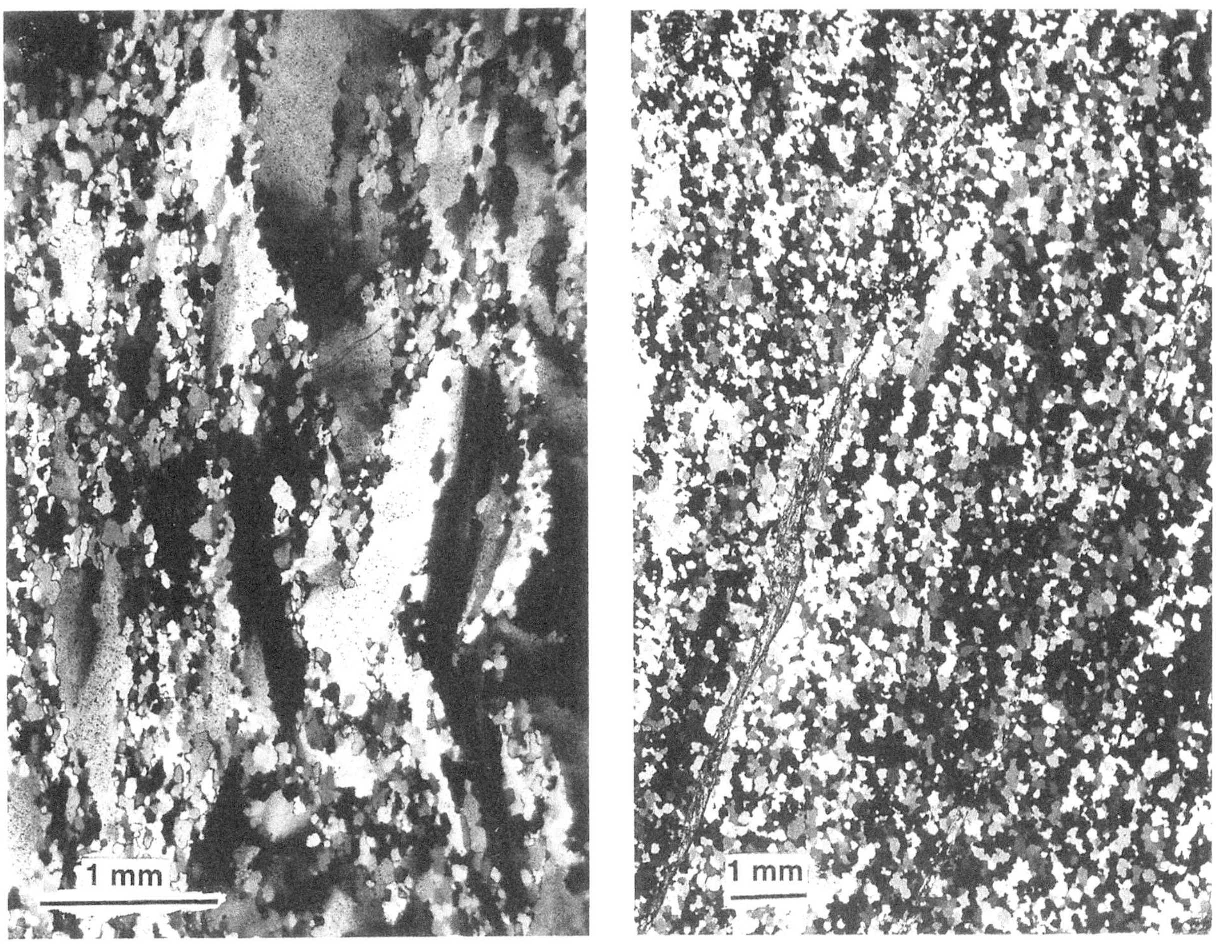


141 C,D

# 142. MYLONITIC FABRIC TRANSITION IN AN EXTENSIONAL TECTONIC SETTING

J. LEE

The northern Snake Range, east-central Nevada, U.S.A. (Figure 142.1), is considered a classic example of a Cenozoic "metamorphic core complex" (e.g., Crittenden and others, 1980), i.e., a region of large-magnitude crustal extension. The most prominent structural feature is the northern Snake Range décollement (NSRD), which is a low-angle detachment fault that juxtaposes an upper plate of complexly normal-faulted Paleozoic and Tertiary strata against a lower plate of ductilely attenuated Proterozoic and Paleozoic metasedimentary rocks and Mesozoic igneous rocks (Miller and others, 1983; Gans and Miller, 1983).

Lower to upper greenschist-facies metamorphism of Tertiary age affected much of the lower plate of the NSRD. This metamorphic event was accompanied by ductile thinning and stretching of lower plate units, resulting in a subhorizontal, bedding-parallel mylonitic foliation and a WNW-ESE-trending mineral elongation lineation. Mesoscopic structures and finite strain measurements indicate a dramatic west-to-east increase in strain across the range, from a low on the west flank of 6:1 (X:Z) (Negro Creek) to a high on the east flank of 600:1 (X:Z) (Hendry's Creek) (Lee and others, 1987). The nature and geometry of quartz microstructures and *c*-axis fabric patterns from mylonitic rocks show a systematic variation with increasing strain across the range. This variation documents a complex extensional strain history of early coaxial strain overprinted by deformation with an eastward increasing component of top-to-the-east noncoaxial strain (Lee and others, 1987).

The photomicrographs and accompanying *c*-axis fabric diagrams illustrate the transition of mylonitic fabrics in lower-plate quartzites from moderate strain and coaxial fabrics on the west flank to high strain and noncoaxial fabrics on the east flank. All photomicrographs are from thin sections cut parallel to the elongation lineation and perpendicular to the foliation. WNW is to the left.

**142A.** Moderate-strain quartzite, Negro Creek, northern Snake Range, Nevada (Lee and others, 1987). This sample comes from a moderate-strain quartzite (~6:1) on the western flank of the range (Figure 142.1). Quartz grains are stretched and flattened into ribbonlike grains, which define a single foliation, $S_1$, parallel to the hand-specimen foliation. These grains exhibit undulatory extinction, deformation lamellae, variable amounts of recrystallization into smaller grains along grain boundaries, and grain-boundary migration. *C*-axes measured from elongate grains yield symmetrical skeletal outlines and density distribution patterns around the finite strain axes (Figure 142.2a). Also present are globular quartz grains (g), around which elongate grains anastomose and flatten and whose measured *c*-axis orientations are at small angles to the Z axis, subvertical to the foliation (Figure 142.2b). Globular grain shapes are preserved during coaxial deformation because of low resolved shear stress along the most easily activated slip planes (basal and prism planes). In contrast, during noncoaxial deformation, globular grains rotate relative to the kinematic framework, and the resolved shear stress varies along slip planes, leading to deformation and elongation. At slightly higher strains (~8:1), quartzites within Negro Creek display the same foliation and contain globular grains with the same crystallographic orientation as that of the lower-strain quartzites (see Figure 142.2b). However, these higher-strain quartzites exhibit *c*-axis patterns that more closely approximate asymmetrical single girdle patterns (Figure 142.2c). Lee and others (1987) interpreted these relations among microstructures, crystallographic orientation, and finite strain axes in Negro Creek quartzites as recording coaxial deformation with a minor noncoaxial overprint. NX

142 A

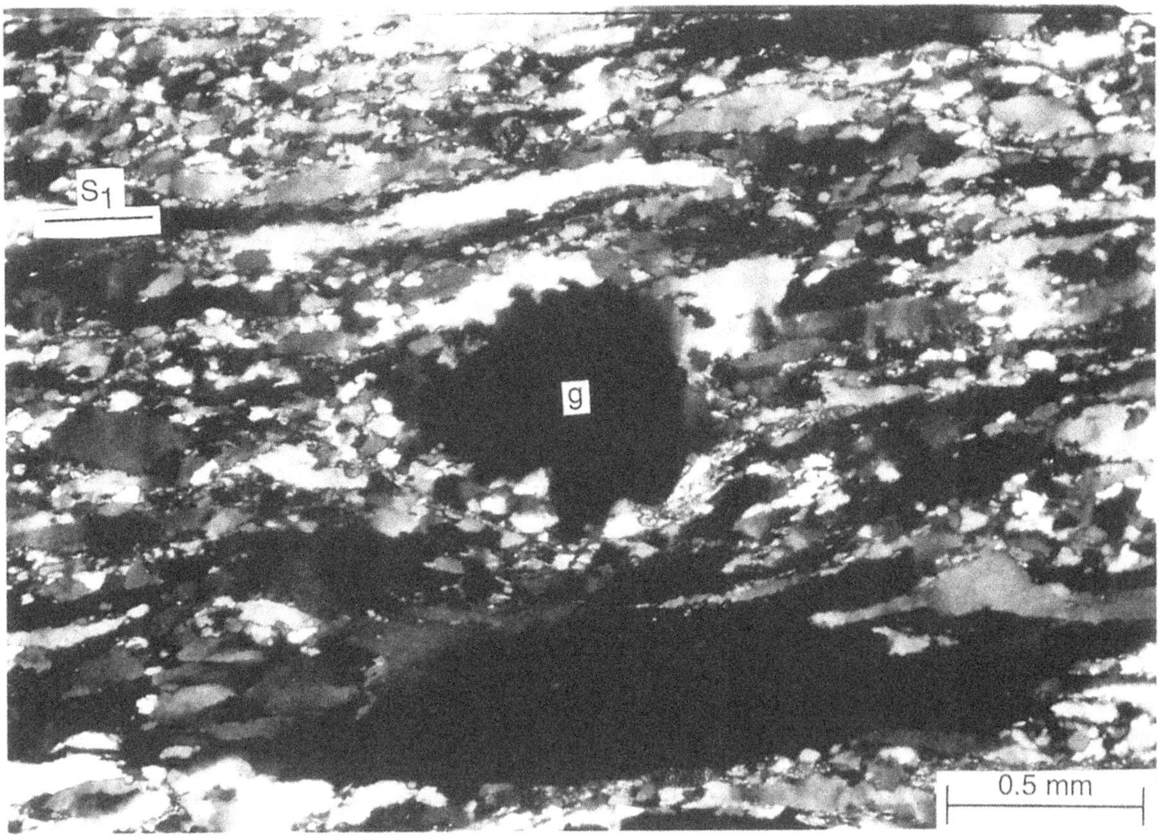


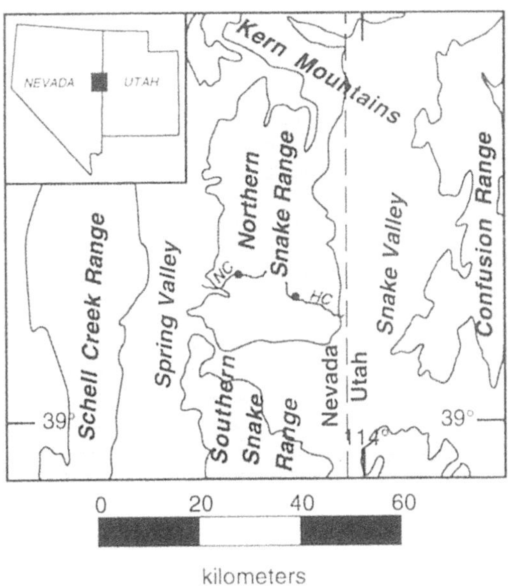


**Figure 142.1.** Index map for east-central Nevada showing locations of the Schell Creek, northern Snake, southern Snake and Confusion Ranges, and Kern Mountains Solid circles show approximate locations of samples discussed in the text NC, Negro Creek, HC, Hendry's Creek

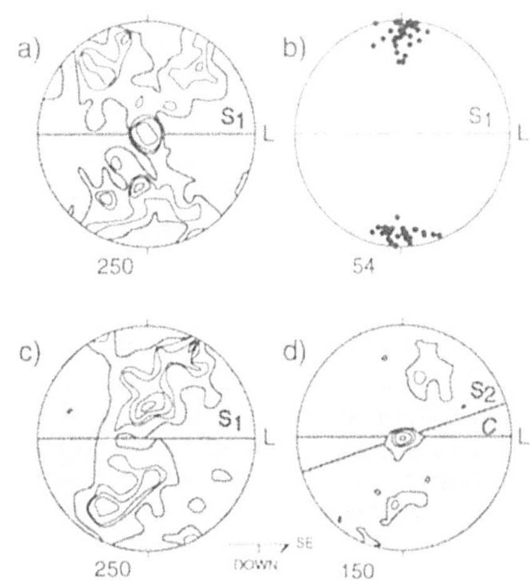


**Figure 142.2.** *C*-axis fabric patterns from Negro and Hendry's creeks (a) *C*-axis fabric pattern from Negro Creek (sample A) (b) Synoptic plot of globular quartz grain *c*-axis measurements from samples in Negro Creek (c) *C*-axis fabric pattern from a slightly higher strain quartzite, Negro Creek (c) *C*-axis fabric pattern from Hendry's Creek (sample B) Foliations are vertical and mineral elongation lineation (L) is horizontal Geographical orientation of stereonets and number of *c*-axes measured are indicated After Lee and others, 1987

**142B.** High-strain mylonitic quartzite, Hendry's Creek, northern Snake Range, Nevada (Lee and others, 1987). This high-strain (~14:1) sample exhibits textures typical of high structural levels at Hendry's Creek (Figure 142.1). At least two foliations are apparent in thin section. The first foliation, $S_1$, is defined by stretched and flattened, ribbonlike, quartz grains with X:Z ratios up to 30:1 that developed subgrains or were variably recrystallized at their boundaries. These ribbons are parallel to the mesoscopic foliation. The second, less well-developed foliation is a discrete, spaced C foliation defined by the strung-out and recrystallized tails of "mica fish" (mf). The C foliation dips more steeply to the east with respect to the mesoscopic foliation.

**142C.** High-strain mylonitic quartzite, Hendry's Creek, northern Snake Range, Nevada (Lee and others, 1987). This high-strain (~14:1), L-S, quartzite is from deeper structural levels than that in 142B above (Figure 142.1). This sample also exhibits two foliations in thin section. The most obvious foliation, in contrast to 142B above, is a discrete, spaced C foliation defined by strung-out and recrystallized tails of mica fish (center of photomicrograph). The C foliation parallels the mesoscopic foliation; the mica fish dip more steeply to the northwest. At deeper structural levels, quartz ribbons have recrystallized dynamically to smaller quartz grains that are equant to slightly elongate (aspect ratios up to 3:1). These grains define a second foliation, $S_2$, that is between and oblique to the C-foliation surfaces, and dips more steeply to the northwest. The asymmetry of the C foliation and the sense of obliquity of the quartz grain-shape foliation indicate top-to-the-southeast noncoaxial strain. No globular quartz grains are present in samples from Hendry's Creek. *C*-axis fabric patterns exhibit a strong Y maximum and a single-girdle "dog-leg" shape that is asymmetric around the finite strain axes, indicating top-to-the-southeast noncoaxial strain (Figure 142.2d).

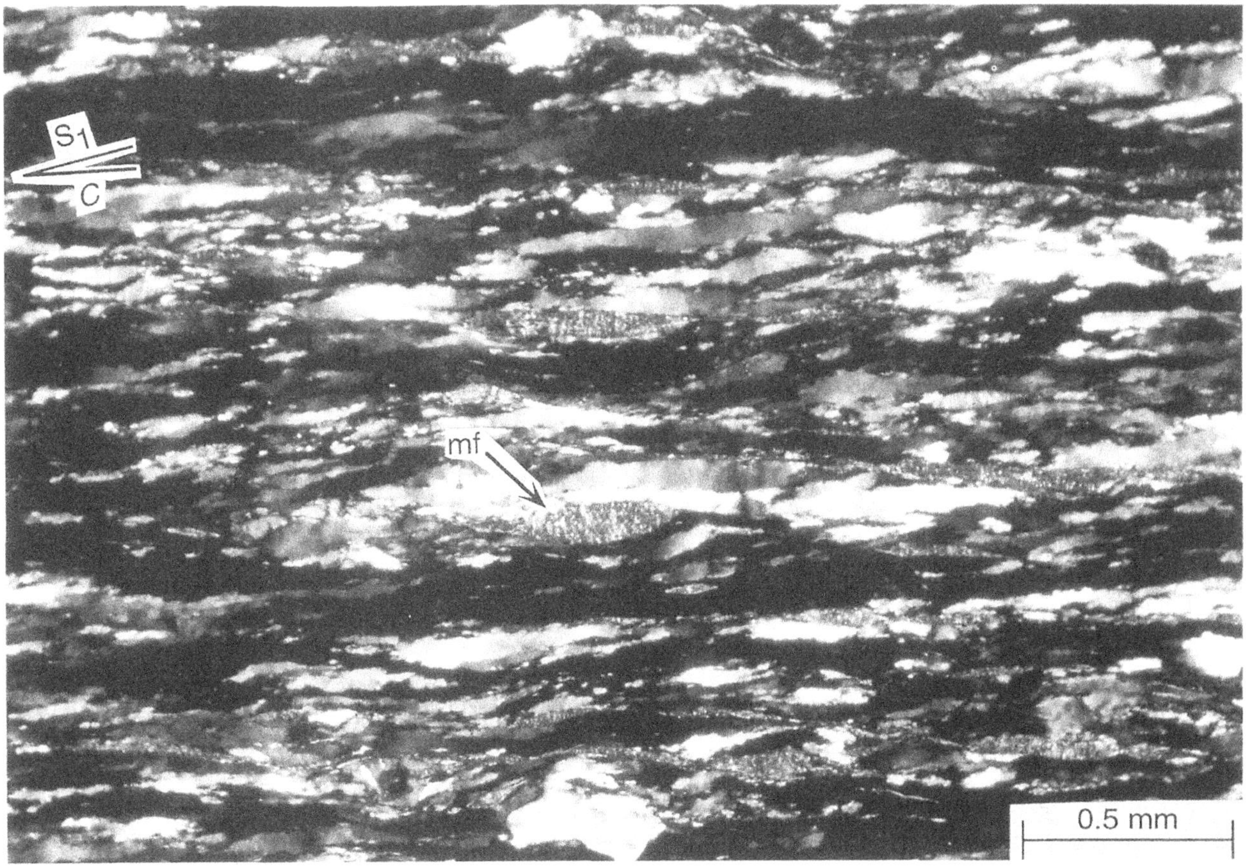

142 B

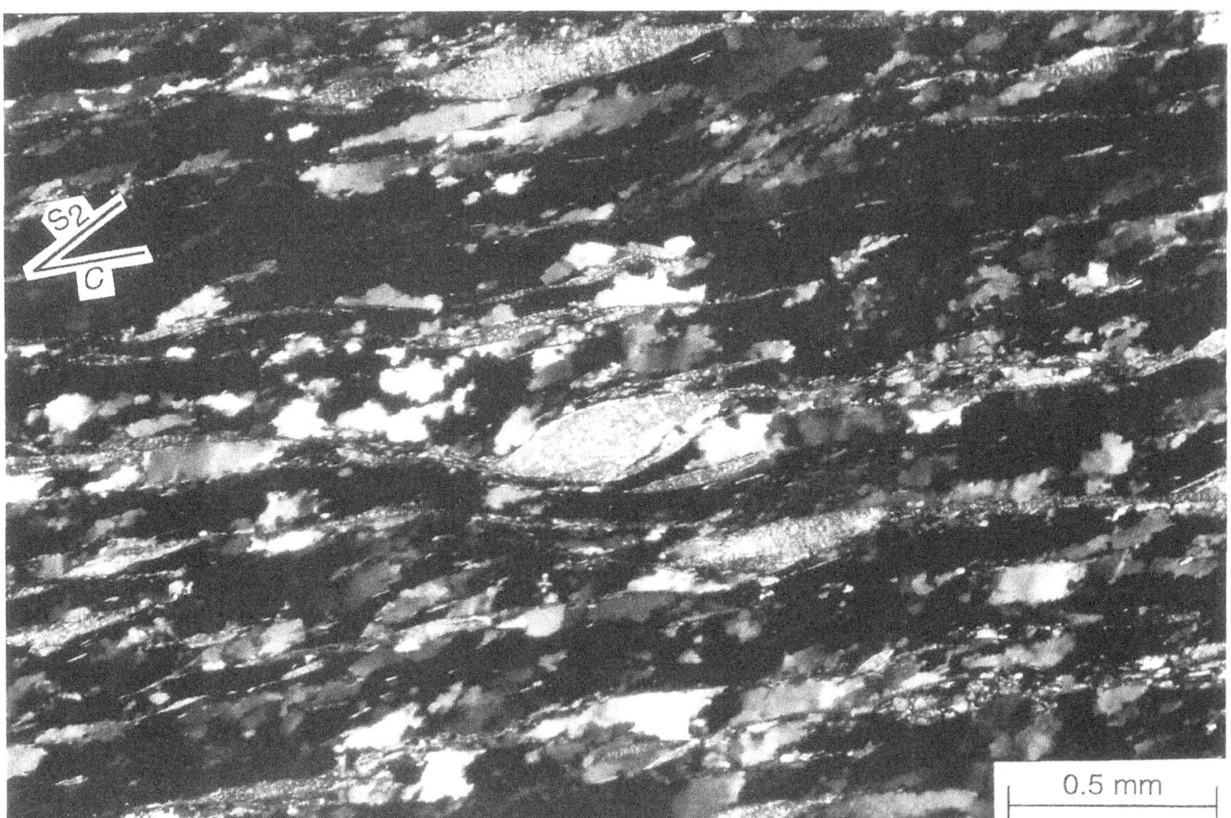

142 C

# 143. QUARTZITE FABRIC TRANSITION IN A CORDILLERAN METAMORPHIC CORE COMPLEX

ALLEN J. MCGREW AND MARTIN CASEY

Photomicrographs 143A–143F record fabric variations in quartzite with increasing structural depth in a >1 km thick, amphibolite-facies, normal-sense shear zone in the East Humboldt metamorphic core complex, Nevada (Figure 143.1). This shear zone and the overlying detachment system unroofed an infrastructure of high-grade, migmatitic gneiss during Oligocene to early Miocene extension (Dallmeyer and others, 1986; Wright and Snoke, 1993; McGrew and Snee, 1994). Thermobarometric constraints from near the base of the mylonitic zone record deformation conditions of 550°–620°C and 300–400 MPa (Hurlow and others, 1991). Sample WBC6 (143A and 143B) characterizes the mylonitic zone, whereas sample 8706-1 (143C and 143D) is transitional in nature, and sample 8727-3 (143E and 143F) represents the infrastructure (Figure 143.1). These three samples record a decreasing contribution from simple shear strain and an actual reversal in shear sense at the deepest structural levels. Inferred variations in quartz deformation mechanisms probably reflect decreasing strain rate and increasing temperature with depth beneath the detachment. We argue that the East Humboldt Range shear system represents a fundamental boundary between an upper crust deforming primarily by rigid block translation and a deeper crust deforming by regional-scale ductile flow.

**143A and 143B** These photomicrographs illustrate quartzite from the upper mylonite zone (sample WBC6, at two different scales), <50 m beneath the East Humboldt Range detachment fault, West Fork of Boulder Creek, East Humboldt Range, Nevada. X = lineation, oriented 282°, 5° in geographic coordinates; Y = transverse direction; Z = pole to foliation, oriented 273°, 29°. Mean quartz grain size is approximately 150 μm. Asymmetric mica and feldspar porphyroclasts record WNW-directed (i.e., dextral) shear. Trails of fine, recrystallized muscovite, biotite, quartz, and feldspar define shear planes spaced at 200–1500 μm, and recrystallized quartz shows a weak grain-shape fabric inclined 25°–35° to this spaced foliation, again implying WNW-directed shear (see inset). NX

X-ray texture results (Figure 143.2) show a kinked single girdle of *c*-axes which is also asymmetric toward the WNW. An extremely strong point maximum of *c*-axes parallel to Y forms the center of a near-single-crystal preferred orientation, with the dominant *a*-axis maximum inclined 25° clockwise from X and with the dominant m pole position inclined 5° anticlockwise from X.

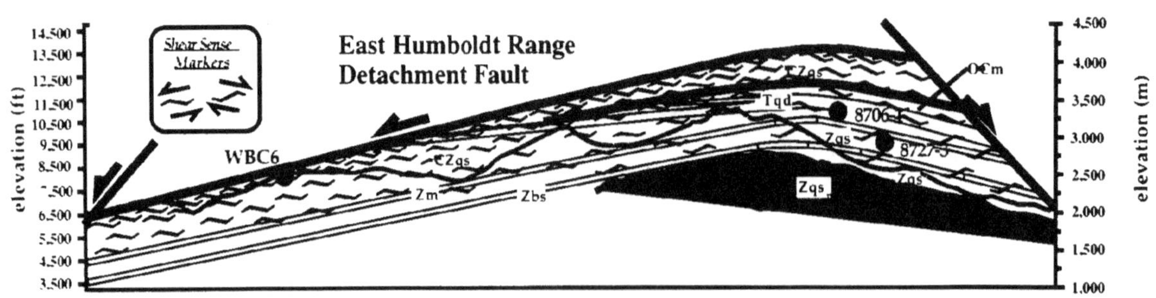


**Figure 143.1.** Schematic cross section through the central East Humboldt Range, Nevada, showing the structural position of samples relative to the detachment fault (Note Sample WBC6 was collected along the line of section but samples 8706-1 and 8727-3 were projected from several kilometers to the north.) Wave pattern indicates shear sense, with closeness of spacing reflecting intensity of deformation

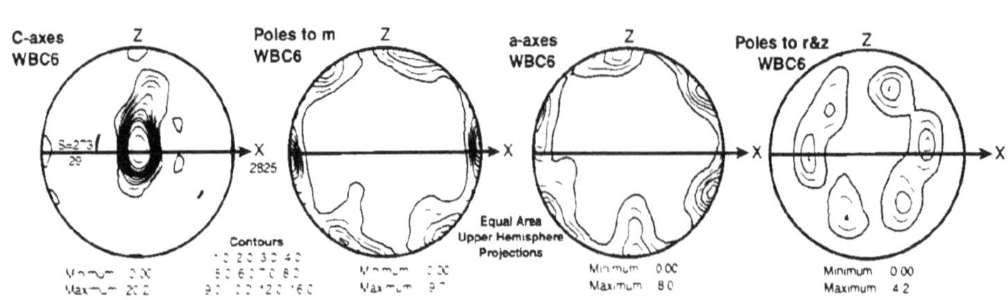


**Figure 143.2.** Selected pole figures for sample WBC6, regenerated from X-ray data and contoured in multiples of uniform distribution

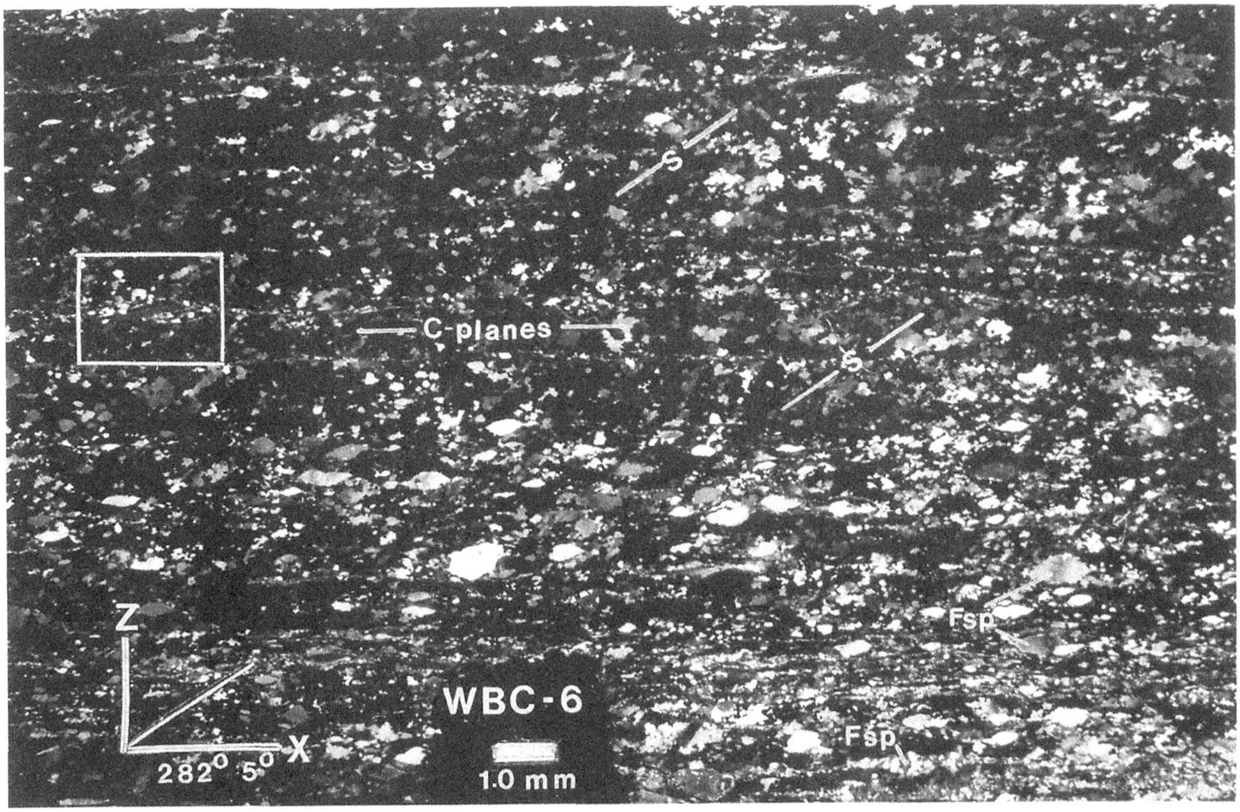


143 A

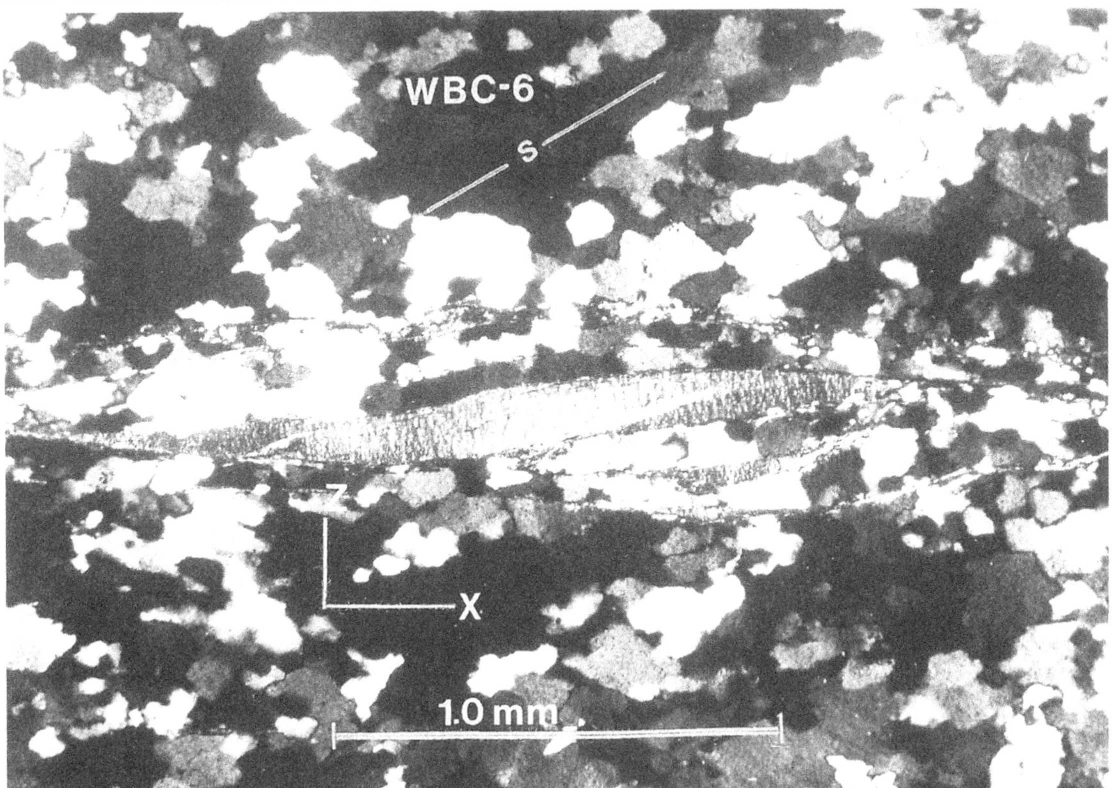


143 B

**143C and 143D.** These photomicrographs illustrate a gneissic quartzite (sample 8706-1, at two different scales), from the boundary between the mylonitic zone and the infrastructure, approximately 750–1000 m beneath the East Humboldt Range detachment fault, East Humboldt Range, Nevada. The rock is transected by thin seams of fine recrystallized grains spaced at 1000–2500 μm parallel to macroscopic foliation. The mean quartz grain size, in the range 500–700 μm (with very coarse grains >1500 μm), contrasts dramatically with that of sample WBC6 from the upper mylonitic zone (143A and 143B). Subgrains and undulose extinction, though present, are far less common than in the mylonites. Cross-hatched microstructure is common; this consists of interlocking cross-mosaics of quartz grains with crudely rectilinear grain boundaries oriented at moderate angles to bulk foliation, and is characteristic of high-temperature deformation (Lister and Dornsiepen, 1982).

Asymmetric microstructures are virtually absent throughout most of the thin section, but the lower part of the field of view (143C) shows a weak, asymmetric grain-shape fabric defined by elongate grains inclined at 15°–20° to bulk foliation, suggesting domainal WNW-directed, noncoaxial strain, as in the overlying mylonites. Optical inspection shows that this domain is also distinguished crystallographically by a higher proportion of grains with *c*-axes lying in the XZ plane and inclined approximately 45° anticlockwise (i.e., ESE-ward) from X (these grains appear bright in 143C and 143D). Grains in this orientation have basal planes dipping 45° clockwise from X, thus placing the *c*-axis and the basal plane parallel to the two preferred grain-boundary orientations of the cross-hatched microstructure. NX

Due to the coarse grain size of this sample, X-ray texture data were combined from four separate scans to improve grain-count statistics and to average out domainal variations. The combined X-ray pole figures (Figure 143.3) show an extremely strong *c*-axis maximum oriented subparallel to Y, with little evidence of the subsidiary grain population with *c*-axes oriented in the XZ plane. Although somewhat resembling the strong Y-maximum fabrics from higher structural levels, this texture shows much higher symmetry, particularly evident in the *m*, *a*, and *r* + *z* pole figures.

Grain-boundary migration probably played a pivotal role in the development of both the texture and the microstructure, favoring growth of crystals in preferred orientations, and thus facilitating both the coarse grain size and the strong point maximum texture. The symmetry of the texture suggests a nearly coaxial strain path, with equal resolved shear stresses (and presumably equal activities) on antithetically paired prism ⟨*a*⟩ slip systems. Because macroscopic fabrics are parallel to those from the mylonitic zone, and because nearby Tertiary orthogneisses show similar fabric relationships, we believe that this texture offers a reasonable estimate of the regional background strain path over which mylonitization developed.

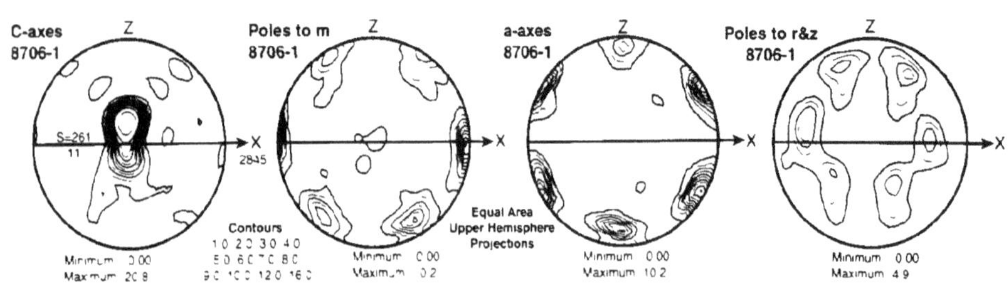


**Figure 143.3.** Selected pole figures for sample 8706-1, regenerated from four X-ray scans and contoured in multiples of uniform distribution

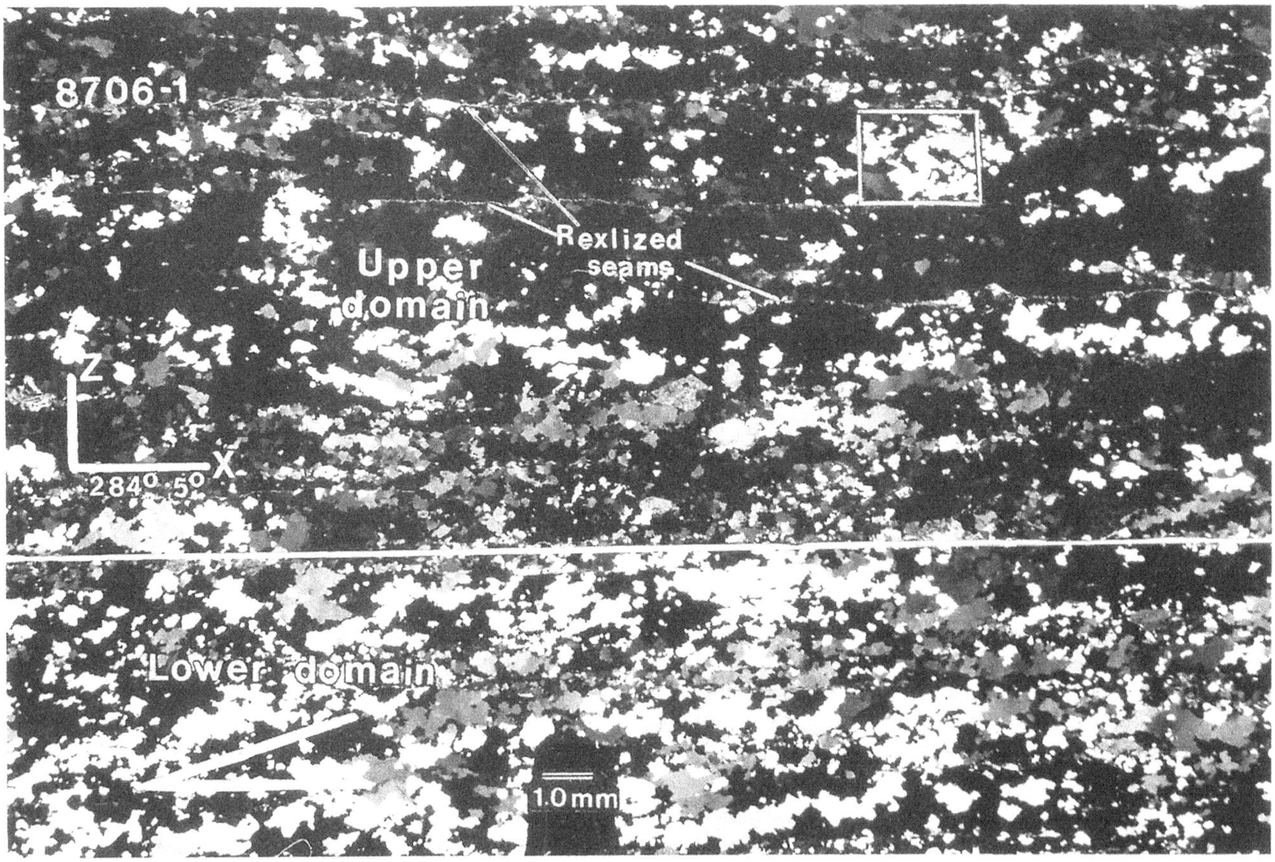
8706-1
Rexlized seams
Upper domain
Z
X
284° 5°
Lower domain
1.0 mm

143 C

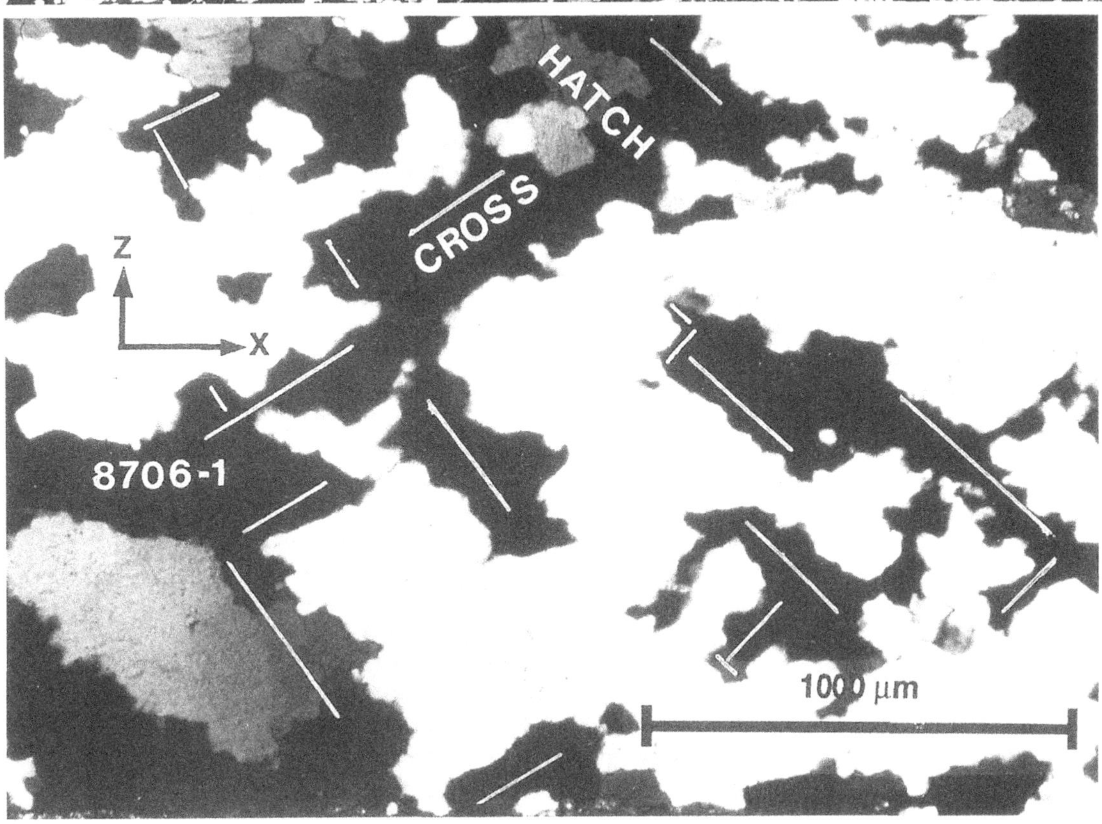
HATCH
CROSS
Z
X
8706-1
1000 μm

143 D

**143E and 143F.** These photomicrographs illustrate a gneissic quartzite (sample 8727-3, at two different scales) from the infrastructure, 1000–1250 m beneath the East Humboldt Range detachment fault, East Humboldt Range, Nevada. Quartz grain size is somewhat domainal in this sample, ranging from approximately 500 μm in the more feldspathic upper domain to >1000 μm in the more biotite-rich, lower domain. Despite the coarse grain size, very fine neoblasts are also present, as are well-developed subgrains, suggesting that the high-temperature deformation characterized by rapid grain-boundary migration was followed by grain-size reduction at lower temperature. Noncoaxial kinematic indicators are uncommon, but some micas show asymmetric shapes, and the more recrystallized upper domain seems to show a subtle quartz grain-shape fabric inclined at 25°–30° to bulk foliation. This suggests weak, domainal noncoaxial strain directed toward the ESE, antithetic to that of the overlying mylonitic zone. Although weakly developed in this sample, ESE-directed kinematic indicators occur locally throughout the infrastructure and suggest a reversal in general shear sense at the deepest structural levels. NX

Due to the coarse grain size of this sample, five X-ray scans were combined to obtain the regenerated pole figures (Figure 143.4), and these agree well with optical results. The crystallographic preferred orientation for this sample consists of three orthogonal maxima, with one maximum near Y and the other two approximately equally inclined to X in or near the XZ plane. Such *c*-axis crystallographic preferred orientations are quite rare and usually occur in granulite-facies terranes (e.g., Lister and Dornsiepen, 1982); complete crystallographic results for such a texture have not been reported previously. The unequal populations of the three maxima impart a weak asymmetry toward the ESE, and the slight misorientation of two of the maxima relative to macroscopic fabric directions imparts a triclinic character to the external symmetry.

An intriguing aspect of these data is the amount of position switching of the *c*-, *m*-, and *a*-axes between the three principal maximum positions. For example, crystals in the central maximum tend to place an *a*-axis and a pole to a first-order prism parallel to each of the other two maxima. Conversely, crystals in the two perimeter maxima tend to place either an *m* pole or an *a*-axis parallel to the other perimeter maximum. Grain-boundary migration probably played an important role in developing this texture by favoring growth of crystals in preferred orientations. If a general, only slightly noncoaxial deformation path is assumed, then *c*-axis positions near Y are well oriented for antithetic slip on paired prism ⟨*a*⟩ systems. Similarly, the *c*-axis maxima positions in the XZ plane are well disposed for paired, antithetic activity on prism ⟨*c*⟩ and basal ⟨*a*⟩ slip systems (Lister and Dornsiepen, 1982). Consequently, we argue for a general, nearly coaxial deformation path with a small contribution from ESE-directed shear.

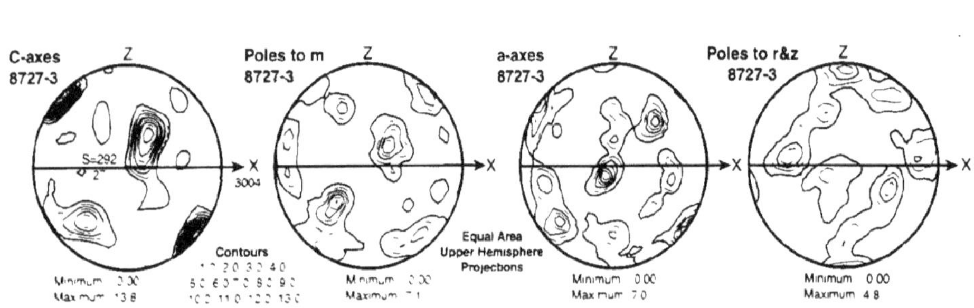


**Figure 143.4.** Selected pole figures for sample 8727-3, regenerated from five X-ray scans and contoured in multiples of uniform distribution

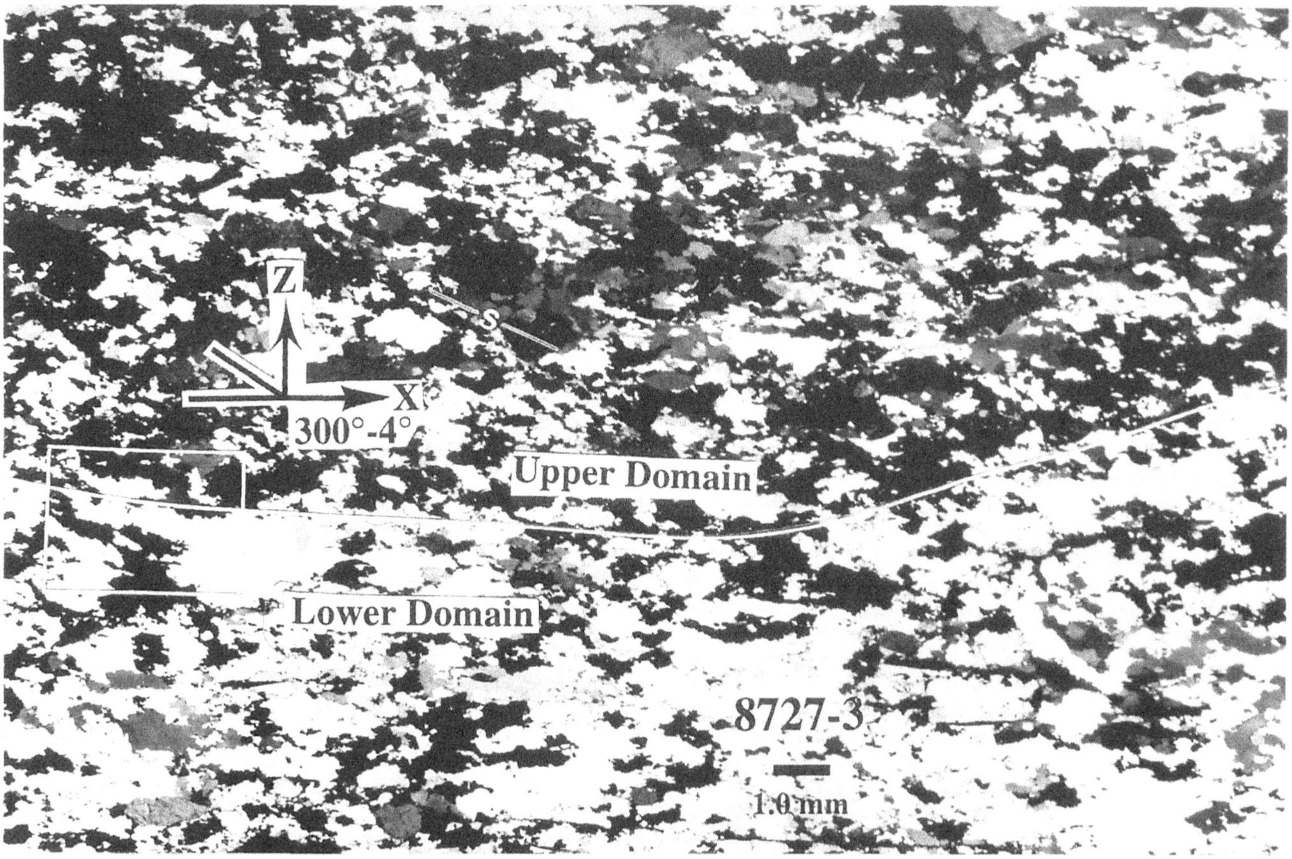
Z
X
300°-4°
Upper Domain
Lower Domain
8727-3
1.0 mm

143 E

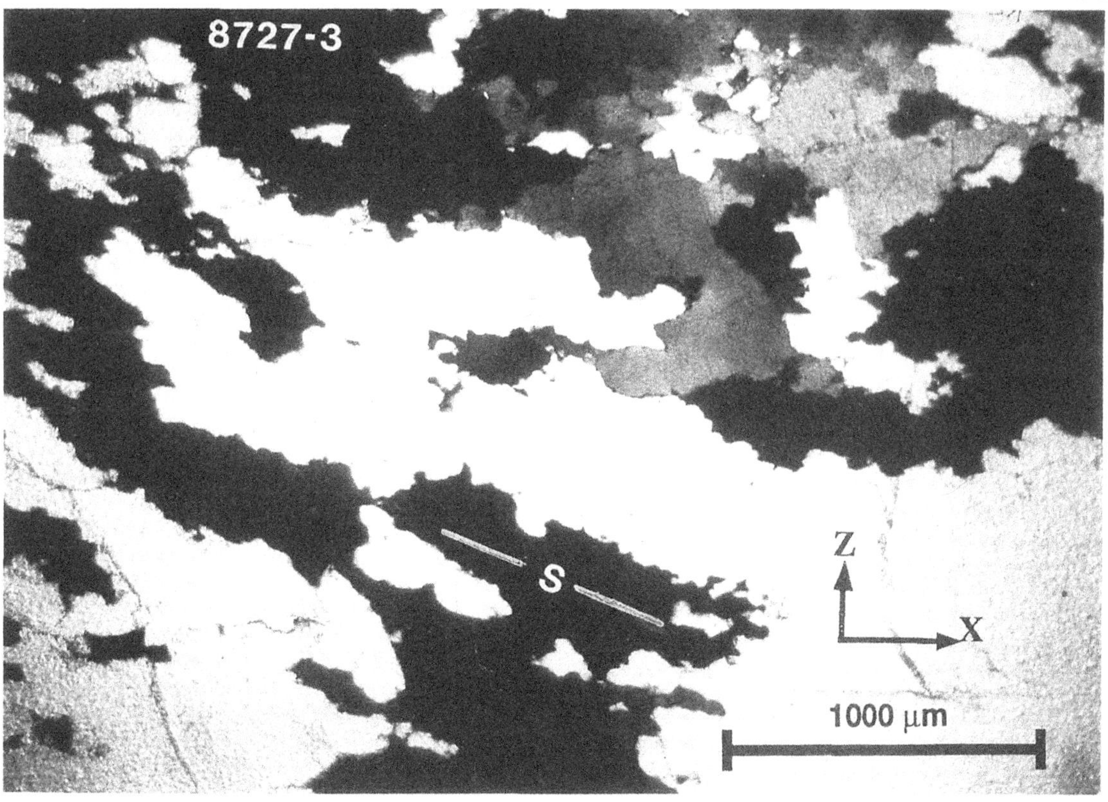
8727-3
S
Z
X
1000 μm

143 F

# 144. QUARTZ MYLONITES FROM THE MOINE THRUST ZONE AT THE STACK OF GLENCOUL, NORTHWEST SCOTLAND

R. D. LAW

The Moine thrust zone consists of a series of west-northwest-directed thrusts of which the Moine thrust is considered to be the oldest. The Moine thrust is structurally the highest thrust; it crops out farthest to the east and carries Moine (late Proterozoic) metasedimentary rocks over the foreland succession contained in the lower levels of the thrust zone (Figure 144.1). The mylonites exposed along the Moine thrust at the Stack of Glencoul are derived from the overriding Moine rocks and the Cambrian quartzites of the footwall. These mylonites are characterized by a strongly developed foliation, which dips gently to the east-southeast, and a grain-shape stretching lineation which plunges down the dip of the foliation planes subparallel to the Moine thrust-zone transport direction. Temperatures of mylonite formation are estimated at 300–350°C (Johnson and others, 1985).

All photographs and micrographs are viewed toward the east-northeast, perpendicular to transport direction.

**144A.** Exposure of the Moine thrust at the Stack of Glencoul (Law and others, 1986). Mylonitic Cambrian quartzite (Q) underlying similarly deformed Moine rocks (M); arrow marks position of Moine thrust. *Since 1986 large sections of this spectacular exposure have been destroyed through thoughtless sample collection by unidentified geologists.*

**144B.** Contact between Moine rocks (M) and mylonitic Cambrian quartzite (Q) with arrow marking the position of the Moine thrust. Polished surface cut perpendicular to foliation and parallel to lineation; note east-southeast-dipping listric extensional fault, which offsets mylonitic foliation in the quartzites (sample SG.1 of Law and others, 1986).

**144C.** Optical micrograph of thin section cut from the polished surface of sample SG.1 illustrated in Plate 144B. Moine thrust plane separates the phyllosilicate-rich Moines (M) of the hanging wall from the mylonitic Cambrian quartzites of the footwall. Note minor listric extensional fault zone (e-e*) with displacement down to the east-southeast, which offsets the mylonitic foliation in the quartzites. This extensional fault zone, which is marked by extremely fine-grained (<10 μm), dynamically recrystallized quartz, predates a quartz vein (V-V*) which truncates the Moine thrust, but is sheared over to the west-northwest in the Moine rocks. The deformation of this vein and the extensional fault zone indicate intermittent changes in the strain path and deformation mechanisms from subhorizontal thrust sheet extension (by dislocation creep and veining) to west-northwest-directed shearing during the late stages of mylonite evolution (Knipe, 1990). NX

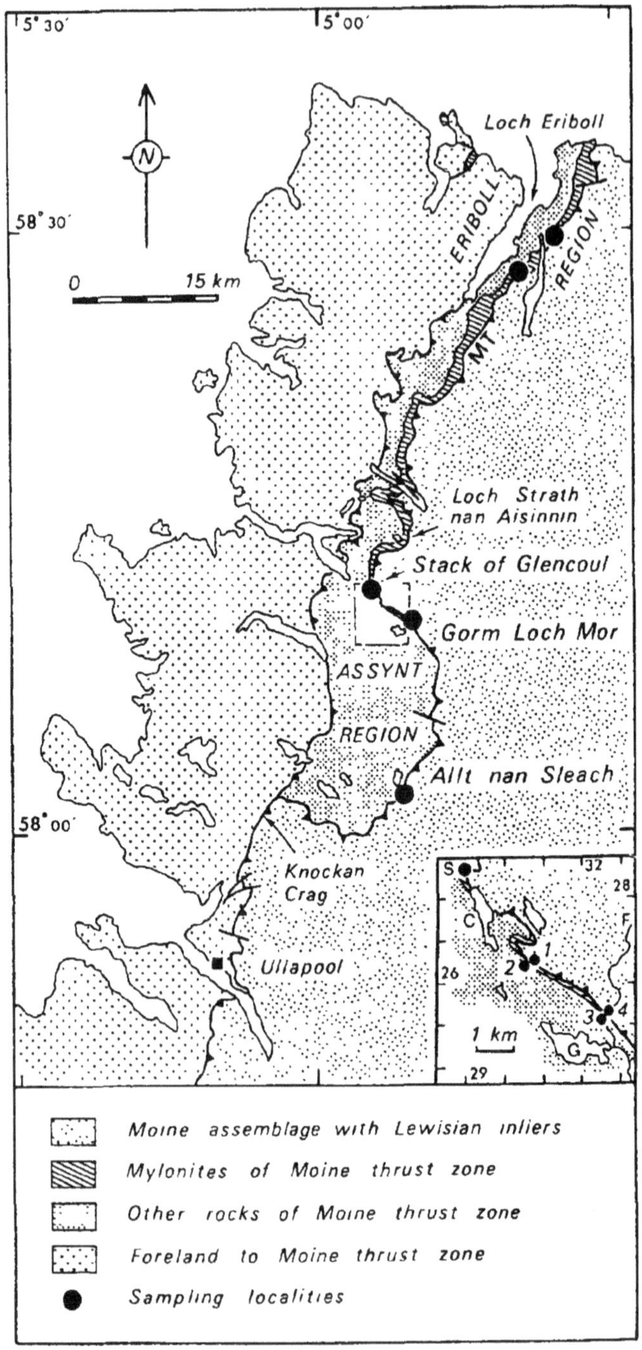


**Figure 144.1.** Simplified geologic map of the northern part of the Moine thrust zone (after Law and others, 1986). Location of exposures of the Moine thrust plane at the Stack of Glencoul and in the Allt nan Sleach stream section (see Plate 145) are indicated.

144 A

144 B

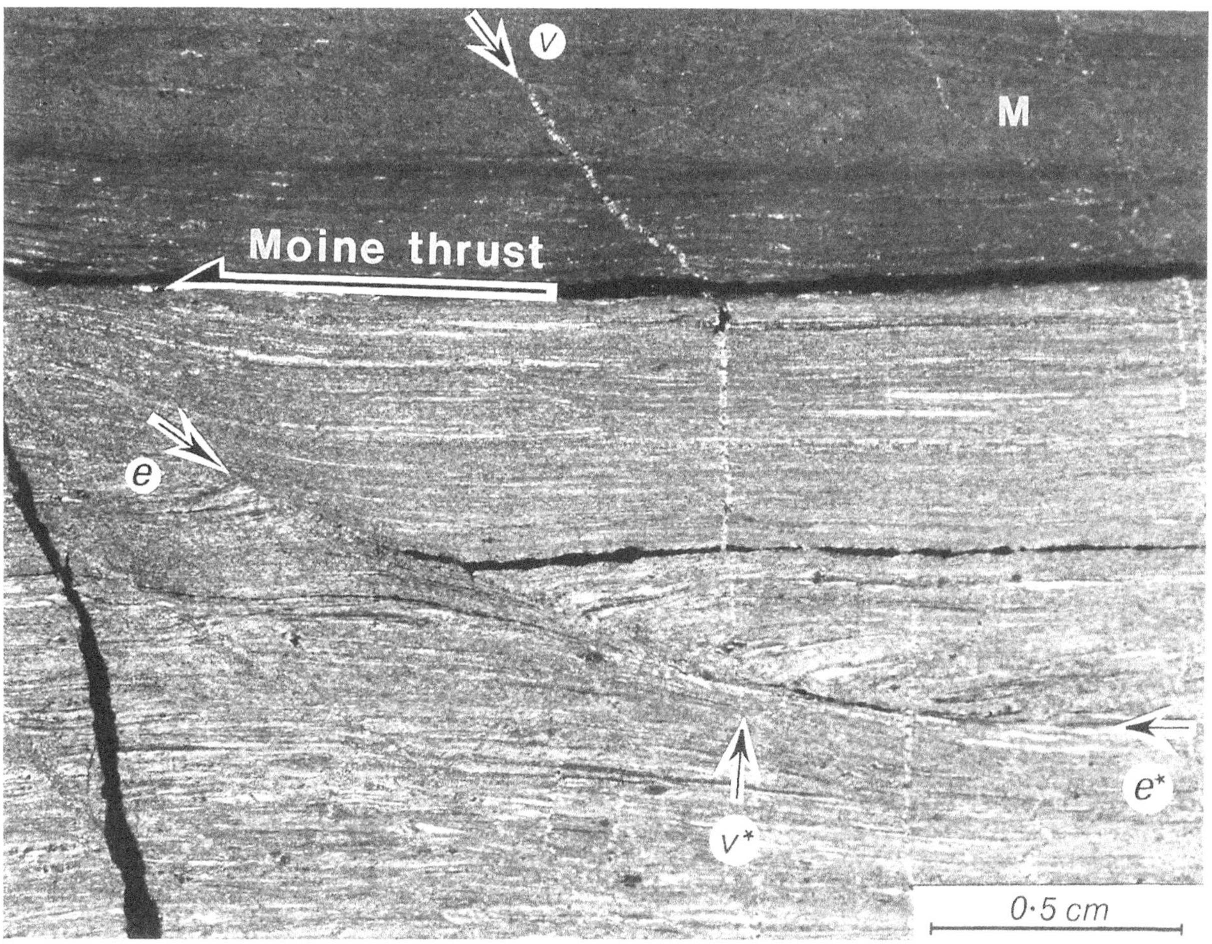


144 C

Cambrian quartzites located at less than 15 cm beneath the Moine thrust plane at the Stack of Glencoul have suffered complete dynamic recrystallization and are characterized by microstructures and asymmetric crystallographic fabrics indicating a noncoaxial deformation associated with west-northwest-directed overthrusting. In contrast, quartzites at greater distances beneath the thrust have only undergone partial dynamic recrystallization and are characterized by microstructures and symmetric crystallographic fabrics indicating approximate coaxial deformation paths (Figure 144.2) (Christie, 1963; Law and others, 1986; Law, 1987). Formation of the asymmetric fabrics (noncoaxial deformation) near the Moine thrust must be either contemporaneous with, or later than, formation of the symmetric fabrics (coaxial deformation) at greater distances beneath the thrust. Flow within the mylonites can be represented by a single value of approximately coaxial flattening and a marked gradient of simple shear which increases towards the thrust surface (Law, 1987).

All thin sections are oriented perpendicular to foliation and parallel to lineation; micrographs and fabric diagrams are viewed toward the north-northeast, a sinistral shear sense being compatible with west-northwest-directed overthrusting.

**144D.** Quartzite at 0.5 cm beneath the thrust. A preferred alignment ($S_B$) of elongate, dynamically recrystallized quartz grains dips more steeply to the east-southeast than the mylonitic foliation ($S_A$) observed in hand sample; both microstructures and crystal fabrics indicate a sinistral shear sense (sample SG.1 of Law and others, 1986; Law, 1987). This oblique, grain-shape alignment is not observed in the phyllosilicate-rich layers (dark zones in micrograph). NX

**144E.** Quartzite at 4.6 m beneath the thrust (sample SG.10 of Law and others, 1986; Law, 1987). Ribbon grains anastomose around a "hard orientation" globular quartz grain (G) of possible detrital origin, suggesting an approximately coaxial bulk deformation history. NX

**144F.** Quartzite at 8.5 m beneath the thrust. Mylonitic foliation ($S_A$) defined by planar zones of recrystallized quartz and ribbonlike, relict detrital grains (sample SG.13 of Law and others, 1986). Large globular quartz grains (G) aligned parallel to foliation are remnants of a single, ribbonlike, parent grain which has been partially destroyed by the processes of lattice kinking, grain-boundary migration, and subgrain rotation. These "hard orientation" globular grains (*c*-axes perpendicular to foliation) suggest that at least the last strain increments may have been coaxially superimposed. NX

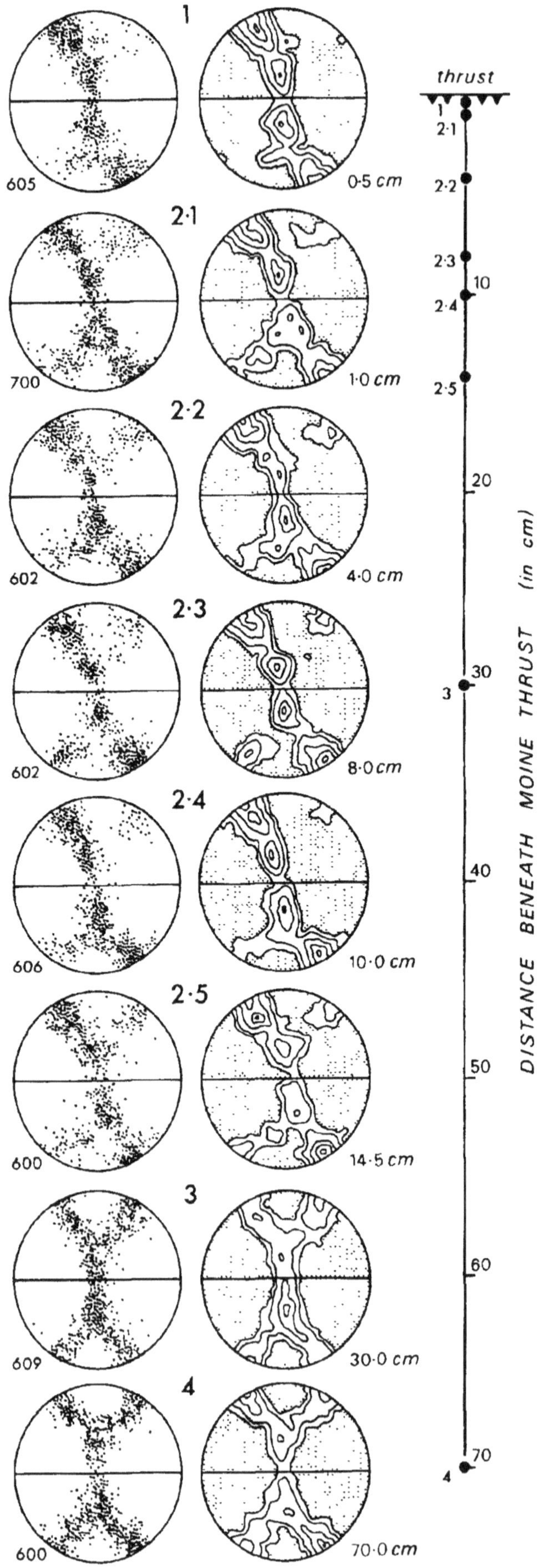


**Figure 144.2.** Transition from asymmetric to symmetric quartz *c*-axis fabrics with increasing distance beneath the Moine thrust (after Law and others, 1986).

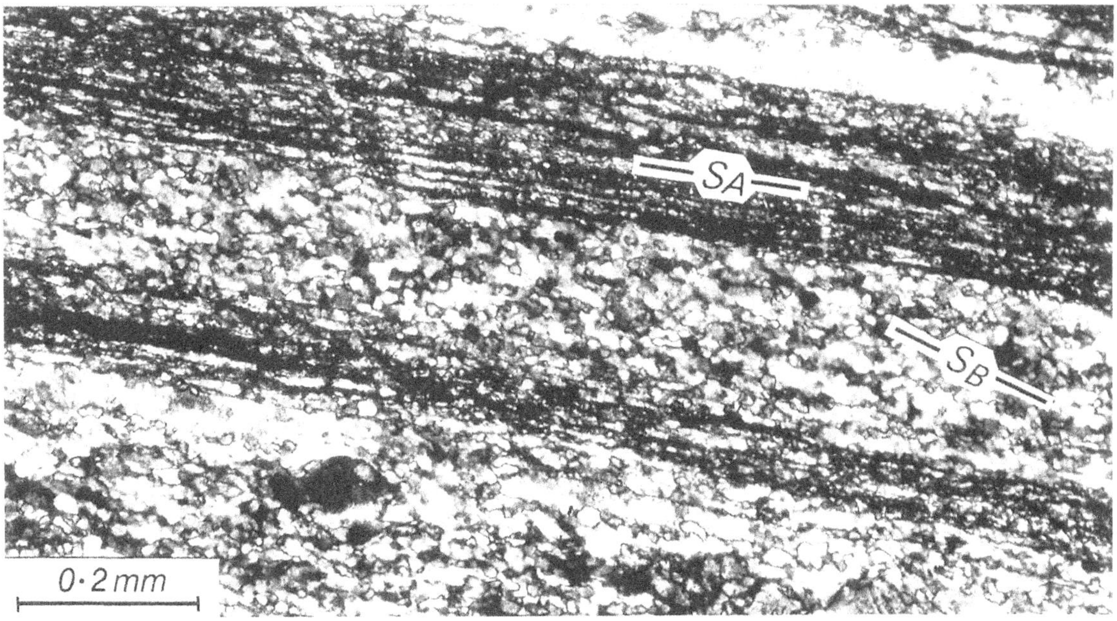


144 D

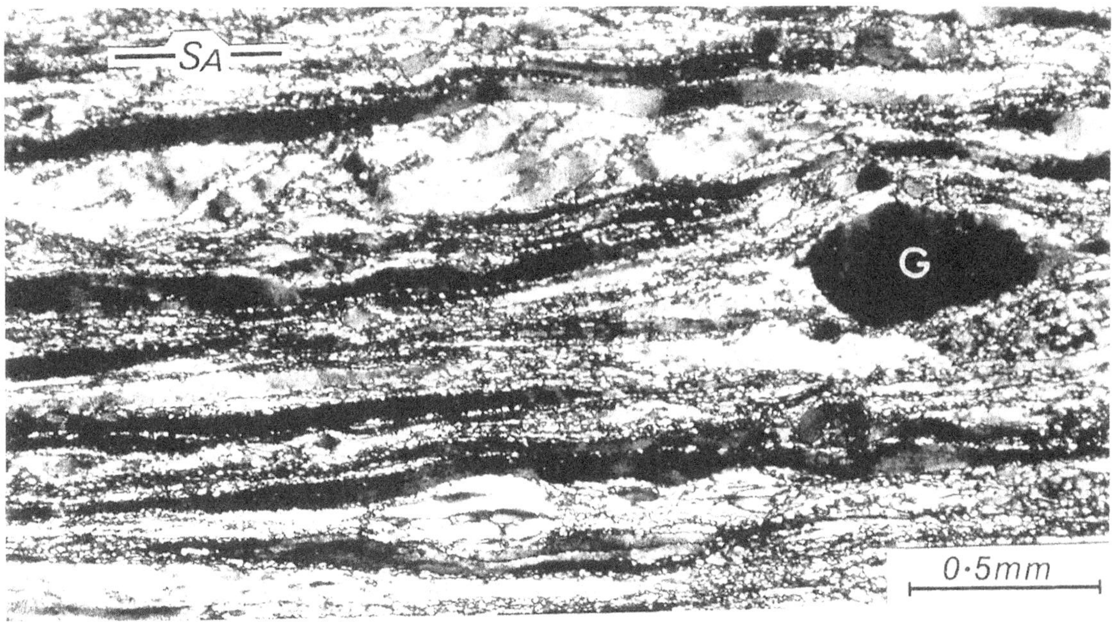


144 E

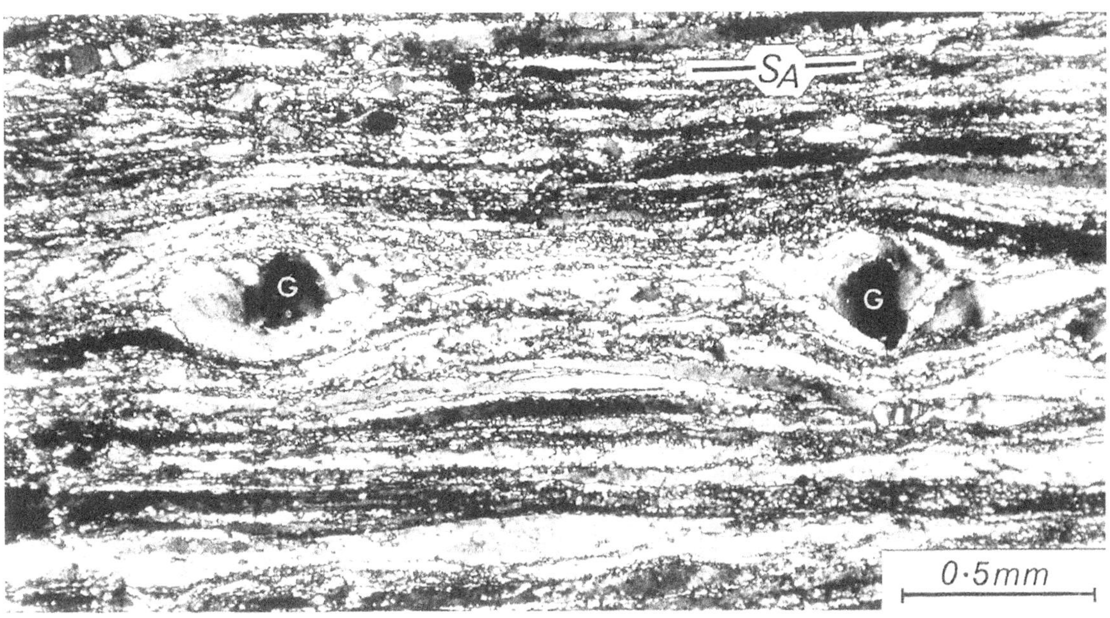


144 F

# 145. QUARTZ MYLONITES FROM THE MOINE THRUST ZONE IN SOUTHERN ASSYNT, NORTHWEST SCOTLAND

R. D. LAW

The Moine thrust plane is exposed in the Allt nan Sleach stream section of southern Assynt (Figure 145.1). Here, the gently dipping Moine thrust may have been breached by a later thrust fault, and a sill of feldspar porphyry is injected along the cross-cutting fault surface.

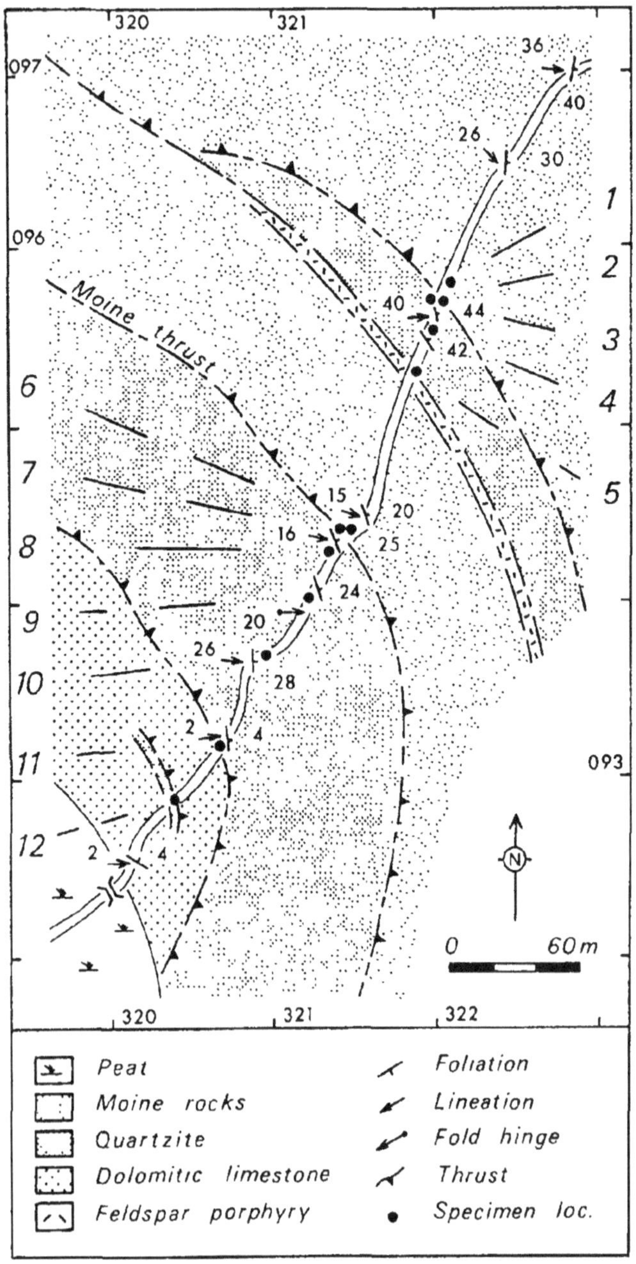


**Figure 145.1.** Geologic map of the Allt nan Sleach stream section (after Law and others, 1986) showing individual sampling locations (AS 1–12) described in that publication.

**145A.** Exposure of the Moine thrust (at arrow tip) viewed toward the northwest (Law and others, 1986); mylonitic Moine rocks (M) overlie similarly deformed Cambrian quartzites (Q). Hand sample AS.7 shown in 145B was collected from the contact (arrowed) separating these rock types.

**145B.** Polished surface (viewed toward the north-northeast) of hand sample AS.7 (Law and others, 1986) cut perpendicular to foliation and parallel to lineation. Mylonitic Moine rocks (dark) overlie similarly deformed Cambrian quartzite (light); the contact between the Proterozoic and underlying Cambrian rock units is taken to mark the position of the Moine thrust. Foliation in both units is parallel to this contact. Note shear bands (dipping toward the west-northwest) within the phyllosilicate-rich Moine rocks; the sense of obliquity between foliation and shear bands is consistent with westward-directed overthrusting of the Moine rocks.

**145C.** Polished surface (viewed toward the north-northeast) of sample AS.5 (Law and others, 1986) cut perpendicular to foliation and parallel to lineation; feldspar porphyry (p) cross-cuts mylonitic foliation in adjacent Cambrian quartzite. Although this porphyry may have been injected along a thrust fault which breaches the Moine thrust, the porphyry is itself deformed, exhibiting a spaced foliation, defined by quartz and phyllosilicate overgrowths on feldspar phenocrysts, which is parallel to the mylonitic foliation in the adjacent Cambrian quartzites.

145 A

145 B

145 C

# 146. DISLOCATION CREEP REGIMES IN EXPERIMENTALLY DEFORMED QUARTZ AGGREGATES

GREG HIRTH AND JAN TULLIS

## Optical microstructures of experimentally deformed samples

Three regimes of dislocation creep occur in experimentally deformed quartz aggregates depending on the relative rates of grain-boundary migration, dislocation climb, and dislocation production (Hirth and Tullis, 1992). The rates of these processes depend on the temperature, strain rate, and amount of water present during deformation. Accordingly, the transitions between dislocation creep regimes occur with decreasing stress from Regime 1 (highest stress) to Regime 3 (lowest stress). Within each regime a distinctive microstructure is produced, due primarily to the operation of different mechanisms of dynamic recrystallization. In the following three plates (Plates 146 and 147), we illustrate the (1) optical and (2) TEM scale microstructures that characterize each regime, and (3) optical micrographs of naturally deformed quartz aggregates that exhibit the same distinctive microstructures.

**146A.** Regime 1. Heavitree quartzite shortened 20% at 700°C and an axial strain rate of $10^{-6}$ $s^{-1}$. The sample was quenched at a differential stress of ~900 MPa. The characteristic microstructure of samples deformed in Regime 1 consists of inhomogeneously flattened original grains that exhibit an irregular and patchy extinction and very fine recrystallized grains along the grain boundaries. The patchy extinction is indicative of high densities of tangled dislocations. Notice the absence of optically visible subgrains in the original grains. At the optical scale, the recrystallized grains are too small to be resolved and the boundaries appear diffuse (e.g., see arrow). The shortening direction is vertical. The horizontal microcracks formed during unloading. NX

**146B.** Regime 2. Heavitree quartzite shortened 60% at 800°C and an axial strain rate of $10^{-6}$ $s^{-1}$. The sample was quenched at a differential stress of ~300 MPa. The distinctive microstructure of samples deformed in Regime 2 consists of homogeneously flattened original grains that exhibit sweeping undulatory extinction, abundant sub-basal deformation lamellae, and optically visible subgrains. Recrystallization occurs by progressive subgrain rotation producing a distinctive core and mantle structure (e.g., White, 1976). This is the only regime of dislocation creep where uniformly flattened original grains are produced at high strains. NX

**146C.** Regime 3. Black Hills quartzite shortened 57% at 900°C and an axial strain rate of $10^{-6}$ $s^{-1}$ with 0.2 wt% added water. The sample was quenched at a differential stress of ~100 MPa. The transition from Regime 2 to Regime 3 occurs due to an enhancement in grain-boundary migration rate. Although dislocation climb remains sufficiently rapid to accommodate recovery, a significant change in the microstructure occurs due to rapid grain-boundary migration. Samples are characterized by having recrystallized grains larger than subgrains and an absence of highly flattened original grains. Since both subgrain rotation and migration recrystallization occur in Regime 3, samples are completely recrystallized at relatively low strains. NX

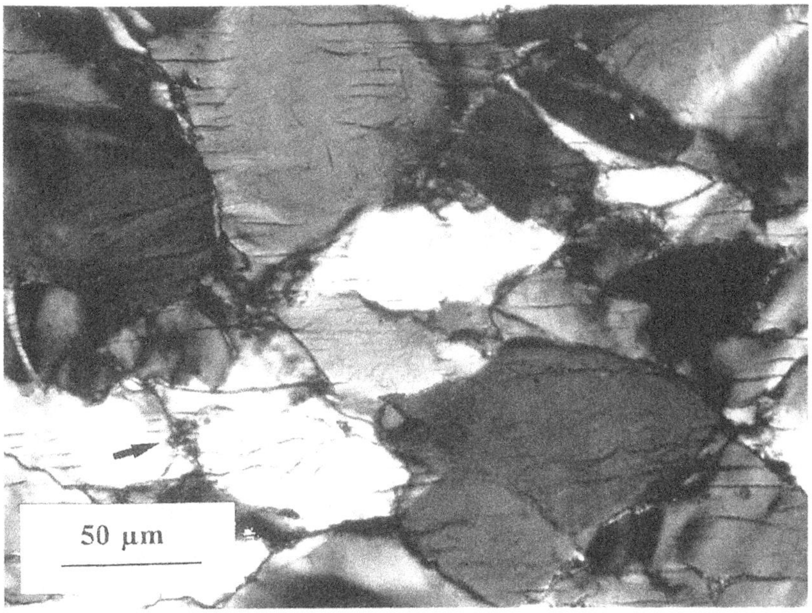


146 A

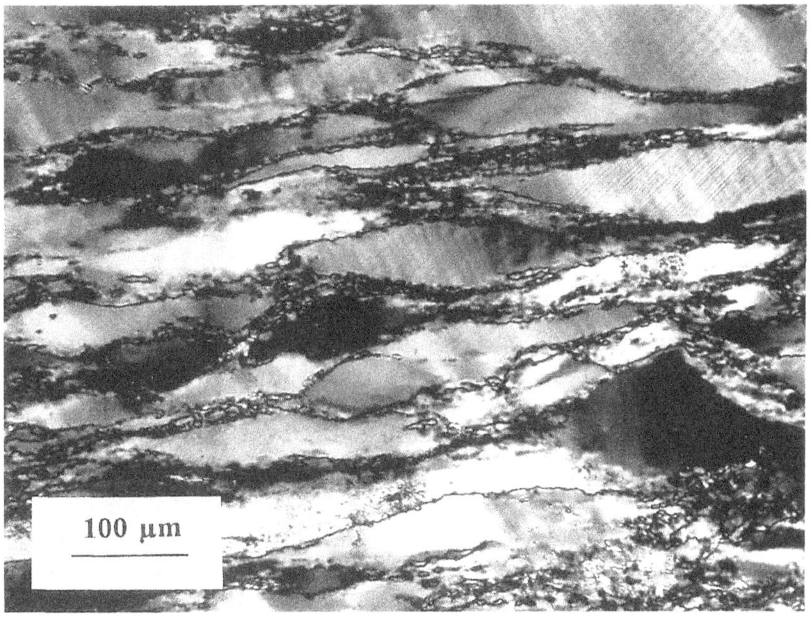


146 B

146 C

### TEM microstructures of experimentally deformed samples

The different mechanisms of dynamic recrystallization that operate during deformation in the dislocation creep regimes are easily illustrated using TEM micrographs. The three micrographs described below were taken from the grain-boundary regions of the samples illustrated in the optical micrographs on the previous plate (146A–146C).

**146D.** In Regime 1, recrystallization occurs by strain-induced grain-boundary migration. The driving force for grain-boundary migration arises from variations in dislocation density between adjacent grains. The micrograph illustrates newly recrystallized grains (indicated by asterisks) replacing an original grain with a high density of tangled dislocations. Tangling of dislocations occurs because the rate of diffusion is too slow for dislocation climb to be an effective recovery mechanism. In Regime 1, recovery is accommodated by the dynamic recrystallization process. This is the only dislocation regime where significant strain weakening is observed. Strain weakening occurs because as a greater percentage of the sample is recrystallized, a correspondingly greater percentage of recovered (i.e., strain-free) grains is produced (e.g., Tullis and Yund, 1985). TEM bright field (BF)

**146E.** In Regime 2, recrystallization occurs predominantly by progressive subgrain rotation. A high-angle grain boundary is formed by this process when a critical number of dislocations of dominantly one sign glide and climb into a low-energy subgrain boundary. Recrystallization by subgrain rotation initiates in the grain-boundary regions due to strain gradients that develop between adjacent grains with different orientations. As illustrated in the micrograph, the free dislocation density in recrystallized grains is the same as that within the original grain from which they were formed. Notice that the recrystallized grains are bounded partly by high-angle (an example is labeled H) and partly by subgrain (an example is labeled S) boundaries. The size of the recrystallized grains is similar to that of the smaller subgrains within the original grain. TEM BF

**146F.** In Regime 3, the recrystallization process is dominated by grain-boundary migration. The driving force for grain-boundary migration arises partly from a difference in the free dislocation density between grains. However, grain boundaries in samples deformed in Regime 3 often form a cusp where they intersect a subgrain boundary. An example of this microstructure is indicated by the arrow in the micrograph. This observation indicates that interfacial energy is an important component of the driving force for grain-boundary migration during the dynamic recrystallization process. Notice that the recrystallized grain is significantly larger than the subgrain within the grain that it is replacing. TEM BF

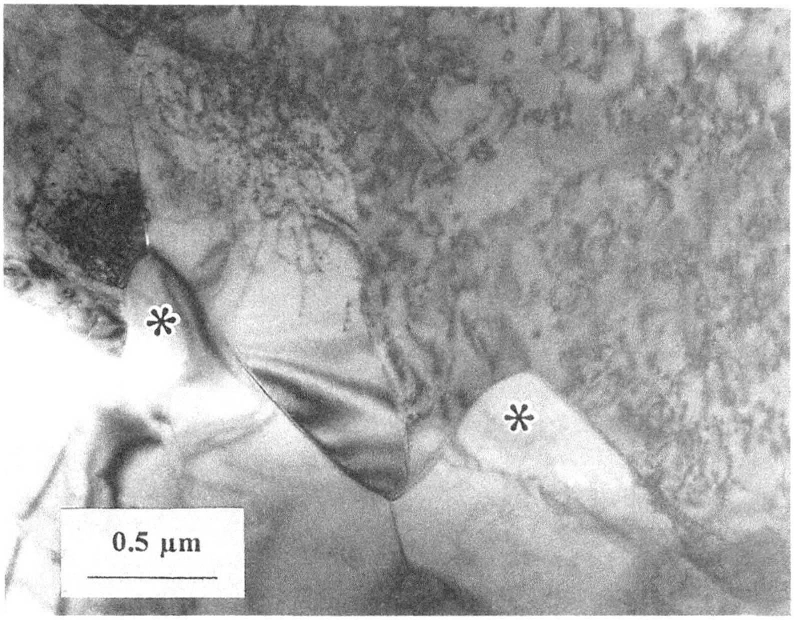

146 D

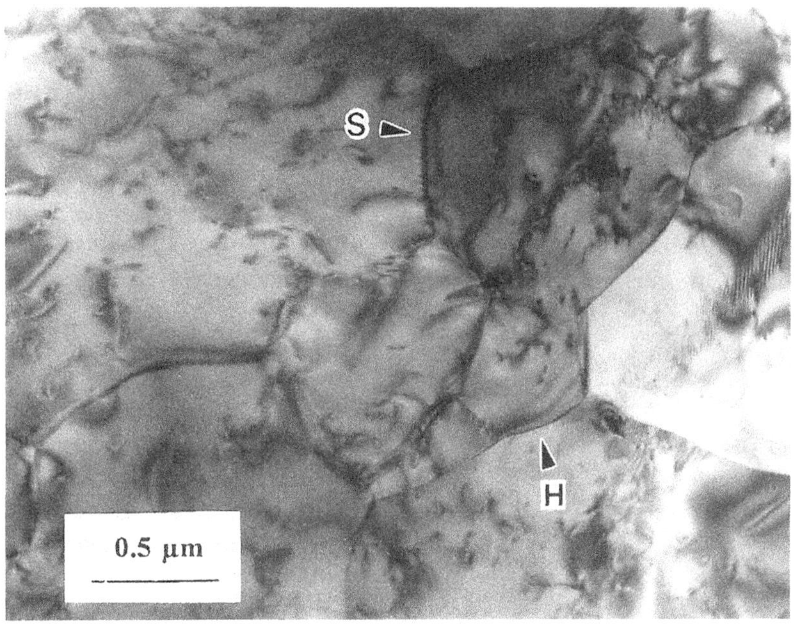

146 E

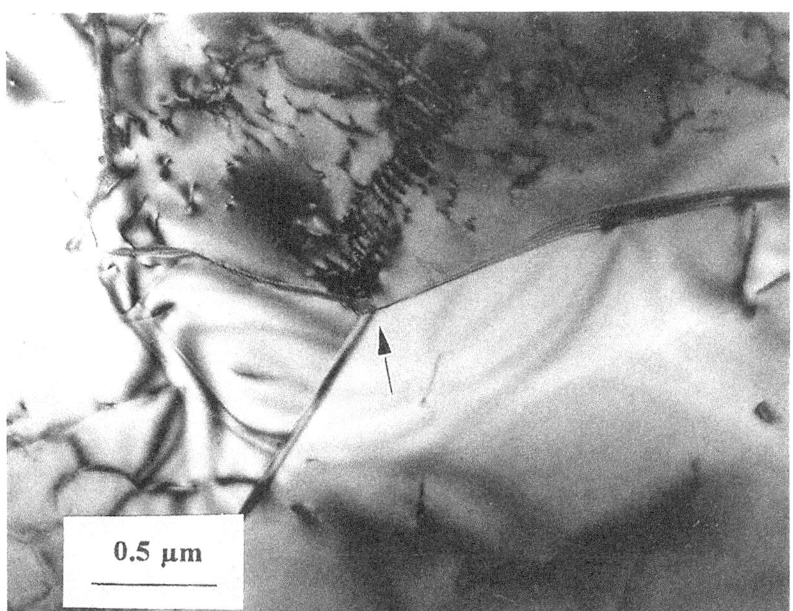

146 F

# 147. DISLOCATION CREEP REGIMES IN NATURALLY DEFORMED QUARTZ AGGREGATES

GREG HIRTH, W. JAMES DUNLAP, AND CHRISTIAN P. TEYSSIER

## Optical microstructures of naturally deformed samples

An analysis of microstructures preserved in naturally deformed Heavitree Quartzite from the Ruby Gap duplex, central Australia, indicates that the three dislocation creep regimes also occur during deformation in the crust. The duplex is comprised of a 3 km thick antiformal stack of six superposed thrust sheets that formed under greenschist-facies conditions. For further details, see Dunlap and others (1991) and Plate 86 by Dunlap and Teyssier in this volume.

**147A.** Regime 1. In sheet 1 of the duplex (the lowest of six sheets that define the thrust system), the original grains exhibit irregular patchy extinction, numerous deformation bands, and extremely fine recrystallized grains along the grain boundaries. These microstructures are characteristic of Regime 1, where dislocation climb is difficult and recovery is accommodated by grain-boundary migration recrystallization. NX

**147B.** Regime 2. Moving upward in the thrust system, abrupt transitions in the quartz microstructure are observed between the major thrust sheets. In sheet 2, the original quartz grains are extremely flattened and exhibit optically visible subgrains, as well as recrystallized grains along their boundaries. These microstructures are characteristic of Regime 2, where recovery is accommodated by dislocation climb and recrystallization occurs dominantly by progressive subgrain rotation. NX

**147C.** Regime 3. In sheets 3 through 6, the quartzite is completely recrystallized, and exhibits an oblique foliation defined by the recrystallized grains (e.g., Lister and Snoke, 1984). This microstructure is characteristic of Regime 3, where recovery is accommodated by dislocation climb, and recrystallization occurs dominantly by grain-boundary migration. NX

The positive correlation between experimentally and naturally produced microstructures makes it possible to constrain the conditions at which deformation in the duplex occurred. In this example, the microstructural analysis indicates that the flow stress was greatest in sheet 1 and lowest in sheet 6, suggesting a variation in either temperature or strain rate during the deformation history.

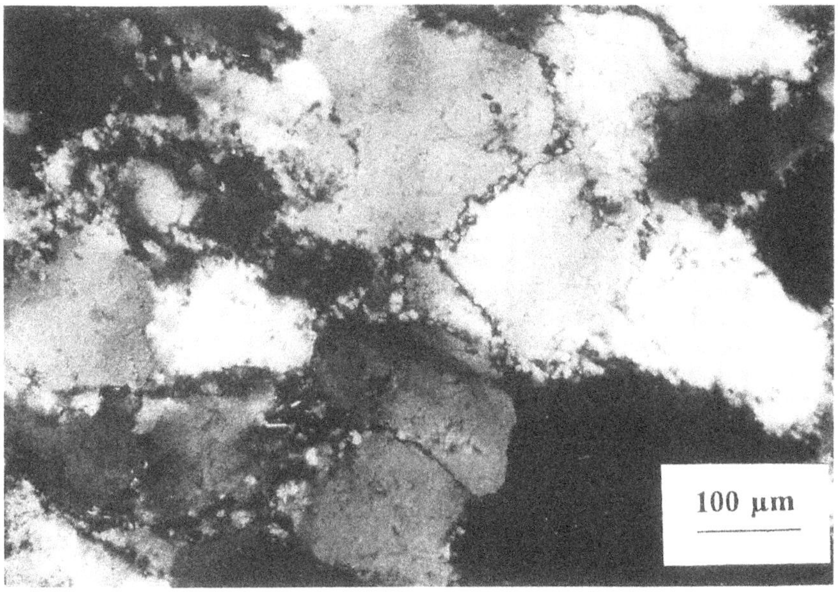


147 A

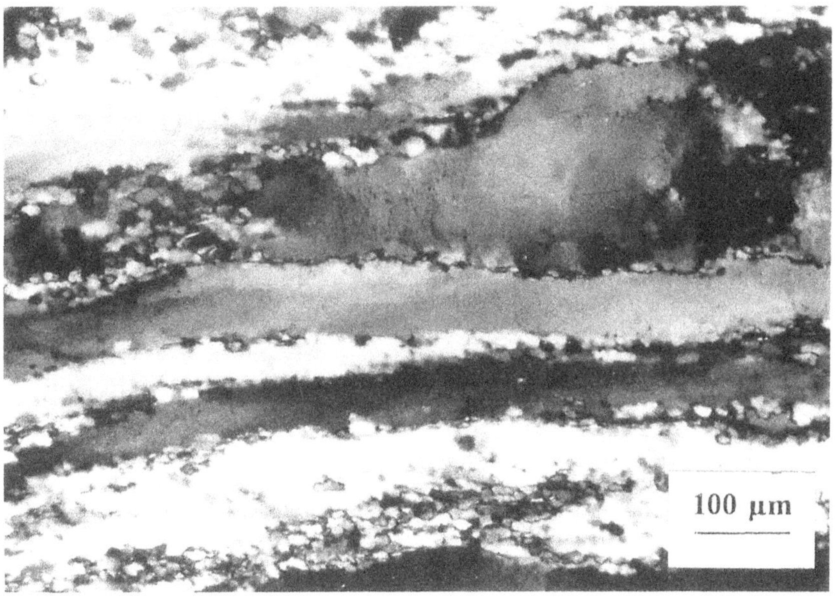


147 B

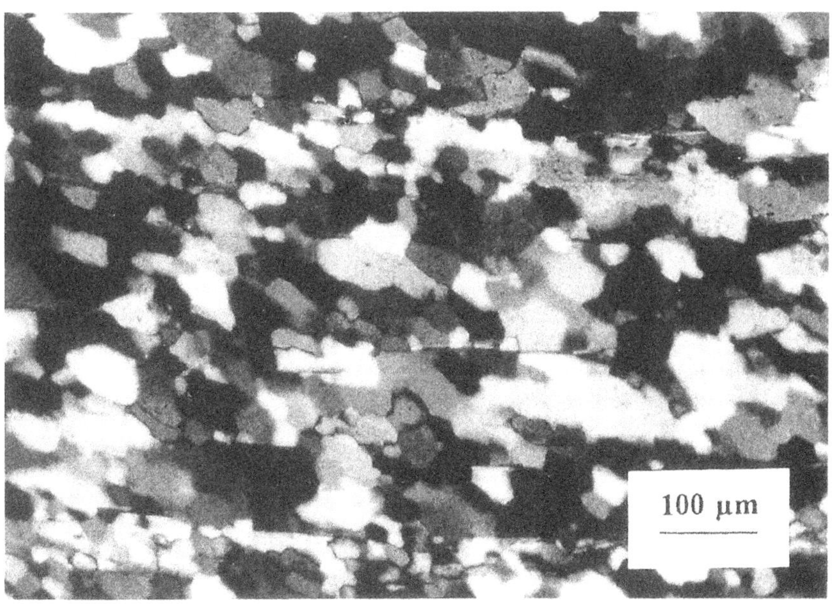


147 C

# Microstructural development of mylonitic rocks of specific composition

## Feldspathic rocks



Common abbreviations used in regard to the plates in this section include PPL = plane polarized light, NX = crossed nicols, SEM = scanning electron microscope, and TEM = transmission electron microscope.

This set of plates illustrates the microstructures developed in feldspars from naturally sheared anorthosites, granites, tonalites, and monzogranites, as well as experimentally deformed albite aggregates. Several of these contributions indicate that feldspars show two of the same dislocation creep regimes identified in quartz, including a lower-temperature regime characterized by grain-boundary migration recrystallization and a higher-temperature regime characterized by subgrain rotation recrystallization. However, the temperatures required for these regimes are significantly higher than those required for the same regimes in quartz; e.g., upper greenschist to lower amphibolite facies for the onset of crystal-plasticity (with grain-boundary migration recrystallization) and upper amphibolite to granulite facies for subgrain rotation recrystallization. Another important difference from quartz is that in feldspars, neomineralization is perhaps more common than simple strain-induced dynamic recrystallization. For example, at high metamorphic grades, myrmekite may form during deformation. At lower metamorphic grades, feldspars are prone to retrograde hydration reactions, which tend to produce a fine-grained mixture of several phases; mutual grain-boundary pinning keeps the grain size small and allows the aggregates to undergo grain-size-sensitive creep.

# 148. PROGRESSIVE DEFORMATION OF PLAGIOCLASE IN DUCTILE SHEAR ZONES

William L. Brown and Jean Macaudière

New textures and mineral parageneses are progressively produced by small ductile shear zones of Proterozoic age, which cut the meta-anorthosite massif of Roineabhal, South Harris, Scotland. Though the shear zones are of variable thickness (millimeter to meter) and orientation, they develop the same textures and parageneses of upper amphibolite facies within the granulite-facies meta-anorthosite. The composition of the meta-anorthosite varies from that of an anorthosite to a gabbro, and that of the undeformed plagioclase from bytownite to calcic andesine, as the mafic index increases. The changes in texture and composition can be related to the amount of shear strain, $\gamma$, calculated either from the rotation of the previously existing foliation (defined by mineral banding and weak orientation of amphibole) or from the angle of the new foliation (defined by strong mineral orientation) and the shear-zone boundary (Figure 148.1 and Brown and others, 1980).

The mineralogical changes include the destruction of clinopyroxene and garnet at $\gamma < 0.5$–$2.5$, the appearance of scapolite, epidote, and biotite, and the recrystallization of amphibole and plagioclase with or without change of composition.

The textural changes are dominated by plagioclase. They include recrystallization of old grains at their boundaries with the progressive formation of new grains surrounding almost undeformed cores. New grains of plagioclase first appear at grain boundaries at $\gamma \approx 0.5$ but relic grains are still present at $\gamma > 10$. The textures vary from protoclastic ($\gamma < 0.5$–$0.7$), through porphyroclastic festoon (148A) or mortar (148B) ($0.7 < \gamma < 10$) to polygonal augen (148C) or homogeneous textures ($\gamma > 10$).

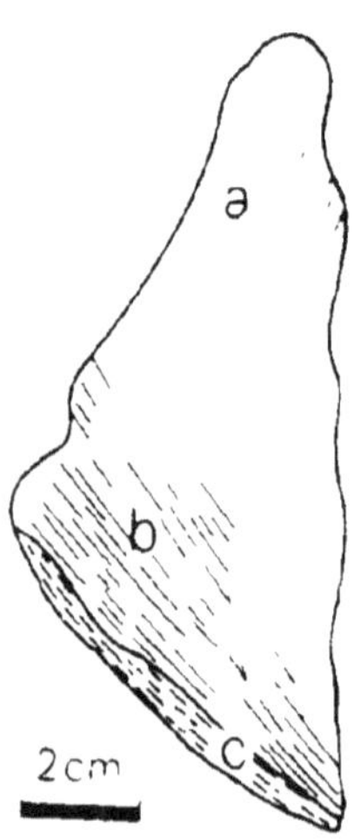


**Figure 148.1** Development of new foliation in a shear zone: a) old foliation; b) new foliation; and c) core of the shear

**148A.** Recrystallized grains along old grain boundaries. New grains develop only along the boundaries of old grains and give rise to a porphyroclastic discontinuous "festoon" texture ($\gamma \approx 1$) near the edge of a small shear zone. Bulges and new grains of plagioclase (40–120 μm) develop along the boundaries of relic grains, which are very homogeneous ($An_{68}$) and show little sign of plastic deformation. The new grains have compositions between $An_{49}$ and $An_{76}$ and are often strongly reverse zoned (a) with a range of up to 20% for a single grain. Bulges have at first the same orientation as the parent plagioclase, but separated new grains become more disorientated and may show rare twins. NX

**148B.** Development of continuous walls of new grains (mortar texture). At larger values $\gamma$ (1–5) in the same shear zone, the new grains are slightly larger (80–300 μm) and form more or less continuous walls or bands of polygonal grains separating very slightly deformed relic grains. As the walls of new grains thicken, the porphyroclastic texture varies from the festoon texture (148A) to a mortar texture. NX

**148C.** Development of a new foliation with isolated relic grains (augen texture). Toward the center of the shear zones, a new foliation develops progressively as $\gamma$ increases. It is defined by the flattened polygonal grains of plagioclase and hornblende (h) in this meta-leucogabbro, in which are preserved a few old rounded plagioclase grains. The angle between the new foliation and the shear-zone boundary is too small to be measured with any precision and corresponds to a $\gamma > 10$. NX

148 A

148 B

148 C

# 149. PLAGIOCLASE PORPHYROCLASTS IN A HIGH-TEMPERATURE SHEAR ZONE

NIELS Ø. OLESEN

Anorthositic gneisses from the Jotun nappe complex, Norway, are transformed to mylonites in shear zones of Precambrian age at Mt. Hornsnipa near Aurlandsvangen (Olesen, 1986).

The gneisses equilibrated under granulite-facies conditions, resulting in a large grain size (i.e., centimeter scale). Mylonitization took place under high amphibolite- to granulite-facies conditions and resulted in a large reduction in grain size in a bimodal distribution. Porphyroclasts range in shape from globular outlines to ribbon shapes with aspect ratios as large as 25:1. A matrix of recrystallized plagioclase displays grain sizes typically of 0.3 mm but <0.05 mm in zones of very high strain. The mylonitic foliation is defined by elongate aggregates of mafic minerals and by the ribbon-shaped porphyroclasts. The compositions of porphyroclasts and neoblasts vary between $An_{35}$ and $An_{60}$ (Olesen, 1987).

**149A.** Plagioclase porphyroclasts. A ribbon-shaped porphyroclast with aspect ratio ~11:1 and smaller porphyroclasts with globular outlines. Garnets (g) are also visible in the matrix of recrystallized plagioclase. Structures in the field demonstrate that this texture is the result of a large strain with a component of sinistral shear. The lattice orientation of these porphyroclasts is such that (010) preferably has an angle to the mylonitic foliation between 15° and 50°, and the porphyroclasts are oriented such that they were able to accommodate strain by twinning. NX

**149B.** Partly recrystallized kink band. For the ribbon-shaped porphyroclasts the angle between (010) and the foliation is commonly between 20° and 40°, displayed here by albite-law twins from upper right to lower left. Furthermore, they are oriented such that they were able to accommodate shear strain by kinking/lattice rotation in bands roughly parallel to (001). Such a band, with a few discontinuous pericline-law twins, is illustrated here (arrows), invaded by a recrystallization "front." The ribbon-shaped porphyroclasts are the result of kinking and preferential recrystallization of bands of high lattice strain, leaving behind the less-strained parts of the original crystal. The aspect ratio of these porphyroclasts is consequently of no use as a measure of strain. NX

**149C.** Mechanism of recrystallization. Lattice rotation is discontinuous. It takes place through a number of sharp lattice bends, ranging from low-angle grain boundaries (lower part of figure) to irregular grain-boundary-like microstructures (central part of figure). There is a complete gradation between these extremes along kink bands, displaying a gradient in lattice rotation. This demonstrates that "rotation recrystallization" was a dominant mechanism of transformation in this geological environment. NX

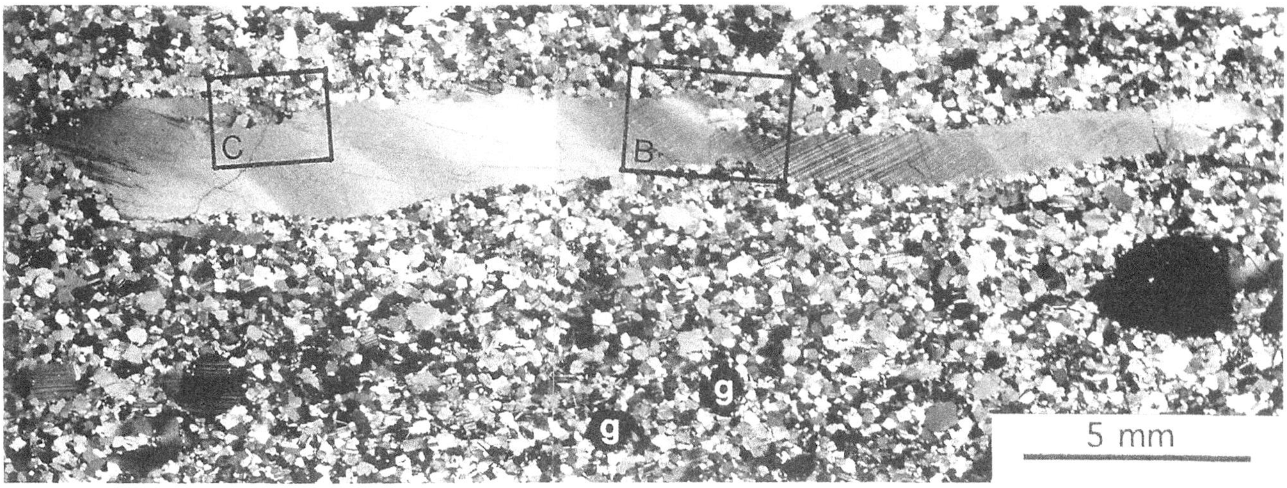


149 A

149 B

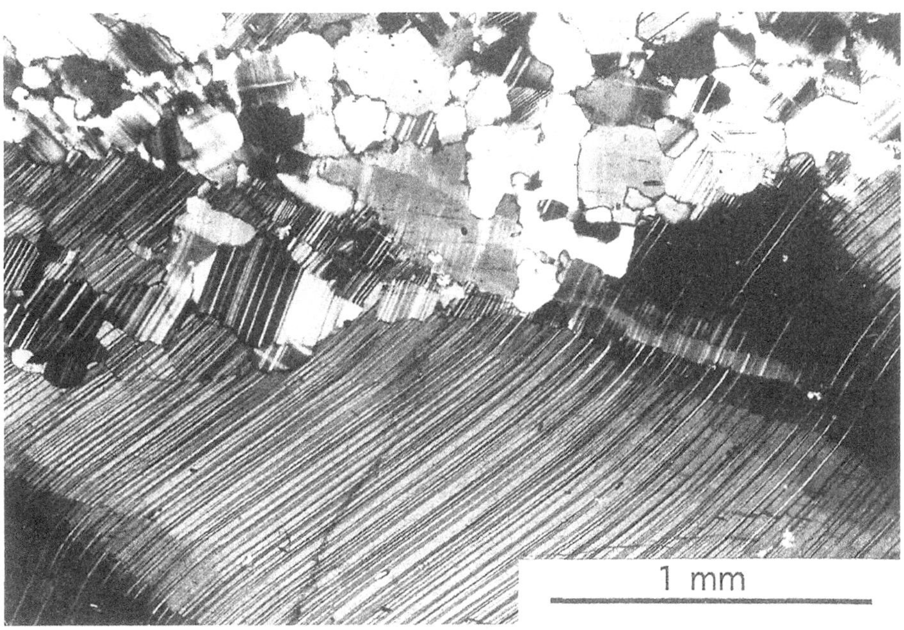


149 C

# 150. EVIDENCE FOR GRANULAR FLOW IN FELDSPAR-DOMINATED LITHOLOGIES

HOLGER STÜNITZ

Feldspars commonly do not exhibit extensive intracrystalline plasticity at temperatures below approximately 550°C, i.e., the beginning of the amphibolite facies (e.g., Tullis, 1983). However, many low-grade mylonites contain fine-grained albite, which results from syntectonic recrystallization or neomineralization of former feldspar clasts (Fitz Gerald and Stünitz, 1993; Stünitz, 1993). Usually, the albite occurs in fine-grained aggregates together with other phases, such as quartz, epidote/clinozoisite, or mica (1 to 30 μm grain diameter). Generally, these fine-grained aggregates are mechanically weaker than pure quartz aggregates. Mechanically weak, very fine-grained aggregates suggest the operation of a dominant deformation mechanism different from intracrystalline plasticity, namely, grain-boundary sliding, i.e., deformation of the grain aggregate as a whole, termed granular flow here. Grain-boundary sliding is difficult to demonstrate other than in experiments (e.g., Schmid and others, 1977). However, it may be possible to find criteria to demonstrate that some examples of nonbrittle, mylonitic deformation did not occur by intracrystalline plasticity as the dominant mechanism, but by granular flow (Stünitz and Fitz Gerald, 1993).

Grain shapes in polyphase aggregates can be indicators for deformation mechanisms. During crystal-plastic deformation of the phases, the stronger phases develop more convex boundary shapes, whereas the weaker phases develop more concave boundary shapes. This shape information can be used in conjunction with TEM investigations on the intracrystalline deformation microstructures. In the cases described here, only quartz shows evidence for extensive crystal-plasticity. If only one of the phases shows evidence for intracrystalline plasticity, but the grain shapes of different phases do not differ systematically, then intracrystalline plasticity is probably not the dominant deformation mechanism of the whole aggregate.

Another criterion for granular flow is a random distribution of different phases in a polyphase aggregate. A nonrandom distribution, i.e., a concentration of weak or strong phases in layers, is a common feature which develops during the crystal-plastic deformation of two-phase aggregates with different strengths of the phases (Jordan, 1987, 1988; Handy, 1990). A further important line of evidence for granular flow would be a weak lattice preferred orientation. However, due to peak overlap in these polyphase aggregates, it is impossible to measure the lattice preferred orientation by X-ray methods. Optical methods cannot be used due to the small grain size. In two examples presented here, the grain-size refinement occurred through recrystallization or neomineralization of the former intermediate plagioclase to albite plus other minerals. The fine grain size is one prerequisite for granular flow and is maintained through the mixture of phases, which inhibits grain growth effectively and thus forms a stable microstructure.

**150A.** Hand specimen of a gabbro mylonite at M. Nery, central Sesia zone, Italian Alps (Stünitz, 1993). The mylonite zones occur predominantly at the margins of gabbro bodies. Deformation took place near 350°C (Stünitz, 1993). The white portions are fine-grained aggregates of predominantly albite + zoisite, which both result from the neomineralization of former intermediate plagioclase. The dark porphyroclasts consist of hornblende/pyroxene intergrowths, which become progressively replaced during deformation by aggregates of actinolite needles (gray streaks extending from porphyroclasts). This replacement represents a solution transfer of material during deformation. The mechanical strength is controlled by the albite + zoisite aggregates (the mechanically weakest component of the rock). The mafic phases show only reactions and dissolution/precipitation. Scale in millimeters.

**150B.** Detail of a typical gabbro mylonite. SEM backscatter image (Ab = albite, Zoi = zoisite, Hbl = hornblende, Act = actinolite). The zoisite needles are well aligned in the orientation of the lineation in mylonites, whereas in weakly deformed portions of the rock zoisite needles form radiating acicular arrangements. The zoisite grains have hypidiomorphic shapes, whereas the albite grains form interstitial material (albite grains display more concave boundaries). However, albite does not accommodate the zoisite shapes by crystal-plastic deformation, because TEM investigations show no evidence for intracrystalline plasticity. Extensive microscale fracturing is also absent in albite and zoisite, so that the shape accommodation of albite must have taken place by diffusive mass transfer (Stünitz, 1993). Albite grains do not show large aspect ratios, so that grain-scale diffusive mass transfer alone cannot have accommodated the high strain, and thus grain-boundary sliding most likely has occurred simultaneously. Thus, the strong shear deformation did not occur by intracrystalline plasticity but most probably by granular flow.

150 A

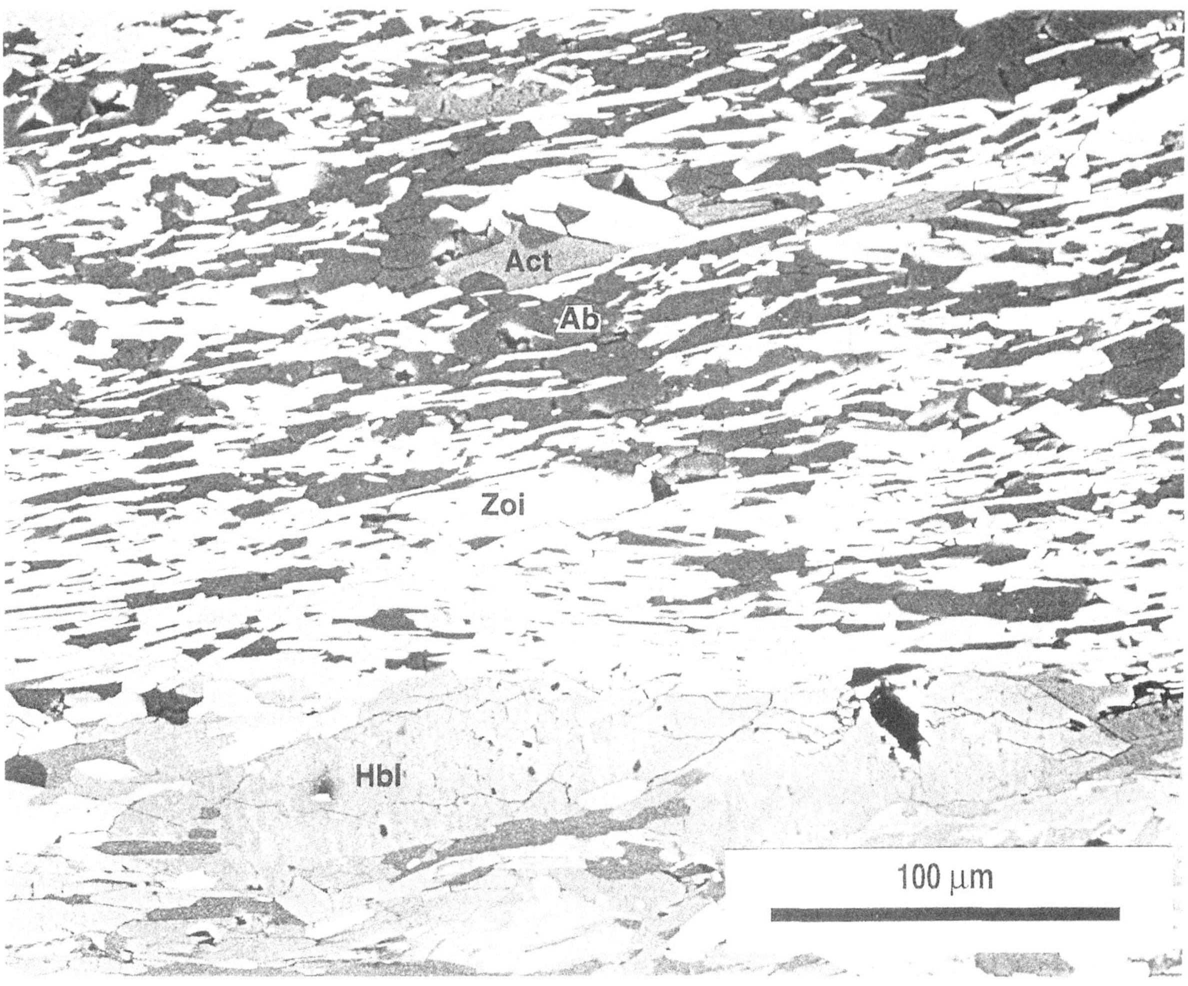


150 B

**150C.** Quartz porphyroclast in a fine-grained albite + quartz + epidote/clinozoisite + mica matrix. Corvatsch mylonite, Corvatsch, Engadine, Switzerland (Stünitz and Fitz Gerald, 1993). The protolith of the mylonite is an equigranular granodiorite, and all high-strain shear deformation is strongly localized. Deformation took place under lower greenschist-facies conditions (300–400°C; Fitz Gerald and Stünitz, 1993). This example illustrates a common feature of low-grade deformation of granitoids, namely, that quartz may be mechanically stronger than a fine-grained polyphase matrix dominated by feldspar. The large quartz grain (Qtz) forms a porphyroclast, which exhibits dynamic recrystallization by rotation recrystallization across the center and at the tails. The recrystallized pure quartz aggregates (Qtz) are recognized by their coarser grain size compared to the polyphase matrix. Pure quartz aggregates show a strong lattice preferred orientation, when inspected with the compensation plate inserted, whereas the polyphase matrix does not. NX

**150D.** Detail of 150C, fine-grained albite + quartz + epidote matrix of Corvatsch mylonite (Stünitz and Fitz Gerald, 1993). The micrograph was taken in reflected light using differential interference contrast on a polished and etched surface. Etching with HF vapor attacks the albite faster than quartz and produces a damaged surface on albite grains, so that they can be distinguished from quartz, which maintains smooth surfaces (Epi = epidote, Ab = albite, Qtz = quartz). Quartz and albite do not show systematic differences in grain size and shape in the polyphase matrix. However, quartz and albite are expected to display contrasting mechanical behavior at these low temperatures, i.e., quartz should be deformed by crystal-plasticity, whereas feldspar crystal-plasticity usually commences above 500 to 550°C (e.g., Tullis, 1983). TEM investigations on these samples confirm that quartz undergoes rotation recrystallization, whereas albite shows only evidence of minor fracturing and no indication of intracrystalline plasticity. The operation of different *intra*crystalline deformation mechanisms, given that the grain shapes of albite and quartz in the aggregates are very similar, indicates that *intra*crystalline deformation was not a dominant deformation mechanism in these polyphase aggregates. This inference is supported by the fact that a mixture of phases containing approximately 50% of a mechanically stronger phase (=albite, in terms of intracrystalline deformation) is *weaker* than a pure aggregate containing only the weaker phase (quartz). Thus, the aggregate probably deformed by granular flow. NX

One observation supporting the operation of grain-boundary sliding in these rocks is that some grain boundaries are aligned in the same orientation, and are connected and subparallel to the shear plane (e.g., arrows). Such boundaries exhibit a favorable orientation and geometry for sliding. Furthermore, albite and quartz are not concentrated in layers but randomly distributed in these aggregates. Such a distribution is consistent with granular flow as the dominant deformation mechanism.

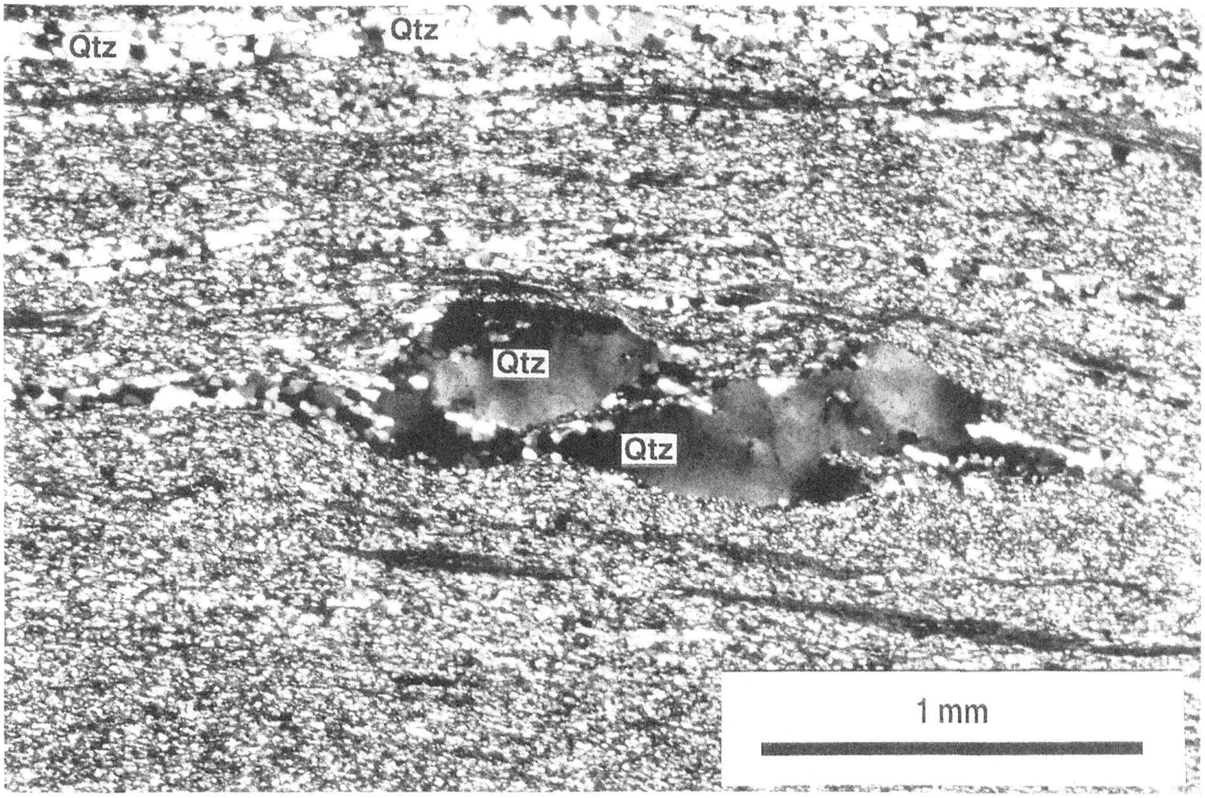


150 C

150 D

# 151. FELDSPAR TRANSFORMATION REACTIONS MEDIATING SUPERPLASTIC FLOW

R. W. KERRICH, I. ALLISON, AND R. L. BARNETT

The photomicrographs opposite illustrate the development of a tectonic fabric at progressively higher states of strain in the Miéville shear zone, Switzerland. At Miéville, located in the Aiguilles Rouges massif, granitic basement rocks of Hercynian age are transected by a major subvertical shear zone, which varies in width from 10 to 500 m, and can be traced laterally for about 15 km. A protracted tectonic history of the shear zone is evident, with initiation in the Permian, followed by varied intensities of Alpine deformation.

The deformation sequence across the shear zone commences with a coarse-grained granodiorite containing sporadic mafic xenoliths. Undeformed granodiorite consists of orthoclase and oligoclase phenocrysts up to 20 mm long (45–60%), quartz (25–30%), biotite (10–30%), and muscovite (<5%). The metamorphic mineral assemblage of deformed counterparts (quartz + albite + epidote + chlorite) corresponds to lower-greenschist facies. Ambient temperatures of deformation are estimated at 280–320°C from $^{18}O/^{16}O$ fractionations between albite and ilmenite and fluid inclusion filling temperatures for quartz occupying microfractures. Overall, deformation of the granodiorite was isochemical (see Kerrich and others, 1980). See Table 151.1 for representative mineral compositions.

**151A.** Granodiorite at a low state of strain in the shear-zone boundary. NX

**151B.** Granodiorite at an intermediate state of strain. Mantle of ultrafine-grained (5–20 μm) orthoclase + albite + oligoclase + quartz (a) surrounding an elliptical orthoclase porphyroclast (b) and developing into a trail (c). NX

**151C.** Granodiorite mylonite at the highest strain intensity. Long trails (a) of ultrafine grained (5–20 μm) equant recrystallized orthoclase + albite + oligoclase + quartz extending away from the remnant orthoclase and oligoclase porphyroclasts (b). Quartz is sporadically distributed in the trails. It appears to have been derived from igneous quartz grains undergoing dynamic recrystallization, and to have been incorporated by grain-boundary sliding. NX

**Table 151.1.** Analyses of plagioclase-feldspar porphyroclasts and ultrafine-grained feldspar in surrounding mantles

| | *center* | *margin* | *recrystallized mantle* | | | | |
|---|---|---|---|---|---|---|---|
| | *1* | *2* | *3* | *4* | *5* | *6* | *7* |
| $SiO_2$ | 60.92 (1.19) | 63.94 (1.14) | 61.37 | 66.32 | 67.89 | 68.00 | 64.24 |
| $Al_2O_3$ | 24.70 (0.52) | 22.63 (1.01) | 24.02 | 21.59 | 20.56 | 20.39 | 18.00 |
| CaO | 5.37 (0.78) | 3.83 (0.39) | 4.78 | 2.92 | 1.43 | 1.39 | 0.04 |
| $Na_2O$ | 8.15 (0.47) | 8.88 (0.08) | 8.62 | 9.14 | 9.97 | 10.42 | 0.33 |
| $K_2O$ | 0.19 (0.06) | 0.13 (0.05) | 0.23 | 0.08 | 0.09 | 0.13 | 16.30 |
| Total | 99.33 (0.44) | 99.43 (0.74) | 99.02 | 100.05 | 99.94 | 100.33 | 98.91 |
| Si | 10.873 | 11.322 | 10.981 | 11.606 | 11.846 | 11.844 | 12.016 |
| Al | 5.195 | 4.722 | 5.065 | 4.452 | 4.228 | 4.185 | 3.968 |
| Na | 1.027 | 0.730 | 0.916 | 0.548 | 0.267 | 0.259 | 0.008 |
| Ca | 2.820 | 3.049 | 2.990 | 3.101 | 3.373 | 3.519 | 0.120 |
| K | 0.043 | 0.029 | 0.052 | 0.018 | 0.020 | 0.029 | 3.889 |
| Sum | 19.958 | 19.853 | 20.005 | 19.725 | 19.734 | 19.836 | 22.001 |
| Ab | 72.49 | 80.05 | 75.53 | 84.58 | 92.15 | 92.43 | 2.98 |
| An | 26.40 | 19.18 | 23.14 | 14.93 | 7.30 | 6.81 | 0.20 |
| Or | 1.11 | 0.77 | 1.33 | 0.49 | 0.55 | 0.76 | 96.82 |
| n[a] | 4 | 4 | 1 | 1 | 1 | 1 | 1 |

[1] = analyses from the center of an oligoclase porphyroclast, specimen [?]. Figures in parentheses represent one standard deviation. [2] = analyses from the margin of the porphyroclast. [3–6] = ultrafine-grained feldspar in mantle surrounding porphyroclast, including oligoclase (3) and progressive reaction by Ca loss to albite (6). [7] = orthoclase. n[a] = the number of determinations.

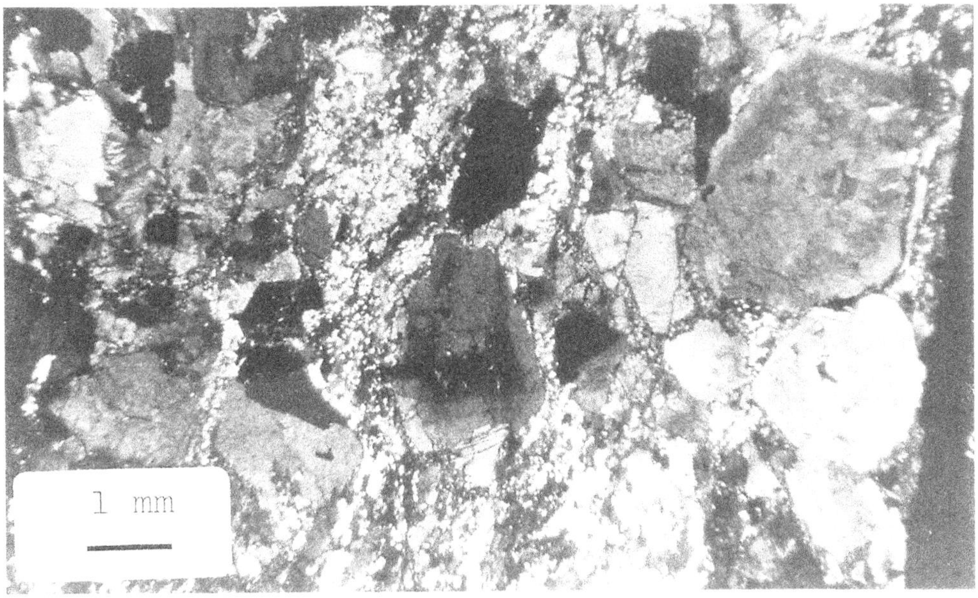


151 A

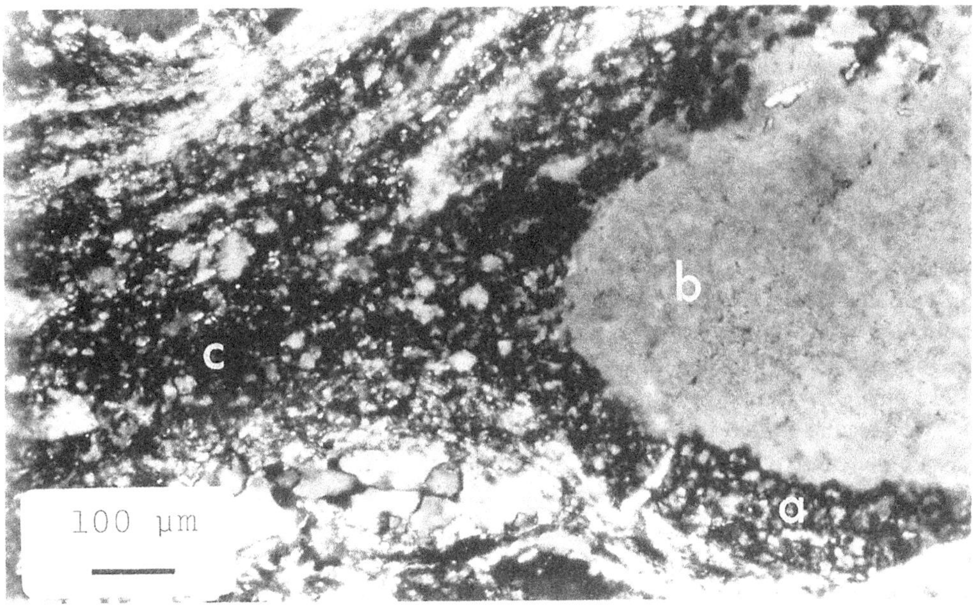


151 B

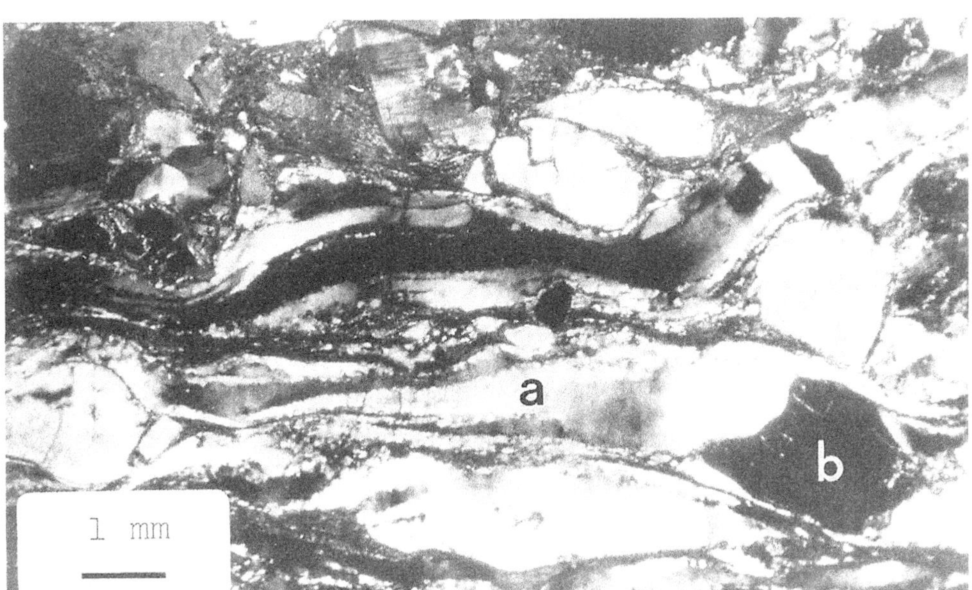


151 C

# 152. ANORTHOSITE MYLONITES PRODUCED BY SUPERPLASTIC FLOW

ANNE-MARIE BOULLIER AND YVES GUÉGUEN

The anorthosite samples belong to the Sogn Jotun nappe, Norway (Lacour, 1971) and have been studied by Boullier and Guéguen (1975). The samples show a well-defined foliation determined by the flattened aggregates of porphyroclasts of plagioclase, amphibole, and garnet. A strong lineation is defined by the elongation of porphyroclasts and mineral aggregates. The rocks are essentially made of plagioclase (PL, 55%), amphibole (AM, 35%), and garnet (GT, 10%). The amphibole and garnet are the results of the complex reactions between olivine and plagioclase (Griffin, 1971). The average grain size is 25 μm; plagioclase is partly (152A) or completely (152B) recrystallized into an equant mosaic of small grains (50 μm); amphibole recrystallized into very small and equant grains (10 μm), forming a band in which they are mixed with plagioclase. The average strain calculated by using amphibole as a marker is around 350% in 152A and larger than 1000% in 152B. The 10 μm amphibole grains which form the dark layers have almost no distinct fabric. Dislocation densities are very low, less than $10^6$ $cm^{-2}$ in small amphibole grains, and no subgrains are observed. The conditions of deformation were close to those of granulite facies (Lacour, 1971), i.e., temperature about 600°C. Thus, deformation took place at a high temperature relative to the amphibole breakdown temperature (900 to 1000°C). The stress and strain-rate conditions are unknown.

Even if the six criteria for superplastic flow (Boullier and Guéguen, 1975) are not all met for these rocks, we suggest that grain-boundary sliding was the dominant mechanism of deformation. The fact that small amphibole grains are partly mixed with the plagioclase matrix indicates that grain-boundary sliding took place between both phases as well as within amphibole. These rocks show that superplastic flow is a very effective process for developing a fine-scale mineral layering.

**152A and 152B.** Two stages of increasing strain. In 152A large porphyroclasts of plagioclase persist in the mosaic of small grains, and fine amphibole grains begin to be disseminated; garnets are fractured and partly chloritized. In 152B, almost all the amphibole porphyroclasts have disappeared; recrystallized, small grains of amphibole and plagioclase are mixed together. At this stage, the mylonitic layering is very fine and is only deflected by rounded and smooth grains of garnet and by some rare amphibole and plagioclase grains.

Foliation is normal to the photographs. Lineation is parallel to the length of the photographs. PPL

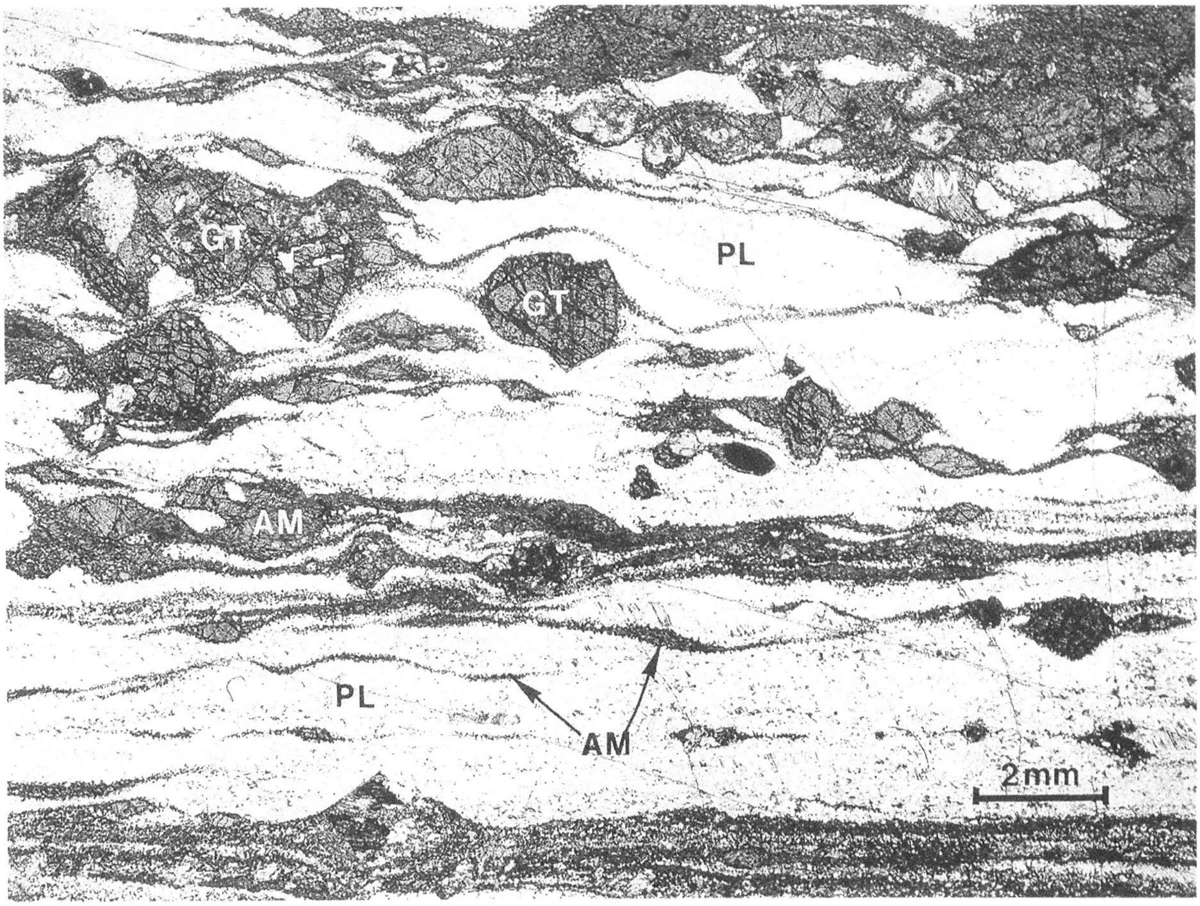


152 A

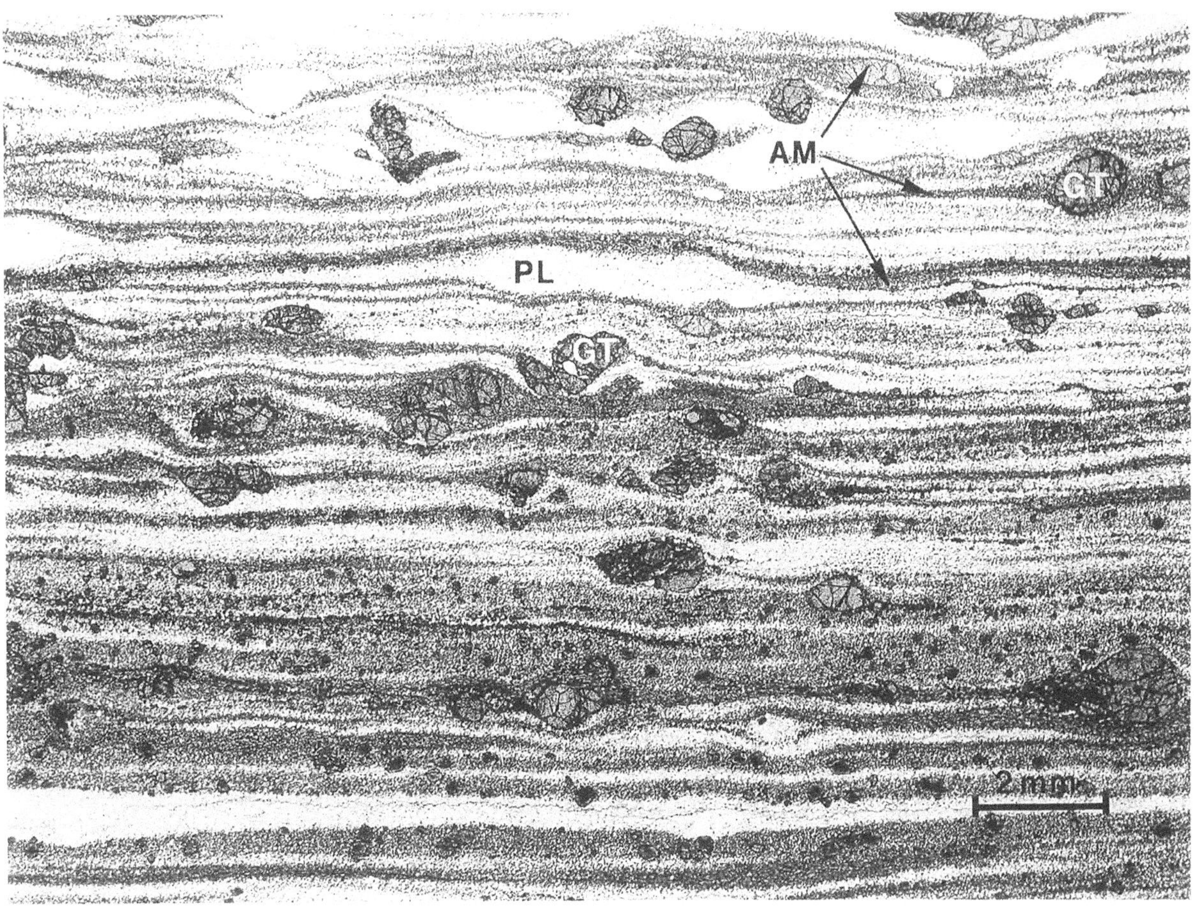


152 B

# 153. FELDSPATHIC MYLONITES

H.-R. WENK

The emplacement of the mostly crystalline Tertiary Bergell granite into the higher Pennine nappes (Figure 153.1) produced pervasive ductile deformation within the granite and surrounding metamorphic wall rocks (Wenk, 1973, 1987). At the time of emplacement (Oligocene-Miocene), metamorphic conditions ranged from granulite- to amphibolite-facies conditions in the southwest to lower greenschist facies in the northeast, as indicated by mineral assemblages, e.g., in pelitic schists (Wenk and others, 1974). Metamorphic grade has controlled the mode of deformation. In the high-grade part, intragranular deformation affected chiefly quartz, whereas stronger feldspars are largely preserved. The macroscopic preferred orientation of alkali-feldspar megacrysts which is pronounced near the external contact of the granite (153A) is likely the result of rotations of rigid anisotropic crystals in a more viscous matrix during deformation. Quartz commonly exhibits undulatory extinction, and in more highly strained samples is recystallized to fine-grained aggregates, as is typical of mylonites (153B and 153C). The reduction in grain size is not caused by cataclastic processes (i.e., "grinding"), but by dynamic recrystallization. Dislocations are arranged in equilibrated networks along subgrain boundaries. Subgrains and grains meet in triple junctions (153D). Dislocation movements and associated slip produced strong preferred orientation with [0001] of quartz lying in the plane of foliation, normal to the lineation, or a triclinic variant of this texture (Figure 153.2) which is attributed to simple shear in heterogeneous local shear zones (Wenk, 1984). While deformation was ductile at high grade and produced pervasive mylonites, brittle conditions dominated in the greenschist-facies zone, and strain was accommodated along discrete faults (e.g., the Maloja fault with a left-lateral displacement of 2–4 km at Maloja Pass) (Wenk, 1984).

**153A.** Preferred orientation of alkali-feldspar megacrysts in Bergell granite. Alpe Trubinasca. NX

**153B.** Mylonitic Bergell granite with quartz recrystallized to a fine-grained matrix and feldspars preserved as porphyroclasts. NX

**153C.** Ultramylonite in rocks of the contact region. Feldspars are rounded and quartz recrystallized to a fine-grained aggregate with strong preferred orientation. NX

**153D.** TEM photograph of quartzo-feldpathic mylonite as in 153C with dislocations recovered in subgrain boundaries and grains meeting in ideal triple junctions.

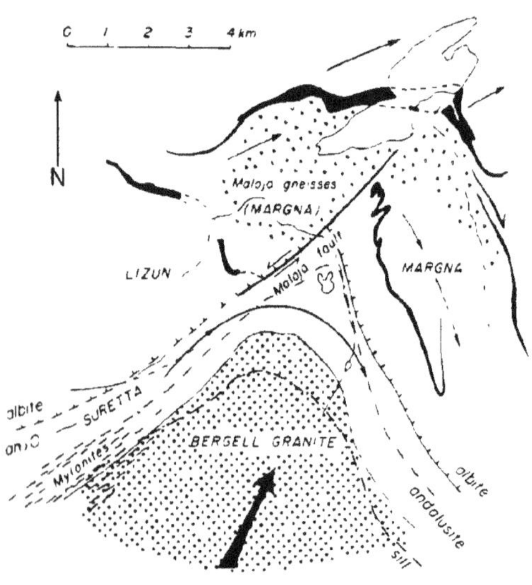


**Figure 153.1.** Geologic sketch map of the northeastern Bergell Alps, Switzerland. Thin arrows indicate linear structures, carbonate marker beds are shown in black, and some metamorphic isograds are also delineated

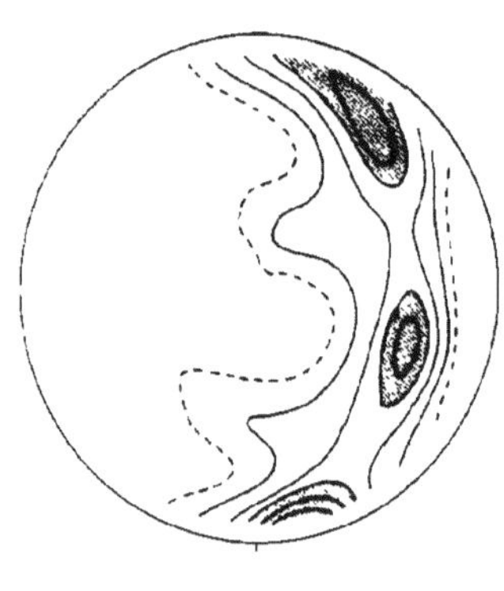

**Figure 153.2.** (11$\bar{2}$0) pole figure of quartz in quartzite mylonite from Boch Tegiola with triclinic symmetry attributed to simple shear. Projected onto the plane of schistosity. l is the lineation. Contours are in multiples of a random distribution [0.5 (dashed), 1, 2, 3, 4, 5; area shaded above 3]

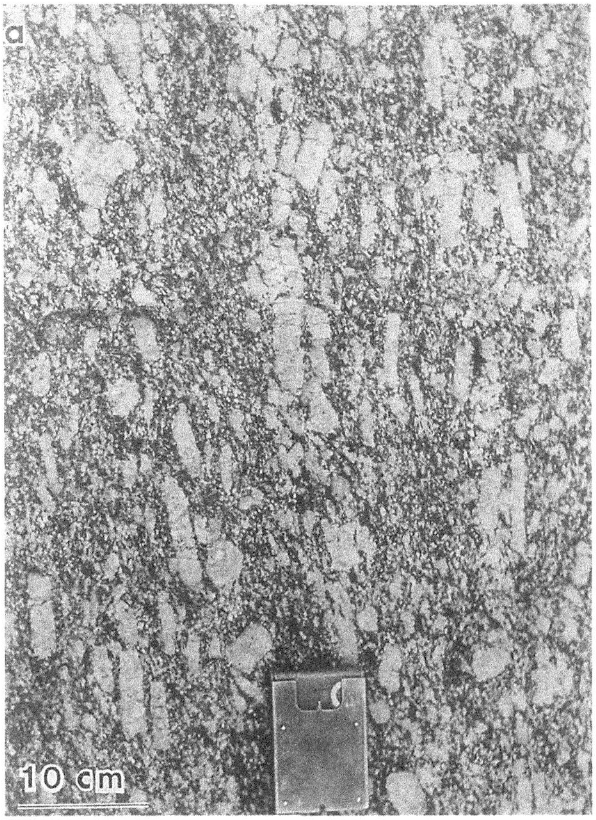

153 A

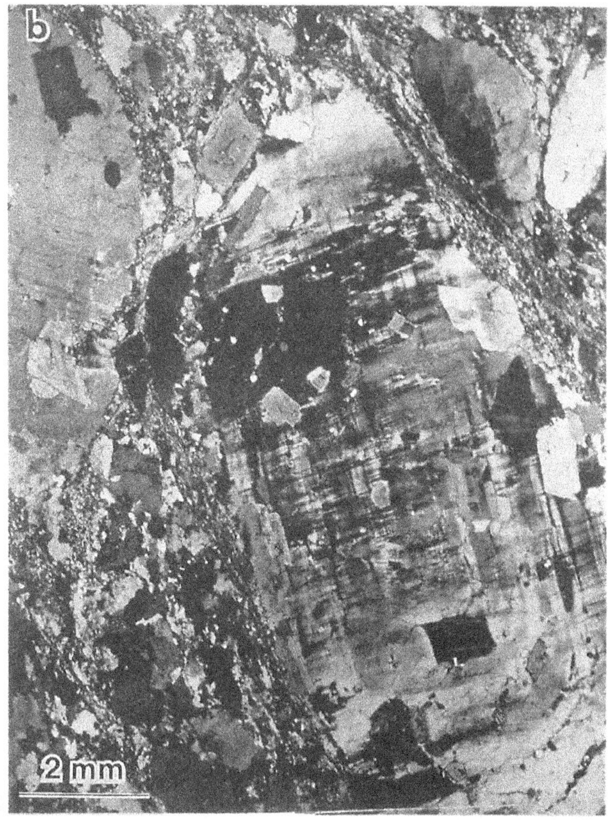

153 B

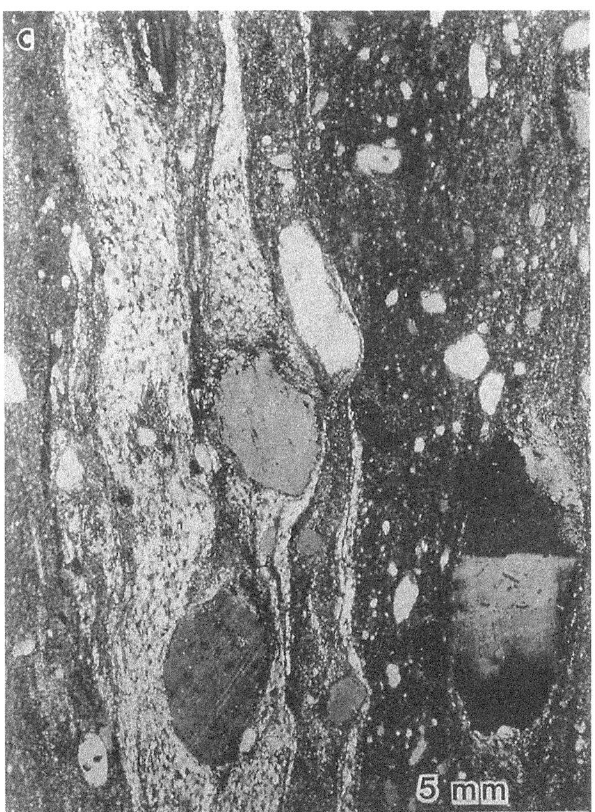

153 C

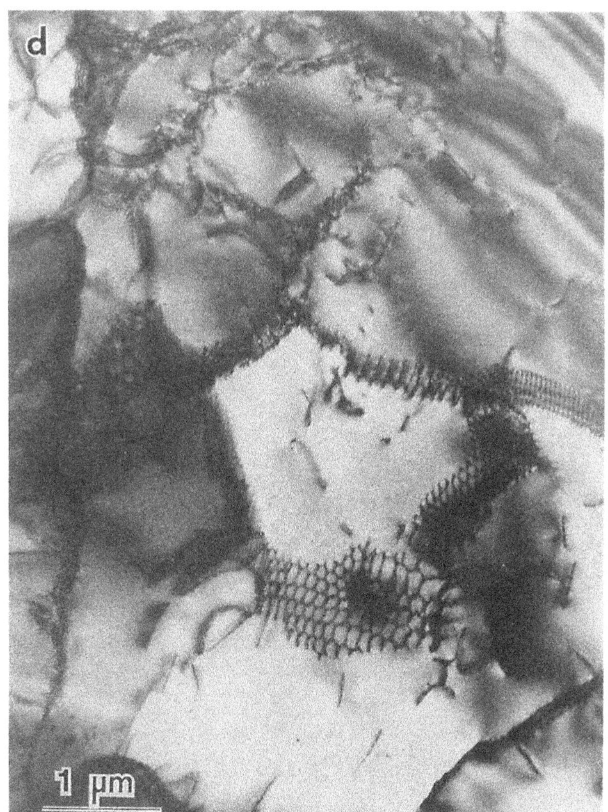

153 D

# 154. TONALITE MYLONITE FORMED DURING CRUSTAL EXHUMATION

Tsuyoshi Toyoshima

A large mass of tonalite mylonite (TM in Figure 154.1) occurs as a north-south-striking, basal zone (western-margin mylonite zone) of the Hidaka metamorphic belt (HMB), Hokkaido (H), northern Japan, and extends nearly the length of the belt. Mylonitic foliation and mineral lineation are subparallel to the general north-south strike of the Hidaka metamorphic belt. The resultant texture in the basal mylonite zone changes gradually toward the west from coarse porphyroclastic texture to phyllitic texture, as shown in Plates 154A and 154B on the facing page. This textural change thus indicates a westward increase in the reduction of grain size in the tonalite. The basal zone developed as an intracrustal basal décollement under amphibolite- to greenschist-facies conditions during the middle Paleogene to Miocene (Toyoshima and others, 1994). Asymmetrical microstructures in the tonalite mylonite indicate a dextral sense-of-shear. Metamorphic and structural studies indicate that the rocks of the Hidaka metamorphic belt were exhumed and tilted steeply eastward during the mylonitization. Thus, the sense-of-shear in the mylonite zone included components of both east-west shortening and dextral strike-slip displacement, i.e., dextral transpression (Toyoshima and others, 1994).

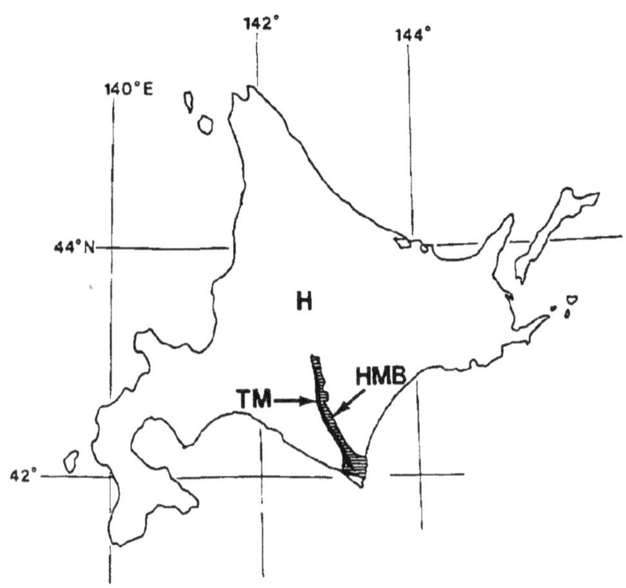


**Figure 154.1.** Sketch map of Hokkaido Japan showing the location of the Hidaka metamorphic belt (HMB) and the distribution of tonalite mylonites (TM) in the belt

**154A.** Tonalite mylonite with S-C-C′ composite planar fabrics, typical of the basal mylonite zone of the Hidaka metamorphic belt. The mylonite contains rounded porphyroclasts of plagioclase (Plpc) and much subordinate porphyroclastic biotite (Btpc). The matrix minerals of the mylonite are chiefly biotite (Bt), muscovite (Ms), plagioclase, and quartz. The mylonitization occurred under lower amphibolite-facies conditions. The mylonite displays asymmetrical structures such as S-C fabric (Berthé and others, 1979a; Lister and Snoke, 1984) and C′ surfaces (Ponce de Leon and Choukroune, 1980) as well as porphyroclasts with asymmetrical tails. S, C, and C′ surfaces are formed by recrystallized, fine-grained quartz, plagioclase, biotite, and small amounts of muscovite. Quartz ribbons (Qrb) also define these surfaces. The asymmetrical tails (At) in the tonalite mylonite consist of a mixture of biotite, plagioclase, and quartz, or of the alternation of biotite-rich layers, plagioclase-rich layers, and quartz ribbons. PPL

**154B.** Ultramylonite with S-C composite planar fabrics. The ultramylonite occurs locally as thin bands within the western part of the basal mylonite zone. The ultramylonite is mostly formed by recrystallized fine-grained biotite (Bt), by a mixture (M) of recrystallized fine-grained biotite, muscovite, plagioclase, and quartz, and by quartz ribbons (Qrb). These minerals define asymmetrical structures such as S-C fabric, foliation boudinage (Platt and Vissers, 1980), or foliation fish (Hanmer, 1986). Metamorphic and structural analysis indicates that the ultramylonite was formed somewhat later and at lower-temperature conditions (greenschist-facies conditions) than the mylonite shown in 154A, during a retrograde metamorphic phase of the Hidaka metamorphic belt (Toyoshima, 1991). Textural and microstructural transitions indicate that the strain and dynamic recrystallization of the mylonitization were successively concentrated into a narrow zone within the basal mylonite zone owing to decreasing temperature. It is also possible that the formation of the ultramylonite resulted from an increase of the strain rate. PPL

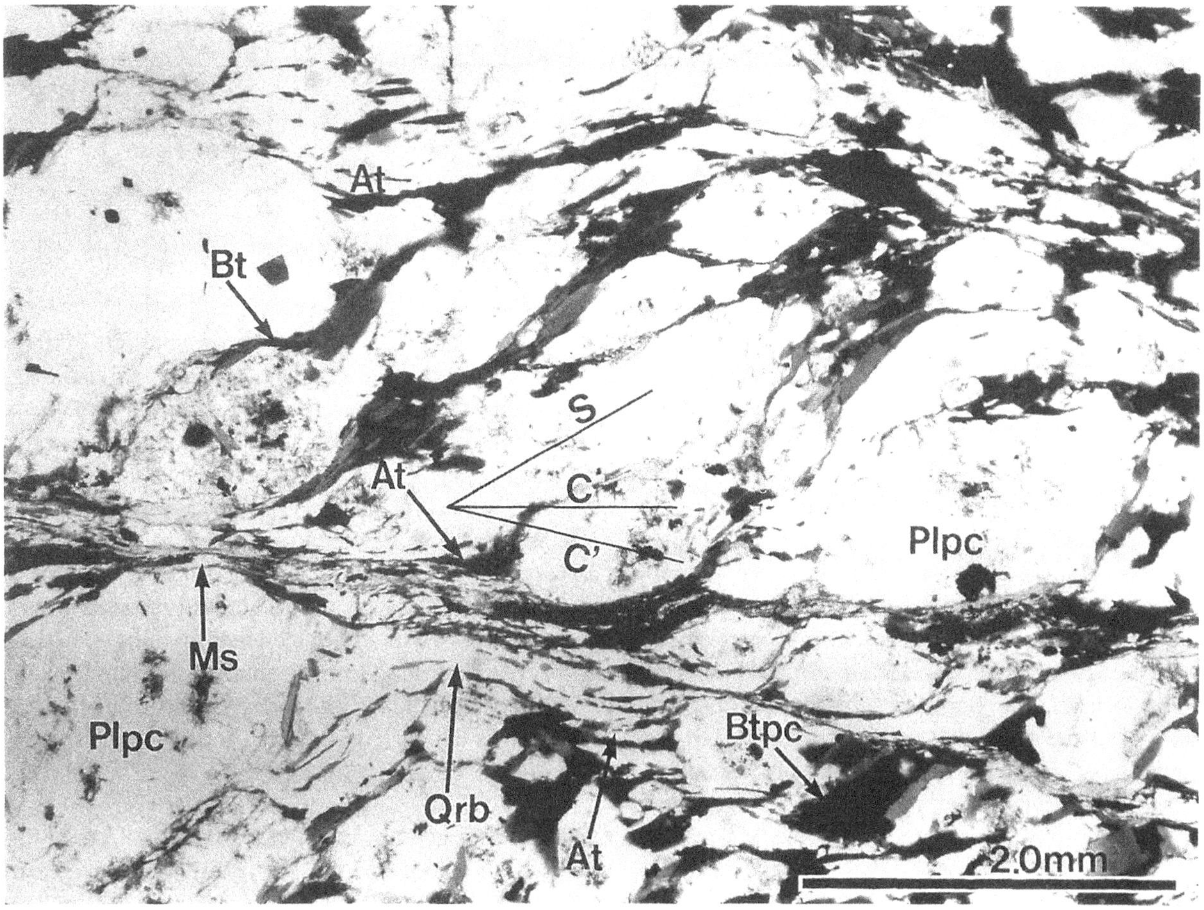


154 A

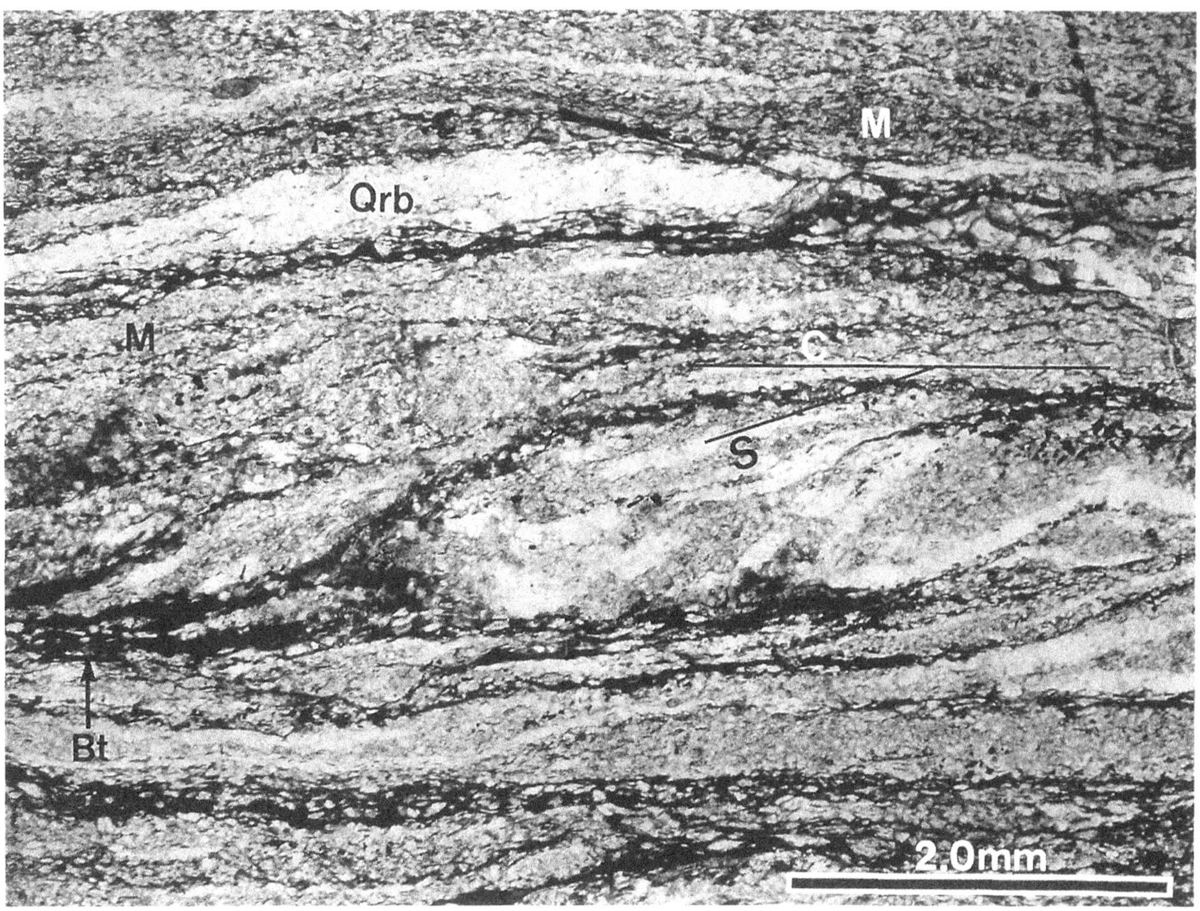


154 B

# 155. MYLONITIZATION AND REMOBILIZATION OF MONZOGRANITE DURING METAMORPHISM AND INTRUSION

VICTORIA R. TODD

Basin and Range normal faults of Pliocene or younger age expose a region in northwest Utah, U.S.A., that was affected by Tertiary heating, extension, and uplift (Raft River–Grouse Creek Mountains metamorphic core complex; Compton and others, 1977) (Figure 155.1). Within the central Grouse Creek Mountains, there is an Oligocene-Miocene culmination (gneiss dome) in which 2.5 Ga monzogranite underwent progressive upward mylonitization and became intercalated with lowermost Proterozoic(?) strata that, elsewhere in the region, overlie the monzogranite nonconformably (Todd, 1980) (Figure 155.2). Three near-horizontal, younger-over-older allochthons developed in Proterozoic(?) through Mesozoic rocks that overlay the gneiss dome during Tertiary peak-grade, upper amphibolite-facies metamorphism, folding, and low-angle faulting. During this deformation, Proterozoic(?) and Paleozoic strata were thinned ductilely and by low-angle faulting to about one-fifth their original thickness. Late- to post-metamorphic low-angle faults carried Paleozoic and Mesozoic strata over Miocene sedimentary rocks in adjoining basins (Todd, 1983).

The intensity of strain and recrystallization in the gneiss dome increases upward as shown by a progressive decrease in grain size and increase in degree of overprinting of magmatic by crystal-plastic textures. Monzogranite showing relict magmatic textures in the structurally deepest exposures of the gneiss dome (~900 m below the projected contact with Proterozoic? cover rocks) grades upward to fine-grained mylonitic gneiss and schist in the uppermost 200 m. Mylonitic rocks have a low-dipping foliation defined by flattened mineral grains and aggregates and by recumbently folded leucocratic schlieren. At the top of the dome, muscovite-rich gneiss and schist are interlayered concordantly with relict beds of the lowermost 5 m of autochthonous Proterozoic(?) Elba Quartzite in a pseudointrusive relation. Elsewhere, the contact between the Precambrian monzogranite and Elba Quartzite is a low-angle fault facilitated by relatively high fluid pressure beneath the impermeable metaquartzite. Remobilization of the uppermost part of the 2.5 Ga autochthonous monzogranite evidently was facilitated by the ductility contrast between the metaquartzite and quartz- and mica-rich mylonitic schist during retrograde metamorphism.

**155A.** Medium-grained monzogranite gneiss from western foot of central Grouse Creek Mountains (Figure 155.2), ~900 m below top of gneiss dome, consists of feldspar porphyroclasts, lenticular quartz aggregates, biotite and muscovite folia, and fine-grained granoblastic matrix composed of the same minerals. The largest feldspar grains are porphyroclasts (relict magmatic grains) whose margins are embayed by new, smaller recrystallized feldspar grains (dynamic recrystallization). Large twinned plagioclase ($An_{20-22}$) porphyroclast near center of view is rimmed by fine-grained mosaic consisting of quartz, plagioclase, potassium feldspar, and mica. Subhedral biotite and muscovite grains project into this porphyroclast, and smaller muscovite plates within it are oriented parallel to cleavage planes, both suggesting significant element mobility. Whereas some relatively intact potassium-feldspar phenocrysts (K) survive, much potassium feldspar is present in complex fine-grained intergrowths with plagioclase (the latter extensively replaced by potassium feldspar) and myrmekite. NX

**155B.** Fine-grained monzogranite gneiss collected ~8 m below top of gneiss dome (Figure 155.2) consists of extensively sericitized plagioclase porphyroclasts (P), relatively clear potassium-feldspar porphyroclasts (K), quartz ribbons, and fine-grained matrix (M). The matrix is composed of quartz, potassium feldspar, muscovite, and plagioclase that has undergone sericitization, albitization, and replacement by potassium feldspar. Some plagioclase ($P_1$) and potassium-feldspar ($K_1$) porphyroclasts contain multiple subgrains in close optical continuity, which suggests upper greenschist- to lower amphibolite-facies metamorphic conditions. Quartz ribbons are interleaved with folia composed of fine-grained muscovite, biotite, and clinozoisite. NX

**155C.** Schistose monzogranite (typically albite–quartz–potassium feldspar–muscovite–biotite schist) collected ~3 m below top of gneiss dome (Figure 155.2). Very fine-grained albitic plagioclase occurs almost exclusively in matrix mosaic with quartz and potassium feldspar; larger grains in this view are microcline-twinned potassium-feldspar porphyroblasts (K), bands of coarse muscovite with very fine-grained chloritized biotite (M), and quartz ribbons typically composed of subrectangular subgrains. Schists of this zone exemplify retrograde reactions that took place throughout the Precambrian monzogranite during intrusion of the Red Butte stock, low-angle faulting, and second metamorphic folding. At this time, monzogranite was converted to muscovite-quartz-albite-chlorite-epidote-sphene assemblages under greenschist-facies conditions. Abundant quartz released in these reactions forms mesoscopic veinlets in schist (not illustrated). Muscovite formed by these reactions is iron- and magnesium-rich, suggesting relatively high water pressures. NX

155 A

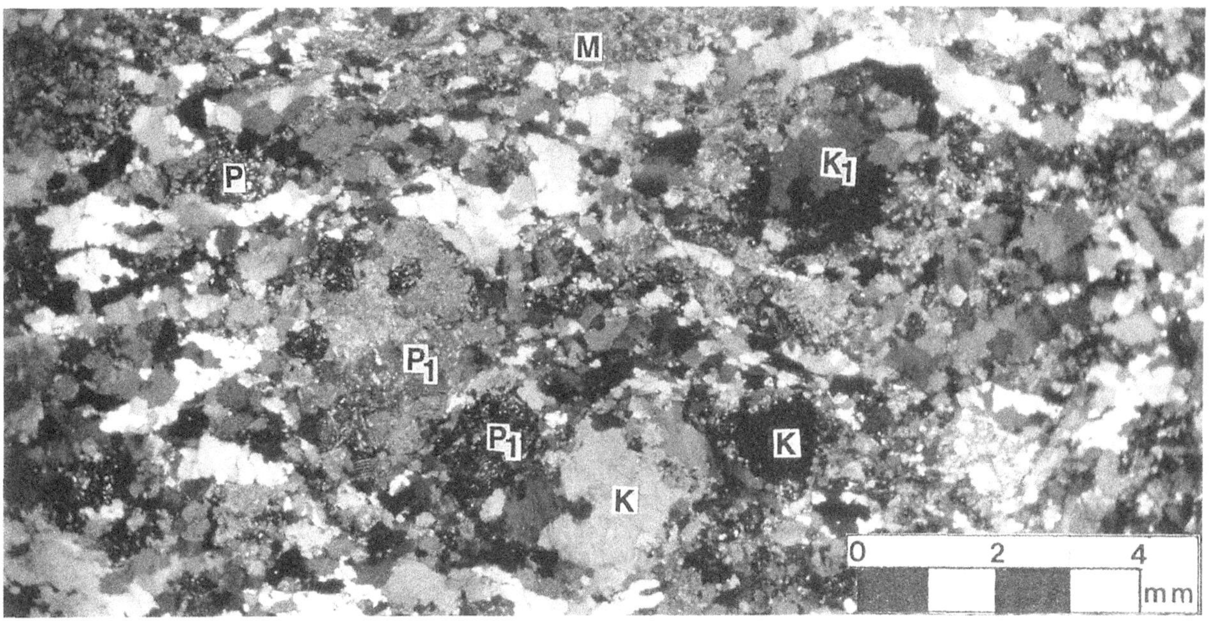


155 B

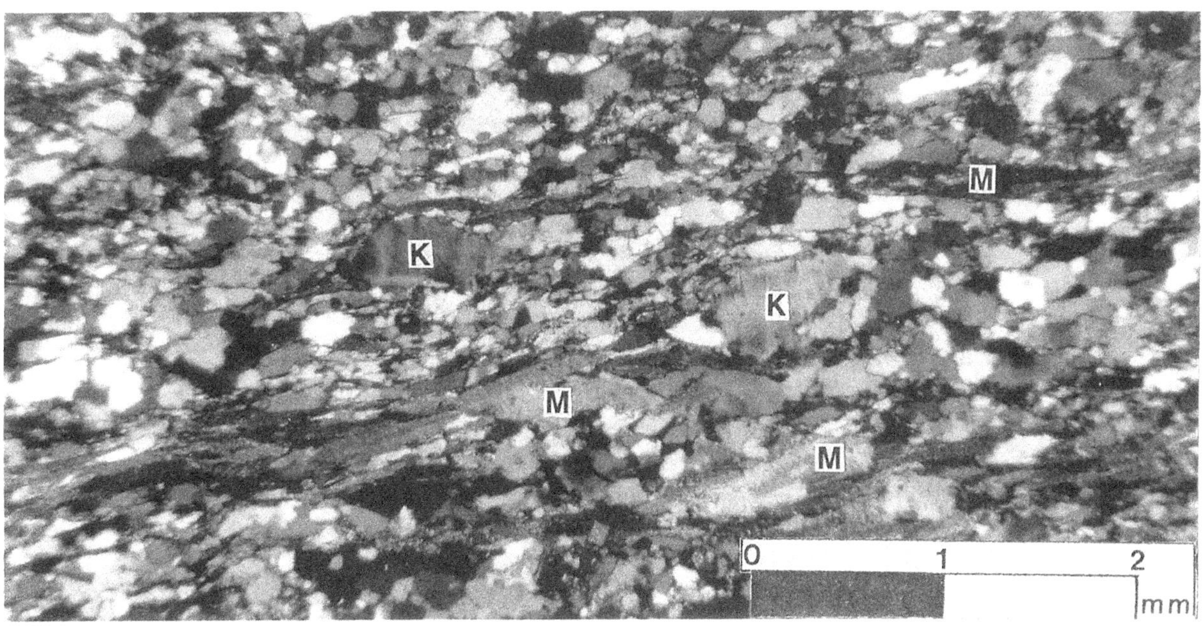


155 C

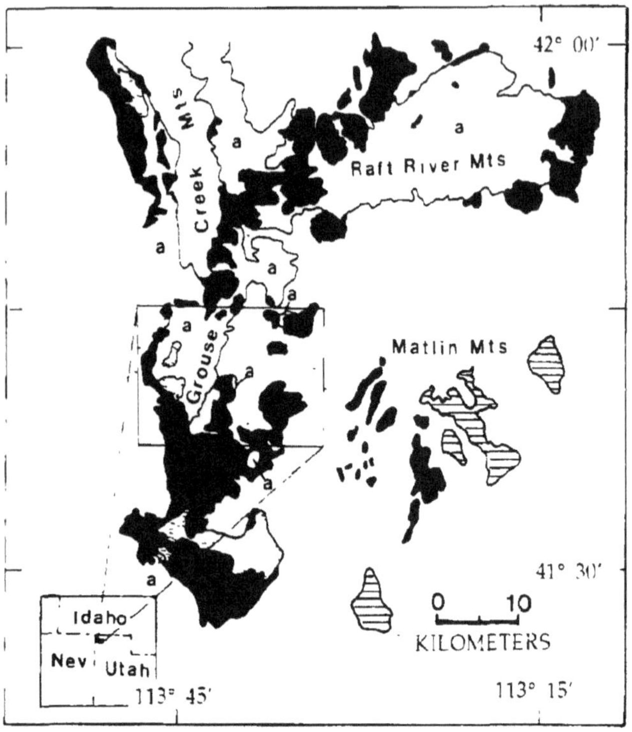


**Figure 155.1.** Generalized geologic map of the Raft River–Grouse Creek Mountains metamorphic core complex, northwest Utah (after Compton and others, 1977). Autochthon, composed mainly of 2.5 Ga monzogranite, a; three stacked allochthonous sheets (Proterozoic?, Paleozoic, and Triassic), black; rooted Pennsylvanian through Triassic carbonate rocks, horizontal lines; Tertiary granite, heavy stipple; Cenozoic deposits, light stipple. Box indicates area of Figure 155.2.

Remobilization of 2.5 Ga monzogranite in the central Grouse Creek Mountains was spatially and temporally related to intrusion of a 24.9 ± 0.5 Ma (Compton and others, 1977) monzogranite pluton (Red Butte stocks; Todd, 1980) (Figure 155.2). The presence of NW- to E-W-striking (second) metamorphic lineation and folds in the outer part of the Oligocene pluton and in the Precambrian monzogranite surrounding it, as well as in the uppermost 200 m of the gneiss dome, indicates a genetic relation between development of the dome and Oligocene intrusion. The upper part of the gneiss dome is also marked by abundant leucogranite gneiss and gneissic pegmatite and by the synkinematic assemblage muscovite (phengite)-quartz-albite-chlorite-sphene, which together indicate relatively high water pressure and greenschist-facies conditions during formation of the dome. Rubidium-strontium biotite whole rock ages of 11.9 and 8.0 Ma for Precambrian monzogranite and fission track ages of 18.3 ± 1.9 Ma for zircon and 18.9 ± 6.3 Ma and 13.7 ± 3.7 Ma for apatite from the Oligocene pluton (Compton and others, 1977) indicate that the Raft River–Grouse Creek Mountains core complex cooled below metamorphic temperatures in the Miocene.

**155D.** Weakly foliated, coarse-grained monzogranite from interior of northern stock (Figure 155.2) shows magmatic texture including euhedral and subhedral phenocrysts of potassium-feldspar (K) and plagioclase (P), large intact quartz grains (Q), and coarse biotite (B). Potassium-feldspar grains display euhedral oscillatory zoning ($K_1$) and Carlsbad twins ($K_2$): both grains contain inclusions of early plagioclase. Mild deformation and alteration in stock interior are indicated by weakly strained, fractured quartz and slightly sericitized plagioclase. NX

**155E.** Mylonitic monzogranite ~3.5 m below middle allochthon, outer part of southern stock (Figure 155.2). Middle allochthon is contact-metamorphosed next to stock and both it and mylonitic monzogranite bear second metamorphic lineation (oriented ~normal to page). Rock consists of plagioclase (P) and potassium-feldspar (K) porphyroclasts surrounded by bands of very fine-grained matrix (M) and slightly coarser folia composed of recrystallized quartz (Q) and chloritized biotite. Large plagioclase porphyroclast ($P_1$) is fractured and segment ($P_2$) is separated from main grain by seam of recrystallized quartz and matrix. Smaller, ovoid and lenticular potassium-feldspar porphyroclasts (dark grains) are strain-shadowed and locally contain deformation lamellae (not shown). NX

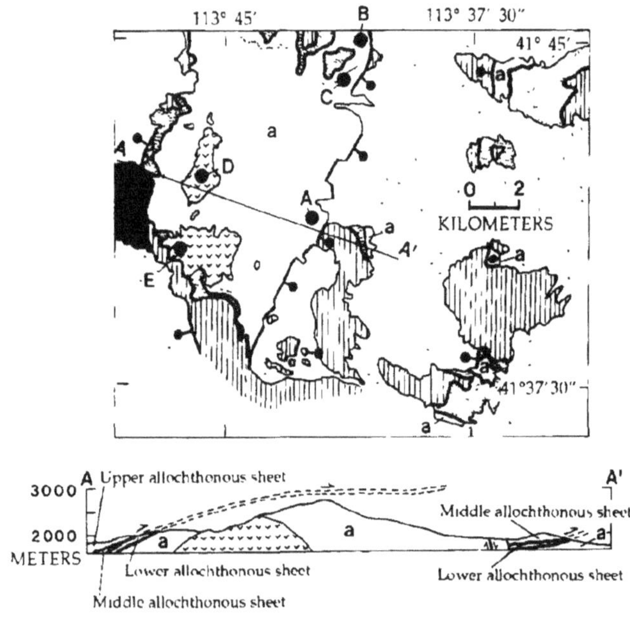


**Figure 155.2.** Generalized geologic map and cross section (A-A′) of central Grouse Creek Mountains, Utah (after Todd, 1980). Autochthon composed chiefly of Precambrian monzogranite (~2.5 Ga), unpatterned areas marked "a", discontinuous dotted line, exposures of base of autochthonous Proterozoic(?) quartzite and schist (top of remobilized Precambrian monzogranite). Lower allochthonous sheet (Proterozoic? through Ordovician), close stipple; middle allochthonous sheet (Mississippian through Permian), vertical lines; upper allochthonous sheet (Permian and Triassic), black. Red Butte stocks (24.9 ± 0.5 Ma), v pattern; small bodies adjacent to stocks are Tertiary alaskite. Cenozoic rocks, light stipple. Light solid line, contact; heavy solid line, low-angle fault (except where bar and bell indicate downthrown block of Basin and Range high-angle fault). Solid circles, locations of samples 155A–155E.

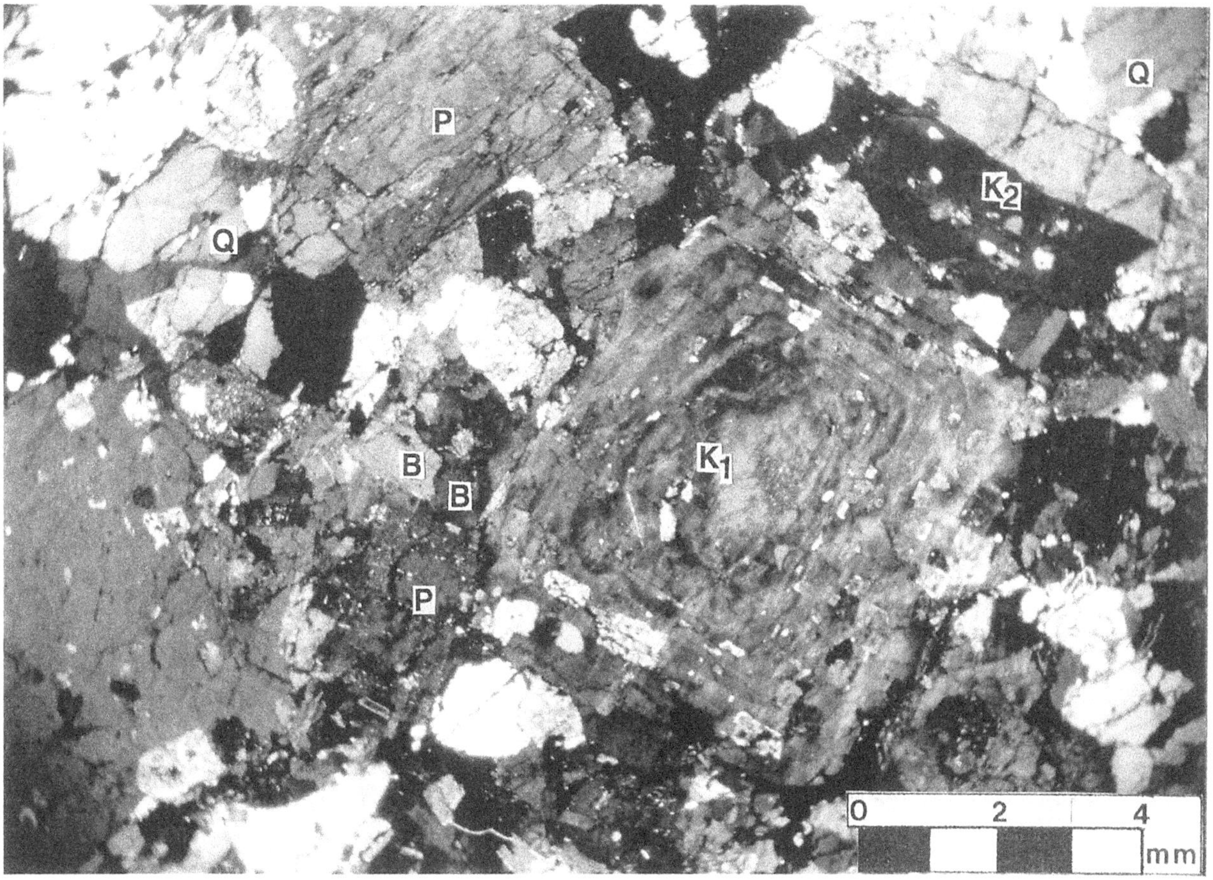


155 D

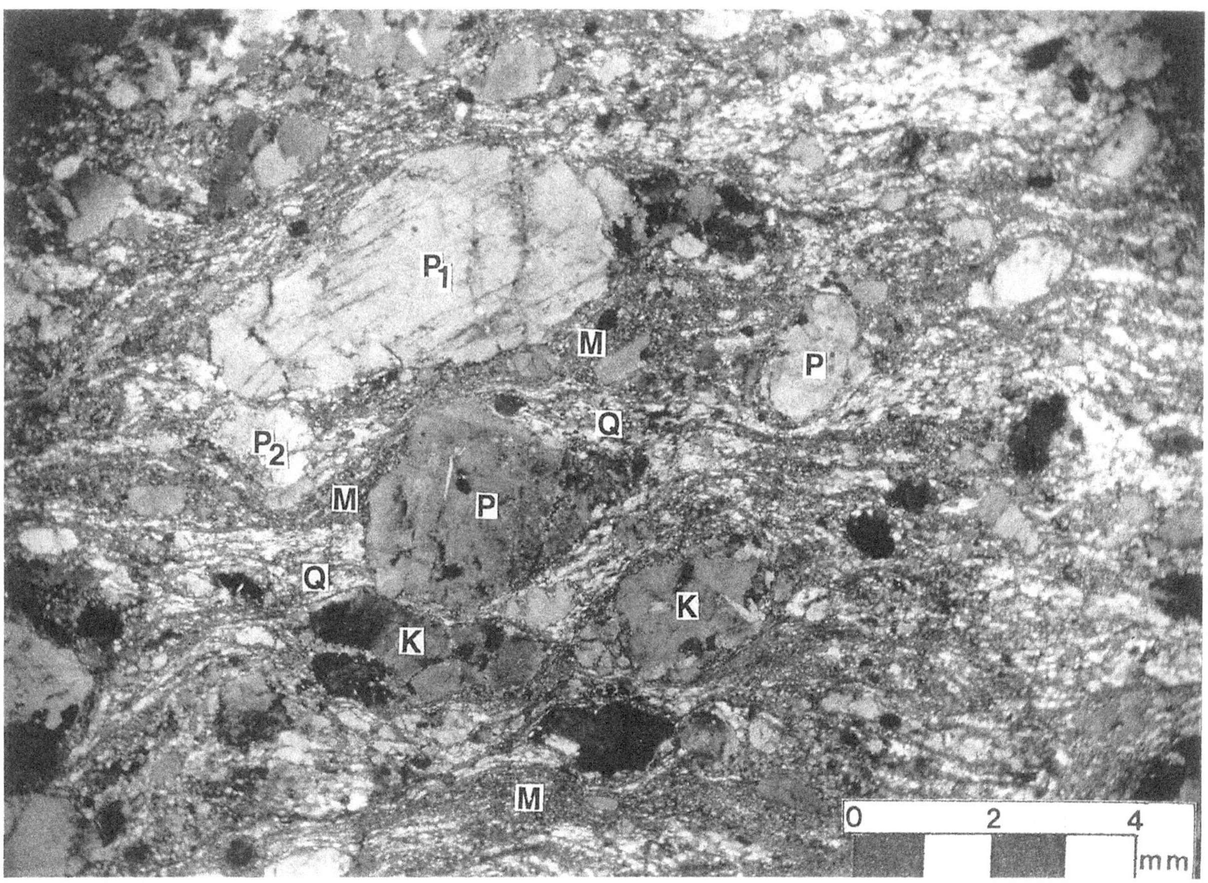


155 E

# 156. INTRUSION AND DUCTILE DEFORMATION IN A LONG-LIVED BATHOLITHIC SHEAR ZONE

VICTORIA R. TODD

The Peninsular Ranges batholith, southern California, U.S.A., and Baja California, Mexico, was emplaced in the Jurassic and Cretaceous across an early Mesozoic lithospheric boundary between North America and oceanic plates. Recent studies indicate that intra-arc ductile deformation played an important role in the evolution of the batholith (Grove, 1987; Todd and others, 1988; Thomson and others, 1994). As the Cretaceous arc migrated eastward, progressively younger, arc-parallel zones of synintrusive and postintrusive deformation developed at, or near, preexisting crustal boundaries. The Early Cretaceous Cuyamaca-Laguna Mountains shear zone (CLMSZ) (Figure 156.1) developed in a 15 to 20 km wide zone of mechanically weak Jurassic S-type granitoid plutons and amphibolite-facies Triassic-Jurassic metasedimentary wall rocks. The NW-elongate shapes and hypabyssal textures of some Cretaceous plutons in the zone suggest that magmas were emplaced initially as dikes into crust that was undergoing NE-SW extension. As intrusion continued and the crust was thermally weakened, Jurassic plutons in the shear zone developed mylonitic fabrics, and Cretaceous plutons developed foliated, gneissic, and protomylonitic fabrics.

Thomson and others (1994; see also Girty and others, this volume, Plate 100) documented an Early Cretaceous episode of NE-SW contraction in the CLMSZ followed by mid- to Late Cretaceous NE-SW extension. The ~118 Ma syntectonic Pine Valley monzogranite-granodiorite pluton was emplaced into the shear zone during mylonitization of an adjacent Jurassic pluton. Microtextures of the Pine Valley pluton and its dikes record magmatic flow, submagmatic to high-temperature solid-state flow, and lower-temperature solid-state flow during an interval of about 10 to 15 m.y.

**156A.** Monzogranite with well-developed magmatic foliation defined by aligned subhedral feldspar grains, lenticular quartz grains, and biotite aggregates is typical of much of the Pine Valley pluton (Figure 156.1). Sparsely perthitic potassium-feldspar ($Or_{?}$) phenocrysts (K) contain Carlsbad twins oriented parallel to rock foliation. Large grain (bottom left) contains early plagioclase inclusions. Small subhedral plagioclase (average $An_{2?}$) grains (P) show oscillatory zoning. Slight submagmatic strain is indicated by recrystallization of large magmatic quartz grains to aggregates of smaller grains with sutured boundaries (Q); marginal embayment of potassium feldspar by fine-grained quartz-feldspar-biotite aggregates and (adjacent to plagioclase) myrmekite (M); and partial recrystallization of magmatic biotite to aggregates of smaller platy grains (B). NX

**156B.** Gneissic foliation, overall finer grain size, and development of fine-grained quartz-feldspar-biotite matrix (M) indicate stronger deformation of granodiorite from western margin of Pine Valley pluton (Figure 156.1). Most quartz occurs in elongate fine-grained aggregates (Q); many feldspar grains (P, K) are ovoid rather than subhedral; and biotite is present mainly in folia of recrystallized grains. Oscillatory zoning in plagioclase (average $An_{25}$) is progressively obliterated with increase in deformation twinning. Strongly deformed plagioclase phenocryst ($P_1$) contains subgrains with similar crystallographic orientations, suggesting dynamic recrystallization at middle to upper amphibolite-facies temperatures. The presence of interstitial melt during deformation is suggested by the occurrence of spidery, irregular potassium-feldspar grains (visible in stained slabs) that replace plagioclase and are partly oblique to foliation. NX

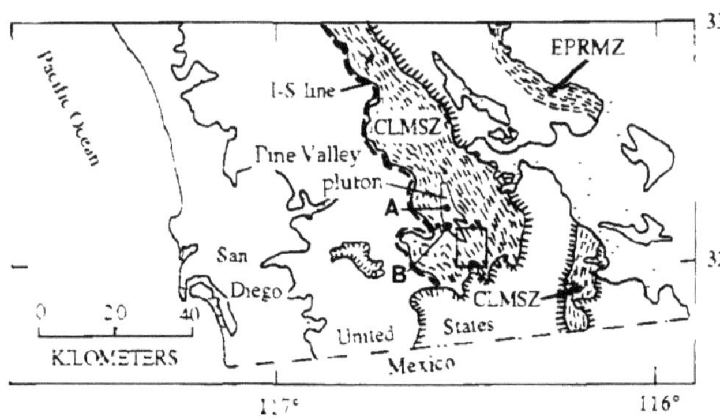


**Figure 156.1.** Schematic map of Peninsular Ranges batholith east of San Diego, California, showing Cretaceous shear zones. Unpatterned areas, Mesozoic batholithic and prebatholithic rocks; heavy dashed line is western limit of Jurassic metagranitoids. Early Cretaceous plutons such as the Pine Valley pluton intruded the Jurassic terrane but are most abundant west of it; hachured line is western boundary of mid- to Late Cretaceous plutons. Cenozoic rocks stippled. Ductile shear zones dashes: Early Cretaceous Cuyamaca-Laguna Mountains shear zone, CLMSZ; Late Cretaceous eastern Peninsular Ranges mylonite zone (Sharp, 1979), EPRMZ. Box indicates area of Figure 156.2; solid circles, locations of samples 156A and 156B.

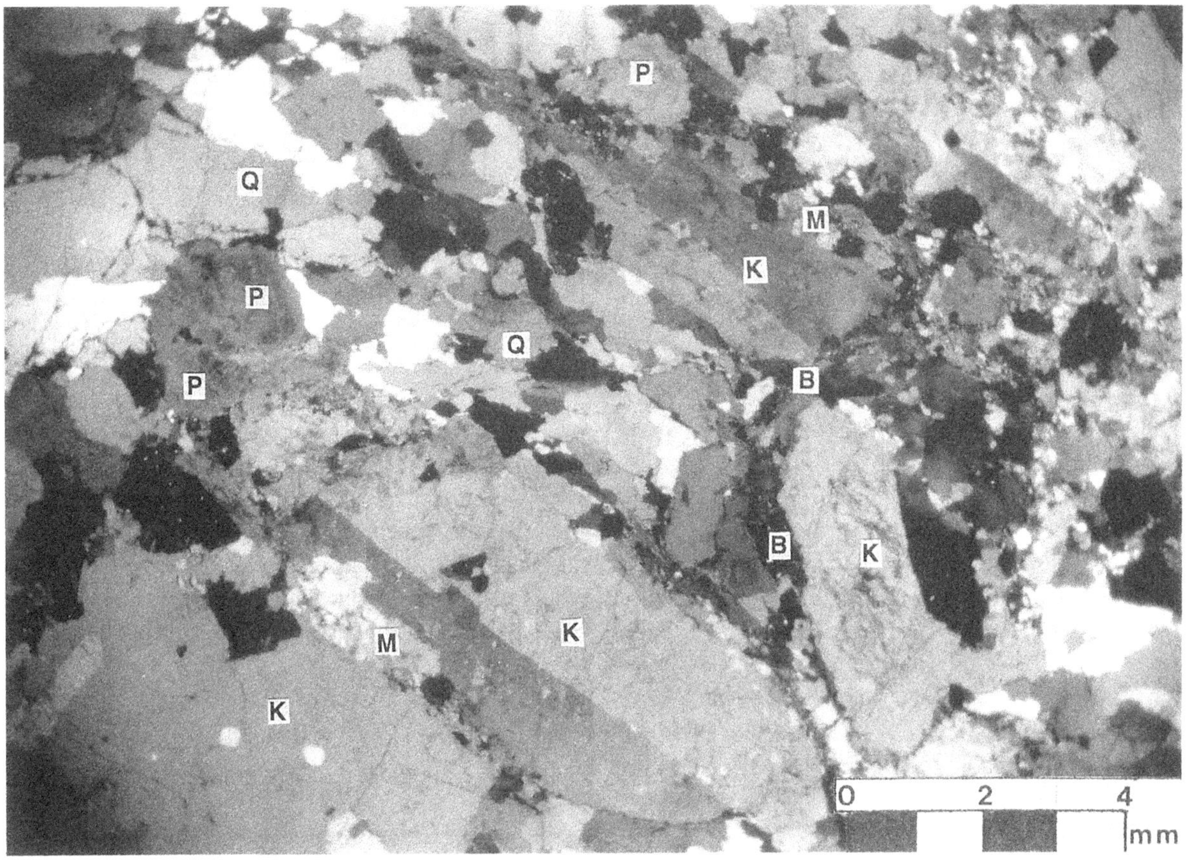

156 A

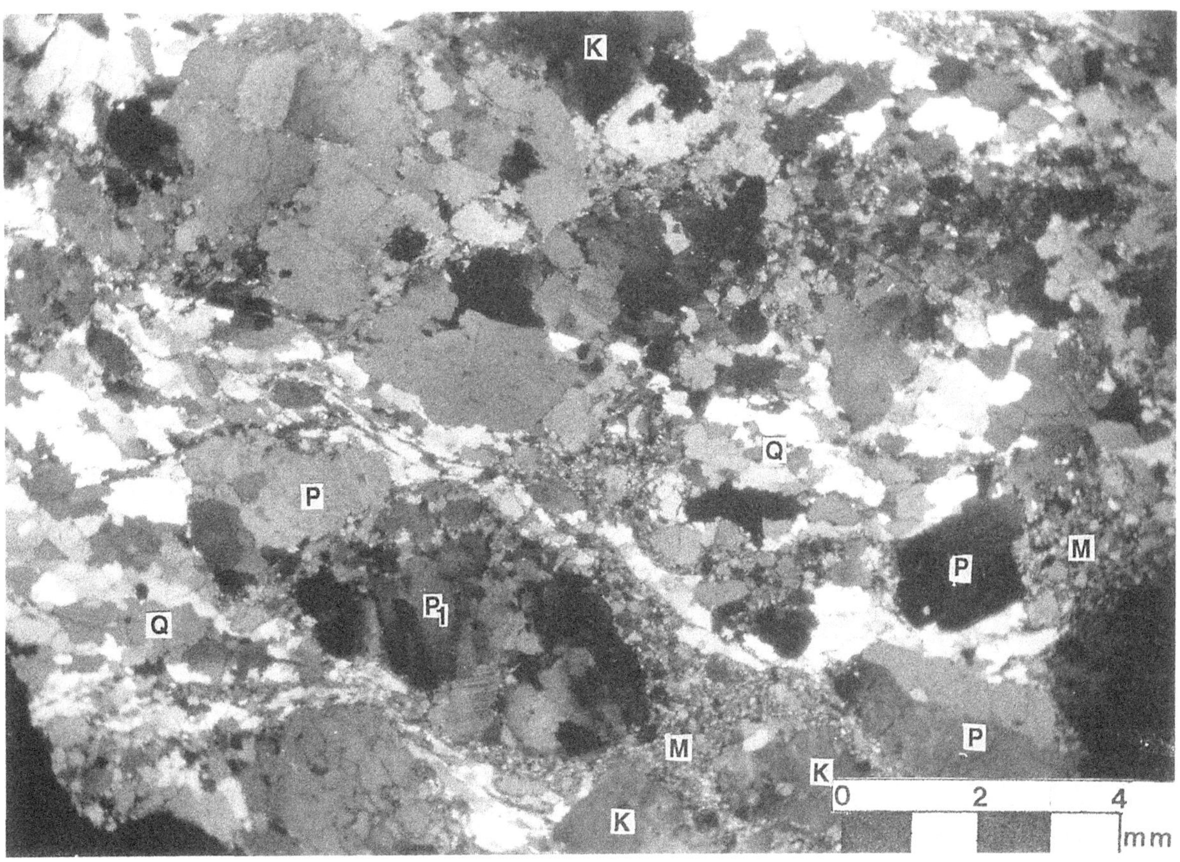

156 B

Plutons in the Cottonwood Valley segment (CVS) of the Cuyamaca-Laguna Mountains shear zone (Figure 156.2) show the effects of submagmatic to post-magmatic deformation and intrusion by younger plutons. Jurassic and Cretaceous granitoid plutons, including the southern end of the Pine Valley pluton and associated leucogranite dikes, were attenuated in the CVS to a dozen or more thin (~50–150 m) sheets that dip ENE from 65° to 85°. The sheeted plutons bear heterogeneous mylonitic and gneissic foliation and steep down-dip lineation. Local development of synkinematic retrograde minerals indicates upper greenschist-facies conditions during the latest preserved solid-state deformation. The CVS was intruded on the east and south by two regionally extensive, mid- to Late Cretaceous tonalite plutons (Figure 156.2), whose margins are strongly foliated to gneissic adjacent to the CVS. The observation that the CVS, the most deformed part of the CLMSZ, lies between (and may be partly underlain by) these two plutons suggests that heat and fluids from nearby crystallizing tonalite magma enhanced deformation and recrystallization. Compositions of hornblende from the tonalite pluton to the south suggest maximum crystallization pressures of ~450 MPa (Hammarstrom, 1992) at ~101 ± 2 Ma (U-Pb zircon data, D. L. Kimbrough, oral commun., 1993). An $^{40}Ar/^{39}Ar$ cooling age for the same pluton indicates that hornblende closed to argon loss at 94.3 ± 0.3 Ma (L. W. Snee, written commun., 1995).

**156C.** Protomylonite from southern "tail" of Pine Valley pluton (Figure 156.2) consists of relict magmatic phenocrysts in a fine-grained matrix of recrystallized quartz, feldspar, and mica (relict quenched texture?). Plagioclase porphyroclasts (P) contain deformation twins and patchy antiperthite. Band (B) consisting of fine-grained recrystallized quartz, biotite (partly altered to chlorite, sericite, and clinozoisite), and matrix is attenuated between large potassium-feldspar porphyroclast (K) (center of view) and smaller plagioclase porphyroclast (P). Shapes of many quartz aggregates and near optical continuity of subgrains within them suggest that aggregates represent recrystallization of late-stage interstitial quartz grains. NX

**156D.** Most intensely sheared rocks in CVS are isoclinally folded and boudinaged, mylonitic garnetiferous leucogranite dikes (Figure 156.2). Although grain size is greatly reduced, scarce large porphyroclasts (relict magmatic grains) survive in a groundmass composed of smaller feldspar porphyroclasts, fine-grained matrix (M), and folia consisting of recrystallized quartz and altered biotite (QB). The large perthitic potassium-feldspar porphyroclast consists of three segments ($K_{1-3}$) in near optical continuity, segments 1 and 2 separated by a seam of quartz and plagioclase. Plagioclase porphyroclasts (P) are sericitized, and fine-grained clinozoisite and chlorite locally replace plagioclase and biotite, indicating postintrusive, lower-temperature deformation. Sigmoidal bending of bands of matrix and recrystallized quartz (Q) implies dextral shear sense. NX

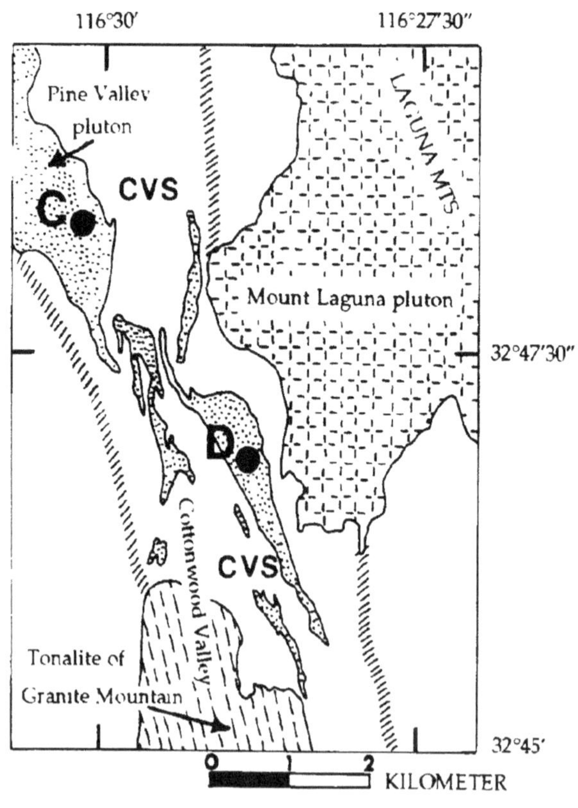


**Figure 156.2.** Simplified geologic map of Cottonwood Valley segment (CVS) of Cuyamaca-Laguna Mountains shear zone. CVS outlined approximately by hachured lines. Late- to post-tectonic tonalite plutons: tonalite of Granite Mountain (~101 Ma), dashes, Mount Laguna pluton (~104 Ma), crosses. Pine Valley pluton (~118 Ma) and related dikes, stippled. Undivided Cretaceous and Jurassic plutonic rocks, unpatterned. Solid circles show locations of samples 156C and 156D. No faults shown.

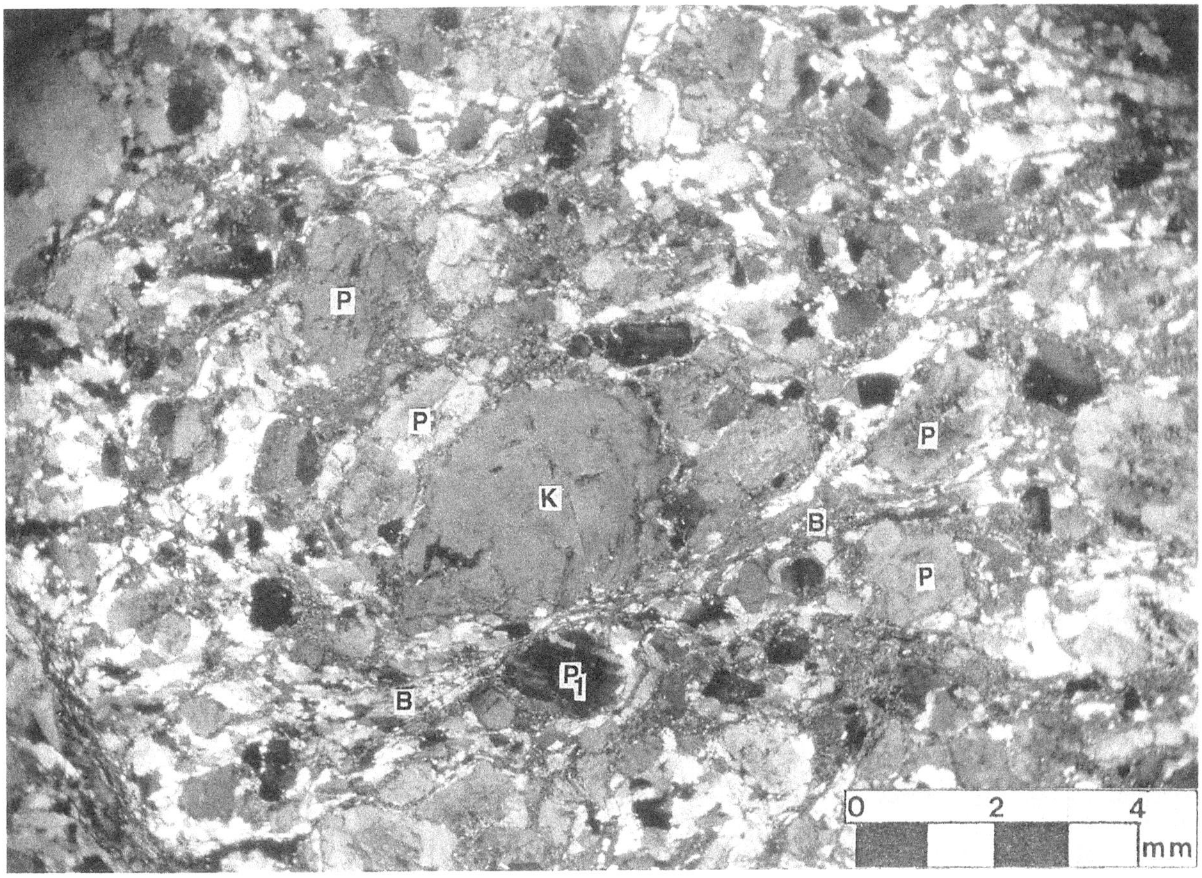


156 C

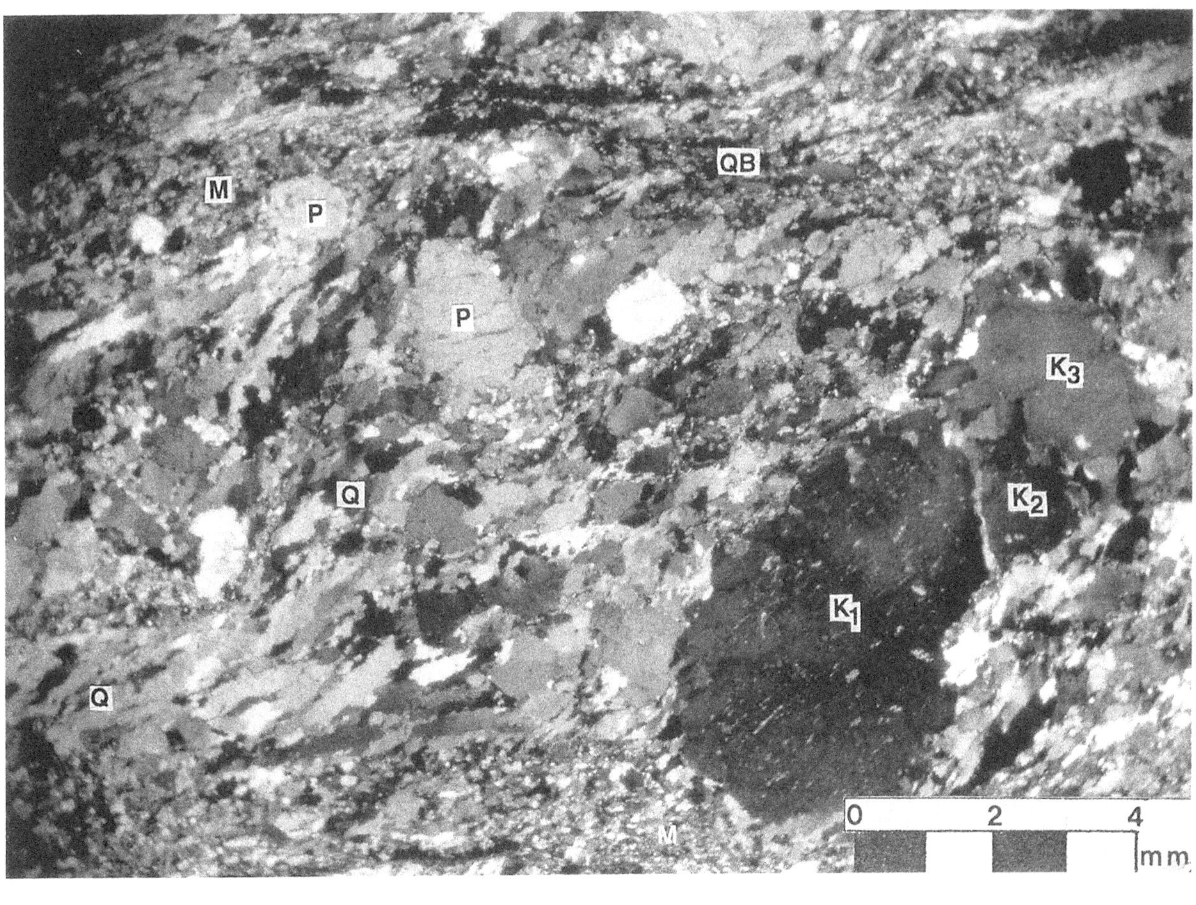


156 D

# 157. RECRYSTALLIZATION-ACCOMMODATED DISLOCATION CREEP IN ALBITE AGGREGATES

JAN TULLIS AND RICHARD A. YUND

These photographs illustrate the microstructural development that occurs during progressive deformation of albite aggregates by recrystallization-accommodated dislocation creep (Regime 1 of Hirth and Tullis, 1992; see also Tullis and Yund, 1985). The samples were subjected to axial compression at a confining pressure of 1500 MPa, a constant strain rate of $10^{-6}$/s, and a temperature of 1100°C. Recrystallization-accommodated dislocation creep occurs in the lower-temperature portion of the dislocation creep regime, where dislocation climb is very difficult. Dislocations tangle and become immobile after gliding only a short distance, leading to work-hardening. Local variations in dislocation density across original grain boundaries allow grain-boundary bulging, and tiny new strain-free recrystallized grains are formed. These can undergo an increment of easy slip before becoming work-hardened and replaced during a new episode of grain-boundary migration. Original grains are not flattened but are gradually reduced in size by progressive recrystallization. The strength of the aggregate decreases continuously from the time recrystallization is initiated until it is complete. This creep mechanism produces aggregates with a very fine (~1 μm) grain size and a strong crystallographic preferred orientation. At natural strain rates, this type of dislocation creep appears to be the dominant deformation mechanism in feldspars from about middle greenschist- through middle to upper amphibolite-facies conditions. It may be important in forming ductile shear zones (Tullis and Yund, 1985).

**157A.** Starting material, Hale albite rock from the border zone of the Hale pegmatite in Middletown, Connecticut, U.S.A. Grains are equant and randomly oriented, with an average diameter of 150–200 μm. NX

**157B.** Sample shortened 50% (N-S). Original grains have separated into several relatively undeformed lozenges bounded by shear zones of extremely fine (~1 μm) recrystallized grains. Same scale as 157A. NX

**157C.** Sample shortened 85% (N-S). Augen are unrecrystallized remnants of original grains; recrystallized matrix has a very strong crystallographic preferred orientation. Same scale as 157A. NX

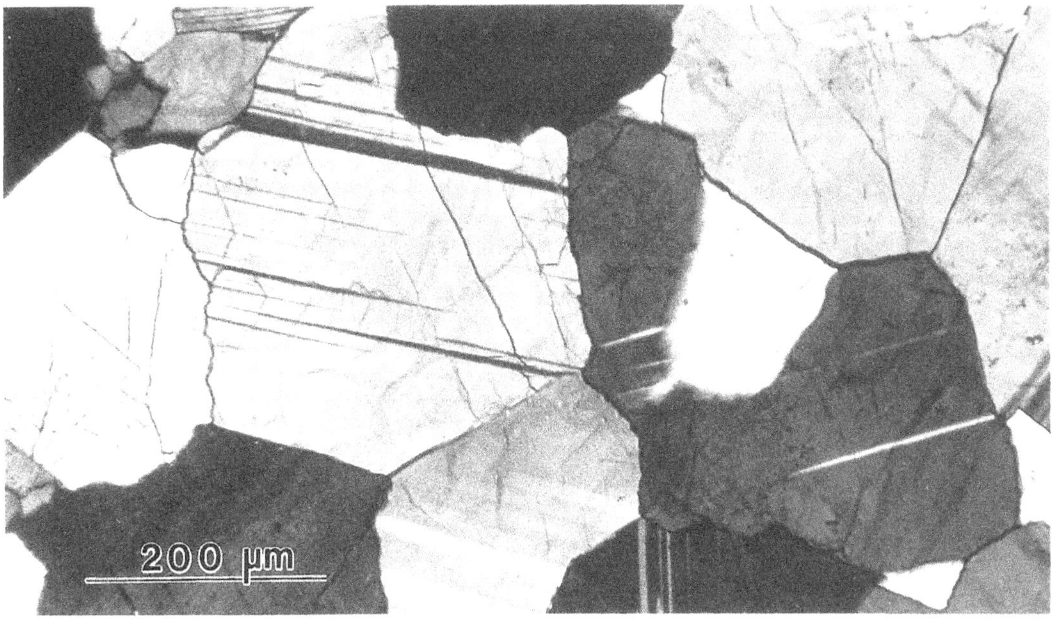

157 A

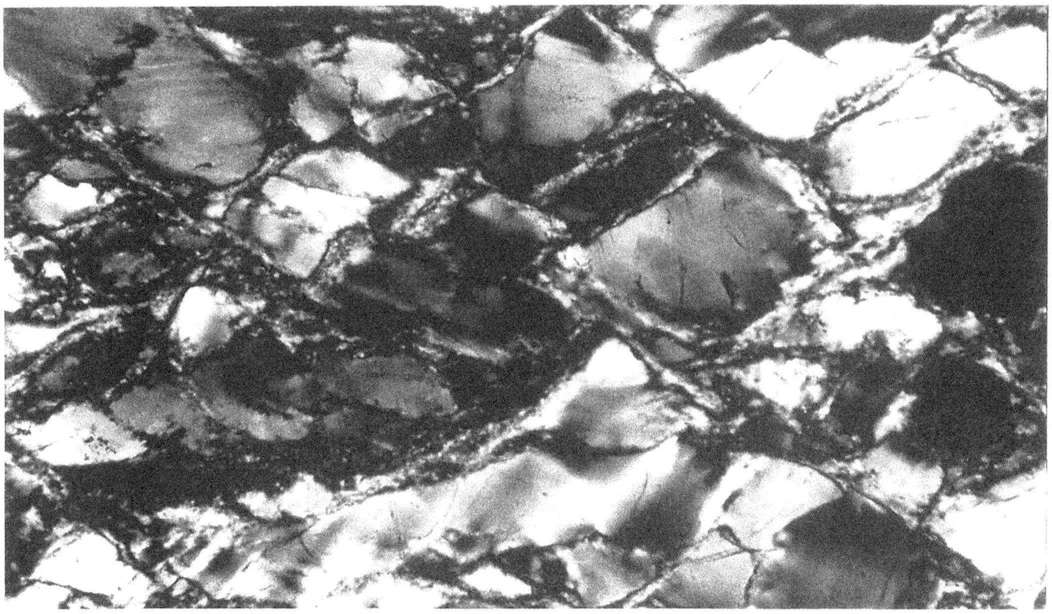
157 B

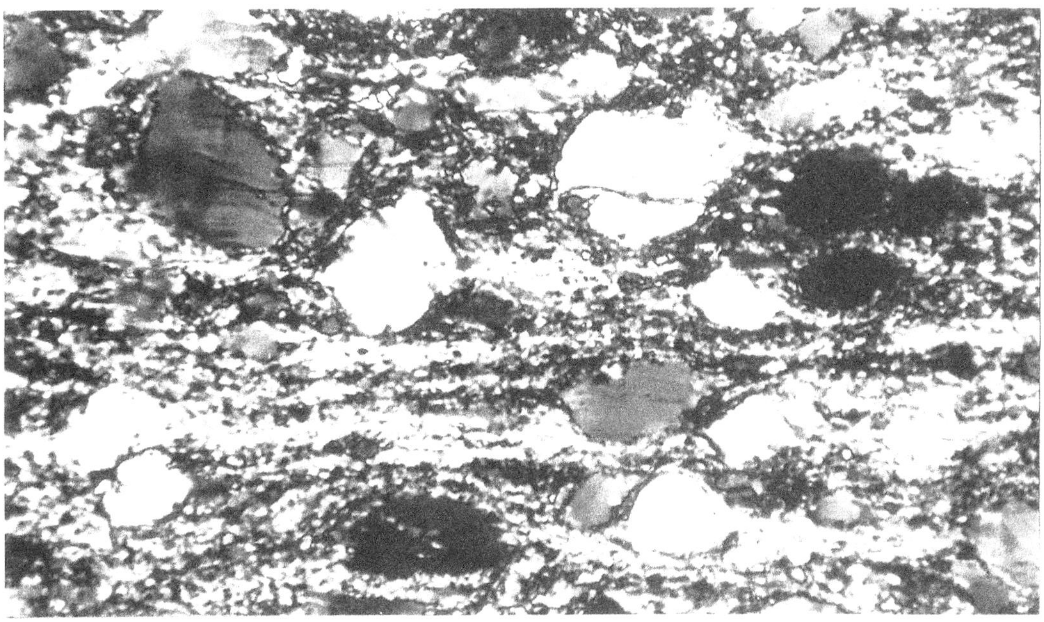
157 C

# DUCTILE BEHAVIOR

## Microstructural development of mylonitic rocks of specific composition

### Carbonate rocks



Common abbreviations used in regard to the plates in this section include PPL = plane polarized light, NX = crossed nicols, SEM = scanning electron microscope, and TEM = transmission electron microscope.

This set of plates illustrates the microstructures and crystallographic preferred orientations (CPOs) developed in experimentally and naturally deformed calcite aggregates, and one naturally deformed dolomite aggregate. Several of the contributions illustrate the important results that have come from experimental studies; for example, identification of different mechanisms of dynamic recrystallization and calibration of a recrystallized grain-size paleopiezometer. In addition, they illustrate the experimentally documented change in microstructures and CPOs accompanying the transition from dislocation creep to grain-size-sensitive creep that occurs with decreasing grain size. This is especially important, because unlike the situation for quartz, pure calcite aggregates appear to undergo grain-size-sensitive creep ("superplastic flow") in nature. Several contributions illustrate natural calcite mylonites, some of which appear to be the result of dislocation creep with partial or complete dynamic recrystallization, and some of which appear to represent grain-size-sensitive creep. Dolomite is significantly stronger than calcite, but there are few experimental data on its strength, deformation mechanisms, and microstructures. The illustrations of naturally deformed dolomite from near a fault zone document the importance of fluids in facilitating dislocation creep and dynamic recrystallization.

# 158. PARTLY RECRYSTALLIZED CALCITE IN NATURALLY DEFORMED MARBLE

R. H. VERNON

Marbles formed in the sillimanite zone of a low-pressure regional metamorphic terrain in the Mount Lofty Ranges, about 50 km east of Adelaide, South Australia, have undergone late deformation in localized mylonite zones under greenschist-facies conditions. The resulting calcite mylonites are composed of deformed, relatively coarse-grained relics of calcite, interspersed with much finer-grained aggregates of recrystallized calcite. Recrystallized calcite also occurs as scattered grains and elongate aggregates inside the deformed relics, many of which show well-developed subgrains and evidence of strain-induced migration of twin and grain boundaries. These features are revealed in ultrathin (2–7 μm) sections, as shown in the four following photomicrographs. Optical microstructural evidence suggests that dynamic recrystallization has taken place, along with some static recrystallization Probable mechanisms for the initiation of recrystallization include bulge nucleation and subgrain rotation (Vernon, 1981).

**158A.** Deformed relics of calcite in a much finer-grained aggregate of recrystallized calcite grains that appear optically to be unstrained, apart from twinning in some grains. The deformed relics contain numerous irregular to ragged, lenticular to bleblike twin remnants, as well as subgrains elongate parallel to the twins (approximately east-west in the photograph). Some subgrain boundaries arranged perpendicular to these elongate subgrains are associated with irregularities in the boundaries of the subgrains (just above center). NX. From Vernon (1981, figure 1).

**158B.** Preferential recrystallization along former grain boundaries (center, right and bottom) and along a former twin boundary (top left) in deformed calcite containing abundant subgrains that are elongate parallel to twin remnants. The vague NNE-SSW-trending streaking is an artifact of the section. NX. From Vernon (1981, figure 3).

158 A

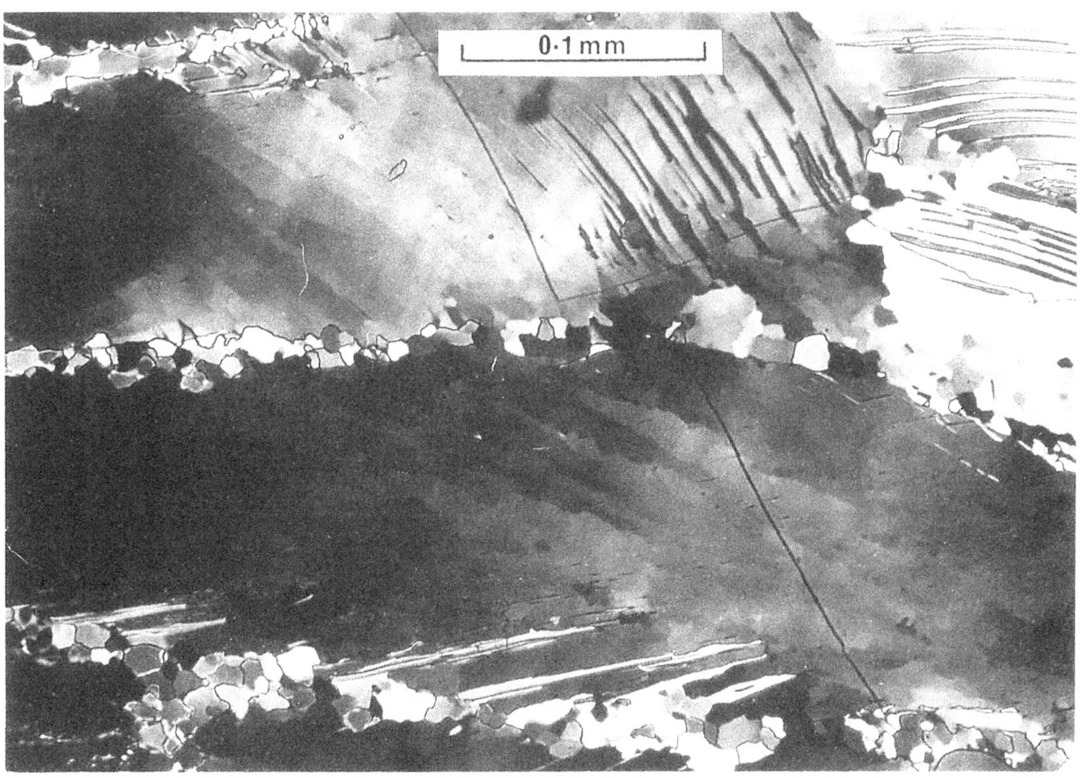


158 B

**158C.** Preferential recrystallization along lines inferred to represent traces of former curved twins or kink bands in a deformed relic of calcite, surrounded by finer-grained, recrystallized calcite. Narrow, smoothly bounded, late-formed twins are present in the deformed relic, and late twins are also present in some recrystallized grains in the matrix. NX. From Vernon (1981, figure 4).

**158D.** Irregularly shaped new (recrystallized) grains scattered through a deformed relic of calcite composed of equant subgrains, together with bulged and bleblike twin remnants. A typical new grain (N) and subgrain (S) are labeled. Some of the new grains also contain twins. The subgrains vary greatly in average diameter, and the new grains vary similarly, so that small and large new grains may have developed from small and large subgrains, respectively, although the larger new grains may have grown considerably since their initiation. Irregularities in the boundaries of the new grains are related to the size of adjacent subgrains. NX. From Vernon (1981, figure 11).

158 C

158 D

# 159. MARBLE MYLONITES

BEN A. VAN DER PLUIJM AND JAY P. BUSCH

The Bancroft shear zone is a <20 m thick mylonite zone that is exposed in the western portion of the Metasedimentary Belt of the deeply eroded, Middle Proterozoic Grenville orogen in southern Ontario, Canada. Sense-of-shear indicators (e.g., rotated clasts, clast-tail relationships, folds) indicate normal-sense displacement for this shallowly dipping zone (Carlson and others, 1990). Geochronologic data show that extension initiated during peak metamorphism, which coincides with the final stages of orogenic contraction (~1045 Ma); however, normal-sense displacement continued intermittently as late as 960 Ma, when regional contractional activity had stopped. Thus, the zone preserves the transition from synorogenic collapse to late-orogenic regional extension in the Grenville orogen (van der Pluijm and others, 1994).

The microstructures that are preserved in the zone vary from protomylonite to ultramylonite (Figure 159.1). The temperature during mylonitization was ~475°C, based on calcite-graphite carbon isotope thermometry of ultramylonite samples (van der Pluijm, 1991). The corresponding crystallographic fabrics of the microstructures are shown in Figure 159.2. Interpretations of the microstructure–crystallographic fabric relationships in the various marble mylonite domains indicate that the competition between deformation mechanisms at constant temperature is highly sensitive to secondary phases, grain size, strain rate, and differential stress (Busch and van der Pluijm, 1995).

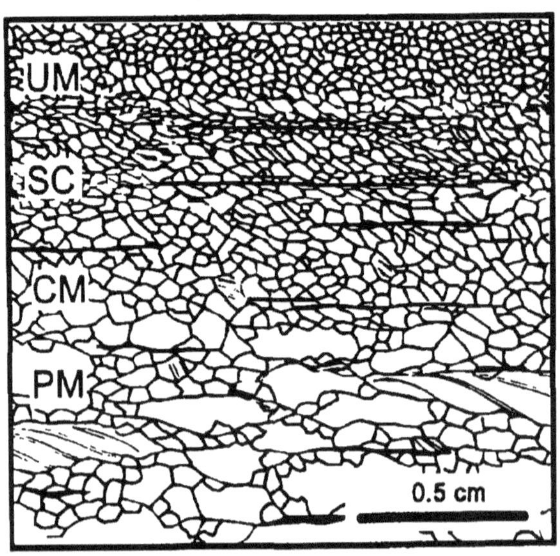


**Figure 159.1.** Schematic illustration of the microstructures in marble protomylonite (PM), coarse mylonite (CM), S-C mylonite (SC), and ultramylonite (UM). Shear sense is top to the left. Modified from Busch and van der Pluijm (1995)

**159A.** Typical outcrops of the shear zone expose branches that are several meters in thickness, in which rapid transitions from coarse marble protolith (light) to fine, foliated mylonite (dark gray) occur. Ultramylonite is very fine-grained and dark-colored, with an irregular foliation ("flow banding") that anastomoses around clasts that are derived from host rock lithologies.

**159B.** The marble *protolith* is dominated by equant, coarse calcite grains of several millimeters diameter. No dimensional preferred orientation (DPO) is evident, and only a weak banding defines the foliation in hand specimen. Grain boundaries are generally serrated and interpenetrating. Grain interiors are dominated by twins that commonly narrow toward grain boundaries or are lensoid in shape, indicative of a mechanical origin. Undulose extinction and deformation bands are restricted to regions surrounding secondary phases. Recrystallization is evident only in a few places along grain boundaries and twin lamellae. There is no obvious crystallographic preferred orientation, and poles of twin lamellae at ca. 26° from the *c*-axis indicate *e*-twinning (Figure 159.2a). Mechanical twin analysis gives a shortening direction consistent with the sense of shear deduced from field evidence. NX

**159C.** The *protomylonite* is composed of millimeter-size calcite porphyroclasts surrounded by equant, smaller grains (50–100 µm) in a core-mantle or mortar texture. A DPO is developed in elongate porphyroclasts that lie parallel or slightly oblique to the dominant foliation (shear plane), with an aspect ratio of ~2.1. A weaker DPO is evident in the domains of recrystallized grains that surround porphyroclasts. The aspect ratio decreases (~1.6), and the orientations of long axes are more dispersed in recrystallized matrix grains. Grain boundaries of porphyroclasts are irregular, whereas matrix grains have straight and curved boundaries. The internal microstructures of porphyroclasts are characterized by undulatory extinction, deformation bands, and subgrains. Matrix grains generally have sharp extinction with little subgrain development. The crystallographic orientations of subgrains at the rims of porphyroclasts are slightly misoriented relative to the core of the grain, indicative of rotation recrystallization (dislocation creep). Porphyroclast *c*-axes tend to be oriented subperpendicular to the shear plane, whereas matrix grain *c*-axes are distributed about a great circle containing the lineation (Figure 159.2b). The crystallographic fabric is similar to those from high-temperature, simple shear experiments on marble, where *r*, *f*, and *c* slip systems were active. NX

159 A

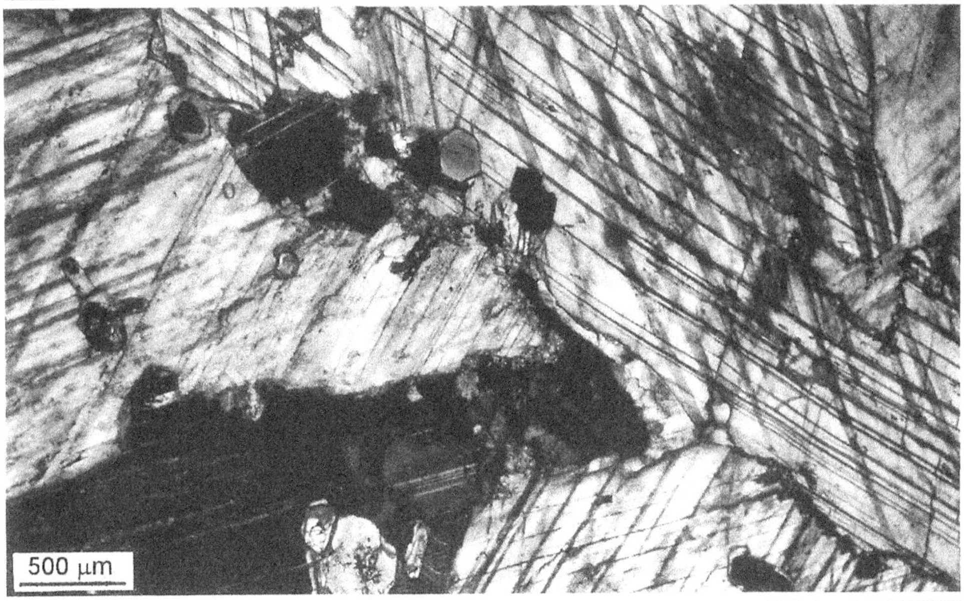


159 B

159 C

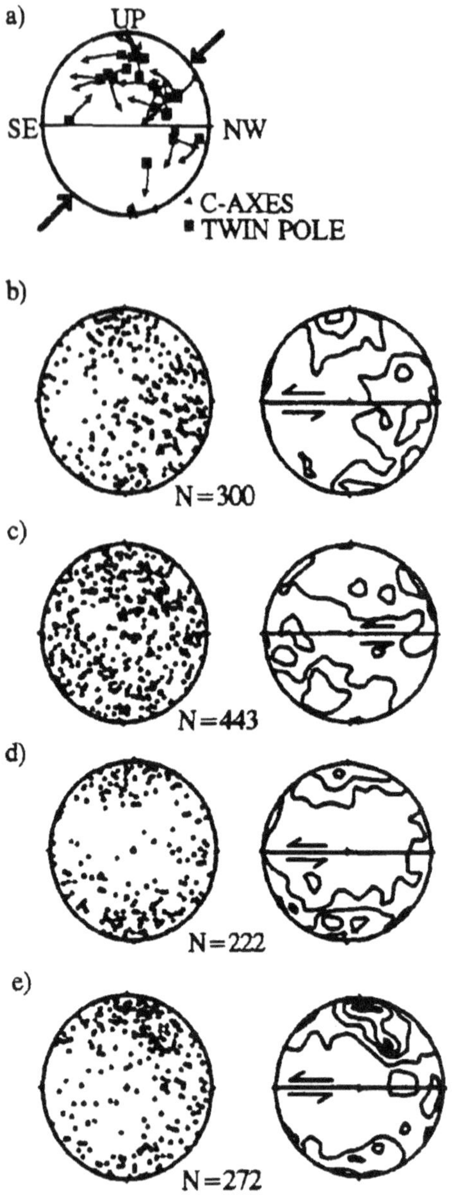


**Figure 159.2.** Crystallographic preferred orientation fabrics for marble protolith (a), protomylonite (b), coarse mylonite (c), S-C mylonite (d), and ultramylonite (e). Data are presented in lower-hemisphere, equal-area projection and corresponding contoured plot (multiples of uniform distribution). Shear sense in all projections is top to the left (i.e., to the SE). N is the number of grains measured using an optical microscope that is equipped with a universal stage. Modified from Busch and van der Pluijm (1995).

**159D.** *Coarse mylonite* domains consist entirely of dynamically recrystallized grains. The grain size (~50 μm) and aspect ratio (~1.7) of these grains are essentially equal to those of the matrix grains in the protomylonite. A weak DPO is illustrated by long-axis orientations aligned oblique to the shear plane. Grains are characterized by sharp extinction and straight to slightly curved boundaries. Larger grains in this domain commonly have subgrains of half the host grain size. The crystallographic fabric of coarse mylonite is nearly random (Figure 159.2c), which is interpreted as evidence for grain-boundary sliding (probably assisted by grain-boundary dissolution and growth) as the dominant deformation mechanism. Several groups of adjacent, similarly oriented grains are present, reflecting an inherited single grain orientation. NX

**159E.** *S-C mylonite* is developed across the textural transition from mylonite to ultramylonite. Relative to adjacent domains the DPO is strong, as indicated by alignment of long axes and greater aspect ratios (~2.0). Elongate grains aligned oblique (~30°) to the shear plane define the S foliation. Bands of finer, more elongate grains, mantled by very fine-grained secondary phases, define the C planes. The shear sense inferred from this microstructure agrees with other indicators. Grain boundaries are generally straight to curved, and most grains are characterized by sharp extinctions. However, subgrain development is common in larger grains. The *c*-axis orientations show a preferential alignment perpendicular to the shear plane and a girdle containing the lineation (Figure 159.2d). Comparison of microstructure and *c*-axis orientation of grains in this domain shows that the most elongate grains tend to dominate the maximum perpendicular to the shear plane; the more equant grains that dominate the S foliation lie throughout the great circle. NX

**159F.** *Ultramylonite* is characterized by decreased grain size and weakening of the DPO. Aspect ratios of grains further decrease (~1.7) and alignment of grain long axes becomes C-parallel with decreasing grain size (<25 μm). Two populations of grains observed in the S-C mylonite are no longer present; instead, the microstructure is homogeneous and secondary phases are disseminated. Grain boundaries remain straight to curved in the finer domains, and grains generally have sharp extinction with subgrains developed in only a few grains. Crystallographic fabrics are characterized by a strong point maximum subperpendicular to the shear plane and a weak girdle (Figure 159.2e). In the finest-grained ultramylonite domain, this maximum becomes deflected in the shear direction, toward the extension direction. Comparison of microstructure and grain orientation shows an essentially random relationship except for occasional clustering of similarly oriented grains. NX

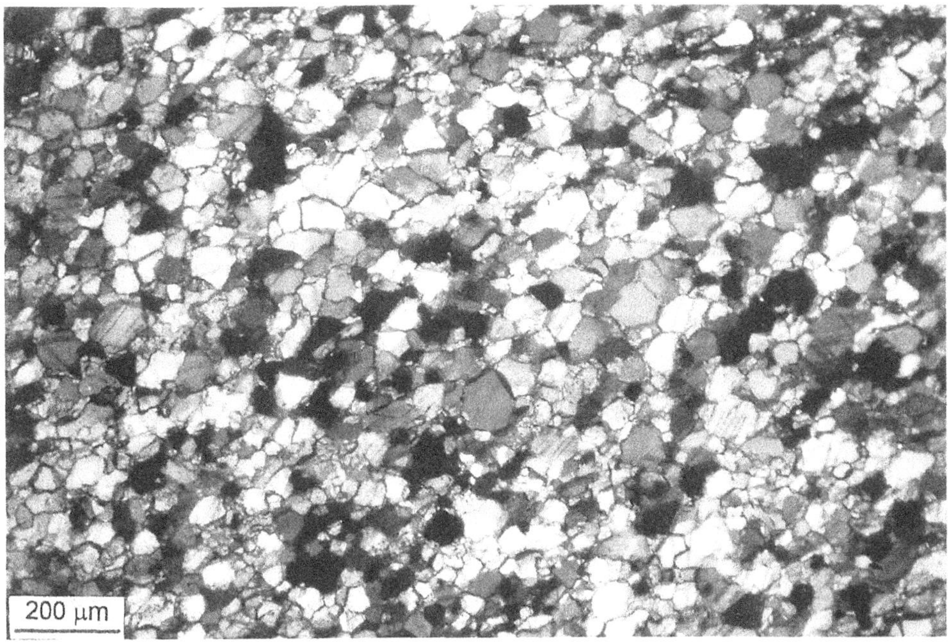

159 D

159 E

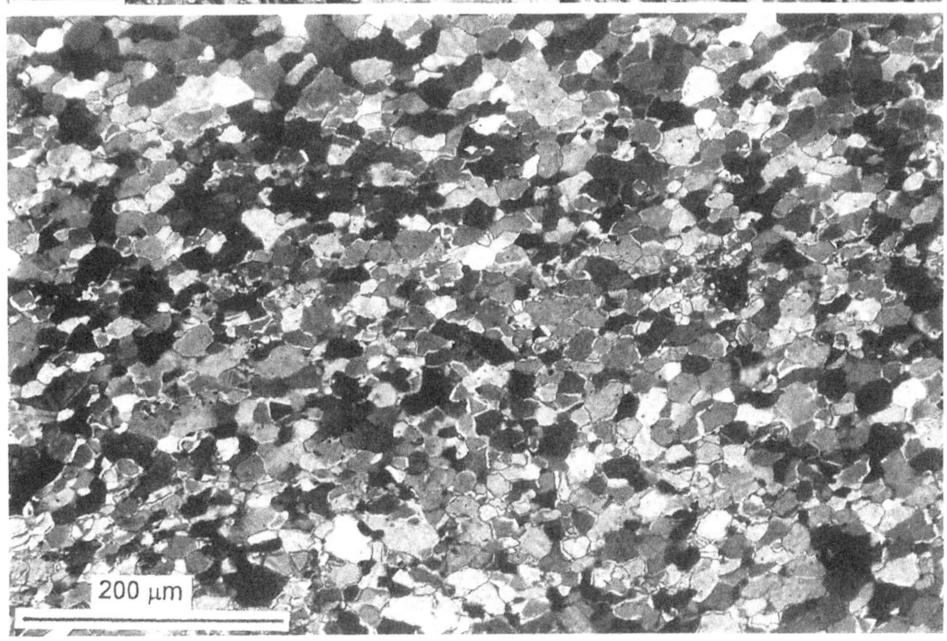

159 F

# 160. DEFORMATION AND DYNAMIC RECRYSTALLIZATION OF DOLOMITE

JULIE NEWMAN AND GAUTAM MITRA

Many natural fault zones show evidence for large volumes of fluids that passed through during deformation (e.g., Beach, 1980; O'Hara, 1988; Losh, 1989; Newman and Mitra, 1993), and the presence of these fluids may dramatically affect the dominant deformation mechanisms in fault rocks. Grain-size reduction in mylonites typically takes place by dislocation creep accompanied by dynamic recrystallization (Bell and Etheridge, 1973; White and others, 1980). These processes may be enhanced by the presence of fluids so that dynamic recrystallization occurs at lower temperatures than in dry minerals (e.g., quartz, see Hirth and Tullis, 1992; dolomite, see Newman and Mitra, 1994; calcite, see Carter and Dworkin, 1990), or in relationship to fractures along which fluids have infiltrated (Newman and Mitra, 1994).

The Pioneer Landing fault zone (Tennessee, U.S.A.) provides an example of concurrent brittle and ductile deformation in dolomite rocks. Regional stratigraphic relations and restorations of balanced cross sections indicate that the rocks were deformed under an overburden of ~9 km (Boyer and Elliott, 1982), suggesting low pressures (~300 MPa) and temperatures (270°C at 30°C/km). The fault-related deformation zone extends for 10–15 m into the hanging wall, defined by the presence of mesoscopic fractures, transgranular microfaults, reactivated stylolites, and thin deformation zones that increase in thickness and density toward the center of the fault zone (Newman and Mitra, 1994). Grain-size reduction is observed across the fault-related deformation zone from 15 m away to the center of the fault zone. The presence of fluids in these rocks is suggested by cathodoluminescence (Newman and Mitra, 1994). Dynamically recrystallized grains are observed adjacent to luminescing veins and stylolites and within luminescing deformation zones, suggesting an association between fluid pathways and dynamically recrystallized grains. Evidence that these recrystallized grains formed by strain-induced processes (not precipitated) is suggested by grains with dimensional preferred orientations indicating glide of dislocations, and by the grain-size reduction observed toward the center of the fault zone.

These micrographs show deformation in rocks ~11 m above the center of the fault zone. They exhibit features representing the transition from undeformed wall rocks to mylonitic fault rocks. Deformation in dolomite rocks is typically difficult to observe due to the mineral's high birefringence. The ultrathin sections pictured here are doubly polished and less than ~5 μm thick. As a result of the lowered birefringence, variations in microstructure resulting from differences in deformation mechanisms are more easily observed.

**160A.** Fine-grained deformation zone (dz) offset by a microfault (F). NX

**160B.** Higher magnification of fine-grained deformation zone in 160A. Subgrains (s) indicate that dislocation creep was accommodated by dynamic recrystallization. Serrated grain boundaries (g) suggest that dynamic recrystallization took place by grain-boundary migration as well as by subgrain rotation. NX

**160C.** Dimensional preferred orientation of grains (d) suggests deformation by dislocation creep. Subgrains (s) and serrated grain boundaries (g) indicate that dislocation creep was accompanied by dynamic recrystallization. This section may represent an early stage in the development of a fine-grained deformation zone. Less deformed dolomite grains can be seen in the top left and bottom right corners of this micrograph. NX

**160D.** Dimensional preferred orientation of grains (d) suggests deformation by dislocation creep. Elongated grains are adjacent to a stylolite (St). Under cathodoluminescent light the stylolite luminesces a similar color as the veins and differently than the host rock, suggesting that fluids moved along the stylolite. NX

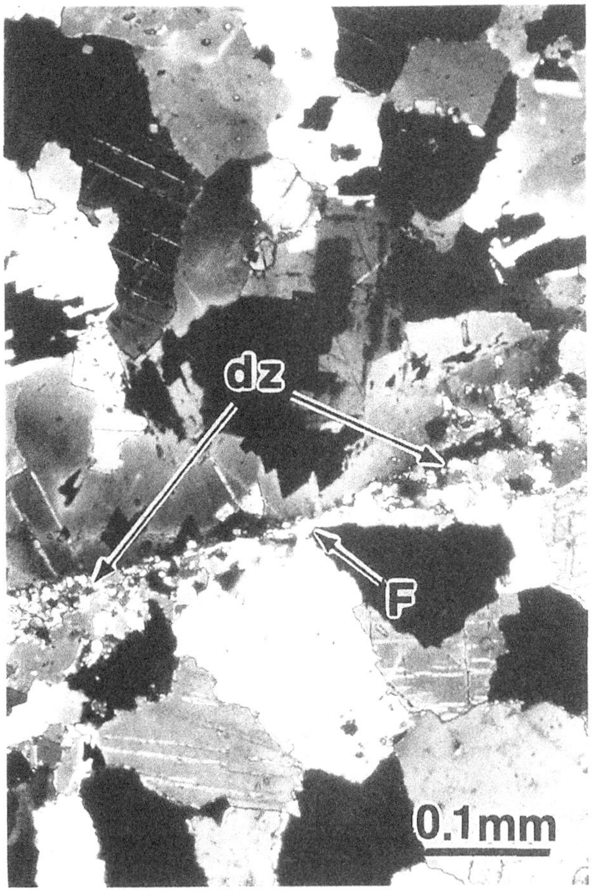


160 A

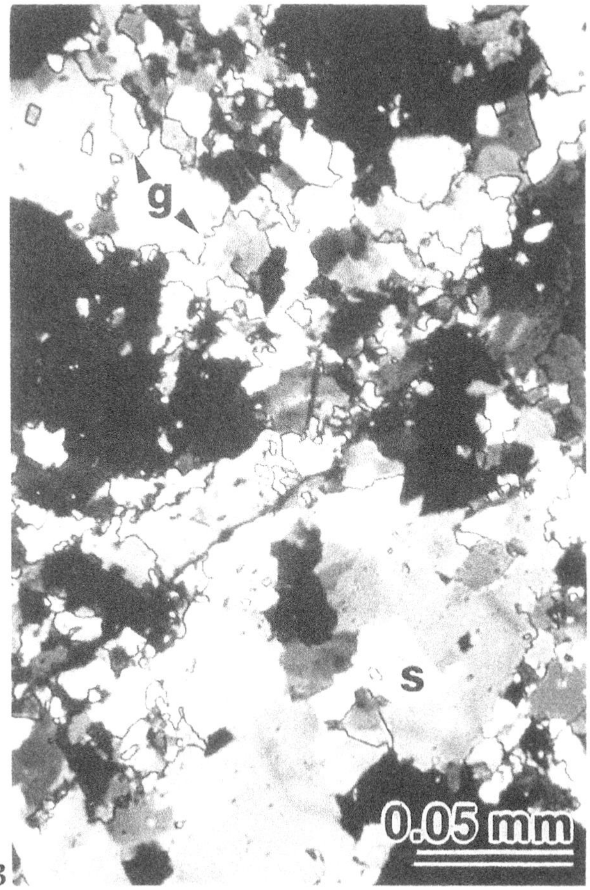


160 B

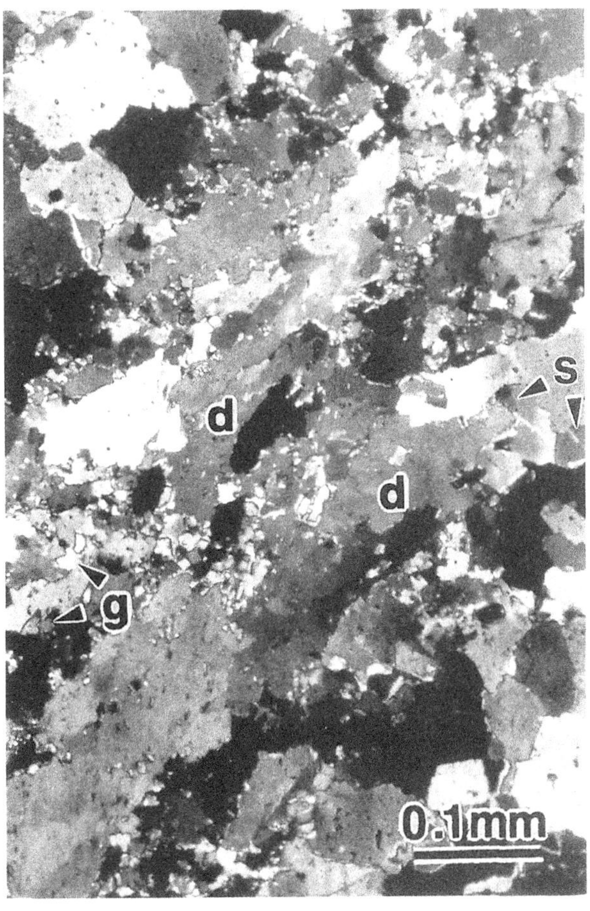


160 C

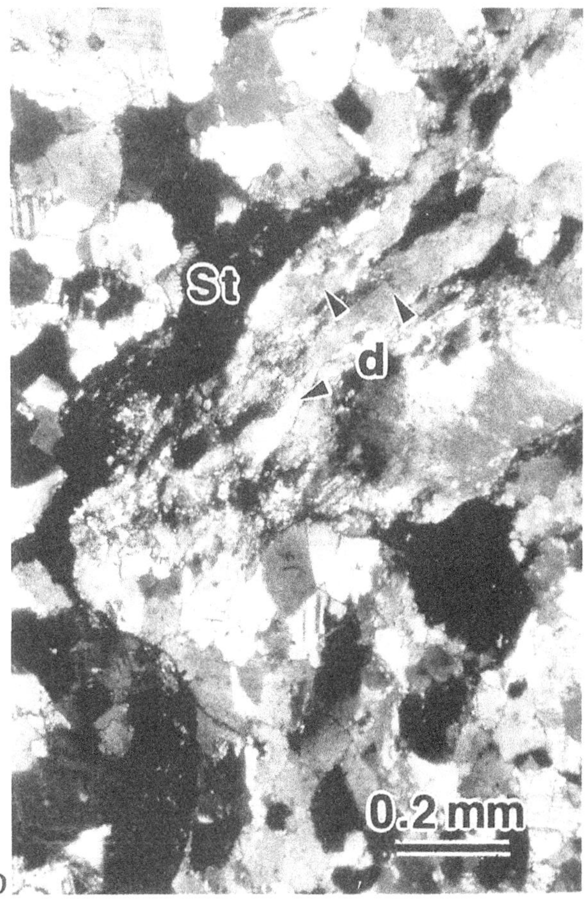


160 D

# 161. EXPERIMENTAL DYNAMIC RECRYSTALLIZATION OF CALCITE ROCKS

E. H. RUTTER

High-strain, plastic deformation of calcite rocks in nature is usually accompanied by dynamic recrystallization, which may result in grain-size reduction. Calcite mylonites are commonly characterized by grain refinement relative to the protolith. Such microstructural change is readily produced during the experimental deformation of calcite rocks at high temperatures, and is favored by combinations of high temperature and strain (Figure 161.1).

**161A** illustrates how dynamic recrystallization develops in two stages in samples deformed in extension by 650% at 700°C and 250 MPa confining pressure. Shortening direction indicated by arrows. NX. First, necklaces of small, new grains form around old, highly elongated grains by a combination of subgrain rotation and grain-boundary bulging [labeled (b)]. These new grains appear to act as nuclei for eventual total recrystallization by grain-boundary migration. The grain size resulting from migration recrystallization is always larger than that for rotation recrystallization [see patch of migration-recrystallized grains labeled (a)]. Both grain sizes vary systematically with differential stress (Figure 161.2).

**161B** shows the upper right corner of a specimen deformed at a higher temperature (800°C, 200 MPa confining pressure, 40% bulk shortening indicated by arrows, NX). In the top left of the photograph the original grain size and shape are seen, protected from strain in the "pressure shadow" immediately beneath the upper loading piston. Toward the lower right there is a marked strain intensification, leading to total migration recrystallization of the rock with concomitant grain-size reduction, resulting in an aggregate of near-equidimensional, strain-free grains.

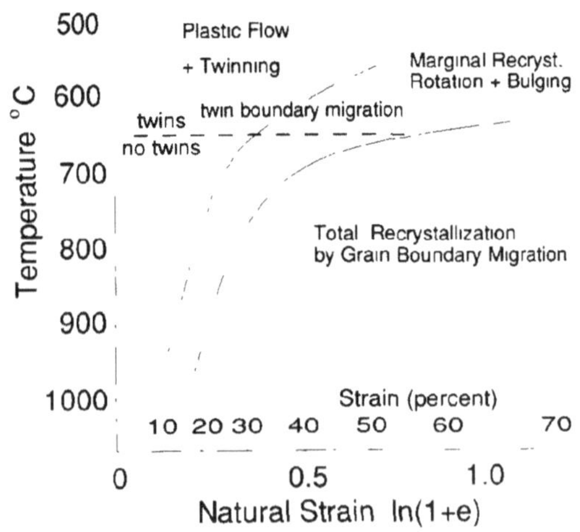


**Figure 161.1.** Regimes of deformation and dynamic recrystallization as a function of temperature and strain from experiments performed on Carrara marble at $10^{-4}$ $s^{-1}$ and 200 MPa confining pressure (Rutter, unpublished data)

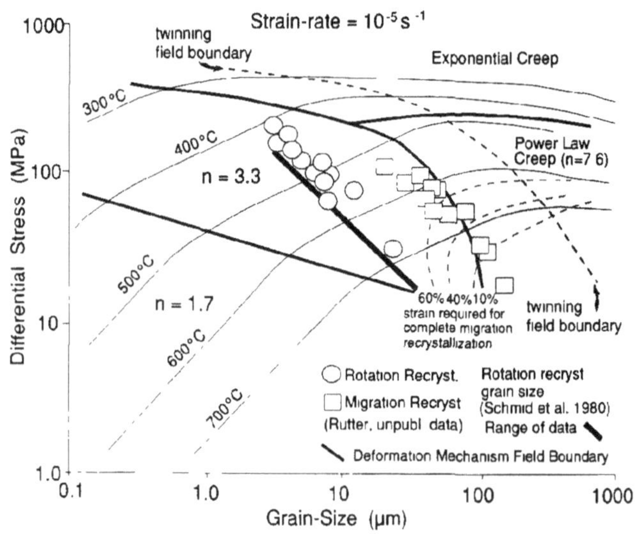


**Figure 161.2.** Deformation mechanism map for calcite rocks deformed dry at a constant (laboratory) strain rate, constructed from data of Rutter (1974), Schmid and others (1980), Walker and others (1990), Rowe and Rutter (1990), and Rutter (unpublished data). Twinning only occurs above the twinning field boundary Stress exponents n of 1.7 and 3 3 characterize the two fields of grain-size-sensitive flow.

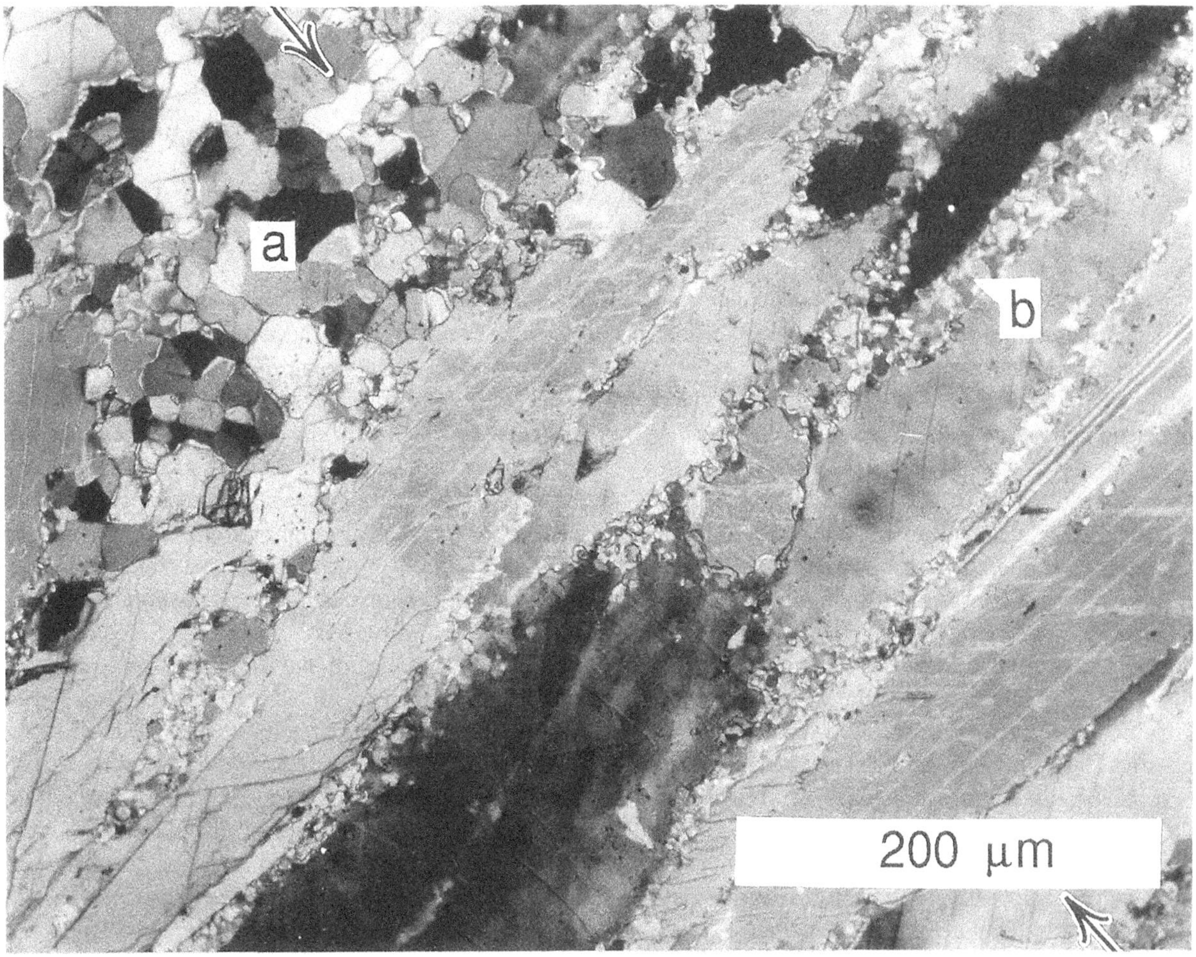


161 A

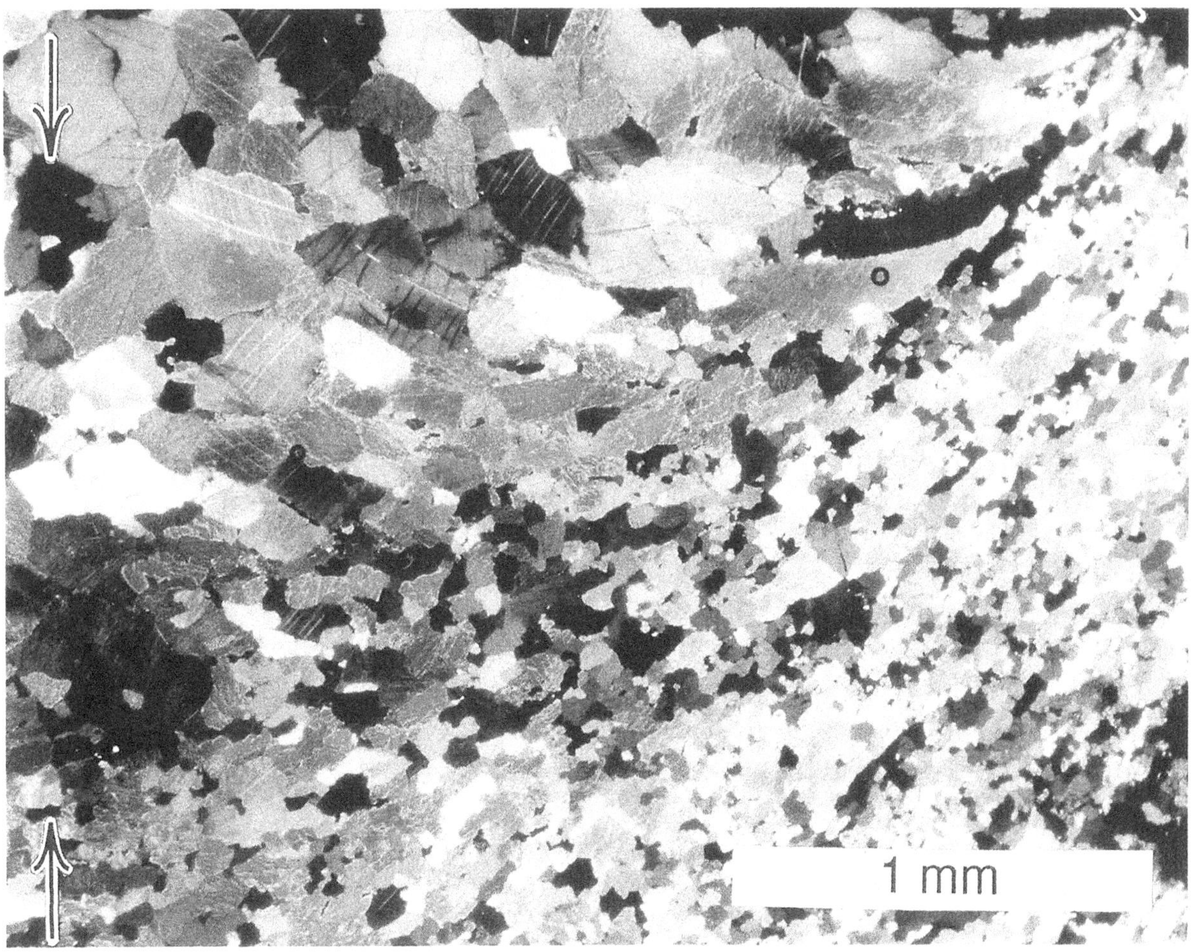


161 B

# 162. COMPARISON OF EXPERIMENTAL AND NATURAL SUPERPLASTIC DEFORMATION OF FINE-GRAINED CALCITE ROCKS

E. H. RUTTER, L. BURLINI, AND M. CASEY

Allowing for differences in temperature and strain rate, the microstructures of naturally and experimentally (dry) deformed, fine-grained calcite rocks correspond closely and probably indicate that in both cases deformation was dominated by the same mechanisms.

**162A.** Optical section (NX) of a synthetic hot-pressed, fine-grained calcite rock stretched experimentally by about 1000% at 700°C, 250 MPa confining pressure, and at a strain rate of about $10^{-3}\,s^{-1}$. Deformation was in the grain-size-sensitive regime (stress exponent n = 3.3; Walker and others, 1990; Rutter and others, 1994). An embedded large grain has deformed plastically, with the grain shape tracking the bulk strain. The finer grains of the matrix, however, have remained almost equidimensional, with a weak grain-shape fabric defining a foliation parallel to the extension direction (indicated by arrows). No grain size changes have occurred.

A marked crystallographic fabric (CPO) was produced in the above sample (Figure 162.1a), despite the importance of the grain-boundary sliding deformation mechanism under these deformation conditions (Walker and others, 1990). After only about 80% extensional strain, normally a large deformation by experimental standards, the CPO was, in contrast, almost undetectable (Rutter and others, 1994). Thus in this deformation regime CPO develops only slowly, and is only significant after a very large strain. At lower flow stresses, CPO is thought to develop even more slowly, and preexisting fabrics may become weakened.

**162B.** Optical microstructure (NX) of a calcite mylonite from the overturned limb of the Morcles nappe (Saillon, Switzerland). The grain size and shape are broadly comparable to those of the experimental sample shown in Plate 162A, and this rock probably deformed by the same mechanism, but it formed at much lower temperatures (<300°C) and at a correspondingly lower "geological" strain rate. Based on the observed ductile thinning of the stratigraphic sequence, the strains in both examples are likely to be comparable. The CPO is represented by the pole figure for poles to the *e*-twin plane shown in Figure 162.1b. From the asymmetry of the pole figure relative to the foliation and lineation, the sense-of-shear can be inferred (Schmid and others, 1981). It can be argued that this CPO developed during the coarser-grained, intracrystalline-plastic protomylonitic stage when flow stresses were higher than would be required to maintain the flow after grain refinement. The CPO probably persists because grain-boundary sliding accommodated by a mixture of plasticity and diffusion tends to produce a similar type of CPO but more slowly (Rutter and others, 1994).

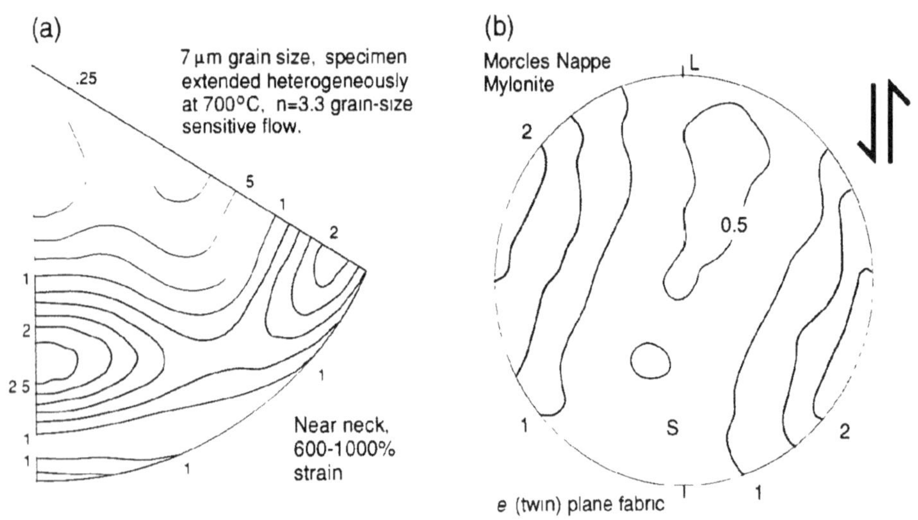


**Figure 162.1.** (a) Inverse pole figure for the stretching direction for the sample illustrated in Plate 162A. The fabric type is the extensional equivalent to that produced in the twinning-absent, intracrystalline-plastic deformation mechanism field. The eccentric shaded ellipse shows the bulk strain shape factor for the whole sample, and the inset unshaded ellipse was measured from the observed grain shape of the fine matrix grains. (b) Pole figure showing the preferred orientation of *e* twins in the Morcles nappe mylonite shown in Plate 162B. The shear sense deduced from geological considerations is indicated. L = lineation, S = foliation.

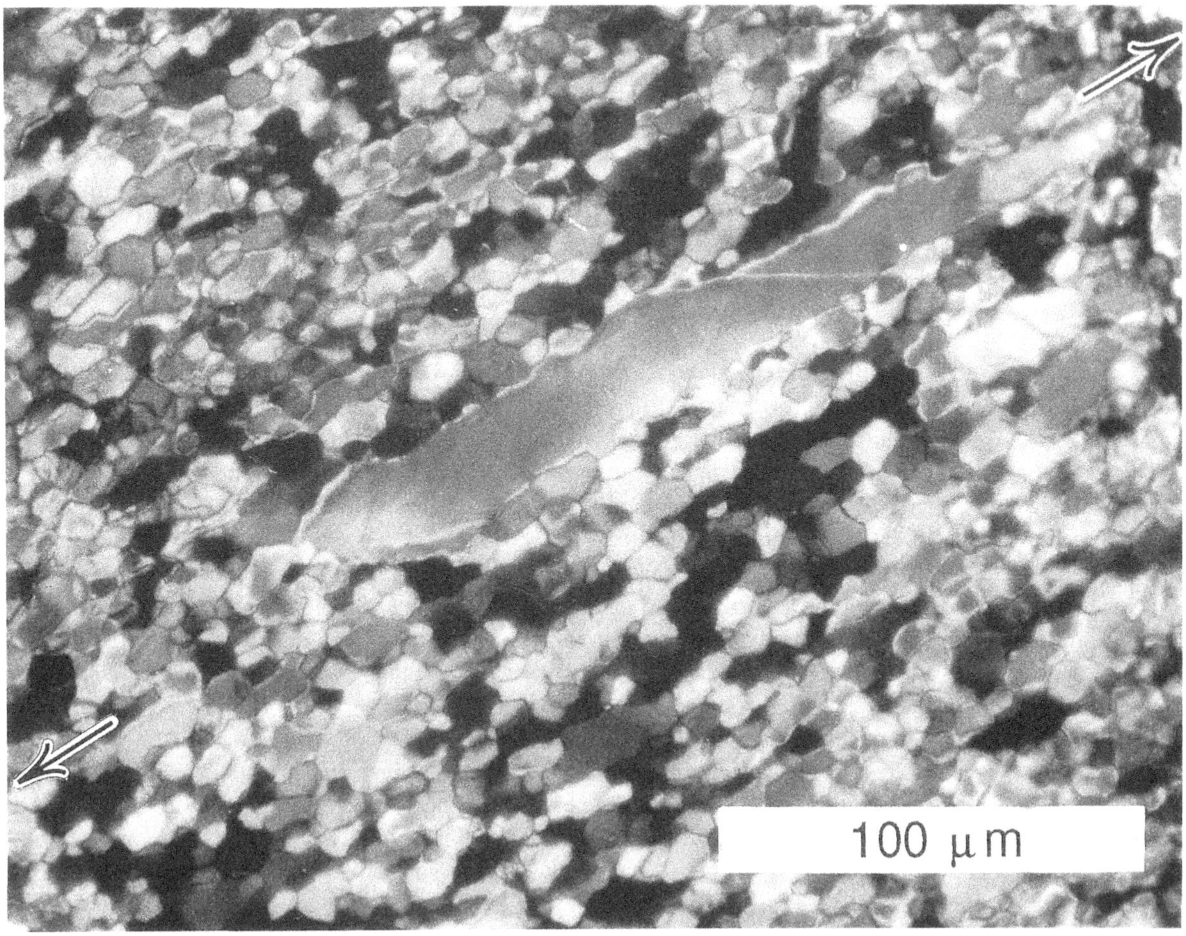

162 A

162 B

# 163. DEFORMATION MECHANISMS AND FABRIC DEVELOPMENT IN FINE-GRAINED CALCITE ROCKS

E. H. RUTTER, L. BURLINI, AND M. CASEY

Recognizing that mylonitic rocks are commonly characterized by marked tectonic grain-size reduction, it is important to inquire how the grain-refined rocks may subsequently "behave." From experimental studies of synthetic calcite rocks prepared by hot isostatic pressing (Walker and others, 1990; Rutter and others, 1994) and natural fine-grained limestone (Schmid and others, 1977, 1987), it is clear that at a given temperature, with decreasing grain size there is a transition from intracrystalline plastic flow involving twinning, through intracrystalline plastic flow without twinning, into a broad regime of grain-size-sensitive flow involving grain-boundary sliding with a mixture of plastic and diffusive accommodation. These changes are accompanied by marked mechanical weakening on entering the grain-size-sensitive flow region, hence fine-grained mylonitic rocks are expected to be weaker than their coarser protoliths.

In the intracrystalline plastic flow regimes, grain shape tends to track applied strain, unless flow is accompanied by dynamic recrystallization. In the grain-size-sensitive regime, grain shape tends to remain equant, irrespective of the applied strain. Plates 163A and 163B illustrate these two mechanism regimes for synthetic, hot-pressed aggregates initially sufficiently fine that dynamic recrystallization does not occur.

**163A.** Sample (NX) shortened experimentally by 30% at 500°C and 200 MPa confining pressure at a strain rate of $3 \times 10^{-4}$ $s^{-1}$ in the twinning-absent intracrystalline plastic flow regime. A marked grain-flattening shape fabric has resulted (arrows indicate shortening direction).

**163B.** Microstructure of equant grains produced during straining by 30% in the grain-size-sensitive flow regime (stress exponent n=1.7) at 700°C and at a strain rate of $3 \times 10^{-6}$ $s^{-1}$. Compression direction vertical, indicated by arrows. NX

Crystallographic preferred orientation (CPO) patterns that are produced reflect the deformation mechanism and the strain geometry. In the intracrystalline plastic flow regimes, strong CPOs are rapidly produced (Figure 163.1). Passing into and through the grain-boundary sliding regime, the rate of CPO development gradually weakens, as the contribution of dislocation motion to the grain-boundary sliding accommodation process gradually gives way to solid-state diffusion (Rutter and others, 1994). The CPO of the sample of Plate 163A is shown in Figure 163.1b, whereas at the same strain the sample of Plate 163B displays no detectable CPO.

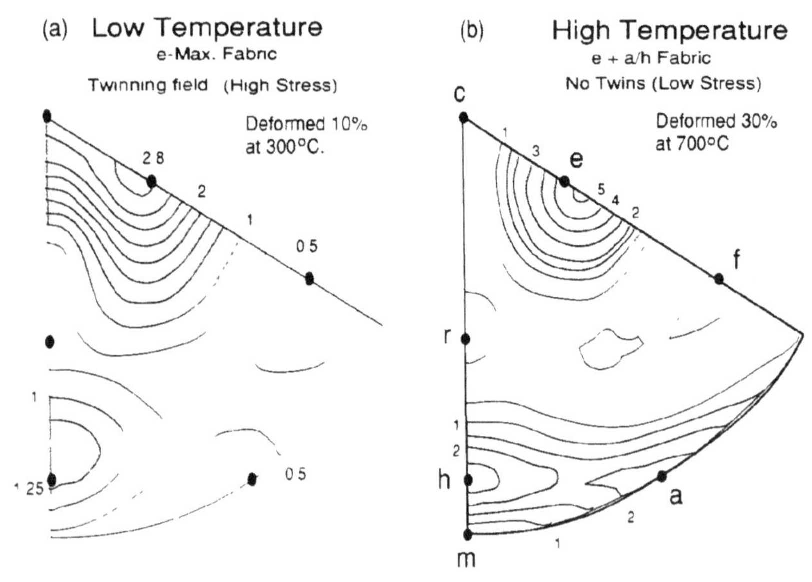


**Figure 163.1.** Inverse pole figures determined by X-ray texture goniometry for axisymmetric shortening of calcite rocks, comparing the preferred orientation patterns characteristic of intracrystalline plastic flow at (a) high flow stress accompanied by twinning activity and (b) lower flow stresses (below the twinning threshhold). Contours are labeled as multiples of a uniform distribution. Poles to crystallographic planes *e*, *r*, *f*, and *h* and axes *a* and *c* are indicated

163 A

163 B

# 164. NATURAL CALCITE MYLONITIC ROCKS

L. BURLINI, E. H. RUTTER, M. CASEY, AND D. GRITTI

These samples were taken from a complex network of shear zones in carbonate rocks from the Khudabad district, Hunza River, Pakistan. Different fault strands preserve different degrees of microstructural development which evolved during a complex movement history.

**164A.** Homogeneous, fully recrystallized, fine-grained calcite mylonite (optical micrograph, NX), in which a weak grain-shape foliation extends from upper left to lower right. In contrast to the Morcles nappe mylonite (Plate 162B), this rock displays no detectable preferred crystallographic orientation. The fine grain size suggests that it recrystallized under a flow stress of about 70 to 100 MPa (Figure 161.2), but it may subsequently have suffered large strain by low-stress, grain-size-sensitive flow that weakened any initial crystallographic fabric. Extrapolation of experimental data for grain-size-sensitive flow (Figure 164.1) suggests a flow temperature at the above stress level of 200°C at a strain rate of $3 \times 10^{-14}$ s$^{-1}$, or a slightly higher temperature if the strain rate were higher or the flow stress lower. This is consistent with geological constraints on the temperature.

**164B.** The microstructure of this sample (optical, NX) is rather different, despite being taken from a nearby locality in the same shear zone. The initial grain size of this rock is not dissimilar to that of Plate 164A. The strain is small, about 50% shortening based on the observed grain-shape fabric. Differential stress levels were high enough to produce twinning (>200 MPa for this grain size, Rowe and Rutter, 1990). The preferred orientation of the twins indicates an *e*-maximum fabric in the rock (cf. the fabric of the Morcles nappe mylonite, Figure 162.1b). Twin boundary migration recrystallization is occurring, and necklaces of small recrystallized grains produced by boundary bulging and subgrain rotation are beginning to appear. The small size of these grains (2–3 μm) again suggests very high flow stresses (>200 MPa). It is possible that the microstructure of Plate 164A represents the total recrystallization of the Plate 164B microstucture by grain-boundary migration. Alternatively, a doubling of the applied stress could convert the microstructure of Plate 164A to that of Plate 164B. It is emphasized that the palaeopiezometric methods referred to here are only likely to record the peak stress levels attained, and do not necessarily correspond to the stresses at which most strain occurred.

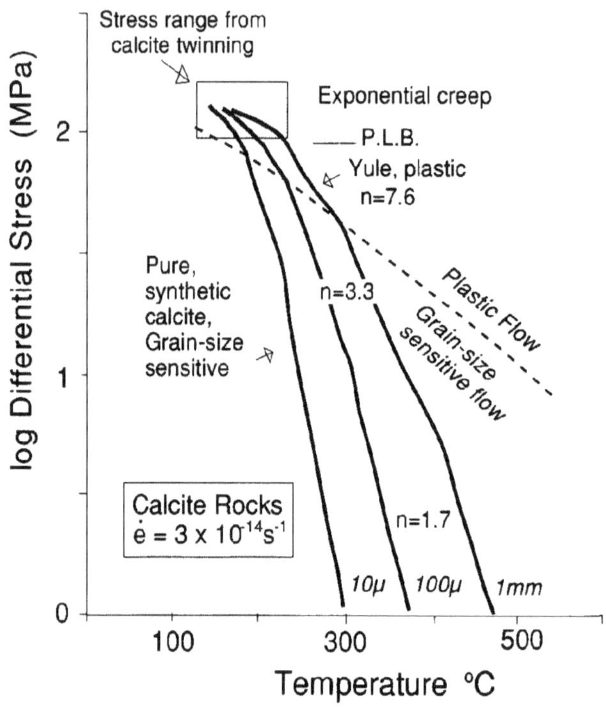


**Figure 164.1.** Graph constructed from experimentally determined flow laws for synthetic, hot-pressed calcite rocks of controlled grain size (Walker and others, 1990) and Yule marble (Heard and Raleigh, 1972, grain size 400 μm), showing log flow stress vs temperature extrapolated to a geological strain rate of $3 \times 10^{-14}$ s$^{-1}$. For grain-size-sensitive flow, three grain sizes are shown. The dashed line separates the regimes of intracrystalline plastic and grain-size-sensitive flow. n values shown are stress exponents for the applicable flow law. Stress ranges estimated from natural calcite twinning of coarse-grained rocks are indicated by the shaded box (after Rowe and Rutter, 1990), but even higher stress levels would apply to the finer-grained, yet twinned, Karakorum mylonites illustrated here. P.L.B. = power-law breakdown stress.

164 A

164 B

# DUCTILE BEHAVIOR

## Microstructural development of mylonitic rocks of specific composition

### Evaporites



Common abbreviations used in regard to the plates in this section include NX = crossed nicols and TEM = transmission electron microscope.

This set of plates illustrates the deformation of evaporites, which are even weaker than carbonates. There have been many studies of halite deformation, due in part to its possible use for nuclear waste repositories. The plates here illustrate mylonitic halite from salt domes and glaciers, in which the halite has undergone a combination of dislocation and solution transfer creep. These halite mylonites, especially when they contain small amounts of harder phases, have microstructures ranging from high-grade "gneisses" to fine-grained mylonite. Dynamic recrystallization is rapid, especially where it is enhanced by fluids (including rain), and this process effectively erases most of the strain history. One contribution documents a transition observed in experimentally deformed anhydrite aggregates, from dislocation creep to grain-size-sensitive diffusion creep. Extrapolation of the experimental results indicates that natural anhydrite aggregates, such as observed along some thrust faults, were probably superplastic.

## 165. LOW-TEMPERATURE MYLONITE IN ROCK SALT

M. P. A. JACKSON

Photograph of a vertical rock face (the image is approximately true scale) which displays subvertical trains of sandstone clasts (pale) in a matrix of rock salt (dark), Avery Island salt mine, Louisiana Gulf Coast, U.S.A. (Jackson, 1985). The rock salt consists mainly of recrystallized halite grains with about 5% anhydrite crystals, and traces of water between and within the grains. The salt is a metamorphic rock. From its origin in a subhorizontal source layer of mid-Jurassic Louann Salt, the salt flowed about 15 km horizontally and at least 10 km vertically, finally forming a plug-shaped salt stock whose sediment-draped crest rises 48 m above the surrounding bayous. During this flow the rock salt passed through a variety of strain regimes (Talbot and Jackson, 1987), and the halite grain size and shape changed by a combination of dislocation and solution transfer creep (e.g., van Keken and others, 1993). Repeated recrystallization continuously erased most of the strain history in the halite.

The pale sandstone clasts have a much more robust strain memory. Country rock of terrigenous clastic sediments—chiefly Oligocene—became incorporated into the rising diapir. The mechanism of inclusion is unclear, a matter of controversy. According to Kupfer (1976), the clastic sedimentary rocks were trapped from above by differential rise of adjacent spines of salt within the diapir. In contrast, Talbot and Jackson (1987, 1989) speculated that the clastic rocks were drawn in from below the bulb of a diapir undergoing internal toroidal circulation. Within the salt diapir, competent inclusions of country rock were drawn out and fragmented into screens of inclusions encased in incompetent rock salt. The clasts and their tails range in size from 2 cm to dust size. As the diapir's contact is approached, such screens, as well as deformed bedding in the salt, become progressively more parallel to the margins of the diapir; additionally, the strains become more flattening and less constrictive.

The difference in competency between sandstone clasts and the halite matrix is roughly analogous to that of feldspar porphyroclasts and recrystallized quartz and mica in normal mylonites, except for two important differences. First, the protolith, in the case of the rock salt, did not consist of one rock type, but rather of foreign inclusions of sandstone within rock salt. Second, under metamorphic conditions speculated to be lower greenschist facies, the clasts in the rock salt may have been comminuted more by cataclasis than by dynamic recrystallization.

165

# 166. ACTIVE MYLONITES IN NEOPROTEROZOIC ROCK SALT

Christopher J. Talbot and Hemin Koyi

This contribution (166A through 166E) illustrates one of the most spectacular of about 120 extrusions of Hormuz salt (Neoproterozoic-Cambrian) in the Zagros mountains of southwest Iran; this particular Kuh-e-Namak or "mountain of salt" is located at 28°30′N, 51°40′ E in Dashti Province (Talbot, 1979, 1993; Talbot and Jarvis, 1984). The salt in this region developed into a pillow during the Jurassic and has extruded episodically since the early Cretaceous, i.e., prior to the still-propagating Zagros orogeny having reached this site. The salt rises ~7 km up the stem of a diapir emplaced at a releasing jog along a major strike-slip fault. Shortening associated with the Zagros orogeny coupled with the loading of dense carbonate overburden results in the extrusion of the salt high into the sky, where it spreads over the surface and flows down the flanks of a Zagros anticline to near sea level (Figure 166.1). Structures and fabrics similar to those shown here and in Plate 167 are exposed in many of the other salt extrusions in the Zagros mountains, and probably simulate allochthonous sheets of salt now spreading in the Gulf of Mexico as well as other salt sheets that spread beneath air, water, or soft sedimentary veneers in many other basins in the past.

**166A.** General view of the deeply dissected southern namakier (= glacier of salt). Transparent colorless to pale pink halite has flowed from the summit (out of sight at 1500 m above sea level, 3000 m to the right) down over Recent sediments (near sea level out of sight, 300 m to the left) with the surface slope of the salt ranging from 11° to 8°. The upper ten meters of salt behaves like an elastic solid with joints that open and close with fluctuations of temperature under dry conditions. However, much of the salt mass flows as a viscous crystalline fluid when its surface is dampened by rain (Talbot and Rogers, 1980). Trains of recumbent flow folds and associated ductile shear zones in low-viscosity salt model small-scale piles of nappes in stiffer silicate tectonites (Talbot, 1979; Talbot and Jackson, 1987). Scale bar = 50 m.

**166B.** Salt emerges near the summit of Kuh-e-Namak (Dashti) as thick beds of coarse-grained (~10 cm), gneissose halite containing pegmatites of transparent halite. Layer-parallel pegmatites, tens of meters long, occur in the strain shadows of boudins of carbonates (not shown). No initial depositional fabrics have survived in this salt. Hammer head is 13 cm long.

**166C and 166D.** In these halite mylonites the dark inclusions are remnant megacrysts or porphyroclasts of halite, survivors of ductile deformation of the coarse-grained, transparent gneissose halite shown in 166B. The salt flows (Talbot and Rogers, 1980) by fluid-assisted, dynamic recrystallization (Urai and others, 1986), a process in which original mother grains (>10 cm) recrystallize around their margins and decrease in size as they disperse in an ever increasing groundmass of finer, daughter grains (<1 mm). Original grains that are ellipsoidal (166D) where the layering is folded recrystallize to subspherical grains where the folds tighten so much that the layering appears to return to planar (166C). Such dynamic recrystallization means that the grain-shape fabric has a short strain memory and maps the streamlines of the last few increments of bulk flow. The foliation is only clear in ductile shear zones, many of which initiate as the axial surfaces of flow folds generated where the streamlines cross the layering because of changes in flow velocity (Talbot, 1979; Talbot and Jackson, 1987). Pipe is 10 cm long.

**166E.** Original grains of clear halite are smaller (than in 166B) and form a lower proportion of a fine-grained mylonite in the snout of the northern namakier. Structures formed during down-slope flow to the right indicate unusually strong competence contrast. Compass is 10 cm long.

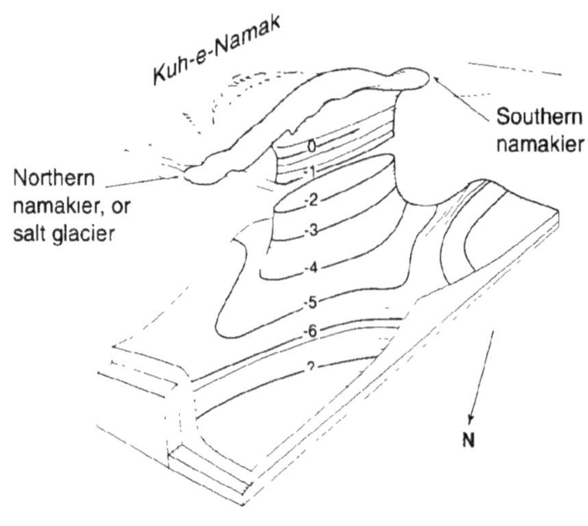


**Figure 166.1.** Block diagram showing extrusive salt diapir at Kuh-e-Namak (Dashti) and its relationship to a strike-slip fault in basement rocks as well as a younger Zagros anticline. Scale is indicated by contours labeled in kilometers below sea level.

166 A

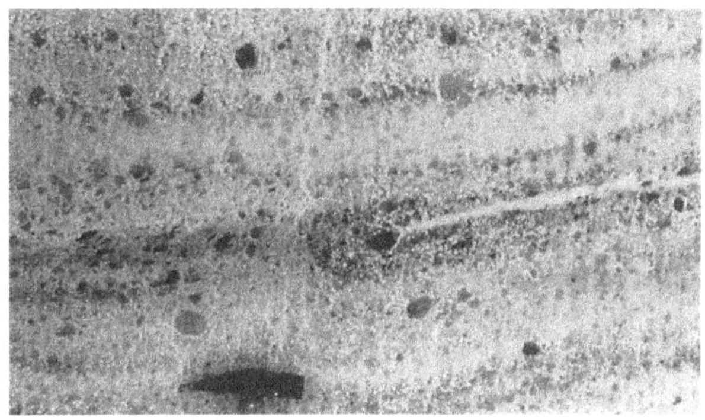

166 C

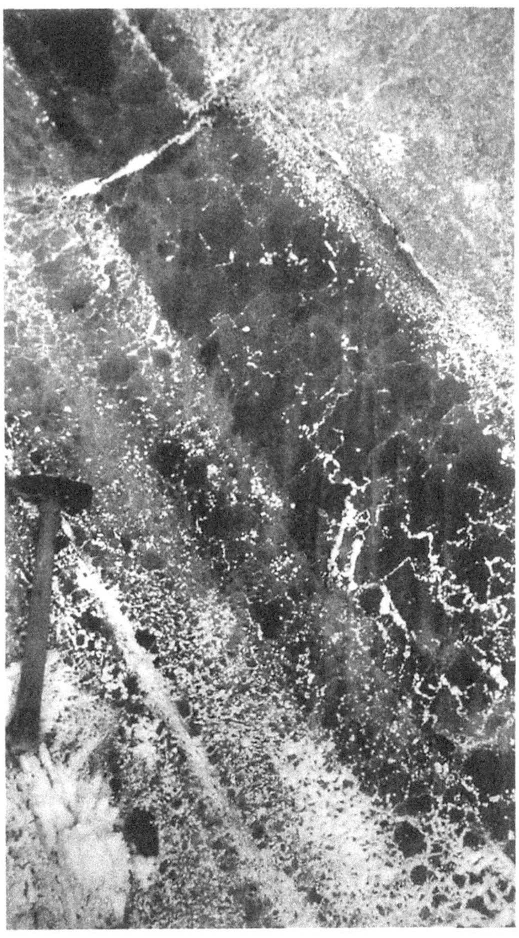

166 B

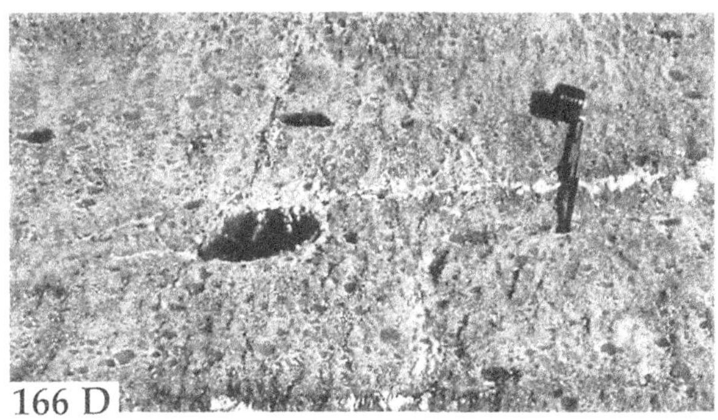

166 D

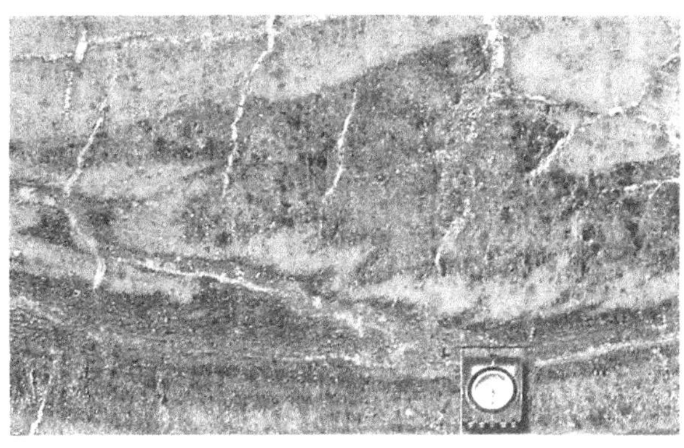

166 E

# 167. AN ACTIVE NAPPE IN NEOPROTEROZOIC ROCK SALT

CHRISTOPHER J. TALBOT AND HEMIN KOYI

The Kuh-e-Namak, or mountain of salt, in Dashti Province, southwestern Iran, is one of many extrusions of Hormuz salt (Neoproterozoic-Cambrian) located in the Zagros mountains (Talbot, 1979, 1993; Talbot and Jarvis, 1984). The northern namakier (salt glacier) of Kuh-e-Namak, which is an active sheathlike nappe of rock salt, approximately 4000 × 3000 × 50 m, is pervasively folded. Layering and flow fabrics are generally subparallel to both the basal and top boundaries of the salt sheet. The salt extrusions of Iran typically start as topographic salt domes that spread due to gravity, thus forming distinctive parabolic profiles of viscous fountains until the deep source is exhausted or sheared off (Talbot, 1979, 1993; Talbot and Jarvis, 1984). The fountainhead then sinks and dissolution begins to predominate over flow effects. Whereas some namakiers have a steep snout overriding a terminal moraine that is typical of advancing glaciers, the southern sheet in Plate 166 has a feather edge characteristic of a retreating glacier. The snout of the northern namakier seen here is probably retreating more slowly than that of the southern namakier.

**167A.** The basic structure of a namakier is that of a recumbent sheath fold that rolls foward with a longitudinal flow profile like a caterpillar or tank-track (Figure 167.1). Similar-style flow folds generate wherever the streamlines (mapped by the foliation) cross the layering because of changes in flow velocity in time or space. The limb thicknesses of the fold in 167A are unusually symmetric.

**167B.** A single bed of magenta-colored salt, about 2 m thick, defines a recumbent fold on the eastern margin of the northern namakier. Rather than eroding its channel, the salt backs up behind obstructions and flows over folded but static salt, infilling irregularities. The unfolded salt that tops the last barrier ridge (pale in color) was probably the last distal salt to flow when it spread like a piedmont glacier on the plain beyond. A short survey line painted onto the marginal salt about 900 m upstream of this view in January 1977 (Talbot and Rogers, 1980) was 53 cm downstream in June 1994, indicating a seventeen-year average flow rate of ≈31 mm/a.

**167C.** In places the tank-track appears to have broken and cascaded forward so that there are piles of isoclinal, similar-type folds with an axial planar foliation. The single magenta bed in 167B is thinned to ~10 cm and repeated at least four times upstream of the last barrier ridge.

**167D.** Stacks of folds like those in 167C can commonly be seen to be sheath folds in profiles perpendicular to the general flow direction. Ductile shears commonly truncate the top and/or bottom of many such "eyes." Basal distal salt is typically a dirty mélange with a flow fabric smeared to conform with the irregular tops of terminal moraines shed over Recent alluvium. Like some other nappes, the namakiers of salt in the Zagros mountains flow over their own debris (Figure 167.1).

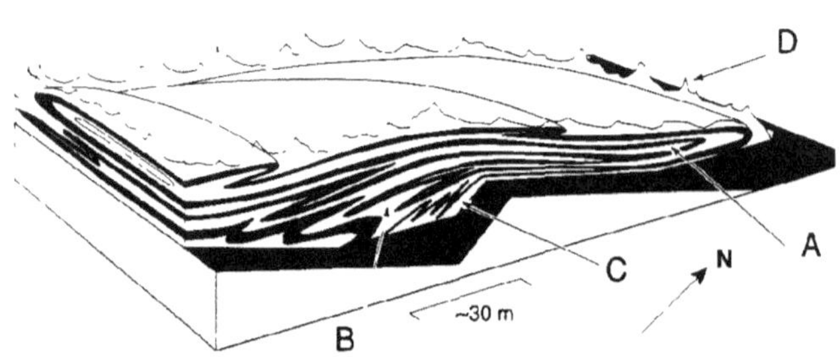


**Figure 167.1.** Schematic diagram of snout of northern namakier with arrows indicating localities and direction of view of Plates 167A–167D.

167 A

167 B

167 C

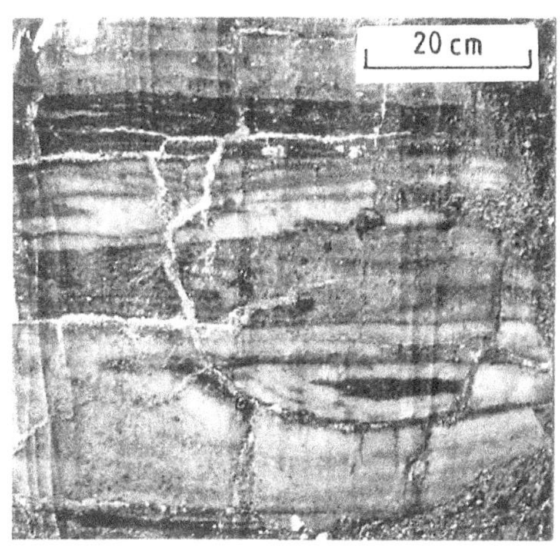


167 D

# 168. DEFORMATION OF ANHYDRITE: DISLOCATION CREEP TO DIFFUSION CREEP

LISA N. DELL'ANGELO AND DAVID L. OLGAARD

High-temperature deformation experiments on fine-grained anhydrite aggregates define two low-stress plastic flow regimes: one indicative of twinning/dislocation creep (I) and one of diffusion creep (II) (Dell'Angelo and Olgaard, 1995). The experiments were performed at 400°–800°C, strain rates of $10^{-3}$–$10^{-6}$/s, and a confining pressure of 300 MPa on synthetic and natural aggregates with grain sizes of 12 and 30 μm, respectively. The two deformation regimes are defined by distinct flow laws (relating stress, grain size, temperature, and strain rate) and diagnostic microstructures. Presented here are examples of the observed microstructures. These microstructures are very similar to those observed in naturally deformed anhydrite rocks by Jordan and others (1990). In order of decreasing stress (or increasing temperature) the dominant deformation processes inferred from the microstructures are regime I, twinning, dislocation glide, grain-boundary migration recrystallization, dislocation climb, and subgrain rotation recrystallization, and regime II, grain-boundary sliding and diffusion. The scale is the same for all optical micrographs (168A, 168B, 168C, and 168F), and the compression direction is horizontal. The two TEM micrographs (168D and 168E) are at the same scale.

**168A.** Regime I, 400°C, 20% strain, and $\sigma = 373$ MPa. This sample was continuously work-hardened. Twins and evidence for grain-boundary migration recrystallization dominate the microstructure. Müller and others (1981) demonstrated that twinning is well established after only a few percent strain, and by ~20% strain virtually all twinning volumetrically possible is accomplished. Further deformation, therefore, must be accommodated by another mechanism. Observations of serrate grain boundaries as well as incipient grain- and twin-boundary recrystallization suggest that grain-boundary migration recrystallization is important at higher strains. NX

**168B.** Regime I, 500°C, 37% strain, and $\sigma = 200$ MPa. Increased grain-boundary mobility is evidenced by highly serrate grain and twin boundaries and extensive grain-size reduction by dynamic recrystallization. The recrystallization mechanism is interpreted as grain-boundary migration recrystallization, which suggests that dislocation climb is limited (e.g., Tullis and Yund, 1985). The remnants of original grains exhibit undulatory extinction and highly serrate grain boundaries. NX

**168C.** Regime I, 600°C, 36% strain, and $\sigma = 40$ MPa. Samples deformed at stresses below ~110 MPa generally exhibit a recovered texture suggesting that dislocation climb is active (e.g., Zeuch, 1982; Tullis and Yund, 1985). Subgrains are common and recrystallized grains form within grains, presumably by progressive subgrain misorientation. The recrystallized grain size is somewhat coarser than at higher stresses (e.g., 168B). Serrate grain boundaries and grain-boundary recrystallization are present but are not as common as at higher stresses (compare with 168B). NX

168 A

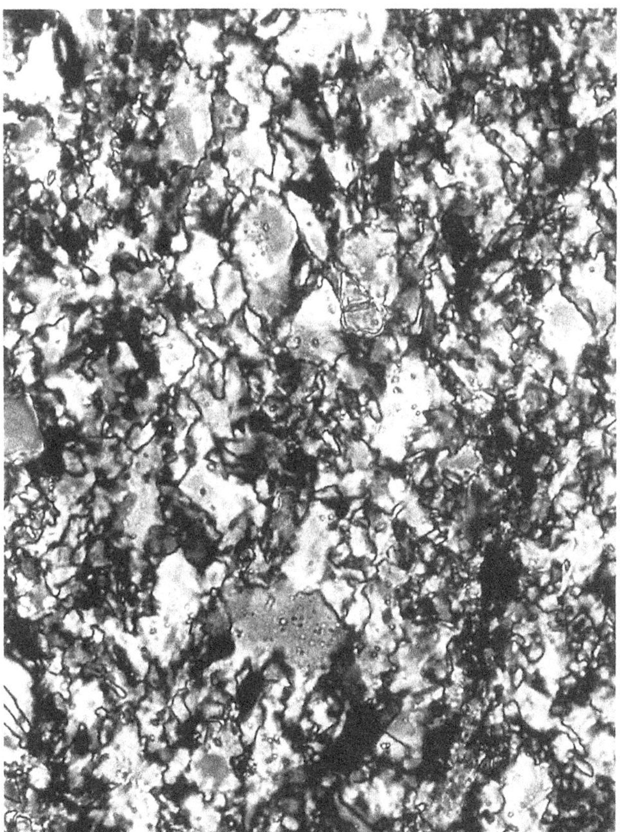

168 B

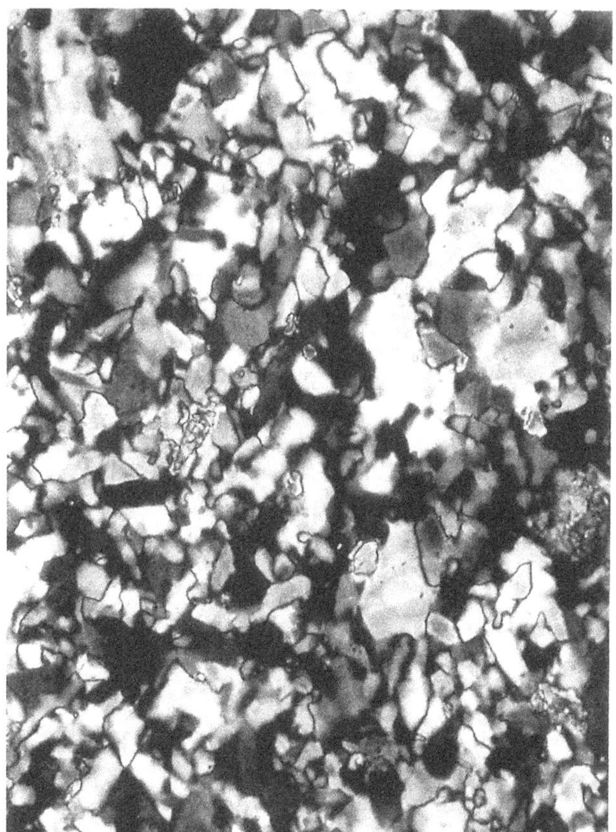

168 C

**168D.** Regime I, 500°C, 37% strain, and σ = 200 MPa (same sample as 168B, previous page). This image is typical of grain-boundary migration recrystallization. Small recrystallized grains (star) with low dislocation densities are located at the grain boundaries of larger original grains, which usually contain high dislocation densities. Dislocations commonly form tangles (arrow) and pileups (not shown), indicating that climb is severely limited. The fine speckles in some grains are electron beam damage. TEM

**168E.** Regime I, 800°C, 47% strain, and σ = 44 MPa. This sample contains similar microstructures as in 168C, previous page. At lower stresses and higher temperatures than 168A, dislocations are arranged into low-energy structures such as dislocation walls (arrow) and intersecting parallel sets (networks) of dislocations (center). The microstructures indicate that dislocation climb is important. Recrystallization is interpreted as forming by progressive subgrain misorientation. TEM

**168F.** Regime II, 800°C, 42% strain, and σ = 8 MPa. The microstructures in the diffusion creep regime are dramatically different from those in the twinning/dislocation creep regime. Despite the high sample strain, the grains are equant and contain no undulatory extinction. Note that most grain boundaries are straight or very gently curved. There is little evidence for recrystallization, although the grain size has increased by ~1.5 times the initial size. Whereas samples deformed in the twinning/dislocation creep regime always exhibit a crystallographic preferred orientation, a random orientation is measured for samples deformed in the diffusion creep regime. These observations strongly suggest that intercrystalline mechanisms such as grain-boundary sliding and diffusion are dominant. NX

Flow laws derived from experimental data can be used to establish the dominant plastic flow regimes in naturally deformed rocks, given estimates of grain size, temperature, and strain rate. In Figure 168.1, the solid line (A) marks the transition between the dislocation creep and diffusion creep regimes for anhydrite (based on the data of Dell'Angelo and Olgaard, 1995) in terms of grain size vs. temperature (T) for a geologic strain rate of $10^{-13}$/s. For comparison to similar regimes for fine-grained calcite rocks, the dotted line (S) marks the transition for Solnhofen limestone (Regimes 2 and 3 of Schmid and others, 1977), and the dashed line (W) for synthetic aggregates (high- and low-stress regimes of Walker and others, 1990). Predicted conditions for deformed anhydrites from the Jura mountains, Switzerland (Jordan and others, 1990) are shown in the box. In general, fine-grained natural anhydrites would deform by diffusion creep under geologic conditions. The transition between the deformation regimes occurs at a larger grain size for anhydrite than calcite.

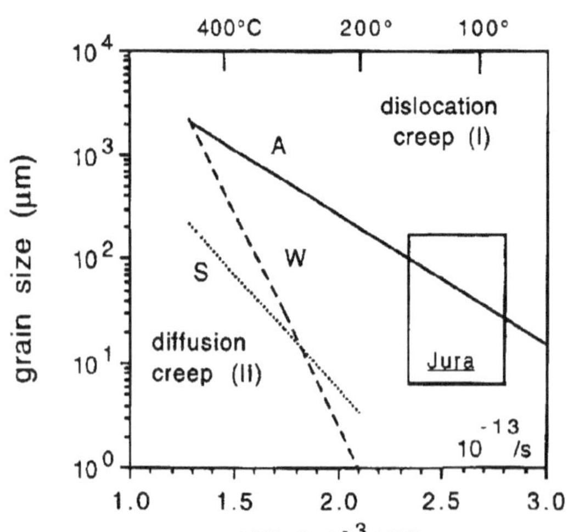


**Figure 168.1.** Plot of log grain size vs. 1/T for a strain rate of $10^{-13}$/s, showing boundaries between dislocation and diffusion creep for anhydrite (A) and for limestone (S and W)

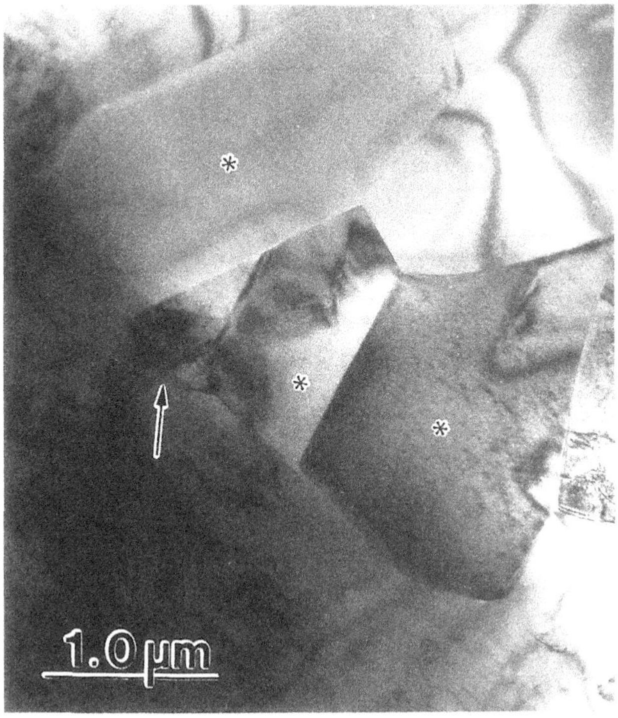


168 D

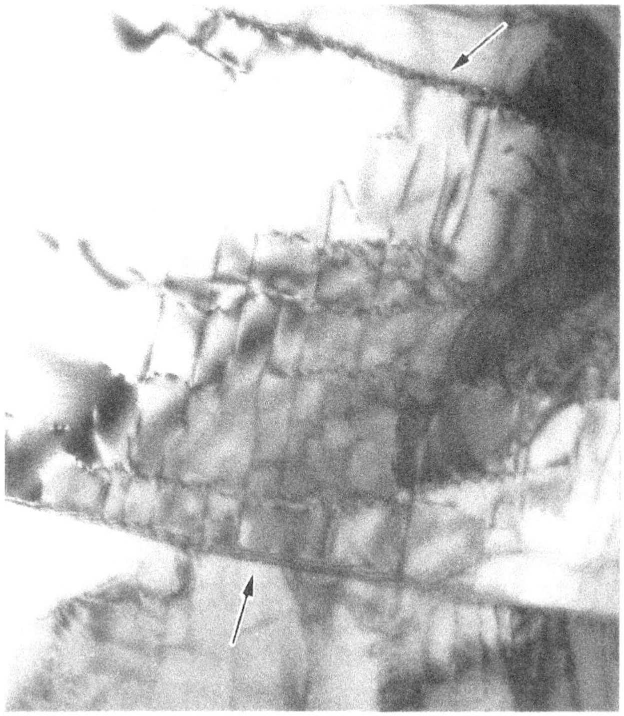

168 E

168 F

# DUCTILE BEHAVIOR

## Microstructural development of mylonitic rocks of specific composition

### Mafic rocks



Common abbreviations used in regard to the plates in this section include PPL = plane polarized light and NX = crossed nicols.

This set of plates illustrates deformation microstructures characteristic of the oceanic crust. These rocks are more difficult to sample, and our knowledge is more limited. Two contributions illustrate deformed gabbros from ophiolites, and two others illustrate deformed gabbros from dredged and cored oceanic ridge environments. These plates show that at higher temperature (granulite-facies) pyroxene and plagioclase undergo slip and dynamic recrystallization, but with decreasing temperature (amphibolite-facies) deformation becomes more localized and strain is partitioned almost entirely into the weaker plagioclase and amphibole. Further decrease in temperature promotes even more localized deformation, accompanying a transition in plagioclase from larger recrystallized grains formed by subgrain rotation to much finer recrystallized grains formed by grain-boundary migration; this is also observed in naturally deformed anorthosites (e.g., Plates 148 and 149, this volume).

BRADLEY R. HACKER

These four photographs illustrate a vertical section through the metamorphic rocks beneath the Samail ophiolite (Figure 169.1), Oman, from the Green Pool section along Wadi Tayin (Ghent and Stout, 1981). These rocks are inferred to represent oceanic sedimentary and igneous rocks accreted to the base of the ophiolite during thrusting over a lower plate of oceanic lithosphere. An extreme inverted metamorphic gradient is present—in excess of 1000°C/km—ranging from partially melted, upper amphibolite (locally granulite) facies immediately beneath the peridotite tectonite of the ophiolite's mantle section, down through lower amphibolite-facies rocks (Searle, 1980). Zircon ages from the crustal plutonic sequence fall chiefly in the range 95.4–94.5 Ma (Tilton and others, 1981), whereas $^{40}$Ar/$^{39}$Ar ages indicate that the metamorphic rocks shown here cooled below ~525° C by 95.7–92.4 Ma.

The Samail ophiolite was probably thrust ≤300 km at a rate of 50–100 km/Ma over a period of 1–3 Ma (Hacker, 1990, 1991, 1994). Some of this displacement occurred within the overlying peridotite tectonite, and some was accommodated at lower temperatures in fault zones structurally beneath the section shown here. If 100 km of displacement was accommodated within the ~200 m thick metamorphic sole, the average shear strain and shear strain rate were on the order of ~500 and $10^{-12}$ s$^{-1}$, respectively.

All photographs were taken in plane polarized light of thin sections cut parallel to the lineation and perpendicular to the foliation. Labels: cpx: clinopyroxene; ep: epidote, grt: garnet; hbl: hornblende; plg: plagioclase.

**169A.** Metasomatic gneiss collected 1 m beneath the peridotite tectonite. Peak upper amphibolite-facies mineral assemblage is 60% ferroan pargasitic hornblende + 10% clinopyroxene (Wo[illegible]-En[illegible]Fs[illegible]) + 20% almandine (Alm[illegible]Grs[illegible]Pyp[illegible]Sps[illegible]) + 5% plagioclase + 5% ilmenite formed at ~500–800 MPa and ~725°C. Pyroxene crystals contain deformation twins and undulatory extinction; hornblende crystals exhibit undulatory extinction and large subgrains bounded by {hk0} walls. Garnet is undeformed and plagioclase deformation features are not visible because of alteration. Subsequent retrogression includes replacement of plagioclase by muscovite + albite, hornblende and clinopyroxene by epidote + pumpellyite + sphene, and ilmenite by hematite; none of the secondary phases are deformed. Sample Oman-153. PPL

**169B.** Metabasaltic amphibolite-facies gneiss collected 48 m beneath the peridotite tectonite. Peak mineral assemblage is 73% tschermakitic hornblende zoned to magnesio-hornblende + 10% epidote + 15% An[illegible] plagioclase + 2% rutile. Hornblende porphyroclasts contain numerous inclusions, are oriented with their *c*-axes at an angle to the foliation, exhibit undulatory extinction, and are recrystallized at their margins to smaller, subequant hornblendes (arrow). The abundant, aligned, smaller hornblendes contain few inclusions or deformation features. Retrogression includes alteration of plagioclase to albite + sphene + muscovite, replacement of hornblende by actinolite + chlorite + hematite, and crystallization of albite + pumpellyite + actinolite veins. None of the alteration phases are deformed. Sample Oman-25. PPL

**169C.** Metabasaltic amphibolite-facies schist collected 124 m beneath the peridotite tectonite. Peak mineral assemblage is 85% tschermakitic hornblende + 4% epidote + 10% An[illegible] plagioclase + 1% rutile. Hornblende porphyroclasts are oriented with their *c*-axes at an angle to the foliation, exhibit undulatory extinction, and are recrystallized at their margins to form σ-type clasts with tails of much smaller, fibrous hornblendes (arrows). Plagioclase crystals also form σ-type clasts, and areas between brittlely extended epidote crystals are filled with chlorite. Subsequent retrogression includes replacement of plagioclase by albite + muscovite, and alteration of hornblende to actinolite + chlorite + sphene; none of these phases are deformed. Sample Oman-135. PPL

**169D.** Metabasaltic amphibolite-facies schist collected 202 m beneath the peridotite tectonite. Peak mineral assemblage is 45% tschermakitic hornblende zoned to magnesiohornblende + 35% plagioclase + 15% epidote + 10% ilmenite. Plagioclase porphyroclasts dot an otherwise uninterrupted fabric of optically undeformed fine-grained hornblende crystals. Retrogressive phases include calcite + chlorite in veins, and albite, muscovite, chlorite, and hematite; none of these phases are deformed. Sample Oman-126. PPL

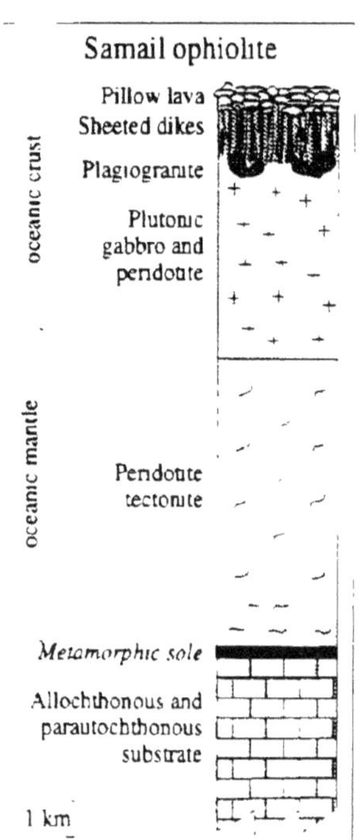


**Figure 169.1.** Schematic pseudostratigraphic column through the Samail ophiolite, Oman

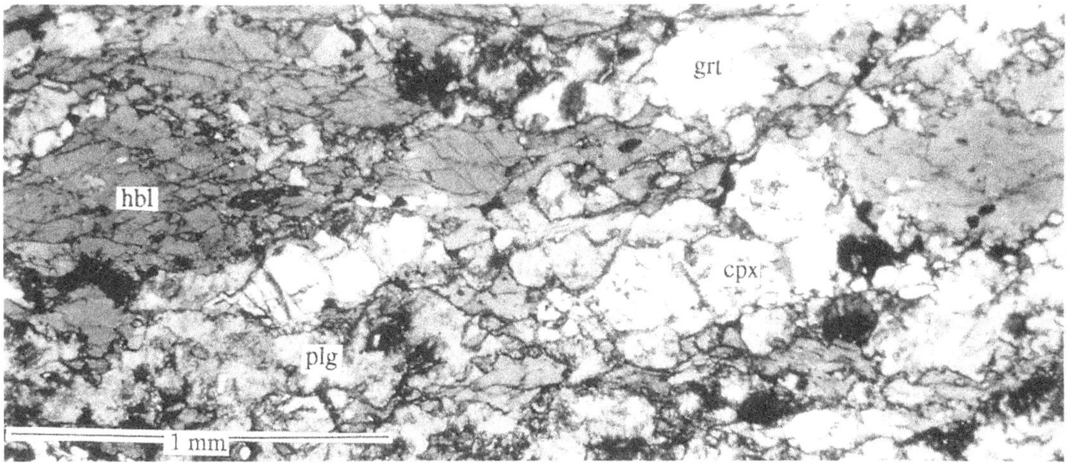

169 A

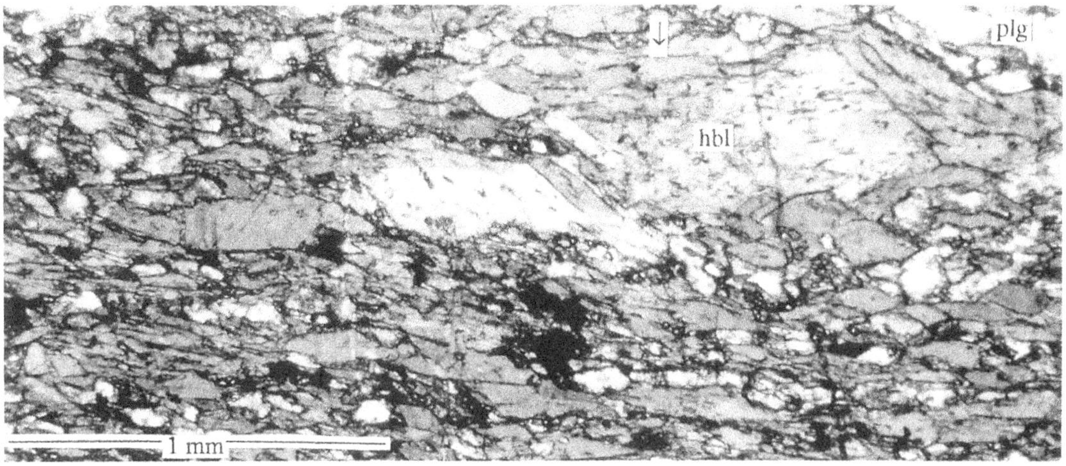

169 B

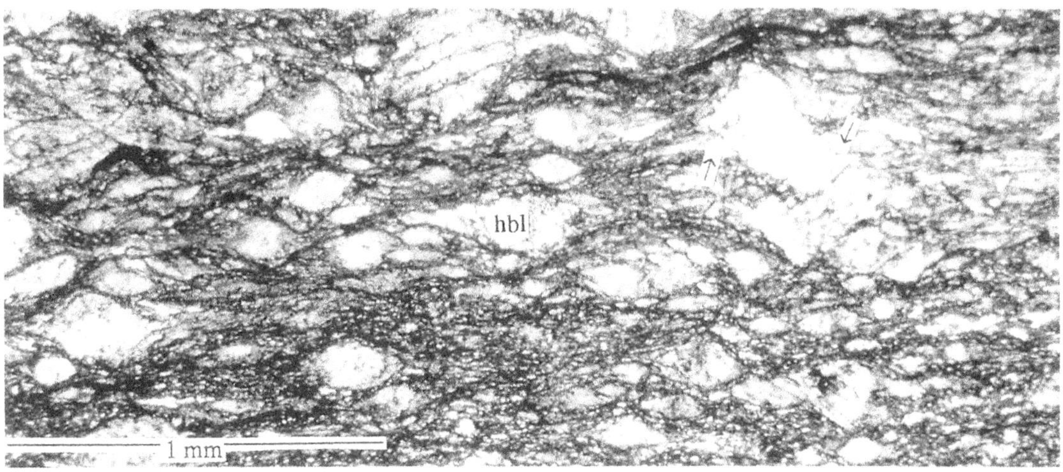

169 C

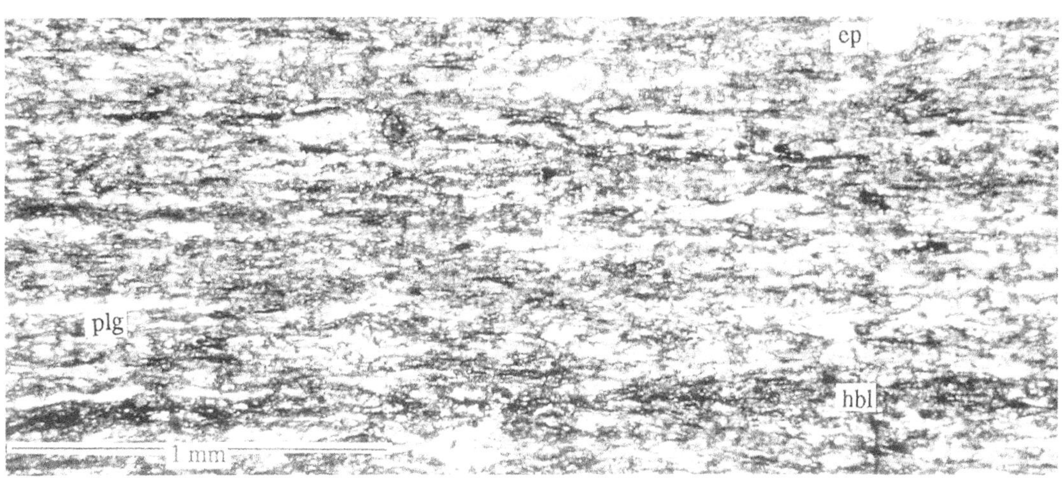

169 D

# 170. MYLONITIC SHEAR ZONES ASSOCIATED WITH DIKE INTRUSION IN AN OPHIOLITE

SUSAN M. AGAR

Observations of fault-rock structures in the oceanic lithosphere are currently restricted to borehole and dredge samples. The three-dimensional structural frameworks and cross-cutting relationships that can be determined in on-land exposures of ophiolites provide a complementary record of deformation at spreading centers. Although the geochemical signatures of ophiolites indicate that they formed in suprasubduction zone environments (Miyashiro, 1973), they still represent valuable mechanical analogs for deformation at active spreading centers.

The distribution and timing of deformation at a spreading center will be strongly dependent on the magmatic and hydrothermal histories that influence local thermal gradients and pore fluid pressures. In the Troodos ophiolite (Figure 170.1), a polyphase intrusive history resulted in multiple generations of dikes sourced by plutons at different crustal levels (Malpas and others, 1989). This has generated complex cross-cutting relationships within the dike-pluton transition, where dikes sourced by deeper-level plutons intrude shallow, older plutons, but some plutons also intrude the sheeted-dike complex. The Troodos ophiolite has not suffered any major structural or metamorphic overprint that might be associated with obduction and exposes a crustal section perpendicular to the spreading direction inferred from the dominant strike of sheeted dikes.

In the Agros region (Figure 170.1), an erosional window provides a three-dimensional exposure of the dike-pluton transition. A strong, low-temperature zeolite alteration dominates the upper plutonic section, but high-temperature (about 370°C) hydrothermal alteration assemblages (quartz, epidote, and sulfides) are found as fracture fills in fault zones that cut the plutonic rocks and replacement phases in plagiogranite intrusions. A peak amphibolite-facies alteration is preserved in a thermal aureole surrounding a late-stage intrusion south of the Agros window (Agar and Klitgord, 1995). Consistent with the model for "isotropic" gabbros in the uppermost layer three of oceanic crust, the upper plutonic section in the Agros window is mainly unaffected by penetrative, crystal-plastic deformation. However, narrow (centimeter-scale) mylonitic shear zones are commonly found along the margins of isolated dikes that intrude the upper plutonic section. The Agros window exposures demonstrate a previously unrecognized association between the magmatic history and localized, high-temperature, crystal-plastic deformation in the Troodos ophiolite (Agar, 1994; Agar and Klitgord, 1995) (Figure 170.2).

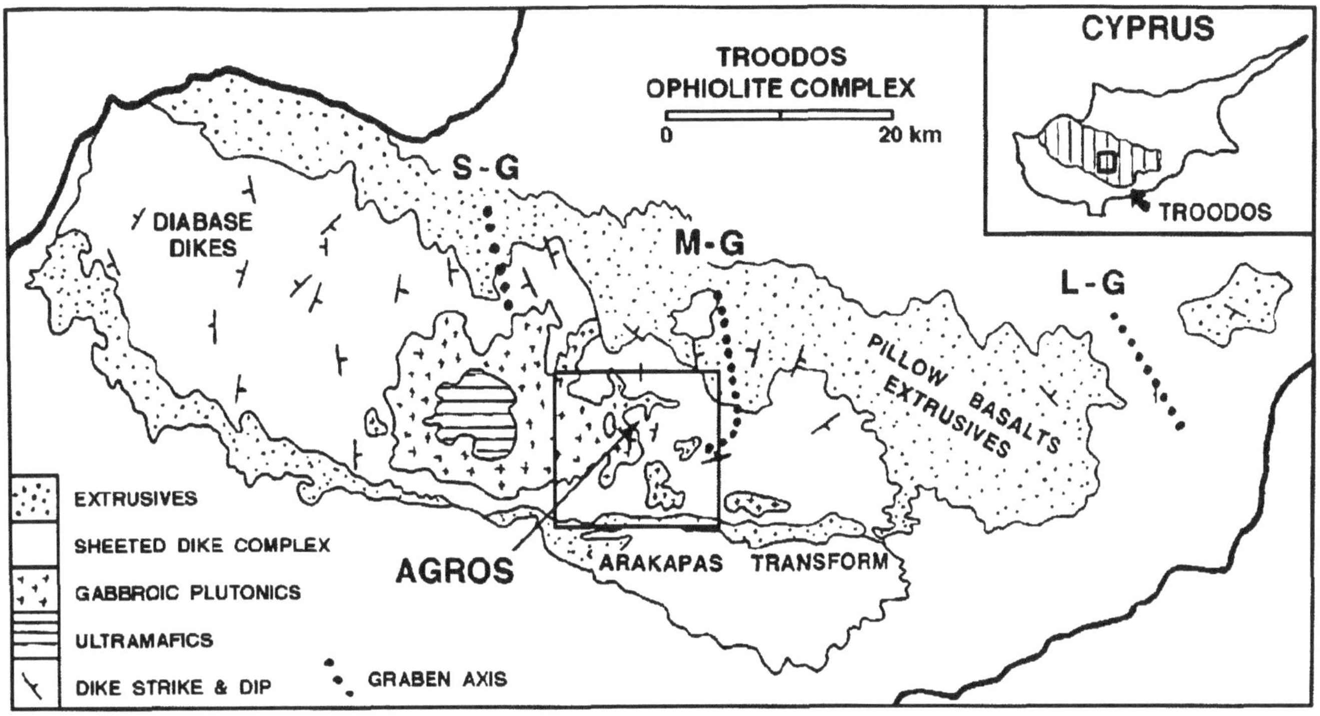


**Figure 170.1.** Location of the Agros window at the dike-pluton boundary in the Troodos ophiolite, Cyprus.

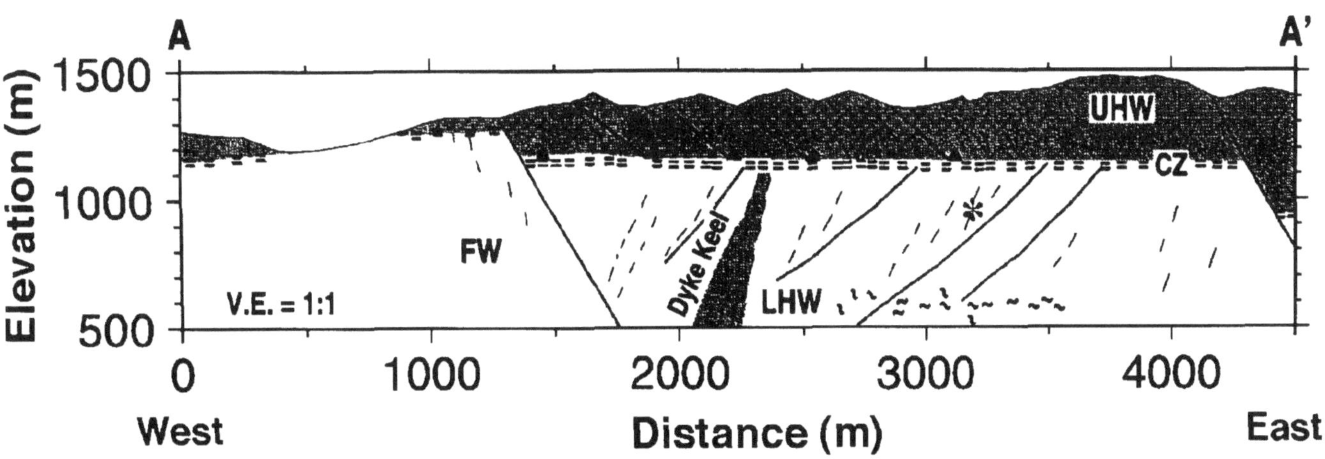


**Figure 170.2.** Cross section of the Agros window showing crustal location (asterisk) for Plates 170A and 170B

**170A.** Field photograph showing a narrow mylonitic shear zone along the margin of a dike intruding an olivine gabbro in the upper-level plutonic unit of the Agros window (Agar, 1994) (Figure 170.3). A mesoscopic shape-preferred orientation of plagioclase and amphibole (replacing pyroxene) crystals and segregations defines a 5–10 cm wide mylonite zone along the chilled dike margin (d, arrow in 170A). In thin section, apophyses of the dike can be seen to cut the mylonite zone, suggesting that at least some dike intrusion postdated shearing (cross-cutting relations are not illustrated here). The segregations of plagioclase and amphibole may have formed by partial melting adjacent to the dike. The close parallelism of the mylonitic foliation to the dike margin orientation and the high-temperature deformation fabrics restricted to the immediate vicinity of the dike suggest that at least some shearing occurred during dike intrusion. It is also possible that a preexisting shear zone generated a weakness that controlled the magma intrusion path.

**170B.** Mylonitic shear zone along dike margin (shown in 170A) (Figure 170.4). The thin section is oriented subhorizontally, perpendicular to the plane of dike margin. The mylonite consists mainly of plagioclase (p), amphibole (a), and Fe-Ti oxide (o). The Fe-Ti oxides are concentrated within the shear zone and form black stringers parallel to the foliation. Backscattered electron images (not shown here) suggest that the oxide phases have dynamically recrystallized. Amphibole, replacing clinopyroxene, occurs as porphyroclasts and neoblasts. Plagioclase is also strongly dynamically recrystallized and is partly altered to albite and chlorite. A very fine-grained, dark matrix of chlorite and clay minerals may represent devitrified glass (formed by melting of the wall rock in the dike margins). Thin, undeformed veins of epidote (e) are oriented perpendicular to and cut the foliation. This suggests that deformation ceased prior to high-temperature (370°C) hydrothermal alteration. The polyphase intrusive history for the Agros window means that these veins are not necessarily synchronous with other epidote veins in the region. The strong L-S fabric of the mylonite zone is disrupted by crenulations. In this example, a tight microfold with crenulations has formed in the hanging wall of a microthrust fault. Extensional shear bands cut across several foliation planes (Figure 170.4). The minor structures indicate a dextral sense-of-shear across the mylonite zone. The folding and faulting of the foliation may have been generated during a single phase of deformation as the result of local strain-rate variations. They could alternatively indicate a reactivation of the mylonite caused by the dike intrusion. PPL

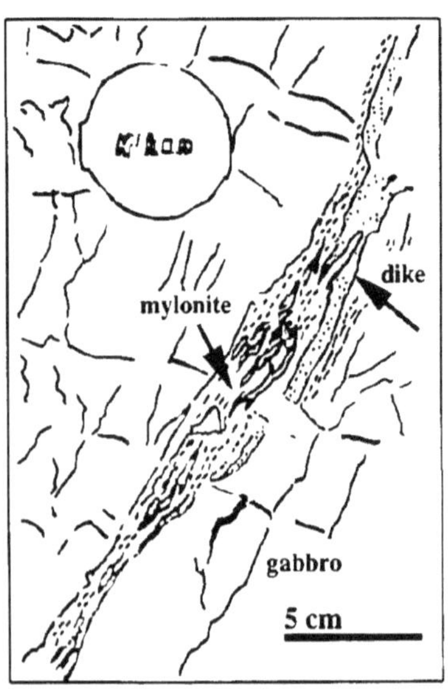


**Figure 170.3.** Line drawing of Plate 170A

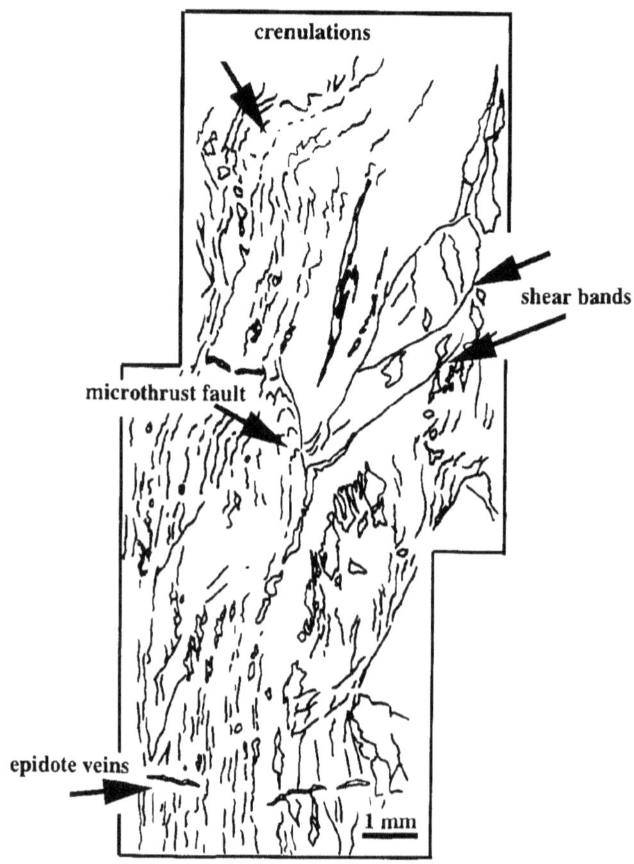


**Figure 170.4.** Line drawing of Plate 170B

170 A

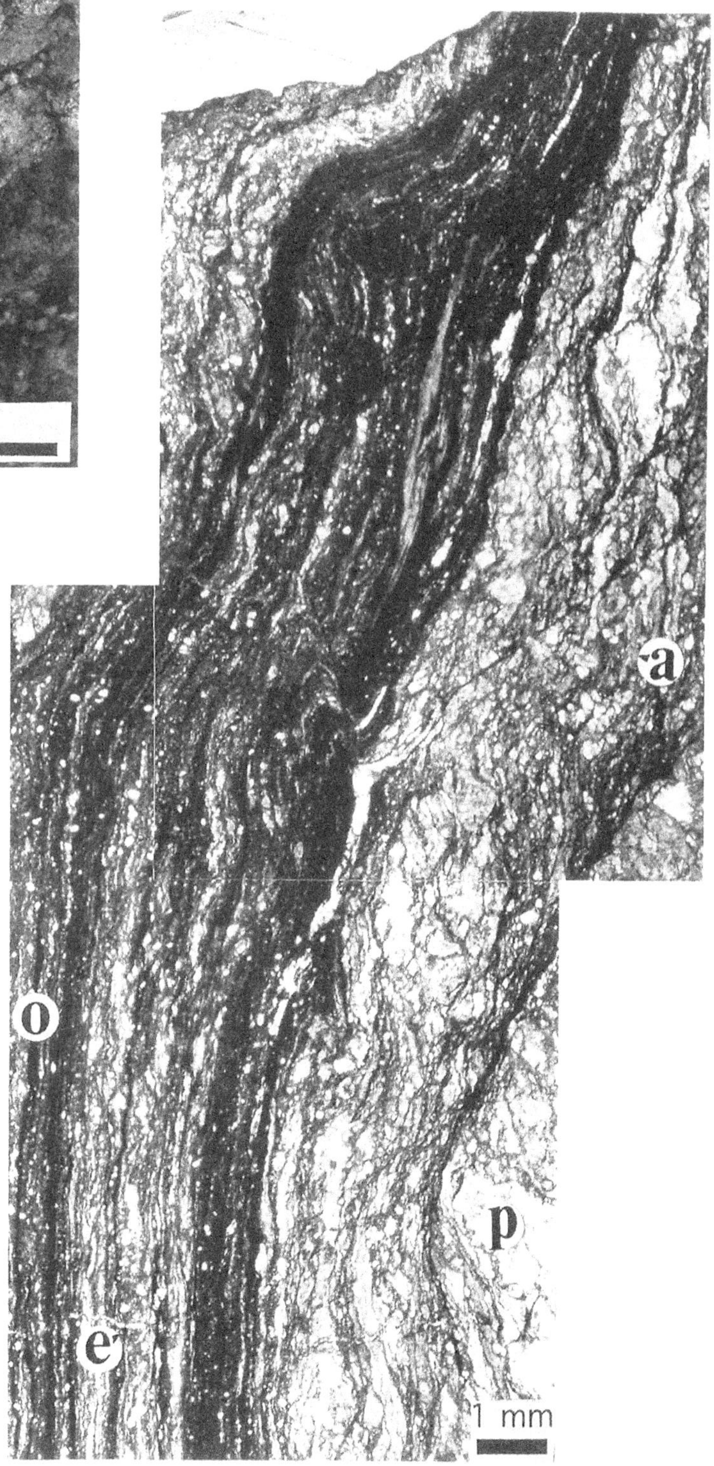

170 B

# 171. DEFORMED OCEANIC PLUTONIC ROCKS

DAVID A. VANKO AND RODEY BATIZA

Coarse-grained gabbroic rocks were dredged from the axial trough walls of the Mathematician Ridge, a failed spreading center in the eastern Pacific Ocean that was abandoned 3.5 m.y.b.p. (Vanko and Batiza, 1982). The coarse-grained plutonic suite consists of (a) undeformed olivine gabbro, gabbro, and two-pyroxene gabbro; (b) deformed gabbroic rocks having primary mineralogy identical to the undeformed rocks, with additional small amounts of hydrated phases such as magnesio-hornblende; (c) foliated metagabbro containing abundant hornblende, actinolitic hornblende, and actinolite, Fe-Ti oxides, chlorite, and albite; and (d) amphibolite with accessory albite, Fe-Ti oxides, sphene, and quartz as well as quartz-epidote veins.

Various stages in the alteration/deformation sequence from fresh plutonic rocks through foliated metagabbros are shown by the accompanying photomicrographs. Original subophitic textures (171A), some with primary igneous foliation owing to slight parallelism of plagioclase (bytownite) tablets, are modified by grain-boundary recrystallization of plagioclase and pyroxene (171B), and by the development of curved, stress-induced polysynthetic twinning in plagioclase. At the same time, pyroxene is usually partly altered to tschermakitic hornblende and magnesio-hornblende. Metamorphic conditions were amphibolite-facies conditions with T > 600°C (e.g., Ito and Anderson, 1983) and $P_{(H_2O)}$ about equal to $P_{(tot)}$. Further metamorphism results in grain-size reduction and a marked increase in modal amphibole and Fe-Ti oxides, with decreasing modal pyroxene and plagioclase. The relict pyroxene retains its augitic composition, whereas the plagioclase has been transformed to andesine. The amphiboles are increasingly variable in composition, with an irregular distribution of magnesio-hornblende and actinolitic hornblende domains within single crystals. The end product is a foliated metagabbro (171C and 171D), many samples of which are further altered at greenschist-facies conditions to actinolite-, chlorite-, and albite-bearing rocks, or to amphibolites with actinolitic hornblende and albite. Both fresh and metamorphosed gabbros were recovered in a single dredge haul, and metamorphosed gabbros were present in a second haul. Some diabases and extrusive rocks were also recovered. We suggest that faulting through the young oceanic crust and hydrothermal circulation of seawater-derived fluids along the faults are responsible for the occurrence of a complete spectrum of fresh through metamorphosed gabbros which was sampled by the dredge.

**171A.** Fresh clinopyroxene gabbro. Bytownite tablets are enclosed subophitically by augite. Less than 1% interstitial opaque oxides are present. This photomicrograph shows a finer-grained portion of the sample. In the coarser-grained portion, plagioclase and pyroxene reach 6 mm in length, and are difficult to photograph. Sample 8-10. Partial NX

**171B.** Metagabbro. Stress-induced polysynthetic twinning and grain-boundary recrystallization (incipient mortar texture) in plagioclase, and amphibole pseudomorphs of original pyroxene are shown. The brown amphiboles are mostly magnesio-hornblende, and the plagioclase grains, including the recrystallized edges, are labradorite-andesine. Coarse interstitial opaque oxide is also present. Sample 7-44. Partial NX

**171C.** Foliated metagabbro. The two amphiboles present are a brown hornblende and a green actinolitic hornblende. Relict plagioclase is andesine, and no relict augite is present. Sample 7-38. Partial NX

**171D.** Foliated metagabbro. The amphiboles are actinolite, the plagioclase is albitic, and relict augite is scarce (not shown in the photograph). Note the reduced grain size. Sample 7-28. PPL

171 A

171 B

171 C

171 D

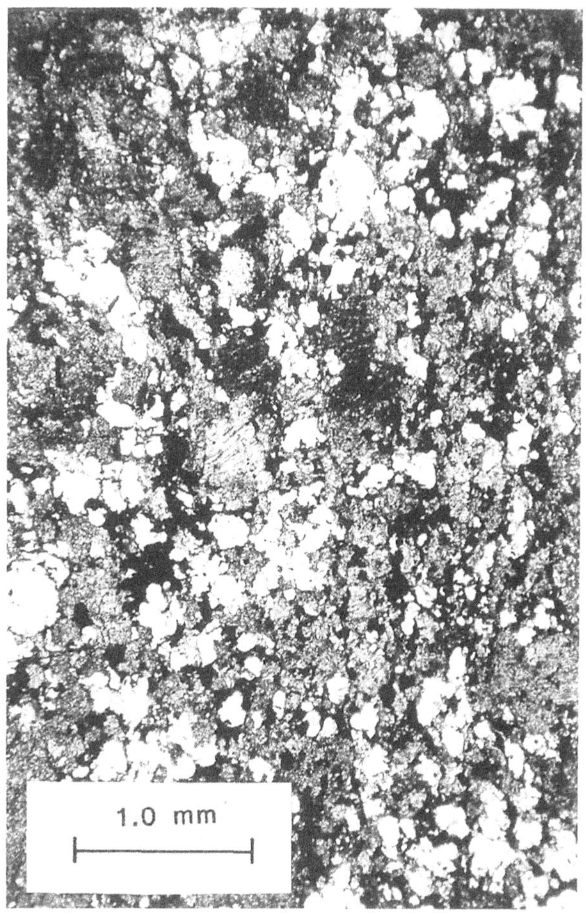

# 172. PLASTICALLY DEFORMED OCEANIC GABBROS

JULIE NEWMAN AND ANDREAS K. KRONENBERG

Oceanic layer 3 gabbros from the Southwest Indian Ridge (available from Ocean Drilling Program core 735B of leg 118) allow us to examine directly rocks that record the early, high-temperature deformation of oceanic crust at a slow spreading ridge. The recovered core contains many zones exhibiting strong mylonitic foliations with orientations that are consistent with normal faulting. The deformation microstructures and compositions of these mylonite zones offer the opportunity to constrain the stresses and temperatures of deformation at the ridge. The different zones display variable textures and mineral assemblages suggesting a range of conditions over which the ductile deformation took place, from granulite through amphibolite facies, reflecting cooling following the original magmatic event. Mylonitic foliations in the gabbros are defined by alternating layers of recrystallized plagioclase feldspar and recrystallized pyroxene and/or amphibole.

Comparison of recrystallized grain sizes from the naturally deformed samples with experimentally deformed clinopyroxenites yields estimates of differential stress at the time of granulite-grade deformation of 35 to 80 MPa. Temperatures at the time of deformation estimated from the compositions of recrystallized clinopyroxene from samples consisting predominantly of pyroxene and plagioclase range from 770 to 970°C.

Deformation at these conditions involved solid-state ductile creep by dislocation processes following igneous activity at the ridge, yet stress estimates are nearly as large as those predicted by Byerlee's law for frictional sliding at depths >5 km. Thus, gabbros recovered from hole 735B represent deep sections of layer 3 crust, and observed pyroxene microstructures record the last stages of high-temperature, plastic deformation as the crust cooled, just prior to the transition to brittle behavior.

**172A.** Mylonite zone from section 14R-01 of the core. The foliation is defined by alternating layers of recrystallized plagioclase (rx pl) and recrystallized pyroxene (px) with clinopyroxene (C) and plagioclase (Pl) augen. The recrystallized pyroxene has been partially replaced by amphibole (am). Opaque minerals (op) are abundant within regions of recrystallized plagioclase and pyroxene; in this sample they form a layer with recrystallized pyroxene. The arrow points toward the top of the core. This mylonite zone is >1 m thick as observed in the core. NX

**172B.** Mylonite zone from section 82R-01 of the core. This thin mylonite zone containing recrystallized plagioclase (rx pl) and pyroxene (px) and augen of clinopyroxene (C) and plagioclase (Pl) cuts through rock that shows predominantly igneous texture (top left and bottom right corners). This sample also contains a layer of opaque minerals (op) and pyroxenes altered to amphibole (am). The arrow points toward the top of the core. NX

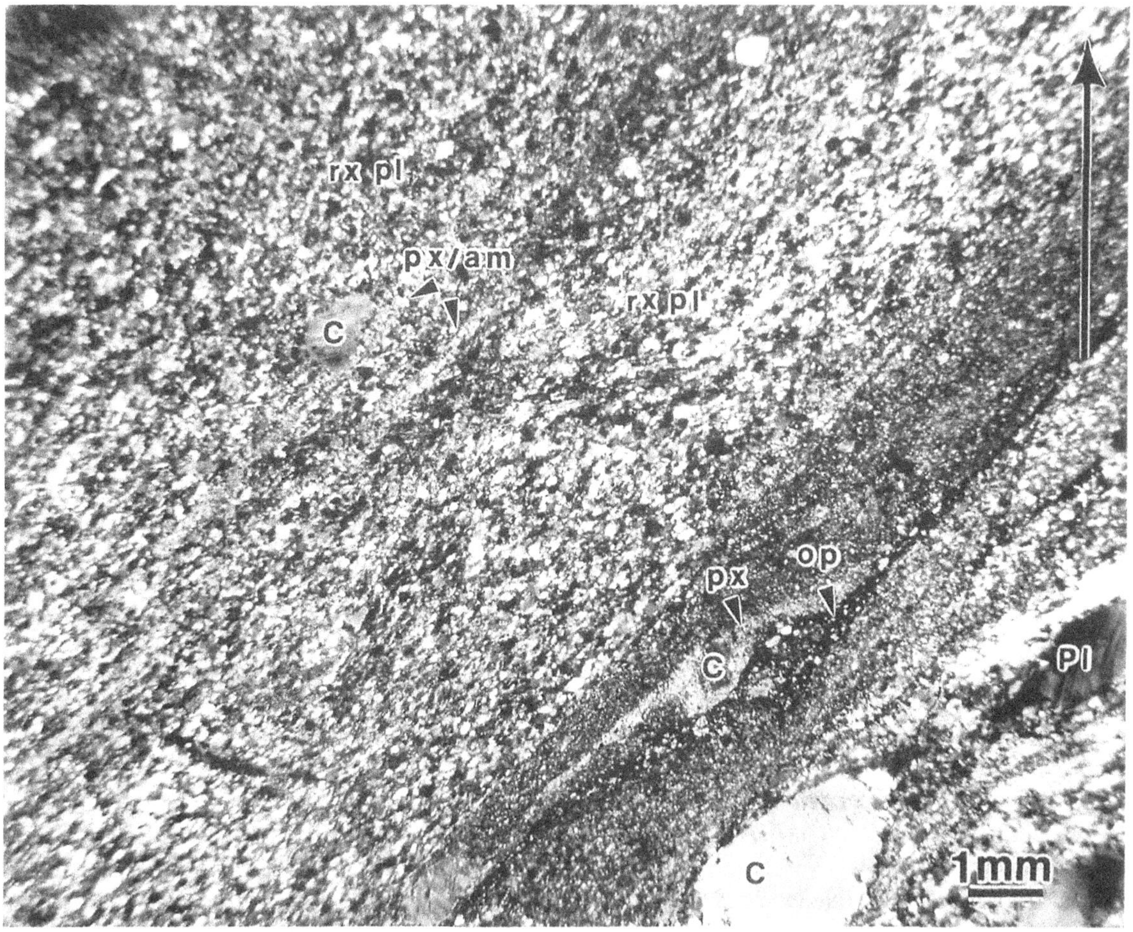


172 A

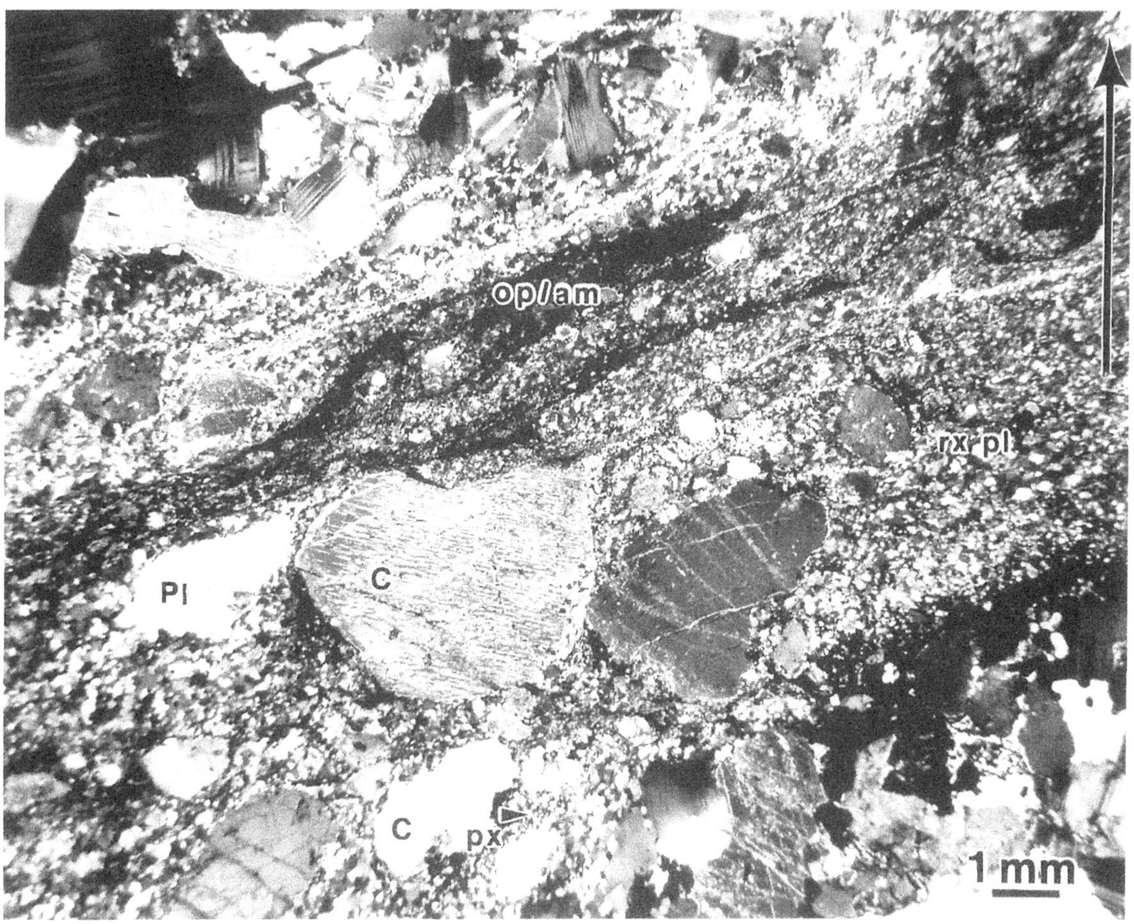


172 B

The textures and mineral assemblages in mylonite zones from core 735B record a range of conditions over which ductile deformation took place, from granulite through amphibolite facies, reflecting cooling following the original magmatic event. Deformation in these zones took place by dislocation creep accompanied by dynamic recrystallization of plagioclase, pyroxene, and amphibole. Thick granulite-facies zones consisting of dynamically recrystallized plagioclase and pyroxene are overprinted by more localized zones in which foliations are defined by layers of fine-grained, dynamically recrystallized plagioclase alternating with layers of pyroxene and amphibole. Pyroxene still exhibits evidence for dynamic recrystallization, but is largely altered to amphibole. Amphibole shows replacement textures, as well as textures suggesting dynamic recrystallization. Within the center of these amphibolite-facies zones, no pyroxene is observed, opaque minerals are common, and the foliation is defined by alternating layers of very fine-grained recrystallized plagioclase and amphibole. The association of amphibole with the later, finer-grained mylonite zones suggests that the development of these localized zones was related to the introduction of fluids to this system, resulting in strain softening and strain localization.

**172C.** Layers of dynamically recrystallized pyroxene (px) with a clinopyroxene augen (C) alternating with layers of dynamically recrystallized plagioclase (Pl) from section 14R-01 of the core. Pyroxene grains are rimmed by alteration to amphibole (am). Opaque minerals are present within layers of both plagioclase and pyroxene. PPL

**172D.** Dynamically recrystallized plagioclase from a localized amphibolite-facies mylonite zone (section 7D-01 of the core). The recrystallized grain size distribution in this sample is homogeneous. NX

**172E.** Dynamically recrystallized plagioclase from section 14R-01 of the core with a heterogeneous (bimodal) grain size distribution. The finer recrystallized grains along the grain boundaries of coarser recrystallized grains may have resulted from later, lower-temperature deformation. Transmission electron microscopy shows internal recovery microstructures and development of subgrain walls. However, sutured grain boundaries (s) indicate that dynamic recrystallization took place predominantly by grain-boundary migration. NX

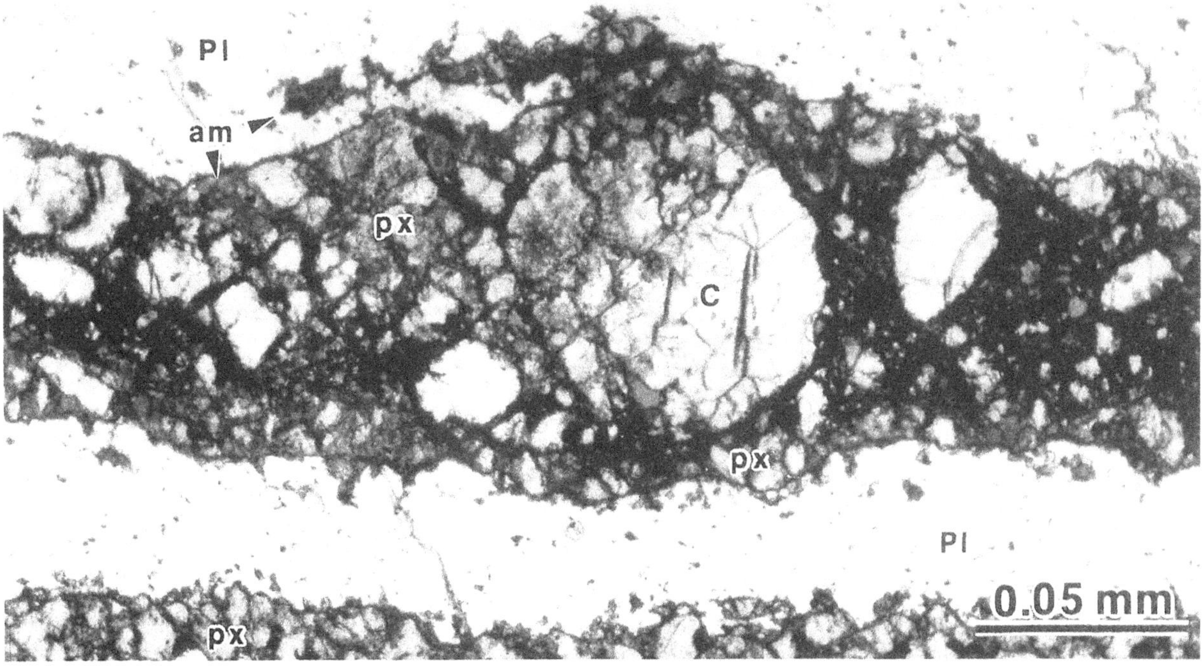

172 C

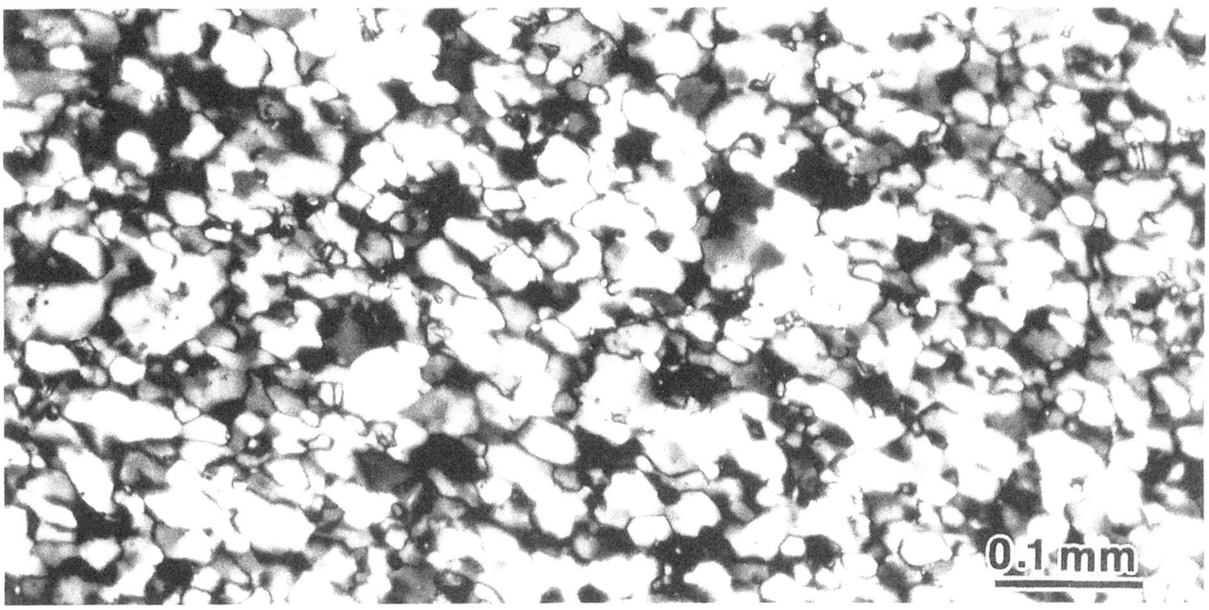

172 D

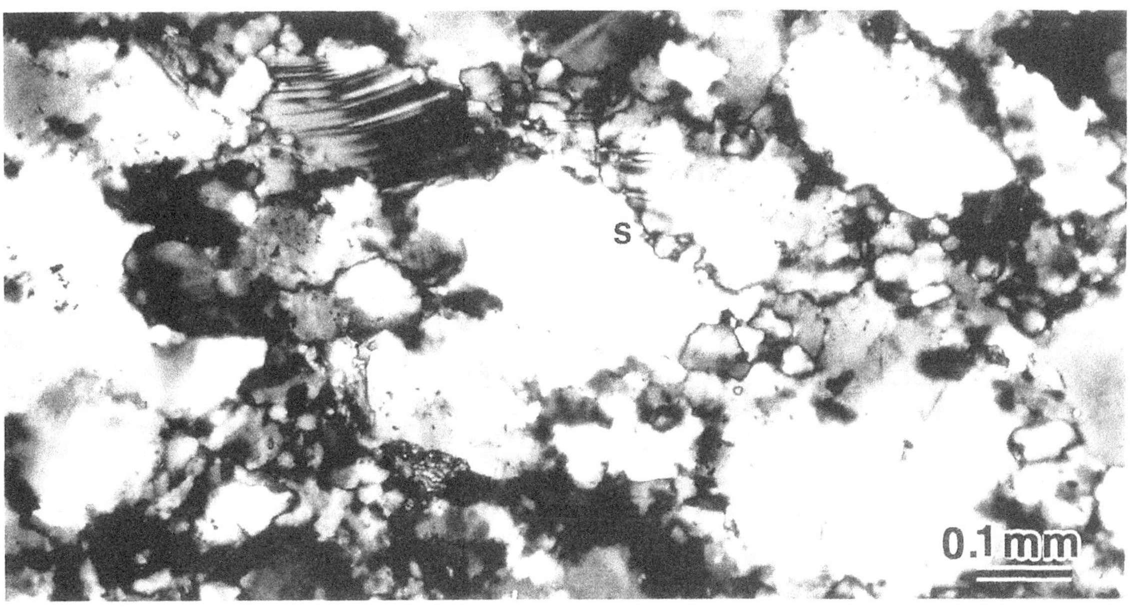

172 E

# Microstructural development of mylonitic rocks of specific composition

## Ultramafic rocks



Common abbreviations used in regard to the plates in this section include PPL = plane polarized light and NX = crossed nicols.

This set of plates illustrates outcrop and thin-section scale structures developed in localized high-strain zones within upper mantle peridotites, involving chiefly olivine, clinopyroxene, and orthopyroxene. Most of the examples come from ophiolites, and one comes from a kimberlite xenolith. Most of these rocks record deformation at extremely high temperatures, characteristic of asthenospheric or lithospheric mantle, although one illustrates the structures developed at lower temperatures where serpentine was stable. The microstructures in these rocks demonstrate that the order of decreasing dislocation creep strength is orthopyroxene, clinopyroxene, olivine. Olivine shows evidence of subgrain rotation recrystallization as well as high-temperature grain-boundary migration, and develops strong crystallographic preferred orientations. These plates illustrate that localization of deformation tends to occur with decreasing temperature, or after a critical strain if grain-size reduction by recrystallization allows a switch to grain-size-sensitive creep. One factor favoring grain-size-sensitive creep in nature is the presence of other phases, because mutual grain-boundary pinning will keep the grain size small. Strain localization may also be enhanced by the introduction of water into shear zones, e.g., causing a reaction of pyroxene to weaker amphibole.

# 173. MICROSTRUCTURES IN LOW- AND HIGH-TEMPERATURE SERPENTINITE MYLONITES

EILARD H. HOOGERDUIJN STRATING AND REINOUD L. M. VISSERS

The microstructures shown in this section are from ductile serpentinite mylonites exposed in the Voltri ophiolite massif, Liguria, northwestern Italy (Scambelluri and others, 1991; Hoogerduijn Strating and Vissers, 1991). Prograde and retrograde, low-temperature serpentinite mylonites are very common in the Voltri massif, but also occur in other ophiolitic complexes (e.g., Josephine complex, Klamath Mountains, U.S.A.; Norrell and others, 1989). Below we also document less common, higher-temperature serpentinite mylonites developed under eclogite-facies conditions (1300–1600 MPa, 500–550°C). The mylonites are developed in shear zones that bound and transect a host rock of massive peridotites with scattered gabbroic and basaltic dikes. These shear zones make up a ductile thrust system formed in an intra-oceanic subduction complex which accommodated Cretaceous/Tertiary convergence between the European and African plates. The serpentinite mylonite zones record a prograde metamorphic history from low-grade to eclogite-facies conditions, followed by unroofing and decompression leading to greenschist-facies assemblages.

**173A.** Low-temperature serpentinite mylonite with well-developed extensional crenulation cleavages indicating dextral shear. These mylonites developed both early in the prograde and late in the retrograde evolution, at temperatures in the range 250–450°C. The dominant wall-rock minerals (olivine, pyroxene, and spinel) are hydrated and transformed into antigorite, chlorite, and magnetite. Synkinematic growth and rotation of the sheet silicates resulted in the development of a very strong, shear-induced, foliation fabric (S). The foliations are generally parallel to the shear-zone boundaries, indicating that considerable shear strains have been accommodated during ductile thrusting. Asymmetric foliation boudinage and extensional crenulation cleavages (E) overprinting the shear foliation are very common, suggesting that ductile deformation in the serpentinite mylonite shear zones involved progressive stretching of the extremely anisotropic serpentinite fabric (Platt and Vissers, 1980). NX

**173B.** Low-temperature serpentinite mylonite with $\sigma$-type clinopyroxene porphyroclast. Most serpentinite mylonites contain porphyroclasts of pyroxene and, to a lesser extent, also of olivine and spinel. The pyroxene porphyroclasts characteristically develop into $\sigma$-type porphyroclast systems (Passchier and Simpson, 1986), with tails of recrystallized antigorite, chlorite, and minor tremolite. The tails of the clinopyroxene crystal shown bend into low-angle shear bands (E), thereby defining an $\sigma_b$-type geometry (Passchier and Simpson, 1986). The porphyroclast is deformed internally by antithetic slip on partially serpentinized [100] lattice planes, accommodating forward rotation of these lattice planes relative to the flow plane. Microfractures perpendicular to the [100] planes contribute to the fragmentation of the porphyroclast. NX

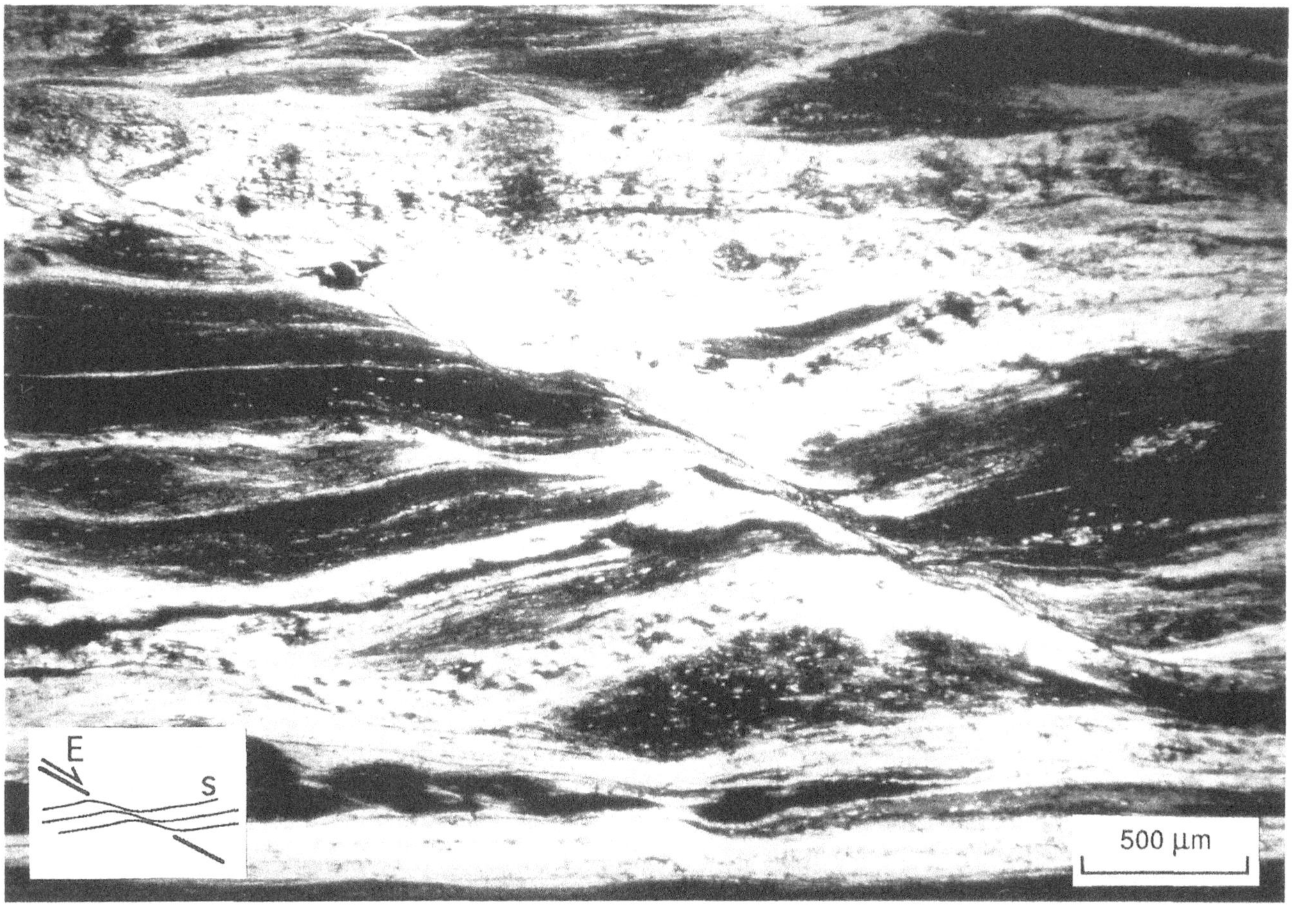


73 A

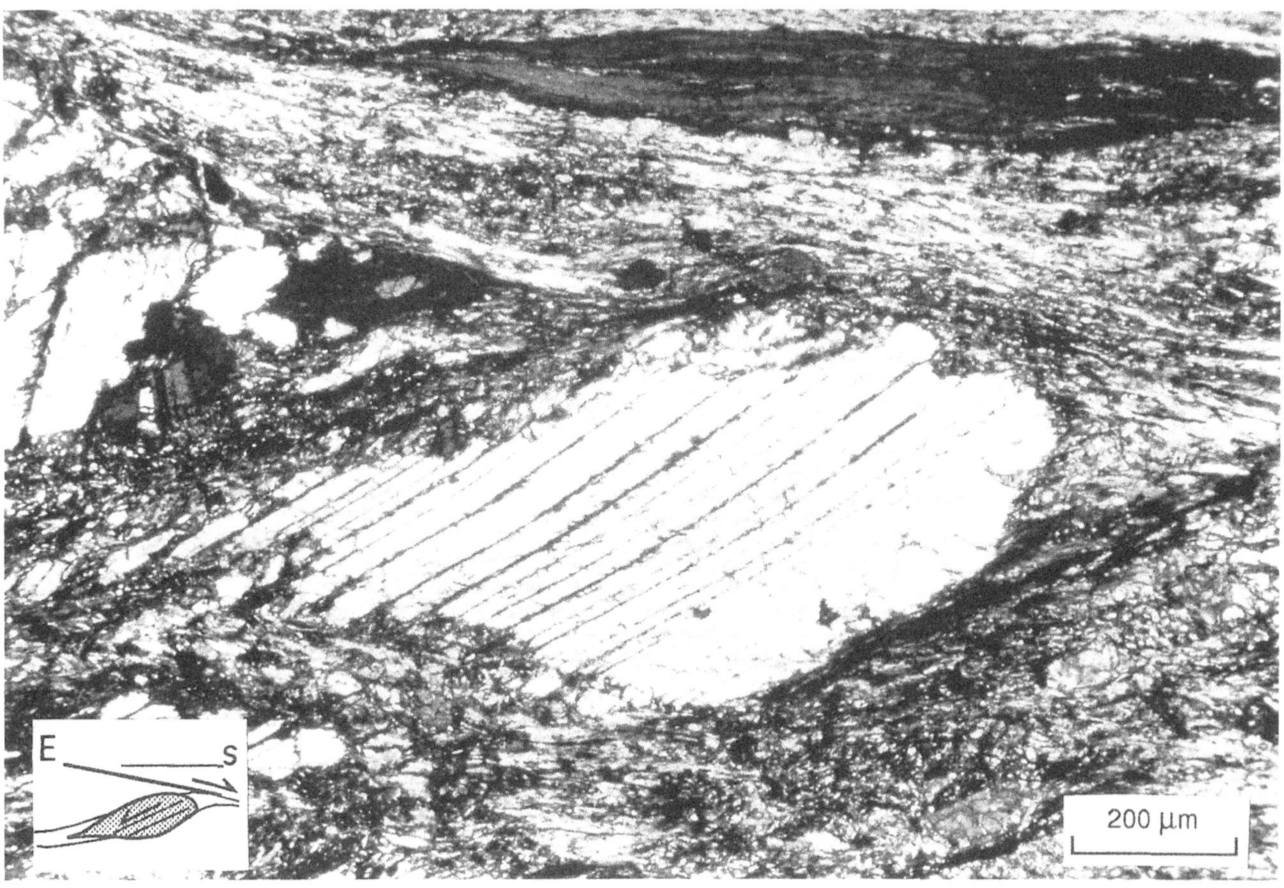


73 B

**173C.** Multiple sets of olivine-bearing, extensional crenulation cleavages. Displacement across the low-temperature serpentinite mylonites in the Voltri massif continued during progressive subduction to higher pressure and temperature conditions in the eclogite stability field. The serpentinite system is rather insensitive to pressure, and the only clear indication of a prograde metamorphic evolution is provided by the transformation of antigorite in the shear bands into a fine-grained olivine- and titanian-clinohumite-bearing assemblage, indicating higher temperatures (>450°C). On weathered surfaces in outcrop, this fine-grained assemblage has a characteristic red-brown color (dark gray bands on the photograph) contrasting against the gray-green serpentinite (lighter gray tones). The photograph shows a complicated geometry defined by multiple sets of extensional crenulation cleavages (ecc's). The low-angle set ($E_1$) dissects the olivine-free, shear foliation (S) at an average angle of about 20°. This composite fabric is overprinted by a younger, high-angle set ($E_2$) oriented at about 40–50° to the foliation.

**173D.** Photomicrograph of the structure shown in 173C. The mylonitic foliation (S) made up of antigorite and minor chlorite is overprinted by a very fine-grained shear band ($E_1$), containing olivine, titanian clinohumite, diopside, chlorite, brucite, and pyrite. This assemblage developed as the result of syntectonic partial dehydration of antigorite along the shear planes at high pressures and temperatures exceeding 450°C. Except for the generally equidimensional olivine grains, all phases show a moderate to strong shape-preferred orientation. The grain size is very small and generally amounts to less than 5 μm. A slightly coarser-grained shear band (20–40 μm grain size), oriented at an angle of 40–45° to S, belongs to the younger set of $E_2$ ecc's (see 173C). Although Scambelluri and others (1991) were the first to report these olivine-bearing shear bands from natural rocks, Rutter and Brodie (1988) had observed similar equidimensional ultrafine-grained olivine decorating microscopic shear zones in their synkinematic dehydration experiments on intact serpentinite and serpentine gouge. The mechanical behavior of the testing material in these experiments and the microstructures produced suggest that the deformation in the fine-grained olivine shear bands was controlled by diffusion-accommodated grain-boundary sliding. There are, as yet, no fabric nor TEM data of the natural samples shown to test whether the deformation in the prograde, high-temperature serpentinite mylonites was also controlled by some form of diffusion creep of fine-grained olivine. PPL

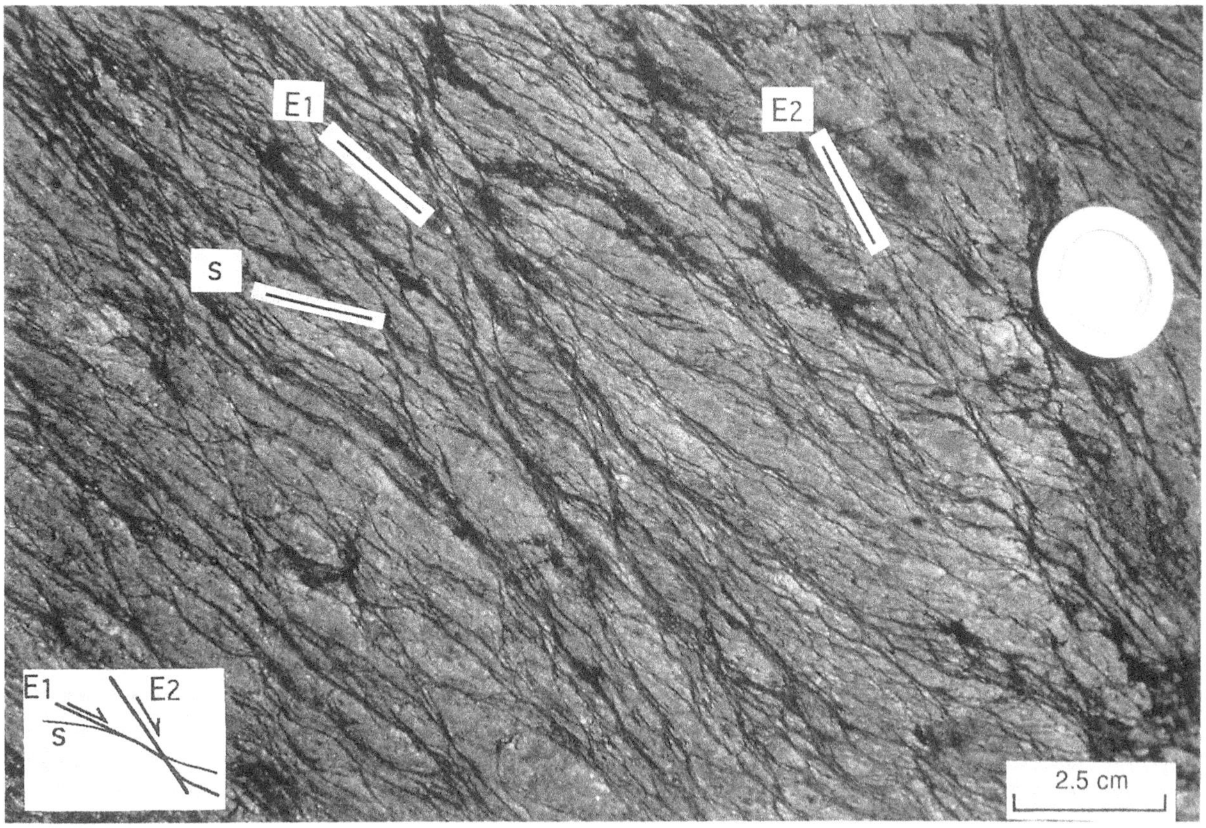


173 C

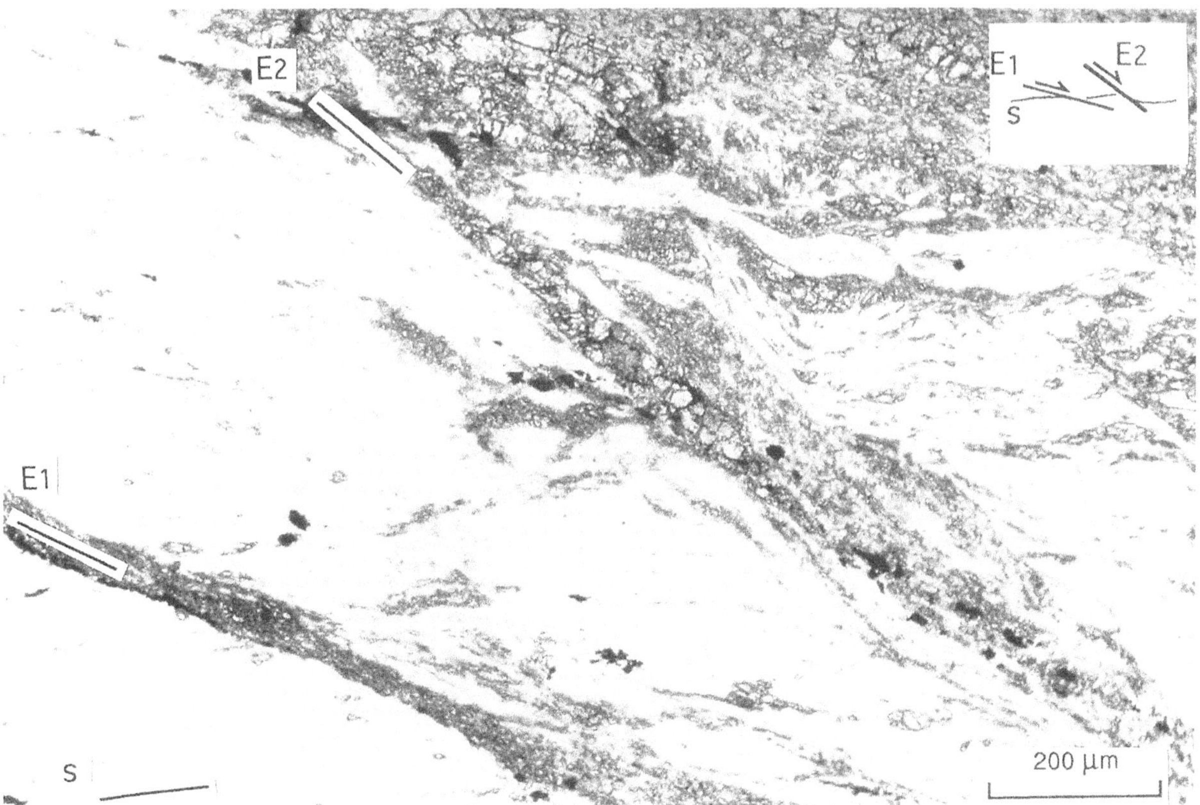


173 D

# 174. MICROSTRUCTURES FORMED DURING SHEAR LOCALIZATION IN THE UPPER MANTLE

R. L. M. VISSERS, E. H. HOOGERDUIJN STRATING, M. R. DRURY, AND D. VAN DER WAL

The microstructures shown represent different degrees of shear localization in upper mantle peridotites. Examples are mainly from the Voltri massif of northwestern Italy, where thrusted fragments of lithospheric mantle, initially exhumed to the ocean floor during Jurassic rifting and continental breakup, were emplaced in the Alpine collisional stack during Tertiary convergence between Africa and Europe. Coherent slices of these mantle rocks contain several deformational structures formed during the early rift evolution, including a transition, over several hundreds of meters, from granular and virtually undeformed spinel-peridotites to porphyroclastic (Mercier and Nicolas, 1975) spinel-peridotite tectonites (Drury and others, 1991; Vissers and others, 1991; Hoogerduijn Strating and others, 1993). This transition is marked by a progressive development of the tectonite foliation and by gradually decreasing angles between this foliation and pyroxenite bands, suggesting increasing strain. In the more intensely deformed rocks, elongate and locally asymmetric pyroxene porphyroclast systems (Passchier and Simpson, 1986) define a stretching lineation. The observed map pattern as well as oblique olivine grain shape and lattice fabrics indicate that the foliated tectonites represent part of a kilometer-scale ductile shear zone. Thermobarometry points to ambient temperatures of over 950°C, in the spinel lherzolite stability field, during deformation in this shear zone.

**174A**. Because a variable but distinct serpentinization in the Voltri tectonites obscures the details of the tectonite microstructures, we illustrate these tectonites with an entirely similar microstructure from the Horoman peridotite of Japan. Note elongate olivine ribbon parallel to a coarse tectonite foliation ($S_{tect}$), with obliquely oriented kink-type subgrains ($S_k$). The ribbon grain is surrounded by equant olivine neoblasts with slightly curved grain boundaries (arrows). The microstructure illustrates extensive rotation recrystallization and grain-boundary migration, leading to grain-size reduction. NX

**174B.** Oblique grain-shape fabric in a Voltri tectonite. The tectonite foliation ($S_{tect}$) defined by stretched pyroxenes and elongate spinel aggregates is horizontal. The oblique fabric ($S_{obl}$) is defined by slightly inequant to highly elongate olivine grains with smoothly curved to serrated grain boundaries. At least some grain boundaries seem to be high-angle, kink-type boundaries. A detailed study of this microstructure (van der Wal and others, 1992) suggests a kink-type glide polygonization process produced by slip on the olivine high-temperature {0kl} slip systems (Avé Lallemant and Carter, 1970), leading to the development of [100] tilt walls such that the subgrain boundaries initiate an oblique fabric as in 173A. With ongoing shear, these tilt walls rotate whilst developing into high-angle, kink-type grain boundaries. The deformation geometry implies that the sense-of-shear, dextral in this example, is the same as that inferred on the basis of associated oblique lattice fabrics (van der Wal and others, 1992). NX

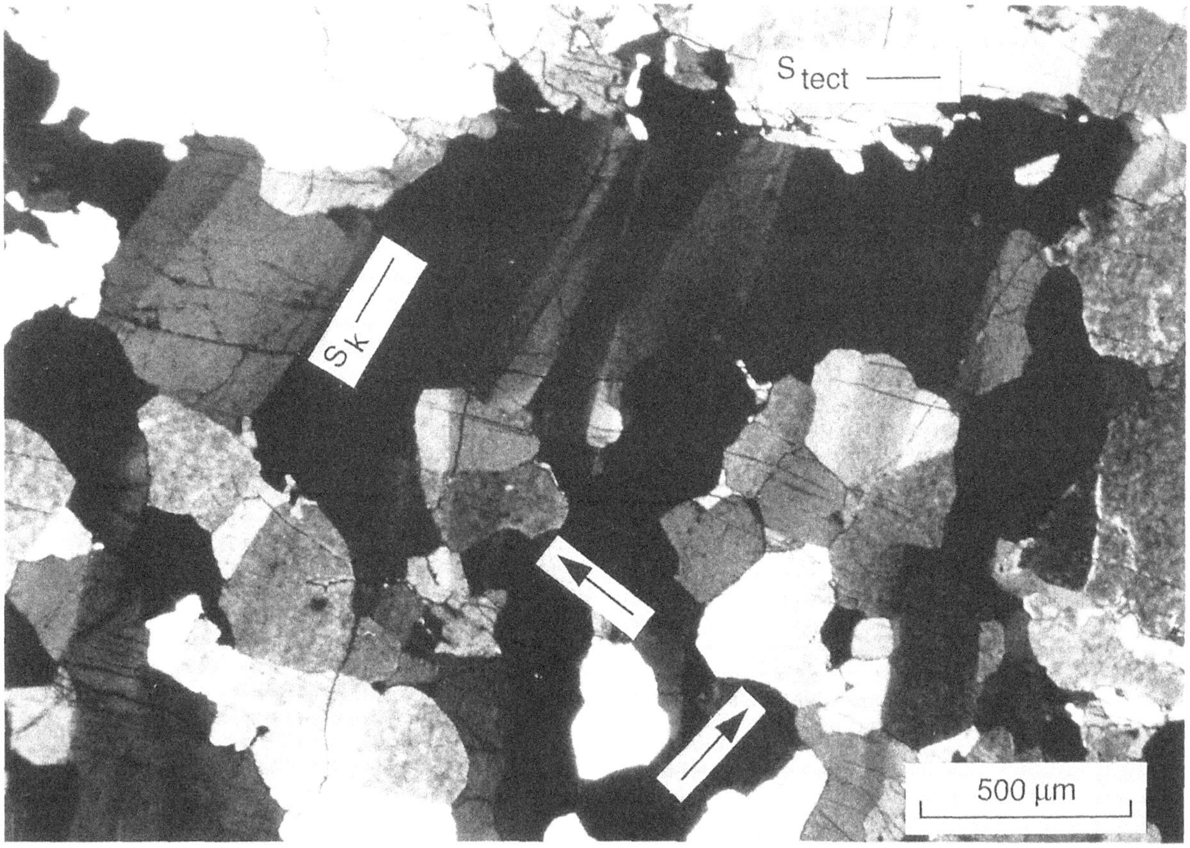

174 A

174 B

Elsewhere in the Voltri massif, kilometer-scale slices of peridotite bounded by Alpine ductile thrusts contain parts of a network of mylonitic shear zones, 0.5–200 m wide, which transect the tectonite fabric and locally also domains of granular, virtually undeformed peridotite. The associated mylonitic foliation is much more intense than the tectonite fabric, and commonly contains a clear stretching lineation of elongate orthopyroxenes. Most mylonites are spinel-bearing mylonites, but there are also plagioclase-bearing mylonites in which spinel is transformed into flattened aggregates of Cr-spinel rimmed by plagioclase. In addition, some mylonites show synkinematic transformation of spinel into trains of magnetite surrounded by rims of strongly oriented chlorite. Most mylonites contain syntectonic pargasitic-edenitic amphibole. The mineral chemistry of recrystallized pyroxene pairs in the spinel- and plagioclase-bearing mylonites yields temperatures between 890 and 1010°C, at pressures between 500 and 800 MPa, consistent with the growth of pargasitic-edenitic hornblende indicating temperatures in the range 800 to 975°C, at ambient pressures around 700 MPa (Hoogerduijn Strating and others, 1993). These temperatures are markedly higher than the temperature range of 550 to 780°C deduced for the chlorite-bearing, peridotite mylonites on the basis of their syntectonic mineral assemblages. All of the mylonitic structures are cut by gabbroic and basaltic dikes with a normal mid-oceanic ridge basalt (MORB) affinity, and we infer that the different mylonites developed progressively during ongoing lithosphere stretching and mantle exhumation at the onset of continental breakup and development of oceanic crust in the Piemonte-Ligurian realm.

**174C.** Detail of an approximately 0.5 m wide mylonitic shear zone, showing progressive bending of a coarse tectonite fabric into the plane of the mylonitic foliation in the shear zone.

**174D.** Microstructure of spinel-peridotite mylonite shown in 174C, with thin, ultrafine-grained (1–10 μm), "fluidal" bands surrounding porphyroclasts of stretched and fragmented orthopyroxenes (Opx), subequant clinopyroxenes (Cpx), and spinel (Spl). The ultrafine-grained bands anastomose around lenticular, pinched domains of a fine-grained (10–150 μm), recrystallized olivine mosaic with subequant porphyroclasts (1–10 mm diameter) mainly of olivine and occasionally of clinopyroxene and spinel. Backscattered and transmission electron microscopy of these mylonites shows that the ultrafine-grained bands consist of an intimately mixed aggregate of equiaxed to slightly elongate olivine grains with variable dislocation densities, pyroxene grains virtually free of dislocations showing a weak preferred orientation, and up to 30 volume percent of pargasitic to edenitic hornblende. The lenticular olivine domains between the ultrafine-grained bands preserve lattice fabrics indicating noncoaxial, crystal-plastic flow dominated by slip on [a]{0kl} and, in some cases, on the lower-temperature [c]{hk0} system (Avé Lallemant and Carter, 1970). PPL

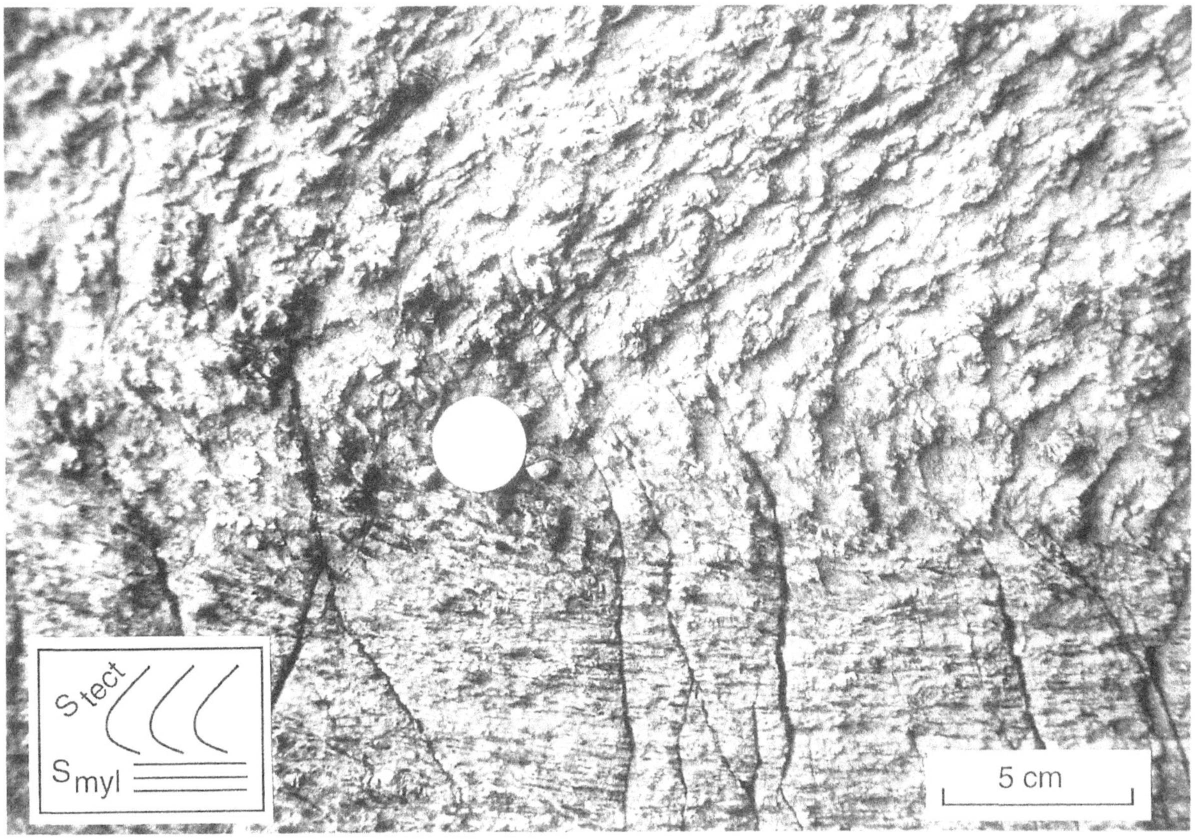

174 C

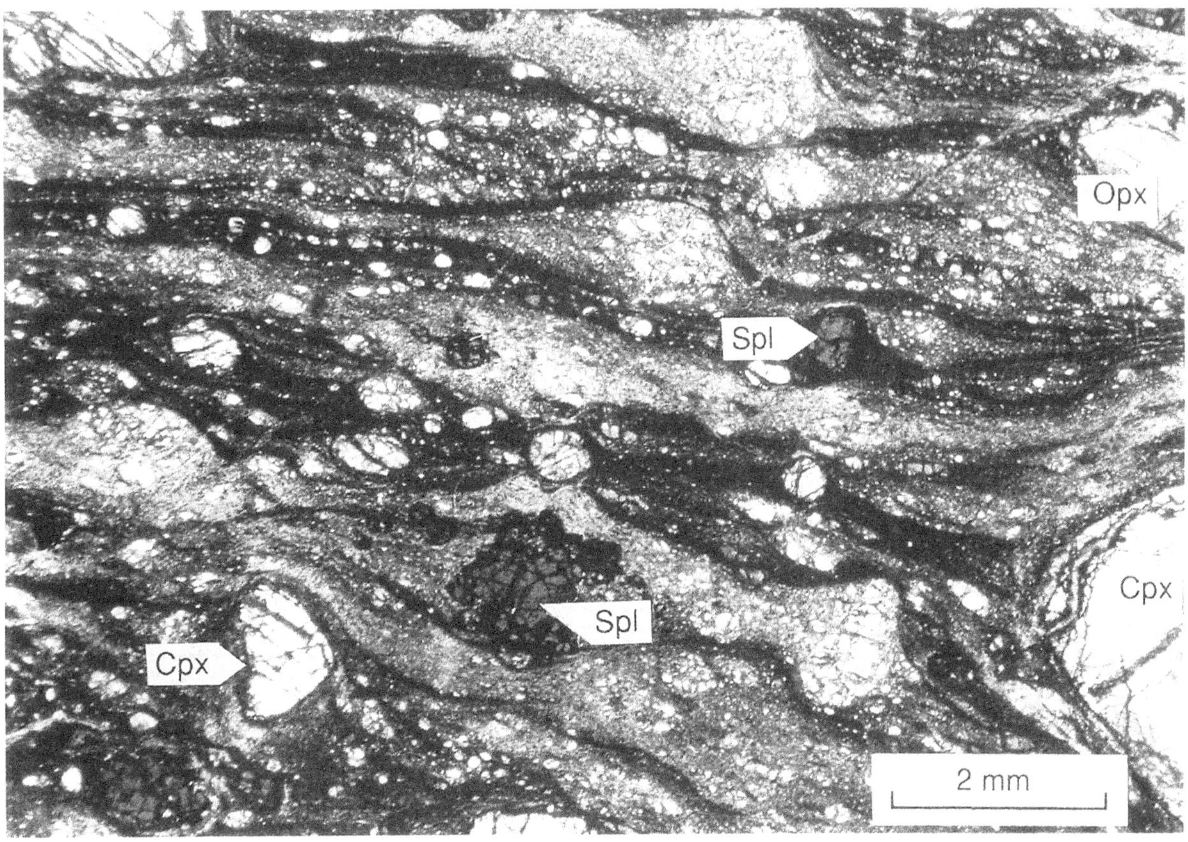

174 D

**174E.** Detail of spinel-peridotite mylonite showing syntectonic transformation of orthopyroxene to pargasitic-edenitic amphibole. These amphiboles developed as fiber growths on a pulled-apart orthopyroxene crystal. Paragasitic-edenitic amphiboles also occur in the anastomosing, very fine-grained bands (shown in 174D) derived as tails from such ortho- and clinopyroxene porphyroclasts. These microstructures are believed to have a rheological significance. The microstructural evidence for coupled, fracture-induced microboudinage plus hydration of pyroxene clasts, occurring as an integral part of the development of the ultrafine-grained bands, lends support to the possibility that the water content and fluid pressure in the mylonitic shear zones were relatively high. The progressive hydration associated with strain localization in the very fine-grained bands may have caused a marked reduction in bulk strength, since experimental evidence (Karato and others, 1986; Rutter and Brodie, 1988) indicates that wet olivine rocks weaken considerably with decreasing grain size. This weakening is ascribed to a change in dominant deformation mechanism, from grain-size-insensitive dislocation creep to diffusion-controlled, grain-size-sensitive mechanisms. Irrespective of the large-scale geometry, the tectonite shear zone and cross-cutting mylonites in the Voltri peridotites indicate that deformation of the Piemonte-Ligurian upper mantle during lithosphere extension and breakup became progressively localized in a network of relatively narrow, fine-grained, and hydrated mylonite zones. It follows that, once this shear-zone network became throughgoing, the bulk strength of the lithospheric mantle may have become controlled by the much weaker rheology of the shear-zone rocks. NX

**174F.** Micrograph of peridotite ultramylonite from the Turon de Técouère, a small peridotite body less than 1 km across exposed along the North Pyrenean fault zone of southern France, showing a δ-type porphyroclast (Passchier and Simpson, 1986) of spinel, embedded in an ultrafine-grained peridotite matrix. Unlike occasional σ-type pyroxene porphyroclasts, such δ-type porphyroclasts have not been observed in the Voltri peridotite mylonites, nor are we aware of any other documented mylonitic or ultramylonitic peridotite containing δ-type porphyroclasts. We speculate that in upper mantle peridotites these microstructures are probably scarce. PPL

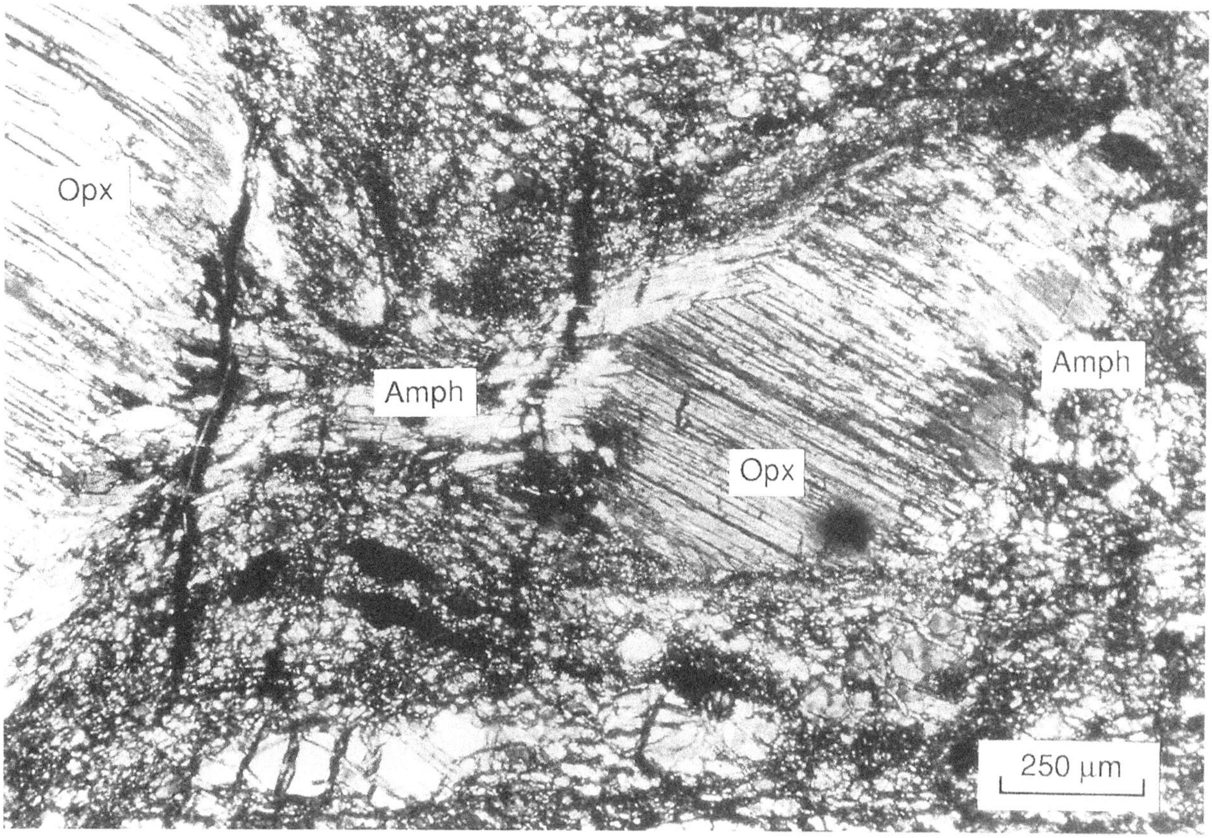


174 E

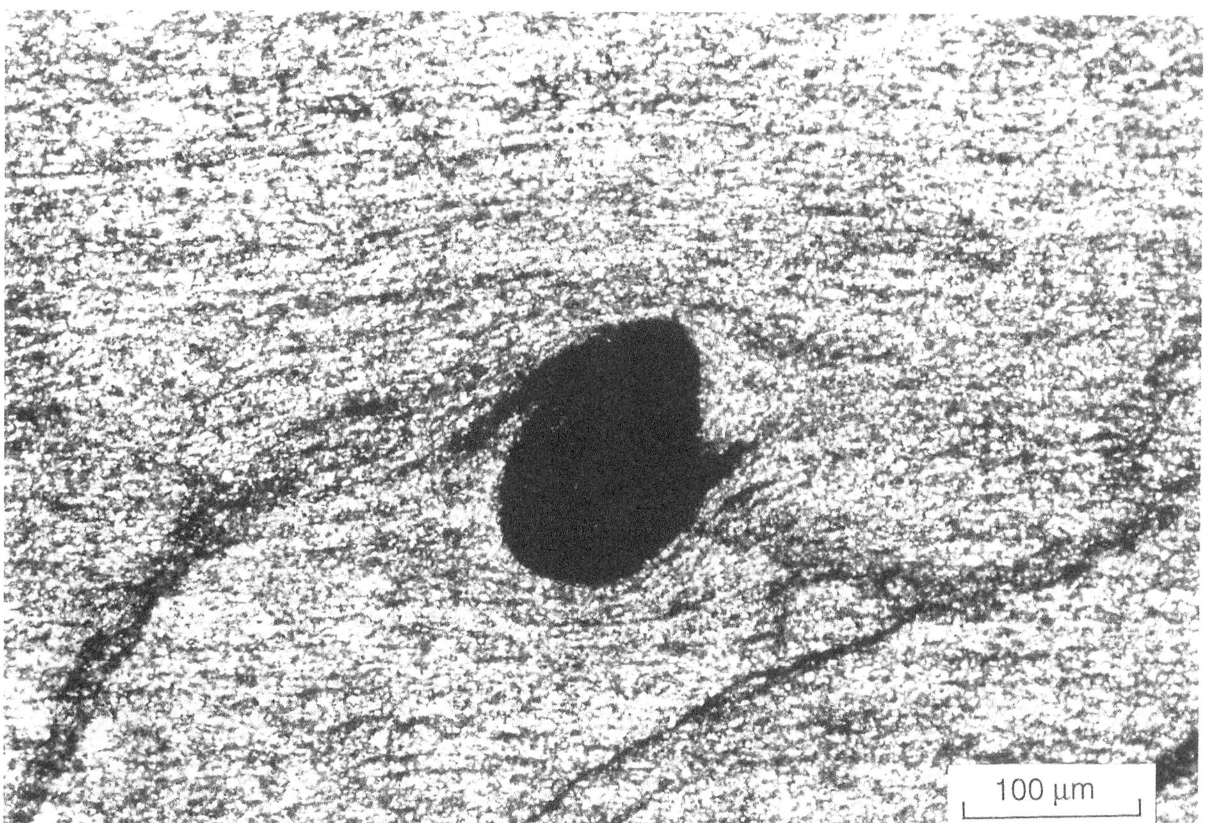


174 F

# 175. ASTHENOSPHERIC DEFORMATION IN OPHIOLITIC PERIDOTITE

Benoît Ildefonse, Françoise Boudier, and Adolphe Nicolas

The textures and structures of deformed ophiolitic peridotite are chiefly a function of the temperature at which these features developed (Nicolas and Poirier, 1976). We illustrate here characteristic structures and textures of ophiolitic peridotite related to high-temperature (≥1200°C) asthenospheric flow. The examples are from the 500 km long Oman ophiolite, which represents a palaeo-fast-spreading ridge (Nicolas, 1989). Continuous deformation produced a remarkable structural fabric (e.g., trajectories of the foliation planes in peridotite) that can be mapped the length of the ophiolite (Nicolas, 1989). The high-temperature structural features were produced by asthenospheric plastic flow within locally partially molten peridotite beneath the ridge while it was still active.

**175A.** High-temperature mantle banding in deformed peridotite from Wadi Sham (Hilti massif, Sultanate of Oman). At outcrop scale, the asthenospheric mantle commonly displays strong compositional banding, defined by alternating layers of harzburgite, dunite, and pyroxenite. This banding may be the result of transposition of either dunite veins and pyroxenite dikes or previous mantle heterogeneities. The high-temperature plastic foliation, best defined by the preferred shape orientation of spinel, is parallel to the banding in this outcrop. However, it may be locally oblique to the banding, as a result of local flow heterogeneities, expressed by meter- to decameter-scale folds. A mineral lineation is then parallel to the fold axis. The hammer is 35 cm long.

**175B.** Photomicrograph of a typical, deformed asthenospheric peridotite from Wadi Sudum (Hilti massif, Sultanate of Oman). Asthenospheric deformation at temperatures ≥1200°C is expressed by a coarse, porphyroclastic texture (Nicolas, 1989) characterized by equant to elongate olivine grains (in the range of 0.5 to 5 mm). Olivine porphyroclasts ($ol_1$) have irregular grain boundaries and commonly contain optically visible subgrain boundaries (sb). High-temperature, hypersolidus to solidus conditions account for the ease of recovery. The development of polygonal olivine neoblasts ($ol_2$) and 120° triple junctions reflects very active grain-boundary migration, possibly enhanced by the presence of partial melt during plastic flowage of the rock. Orthopyroxene also is present as porphyroclasts (opx) and, at the highest temperatures, displays polygonal shapes indicative of dynamic recrystallization. Lattice fabrics for olivine in this type of layered peridotite reveal high strain, dominated by simple shear; they result from high-temperature intracrystalline gliding on [100](0kl) (Boudier and Coleman, 1981). NX

175 A

175 B

# 176. LITHOSPHERIC DEFORMATION IN OPHIOLITIC PERIDOTITE

Benoît Ildefonse, Françoise Boudier, and Adolphe Nicolas

In the Oman ophiolite, the conditions for deformation lowered from asthenospheric (see Plate 175) to lithospheric (<1000°C) during the early and fast (one million years) detachment of the nappe (Boudier and others, 1988; Hacker, 1994), which took place along a Neo-Tethys ridge in the Cenomanian (Nicolas, 1989). As the temperature decreased, the mantle peridotite underwent more localized deformation. We illustrate here examples of mylonitic mesostructures and microstructures that are typically associated with deformation at lowermost lithospheric conditions (800–1000°C), which in the present-day situation (Figure 176.1) is encountered at the base of the nappe and along steep shear zones (Boudier and others, 1988).

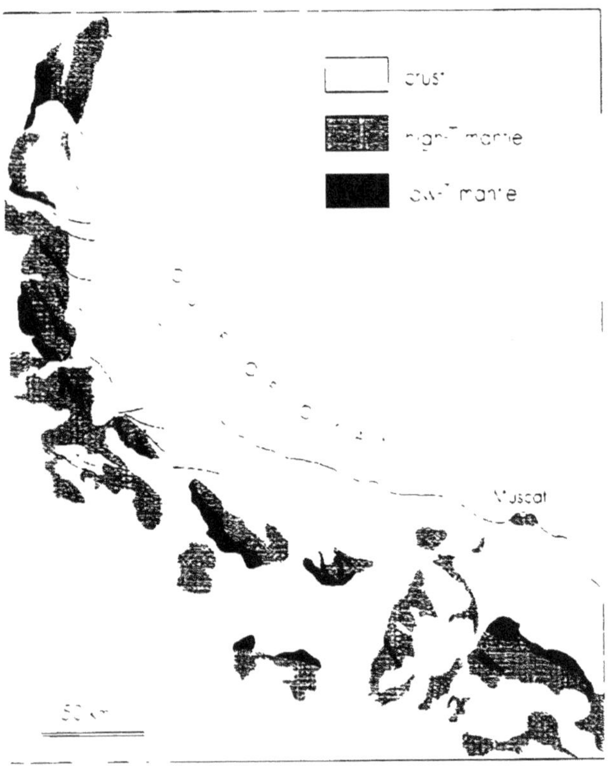

**Figure 176.1.** Generalized geologic map of Oman ophiolite

**176A.** Low-temperature (about 800–1000°C) mylonitic foliation in peridotite from Wadi Hayl (Fizh massif, Sultanate of Oman). Lithospheric deformation recorded in peridotites within the basal thrust and along transcurrent shear zones, the stretching lineation in the latter being typically parallel to the one in the base of the nappe. The outcrop is located in a kilometer-scale, nearly vertical shear zone. The foliation strikes 150° (parallel to the photograph) and dips 60° to the west. The scale bar is 10 cm long.

**176B.** Photomicrograph of a typical lithospheric mylonite in peridotite from Wadi Tayin area (Wadi Tayin massif, Sultanate of Oman). Lithospheric conditions (800–1000°C) are manifested by porphyroclastic microstructures (Nicolas, 1989) with characteristic bimodal grain-size distribution, grading with increasing deformation to mylonitic microstructures in shear zones such as that of 176A. Olivine grains are typically much smaller than in the high-temperature peridotites (see Plate 175); they are elongated, contain abundant subgrain boundaries, and display undulose extinction. The grain-size reduction and mosaic-like texture are due to dynamic recrystallization which, together with grain-boundary sliding, results in weakening of the lattice fabric (Boudier and Coleman, 1981). Orthopyroxene appears as coarse porphyroclasts (opx) that are commonly bent and contain clinopyroxene exsolution lamellae. They can become lamellar by slip with an elongation ratio of 100/1. This sample is from the sole of the Wadi Tayin massif, which is in contact with metamorphic rocks whose peak temperatures were estimated as 850 to 900°C (Ghent and Stout, 1981). NX

176 A

176 B

# 177. PERIDOTITE MYLONITE PRODUCED BY SUPERPLASTIC FLOW

Anne-Marie Boullier and Yves Guéguen

This peridotite is a xenolith from the kimberlite of Kao pipe (Lesotho); its mosaic fluidal texture has been described as representing the most highly deformed state in xenoliths from kimberlites (Boullier and Nicolas, 1975; Boullier, 1977). A detailed study of this sample has been carried out by Guéguen and Boullier (1976). The rock shows a strong foliation bearing a lineation defined by aggregates of small orthopyroxenes. The rock contains 65% olivine (OL), 28% orthopyroxene (OPX), 3.5% clinopyroxene (CPX), and 3.5% garnet (GT) which is surrounded by a kelyphitic rim. Olivine is almost completely recrystallized to a mosaic of small grains (OLM, 0.07 mm in average); however, one porphyroclast (OLP) remains, and its retorted shape gives the sense-of-shear without ambiguity (double arrow; Etchecopar, 1977). Orthopyroxene is less recrystallized, but it must be noted that orthopyroxene recrystallization is rare in other peridotite mylonites. The recrystallized orthopyroxene grains are very small (less than 10 μm), almost equant, and form stripes which can be seen on the photographs. The following important observation may be added: the orthopyroxene stripes are prolonged within the olivine matrix by fracture lines (arrows). This implies that the deformation was controlled by the small orthopyroxenes and that olivine could not keep up with it, and therefore responded by fracturing. The average percentage of strain that has been achieved without necking has been calculated by using orthopyroxene as a marker; it is $\varepsilon = \Delta L/L = 840\%$. The fine-grained orthopyroxene grains that form the stripes show no crystallographic fabric, although the orthopyroxene porphyroclasts exhibit a strong preferred orientation (Boullier, 1977). The olivine porphyroclasts and mosaic grains have the preferred orientation usually observed in deformed peridotites (Nicolas and Poirier, 1976). The small orthopyroxene grains contain no subgrains and their dislocation density is moderate ($5 \times 10^7$ cm$^{-2}$, Guéguen and Boullier, 1976). In olivine porphyroclasts, except for some locally high dislocation densities in the olivine porphyroclasts ($5–8 \times 10^8$ cm$^{-2}$; Green, 1976), the average value is low ($10^7$ cm$^{-2}$; Guéguen, 1979).

Geochemical data show that such xenoliths originated at a depth of 200 km and a temperature of 1400°C. That implies that the deformation was achieved at high temperature relative to the melting temperature. The $(\sigma, \dot{\varepsilon})$ domain in which the deformation took place cannot be accurately determined, but the low dislocation density in olivine suggests a moderate stress (2.5 to 14 MPa, Guéguen, 1979). Strain-rate sensitivity and activation energy are unknown, but an interpretive model based on diapiric upwelling in the upper mantle (Green and Guéguen, 1974) concludes that $\dot{\varepsilon}$ was faster than $10^{-13}$ s$^{-1}$, which is fast by geological standards.

For the strain rate to be faster than usual and controlled by the less ductile mineral, given that olivine is ordinarily more plastic than orthopyroxene, a completely different flow mechanism has to be operative. The six criteria given by Boullier and Guéguen (1975) from comparison with superplastic, fine-grained alloys are met by the small orthopyroxene grains. This suggests that the mylonitic texture of this rock is the result of superplastic flow in the orthopyroxene. Metallurgical experiments show that grain-boundary sliding must be the dominant mechanism of superplasticity (Hayden and others, 1972; Ashby and Verall, 1973). Consequently, the small orthopyroxene grains act as a "lubricating layer" or "ball-bearings" in the olivine matrix.

The photograph is normal to foliation, and lineation is parallel to the length of the photograph. PPL

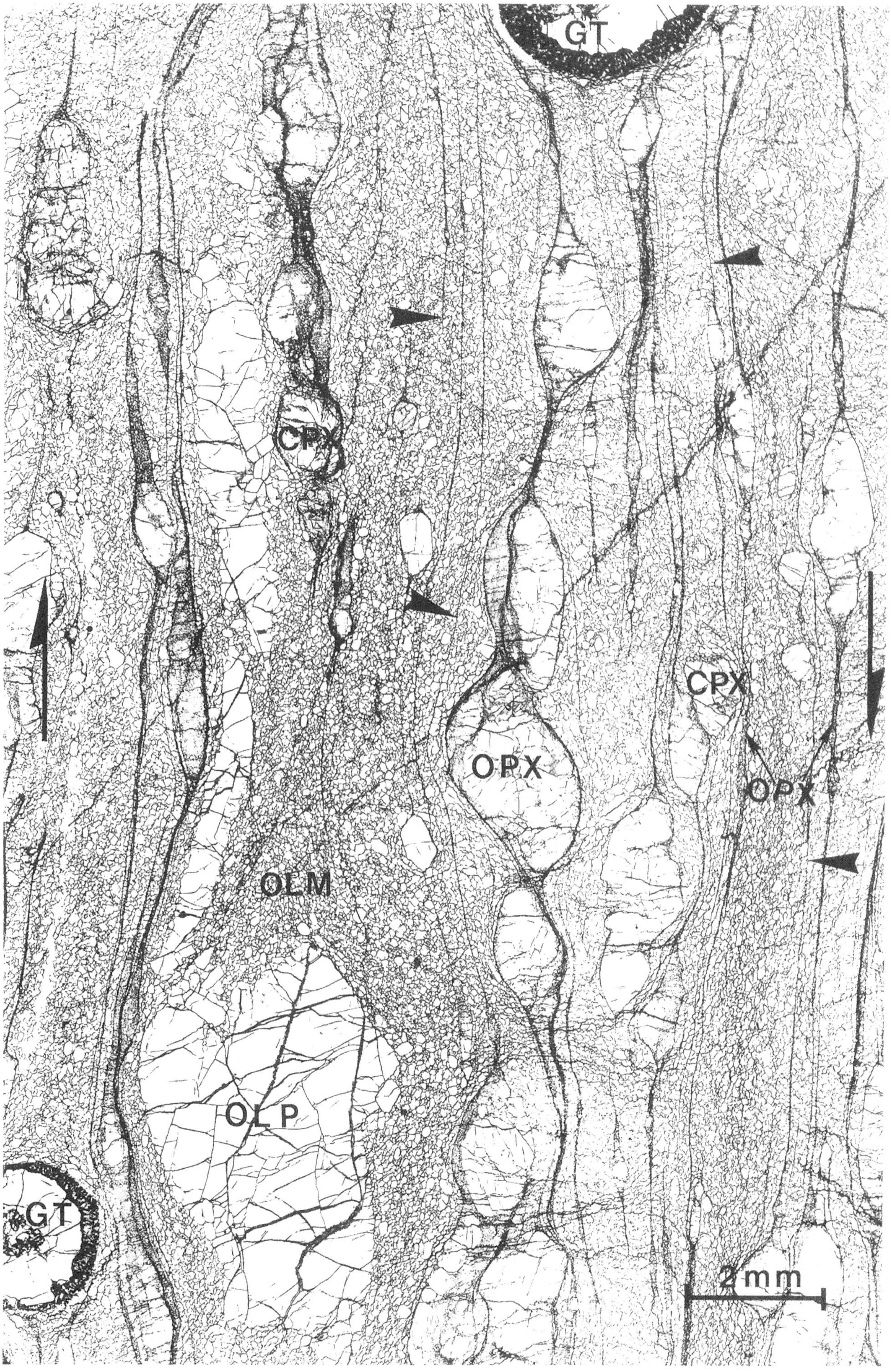
GT
CPX
CPX
OPX
OPX
OLM
OLP
GT
2 mm

# References cited

Agar, S. M., 1990, Fracture evolution in the upper ocean crust: evidence from DSDP hole 504B, *in* Knipe, R. J. and Rutter, E. H., eds., Deformation mechanisms, rheology and tectonics: Geological Society of London, Special Publication No. 54, p. 41–50.

Agar, S. M., 1991, Microstructural evolution of a deformation zone in the upper ocean crust: evidence from DSDP Hole 540B: Journal of Geodynamics, v. 13, p. 119–140.

Agar, S. M., 1994, Rheological evolution of the ocean crust: a microstructural view: Journal of Geophysical Research, v. 99, p. 3175–3200.

Agar, S. M., and Klitgord, K. D., 1995, A mechanism for decoupling within the oceanic lithosphere revealed by the Troodos ophiolite: Nature, v. 374, p. 232–238.

Allison, I., and La Tour, T. E., 1977, Brittle deformation of hornblende in a mylonite: a direct geometrical analogue of ductile deformation by translation gliding: Canadian Journal of Earth Sciences, v. 14, p 1953–1958

Alt, J. C., Honnorez, J., Laverne, C., and Emmerman, R., 1986, Hydrothermal alteration of a 1 km section through the upper oceanic crust, Deep Sea Drilling Project Hole 504B, mineralogy, chemistry, and evolution of seawater-basalt interactions: Journal of Geophysical Research, v. 91, p. 10,309–10,335.

Alt, J. C., and others, 1993, Proceedings of the Ocean Drilling Program, Initial Reports: College Station, Texas, Ocean Drilling Program, Texas A&M University, v. 148, 352 p.

Ashby, M. F., and Verall, R. A., 1973, Diffusion accommodated flow and superplasticity: Acta Metallurgica, v. 21, p. 149–163.

Atkinson, B K., 1984, Subcritical crack growth in geological materials: Journal of Geophysical Research, v. 89, p. 4077–4114.

Austrheim, H., 1986/87, Eclogitization of lower crustal granulites by fluid migration through shear zones: Earth and Planetary Science Letters, v. 81, p. 221–232.

Austrheim, H., and Boundy, T. M., 1994, Pseudotachylytes generated during seismic faulting and eclogitization of the deep crust. Science, v. 265, p. 82–83.

Austrheim, H., and Griffin, W L., 1985, Shear deformation and eclogite formation within the granulite-facies anorthosites of the Bergen Arcs, western Norway: Chemical Geology, v. 50, p. 267–281.

Avé Lallemant, H. G., and Carter, N. L., 1970, Syntectonic recrystallization of olivine and modes of flow in the upper mantle Geological Society of America Bulletin, v. 81, p. 2203–2220.

Aydin, A., and Reches, Z., 1982, Number and orientation of fault sets in the field and in experiments: Geology, v 10, p. 107–112.

Babaie, H. A., and La Tour, T E., 1994, Semibrittle and cataclastic deformation of hornblende-quartz rocks in a ductile shear zone. Tectonophysics, v. 229, p. 19–30.

Bailey, C. M., and Simpson, C., 1993, Extensional and contractional deformation in the Blue Ridge province, Virginia: Geological Society of America Bulletin, v. 105, p. 411–422.

Bailey, C. M , Simpson, C., and De Paor, D. G., 1994, Volume loss and tectonic flattening strain in granitic mylonites from the Blue Ridge province, central Appalachians: Journal of Structural Geology, v. 16, p. 1403–1416.

Barnett, R. L., and Kerrich, R., 1980, Stress corrosion cracking of biotite and feldspar. Nature, v. 283, p. 185–187.

Bartholomew, M. J., 1981, Geology of the Roanoke and Stewartsville quadrangles, Virginia: Virginia Division of Mineral Resources, Publication 34, 23 p.

Bartlett, W. L., Friedman, M., and Logan, J. M., 1981, Experimental folding and faulting of rocks under confining pressure. Part IX. Wrench faults in limestone layers: Tectonophysics, v. 79, p. 255–277.

Bayly, M. B., Borradaile, G. J., and Powell, C., McA., eds., 1977, Atlas of rock cleavage—provisional edition: Hobart, University of Tasmania, 110 plates + 26 p.

Beach, A., 1979, Pressure solution as a metamorphic process in deformed terriginous sedimentary rocks: Lithos, v. 12, p. 51–59.

Beach, A., 1980, Retrogressive metamorphic processes in shear zones with special reference to the Lewisian complex: Journal of Structural Geology, v. 2, p. 257–263.

Becker, K., and others, 1988, Proceedings of the Ocean Drilling Program, Initial Reports, v. 111, College Station, Texas.

Beeler, N. M., Tullis, T. E., Blanpied, M. L., and Weeks, J. D., 1995, Frictional behavior of large displacement faults: Journal of Geophysical Research, v. 101, p. 8697–8716.

Bell, T. H., 1978a, Progressive deformation and reorientation of fold axes in a ductile mylonite zone: The Woodroffe thrust Tectonophysics, v. 44, p. 285–320.

Bell, T. H , 1978b, Syntectonic nucleation of new grains in deformed mica: Tectonophysics, v. 51, p. T31–T37.

Bell, T. H., 1979, The deformation and recrystallization of biotite in the Woodroffe thrust mylonite zone: Tectonophysics, v. 58, p. 139–158.

Bell, T. H., 1981, Foliation development—the contribution, geometry and significance of progressive, bulk, inhomogeneous shortening. Tectonophysics, v. 75, p. 273–296.

Bell, T. H., and Etheridge, M. A., 1973, Microstructures of mylonites and their descriptive terminology: Lithos, v. 6, p. 337–348.

Berthé, D., Choukroune, P., and Jégouzo, P., 1979a, Orthogneiss, mylonite and noncoaxial deformation of granites: the example of the South Armorican shear zone· Journal of Structural Geology, v 1, p. 31–42.

Berthé, D., Choukroune, P., and Gapais, D., 1979b, Orientations préférentielles du quartz et orthogneissification progressive en régime cisaillant: l'exemple du cisaillement sud-armoricain: Bulletin de Minéralogie, v. 102, p. 265–272.

Beutner, E. C., and Diegel, F. A., 1985, Determination of fold kinematics from syntectonic fibers in pressure shadows, Martinsburg slate, N.J.: American Journal of Science, v. 285, p. 16–50.

Bisschoff, A. A., 1962, The pseudotachylite of the Vredefort Dome: Transactions of the Geological Society of South Africa, v. 65, p. 207–225.

Blanpied, M. L., Lockner, D. A., and Byerlee, J. D., 1992, An earthquake mechanism based on rapid sealing of faults: Nature, v. 358, p. 574–576.

Blenkinsop, T. G., 1987, Mechanisms and conditions of deformation in quartzites from the Cantabrian and West Asturian-Leonese zones, NW Spain [Ph.D dissertation] Keele, Staffordshire, England, Keele University, 427 p

Blenkinsop, T G., and Drury, M R., 1988, Stress estimates and fault history from quartz microstructures, Journal of Structural Geology, v 10, p 673–684.

Blenkinsop, T G., and Rutter, E. H., 1986, Cataclastic deformation of quartzite in the Moine thrust zone Journal of Structural Geology, v 8, p 669–681.

Bobyarchick, A. R., 1976, Tectogenesis of the Hylas zone and eastern Piedmont near Richmond, Virginia [M.S. thesis] Blacksburg, Virginia, U.S.A., Virginia Polytechnic Institute and State University, 168 p.

Bobyarchick, A. R., 1981, The Eastern Piedmont fault system and its relationship to Alleghanian tectonics in the southern Appalachians: Journal of Geology, v. 89, p. 335–347.

Bobyarchick, A. R., 1983, Structure of the Brevard zone and Blue Ridge near Lenoir, North Carolina with observations on oblique crenulation cleavage and a preliminary theory for irrotational structures in shear zones [Ph D. dissertation]: Albany, New York, U.S.A., State University of New York at Albany, 306 p.

Bobyarchick, A R., and Glover, L., III, 1979, Deformation and metamorphism in the Hylas zone and adjacent parts of the eastern Piedmont in Virginia: Geological Society of America Bulletin, v. 90, p. 739–752.

Boeckeler, A. J., 1994, Isotopic ages of pseudotachylite veins from coastal New Hampshire and SW Maine: evidence for post-Acadian strike-slip motion: Geological Society of America Abstracts with Programs, v. 26, no. 3, p. 7.

Bons, A.-J., 1988, Deformation of chlorite in naturally deformed low-grade rocks Tectonophysics, v. 154, p. 149–165.

Bons, P D., 1993, Experimental deformation of polyphase rock analogues [Ph.D. dissertation]: Utrecht, The Netherlands: Geologica Ultraiectina, v 110, 207 p.

Bons, P. D., and Cox, S. J. D., 1994, Analogue experiments and numerical modelling on the relation between microgeometry and flow properties of polyphase materials: Materials Science and Engineering, v. A175, p. 237–246.

Bons, P. D., and Urai, J. L., 1994, Experimental deformation of two-phase rock analogues: Materials Science and Engineering, v. A175, p. 221–230

Boon, J. D., and Albritton, C C., 1936, The impact of large meteorites Field laboratory, v. 5, p. 1–9.

Borg, I., and Handin, J., 1966, Experimental deformation of crystalline rocks Tectonophysics, v. 3, p. 249–368.

Borradaile, G. J., Bayly, M B., and Powell, C McA., eds., 1982, Atlas of deformational and metamorphic rock fabrics: Berlin, Springer-Verlag, 551 p.

Bouchez, J. L., and Pêcher, A., 1981, The Himalayan Main Central Thrust pile and its quartz-rich tectonites in central Nepal: Tectonophysics, v 78, p 23–50.

Bouchez, J L., Lister, G. S., and Nicolas, A., 1983, Fabric asymmetry and shear sense in movement zones: Geologische Rundschau, v 72, p 401–419

Boudier, F., and Coleman, R. G., 1981, Cross section through the peridotite in the Samail ophiolite, southeastern Oman Mountains: Journal of Geophysical Research, v. 86, p 2573–2592.

Boudier, F., Nicolas, A., and Bouchez, J. L., 1982, Kinematics of oceanic thrusting and subduction from basal sections of ophiolites Nature, v 296, p. 825–828.

Boudier, F., Ceuleneer, G., and Nicolas, A., 1988, Shear zones, thrusts and related magmatism in the Oman ophiolite: initiation of thrusting on an oceanic ridge: Tectonophysics, v. 151, p. 275–296.

Boullier, A.-M., 1977, Structure des péridotites en enclaves dans les kimberlites d'Afrique du Sud. Conséquences sur la constitution du manteau supérieur: Bulletin de la Société française Minéralogie et de Cristallographie, v. 100, p. 214–229.

Boullier, A.-M., and Guéguen, Y., 1975, SP mylonites: origin of some mylonites by superplastic flow: Contributions to Mineralogy and Petrology, v. 50, p. 93–104.

Boullier, A.-M., and Nicolas, A., 1975, Classification of textures and fabrics of peridotite xenoliths from South African kimberlites, *in* Ahrens, L. H., ed., Physics and chemistry of the Earth: Oxford, Pergamon Press, v. 9, p. 97–105.

Boullier, A.-M., and Robert, F., 1992, Palaeoseismic events recorded in Archaean gold-quartz vein networks, Val d'Or, Abitibi, Quebec, Canada: Journal of Structural Geology, v. 14, p. 161–179.

Boundy, T. M., Fountain, D. M., and Austrheim, H., 1992, Structural development and petrofabrics of eclogite facies shear zones in the granulite facies anorthosite complex, Bergen Arcs, western Norway: implications for deep crustal deformational processes: Journal of Metamorphic Geology, v. 10, p 127–146.

Boyer, S. E., and Elliott, D., 1982, Thrust systems: American Association of Petroleum Geologists Bulletin, v. 66, p. 1196–1230.

Brodie, K. H., 1995, The development of orientated symplectites during deformation. Journal of Metamorphic Geology, v. 13, p. 499–508

Brodie, K. H., and Rutter, E. H., 1987, Deep crustal extensional faulting in the Ivrea Zone of Northern Italy: Tectonophysics, v. 140, p. 193–212.

Brodie, K. H., Rutter, E. H., and Evans, P., 1992, On the structure of the Ivrea-Verbano Zone (northern Italy) and its implications for present-day lower continental crust geometry: Terra Nova, v. 4, p. 34–40.

Brown, R. L., Journeay, J. M., Lane, L. S., Murphy, D. C., and Rees, C. J., 1986, Obduction, backfolding and piggyback thrusting in the metamorphic hinterland of the southern Canadian Cordillera: Journal of Structural Geology, v. 8, p 255–268.

Brown, R. L., Carr, S D., Johnson, B J., Coleman, V. J., Cook, F. A., and Varsek, J. L., 1992, The Monashee decollement of the Canadian Cordillera: a crustal-scale shear zone linking the Rocky Mountain Foreland belt to lower crust beneath accreted terranes, *in* McClay, K. R., ed., Thrust tectonics: London, Chapman & Hall, p. 357–364.

Brown, W. L., Macaudière, J., Ohnenstetter, D., and M., 1980, Ductile shear zones in a meta-anorthosite from Harris, Scotland: textural and compositional changes in plagioclase: Journal of Structural Geology, v. 2, p. 281–287.

Bulau, J. R., Waff, H. S., and Tyburczy, J. A., 1979, Mechanical and thermodynamical constraints on fluid distribution in partial melts: Journal of Geophysical Research, v. 84, p. 6102–6108.

Burkhard, M., 1993, Calcite twins: their geometry, appearance and significance as stress-strain markers and indicators of tectonic regime: a review: Journal of Structural Geology, v. 15, p. 351–368

Busch, J P., and van der Pluijm, B. A., 1995, Calcite textures, microstructures and rheological properties of marble mylonites in the Bancroft shear zone, Ontario, Canada: Journal of Structural Geology, v. 17, p. 677–688.

Byerlee, J D., 1978, Friction of rocks: Pure and Applied Geophysics, v 116, p 615–626.

Cannat, M., and others, 1995, Proceedings of the Ocean Drilling Program, Initial Reports: College Station, Texas, Ocean Drilling Program, Texas A&M University, v. 153, 798 p.

Carlson, K. A., van der Pluijm, B. A., and Hanmer, S., 1990, Marble mylonites of the Bancroft shear zone: Evidence for extension in the Canadian Grenville: Geological Society of America Bulletin, v. 102, p. 174–181.

Carmichael, D. M., 1969, On the mechanism of prograde metamorphic reactions in quartz-bearing pelitic rocks: Contributions to Mineralogy and Petrology, v. 20, p. 244–267.

Carreras, J., and Casas, J. M., 1987, On folding and shear zone development: a mesoscale structural study on the transition between two different tectonic styles: Tectonophysics, v. 135, p. 87–98.

Carreras, J., and Druguet, E., 1994, Structural zonation as a result of inhomogeneous non-coaxial deformation and its control on syntectonic intrusions: an example from the Cap de Creus area (eastern-Pyrenees): Journal of Structural Geology, v. 16, p. 1525–1534.

Carreras, J., and Garcia-Celma, A., 1982, Quartz *c*-axis fabric variation at the margins of a shear zone developed in schists from Cap de Creus (Spain): Acta Geologica Hispanica, v. 17, p. 137–149.

Carreras, J., and Losantos, M., 1982, Geological setting of the Roses granodiorite, (E-Pyrenees, Spain): Acta Geologica Hispanica, v. 17, p. 211–217.

Carreras, J., Estrada, A., and White, S., 1977, The effects of folding on the *c*-axis fabrics of a quartz-mylonite: Tectonophysics, v. 39, p. 3–24.

Carter, K. E., and Dworkin, S. I., 1990, Channelized fluid flow through shear zones during fluid-assisted dynamic recrystallization, Northern Apennines, Italy: Geology, v. 18, p. 720–723.

Carter, N. L., and Tsenn, M. C., 1987, Flow properties of continental lithosphere: Tectonophysics, v. 136, p. 27–63.

Carter, N. L., Christie, J. M., and Griggs, D. T., 1964, Experimental deformation and recrystallization of quartz: Journal of Geology, v. 72, p. 687–733.

Chen, W.-P., and Molnar, P., 1983, Focal depths of intracontinental and intraplate earthquakes and their implications for the thermal and mechanical properties of the lithosphere: Journal of Geophysical Research, v. 88, p. 4183–4214.

Chester, F. M., 1988, The brittle-ductile transition in a deformation-mechanism map for halite: Tectonophysics, v. 154, p. 125–136.

Chester, F. M., 1989, Dynamic recrystallization in semi-brittle faults: Journal of Structural Geology, v. 11, p. 847–858.

Chester, F. M., and Logan, J. M., 1986, Implications for mechanical properties of brittle faults from observations of the Punchbowl fault zone, California: Pure and Applied Geophysics, v. 124, p. 79–106.

Chester, F. M., and Logan, J. M., 1987, Composite planar fabric of gouge from the Punchbowl fault, California: Journal of Structural Geology, v. 9, p. 621–634.

Chester, F. M., and Logan, J. M., 1990, Frictional faulting in polycrystalline halite: correlation of microstructure, mechanisms of slip and constitutive behavior, *in* Duba, A. G., Durham, W. B., Handin, J. W., and Wang, H. F., eds., The brittle-ductile transition in rocks: The Heard volume: Washington, D.C., American Geophysical Union, Geophysical Monograph 56, p. 49–65.

Chester, F. M., Friedman, M., and Logan, J. M., 1985, Foliated cataclasites: Tectonophysics, v. 111, p. 139–146.

Chester, F. M., Evans, J. P., and Biegel, R. L., 1993, Internal structure and weakening mechanisms of the San Andreas fault: Journal of Geophysical Research, v. 98, p. 771–786.

Christie, J. M., 1960, Mylonitic rocks of the Moine Thrust-Zone in the Assynt region, north-west Scotland: Transactions of the Edinburgh Geological Society, v. 18, p. 79–93.

Christie, J. M., 1963, The Moine thrust zone in the Assynt region, northwest Scotland: University of California Publications in Geological Sciences, v. 40, p. 345–440.

Clarke, G. L., and Norman, A. R., 1993, Generation of pseudotachylite under granulite facies conditions, and its preservation during cooling: Journal of Metamorphic Geology, v. 11, p. 319–335.

Compton, R. R., Todd, V. R., Zartman, R. E., and Naeser, C. W., 1977, Oligocene and Miocene metamorphism, folding, and low-angle faulting in northwestern Utah: Geological Society of America Bulletin, v. 88, p. 1237–1250

Coney, P. J., 1979, Tertiary evolution of Cordilleran metamorphic core complexes, *in* Armentrout, J. W., Cole, M. R., and TerBest, H., Jr., eds., Cenozoic paleogeography of the western United States, Pacific Coast Paleogeography Symposium 3: Los Angeles, Calif. Society of Economic Paleontologists and Mineralogists, The Pacific Section, p. 15–28.

Conrad, R. E., and Friedman, M., 1976, Microscopic feather fractures in the faulting process: Tectonophysics, v. 33, p. 187–198.

Coulton, A. J., Harper, G. D., and O'Hanley, D. S., 1995, Oceanic versus emplacement age serpentinization in the Josephine ophiolite: Implications for the nature of the Moho at intermediate and slow spreading ridges: Journal of Geophysical Research, v. 100, p. 22,245–22,260.

Coward, M. P., 1984, Major shear zones in the Precambrian crust: examples from NW Scotland and southern Africa and their significance, *in* Kröner, A., and Greiling, R., eds., Precambrian tectonics illustrated: Stuttgart, e., Schweizerbart'sche Verlagsbuchhandlung (Nägele u. Obermiller), p. 207–235.

Coward, M. P., and McClay, K. R., 1979, The geology of NW Scotland and the Moine thrust belt: Field guide: Conference on Thrust and Nappe Tectonics, London.

Cox, H. L., 1952, The elasticity and strength of paper and other fibrous materials: British Journal of Applied Physics, v. 3, p. 72–79.

Cox, S. F., 1981, The stratigraphic and structural setting of the Mt. Lyell volcanic-hosted sulfide deposits: Economic Geology, v. 76, p. 231–245.

Cox, S. F., 1987, Antitaxial crack-seal vein microstructures and their relationships to displacement paths: Journal of Structural Geology, v. 9, p. 779–787.

Cox, S. F., 1995, Faulting processes at high fluid pressures: An example of fault valve behavior from the Wattle Gully fault, Victoria, Australia: Journal of Geophysical Research, v. 100, p. 12,841–12,859.

Cox, S. F., and Etheridge, M. A., 1983, Crack-seal fibre growth mechanisms and their significance in the development of oriented layer silicate microstructures: Tectonophysics, v. 92, p. 147–170

Cox, S. F., and Etheridge, M. A., 1984, Deformation microfabric development in chalcopyrite in fault zones, Mt. Lyell, Tasmania: Journal of Structural Geology, v. 6, p. 167–182.

Cox, S. F., and Paterson, M. S., 1991, Experimental dissolution-precipitation creep in quartz aggregates at high temperature: Geophysical Research Letters, v. 18, p. 1401–1404.

Cox, S. F., Wall, V. J., Etheridge, M. A., and Potter, T. F., 1991a, Deformational and metamorphic processes in the formation of mesothermal, vein-hosted gold deposits: examples from the Lachlan fold belt in central Victoria, Australia: Ore Geology Reviews, v. 6, p. 391–423.

Cox, S. F., Wall, V. J., Etheridge, M. A., and Potter, T. F., 1991b, Deformational style of the Castlemaine area, Bendigo-Ballarat Zone: Implications for evolution of crustal structure in central Victoria: Australian Journal of Earth Sciences, v. 38, p. 151–170.

Crittenden, M. D., Jr., Coney, P. J., and Davis, G. H., eds., 1980, Cordilleran metamorphic core complexes: Boulder, Colo., Geological Society of America Memoir 153, 490 p.

Dallmeyer, R. D., Snoke, A. W., and McKee, E. H., 1986, The Mesozoic-Cenozoic tectonothermal evolution of the Ruby Mountains–East Humboldt Range, Nevada: A Cordilleran metamorphic core complex: Tectonics, v. 5, p. 931–954.

Daly, R. A., 1947, The Vredefort ring-structure of South Africa: Journal of Geology, v. 55, p. 125–145.

Davidson, A., Culshaw, N. G., and Nadeau, L., 1982, A tectono-metamorphic framework for part of the Grenville Province, Parry Sound region, Ontario: Geological Survey of Canada Paper 82–1B, p. 195–207.

Davidson, C., Schmid, S. M., and Hollister, L. S., 1994, Role of melt during deformation in the deep crust: Terra Nova, v. 6, p. 133–142.

de Bresseur, H. P., 1996, Steady state dislocation densities in experimentally deformed calcite materials: Single crystals versus polycrystals: Journal of Geophysical Research, v. 101, p. 22, 189–22, 202.

Dell'Angelo, L. N., and Olgaard, D. L., 1995, Deformation of fine-grained anhydrite: Evidence for dislocation and diffusion creep: Journal of Geophysical Research, v. 100, p. 15,424–15,440.

Dell'Angelo, L. N., and Tullis, J., 1988, Experimental deformation of partially melted granitic aggregates: Journal of Metamorphic Geology, v. 6, p. 495–515.

Dell'Angelo, L. N., and Tullis, J., 1989, Fabric development in experimentally sheared quartzites: Tectonophysics, v. 169, p. 1–21.

Dell'Angelo, L. N., and Tullis, J., 1996, Textural and mechanical evolution with progressive strain in experimentally deformed aplite: Tectonophysics, v. 256, p. 57–82.

Den Brok, B., 1992, An experimental investigation into the effect of water on the flow of quartzite: Geologica Ultraiectina, v. 95, 178 p.

Dennis, A. J., and Secor, D. T., Jr., 1987, A model for the development of crenulations in shear zones with applications from the southern Appalachian Piedmont: Journal of Structural Geology, v. 9, p. 621–634.

Derby, B., 1991, The dependence of grain size on stress during dynamic recrystallization: Acta Metallurgica et Materialia, v. 39, p. 955–962.

Dewers, T., and Ortoleva, P., 1991, Influences of clay minerals on sandstone cementation and pressure solution: Geology, v. 19, p. 1045–1048.

Dewey, J. F., Shackleton, R. M., Chang, C., and Sun, Y., 1988, The tectonic evolution of the Tibetan plateau: Philosophical Transactions of the Royal Society of London, Series A, v. 327, p. 379–413.

Dieterich, J. H., 1979, Modeling of rock friction 1. Experimental results and constitutive equations: Journal of Geophysical Research, v. 84, p. 2161–2168.

Dietrich, R. V., Fullagar, P. D., and Bottino, M. L., 1969, K/Ar and Rb/Sr dating of tectonic events in the Appalachians of southwestern Virginia: Geological Society of America Bulletin, v. 80, p. 307–314.

Dietz, R. S., 1961, Vredefort ring structure: meteorite impact scar?: Journal of Geology, v. 69, p. 499–516.

Doherty, R., 1980, Dendritic growth, *in* Hargraves, R. B., ed., Physics of magmatic processes: Princeton, N.J., Princeton University Press, p. 576–600.

Drury, M. R., 1993, Deformation lamellae in metals and minerals, *in* Boland, J. N., and Fitz Gerald, J. D., eds., Defects and processes in the solid state: geoscience applications: The McLaren volume: Amsterdam, The Netherlands, Elsevier, p. 195–212.

Drury, M. R., and Humphreys, F. J., 1986, The development of microstructure in Al 15% Mg during high temperature deformation: Acta Metallurgica, v. 34, p. 2259–2271.

Drury, M. R., Vissers, R. L. M., van der Wal, D., and Hoogerduijn Strating, E. H., 1991, Shear localization in upper mantle peridotites, *in* Ord, A., and Hobbs, B. E., eds., Shear localization in rocks and metals: Pure and Applied Geophysics, v. 137, p. 439–460.

Duebendorfer, E. M., and Christensen, C., 1995, Synkinematic(?) intrusion of the "anorogenic" 1425 Ma Beer Bottle Pass pluton, southern Nevada: Tectonics, v. 14, p. 168–184.

Dunlap, W. J., 1992, Structure, kinematics, and cooling history of the Arltunga Nappe Complex, central Australia [Ph.D. dissertation]: Minneapolis, Minn., U.S.A., University of Minnesota, 276 p.

Dunlap, W. J., Teyssier, C., McDougall, I., and Baldwin, S., 1991, Ages of deformation from K/Ar and $^{40}$Ar/$^{39}$Ar dating of white micas: Geology, v. 19, p. 1213–1216.

Dunlap, W. J., Teyssier, C., McDougall, I., and Baldwin, S., 1995, Thermal and structural evolution of the intracratonic Arltunga Nappe Complex, central Australia: Tectonics, v. 14, p. 1182–1204.

Durney, D. W., and Ramsay, J. G., 1973, Incremental strains measured by syntectonic crystal growth, *in* De Jong, K. A., and Scholten, R., eds., Gravity and tectonics: New York, John Wiley & Sons, p. 67–96.

Eisbacher, G. H., 1970, Deformation mechanics of mylonitic rocks and fractured granites in Cobequid Mountains, Nova Scotia, Canada: Geological Society of America Bulletin, v. 81, p. 2009–2020.

Erskine, B. G., and Wenk, H.-R., 1985, Evidence for Late Cretaceous crustal thinning in the Santa Rosa mylonite zone, southern California: Geology, v. 13, p. 274–277.

Erskine B. G., Heidelbach, F., and Wenk, H.-R., 1993, Lattice preferred orientations and microstructures of deformed Cordilleran marbles: correlation of shear indicators and determination of strain path: Journal of Structural Geology, v. 15, p. 1189–1205.

Etchecopar, A., 1977, A plane kinematic model of progressive deformation in a polycrystalline aggregate: Tectonophysics, v. 39, p. 121–142.

Etheridge, M. A., and Wilkie, J. C., 1981, An assessment of dynamically recrystallized grain size as a paleopiezometer in quartz-bearing mylonite zones: Tectonophysics, v. 78, p. 475–508.

Etheridge, M. A., Wall, V. J., and Vernon, R. H., 1983, The role of the fluid phase during regional metamorphism and deformation: Journal of Metamorphic Geology, v. 1, p. 205–226.

Evans, J. P., 1988, Deformation mechanisms in granitic rocks at shallow crustal levels: Journal of Structural Geology, v. 10, p. 437–443.

Evans, J. P., 1990, Textures, deformation mechanisms, and the role of fluids in cataclastic deformation of granitic rocks, *in* Knipe, R. J., and Rutter, E. H., eds., Deformation mechanisms, rheology and tectonics: Oxford, England, Geological Society of London, Special Publication No. 54, p. 29–39.

Evans, J. P., 1993, Deformation mechanisms and kinematics of a crystalline-cored thrust sheet: The EA thrust system, Wyoming, *in* Schmidt, C. J., Chase, R. B., and Erslev, E. A., eds., Laramide basement deformation in the Rocky Mountain foreland of the western United States: Boulder, Colo., Geological Society of America, Special Paper 280, p. 147–161.

Evans, J. P., and Chester, F. M., 1995, Fluid-rock interaction and weakening of faults of the San Andreas system: Inferences from San Gabriel fault rock geochemistry and microstructures: Journal of Geophysical Research, v. 100, p. 13,007–13,020.

Evans, P., 1995, Structural geometry of parts of the Ivrea-Verbano lower crustal section, northern Italy [Ph.D. dissertation]: Manchester, England, University of Manchester.

Farver, J. R., and Yund, R. A., 1991, Measurement of oxygen grain boundary diffusion in natural, fine-grained, quartz aggregates: Geochimica et Cosmochimica Acta, v. 55, p. 1597–1607.

Farver, J. R., and Yund, R. A., 1996, Volume and grain boundary diffusion of calcium in natural and hot-pressed calcite aggregates: Contributions to Mineralogy and Petrology, v. 123, p. 77–91

Fitz Gerald, J. D., and Stünitz, H., 1993, Deformation of granitoids at low metamorphic grade. I: Reactions and grain size reduction: Tectonophysics, v. 221, p. 269–297.

Fitzgerald, P. G., Reynolds, S. J., Stump, E., Foster, D. A., and Gleadow, A. J. W., 1993, Thermochronologic evidence for timing of denudation and rate of crustal extension of the South Mountains metamorphic core complex and Sierra Estrella, Arizona: Nuclear Tracks Radiation Measurements, v. 21, p. 555–563.

Fossen, H., and Holst, T. B., 1995, Northwest-verging folds and the northwestward movement of the Caledonian Jotun nappe, Norway: Journal of Structural Geology, v. 17, p. 3–15.

Fountain, D. M., Boundy, T. M., Austrheim, H., and Rey, P., 1994, Eclogite-facies shear zones—deep crustal reflectors?: Tectonophysics, v. 232, p. 411–424.

Fredrich, J., and Evans, B., 1992, Strength recovery along simulated faults by solution transfer processes, *in* Tillerson, J. R., and Wawersik, W. R., eds., Rock mechanics: Rotterdam, A. A. Balkema, p. 121–130.

Friedman, M., and Higgs, N. G., 1981, Calcite fabrics in experimental shear zones, *in* Carter, N. L., Friedman, M., Logan, J. M., and Stearns, D. W., eds., Mechanical behavior of crustal rocks: The Handin volume: Washington, D.C., American Geophysical Union, Geophysical Monograph, no. 24, p. 11–27.

Frost, H. J., and Ashby, M. F., 1982, Deformation-mechanism maps: Oxford, Pergamon Press, 166 p.

Fry, N., 1979, Random point distributions and strain-measurements in rocks: Tectonophysics, v. 60, p. 89–105.

Fullagar, P. D., 1992, Geochronological studies of fault-related rocks, *in* Bartholomew, M. J., Hyndman, D. W., Mogk, D. W., and Mason, R., eds., Basement Tectonics 8: Characterization and comparison of ancient and Mesozoic continental margins—Proceedings of the 8th international conference on basement tectonics (Butte, Montana, 1988): Dordrecht, The Netherlands, Kluwer Academic Publishers, p. 37–50.

Gans, P. B., and Miller, E. L., 1983, Style of mid-Tertiary extension in east-central Nevada, *in* Gurgel, K. D., ed., Geologic excursions in the overthrust belt and metamorphic core complexes of the Intermountain region. Guidebook—Part 1 (Geological Society of America Rocky Mountain and Cordilleran sections meeting, Salt Lake City, Utah): Salt Lake City, Utah Geological and Mineral Survey Special Studies 59, p. 107–160.

Gapais, D., 1989, Shear structures in deformed granites: mechanical and thermal indicators: Geology, v. 17, p. 1144–1147.

Garcia-Celma, A., 1983, *C*-axis- and shape-fabrics in quartz mylonites of the Cap de Creus (Spain): their properties and development [Ph.D. dissertation]: Utrecht, The Netherlands, Utrecht University.

Gastil, G., Wracher, M., Strand, G., Kear, L. L., Ely, D., Chapman, D., and Anderson, C., 1992, The tectonic history of the southwestern United States and Sonora, Mexico, during the past 100 m.y.: Tectonics, v. 11, p. 990–997.

Ghent, E. D., and Stout, M. Z., 1981, Metamorphism at the base of the Samail ophiolite, southeastern Oman Mountains: Journal of Geophysical Research, v. 86, p. 2557–2571.

Giletti, B. J., and Yund, R. A., 1984, Oxygen diffusion in quartz: Journal of Geophysical Research, v. 89, p. 4039–4046

Gilotti, J. A., 1989, Reaction progress during mylonitization of basaltic dikes along the Särv thrust, Swedish Caledonides: Contributions to Mineralogy and Petrology, v. 101, p. 30–45.

Gilotti, J. A., 1992, The rheologically critical matrix in arkosic mylonites along the Särv thrust, Swedish Caledonides, *in* Mitra, S. and Fisher, G., eds., Structural geology of fold and thrust belts. Baltimore, Johns Hopkins University Press, p. 145–160.

Gilotti, J. A., and Hull, J. M., 1990, Phenomenological superplasticity in rocks, *in* Knipe, R. J. and Rutter, E. H., eds., Deformation mechanisms, rheology and tectonics: Geological Society of London, Special Publication No. 54, p. 229–240.

Gilotti, J. A., and Kumpulainen, R., 1986, Strain softening induced flow in the Särv thrust sheet, Scandinavian Caledonides: Journal of Structural Geology, v. 8, p. 441–455.

Gleason, G. C., and Tullis, J., 1995, A flow law for dislocation creep of quartz aggregates determined with the molten salt cell: Tectonophysics, v. 247, p. 1–23.

Glover, L., III, Mose, D. G., Poland, F. B., Bobyarchick, A. R., and Bourland, W. C., 1978, Grenville basement in the eastern Piedmont of Virginia: implications for orogenic models: Geological Society of America Abstracts with Programs, v. 10, no. 4, p. 169.

Goetze, C., and Evans, B., 1979, Stress and temperature in the bending lithosphere as constrained by experimental rock mechanics: Geophysical Journal of the Royal Astronomical Society, v. 59, p. 463–478.

Goodwin, L. B., and Renne, P. E., 1991, Effects of progressive mylonitization on Ar retention in biotites from the Santa Rosa mylonite zone, California, and thermochronologic implications: Contributions to Mineralogy and Petrology, v. 108, p. 283–297.

Goodwin, L. B., and Wenk, H.-R., 1990, Intracrystalline folding and cataclasis in biotite of the Santa Rosa mylonite zone: HVEM and TEM observations: Tectonophysics, v. 172, p. 201–214.

Goodwin, L. B., and Wenk, H.-R., 1995, Development of phyllonite from granodiorite: Mechanisms of grain-size reduction in the Santa Rosa mylonite zone, California: Journal of Structural Geology, v. 17, p. 689–707.

Gottschalk, R. R., Kronenberg, A. K., Russell, J. E., and Handin, J., 1990, Mechanical anisotropy of gneiss. Failure criterion and textural sources of directional behavior: Journal of Geophysical Research, v. 95, p. 21,613–21,634.

Gower, R. J. W., and Simpson, C., 1992, Phase boundary mobility in naturally deformed, high-grade quartzofeldspathic rocks: evidence for diffusional creep: Journal of Structural Geology, v. 14, p. 301–313.

Grady, K. A., 1990, Geology and structure of the rocks associated with the basal (Madstone) thrust of the Josephine ophiolite in southwestern Oregon: evidence for a metamorphic sole [M.S. thesis]: Albany, N.Y., U.S.A., State University of New York at Albany, 160 p.

Green, H. W., 1976, Plasticity of olivine in peridotites, *in* Wenk, H.-R., ed., Electron microscopy in mineralogy: Berlin, Springer-Verlag, p. 443–464.

Green, H. W., and Guéguen, Y., 1974, Origin of kimberlite pipes by diapiric upwelling in the upper mantle: Nature, v. 249, p. 617–620.

Griffin, W. L., 1971, Origin of coronas in anorthosites of the upper Jotun nappe, Indre Sogn, Norway: Journal of Petrology, v. 12, p. 219–243.

Griggs, D. T., 1967, Hydrolytic weakening of quartz and other silicates: Geophysical Journal of the Royal Astronomical Society, v. 14, p. 19–31.

Grocott, J., 1981, Fracture geometry of pseudotachylyte generation zones: a study of shear fractures formed during seismic events: Journal of Structural Geology, v. 3, p. 169–178.

Groshong, R. H., Teufel, L. W., and Gasteiger C., 1984, Precision and accuracy of the calcite strain-gauge technique: Geological Society of America Bulletin, v. 95, p. 357–363.

Grove, M., 1987, Metamorphism and deformation of prebatholithic rocks in the Box Canyon area, eastern Peninsular Ranges, San Diego County, California [M.S. thesis]: Los Angeles, Calif., U.S.A., University of California, 174 p.

Grubenmann, U., and Niggli, P., 1924, Die Gesteinsmetamorphose. Berlin, Borntraeger, p. 219–223.

Gu, Y., and Wong, T.-f., 1994, Development of shear localization in simulated quartz gouge: Effect of cumulative slip and gouge particle size, *in* Marone, C. J., and Blanpied, M. L., eds., Special Issue. Faulting, friction, and earthquake mechanics: Pure and Applied Geophysics, v. 143, p. 387–424.

Guéguen, Y., 1979, Les dislocations dans l'olivine des péridotites [Thèse Doctorate Etate]: Nantes, France, Université Nantes, 193 p.

Guéguen, Y., and Boullier, A.-M., 1976, Evidence of superplasticity in mantle peridotites, *in* Stens, R. G. J., ed., The physics and chemistry of minerals and rocks: New York, John Wiley & Sons, p. 19–33.

Guillope, M., and Poirier, J. P., 1979, Dynamic recrystallization during creep of single-crystalline halite: an experimental study: Journal of Geophysical Research, v. 84, p. 5557–5567.

Hacker, B. R., 1990, Simulation of the metamorphic and deformational history of the metamorphic sole of the Oman ophiolite: Journal of Geophysical Research, v. 95, p. 4895–4907.

Hacker, B. R., 1991, The role of deformation in the formation of metamorphic field gradients. Ridge subduction beneath the Oman ophiolite: Tectonics, v. 10, p. 455–473.

Hacker, B. R., 1994, Rapid emplacement of young oceanic lithosphere: Science, v. 265, p. 1563–1565.

Hacker, B. R., Yin, A., Christie, J. M., and Snoke, A. W., 1990, Differential stress, strain rate, and temperature of mylonitization in the Ruby Mountains, Nevada: Implications for the rate and duration of uplift: Journal of Geophysical Research, v. 95, p. 8569–8580.

Hallbauer, D. K., Wagner, H., and Cook, N. G. W., 1973, Some observations concerning the microscopic and mechanical behavior of quartzite specimens in stiff, triaxial compression tests: International Journal of Rock Mechanics, Mining Science and Geomechanical Abstracts, v. 10, p. 713–726.

Hammarstrom, J. M., 1992, Mineral chemistry of Cretaceous plutons: hornblende geobarometry in southern California and southeastern Alaska: Geological Society of America Abstracts with Programs, v. 24, no. 5, p. 30.

Handy, M. R., 1987, The structure, age and kinematics of the Pogallo Fault Zone; Southern Alps, northwestern Italy: Eclogae geologicae Helvetiae, v. 80, p. 593–632.

Handy, M. R., 1990, The solid-state flow of polymineralic rocks: Journal of Geophysical Research, v. 95, p. 8647–8661.

Handy, M. R., 1994a, Flow laws for rocks containing two non-linear viscous phases: a phenomenological approach: Journal of Structural Geology, v. 16, p. 287–301.

Handy, M. R., 1994b, The energetics of steady state heterogeneous shear in mylonitic rocks: Materials Science and Engineering, v. A175, p. 261–272.

Handy, M. R., and Zingg, A., 1991, The tectonic and rheologic evolution of an attenuated cross section of the continental crust: Ivrea crustal section, southern Alps, northwestern Italy and southern Switzerland: Geological Society of America Bulletin, v. 103, p. 236–253.

Hanmer, S., 1984, The potential use of planar and elliptical structures as indicators of strain regime and kinematics of tectonic flow: Geological Survey of Canada Paper 84–1B, p. 133–142.

Hanmer, S., 1986, Asymmetrical pull-aparts and foliation fish as kinematic indicators: Journal of Structural Geology, v. 8, p. 111–122.

Hanmer, S., and Passchier, C. W., 1991, Shear-sense indicators: a review: Geological Survey of Canada Paper 90–17, 72 p.

Hanmer, S., Darrach, M., and Kopf, C., 1992, The East Athabasca mylonite zone: an Archean segment of the Snowbird tectonic zone in northern Saskatchewan: Geological Survey of Canada Paper 92-1C, p. 19–29.

Hanmer, S., Parrish, R., Williams, M., and Kopf, C., 1994, Striding-Athabasca mylonite zone: complex Archean deep-crustal deformation in the East Athabasca mylonite triangle, northern Saskatchewan: Canadian Journal of Earth Sciences, v. 31, p. 1287–1300.

Harley, S. L., 1988, Proterozoic granulites from the Rauer Group, East Antarctica. I. Decompressional pressure-temperature paths deduced from mafic and felsic gneisses: Journal of Petrology, v. 29, p. 1059–1095.

Harley, S. L., and Fitzsimons, I. C. W., 1991, Pressure-temperature evolution of metapelitic granulites in a polymetamorphic terrane, the Rauer Group, East Antarctica: Journal of Metamorphic Geology, v. 9, p. 231–243.

Harper, G. D., Grady, K., and Wakabayashi, J., 1990, A structural study of a metamorphic sole beneath the Josephine ophiolite, western Klamath terrane, California-Oregon, *in* Harwood, D. S., and Miller, M. M., eds., Paleozoic and early Mesozoic paleogeographic relations; Sierra Nevada, Klamath Mountains and related terranes: Boulder, Colorado, Geological Society of America Special Paper 255, p. 379–396.

Hatcher, R. D., Jr., 1978, Eastern Piedmont fault system. Speculations on its extent: Reply: Geology, v. 6, p. 580–582.

Hatcher, R. D., Jr., and Hooper, R. J., 1981, Controls of mylonitization processes: relationships to orogenic thermal/metamorphic peaks: Geological Society of America Abstracts with Programs, v. 13, no. 7, p. 469.

Hatcher, R. D., Jr., Howell, D. E., and Talwani, P., 1977, Eastern Piedmont fault system: Speculations on its extent: Geology, v. 5, p. 636–640.

Hayden, H. W., Floreen, S., and Goodell, P. D., 1972, The deformation mechanism of superplasticity. Metallurgical Transactions, v. 3, p. 833–842.

Heard, H. C., and Raleigh, C. B., 1972, Steady-state flow in marble at 500 to 800°C: Geological Society of America Bulletin, v. 83, p. 935–956.

Heitzmann, P., 1985, Kakrite, kataklasite, mylonite—zur nomenklatur der metamorphite mit verformungsgefügen: Eclogae geologicae Helvetiae, v. 78, p. 273–286.

Hensen, B. J., and Harley, S. L., 1990, Graphical analysis of P-T-X relations in granulite facies metapelites, *in* Ashworth, J. R., and Brown, M., eds., High-temperature metamorphism and crustal anatexis: London, Unwin Hyman, p. 320–370.

Herwegh, M., and Handy, M. R., 1996, The evolution of high-temperature mylonitic microfabrics: evidence from simple shearing of a quartz analogue (norcamphor): Journal of Structural Geology, v. 18, p. 689–710.

Heyes, A. J., 1978, A metamorphic structural and geochemical study of the South Harris anorthosite, Outer Hebrides [Ph.D. dissertation]: London, England, University of London.

Hickman, S. H., and Evans, B., 1995, The kinetics of pressure solution at halite-silica interfaces and intergranular clay films: Journal of Geophysical Research, v. 100, p. 13,113–13,132.

Hickman, S., Sibson, R., and Bruhn, R., 1995, Introduction to special section: Mechanical involvement of fluids in faulting: Journal of Geophysical Research, v. 100, p. 12,831–12,840.

Higgins, M. W., 1971, Cataclastic rocks: Washington, D.C., U.S. Geological Survey Professional Paper 687, 97 p.

Hill, D. P., Eaton, J. P., and Jones, L. M., 1990, Seismicity, 1980–86, *in* Wallace, R. E., ed., The San Andreas fault system, California: Washington, D.C., U.S. Geological Survey Professional Paper 1515, p 115–151

Hippertt, J. F. M., 1993, "V"-pull-apart microstructures: a new shear-sense indicator: Journal of Structural Geology, v. 15, p. 1393–1403.

Hippertt, J. F. M., 1994a, Grain boundary microstructures in micaceous quartzite. significance for fluid movement and deformation processes in low metamorphic grade shear zones: Journal of Geology, v. 102, p. 331–348.

Hippertt, J. F., 1994b, Structures indicative of helicoidal flow in a migmatitic diapir (Bação Complex, southeastern Brazil): Tectonophysics, v. 234, p. 169–196.

Hippertt, J. F., 1994c, Direct observation of porosity in quartzite and phyllonite: Neues Jahrbuch für Mineralogie Abhandlung, v. 166, p. 239–259.

Hippertt, J. F., 1994d, Microstructures and *c*-axis fabrics indicative of quartz dissolution in sheared quartzites and phyllonites: Tectonophysics, v. 229, p. 141–163.

Hippertt, J. F. M., and Borba, R. P., 1992, Quartz *c*-axis fabric differences between porphyroclasts and recrystallized grains: Discussion: Journal of Structural Geology, v. 14, p. 627–630.

Hirth, G., and Tullis, J., 1992, Dislocation creep regimes in quartz aggregates: Journal of Structural Geology, v. 14, p. 145–159.

Hirth, G., and Tullis, J., 1994, The brittle-plastic transition in experimentally deformed quartz aggregates. Journal of Geophysical Research, v. 99, p. 11,731–11,747.

Hobbs, B. E., and Ord, A., 1988, Plastic instabilities: implications for the origin of intermediate and deep focus earthquakes: Journal of Geophysical Research, v. 93, p. 10,521–10,540.

Hobbs, B. E., Means, W. D , and Williams, P F., 1976, An outline of structural geology: New York, John Wiley & Sons, 571 p.

Hobbs, B. E., Ord, A., and Teyssier, C., 1986, Earthquakes in a ductile regime?: Pure and Applied Geophysics, v. 124, p. 309–336

Hollister, L. S., and Crawford, M. L., 1986, Melt-enhanced deformation: A major tectonic process: Geology, v. 14, p. 558–561.

Honnorez, J., Laverne, C., Hubberten, H W., Emmerman, R., and Meuhlenbachs, K., 1983, Alteration processes in layer 2 basalts from deep sea drilling project hole 504B, Costa Rica Rift: Initial Reports of Deep Sea Drilling Project, Washington, D.C., U.S. Govt Printing Office, v. 69, p. 509–546.

Hoogerduijn Strating, E. H., and Vissers, R. L. M., 1991, Dehydration-induced fracturing of eclogite facies peridotites—implications for the mechanical behaviour of subducting oceanic lithosphere: Tectonophysics, v. 200, p. 187–198.

Hoogerduijn Strating, E. H., and Vissers, R.L.M., 1994, Structures in natural serpentinite gouges: Journal of Structural Geology, v. 16, p. 1205–1216.

Hoogerduijn Strating, E. H., Rampone, E. H., Piccardo, G. B., Drury, M., and Vissers, R.L.M., 1993, Subsolidus emplacement of mantle peridotites during incipient oceanic rifting of the Mesozoic Tethys (Voltri Massif, NW Italy): Journal of Petrology, v. 34, p. 901–927

Hooper, R. J., and Hatcher, R. D., Jr., 1988a, The Pine Mountain terrane, a complex window in the Georgia and Alabama Piedmont: Evidence from the eastern termination. Geology, v. 16, p. 307–310.

Hooper, R. J., and Hatcher, R. D., Jr., 1988b, Mylonites from the Towaliga fault zone, central Georgia: Products of heterogeneous non-coaxial deformation: Tectonophysics, v. 44, p. 1–17.

Houseknecht, D. W., 1988, Intergranular pressure solution in four quartzose sandstones: Journal of Sedimentary Petrology, v. 58, p. 228–246.

Howard, K. A., 1968, Flow direction in triclinic folded rocks: American Journal of Science, v. 266, p. 758–765.

Hsu, K. J., 1955, Granulites and mylonites of the region about Cucamonga and San Antonio canyons, San Gabriel Mountains, California: University of California Publications in Geological Sciences, v. 30, no. 4, p. 223–352.

Hubbert, M. K., and Rubey, W. W., 1959, Role of fluid pressure in mechanics of overthrust faulting, I. Mechanics of fluid-filled porous solids and its application to overthrust faulting: Geological Society of America Bulletin, v. 70, p. 115–166.

Hudleston, P. J., 1980, The progressive development of inhomogeneous shear and crystallographic fabric in glacial ice: Journal of Structural Geology, v. 2, p. 189–196.

Hurlow, H. A., Snoke, A. W., and Hodges, K. V., 1991, Temperature and pressure of mylonitization in a Tertiary extensional shear zone, Ruby Mountains–East Humboldt Range, Nevada: Tectonic implications: Geology, v. 19, p. 82–86.

Ito, E., and Anderson, A. T., Jr., 1983, Submarine metamorphism of gabbros from the Mid-Cayman Rise: petrographic and mineralogic constraints on hydrothermal processes at slow-spreading ridges: Contributions to Mineralogy and Petrology, v 82, p 371–388.

Jackson, M. P. A., 1985, Natural strain in diapiric and glacial rock salt, with emphasis on Oakwood Dome, east Texas: University of Texas at Austin Bureau of Economic Geology Report of Investigations 143, 74 p.

Jamtveit, B., Bucher-Nurminen, K., and Austrheim, H., 1990, Fluid controlled eclogitization of granulites in deep crustal shear zones, Bergen arcs, western Norway: Contributions to Mineralogy and Petrology, v. 104, p. 184–193.

Janecke, S. U., and Evans, J. P., 1988, Feldspar-influenced rock rheologies: Geology, v. 16, p. 1064–1067.

Jégouzo, P., 1980, The South Armorican shear zone: Journal of Structural Geology, v 2, p 39–47.

Jehu, T. J., and Craig, R. M., 1927, Geology of the Outer Hebrides—IV South Harris: Transactions of the Royal Society of Edinburgh, v. 55, p. 457–488.

Jessell, M. W., 1986, Grain boundary migration and fabric development in experimentally deformed octachloropropane: Journal of Structural Geology, v. 8, p. 527–542.

Jessell, M. W., 1988, Simulation of fabric development in recrystallizing aggregates—II. Example model runs: Journal of Structural Geology, v. 10, p. 779–793.

Jessell, M. W., and Lister, G. S., 1990, A simulation of the temperature dependence of quartz fabrics, *in* Knipe, R. J., and Rutter, E. H., eds., Deformation mechanisms, rheology and tectonics, Geological Society of London, Special Publication No. 54, p. 353–362.

Ji, S., and Mainprice, J. M., 1988, Natural deformation fabrics of plagioclase: implications for slip systems and seismic anisotropy: Tectonophysics, v. 147, p. 145–163.

Ji, S., and Mainprice, J. M., 1990, Recrystallization and fabric development in plagioclase: Journal of Geology, v. 98, p. 65–79.

Ji, S., and Martignole, J., 1994, Ductility of garnet as an indicator of extremely high temperature deformation: Journal of Structural Geology, v. 16, p. 985–996.

Ji, S., and Zhao, P., 1993, Location of tensile fracture within rigid-brittle inclusions in ductilely flowing matrix: Tectonophysics, v. 220, p. 23–31.

Jiang, D., 1994, Vorticity determination, distribution, partitioning and the heterogeneity and non-steadiness of natural deformations: Journal of Structural Geology, v. 16, p. 121–130.

Jin, Z.-M., Green, H. W., and Borch, R. S., 1989, Microstructures of olivine and stresses in the upper mantle beneath eastern China: Tectonophysics, v. 169, p. 23–50.

Jin, Z.-M., Green, H. W., and Zhou, Y., 1994, Melt topology during dynamic partial melting of peridotite: Nature, v. 372, p. 164–167.

John, B. E., and Howard, K. A., 1995, Rapid extension recorded by cooling-age patterns and brittle deformation, Naxos, Greece: Journal of Geophysical Research, v. 100, p. 9969–9979.

Johnson, M. R. W., Kelly, S. P., Oliver, G. J. H., and Winter, D. A., 1985, Thermal effects and timing of thrusting in the Moine thrust zone: Journal of the Geological Society (London), v. 142, p. 863–874.

Jordan, P. G., 1987, The deformational behaviour of bimineralic limestone-halite aggregates: Tectonophysics, v. 135, p. 185–197.

Jordan, P. G., 1988, The rheology of polymineralic rocks: An approach: Geologisches Rundschau, v. 77, p. 285–294.

Jordan, P., Noak, T., and Widmer, D., 1990, The evaporite shear zone of the Jura Boundary Thrust—new evidence from the Wisen well (Switzerland): Eclogae geologicae Helvetiae, v. 83, p. 525–542.

Kalakay, T. J., and Snoke, A. W., 1995, Anatectic melt generation and migration in a shear zone associated with a lower crustal mafic intrusion, Ivrea-Verbano Zone, Italy, *in* Brown, M., and Piccoli, P. M., eds., The origin of granites and related rocks: Washington, D.C., U.S. Geological Survey Circular 1129, p. 77–78.

Karato, S.-I., 1984, Grain-size distribution and rheology of the upper mantle: Tectonophysics, v. 104, p. 155–176.

Karato, S.-I., and Wu, P., 1993, Rheology of the upper mantle: A synthesis: Science, v. 260, p. 771–778.

Karato, S.-I., Paterson, M. S., and Fitz Gerald, J. D., 1986, Rheology of synthetic olivine aggregates: Influence of grain size and water: Journal of Geophysical Research, v. 91, p. 8151–8176.

Karson, J. A., 1990, Seafloor spreading on the Mid-Atlantic Ridge: Implications for the structure of ophiolites and oceanic lithosphere produced in slow-spreading environments, *in* Malpas, J., Moores, E. M., Panayiotou, A., and Xenophontos, C., eds., Oceanic crustal analogues, Proceedings of the symposium "TROODOS 1987": Nicosia, Cyprus, Geological Survey Department, p. 547–555.

Karson, J. A., and Dick, H. J. B., 1983, Tectonics of ridge-transform intersections at the Kane Fracture Zone: Marine Geophysical Researches, v. 6, p. 51–98.

Karson, J. A., Thompson, G., Humphris, S. E., Edmond, J. M., Bryan, W. B., Brown, J. R., Winters, A. T., Pockalny, R. A., Casey, J. F., Campbell, A. C., Klinkhammer, G., Palmer, M. R., Kinzler, R. J., and Sulanowska, M. M., 1987, Along-axis variations in seafloor spreading in the MARK area: Nature, v. 328, p. 681–685.

Keleman, P. B., and Dick, H. J. B., 1995, Focused melt flow and localized deformation in the upper mantle: Juxtaposition of replacive dunite and ductile shear zones in the Josephine peridotite, SW Oregon: Journal of Geophysical Research, v. 100, p. 423–438.

Kelley, S. P., Reddy, S. M., and Maddock, R. H., 1994, Laser-probe $^{40}Ar/^{39}Ar$ investigation of a pseudotachylyte and its host rock from the Outer Isles thrust, Scotland: Geology, v. 22, p. 443–446.

Kern, H., and Wenk, H.-R., 1983, Texture development in experimentally induced ductile shear zones: Contributions to Mineralogy and Petrology, v. 83, p. 231–236.

Kerrich, R., 1986, Fluid infiltration into fault zones: chemical, isotopic, and mechanical effects: Pure and Applied Geophysics, v. 124, p. 226–268.

Kerrich, R., Allison, I., Barnett, R. L., Moss, S., and Starkey, J., 1980, Microstructural and chemical transformations accompanying deformation of granite in a shear zone at Miéville, Switzerland, with implications for stress corrosion cracking and superplastic flow: Contributions to Mineralogy and Petrology, v. 73, p. 221–242.

Kerrich, R. W., La Tour, T. E., and Barnett, R. L., 1981, Mineral reactions participating in intragranular fracture propagation: implications for stress corrosion cracking: Journal of Structural Geology, v. 3, p. 77–87.

Kerrich, R., La Tour, T. E., and Willmore, L., 1984, Fluid participation in deep fault zones, evidence from geological, geochemical, and $^{18}O/^{16}O$ relations: Journal of Geophysical Research, v. 89, p. 4331–4343.

Killick, A. M., and Reimold, W. U., 1990, A review of the pseudotachylites in and around the Vredefort Dome, South African Journal of Geology, v. 93, p. 350–365.

Kirby, S. H., 1983, Rheology of the lithosphere: Reviews of Geophysics and Space Physics, v. 21, p. 1458–1487.

Kirby, S. H., and Kronenberg, A. K., 1987, Rheology of the lithosphere: Selected topics: Reviews of Geophysics, v. 25, p. 1219–1244.

Kirschner, D. L., and Teyssier, C., 1992a, Deformation history of the White Range duplex, central Australia, with implications for fold reorientation: Australian Journal of Earth Sciences, v. 39, p. 441–456.

Kirschner, D. L., and Teyssier, C., 1992b, Quartz *c*-axis fabric differences between porphyroclasts and recrystallized grains: Reply: Journal of Structural Geology, v. 14, p. 631–634.

Kirschner, D. L., Sharp, Z. D., and Teyssier, C., 1993, Vein growth mechanisms and fluid sources revealed by oxygen isotope laser microprobe: Geology, v. 21, p. 85–88.

Kirschner, D. L., Teyssier, C., Gregory, R., and Sharp, Z. D., 1995, Effect of deformation on oxygen isotope exchange in the Heavitree Quartzite, Ruby Gap duplex, central Australia: Journal of Structural Geology, v. 17, p. 1407–1423.

Knipe, R. J., 1989, Deformation mechanisms—recognition from natural tectonites: Journal of Strucural Geology, v. 11, p. 127–146.

Knipe, R. J., 1990, Microstructural analysis and tectonic evolution in thrust systems: examples from the Assynt region of the Moine thrust zone, Scotland, *in* Barber, D. J., and Meredith, P. G., eds., Deformation processes in minerals, ceramics and rocks: London, Unwin Hyman, p. 228–261.

Knipe, R. J., and Lloyd, G. E., 1994, Microstructural analysis of faulting in quartzite, Assynt, NW Scotland: Implications for fault zone evolution, *in* Marone, C. J., and Blanpied, M. L., eds., Special Issue: Faulting, friction, and earthquake mechanics: Pure and Applied Geophysics, v. 143, p. 229–254.

Knipe, R. J., and McCaig, A. M., 1994, Microstructural and microchemical consequences of fluid flow in deforming rocks, *in* Parnell, J., ed., Geofluids: origin, migration and evolution of fluids in sedimentary basins: Geological Society of London, Special Publication No. 78, p. 99–111.

Knipe, R. J., and Wintsch, R. P., 1985, Heterogeneous deformation, foliation development, and metamorphic processes in a polyphase mylonite, *in* Thompson, A. B., and Rubie, D. C., eds., Metamorphic reactions: kinetics, textures and deformation: New York, Springer-Verlag, Advances in Physical Geochemistry, v. 4, p. 180–210.

Koch, P. S., and Christie, J. M., 1981, Spacing of deformation lamellae as a paleopiezometer: EOS, Transactions of the American Geophysical Union, v. 62, p. 1030.

Kohlstedt, D. L., Evans, B., and Mackwell, S. J., 1995, Strength of the lithosphere: Constraints imposed by laboratory experiments: Journal of Geophysical Research, v. 100, p. 17,587–17,602.

Komatsu, M., Toyoshima, T., Osanai, Y., and Arai, M., 1994, Prograde and anatectic reactions in the deep arc crust exposed in the Hidaka metamorphic belt, Hokkaido, Japan: Lithos, v. 33, p. 31–49.

Kronenberg, A. K., 1981, Quartz preferred orientations within a deformed pebble conglomerate from New Hampshire, U.S.A.: Tectonophysics, v. 79, p. T7-T15.

Kronenberg, A. K., 1994, Hydrogen speciation and chemical weakening of quartz, *in* Heaney, P. J., Prewitt, C. T., and Gibbs, G. V., eds., Silica: Mineralogical Society of America, Reviews in Mineralogy, v. 29, p. 123–176.

Kronenberg, A. K., Segall, P., and Wolf, G. H., 1990, Hydrolytic weakening and penetrative deformation within a natural shear zone, *in* Duba, A. G., Durham, W. B., Handin, J. W., and Wang, H. F., eds., The brittle-ductile transition in rocks: Washington, D.C., American Geophysical Union, Geophysical Monograph 56, p. 21–36.

Kumpulainen, R., 1980, Upper Proterozoic stratigraphy and depositional environments of the Tossåsfjallet Group, Särv Nappe, southern Swedish Caledonides: Geologiska Föreningens i Stockholm Förhandlingar, v. 102, p. 531–550.

Kupfer, D. H., 1976, Shear zones inside Gulf Coast salt stocks help to delineate spines of movement: American Association of Petroleum Geologists Bulletin, v. 60, p. 1434–1447.

Lachenbruch, A. H., and Sass, J. H., 1980, Heat flow and energetics of the San Andreas fault zone: Journal of Geophysical Research, v. 85, p. 6185–6222.

Lacour, A., 1971, Observations sur la mylonitisation des roches de la nappe du Jotun dans la région de Sogndal, Norvège centro-méridional: C.R. somm. Societe Géologique de France, v. 9, p. 150–151.

Lafrance, B., and Vernon, R. H., 1993, Mass transfer and microfracturing in gabbroic mylonites of the Guadalupe igneous complex, California, *in* Boland, J. N., and Fitz Gerald, J.D., eds., Defects and processes in the solid state: geoscience applications, The McLaren volume: The Netherlands, Elsevier, p. 151–167.

Lailey, M., Stein, A. M., and Reston, T. J., 1989, The Outer Hebrides fault: a major Proterozoic structure in NW Britain: Journal of the Geological Society (London), v. 146, p. 253–259.

Lambert, P., 1981, Breccia dikes: geological constraints on the formation of complex craters, *in* Schultz, D. H. and Merrill, R. B., eds., Multi-ring basins: Proceedings of Lunar and Planetary Science, v. 12A, p. 58–78.

Lane, L. S., 1984, Brittle deformation in the Columbia River fault zone near Revelstoke, southeastern British Coumbia: Canadian Journal of Earth Sciences, v. 21, pp. 584–598.

Lane, L. S., and Brown, R. L., 1987, Structural relationships on the eastern margin of the Shuswap metamorphic complex near Revelstoke, British Coumbia, Canada, *in* Hill, M. L., ed., Geological Society of America Centennial Field Guide, v. 1, Cordilleran Section, p. 413–418.

Lane, L. S., Ghent, E. D., Stout, M. Z., and Brown, R. L., 1989, PT history and kinematics of the Monashee Décollement near Revelstoke, British Columbia: Canadian Journal of Earth Sciences, v. 26, p. 231–243.

Langseth, M. G., Mottl, M. J., Hobart, M., and Fischer, A., 1988, The distribution of geothermal and geochemical gradients near Site 501/504: implications for hydrothermal circulation in the oceanic crust, *in* Becker, K., Sakai, H. et al., Proceedings of the Ocean Drilling Program, Initial Reports: College Station, Texas, v. 111: p. 23–32.

Lapworth, C., 1885, The highland controversy in British geology; its causes, course and consequences: Nature, v. 32, p. 558–559.

La Tour, T. E., 1981, Metamorphism and geothermometry near Coniston, Ontario: a clue to the tectonic evolution of the Grenville front: Canadian Journal of Earth Sciences, v. 18, p. 884–898.

Law, R. D., 1986, Relationships between strain and quartz crystallographic fabrics in the Roche Maurice quartzites of Plougastel, western Brittany: Journal of Structural Geology, v. 8, p. 493–515.

Law, R. D., 1987, Heterogeneous deformation and quartz crystallographic fabric transitions: natural examples from the Moine thrust zone at the Stack of Glencoul, northern Assynt: Journal of Structural Geology, v. 9, p. 819–833.

Law, R. D., 1990, Crystallographic fabrics: a selective review of applications to research in structural geology, *in* Knipe, R. J., and Rutter, E. H., eds., Deformation mechanisms, rheology and tectonics: Geological Society of London Special Publication No. 54, p. 335–352.

Law, R. D., Casey, M., and Knipe, R. J., 1986, Kinematic and tectonic significance of microstructures and crystallographic fabrics within quartz mylonites from the Assynt and Eriboll regions of the Moine thrust zone, NW Scotland: Transactions of the Royal Society of Edinburgh: Earth Sciences, v. 77, p. 99–125.

Law, R. D., Schmid, S. M., and Wheeler, J., 1990, Simple shear deformation and quartz crystallographic fabrics: a possible natural example from the Torridon area of NW Scotland: Journal of Structural Geology, v. 12, p. 29–45.

Lee, J., Miller, E. L., and Sutter, J. F., 1987, Ductile strain and metamorphism in an extensional tectonic setting: a case study from northern Snake Range, Nevada, USA, *in* Coward, M. P., Dewey, J. F., and Hancock, P. L., eds., Continental extensional tectonics:

Geological Society of London, Special Publication No. 28, p. 267–298.

Li, J.-L., 1988, Tectonic evolution of the Three-River belt in the western Yunnan, Scientia Geologica Sinica, v. 4, p. 337–346 (in Chinese).

Lin, A., 1991, Origin of fault-generated pseudotachylites [Ph.D. dissertation]: Tokyo, Japan, University of Tokyo, 108 p.

Lin, A., 1994a, Glassy pseudotachylytes from the Fuyun fault zone, northwest China: Journal of Structural Geology, v. 16, p. 71–83.

Lin, A., 1994b, Microlite morphology and chemistry in pseudotachylyte from the Fuyun fault zone, China: Journal of Geology, v 102, p. 317–329.

Lin, C. Y., and Fan, F. T., 1984, Microstructural analysis of Fuyun fault zone, Xingjiang, China: Journal of Seismology, v. 2, p. 1–8 (in Chinese).

Lister, G. S., and Dornsiepen, U. F., 1982, Fabric transitions in the Saxony granulite terrain, Journal of Structural Geology, v. 4, p. 81–92.

Lister, G. S., and Price, G. P., 1978, Fabric development in a quartz-feldspar mylonite: Tectonophysics, v. 49, p. 37–78.

Lister, G. S., and Snoke, A. W., 1984, S-C mylonites: Journal of Structural Geology, v. 6, p. 617–638.

Lister, G. S., and Williams, P. F., 1983, The partitioning of deformation in flowing rock masses: Tectonophysics, v. 92, p. 1–33.

Lloyd, G. E., 1985, Review of instrumentation, techniques and applications of SEM in mineralogy, *in* White, J. C., ed., Applications of electron microscopy in the earth sciences: Mineralogical Association of Canada Short Course, v. 11, p. 151–188.

Lloyd, G. E., 1987, Atomic number and crystallographic contrast images with the SEM, a review of backscattered electron techniques: Mineralogical Magazine, v. 51, p. 3–19.

Lloyd, G. E., 1994, An appreciation of the SEM electron channelling technique for petrofabric and microstructural analysis of geological materials, *in* Bunge, H. J., Siegesmund, S., Skrotzki, W., and Weber, K., eds., Textures of geological materials: Oberursel, Germany, German Society of Materials Science, DGM Informationsgesellschaft, Verlag, p. 109–125.

Lloyd, G. E., and Knipe, R. J., 1992, Deformation mechanisms accommodating faulting of quartzite under upper crustal conditions: Journal of Structural Geology, v. 14, p. 127–143.

Lloyd, G. E., Schmidt, N.-H., Mainprice, D., and Prior, D. J., 1991, Crystallographic textures: Mineralogical Magazine, v. 55, p. 331–345.

Lloyd, G. E., Mainprice, D., Law, R. D., and Wheeler, J., 1992, Microstructural and crystal fabric evolution during shear zone formation: Journal of Structural Geology, v. 14, p. 1079–1100.

Lockwood, J. P., and Lydon, P. A., 1975, Geologic map of the Mount Abbot quadrangle, central Sierra Nevada, California: Washington, D.C., U.S. Geological Survey Geological Quadrangle Map GQ-1155, scale 1:62,500.

Lofgren, G., 1971a, Spherulitic structures in glassy and crystalline rocks: Journal of Geophysical Research, v. 76, p. 5635–5648.

Lofgren, G. E., 1971b, Experimentally produced devifrication textures in natural rhyolite glass: Geological Society of America Bulletin, v. 82, p. 111–124.

Lofgren, G. E., 1974, An experimental study of plagioclase crystal morphology: isothermal crystallization: American Journal of Science, v. 274, p. 243–273.

Lofgren, G. E., 1980, Experimental studies on the dynamic crystallization of silica melts, *in* Hargraves, R. B., ed., Physics of magmatic processes: Princeton, N.J., Princeton University Press, p 487–551.

Logan, J. M., Friedman, M., Higgs, N. G., Dengo, C., and Shimamoto, T., 1979, Experimental studies of simulated gouge and their application to studies of natural fault zones: Proceedings of Conference VIII: Analysis of actual fault zones in bedrock: Washington, D.C., U.S. Geological Survey Open File Report 70–1239, p. 305–343.

Logan, J. M., Dengo, C. A., Higgs, N. G., and Wang, Z. Z., 1992, Fabrics of experimental fault zones. Their development and relationship to mechanical behavior, *in* B. Evans and T.-f. Wong, eds., Fault mechanics and transport properties of rocks: New York, Academic Press, p. 33–68.

Lonsdale, P., and Klitgord, K.D., 1978, Structure and tectonic history of the eastern Panama Basin: Geological Society of America Bulletin, v. 89, p. 981–999.

Losh, S., 1989, Fluid-rock interaction in evolving ductile shear zone and across the brittle-ductile transition, central Pyrenees, France: American Journal of Science, v. 289, p. 600–648.

Lovera, O. M., Richter, F. M., and Harrison, T. M., 1989, The $^{40}Ar/^{39}Ar$ thermochronometry for slowly cooled samples having a distribution of diffusion domain sizes: Journal of Geophysical Research, v. 94, p. 17,917–17,935.

Lundgren, L., and Eblin, C., 1972, Honey Hill fault in eastern Connecticut: regional relations: Geological Society of America Bulletin, v. 83, p. 2773–2794.

Macaudière, J., and Brown, W. L., 1982, Transcrystalline shear fracturing and pseudotachylyte generation in a meta-anorthosite (Harris, Scotland): Journal of Structural Geology, v. 4, p. 399–406.

Macdonald, K. C., Sheirer, D. S., and Carbotte, S. M., 1991, Mid-ocean ridges: discontinuities, segments and giant cracks: Science, v. 253, p. 986–994.

Maddock, R. H., 1983, Melt origin of fault-generated pseudotachylytes demonstrated by textures: Geology, v. 11, p. 105–108.

Maddock, R. H., 1986, Partial melting of lithic porphyroclasts in fault-generated pseudotachylytes: Neues Jahrbuch für Mineralogie Abhandlung, v. 155, p. 1–14.

Maddock, R. H., 1992, Effects of lithology, cataclasis and melting on the composition of fault-generated pseudotachylytes in Lewisian gneiss, Scotland: Tectonophysics, v. 204, p. 261–278.

Maddock, R. H., Grocott, J., and Van Nes, M., 1987, Vesicles, amygdales and similar structures in fault-generated pseudotachylytes: Lithos, v. 20, p. 419–432.

Magloughlin, J. F., 1989, The nature and significance of pseudotachylite from the Nason terrane, North Cascade Mountains, Washington: Journal of Structural Geology, v. 11, p. 907–917.

Magloughlin, J. F., 1992, Microstructural and chemical changes associated with cataclasis and frictional melting at shallow crustal levels: the cataclasite-pseudotachylyte connection: Tectonophysics, v. 204, p. 243–260

Magloughlin, J. F., 1993, A Nason terrane trilogy: I. Nature and significance of pseudotachylyte; II. Summary of the structural and tectonic history; III. Major and trace element geochemistry, and strontium and neodymium isotope geochemistry of the Chiwaukum Schist, amphibolite, and metatonalite gneiss of the Nason terrane [Ph.D. dissertation]: Minneapolis, Minn., U.S.A., University of Minnesota, 309 p.

Magloughlin, J. F., and McEwan, R. F., 1988, The White River Shear Zone: a major North Cascades terrane boundary and compressional structure: Geological Society of America Abstracts with Programs, v. 20, no. 7, p. A395–396.

Magloughlin, J. F., and Spray, J. G., 1992, Frictional melting processes and products in geological materials: introduction and discussion: Tectonophysics, v. 204, p. 197–204.

Malpas, J., Case, G., and Moore, P., 1989, The geology of the area immediately surrounding the CY-4 borehole of the Cyprus Crustal Study: Geological Survey of Canada Paper 88-9, p. 31–37.

Mandl, G., DeJong, L. N. J., and Maltha, A., 1977, Shear zones in granular media: an experimental study of their structure and mechanical genesis: Rock Mechanics, v. 9, p. 95–144.

Marshak, S., and Alkmin, F. F., 1989, Proterozoic contraction/extension tectonics of the southern São Francisco craton region, Minas Gerais, Brazil: Tectonics, v. 8, p. 555–571.

Martini, J. E. J., 1978, Coesite and stishovite in the Vredefort Dome, South Africa: Nature, v. 278, p. 715–717.

Martini, J. E. J., 1991, The nature, distribution and genesis of the coesite and stishovite associated with pseudotachlylite of the Vredefort Dome, South Africa: Earth and Planetary Science Letters, v. 103, p. 285–300.

Martini, J. E. J., 1992, The metamorphic history of the Vredefort dome at approximately 2 Ga as revealed by coesite-stishovite-bearing pseudotachylites: Journal of Metamorphic Geology, v. 10, p. 517–527.

Masch, L., Wenk, H.-R., and Preuss, E., 1985, Electron microscopy study of hyalomylonites—evidence for frictional melting in landslides: Tectonophysics, v. 115, p. 131–160.

Masuda, T., and Ando, S., 1988, Viscous flow around a rigid spherical body: A hydrodynamical approach: Tectonophysics, v. 148, p. 337–346.

Mawer, C. K., 1983, State of strain in a quartzite mylonite, central Australia: Journal of Structural Geology, v. 5, p. 401–409.

Mawer, C. K., 1986, What is a mylonite?: Geoscience Canada, v. 13, p. 33–34.

Mawer, C. K., and Fitz Gerald, J. D., 1993, Microstructure of kink band boundaries in naturally deformed Chewings Range Quartzite, *in* Boland, J. N., and Fitz Gerald, J. D., eds., Defects and processes in the solid state: geoscience applications, The McLaren volume: Amsterdam, The Netherlands, Elsevier, p.

McCaig, A. M., and Knipe, R. J., 1990, Mass-transport mechanisms in deforming rocks: Recognition using microstructural and microchemical criteria: Geology, v. 18, p. 824–827.

McCaig, A. M., Wickham, S. M., and Taylor, H. P., Jr., 1990, Deep fluid circulation in alpine shear zones, Pyrenees, France: field and oxygen isotope studies: Contributions to Mineralogy and Petrology, v. 106, p. 41–60.

McCaig, A. M., Wayne, D. M., Marshall, J. D., Banks, D., and Henderson, I., 1995, Isotopic and fluid inclusion studies of fluid movement along the Gavarnie thrust, central Pyrenees: Reaction fronts in carbonate mylonites: American Journal of Science, v. 295, p. 309–343.

McClay, K. R., and Coward, M. P., 1981, The Moine thrust: An overview, *in* McClay, K. R., and Price, N. J., eds., Thrust and nappe tectonics: Geological Society of London, Special Publication No. 9, p. 241–260.

McCormick, J. W., 1977, Transmission electron microscope investigation of some naturally deformed quartzites [Ph.D. dissertation]: Los Angeles, Calif., U.S.A., University of California, 171 p.

McGrew, A. J., and Snee, L. W., 1994, $^{40}Ar/^{39}Ar$ thermochronologic constraints on the tectonothermal evolution of the northern East Humboldt Range metamorphic core complex, Nevada: Tectonophysics, v. 238, p. 425–450.

Means, W. D., 1981, The concept of a steady state foliation: Tectonophysics, v. 78, p. 179–199.

Means, W. D., 1983, Microstructure and micromotion in recrystallization flow of octachloropropane: a first look: Geologische Rundschau, v. 72, p. 511–528

Means, W. D., 1989, Synkinematic microscopy of transparent polycrystals: Journal of Structural Geology, v. 11, p. 163–174.

Mercier, J.-C. C., and Nicolas, A., 1975, Textures and fabrics of upper-mantle peridotites as illustrated by xenoliths from basalts: Journal of Petrology, v. 16, p. 454–487.

Mével, C., Cannat, M., Gente, P., Marion, E., Auzende, J. M., and Karson, J. A., 1991, Emplacement of deep crustal and mantle rocks on the West Median Valley Wall of the MARK Area (MAR 23°N): Tectonophysics, v. 190, p. 31–53.

Miller, E. L., Gans, P. B., and Garing, J. D., 1983, The Snake Range décollement: An exhumed mid-Tertiary ductile-brittle transition: Tectonics, v. 2, p. 239–263.

Miller, R. B., and Paterson, S. R., 1992, Tectonic implications of syn- and post-emplacement deformation of the Mount Stuart batholith for mid-Cretaceous orogenesis in the North Cascades: Canadian Journal of Earth Sciences, v. 29, p. 479–485.

Miller, R. B., and Paterson, S. R., 1994, The transition from magmatic to high-temperature solid-state deformation: implications from the Mount Stuart batholith, Washington: Journal of Structural Geology, v. 16, p. 853–865.

Milnes, A. G., and Koestler, A. G., 1985, Geological structure of Jotunheimen, southern Norway (Sognefjell-Valdres cross-section), *in* Gee, D. G., and Sturt, B. A., eds., The Caledonide orogen—Scandinavia and related areas: Chichester, John Wiley & Sons, p. 457–474.

Mitra, G., 1978, Ductile deformation zones and mylonites: the mechanical processes involved in the deformation of crystalline basement rocks: American Journal of Science, v. 278, p. 1075–1084.

Mitra, G., 1979, Ductile deformation zones in Blue Ridge basement rocks and estimation of finite strains: Geological Society of America Bulletin, v. 90, p. 935–951.

Mitra, G., 1984, Brittle to ductile transition due to large strains along the White Rock thrust, Wind River Mountains, Wyoming. Journal of Structural Geology, v. 6, p. 51–61.

Mitra, G., 1992, Deformation in granitic basement rocks along fault zones at shallow to intermediate crustal levels, *in* Mitra, S., and Fisher, G.W., eds., Structural geology of fold and thrust belts: Baltimore, Maryland, The Johns Hopkins University Press, p. 123–144.

Mitra, G., 1993, Deformation processes in brittle deformation zones in granitic basement rocks: A case study from the Torrey Creek area, Wind River Mountains, *in* Schmidt, C. J., Chase, R. B., and Erslev, E. A., eds., Laramide basement deformation in the Rocky Mountain foreland of the western United States: Geological Society of America, Special Paper 280, p. 177–195.

Mitra, G., and Elliott, D., 1979, Development of the South Mountain cleavage and deformation of basement in the Blue Ridge, *in* Wones, D. R., ed., Proceedings of "the Caledonides in the USA," I.G.C.P. project 27, Blacksburg, Va., Department of Geological Sciences, Virginia Polytechnic Institute and State University, p. A16.

Mitra, G., and Frost, B. R., 1981, Mechanisms of deformation within Laramide and Precambrian deformation zones in basement rocks of the Wind River Mountains: University of Wyoming Contributions to Geology, v. 19, p. 161–173.

Mitra, G., Hull, J. M., Yonkee, W. A., and Protzman, G. M., 1988, Comparison of mesoscopic and microscopic deformational styles in the Idaho-Wyoming thrust belt and the Rocky Mountain foreland, *in* Schmidt, C. J., and Perry, W. J., Jr., eds., Interaction of the Rocky Mountain foreland and Cordilleran thrust belt: Geological Society of America Memoir 171, p. 119–141.

Miyashiro, A., 1973, The Troodos ophiolitic complex was probably formed in an island arc: Earth and Planetary Science Letters, v. 19, p. 218–224.

Molnar, P., 1988, Continental tectonics in the aftermath of plate tectonics: Nature, v. 335, p. 131–137.

Moore, D. E., and Byerlee, J. D., 1991, Comparative geometry of the San Andreas fault, California, and laboratory fault zones: Geological Society of America Bulletin, v. 103, p. 762–774.

Moore, D. E., and Lockner, D. A., 1995, The role of microcracking in shear-fracture propagation in granite: Journal of Structural Geology, v. 17, p. 95–114.

Moore, D. E., Lockner, D. A., Summers, R., Shengli, M., and Byerlee, J. D., 1996, Strength of chrysotile-serpentine gouge under hydrothermal conditions: Can it explain a weak San Andreas fault?: Geology, v. 24, p. 1041–1044.

Muller, W. H., Schmid, S. M., and Briegel, U., 1981, Deformation experiments on anhydrite rocks of different grain sizes. Rheology and microfabrics: Tectonophysics, v. 78, p. 527–543.

Mutter, J. C., and Karson, J. A., 1992, Structural processes at slow-spreading ridges: Science, v. 257, p. 627–634.

Naylor, R. S., 1971, Geology of the Mascoma mantled gneiss dome near Hanover, New Hampshire: 63rd Annual New England Intercollegiate Geological Conference, Trip A-4, p. 28–37.

Newman, J., and Mitra, G., 1993, Lateral variations in mylonite zone thickness as influenced by fluid-rock interaction, Linville Falls fault, North Carolina: Journal of Structural Geology, v. 15, p. 849–863.

Newman, J., and Mitra, G., 1994, Fluid-influenced deformation and recrystallization of dolomite at low temperatures along a natural fault zone, Mountain City window, Tennessee: Geological Society of America Bulletin, v. 106, p. 1267–1280.

Nicolas, A., 1989, Structures in ophiolites and dynamics of oceanic lithosphere: Dordrecht, The Netherlands, Kluwer Academic Publishers, 367 p.

Nicolas, A., and Le Pichon, X., 1980, Thrusting of young lithosphere in subduction zones with special reference to structures in ophiolitic peridotites: Earth and Planetary Science Letters, v. 46, p. 397–406.

Nicolas, A., and Poirier, J.-P., 1976, Crystalline plasticity and solid state flow in metamorphic rocks: London, John Wiley & Sons, 444 p.

Nicolaysen, L. O., Burger, A. J., and Van Neikerk, C. B., 1963, The origin of the Vredefort dome structure in the light of new isotopic data [abstract]: 13th General Assembly of the International Union of Geodynamics and Geophysics (Berkeley, California).

Norrell, G. T., 1989, Structural geology of the Josephine peridotite, northern California: Implications for structural processes at slow-spreading centers [Ph.D. dissertation]: Albany, N.Y., U.S.A., State University of New York at Albany, 234 p.

Norrell, G. T., and Harper, G. D., 1988, Detachment faulting and amagmatic extension at mid-ocean ridges: The Josephine ophiolite as an example: Geology, v. 16, p. 827–830.

Norrell, G. T., and Harper, G. D., 1990, Ultramafic fault-rocks and oceanic detachment faulting, *in* Malpas, J., Moores, E. M., Panayiotou, A., and Xenophontos, C., eds., Ophiolites, Proceedings of the symposium "TROODOS 1987": Republic of Cyprus, Geological Survey Department, p. 521–534.

Norrell, G. T., Teixell, A., and Harper, G. D., 1989, Microstructure of serpentinite mylonites from the Josephine ophiolite and serpentinization in retrogressive shear zones, California: Geological Society of America Bulletin, v. 101, p. 673–682.

Nyman, M. W., and Tracy, R. J., 1993 Petrological evolution of amphibolite shear zones, Cheyenne Belt, south-eastern Wyoming, USA: Journal of Metamorphic Geology, v. 11, p. 757–773.

Nyman, M. W., Law, R. D., and Smelik, E. A., 1992, Cataclastic deformation mechanism for the development of core-mantle structures in amphibole: Geology, v. 20, p. 455–458.

O'Brien, D. K., Wenk, H.-R., Ratschbacher, L., and You, Z., 1987, Preferred orientation of phyllosilicates in phyllonites and ultramylonites: Journal of Structural Geology, v. 9, p. 719–730.

O'Hanley, D. S., Chernosky, J. V., Jr., and Wicks, F. J., 1989, The stability of lizardite and chrysotile: Canadian Mineralogist, v. 27, p. 483–493.

O'Hara, K., 1988, Fluid flow and volume loss during mylonitization: an origin for phyllonite in an overthrust setting, North Carolina, U.S.A.: Tectonophysics, v. 156, p. 21–36.

Okada, A., 1968, Strike-slip faulting of late Quaternary along the Median Dislocation Line in the surroundings of Awa-Ikeda, northeastern Shikoku: Quaternary Research, v. 7, p. 15–26.

Olesen, N. Ø., 1986, Three thrust sheets on Hornsnipa, Jotun nappe complex, west Norway: Norges Geologiske Indersokelse Bulletin, v. 404, p. 55–66.

Olesen, N. Ø., 1987, Plagioclase fabric development in a high-grade shear zone, Jotunheimen, Norway, Tectonophysics v. 142, p. 291–308.

Olsen, T. S., and Kohlstedt, D. L., 1985, Natural deformation and recrystallization of some intermediate plagioclase feldspars: Tectonophysics v. 111, p. 107–131.

Parks, G. A., 1984, Surface and interfacial free energies of quartz. Journal of Geophysical Research, v. 89, p. 3997–4008.

Passchier, C. W., 1982a, Mylonitic deformation in the Saint-Barthélemy massif, French Pyrenees, with emphasis on the genetic relationship between ultramylonite and pseudotachylyte: GUA papers of geology, City University of Amsterdam, Series 1, v. 16, p. 1–173.

Passchier, C. W., 1982b, Pseudotachylite and the development of ultramylonite bands in the Saint-Barthélemy Massif, French Pyrenees: Journal of Structural Geology, v. 4, p. 69–79.

Passchier, C. W., 1984, Mylonite-dominated footwall geometry in a shear zone, central Pyrenees, Geological Magazine, v. 121, p. 429–436.

Passchier, C. W., 1994, Mixing in flow perturbations: a model for the development of mantled porphyroclasts in mylonites: Journal of Structural Geology, v. 16, p. 733–736.

Passchier, C. W., and Simpson, C., 1986, Porphyroclast systems as kinematic indicators: Journal of Structural Geology, v. 8, p. 831–843.

Passchier, C. W., and Sokoutis, D., 1993, Experimental modelling of mantled porphyroclasts: Journal of Structural Geology, v. 15, p. 895–910.

Passchier, C. W., and Trouw, R. A. J., 1996, Microtectonics, Berlin, Springer-Verlag, 289 p.

Passchier, C. W., ten Brink, C. E., Bons, P. D., and Sokoutis, D., 1993, δ-objects as a gauge for stress sensitivity of strain rate in mylonites: Earth and Planetary Science Letters, v. 120, p. 239–245.

Paterson, M. S., 1989, The interaction of water with quartz and its influence on dislocation flow—an overview, *in* Karato, S., and

Tonumi, M., eds., Rheology of solids and of the earth: New York, Oxford University Press, p. 171–194.

Petit, J. P., 1987, Criteria for the sense of movement on fault surfaces in brittle fault zones· Journal of Structural Geology, v. 9, p. 597–608.

Pike, R. J., 1985, Some morphologic systematics of complex ring structures: Meteoritics, v. 20, p. 49–68.

Platt, J. P., and Vissers, R. L. M., 1980, Extensional structures in anisotropic rocks: Journal of Structural Geology, v. 2, p. 397–410.

Poirier, J.-P., 1985, Creep of crystals: Cambridge, England, Cambridge University Press, 260 p.

Ponce de Leon, M. I., and Choukroune, P., 1980, Shear zones in the Iberian Arc: Journal of Structural Geology, v. 2, p. 63–68.

Post, A., Tullis, J., and Yund, R. A., 1996, Experimental deformation of quartzite in a controlled chemical environment: Journal of Geophysical Research, v. 101, p. 22, 143–22, 155.

Power, W. L., and Tullis, T. E., 1989, The relationship between slickenside surfaces in fine-grained quartz and the seismic cycle: Journal of Structural Geology, v. 11, p. 879–893.

Price, G. P., 1985, Preferred orientations in quartzites, *in* Wenk, H.-R., ed., Preferred orientation in deformed metals and rocks: An introduction to modern texture analysis: Orlando, Academic Press, p. 385–406.

Prior, D. J., Knipe, R. J., and Handy, M. R., 1990, Estimates of the rates of microstructural change in mylonites, *in* Knipe, R. J., and Rutter, E. H., eds., Deformation mechanisms, rheology and tectonics: Geological Society of London, Special Publication No. 54, p. 309–319.

Pryer, L. L., 1993, Microstructures in feldspars from a major crustal thrust zone: the Grenville Front, Ontario, Canada: Journal of Structural Geology, v. 15, p. 21–36.

Quick, J. E., Sinigoi, S., and Mayer, A., 1994, Emplacement dynamics of a large mafic intrusion in the lower crust, Ivrea-Verbano Zone, northern Italy: Journal of Geophysical Research, v. 99, p. 21,559–21,573.

Raleigh, C. B., Kirby, S. H., Carter, N. L., and Avé Lallemant, H. G., 1971, Slip and the clinoenstatite transformation as competing rate processes in enstatite: Journal of Geophysical Research, v. 76, p. 4011–4022.

Ramsay, J. G., 1980a, Shear zone geometry: a review: Journal of Structural Geology, v. 2, p. 83–99.

Ramsay, J. G., 1980b, The crack-seal mechanism of rock deformation: Nature, v. 284, p. 135–139.

Ramsay, J G., and Graham, R. H., 1970, Strain variations in shear belts: Canadian Journal of Earth Sciences, v. 7, p. 786–813.

Ranalli, G., and Murphy, D. C., 1987, Rheological stratification of the lithosphere: Tectonophysics, v. 132, p. 281–295.

Rankin, D. W., Espenshade, G. H., and Shaw, K. W., 1973, Stratigraphy and structure of the metamorphic belt in northwestern North Carolina and southwestern Virginia: A study from the Blue Ridge across the Brevard fault zone to the Sauratown Mountains anticlinorium: American Journal of Science, v. 273-A (Cooper Volume), p. 1–40.

Ree, J.-H., 1991, An experimental steady-state foliation: Journal of Structural Geology, v. 13, p. 1001–1011.

Ree, J.-H., 1994, Grain boundary sliding and development of grain boundary openings in experimentally deformed octachloropropane: Journal of Structural Geology, v. 16, p. 403–418.

Reimold, W. U., and Colliston, W. P., 1994, Pseudotachylites of the Vredefort Dome and the surrounding Witwatersrand Basin, South Africa, *in* Dressler, B. O., Grieve, R. A. F., and Sharpton, V. L., eds., Large meteorite impacts and planetary evolution: Boulder, Colorado, Geological Society of America, Special Paper 293, p. 177–196.

Reinen, L. A., Weeks, J. D., and Tullis, T. E., 1991, The frictional behavior of serpentinite: implications for aseismic creep on shallow crustal faults: Geophysical Research Letters, v 18, p. 1921–1927.

Reynolds, S. J., 1985, Geology of the South Mountains, central Arizona: Arizona Bureau of Geology and Mineral Technology Bulletin 195, 61 p.

Reynolds, S. J., and Lister, G. S., 1987, Structural aspects of fluid-rock interactions in detachment zones: Geology, v. 15, p. 362–366.

Reynolds, S. J. and Lister, G. S., 1990, Folding of mylonite zones in Cordilleran metamorphic core complexes; evidence from near the mylonitic front: Geology, v. 18, p. 216–219.

Reynolds, S. J., and Rehrig, W. A., 1980, Mid-Tertiary plutonism and mylonitization, South Mountains, central Arizona, *in* Crittenden, M. D., Coney, P. J., and Davis, G. H., eds., Cordilleran metamorphic core complexes: Geological Society of America Memoir 153, p. 159–174.

Reynolds, S. J , Shafiqullah, M., Damon, P. E., and DeWitt, E., 1986, Early Miocene mylonitization and detachment faulting, South Mountain, central Arizona: Geology, v. 14, p 283–286.

Rice, J. R., 1992, Fault stress states, pore pressure distributions, and the weakness of the San Andreas fault, *in* Evans, B., and T.-f. Wong, eds., Fault mechanics and transport properties of rocks: New York, Academic Press, p. 475–504.

Robbins, G. A., 1972, Radiogenic argon diffusion in muscovite under hydrothermal conditions [M.S. thesis]: Providence, Rhode Island, U.S.A., Brown University, 47 p.

Robin, P.-Y., 1978, Pressure solution at grain-to-grain contacts: Geochimica et Cosmochimica Acta, v. 42, p 1383–1389.

Roddy, D. J., and Davis, L. K., 1977, Shatter cones formed in large-scale experimental explosion craters, *in* Roddy, D. J., Pepin, R. O., and Merrill, R. B., eds., Impact and explosion cratering: New York, Pergamon Press, p. 715–750.

Roper, P. J., 1972, Structural significance of "button" or "fish scale" texture in the phyllonitic schist of the Brevard zone: Geological Society of America Bulletin, v. 83, p. 853–860.

Rowe, K. J., and Rutter, E. H., 1990, Palaeostress estimation using calcite twinning: experimental calibration and application to nature: Journal of Structural Geology, v. 12, p. 1–17.

Royse, F, Jr., 1993, An overview of the geologic structure of the thrust belt in Wyoming, northern Utah, and eastern Idaho, *in* Snoke, A. W., Steidtmann, J. R., and Roberts, S. M., eds., Geology of Wyoming: Geological Survey of Wyoming Memoir No. 5, p. 272–311.

Royse, F., Jr., Warner, M. A., and Reese, D. L., 1975, Thrust belt structural geometry and related stratigraphic problems, Wyoming-Idaho-northern Utah, *in* Bolyard, D. W., ed., Deep drilling frontiers of the central Rocky Mountains: Denver, Colo., Rocky Mountain Association of Geologists Symposium, p. 41–54.

Rutter, E. H., 1974, The influence of temperature, strain rate and interstitial water in the experimental deformation of calcite rocks Tectonophysics, v. 22, p. 311–334.

Rutter, E H., 1983, Pressure solution in nature, theory, and experiment: Journal of the Geological Society (London), v. 140, p. 725–740.

Rutter, E. H., 1986, On the nomenclature of mode of failure transitions in rocks: Tectonophysics, v 122, p. 381–387.

Rutter, E. H., 1995, Experimental study of the influence of stress, temperature, and strain on the dynamic recrystallization of Carrara marble: Journal of Geophysical Research, v. 100, p. 24,651–24,663.

Rutter, E. H., and Brodie, K. H., 1988, The role of tectonic grain size reduction in the rheological stratification of the lithosphere: Geologische Rundschau, v. 77, p. 295–308.

Rutter, E. H., and Brodie, K. H., 1991, Lithosphere rheology—a note of caution: Journal of Structural Geology, v. 13, p. 363–367.

Rutter, E. H., and Brodie, K. H., 1992, Rheology of the lower crust, *in* Fountain, D. M., Arculus, R., and Kay, R. W., eds., Continental lower crust: Amsterdam, Elsevier, p. 201–267.

Rutter, E. H., Maddock, R. H., Hall, S. H., and White S. H., 1986, Comparative microstructures of natural and experimentally produced clay-bearing fault gouges: Pure and Applied Geophysics, v. 124, p. 3–30.

Rutter, E. H., Brodie, K. H., and Evans, P., 1993, Structural geometry, lower crustal magmatic underplating and lithospheric stretching in the Ivrea-Verbano zone, northern Italy: Journal of Structural Geology, v. 15, p. 647–662.

Rutter, E. H., Casey, M., and Burlini, L., 1994, Preferred crystallographic orientation development during the plastic and superplastic flow of calcite rocks: Journal of Structural Geology, v. 16, p. 1431–1446.

Scambelluri, M., Hoogerduijn Strating, E. H., Piccardo, G. B., Vissers, R. L. M., and Rampone, E., 1991, Alpine olivine- and titanian clinohumite-bearing assemblages in the Erro-Tobbio peridotite (Voltri Massif, NW Italy), Journal of Metamorphic Geology, v. 9, p. 79–91.

Schmid, S. M., 1975, The Glarus overthrust: Field evidence and mechanical model: Eclogae geologae Helvetiae, v 68, p. 247–280.

Schmid, S. M., 1983, Microfabric studies as indicators of deformation mechanisms and flow laws operative in mountain building, *in* Hsü, K. J., ed., Mountain building processes New York, Academic Press, p. 95–110.

Schmid, S. M., 1993, Ivrea Zone and adjacent Southern Alpine basement, *in* von Raumer, J. F., and Neubauer, F., eds., Pre-Mesozoic geology in the Alps: Berlin, Springer-Verlag, p. 567–583.

Schmid, S. M., 1994, Textures of geological materials: computer model predictions versus empirical interpretations based on rock deformation experiments and field studies, *in* Bunge, H.-J., Siegesmund, S., Skrotzki, W., and Weber, K., eds., Textures of geological materials: German Society of Materials Science, DGM Informationsgesellschaft, Verlag, Germany, p. 279–301.

Schmid, S. M., and Casey, M., 1986, Complete fabric analysis of some commonly observed quartz *c*-axis patterns, *in* Hobbs, B. E., and Heard, H. C., eds., Mineral and rock deformation: Laboratory studies (Paterson volume): Washington, D.C., American Geophysical Union, Monograph Series 36, p. 263–286.

Schmid, S. M., and Handy, M. R., 1991, Towards a genetic classification of fault rocks: geological usage and tectonophysical implications, *in* Muller, D. W., McKenzie, J., and Weissert, H., eds., Controversies in modern geology: New York, Academic Press, p. 339–361.

Schmid, S. M., Boland, J. N., and Paterson, M. S., 1977, Superplastic flow in fine-grained limestone, Tectonophysics, v. 43, p. 257–291.

Schmid, S. M., Paterson, M. S., and Boland, J. N., 1980, High temperature flow and dynamic recrystallization in Carrara marble: Tectonophysics, v 65, p. 245–280.

Schmid, S. M., Casey, M., and Starkey, J., 1981, The microfabric of calcite tectonites from the Helvetic nappes (Swiss Alps): Geological Society of London, Special Publication No. 9, p. 151–158.

Schmid, S. M., Pannozzo, R., and Bauer, S., 1987, Simple shear experiments on calcite rocks: rheology and microfabric: Journal of Structural Geology, v. 9, p. 747–778.

Schmidt, C. J., Evans, J. P., Harlan, S. S., Weberg, E. D., Brown, J. S., Batatian, D., Derr, D. N., Malizzi, L., McDowell, R. J., Nelson, G. C., Parke, M., and Genovese, P. W., 1993, Mechanical behavior of basement rocks during movement of the Scarface thrust, central Madison Range, Montana, *in* Schmidt, C. J., Chase, R. B., and Erslev, E. A., eds., Laramide basement deformation in the Rocky Mountain foreland of the western United States: Geological Society of America Special Paper 280, p. 89–105.

Scholz, C. H., 1988, The brittle-plastic transition and the depth of seismic faulting: Geologische Rundschau, v. 77, p. 319–328.

Scholz, C. H., 1990, The mechanics of earthquakes and faulting: New York, Cambridge University Press, 439 p.

Sclar, C. B., Young, A P., Carrison, L. C, and Schwartz, C. M., 1962, Synthesis and optical crystallography of stishovite, a very high pressure polymorph of $SiO_2$: Journal of Geophysical Research, v. 67, p. 4049–4054.

Searle, M. P., 1980, The metamorphic sheet and underlying volcanic rocks beneath the Semail ophiolite in the northern Oman Mountains of Arabia, [Ph.D. dissertation]: Open University, United Kingdom.

Secor, D. T., Jr., 1965, Role of fluid pressure in jointing: American Journal of Science, v. 263, p. 633–646.

Segall, P., and Pollard, D. D., 1980, Mechanics of discontinuous faults: Journal of Geophysical Research, v. 85, p. 4337–4350.

Segall, P., and Pollard, D. D., 1983, Nucleation and growth of strike-slip faults in granite: Journal of Geophysical Research, v. 88, p. 555–568.

Segall, P., and Simpson, C., 1986, Nucleation of ductile shear zones on dilatant fractures: Geology, v. 14, p. 56–59.

Shand, S. J., 1916, The pseudotachylyte of Parijs (Orange Free State) and its relation to "trap-shotten gneiss" and "flinty crush-rock": The Quarterly Journal of the Geological Society of London, v. 72, p. 198–221.

Sharp, R. V., 1979, Some characteristics of the eastern Peninsular Ranges mylonite zone, *in* Proceedings of Conference VIII: Analysis of actual fault zones in bedrock: Washington, D.C., U.S. Geological Survey Open-file Report 79–1239, p. 258–267.

Shibata, K., and Takagi, H., 1988, Isotopic ages of rocks and intrafault materials along the Median Tectonic Line—An example in the Bunguitoge area, Nagano Prefecture: Journal of the Geological Society of Japan, v. 94, p. 35–50.

Shimamoto, T., 1989, The origin of S-C mylonites and a new fault-zone model. Journal of Structural Geology, v. 11, p. 51–64.

Shimamoto, T., and Logan, J. M., 1981, Effects of simulated fault gouge on the sliding behavior of Tennessee sandstone, nonclay gouges: Journal of Geophysical Research, v. 86, p. 2902–2914.

Sibson, R. H., 1975, Generation of pseudotachylyte by ancient seismic faulting: Geophysical Journal of the Royal Astronomical Society, v. 43, p. 775–794.

Sibson, R. H., 1977, Fault rocks and fault mechanisms: Journal of the Geological Society (London), v. 133, p. 191–213.

Sibson, R. H., 1980, Transient discontinuities in ductile shear zones: Journal of Structural Geology, v. 2, p. 165–171.

Sibson, R. H., 1982, Fault zone models, heat flow, and the depth distribution of earthquakes in the continental crust of the United

States: Bulletin of the Seismological Society of America, v. 72, p. 151–163.
Sibson, R. H., 1984, Roughness at the base of the seismogenic zone: Contributing factors: Journal of Geophysical Research, v. 89, p. 5791–5799.
Sibson, R. H., 1985, A note on fault reactivation: Journal of Structural Geology, v. 7, p. 751–754.
Sibson, R. H., 1986, Earthquakes and rock deformation in crustal fault zones: Annual Reviews of Earth and Planetary Sciences, v. 14, p. 149–175.
Sibson, R. H., 1987, Earthquake rupturing as a mineralizing agent in hydrothermal systems: Geology, v. 15, p. 701–704.
Sibson, R. H., 1989, Earthquake faulting as a structural process: Journal of Structural Geology, v. 11, p. 1–14.
Sibson, R. H., 1990, Conditions for fault-valve behavior, *in* Knipe, R. J., and Rutter, E. H., eds., Deformation mechanisms, rheology and tectonics: Geological Society of London, Special Publication No. 54, p. 15–28.
Sibson, R. H., 1992, Implications of fault-valve behaviour for rupture nucleation: Tectonophysics, v. 211, p. 283–293.
Sibson, R. H., Robert, F., and Poulsen, K. H., 1988, High-angle reverse faults, fluid-pressure cycling, and mesothermal gold-quartz deposits: Geology, v. 16, p. 551–555.
Simpson, C., 1981, Ductile shear zones: a mechanism of rock deformation in the orthogneiss of the Maggia nappe, Ticino [Ph.D. dissertation]: Zürich, Switzerland, ETH Zürich, 265 p.
Simpson, C., 1983, Displacement and strain patterns from naturally occurring shear zone terminations: Journal of Structural Geology, v. 5, p. 497–506.
Simpson, C., 1984, Borrego Springs–Santa Rosa mylonite zone: A Late Cretaceous west-directed thrust zone in southern California: Geology, v. 12, p. 8–11.
Simpson, C., 1985, Deformation of granitic rocks across the brittle-ductile transition: Journal of Structural Geology, v. 7, p. 503–511.
Simpson, C., 1986, Fabric development in brittle-to-ductile shear zones: Pure and Applied Geophysics, v. 124, p. 269–288.
Simpson, C., 1990, Microstructural evidence for NE-directed movement on the Chocolate Mountain fault system, southeastern California: Journal of Geophysical Research, v. 95, p. 529–537.
Simpson, C., and De Paor, D. G., 1993, Strain and kinematic analysis in general shear zones: Journal of Structural Geology, v. 15, p. 1–20.
Simpson, C., and Kalaghan, T., 1989, Late Precambrian crustal extension preserved in the Fries fault zone mylonites, southern Appalachians: Geology, v. 17, p. 148–151.
Simpson, C., and Schmid, S. M., 1983, An evaluation of criteria to deduce the sense of movement in sheared rocks: Geological Society of America Bulletin, v. 94, p. 1281–1288.
Simpson, C., and Wintsch, R. P., 1989, Evidence for deformation-induced K-feldspar replacement by myrmekite: Journal of Metamorphic Geology, v. 7, p. 261–275.
Simpson, C., Carreras, J., and Losantos, M., 1982, Inhomogeneous deformation in Rosas granodiorite (NE Spain): Acta Geologica Hispanica, v. 17, p. 219–226.
Sims, J. P., Dirks, P. H. G. M., Carson, C. J., and Wilson, C. J. L., 1995, The structural evolution of the Rauer Group, East Antarctica: mafic dykes as passive markers in a composite Proterozoic terrain: Antarctic Science, v. 6, p. 379–394.
Sleep, N. H., and Blanpied, M. L., 1992, Creep, compaction, and the weak rheology of major faults: Nature, v. 359, p 687–692.
Smith, B. M., Reynolds, S. J., Day, H. W., and Bodnar, R. J., 1991, Deep-seated fluid involvement in ductile-brittle deformation and mineralization, South Mountains metamorphic core complex, Arizona: Geological Society of America Bulletin, v. 103, p. 559–569.
Smith, R. B., and Bruhn, R. L., 1984, Intraplate extensional tectonics of the eastern Basin-Range: Inferences on structural style from seismic reflection data, regional tectonics, and thermal-mechanical models of brittle-ductile deformation: Journal of Geophysical Research, v. 89, p. 5733–5762.
Snoke, A. W., 1980, Transition from infrastructure to suprastructure in the northern Ruby Mountains, Nevada, *in* Crittenden, M. D., Jr., Coney, P. J., and Davis, G. H., eds., Cordilleran metamorphic core complexes: Geological Society of America Memoir 153, p. 287–333.
Snoke, A. W., and Lush, A. P., 1984, Polyphase Mesozoic-Cenozoic deformational history of the northern Ruby Mountains-East Humboldt Range, Nevada, *in* Lintz, J., Jr., ed., Western geological excursions (Geological Society of America 1984 Annual Meeting field trip guidebook): Reno, Nevada, MacKay School of Mines, v. 4, p. 232–260.
Snoke, A. W., Hudec, M. R., Hurlow, H. A., and McGrew, A. J., 1990a, The anatomy of a Tertiary extensional shear zone, Ruby Mountains–East Humboldt Range, Nevada: Geological Society of America Abstracts with Programs, v. 22, no. 3, p. 85.
Snoke, A. W., McGrew, A. J., Valasek, P. A., and Smithson, S. B., 1990b, A crustal cross-section for a terrain of superimposed shortening and extension: Ruby Mountains–East Humboldt Range metamorphic core complex, Nevada, *in* Salisbury, M. H., and Fountain, D. M., eds., Exposed cross-sections of the continental crust (NATO ASI series): Dordrecht, The Netherlands, Kluwer Academic Publishers, p. 103–135.
Spray, J G., 1992, A physical basis for the frictional melting of some rock forming minerals: Tectonophysics, v. 204, p. 205–221.
Spray, J. G., 1993, Viscosity determinations of some frictionally generated silicate melts: implications for fault zone rheology at high strain rates: Journal of Geophysical Research, v. 98, p. 8053–8068.
Spray, J. G., 1995, Pseudotachylyte controversy: Fact or friction?: Geology, v. 23, p. 1119–1122.
Spray, J. G., and Thompson, L. M., 1995, Friction melt distribution in a terrestrial multi-ring impact basin: Nature, v. 373, p. 130–132.
Spray, J. G., Kelley, S. P., and Reimold, W. U., 1995, Laser probe $^{40}Ar/^{39}Ar$ dating of coesite- and stishovite-bearing pseudotachylytes and the age of the Vredefort impact event: Meteoritics, v. 30, p. 335–343.
Spry, A., 1969, Metamorphic textures: Oxford, Pergamon Press, 350 p.
Spudis, P. D., 1993, The geology of multi-ring impact basins: New York, Cambridge University Press, 263 p.
Stauffer, M R., and Lewry, J. F., 1993, Regional setting and kinematic features of the Needle Falls Shear Zone, Trans-Hudson orogen: Canadian Journal of Earth Sciences, v. 30, p. 1338–1354.
Stewart, I. S., and Hancock, P., 1994, Neotectonics, *in* Hancock, P., ed., Continental deformation: Oxford, Pergamon Press, p. 370–409.
Stöffler, D., 1971, Coesite and stishovite in shocked crystalline rocks: Journal of Geophysical Research, v. 76, p. 5474–5488.
Student, J. J., and Sinha, A. K., 1992, Carboniferous U-Pb ages of zircons from the Box Ankle and Ocmulgee faults, central Georgia: implications for accretionary models: Geological Society of America Abstracts with Programs, v. 24, no. 2, p. 69.

Stünitz, H., 1993, Transition from fracturing to viscous flow in a naturally deformed metagabbro, *in* Boland, J. N., and Fitz Gerald, J. D., eds., Defects and processes in the solid state: geoscience applications: Developments in Petrology, v. 14, p. 121–150.

Stünitz, H., and Fitz Gerald, J D., 1993, Deformation of granitoids at low metamorphic grade. II: Granular flow in albite-rich mylonites Tectonophysics, v. 221, p. 299–324

Subimal, S. R., 1977, Deformation and chemical processes in mylonite genesis Geologiska Foreningens i Stockholm Forhandlingar, v. 99, p. 3–9.

Swanson, M. T., 1988, Pseudotachylyte-bearing strike-slip duplex structures in the Fort Foster Brittle Zone of southern Maine: Journal of Structural Geology, v. 10, p. 813–828.

Swanson, M T., 1989, Sidewall ripouts in strike-slip faults: Journal of Structural Geology, v. 11, p. 933–948.

Swanson, M. T., 1992a, Fault structure, wear mechanisms and rupture processes in pseudotachylyte generation: Tectonophysics, v. 204, p. 223–242.

Swanson, M. T., 1992b, Late Acadian-Alleghenian transpressional deformation: evidence from asymmetric boudinage in the Casco Bay area, coastal Maine: Journal of Structural Geology, v. 14, p. 323–341.

Swanson, M. T., 1994, Minimum dextral shear strain estimates in the Casco Bay area of coastal Maine from vein reorientation and elongation: Geological Society of America Abstracts with Programs, v. 26, no. 3, p 75.

Tabor, R. W., Frizzell, V. A., Jr., Whetten, J. T., Waitt, R. B., Swanson, D. A., Byerly, G. R., Booth, D. B., Hetherington, M. J., and Zartman, R. E., 1987, Geologic map of the Chelan 30-minute by 60-minute quadrangle: Washington, D.C., U.S. Geological Survey Miscellaneous Investigations Series Map I-1661, scale: 100,000.

Takagi, H., 1984, Mylonitic rocks along the Median Tectonic Line in Takago-Ichinose area, Nagano Prefecture: Journal of the Geological Society of Japan, v. 90, p. 81–100.

Takagi, H., 1986, Implications of mylonitic microstructures for the geotectonic evolution of the Median Tectonic Line, central Japan: Journal of Structural Geology, v. 8, p. 3–14.

Takagi, H , 1992, Development of composite planar fabric in mylonites along the Median Tectonic Line, southwest Japan: The Island Arc, v. 1, p. 92–102

Takagi, H., and Ito, M., 1988, The use of asymmetric pressure shadows in mylonites to determine the sense of shear, Journal of Structural Geology, v. 10, p. 347–360.

Takagi, H, Takeshita, T., Shibata, K., Uchiumi, S., and Inoue, M., 1992, Middle Miocene normal faulting along the Tobe thrust in western Shikoku: Journal of the Geological Society of Japan, v. 98, p. 1069–1072.

Takeshita, T., and Wenk, H.-R., 1988, Plastic anisotropy and geometric hardening in quartzites, Tectonophysics, v. 149, p. 345–361.

Talbot, C. J., 1979, Flow folds in a glacier of salt in southern Iran: Journal of Structural Geology, v. 1, p. 5–18.

Talbot, C. J., 1981, Sliding and other deformation mechanisms in a glacier of salt, S Iran, *in* McClay, K. R., and Price, N. J , eds., Thrust and nappe tectonics Geological Society of London, Special Publication No 9, p. 173–183.

Talbot, C. J., 1993, Spreading of salt structures in the Gulf of Mexico Tectonophysics, v. 228, p. 151–166.

Talbot, C. J., and Jackson, M. P. A., 1987, Internal kinematics of salt diapirs: American Association of Petroleum Geologists Bulletin, v. 71, p. 1068–1093.

Talbot, C J., and Jackson, M. P. A., 1989, Internal kinematics of salt diapirs. Reply: American Association of Petroleum Geologists Bulletin, v. 73, p. 946–950.

Talbot, C. J., and Jarvis, R. J., 1984, Age, budget and dynamics of an active salt extrusion in Iran: Journal of Structural Geology, v. 6, p. 521–533.

Talbot, C J., and Rogers, E. A., 1980, Seasonal movements in a salt glacier in Iran: Science, v. 208, p. 395–397.

Tanaka, H., 1992, Cataclastic lineations: Journal of Structural Geology, v. 14, p. 1239–1252.

Tapponnier, P., Peltzer, G., and Armijo, R., 1986, On the mechanics of the collision between India and Asia, *in* Coward, M. P., and Ries, A. C., eds., Collision tectonics: Geological Society of London, Special Publication No. 19, p. 115–157.

Tapponnier, P., Lacassin, R., Leloup, P. H., Schärer, U., Zhong, D., Wu, H., Liu, X., Ji, S., Zhang, L., and Zhong, J., 1990, The Ailao Shan/Red River metamorphic belt: Tertiary left-lateral shear between Indochina and South China, Nature, v. 343, p. 431–437.

Tazaki, K., Takahashi, J., Itaya, T., and Kashima, N., 1990, K-Ar ages of the andesites intruding along the Median Tectonic Line in northwestern Shikoku, Japan: Journal of the Japanese Association of Mineralogists, Petrologists and Economic Geologists, v. 85, p 155–160.

Tchalenko, J. S., 1970, Similarities between shear zones of different magnitudes: Geological Society of America Bulletin, v. 81, p. 1625–1640

Tchalenko, J. S., and Ambraseys, N. N., 1970, Structural analysis of the Dasht-e Bayaz (Iran) earthquake fractures: Geological Society of America Bulletin, v. 81, p. 41–60.

Teall, J. J. H., 1918, Dynamic metamorphism, a review, mainly personal: Proceedings of the Geological Association (London), v. 29, p. 1–15.

Techmer, K., 1989, Die Genese von Pseudotachyliten und ihre Bedeutung für die geotektonische Stellung der Ivrae-Verbano Zone [Ph.D. dissertation]: Göttingen, Germany, University of Göttingen, 144 p.

Techmer, K. S., Ahrendt, H., and Weber, K., 1992, The development of pseudotachylyte in the Ivrea-Verbano Zone of Italian Alps: Tectonophysics, v. 204, p. 307–322.

ten Brink, C. E., and Passchier, C. W., 1995, Modelling of mantled porphyroclasts using non-Newtonian rock analogue materials: Journal of Structural Geology, v. 17, p. 131–146.

Theodore, T. G., and Christie, J. M., 1969, Mylonites and mylonitic gneisses. Correct use of the terms: Geological Society of America Abstracts with Programs, v. 1, no. 3, p 69–70.

Therriault, A. M., Reid, A. M., and Reimold, W. U., 1993, Original size of the Vredefort Structure, South Africa: Lunar and Planetary Science, v. 24, p. 1419–1420.

Thompson, L. M., and Spray, J. G., 1994, Pseudotachylytic rock distribution and genesis within the Sudbury impact structure, *in* Dressler, B. O., Grieve, R.A.F., and Sharpton, V. L., eds., Large meteorite impacts and planetary evolution: Boulder, Colorado, Geological Society of America, Special Paper 293, p. 275–287.

Thomson, C. N., and Girty, G. H., 1994, Early Cretaceous intra-arc ductile strain in Triassic-Jurassic and Cretaceous continental margin arc rocks, Peninsular Ranges, California: Tectonics, v. 13, p. 1108–1119.

Thomson, C. N., Girty, G. H., and Bracchi, K. A., 1994, Jurassic continental-arc magmatism and the record of Early Cretaceous normal convergence between the North American and Farallon plates: Cuyamaca-Laguna Mountains shear zone (CLMSZ), Peninsular Ranges batholith (PRB), southern California: Geological Society of America Abstracts with Programs, v. 26, no. 2, p. 98.

Tilton, G. R., Hopson, C. A., and Wright, J. E., 1981, Uranium-lead isotopic ages of the Samail ophiolite, Oman, with applications to Tethyan ocean ridge tectonics: Journal of Geophysical Research, v. 86, p. 2763–2775.

Todd, V. R., 1980, Structure and petrology of a Tertiary gneiss complex in northwestern Utah, *in* Crittenden, M. D., Jr., Coney, P. J., and Davis, G. H., eds., Cordilleran metamorphic core complexes: Geological Society of America Memoir 153, p. 349–383.

Todd, V. R., 1983, Late Miocene displacement of pre-Tertiary rocks in the Matlin Mountains, northwestern Utah, *in* Miller, D. M., Todd, V. R., and Howard, K. A., eds., Tectonic and stratigraphic studies in the eastern Great Basin: Geological Society of America Memoir 157, p. 239–270.

Todd, V. R., Erskine, B. G., and Morton, D. M., 1988, Metamorphic and tectonic evolution of the northern Peninsular Ranges batholith, southern California, *in* Ernst, W.G., ed., Metamorphism and crustal evolution of the western United States: Rubey Volume VII: Englewood Cliffs, New Jersey, Prentice-Hall, p. 894–937.

Toyoshima, T., 1991, Tectonic evolution of the Hidaka metamorphic belt and its implication in late Cretaceous–middle Tertiary tectonics of Hokkaido, Japan. Journal of the Faculty of Science, Niigata University, Series E, no. 8, p. 1–107.

Toyoshima, T., Komatsu, M., and Shimura, T., 1994, Tectonic evolution of lower crustal rocks in an exposed magmatic arc section in the Hidaka metamorphic belt, Hokkaido, northern Japan: The Island Arc, v. 3, p. 182–198.

Treloar, P. J., Brodie, K. H., Coward, M. P., Jan, M. Q., Khan, M. A., Knipe, R. J., Rex, D. C., and Williams, M. P, 1990, The evolution of the Kamila shear zone, Kohistan, Pakistan, *in* Salisbury, M. H., and Fountain, D. M., eds., Exposed cross-sections of the continental crust (NATO ASI Series): Dordrecht, The Netherlands, Kluwer Academic Publishers, p. 175–214.

Truman, W E., 1976, Geology of the Blue Ridge Front near Riner, Virginia [M S thesis]: Blacksburg, Virginia, U.S.A., Virginia Polytechnic Institute and State University, 102 p.

Tse, S. T., and Rice, J. R., 1986, Crustal earthquake instability in relation to the depth variation of frictional slip properties: Journal of Geophysical Research, v. 91, p. 9452–9472.

Tullis, J., 1983, Deformation of feldspars, *in* Ribbe, P. H., ed., Feldspar mineralogy: Mineralogical Society of America, Reviews in Mineralogy, v. 2, p. 297–323.

Tullis, J., 1990, Experimental studies of deformation mechanisms and microstructures in quartzo-feldspathic rocks, *in* Barber, D. J., and Meridith, P. G., eds., Deformation processes in minerals, ceramics and rocks: London, Unwin Hyman, p. 190–227.

Tullis, J., and Yund, R. A., 1985, Dynamic recrystallization of feldspar: A mechanism for ductile shear zone formation: Geology, v. 13, p. 238–241.

Tullis, J., and Yund, R. A., 1987, Transition from cataclastic flow to dislocation creep of feldspar: Mechanisms and microstructures: Geology, v. 15, p. 606–609.

Tullis, J., and Yund, R. A., 1991. Diffusion creep in feldspar aggregates: experimental evidence: Journal of Structural Geology, v. 13, p. 987–1000.

Tullis, J., and Yund, R. A., 1992, The brittle-ductile transition in feldspar aggregates: An experimental study, *in* Evans, B., and Wong, T.-f., eds., Fault mechanics and transport properties of rocks: Orlando, Academic Press, p. 89–117

Tullis, J., Christie, J. M., and Griggs, D. T., 1973, Microstructures and preferred orientations of experimentally deformed quartzites: Geological Society of America Bulletin, v. 84, p. 297–314.

Tullis, J., Snoke, A. W., and Todd, V. R., 1982, Penrose Conference Report: Significance and petrogenesis of mylonitic rocks: Geology, v. 10, p. 227–330.

Tullis, J., Dell'Angelo, L. N., and Yund, R. A., 1990, Ductile shear zones from brittle precursors in feldspathic rocks: the role of dynamic recrystallization, *in* Duba, A. G, Durham, W. B., Handin, J. W., and Wang, H. F., eds., The brittle-ductile transition in rocks: Washington, D.C., American Geophysical Union, Geophysical Monograph 56, p 67–82.

Tullis, J., Yund, R., and Farver, J., 1996, Deformation-enhanced fluid distribution in feldspar aggregates: Geology, v. 24, p. 63–66.

Tullis, T. E., 1980, The use of mechanical twinning in minerals as a measure of shear stress magnitudes: Journal of Geophysical Research, v. 85, p. 6263–6268

Turner, F. J., 1962, "Compression" and "tension" axes determined from {0112} twinning in calcite [abstract]. Journal of Geophysical Research, v. 67, p. 1660.

Twiss, R. J., 1977, Theory and applicability of a recrystallized grain size paleopiezometer, Pure and Applied Geophysics, v. 115, p. 227–244.

Twiss, R. J., and Moores, E. M., 1992, Structural geology: New York, W. H. Freeman and Company, 532 p.

Urai, J. L., Humphreys, F. J., and Burrows, S. E., 1980, In-situ studies of the deformation and dynamic recrystallization of rhombohedral camphor: Journal of Materials Science, v. 15, p. 1231–1240.

Urai, J. L., Means, W. D., and Lister, G. S., 1986, Dynamic recrystallization of minerals, *in* Hobbs, B E, and Heard, H C, eds., Mineral and rock deformation: Laboratory studies: American Geophysical Union, Geophysical Monograph 36, p. 161–199.

Urai, J. L., Spiers, C. J., Zwart, H. J., and Lister, G. S., 1986, Weakening of rock salt by water during long term creep: Nature, v. 324, p. 554–557.

van den Driessche, J., and Brun, J.-P., 1987, Rolling structures at large shear strain: Journal of Structural Geology, v. 9, p. 691–704.

van der Pluijm, B. A., 1991, Marble mylonites in the Bancroft shear zone, Ontario, Canada: microstructures and deformation mechanisms: Journal of Structural Geology, v. 13, p. 1125–1135.

van der Pluijm, B. A., Mezger, K., Cosca, M. A, and Essene, E J., 1994, Determining the significance of high-grade shear zones by using temperature-time paths, with examples from the Grenville orogen: Geology, v. 22, p. 743–746.

van der Wal, D., Vissers, R. L. M., Drury, M. R., and Hoogerduijn Strating, E. H., 1992, Oblique fabrics in porphyroclastic Alpine-type peridotites: a shear-sense indicator for upper mantle flow: Journal of Structural Geology, v. 14, p 839–846.

van der Wal, D., Chopra, P., Drury, M., and Fitz Gerald, J., 1993, Relationships between dynamically recrystallized grain size and deformation conditions in experimentally deformed olivine rocks: Geophysical Research Letters, v. 20, p. 1479–1482.

van Keken, P. E., Spiers, C. J., van den Berg, A. P., and Muyzert, E J., 1993, The effective viscosity of rocksalt implementation of steady-state creep laws in numerical models of salt diapirism: Tectonophysics, v. 225, p. 457–476.

Vanko, D. A., and Batiza, R., 1982, Gabbroic rocks from the Mathematician Ridge failed rift: Nature, v. 300, p. 742–744.

Vernon, R. H., 1977, Relationships between microstructures and metamorphic assemblages Tectonophysics, v. 39, p. 439–452.

Vernon, R. H., 1981, Optical microstructure of partly recrystallized calcite in some naturally deformed marbles: Tectonophysics, v. 78, p. 601–612.

Vernon, R. H., Williams, V. A., and D'Arcy, W. F., 1983, Grain-size reduction and foliation development in a deformed granitoid batholith: Tectonophysics, v. 92, p. 123–145.

Vidal, J. L., Kubin, L., Debat, P., and Soula, J. C., 1980, Deformation and dynamic recrystallization of K-feldspar augen in orthogneiss from Montagne Noire, Occitania, southern France: Lithos, v. 13, p. 247–255.

Vissers, R. L. M., Drury, M. R., Hoogerduijn Strating, E. H., and van der Wal, D., 1991, Mantle shear zones—a case study in an Alpine lherzolite massif: Geology, v. 19, p. 990–993.

Walker, A. N., Rutter, E. H., and Brodie, K. H., 1990, Experimental study of grain-size sensitive flow of synthetic, hot-pressed calcite rocks, *in* Knipe, R. J., and Rutter, E. H., eds., Deformation mechanisms, rheology and tectonics, Geological Society of London, Special Publication No. 54, p. 259–284.

Warner, R. H., 1961, Structural geology of Carboniferous rocks near Marble Falls, Burnett County, Texas [M.S. thesis]: Austin, Texas, U.S.A., University of Texas, 72 p.

Waters, A. C. and Campbell, C. D., 1935, Mylonites from the San Andreas fault zone: American Journal of Science, Fifth Series, v. 29, no. 174, p. 473–503.

Weathers, M. S., Bird, J. M., Cooper, R. F., and Kohlstedt, D., 1979, Differential stress determined from deformation induced microstructures of the Moine thrust zone: Journal of Geophysical Research, v. 84, p. 7495–7509.

Weber, K., Ahrendt, H., Dornbusch, J., and Techmer, K., 1989, The ductile-brittle strain path along the northwestern rim of the Ivrea-Verbano zone: Conference on the Lower Continental Crust, Varallo, Italy, May 17–20, 1989, p. 20.

Wenk, H.-R., 1973, The structure of the Bergell Alps: Eclogae geologicae Helvetiae, v. 66, p. 255–291.

Wenk, H.-R., 1978, Are pseudotachylites products of fracture or fusion?: Geology, v. 6, p. 507–511.

Wenk, H.-R., 1984, Brittle-ductile transition zone in the northern Bergell Alps: Geologische Rundschau, v. 73, p. 419–431.

Wenk, H.-R., ed., 1985, Preferred orientation in deformed metals and rocks: An introduction to modern textural analysis: New York, Academic Press, 610 p.

Wenk, H.-R., 1987, Introduction to the geology of the Bergell Alps with guide for excursions: Jahrbuch Naturforschungen des Graubünden, v. 103, p. 29–90.

Wenk, H.-R., and Christie, J. M., 1991, Comments on the interpretation of deformation textures in rocks: Journal of Strucural Geology, v. 13, p. 1091–1110.

Wenk, H.-R., and Pannetier, J., 1990, Texture development in deformed granodiorites from the Santa Rosa mylonite zone, southern California: Journal of Structural Geology, v. 12, p. 177–184.

Wenk, H.-R., Wenk, E., and Wallace, J., 1974, Metamorphic mineral assemblages in pelitic rocks of the Bergell Alps, *in* Alpine metamorphism: Schweizerisches Mineralogisches und Petrographisches Mitteilungen, v. 54, p. 507–554.

Wenk, H.-R., Kern, H., Van Houtte, P., and Wagner, F., 1986, Heterogeneous strain in axial deformation of limestone, textural evidence, *in* Hobbs, B. E., and Heard, H. C., eds., Mineral and rock deformation: Laboratory studies: Washington, D.C., American Geophysical Union, Geophysical Monograph 36, p. 287–295.

Wenk, H.-R., Takeshita, T., Bechler, E., Erskine, B. G., and Matthies, S., 1987, Pure shear and simple shear calcite textures. Comparison of experimental, theoretical and natural data: Journal of Structural Geology, v. 9, p. 731–745.

Wernicke, B., Walker, J. D., and Beaufait, M. S., 1985, Structural discordance between Neogene detachments and frontal Sevier thrusts, central Mormon Mountains, southern Nevada: Tectonics, v. 4, p. 213–246.

White, J. C., 1982, Quartz deformation and the recognition of recrystallization regimes in the Flinton Group conglomerates, Ontario: Canadian Journal of Earth Sciences, v. 19, p. 81–93.

White, J. C., and Mawer, C. K., 1986, Extreme ductility of feldspars from a mylonite, Parry Sound, Canada: Journal of Structural Geology, v. 8, p. 133–143.

White, J. C., and Mawer, C. K., 1988, Dynamic recrystallization and associated exsolution in perthites: evidence of deep crustal thrusting: Journal of Geophysical Research, v. 93, p. 325–337.

White, J. C., and Mawer, C. K., 1992, Deep-crustal deformation textures along megathrusts from Newfoundland and Ontario: implications for microstructural preservation, strain rates, and strength of the lithosphere: Canadian Journal of Earth Sciences, v. 29, p. 328–337.

White, S. H., 1973a, The dislocation structures responsible for the optical effects in some naturally deformed quartzites: Journal of Material Science, v. 8, p. 490–499.

White, S. H., 1973b, Syntectonic recrystallization and texture development in quartz: Nature, v. 244, p. 276–278.

White, S. H., 1976, The effects of strain on the microstructures, fabrics and deformation mechanisms in quartzite: Philosophical Transactions of the Royal Society of London, Ser. A, v. 283, p. 69–86.

White, S. H., 1979, Difficulties associated with paleo-stress estimates: Bulletin de Minéralogie, v. 102, p. 210–215.

White, S. H., 1982, Fault rocks of the Moine Thrust Zone: a guide to their nomenclature: Textures and Microstructures, v. 4, p. 211–221.

White, S. H., and Glasser, J., 1987, The Outer Hebrides fault zone, *in* Park, R. G., and Tarney, J. eds., The Lewisian: Geological Society of London, Special Publication No. 27, p. 125–183.

White, S. H., and Knipe, R. J., 1978, Transformation and reaction enhanced ductility in rocks: Journal of the Geological Society (London), v. 135, p. 513–516.

White, S. H., Burrows, S. E., Carreras, J., Shaw, N. D., and Humphreys, F. J., 1980, On mylonites in ductile shear zones: Journal of Structural Geology, v. 2, p. 175–187.

White, S. H., Evans, D. J., and Zhong, D.-L., 1982, Fault rocks of the Moine Thrust Zone: Microstructures and textures of selected mylonites: Textures and Microstructures, v. 5, p. 33–61.

Williams, P. F., Goodwin, L. B., and Raiser, S., 1994, Ductile deformation processes, *in* Hancock, P. L., ed., Continental deformation: Oxford, Pergamon Press, p. 1–27.

Wilshire, H. G., 1971, Pseudotachylite from the Vredefort Ring, South Africa: Journal of Geology, v. 79, p. 195–206.

Wilshire, H. G., and Kirby, S. H., 1989, Dikes, joints, and faults in the upper mantle: Tectonophysics, v. 161, p. 23–31.

Wintsch, R. P., and Ambers, C. P., 1989, Core description of the 1984 Gilette Castle Well, Connecticut: Hartford, Conn., U.S.A., Connecticut Geological and Natural History Survey, Open-file Report OF-89-2, 152 p.

Wintsch, R. P. and Knipe, R. J., 1983, Growth of a zoned plagioclase porphyroblast in a mylonite: Geology, v. 11, p. 360–363.

Wintsch, R. P., Sutter, J. F., Kunk, M. J., Aleinikoff, J. N., and Dorais, M. J., 1992, Contrasting P-T-t paths: Thermochronologic evidence for a late Paleozoic final assembly of the Avalon composite terrane in the New England Appalachians: Tectonics, v. 11, p. 672–689.

Wintsch, R. P., Sutter, J. F., Kunk, M. J., Aleinikoff, J. N., Boyd, J. L., and Hellickson, B. P., 1993, Geochronological evidence for an Alleghanian assembly age of Proterozoic and Paleozoic lithotectonic terranes in south-central New England: Geological Society of America Abstracts with Programs, v. 25, no. 6 p. A-478–A-479.

Wintsch, R. P., Christoffersen, R., and Kronenberg, A. K., 1995, Fluid-rock reaction weakening of fault zones: Journal of Geophysical Research, v. 100, p. 13,021–13,032.

Wise, D. U., Dunn, D. E., Engelder, J. T., Geiser, P. A., Hatcher, R. D., Kish, S. A., Odom, A. L., and Schamel, S., 1984, Fault-related rocks: Suggestions for terminology: Geology, v. 12, p. 391–394.

Witty, G. J., 1975, The geochemistry of the Roneval anorthosite, South Harris, Scotland [Ph.D. dissertation]: London, England, University of London.

Wright, J. E., and Snoke, A. W., 1993, Tertiary magmatism and mylonitization in the Ruby–East Humboldt metamorphic core complex, northeastern Nevada: U-Pb geochronology and Sr, Nd, Pb isotope geochemistry: Geological Society of America Bulletin, v. 105, p. 935–952.

Wright, T. O., and Platt, L. B., 1982, Pressure dissolution and cleavage in the Matinsburg Shale: American Journal of Science, v. 282, p. 122–135.

Yonkee, W. A., 1990, Geometry and mechanics of basement and cover deformation, Farmington Canyon complex, Sevier orogenic belt, Utah [Ph.D. dissertation]: Salt Lake City, Utah, U.S.A., University of Utah, 225 p.

Yonkee, W. A., 1992, Basement-cover relations, Sevier orogenic belt, northern Utah: Geological Society of America Bulletin, v. 104, p. 280–322.

Yonkee, W. A., and Mitra, G., 1993, Comparison of basement deformation styles in parts of the Rocky Mountain foreland, Wyoming, and Sevier orogenic belt, northern Utah, *in* Schmidt, C. J., Chase, R. B., and Erslev, E. A., eds., Laramide basement deformation in the Rocky Mountain foreland of the western United States: Geological Society of America Special Paper 280, p. 197–228.

Yonkee, W. A., Parry, W. T., Bruhn, R. L., and Cashman, P. C., 1989, Thermal models of thrust faulting: Constraints from fluid inclusion observations, Willard thrust sheet, Idaho-Utah-Wyoming thrust belt: Geological Society of America Bulletin, v. 101, p. 304–313.

Yund, R. A., and Tullis, J., 1991, Compositional changes of minerals associated with dynamic recrystallization: Contributions to Mineralogy and Petrology, v. 108, p. 346–355.

Yund, R. A., Blanpied, M. L., Tullis, T. E., and Weeks, J. D., 1990, Amorphous material in high strain experimental fault gouge: Journal of Geophysical Research, v. 95, p. 15,589–15,602.

Zeck, H. P., 1974, Cataclastites, hemiclastites, holoclastites, blasto-ditto, and myloblastites—cataclastic rocks: American Journal of Science, v. 274, p. 1064–1073.

Zeuch, D. H., 1982, Ductile faulting, dynamic recrystallization, and grain size sensitive flow of olivine: Tectonophysics, v. 83, p. 293–308.

Zingg, A., 1983, The Ivrea and Strona-Ceneri Zones (Southern Alps, Ticino and N-Italy)—a Review: Schweizerisches Mineralogisches und Petrographisches Mitteillungen, v. 63, p. 361–392.

Zingg, A., Handy, M. R., Hunziker, J. C., and Schmid, S. M., 1990, Tectonometamorphic history of the Ivrea Zone and its relationship to the crustal evolution of the Southern Alps: Tectonophysics, v. 182, p. 169–192.

Zoback, M. D., Zoback, M. L., Mount, V. S., Suppe, J., Eaton, J. P., Healy, J. H., Oppenheimer, D., Reasenberg, P., Jones, L., Raleigh, C. B., Wong, I. G., Scotti, O., and Wentworth, C., 1987, New evidence on the state of stress of the San Andreas fault system: Science, v. 238, p. 1105–1111.

# Index

GPSR Authorized Representative: Easy Access System Europe - Mustamäe tee 50, 10621 Tallinn, Estonia, gpsr.requests@easproject.com

www.ingramcontent.com/pod-product-compliance
Lightning Source LLC
LaVergne TN
LVHW080333021026
849160LV00028B/1075

* 9 7 8 0 6 9 1 6 0 0 7 3 4 *